Modern Carpentry

Essential Skills for the Building Trades

Willis H. Wagner
Professor Emeritus
Industrial Technology
University of Northern Iowa, Cedar Falls

Howard Bud Smith
Technical Author and Editor
Lee Howard Associates
Black River Falls, WI

Consultants

Jim Maddox
St. Paul Technical College
St. Paul, MN

John M. Murphy
Master Builder
Milford, MI

Publisher
The Goodheart-Willcox Company, Inc.
Tinley Park, Illinois
www.g-w.com

Library of Congress Catalog Card Number 2005056701

ISBN-13: 978-1-59070-648-0
ISBN-10:1-59070-648-X

3 4 5 6 7 8 9 – 08 – 12 11 10 09

Library of Congress Cataloging-in-Publication Data

Wagner, Willis H.
Modern carpentry: building construction details in easy-to-understand form / Willis H. Wagner, Howard Bud Smith.
    p. cm.
  Includes index.
  ISBN 1-59070-648-X
  1. Carpentry.   I. Smith, Howard Bud   II. Title.
TH5604.W34 2006
694--dc22   2005056701

Introduction

Modern Carpentry is a colorful, easy-to-understand source of authoritative and up-to-date information on building materials, tools, and construction methods. It provides detailed coverage of all aspects of light construction. Included are site development, print reading, site layout, excavating, foundation work, framing, sheathing, insulating, roofing, window and door installation, exterior finishing, interior finishing, and mechanical systems.

Chapters are arranged in a logical sequence similar to the order builders use to complete various phases of home construction. Special emphasis is given to the carpenter's workplace and what is expected of a new carpenter on a job site. The workplace chapter covers basic skills, attitudes, and relationship to coworkers. It strongly focuses on safe work habits in handling tools, materials, and prefabricated components.

An introductory chapter discusses the entire scope of materials—old and new—that are available. This chapter includes size and grade descriptions as well as technical information covering physical properties and other important characteristics. Scientific and technical discoveries have led to the development of many new materials including engineered lumber, better insulation, and more-energy-efficient doors and windows.

The proper use and application of these materials depends on craftspersons with considerable knowledge of the products and materials and how those materials should function in a completed structure. Use of prefabricated components, or even entire prefabricated structures, is having a significant impact on how homes are constructed on the building site. The market for housing partly or entirely built in a factory is expected to show continued growth beyond the 30% market share witnessed in recent years.

Beyond a solid discussion of building basics, **Modern Carpentry** covers special topics such as constructing stairs, chimneys, fireplaces, and decks. Systems-built (prefabricated) houses, remodeling, cabinet installation, solar construction, and painting and decorating are also covered.

Modern Carpentry also provides chapters introducing mechanical systems: electrical wiring, plumbing, and heating and cooling (HVAC). More than cursory discussions of principles, these chapters explain how to install such systems.

Supplementing the text materials are more than 1600 photographs and line art illustrations. These are coordinated with step-by-step instructions and clear descriptions.

Modern Carpentry is designed as a comprehensive text for carpentry students at various levels, including high schools, vocational-technical schools, colleges, and apprentice training programs. It also serves as a valuable reference for students in architectural drafting and for journeymen carpenters and construction supervisors. The book will enable do-it-yourselfers to handle many construction projects they may otherwise be reluctant to undertake. This edition incorporates input of consultants in both residential and light commercial construction.

Willis H. Wagner
Howard Bud Smith

Modern Carpentry Chapter Components

Power Tools 5

Learning Objectives

After studying this chapter, you will be able to:
- Identify common power tools.
- Explain the function and operation of the carpenter's principal power tools.
- Identify the major parts of common power tools.
- Apply power tool safety rules.

Technical Vocabulary

Belt sander
Chain saw
Continuity monitor
Drywall screw
  shooter
Electric drill
Finishing sander
Frame and trim saw
Ground fault circuit
  interrupter (GFCI)
Jointer
Magnetic starters
Nailers
Panel saw

Plate joiner
Portable tools
Powder-actuated tools
Power block plane
Power miter saw
Power plane
Power stapler
Radial arm saw
Reciprocating saw
Rotary hammer drill
Router
Saber saw
Stationary tools
Table saw

Power tools greatly reduce the time required to perform many of the operations in carpentry work. The carpenter can plane, route, bore, and saw in less time with less exertion. In addition, when proper tools are used in the correct way, a high level of accuracy can easily be achieved. There are two general types of power tools:
- *Portable tools* are lightweight and intended to be carried around by the carpenter. They may be used anywhere on the construction site or on any part of the structure.
- *Stationary tools*, usually called machines, are heavy equipment mounted on benches or stands. The workpiece is brought to them. Benches and stands rest firmly on the floor or ground. The machines must be level when used on the job site.

This chapter provides only brief descriptions of the kinds of power tools most commonly associated with carpentry work. It should be supplemented with woodworking textbooks and reference books devoted to power tool operation. Manufacturer bulletins and operator's manuals are also good sources of information.

Portable tools: Lightweight tools that are intended to be carried around by the carpenter.
Stationary tools: Heavy equipment mounted on benches or stands. Usually called machines.

121

Objectives. Provide an overview of the chapter content and explain what should be understood on completion of the chapter.

Technical Vocabulary. List of important technical terms introduced in the chapter. The terms in this list are displayed in **bold italic type** when they first appear in the chapter text.

Running Glossary. Definitions of important terms introduced on the page. The running glossary reinforces the terminology carpenters will encounter on the job.

Working Knowledge. Supplemental information and hints related to the components or procedures discussed in the chapter.

Procedures. Detailed, step-by-step instructions for accomplishing specific carpentry tasks. The procedures promote a logical, organized approach to completing common jobs.

Chapter 6 Building Layout   155

The rod shown in **Figure 6-8** has graduations in feet and decimal parts of a foot. This is the type used for regular surveying work. Rods are also available with graduations in feet and inches.

When sighting short distances (100' or less), a regular wooden or metal folding rule can be held against a wood stake and read through the instrument. This procedure will be satisfactory for jobs such as setting grade stakes for a footing. Always be sure to hold the stake and rule in a vertical position.

Working Knowledge

Leveling instruments and equipment will vary somewhat, depending on the manufacturer. Always carefully read and study the instructions for a given brand.

6.4.2 Care of Leveling Instruments

Leveling instruments are more delicate than most other carpentry tools and equipment. Special precautions must be followed in their use so they will continue to provide accurate readings over a long period of time. Some suggestions follow:
- Keep the instrument clean and dry. Store it in its carrying case when not in use.
- When the instrument is set up, have a plastic bag or cover handy to use in case of rain. If the instrument becomes wet, dry it before storing.
- When moving the instrument from its case to the tripod, grip it by the base.
- Never leave the instrument unattended when it is set up near moving equipment.
- When moving a tripod-mounted instrument, handle it with care. Hold the instrument upright; never carry it in a horizontal position.
- Never over-tighten leveling screws or any of the other adjusting screws or clamps.

- Always set the tripod on firm ground with the legs spread well apart. When it is set up on a floor or pavement, take extra precautions to ensure that the legs will not slip.
- For precision work, permit the instrument to reach ambient (air) temperature before making readings.
- When the lenses collect dust and dirt, clean them with a camel hair brush or special lens paper.
- Never use force on any of the adjustments. They should easily turn by hand.
- Have the instrument cleaned, lubricated, and checked yearly by a qualified repair station or by the manufacturer.

PROCEDURE

Setting up the instrument

Use the following procedure to set up a tripod-mounted instrument.
1. Place the tripod so it will be a firm and stable base for the instrument. The base of the legs should be about 3'-6" apart. Make sure the points are well into the ground and the head is fairly level.
2. Check the wing nuts on the adjustable legs. They should be tight enough to carry the weight of the instrument without collapsing or sinking. Tighten the hex nuts holding the legs to the head to the desired tension.
3. Carefully lift the instrument from its case by the base plate. Before mounting the instrument, loosen the clamp screws. On some instruments, the leveling screws must be turned up so the tripod cup assembly can be hand-tightened to the instrument mounting stud. Set the telescope lock lever of the transit in the closed position.
4. Attach the instrument to the tripod. If it is to be located over an exact point, such as a benchmark, attach the plumb bob and move the instrument over the spot. Do this before the final leveling.

---

800   Section 5   Special Construction

Safety Note

Dealing with hazardous materials is a concern in every remodeling project. In a house built before 1978, painted surfaces should be tested for lead content. Lead-based paint was banned in 1978. Kits for testing lead content in paint are available. Lead-contaminated materials should be professionally stripped before use. Some experts discourage reuse of such material and suggest it be disposed of in an approved landfill. Other possible hazards include chlordane contaminated lumber, lead solder in old pipes, and asbestos in pipe insulation, old flooring tile, and exterior siding.

26.11.1 Fall Protection

OSHA fall protection rules allow residential carpenters and contractors to write their own fall protection plans. The plan has to meet a set of guidelines supplied by OSHA.

The residential plan, which applies at heights of 6' or more, must be specific to the site, properly supervised, and only applied in situations where conventional fall protection procedures would cause a greater hazard than the work itself.

Safety Note

The OSHA regulations covering actions on scaffolding, stairs, ladders, roofs, and other high surfaces, such as top plates, are important to remodelers. Contact a local OSHA office or write Superintendent of Documents, U. S. Government Printing Office, Washington DC 20402. Ask for Code of Federal Regulations (CFR) Title 29, Part 1926.500. The title is *Safety Standards for Fall Protection in the Construction Industry; Final Rule.*

ON THE JOB

Asbestos Removal Technician

For many years, the mineral fiber asbestos was used for fireproofing, pipe insulation, and other applications in buildings. It is no longer used in construction since airborne particles of the material have been shown to cause cancer and other diseases of the lungs. Many older buildings have large amounts of asbestos in them, which poses danger to workers involved in demolition or remodeling work. This material must be removed for health and safety of workers and occupants. Asbestos contamination has been a particular concern in school buildings where young children can be exposed to asbestos particles.

These concerns have created a new building trades specialty, the asbestos removal technician. A related field is lead removal, since old lead-based paints are a health hazard as well. Many people working in this field are trained in both lead and asbestos removal. Older structures frequently have both types of material.

Since buildings usually cannot be occupied while asbestos is being removed, asbestos removal technicians typically work in teams to complete the work in as short a time as possible. The building or portion of the structure where the work is being done is sealed to prevent the escape of airborne asbestos particles. To protect their health, asbestos removal technicians (and, often, lead removal workers) wear full body suits with an air supply. They physically remove the asbestos materials from the structure, enclosing all debris in sealed containers for safe disposal. Specialized vacuum cleaners with highly efficient filters are used to clean up dust resulting from the removal activities.

Both asbestos and lead removal work can be dangerous, although strict preventive measures are taken to prevent the breathing of toxic dust. As in most construction occupations, working with hand and power tools presents dangers, as do the necessary climbing, lifting, and other physical activities. Working inside full body suits is hot and uncomfortable and may cause claustrophobia in some people. Work hours may often be long, since there is usually time pressure to complete the work. These specialty technicians are generally well-paid, however, ranking with the more well-compensated members of the traditional building trades.

Workers in both asbestos and lead removal must be licensed. Although formal education beyond high school is not needed, licensing requires successful completion of a formal 32-to-40 hour training program. Employment in these fields is expected to continue to grow as more and more aging buildings containing asbestos and lead are demolished or renovated.

Safety Notes. Identify situations or practices that could result in personal injury or damage if proper safety procedures are not followed.

On the Job. Brief profiles of the job specialties found in the construction field. Provides exposure to different career opportunities.

Test Your Knowledge Questions. Designed to reinforce the material covered in the chapter.

Curricular Connections. Activities designed to relate the book's content to other curriculum, such as language, mathematics, science, and social studies.

Outside Assignments. Various tasks and projects designed to help the student expand their knowledge of the chapter content or gain competence in the subject area.

---

120   Section 1   Preparing to Build

Summary

Hand tools are essential to the work of the carpenter. As a skilled worker, the carpenter must carefully select the kind, type, and size of tools that best suit personal requirements. The carpenter works with measuring and layout tools, such as rules and squares. The most common carpentry tool is the claw hammer used to nail wooden building components. The carpenter also uses hand saws; smoothing tools, such as planes; prying tools; and clamping tools. Experienced carpenters take pride in their tools and keep them in good working condition. This may involve cleaning, adjusting, sharpening, and repairing.

Test Your Knowledge

Answer the following questions on a separate piece of paper. Do not write in this book.

1. A standard folding wood rule is _____ feet long.
2. List five uses of the framing or rafter square.
3. The outside edge of the back side of a framing square is graduated in what fractions of an inch?
   A. 1/10"
   B. 1/12"
   C. 1/16"
4. Where would you find the rafter table on a framing square? What would you use it for?
5. When checking structural members to see if they are horizontal or vertical (plumb), what tool(s) should be used?
6. A backsaw is used for fine work. It usually has _____.
   A. 8 points per inch.
   B. 8–10 points per inch.
   C. 12–14 points per inch.
   D. 14–16 points per inch.
7. The bevel of a block plane blade is turned _____.
   A. up
   B. down
8. The size of a claw hammer is determined by the _____.
   A. length of the handle
   B. length of the head
   C. weight of the entire hammer
   D. weight of the head
9. Name the different types of claw hammers and explain the purpose of each.
10. *True or False?* Never strike a wood chisel with a mallet.
11. Where would a carpenter use tin snips?
12. How are sizes of slotted screwdrivers and Phillips screwdrivers determined?
13. The operation of filing off the points of saw teeth until all are level is called _____.

Curricular Connections

Social Studies. Prepare a report on the historical development of woodworking tools used by the carpenter. Reference books on woodworking and encyclopedias will contain helpful material. Make a report to your class using visual aids to show students pictures and drawings of early tools. If possible, secure actual examples from a collector.

Mathematics. Using reference books and manufacturers' catalogs, prepare a list of hand tools you believe the carpenter will need for rough framing and exterior finish of a typical residential structure. Determine the cost of equipping a carpenter with a complete hand tool kit based on your list. Take the list to a hardware store or home improvement center and find the cost of each tool. Add up the costs and report to the class.

Outside Assignments

1. Claw hammers are available in different weights and constructions (one piece, wooden handle, fiberglass handle). Find a carpenter or a knowledgeable clerk at a hardware store or home improvement center. Research the uses and the advantages/disadvantages of each type. Based on your research, choose a hammer with a weight and construction that would be suitable for you to use in framing a house.
2. Research the proper method for cutting a mortise in a door jamb, using a wood chisel and mallet. Make a poster showing the step-by-step procedure.

6

Brief Contents

Contents

8

Section 3 Closing In

Section 4 Finishing

Section 5 Special Construction

Section 6 Mechanical Systems

Acknowledgments

The authors wish to thank the individuals and organizations listed below for the valuable information, photographs, and line illustrations they so willingly provided.

ABTco, Inc., Troy, MI
Ace Hardware, Sister Bay, WI
Acoustical and Board Products Assn., Palatine, IL
Agricultural Extension Service, University of
　　Minnesota, Minneapolis, MN
Gene Ahnen, Bayfield, WI
Alcoa Building Products, Pittsburgh, PA
Alum-A-Pole Corp., Scranton, PA
American Building Components, Omaha, NE
American Institute of Timber Construction,
　　Englewood, CO
American Olean Tile Co., Lansdale, PA
American Standard, Chicago, IL
American Tool, Dewitt, NE
Amerock Corp., Rockford, IL
Amoco Foam Products Co., Atlanta, GA
Andersen Corp., Bayport, ME
APA-The Engineered Wood Association,
　　Tacoma, WA
Architectural Woodwork Institute, Arlington, VA
Arcways, Inc., Neenah, WI
Ark Seal, Inc., Denver, CO
Armstrong World Industries, Inc., Lancaster, PA
Arxx Building Products, Cobourg, ON, Canada
Asphalt Roofing Manufacturers Assn.,
　　New York, NY
Baldwin Hardware Corp., Reading, PA
Bayfield Lumber Company, Bayfield, WI
Benjamin Obdyke, Inc., Warminster, PA
Bird Division, CertainTeed Corp., Norwood, MA
Harry Black, Animosa, IA
Black & Decker, Hunt Valley, MD
Blandin Wood Products Co., Grand Rapids, MI
Boen Hardwood Flooring, Inc., Martinsville, VA
Boise-Cascade Corp., Boise, ID
Borden, Inc., New York, NY
Bosch Power Tool Corp., Mount Prospect, IL
Brammer Mfg. Co., Chicago, IL
Bruce Hardwood Floors, Dallas, TX
Bullard-Haven Technical School, Bridgeport, CT
Calculated Industries, Inc., Yorba Linda, CA

Caradon Peachtree, Inc., Duluth, GA
Cardinal Industries, Inc., Columbus, OH
Carol Electric, Los Alamitos, CA
Cascade Precast Concrete Products, Cascade, IA
C-E Morgan, Oshkosh, WI
Cedar Shake and Shingle Bureau, Sumas, WA
Cella Barr Associates, Tucson, AZ
Cemco, Louisville, KY
CertainTeed Corp., Valley Forge, PA
Colonial Stair and Woodwork Co.,
　　Jeffersonville, OH
Columns, Inc., Pearland, TX
Construction Training School, St. Louis, MO
Council of Forest Industries of British Columbian
　　North Vancouver, BC, Canada
Crown Aluminum Industries, Pittsburgh, PA
Curran & LaPointe, Bayfield, WI
Dal-Tile Corp., Dallas, TX
Daniel Woodhead Co., Northbrook, IL
David White Instruments, Watseka, IL
Deckmaster
Deft, Inc., Irvine, CA
Del Webb's Sun City, Tucson, AZ
Delta International Machinery Corp.,
　　Pittsburgh, PA
Dennis C. Christensen and Son
Des Moines, Iowa, Public Schools, Des Moines, IA
DeVilbiss Co., Toledo, OH
DeWALT Industrial Tool Co., Baltimore, MD
Dexter Industries, Inc., Windsor Locks, CT
Dickinson Homes, Inc., Kingsford, MI
District 1 Technical Institute, Eau Claire, WI
Duo-Fast Corp., Franklin Park, IL
Dutcher Glass and Paint, Cedar Falls, IA
E.I. duPont de Nemours & Co., Wilmington, DE
Eaton Yale & Towne, Inc., White Plains, NY
Ekco Building Products Co., Canton, OH
Enercept, Inc., Watertown, SD
Estwing Mfg. Co., Rockford, IL
Forest Products Laboratory, Madison, WI
Formica Corp., Cincinnati, OH

Forslund Bldg. Supply, Inc., Ashland, WI
Frank Paxton Lumber Co., Des Moines, IA
GAF Building Materials Corp., Wayne, NJ
Gang-Nail Systems, Inc., Miami, FL
The Garlinghouse Company, Chantilly, VA
GB Electrical, Inc., Milwaukee, WI
General Products, Inc., Fredericksburg, VA
Georgia-Pacific Corp., Atlanta, GA
Gold Bond Building Products, Charlotte, NC
Goldblatt Tool Co., Kansas City, KS
Gorrell Windows and Doors
Gory Associated Industries, Inc., North Miami, FL
Greco Painting, Tucson, AZ
Henry Brothers Co.
Grimes-Port-Jones-Schwerdtfeger Architects, Inc.,
    Waterloo, IA
Grosse Steel Co., Cedar Falls, IA
Gypsum Assn., Evanston, IL
H.L. Stud Corporation, Columbus, OH
Haas Cabinet Co., Inc., Sellersburg, IN
Hadley-Hobley Construction, Kansas City, MO
Harrington, Los Angeles, CA
Headlee Roofing, Phoenix, AZ
Henry Brothers Co.
Homewood Scavenger Services, Inc.,
    East Hazelcrest, IL
Ideal Co., Waco, TX
Independent Nail and Packing Co.,
    Bridgewater, MA
Insulspan, Nashville, TN
International Code Council, Falls Church, VA
Iowa Energy Policy Council, Des Moines, IA
ITW Paslode, Lincolnshire, IL
I-XL Furniture Co. Goshen, IN
James Hardie Building Products, Fontana, CA
Jeld-Wen, Inc., Klamath Falls, OR
Johns Manville, Denver, CO
Johnson-Manley Lumber Co., Tucson, AZ
Journal of Light Construction, Williston, VT
Kasten-Weiler Construction, Fish Creek, WI
KCPL, Kansas City, MO
Kentron Div., North American Reiss,
    Belle Mead, NJ
Klein Tools, Chicago, IL
KraftMaid Cabinetry, Inc., Cleveland, OH
Kunkle Valve Co., Inc., Fort Wayne, IN
Kwikset, Lake Forest, CA
L. J. Smith, Inc., Bowerstown, OH
L. S. Starrett Co., Athol, MA
Lennox Industries, Inc., Dallas, TX
Libby-Owens-Ford Co., Glass Div., Toledo, OH
Lifetile, Irvine, CA
Lite-Form International, Sioux City, IA
Loren LaPointe Drywalling, Bayfield, WI
Louisiana-Pacific Corp., Portland, OR
LTL Home Products, Inc., Schuylkill, PA

Luxaire Heating and Air Conditioning, York, PA
Maclanburg-Duncan, Oklahoma City, OK
Majestic Co., Huntington, IN
Makita USA, Inc., La Mirada, CA
Malm Fireplaces, Inc., Santa Rosa, CA
Manufactured Housing Institute, Arlington, VA
Manville Building Products, Inc., Denver, CO
Marshfield Homes, Inc., Marshfield, WI
Marvin Windows and Doors, Warroad, MN
Masonite Corp., Chicago, IL
McDaniels Construction Co., Inc. Columbus, OH
McRae True Value Hardware, Ashland, WI
Mellin Well Service, Ashland, WI
Memphis Hardwood Flooring Co., Memphis, TN
Merillat Industries, Inc., Adrian, MI
Met-Tile Inc., Ontario, CA
Miller Electric Mfg. Co., Appleton, WI
Milwaukee Electric Tool Corp., Brookfield, WI
Moen Incorporated, North Olmstead, OH
Monday Construction, Dauphin Island, AL
Monier Roof Tile, Irvine, CA
Montachusett Regional Vo-Tech School,
    Fitchburg, MA
National Concrete Masonry Assn., Herndon, VA
National Decorating Products Assn., St. Louis, MO
National Forest Products Assn., Washington, DC
National Gypsum Co., Dallas, TX
National Lock Hardware, Milwaukee, WI
National Solar Heating and Cooling Center,
    Rockville, MD
North Bennett Street School, Boston. MA
Oak Flooring Institute, Memphis, TN
Omni Products, Addison, IL
Orem Research, Hinsdale, IL
Owens-Corning, Toledo, OH
Pacific International Tool and Shear, Kingston, WA
Panel Clip Co., Farmington, MI
Paslode Co., Div. of Signode, Vernon Hills, IL
Patent Scaffolding Co., Long Island City, NY
Pease Industries, Inc., Fairfield, OH
Pella Corp., Pella, IA
Pemko Manufacturing, Inc., Ventura, CA
Pergo, Inc., Raleigh, NC
Perlite Institute, Inc., New York, NY
Pierce Construction, Ltd., Green Valley, AZ
Pittsburgh Corning Corp., Pittsburgh, PA
Porter-Cable Corp., Jackson, TN
Portland Cement Assn., Skokie, IL
PPG Industries, Pittsburgh, PA
Preway, Inc., Wisconsin Rapids, WI
RECON —Reconstruction Unlimited, Inc.,
    Columbus, OH
Redman Industries Inc., Dallas, TX
Reed Manufacturing Co., Erie, PA
Rick Pomerlou
Ridge Tool Company, Elyria, OH

Riviera Kitchens, Evans Products Co.,
    Colorado Springs, CO
Riviera Marmitec, Dallas, TX
Robbins/Sykes, Warren, AR
Roberto's Roofing, Tucson, AZ
Rock Island Millwork, Waterloo, IA
Rollex Corp., Elk Grove Village, IL
Ronthor Plastics Div., U.S. Mfg. Corp.,
    Social Circle, GA
Santa Rita High School, Tucson, AZ
Schlage Lock Company, Olathe, KS
Schulte Distinctive Storage, Cincinnati, OH
Senco Fastening Systems, Cincinnati, OH
Shakertown Corp., Cleveland, OH
Sherwin-Williams Co., Cleveland, OH
Ship and Shore General Store, Dauphin Island, GA
Simplex Product Division, Adrian, MI
Simpson Strong-Tie Company, Inc.,
    San Leandro, CA
SkillsUSA, Leesburg, VA
Slant/Fin Corp., Greenvale, NY
Solatube North America, Ltd., Vista, CA
Southern Forest Products Assn., New Orleans, LA
Southern Pine Council
Spectra-Physics Laserplane, Inc., Dayton, OH
Speed Cut, Inc. Corvallis, OR
St. Paul Technical College, St. Paul, MN
Stan Greer Millwork, Sierra Vista, AZ
Stanley Door Systems, Farmington, MI
Stanley Tools, Covington, GA
Sto Finish Systems Division, Atlanta, GA
Structural Board Association, Markham,
    ON, Canada
Superior Fireplace Co., Fullerton, CA
T. W. Lewis Construction Company, Tempe, AZ
Tapco International, Plymouth, MI
TECO, Hayward, CA
The Burke Co., San Mateo, CA
The Flintkote Co., New York, NY
Therma-Tru, Bowling Green, OH
Tibbias Flooring Inc., Oneida, TN

Tilley Ladder Co., Davenport, IA
Timber Engineering Co., Washington, DC
Timberpeg, Claremont, NH
Tony's Construction, Inc., Tucson, AZ
Trex Company, Inc.
Trudeau Construction Co., Bayfield, WI
Truecraft Tools, Somerset, NJ
Trus Joist MacMillan, Boise, ID
Truss Plate Institute, Madison, WI
Trussworks, Inc., Hayward, WI
TrusWal Systems Corp., Troy, MI
U.S. Department of Housing and Urban Development, Washington, DC
USG Acoustical Products Co.
U.S. Gypsum Co., Chicago, IL
United Brotherhood of Carpenters and Joiners of America, Washington, DC
United States Steel Corp., Pittsburgh, PA
United Technologies Carrier, Indianapolis, IN
Universal Form Clamp Co., Chicago, IL
Vermiculite Institute Minneapolis, MN
Vermont American Tool Co., Lincolnton, NC
Village of Flossmoor, Flossmoor, IL
Visador Co., Jacksonville, FL
Waterloo-Cedar Falls Daily Courier, Waterloo, IA
Wausau Homes, Inc., Wausau, WI
Western Forms, Kansas City, MO
Western Wood Products Assn., Portland, OR
Weyerhaeuser Co., Tacoma, WA
Whirlpool Corp., Benton Harbor, MI
William Powell Co., Cincinnati, OH
Wiss, Division of Cooper Tools, Raleigh, NC
Witkowski Construction, Bayfield, WI
Wolmanized Wood, Smyrna, GA
Wood Conversion Co., St. Paul, MN
Wood Flooring Manufacturers Assn. (NOFMA), Memphis, TN
Wood Frame House Construction, U.S. Dept. of Agriculture, Washington, DC
Yankee Barn Homes, Inc., Grantham, NH

Section 1

Preparing to Build

Henry Bros. Co. Photography by Larry Morris

Building Materials

Learning Objectives

After studying this chapter, you will be able to:

- Describe the hardwood and softwood classifications.
- Define moisture content (M.C. and E.M.C.).
- Identify common defects in lumber.
- Define lumber grading terms.
- Calculate lumber sizes according to established industry standards.
- Explain plywood, hardboard, and particleboard grades and uses.
- Identify nail types and sizing units.
- List precautions to observe while working with treated lumber.
- Identify types of engineered lumber and list their uses and advantages.
- Discuss the uses of metal structural materials and describe their advantages/disadvantages.
- Identify a variety of metal framing connectors and indicate where each is used.

Technical Vocabulary

Annular rings
Blue stain
Board foot
Cambium
Casein glue
Composite board
Composite panels
Conifers
Contact cement
Core
Cross-bands
Decay
Deciduous
Edge-grained
Engineered lumber products (ELP)
Equilibrium moisture content (E.M.C.)
Faces
Factory and shop lumber
FAS (firsts and seconds)
Fiber saturation point
Flat-grained
Glue-laminated beams
Glulams
Hardboard
Heartwood
Holes
Honeycombing
Kiln-dried
Knots
Laminated-strand lumber (LSL)
Laminated-veneer lumber (LVL)
Lignin
Lumber
Lumber core
Mastics
Moisture content (M.C.)
No. 1 common
Open-grain woods
Open time
Oriented strand board (OSB)
Parallel lamination
Parallel-strand lumber (PSL)
Particleboard
Phloem
Photosynthesis
Pitch pockets
Plain-sawed
Polyvinyl resin emulsion adhesive
Polyurethane adhesives
Premium grade
Quarter-sawed
Sapwood
Selects
Shakes
Span ratings
Splits and checks
Steel framing members
Tension bridging
Thermoplastic
Thermoset adhesives
Tracheids
Urea-formaldehyde resin glue

Veneer core
Wane
Warp

Wood I-beams
Xylem

Many different types of materials go into the construction of a modern residence, **Figure 1-1.** A carpenter should be familiar with all of these materials. Each has special properties that make it suited to certain building applications.

Before being used in construction, some of these materials are treated or engineered to improve their qualities. Some are composites of different materials, which are designed to do a certain job better than a natural material. Construction materials include:

- Sawed lumber and logs.
- Plywood.
- Composition panels (composite board, particleboard, hardboard, oriented strand board, and fiberboard).
- Engineered lumber.
- Wood and nonwood materials for shingles and flooring.
- Steel and aluminum.
- Metal fasteners.
- Concrete and masonry units.
- Mortar.
- Adhesives and sealers.
- Vapor barriers.
- Gypsum board and other interior coverings.

Figure 1-1. The frame of a house is usually constructed of wood, though other materials are sometimes used. As construction is completed, many different materials will be used. This chapter describes these materials.

1.1 Lumber

Wood is one of our greatest natural resources. When cut into pieces uniform in thickness, width, and length, it becomes lumber. This material has always been widely used for residential construction.

Lumber is the name given to natural or engineered products of the sawmill. Lumber includes:

- Boards used for flooring, sheathing, paneling, and trim.
- Dimension lumber used for sills, plates, studs, joists, rafters, and other framing members.
- Timbers used for posts, beams, and heavy stringers.
- Numerous specialty items.

Carpenters must have a good working knowledge of lumber. They must be familiar with type, grade, size, and other details to properly select and use lumber. To understand how to handle and treat wood, carpenters should also know something about the growth, structure, and characteristics of wood.

1.2 Wood Structure and Growth

Wood is made up of long narrow tubes or cells called fibers or *tracheids.* The cells are no larger around than human hair. Their length varies from about 1/25″ for hardwoods to approximately 1/8″ for softwoods. Tiny strands of cellulose make up the cell walls. The cells are held together with a natural cement called *lignin.* This cellular structure makes it possible to drive nails and screws into the wood. It also

---

Lumber: Logs that have been sawed into usable boards, planks, and timbers. Some planing and matching of ends and edges may be included.

Tracheids: Long narrow tubes or cells that make up wood. Also called fibers.

Lignin: Substance in wood that binds cell walls together.

accounts for the light weight, low heat transmission, and sound absorption qualities of wood. The growing parts of a tree are:
- The tips of the roots.
- The leaves.
- A layer of cells just inside the bark called the *cambium.*

Water absorbed by the roots travels through the sapwood to the leaves, where it is combined with carbon dioxide from the air. Through the chemical process of *photosynthesis,* sunlight changes these elements to a food called carbohydrates. The sap carries this food back to the various parts of the tree.

Figure 1-2 shows the location of the cambium layer, where new cells are formed. The inside area of the cambium layer is called *xylem.* It develops new wood cells. The outside area, known as *phloem,* develops cells that form the bark.

1.2.1 Annular Rings

Growth in the cambium layer takes place in the spring and summer. Separate layers form each season. These layers are called *annular rings.* Each ring is composed of two layers: springwood and summerwood.

Cambium: The layer of wood cells located beneath the bark of a tree.

Photosynthesis: The chemical process that plants use to store the sun's energy. Trees use sunlight to convert carbon dioxide and water into leaves and wood.

Xylem: Inside area of the cambium layer that develops new wood cells.

Phloem: Outside area of the cambium layer of a tree. It develops cells that form the bark.

Annular rings: Rings or layers of wood that represent one growth period of a tree. In cross-section, the rings may indicate the age of the tree.

Sapwood: The layers of wood next to the bark, usually lighter in color than the heartwood, that are actively involved in the life processes of the tree. More susceptible to decay than heartwood, sapwood is not necessarily weaker or stronger than heartwood of the same species.

Heartwood: The wood extending from the pith or center of the tree to the sapwood. Heartwood cells no longer participate in the life processes of the tree.

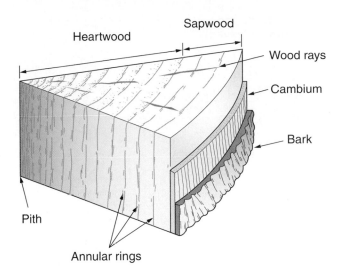

Figure 1-2. A tree is made up of many different layers. Growth takes place in the cambium layer.

In the spring, trees grow rapidly and the cells produced are large and thin-walled. As growth slows down during summer months, the cells produced are smaller, thick-walled, and darker in color. These annual growth rings are largely responsible for the grain patterns that are seen on the surface of boards.

Sapwood is located inside the cambium layer. It contains living cells and may be several inches thick. Sapwood carries sap to the leaves. The *heartwood* of the tree is formed as the sapwood becomes inactive. Usually, it turns darker in color because of the presence of gums and resins. In some woods such as hemlock, spruce, and basswood, there is little or no difference in the appearance of sapwood and heartwood. Sapwood is as strong and heavy as heartwood, but is not as durable when exposed to weather.

1.3 Kinds of Wood

Lumber is either softwood or hardwood. Softwoods come from evergreen, or needle-bearing, trees. These are called *conifers* because many of them bear cones. Hardwoods come from broadleaf (*deciduous*) trees that shed their leaves at the end of the growing season. See **Figure 1-3.** This classification is somewhat confusing because many of

**Hardwood
(broad-leaved)**

**Softwood
(conifers)**

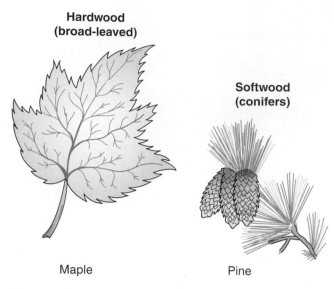

Maple

Pine

Figure 1-3. Woods are classified as either hardwoods or softwoods. Hardwoods have broad leaves; softwoods have needles.

the hardwood trees produce a softer wood than some of the so-called softwood trees. A number of the more common kinds of commercial softwoods and hardwoods are listed in **Figure 1-4.**

A number of hardwoods have large pores in the cellular structure and are called *open-grain woods.* They require additional operations

Softwoods	Hardwoods
Douglas fir Southern pine Western larch	Basswood Willow American elm
Hemlock White fir Spruce	*Mahogany Sweet gum *White ash
Ponderosa pine Western red cedar Redwood	Beech Birch Cherry
Cypress White pine Sugar pine	Maple *Oak *Walnut

*Open-grain wood

Figure 1-4. A list of woods popular for residential use. Usually, softwoods are used as framing lumber.

during finishing. Different kinds of wood will also vary in weight, strength, workability, color, texture, grain pattern, and odor.

Study the full-color wood samples at the end of this chapter as a first step in wood identification. To further develop your ability to identify woods, study actual specimens. Several of the softwoods used in construction work are similar in appearance. Considerable experience is required to make accurate identification.

Most of the samples shown in this text were cut from plain-sawed or flat-grained stock. Edge-grained views look different.

Availability of different species (kinds) of lumber varies from one part of the country to another. This is especially true of framing lumber, which is expensive to transport long distances. It is usually more economical to select building materials found in the area.

1.4 Cutting Methods

Most lumber is cut so that the annular rings form an angle of less than 45° with the surface of the board. This produces lumber called *flat-grained* if it is softwood, or *plain-sawed* if it is hardwood. This method produces the least waste and more desirable grain patterns.

Lumber can also be cut so the annular rings form an angle of more than 45° with the surface of the board, **Figure 1-5.** This method produces lumber called *edge-grained* if it is softwood, and

---

Conifers: Needle-bearing trees, such as evergreens. Softwoods come from these trees.

Deciduous: Broad leaf trees that lose their leaves at the end of the growing season.

Open-grain woods: Woods with large pores, such as oak, ash, chestnut, and walnut.

Flat-grained: Softwood lumber cut so the annular rings form an angle of less than 45° with the board's surface.

Plain-sawed: Hardwood lumber that is cut so that the annular rings form an angle of less than 45° with the surface of the board.

Edge-grained: Boards sawed so that the annular rings form an angle with the face of greater than 45°.

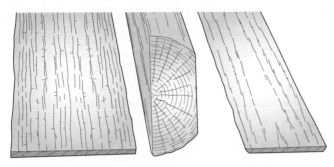

Figure 1-5. Lumber may be sawed in different ways. Left—Board is edge-grained or quarter-sawed. Saw cut was made roughly parallel to a line running through center of log. Right—Flat-grained or plain-sawed board.

quarter-sawed if it is hardwood. It is more difficult and expensive to use this method. However, it produces lumber that swells and shrinks less across its width and is not as likely to warp.

1.5 Moisture Content and Shrinkage

Before wood can be commercially used, a large part of the moisture (sap) must be removed. When a living tree is cut, more than half its weight may be moisture. Lumber used for framing and outside finish should be dried to a moisture content of 15%. Cabinet and furniture woods are dried to a moisture content of 7% to 10%.

The amount of moisture, or *moisture content (M.C.)*, in wood is given as a percent of the oven-dry weight. To determine the moisture content, a sample is first weighed. It is then placed in an oven and dried at a temperature of about 212°F (100°C). The drying is continued until the wood no longer loses weight. The sample is weighed again and this oven-dry weight is subtracted from the initial weight. The difference is then divided by the oven-dry weight, **Figure 1-6.**

Moisture contained in the cell cavities is called *free water.* Moisture contained in the cell walls is called *bound water.* As the wood dries, moisture first leaves the cell cavities. When the cells are empty, but the cell walls are still full of moisture, the wood has reached a condition called the *fiber saturation point.* For most woods, this is about 30%, **Figure 1-7.** The fiber saturation point is important because wood does not start to shrink until this point is reached. As the M.C. drops below 30%, moisture is removed from the cell walls and they shrink. **Figure 1-8** shows the actual shrinkage in a 2 × 10 joist.

Wood shrinks the most along the direction of the annual rings (tangentially) and about one-half as much across these rings. There is little shrinkage in the length. How this shrinkage affects lumber cut from different parts of a log is shown in **Figure 1-9.** As wood takes on moisture, it swells in the same proportion as the shrinkage that took place.

Quarter-sawed: Hardwood cut so the annular rings form an angle of more than 45° with the surface of the board.

Moisture content (M.C.): The amount of water contained in wood, expressed as a percentage of the weight of oven-dry wood.

Fiber saturation point: The stage in the drying or wetting of wood at which the cell walls are saturated and the cell cavities are free from water. It is assumed to be 30% moisture content, based on oven-dry weight, and is the point below which shrinkage occurs.

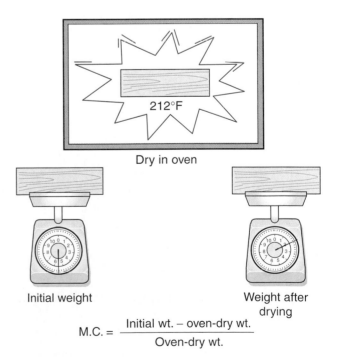

$$M.C. = \frac{\text{Initial wt.} - \text{oven-dry wt.}}{\text{Oven-dry wt.}}$$

Figure 1-6. The moisture content of wood can be determined by weighing it before and after it is dried.

1.5.1 Equilibrium Moisture Content

A piece of wood gives off or takes on moisture from the air around it until the moisture in the wood is balanced with that in the air. At this point the wood is said to be at *equilibrium moisture content (E.M.C.)*.

Since wood is exposed to daily and seasonal changes in the relative humidity of the air, its moisture content is always changing. Therefore, its dimensions are also changing. This is the reason doors and drawers often stick during humid weather.

Ideally, a wood structure should be framed with lumber at an M.C. equal to that of the environment it is located in. This is not practical. Lumber with such low moisture content is seldom available and would likely gain moisture during construction. Standard practice is to use

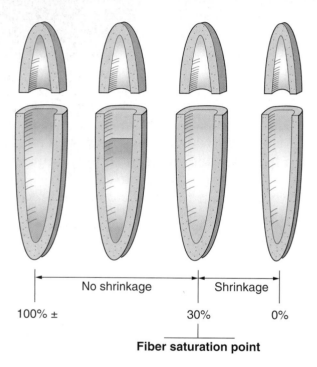

Figure 1-7. How a wood cell dries. First, the free water in the cell cavity is removed. Then, the cell wall dries and shrinks.

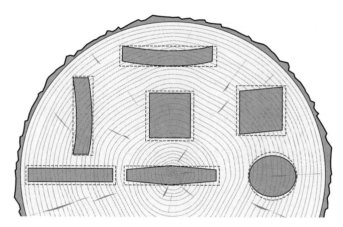

Figure 1-9. Shrinkage and distortion of flat, square, and round pieces is affected by the direction of the annual rings in relation to the lumber piece.

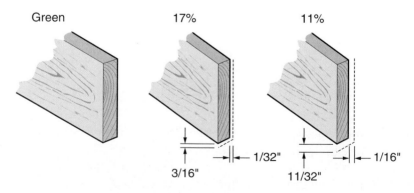

Figure 1-8. A 2 × 10 joist may shrink 1/16″ across its shortest dimension.

Equilibrium moisture content (E.M.C.): The moisture content at which wood neither gains nor loses moisture when surrounded by air at a given relative humidity and temperature.

lumber with a moisture content in the range of 15%–19%. In heated structures, it will eventually reach a level of about 8%. However, this will vary in different geographical areas, **Figure 1-10.**

Carpenters understand that shrinkage is inevitable. They make allowances where needed to reduce the effects of shrinkage. The first, and by far the greatest, change in moisture content occurs during the first year after construction, particularly during the first heating season. When green lumber (more than 20% M.C.) is used, shrinkage is excessive. Warping, plaster cracks, nail pops, squeaky floors, and other difficulties are almost impossible to prevent.

1.5.2 Seasoning Lumber

Seasoning is reducing the moisture content of lumber to the required level specified for its grade and use. In air drying, the lumber is simply exposed to the outside air. It is carefully stacked with *stickers* (wood strips) between layers so air can circulate through the pile. Boards are also spaced within the layers so air can reach edges. Air-drying is a slower process than kiln drying. It often creates additional defects in the wood.

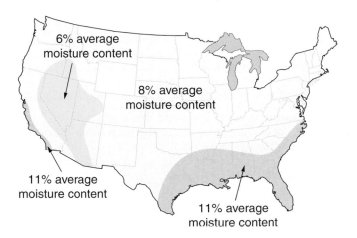

Figure 1-10. This map shows average moisture content of interior woodwork for various regions of the United States.

Kiln-dried: Wood seasoned in a kiln by means of artificial heat, controlled humidity, and air circulation.

Lumber is *kiln dried* by placing it in huge ovens where the temperature and humidity can be carefully controlled. When the green lumber is first placed in the kiln, steam is used to keep the humidity high. The temperature, meanwhile, is kept at a low level. Gradually the temperature is raised while the humidity is reduced. Fans keep the air in constant circulation. Bundles of lumber may carry a stamp to indicate that they have been kiln dried. The letters k.d. mean "kiln dried," while p.k.d. stands for "partly kiln dried."

1.5.3 Moisture Meters

The moisture content of wood can be determined by:
- Oven drying a sample as previously described.
- Using an electronic moisture meter.

Although the oven drying method is more accurate, meters are often used because readings can be secured rapidly and conveniently. The meters are usually calibrated to cover a range from 7% to 25%. Accuracy is plus or minus 1% of the moisture content.

There are two basic types of moisture meters. One determines the moisture content by measuring the electrical resistance between two pin-type electrodes that are driven into the wood. The other type measures the capacity of a condenser in a high-frequency circuit where the wood serves as the *dielectric* (nonconducting) material of the condenser. See **Figure 1-11.**

1.6 Lumber Defects

A *defect* is an irregularity occurring in or on wood that reduces its strength, durability, or usefulness. It may also detract from or improve the wood's appearance. For example, knots are commonly considered a defect, but may add to the appearance of pine paneling. Lumber with certain defects, particularly knots, should not be used where strength is important. An imperfection that impairs only the appearance of wood is called a *blemish*. Some common defects include:

Figure 1-11. An electronic moisture meter uses LEDs to indicate moisture content. This model has probes that are inserted into the cut end. (Forestry Suppliers, Inc.)

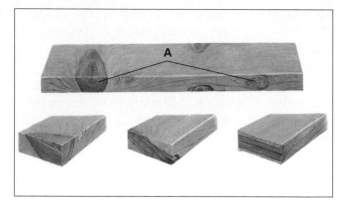

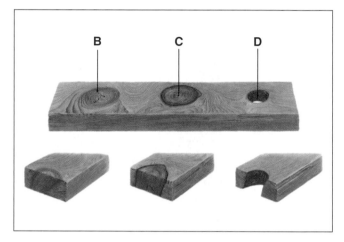

Figure 1-12. Types of knots. A—Sound intergrown through two faces. B—Sound star-checked, intergrown through two wide faces. C—Sound encased through two wide faces. D—Round knothole through two wide faces. (Western Wood Products Assn.)

- *Knots*—Caused by an embedded branch or limb. Knots reduce the strength of the wood. The extent of damage depends on the type of knot, its size, and its location. See **Figures 1-12** and **1-13**.

- *Splits and checks*—Separations of the wood fibers that run along the grain and across the annular growth rings. Splits usually occur at the ends of lumber that has been unevenly seasoned.

- *Shakes*—Separations along the grain and between the annular growth rings. Shakes are likely to occur only in species with abrupt change from spring to summer growth.

- *Pitch pockets*—Cavities that contain or have contained pitch in solid or liquid form.

- *Honeycombing*—Separation of the wood fibers inside the tree. Honeycombing may not be visible on the board's surface.

- *Wane*—The presence of bark or the absence of wood along the edge of a board; wane forms a bevel and/or reduces width.

- *Blue stain*—Caused by a mold-like fungus. Blue stain discoloration has little or no effect on the strength of the wood. However, it is objectionable for appearance in some grades of lumber.

- *Decay*—Disintegration of wood fibers due to fungi. Decay is often difficult to recognize

Knot: Lumber defect caused by a branch or limb embedded in the tree and cut through during lumber manufacture.

Splits and checks: Separations that run along the grain and across the annular growth rings.

Shakes: Handsplit shingles.

Pitch pockets: Cavities that contain or have contained pitch in solid or liquid form.

Honeycombing: Separation of wood fibers inside the tree, but not always visible in the lumber.

Wane: Bark or absence of wood on the edge of lumber.

Blue stain: A stain caused by a fungus growth in unseasoned lumber, especially pine. It does not affect the strength of the wood.

Decay: Disintegration of wood substance due to action of wood-destroying fungi.

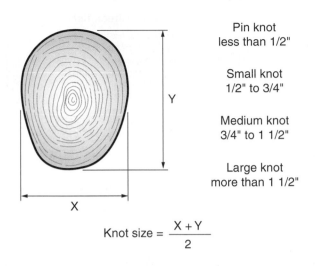

Pin knot
less than 1/2"

Small knot
1/2" to 3/4"

Medium knot
3/4" to 1 1/2"

Large knot
more than 1 1/2"

$$\text{Knot size} = \frac{X + Y}{2}$$

Figure 1-13. Knot size is determined by adding together the width and length of the knot and dividing by two.

A

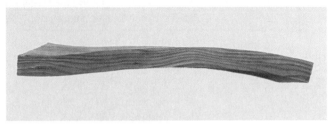

B

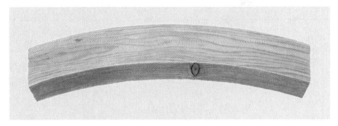

C

Figure 1-14. Warping affects quality, strength, and appearance. A—Crook. B—Twist or wind. C—Bow. (Western Wood Products Assn.)

in its early stages. In advanced stages, wood is soft, spongy, and easily crumbles.

- *Holes*—Openings caused by handling equipment or boring insects and worms. These lower the lumber grade.
- *Warp*—Any variation from true or plane surface. Warp may include any one or combinations of the following: cup, bow, crook, and twist (or wind). See **Figure 1-14.**

1.7 Softwood Grades

The U.S. Department of Commerce publishes basic principles of grading lumber, which are set down by the American Lumber Standards

Holes: Openings in lumber caused by handling equipment or boring insects and worms.

Warp: Any variation from true or plane surface, warp may include any one or combinations of the following: cup, bow, crook, and twist (or wind).

Factory and shop lumber: A grade of lumber intended to be cut up for use in further manufacture. It is graded on the basis of the percentage of the area that will produce a limited number of cuttings of a specified size and quality.

Committee. Various associations of lumber producers—Western Wood Products Association, Southern Forest Products Association, California Redwood Association, and similar groups—develop and apply detailed rules. These agencies publish grading rules for the species of lumber produced in their regions. They also have qualified personnel who supervise grading standards at sawmills.

Basic classifications of softwood grading include boards, dimension, and timbers. The grades within these classifications are shown in **Figure 1-15.**

Another classification, called *factory and shop lumber,* is graded primarily for remanufacturing purposes. Millwork plants use it to fabricate windows, doors, moldings, and other trim items.

Boards

Appearance grades	Selects	B & Better (IWP – supreme) C select (IWP – choice) D select (IWP – quality)	**Specification check list** ☐ Grades listed in order of quality. ☐ Include all species suited to project. ☐ For economy, specify lowest grade that will satisfy job requirement. ☐ Specify surface texture desired. ☐ Specify moisture content suited to project. ☐ Specify Ⓦ grade stamp. For finish and exposed pieces, specify stamp on back or ends.
	Finish	Superior Prime E	
	Paneling	Clear (any select or finish grade) No. 2 common selected for knotty paneling No. 3 common selected for knotty paneling	
	Siding (bevel, bungalow)	Superior Prime	**Western red cedar**

Boards sheathing		No. 1 common (IWP – colonial) **Alternate board grades**
		No. 2 common (IWP – sterling) Select merchantable
		No. 3 common (IWP – standard) Construction
		No. 4 common (IWP – utility) Standard
		                                              Utility
			Finish **Clear heart**
			Paneling **A**
			and ceiling **B**
			
			Bevel Clear – V.G. heart
			siding A – Bevel siding
			            B – Bevel siding
			            C – Bevel siding

Dimension

Light framing 2" to 4" thick 2" to 4" wide	Construction Standard Utility Economy	This category for use where high strength values are NOT required; such as studs, plates, sills, cripples, blocking, etc.
	Stud Economy stud	An optional all-purpose grade limited to 10 feet and shorter. Characteristics affecting strength and stiffness values are limited so that the "Stud" grade is suitable for all stud uses, including load bearing walls.
Structural light framing 2" to 4" thick 2" to 4" wide	Select structural No. 1 No. 2 No. 3 Economy	These grades are designed to fit those engineering applications where higher bending strength ratios are needed in light framing sizes. Typical uses would be for trusses, concrete pier wall forms, etc.
Structural joists and planks 2" to 4" thick 6" and wider	Select structural No. 1 No. 2 No. 3 Economy	These grades are designed especially to fit in engineering applications for lumber six inches and wider, such as joists, rafters, and general framing uses.

Timbers

Beams and stringers	Select structural No. 1 No. 2 (No. 1 mining) No. 3 (No. 2 mining)	Posts and timbers	Select structural No. 1 No. 2 (No. 1 mining) No. 3 (No. 2 mining)

Figure 1-15. Softwood lumber classifications and grades. Names of grades and their specifications will vary among lumber manufacturers' associations and among regions producing lumber. The grade listed on top is better than the grades below it for each classification.

The carpenter must understand that quality construction does not require that all lumber be of the best grade. Today, lumber is graded for specific uses. In a given structure, several grades may be appropriate. The key to good economical construction is matching the proper grade with its function.

1.8 Hardwood Grades

The National Hardwood Lumber Association establishes grades for hardwood lumber. *FAS (firsts and seconds)* is the best grade. It specifies that pieces be no less than 6″ wide by 8′ long and yield at least 83 1/3% clear cuttings. The next lower grade, *selects,* permits pieces 4″ wide by 6′ long. A still lower grade is *No. 1 common.* Lumber in this group is expected to yield 66 2/3% clear cuttings.

1.9 Lumber Stress Values

In softwood lumber, all dimension and timber grades, except economy and mining grades, are assigned stress values. Slope of grain, knot sizes, and knot locations are critical considerations.

There are two methods of assigning stress values:
- Visual.
- Machine rated.

FAS (firsts and seconds): Best grade of hardwood lumber. It specifies that pieces be no less than 6″ wide by 8′ long and yield at least 83 1/3% clear cuttings.

Selects: The hardwood grade below first and seconds.

No. 1 common: Low grade of lumber which is expected to yield 66 2/3% clear cuttings.

Board foot: The equivalent of a board 1′ square and 1″ thick.

In the machine-rated method, lumber is fed into a special machine and subjected to bending forces. The stiffness of each piece (modulus of elasticity, E) is measured and marked on each piece. Machine stress-rated lumber (MSR) must also meet certain visual requirements.

1.10 Lumber Sizes

When listing and calculating the size and amount of lumber, the nominal dimension is always used. **Figure 1-16** lists the nominal and dressed sizes for various classifications of lumber. Note that nominal sizes are sometimes listed in quarters. For example: 1 1/4″ material is given as 5/4. This nominal dimension is its rough unfinished measurement (width and thickness), **Figure 1-17.** The dressed size is less than the nominal size as a result of seasoning and surfacing. Dressed sizes of lumber, established by the American Lumber Standards, are consistently applied throughout the industry.

1.11 Calculating Board Footage

The unit of measure for lumber is the *board foot.* This is a piece 1″ thick and 12″ square or its equivalent (144 cu. in.). Standard-size pieces can be quickly calculated by visualizing the board feet included. For example: a board 1″ × 12″ and 10′ long contains 10 bd. ft. If the piece is only 6″ wide, it contains 5 bd. ft. If the original board is 2″ thick, it contains 20 bd. ft.

The following formula can be applied to calculate board feet. *T* represents the thickness of the board (in.), *W* is the width (in.), and *L* is the length (ft.):

$$\text{bd. ft.} = \frac{\text{No. pcs.} \times T \times W \times L}{12}$$

An example of the application of the formula follows: find the number of board feet in six pieces of lumber that measure 1″ × 8″ × 14′:

Board Measure

The term "board measure" indicates that a board foot is the unit for measuring lumber. A board foot is one inch thick and 12 inches square.

The number of board feet in a piece is obtained by multiplying the normal thickness in inches by the nominal width in inches by the length in feet and dividing by 12:

$$\frac{(T \times W \times L)}{12}$$

Lumber less than one inch in thickness is figured as one-inch.

Product classification

	Thickness	Width		Thickness	Width
Board lumber	1"	2" or more	Beams and stringers	5" and thicker	more than 2" greater than thickness
Light framing	2" to 4"	2" to 4"	Posts and timbers	5" x 5" and larger	not more than 2" greater than thickness
Studs	2" to 4"	2" to 4" 10' and shorter	Decking	2" to 4"	4" to 12" wide
Structural light framing	2" to 4"	2" to 4"	Siding	thickness expressed by dimension of butt edge	
Joist and planks	2" to 4"	6" and wider	Mouldings	size at thickest and widest points	
Lengths of lumber generally are 6 feet and longer in multiples of 2'					

Standard Lumber Sizes/Nominal, Dressed, Based on WWPA Rules

Product	Description	Nominal Size		Dressed Dimensions		Lengths feet
				Thickness and Widths inches		
		Thickness inches	Width inches	Surfaced Dry	Surfaced Unseasoned	
Framing	S4S...............	2	2	1-1/2	1-9/16	6' and longer in multiples of 1'
		3	3	2-1/2	2-9/16	
		4	4	3-1/2	3-9/16	
			6	5-1/2	5-5/8	
			8	7-1/4	7-1/2	
			10	9-1/4	9-1/2	
			12	11-1/4	11-1/2	
			Over 12	Off 3/4	Off 1/2	
Timbers	Rough or S4S...............	5 and larger		Thickness inches	Width inches	Same
				1/2 off nominal		

| | | Nominal Size | | Dressed Dimensions Surfaced Dry | | |
		Thickness inches	Width inches	Thickness inches	Width inches	Lengths feet
Decking Decking is usually surfaced to single T&G in 2" thickness and double T&G in 3" and 4" thicknesses	2" single T&G...............	2	6	1 1/2	5	6' and longer in multiples of 1'
			8		6 3/4	
			10		8 3/4	
			12		10 3/4	
	3" and 4" double T&G...............	3	6	2 1/2	5 1/4	
		4		3 1/2		
Flooring	(D & M), (S2S and CM)...............	3/8	2	5/16	1 1/8	4' and longer in multiples of 1'
		1/2	3	7/16	2 1/8	
		5/8	4	9/16	3 1/8	
		1	5	3/4	4 1/8	
		1 1/4	6	1	5 1/8	
		1 1/2		1 1/4		
Ceiling and partition	(S2S and CM)...............	3/8	3	5/16	2 1/8	4' and longer in multiples of 1'
		1/2	4	7/16	3 1/8	
		5/8	5	9/16	4 1/8	
		3/4	6	11/16	5 1/8	
Factory and shop lumber	S2S...............	1 (4/4)	5 and wider (4" and wider in 4/4 No. 1 Shop and 4/4 No. 2 Shop)	25/22 (4/4)	Usually sold random width	4' and longer in multiples of 1'
		1 1/4 (5/4)		1 5/22 (5/4)		
		1 1/2 (6/4)		1 13/32 (6/4)		
		1 3/4 (7/4)		1 19/32 (7/4)		
		2 (8/4)		1 13/16 (8/4)		
		2 1/2 (10/4)		2 3/8 (10/4)		
		3 (12/4)		2 3/4 (12/4)		
		4 (16/4)		3 3/4 (16/4)		

Abbreviations

Abbreviated descriptions appearing in the size table are explained below.
S1S – Surface one side.
S2S – Surfaced two sides.

S4S – Surfaced four sides.
S1S1E – Surfaced one side, one edge.
S1S2E – Surfaced one side, two edges.
CM – Center matched.

D & M – Dressed and matched.
T & G – Tongue and grooved.
EV1S – Edge vee on one side.
S1E – Surfaced one edge.

Figure 1-16. Standard lumber sizes are set by government agencies and lumber associations. (Western Wood Products Assn.)

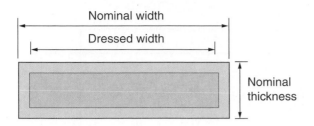

Figure 1-17. Nominal size is greater than dressed size.

$$\text{bd. ft.} = \frac{6 \times 1 \times 8 \times 14}{12}$$

$$\text{bd. ft.} = \frac{\overset{1}{\cancel{6}} \times 1 \times 8 \times 14}{\underset{2}{\cancel{12}}}$$

$$\text{bd. ft.} = \frac{1 \times 1 \times \overset{4}{\cancel{8}} \times 14}{\underset{1}{\cancel{2}}}$$

$$\text{bd. ft.} = \frac{1 \times 1 \times 4 \times 14}{1}$$

$$= 56 \text{ bd. ft.}$$

Stock less than 1″ thick is figured as though it is 1″. When the stock is thicker than 1″, the nominal size is used. When this size contains a mixed fraction, such as 1 1/4, change it to an improper fraction (5/4) and place the numerator above the formula line and the denominator below. For example: find the board feet in two pieces of lumber that measure 1 1/4″ × 10″ × 8′:

$$\text{bd. ft.} = \frac{2 \times 5 \times 10 \times 8}{4 \times 12}$$

$$\text{bd. ft.} = \frac{2 \times 5 \times \overset{5}{\cancel{10}} \times \overset{2}{\cancel{8}}}{\underset{1}{\cancel{4}} \times \underset{6}{\cancel{12}}}$$

$$\text{bd. ft.} = \frac{\overset{1}{\cancel{2}} \times 5 \times 5 \times 2}{1 \times \underset{3}{\cancel{6}}}$$

$$\text{bd. ft.} = \frac{1 \times 5 \times 5 \times 2}{1 \times 3}$$

$$= \frac{50}{3}$$

$$= 16 \text{ 2/3 bd. ft.}$$

Use the nominal size of the material when figuring the footage. Items such as moldings, furring strips, and rounds are priced and sold by the lineal foot. Thickness and width are disregarded.

1.12 Metric Lumber Measure

Metric-size lumber gives thickness and width in millimeters (mm) and length in meters (m). There is little difference between metric and conventional dimensions for common sizes of lumber. For example, the common 1 × 4 board is about 25 mm × 100 mm. Visually, they appear to be about the same size. Metric lumber lengths start at 1.8 m (about 6′) and increase in steps of 300 mm (about a foot) to 6.3 m. This is a little more than 20′. See **Figure 1-18** for a chart of standard sizes. Metric lumber is sold by the cubic meter (m³). See Appendix B, **Technical Information** for more details.

1.13 Panel Materials

Wood panels for construction are manufactured in several different ways:
- Plywood. Thin sheets are laminated to various thicknesses.
- Composite plywood. Veneer faces are bonded to different kinds of wood cores.
- Nonveneered panels. Includes particleboard, fiberboard, and oriented strand board.

1.13.1 Plywood

Plywood is constructed by gluing together a number of thin layers (plies) of wood with the grain direction turned at right angles in each successive layer. An odd number (3, 5, 7) of plies is used so they are balanced on either side of

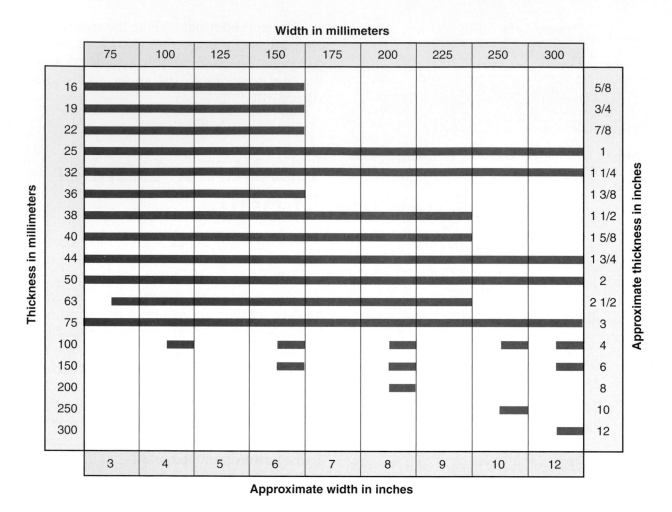

Figure 1-18. This chart compares metric and inch lumber sizes. Dimensions of metric lumber are given in millimeters. Lengths are always given in meters and range from 1.8 m to 6.3 m in increments of 0.3 m (about 1′).

a center core and so the grain of the outside layers run in the same direction. The outer plies are called *faces* or face and back. The next layers under these are called *cross-bands;* the other inside layer or layers are called the *core.* See **Figure 1-19.** A thin plywood panel made of three layers consists of two faces and a core. There are two basic types of plywood:

- Exterior plywood is bonded with waterproof glues. It can be used for siding, concrete forms, and other construction where it will be exposed to the weather or excessive moisture.

- Interior plywood may be manufactured with either interior or exterior glue. Exterior glue is generally used. The grade of the panel is mainly determined by the quality of the layers. Plywood is made in thicknesses from 1/8″ to more than 1″. Common sizes are 1/4″, 3/8″, 1/2″, and 3/4″. Thicknesses of 15/32″, 19/32″, and 23/32″ are also available. A standard panel size is 4′ wide by 8′ long.

Metric plywood panels are 1200 mm × 2440 mm, slightly smaller than the standard 4′ × 8′ panel. **Figure 1-20** compares metric and conventional panel sizes.

Faces: Outer plies of plywood.

Cross-bands: Inner layers of plywood between the face plies and the core.

Core: Innermost layer(s) of plywood.

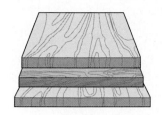

Plywood: This staple of the construction industry is made by peeling logs and laying up the veneers at right angles to each other for rigidity and strength.

A

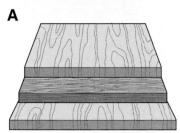

Composite panel: This structural panel closely resembles plywood, except the middle "ply" is a core made of oriented wood fibers.

B

Figure 1-19. Panel construction. A—Standard plywood construction. B—Composite panel plywood construction. (Georgia-Pacific Corp.)

Softwood plywood for general construction is manufactured in accordance with *U.S. Product Standard PS 1-95 for Construction and Industrial Plywood.* This provides for designating species, strength, type of glue, and appearance.

Many species of softwood are used in making plywood. There are five separate plywood groups based on stiffness and strength. Group 1 includes the stiffest and strongest woods. See **Figure 1-21.**

1.13.2 Grade-Trademark Stamp

Structural plywood panels are quality graded in two ways:

- A lettering system indicates the veneer quality for the face and back of the panel, **Figure 1-22.**
- A name indicates the intended use or "performance rating."

The APA-The Engineered Wood Association has a rigid testing program based upon standard PS 1-95. Mills that are members of the association may use the official grade-trademark. It is stamped on each piece of plywood. **Figure 1-23** lists some typical engineered grades of plywood. Included are descriptions and the most common uses.

Panel nomimal thickness

in.	mm
1/4	6.4
5/16	7.9
11/32	8.7
3/8	9.5
7/16	11.1
15/32	11.9
1/2	12.7
19/32	15.1
5/8	15.9
23/32	18.3
3/4	19.1
7/8	22.2
1	25.4
1 3/32	27.8
1 1/8	28.6

A

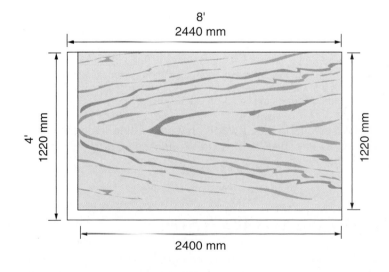

B

Figure 1-20. Comparisons of conventional and metric thicknesses/sizes of plywood panels. A—Panel nominal thicknesses. B—Sizes for a standard plywood panel.

Group 1	Group 2	Group 3	Group 4	Group 5
Apitong	Cedar, Port	Alder, Red	Aspen	Basswood
Beech,	Orford	Birch, Paper	Bigtooth	Poplar,
American	Cypress	Cedar, Alaska	Quaking	Balsam
Birch	Douglas	Fir,	Cativo	
Sweet	Fir 2[a]	Subalpine	Cedar	
Yellow	Fir	Hemlock,	Incense	
Douglas	Balsam	Eastern	Western	
Fir 1[a]	California	Maple,	Red	
Kapur	Red	Bigleaf	Cottonwood	
Keruing	Grand	Pine	Eastern	
Larch,	Noble	Jack	Black	
Western	Pacific	Lodgepole	(Western	
Maple, Sugar	Silver	Ponderosa	Poplar)	
Pine	White	Spruce	Pine	
Caribbean	Hemlock,	Redwood	Eastern	
Ocote	Western	Spruce	White	
Pine, Southern	Lauan	Englemann	Sugar	
Loblolly	Almon	White		
Longleaf	Bagtikan			
Shortleaf	Mayapis			
Slash	Red Lauan			
Tanoak	Tangile			
	White Lauan			
	Maple, Black			
	Mengkulang			
	Meranti, Red[b]			
	Mersawa			
	Pine			
	Pond			
	Red			
	Virginia			
	Western			
	White			
	Spruce			
	Red			
	Sitka			
	Sweetgum			
	Tamarack			
	Yellow poplar			

(a) Refers to Douglas fir from the states of Washington, Oregon, California, Idaho, Montana, and Wyoming, as well as from Alberta and British Columbia shall be classified as Douglas Fir No. 1. Those grown in Nevada, Utah, Colorado, Arizona, and New Mexico shall be classified as Douglas Fir No. 2. (b) Red Meranti shall be limited to species with a specific gravity of 0.41 or more, based on green volume and oven-dry weight.

Figure 1-21. Classification of softwood plywood, rating species for strength and stiffness. Group 1 represents the strongest woods. (APA-The Engineered Wood Association)

1.13.3 Exposure Ratings

The grade-trademark stamp gives an "exposure durability" classification to plywood. There are two basic categories:
- Exterior type, which has 100% waterproof glueline.
- Interior type, with moisture-resistant glueline.

However, panels can be manufactured in three exposure durability classifications:
- Exterior—Fully waterproof adhesives, for use where subject to permanent exposure to weather or moisture.
- Exposure 1—Waterproof bond designed for use where long construction delays are expected before providing protection.

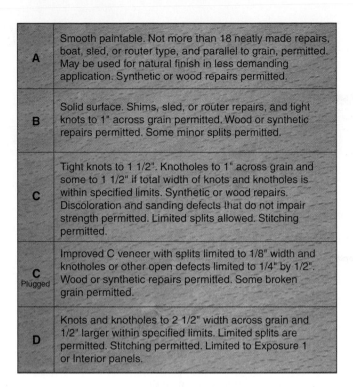

A	Smooth paintable. Not more than 18 neatly made repairs, boat, sled, or router type, and parallel to grain, permitted. May be used for natural finish in less demanding application. Synthetic or wood repairs permitted.
B	Solid surface. Shims, sled, or router repairs, and tight knots to 1" across grain permitted. Wood or synthetic repairs permitted. Some minor splits permitted.
C	Tight knots to 1 1/2". Knotholes to 1" across grain and some to 1 1/2" if total width of knots and knotholes is within specified limits. Synthetic or wood repairs. Discoloration and sanding defects that do not impair strength permitted. Limited splits allowed. Stitching permitted.
C Plugged	Improved C veneer with splits limited to 1/8" width and knotholes or other open defects limited to 1/4" by 1/2". Wood or synthetic repairs permitted. Some broken grain permitted.
D	Knots and knotholes to 2 1/2" width across grain and 1/2" larger within specified limits. Limited splits are permitted. Stitching permitted. Limited to Exposure 1 or Interior panels.

Figure 1-22. Description of veneer grades for softwood plywood. (APA-The Engineered Wood Association)

- Exposure 2—Identified as interior type with intermediate glue for protected applications in which high humidity or water leaks may exist.

Most plywood is manufactured with waterproof exterior glue. However, interior panels may be manufactured with intermediate or interior glue.

1.13.4 Span Ratings of Plywood

In 1980, the APA issued APA PRP-108, *Performance Standards and Policies for Structural Use Panels.* This standard was a response to a need to conserve lumber without affecting the strength of structural plywood. At that time, APA also began placing its trademark on

---

Span rating: Maximum distance a structural member can extend without support when the long dimension is at right angles to its supports.

panel thicknesses of 15/32", 19/32", and 23/32". Today, the trademark is placed on sheathing, floor underlayment, and siding panels with numbers called *span ratings.* This rating is the maximum recommended center-to-center distance (in inches) between supports when the long dimension of the panel is at a right angle to the supports.

A typical stamp for an engineered grade of structural panel is shown in **Figure 1-24.** On the right, a fraction indicates the thickness of the panel (for example, 7/16 means the panel is seven-sixteenths of an inch thick). To the left is a pair of numbers separated by a slash mark (/). The number on the left indicates the maximum recommended span in inches when the plywood is used as roof decking (sheathing). The number on the right indicates the maximum recommended span when the plywood is used as subflooring. The rating applies only when the sheet is placed with its long dimension spanning three or more supports. Generally, the larger the span rating, the greater the panel's stiffness. Panel products intended strictly for use as wall sheathing have only one number.

APA-rated sheathing is typically manufactured with span ratings of 24/0, 24/16, 40/20, or 48/24. Typical Sturd-I-Floor ratings are 16", 20", 24", 32", and 48". This single number is the recommended center-to-center spacing of floor joists when panels are installed with the long dimension at a right angle to the supports.

Siding is manufactured with ratings of 16" and 24". Panel and lap siding rated at 16" may be applied directly to studs spaced at 16" O.C. Panel and lap siding rated at 24" can be applied to studs spaced 24" O.C. When applied over nailable structural sheathing, the span rating indicates the maximum recommended spacing of vertical rows of nails.

1.13.5 HDO and MDO Plywood

HDO and MDO plywood combine the properties of exterior type plywood and plywood with the superior wearability of overlaid surfaces. MDO (medium density overlay) is an exterior-type panel manufactured with 100%

Panel name	Typical trademarks	Descriptions/sizes/uses
APA RATED SHEATHING Typical Trademark	**APA** THE ENGINEERED WOOD ASSOCIATION — RATED SHEATHING 40/20 19/32 INCH SIZED FOR SPACING EXPOSURE 1 — 000 — PS-2-92 SHEATHING PRP-108 HUD-UM-40C · **APA** THE ENGINEERED WOOD ASSOCIATION — RATED SHEATHING 24/16 7/16 INCH SIZED FOR SPACING EXPOSURE 1 — 000 — PRP-108 HUD-UM-40C	Specially designed for subflooring and wall and roof sheathing. Also good for a broad range of other construction and industrial applications. Can be manufactured as plywood, as a composite, or as OSB. EXPOSURE DURABILITY CLASSIFICATION: Exterior, Expose 1, Exposure 2. COMMON THICKNESSES: 5/16, 3/8, 7/16, 15/32, 1/2, 19/32, 5/8, 23/32, 3/4.
APA STRUCTURAL I RATED SHEATHING[a] Typical Trademark	**APA** THE ENGINEERED WOOD ASSOCIATION — RATED SHEATHING STRUCTURAL I 32/16 15/32 INCH SIZED FOR SPACING EXPOSURE 1 — 000 — PS 1-95 C-D PRP-108 · **APA** THE ENGINEERED WOOD ASSOCIATION — RATED SHEATHING 32/16 15/32 INCH SIZED FOR SPACING EXPOSURE 1 — 000 — STRUCTURAL I RATED DIAPHRAGMS-SHEAR WALLS PANELIZED ROOFS — PRP-108 HUD-UM-40C	Unsanded grade for use where shear and cross-panel strength properties are of maximum importance, such as panalized roofs and diaphragms. Can be manufactured as plywood, as a composite, or as OSB. EXPOSURE DURABILITY CLASSIFICATION: Exterior, Expose 1, COMMON THICKNESSES: 5/16, 3/8, 7/16, 15/32, 1/2, 19/32, 5/8, 23/32, 3/4.
APA RATED STURD-I-FLOOR Typical Trademark	**APA** THE ENGINEERED WOOD ASSOCIATION — RATED SHEATHING 24 OC 23/32 INCH SIZED FOR SPACING T&G NET WIDTH 47-1/2 EXPOSURE 1 — 000 — PS-2-92 SINGLE FLOOR PRP-108 HUD-UM-40C · **APA** THE ENGINEERED WOOD ASSOCIATION — RATED SHEATHING 20 OC 19/32 INCH SIZED FOR SPACING T&G NET WIDTH 47-1/2 EXPOSURE 1 — 000 — PRP-108 HUD-UM-40C	Specially designed as combination subfloor-underlayment. Provides smooth surface for application of carpet and pad and possesses high concentrated and impact load resistance. Can be manufactured as plywood, as a composite, or as OSB. EXPOSURE DURABILITY CLASSIFICATION: Exterior, Expose 1, Exposure 2. COMMON THICKNESSES: 19/32, 5/8, 23/32, 3/4, 1, 1-1/8.
APA RATED SIDING Typical Trademark	**APA** THE ENGINEERED WOOD ASSOCIATION — RATED SIDING 24 OC 19/32 INCH SIZED FOR SPACING EXPOSURE 1 — 000 — PRP-108 HUD-UM-40C · **APA** THE ENGINEERED WOOD ASSOCIATION — RATED SIDING 303-18-S/W 16 OC 11/32 INCH GROUP 1 SIZED FOR SPACING EXTERIOR — 000 — PS 1-95 PRP-108 FHA-UM-64	For exterior siding, fencing, etc. Can be manufactured as plywood, as a composite or as an overlaid OSB. Both panel and lap siding available. Special surface treatment such as V-groove, channel groove, deep groove (such as APA Texture 1-11), brushed, rough sawn and overlaid (MDO) with smooth-or texture-embossed face. Span Rating (stud spacing for siding qualified for APA Sturd-I-Wall applications) and face grade classification (for veneer-faced siding) indicated in trademark. EXPOSURE DURABILITY CLASSIFICATION: Exterior, COMMON THICKNESSES: 11/32, 3/8, 7/16, 15/32, 1/2, 19/32, 5/8.

Specific grades, thickness and exposure durability classifications may be in limited supply in some areas. Check with your supplier before specifying.

Specify Performance Rated Panels by thickness and Span Rating. Span Ratings are based on panel strength and stiffness. Since these properties are a function of panel composition and configuration as well as thickness, the same Span Rating may appear on panels of different thickness. Conversely, panels of the same thickness may be marked with different Span Ratings.

(a) All plies in Structural I plywood panels are special improved grades and panels marked PS I are limited to Group I species. Other panels marked Structural I Rated qualify through special performance testing. Structural II plywood panels are also provided for, but rarely manufactured. Application recommendations for Structural II plywood are identical to those for APA RATED SHEATHING plywood.

Figure 1-23. Table shows representative engineered softwood plywood panels manufactured to APA standards. Panels are stamped with the APA trademark. Note the span ratings on the trademarks. (APA-The Engineered Wood Association) waterproof adhesive. It readily accepts paint and is suited for structural siding, exterior color accented panels, soffits, and other applications where long-lasting paint or coating performance is required. See **Figure 1-25.**

MDO panels that are to be used outside should be edge-sealed as soon as possible. One or two coats of high-quality exterior house paint primer formulated for wood should be used. Edges are more easily sealed while the panels are in a stack.

HDO (high-density overlay) plywood is manufactured with a thermosetting resin-impregnated fiber surface that is bonded under heat and pressure to both sides of the panel. It is more rugged than MDO and suited for such punishing applications as concrete forming. The resin overlay requires no additional finish and resists abrasion, moisture penetration, and damage from common chemicals and solvents.

1.13.6 Hardwood Plywood Grades

The Hardwood Plywood Institute uses a numbering system for grading the faces and backs of panels. A grading specification of 1-2 indicates a good face with grain carefully matched and a good back, but without grain carefully matched. A No. 3 back permits noticeable defects and patching, but is generally sound. A special or *premium grade* of hardwood is known as "architectural" or "sequence-matched." This usually requires an order to a plywood mill for a series of matched plywood panels.

Premium grade: A special grade of hardwood known as "architectural" or "sequence-matched." It usually requires an order to a plywood mill for a series of matched plywood panels.

Veneer core: Hardwood plywood core construction that is inexpensive, fairly stable, and warp resistant.

Lumber core: Hardwood plywood core that is easier to cut, the edges are better for shaping and finishing, and they hold nails and screws better than a veneer core.

For either softwood or hardwood plywood, it is common practice to designate the grade by a symbol. G2S means "good two sides." G1S means "good one side."

In addition to the various types and grades, hardwood plywood is made with different core constructions. The two most common are the *veneer core* and the *lumber core.* See **Figure 1-26.** Veneer cores are the least expensive, are fairly stable, and are warp-resistant. Lumber cores are easier to cut, the edges are better for shaping and finishing, and they

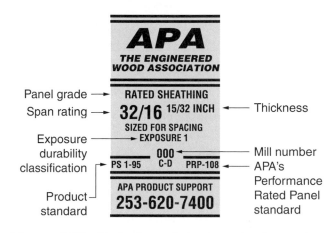

Figure 1-24. Typical grade-trademark stamped on all plywood panels manufactured in compliance with APA standard PS 1-95. (APA-The Engineered Wood Association)

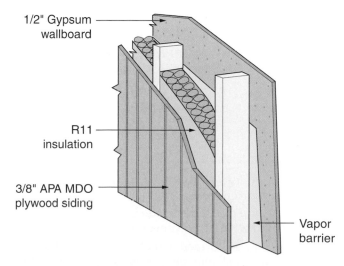

Figure 1-25. This sketch shows the makeup of APA Sturd-I-Wall, which uses MDO plywood as an exterior finish. This wall has an R-value of approximately 12.77. (APA-The Engineered Wood Association)

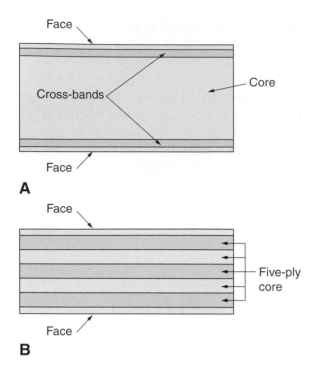

Figure 1-26. Hardwood plywood is of two types. A—Lumber core. B—Veneer core.

hold nails and screws better. Plywood is also manufactured with a particleboard core. It is made by gluing veneers directly to a particleboard surface.

1.13.7 Composite Board

Panels made up of a core of reconstituted wood with a thin veneer on either side are called *composite board* or *composite panels.* These materials are widely used in modern construc-

---

Composite board: A board consisting of a core of wood between veneered surfaces.

Composite panels: Panels made up of a core of reconstituted wood with a thin veneer on either side.

Particleboard: A formed panel consisting of wood flakes, shavings, slivers, etc., bonded together with a synthetic resin or other added binder.

Hardboard: A board material manufactured of wood fiber, formed into a panel with a density of approximately 50–80 lb. per cu. ft.

tion. They are good as sheathing, subflooring, siding, and interior wall surfaces.

1.13.8 Particleboard

In cabinetwork, hardboard and particleboard serve as appropriate materials for drawer bottoms and concealed panels in cases, cabinets, and chests. They are manufactured by many different companies and sold under various trade names.

Particleboard is made of wood flakes, chips, and shavings bonded together with resins or adhesives. It is not as heavy as hardboard (about 40 lb. per cu. ft. compared to 50–80 lb. per cu. ft.) and is available in thicker panels. Particleboard may be constructed of layers made of different-size wood particles. Large ones in the center provide strength. Fine ones at the surface provide smoothness. Particleboard is available in thicknesses ranging from 1/4" to 17/16". The most common panel size is 4' × 8'.

Particleboard is used as a base for veneers and laminates. It is important in the construction of countertops, cabinets, drawers and shelving, many types of folding and sliding doors, room dividers, and a variety of built-ins.

It is popular because it has a smooth, grain-free surface and is stable. Its surface qualities make it a popular choice as a base for laminates. Particleboard doors do not warp and require little adjustment after installation.

1.13.9 Hardboard

Hardboard is made of refined wood fibers pressed together to form a hard, dense material (50–80 lb. per cu. ft.). There are two types: standard and tempered.

Tempered hardboard is impregnated (filled) with oils and resins. These materials make it harder, slightly heavier, more water resistant, and darker in appearance. Hardboard is manufactured with one side smooth (S1S) or both sides smooth (S2S). It is available in thicknesses from 1/12" to 5/16". The most common thicknesses are 1/8", 3/16", and 1/4". Panels are 4' wide and come in standard lengths of 8', 10', 12', and 16'.

1.13.10 Oriented Strand Board

Oriented strand board (OSB) is made up of individual flakes of wood that are laid down like a blanket and adhered to each other with suitable resins and glues. The fibers are put down in successive layers arranged at right angles to one another. It is generally manufactured from aspen poplar in the northern part of North America and from southern yellow pine in the south. The panels are bonded under heat and pressure with thermosetting phenol formaldehyde or isocyanate binders. Panels range in thickness from 1/4" to 1 1/8". All model building codes in the United States and Canada have approved oriented strand board for various structural applications. Refer to the table in **Figure 1-27**. The applications of OSB are many:
- A combination subfloor and underlayment.
- Sheathing for roofs and exterior walls. Oriented strand board provides an excellent nailing base for siding. The large size of the panels also reduces air infiltration through the building's walls.
- APA-rated siding. When special surface treatments are used, oriented strand board is suitable as an exterior finish.
- Soffits.

1.14 Wood Treatments

Wood and wood products that will be exposed to high levels of moisture should be protected from attack by fungi, insects, and borers. Millwork plants employ extensive treatment processes in the manufacture of items such as door frames and window units.

Structures that are continually exposed to weather—outside stairs, fences, decks, and furniture—should be constructed from pressure-treated lumber that has long-lasting resistance to termites and fungal decay. In addition, the lumber should be pressure-treated with a liquid repellent that slows the rate at which the moisture is absorbed and released.

When wood is treated under pressure and with controlled conditions, the treatment deeply penetrates into the wood's cellular structure. Thus treated, the wood resists rot, decay, and termites. Even when exposed to severe conditions, the wood provides excellent service.

In recent years, however, concern has grown about the potential health hazards of exposure to the most common preservative material, chromated copper arsenate (CCA). The EPA ordered the phasing-out of CCA-treated lumber for residential applications by the end of 2003.

Safety Note

The Environmental Protection Agency (EPA) warns that treated wood should *never* be used where the waterborne arsenical preservatives in the treatment may become a component of food, animal feed, or drinking water.

Replacements for CCA-treated material eliminate the arsenic hazard, but are somewhat more expensive. The most widely available replacement is lumber treated with an ammonia/copper preservative known as ACQ. Also available are copper azole–treated wood and borate-treated lumber. Borate preservatives are not toxic to animals or humans, but cannot be used in applications where moisture is constantly present.

The EPA has produced a consumer information sheet that outlines uses and handling precautions for treated wood products. Copies of the sheet are available at home improvement centers and lumberyards.

There are three major types of liquid preservatives:
- Waterborne—These preservatives are used for residential, commercial, recreational, marine, agricultural, and industrial applications. Waterborne treatments leave the wood

---

Oriented strand board (OSB): A formed panel consisting of layers of compressed strand-like wood particles, arranged at right angles to each other.

Use	Conditions	Panel thickness (inches)	Maximum support spacing (inches)	Building Code[1]			
				National (BOCA)	1 & 2 family (CABO)	Standard (SBCC)	Uniform (ICBO)
Subfloors		1/2 5/8 3/4	16 19.2/20.0 24	x x x	x x x	x x x	x x x
Single layer floors		5/8 3/4	16 24	x x	x x	x x	x x
Wall sheathing	Under siding nailed to studs	5/16 3/8	16 24	x x	x x	x x	x x
	Under siding nailed to sheathing	5/8 3/4	16 24	x x	x x	x x	x x
Roof[2] sheathing	With edge clips	3/8 7/16	16 24	x x	x x	x x	x x
	With no edge support	7/16 1/2	16 24	x x	x x	x x	x x
Siding	Over sheathing	5/16 3/8	cont. cont.	x x	x x	x x	x x
	Applied directly to studs	3/8 1/2 5/8	16 24 16	x x 	 x	x x 	x x
Soffits		5/16 3/8	16 24	x x	N/C o N/C o	x x	x x
Horizontal diaphragms		Various		N/C	N/C o	x	x
Shear walls		Various		N/C	N/C o	x	x
Underlayment		1/4S		x	x	x	x

(1) — Current edition

(2) — Special fastener spacing is required in high wind areas and thicker panels may also be required. Check with your local building inspector.

x — Specified in the building code and amendments

o — Specified in the agency listings - not embodied in the code

N/C — Not cited in building code

S — Sanded

Notes: Local building codes may vary. Check with your local building inspector if you have a question.

Figure 1-27. Oriented strand board is approved by all building code organizations for a variety of construction projects. (Structural Board Association)

clear, odorless, and paintable. Waterborne preservatives (in addition to the phased-out CCA), include ACQ, borate, copper azole, acid copper chromate (ACC), and chromated zinc chloride (CZC).

- Oilborne—The best-known oilborne preservative is pentachlorophenol, which can be mixed with various solvents. It is highly toxic to insects and fungi. However, treated wood may become discolored. It is used on farms, around utility poles, and in industry.

- Creosote—This is a mixture of creosote and coal tar in heavy oil. It is suitable for treatment of pilings, utility poles, and railroad ties. Creosoted wood cannot be painted.

Often, treated lumber carries an identifying end tag or stamp, **Figure 1-28.** This tag or stamp indicates the type of preservative, proper expo-sure conditions, and other information. More information on treated lumber can be obtained by contacting:

American Wood Preservers' Association
P.O. Box 286
Woodstock, MD 21163-0286
www.awpa.com

Canadian Institute of Treated Wood
202-2141 Thurston Drive
Ottawa, Ontario, Canada K1G 6C9
www.citw.org

Southern Forest Products Association
P.O. Box 641700
Kenner, LA 70064-1700
www.sfpa.org

Southern Pine Inspection Bureau
4709 Scenic Highway
Pensacola, FL 32504
www.spib.org

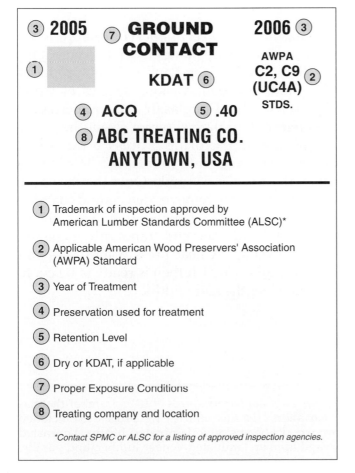

Figure 1-28. An end tag identifies treated lumber for ground-contact use. Other tags denote lumber suitable for above ground use. (Southern Pine Council)

1.15 Handling and Storing Lumber

Building materials are expensive and should be maintained in the best condition that is practical. After they are delivered to the construction site, this becomes the responsibility of the carpenter. Piles of framing lumber and sheathing should be laid on level skids raised at least 6″ above the ground. Be sure all pieces are supported and lying straight. Cover the material with canvas or waterproof paper. Polyethylene film also provides a watertight covering.

If moisture absorption is likely, cut the steel banding on panel materials to prevent edge damage when the fibers expand. Keep coverings open and away from the sides and bottom of lumber stacks to promote good ventilation. Tight coverings promote mold growth.

Exterior finish materials, door frames, and window units should not be delivered until the structure is partially enclosed and the roof surfaced. In cold weather, the entire structure should be enclosed and heated before interior finish and cabinetwork are delivered and stored.

APA structural wood panels delivered on site should be properly stored and handled. Protect ends and edges from damage. Whenever possible, store panels under a cover. Keep sanded panels and other appearance grades away from high traffic areas. Weigh down the top panel to prevent warpage.

Panels stored outdoors should be stacked on a level surface atop 4 × 4 stringers or other blocking. Use at least three stringers. Never allow panels to touch the ground. Keep panels well ventilated by separating them with stickers (wood strips) to prevent mildew growth.

When finish lumber is received at a moisture content different from that it will attain in the structure, it should be open-stacked with wood strips so air can circulate around each piece. Plywood, especially the fine hardwoods, must be handled with care. Sanded faces can become soiled and scarred if not protected. While stored, the panels should lie flat.

1.16 Engineered Lumber

Builders have long relied on traditional wood products for residential and light commercial framing. Modern tree-farming technology has assured that this material continues to be available, keeping up with the increased demand for wood structures. It once took more than 80 years to grow a tree to sawmill size; tree farming has shortened the span to 27 years. In the process, unfortunately, the tree's annular rings, which determine its tensile and comprehensive strength, are much farther apart, resulting in weaker dimension lumber. The industry has found a way to strengthen lumber through the development of *engineered lumber products (ELP).*

Engineered lumber products include those wood structural units that have been altered through manufacturing processes to make them stronger, straighter, and more dimensionally stable than sawn lumber. These processes produce more lumber products from less timber and can utilize smaller trees that are unsuitable as sawn lumber. Where sawn lumber typically

uses only 40% of a log, engineered materials use 75% of the log for structural lumber. At the same time, the manufacturing processes require less energy than those used for solid lumber.

Components are glued together in different configurations—some are solid, some are shaped like steel I-beams, and some are in truss form. These products include:

- Laminated-veneer lumber (LVL), including I-beams.
- Glue-laminated beams (commonly known as "glulams").
- Wood I-beams.
- Open-web trusses.
- Parallel-strand lumber (PSL).
- Laminated-strand lumber (LSL).

1.16.1 Laminated-Veneer Lumber

Laminated-veneer lumber (LVL) is produced under various trade names (such as Microlam®, Parallam®, and Gang-Lam™). It is produced much like plywood. Various species of wood may be used, but Douglas fir and southern pine are preferred for their superior strength.

In the manufacturing process, a log is first debarked, then cut into thin (1/10"–3/16" thick) sheets that are 27"–54" wide. Once the sheets are dried, the next step removes defects and coats the veneer sheets with phenolic glue. Next, sheets are stacked with their ends randomly staggered so they overlap. A heat press compresses and cures the layup, which then is ready to be cut to specified lengths and widths.

Engineered lumber products (ELP): Lumber that has been altered by manufacturing processes. Included are wood I-beams, glue laminated beams, laminated strand lumber, laminated veneer lumber, and parallel strand lumber.

Laminated-veneer lumber (LVL): Lumber produced much like plywood, except all of the veneer panels have their grain running in the same direction.

Another widely used process cuts the veneer into 1/2" wide strips. These strips are coated with waterproof phenolic glue and fed into a series of automated machines. The resulting billet comes out measuring 12" × 17" × 66'. The billet can be cut into sizes commonly used by the construction industry. The laminate material is used as headers, beams, and columns.

LVL products are dimensionally stable, uniform in size throughout, predictable in performance, and can be nailed, drilled, and cut with ordinary construction tools. LVL products easily take stain and can be left exposed, where they become a natural part of the interior decor.

Although a cross section of LVL looks somewhat like plywood, there is an important difference—grain orientation. In conventional plywood, each veneer layer is placed at a 90° angle to the previous layer. In LVL, all of the veneer panels have their grain running in the same direction. Called *parallel lamination*, this provides greater strength, allowing the member to carry the heavy loads required of beams, headers, and truss components. Standard laminated beams are 1 3/4" thick with depths of 5 1/2", 7 1/4", 9 1/2", 11 7/8", 14", 16", and 18".

1.16.2 Glue-Laminated Beams

Glue-laminated, also called *glulams*, are made by gluing and then applying heavy pressure to a stack of four or more layers of 1 1/2" thick stock. The result is a beam that can be up to 30" deep. The layers, called lams, are arranged to produce maximum strength. Under load, the top of a horizontal beam is compressed while the bottom is in tension. The manufacturer places high-grade compression layers at the top and high-grade tension layers at the bottom.

Two types

Manufacturers produce two types of glulams: balanced and unbalanced. Glue-laminated beams always come with one edge marked "top." This edge must be installed upward.

Glulams are normally used (either exposed or hidden) for rafters, floor-support beams, and stair stringers. Manufacturers also produce glulams in curved shapes for supporting arched roofs such as those found in churches or other public buildings. Three grades are available:
- Industrial appearance, designed for use where appearance is not important.
- Architectural appearance, where appearance is important.
- Premium appearance, when appearance is critical.

Regardless of grade, there is no difference in strength. Those with the same design values also have the same strength.

Fire resistance is a significant advantage of glue-laminated beams. They are slow to ignite and burn slowly.

Sizes

Glulams are sold in both custom and stock sizes. Manufacturers produce to customer specifications where there are long spans, unusually heavy loads, or other conditions to be considered. Stock beams are manufactured in commonly-used dimensions. Distributors and dealers cut them to order for customers. Typical stock widths are 3 1/8", 3 1/2", 5 1/8", 5 1/2", and 4 3/4". These meet the requirements of most residential applications.

Another engineered timber, called Reid-Lam, combines elements of both glue-laminates and laminated veneer lumber. The core and the compression face (top of the timber) is a standard glue-laminated beam. The bottom quarter, the tension face, is laminated veneer lumber. Sizes are the same as a standard glue-laminated beam.

Parallel lamination: All of the veneer panels have their grain running in the same direction.

Glue-laminated beams: Large beams fabricated by gluing up small dimension lumber.

Glulams: Beams made by gluing and then applying heavy pressure to a stack of four or more layers of 1 1/2" thick stock.

1.16.3 Wood I-Beams

Wood I-beams are shaped similar to steel I-beams. They consist of flanges (also called *chords*) of structural composite lumber, such as laminated veneer lumber, or sawn lumber and a vertical web of plywood or oriented strand board, 3/8″ or 7/16″ thick. The web is fastened into grooves in the flanges using waterproof glue. See **Figure 1-29.**

Wood I-beams are light, straight, and strong. They are designed to span long distances. Though mostly used as joists and rafters, wood I-beams may be used as headers, too. Most mills deliver them in lengths up to 60′. Distributors and dealers cut them to frequently-used lengths (16′ to 36′) specified by builders. Most common depths are 9 1/2″, 11 1/2″, 14″, and 16″.

An added feature of the wood I-beam is that it is often manufactured with knockouts scored into the web to allow access for plumbing and electrical services. To preserve the strength engineered into wood I-beams, the APA specifies the size and location of holes cut into the web. These are found in Appendix B, **Technical Information**. A table of dimensions and span tables is also shown in Appendix B.

Figure 1-29. Engineered wood I-beam with top and bottom flanges consisting of either laminated veneer lumber or solid finger-jointed lumber and solid webbing of plywood or OSB. To avoid damage, always store wood I-beams with webs upright.

Wood I-beams manufactured to APA specifications carry a trademark indicating conformance to their PRI Standard 400. **Figure 1-30** is a sample of this trademark with an explanation of its markings.

1.16.4 Open-Web Trusses

Open-web trusses are most often used in place of floor joists, **Figure 1-31.** This is especially true when dealing with long spans. These trusses are fabricated in factories from solid 2 × 4 lumber. Sometimes, steel webs are used instead of wood. Depths of 14″ and 16″ are most common. The open webs of the trusses allow installation of pipes, wiring, drains, and other mechanical systems without having to cut openings, as in other types of joists.

1.16.5 Parallel-Strand Lumber

To make *parallel-strand lumber (PSL)* the manufacturer lays up 1″ wide by 8′ long strands of veneer that have been soaked with adhesive. Bonded under pressure, the veneer strips form billets up to 11″ thick by 19″ deep. Billets can be extruded to any length, but 66′ is usual. Lumber yards cut them to specified lengths for customers. There are five common thicknesses: 1 3/4″, 2 11/16″, 3 1/2″, 5 1/4″, and 7″. PSL products are used as exposed posts and beams. When pressure-treated, they are suitable for outdoor use in porches, decks, and gazebos.

Wood I-beams: Beams consisting of flanges (also called *chords*) of structural composite lumber, such as laminated veneer lumber, or sawn lumber, and a vertical web of plywood or oriented strand board, 3/8″ or 7/16″ thick. Also called solid-web trusses.

Parallel-strand lumber (PSL): Engineered lumber made up of strands of wood up to 8′ long that are bonded together using adhesives, heat, and pressure.

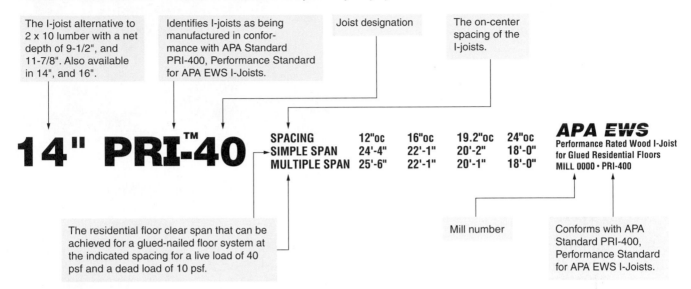

SAMPLE TRADEMARK – Position of trademark on I-joist may vary by manufacturer

The I-joist alternative to 2 x 10 lumber with a net depth of 9-1/2", and 11-7/8". Also available in 14", and 16".

Identifies I-joists as being manufactured in conformance with APA Standard PRI-400, Performance Standard for APA EWS I-Joists.

Joist designation

The on-center spacing of the I-joists.

14" PRI™-40

SPACING	12"oc	16"oc	19.2"oc	24"oc
SIMPLE SPAN	24'-4"	22'-1"	20'-2"	18'-0"
MULTIPLE SPAN	25'-6"	22'-1"	20'-1"	18'-0"

APA EWS
Performance Rated Wood I-Joist for Glued Residential Floors
MILL 0000 • PRI-400

The residential floor clear span that can be achieved for a glued-nailed floor system at the indicated spacing for a live load of 40 psf and a dead load of 10 psf.

Mill number

Conforms with APA Standard PRI-400, Performance Standard for APA EWS I-Joists.

Figure 1-30. This trademark appears on wood I-beams manufactured to APA specifications. Spacing and span information save builders' time. (APA-The Engineered Wood Association)

Figure 1-31. Open-web trusses as shipped from the factory. Webs and chords are fastened with metal plate connectors. Typically, such trusses are stronger than solid joists of the same depth. The arrow points to offsets that are designed to receive a ribbon. This ribbon ties joists together and holds them upright.

Laminated-strand lumber (LSL): Lumber laid up from 1/32"×1"×12" strands of wood bonded with polyurethane adhesive. It is available in two thicknesses: 1 1/4" and 3 1/2". Depths vary up to a maximum 16" and lengths vary up to 35'. Uses include: door and window headers, rim joists in floors, and core stock for flush doors with veneer overlays.

1.16.6 Laminated-Strand Lumber

Laminated-strand lumber (LSL) is laid up from 1/32" × 1" × 12" strands of wood bonded with polyurethane adhesive. It is available in two thicknesses: 1 1/4" and 3 1/2". Depths vary up to a maximum 16" and lengths vary up to 35'. Uses include: door and window headers, rim joists in floors, and core stock for flush doors with veneer overlays.

1.17 Nonwood Materials

The carpenter works with a number of materials other than lumber and wood-based products. Some of the more common items include:
- Metal framing members, especially joists and studs.
- Gypsum and metal lath.
- Wallboard and sheathing.
- Insulating boards, batts, and loose insulation.

- Shingles of asphalt, metal, fiberglass, tile, and concrete, **Figure 1-32.**
- Metal flashing material.
- Caulking materials.
- Resilient flooring materials and carpeting.
- Bonding materials.

1.18 Metal Structural Materials

Steel framing is often found in light commercial buildings, **Figure 1-33.** For reasons of economy, it is increasingly being used in residential construction. *Steel framing members* are manufactured in various widths and gages and are used as studs and joists.

The studs are attached to base and ceiling channels by welding or with screws or clips. A typical stud consists of a metal channel with openings through which electrical and plumbing lines can be installed.

Wall surface material, such as drywall or paneling, is attached to the metal stud with self-tapping drywall screws. Metal stud systems are most used for nonloadbearing walls and partitions. Metal framing materials are covered in greater detail in Chapter 11, **Framing with Steel**.

Figure 1-33. Steel framing units as delivered to a building site. Due to the rising costs of lumber, steel framing is finding greater use in modern residential and light commercial buildings.

1.18.1 Metal Framing Connectors

At one time, all wood-to-wood connections in a wood frame were made with nails. While this practice still continues and produces adequate strength for the structure, metal connectors are faster to install and improve uniformity in strength. They may be required by code in areas that experience high winds.

Basically, metal framing connectors are stamped brackets or strapping designed to make

Figure 1-32. In mild climates, fiber-cement shingles are popular and durable.

Steel framing members: Manufactured in various widths and gages and are used as studs, joists, and truss rafters.

wood-to-wood, wood-to-masonry, or wood-to-concrete connections. Unless the connector is designed to be exposed and decorative, it is made from galvanized metal in various gages (thicknesses).

1.18.2 Strapping or Ties

Strapping or ties are designed to hold parts of a frame together, **Figure 1-34.** They are often used to "quake-proof" structures by tying frames to foundations and roofs to walls. Straps are perforated so that they can be fastened with nails without first drilling holes.

1.18.3 Hangers

Hangers connect the end of one framing member to another. Hangers are used where floor or ceiling joists intersect another framing member, such as a beam, **Figure 1-35.** Designs are available for either solid lumber or laminated lumber.

1.18.4 Other Connector Types

Other connectors are designed for special purposes. *Tension bridging* can be used instead of solid wood or wood braces to transfer loads from joist to joist. They can be used on either solid lumber joists or on wood I-beams. Metal corner braces are another type of tension connector. See **Figure 1-36.**

1.18.5 Fastener Strength

Fasteners used for metal connectors should be able to withstand shear stresses placed on them by the connectors. Bolts may sometimes

Tension bridging: Use of metal connectors to transfer load from one joist to another.

be used, but nails or screws are most common. Some builders secure the fasteners with drywall screws, which penetrate wood quickly and cannot be easily withdrawn. However, drywall screws have little shear strength and are not recommended. Generally, only nails should be used, **Figure 1-37.** Nail length varies with the type of connector. Manufacturer's literature should be consulted.

1.18.6 Metal Lath

Where stucco is the exterior wall finish, metal lath is attached to the sheathing as a base for the stucco plaster. The lath comes in rolls and is attached with staples.

1.18.7 Metal Fasteners

Nails, the metal fasteners commonly used by carpenters, are available in a wide range of types and sizes. Basic kinds are illustrated in **Figure 1-38.** Nails for use in power nailers are designed for automatic feeding through the nailer. They are provided in flat blanks called clips and in round coils.

The common nail has a heavy cross section and is designed for rough framing. The thinner box nail is used for toenailing in frame construction and light work. The casing nail is the same weight as the box nail and used in finish carpentry work to attach door and window casings and other wood trim. Finishing nails and brads are quite similar and have the thinnest cross section and smallest head.

The nail size unit is called a *penny* and is abbreviated with the lowercase letter d. It indicates the length of the nail. A 2d (2 penny) nail is 1" long. A 6d (6 penny) nail is 2" long. See **Figure 1-39.** This measurement applies to common, box, casing, and finish nails. Brads and small box nails are specified by their actual length and gage number.

A few of the many nails designed for special purposes are shown in **Figure 1-40.** Annular or spiral threads greatly increase holding power. Some nails have special coatings of zinc, cement, or resin. Coating or threading increases nail-

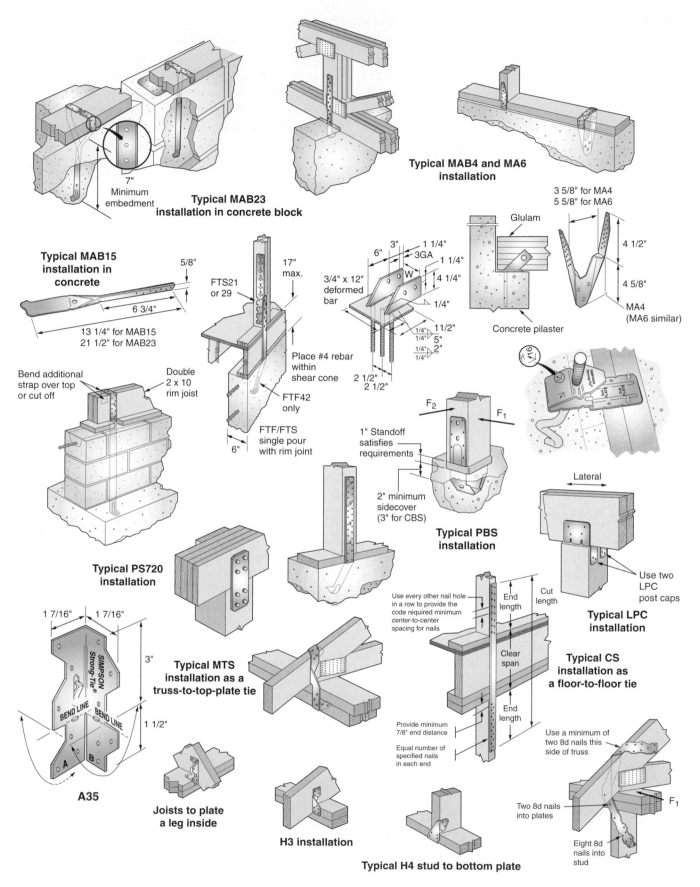

Typical MAB23 installation in concrete block

7"
Minimum embedment

Typical MAB4 and MA6 installation

Typical MAB15 installation in concrete

5/8"

6 3/4"

13 1/4" for MAB15
21 1/2" for MAB23

3 5/8" for MA4
5 5/8" for MA6

4 1/2"

4 5/8"

MA4
(MA6 similar)

FTS21 or 29

17" max.

3/4" x 12" deformed bar

6" 3" 1 1/4"
3GA
1 1/4"
W
4 1/4"
1/4"

11/2"
1/4" 5"
1/4" 2"
1/4"

2 1/2"
2 1/2"

Glulam

Concrete pilaster

Bend additional strap over top or cut off

Double 2 x 10 rim joist

Place #4 rebar within shear cone

FTF42 only

FTF/FTS single pour with rim joint

6"

F₂   F₁

1" Standoff satisfies requirements

2" minimum sidecover (3" for CBS)

Typical PBS installation

Lateral

Use two LPC post caps

Typical LPC installation

Typical PS720 installation

1 7/16" 1 7/16"

SIMPSON Strong-Tie®

BEND LINE BEND LINE

A   B

3"

1 1/2"

A35

Typical MTS installation as a truss-to-top-plate tie

Joists to plate
a leg inside

H3 installation

Use every other nail hole in a row to provide the code required minimum center-to-center spacing for nails

Provide minimum 7/8" end distance

Equal number of specified nails in each end

End length

Clear span

End length

Cut length

Typical CS installation as a floor-to-floor tie

Use a minimum of two 8d nails this side of truss

Two 8d nails into plates

Eight 8d nails into stud

F₁

Typical H4 stud to bottom plate

Figure 1-34. Framing connectors and ties are made in many different configurations to secure walls to foundations and roofs to walls. (©Simpson Strong-Tie Company, Inc.)

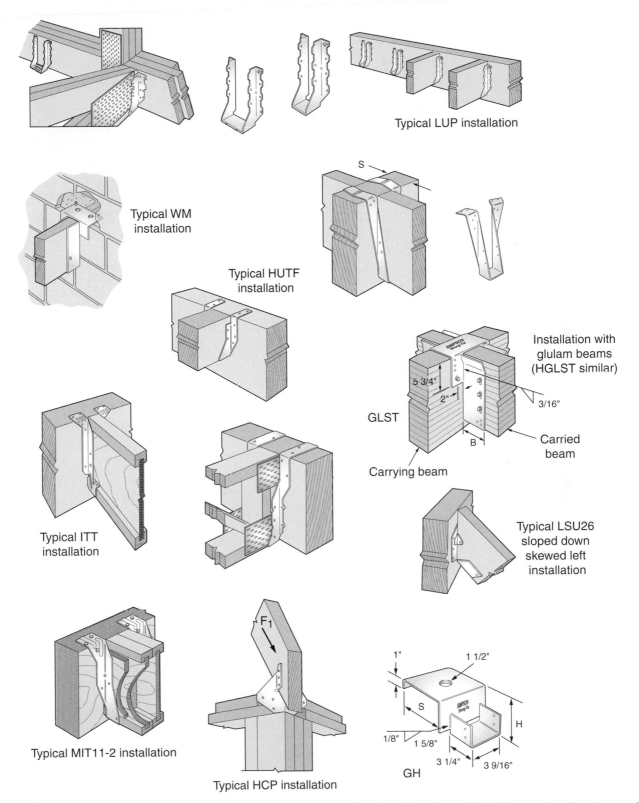

Figure 1-35. Wood-framing hangers can be used at various framing joints in residential construction to provide support of loads and stresses placed on joists and beams. (©Simpson Strong-Tie Company, Inc.)

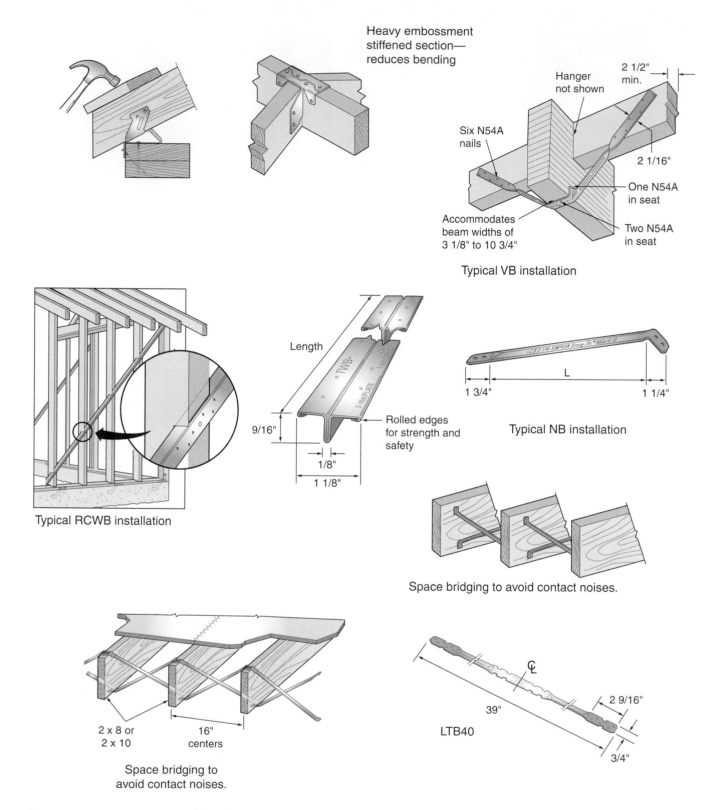

Heavy embossment
stiffened section—
reduces bending

Hanger
not shown

2 1/2"
min.

Six N54A
nails

2 1/16"

One N54A
in seat

Accommodates
beam widths of
3 1/8" to 10 3/4"

Two N54A
in seat

Typical VB installation

Length

Rolled edges
for strength and
safety

9/16"

1/8"

1 1/8"

Typical RCWB installation

1 3/4"

L

1 1/4"

Typical NB installation

Space bridging to avoid contact noises.

2 x 8 or
2 x 10

16"
centers

Space bridging to
avoid contact noises.

39"

2 9/16"

3/4"

LTB40

Figure 1-36. Some types of strapping and ties are designed to work under tension to prevent joists from tipping or building frames from racking. (©Simpson Strong-Tie Company, Inc.)

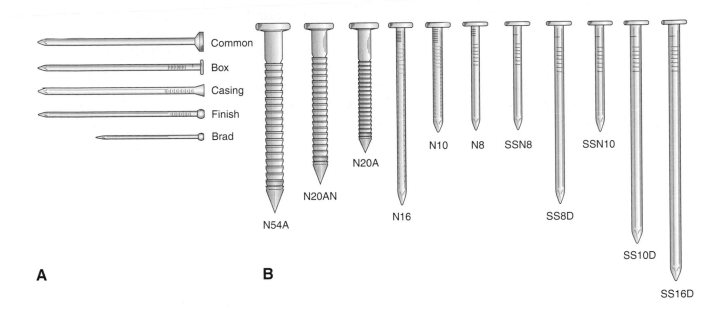

Model No.	Canadian equivalent	Metric equivalent (mm)	Description	Finish	Allowable loads				Nails per cwt
					Light Gage		3 Gage		
					Shear (100)	Gage	Shear (100)		
N8	1 1/2" x 10 1/4 ga	3.3 x 38.1	(8d) 10 1/4 ga x 1 1/2" Smooth shank	HDG	86	14	105		15200
SSN8	1 1/2" x 10 1/4 ga	3.3 x 38.1	(8d) 10 1/4 ga x 1 1/2" Smooth shank	SS	86	14	105		15200
SS8D	2 1/2" COMMON	3.3 x 63.5	(8d) 10 1/4 ga x 2 1/2" Smooth shank	SS	92	20	131		9400
SD8x1.25	–	4.1 x 31.7	#8 x 1 1/4" Screw	EG	76	18	–		9926
N10	1 1/2" x 9 ga	3.8 x 38.1	(10d) 9 ga x 1 1/2" Smooth shank	HDG	92	14	112		11900
SSN10	1 1/2" x 9 ga	3.8 x 38.1	(10d) 9 ga x 1 1/2" Smooth shank	SS	92	14	112		12200
SS10D	3" COMMON	3.8 x 76.2	(10d) 9 ga x 3" Smooth shank	SS	112	18	158		6700
N16	2 1/2" x 8 ga	4.1 x 63.5	(16d) 8 ga x 2 1/2" Smooth shank	BRIGHT	134	18	187		6300
SS16D	3 1/2" COMMON	4.1 x 88.9	(16d) 8 ga x 3 1/2" Smooth shank	SS	134	18	187		4400
N20A	1 3/4" x 6 ga	4.9 x 44.5	(20d) .192 x 1 3/4" Annular ring	BRIGHT	119	14	140		6300
N20AN	2 1/8" x 6 ga	4.9 x 54.0	(20d) .192 x 2 1/8" Annular ring	BRIGHT	145	14	174		5500
N54A	2 1/2" x 3 ga	6.4 x 63.5	.250 x 2 1/2" Annular ring	BRIGHT	167	14	188		2700

1. Allowable loads are based on the 1991 NDS. Adjustments are made for use with metal side plates, F_{es} = 45 ksi. Loads under light gage are for gages listed through 22 gage.

Allowable loads for gages not indicated must be calculated according to the code. Contact the factory for more detail if required.

2. N16, N20, N20AN and N54A fasteners may be ordered galvanized; specify EG; for example N16EG.

3. Metric equivalents are listed by Diameter x Length.

C

Figure 1-37. The most common fastener is a nail. A—Carpenters generally use these five basic types of nails. B—Various sizes of stainless steel nails. C—Nail types and specifications for securing connectors and hangers. (©Simpson Strong-Tie Company, Inc.)

holding power. Nails are made from such materials as iron, steel, copper, bronze, aluminum, and stainless steel.

Wood screws have greater holding power than nails and are often used for interior construction and cabinetwork. The length and diameter (gage number) determine their size. Screws are classified according to the shape of head, surface finish, and the material from which they are made. See **Figure 1-41.**

Wood screws are available in lengths from 1/4" to 6" and in gage numbers from 0 to 24.

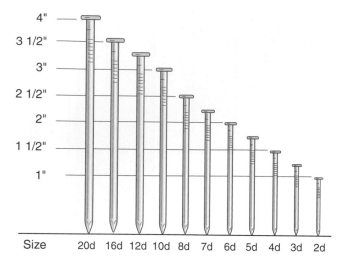

Figure 1-38. Nail sizes are given in a unit called the "penny." It is written as d. (United States Steel)

The gage number can vary for a given length of screw. For example, a 3/4" screw is available in gage numbers of 4 through 12. The No. 4 is a thin screw, while the No. 12 has a large diameter. From one gage number to the next, the size of the wood screw changes by 13 thousandths (.013) of an inch.

Most wood screws used today are made of mild steel with a zinc chromate finish. Nickel and chromium plated screws, as well as screws made of brass or stainless steel, are available for special work. Wood screws are usually priced and sold by the box.

Additional useful fasteners include lag screws, hanger bolts, carriage bolts (specially designed for woodwork), corrugated fasteners, and metal splines. Specialized metal fasteners are described in other sections of this book.

Size	Length"	Common Diam."	Common No./lb.	Box Diam."	Box No./lb.
4d	1 1/2	.102	316	.083	473
5d	1 3/4	.102	271	.083	406
6d	2	.115	181	.102	236
7d	2 1/4	.115	161	.102	210
8d	2 1/2	.131	106	.115	145
10d	3	.148	69	.127	94
12d	3 1/4	.148	63	.127	88
16d	3 1/2	.165	49	.134	71
20d	4	.203	31	.148	52
30d	4 1/2	.220	24	.148	46
40d	5	.238	18	.165	35

Figure 1-39. Nail sizes and approximate number in a pound are shown on this chart. (Georgia-Pacific Corp.)

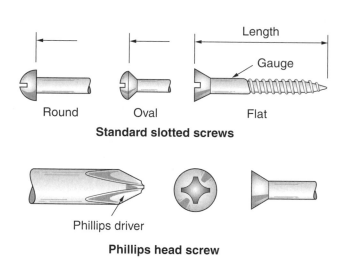

Round   Oval   Flat
Standard slotted screws

Phillips driver
Phillips head screw

Figure 1-41. Kinds of heads found on common wood screws.

Nail for general use
Nail for general use
Trussed rafter nail
Pole-construction nail
Flooring nail
Underlay floor nail

Drywall nail
Roofing nail with neoprene washer
Roofing nail with neoprene washer
Asphalt shingle nail
Asphalt shingle nail

Figure 1-40. Annular- and spiral-threaded nails are designed for special purposes. (Independent Nail and Packing Co.)

1.19 Concrete

Concrete, undoubtedly the most-used material in modern construction, is covered more extensively in Chapter 7, **Footings and Foundations**. It is the material of choice in most foundations and all slab-on-grade construction. Blocks, which are widely used in foundation walls and even for entire building walls, are entirely of concrete. Frames of some commercial buildings are of poured or prefabricated concrete.

Concrete is a combination of several materials: cement, sand, gravel or crushed stone, and water. Sometimes additives are included to provide desired characteristics. Cement is a manufactured product made from limestone mixed with shale, clay, or marl. These materials are combined in carefully measured proportions. When water is added, the mixture becomes plastic and is easily shaped in forms. The water also sets in motion a chemical reaction called hydration. It is this action, not drying, that cures concrete and imparts the desired strength.

1.20 Adhesive Bonding Agents

Bonding materials that carpenters use include glues, adhesives, cements, and mastics. Research and development have created many new products in this area. Glues come from natural materials; adhesives are developed from synthetic materials; cements and mastics are rubber-based. Some are highly specialized, designed for use with a specific material and/or application. Brief descriptions of several of the commonly used bonding materials follow.

Polyvinyl resin emulsion adhesive (generally called polyvinyl or white glue) is excellent for interior construction. It comes ready to use in plastic squeeze bottles and is easily applied. This adhesive sets up rapidly, does not dull tools or stain the wood, and securely holds wood parts.

Polyvinyls harden when their moisture is absorbed by the wood or evaporates. They are not waterproof, so they should not be used for assemblies that will be subjected to high humidity or moisture. The vinyl-acetate resins used in the adhesive are *thermoplastic.* Under heat they will soften. They should not be used in situations where the temperature may rise above 165°F (75°C).

Yellow glues are somewhat stronger than white and more resistant to lacquers and solvents. Since they are more heat-resistant, they do not clog sandpaper. Newer yellow glues are cross-linked polymers. This makes them even stronger with much better water resistance. They are also tackier with faster drying time. They still can be cleaned up with water.

Thermoset adhesives are more water- and heat-resistant than the polyvinyl adhesives. In this category are polyurethanes, urea formaldehydes, and resorcinal formaldehydes. The *polyurethane adhesives* are known as single-component, moisture-catalyzed adhesives. They use moisture from the air or wood to set. As they set up, they foam and spread themselves. This action also fills any gaps.

Polyurethanes have few or no solvents, are completely waterproof, and will take stain. They can be applied at temperatures ranging from 40°F (5°C) to 90°F (30°C) and will bond with wood having up to 25% moisture content.

Surfaces must fit snugly and be free of dust, oil, wax, old glue, or paint. On wood, it is good practice to wipe each surface with a damp cloth about a minute before gluing.

To wipe up excess polyurethane adhesive, use mineral spirits or acetone. If this is not practical (since the adhesive may foam for up to an

---

Polyvinyl resin emulsion adhesive: Wood adhesive intended for interiors. Made from polyvinyl acetates which are thermoplastic and not suited for temperatures over 165°F (75°C). Also called *white glue.*

Thermoplastic: Material that softens under heat.

Thermoset adhesives: Adhesives that are more water- and heat-resistant than the polyvinyl adhesives. In this category are polyurethanes, urea formaldehydes, and resorcinal formaldehydes.

Polyurethane adhesive: A single component, moisture-catalyzed adhesive that will bond with wood of up to 25% moisture content.

hour) trim off excess with a chisel or scraper after the material has set for several hours.

Urea-formaldehyde resin glue (usually called urea resin) is available in a dry powder form that contains the hardening agent, or catalyst. It is mixed with water to a thick, creamy consistency before use. Joints should be clamped securely during curing time.

Urea resin is moisture-resistant, dries to a light brown color, and securely holds wood. It hardens through chemical action when water is added and sets in four to eight hours, depending on room temperature.

Contact cement, shown in **Figure 1-42**, is made with a neoprene rubber base. It is an excellent adhesive for applying plastic laminates or joining parts that cannot be easily clamped together. It works well for applying thin veneer strips to plywood edges and can also be used to join combinations of wood, cloth, leather, rubber, sheet foams, and plastics.

To use contact cement, apply it to each surface with a brush, short nap roller, or finishing trowel. Allow to dry until a piece of paper will not stick to the film. Surfaces must never be allowed to touch before they are in perfect alignment. Carefully align cemented surfaces and firmly press them together. Bonding takes place immediately. The pieces must be carefully aligned for the initial contact, because they cannot be moved after they touch. The bonding time is not critical and can usually be performed anytime within one hour.

Contact cement usually contains volatile, flammable solvents. Thoroughly ventilate the work area where it is applied. Carefully read directions on the container for additional handling instructions.

Casein glue is made from milk curd, hydrated lime, and sodium hydroxide. It is supplied in powder form and is mixed with cold water for use. After mixing, it should sit for about 15 minutes before it is applied. It is classified as a water-resistant glue.

Casein glue is used for structural laminating and works well with wood that has high moisture content. It has good joint filling qualities and, therefore, is often used on materials that have not been carefully surfaced. Casein is used for gluing oily woods such as teak, padouk, and lemon wood. Its main disadvantages are that it stains the wood, especially such species as oak, maple, and redwood, and has an abrasive effect on tool edges.

1.20.1 Mastics

Mastics are heavy, pasty adhesives that have revolutionized the methods used in the application of wallboards, wood paneling, and some types of floors. See **Figure 1-43.** They vary in their characteristics and application methods, and are usually designed for a specific type of material. Some are waterproof; others must be used where there is no excessive moisture.

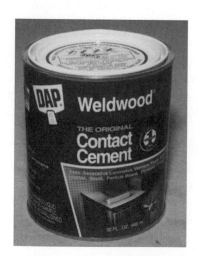

Figure 1-42. Contact cement forms an instant bond to two surfaces. It is especially useful where clamping would be hard or impossible. (Bayfield Ace Hardware)

Urea-formaldehyde resin glue: Moisture-resistant glue that hardens through chemical action when water is added to the powdered resin.

Contact cement: Neoprene-rubber-based adhesive that bonds instantly upon contact of parts being fastened.

Casein glue: An adhesive of casein and hydrated lime that is suitable for gluing oily woods and for laminating wood with a high moisture content.

Mastic: Thick adhesive used in construction as a bonding agent for tile and paneling.

One application method consists of placing several gobs on the surface of the material and then pressing the unit firmly in place. This causes the mastic to spread over a wider area. Some mastics are spread over the surface with a notched trowel. Still others are designed for caulking gun application.

Mastics are usually packaged in metal or plastic containers or in gun cartridges and are ready for application. Always follow the directions of the manufacturer.

1.20.2 Bonding Basics for Adhesives

In the U.S., the construction industry classifies glue's resistance to water this way:
- *Type I* includes those adhesives that can be submerged in water indefinitely.
- *Type II* designates adhesives that must be tested by soaking in water for four hours, followed by 19 hours of drying at 120°F (50°C) before they can be called water resistant.

1.20.3 Open time and other variables

Also called "setting time," **open time** is the amount of time between spreading the material and when the parts must be clamped. This is not important on small parts like casing or baseboard miters, but is critical when large surfaces must be glued. Setting time varies with temperature and humidity.

Figure 1-43. Today's carpentry makes heavy use of various adhesives and mastics. (Bayfield Ace Hardware)

Temperature is closely related to open time. Usually, the range is between 65°F (20°C) and 75°F (25°C). At these temperatures, adhesives easily spread and have open times as indicated by the manufacturer. Above or below this temperature range, viscosity and open time change.

Squeeze-out of glue can create extra work for finishers if it is not promptly cleaned off after clamping. Some glues can be cleaned with a damp rag, while others require mineral spirits or thinner. Some finishers prefer that glues be allowed to dry before the excess is removed with a chisel. It is always best to ask which is preferred.

Open time: Amount of time between spreading an adhesive and when the parts must be clamped.

ON THE JOB

Estimator

Accurate estimates of cost are vital to the success of a construction project. Developing such estimates in great detail is the responsibility of the construction estimator. Usually employed by a general contractor, the construction estimator factors in everything from the price of lumber to the cost of renting a large crane for a day, the labor pay rates to insurance coverage, and expected loss of time resulting from bad weather. The resulting estimate is used as the basis for the contractor to competitively bid on the construction contract. For firms building single homes or residential developments, such an estimate is used to set the selling price of the homes.

Construction estimators usually begin with a set of detailed building plans that provide dimensions and other information needed to begin the estimate. In most cases, the estimator also visits the site. This visit provides information on terrain, site access, drainage, and the availability of utilities.

Next, careful analyses of material and labor requirements are done. These are called *takeoffs* and are entered on standard estimating forms. The estimator must identify and enter on the forms the exact quantities of each item that the contractor must provide. If portions of the work (such as plumbing or electrical) are subcontracted, the estimator must usually analyze the bids submitted by the subcontractors. Decisions involving sequencing of operations must be made, since improperly scheduled equipment, materials, or work crews may cause delays and increased costs. Allowances must be made for shipping delays, waste and damage to materials, weather problems, and so on.

While most construction estimators are employed by building contractors, some work for large architectural or engineering firms. Estimators may also operate their own consulting businesses, providing fee-based services to contractors, government agencies, and building owners or financing organizations.

Educational requirements for construction estimators usually consist of a degree in a field such as construction management, engineering, or architecture. A strong aptitude for mathematics and proficiency with computer software are also valuable assets. Many estimators also have practical experience in construction work, giving them familiarity with materials, job practices, and the various specialty trades. Additional on-the-job training is usually provided, since each company has its own specific way of preparing and presenting estimates.

Professional certification is voluntary, but beneficial. Such certification is administered by two professional organizations—the Association for the Advancement of Cost Engineering and the Society of Cost Estimating and Analysis. The requirements vary, but include a number of years of professional experience and passing an examination.

Summary

Many kinds of materials are used in building construction, but the most common is wood. All woods can be classified as either hardwoods or softwoods. Wood is used in construction in the form of dimension lumber or panels. Panels may be made as laminated layers (plywood) or composite material of wood particles or chips and a resin binder. Traditional dimension lumber is solid wood sawn to size, but engineered lumber is becoming more common. Engineered lumber consists of layers of material laminated together with adhesives or may be wooden I-beams combining solid and plywood components. Engineered lumber is typically stronger and lighter than traditional dimension lumber. Steel joist hangers and other metal framing connectors improve joint strength in lumber construction. Metal framing materials, fabricated from thin steel, are widely used in commercial construction, but are becoming more common in residential use. They are fastened together with screws or welded joints. Various types of construction adhesives have been introduced in recent years.

Test Your Knowledge

Answer the following questions on a separate piece of paper. Do not write in this book.

1. The natural cement that holds wood cells together is called _____.
2. New wood cells are formed in which layer?
3. Which of the following kinds of wood are classified as hardwoods? Hemlock, redwood, willow, spruce.
4. When a softwood log is cut so the annular rings form an angle greater than 45° with the surface of the boards, the lumber is called _____.
5. What is the moisture content of a board if a test sample that originally weighed 11.5 oz. was found to weigh 10 oz. after oven drying?
6. The fiber saturation point is about _____ M.C. for nearly all kinds of wood.
7. The letters E.M.C. are an abbreviation for the term _____ moisture content.
8. A large knot is defined as one that is over _____ inches in size.
9. The best grade of "selects" softwood lumber is _____.
10. The best available grade of hardwood lumber is _____.
11. How many board feet of lumber are contained in a stack of 24 pieces of 2" × 4" × 8'?
12. Where should a plywood panel marked "Exposure 1" be used?
13. What is a span rating for plywood?
14. How do polyvinyl glues cure and how does temperature affect them?
15. Explain how moisture-catalyzed adhesives are cured (hardened).

Curricular Connections

Social Studies. Before barbed wire was introduced in the mid-1800s, farmers used locally available materials to build fences around their fields and pastures. Do research in the library or on the Internet to identify the materials used in different areas of North America. Try to determine why a particular material (such as field-stone) was used in one locality, but not in others. If possible, find pictures or create sketches of fences made from various materials. Prepare a display.

Mathematics. Visit a local building supply center and secure information and literature concerning the various grades and species of lumber normally carried in stock. Prepare written descriptions of the defects permitted in several of the grades commonly selected by builders in the area. Obtain list prices of these grades to gain some understanding of the savings that can be gained by using a lower classification. Work out the difference in cost between two grades of 2 × 4s if you were purchasing 50 eight-foot boards. Make a summary report to your class on your findings and conclusions.

Outside Assignments

1. Prepare a visual aid that shows various metal fasteners used in carpentry. Include nails, brads, screws, carriage bolts, staples (such as those used in power staplers), and other items. Include spiral, ring groove, and coated nails. Label each item, giving the correct name, size, and other information.
2. Gather softwood samples representative of the species of lumber used in your locality. Instead of writing the proper name on each piece, use a number that corresponds with your master list. This will permit you to give a wood identification quiz to members of your class. As the samples are passed around, the students can record the numbers and their answers on a sheet of paper.
3. After a study of reference materials, prepare a paper on stress-rated lumber. Include information on grade stamps and their interpretation. Also define the f (fiber stress in bending) and the corresponding E (stiffness) rating. Try to include a description of modern equipment and/or new techniques used in grading.

Wood Identification

A key element in woodworking and in carpentry is the proper identification of the wood species used by carpenters. These samples are intended as a guide and aid to the student carpenter in learning to identify various woods. Shown are typical color and grain characteristics for 16 different species.

Ash, white. The heartwood of the white ash is brown and the sapwood is light-colored or nearly white. Wood from second-growth trees is heavy, strong, hard, and stiff and it has high resistance to shock. White ash is used principally for nonstriking tool handles, oars, baseball bats, and other sporting and athletic goods. Additional uses are decorative veneer, cabinets, furniture, flooring, millwork, and crates.

Beech, American. Only one species of beech, American beech is native to the United States. It grows in the eastern one-third of the United States and adjacent Canadian provinces. The wood has little figure and is of close, uniform texture. It has no characteristic taste or odor. The wood of beech is classed as heavy, hard, and strong, but shrinks substantially and therefore requires careful drying. It machines smoothly, is an excellent wood for turning, and wears well. Most beech is used for flooring, furniture, handles, veneer, containers, and cooperage.

Birch. A hard, strong wood (47 lb. per cu. ft.). Easily machine-worked with excellent finishing characteristics. Heartwood is reddish-brown with white sapwood. Fine grain and texture. Used extensively for quality furniture, cabinetwork, doors, interior trim, and plywood. Also used for dowels, spools, toothpicks, and clothespins.

Butternut. Also called white walnut, butternut (*Juglans cinerea*) is a hardwood that grows from southern New Brunswick and Maine west to Minnesota. The wood is moderately light in weight (about the same as eastern white pine), rather coarse textured and moderately soft. It machines easily and finishes well. Used for furniture, cabinets, paneling, and interior woodwork.

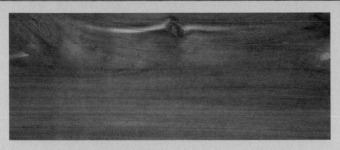

Cedar, red, eastern. The heartwood of eastern red cedar is bright or dull red and the narrow sapwood is nearly white. The wood is moderately heavy and is dimensionally stable after drying. The texture is fine and uniform and the wood has a distinctive aroma that is reputed to repel moths, but this claim has not been supported by research. Lumber is manufactured into chests, wardrobes, and closet lining. Other uses include flooring, novelties, pencils, scientific instruments, and small boats. The greatest quantity of eastern red cedar is used for fence posts.

Mahogany, Honduran. Also known as American mahogany, Honduran mahogany is an attractive and dimensionally stable wood. It is grown on plantations established in its natural range in Central and South America. The heartwood varies from pale pink or salmon colored to dark reddish brown. The grain is generally straighter than that of African mahogany. The wood is very easy to work with hand and machine tools and is often sliced into fine veneer. The principal uses for mahogany are fine furniture and cabinets, interior woodwork, pattern woodwork, boat construction, musical instruments, and paneling.

Cherry, black. The heartwood of black cherry varies from light to dark reddish brown and has a distinctive luster. The wood has a fairly uniform texture and very good machining properties. It is moderately heavy, strong, stiff, and moderately hard; it is very dimensionally stable after drying. Black cherry is used principally for furniture, fine veneer panels, and architectural woodwork.

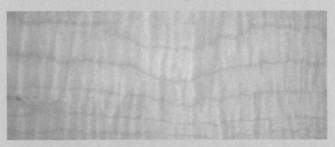

Maple, sugar. Also called hard maple, it is hard, strong, and heavy (44 lb. per cu. ft.) with a fine texture and grain pattern. Light tan color, with occasional dark streaks, the wood is hard to work with hand tools. However, it machines easily and is an excellent turning wood. Used for floors, bowling alleys, woodenware, handles, and quality furniture.

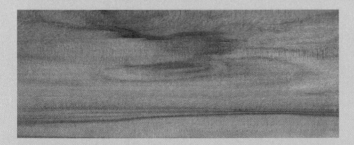

Gum, sweet. Also called red gum. Fairly hard and strong (36 lb. per cu. ft.). A close-grained wood that machines well but has a tendency to warp. Heartwood is reddish-brown and may be highly figured. Used extensively in furniture and cabinetmaking. Stains well; often used in combination with more expensive woods.

Oak, white. Heavy (47 lb. per cu. ft.), very hard, durable, and strong. Works best with power tools. Heartwood is grayish-brown with open pores that are distinct and plugged with a hairlike growth called tyloses. Sawing or slicing oak in a radial direction results in a striking pattern as shown. The "flakes" are formed by large wood rays that reflect light. Used where dramatic wood grain effects are desired. Used for high quality millwork, interior finish, furniture, carvings, boat structures, barrels, and kegs.

Oak, red. Heavy (45 lb. per cu. ft.) and hard with the same general characteristics as white oak. Heartwood is reddish-brown in color. No tyloses in wood pores. Difficult to work with hand tools. Used for flooring, millwork, and inside trim.

Poplar. Poplar is also known as yellow-poplar, tulip-poplar, and tulipwood. The greatest commercial production of poplar lumber is in the South and Southeast. The wood is generally straight grained and comparatively uniform in texture. The lumber is used primarily for furniture, interior moulding, siding, cabinets, musical instruments, and structural components. Boxes, pallets, and crates are made from lower grade stock. Poplar is also made into plywood for paneling, furniture, piano cases, and various other special products.

Pecan. A member of the hickory family, pecan grows from central Texas and Louisiana to Missouri and Indiana. The sapwood of this group is white or nearly white, while the heartwood is somewhat darker. The wood is heavy, and is used for tool and implement handles. Higher grade logs are sliced to provide veneer for furniture and decorative paneling.

Teak. Teak is a hardwood that is plantation-grown in commercial quantities in Asia, Latin America, and Africa. The heartwood varies from yellow-brown to dark golden-brown and eventually turns a rich brown on exposure to air. Teakwood is usually straight-grained and has a distinctly oily feel. The heartwood has excellent dimensional stability and a very high degree of natural durability. Teak is one of the most valuable woods, but its use is limited by scarcity and high cost. Because teak does not cause rust or corrosion when in contact with metal, it is extremely useful in the shipbuilding industry and similar applications. It is used in the construction of boats, furniture, flooring, decorative objects, and decorative veneer.

Pine, white. Soft, light (28 lb. per cu. ft.), and even textured. Cream colored with some resin canals but not as prevalent as sugar pine. Works easily with hand or machine tools. Used for interior and exterior trim and millwork item. Knotty grades often used for wall paneling.

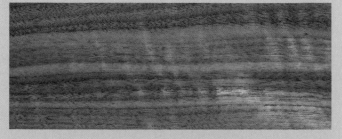

Walnut, black. Fairly dense and hard. Very strong in comparison to its weight (38 lb. per cu. ft.). Excellent machining and finishing properties. A fine textured, open-grain wood with beautiful grain patterns. Heartwood is a chocolate brown with sapwood near white. Used on quality furniture, gunstocks, and fine cabinetwork.

The Carpenter's Workplace

Learning Objectives

After studying this chapter, you will be able to:
- Describe the qualities employers value in a beginning carpenter.
- List advantages of maintaining a proper relationship with other members of the crew.
- List and describe clothing safety as it applies to carpenters.
- List other personal safety equipment recommended for carpenters to use.
- Cite safety measures relating to hand and power tools.
- Explain housekeeping measures that promote safe working conditions.
- List safety measures relating to shoring and scaffolding.
- Describe proper methods of lifting and carrying to avoid personal injury.
- Describe the classes of fires.

Technical Vocabulary

Angle of repose
Class A fires
Class B fires
Class C fires
Dust mask
Ground fault circuit
  interrupter (GFCI)
Hard hat
Pressure-treated
  lumber
Safety factor
Safety glasses
Sexual harassment

As in other occupations, carpentry requires a level of knowledge and skill that allows the worker to turn out good work in a timely manner. Appropriate personal conduct on the job site is important. Safety in areas such as dress, moving about the site, and proper handling of potentially dangerous tools is also an important job factor.

2.1 Skill Development and Job Competency

Obviously, good reading and comprehension skills are important for carpenters. They are crucial for following construction drawings, building codes, and specifications. Not so obvious is the need to take the time to carefully read instructions. Though any construction worker faces time constraints on a daily basis, paying close attention to manufacturers' instructions on application of their products actually saves time. The old adage applies: "If you don't have time to do it right the first time, how will you find the time to do it over?"

Doing it right the first time has other benefits for the carpenter and his/her employer, **Figure 2-1.** When the job is done right, it reduces the amount of warranty work the contractor must do and promotes greater customer satisfaction. Lawsuits resulting from improper

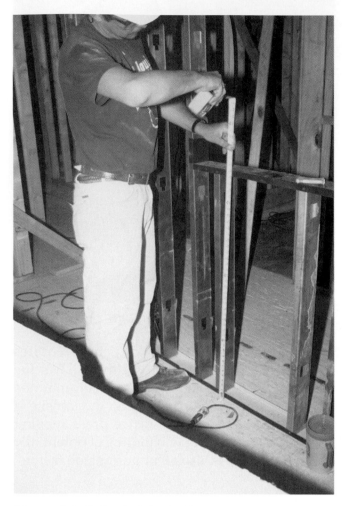

Figure 2-1. Doing it right the first time demands accuracy in taking measurements. Carpenters "measure twice and cut once." (Construction Training School, St. Louis, MO)

application of a manufacturer's siding materials are a case in point. If it is found that a carpenter improperly installed a product that later failed, the contractor who employs the carpenter—not the manufacturer—is liable. The contractor must make it right. This failure to read and follow instructions affects the contractor's profit margin as well as the carpenter's reputation.

2.1.1 What Employers Expect of Beginning Carpenters

Results of a recent survey of union and nonunion contractors in a large Midwest metro-

politan area indicate that what construction contractors most value in beginning carpenters is not always a high level of knowledge or even well-honed carpentry skills. The qualities were separated into three categories or behaviors:

- Affective behavior—Attitude and interpersonal relationships.
- Cognitive behavior—Knowledge of any sort; in this case, carpentry work.
- Psychomotor skills—Eye and hand coordination, meaning how well the worker uses tools and the quality of his/her work.

Responding contractors were asked to rate 54 qualities on a scale of 1 to 5, with 1 being of little or no value for success and 5 being crucial for success. **Figure 2-2** shows the qualities contractors most value in beginning carpenters.

2.2 Conduct on the Job

Maintaining a proper relationship with others on the building project is important for the construction worker. It avoids potential problems in meeting production schedules and needless friction between workers.

Federal law protects workers from various forms of discrimination and abuse (such as sexual harassment) in the workplace. Sexual harassment is considered a type of discrimination and as such is covered by Title VII of the Civil Rights Act of 1964. Title VII says that *sexual harassment* is any unwelcome sexual advances or requests for sexual favors. It also includes any verbal or physical conduct of a sexual nature under specific conditions.

Appropriate job site behavior includes avoiding practical jokes or horseplay that could cause injury. Avoid distracting other workers with casual conversation—except during breaks and lunch periods, communications should be brief and related to the job at hand.

Sexual harassment: Any unwelcome sexual advances or requests for sexual favors. It also includes any verbal or physical conduct of a sexual nature under specific conditions.

Skills/Behaviors Contractors Value Most in Beginning Carpenters		
Name of skill or behavior	**Value** (5 = crucial; 4 & 3 important)	**Type**
Have respect for customers and owners	4.77	Affective
Able to measure material accurately	4.73	Cognitive
Carries basic tools (pencil, tape, knife)	4.68	Affective
Notify superior when unable to report for work due to personal emergency	4.62	Affective
Cooperate with coworkers	4.59	Affective
Observe safety rules	4.58	Affective
Use safe lifting/carrying techniques	4.49	Affective
Take proper care of power tools	4.49	Affective
Follow company rules/procedures	4.49	Affective
Operate skilsaw	4.49	Psychomotor
Report for work properly rested/ready to work	4.49	Affective
Use leveling instruments	4.25	Cognitive
Keep work area clean	4.24	Affective
Attend to details	4.23	Affective
Operate pneumatic nailer	4.07	Psychomotor
Exhibit fall protection procedures	4.04	Cognitive
Use the right nail for the application	3.97	Cognitive
Report for work 10 or more minutes early	3.91	Affective
Wear suitable clothing	3.88	Affective
Recognize fire hazards	3.87	Cognitive
Perform first aid	3.86	Psychomotor
Volunteer for undesirable jobs	3.81	Affective
Record number of hours on the job	3.80	Cognitive
Absent from work no more than four days per year	3.80	Affective
Care for electrical extension cords	3.75	Affective
Toe-nail with minimum splitting of lumber	3.70	Psychomotor
Read blueprints	3.65	Cognitive
Operate screw gun	3.62	Psychomotor
Use telephone etiquette	3.54	Affective
Apply adhesives	3.51	Psychomotor
Scale dimensions from a blueprint	3.48	Cognitive
Operate powder-actuated tool	3.44	Psychomotor
Use job site sanitation facilities	3.41	Affective
Apply caulk	3.41	Psychomotor
Install door hardware	3.35	Psychomotor
Install doors	3.25	Psychomotor
Install moulding	3.17	Psychomotor
Operate radial arm saw	3.17	Psychomotor
Establish building lines	3.15	Cognitive
Install cabinets	3.13	Psychomotor
Lay out stairways	3.07	Cognitive
Install metal studs	3.01	Psychomotor

Figure 2-2. Results of a survey taken among construction companies in a large Midwestern city. Contractors listed these 42 items of attitude, knowledge, or skill as most valued in a beginning carpenter. Of the 42, almost half (18) had to do with attitude and personal relationships, 14 with skills in carpentry, and 10 with knowledge of carpentry. Other qualities were rated, but these were most frequently mentioned.

2.3 General Safety Rules

Good carpenters recognize that safety is an important part of the job. They know that accidents can easily occur in building construction, often resulting in partial or total disability. Even minor cuts and bruises can be painful.

Safety is based on knowledge, skill, and an attitude of care and concern. Carpenters should know correct and proper procedures for performing the work. They should also be familiar with the potential hazards and how to minimize or eliminate hazards. Other sections of this book, especially those dealing with hand and power tools, stress proper safety rules. Read and follow them.

A good attitude toward safety is important. This includes belief in the importance of safety and willingness to give time and effort to a continuous study of the safest ways to perform work. It means working carefully and following the rules.

2.3.1 Clothing

Wear clothing appropriate for the work and weather conditions. Trousers or overalls should properly fit and have legs without cuffs. Avoid loose-fitting clothing that can catch on nails or draw hands or other part of the body into cutting tools. Keep shirts and jackets buttoned. Sleeves should also be buttoned or rolled up. Never wear loose or ragged clothing, especially around moving machinery.

Shoes should be sturdy, with thick soles to protect feet from protruding nails. Tennis or lightweight canvas shoes are not satisfactory. Never wear shoes with leather soles. They will not provide satisfactory traction on smooth wood surfaces or on sloping roofs.

Headgear should provide the necessary protection, be comfortable, permit good visibility, and shade your eyes. All clothing should be maintained in a good state of repair and laundered when soiled.

2.3.2 Personal Protective Equipment

Safety glasses should be worn whenever work involves even the slightest hazard to your eyes. Standard specifications state that a safety lens must withstand the blow of a 1/8″ diameter steel ball dropped from a height of 50″.

Safety boots and shoes are required on heavy construction jobs. They consist of special reinforced toes that will withstand a load of 2500 lb. A *hard hat* should be worn whenever you are exposed to any possibility of falling objects. Since carpenters must often work outside during cold weather, warmer protective headgear such as winter liners should be available. See **Figure 2-3.** Standard specifications require that hard hats withstand a certain degree of denting. They must be able to resist breaking when struck with an 80 lb. ball dropped from a 5′ height.

Wear gloves of an appropriate type when handling rough materials. Use a respirator when

Figure 2-3. Hard hats are required wherever there is danger from falling objects. In cold weather, a winter liner can be worn under the hard hat.

Safety glasses: Eye protection with a safety lens that must withstand the blow of a 1/8″ diameter steel ball dropped from a height of 50″.

Hard hat: Protective headgear that is able to resist breaking when struck with an 80 lb. ball dropped from a 5′ height.

working in dusty areas, while installing insulation, or where finishing materials are being sprayed.

2.3.3 Hand Tools

Always select the correct type and size of tool for your work. Be sure it is sharp and properly adjusted. Guard against using any tool if the handle is loose or in poor condition. Dull tools are hazardous to use because additional force must be applied to make them cut. Oil or dirt on a tool may cause it to slip, resulting in injury.

When using a tool, correctly hold it. Most edge tools should be held in both hands with the cutting action away from your body. Be careful when using your hand or fingers as a guide to start a cut with a handsaw. Carefully handle and carry tools. Keep edged and pointed tools turned downward. Carry only a few tools at one time unless they are mounted in a special holder, **Figure 2-4.** Do not carry sharp tools in pockets of your clothing. When not in use, tools should be kept in special boxes, chests, or cabinets.

Figure 2-4. A tool belt is a safe, handy method of keeping hand tools close by.

2.3.4 Electrical Power

Electrical power is almost always essential on a construction site. Its source may be an alternating current generator or a power pole, **Figure 2-5.** Safe use of this power is important: even a small amount of electrical current is capable of causing serious injury or death.

A

B

Figure 2-5. Power sources. A—Typical building site power pole. B—When there is no electrical utility available, portable diesel- or gasoline-powered generators are used. This construction site generator supplies power for welding and for the operation of lights and other tools. (Miller Electric Mfg. Co.)

Electrical power travels through power cords on the job site. From its source—either the generator or power pole—the current travels through one conductor, known as the hot wire, to the tool. It then returns to the source through a second conductor, known as the neutral. Insulation on hot conductors is always either red or black. Insulation on neutral conductors is always white. A third conductor in the power cord is known as the ground and usually has green insulation. Its purpose is to safely carry away the current from accidental grounding.

Moisture can turn many materials, including soil, into conductors of electrical current. A worker can become a part of the current-conducting loop by coming in contact with both current-carrying conductors or with the hot conductor and the ground. Power tools made of metal are insulated against accidental grounding. This reduces the danger, but does not eliminate it.

Make sure that the tool you are using is grounded. The electrical system should also be checked for proper ground. Any break in the grounding wire makes the grounding system inoperative. The integrity of a grounding wire of a power cord can be checked with an ohmmeter or a continuity tester.

Circuit breakers and fuses provide some protection against shock. They are designed to open the circuit if a short should occur. A much more efficient protective device, shown in **Figure 2-6,** is a *ground fault circuit interrupter (GFCI)*. It detects tiny amounts of current and opens the circuit before shock can occur. One of these devices can be attached to a power cord supplying electricity to a power tool. It is an especially important safety measure whenever moisture is present on the job site.

2.3.5 Power Tools

Before operating any power tool or machine, you must be thoroughly familiar with the way it works and the correct procedures to follow. Read the directions in the owner's manual, if there is one. Make sure you understand them. Be alert and, above all, use common sense as you work. In general, when you learn to use equipment the

A

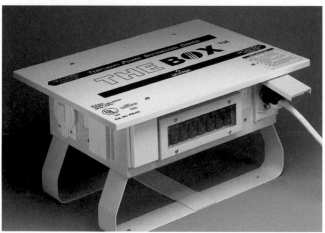

B

Figure 2-6. A GFCI provides protection against electrical shock. A—The power pole outlets often include a GFCI. B—If the outlet is not GFCI-protected, a portable power distribution center can be installed to provide a number of GFCI outlets. (Leviton)

Ground fault circuit interrupter (GFCI): An electrical safety device that can be installed either in an electrical circuit or at an outlet. It is able to detect a short circuit and shut off power automatically. Used as a protection against electrical shock.

correct way, you also learn to use it the safe way, **Figure 2-7.**

There are a number of general safety rules that apply to power equipment. In addition, special safety rules must be observed in the operation of each individual tool or machine. Those that apply directly to the power tools commonly used for carpentry are listed in Chapter 5, **Power Tools.** Study the rules and follow them carefully.

2.3.6 Good Housekeeping

Housekeeping refers to the neatness and good order of the construction site. Maintaining a clean site helps you efficiently work and is an important factor in the prevention of accidents.

Store building materials and supplies in neat piles. Locate them to allow adequate aisles and walkways. Rubbish and scrap should be placed in containers for proper disposal, **Figure 2-8.** Do not permit blocks of wood, nails, bolts, empty cans, or pieces of wire to accumulate. They interfere with your work and are a tripping hazard.

Remove nails from lumber scraps before discarding them. Failure to do so could result in injury from a puncture and lead to lost time on the job. Chapter 4, **Hand Tools**, shows proper tools and methods for pulling nails.

Keep tools and equipment in panels or chests when they are not being used. This provides protection for the tools, as well as for workers on the job site. In addition to improving efficiency and safety, good housekeeping helps maintain a better appearance at the construction project. This will, in turn, contribute to the good morale of all workers.

Figure 2-8. An on-site dumpster provides safe disposal for construction scrap. Properly used, it will help to keep the construction site clear of debris (like the scrap material left on the ground here) that could cause accidents and injuries.

A

B

Figure 2-7. Using a radial arm saw and a portable power tool the safe, correct way. Always wear safety glasses. Hard hats should be worn wherever there is danger of sustaining a head injury. A—When operating a radial arm saw, keep hands at least 6″ away from the saw blade. Make sure that all guards are in place. B—Carefully check hoses for damage and proper connection before using pneumatic nailers and other air-operated tools. (Montachusett Regional Vo-Tech School; Des Moines, Iowa, Public Schools)

2.4 Decks and Floors

To perform an operation safely, either with hand or power tools, the carpenter should stand on a firm, solid base. The surface should be smooth, but not slippery. Do not attempt to work over rough piles of earth or on stacks of material that are unstable. Whenever possible, stay well away from floor openings, floor edges, and excavations. Where this cannot be done, install adequate guardrails or barricades. In cold weather, remove ice, cover it with sand, or apply calcium chloride (salt).

2.5 Excavations

Shoring and adequate bracing must be placed across the face of any excavation where the ground is cracked or a cave-in is likely to occur. Inspect the excavation and shoring daily, especially after rain. Follow state and local regulations. Never climb into an open trench until proper reinforcement against cave-in has been installed or until the sides have been sloped to the *angle of repose* of the material being excavated.

Before beginning an excavation, determine whether there are underground utilities in the area. If so, locate and arrange protection for them during excavation operations. Excavated soil and rock must be stored at least 2′ away from the edge of an excavation. See **Figure 2-9.** Use ladders or steps to enter trenches that are more than 4′ deep.

2.6 Scaffolds and Ladders

Scaffolds should have a minimum *safety factor* of four. This means that the scaffold is capable of carrying a load four times greater than the maximum load it is required to support. Inspections should be made daily before the scaffold is used.

Figure 2-9. Excavated soil should be stockpiled at least 2′ away from a trench to prevent collapse.

Ladders should be inspected for damage at frequent and regular intervals. Their use should be limited to climbing from one level to another. Working while being supported on a ladder is hazardous and should be kept to a minimum.

There are many more safety rules that must be observed in the use of scaffolds and ladders. Take special care to protect against possible electrocution resulting from contact with overhead power lines while erecting, moving, or working from metal or conductive ladders and scaffolds. Refer to Chapter 31, **Scaffolds and Ladders** for more specific rules.

Angle of repose: The greatest slope at which loose material, such as excavated soil, will stand without sliding.

Safety factor: Capable of carrying a load a specified number of times (for example, four) in excess of the expected maximum load.

2.7 Fall Protection

Construction contractors are required to provide fall protection on the job site. At lower heights, this is fairly simple. Carpenters must install guardrails around openings in floors and across wall openings when the fall distance is 6′ or more and on scaffolds more than 10′ off the ground.

An acceptable guardrail will have a top rail strong enough to support a 200 lb. load and is at a height of about 42″. A midrail should be located at half that distance.

In addition:

- Floor openings larger than 2′ square must be covered with material that can safely support the working load.

- Roof openings, including skylights, must be covered securely or protected by guardrails.

- At roof pitches between 4-in-12 and 6-in-12, slide rails must be installed along roof eaves. These should be placed above the first three courses of roofing material.

- At roof pitches between 6-in-12 and 8-in-12, additional slide guards must be installed every 8′ up the roof.

- When the roof pitch is steeper than 8-in-12, or if the height from ground to eaves is more than 25′, a safety-harness system with a solid anchor point must be used. Refer to **Figure 2-10.**

- Hazards that could impale a worker must be properly guarded or removed.

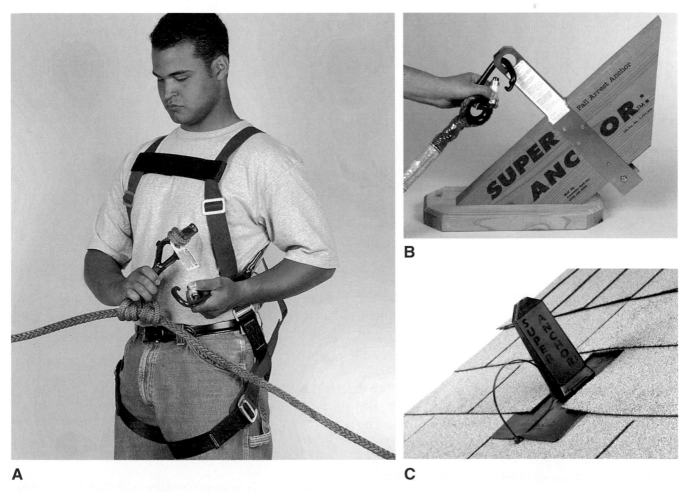

A

B

C

Figure 2-10. OSHA regulations require fall protection for carpenters and other construction workers at heights over 25′. This harness and anchor system is designed to provide such protection without unduly restricting movement. A—Worker wears a harness that allows full range of movement. A clip on the harness attaches to the lifeline. B—During framing, lifeline anchors are fastened to rafters. C—During the shingling stage, an anchor can be fastened to roof sheathing. (Super Anchor Manufacturing)

2.8 Falling Objects

When working on upper levels of a structure, you should be especially cautious in handling tools and materials so there is no chance of them falling on workers below. Do not place tools on the edge of scaffolds, stepladders, windowsills, or on any other surface where they might be knocked off.

If long pieces of lumber must be placed temporarily on end and leaned against the side of the structure, be sure they will not fall sideways. When moving through a building under construction, be aware of overhead work, and, wherever possible, avoid passing directly beneath other workers. Stay clear of materials being hoisted. Wear an approved hard hat whenever there is a possibility of falling objects.

2.9 Handling Hazardous Materials

Pressure-treated lumber, especially wood treated with chromated copper arsenic (CCA) preservative, requires special care in handling for safety of the construction worker. Although wood treated with CCA has been phased out in favor of less dangerous preservatives, precautions are necessary to protect your health. Avoid prolonged breathing in of sawdust particles. Sawing should be done outdoors while wearing a *dust mask.* Wear safety goggles when power sawing or machining. To minimize skin contact, wear gloves when handling any pressure-treated materials.

Before eating any food, carefully wash any skin that has come into contact with treated wood. Clothing soiled by contact with the wood should be laundered before reuse. The clothing should be washed separately from other clothing.

Never burn scraps of treated wood. Rather, bury them or place them in an ordinary trash collection bin or dumpster placed on site during construction.

Use great care when spraying paints, stains, or similar materials. Use an approved respirator and protect exposed skin by covering it with clothing, **Figure 2-11.**

Figure 2-11. Avoid breathing in fumes from paint spraying. Use a mask or respirator and cover exposed skin when necessary. (Greco Painting)

Pressure-treated lumber: Wood that has been treated with chemical preservatives to protect it from rot and insects.

Dust mask: A woven covering worn over the mouth and nose to filter out dust and larger particles of airborne debris. Dust masks are not effective in trapping very small particles, such as paint mists.

2.10 Lifting and Carrying

Improper lifting and carrying of heavy objects may cause injuries. When lifting, stand close to the load, bend your knees, and grasp the object firmly. Then, lift by straightening your legs and keeping your body as close to vertical as possible. To lower the object, reverse the procedure.

When carrying a heavy load, do not turn or twist your body, but make adjustments in position by shifting your feet. If the load is heavy or bulky, have others help. Never underestimate the weight to be moved or overestimate your own ability. Always seek assistance before attempting to carry long pieces of lumber.

Stretching exercises are a precaution against muscle strain. Many company safety programs include stretch training.

2.11 Fire Protection

Carpenters should have a good understanding of fire hazards. They must know the causes of fires and methods of controlling them.

Class A fires: Fires that result from burning wood and debris.
Class B fires: Fires that involve highly volatile materials such as gasoline, oil, paints, and oil-soaked rags.
Class C fires: Fires that result from faulty electrical wiring and equipment.

Class A fires result from burning wood and debris. *Class B fires* involve highly volatile materials such as gasoline, oil, paints, and oil-soaked rags. *Class C fires* result from faulty electrical wiring and equipment. Any of these fires can occur on a typical construction site.

Approved fire prevention practices should be followed throughout the construction project. Good housekeeping is an important aspect. Special precautions should be taken during the final stages of construction when heating equipment and wiring are being installed and when highly flammable surface finishes are being applied. Always keep containers of flammable materials closed when not in use. Dispose of oily rags and combustible materials promptly.

Fire extinguishers should be available on the construction site. Be sure to use the proper kind for each type of fire, **Figure 2-12.** Study and follow local regulations.

2.12 First Aid

Knowledge of first aid is important. You should understand approved procedures and be able to exercise good judgment in applying them. Remember that an accident victim may receive additional injury from unskilled treatment. Information of this nature can be secured from your local Red Cross.

As a preventative measure against infection, keep an approved first aid kit on the job site. Because of the nature of the material being handled and the dirty conditions of the work area, even superficial wounds should be treated promptly. Clean, sterilize, and bandage all cuts and nicks. As a precaution, it is important to maintain a current tetanus shot as a guard against infection.

Fire Extinguishers and Fire Classifications

Fires	Type	Use	Operation
Class A Fires Ordinary Combustibles (Materials such as wood, paper, textiles.) *Requires... cooling-quenching.* Old New	**Soda-acid** Bicarbonate of soda solution and sulfuric acid	Okay for use on **A** Not for use on **B C D**	Direct stream at base of flame.
Class B Fires Flammable Liquids (Liquids such as grease, gasoline, oils, and paints.) *Requires...blanketing or smothering.* Old New	**Pressurized Water** Water under pressure	Okay for use on **A** Not for use on **B C D**	Direct stream at base of flame.
Class C Fires Electrical Equipment (Motors, switches, and so forth.) *Requires... a nonconducting agent.* Old New	**Carbon Dioxide (CO_2)** Carbon dioxide (CO_2) gas under pressure	Okay for use on **B C** Not for use on **A D**	Direct discharge as close to fire as possible, first at edge of flames and gradually forward and upward.
Class D Fires Combustible Metals (Flammable metals such as magnesium and lithium.) *Requires...blanketing or smothering.* **D**	**Foam** Solution of aluminum sulfate and bicarbonate of soda	Okay for use on **A B** Not for use on **C D**	Direct stream into the burning material or liquid. Allow foam to fall lightly on fire.

		Multi-purpose type	Ordinary BC type	
	Dry Chemical	Okay for **A B C**	Okay for **B C**	Direct stream at base of flames. Use rapid left-to-right motion toward flames.
		Not okay for **D**	Not okay for **A D**	

	Dry Chemical Granular type material	Okay for use on **D** Not for use on **A B C**	Smother flames by scooping granular material from bucket onto burning metal.

Figure 2-12. Not every fire extinguisher will put out every kind of fire. Check the label on the extinguisher. Using the wrong extinguisher could electrocute you or produce toxic fumes.

ON THE JOB

Construction Laborer

The construction laborer is a generalist, involved with virtually every aspect of the construction process. Once considered a member of an unskilled occupation, the construction laborer today must learn and make use of a wide variety of skills. He or she may be called on to help clear a job site, dig a trench, unload and distribute building materials, mix or place concrete, read construction prints, operate various machines on the job site, or assist any of the skilled trades workers, such as carpenters, electricians, plumbers, or masons.

Tasks performed by construction laborers are usually the most physically demanding of all of the jobs in the construction field. Such tasks as operating a jackhammer, carrying and stacking lumber, moving wheelbarrow loads of earth or concrete, or clearing brush and debris from a building site require physical strength and a high level of physical fitness.

Much of the laborer's day is spent outdoors working under weather conditions that may be severe.

Almost all laborers are involved in construction work, including building construction and capital projects (roads, bridges, airports). About 1/3 of all laborers work for specialty contractors, the rest for general contractors. About 15% of construction laborers are self employed.

Most construction laborers acquire their skills through informal on-the-job training, learning from experienced laborers, as well as from workers in the skilled trades. Some complete a formal apprenticeship that is typically completed in 2–4 years. It requires at least 4000 hours of supervised on-the-job training and 400 hours of classroom work. Because they work in all areas of a construction project, laborers are in an ideal position to advance to supervisory work or to move into one of the more specialized trades.

Carpenters often build temporary racks like the one shown to speed up work. This rack was constructed and placed so plywood panels could be accessible to workers on an upper level or a roof.

Summary

A carpenter must have the knowledge and skill to turn out good work in a timely manner. Important job factors include appropriate personal conduct on the job site and careful attention to safety. Safety rules must be observed in dress, moving about the worksite, and proper handling of potentially dangerous tools and equipment. Good housekeeping on the site is important to preventing accidents. Excavations must be shored and braced to prevent collapse. Where needed, scaffolds and ladder should be used and able to safely and securely support loads. When working above certain heights, especially on roofs, fall protection devices must be used. To prevent injury, correct lifting and carrying procedures must be observed. Carpenters should have a good understanding of the causes of fires and methods of controlling them. A first aid kit should always be available on the job site.

Test Your Knowledge

Answer the following questions on a separate piece of paper. Do not write in this book.

1. Which of these behaviors in beginning carpenters is rated highest by employers?
   A. Observe safety rules.
   B. Have respect for customers and owners.
   C. Cooperate with coworkers.
   D. Volunteer for undesirable jobs.
2. Indicate the importance of maintaining a proper relationship with other carpentry crew members (select all correct answers):
   A. Work is more likely to get done on time.
   B. It promotes good attitudes among crew members.
   C. The boss leaves you alone to do your job.
3. Which of the following is considered safe clothing for a carpenter on the job site?
   A. Unbuttoned sleeves.
   B. Trousers or overalls without cuffs.
   C. Sandals.
   D. Necktie.
   E. All of the above.
4. When should safety glasses be worn?
5. When should a hard hat be worn?
   A. After a bad haircut.
   B. When the sun is bright.
   C. Wherever there is a danger of falling objects.
   D. All of the above.
   E. None of the above.
6. Safety shoes are required to withstand a load of _____.
   A. 50 lb.
   B. 1500 lb.
   C. 2500 lb.
   D. 3500 lb.
   E. 5000 lb.
7. *True or False?* A tool with a dull edge is more dangerous than one with a sharp edge.
8. When not being used, tools should be _____.
   A. kept near at hand for the next use
   B. stored on a shelf, out of reach
   C. thrown in the back of a truck
   D. stored in panels or chests
   E. None of the above.

9. Scaffolding should have a minimum safety factor of _____.
10. When carrying a heavy load, do not _____ your body; instead, shift your feet.
11. *True or False?* Burning wood is considered a Class C fire.

Curricular Connections

Language Arts. Visit a residential construction site and observe the work in progress. Do not enter the site without permission. Write a report on the safety measures being used, as well as any unsafe practices you saw.

Science. Do research to find out how the user can get a shock from an electrically powered tool that is improperly grounded and the effects of electrical shock on the human body. Develop a list of safety measures appropriate to working with electrically powered tools.

Outside Assignments

1. Obtain a catalog from a safety equipment company. Find five different types of personal protective equipment (safety glasses, hard hats, etc.) and identify the number of choices offered in each category. Compare the highest-priced and lowest-priced items in each category. Try to determine why they are different.
2. Research safety measures appropriate to working with air-powered (pneumatic) tools.
3. If you have a large manufacturing plant or utility firm in your town, contact the person responsible for safety at the plant or utility. Ask them to provide examples of safety materials that they distribute to workers.

Plans, Specifications, and Codes

Learning Objectives

After studying this chapter, you will be able to:
- Identify the elements commonly included in a set of house plans.
- Demonstrate the use of scale in architectural drawings.
- Identify architectural symbols.
- Explain the use of building specifications.
- Summarize the concept of modular construction.
- Describe the application of building codes, standards, and permits.

Technical Vocabulary

Architectural
  drawings
Bill of materials
Blueprints
Building code
Building permit
CADD-CAM
Computer-aided
  drafting and design
  (CADD)
Computer-aided
  manufacturing
  (CAM)
Detail drawing
Dimension lines

Drawn to scale
Elevations
Floor plans
Footprint
Foundation plans
Framing plans
Lumber list
Mill list
Model codes
Modular construction
Pictorial sketch
Plot plan
Prints
Scale
Schedule

Section drawing
Section view
Set of plans
Setback

Specifications (specs)
Stock plan
Symbols
Unicom

A good plan and well-defined contract are important in building construction. The old proverb "Early understandings make long friendships" holds true in the construction industry.

Every carpenter must know how to read and understand *architectural drawings* (plans) and correctly interpret the information found in written specifications. Simply put, the plans tell you *how* to build and the specifications tell you *what* materials must be used. See **Figure 3-1.** It has been said that "blueprints are the language of the construction industry."

Copies of the architect's original drawings are usually called blueprints, or simply *prints.* At one time, all drawing copies were chemically made on treated paper that displayed white lines on a blue background. Today, prints are usually reproduced with dark lines on a white background. Although the term *blueprint* is still used, *print* is the preferred term.

Architectural drawings: Plans that show the shape, size, and features of a building.

Prints: Copies of the architect's original drawings. Preferred term over *blueprints.*

Figure 3-1. Carpenters must frequently study plans as they construct a building. (Tony's Construction, Inc.)

Figure 3-2. A contractor (at right) may discuss a building plan with the architect before submitting a bid. (Tony's Construction, Inc.)

3.1 Set of Plans

Vast amounts of information are needed to build a house or light commercial structure. Since a single drawing could not possibly hold all that information, many drawings are needed. When printed and bound together, these sheets are known as a *set of plans.*

Usually, plans are drawn to several different scales and show different things. A person accustomed to reading plans is familiar with the kinds of information present in each. This narrows the search as the contractor determines what the architect and owner want. For example, the dimensions of a room can be found only on a floor plan. The size of a fastener can be shown only in a detail. Many things can be shown on a section drawing: materials used, dimensions, and elevations.

Builders usually look at the elevation drawings first. These offer the best representation of what is to be built. Carpenters who specialize in the installation of acoustical ceilings look at a plan view. Since plan views are always drawn to show a view looking down from above, they look down at the floor as though it were a photograph. A contractor uses a set of plans to bid on the construction of a building, **Figure 3-2.** An estimator studies them in preparing a bid.

The carpenter and building contractor are not the only persons needing a set of plans. The owner receives a set as assurance that the plan reflects his or her wishes. Tradespeople, such as

the electrician, plumber, and heating contractor, need sets so they can install their systems. Lumber dealers and other suppliers use plans to determine the quantity of materials needed. The carpenter uses only parts of the plan.

3.1.1 Drawings in a Set of Plans

A set of house plans usually includes the following drawings. Each type is explained later.
- Plans—There are several: plot plan, foundation or basement plan, and floor plans. These are bird's eye views.
- Drawings of the mechanical systems— Essentially, these are floor plans that describe and give the location of electrical, plumbing, heating, and air conditioning components.
- Elevation drawings—These show the front, rear, and sides of the building.
- Section drawings.
- Detail drawings.

Set of plans: Many sheets of drawings bound together. They include various plans (such as the floor plan), drawings of the mechanical systems, elevation drawings, section drawings, and detail drawings.

3.1.2 Stock Plans

When a set of plans is mass-produced to be offered for sale to many clients, it is called a *stock plan.* Today, a wide range of such ready-drawn plans is available for people who want to build. This chapter shows a number of drawings taken from a set of stock plans (Plan # 20161) from The Garlinghouse Company, Chantilly, VA. See **Figure 3-3.** The drawings referenced as plans are discussed in more detail.

3.1.3 Scale

Drawings must be reduced so they fit on the drawing sheet. This must always be done in such a way that the reduced drawing is in exact proportion to the actual size. Such drawings are said to be *drawn to scale.* Residential plan views are generally drawn to 1/4″ scale (1/4″ = 1′-0″). This means that for each 1/4″ on the plan, the building dimension is 1′. When certain parts of the structure need to be shown in greater detail, they are drawn to a larger scale, such as 1″ = 1′-0″. Others, such as framing plans, are often drawn to a smaller scale (1/8″ = 1′-0″). **Figure 3-4** is a partial list of conventional inch-foot scales.

While carpenters are not responsible for making drawings, it is helpful to be able to make a simple sketch for a building or remodeling job. The sketch is often good enough to be used as a guide during the construction. A *pictorial sketch* is a drawing showing three dimensions, much like a photograph. It is often needed so the customer can visualize the completed job, **Figure 3-5.**

Stock plan: A set of plans that is mass produced to be offered for sale to many clients.

Drawn to scale: The reduced drawing is in exact proportion to the actual size.

Pictorial sketch: Three-dimensional drawing much like a photo that shows how a project will look when built.

Floor plans: Line drawings that show the size and outline of the building and its rooms, including many dimensions to show the location and size of inside partitions, doors, windows, and stairs.

3.1.4 Floor and Foundation Plans

Floor plans are drawings that show the size and outline of the building and its rooms. They

Figure 3-3. Full-color renderings help the buyer of a stock plan visualize how a residence will look. Plans for this single-story house are shown throughout this chapter. (©The Garlinghouse Company. The drawings in this chapter cannot be reproduced or built. If copies of the plans or more information is required, contact The Garlinghouse Company at 1-800-235-5700, or go to their Web site at www.garlinghouse.com.)

Scale	Where used	Ratio	Relationship to actual size
1/32″	Site plans	1:384	1/32″ = 1′0″
1/16″	Plot plan	1:192	1/16″ = 1′0″
1/8″	Plot plan	1:96	1/8″ = 1′0″
1/4″	Elevation plans	1:48	1/4″ = 1′0″
3/8″	Construction details	1:32	3/8″ = 1′0″
1/2″	Construction details	1:24	1/2″ = 1′0″
3/4″	Construction details	1:16	3/4″ = 1′0″
1″	Construction details	1:12	1″ = 1′0″

Figure 3-4. Common inch-foot scales used in residential drawings. Construction details may also be shown in 1 1/4″ scale.

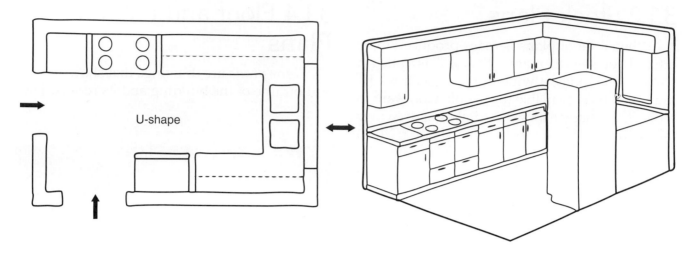

Figure 3-5. It is easier to visualize how the kitchen will look if shown in a picture-like sketch, such as the one at right. This can easily be done using isometric sketch paper.

also give additional information useful to the carpenter and other workers in the construction trades. Floor plans have many dimensions to show the location and size of inside partitions, doors, windows, and stairs. Dimensions are explained later. Plumbing fixtures, as well as appliance and utility installations, can also be shown in this view. **Figure 3-6** is the floor plan of the residence shown in **Figure 3-3.**

Foundation plans are similar to floor plans. They are often combined with basement plans. When shown, the footings are represented as a dashed line. It is assumed that the basement floor is in place and that the grade (ground) covers the footings on the outside. See **Figure 3-7.**

Most stock plans include a *plot plan.* This drawing shows the *footprint* (size and shape) of the building and its location on the site (either a lot or acreage). A stock plot plan may also include the roof plan, indicating ridges of the building and space for filling in setbacks. See **Figure 3-8.** *Setback* is the distance between the building and the property lines. Allowable setbacks vary according to local zoning restrictions.

3.1.5 Elevations

Elevations are drawings showing the outside walls of the structure. These drawings are scaled so that all elements appear in their true relationship. Generally, the various elevations are oriented to the site by labeling them according to the direction they face. However, when plans are not designed for a specific location, the names "front," "rear," "left side," and "right side" are used. **Figures 3-9** and **3-10** are typical elevation drawings. By studying the elevation views, the carpenter can determine:

- Floor levels.
- Grade lines.
- Window and door heights.
- Roof slopes.
- Kinds of materials used on wall and roof surfaces.

Foundation plans: Similar to floor plans and are often combined with basement plans.

Plot plan: A drawing of the view from above a building site. The plan shows distances from a structure to property lines. Sometimes called a *site plan.*

Footprint: Size and shape of a building shown on a plot plan.

Setback: The distance between the building and the property lines.

Elevations: Drawings showing the outside walls of the structure. These drawings are scaled so that all elements appear in their true relationship.

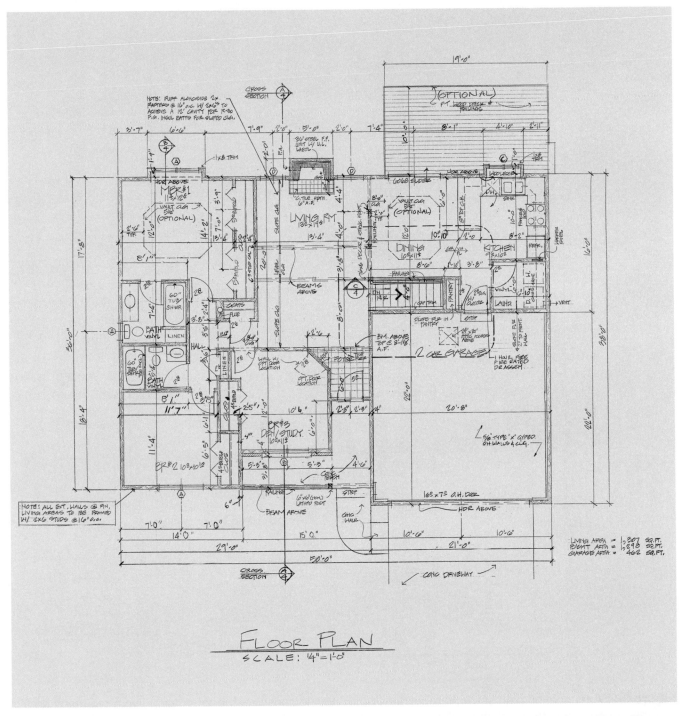

Figure 3-6. The floor plan for one-story house shown in Figure 3-3. Floor plans give overall dimensions, shapes, and sizes of rooms; placement of mechanical systems, cabinetry; and other built-in features of the building. (©The Garlinghouse Company)

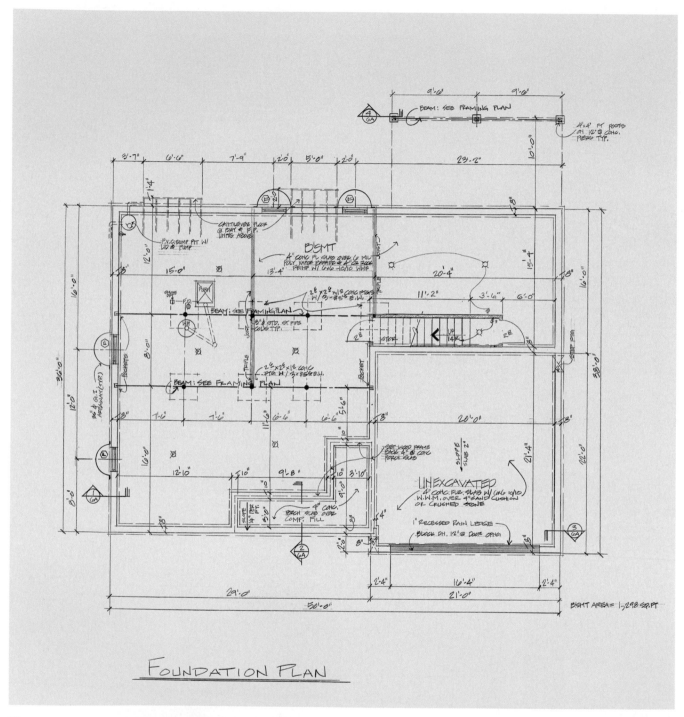

FOUNDATION PLAN

Figure 3-7. A foundation plan shows the builder exactly how to construct the foundation. Note the many details included in such a plan: location of all features, including supporting posts and beams, pockets for supporting beams, window openings, sump, furnace, and electrical boxes. Often, details of special features are also shown. (©The Garlinghouse Company)

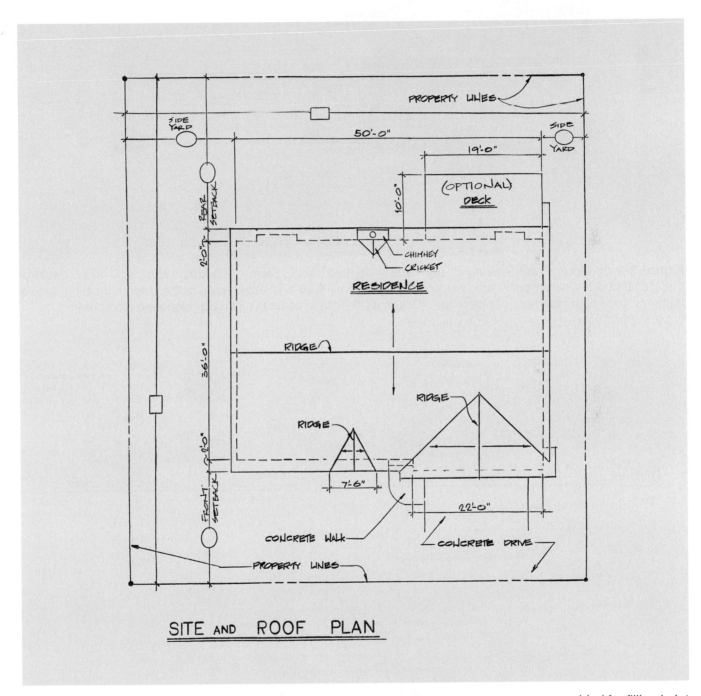

Figure 3-8. Example of a site or plot plan that also includes a roof plan. Blank spaces are provided for filling in lot dimensions and building setbacks. (©The Garlinghouse Company)

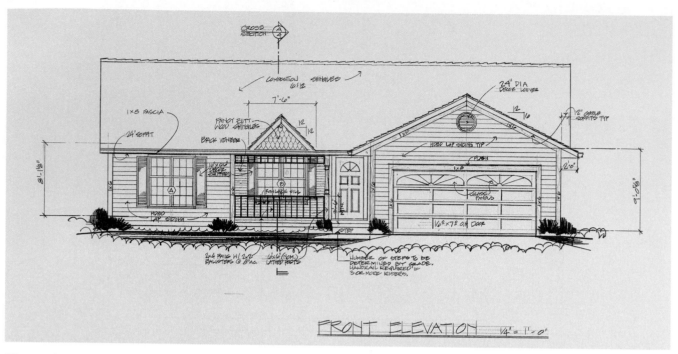

Figure 3-9. In stock plans, elevations are usually marked "front," "rear," "left side," and "right side" because the architect does not know which direction the house will face. Elevations may carry notes indicating special features, code requirements, or reminders. (Drawings are not to scale.) (©The Garlinghouse Company)

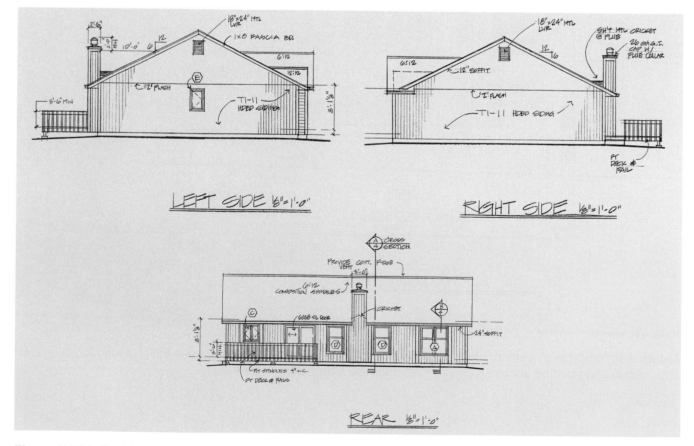

Figure 3-10. Left, right, and rear elevation drawings. All parts appear in true relationship to one another. Note that the slope of the roof is given. See Chapter 10, **Roof Framing**, for more information. (©The Garlinghouse Company)

3.1.6 Framing Plans

Sometimes a house plan also has drawings showing the size, number, and location of the structural members of the building's frame. These are known as *framing plans*. See **Figures 3-11** through **3-13**. Separate plans may be drawn for the floors, ceilings, walls, and roof. These plans specify the size and spacing of the framing members. The members are drawn as they will be positioned in the building. Openings needed for chimneys, windows, and doors are shown. Dimensions are not used, but the drawings are made to scale, as with other plan drawings.

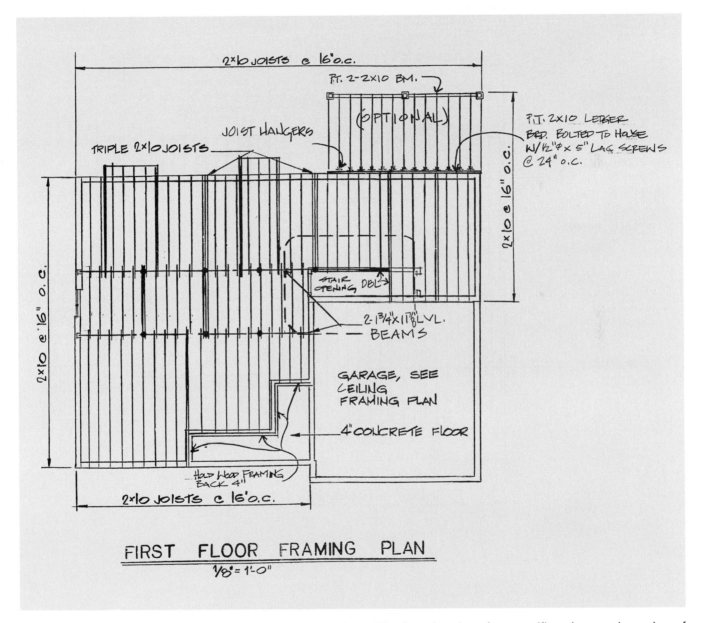

Figure 3-11. House plans usually include framing drawings. The floor framing plan specifies sizes and spacing of joists, girders, and support columns. (©The Garlinghouse Company)

Framing plans: Drawings showing the size, number, and location of the structural members of a building's frame.

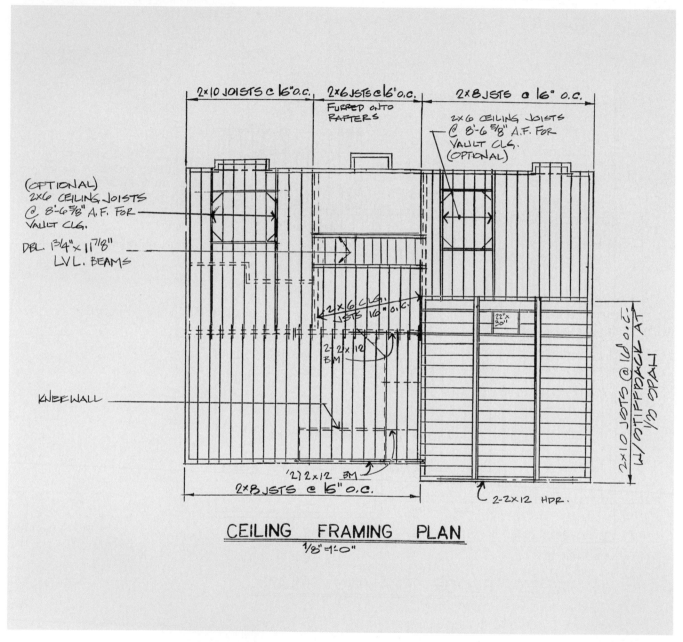

Figure 3-12. The ceiling framing plan shows placement, size, and spacing of joists. (©The Garlinghouse Company)

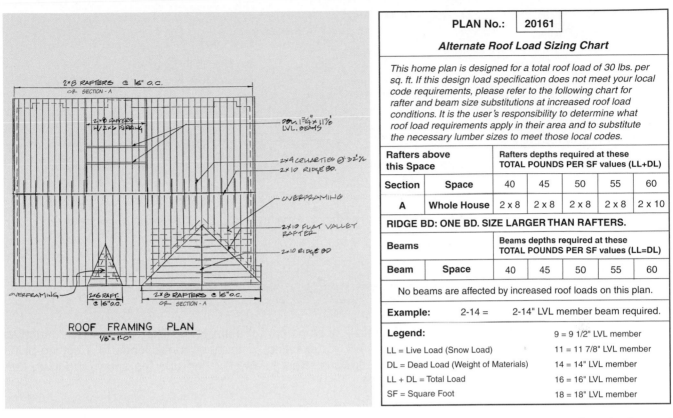

Figure 3-13. A rafter framing plan helps the carpenter visualize layout of rafters and where openings in the roof require special framing. A sizing chart (right) is sometimes included as a guide when heavy roof loads are likely. (©The Garlinghouse Company)

3.1.7 Section and Detail Drawings

A floor plan or elevation drawing does not show small parts of the structure or how the parts fit into the total structure. For this, the carpenter needs to consult drawings called sections and details.

A *section drawing,* or *section view,* gives important information about size, materials, fastening, and support systems, as well as concealed features. Parts of the structure likely to have a section drawing include walls, window and door frames, footings, and foundations.

The section shows how a part of the structure looks when cut by a vertical plane. Imagine you are looking at the cut edge of the part after it has been sawed in two. Because of the need to show many details, section drawings are made to a large scale.

Figures 3-14 and **3-15** are typical section views. They show how a structure looks when vertically cut at a specific point. Specific attention is given to details of materials and size. A complex structure may need many section drawings to show important details of construction.

Like sections, *detail drawings* show the carpenter any important and complex construction that cannot be included in plan drawings.

Section drawing: A type of drawing that shows how a part of a structure would look if cut along a given plane. Also called *section view.*

Detail drawing: In working drawings, an enlarged view of a part of the structure that cannot be clearly explained in elevation views or general plan drawings.

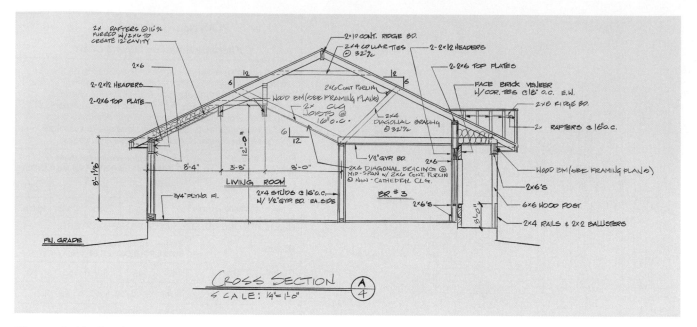

Figure 3-14. Section views are a way to show details of some part of a structure. They carry a label, such as a double letter or a letter and a number (A-A or A-4, for example). The label corresponds to a cutting-plane line shown on an elevation drawing. This section drawing refers to an elevation drawing shown in Figure 3-9. (©The Garlinghouse Company)

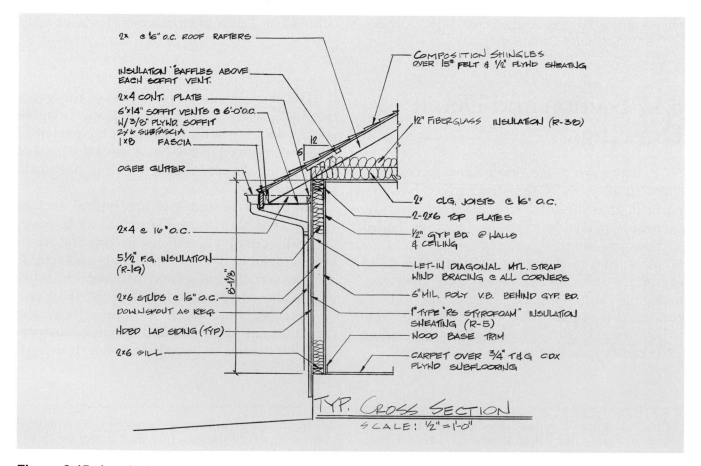

Figure 3-15. A typical section view shows a smaller element of a structure. Note the amount of detail covered, compared to the section view in Figure 3-14. (©The Garlinghouse Company)

They are also large scale and show how various parts are to be located and connected. Fireplaces, stairs, and built-in cabinets are examples of constructions shown as detail drawings. See **Figure 3-16.** Some detail drawings are shown at or near full size. See **Figure 3-17.**

Some stock plans include drawings for mechanical systems—electrical wiring, plumbing, and HVAC. **Figure 3-18** is an electrical plan. It shows the location of every part of an electrical system and diagrams each branch circuit. By following this plan, an electrician knows where to place electrical conductors, receptacles, fixture boxes, and switches.

3.1.8 Dimensions

Dimensions show distance and size. In general, all dimensions greater than one foot are written in feet and inches. For example, standard ceiling height is given as 8'-0" rather than 96".

Carpenters using conventional measure prefer to work with feet and inches, since measurement in this form is easier to visualize. When laying out various distances, they often need to add or subtract dimensions. Steps for making calculations are:

Addition:

$$
\begin{array}{r}
6'\text{-}8" \\
+4'\text{-}6" \\
+2'\text{-}4" \\
+1'\text{-}2" \\
\hline
= 13'\text{-}20" \text{ or } 14'\text{-}8"
\end{array}
$$

Subtraction

$$
\begin{array}{r}
8'\text{-}4" \\
-6'\text{-}10"
\end{array}
$$

(Since 10 cannot be subtracted from 4, borrow 12" from 8'.) Thus:

$$
\begin{array}{r}
7'\text{-}16" \\
-6'\text{-}10" \\
\hline
= 1'\text{-}6"
\end{array}
$$

Dimensions: Lines on a drawing that show distance and size.

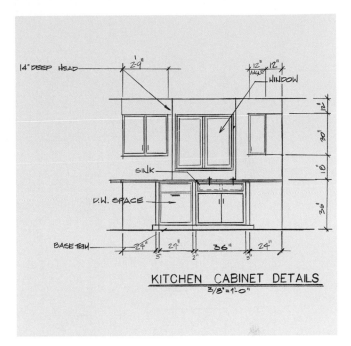

Figure 3-16. This detail drawing of kitchen cabinets is typical. Notice the careful attention given to dimensions and to naming of parts. (©The Garlinghouse Company)

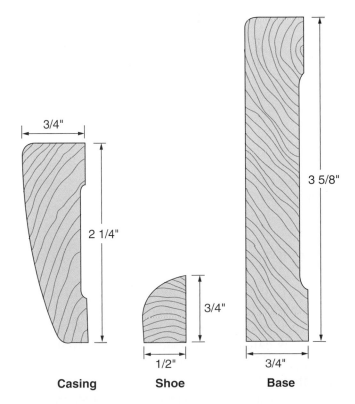

Figure 3-17. Some detail drawings are made at or near full size. Shown are cross-sections of special millwork needed for finishing a house.

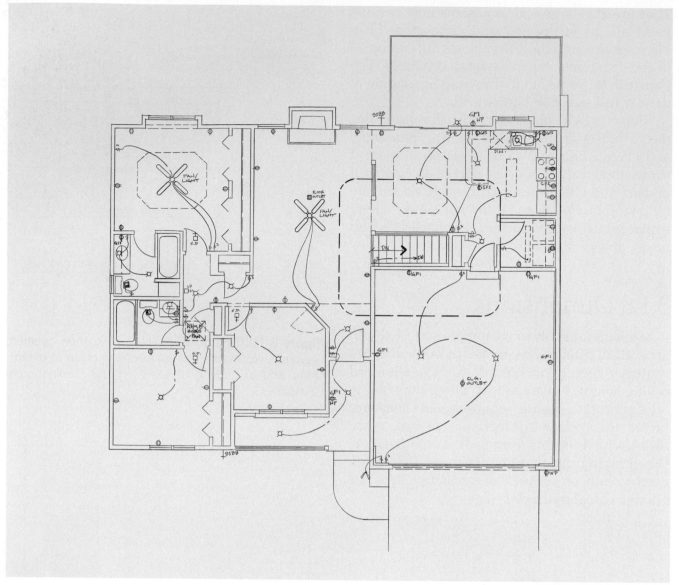

Figure 3-18. The electrical plan is similar to a floor plan, but is intended specifically as a guide for the electrician in making calculations about size of service, switching arrangements, and number of circuits needed. (©The Garlinghouse Company)

3.1.9 Lists of Materials

Sets of plans also include a *list of materials,* **Figure 3-19.** It is known by other names as well: *bill of materials, lumber list,* or *mill list.* Whatever the list is called, it includes all of the materials and assemblies needed to build the structure. A materials list usually includes the number of the item and its name, description, size, and the material of which it is made. Built-in items, such as cabinets, are included.

Another part of the materials list is the window and door *schedule.* This gives the quantity needed, size of the rough openings, and descriptions. Sometimes the manufacturer is also specified. **Figure 3-20** shows a door schedule taken from a materials list.

List of materials: A listing of all the materials and assemblies needed to build the structure. Also called a *bill of materials, lumber list,* or *mill list.*

Schedule: A part of the bill of materials that covers windows and doors as to quantity, size of opening, description, and sometimes the name of the manufacturer.

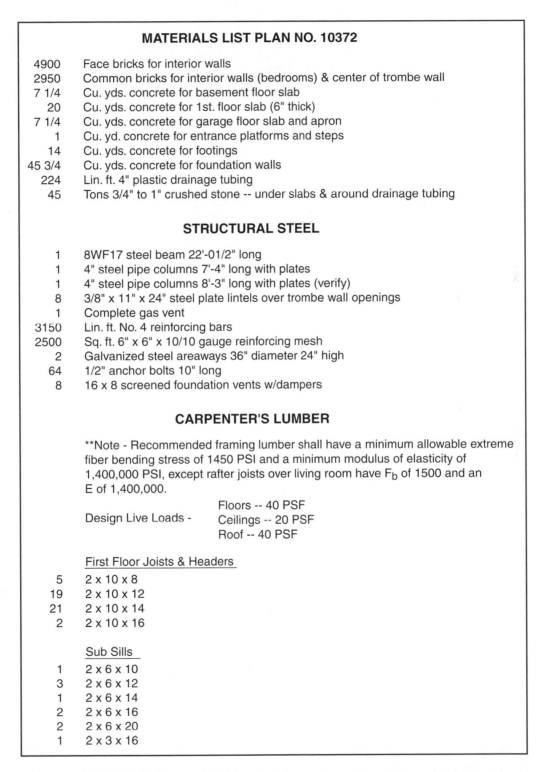

MATERIALS LIST PLAN NO. 10372

4900	Face bricks for interior walls
2950	Common bricks for interior walls (bedrooms) & center of trombe wall
7 1/4	Cu. yds. concrete for basement floor slab
20	Cu. yds. concrete for 1st. floor slab (6" thick)
7 1/4	Cu. yds. concrete for garage floor slab and apron
1	Cu. yd. concrete for entrance platforms and steps
14	Cu. yds. concrete for footings
45 3/4	Cu. yds. concrete for foundation walls
224	Lin. ft. 4" plastic drainage tubing
45	Tons 3/4" to 1" crushed stone -- under slabs & around drainage tubing

STRUCTURAL STEEL

1	8WF17 steel beam 22'-01/2" long
1	4" steel pipe columns 7'-4" long with plates
1	4" steel pipe columns 8'-3" long with plates (verify)
8	3/8" x 11" x 24" steel plate lintels over trombe wall openings
1	Complete gas vent
3150	Lin. ft. No. 4 reinforcing bars
2500	Sq. ft. 6" x 6" x 10/10 gauge reinforcing mesh
2	Galvanized steel areaways 36" diameter 24" high
64	1/2" anchor bolts 10" long
8	16 x 8 screened foundation vents w/dampers

CARPENTER'S LUMBER

**Note - Recommended framing lumber shall have a minimum allowable extreme fiber bending stress of 1450 PSI and a minimum modulus of elasticity of 1,400,000 PSI, except rafter joists over living room have F_b of 1500 and an E of 1,400,000.

Design Live Loads -   Floors -- 40 PSF   Ceilings -- 20 PSF   Roof -- 40 PSF

First Floor Joists & Headers

5	2 x 10 x 8
19	2 x 10 x 12
21	2 x 10 x 14
2	2 x 10 x 16

Sub Sills

1	2 x 6 x 10
3	2 x 6 x 12
1	2 x 6 x 14
2	2 x 6 x 16
2	2 x 6 x 20
1	2 x 3 x 16

Figure 3-19. One page of a typical list of materials. Such a list includes all the materials and assemblies (for example, doors, windows, and cabinets) needed for construction of the building. (This list is for a structure other than the one shown in previous drawings.)

Doors:	Frames	Openings	Jambs
1	Outside entrance	3-0 x 6-8 x 1 3/4	
1	Outside solid core service	2-8 x 6-8 x 1 3/4	
1	Outside service	2-8 x 6-8 x 1 3/4	
1	Inside	2-8 x 6-8 x 1 3/8	
5	Inside	2-6 x 6-8 x 1 3/8	
2	Inside	2-0 x 6-8 x 1 3/8	
1	Inside	1-6 x 6-8 x 1 3/8	
2	Garage	9-0 x 7-0 x 1 3/8	
1	Combination	2-8 x 6-8 x 1 1/8	
1	Combination	3-0 x 6-8 x 1 1/8	
1	Bi-fold door	5-0 x 6-8 x 1 3/8	
1	Bi-fold door	4-0 x 6-8 x 1 3/8	
2	Bi-fold doors	3-0 x 6-8 x 1 3/8	
2	Sides of door trim	5-0 x 6-8, 1/2 x 2 1/4	
4	Sides of door trim	4-0 x 6-8, 1/2 x 2 1/4	
7	Sides of door trim	3-0 x 6-8, 1/2 x 2 1/4	
5	Sides of door trim	2-8 x 6-8, 1/2 x 2 1/4	
10	Sides of door trim	2-6 x 6-8, 1/2 x 2 1/4	
4	Sides of door trim	2-0 x 6-8, 1/2 x 2 1/4	
2	Sides of door trim	1-6 x 6-8, 1/2 x 2 1/4	

Windows: All *Andersen Perma-shield casement and awning windows are to be complete with frames, sash, interior trim, exterior trim, screens, storm sash and hardware.

Quan.	No.		
1	C16-3	Perma-shield casement window	
5	C25	Perma-shield casement window	
1	C16	Perma-shield casement window	*Andersen Corporation
1	C135	Perma-shield casement window	Bayport, MN 55003
1	AN41-22	Perma-shield awning window	
5	A42	Perma-shield awning window	
2	A32	Perma-shield awning window	
2	A31	Perma-shield awning window	
2	Steel basement units 2 lites 15 x 20 with screens		
	Custom Window Material - Trombe Wall		
50	Lin. ft. 2 x 10 Surround material		
15	Lin. ft. 2 x 8 Surround material		
15	Lin. ft. 2 x 4 Blocking		
165	Lin. ft. 2 x 6 Mullion material		
588	Lin. ft. 3/4" x 3/4" Stop material		
294	Lin. ft. 3/4" x 1" Stop material		
36	16 GA 1 1/2" x 8" Galvanized wall ties		
500	Sq. ft. 1/4" Plate glass		

Figure 3-20. One page of a door and window schedule from a list of materials. A manufacturer is often specified. (Not for the plan used in this chapter.)

3.1.10 Symbols

Since architectural plans are drawn to a small scale, materials and construction particulars can seldom be shown as they actually appear. Also, it would require too much time to produce drawings of this nature. The architect, therefore, uses **symbols** to represent materials and other items and certain approved shortcuts (called *conventional representations*). These simplify the illustration of assemblies and other elements of the structure. Generally accepted symbols are illustrated in **Figures 3-21** through **3-24.** Abbreviations are

Symbols: Used in plans to represent materials and other items.

Material	Plan	Elevation	Section
Wood	Floor areas left blank	Siding Panel	Framing Finish
Brick	Face / Common	Face or common	Same as plan view
Stone	Cut / Rubble	Cut Rubble	Cut Rubble
Concrete			Same as plan view
Concrete block			Same as plan view
Earth	None	None	
Glass			Large scale / Small scale
Insulation	Same as section	Insulation	Loose fill or batt / Board
Plaster	Same as section	Plaster	Stud / Lath and plaster
Structural steel		Indicated by note	
Sheet metal flashing	Indicated by note		Show contour
Tile	Floor	Wall	

Figure 3-21. Symbols are used to represent things that are impractical to draw.

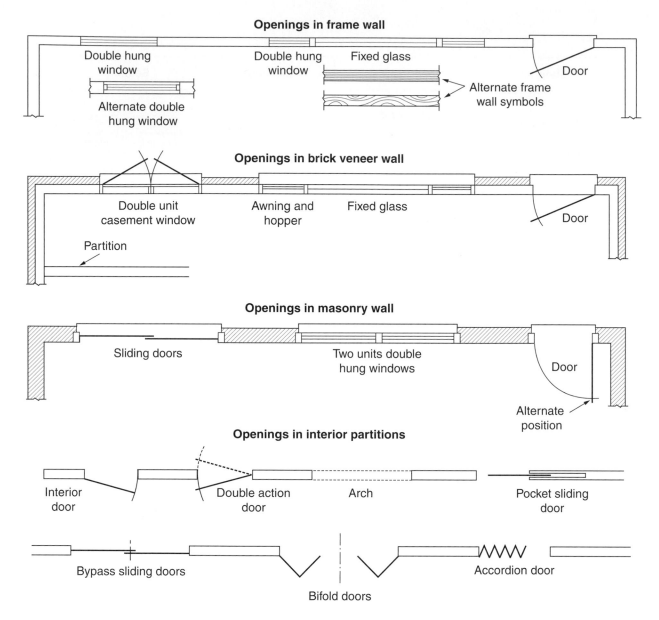

Figure 3-22. How openings in walls are presented in plan views. Handedness of doors (which way they swing) is always shown.

commonly used on plans to save space. Refer to Appendix B, **Technical Information**, for a listing of common abbreviations.

3.2 How to Scale a Drawing

Architectural plans include dimensions that show many distances and sizes. Still, carpenters may require a dimension that is not shown.

To get this dimension, they need to *scale* the drawing. An architect's scale may be used for this purpose, **Figure 3-25**. Each division of an architect's scale represents 1′. The foot division is subdivided into major parts equal to inches and

---

Scale: A term that specifies the size of a reduced-size drawing. For example, a plan is drawn to 1/4″ scale if every 1/4″ represents 1′ on the real structure. Also, the process of measuring a dimension on a plan with an architect's scale.

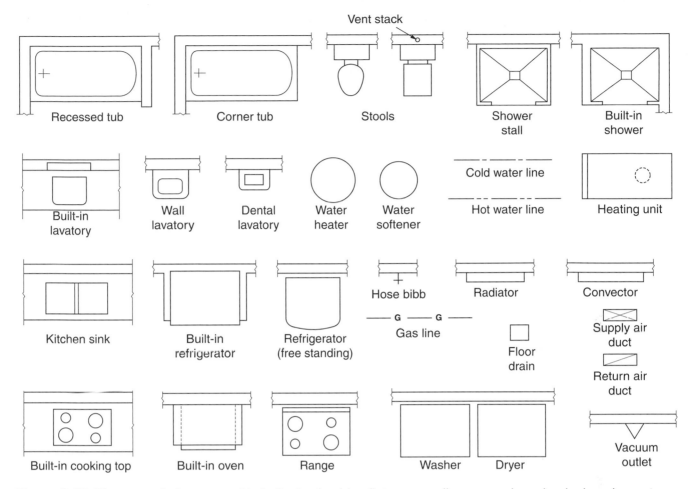

Figure 3-23. These symbols are used to indicate plumbing fixtures, appliances, and mechanical equipment.

Electrical symbols			
Ceiling outlets for fixtures		Lighting panel	
		Power panel	
Wall fixture outlet		S	Single-pole switch
Ceiling outlet with pull switch (PS)		S₂	Double-pole switch
Wall outlet with pull switch (PS)		S₃	Three-way switch
Duplex convenience outlet		S₄	Four-way switch
Waterproof convenience outlet (WP)		Sₚ	Switch with pilot light
Convenience outlet 1 = single 3 = triple (1, 3,)			Push button
Range outlet (R)			Bell

Figure 3-24. Some electrical symbols. Although carpenters are not responsible for wiring, they should recognize the symbols.

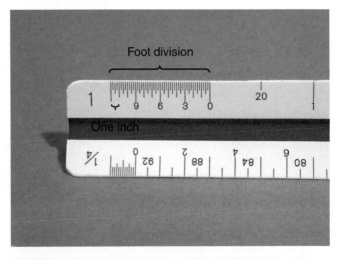

Figure 3-25. An architect's scale may be used to check the dimensions on a scaled drawing. Each division represents 1′. Fractions of a foot are checked against the foot division.

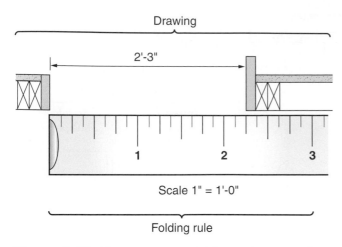

Figure 3-26. How to use a folding rule to scale a plan. Do you know how the correct dimension was determined?

parts of an inch. Another way to scale a plan is to use a regular folding rule. See **Figure 3-26.**

3.3 Changing Plans

Sometimes, an owner wants changes in plans as the job progresses. If the changes are minor—such as changing the size or location of a window or making a revision in the design of a built-in cabinet—the carpenter can usually handle them. Sketches or notations should be recorded on each set of plans so there is no misunderstandings. Major changes, such as the relocation of a load-bearing wall or stairs, may generate a chain reaction of problems. These should be undertaken only after an architect has made the necessary plan changes and the owner has approved them.

3.4 Specifications

Although the working drawings show many of the requirements for a structure, certain supplementary information is best presented in written form as *specifications,* commonly called *specs.* See **Figure 3-27.** The carpenter should carefully check and follow the specs. Specifications for a residential structure generally include these headings:

- General Requirements, Conditions, and Basic Information.
- Excavating and Grading.
- Masonry and Concrete Work.
- Sheet Metal Work.
- Rough Carpentry and Roofing.
- Finish Carpentry and Millwork.
- Insulation, Caulking, and Glazing.
- Lath and Plaster or Drywall.
- Schedule for Room Finishes.
- Painting and Finishing.
- Tile Work.
- Electrical Work.
- Plumbing.
- Heating and Air Conditioning.
- Landscaping.

Under each heading, the content is usually divided into the following sections: scope of work, specifications of materials to be used, application methods and procedures, and guarantee of quality and performance.

Carefully prepared specifications are valuable to the contractor, estimator, tradespeople, and building supply dealer. They protect the owner and help to ensure quality work. In addition to the previously described items, specifications may include information and requirements regarding building permits, contract payment provisions, insurance and bonding, and provisions for making changes in the original plans.

3.5 Modular Construction

The *modular construction* concept is based on the use of a standard grid divided into 4″ squares. Refer to **Figure 3-28.** Consider each

---

Specifications: A written document stipulating the type, quality, and sometimes the quantity of materials and work required for a construction job. Also called *specs.*

Modular construction: Use of a standard grid (based on 4″ squares) for construction.

DIVISION 5	PLAN NO. 12275

CARPENTRY, MILLWORK, & HARDWARE

General Conditions:

This contractor shall read the General Conditions and Supplementary General Conditions which are a part of these Specifications.

Scope of Work:

Furnish and install all rough lumber, millwork, rough and finished hardware, including all grounds, furring, and blocking, frames and doors, etc., as shown on the drawings or hereinafter specified.

Rough Lumber:

All lumber used for grounds, furring, blocking, etc., shall be #1 Common Douglas Fir or Pine.

Grounds and Blocking:

Furnish and install suitable grounds wherever required around all openings for nailing metal or wood trim such as casings, wainscoting, cap, shelving, etc.; also building into masonry all blocking as required. All wood grounds must be perfectly true and level and securely fastened in place.

Doors:

Interior doors to be plain sliced red oak for stain finish. Solid core doors to be laminated 1 3/4" thick, or as called for in the door schedule. See plan for Label doors.

Temporary Doors and Enclosures:

Provide temporary board enclosures and batten doors for all entrance openings. Provide suitable hardware and locks to prevent access to work by unauthorized persons.

At any openings used for ingress and egress of material, provide and maintain protection at jambs and sills as long as openings are so used.

Rough Hardware:

All hardware such as nails, spikes, anchors, bolts, rods, etc., in connection with rough car-
All _____ by this contractor. Use aluminum non-staining nails for all exterior wood.
around openings
edges of trim shall be lightly _____
smoothed and machine sanded at the factory
carpenters at the building in a first class manner before _____
applied.

Finishing Hardware:

An allowance of $1,000.00 shall be included for finishing hardware, which will be selected by the architect and installed by this contractor. Locksets shall match existing design and finish. Adjustment of cost above or below this allowance shall be made according to authentic invoice for hardware for this building from hardware supplier. Hardware supplier shall furnish schedule and templates to frame manufacturer.

Caulking:

Caulk all exterior windows, doors, etc., and openings throughout building with polysulfide based sealant, see specifications under cut stone.

Wood Paneling:

All wood paneling shall be Weyerhauser Forestglo, Orleans Oak 1/4" prefinished. Apply over 1/2" drywall on steel studs and furring.

Figure 3-27. A portion of the specifications that are included with a set of house plans. Other topics included in specifications are plumbing, heating, and wiring.

individual square (module) as the base of a cube so the module can be applied to elevations as well as horizontal planes.

All dimensions are based on multiples of 4″, including 16″, 24″, and 48″. The last two dimensions (24″ and 48″) are sometimes called the minor and major modules. Many building materials and fabricated units are manufactured to coordinate with modular dimensions. This helps to eliminate costly cutting and fitting during construction. A good example of this system is illustrated in standard concrete blocks. They are manufactured in nominal sizes of 8″ × 8″ × 16″. The actual size is 3/8″ less in each dimension to allow for bonding (mortar joints). See Chapter 7, **Footings and Foundations**.

The National Lumber Manufacturers Association has developed modular dimension standards for components. The system is called

Unicom, an acronym for *uniform manufacture of components.* This system helps manufacturers apply modern mass-production methods to building construction. For example, a sheet of plywood is 48″ × 96″ and exactly fits the module.

A modular system also exists in the SI metric system. It is based on a grid made up of 100 mm squares. The 100 mm module is nearly, but not exactly, 4″. The ISO standard also recommends that the submultiples of 25, 50, and 75 mm be used, as well as the multiples of 300, 400, 600, 800, and 1200 mm. The 600 and 1200 multiples become the minor and major modules. See the reference to metric measurements in Appendix B, **Technical Information**.

Computers are frequently used to create a set of plans and even perspective drawings for houses. This is known as *computer-aided drafting and design (CADD).* Working with a variety of archi-

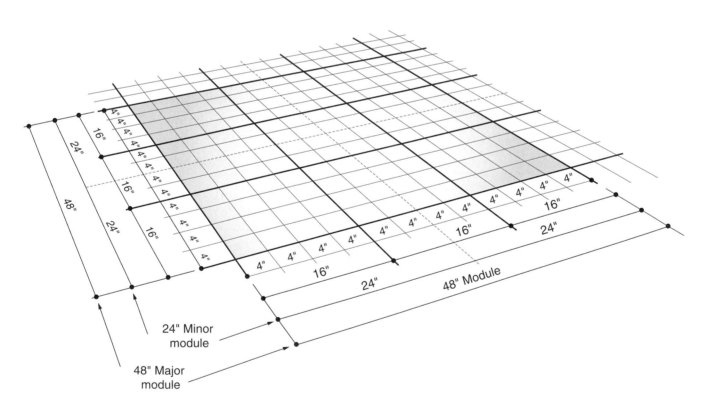

Figure 3-28. Plans for modular structures to be built with components or by conventional framing methods are designed to exact grid sizes. (National Forest Products Assn.)

Unicom system: A term taken from "uniform manufacture of components." The system uses modules with sizes that are multiples of a standard unit.

Computer-aided drafting and design (CADD): Using computers to create a set of plans and perspective drawings for houses.

tectural software, the drafter creates drawings on the computer. Then, the drawings are sent to a printer or plotter to be duplicated on paper. A plotter is capable of producing drawings in larger sizes than most printers. It has pens or inkjet cartridges of various colors to produce a drawing in several colors. An advantage of CADD is the ability to quickly revise a drawing without the erasures necessary when the drawing is first made on paper. Some software can automatically produce materials lists.

Today, the manufacture of construction components, such as truss joists and rafters or even whole houses, can be computer controlled. The combination of design and manufacture controlled by computer is called *CADD-CAM.* As you might imagine, CAM stands for ***computer-aided manufacturing.***

3.6 Metric Measurement

While the United States still uses the customary measurement system inherited from England, most nations, including Canada, use the SI metric system of measurement. Carpenters should be familiar with the metric system because a metric specification may be encountered. The entire metric system is modular, since it is based on units of ten. Thus, the millimeter is 1/1000 of a meter and a centimeter is 1/100 of a meter. A kilometer is equal to 1000 meters. Countries that use metric measure commonly use 300 millimeters as a standard module. This is equal to about 1'. In architectural drawings dimensioned in metric units, measurements of buildings are given in millimeters, while site measurements are in meters.

Computer-aided manufacturing (CAM): A combination of design and manufacturing activities controlled by a computer.

Building code: A collection of rules and regulations for construction established by organizations and based on experience and experiment.

3.7 Building Codes

A ***building code*** is a collection of laws that apply to a given community. See **Figure 3-29.** The code is usually presented in booklet form. It covers all important aspects of the erection of a new building, as well as the alteration, repair, and demolition of existing buildings. The basic purpose is to provide for the health, safety, and general welfare of building occupants and other people in the community.

A code sets the minimum standards that are acceptable in a community for design, quality of materials, and quality of construction. It also sets requirements concerning such design factors as:

- Size, height, and bulk of buildings.
- Room sizes.
- Ceiling heights.
- Lighting and ventilation.

Many local codes contain detailed directions and standards for installation of a building's systems. These govern the methods and materials used when installing plumbing, wiring, and heating systems.

A carpenter should be aware that building codes might vary from community to community. What is common practice in one area may not be allowed in another. For example, some communities' codes will require one or more sumps in basement floors and perimeter drainage around the footings. Another community may not mention them. These differences should be carefully noted.

Some items in a code must be adjusted to local conditions. In northern climates, footings need to be deeper than in southern states. Structures in hurricane belts require extra bracing and strapping.

Working Knowledge

In doing any kind of carpentry work, the importance of closely following all building codes cannot be overemphasized. If a building inspector finds work that does not conform to code, it must be done over. This can considerably add to the expense of construction.

13.1.2   **Required Room Sizes**

No dwelling unit shall be erected or constructed which does not comply with the following minimum room sizes:

(a) Living Room:	250 sq. feet
(b) Dining Room:	100 sq. feet
(c) Kitchen:	90 sq. feet
(d) Bathroom (Three Fixtures)	40 sq. feet
(e) Powder Room (Two Fixtures)	24 sq. feet
(f) First Bedroom:	150 sq. feet
(g) Each Additional Bedroom:	110 sq. feet
(h) Den or Library, etc.	100 sq. feet

(Permitted only when dwelling unit includes <u>two</u> (2) or more bedrooms of the required sizes.)

(i) Garage:	253 sq. feet

(Minimum dimensions shall be 11 feet x 23 feet.)

13.1.3   **Required Rooms**

No dwelling unit shall be erected or constructed which does not contain one each of the following rooms: Living Room, Dining Room, Kitchen, Bathroom, Bedroom and Garage. Each room is to be separated from each other room by full height partitioning and doors, except that Living Room, Dining Room and Kitchen need not be fully separated from each other, and the Garage may be a detached building. Required rooms and all additional bedrooms, den and library shall not be located in a basement.

Figure 3-29. Sample of a local building code. Codes vary from community to community. (Village of Flossmoor, Illinois)

3.7.1 Model Codes

Continual improvements in construction methods and materials make the updating of local building codes difficult and costly. Because of this, many communities have adopted *model codes.* One of the first model codes was introduced by The American Insurance Association (successor to the National Board of Fire Underwriters). This publication is known as the National Building Code. An abbreviated edition is also available.

The International Code Council introduced the *International Residential Code for One- and Two-Family Dwellings* in 2000. This model code was developed from the codes published by the Building Officials and Code Administrators International, Inc. (BOCA), the Uniform Codes published by the International Conference of Building Officials (ICBO), and the Standard Codes published by the Southern Building Code Congress International, Inc. (SBCCI). The code is updated every three years and is intended to replace older codes. Although BOCA, ICBO, and SBCCI ceased to exist as separate organizations in 2002, their codes are still in use in a number of states and municipalities.

In addition to observing building codes adopted by local communities (cities, towns, counties), the carpenter must be aware of certain laws at the state level that govern buildings. Several states have developed building codes for adoption by their local communities. However, for the most part, state codes deal mainly with fire protection and special needs for public buildings.

Model codes: A single, national family of building codes.

A carefully prepared and up-to-date code is not sufficient in itself to ensure safe and adequate buildings. All codes must be properly administered by officials who are experts in the field. Under these conditions, the owner can be assured of a well-constructed building. Also, the carpenter is protected against the unfair competition of those who are willing to sacrifice quality for an excessive margin of profit.

Communities have inspectors who enforce the building code. They make periodic inspections during construction or remodeling. The inspectors are persons who have worked in the construction trades or who are otherwise knowledgeable about construction.

There is a movement to adopt a national residential building code for single- and multi-family residences under three stories. Some states have already adopted it. Builders whose operation extends to several states favor such standardization.

3.7.2 Standards

Building codes are based on standards developed by manufacturers, trade associations, government agencies, professionals, and tradespeople. All are looking for a desirable level of quality through efficient means. A particular material, method, or procedure is technically described through specifications. Specifications become *standards* when their use is formally adopted by broad groups of manufacturers and builders and/or recognized agencies and associations. Organizations devoted to the establishment of standards, many of which are directly related to the field of construction, include:

- The American Society for Testing and Materials (ASTM).
- American National Standards Institute (ANSI).
- Underwriters Laboratories, Inc. (UL).

The Commodity Standards Division of the U.S. Department of Commerce develops commercial standards. The chief purpose of the agency is to establish quality requirements and approved methods of testing, rating, and labeling. These standards are designated by the initials CS, followed by a code number and the year of the latest revision.

3.7.3 Building Permits and Inspections

A *building permit* is issued by the community's building officials stating that construction can begin, **Figure 3-30.** Steps for securing a building permit vary from one community to another. Usually, the contractor or building owner files a formal application with the appropriate local agency. This may be a village clerk, city clerk, or a county building department. The application is given to the agency with one or two sets of plans, **Figure 3-31.** Usually, the submitted drawings must include:

- Floor plans.
- Specifications.

Figure 3-30. Construction cannot begin until a building permit is obtained from the community's building officials. The permit and inspection card must be displayed at the building site. This permit is combined with the card.

Building permit: A document issued by building officials certifying that plans meet the requirements of the local building code. The issuing of a permit allows construction to begin.

- Site plan.
- Elevation drawings.

Sometimes a filing fee and plan review fee are required. These are in addition to the fee for the building permit itself.

The plans are examined by building officials to determine if they meet the requirements of the local code. Some communities have an architectural committee that determines if the plans are satisfactory.

It is sometimes necessary for the builder or owner to submit supporting data to show how the building design meets the code. When the plan meets all of the requirements of the building code, a building permit is issued. Permit fees are based on cost of construction, and can amount to hundreds of dollars for large structures.

When construction begins, the building permit and an inspection card are posted on the building site. Sometimes the two are combined, as in **Figure 3-30.** As work progresses, the building inspector makes inspections and fills out the inspection card for approval of work completed. It is important that the permit and card always be attached to the building or somewhere on the construction site.

Work on the structure should *not* proceed beyond the point indicated in each successive inspection. Carpenters on the job must pay close attention to this record. Mechanical work (heating, plumbing, and electrical wiring) may *never* be enclosed before the building inspector has approved the installations. In some communities, a final inspection must be made and an occupancy permit issued. Until then, the building may not be occupied.

VILLAGE OF FLOSSMOOR
APPLICATION FOR BUILDING PERMIT

Type of Work:
Erection ____ Remodel ____ Addition ____ Repair ____ Demolish ____
Construction: Brick ____ Frame ____ Other ____
Color: Roof ____ Brick ____ Trim ____ Siding ____
Proposed use ____
Livable floor area ____

Required Service Entrance Conductors – Full Rated
100 Amp. Service and 100 Amp. Service Switch - .... Wire Size ....
150 Amp. Service and 150 Amp. Service Switch - .... Wire Size ....
200 Amp. Service and 200 Amp. Service Switch - .... Wire Size ....
400 Amp. or Larger Service and 400 Amp. Service Switch - ....
Wire Size ....
Underground Service ☐

MAJOR APPLIANCE CIRCUITS REQUIRED BY CODE

☐ Range .... Amps.	☐ 2 Laundry Circuits-20 .... Amps.
☐ Built-in Oven .... Amps.	☐ Heating Plant .... Amps.
☐ Water Heater .... Amps.	☐ Lighting Circuits .... Amps.
☐ Dish Washer .... Amps.	ELECTRICAL HEATING CIRCUITS
☐ Garbage Disposal .... Amps.	☐ Cable .... Amps.
☐ Sump Pump .... Amps.	☐ Baseboard Units .... Amps.
☐ Clothes Dryer .... Amps.	☐ Electrical Furnace .... Amps.
☐ Bathroom Heater .... Amps.	COMMERCIAL WIRING
☐ Fixed Air Conditioner .... Amps.	☐ Lighting .... Amps.
☐ Window Air Conditioner .... Amps.	☐ Motors .... Amps.
☐ Electric Door Opener .... Amps.	☐ Appliances .... Amps.
☐ 2 Kitchen Circuits–20 .... Amps.	☐ Other .... Amps.

*NOTE – No Lighting or Other Current Consuming Device Shall Be Connected to the 2 Kitchen Circuits, the 2 Laundry Circuits, Heating Plant, Sump Pump or Air Conditioner Circuits.

Street Address: ____
Date ____ Township ____
Real Estate Index No. ____
Lot: ____ Block ____
Subdivision: ____

Property Owner: ____
Present Address ____
Telephone ____
Architect: ____
Address ____
Telephone ____
General Contractor: ____
Address ____
Telephone ____ Bond Expir. Date ____
Air Conditioning & Heating Contractor: ____
Address ____
Telephone ____
Electrical Contractor: ____
Address ____
Telephone ____ Licensed ____
Plumbing Contractor: ____
Address ____
Telephone ____ Bond Expir. Date ____
State License No. ____

As owner of the property, for which this permit is issued and/or as the applicant for this permit, I expressly agree to conform to all applicable ordinances, rules and regulations of the Village of Flossmoor.

Contractor or Owners Signature: ____

Estimated cost of building complete, including all materials and labor $ ____

Building Permit No. ____
Electrical Permit No. ____
Plumbing Permit No. ____

Figure 3-31. An application for a building permit must be accompanied by information about the structure that is to be built. In addition to the information given on the form, plans must be submitted. (Village of Flossmoor, Illinois)

ON THE JOB

Inspector

There are many different kinds of inspectors whose work touches on the construction industry. Many of these can be grouped under the general title of *building inspector* since they deal with some aspect of the building process, such as electrical, plumbing, or mechanical systems.

Approximately half of all inspectors are employed by local units of government. These inspectors perform the important task of making sure that a building and its systems conform to codes, standards, and local regulations. The inspection process usually begins even before construction starts when plans are submitted to the building inspector's department for review. The review may result in approval or may require changes to bring the plan into compliance with codes and regulations. Once construction gets underway, inspections are made at a number of stages—foundation, framing, electrical, plumbing, and mechanical system (such as HVAC) installations. Once all work is completed to the satisfaction of the inspector, a final approval, often called an occupancy permit, can be issued.

People employed as building inspectors and in the various specialties, such as electrical or plumbing inspectors, should have a thorough knowledge of all codes and local regulations that the builder must meet. A good background in the building trades is very desirable. Many tradespeople "retire" to working as building inspectors. Some inspectors have academic preparation in such areas as construction technology or construction management. A degree in engineering or architecture is helpful in advancing to supervisory duties.

Working conditions are generally the same as most building trades jobs, although less physically demanding. Inspectors must be able to climb ladders, work in confined spaces, and carefully navigate through areas cluttered with materials and tools.

In addition to the inspectors employed by government, a significant number work for engineering or architectural service firms. A growing number are self employed, primarily in the field of home inspection where they are hired by a buyer or seller to examine a property being offered for sale.

Summary

Carpenters must be able to read and understand building plans. Included in a typical set of plans are site, foundation, and floor plans; framing, electrical, plumbing, and HVAC plans; elevations; section drawings; detail drawings; and lists of materials. In addition to plans, there often are written specifications, which must be carefully followed. Dimensioning may be U.S. customary or metric units. Local building codes often govern how a structure may be built. Permits for building must be obtained from local government units and periodic inspections must be done to approve the work.

Test Your Knowledge

Answer the following questions on a separate piece of paper. Do not write in this book.

1. A set of house plans usually includes which drawings?
2. Residential plan views are usually drawn to a scale of _____" = _____'-_____".
3. Floor plans show the _____ and outline of the building and its rooms.
4. The plot plan shows the _____.
5. Elevation drawings show the _____ walls of the structure.
6. A section view shows how a part of a structure looks when _____.
7. Dimensions show _____ and size.
8. Draw symbols that represent these materials and items:
    A. Concrete.
    B. Double hung window.
    C. Interior door.
    D. Refrigerator.
    E. Wall lavatory.
    F. Three-way switch.
    G. Range outlet.
    H. Wall fixture outlet.
9. To obtain a plan dimension not shown, a(n) _____ scale may be used.
10. Working drawings (plans) provide much information required by the builder. Supplementary information is supplied by written _____.

11. The modular construction concept is based on the use of a standard grid divided into _____" squares.
12. What does *CADD-CAM* stand for?
13. What standard unit is used for building measurements in the metric system?
14. *True or False?* A building code covers all important aspects of the erection of a building.
15. Describe the usual procedure for securing a building permit.
16. *True or False?* It is permissible to cover up mechanical systems prior to an inspection, provided you first take photos to prove how the work was done.

Curricular Connections

Mathematics. Draw a simple floor plan for a house at a scale of 1/4" = 1'. Use an 8 1/2" × 11" sheet of plain paper and an ordinary ruler. The house should be rectangular with outside dimensions of 48' × 24'. It should have five rooms: living room, kitchen with dining area, bathroom, and two bedrooms. Be sure to note the scale on your drawing.

Social Studies. Why are businesses, homes, and manufacturing plants usually in separate areas of a community? Interview the person in your community who is in charge of planning or zoning. Find out why zoning is considered important for the community and how a property owner can seek a zoning change. Obtain a copy of the community's zoning map and use it for a visual aid as you report your findings to the class.

Outside Assignments

1. Obtain a complete set of plans for an average-size residence. Try to borrow a set from a local builder. Carefully study the views shown. Make a list of symbols, notes, and abbreviations that you do not understand. Then, go to reference books and architectural standards books to find the informa-

tion. Ask your instructor for help if you have difficulty with some of the views.

2. Study the building code in your community. Become familiar with the various sections that are covered and especially note requirements that apply to residential work. Submit a general outline of the material you feel is most important.

3. Make a trip to offices of the agency that issues building permits in your community. If possible, visit with the commissioner or director of building. Check with your instructor first and be sure to call for an appointment in advance. During your visit, obtain information concerning building permits and inspection procedures. Learn the cost for a permit and what plans and specifications need to be submitted. Prepare carefully organized notes and make an oral report to your class.

Material Safety Data Sheet

Preparation/Revision Date:

ACME Chemical Company

24 hour Emergency Phone: Chemtrec: 1-800-424-9300
Outside United States: 1-202-483-7616

Trade Name/Syn: **DICHLOROMETHANE * METHYLENE DICHLORIDE**
Chem Name/Syn: **METHYLENE CHLORIDE**
CAS Number: **75-09-2**
Formula: Ch_2Cl_2

NFPA Rating:

Health   4
Flammability   1
Reactivity   1

Statement of Hazard:

Possible cancer hazard. May cause cancer based on animal data. Harmful if swallowed or inhaled. Vapor irritating. May cause eye injury and/or skin irritation. May cause damage to liver, kidneys, blood, and central nervous system.

Effects of Overexposure-Toxicity-Route of Entry:

Toxic by ingestion and inhalation. Irritating on contact with skin, eyes, or mucous membranes. May cause eye injury. Inhalation of high vapor concentrations causes dizziness, nausea, headache, narcosis, irregular heartbeats, coma, and death. If vomiting occurs, methylene chloride can be aspirated into the lungs, which can cause chemical pneumonia and systemic effects. Medical conditions aggravated by exposure: heart, kidney, and liver conditions. Routes of entry: inhalation, ingestion.

Hazardous Decomposition Products:

HCL, phosgene, chlorine.

Will Hazardous Polymerization Occur?

Will not occur under normal conditions.

Is the Product Stable?

Product is normally stable.

Conditions to Avoid:

Contact with open flame, welding arcs, and hot surfaces.

Spill Procedures, Disposal Requirements/Methods:

Evacuate the area of all unnecessary personnel. Wear suitable protective equipment. Eliminate any ignition sources until the area is determined to be free from explosion or fire hazard. Contain the release with a suitable absorbent. Place in a suitable container for disposal. Dispose of in accordance with all applicable federal, state, and local regulations.

Ventilation:

Local exhaust: Recommended
Mechanical (Gen):   Recommended
Special: NA
Other: None

Respiratory Protection:

NIOSH/MSHA air supplied respirator.

Protective Gloves:

Viton, PVA, or equivalent to prevent skin contact.

Other Protective Equipment:

Safety glasses with side shields must be worn at all times; eyewash; fume hood.

Material data safety sheets are available from the manufacturers of hazardous materials and chemicals. These sheets contain information on the potential dangers of a specific hazardous material, the characteristics of the material, and protective measures to follow when working with the material. A typical material data sheet (MSDS) is shown above.

Hand Tools

Learning Objectives

After studying this chapter, you will be able to:
- Identify the most common hand tools.
- Select the proper hand tool for a given job.
- Identify the main parts of each major hand tool.
- Explain proper methods of tool maintenance and storage.

Technical Vocabulary

All-hard blades	Hacksaw
Backsaw	Hand screws
Bar or pipe clamps	Jack plane
Block plane	Jointing
Burnisher	Kerf
Butt gauge	Level
C-clamp	Marking gauge
Chalk line	Nail claw
Claw hammer	Nailers
Combination square	Nail puller
Coping saw	Nail set
Crosscut saw	Oilstone
Dados	Pencil compass
Flexible-back blades	Plumb bob
Flexible measuring tape	Pry bar
Folding wood rule	Rabbet
Fore and jointer plane	Rabbet plane
Framing square	Rafter square
	Rasps

Ripping bar	Speed square
Ripsaw	Tackers
Router plane	T-bevel
Scraper	Tin snips
Scratch awl	Try square
Sheetrock-drywall saw	Utility knives
Smooth plane	Wing dividers
Staplers	Wood chisel
	Wood clamp

Hand tools are essential to every aspect of carpentry work. A great variety of tools is required because residential and light commercial construction covers a broad range of activities.

The carpenter, a skilled worker, carefully selects the types and sizes of tools that best suit personal requirements. Tools become an important part of the carpenter's life, helping him or her perform the various tasks of the trade with speed and accuracy.

Although the basic design of common woodworking tools has changed little over many years, modern technology and industrial know-how have brought numerous improvements. Special tools have been developed to do specific jobs. Experienced carpenters appreciate the importance of having good tools and select those that are accurately made from quality materials. They know that such tools last longer and will enable them to do better work.

A detailed study of the selection, care, and use of all the hand tools available is not practical for this text. Instead, a general description of the tools most commonly used by the carpenter is provided.

4.1 Measuring and Layout Tools

Measuring tools must be handled with considerable care and kept clean. Only then can a high level of accuracy be assured. **Figure 4-1** shows tools used for measuring and layout.

4.1.1 Tapes and Rules

A *folding wood rule* and a *flexible measuring tape* are indispensable. Standard length for a folding rule is 6'. Some have a ruled metal slide for taking inside measurements. Measuring tapes can be made of fabric or steel. Both types are flexible and easily carried. Steel tapes are commonly found in 6'–25' lengths. Tapes of 50' or longer are also available. Inch-wide, metal-bladed rules are stiffer than narrower ones, making them easier to use. See **Figure 4-2.**

4.1.2 Squares

The *framing square*, also called a *rafter square,* is specially designed for the carpenter. The blade (or body) is 2" wide and 24" long. The

tongue is at a right angle to the blade. It is 1 1/2" wide and 16" long. The corner of the square is called the *heel.* One side of the square, with the tongue to the right and toward you, is called the *face.*

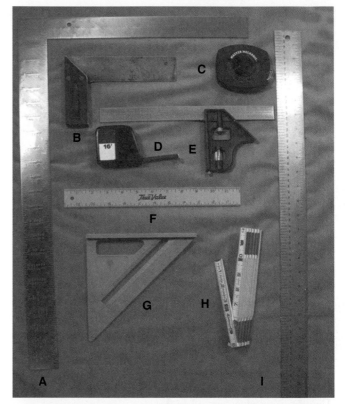

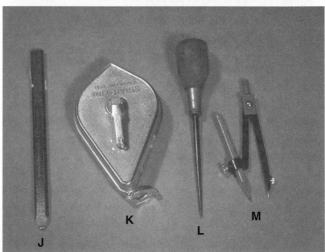

Figure 4-1. Carpenters often use these measurement and layout tools. A—Framing square (also called a "rafter" square). B—Try square. C—Long tape measure. D—Layout tape rule. E—Combination square. F—Steel rule. G—Speed square. H—Folding wood rule. I—4' steel rule. J—Carpenter's pencil. K—Chalk line. L—Awl. M—Pencil compass.

Folding wood rule: Measuring tool with a standard length of 6'. Some have a ruled metal slide for taking inside measurements.

Flexible measuring tape: Easily carried measuring tape made of fabric or steel.

Framing square: A square designed for carpenters with a short section (called the tongue) attached at a right angle to the longer section (called the blade). A number of tables are imprinted on its sides.

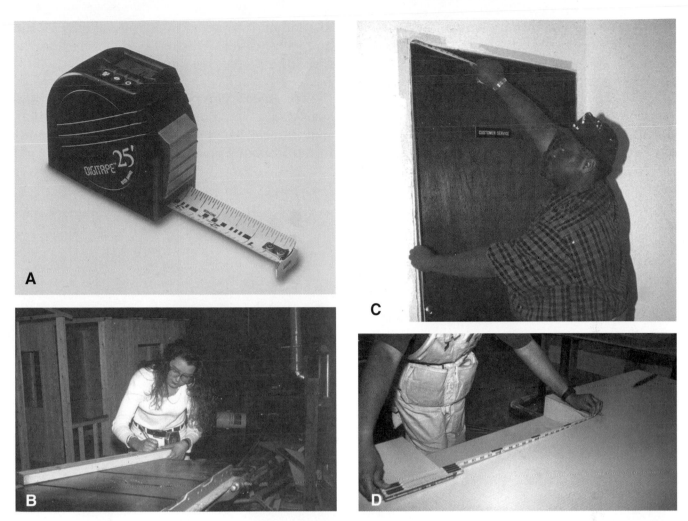

Figure 4-2. Steel tape rules are available in lengths from 6'–25'. A—Newer models have direct digital readouts. B—Rules are equipped with a hook that grips the end of the workpiece as a measurement is taken. C—Since it is flexible, the tape easily bends around corners. D—Using a folding wood rule to take a measurement. (L.S. Starrett Co.; Santa Rita High School, Tucson; McDaniels Construction Co., Inc.; Bullard-Haven Technical School)

Framing squares have a number of tables printed on them. The face has rafter tables printed on the body. These can be used in determining the length of rafters. The tongue carries the octagon scale. This scale is used to lay out the angles of an octagon. The opposite side of the square has the Essex Board Measure Table and the Brace Measure Table. These tables are used very little, compared to the rafter table.

The outside edge of the face is divided into 1/16" graduations, while the inside edge has 1/8" graduations. The outside edge of the opposite side of the square has each inch graduated in twelfths of an inch. This is useful in making a scale drawing. Each inch represents a foot and each 1/12" represents an inch. The inside edge of the tongue is graduated in tenths of an inch.

The framing square is a versatile tool with many uses. Among other tasks, the carpenter can use it to:

- Check or square lines across boards.
- Check the squareness of corners.
- Make a 45° angle across wide boards.
- Find the length and angles of rafters.
- Lay out stairs.

An explanation of how the square is used in framing rafters is included in Chapter 10,

Roof Framing. The framing square is available in aluminum or in steel with a copper-clad or blued finish.

Try squares are available with blades 6″ to 12″ long. Handles are made of wood or metal. These are used to check the squareness of surfaces and edges. They are also used to lay out lines perpendicular to an edge, **Figure 4-3.**

The *combination square* serves a similar purpose as a try square. It is also used to lay out miter joints. Its adjustable, sliding blade allows it to be conveniently used as a gauging tool. It can be used as a marking gauge to make parallel lines and as a square for 90° and 45° angles. Refer to **Figure 4-4.**

The *T-bevel* has an adjustable blade making it possible to transfer an angle from one place to another. It is useful in laying out cuts for hip and valley rafters. **Figure 4-5** illustrates how the framing square is used to set the T-bevel at a 45° angle.

The *pencil compass* is available in several sizes and serves a number of purposes. Dimensions can be "stepped off" along a layout or transferred from one position to another. Circles and

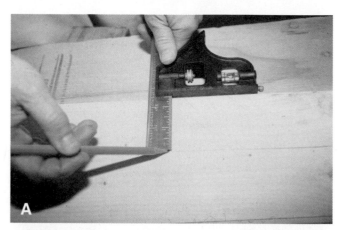

Figure 4-3. The try square is handy for laying out lines perpendicular to an edge. A scratch awl is sometimes used to mark lines.

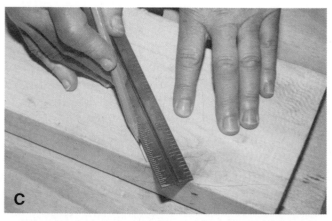

Figure 4-4. A combination square has several uses. A—Drawing a line parallel to the edge of a board. B—Marking a square cut. C—Marking a 45° cut.

Try square: Square available with blades 6″ to 12″ long. Handles are made of wood or metal. These are used to check the squareness of surfaces and edges and to lay out lines perpendicular to an edge.

Combination square: Tool that serves a similar purpose to a try square and is also used to lay out miter joints. Its adjustable sliding blade allows it to be conveniently used as a gauging tool. It can be used as a marking gauge to make parallel lines and as a square for 90° and 45° angles.

T-bevel: Tool with an adjustable blade, making it possible to transfer an angle from one place to another. It is useful in laying out cuts for hip and valley rafters.

Pencil compass: Tool used to draw arcs and circles or to step off short distances.

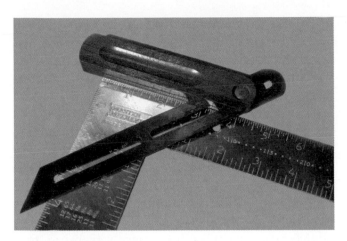

Figure 4-5. The framing square can be used to set a T-bevel at a 45° angle.

Figure 4-6. A pencil compass may be used to draw arcs and circles or to step off short distances.

arcs can also be scribed on surfaces, **Figure 4-6.** Similar to pencil compasses are *wing dividers.* They differ from the pencil compass in that both legs have a point. Legs can be locked to prevent movement that would change the measurement being transferred.

4.1.3 Marking Gauge

A *marking gauge* is used to lay out parallel lines along the edges of material. It may also be used to:

* Transfer a dimension from one place to another.
* Check sizes of material.

The *butt gauge* can be used for the same purposes. However, it is normally used to lay out the gain (recess) for hinges.

In layout work, a *scratch awl* is used to scribe lines on the surface of the material. It is also used to mark points or form starter holes for small screws or nails.

Using a *chalk line* is an easy way to mark long, straight lines. The line is a thin strong cord that is covered with powdered chalk. It is held tight and close to the surface. Then, it is snapped, **Figure 4-7.** This action drives the chalk onto the surface forming a distinct mark. A special reel within the case rechalks the line each time it is wound.

Wing dividers: Tools that are similar to a pencil compass, except with points attached to both legs. The legs can be locked to prevent movement, which would change the measurement being transferred.

Marking gauge: Device used to lay out parallel lines along the edges of material. It may also be used to transfer a dimension from one place to another and check sizes of material.

Butt gauge: Layout tool used to lay out the gain (recess) for hinges.

Scratch awl: Tool used to scribe lines on the surface of the material. It is also used to mark points and to form starter holes for small screws or nails.

Chalk line: Tool used to mark long, straight lines. A thin strong cord that is covered with powdered chalk is held tight and close to the surface. Then it is snapped, driving the chalk onto the surface and forming a distinct mark.

Figure 4-7. Snapping a chalk line is a fast, simple way to lay down straight lines over long distances. (APA-The Engineered Wood Association)

The *level,* **Figure 4-8,** and plumb bob are important devices for laying out vertical and horizontal lines. A standard level is 24″ long. The body may be made of wood, aluminum, or special lightweight alloys. It is often used in connection with a straightedge when the span or height of the work is greater than the length of the level. A *straightedge* is a straight strip of wood, usually laminated for greater stability. Some carpenters prefer a long level that is 4′–6′ in length. It eliminates the need for a straightedge on such work as installing door and window frames.

The *plumb bob* establishes a vertical line when attached to and suspended from a line. Its weight pulls the line in a true vertical position for layout and checking. The point of the plumb bob is always directly below the point from which it hangs.

Many carpenters now use a framing tool called a **speed square,** *quick square,* or *super square.* It is essentially a variation of the framing square, performing most of the same functions. This tool can be used to mark any angle for rafter cuts by aligning the desired degree mark on the tool with the edge of the rafter being cut. See **Figure 4-9.**

Figure 4-9. The speed square can be used for marking square cuts and angle cuts. (Construction Training School, St. Louis)

Level: Device used for laying out vertical and horizontal lines.

Plumb bob: Device that establishes a vertical line when attached to and suspended from a line. Its weight pulls the line in a true vertical position for layout and checking.

Speed square: Triangular measuring tool that can be used to mark any angle for rafter cuts by aligning the desired degree mark on the tool with the edge of the rafter being cut. Also called a *quick square* or *super square.*

A

B

Figure 4-8. Levels are needed to make sure materials are horizontal or vertical. A—The level can be used to lay out level lines, as well as check building frames for level and plumb by centering a bubble. B—Checking a wall frame for plumb. Over longer distances, a straightedge is used in connection with the level. (The Bennett Street School)

4.2 Saws

The principal types of hand saws used by the carpenter are illustrated in **Figure 4-10.** These are available in several different lengths as well as various tooth sizes (given as teeth per inch or points per inch). Each type of saw has a specific purpose.

4.2.1 Saw Types

Crosscut saws, as the name implies, are designed to cut across the wood grain. Their teeth are pointed, **Figure 4-11.** Every carpenter needs a good crosscut saw with a tooth size

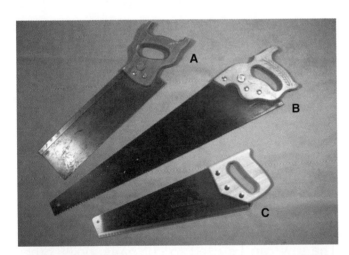

Figure 4-10. Carpenters may use several different kinds of handsaws in the course of their work. A—Backsaw. B—Crosscut saw. C—Drywall saw. D—Ripsaw. (McRae True Value Hardware)

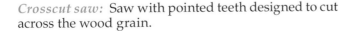

Crosscut saw: Saw with pointed teeth designed to cut across the wood grain.

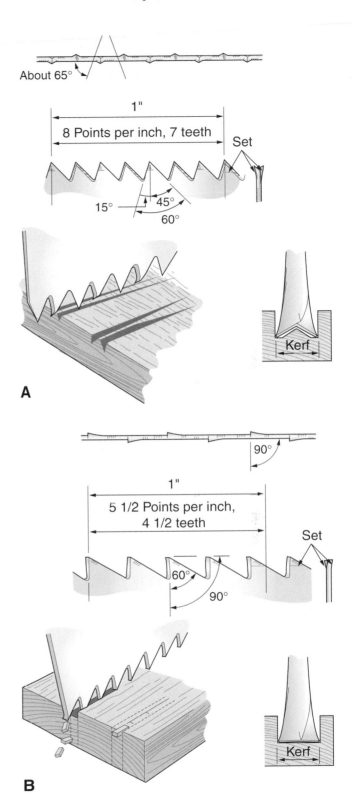

Figure 4-11. The teeth of crosscut saws and ripsaws are quite different. A—Crosscut teeth cut like a knife. The top illustration shows the shape and angle of teeth. The bottom illustration shows how teeth cut. B—Ripsaw teeth cut like a chisel. The top illustration shows shape and angle of teeth. The angle is often increased from 90° to give a negative rake. The bottom illustration shows how rip teeth cut.

ranging from 8 to 11 points. A crosscut saw for general use has 8 teeth per inch. A finishing saw used for fine cutting has 10 or 11 teeth per inch. Using the thumb as a guide will assure getting a crosscut started on the line, as shown in **Figure 4-12.**

Ripsaws have chisel-shaped teeth, which cut best along the grain. Refer to **Figure 4-11.** Note that the teeth are set (bent) alternately from side to side. This is so the *kerf* (cut in the wood) will be large enough for the blade to freely run. Some saws have a taper ground from the toothed edge to the back edge. It eliminates the need for a large set on the teeth. Most ripping operations are now performed with power saws, making the ripsaw an optional tool in the carpenter's arsenal.

A saw occasionally used for cutting curves is the *coping saw,* **Figure 4-13.** This saw has a thin, flexible blade that is pulled tight by the saw frame. With a blade that has 15 teeth per inch, the coping saw can make very fine cuts.

The *backsaw* has a thin blade reinforced with a heavier steel strip along the back edge. The teeth are small (14–16 points). Thus, the cuts produced are fine. It is used mostly for interior finish work.

Many *compass* and *keyhole* saws have a quick-change feature that permits the use of different sizes and kinds of blades. Although originally designed to cut keyholes, they now serve as general-purpose saws for irregular cuts or where space is limited.

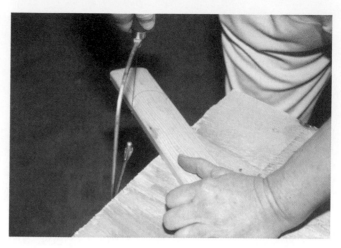

Figure 4-13. A coping saw is designed to cut along curves. Its thin, flexible blade is supported under tension by the saw's frame. The blade has about 15 teeth per inch.

The *sheetrock-drywall saw* has large, teeth specially designed for cutting through paper facings, backings, and the gypsum core. Gullets (spaces between the saw teeth) are rounded to prevent clogging with the gypsum material.

4.2.2 Hacksaw Blades

Due to the wide range of work that the carpenter must be prepared to handle, a *hacksaw* should be included in the hand tool assortment. The saw can be used to cut nails, bolts, other

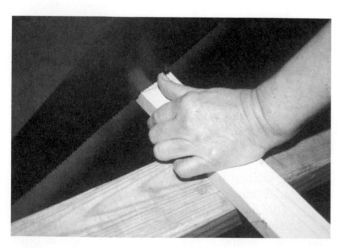

Figure 4-12. Use your thumb as a guide against the saw blade when starting a cut with the hand saw.

Ripsaw: Saw with chisel-shaped teeth, which cut best along the grain.

Kerf: The width of cut made by a saw blade.

Coping saw: Saw with a thin, flexible blade that is pulled tight by the saw frame. With a blade that has 15 teeth per inch, this saw can make very fine cuts. It is occasionally used for cutting curves.

Backsaw: Saw with a thin blade reinforced with a heavier steel strip along the back edge. The teeth are small (14–16 points), producing fine cuts. It is mostly used for interior finish work.

Sheetrock-drywall saw: Saw with large, specially designed teeth for cutting through paper facings, backings, and the gypsum core.

Hacksaw: Saw that can be used to cut nails, bolts, other metal fasteners, and metal trim. Most have an adjustable frame, permitting the use of several sizes of blades.

metal fasteners, and metal trim. Most hacksaws have an adjustable frame, permitting the use of several lengths of blades.

Hacksaw blades are made of high-speed steel, tungsten alloy steel, molybdenum steel, and other special alloys. *All-hard blades* are heat-treated. This makes them very brittle and easily broken if misused. *Flexible-back blades* are hardened only around the teeth. They are usually preferred.

The blade's cutting edge may have anywhere from 14 to 32 teeth per inch. As a rule, you should use a blade with 14 teeth per inch for brass, aluminum, cast iron, and soft iron. For drill rod, mild steel, tool steel, and general work, a blade with 18 teeth per inch is recommended. For tubing and pipe, a blade with 24 teeth per inch is best. Generally, the thinner the metal being cut, the finer (more teeth per inch) the blade should be. At least two or three teeth should rest on the metal. Otherwise, the sawing motion may shear off teeth.

Like woodcutting saws, hacksaw blades have a set. This provides clearance for the blade to slide through the cut. Edge views showing different types of set are illustrated in **Figure 4-14.**

Blades should be installed with teeth pointing forward, away from the handle. Use both hands to operate the saw. Apply enough

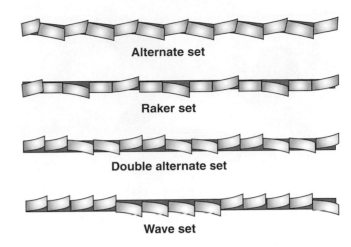

Figure 4-14. This enlarged view of hacksaw blades shows different types of set.

pressure on the forward stroke to allow each tooth to remove a small amount of metal. Release pressure on the return stroke to reduce wear on the blade. Saw with long steady strokes paced at 40 to 50 per minute.

4.3 Planing, Smoothing, and Shaping Tools

The most important tools in this group are the planes and chisels. Several other tools that depend on a cutting edge to perform the work are also shown in **Figure 4-15.**

Standard surfacing planes include the *smooth plane* (8"–9"long), *jack plane* (14" long), and *fore and jointer plane* (18"–24" long). **Figure 4-16** illustrates the parts of a plane and how they fit together. The jack plane is usually selected for general-purpose work.

Another surfacing plane, the **block plane,** is especially useful. It is small (6"–7" long) and can be used with one hand. The blade is mounted at a low angle and the bevel of the cutter is turned up. This plane produces a fine, smooth cut, making it suitable for fitting and trimming work.

The **router plane** and **rabbet plane** are designed to form **dados,** grooves, and **rabbets.** The need for these specialized hand planes has

All-hard blades: Heat-treated hacksaw blades that are very brittle and easily broken if misused.
Flexible-back blades: Hacksaw blades that are hardened only around the teeth.
Smooth plane: 8"–9"long surfacing plane.
Jack plane: 14" long surfacing plane.
Fore and jointer plane: 18"–24" long standard surfacing plane.
Block plane: A small surfacing plane that produces a fine, smooth cut, making it suitable for fitting and trimming work. The blade is mounted at a low angle and the bevel of the cutter is turned up.
Router plane: Plane designed to form dados, grooves, and rabbets.
Rabbet plane: Plane designed to form dados, grooves, and rabbets.
Dado: A rectangular groove cut in wood.
Rabbet: A rectangular shape consisting of two surfaces cut along the edge or end of a board.

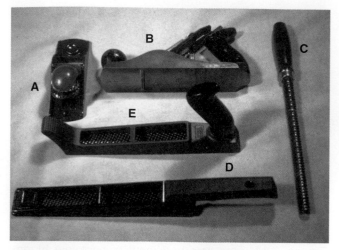
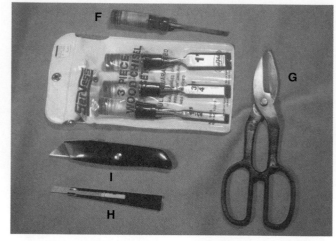
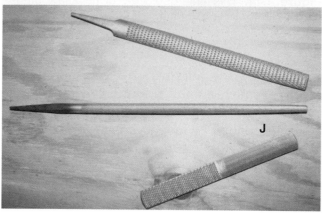

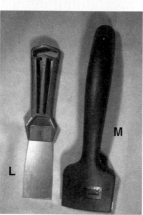

Figure 4-15. Edge tools are used in carpentry for planing, smoothing, and shaping wood. A—Block plane. B—Jack plane. C—Surform (round). D—Surform file. E—Surform (plane type). F—Wood chisels. G—Tin snips. H, I—Utility knives. J—Rasps. K—Flooring chisel. L—Putty knife. M—Scraper. (McRae True Value Hardware)

diminished in recent years due to the widespread use of power-driven carpentry equipment.

Wood chisels are used to trim and cut away wood or composition materials to form joints or recesses, **Figure 4-17.** They are also helpful in paring and smoothing small, interior surfaces that are inaccessible for other edge tools. Chisel widths range from 1/8″–2″. For general work, the carpenter usually selects 3/8″, 1/2″, 3/4″, and 1 1/4″ sizes. A soft-face hammer or mallet should always be used to drive the chisel when making deep cuts.

Tin snips are used to cut asphalt shingles and light sheet metals, such as flashing. *Utility*

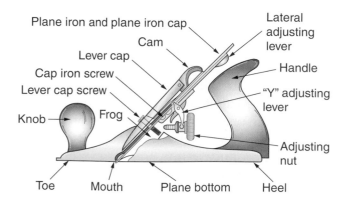

Figure 4-16. A standard plane is made up of many parts.

Plane iron and plane iron cap
Cam
Lever cap
Cap iron screw
Lever cap screw
Knob
Frog
Toe
Mouth
Plane bottom
Heel
Lateral adjusting lever
Handle
"Y" adjusting lever
Adjusting nut

Wood chisel: Tool used to trim and cut away wood or composition materials to form joints or recesses.

Tin snips: Tool used to cut asphalt shingles and light sheet metals, such as flashing.

Utility knives: Tools with very sharp blades. They are useful for trimming wood and cutting veneer, hardboard, particle board, vapor barrier, tar paper, and house wrap. They are also used to cut batt insulation and for accurate layout.

Figure 4-17. Cutting a recess with a wood chisel. Handles should be able to withstand light hammer or mallet blows. Note that saw cuts form the sides of the recess.

knives have very sharp blades and are useful for trimming wood and cutting veneer, hardboard, particle board, vapor barrier, tar paper, and house wrap. They are also used to cut batt insulation and for accurate layout.

Rasps are used for trimming, shaping, and smoothing wood. A round rasp is useful for forming inside curves and shaping or enlarging holes.

The edge of the *scraper* is hooked. It produces thin, shaving-like cuttings. The edge is usually filed, and then a tool called a *burnisher* is used to form the hook. There are a number of different types of scrapers. Most carpenters generally prefer one with a blade that can be quickly replaced.

Two tools used for smoothing and shaping are the tungsten carbide–coated file and the multiblade forming tool (Surform®). Both rapidly cut and will do a considerable amount of work before becoming dull. The surface produced is rough and requires sanding to remove the tool marks.

Rasp: Tool used for trimming, shaping, and smoothing wood.
Scraper: Tool with a hooked edge that produces thin shaving-like cuttings.
Burnisher: Tool used to form the hook on the edge of a scraper.
Claw hammer: A hammer having a head with one end forked for removing nails.
Nail set: Tools designed to drive the heads of casing and finishing nails below the surface of the wood. Tips range in diameter from 1/32"–5/32" in increments of 1/32".

4.4 Fastening Tools

Much of carpentry work consists of fastening parts together. Nails, screws, bolts, and other types of connectors are used. **Figure 4-18** shows a group of hand tools commonly used to perform operations in this area. Of the tools shown, the *claw hammer* is used most often. Carpenters usually carry a hammer in a holster attached to their belts. Two common shapes of hammer head are:
• Curved claw.
• Ripping (straight) claw.

The curved claw is the most common and is best suited for pulling nails. The ripping claw, which may be driven between fastened pieces, is used somewhat like a chisel to pry them apart.

Parts of a ripping claw hammer are shown in **Figure 4-19.** The face can be either flat or slightly rounded (bell face). The bell face is used most often. It will drive nails flush with the surface without leaving hammer marks on the wood. The hammer head is forged from high-quality steel and is heat treated to give the poll and face extra hardness.

The size of a claw hammer is determined by the weight of its head. They are available in sizes from 7 oz.–20 oz. The 13 oz. size is the most popular for general-purpose work. Carpenters generally use either the 16 oz. or 20 oz. size for rough framing.

A hammer should be given good care. It is especially important to keep the handle tight and the face clean. If a wooden handle becomes loose, it can be tightened by driving the wedges deeper or by installing a new one. **Figure 4-20** illustrates the procedure for pulling nails to relieve strain on the hammer handle and protect the work.

Hatchets are used for rough work on such jobs as making stakes and building concrete forms. Some are designed for wood shingle work, especially wood shakes, **Figure 4-21.** See also Chapter 12, **Roofing Materials and Methods**.

The *nail set* is designed to drive the heads of casing and finishing nails below the surface of the wood. Tips range in diameter from 1/32"–5/32" by increments of 1/32". Overall length is usually 4".

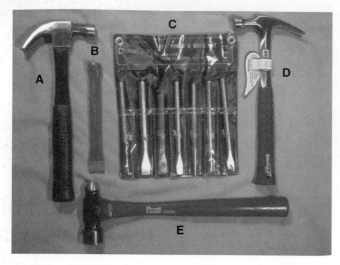

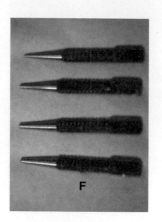

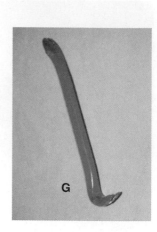

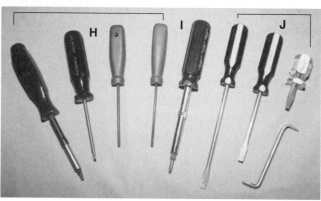

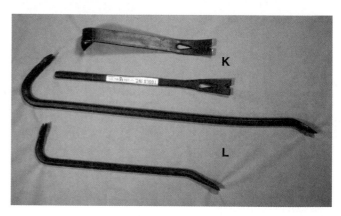

Figure 4-18. These tools are used for fastening and prying apart wood building components. A—Curved claw hammer. B—Large cold chisel. C—Cold chisels in various sizes. D—Straight or ripping claw hammer. E—Ball peen hammer. F—Nail sets. G— Nail claw. H—Torx screwdrivers. I—Phillips screwdriver. J—Slotted or flat blade screwdrivers. K—Ripping chisels. L— Pry bars or crowbars. (Estwing Mfg. Co. and McRae True Value Hardware)

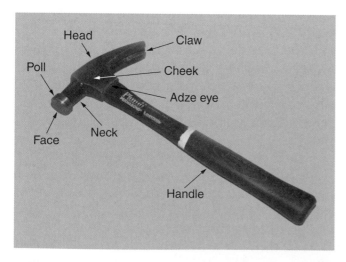

Figure 4-19. Parts of a ripping claw hammer. (McRae True Value Hardware)

Screwdrivers are available in a number of sizes and styles. Sizes are specified by giving the length of the blade, measuring from the ferrule to the tip. The most common sizes for the carpenter are 3″, 4″, 6″, and 8″. The size of a Phillips screwdriver is given as a point number ranging from a No. 0 (the smallest), to a No. 4 (the largest). Size numbers 1, 2, and 3 will fit most of the screws used in carpentry work.

Tips of screwdrivers must be carefully picked for the job. For a slotted screw, tips must be square, have the correct width, and snugly fit into the screw slot. The width of the tip should be equal to the length of the bottom of the screw slot. The sides of the screwdriver tip should be carefully ground to an included angle of not

Figure 4-20. Proper methods for pulling nails and separating two wood pieces. A—Place a putty knife under the hammer to protect the wood surface. B—Use a block of wood to increase leverage and protect the surface. C—To separate two nailed surfaces, a rip claw hammer works best. The rip claw can also be used to split short boards.

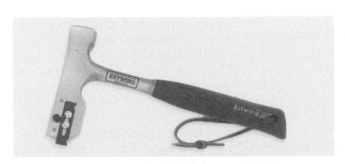

Figure 4-21. A shingling hatchet is designed for installing wood shingles and shakes. The stops on the blade are for measuring the exposure on shingles. The stops are adjustable. (Estwing Mfg. Co.)

more than 8° and to a thickness that will fit the screw slot.

Today, mechanical *tackers, staplers,* and *nailers* perform a variety of operations formerly accomplished by hand nailing. They provide an efficient method of attaching insulation, roofing material, underlayment, ceiling tile, and many other products. An advantage of the regular stapler is that it leaves one hand free to hold the material.

4.5 Prying Tools

Sometimes, particularly when remodeling, it is necessary to take apart nailed structural members. The same is true when forms or temporary braces are removed from a building under construction. Pry bars and ripping chisels are shown in **Figure 4-18.** Lengths range from 12″ for ripping chisels to 30″ for pry bars. See **Figure 4-22** for information on the use of prying tools.

Ripping chisels are easily recognized by the small teardrop slot at one end. The purpose of this slot is to grip and pull nails while in confined spaces where a larger tool cannot be used. The ripping chisel may also be used for light prying.

Ripping bars, also called *wrecking bars,* vary in length from 12″–36″. They are used to strip concrete forms, disassemble scaffolding, or other rough work involving prying, scraping, and nail pulling.

Tackers: Tool that performs a variety of operations formerly accomplished by hand nailing, such as attaching insulation, roofing material, underlayment, and ceiling tile.

Staplers: Tool that performs a variety of operations formerly accomplished by hand nailing, such as attaching insulation, roofing material, underlayment, ceiling tile, and many other products.

Nailer: Tool used for driving nails.

Ripping bar: Tool used to strip concrete forms, disassemble scaffolding, or other rough work involving prying, scraping, and nail pulling. Also called *wrecking bar.*

Pry bars are designed for removing larger nails or spikes from lumber of larger dimensions. The 30″ bar is most-used because it fits into a standard carpenter's toolbox.

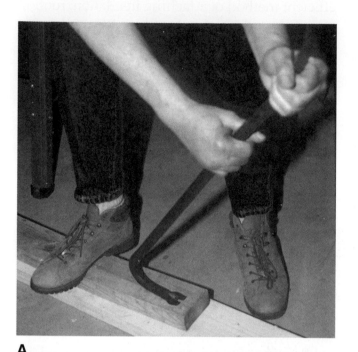

A

B

Figure 4-22. Using prying tools. A—The pry bar is used to remove larger fasteners. Its long handle provides considerable leverage. B—Nail puller claws are first driven under the nail with a weight that slides along the handle. A small fulcrum attached to one claw tightens the puller's grip on the nail.

A *nail claw,* which is much smaller than a pry bar, is only used to pull a nail above the surface of the wood. A hammer is used to drive the sharpened claw under the nail head, raising it sufficiently to be grasped by the hammer or a pry bar.

A *nail puller* is a specialty tool with a fixed jaw and a movable jaw and lever. A sliding driver on the end of the handle drives the jaws under the nail head. Leverage tightens the jaw's grip on the nail, raising it above the wood's surface.

Safety Note
There are certain safety precautions that should be observed while using ripping bars. Be sure to have good balance before applying force to the bar. Grip the bar in such a way that fingers are not bruised by coming into contact with any part of the building structure as the fastener or piece comes free.

4.6 Gripping and Clamping Tools

Carpenters find that certain gripping and clamping tools are helpful, **Figure 4-23.** *Wood clamps* and *C-clamps* are especially useful and

Pry bar: Tool designed for removing larger nails or spikes from lumber of larger dimensions.

Nail claw: Tool used to pull a nail above the surface of the wood where it can be gripped easily.

Nail puller: A specialty tool with a fixed jaw and a movable jaw and lever. A sliding driver on the end of the handle drives the jaws under the nail head. Leverage tightens the jaw's grip on the nail, raising it above the wood's surface.

Wood clamps: Clamps with broad jaws that distribute the pressure over a wide area. Sizes (designated by the width of the jaw opening) range from 4″–24″. Also called hand screws.

C-clamp: Clamping tool that adapts to a wide range of assemblies where parts need to be held together. They are available in sizes from 1″–12″.

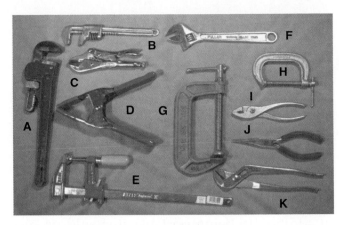

Figure 4-23. Clamping and gripping tools are helpful in carpentry work. A—Straight pipe wrench. B—Monkey wrench. C—Clamping pliers. D—Spring clamp. E—Bar clamp. F—Adjustable wrench. G—Large C-clamp. H—Small C-clamp. I—Slip joint pliers. J—Long-nosed pliers. K—Tongue-and-groove pliers. (McRae True Value Hardware)

adapt to a wide range of assemblies where parts need to be held together:

- While metal fasteners are being attached.
- While adhesives are setting.
- While wood parts are being worked on, **Figure 4-24.**

Sometimes clamps are used to hold jigs or fixtures to a machine for some special setup.

C-clamps are available in sizes from 1″–12″. The clamp size represents the tool's largest opening.

Wood clamps, also known as *hand screws,* are ideal for woodworking because the jaws are broad and distribute the pressure over a wide area. Sizes are designated by the width of the jaw opening and range from 4″–24″.

Bar or *pipe clamps* are useful for holding larger pieces while glue is drying or receiving mechanical fasteners. Some clamps are adjustable at both jaws. Newer types, like the one shown in **Figure 4-25,** feature one-handed clamping mechanisms.

Hand screws: Wood clamps with broad jaws that distribute the pressure over a wide area. Sizes (designated by the width of the jaw opening) range from 4″–24″.

Bar clamps: Clamps useful for holding larger pieces while they are receiving glue or mechanical fasteners. Also called *pipe clamps.*

A

B

Figure 4-24. Clamps can be used to provide temporary pressure on parts being shaped or glued. A— Hand screw or wood clamp. B—A carriage clamp, one type of C-clamp. (Montachusett Regional Vo-Tech School)

Figure 4-25. Newer types of bar clamps have a pistol-grip mechanism that allows quick size adjustment and one-handed clamping. (American Tool)

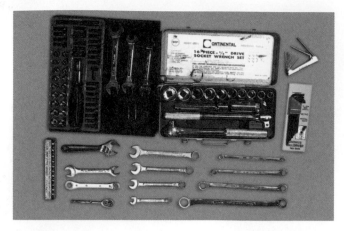

Figure 4-26. A variety of wrenches is useful to the carpenter in maintaining tools and other equipment.

Clamping pliers are an adjustable, all-purpose tool with a lever action that locks the jaws when the handles are fully closed. The pliers can be substituted for a vise or can be used in place of a pipe wrench, open-end wrench, or adjustable wrench.

Figure 4-26 shows various types of wrenches that the carpenter may use to service and adjust equipment. Tool maintenance is an important responsibility. Well-maintained tools are safe tools.

4.7 Tool Belts and Storage

Many carpenters use tool belts and tool holders that hang from the waist. These items keep tools close and ready for use, while freeing hands for other tasks when the tools are not in use. See **Figure 4-27A.**

Some type of container is needed to store and transport tools to and from the job site. In addition to being portable, it should protect the tools from weather damage or loss. At one time, folding tool panels met this need. Today's carpenter prefers to carry tools in a five-gallon bucket fitted with a fabric holder full of pockets. The one shown in **Figure 4-27B** is typical.

4.8 Care and Maintenance of Tools

Experienced carpenters take pride in their tools and keep them in good working condition. They know that even high-quality tools will not

A

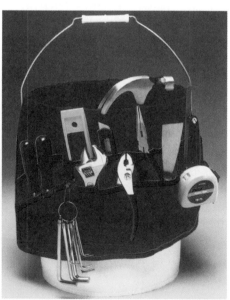

B

Figure 4-27. A—The tool belt can increase a carpenter's efficiency by keeping needed items in easy reach. B—A fabric bucket liner with many pockets is handy for organizing and transporting tools. The liner fits any 5 gallon plastic bucket. (Pierce Construction, Ltd.; Truecraft Tools)

perform satisfactorily if they are dull or out of adjustment.

Tools should be wiped clean after being used. Occasionally wet the wiping cloth lightly with oil. Some carpenters use a lemon oil furniture polish. It has a slight cleaning action and also leaves a thin oil film that protects metal surfaces from rust.

Keep handles on all tools tight. If handles or fittings are broken, they should be replaced.

It is a simple matter to hone edge tools on an *oilstone*, **Figure 4-28.** For tools with single-bevel edges, like planes and chisels, place the tool on the stone with the bevel flat on the surface. Raise the back edge of the tool a few degrees so only the cutting edge is in contact. Then move the tool back and forth until a fine wire edge can be detected by pulling the finger over the edge. Now place the back of the tool flat on the oilstone and stroke lightly several times. Turn the tool over and again stroke the beveled side lightly. Repeat the total operation several times until the wire edge has disappeared from the cutting edge.

When the bevel becomes blunt, reshape it by grinding. An edge can be ground many times. If a cutting edge is not damaged, it can be honed

several times before grinding is required. The grinding angle will vary somewhat depending on the work for which the tool is used. **Figure 4-29** shows grinding and honing angles recommended for a plane iron.

Some tools may be sharpened with a file. Saws require filing as well as setting. Before filing, they often require *jointing*. In this operation the height of the teeth is evenly struck off. Filing a saw is a tedious operation. Most carpenters prefer to send their saws to a shop where they can be machine sharpened by an expert.

When a saw is only slightly dull, the teeth can often be sharpened with a few strokes, using a triangular saw file. Be sure to match the original angle of the teeth. File the back of one tooth and the front of an adjacent tooth in a single stroke. **Figure 4-30** shows how correctly filed teeth should appear.

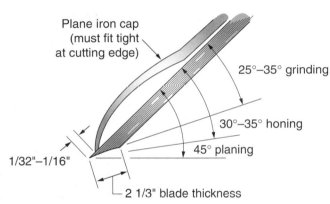

Figure 4-29. Follow these grinding and honing angles for a plane iron.

Figure 4-28. Honing a chisel on an oilstone. Place a few drops of oil on the stone first.

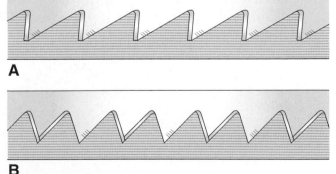

Figure 4-30. Correctly filed saw teeth. A—Rip teeth. B—Crosscut teeth.

Oilstone: Stone lubricated with oil and used for honing edge tools.

Jointing: The height of the saw teeth is struck off evenly.

Summary

Hand tools are essential to the work of the carpenter. As a skilled worker, the carpenter must carefully select the kind, type, and size of tools that best suit personal requirements. The carpenter works with measuring and layout tools, such as rules and squares. The most common carpentry tool is the claw hammer used to nail wooden building components. The carpenter also uses hand saws; smoothing tools, such as planes; prying tools; and clamping tools. Experienced carpenters take pride in their tools and keep them in good working condition. This may involve cleaning, adjusting, sharpening, and repairing.

Test Your Knowledge

Answer the following questions on a separate piece of paper. Do not write in this book.

1. A standard folding wood rule is _____ feet long.
2. List five uses of the framing or rafter square.
3. The outside edge of the back side of a framing square is graduated in what fractions of an inch?
   A. 1/10"
   B. 1/12"
   C. 1/16"
4. Where would you find the rafter table on a framing square? What would you use it for?
5. When checking structural members to see if they are horizontal or vertical (plumb), what tool(s) should be used?
6. A backsaw is used for fine work. It usually has _____.
   A. 8 points per inch.
   B. 8–10 points per inch.
   C. 12–14 points per inch.
   D. 14–16 points per inch.
7. The bevel of a block plane blade is turned _____.
   A. up
   B. down
8. The size of a claw hammer is determined by the _____.
   A. length of the handle
   B. length of the head
   C. weight of the entire hammer
   D. weight of the head
9. Name the different types of claw hammers and explain the purpose of each.
10. *True or False?* Never strike a wood chisel with a mallet.
11. Where would a carpenter use tin snips?
12. How are sizes of slotted screwdrivers and Phillips screwdrivers determined?
13. The operation of filing off the points of saw teeth until all are level is called _____.

Curricular Connections

Social Studies. Prepare a report on the historical development of woodworking tools used by the carpenter. Reference books on woodworking and encyclopedias will contain helpful material. Make a report to your class using visual aids to show students pictures and drawings of early tools. If possible, secure actual examples from a collector.

Mathematics. Using reference books and manufacturers' catalogs, prepare a list of hand tools you believe the carpenter will need for rough framing and exterior finish of a typical residential structure. Determine the cost of equipping a carpenter with a complete hand tool kit based on your list. Take the list to a hardware store or home improvement center and find the cost of each tool. Add up the costs and report to the class.

Outside Assignments

1. Claw hammers are available in different weights and constructions (one piece, wooden handle, fiberglass handle). Find a carpenter or a knowledgeable clerk at a hardware store or home improvement center. Research the uses and the advantages/disadvantages of each type. Based on your research, choose a hammer with a weight and construction that would be suitable for you to use in framing a house.
2. Research the proper method for cutting a mortise in a door jamb, using a wood chisel and mallet. Make a poster showing the step-by-step procedure.

Power Tools

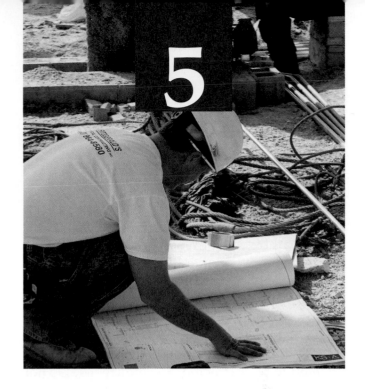

Learning Objectives

After studying this chapter, you will be able to:
- Identify common power tools.
- Explain the function and operation of the carpenter's principal power tools.
- Identify the major parts of common power tools.
- Apply power tool safety rules.

Technical Vocabulary

Belt sander
Chain saw
Continuity monitor
Drywall screw
  shooter
Electric drill
Finishing sander
Frame and trim saw
Ground fault circuit
  interrupter (GFCI)
Jointer
Magnetic starters
Nailers
Panel saw

Plate joiner
Portable tools
Powder-actuated tools
Power block plane
Power miter saw
Power plane
Power stapler
Radial arm saw
Reciprocating saw
Rotary hammer drill
Router
Saber saw
Stationary tools
Table saw

Power tools greatly reduce the time required to perform many of the operations in carpentry work. The carpenter can plane, route, bore, and saw in less time with less exertion. In addition, when proper tools are used in the correct way, a high level of accuracy can easily be achieved. There are two general types of power tools:

- *Portable tools* are lightweight and intended to be carried around by the carpenter. They may be used anywhere on the construction site or on any part of the structure.

- *Stationary tools,* usually called machines, are heavy equipment mounted on benches or stands. The workpiece is brought to them. Benches and stands rest firmly on the floor or ground. The machines must be level when used on the job site.

This chapter provides only brief descriptions of the kinds of power tools most commonly associated with carpentry work. It should be supplemented with woodworking textbooks and reference books devoted to power tool operation. Manufacturer bulletins and operator's manuals are also good sources of information.

Portable tools: Lightweight tools that are intended to be carried around by the carpenter.

Stationary tools: Heavy equipment mounted on benches or stands. Usually called machines.

5.1 Power Tool Safety

Safety must be constantly practiced. Before operating any power tool, you must become thoroughly familiar with:

- The way it works.
- The correct way to use it.

You must be wide-awake and alert. *Never* operate a power tool when you are tired or ill. Think through the operation before performing it. Know what you are going to do and what the tool will do. Make all adjustments before turning on the power. Be sure blades and cutters are sharp and are of the correct type for the work.

While operating a power tool, do not allow yourself to be distracted. See **Figure 5-1.** Do not distract the attention of others while they are operating power tools. Keep all safety guards in place and wear safety glasses. See **Figure 5-2.**

Carefully feed the work and only as fast as the tool will easily cut. Overloading is hazardous to the operator and likely to damage the tool or work. When the operation is complete, turn off the power. Wait until the moving parts have stopped before leaving the machine.

5.1.1 Electrical Safety

Always make sure that the electric power source is of the correct voltage and that the tool switch is in the *off* position before plugging the tool into an electrical outlet. Power tools operate

Figure 5-2. Always wear safety glasses and make sure all guards are in place when working with power equipment. Note that the worker is using a push board for added safety while using the jointer. (Bullard-Haven Technical School)

on 120 V or 240 V electric power. This information is usually shown on a plate attached to the power tool's case. If in doubt about the power source voltage, examine the receptacle, **Figure 5-3.**

Safety Note

Never carry a portable power tool by its cord or disconnect it from the receptacle by pulling on the cord. The insulation and electrical connections inside of the plug or the tool can be damaged by such action, creating a potential shock hazard.

Stationary power tools are factory equipped with **magnetic starters.** These are safety devices that automatically turn the switch to the *off* position in case of power failure. This feature is important because personal injury or damage

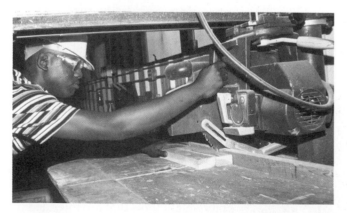

Figure 5-1. When operating a power tool, give your full attention to the work. (Photo © Des Moines, Iowa Public Schools)

Magnetic starter: Safety device on a power tool that automatically turns the switch to the *off* position in case of a power failure.

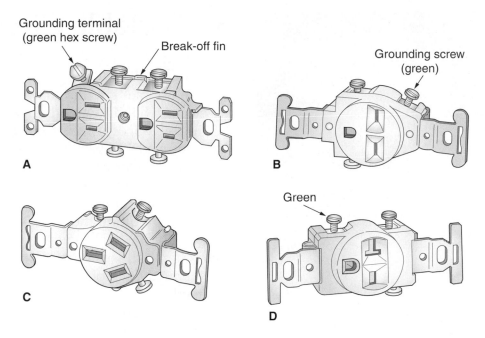

Grounding terminal
(green hex screw)

Break-off fin

A

Grounding screw
(green)

B

C

Green

D

Figure 5-3. Different types of receptacles. A—Receptacles for 120 V power source always have this configuration. B—A 240 V receptacle with tandem blades and U-shaped ground prong. C—A 240 V receptacle for three-bladed plug. D—A 240 V receptacle for horizontal and vertical blades and U-shaped ground prong.

to the equipment could result if power is re-established with the switch in the *on* position. *Always* make sure these switches are in good working order.

The electrical cord and plug must be in good condition and must provide a ground for the tool. This means that extension cords should be the three-wire type. Make sure the conducting wire is large enough to prevent excessive voltage drop.

Be careful in stringing electrical extension cords around the work site. Place them where they will not be damaged or interfere with other workers. Make certain that extension cords maintain *ground continuity*. This means that the third conductor (wire) in a cord is providing an unbroken path for current back to a grounded

terminal. Thus, in case of a short in the tool case, current harmlessly bleeds off to ground. Continuity can be checked with a neon tester or a *continuity monitor.*

5.1.2 Shock Protection

Electric shock is one of the potential hazards of working with power tools. Always be sure that proper grounding is provided. Receptacles should be of the concealed-contact type with a grounding terminal for continuous ground. Plugs and cords should be of an approved type.

Portable power tools should be double insulated or otherwise grounded to protect the worker from dangerous electrical shock. Even though a circuit may be grounded, the operator of a portable power tool could be electrocuted should a bare conductor ground on the metal tool case. A *ground fault circuit interrupter (GFCI)* should be used on all construction job sites. These units can be installed in a circuit or plugged into an outlet that is grounded. These units sense when a short has occurred and turn off power to the tool. See **Figure 5-4.**

Continuity monitor: A device that checks whether grounding conductors in electrical cords are connected to ground (have continuity).

Ground fault circuit interrupter (GFCI): An electrical safety device that can be installed either in an electrical circuit or at an outlet. It is able to detect a short circuit and automatically shut off power. Used as a protection against electrical shock.

Figure 5-4. A portable ground fault circuit interrupter is used where permanent ground fault protection is not available. A power tool is plugged into it to protect the operator. Should a ground occur in the tool, the GFCI trips at around 5 milliamperes, turning off power to the grounded tool. (Carol Electric)

Safety Note

Under certain circumstances, even a double-insulated tool can provide a serious or even fatal shock. If moisture is allowed to get inside of the tool's housing, it can provide an electrical path from energized parts to the outside.

5.2 Portable Circular Saws

A portable circular saw is also called an electric handsaw or builder's saw. Refer to **Figure 5-5.** Its size is determined by the diameter of the largest blade it will take. Most carpenters prefer a 7″ or 8″ saw. The depth of cut is adjusted by raising or lowering the base, or shoe. On many saws, it is possible to make bevel cuts by tilting the shoe. The portable saw may be used to make cuts in assembled work. For example, flooring and roofing boards are often nailed into place before their ends are trimmed.

Portable saws are often guided along the layout line "free hand." Therefore, extra clearance in the saw kerf is required. To provide this clearance, teeth usually have a wide set. **Figure 5-6** shows standard types of blades. The combination blade is popular because it is suitable for both ripping and crosscutting. Some carpenters prefer carbide-tipped saw teeth because they usually stay sharp longer than the teeth of a standard blade.

5.2.1 Making a Cut

To use a portable saw, firmly grasp the handle in one hand with the forefinger ready to operate the trigger switch. Place your other hand on the stock, well away from the cutting line. Some saws require both hands on the machine.

Rest the base of the saw on the work and align the guide mark with the layout line. Turn on the switch, allow the motor to reach full speed, then smoothly feed the blade into the stock, as shown in **Figure 5-7.** Release the switch as soon as the cut is finished. Hold the saw until the blade stops.

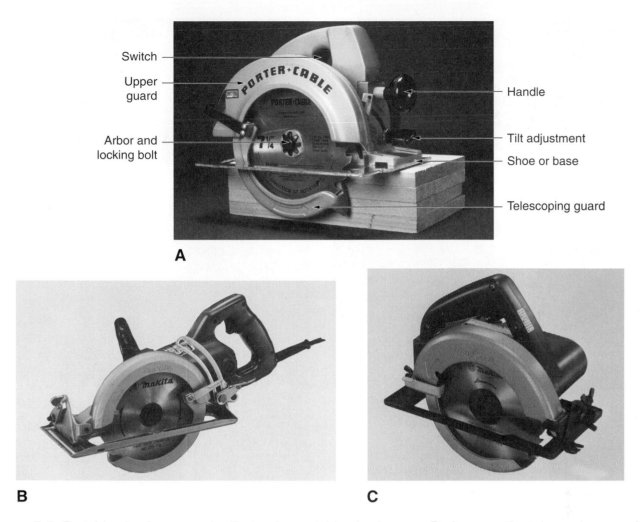

Switch
Upper guard
Arbor and locking bolt
Handle
Tilt adjustment
Shoe or base
Telescoping guard

A

B

C

Figure 5-5. Portable circular saws. A—Parts of a portable circular saw. During use, the telescoping guard is pushed back by the stock. A spring returns the guard when the cut is completed. This type saw is also known as a "sidewinder." B—This style of portable saw is known as a hypoid or worm-drive circular saw. C—A cordless circular saw is practical where electric service is not available. (Porter-Cable Corp., Makita USA, Inc.)

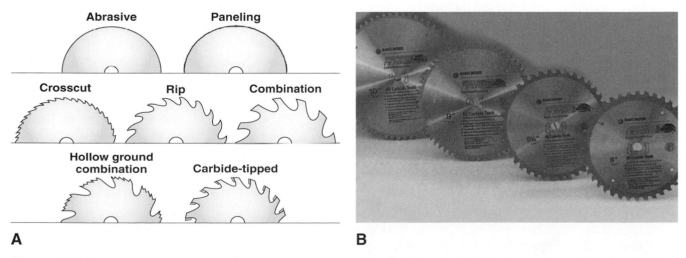

Abrasive   Paneling

Crosscut   Rip   Combination

Hollow ground combination   Carbide-tipped

A

B

Figure 5-6. Circular saw blades. A—Standard blade types. B—Carbide-tipped blades are popular because they stay sharp longer, but they must be carefully handled to prevent damage to the points. (Owner/Builder Directory, Black & Decker)

A

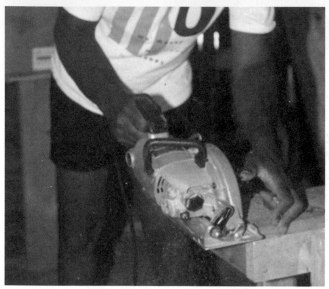

B

Figure 5-7. Circular saws in use. A—Using a portable circular saw to cut a piece of light plywood. The good side of the plywood is placed down because the saw blade rotates upward. Note the workpiece is well-supported during cutting so the saw will not bind. B—Preparing to make a cut on a job site. The carpenter's left thumb is lifting the telescoping guard so it will clear the workpiece. Once the cut has begun, the left hand will be shifted to the second handle as a good safety practice. (Hadley-Hobley Construction)

Safety Note

Follow these safety rules when working with portable circular saws.

- Unplug the saw to replace a saw blade. Some saws have a control that locks the arbor during this operation, **Figure 5-8.**
- If the saw has two handles, keep both hands on them as much as possible during the cut. See **Figure 5-9.** Start the cut using one hand to raise the guard (when needed, usually at the beginning of a compound

cut). When the cut is completed, make sure the guard returns to the closed position. Never block the guard so that it stays open—doing so will severely compromise your safety.

- Support stock being cut so that the kerf will not close up and bind the blade during or at the end of the cut.
- Support thin materials near the cut. Small pieces should be clamped to a bench top or sawhorse.
- Be careful not to cut into the sawhorse or other supporting device.
- Before starting the saw, adjust the depth of cut to the thickness of the stock, plus about 1/8″.
- Check the base and angle adjustments to be sure they are tight.
- Plug the cord into a grounded outlet and be sure it will not become tangled in the work.
- Always place the saw base on the stock with the blade clear before turning on the switch.
- During the cut, stand to one side of the cutting line. Never reach under the workpiece during a cut.
- Never pull back on a running saw once the cut is started—to do so will cause a kickback.
- If saw blade strays off the cutting line, stop the saw, pull back, and start the cut over. Otherwise, the blade may bind and cause an injury.
- Keep hands clear of the cutting line.
- Have the saw at full power and up to speed before beginning a cut.
- Always use a sharp blade of the proper type for the material and cut.
- Never force the saw through the material.
- Since they are light and portable, hand power saws can be carried anywhere on the job site. Use care to avoid damage or injury.
- Dropping the saw or allowing it to fall can damage or destroy it. Never use a damaged saw.
- Keep the saw guard clean and lubricated, so that it functions as designed.
- Immediately remove from service any saw that does not properly function.

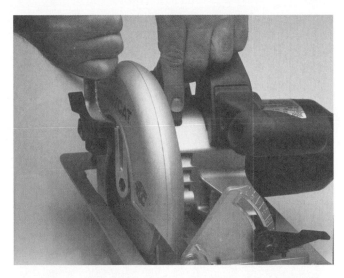

Figure 5-8. Always disconnect the saw from the power source while adjusting or changing blades. On some saws, pressing a plunger locks the arbor during blade change. (Black & Decker)

Figure 5- 9. If a portable saw has two handles, firmly grasp the saw with both hands while cutting. (Black & Decker)

Saber saw: Portable power saw useful for a wide range of light work. Also called a portable jig saw.

5.3 Saber Saws

The *saber saw* is also called a portable jig saw. It is useful for a wide range of light work. Carpenters, cabinetmakers, electricians, and home craftspeople use it. A standard model is shown in **Figure 5-10.** The stroke of the blade is about 1/2″. The saw operates at a speed of approximately 2500 strokes per minute, but some have a variable speed control.

Blades for wood cutting have 6–12 teeth per inch. For general purpose work, a blade with 10 teeth per inch is satisfactory. Always select a blade that will have at least two teeth in contact with the material being cut.

Saber saws vary in the way the blade is mounted in the chuck. Follow the directions in the manufacturer's manual. Also, follow the lubrication schedule specified in this manual.

The saber saw can be used to make straight or bevel cuts, as shown in **Figure 5-11.** Curves are usually cut by guiding the saw along a layout line. However, circular cuts may be more accurately made with a special guide or attachment.

Since the blade cuts on the upstroke, splintering takes place on the top side of the work. This must be considered when making finished cuts, especially in fine hardwood plywood. Always firmly hold the base of the saw against the surface of the material being cut.

Figure 5-10. A saber saw cuts with an action much like a handsaw. This particular model has variable speed control and four-position orbital action. (Black & Decker)

A   **B**

Figure 5-11. Using a saber saw. A—Straight cutting using a quick square as a guide. B—Angle cutting.

Figure 5-12. On internal cuts, a saber saw can make its own opening. Be sure to rest the base on the workpiece before starting the cut.

When cutting internal openings, a starting hole can be drilled in the waste stock or the saw can be held on end so the blade cuts its own opening. See **Figure 5-12.** This is called *plunge cutting* and must be done with considerable care. Firmly rest the toe of the base on the work and turn on the motor. Then, slowly lower the blade into the stock.

Another portable power tool is the ***reciprocating saw,*** shown in **Figure 5-13.** Operation of this saw is similar to that of the saber saw. Like the saber saw, its cutting action is linear (back-and-forth), not circular. It is useful under conditions where a circular saw is not safe or practical. Carpenters use it in remodeling work where sections of framing, sheathing, or inside

Reciprocating saw: Portable power tool that has linear (back-and-forth) cutting action.

A

B

C

Figure 5-13. A reciprocating saw. A—This type of saw has a blade that moves forward and back (reciprocates) through a stroke distance of about 1 1/8″. It can be fitted with a variety of blades to suit all purposes. B—Typical reciprocating saw blades. C—The reciprocating saw is used in situations where space does not permit use of a circular saw. (Milwaukee Electric Tool Corp., Bosch Power Tool Corp., Black & Decker) walls must be removed. Blades are available to cut through plaster, fiberglass, and all kinds of metals.

5.4 Chain Saws

Occasionally, a *chain saw* may be needed to cut heavy timbers, posts, or pilings. See **Figure 5-14.** Both gas and electric models are available, but gasoline-powered types are more powerful. Blade lengths vary from about 10"–20". Chain saws are also useful for demolition during remodeling jobs. An 18" blade is recommended for the carpenter.

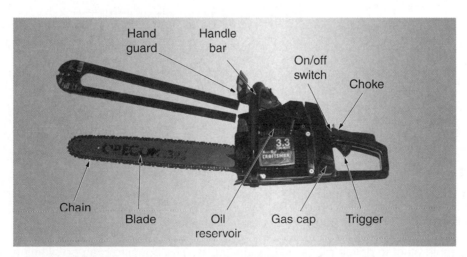

Figure 5-14. A chain saw is suitable for cutting heavy timbers.

Chain saw: Gasoline- or electric-powered saw used to cut heavy timbers, posts, or pilings. Blade lengths vary from about 10"–20".

(Continued)
- Stand with your weight evenly balanced on both feet.
- Do not overreach or cut above shoulder height.
- Wear protective clothing and goggles or a face shield. Earplugs protect you from excessive noise.
- Stop the engine before setting the saw down.
- Never use a chain saw to cut nonwood materials.

5.5 Portable Electric Drills

Portable *electric drills* come in a wide range of types and sizes. The size is determined by the chuck capacity; 1/4″ and 3/8″ chucks are the sizes normally used by the carpenter. **Figure 5-15** illustrates the basic parts and reduction gears of a typical model. Speeds of about 1000 rpm are best for woodworking. The speed of a variable speed drill is adjusted by the trigger. Bits designed for use in a portable electric drill are shown in **Figure 5-16.**

Cordless portable drills are handy for many jobs, **Figure 5-17.** Power is supplied by a rechargeable nickel-cadmium battery. Most carpenters prefer cordless drills.

Safety Note

Follow these safety rules when working with portable drills.
- Select the correct drill or bit for your work and securely mount it in the chuck.
- Be sure stock is firmly held so it does not move during the operation.
- With switch in the *off* position, connect the drill to a properly grounded outlet.
- Turn on the switch for a moment to see if the bit is properly centered and running true.

(Continued)

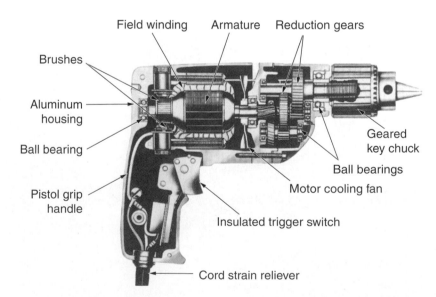

Figure 5-15. A portable electric drill is made up of many parts. The housing may be aluminum or plastic. Many models are cordless.

Electric drill: A portable tool powered by electricity and used primarily for drilling holes. Both corded and cordless (battery-powered) models are available.

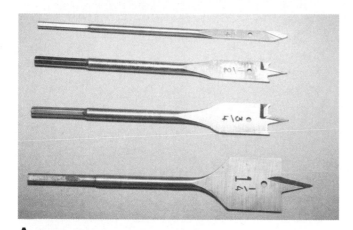

A

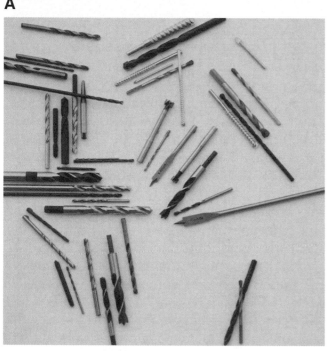

B

Figure 5-16. Bits for portable electric drills come in various types. A—Spade bits for light duty. B—High-speed twist drills. (Black & Decker)

Figure 5-17. Cordless drills are preferred by most carpenters. (Makita U.S.A.)

(Continued)

- Turn on the switch and feed the drill into the work. The pressure required will vary with the size of the drill and the kind of material being drilled.

- During operation, keep the drill aligned with the desired direction of the hole.

- When drilling deep holes, especially with a twist drill, withdraw the drill bit from the hole several times during the process to clear the cuttings.

- Always turn off the power and remove the bit from the drill as soon as you have completed your work.

(Continued)

- With the switch off, place the point of the bit in the punched layout hole.

- Firmly hold the drill at the correct drilling angle, using one or both hands as needed.

(Continued)

Rotary hammer drill: Heavy duty tool for drilling large holes in concrete and other masonry materials.

5.6 Rotary Hammer Drills

Rotary hammer drills are heavy-duty tools used to drill large holes in concrete and other masonry materials. See **Figure 5-18.** Anchor devices are then placed in the holes to receive bolts or screws that fasten materials to floors, walls, or ceilings. Various bit types are available. Wood bits include a flat boring bit, auger bit, ship

Figure 5-19. Portable power planes are replacing hand planes because they are faster and more accurate. (Makita USA, Inc.)

Figure 5-18. A 1 1/2″ rotary hammer fitted with a spine shank four-cutter bit for fast rotary cutting. The depth rod allows drilling holes to a preset depth. A lever disengages the rotary action when hammer action only is needed. (Milwaukee Electric Tool Corp.)

auger bit, and hole saw. Concrete bits include a carbide tip bit. Carbide-tip bits are also available for faster cutting and greater durability in drilling concrete and masonry. Twist drills are also offered for use in drilling steel. Depending on the type of bit used, a rotary hammer can drill holes from 3/16″ to 6″ in diameter.

Safety Note

Follow these safety rules when working with rotary hammer drills.
- Maintain good balance and a tight grip on the handles. Bits can bind in the hole, tearing the tool from your grasp and possibly knocking you off balance.
- Steadily push into the workpiece, pulling back frequently to clear away chips.
- Keep bits sharp.
- Disconnect the drill from power while installing or removing bits.
- Never lock the trigger in the *on* position.
- Check that the switch is in the *off* position before plugging it in.

5.7 Power Planes

The *power plane* produces finished wood surfaces with speed and accuracy, **Figure 5-19.** The motor, which operates at a speed of about 20,000 rpm, drives a spiral cutter. The depth of cut is adjusted by raising or lowering the front shoe. The rear shoe (main bed) must be kept level with the cutting edge of the cutter head. The power plane is equipped with an adjustable fence for planing bevels and chamfers. For surfacing operations, the fence is removed.

Hold and operate the power plane in about the same manner as you would a hand plane. Support the work so that it cannot move, while still permitting easy operation. Start the cut with the front shoe firmly resting on the work and the cutter head slightly behind the surface. Be sure that the electric cord is kept clear to prevent damage. Start the motor before the cutter head engages the stock. Move the plane forward with smooth, even pressure on the work. When finishing the cut, apply heavier pressure on the rear shoe.

The *power block plane* can be used on small surfaces. It has features and adjustments similar to the regular power plane. Since it is small,

---

Power plane: Tool that produces finished wood surfaces with speed and accuracy. The motor, which operates at a speed of about 20,000 rpm, drives a spiral cutter.

Power block plane: Small plane that can be used on small surfaces and operated with one hand.

this plane is designed to be operated with one hand. When using the tool, the work should be securely held or clamped in place. In planing small stock, kickbacks may occur. Be sure your free hand is kept well out of the way.

5.8 Portable Routers

Routers are used to cut irregular shapes and to form various contours on edges, **Figure 5-20.** When equipped with special guides, routers can be used to cut dados, grooves, mortises, and dovetail joints. Important uses in carpentry include the cutting of gains for hinges when hanging passage doors and routing housed stringers for stair construction. See Chapters 18 and 19.

A collet-type chuck is typically used to hold a standard bit. When changing bits on the router, it may be necessary to remove the base. However, it is usually possible to change bits through openings in the base. See **Figure 5-21.**

Router: Tool used to cut irregular shapes and to form various contours on edges.

Figure 5-20. A portable router has a motor mounted in an adjustable base that rests on the workpiece. The motor revolves in a clockwise direction. Always use two hands when routing. (Montachusett Regional Vo-Tech School)

PROCEDURE

Changing a router bit

1. Disconnect the router cord from power.
2. Lock the shaft or hold it with a wrench, depending on the router.
3. Loosen the chuck with a wrench.
4. Remove the bit.
5. Select the new bit. Some routers have a sleeve that is part of the bit assembly.
6. Install the bit on the arbor.
7. Tighten the chuck with a wrench.

Straight bits are used when cutting dados and grooves. Some bits for shaping and forming edges have a pilot tip that guides the router. The router motor revolves in a clockwise direction (when viewed from above) and should be fed from left to right when making a cut along an edge. When cutting around the outside of rectangular or circular pieces, always move in a counterclockwise direction.

Fixtures and templates are available that guide the router through various decorative or blanking cuts. **Figure 5-22** shows several router attachments that improve both the quality and efficiency of work.

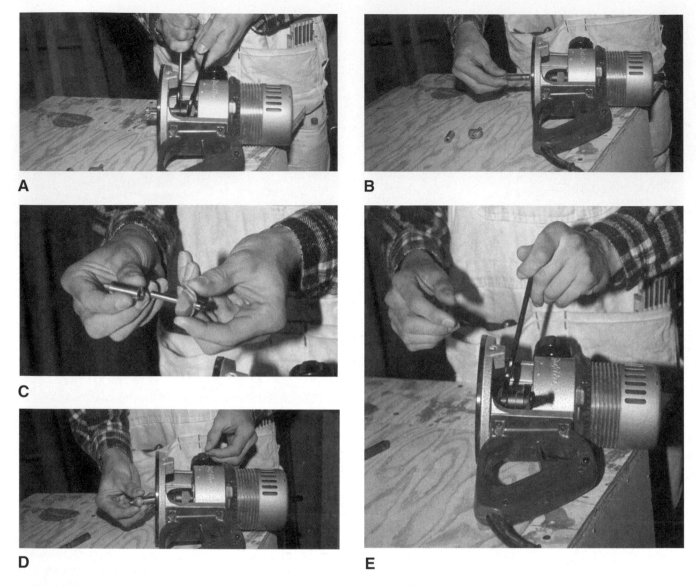

Figure 5-21. Changing a router bit. A—Use a wrench to loosen the chuck. B—Remove the bit being replaced. C—Select a new bit. D—Slide the new bit onto the arbor. E—Tighten the chuck and adjust the cutting depth. (Bullard-Haven Technical School)

Safety Note

Follow these safety rules when working with routers.
- Securely mount the bit in the chuck and make sure the base is tight.
- Be sure the motor is properly grounded.
- Wear eye protection.
- Be certain the work is securely clamped so it will remain stationary during the routing operation.
- Place the router base on the work, template, or guide—with the bit clear of the wood—before turning on the power. Firmly hold the router when turning on the motor. Starting torque could wrench the tool from your grasp.
- Hold the router with both hands and smoothly feed it through the cut in the correct direction.
- When the cut is complete, turn off the motor. Do not lift the machine from the work until the motor has stopped.
- Always unplug the router when mounting bits or making adjustments.

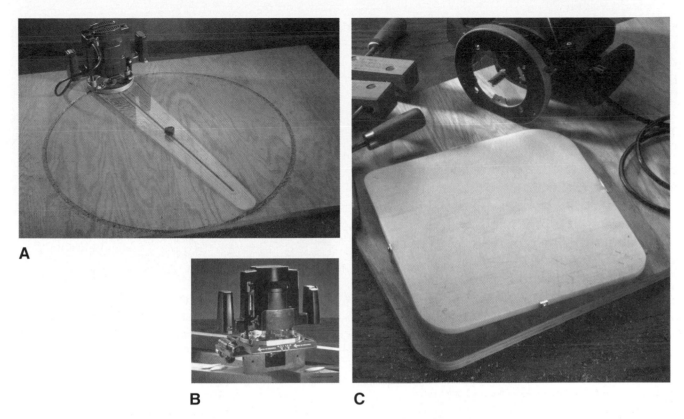

A

B   **C**

Figure 5-22. Router attachments. A—This circle cutting attachment produces circles from 8″–48″ in diameter. B—A converter allows the router to act as a biscuit joiner. C—A radius edge guide is used to cut smooth rounded corners. (Vermont American)

5.9 Portable Sanders

There are three basic types of portable sanders:
- Belt.
- Disc.
- Finish.

The types vary widely in size and design. Carefully follow the manufacturer's instructions when mounting abrasive belts, discs, and sheets, **Figure 5-23.** Be sure to follow the manufacturer's lubrication schedule, as well.

5.9.1 Belt Sander

The size of a ***belt sander*** is determined by the width of the belt. Using the sander takes some

Figure 5-23. When changing sandpaper on a power sander, always follow the manufacturer's instructions. Disconnect the tool from power before attempting to replace sandpaper. (Montachusett Regional Vo-Tech School)

Belt sander: Type of portable sander. Its size is determined by the width of the belt.

skill. Firmly support the stock. The switch must always be in the *off* position before plugging in the electric cord. Like all portable power tools, the sander should be properly grounded. Check the belt and make sure it is properly tracking.

Hold the sander over the work. Start the motor. Then, carefully and evenly lower the sander onto the surface. When using belt and finish sanders, be sure to travel with the grain. Move the sander forward and backward over the surface in even strokes. At the end of each stroke, shift it sideways about one-half the width of the belt.

Continue over the entire surface, holding the sander level. Sand each area the same amount. Do not press down on the sander; its weight is sufficient to provide the proper pressure for the cutting action. When work is completed, raise the sander from the surface and turn off the motor.

Figure 5-24. Using an orbital sander to perform finish sanding on cabinetwork. Some orbital finishing sanders are equipped with a dust bag. (Montachusett Regional Vo-Tech School)

5.9.2 Finishing Sanders

Finishing sanders are used for final sanding where only a small amount of material needs to be removed. They are also used for cutting down and rubbing finishing coats. There are two general types:

- Orbital sander, **Figure 5-24.**
- Oscillating sander.

Also called a *pad sander,* the finishing sander has a pad backing up the sanding paper. Unlike belt sanders, where the sanding belt moves, the paper is held stationary on the pad by mechanical grippers or is attached to the pad with a strong adhesive. The pad moves, carrying the paper.

Figure 5-25. A pneumatic stapler that drives 15 gage staples from 1 1/4″–2 1/2″long. (Senco Fastening Systems)

5.10 Staplers and Nailers

A wide variety of **power staplers** and **nailers** is available. Most of them are *pneumatic* (air powered), **Figure 5-25.** Those that are electrically operated should be properly grounded.

Using power nailers and staplers speeds up work and also allows the carpenter to work in

Finishing sander: Tool used for final sanding where only a small amount of material needs to be removed. It is also used for cutting down and rubbing finishing coats.

Power stapler: Electrically or pneumatically powered tool that drives staples.

Nailer: Tool used for driving nails.

tight places where swinging a hammer might be difficult or impossible. Staplers offer an alternative to nails, especially for securing sheathing and shingles. Nailer types include:

- **Strip-fed.** These hold a strip of nails in a long magazine located under the handle.
- **Coil-fed.** This type carries nails in a round canister, usually located ahead of a trigger handle.

To operate a nailer, first press the nose against the work to disarm the safety device that keeps it from driving the fastener. Then, press the trigger to drive the nail, as shown in **Figure 5-26.**

Most air compressors used on job sites are portable, **Figure 5-27.** They may be driven by either electric motors or gas engines. Manufacturers of pneumatic tools can assist carpenters in selecting a compressor large enough to handle air needs on the job.

Safety Note

Follow these safety rules when working with power staplers and nailers.

- Study the manufacturer's operating directions and carefully follow them.
- Use the correct type and size of fastener recommended by the manufacturer.
- For air-powered nailers, always use the correct pressure (seldom over 90 psi). Be sure the compressed air is free of dust and excessive moisture.
- Always keep the nose of the stapler or nailer pointed toward the work. Never aim it toward yourself or others.
- Check all safety features and be sure they are working. Make a test by driving the staples or nails into a block of scrap wood.

(Continued)

A

B

C

Figure 5-26. Pneumatic nailers speed work and can be used in close quarters where using a hammer may be difficult. A—This finish nailer requires no lubrication and drives 1-2″ brads. B—Framing nailer drives 2″–3 1/2″ smooth shank nails, 2″–3″ screw-shank nails, and 2″ and 2 3/8″ ring-shanked nails. C—A pneumatic nailer is being used to fabricate truss rafters. (Hadley-Hobley Construction; Trussworks, Inc.)

A

B

Figure 5-27. An air compressor suitable for operating pneumatic tools must deliver up to 110 pounds per square inch (psi) of pressure. A—A typical portable compressor. B—This generator and compressor on a job site power electrical as well as pneumatic tools. (Johnson-Manley Lumber Co.)

(Continued)
- During use on the job, firmly hold the tool's nose against the surface being stapled or nailed.
- Always disconnect the tool from the air or electrical power supply when it is not being used or before you attempt to clear a jam.
- Always wear eye protection while using a power nailer or stapler. Hearing protection is also recommended.
- Remove your finger from the trigger when carrying a nailer or stapler. An accidental discharge could cause injury.
- Never use bottled gas to operate a pneumatic nailer. Sparks could cause a fire and many gases are bottled at extremely high pressures.

(Continued)

(Continued)
- Keep hoses in good condition.
- Keep your free hand out of the way of the tool.
- Never operate power nailers or staplers around flammable materials. Sparks may cause a fire.
- Always move forward when nailing or stapling on a roof. This helps prevent backing off the roof.
- Secure the hose of a pneumatic tool with rope or other tie when working on scaffolding. This prevents the tool from falling on other workers.

5.11 Radial Arm Saws

The motor and blade of the *radial arm saw* are carried by an overhead arm. The stock being cut is supported on a stationary table. The arm is attached to and supported by a column at the back of the table. By raising or lowering the overhead arm, the depth of cut can be controlled. **Figure 5-28** shows the parts of a typical radial arm saw.

The motor is mounted in a yoke and may be tilted for angle cuts. The yoke is suspended from the arm on a pivot, which permits the motor to be rotated in a horizontal plane. Adjustments make it possible to perform many sawing operations.

While crosscutting, mitering, beveling, or dadoing, the work is firmly held on the table. The saw is pulled through the workpiece, **Figure 5-29**. For ripping and grooving, the blade is turned parallel to the table and locked in this position. Stock is then fed into the blade in a manner similar to feeding a table saw.

For regular crosscuts and miters, first be sure the saw is resting against the column.

Radial arm saw: The motor and blade are carried by an overhead arm. The stock being cut is supported on a stationary table. The arm is attached to and supported by a column at the back of the table. By raising or lowering the overhead arm, the depth of cut can be controlled.

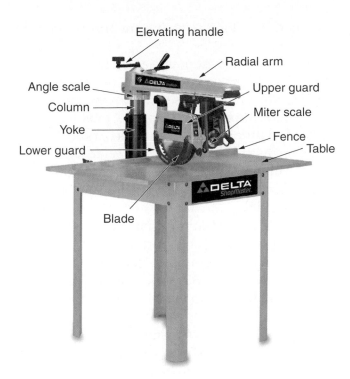

Figure 5-28. Radial arm saws make most cuts with the material held stationary. (Delta International Machine Corp.)

Figure 5-29. Crosscutting on a radial arm saw. Note that the lower guard is in place. (Montachusett Regional Vo-Tech School)

Then, place your work on the table and align the cut. Firmly hold the stock against the table fence with your hand at least 6″ away from the path of the saw blade. Turn on the motor. Grasp the saw handle and firmly and slowly pull the saw through the cut. See **Figure 5-30.** The radial arm saw may tend to "feed itself." You must control the rate of feed. When the cut is completed,

return the saw to the rear of the table and shut off the motor.

The radial arm saw is especially useful in cutting compound miters. It is also a good tool for cutting larger dimension lumber. Because of its weight, such lumber is difficult to slide across the table of a conventional table saw. See **Figure 5-31.**

Safety Note

Follow these safety rules when working with radial arm saws.
*   Stock must be firmly held on the table and against the fence for all crosscutting operations. The ends of long boards must be supported level with the table.

(Continued)

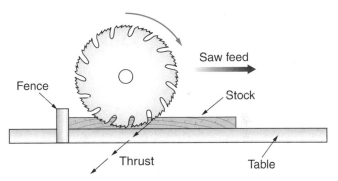

Figure 5-30. On a radial arm saw, blade rotation is in the direction of the saw feed.

Figure 5-31. Using a radial arm saw to cut dimension lumber. Materials of this size would be difficult to slide through a table saw. (Delta International Machinery Corp.)

(Continued)

- Before turning on the motor, be sure clamps and locking devices are tight. Check the depth of cut and table slope. The table must be slightly lower at the back than at the front to prevent the blade from "running" forward.

- Keep the guard and anti-kickback device in position.

- Always return the saw to the rear of the table after completing a crosscut or a miter cut. Never remove stock from the table until the saw has been returned.

- Maintain a 6″ margin of safety. Keep your hands this distance away from the path of the saw blade.

- Shut off the motor and wait for the blade to stop before making any adjustments.

- Do not leave the machine before the blade has stopped.

- Keep the table clean and free of scrap pieces and excessive amounts of sawdust. Do not attempt to clean off the table while the saw is running.

- In crosscutting, always pull the blade toward you.

- Stock to be ripped must be flat with one straight edge to guide it along the fence.

- When ripping, always feed stock into the blade so that the bottom teeth are turning toward you. This is the side opposite of the anti-kickback fingers.

5.12 Table Saws

Table saws are basic machines used in finish work, such as interior trim. They are frequently used by carpenters on projects that include on-the-job-built cabinets and other built-ins. Carpenters use them, to some extent, for cutting and fitting moldings and other inside trim work. When used for carpentry, the smaller sizes (8″–10″) are usually selected because they can be easily moved from one job to another. This book provides a brief introduction to this equipment. Woodworking textbooks provide a complete description of the wide variety of work these

saws can do, along with instruction on how to use them.

The table saw, sometimes called a circular saw, is used for ripping stock to width and cutting it to length. It also cuts bevels, chamfers, and tapers. Properly set up, the table saw can be used to produce grooves, dados, rabbets, and other forms that are basic to a wide variety of joints. The size of the saw is determined by the largest blade it will take. **Figure 5-32** shows a typical model.

Stock to be ripped must have at least one flat face to rest on the table and one straight edge to run along the fence, thus serving as a guide for the cut. **Figure 5-33** shows the correct procedure for making a rip cut. Be sure to follow the safety rules for sawing.

Figure 5-34 shows a standard crosscutting operation. A line is squared across the stock to show where the cut is to be located. For accurate work, make a check mark on the side of the line where the saw kerf will be located. Since the guard tends to hide the saw blade, it is helpful

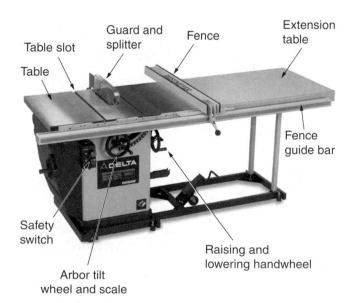

Figure 5-32. Table saws are most useful for finish work, such as cutting moldings and components for built-ins. (Delta International Machinery Corp.)

Table saw: Basic machine used in finish work.

Figure 5-33. Ripping stock with a table saw. The guard should always be in place for safe operation. (Des Moines Public Schools)

Figure 5-35. A line scribed on the saw table will help align stock for more accurate cutting to a line. This mark should align with left side of the saw kerf.

to use a line scribed in the table surface when aligning the cut. Since most of the work will be located to the left, it should extend back from the left side of the blade, as shown in **Figure 5-35.**

Figure 5-36 shows a compound angle being cut on a table saw. To make these cuts, both the miter gage and the saw blade are set at angles. Tables of compound angles are available to determine the proper settings.

Figure 5-36. Cutting a compound angle on 2 × 4 stock. To make these cuts, both the blade and miter gage are set at an angle.

A

B

Figure 5-34. Crosscutting. A—Squaring stock to a marked line. Securely hold the stock against the miter gage. B—The guard has been lifted out of the way to show operation.

Safety Note

Follow these safety rules when working with table saws.

- Be certain the blade is sharp and right for the job at hand.

- Make sure the saw is equipped with a guard and use it.

- Set the blade so it extends about 1/4" above the stock to be cut.

- Stand to one side of the operating blade and do not reach across it.

- Maintain a 4" margin of safety. Do not let your hands come closer than 4" to the operating blade, even though the guard is in position.

- Stock should be surfaced and at least one edge jointed before the piece is cut on the saw.

- Use the fence or miter gage to control the stock. Do not cut stock freehand.

- Always use push sticks when ripping short, narrow pieces.

- Stop the saw before making adjustments.

- Do not let small, scrap cuttings accumulate around the saw blade. Use a push stick to move them away.

- Setups must be carefully made and checked before the power is turned on.

- Remove the dado head or any special blades after use.

- Workers helping to "tail-off" the saw should not push or pull on the stock, only support it. The operator must control the feed and direction of the cut.

- Once work is complete, turn off the machine and remain until the blade has stopped. Clear the saw table and place waste in a scrap box.

The motor and blade of the power miter saw are supported on a pivot. See **Figure 5-37.** To operate the saw, the carpenter sets the angle from a scale marked off in degrees. The cut is made by pulling downward on the handle. A

A

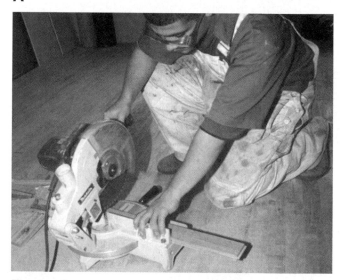

B

Figure 5-37. The miter saw. A—Miter saw makes precise mitered cuts to 46 1/2° left and right. B—Using a power miter to make a 90° cut on a piece of flooring. (Porter-Cable Corp.; Bullard-Haven Technical School)

5.13 Special Saws

Special saws have been developed for working with wood construction. If lightweight and compact size are important factors, carpenters may use a *power miter saw* or a *frame and trim saw* for accurate crosscuts and mitering.

Power miter saw: The motor and blade of this saw are supported on a pivot. The carpenter sets the angle from a scale marked off in degrees, then the cut is made by pulling downward on the handle.

Frame and trim saw: A saw that is supported on a pair of overhead shafts or guides. The support rotates left and right a little more than 45° to make crosscuts and miter cuts.

trigger control in the handle turns the motor on and off.

The frame and trim saw is supported on a pair of overhead shafts or guides. The support rotates left and right a little more than 45° to make crosscuts and miter cuts. It is capable of all sawing operations except ripping. The saw's capacity is 16" width on crosscuts and 12" on miter cuts of 45°. An extension table allows one worker to cut long stock alone

Safety Note

Follow these safety rules when working with all types of power saws.
- Keep guards in place while operating.
- Wear safety glasses or a face shield to protect your eyes from sawdust and other flying debris.
- Securely lock the saw at the angle of the cut.
- Firmly hold stock against the fence.
- Keep your free hand clear of the cutting area.
- Work only with a sharp saw blade.

5.14 Jointers

Jointers are commonly used to dress the edges and ends of boards. They may also be used for planing a face or for cutting a rabbet, bevel, chamfer, or taper. The cutter head is cylindrical and usually has three or four knives. As the cylinder rotates at high speed, the knives cut away small chips. This cutting action produces a smooth surface on the wooden workpiece.

Principal parts of a jointer are shown in **Figure 5-38.** The cutter head revolves at a speed of about 4500 rpm. The size of the jointer is

Jointer: Tool used to dress the edges and ends of boards. They may also be used for planing a face or for cutting a rabbet, bevel, chamfer, or taper. The cutter head is cylindrical and usually has three or four knives. As the cylinder rotates at high speed, the knives cut away small chips.

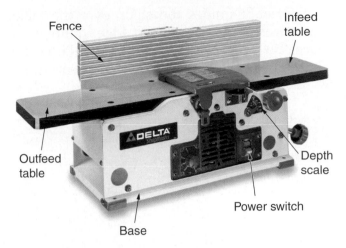

Figure 5-38. Principal parts of a jointer. Small 4" to 6" jointers like this one can be easily moved from one job site to another. (Delta International Machinery Corp.)

determined by the length of the knives. The three main adjustable parts are:
- Infeed table.
- Outfeed table.
- Fence.

The outfeed table must be the same height as the knife edges at their highest point of rotation. This is a critical adjustment. See **Figure 5-39.** If the outfeed table is too high, the stock will gradually rise out of the cut and a slight taper will be formed. If it is too low, the tail end of the stock will drop as it leaves the infeed table and cause a "bite" in the surface or edge.

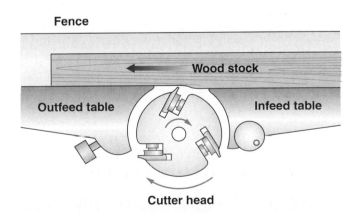

Figure 5-39. Operation of a jointer. Note the direction of the wood grain. Avoid working against the grain. The outfeed table must be set at the exact height of the cutter head blades.

The fence guides the stock over the table and knives. When jointing an edge square with a face, the fence should be perpendicular to the table surface, **Figure 5-40.** The fence is tilted when cutting chamfers or bevels.

Figure 5-40. Hold the board perpendicular to the table surface when jointing an edge. Keep hands at least 4″ away from the cutter head. Here, a worker uses a push stick to hold down the work. (Photo © Des Moines Public Schools)

Safety Note

Follow these safety rules when working with jointers.
- Before turning on the machine, make adjustments for depth of cut and position of fence.
- Be sure the guard is in place and is operating properly.
- The maximum cut for jointing on a small jointer is 1/8″ for an edge and 1/16″ for a flat surface.
- Stock must be at least 12″ long. Stock to be surfaced must be at least 3/8″ thick unless a special feather board is used.
- Feed the work so the knives will cut with the grain. Use stock that is free from knots, splits, and checks.
- Keep your hands away from the cutter head, even though the guard is in position. Maintain at least a 4″ margin of safety.
- Use a push block when planing a flat surface. Do not apply pressure directly over the knives with your hand.
- Do not plane end grain unless the board is at least 12″ wide.
- The jointer knives must be sharp. Dull knives vibrate the stock and may cause a kickback.
- When work is complete, turn off the machine. Stand by until the cutter head has stopped.

---

control the depth below the face of the drywall. Usually a belt clip attached to the shooter's housing allows the gun to be carried on a tool belt.

A ***plate joiner,*** often referred to as a *biscuit cutter,* is a portable tool for cutting half-moon-shaped slots in the edges of lumber for insertion of wood plates or biscuits. The plates strengthen the wood joints used in cabinetmaking and other home-building projects, just like dowels.

5.15 Specialty Tools

Drywall screw shooters (also called *screw guns*) speed up drywall installation and provide better connections, all but eliminating "nail popping." See **Figure 5-41.** Drywall screws are inserted one at a time into the nose. Locators

Drywall screw shooter: Tool used for drywall installation. Drywall screws are inserted one at a time into the nose. Locators control depth below the face of the drywall. Also called screw gun

Plate joiner: A portable tool for cutting half-moon slots in the edges of lumber for insertion of wood plates or biscuits. Often referred to as a biscuit cutter.

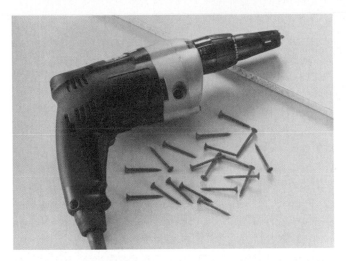

Figure 5-41. This drywall screw gun has variable speeds from 0–2500 rpm and trigger control reversing. (Milwaukee Electric Tool Corp.)

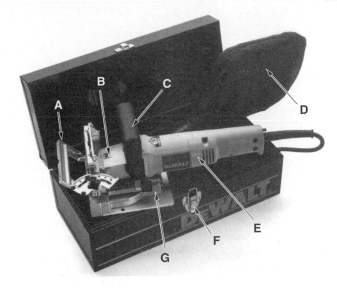

Figure 5-42. A plate joiner cuts crescent-shaped slots in matching parts of wood stock to be edge-joined. Wood splines called "plates" or "biscuits" are then inserted in the slots to strengthen the joint. A—Adjustable fence. B—Switch. C—Handle. D—Dust bag. E—Electric motor and housing. F—Base plate. G—Depth adjustment. (DeWALT Industrial Tool Co.)

Trim carpenters sometimes use plate joiners when assembling casing lumber and joining shelves and facings of cabinets built on site. The tool is also useful anywhere the carpenter needs to butt-join wood flooring or other wood members. **Figure 5-42** shows the major parts of the plate joiner.

A *panel saw* is useful for cutting large sheets of paneling or plywood. It holds a sheet in a vertical position while a cut is made from top to bottom. A typical panel saw is shown in **Figure 5-43.**

Powder-actuated tools are used to drive fasteners into concrete and structural steel. See **Figure 5-44.** The tool may be either direct acting or indirect acting. Both types depend on the expanding gases from an exploded cartridge to drive the fastener. Both types have a triggering mechanism similar to that of a firearm. To use the tool, it must be firmly pressed against the material being fastened. Then, pull the trigger to activate the tool. Expanding gases from the exploding powder act on either the fastener or a piston. The fastener is driven into the concrete or metal.

Panel saw: Saw used for cutting large sheets of paneling or plywood.

Powder-actuated tool: A striking tool that uses the force of an exploding cartridge to drive fasteners into concrete and steel.

Figure 5-43. A panel saw is usually found in a woodworking shop, but some carpenters and contractors use them on residential building projects. These saws are used to cut large sheets pof plywood or paneling. (Milwaukee Electric Tool Corp.)

Figure 5-44. A powder-actuated tool with fasteners. The gun ignites a blasting powder cartridge to drive fasteners into tough material like concrete and steel. (Forslund Bldg. Supply, Inc.)

5.16 Power Tool Care and Maintenance

Care of power tools is especially important if they are to properly function and give long service. Sharp blades and cutters ensure accurate work and make the tool much safer to operate. Good carpenters take pride in the condition and appearance of their tools.

Most power tools are equipped with sealed bearings that seldom need attention. Follow the manufacturer's recommendations for lubrication schedules. Gear mechanisms for portable power tools usually require a special lubricant. All equipment requires a few drops of oil on controls and the adjustment of bearings from time to time.

Clean and polish bare metal surfaces with 600 grit wet-or-dry abrasive paper when required. These surfaces can be kept smooth and clean by occasionally wiping them with light oil or furniture polish. Some carpenters apply a coat of paste wax to protect the surface and reduce friction.

Some power tools, especially those with a number of accessories, can be purchased with a case. In addition to making transport easier, such cases keep the accessories organized and protected.

Cutters and blades require periodic sharpening. Most carpenters are too busy and usually do not have the equipment to accurately grind cutters or completely fit saw blades. They usually send these items to a shop for an expert sharpening job. Still, carpenters may lightly hone cutters before grinding operations are needed. Also, they may prefer to file saw blades several times before sending them in for a complete fitting. A fitting includes jointing, gumming, setting, and filing. Carbide-tipped tools stay sharp about ten times longer than those with regular steel teeth or edges. A special diamond grinding wheel is used to sharpen carbide edges.

A modern high-production cutoff saw. A foot pedal controls the hydraulic power feed that moves the cutter head through a workpiece. (Speed Cut, Inc., Corvallis, OR)

Summary

Power tools greatly reduce the time needed to do many carpentry tasks. There are two categories of power tools used by carpenters: portable and stationary. Portable tools are lightweight and intended to be carried around by the worker. They include such items as drills, saws, power planes, routers, sander, screw shooters, and nailers. Stationary power tools are larger machines mounted on benches or stands. They are set up on the job site and work is brought to them. Stationary power tools include various types of saws, jointers, and planers. Observing safety rules and wearing proper protective devices are vital when working with power tools. Like hand tools, power tools must be properly cleaned and maintained for safe use.

Test Your Knowledge

Answer the following questions on a separate piece of paper. Do not write in this book.

1. How can you determine the voltage required by a portable power tool?
2. Give the meaning of the term *ground continuity* and explain why it is important.
3. What is a ground fault circuit interrupter and why should it be used in carpentry?
4. The size of a portable circular saw is determined by the _____.
5. For general-purpose work, a saber saw blade should have about _____ teeth per inch.
6. The blade of a saber saw cuts on the _____ stroke.
7. *True or False?* When drilling deep holes, do not withdraw a twist drill until the hole is completed.
8. The depth of cut of a power plane is adjusted by raising or lowering the _____.
9. Typically, a standard router bit is held in a _____-type chuck.
10. The size of a belt sander is determined by the _____.
11. Name two advantages of a power nailer.
12. *True or False?* It is safe to use bottled oxygen gas to power a pneumatic stapler.
13. To adjust the depth of cut of a radial arm saw, the _____ is raised or lowered.
14. When crosscutting with the radial arm saw, the blade is moved _____ the operator.

15. For regular work, the _____ of the jointer should be perfectly aligned with the knife edges at their highest point of rotation.

Curricular Connections

Science. Power tools often combine the principles of one or more of the six simple machines (lever, screw, wheel and axle, pulley, wedge, and inclined plane). Observe the operation of different types of power tools and try to identify which simple machines are involved. Example: a circular saw combines the wheel and axle (rotating blade) with the wedge (cutting teeth). Try to list as many tools as possible along with the simple machines used.

Language Arts. Make an oral presentation to the class on the proper use of a jointer. Demonstrate the proper setting of the outfeed table and explain what problems can occur if the table is set too low or too high. Be sure to stress safety in your presentation.

Outside Assignments

1. Visit a builders' supply center and study the various portable circular saws on display. Also, obtain descriptive literature concerning the various sizes. After careful consideration, select a brand, model, and size that you believe would be best for rough framing and sheathing work on residential structures. Give your reasons and report to your class. Include specifications and prices of your selection.
2. Visit with a carpenter in your locality and learn what procedures are followed in maintaining and sharpening tools. If he or she uses standard saw blades, learn how they are kept sharp. Ask about the use of hardened-tooth and carbide-tipped blades. If he or she has some tools sharpened at a saw shop, find approximate prices. If a tool maintenance center is located in your area, find out what services are available and what they cost. Carefully prepare your notes, then make a report to your class.
3. If a powder-actuated tool is available, demonstrate its safe and proper use.

Much planning goes into a building. This series of town homes was accurately laid out based on the architect's plans before the first footing was dug.

Building Layout

Learning Objectives

After studying this chapter, you will be able to:

- Explain the operation of the builder's level and level-transit.
- Explain the basic operation of a laser level system.
- Demonstrate proper setup, sighting, and leveling procedures.
- Measure and lay out angles using leveling equipment.
- Read the vernier scale.
- Use a plumb line.

Technical Vocabulary

Benchmark	Laser level
Builder's level	Leveling rod
Building lines	Line of sight
Building site	Lots
Carpenter's level	Measuring tape
Cut	Optical level
Datum	Plumb
Dumpy level	Property line
Elevation	Station mark
Fill	Steel tape
Grade	Transit
Grade leveling	Vernier scale

Before construction of a foundation or slab for a building can begin, the carpenter must know where the structure will be located on the property. This requires a series of steps involving the use of measuring tapes and building layout instruments. The place where a building is to be constructed is called a *building site*. Often, these sites are in communities where there are streets and where small tracts of land are broken up into smaller parcels called *lots*.

If the building site is a city lot, placing the building requires more steps. Most communities have strict requirements—buildings must be set back a certain distance from the street and maintain minimum clearances from adjoining properties. The local code must be carefully checked for these requirements before layout begins.

6.1 Plot Plan

Most, if not all, communities require the builder or owner to furnish a plot plan before

---

Building site: The place where a building is to be constructed.

Lots: Small tracts of land broken up into smaller parcels.

they will issue a building permit, **Figure 6-1.** A surveyor provides a survey of the site and locates stakes to indicate the boundaries, also called *property lines.* If there is a plot plan, it indicates the location of the structure and distances to property lines on all sides. Surveyors should always locate the property lines. They should also draw the plot plan, if one is required.

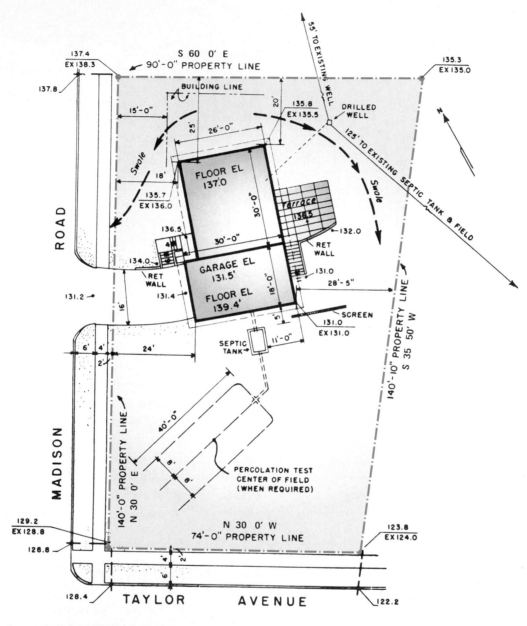

LOT 29. BLK. 2
HOMESTEAD ADDITION, LINCOLN COUNTY, COLUMBIA

Figure 6-1. Providing a plot plan is the responsibility of the architect or owner. The plan shows the property boundaries, along with the building lines for the proposed building. Many communities require a plot plan before issuing a building permit.

Property lines: The boundaries of a site.

The work of an engineer or surveyor protects the owner and builder from costly errors in measurement.

6.2 Measuring Tapes

For measurements and layouts involving long distances, *steel tapes,* usually called *measuring tapes,* may be used, **Figure 6-2.** Tapes are available in lengths of 50′ to 300′. There are various types with differing graduations. A carpenter will usually select one that is marked in feet, inches, and eighths of an inch. Surveyors, on the other hand, require a tape graduated in feet and decimal parts of a foot.

6.3 Establishing Building Lines

Laying out a building means locating the outside corners of its foundation, then marking

A

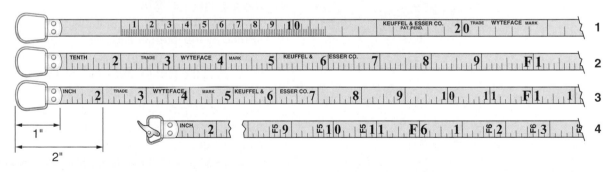

B

Figure 6-2. Measuring tape. A—Long tapes for layout use are made in lengths of up to 300′. B—Measuring tapes are made with different systems and graduations. 1. Metric. 2. Feet and decimal graduations. 3. Feet, inch, and eighth-inch graduations. 4. Feet and inches with feet repeated at each inch mark. (The Stanley Co.; Keuffel & Esser)

Steel tape: Tape that comes in lengths of 50′ to 300′. It is used for measurements and layouts involving long distances. Also called *measuring tape.*

Measuring tape: Tape that comes in lengths of 50′ to 300′. It is used for measurements and layouts involving long distances. Also called *steel tape.*

Building lines: The lines marking where the walls of a structure will be located.

them with stakes. *Building lines* are the lines marking where the walls of the structure will be. These lines must conform to code requirements on distance of the structure from boundary lines of the property.

Once the property lines are known and marked by the surveyor, the building lines can be found by measuring distances with a tape. See **Figure 6-3.** Be sure to observe proper setbacks and clearances. Carefully check the local code to ensure compliance. To start:
1. Locate the surveyor's stakes marking the lot corners. Often, these are 2 × 2 wood stakes driven flush with the surface of the ground. A small nail driven in the top of the stake marks the exact corner of the property.
2. Measure the setback to locate the front building line. Use a tape to measure the distance. To be accurate, the tape measurement must be perpendicular to the building line. String a line from stake-to-stake as needed.
3. Locate and mark the setbacks on the other three sides of the lot. Building lines must always be within these boundaries.

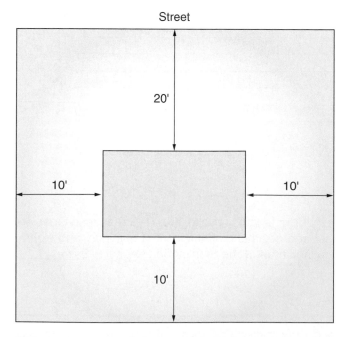

Figure 6-3. A simple rectangular structure can be laid out by taking measurements, using lot lines as reference points.

PROCEDURE

Using measuring tapes to establish building lines

In the absence of optical instruments (which are described later), it is possible to lay out building lines for small structures with measuring tapes. To start:
1. Drive a 2 × 2 stake marking one corner of the building. The exact location should never be closer to the lot lines than the intersection of two setback lines. Drive a nail in the top of the stake to mark the building line.
2. With a tape, measure off one side of the building dimension along the setback.
3. Drive another corner stake at this point and drive a nail in the top to mark the dimension. This establishes the beginning point of the second side of the building.
4. From the second stake, take a measurement for the intersecting building line. Use the 6–8–10 method shown in **Figure 6-4** to establish a square corner.
5. In the 6–8–10 method, measure and mark 6' along one building line already established. Next, measure 8' on the intersecting building line and mark that location. Then, measure the diagonal distance from one mark to the other. Adjust the building line as needed until the diagonal line measures 10' long. At that point, the corner formed by the intersecting lines is square.
6. Establish the remaining two building lines by measuring the length and squaring the corners.
7. Test the accuracy of the layout by diagonally measuring from corner to corner. If these measurements are equal, the building lines are laid out square.

6.4 Laying Out with Leveling Instruments

In residential construction, it is important that building lines be accurately established in relation to lot lines. It is also important that footings and foundation walls be level, square, and the correct size.

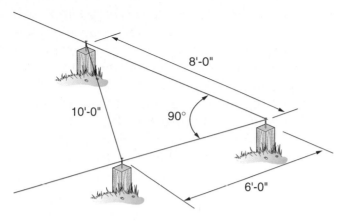

Figure 6-4. To check whether a corner is square, mark 6′ along one building line and 8′ on the intersecting building line. If the distance between the marks is 10′, the corner is square.

If the building is small, such as a 24′ garage, a *carpenter's level,* line level, framing square, and rule are accurate enough for laying out and checking the building lines. But, as size increases, special leveling instruments are needed for greater accuracy and efficiency.

6.4.1 Leveling Instruments

The *builder's level* and *transit* are more accurate instruments for building layout. These leveling instruments are frequently used in construction work. They are basically telescopes with accurate spirit levels and must be mounted and leveled on a base that can be rotated. When the job is too large for the chalk line, straight-edge, level, and square, then leveling instruments should be used.

The optical device of these instruments operates on the basic principle that a *line of sight* is a straight line that does not dip, sag, or curve. Any point along a level line of sight will be at the same height as any other point. Through the use of these instruments, the line of sight replaces the chalk line, line level, and straightedge.

The builder's level, also called a *dumpy level* or *optical level,* is shown in **Figure 6-5.** It consists of an accurate spirit level and a telescope assembly attached to a circular base that swivels 360°. Leveling screws are used to adjust the base after the instrument has been mounted on a tripod. The telescope is fixed so that it does not move up or down, but it rotates on the base. This permits any angle in a horizontal plane to be laid out or measured.

The transit is like the builder's level in most respects, **Figure 6-6.** However, the telescope can be pivoted up and down 45° in each direction. Using this instrument, it is possible to accurately measure vertical angles. The transit is also used to determine if a wall is perfectly *plumb* (vertical) by sighting the vertical cross-hair while pivoting the scope up and down. Its vertical movement also simplifies the operation of aligning a row of stakes, especially when they vary in height.

In use, both the builder's level and transit are mounted on tripods, **Figure 6-7.** Some models,

Carpenter's level: Tool used for laying out and checking the building lines.

Builder's level: An optical device used by builders to determine grade levels and angles for laying out buildings on a site. Also called a *dumpy level* or an *optical level.*

Transit: Optical leveling instrument commonly used for building layout.

Line of sight: A straight line that does not dip, sag, or curve. Any point along a level line of sight is at the same height as any other point.

Plumb: Exactly perpendicular or vertical; at a right angle to the horizon or floor.

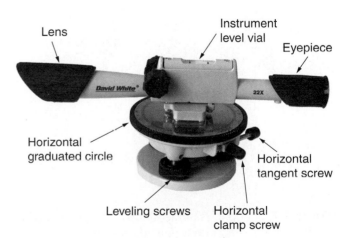

Figure 6-5. The builder's level is used to sight level lines and lay out or measure horizontal lines. (David White)

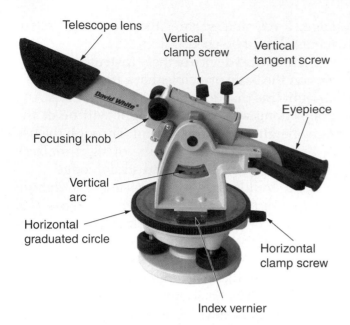

Figure 6-6. A transit can be used to lay out or check both level and plumb lines. It can also be used to measure angles in either the horizontal or vertical planes. (David White)

like the one shown, have legs whose length can be independently adjusted. This feature makes the tripod easier to level on sloping ground. It also permits the legs to be shortened for easier handling and storing.

When sighting over long distances, use a *leveling rod*. See **Figure 6-8.** It is a long rod marked off with numbered graduations. It allows differences in elevation between the position of the level and various positions where the rod is held to be easily read. The rod is especially useful for surveying. The person operating the level can make the readings or the target can be adjusted up and down to the line of sight and then the person holding the rod (rod holder) can make the reading.

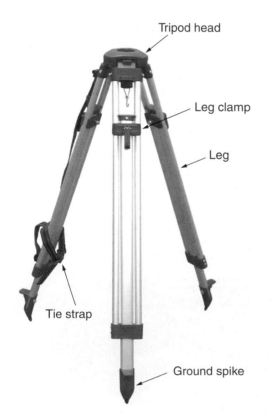

Figure 6-7. Tripod legs hinge at top and are adjustable for use on uneven terrain. (David White)

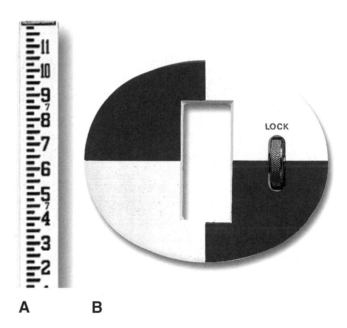

A   B

Figure 6-8. Leveling rod with target. The target can be moved up or down to match the line of sight of the transit. (David White)

Leveling rod: A long rod marked off with numbered graduations that is used, along with a builder's level or transit, to sight differences in elevation over long distances.

The rod shown in **Figure 6-8** has graduations in feet and decimal parts of a foot. This is the type used for regular surveying work. Rods are also available with graduations in feet and inches.

When sighting short distances (100′ or less), a regular wooden or metal folding rule can be held against a wood stake and read through the instrument. This procedure will be satisfactory for jobs such as setting grade stakes for a footing. Always be sure to hold the stake and rule in a vertical position.

Working Knowledge

Leveling instruments and equipment will vary somewhat, depending on the manufacturer. Always carefully read and study the instructions for a given brand.

6.4.2 Care of Leveling Instruments

Leveling instruments are more delicate than most other carpentry tools and equipment. Special precautions must be followed in their use so they will continue to provide accurate readings over a long period of time. Some suggestions follow:

- Keep the instrument clean and dry. Store it in its carrying case when not in use.

- When the instrument is set up, have a plastic bag or cover handy to use in case of rain. If the instrument becomes wet, dry it before storing.

- When moving the instrument from its case to the tripod, grip it by the base.

- Never leave the instrument unattended when it is set up near moving equipment.

- When moving a tripod-mounted instrument, handle it with care. Hold the instrument upright; never carry it in a horizontal position.

- Never over-tighten leveling screws or any of the other adjusting screws or clamps.

- Always set the tripod on firm ground with the legs spread well apart. When it is set up on a floor or pavement, take extra precautions to ensure that the legs will not slip.

- For precision work, permit the instrument to reach ambient (air) temperature before making readings.

- When the lenses collect dust and dirt, clean them with a camel hair brush or special lens paper.

- Never use force on any of the adjustments. They should easily turn by hand.

- Have the instrument cleaned, lubricated, and checked yearly by a qualified repair station or by the manufacturer.

PROCEDURE

Setting up the instrument

Use the following procedure to set up a tripod-mounted instrument.

1. Place the tripod so it will be a firm and stable base for the instrument. The base of the legs should be about 3′-6″ apart. Make sure the points are well into the ground and the head is fairly level.

2. Check the wing nuts on the adjustable legs. They should be tight enough to carry the weight of the instrument without collapsing or sinking. Tighten the hex nuts holding the legs to the head to the desired tension.

3. Carefully lift the instrument from its case by the base plate. Before mounting the instrument, loosen the clamp screws. On some instruments, the leveling screws must be turned up so the tripod cup assembly can be hand-tightened to the instrument mounting stud. Set the telescope lock lever of the transit in the closed position.

4. Attach the instrument to the tripod. If it is to be located over an exact point, such as a benchmark, attach the plumb bob and move the instrument over the spot. Do this before the final leveling.

PROCEDURE

Leveling the instrument

Leveling the instrument is the most important operation in preparing it for use. None of the readings taken or levels sighted will be accurate unless the instrument is level throughout the work. To level the instrument:

1. Release the horizontal clamp screw and line up the telescope so it is directly over a pair of the leveling screws.
2. Grasp the two screws between the thumb and forefinger, as shown in **Figure 6-9.** Uniformly turn both screws with your thumbs moving toward each other or away from each other.
3. Keep turning until the bubble of the level vial is centered between the graduations. You will find on most instruments that the bubble travels in the direction your left thumb moves. See **Figure 6-10.** Leveling screws should bear firmly on the base plate. Never tighten the screws so much that they bind.
4. When the bubble is centered, rotate the telescope 90° (so it is over the other pair of leveling screws) and repeat the leveling operation.
5. Recheck the instrument over each pair of screws. When the instrument is level at both positions, the telescope can be turned in a complete circle without any change in the bubble.

PROCEDURE

Sighting

Most builder's levels have a telescope with a power of about 20×. This means that the object being sighted appears to be 20 times closer than it actually is. The procedure for sighting is easy to learn:

1. Line up the telescope by sighting along the barrel and then look into the eyepiece, **Figure 6-11.**
2. Adjust the focusing knob until the image is clear and sharp.
3. When the crosshairs are in approximate position on the object, **Figure 6-12**, tighten the horizontal-motion clamp.
4. Make the final alignment by turning the tangent screw.

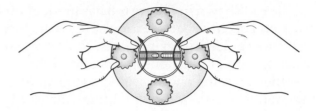

Figure 6-9. Adjust leveling screws to center the bubble in the level vial.

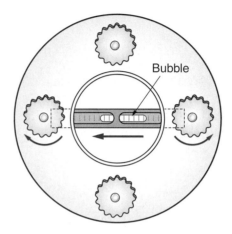

Figure 6-10. The bubble of the level vial will generally move in the same direction as the left thumb. This bubble needs to move left.

Figure 6-11. Sighting a level line with a builder's level. Both eyes are kept open during sighting. This reduces eyestrain and provides the best view. Hand signals tell the rod holder whether to raise or lower the target on the leveling rod. (Kasten-Weiler Construction)

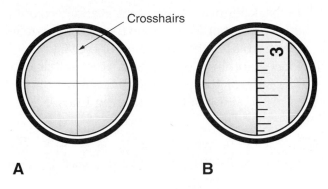

Figure 6-12. View through the telescope. A—Crosshairs vertically and horizontally split the image area in half. B—The object in the view should be centered on the crosshairs.

6.5 Using the Instruments

A carpenter can use leveling instruments to prepare the building site for excavation and grade leveling. Jobs that can be done with leveling instruments include:

- Locating the building lines and laying out horizontal angles (square corners).
- Finding grade levels and elevations.
- Determining plumb (vertical) lines.

For layout, the builder's level or transit must start from a reference point. This can be a stone marker in the ground, point on a manhole cover, or mark on a permanent structure nearby. The point where the instrument is located is called the *station mark*. It may be the *benchmark*, the corner of the property, or a previously marked point that is to be a corner of the building. This

Station mark: The reference point where a builder's level or transit is set up when laying out building lines or finding grade levels and elevations. It must be a type of marker that will not be disturbed by any construction activity.

Benchmark: A mark on a permanent object fixed to the ground from which grade levels and elevations are taken for construction of a building. Sometimes officially established by government survey.

Vernier scale: Device on a transit that measures minute portions of an angle.

might be a stake with a nail in it or the intersection of two string lines marking the corner of a building line.

6.5.1 The Horizontal Graduated Circle

Laying out corners with the transit requires an understanding of how the horizontal graduated circle is marked. It is divided into spaces of 1°, **Figure 6-13.** When you swing (rotate) the telescope of the builder's level or transit, the graduated circle remains stationary, but another scale, called the *vernier scale,* moves. It is marked off in 15-minute intervals. When laying out or measuring angles where there are fractions of degrees involved, you will use this vernier scale.

Figure 6-14A shows a section of the graduated circle and the scale. It reads 75°. Notice that the zero mark on the vernier exactly lines up with the 75° mark. Now look at **Figure 6-14B.** The zero mark has moved past the mark for 75°, but is not on 76°. You need to read along the vernier scale until you find a mark that is closest to being directly over a degree mark on

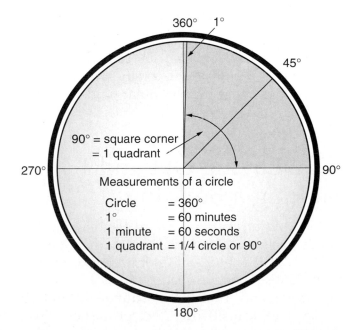

Figure 6-13. The graduated circle of a transit corresponds to the 360° of a full circle; 90° represents a quadrant, which would give you a square corner for a building.

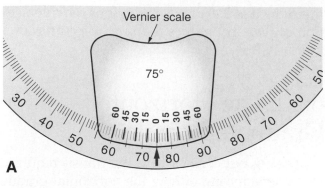

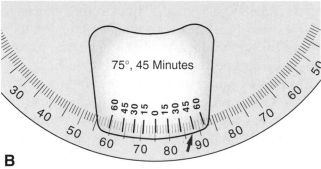

Figure 6-14. Reading the horizontal circle of a transit. A—When the zero mark of the vernier is exactly on a degree mark, the reading is an even degree. In this case, the reading is 75°. B—When the zero mark is between degrees, read across the vernier to find the minute mark that aligns with a degree mark. This reading is 75°, 45′. (David White)

the circle. That number is 45. The reading is 75°, plus the number on the vernier, 45 minutes.

Vernier scales will not be the same on all instruments. Study the operator's manual for instructions about the particular model you are using.

6.5.2 Laying Out and Staking a Building

Staking out is done before establishing the grade level. It begins with locating the lot lines. Corners of the lot should normally be marked with stakes. Then, proceed:

1. Center and level the instrument (builder's level or transit) over the lot corner stake. Measure the setback called for by local codes. Sight across to the opposite corner stake.
2. Drive a 2 × 2 stake at the setback in line with the lot stakes. Use the transit or builder's level

to check alignment. The vertical and horizontal crosshairs should center on the top of the stake. Drive a nail in the top-center of the stake.
3. Place another stake at the correct setback for one side of the property line. You are now ready to stake out the building lines.

PROCEDURE

Staking a building

Staking out building lines requires two persons. When a builder's level is used, the second person will use a rod that must be plumbed along the line of sight. Since a transit can pivot up and down, the second person uses a stake to locate corners along the building line.

1. Attach a plumb bob to the center screw or hook on the underside of the instrument. Some instruments have an optical plumb for zeroing in over a point. Shift the tripod until the point of the plumb bob is directly over the point marking the corner of the building lines. This is at point A on line AB, as shown in **Figure 6-15.**
2. Level the instrument before proceeding further. Recheck for plumb.
3. From point A (or station A), turn the telescope so the vertical crosshair is directly in line with the edge of a stake or rod held at point B. When using a transit, sight the telescope on the stake.
4. Use a measuring tape along line AB to locate distance to the corner. Drive a corner stake at this point.
5. Set the horizontal circle on the instrument at zero to align with the vernier zero and swing the instrument 90° (or any other required angle).
6. Position the rod or stake along line AC so it aligns with the crosshairs.
7. Locate the other corner along line AC using a measuring tape.
8. Move the instrument to point C, sight back to point A and then turn 90° to locate the line of sight to point D.
9. Measure the distance to point D and place a stake.
10. Use a measuring tape to check the diagonal distances. If these are equal, the building line is square.

If the resulting figure is a rectangle or square, you have completed the layout. However, you may want to move the instrument to point D to check your work.

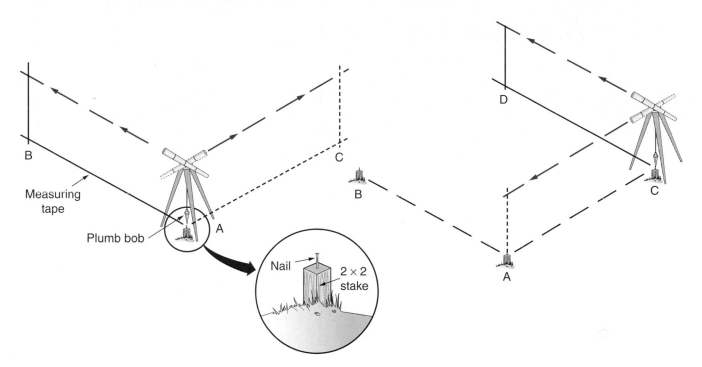

Figure 6-15. Steps for laying out building lines. Left—Locate the instrument over a stake marking a corner. Line up 0 on the instrument circle with the building line AB. Swing the instrument 90° to establish line AC. Right—Move the instrument to point C to establish point D. Rod must be used when the instrument is a builder's level. Rod must be held plumb using a plumb line or carpenter's level. A transit is a much better instrument in this operation since it is not necessary to use the rod. Simply swivel the transit telescope and sight on the corner stake.

In practice, you will find that it is difficult to locate a stake in a single operation. This is especially true when using a builder's level, where the line of sight must be "dropped" to ground level with a plumbed rod or straightedge. Usually, it is best to set a temporary stake, as in **Figure 6-16.** Mark it with a line sighted from the instrument. Then, with the measuring tape pulled taut and aligned with the mark, drive the permanent stake and locate the exact point as shown.

All major rectangles and squares of a building line can be laid out using leveling instruments in the manner just described. After batter boards are set and lines attached, the carpenter's level and square can be used to locate stakes for small projections and irregular shapes.

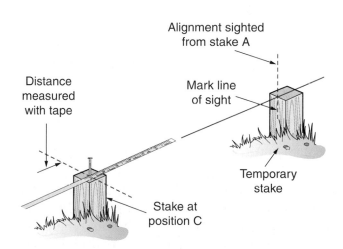

Figure 6-16. A temporary stake may be used to establish an exact point. First, set the temporary stake and mark the line of sight on it. Drive a second stake and transfer the mark from stake A.

6.5.3 Finding Grade Level

Many points on the building site need to be set at certain elevations or grade levels. These points might include:

*   The depth of excavation, such as for a basement.
*   The finished height (elevation) of the foundation footings.
*   The height (elevation) of foundation walls.
*   The elevation of floors.
*   Site features, such as proper grading to ensure that surface water is directed away from the building.
*   Bearing elevation (point at which footings contact the earth) for footings in order to ensure adequate frost protection of the foundation.
*   Establishing rise and run of steps and walkways as part of the exterior of the structure.

Not all building sites are level. Finding the difference in the grade level between several points or transferring the same level from one point to another is called *grade leveling.* This operation is immediately useful to the excavator, who must determine how much earth must be removed to excavate a basement or trench a foundation footing. Grade leveling is also useful to determine how much earth must be deposited in a particular area to achieve a desired height or elevation at that location.

When the leveling instrument has been set level, the line of sight will also be level. The readings can be used to calculate the difference in elevation, **Figure 6-17.** If the building site has a large slope, the instrument may need to be set up more than once between the points where you want to take readings. The first reading is taken with the rod in one position. Then, the instrument is carefully rotated 180° to get the reading at a second rod position. This position may be higher or lower than the first position. From a practical standpoint, it is simpler to work from a higher point on the site than a lower point. Depending on the actual slope, this measurement can often be made with a single reading.

The term *grade* means the level of the ground. *Elevation* refers to the major structural levels

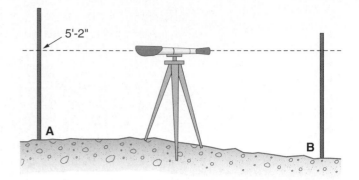

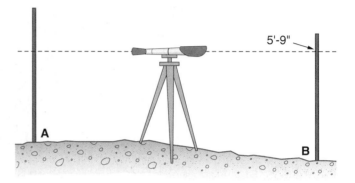

Figure 6-17. Establishing a grade level and finding the difference between two points on a building lot. Top—Level the instrument and take a line of sight reading on point A. Mark down the rod reading. Bottom—Swing the telescope 180° and take a line of sight reading on the rod. Compare the two elevations. In this example, point A is 7″ higher than point B.

of the building. More specifically, these levels include the top of the footing, top of the foundation wall, and finish height of the first floor.

There should be a reference point (level) for all elevations established on the building site. It is called the benchmark, *datum,* or simply the beginning point. This point must remain

---

Grade leveling: Finding the difference in the grade between several points or transferring the same level from one point to another.

Grade: Quality of lumber. Also, the height or level of a building site.

Elevation: The height of an object above grade. Also, a type of drawing that shows the front, rear, and sides of a building.

Datum: An established reference point for determining elevations in an area. Also called *benchmark* or simply the beginning point.

undisturbed during the construction of the building project. A stake driven at one corner of the lot or building site, or even a mark chiseled into concrete curbing, often serves as the benchmark.

PROCEDURE

Checking grade

Building sites are rarely perfectly level. All have high and low points. These highs and lows need to be determined before the height of the foundation is established. This job is easier when the site is fairly level, as in **Figure 6-17.**

1. Locate the instrument midway into the site, then level the instrument.
2. Take a line-of-sight reading on a rod held at one edge of the site.
3. From the position of the target on the rod, note the elevation (5'-2" in **Figure 6-17**) and record it. Surveyors keep a notebook for such recording, since many people may need to refer to the readings. Carpenters generally jot readings on a piece of scrap lumber. Usually, they are the only ones needing to refer to the readings.
4. Rotate the telescope 180°. Take a line-of-sight reading on the rod located at the opposite side of the site.
5. Note the elevation at the target position on the rod (5'-9" in **Figure 6-17**). Record it as before.
6. Subtract the lower elevation from the higher one to find the difference (7"). This number is the actual vertical increase or decrease from one known point to another.

When setting grade stakes for a footing or erecting batter boards, set the instrument in a central location on the site, as shown in **Figure 6-18.** The distances to the target (or rod) will be roughly equal. This will improve the accuracy of the readings taken for each corner. An elevation established at one corner can be quickly transferred to other corners or points in between.

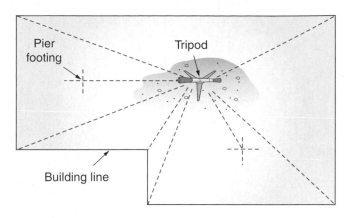

Figure 6-18. A central location for the instrument will make finding and setting grade stakes easier and more accurate.

6.5.4 Setting Footing Stakes

Grade stakes for footings are usually first set to the approximate level "by eye." They are then carefully checked with the rod and level as they are driven deeper. The top of each stake should be driven to the required elevation.

Sometimes, reference lines are marked on construction members, stakes, or other objects near the work. The carpenter then transfers them to the formwork with a carpenter's level and rule as needed. This eliminates the need to repeatedly establish the same elevation.

There may be situations where the existing grade will not permit the setting of a stake or reference mark at the actual level of the grade. In such cases, a mark is made on the stake with the information on how much fill to add or remove. The letters *C* and *F*, standing for *cut* and *fill,* are generally used. See **Figure 6-19** for an example of how stakes are marked.

Cut: The process of removing material to achieve the desired grade.

Fill: The process of adding material to achieve the desired grade.

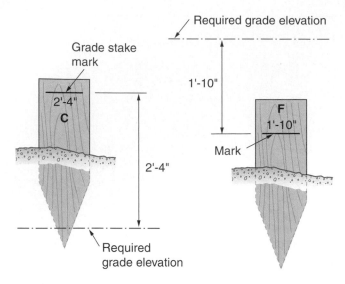

Required grade elevation

Grade stake mark

2'-4"

C

1'-10"

F

1'-10"

Mark

2'-4"

Required grade elevation

Figure 6-19. Cut and fill stakes are used to tell the excavator how much material must be removed or added to reach grade level. The letter *C* means "cut" (remove) and the letter *F* means "fill" (add).

P R O C E D U R E

Using multiple readings

When laying out steeply sloped building plots or carrying a benchmark to the building site, it will likely be necessary to set up the instrument in several locations. **Figure 6-20** shows how reading from two positions is used to calculate, establish, or determine differences in grade at various locations on the plot.

1. Set up the instrument midway between two points on the plot. In **Figure 6-20,** these are identified as points A and B.
2. Take a line-of-sight reading at point A (or station A). Record the reading in the notebook. In this case, the reading is 5'-8".
3. With the rod at point B and the instrument still at setup #1, take a second reading. Record the reading in the notebook. In this instance, the reading is 2'-6".
4. Move the instrument to a point midway between points B and C. Level it as before.
5. Take a second reading on point B and record it. In this example, the reading is 6'-0".
6. With the rod located at point C, take a reading by rotating the telescope 180°. Note and record the elevation. In this case, the reading is 2'-4".
7. Calculate the differences in grade level for each pair of stakes, as shown in **Figure 6-20.**
8. Add the resulting distances and subtract the "minus" sum from the "plus" sum to find the difference in grade from one edge of the plot to the other. The result, in this example, is 6'-10".

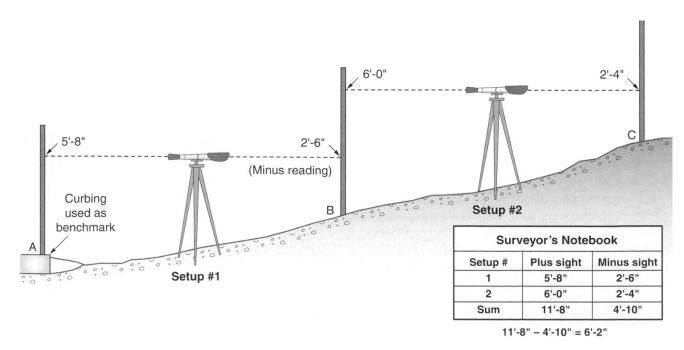

Surveyor's Notebook		
Setup #	Plus sight	Minus sight
1	5'-8"	2'-6"
2	6'-0"	2'-4"
Sum	11'-8"	4'-10"

11'-8" − 4'-10" = 6'-2"

Figure 6-20. When there is a large slope on the property or when long distances are involved, the transit will need to be set up in two or more locations.

6.5.5 Running Straight Lines with a Transit

Although the builder's level can be used to line up stakes, fence posts, poles, and roadways, the transit is more accurate for these tasks, especially when different elevations are involved. Set the instrument directly over the reference point. Level the instrument and then release the lock that holds the telescope in the level position. Swing the instrument to the required direction or until a stake is aligned with the vertical crosshair. Tighten the horizontal circle clamp so the telescope can move only in a vertical plane. Now, by pointing the telescope up or down, any number of points can be located in a perfectly straight line. See **Figure 6-21.**

6.5.6 Vertical Planes and Lines

Beyond the leveling tasks just mentioned, the transit is a good instrument for:

- Measuring vertical angles above or below the line of sight.
- Plumbing building walls, columns, and posts.

PROCEDURE

Measuring vertical angles

1. Position the instrument near the structure. Level the instrument.
2. Release the lever that holds the telescope in a horizontal position.
3. Swing the instrument vertically.
4. Set the horizontal crosshair at the point you wish to measure.
5. Tighten the vertical clamp.
6. Make a final, fine adjustment with the tangent screw to locate the horizontal crosshair exactly on the point.
7. Read the vertical angle on the vertical arc scale and the vernier.

PROCEDURE

Establishing plumb lines

Plumb lines can be checked or established by first operating the instrument as shown in **Figure 6-22.** As you tilt the telescope up and down, all of the sighted points are located in the same vertical plane. To plumb structures, such as posts or walls, follow these steps and refer to **Figure 6-23:**

1. Set up the transit at a distance from the object equal to at least equal to the height of the object. Tilt the telescope to sight on the base.
2. Loosen the horizontal clamp and line up the vertical crosshair with the base of the object.
3. Tighten the horizontal clamp.
4. Tilt the telescope upward to the top of the object. If the object is plumb, the crosshair will be on the same plane as it is at the base.
5. If object is not plumb, adjust the brace to bring the object into plumb.
6. Move the transit to a second position, preferably 90° either to the right or left, and repeat the procedure.

A plumb bob and line may often be the most practical way to check vertical planes and lines. For layouts inside a structure, where a regular builder's level or transit is impractical, use a plumb line.

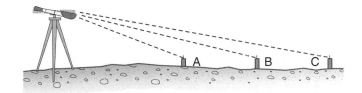

Figure 6-21. How to use the transit to align a row of stakes.

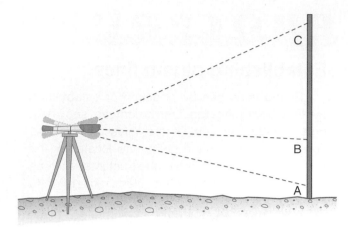

Figure 6-22. A transit can be used to lay out or check points in a vertical plane.

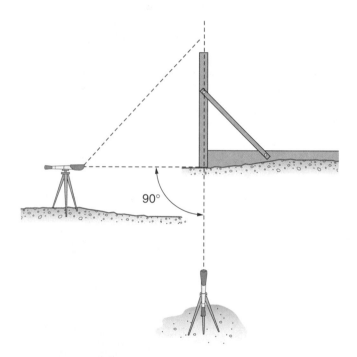

90°

Figure 6-23. Using the transit to plumb an object. First, align the vertical crosshair with the base. Then, swing the telescope to or near the top. Adjust the object for plumb as needed. Reposition the instrument at a 90° angle to the first line and repeat the process.

6.6 Laser Systems

The laser has been integrated into the building industry. Manufacturers have been able to produce low-power, visible lasers in small, inexpensive units. Using various names—laser plane, construction laser, laser transit, or simply *laser level*—the operating principle of the various units is essentially the same. They perform most of the functions of a conventional transit. However, they need only one person to carry out any layout operation. Some are designed to rotate 360°, allowing the operator to establish level lines throughout the structure without moving the instrument.

Safety Note

For safety, power output of the laser is controlled. Federal regulations limit Class II construction lasers to 1 milliwatt and Class IIIa units to less than 5 milliwatts. These levels are safe for the eyes so long as the user does not stare into the beam. Manufacturer's cautions should be carefully observed.

A laser level can be set to emit a laser beam for a full 360° without being tended by an instrument person. **Figure 6-24** shows the laser transmitter and its receiver. During use, the transmitter is untended. The receiver can be tended by a rod

Figure 6-24. Laser level and receiver. The transmitter is at the right and the receiver is at the left. (Spectra-Physics Laserplane, Inc.)

Laser level: Leveling instrument that emits a level laser beam over 360° without being attended. A receiver attached to a rod and attended by a rod holder makes a sound when the laser beam strikes it.

holder, **Figure 6-25,** or it can be attached to an excavator or backhoe, **Figure 6-26.** Operators of excavating equipment can work the entire job site without stopping to check grade levels. The laser level continuously monitors grade, eliminating excess excavation.

Figure 6-26. A laser receiver can be attached to an excavating machine, speeding up the excavation process. (Spectra-Physics Laserplane, Inc.)

Figure 6-25. A rod holder moves the receiver up or down until it is on line of sight with the laser level. (Spectra-Physics Laserplane, Inc.)

The operator places the bucket or blade cutting edge on the benchmark or finished elevation. Then, the receiver is adjusted up or down on the machine and the operator tightens the clamp when the "on-grade" point is reached. The receiver "catches" the laser beam from the rotating transmitter and signals the operator whether measured surface is above, below, or on grade.

The laser level is also adapted to other construction tasks to establish either level or plumb lines. Horizontal operations for which it might be used include leveling suspended ceiling grids and leveling raised-access computer floors. Vertical operations include plumbing partitions, curtain walls, posts, columns, elevator shafts, or any operation requiring a vertical reference point.

ON THE JOB

Surveyor/Surveying Technician

Surveyors and surveying technicians establish official land boundaries and the exact location and extent of building sites and other land uses. Survey parties, usually consisting of several surveying technicians under the direction of a party chief, traditionally use a variety of instruments to establish property boundaries and elevations. A licensed surveyor uses this information, combined with research in legal records and other sources, to prepare an official map called a *plat of survey* that establishes the official location of the property. Such a survey is often required before a deed can be issued in a property transfer.

Surveyors are professionals who must meet educational and experience requirements and pass licensing examination. In the past, surveyors could be licensed by gaining experience on a survey party, then passing a licensing exam. However, most states today require a four-year degree in surveying or a related field, such as civil engineering, in addition to the exam and field experience.

Surveying technicians often have some postsecondary training, usually in a community college certificate or associate degree program. High school courses in algebra, geometry, trigonometry, and drafting are good preparation. Technicians are usually responsible for using optical, physical, and electronic tools to make the needed measurements in the field. While traditional tools such as theodolites and measuring tapes are still used, more and more survey parties are working with electronic distance-measuring devices and global positioning system receivers that use satellite data for precise location-finding. Advancement to party chief is possible with additional formal training and experience.

Working conditions for survey parties can be extreme, since they are exposed to all kinds of weather conditions and may have to carry equipment for long distances in rugged terrain. Licensed surveyors often work in the field, but also spend time indoors doing research, data analysis, and report writing.

Approximately 2/3 of all surveyors and surveying technicians are employed by architectural and engineering firms or companies providing related services. Most of the remaining employment is provided by government at all levels, ranging from federal agencies, such as the U.S. Forest Service, to state and local planning departments and highway agencies. A relatively small number of surveyors are self employed.

Summary

The correct location on the building site must be identified before the foundation or slab can be constructed. Laying out a building involves locating the outside corners of the foundation, driving stakes, and then stretching building lines between stakes to mark where the walls will be. Measuring tapes can be used for layout work, but leveling instruments are more precise. Leveling instruments are also used to establish grades and elevations. Laser levels are replacing conventional transits and levels in building site layout. Their biggest advantage is requiring only one person to take measurements, rather than two.

Test Your Knowledge

Answer the following questions on a separate piece of paper. Do not write in this book.
1. For surveying work, a measuring tape with graduations reading in feet and _____ is required.
2. What are *building lines?*
3. Explain how to check whether corners of a building layout are square.
4. In the use of leveling instruments, the _____ replaces the chalk line and straightedge.
5. The builder's level consists of a telescope assembly that is mounted on a _____ base.
6. The most important operation in setting up a builder's level or transit is _____ the instrument.
7. When sighting through the telescope, you should adjust the _____ until the image is sharp and clear.
8. *True or False?* The vernier scale is used to measure angles in fractions of a degree.
9. To position a leveling instrument without an optical plumb directly over a given point, a _____ is used.
10. When setting grade stakes for a building footing, the instrument should be set up in a _____ location.
11. *True or False?* A laser level can check the level of an entire structure from one position.

Curricular Connections

Mathematics. Make a study of the procedures you would follow and calculations you would make to determine the height of a flagpole, tall building, or mountain, using a transit and trigonometric functions. Use the information to determine the height of at least one tall object in your community. Prepare a report for your class explaining the theory and how you proceeded.

Outside Assignments

1. Study the catalog of a supplier or manufacturer and develop a set of specifications for a builder's level. Be sure it includes a good carrying case. Also select a suitable tripod and measuring tape. Secure prices for all of the items.
2. Through drawings and a written description, tell how you would proceed to lay out a baseball diamond using a transit.

Section 2

Footings,
Foundations,
and Framing

Jack Klasey

Footings and Foundations

Learning Objectives

After studying this chapter, you will be able to:

- Lay out building lines and set up batter boards.
- Describe excavation procedures.
- Explain footing requirements and how to build footing forms.
- Define the terms *concrete, cement,* and *aggregate.*
- Describe the building, erecting, and use of forms for poured foundation walls.
- Discuss the types of foundation systems used for residential buildings.
- List steps and professional practices for laying up concrete block foundation walls.
- Explain foundation insulating and waterproofing procedures.
- Discuss design factors that apply to sidewalks and driveways.
- Estimate concrete materials required for a specific area.

Technical Vocabulary

Admixtures
Aggregate
Anchor bolts
Anchor straps
Backfilling
Batter boards
Bucks
Cement
Chairs
Closure block
Concrete blocks
Control point
Corbel blocks
Course pole
Crawl-space foundation
Curing
Drywood termites
Fixed anchor
Flat ICF wall
Floating
Form ties
Grade beam foundation
Ground-supported slab
Head joints
Hydration
Ledger boards
Lintel
Mason's line
Monolithic pour
Mortar
Nailing strips
Permanent wood foundation (PWF)
Pilaster
Piling and girder foundations
Plain footings
Post-and-beam ICF wall
Rebar
Reinforced footings
Screeding
Screen-grid ICF wall
Slab foundations
Slab-on-foundation
Slab-on-grade
Spread foundation
Stepped footing
Structurally supported slab
Subterranean termites
Troweling
Waffle-grid ICF wall
Walers
Wall pocket

In the construction of single-family dwellings and other structures, carpenters must work with many people in other construction trades. They also must closely work with the architect and owner in carrying out the total building plan.

On some jobs, carpenters may be required to lay out the building lines and supervise the excavation. They may also build forms for footings and poured foundation walls. Anyone working in the carpentry trade needs a working knowledge of standards and practices in concrete work. In this chapter, some of the material also deals with masonry. It is included because of the close relationship to carpentry.

7.1 Clearing the Site

Preparation of the building site may require removal of trees, boulders, and other obstructions, **Figure 7-1.** Grading may be needed before the building lines are laid out. This may require the placement of grade level stakes. The proper establishing of grade is explained in Chapter 6, **Building Layout**. It usually requires the use of a transit or laser level.

If the property is wooded, use care in deciding which trees are to be removed. Much depends on where the trees are located and their types. In general, evergreens should be used as protection against the cold winter winds. Deciduous (leaf-dropping) trees are best used as shade from the hot summer sun. Try to place the house to take advantage of the protection offered by trees already on the property.

Trees that are to be taken down should be clearly marked so the person responsible for their removal takes the right ones. Avoid digging trenches through the root system of trees being retained. It could cause them to die. Likewise, trees usually will not tolerate more than a foot of additional fill over their root systems.

7.2 Laying Out Building Lines

After the site is cleared, the next step is to locate and mark lot lines. This must be carefully and accurately done. Errors in establishing boundaries of the property can lead to locating the building too close to lot lines or even

Figure 7-1. Site preparation may involve removing stumps and boulders, as well as leveling and stockpiling topsoil.

encroachment on adjoining property. Therefore, to protect the owner and builder, establishing boundaries is best done by or with the help of a registered engineer or licensed surveyor, **Figure 7-2.** Such help may include establishing building lines and grade levels. However, once property lines are located, the carpenter can do this. Therefore, be familiar with the setback requirements of local building codes.

It is best to locate building lines with a builder's level, transit, or theodolite. Follow the procedures described in Chapter 6, **Building Layout**. Lines can, however, be transferred from lot markers. In such cases, it is important that

Figure 7-2. Establishing property or lot lines is a task best left to an engineer or surveyor. This crew is using a total station for the job. (Cella Barr Associates)

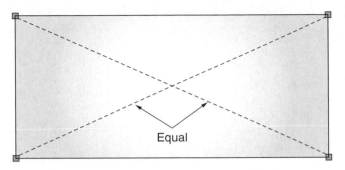

Figure 7-3. The diagonals of a perfect square or rectangle are equal. Always measure diagonals to be sure corners are square.

distances be laid out perpendicular to existing lines. Building lines also must be square. To establish a right angle, the 6–8–10 method can be used. Refer to **Figure 6-4** in Chapter 6.

Following the procedures described in Chapter 6, locate corners formed by the intersection of the outside edges of the foundation walls. Mark the positions by driving stakes. Drive a nail in each stake top at the exact spot marking the corners of the foundation or slab.

After locating all building lines, carefully check them. Measure their length and, even though they were laid out with an optical instrument, measure the diagonals of squares and rectangles, **Figure 7-3.** An out-of-square foundation causes problems throughout construction.

7.2.1 Batter Boards

Batter boards consist of horizontal boards and stakes. They are set up well away from where each corner of the new building will be. Use 2 × 4s for the stakes and 1 × 6s or wider boards for the horizontal pieces, **Figure 7-4.** The horizontal boards are called *ledger boards.* Nail the ledger boards to the stakes. Check that they

---

Batter boards: A temporary framework of stakes and ledger boards used to locate corners and building lines when laying out a foundation.

Ledger board: Horizontal components of batter boards that support the lines set up to locate building lines and corners.

are level and at a convenient working height, preferably slightly above the top of the foundation. The batter boards should be roughly level with each other. Sometimes a straight batter board is used when it will not be located at a corner. See **Figure 7-5.**

Most excavations are sloped at a 45° angle. Therefore, batter boards must be set back far enough so the excavator does not disturb them. A garage with a slab-on-grade foundation might require 3′–4′ of clearance. A residence may need as much as 8′–10′ of clearance and a commercial building much more.

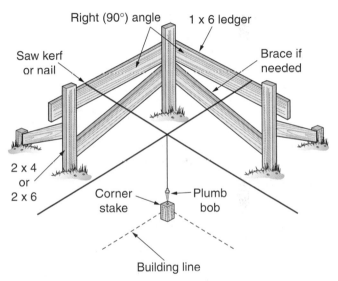

A

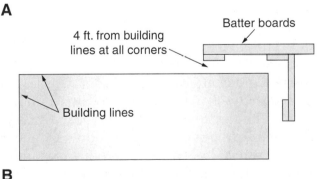

B

Figure 7-4. Using batter boards and plumb line to establish building lines. A—Drop the plumb where building lines intersect. Drive a tack in the corner stake to mark the exact corner. B—Set up batter boards at least 4′ from building lines on all four corners. Allow more room if heavy equipment is likely to disturb boards. In loose soil, or when boards are more than 3′ off the ground, use braces.

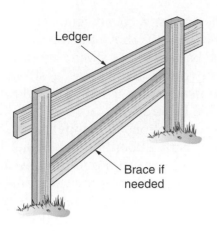

Figure 7-5. A straight batter board is used in some situations to mark offsets or extensions of a foundation. An example is shown in **Figure 7-6.**

Using strong cord (such as mason's line) and a plumb bob or laser level, locate the lines so they intersect over the layout stakes. Mark the tops of the ledger boards where the intersecting lines rest. Make a shallow saw kerf or drive a nail at this mark. Pull the lines tight and fasten them. If you are using saw kerfs, drive nails into the backs of the ledger boards to fasten the lines. You may prefer to wrap lines around the ledger several times, running them through the saw kerf.

7.3 Excavation

Building sites on steep slopes or in rugged terrain should be rough graded before the building is laid out. Topsoil should be removed and piled where it will not interfere with construction. This soil can be used for the finished grade after the building is completed.

Figure 7-6 shows a typical residential building layout. The carpenter erects batter boards at least 4′ away from the intended excavation. This allows the excavation to proceed without disturbing the batter boards. Note the use of straight batter boards where there are no corners.

Next, the carpenter drives stakes marking the outer edge of the rough excavation. The lines are then removed from the batter boards so they will not interfere with the excavation.

For regular basement foundations, the excavation should extend at least 2′ beyond the building lines to allow clearance for formwork.

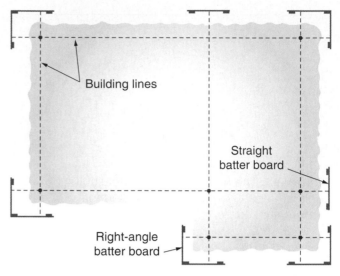

Figure 7-6. Starting the foundation for a home. Building lines (dotted) indicate the area to be excavated. The actual excavation (shown in gray) should be at least 2′ outside of the building lines to provide working room for forming and placing the foundation.

Foundations for structures with a slab floor or crawl space need little excavating beyond the trench for footings and walls.

The depth of the excavation can be calculated from a study of the vertical section views of the architectural plans. In cold climates, it is important to locate foundations below the frost line or frost-proof them with rigid insulation. Local building codes usually cover these requirements. If footings are set too shallow or are not insulated, moisture in the soil under the footing may freeze. This could force the foundation wall upward, a condition called *heaving.* Heaving can cause cracks and serious damage.

It is common practice to establish both the depth of the excavation and the height of the foundation by using the highest elevation on the perimeter of the excavation. This is known as the **control point** or *high point,* **Figure 7-7.** This practice is followed whether the site is graded or not.

Control point: A reference point for determining the elevation of footings, floors, and other parts of a building. Also, the highest elevation on the perimeter of an excavation (also called *high point*).

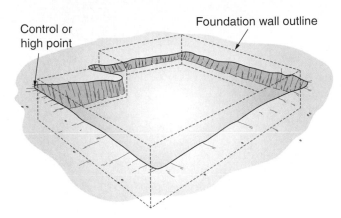

Figure 7-7. The control point is the highest elevation outside of the excavation and used to establish depth of excavation.

Foundations should extend about 8" above the finished grade. At this height, wood framing and finish members are protected from moisture. The finished grade should slope away from all sides of the structure so surface water drains away from the foundation.

The depth of the excavation may also be affected by the elevation of the site. It may be higher or lower than the street or adjacent property. The level of sewer lines also has an effect. Normally, solving these problems is the responsibility of the architect. Information on grade, foundation, and floor levels is usually included in the working drawings.

Spread foundation: Popular in light construction, a poured or block foundation resting on a footing twice as wide as the foundation wall. The footing transfers the load received from walls, pilasters, columns, and piers to the soil below.

Crawl-space foundation: Similar to a spread foundation used for a home with a basement, but less deep. The space between the ground and the floor joists is only a few feet.

Slab-on-foundation: Foundation type usually employed in commercial construction.

Slab-on-grade: Slab that is supported directly on top of the ground. Also called *ground-supported slab.*

Piling and girder foundations: Foundation system usually found in warm climates where freezing of supply piping and drains is not a problem.

Safety Note

Follow these safety rules whenever the project involves excavation.

- Contact local utilities to determine the locations of any underground utility lines.

- If soil or moisture conditions make the sides of the excavation potentially unstable, provide adequate shoring.

- Pile all excavated material a minimum of 2' outside the excavation and be sure it rests at a stable angle.

- Check air quality in the excavation, especially if gasoline-engine-powered equipment is being used. Provide additional ventilation if needed.

- When necessary, pump water seepage or rainwater accumulation out of the excavation.

- Provide an adequate number of ladders or other means of safe exit from the excavation.

7.4 Foundation Systems

All structures settle. A properly designed and constructed foundation distributes the weight to the ground in such a way that the settling is negligible or at least uniform. **Figure 7-8** presents several foundation types, shown in simplified form.

For light construction, such as residential, the *spread foundation* is most common. It transmits the load through the walls, pilasters, columns, or piers. These elements rest on a footing that transmits the load to the soil beneath. This type is often used when the owner needs a basement. The *crawl-space foundation* is similar in construction.

The *slab-on-foundation* is usually used in commercial construction. The *slab-on-grade* type is popular in warm climates and is also used for smaller structures, such as detached garages.

Piling and girder foundations are usually found in warm climates. There, freezing of supply piping and drains is not a problem.

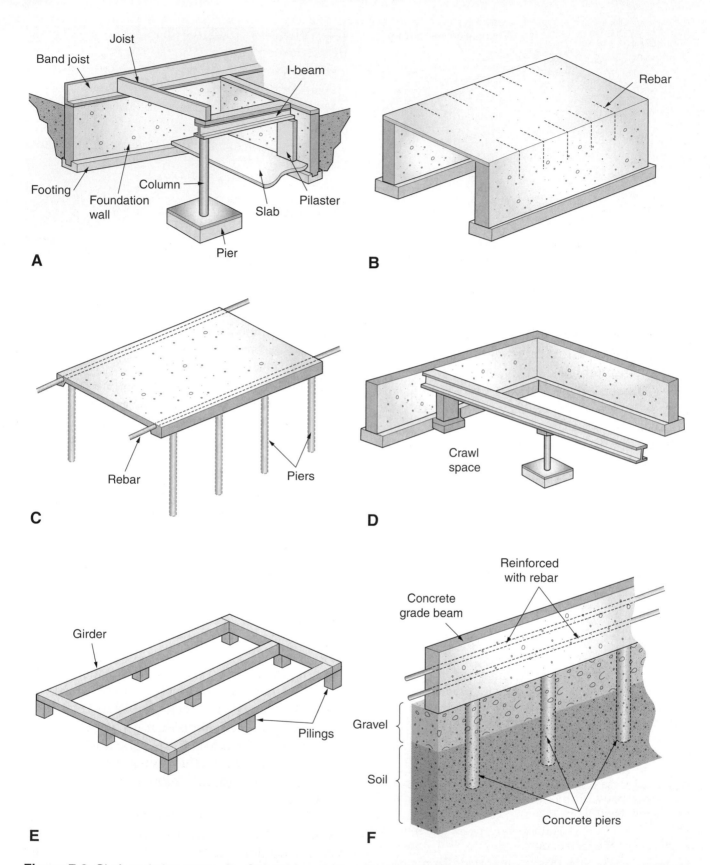

Figure 7-8. Six foundation types. A—Spread foundation. B—Slab-on-foundation. C—Slab-on-grade foundation. Piers (dashed lines) can be used as additional support in unstable soils. D—Crawl-space foundation (similar to a spread foundation). E—The piling-and-girder foundation is used in warm climates where water pipes and drain pipes are not in danger of freezing. F—Reinforced grade beam foundation.

A

Wood-framed wall

Flash

Concrete slab

Rigid-polystyrene
insulation

0°F

10°F

Heat flow

Vapor barrier   Soil

B

Figure 7-9. Frost-protected shallow foundations. A—This type of foundation looks like any other except for the visible rigid insulation around the perimeter. B—Rigid polystyrene insulation placed around perimeter keeps the soil from freezing.

The *grade beam foundation* is used where soil has poor load-bearing qualities. This foundation is similar to the footing of a spread foundation. It differs in that the footing is reinforced with steel reinforcing bar (*rebar*) and rests on concrete piers. A machine-operated auger can be used to dig the holes for the piers. In firm soil, concrete can be poured directly into the

---

Grade beam foundation: Similar to a spread foundation footing but differs in that it is reinforced with rebar and rests on piling.

Rebar: Steel reinforcing rod placed in concrete.

Plain footings: Footings that carry light loads and usually do not need reinforcing.

Reinforced footings: Footings containing steel rebar for added strength.

Stepped footing: A footing that changes grade levels at intervals to accommodate a sloping site.

holes. In less stable or soft soil, round fiber forms are placed in the holes and filled with concrete. Rebar is placed in the holes or forms before the pour. This reinforcement extends above the ground and ties into other rebar laid horizontally in the beam.

7.4.1 Frost-Protected Shallow Foundations

The Council of American Building Officials (CABO) has modified its model energy code to allow the use of shallow foundations if they are frost-protected with rigid polystyrene. The method allows footings in extremely cold climates, such as in North Dakota, to be only 18" below grade. The change affects CABO's one- and two-family dwelling code.

This method has been successfully applied in Scandinavian countries. The rigid insulation is placed both vertically and flat around the foundation perimeter along the slab edges and foundation corners. This insulation prevents frost heave even where footings do not extend below the frostline. One strip is vertically installed to protect the edge of the foundation. Another strip, usually about 4' wide, is placed flat around all sides. This is covered with several inches of soil, **Figure 7-9**.

7.4.2 Footings

Footings must be strong enough to carry the load placed on the foundation and transfer this load to the soil beneath, **Figure 7-10**. *Plain footings* carry light loads and usually do not need reinforcing. *Reinforced footings* have steel rebar embedded in them for added strength against cracking. They are used when the load must be spread over a large area or bridged over a weak spot, such as excavations or sewer lines.

A *stepped footing* is one that changes grade levels at intervals to accommodate a sloping lot. Vertical sections should be at least 6" thick. Horizontal sections between steps should be at least 2' long. If masonry units are to be used over the footing, distances should fit standard brick or block modules.

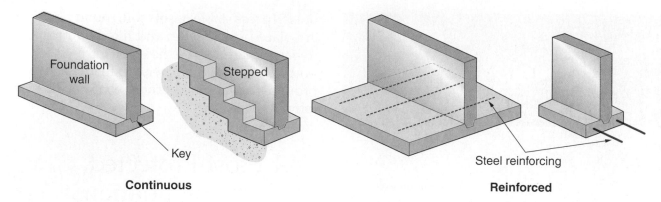

Figure 7-10. Different types of footings are needed for different slope and soil conditions. Vertical runs of a stepped footing should not exceed 3/4 of the horizontal run between steps.

Footings must be wide enough to spread the load over a sufficient area. Load-bearing capacities of soils vary considerably. See **Figure 7-11.** In residential and smaller building construction, the usual practice is to make the footing twice as wide as the foundation wall, **Figure 7-12.** The average thickness of a footing is about 8″. Footings that must support cast-in-place concrete walls may include a recess or groove forming a keyed joint. Steel dowels also can be installed to secure a poured wall.

Footings under columns and posts carry heavy, concentrated loads and are usually from 2′ to 3′ square. The thickness should be about 1 1/2 times the distance from the face of the column to the edge of the footing.

Reinforced footings are used in:

- Regions subject to earthquakes.
- Situations where the footings must extend over soils containing poor load-bearing material.

Some structural designs may also require the use of reinforcing. The common practice is to use two No. 5 (5/8″) rebar for 12″ × 24″ footings. At least 3″ of concrete should cover the reinforcing at all points.

In a single-story dwelling where the chimney footings are independent of other footings, they should have a minimum projection of 4″ on each side. For a two-story house, chimney footings

Load-bearing capacities of different soils	
Soil type	Capacity (lb./square foot)*
Crystalline bedrock	12,000
Sedimentary and layered rock	4,000
Sandy gravel or gravel	3,000
Sand, silty sand, clayey sand, silty gravel and clayey gravel	2,000
Clay, sandy clay, silty clay, clayey silt, silt, and sandy silt	1,500

*One-third increase permitted when considering load combinations, including wind and earthquake loads.

Figure 7-11. The load-bearing capacity of different soil types varies considerably.

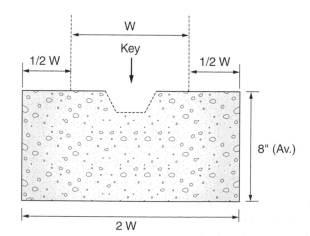

Figure 7-12. Standard footing design for residential construction. It should be twice the width of the foundation wall that will rest on it. The thickness can vary. Sometimes reinforcement is added. The key is designed to anchor the wall. It is often replaced with dowels of steel rebar.

Figure 7-13. A slab-on-grade foundation "floats" on the soil under it. Reinforcing steel is often placed around the perimeter. (Pierce Construction, Ltd.)

should have a minimum thickness of 12″ and a minimum projection of 6″ on each side. Exact dimensions will vary according to the weight of the chimney and the nature of the soil. Where chimneys are a part of outside walls or inside bearing walls, chimney footings should be constructed as part of the wall footing. Concrete for both chimney and wall footings should be placed at the same time.

7.4.3 Slabs

Slab foundations take several forms. The slab can be used with elements such as walls, piers, and footings, as shown in **Figure 7-8C.** This is called a *structurally supported slab.* A second type is supported directly by the ground, like the one shown in **Figure 7-13.** These are referred to as a *ground-supported slab* or *slab-on-grade.*

Slab foundation: A poured concrete foundation supported by the soil. When constructed in one continuous session, it is called a *monolithic pour.*

Structurally supported slab: Slab that can be used with other elements such as walls, piers, and footings.

Ground-supported slab: Slab that is supported directly on top of the ground. Also called *slab-on-grade.*

Monolithic pour: Term used for concrete construction poured and cast in one unit without joints.

Some slabs are constructed in one continuous pour. There are no joints or separately poured sections. This is called a *monolithic pour.* See **Figure 7-14.** This type of slab is constructed particularly in warmer climates.

Safety Note

When using wire mesh for slab reinforcement, take precautions to prevent the ends of the unrolled mesh from springing up and causing injury. Secure both ends of the mesh or reverse the material so it curves downward as it is unrolled.

7.4.4 Forms for Footings

After the excavation is completed, carefully check the batter boards. They may have been disturbed by the excavating equipment. Make necessary adjustments before proceeding with layout of the footings and construction of the forms. Forms contain the concrete until it is cured. Replace the lines on the batter boards. You need them to locate corners of the footing form. Drop a plumb bob from the intersection of the lines to the bottom of the excavation. Then proceed as follows.

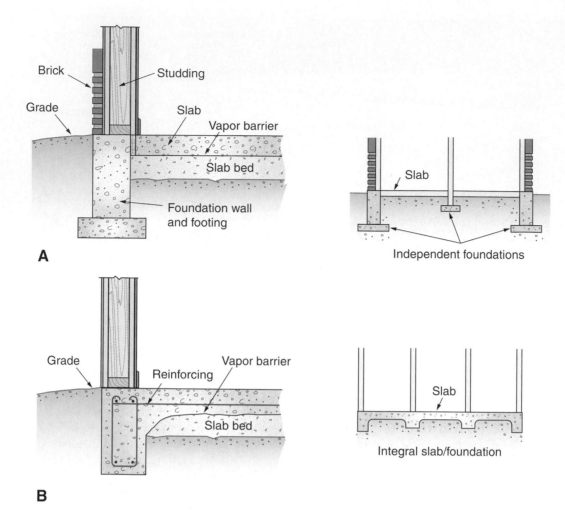

Figure 7-14. Variations of the slab-on-grade foundation. A—Slab is independent of foundation walls. The left view shows only part of the slab. The total foundation is shown at the right. B—This slab has a monolithic structure incorporating a grade beam and is used where soil provides poor support. The right view shows the entire foundation.

PROCEDURE

Constructing footing forms

1. Drive stakes and establish points at the corners of the foundation walls.
2. Set up a builder's level at a central point of the excavation. Drive a number of grade stakes level with the top of the footing along the footing line and at approximate points where column footings are required. Corner stakes can also be driven to the exact height of the top of the footing.
3. Connect the corner stakes with lines tied to nails in the top of the stakes.
4. Working from these building lines, construct the outside form for the footing. The form boards are located outside of the building lines by a distance equal to the footing extension beyond the building line. This varies according to the width of the foundation wall. Footings should be twice the width of the wall. For example, the footing for an 8″ foundation wall is usually 16″. Thus, it should extend 4″ on each side of the finished wall. The top edge of the form boards must be level with the grade stake. See **Figure 7-15.**
5. With the outside form boards nailed in place, it is easy to set and level the inside form using a folding rule or steel tape and a level. **Figure 7-16** illustrates this method.

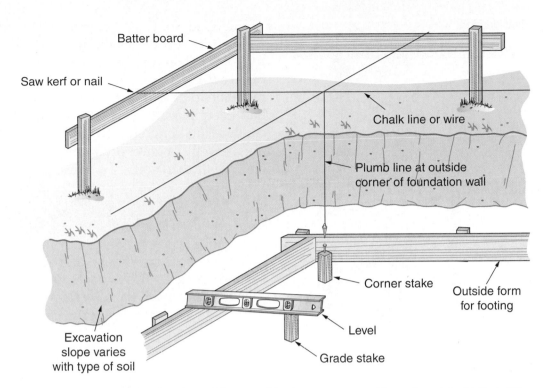

Batter board

Saw kerf or nail

Chalk line or wire

Plumb line at outside corner of foundation wall

Corner stake

Outside form for footing

Excavation slope varies with type of soil

Level

Grade stake

Figure 7-15. Laying out forms for footings. The outside forms are positioned and built first. The corner stake indicates the building line. Footings should be twice the width of the wall. The top of the grade stake is the height of the footing. Place one end of a level on the grade stake and the other end on the outer form board. The form boards must be level with the grade stake.

Normally, forms are constructed of 2" lumber. Support stakes are placed at 4' intervals when 2" material is used. If 1" lumber is substituted, it should be supported with stakes placed 2' apart. Use precut spreaders (also called *spacers*) between the inside and outside form boards so you do not have to measure the distance each time, **Figure 7-17.** Strips of 1" lumber called *ties* are nailed across the top of the form boards at intervals to strengthen the form during the pour. Remove them as you place the concrete. Bracing of footing forms may sometimes be desirable. Usually it is necessary only if 1" lumber is used for forming. Attach the brace to the top of a form stake and to the bottom of a brace stake located about 1' away.

Another method of building footing forms is shown in **Figure 7-18.** Plastic channel with preformed metal spreaders can be quickly installed using only a level and an occasional stake to keep the hollow form from shifting as concrete is poured. The form has perforations along its outer edges. These allow it to become a permanent drain that removes water from around the foundation. The plastic form is not removed.

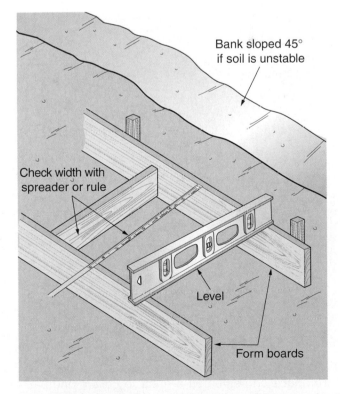

Bank sloped 45° if soil is unstable

Check width with spreader or rule

Level

Form boards

Figure 7-16. Setting forms. Measure the distance and set the inside form board. Use a precut spreader or a rule and a level to space and level the inside form boards.

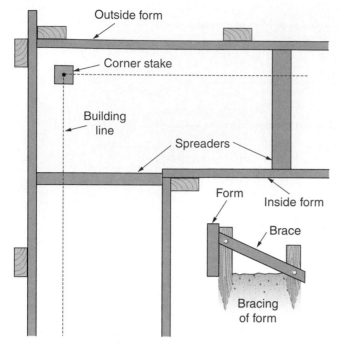

Figure 7-17. Properly located footing form. Note that the outer form extends beyond corner stake and building line for a distance determined by the thickness of the foundation wall. Precut spreaders are used to locate the inside form boards. Ties may be fastened across the tops of the forms to keep the forms from spreading.

Figure 7-18. Form-a-Drain® is a patented plastic footing form that also serves as a drain to collect and carry away water. The form remains in place as part of the foundation. Preformed metal spreaders also act as ties. (Kasten-Weiler Construction)

Stepped footings require some additional formwork. Vertical blocking must be nailed to the form to contain the concrete until it sets. See **Figure 7-19.** Be careful not to install any form members that will be trapped after the concrete has been placed. The soil should be able to contain the concrete at these points. All footings, except in rare situations, must be placed on undisturbed soil.

Working Knowledge

Loose dirt and debris must always be removed from the ground that will be located under a footing. This is necessary even though the resulting depth will be greater than required.

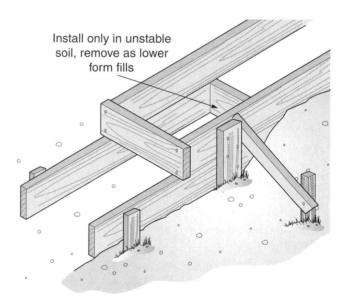

Install only in unstable soil, remove as lower form fills

A

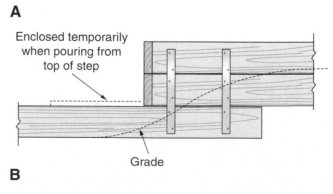

Enclosed temporarily when pouring from top of step

Grade

B

Figure 7-19. Form constructed for a vertical section in a stepped footing. The lower level is placed first and usually allowed to set up slightly before the step is poured. A—Basic form. B—Alternate design when more height is needed.

Column footings are pads of concrete designed to support columns by distributing the load to the ground. They carry weight transmitted to the column from the beams, stringers, and joists of a building, **Figure 7-20.**

Forms for column footings are usually set after the wall footing forms are complete. These are located by taking measurements from the building lines. Forms are leveled to previously set grade level stakes.

Some hand digging and leveling of the excavation will probably be necessary as forms are set. The top of the footing must be level. The bottom may vary as long as the minimum thickness is maintained.

Proper location of nails is necessary to avoid problems in form removal. It is important that they be only temporarily nailed (always from

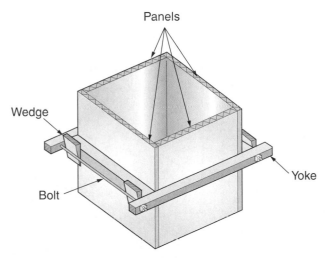

A

B

Figure 7-20. Column footing. A—A typical reusable column footing form. B—An actual footing.

the outside) to stakes and to each other. Duplex (double-headed) nails may be used. If regular nails are used, they should be driven only partway into the wood so they can be easily removed.

When the forms are completed, check for sturdiness and accuracy. Remove the line, line stakes, and grade stakes. The concrete can now be placed.

7.5 Erecting Wall Forms

Many different types of wall forming systems are available. There are certain basic considerations that should be understood and applied to all systems. For quality work, the forms must be tight, smooth, defect free, and properly aligned. Joints between form boards or panels should be tight. This prevents loss of the cement paste, which tends to weaken the concrete and cause honeycombing.

Wall forms must be strong and well braced to resist the side pressure created by the plastic concrete. This pressure tremendously increases as the height of the wall is increased. Regular concrete weighs about 150 lb. per cu. ft. If it is immediately poured into a form 8′ high, it creates a pressure of about 1200 lb. per sq. ft. along the bottom edges of the form. In practice, this pressure is reduced through compaction and hardening of the concrete. It tends to support itself. Thus, the lateral pressure is related to:

- The amount of concrete placed per hour.
- The outside temperature.
- The amount of mechanical vibration.

7.5.1 Re-establishing the Building Line

Before setting up the outside foundation wall form, you will need to find and mark the building line on top of the footing. To do this, set up your lines on the batter boards once more. Then drop a plumb line from the intersections (corners) of the building lines to the footing.

Mark the corners on the footing. Snap a chalk line from corner to corner on the footing. This line serves as a guide, marking the outside face of the foundation wall. As you set up the foundation forms, align the inner face of the outside form with the chalk line, **Figure 7-21.**

Low wall forms, up to about 3′ in height, can be assembled from 1″ sheathing boards or 3/4″ plywood supported by 2 × 4 studs spaced 2′ apart. The height can be increased somewhat if the studs are closer together. See **Figure 7-22A.**

For walls over 4′ in height, the studs should be backed with *walers*. These are horizontal stiffeners added to provide greater strength to the form, **Figure 7-22B.** Walers can be 4 × 4 lumber bored to accept metal form ties or doubled 2 × 4s

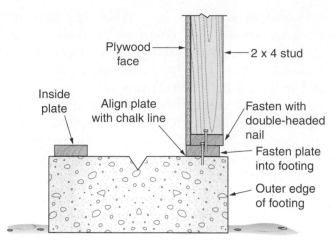

Figure 7-21. One method of locating and fastening an outside wall form to the footing. The plywood face of the form must align with the building line.

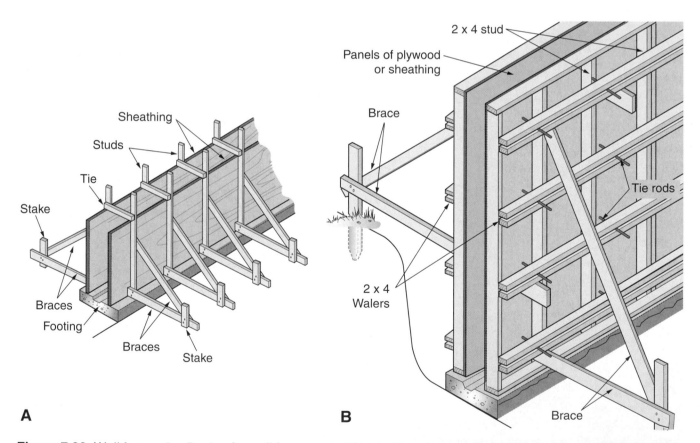

A   **B**

Figure 7-22. Wall forms. A—Design for wall forms up to 3′ high. Use plywood sheathing and space studs 2′ apart. B—Prefabricated wooden panels for a wall form. For walls over 4′ high, walers provide greater strength to the form. Add bracing to the top and bottom.

Waler: A horizontal member used in concrete form construction to stiffen and support the walls of the form; used to keep the form walls from bending outward under the pressure of poured concrete. Also called *wale.*

with space between them to run the ties. Spacing of the walers depends on the amount of pressure the concrete exerts on the form.

7.5.2 Form Hardware

Wire ties and wooden spreaders have been largely replaced with various manufactured devices. **Figure 7-23** shows three types of form ties. The rods go through small holes in the sheathing and studs. Holes through the walers can be larger for easy assembly or the walers can be doubled as shown. The snap tie is designed so that a portion of it remains in the

wall. Release solution or another type of lubricant is applied to tapered ties prior to pouring the concrete. These solutions allow the ties to be removed after the concrete sets. Bolts in the coil-type ties should also be coated with release solution. Corner locks can be used to secure the corners of concrete forms. Two types are shown in **Figure 7-24.**

After the concrete has set, the clamps can be quickly removed and the forms stripped away. To break off the outer sections of the snap tie rod, a special wrench is used. The rod breaks at a small indentation located about 1″ beneath the concrete surface. The hole in the concrete is later patched with grout or mortar.

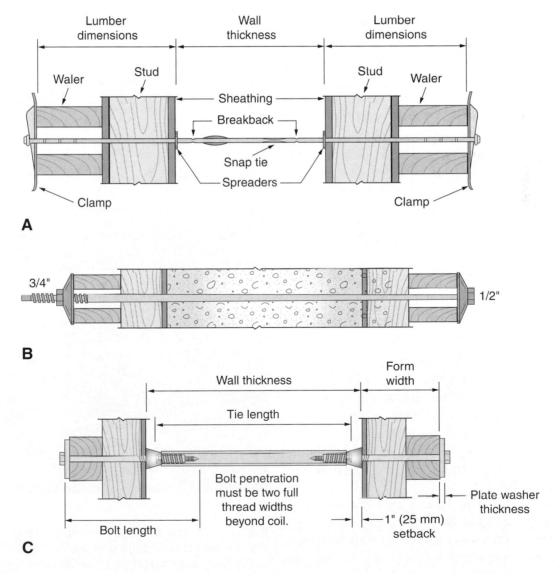

Figure 7-23. Some form ties combine the functions of both ties and spreaders. A—Snap tie. B—Reusable taper tie. C—Heavy-duty coil tie. (Universal Form Clamp Co., The Burke Co.)

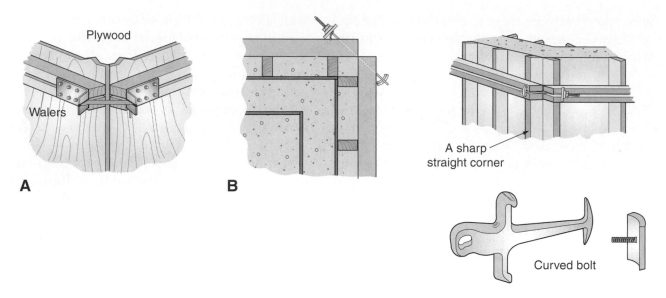

Figure 7-24. Corners of wall forms must be carefully fastened so they can withstand the pressure from the poured concrete. A—A high-speed corner lock. Clamps are attached with 6d–8d duplex nails and lag screws. B—A corner lock requiring no nails or screws. (The Burke Co.)

Referring to **Figure 7-24,** the coil-type tie is assembled with a wood, plastic, or metal cone and a pair of lag bolts. The cone provides smooth contact with the form and leaves a void that is easily patched. Taper ties are threaded at both ends for easy removal. Take off the threaded plate from the small end and pull the rod free from the other end.

7.5.3 Panel Forms

Carpenters may build their own wooden panel forms using 3/4″ plywood and 2 × 4 studs to form 2′ × 8′ or 4′ × 8′ units. However, prefabricated panels are now used for most poured concrete wall forms, **Figure 7-25.** Some panels are made from a special grade of plywood attached to a wood or metal frame. Other panels are steel or aluminum. A new type, called insulating concrete forms (ICFs), is discussed later in this unit. These forms become a permanent part of the foundation.

If you are erecting the forms, be sure to add walers as needed for stiffening. Brace the forms on each side as needed to straighten them and provide support. Straightening is easy to do by sighting along the top edge of the form from corner to corner while another carpenter

A

B

Figure 7-25. Prefabricated wall form panels A—Most wall forming is now done with prefabricated forms that are quickly assembled and dismantled. B—Aluminum forms are light and easy to assemble. Note the hooks for installation of walers. (Western Forms)

secures the bracing. If working alone, stretch a line from corner to corner. Insert blocks of equal thickness at each end to hold the line away from the form. Use a test block of the same thickness as the wall is straightened. The form is straight when the test block slips between the form and the line. Secure the form with a brace at each checkpoint.

> **Working Knowledge**
>
> Carefully recheck formwork before placing the concrete. A form that fails during the pouring wastes material and causes extra work.

Since panel forms are designed to be used many times, they should be treated to prevent the concrete from sticking to the surfaces. Use special form release coatings that are commercially available, **Figure 7-26**. After each removal, thoroughly clean form components. Carefully sort and stack them for the next job or storage.

Manufacturers have developed many forming systems to replace or supplement panel forms built by the carpenter. The panel units are light for easy handling and transporting from one building site to another. Specially designed devices are used to quickly and accurately assemble and space the components. For residential work, these systems may be made from aluminum or from steel frames and exterior grade plywood panels. Lightweight polystyrene forms are designed to remain as insulation for the structure. Sometimes plywood forms are coated with a special plastic material to create a smooth finish on the concrete and prevent it from sticking to the surface.

Buck: A temporary wooden form installed within the forms for poured concrete walls. Its purpose is to provide an opening or void for later installation of features such as windows or doors.

Nailing strips: Wooden strips cast into a concrete foundation around a window or door opening and used to attach the door or window frame.

Figure 7-26. New aluminum forms should be seasoned with special compounds before use. This reduces "worming" and surface voids in the concrete. After each use, forms should be carefully cleaned with a compound designed for that purpose. Wooden forms usually are treated with a release compound before use. (Western Forms)

7.5.4 Wall Openings

Several procedures are followed to form openings in foundation walls for doors, windows, and other voids. In poured walls, temporary frames called *bucks* or stops are built and fastened into the regular forms. Beveled keys or *nailing strips* may be attached to the buck and cast into the concrete. Frames are then secured to these strips after the bucks are removed. **Figure 7-27** shows two methods of framing openings. Use duplex nails to fasten the buck to the inside face of the form wall. Drive duplex nails through the outside of the form wall into the buck.

Special framing must also be attached inside the form for pipes or voids for carrying beams. As with windows and doors, formwork for these structures must be attached to either the inside or outside wall form before the other form is erected.

Tubes of fiber, plastic, or metal can be used for small openings. Hold them in place with wood blocks or plastic fasteners attached to the form. Larger forms made of wood can be attached with duplex nails driven through the form from the outside.

In concrete block construction, door and window frames are first set in place. The masonry units are constructed around the sides. The outside surface of the frames has grooves into which the mortar flows, forming a key. Basement windows are often located level with the top of the foundation wall. The sill carries the weight of the structure across the opening, **Figure 7-28.**

7.5.5 Forming a Wall Pocket

A *wall pocket* is a void in a concrete wall used to receive a large framing member that

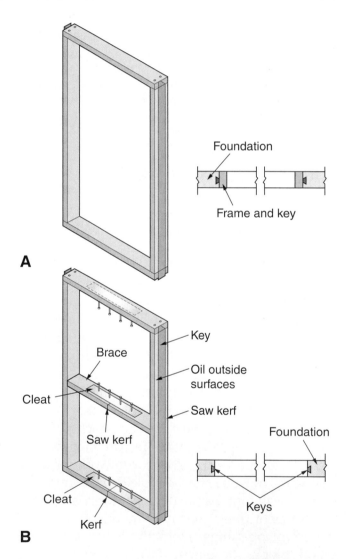

A

B

Figure 7-27. Wood forms are used to frame openings in foundation walls. Carpenters often construct them on site. A—Permanent frame that will be left in the wall. B—This frame, called a buck, is designed to be removed. Members are cut partway through for easy removal. Cleats and braces reinforce the members at saw cuts. Bucks and frames are nailed to the form with duplex nails driven from the outside.

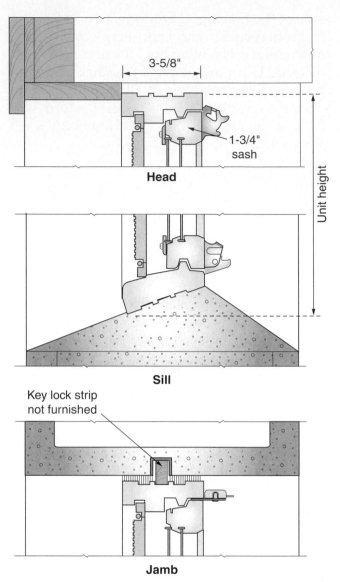

Head

Sill

Jamb

Figure 7-28. Detail of basement window unit that can be placed in a concrete or masonry foundation wall.

supports a floor. Normally, a pocket is required to receive a steel or wood beam. It should be large enough to provide 1/2″ of clearance on the sides and end for proper ventilation of the beam. As with other voids in poured concrete walls, a buck must be installed on the inside form. Never allow the pocket to go all the way through the wall. This exposes the beam to the weather. Also, there is air leakage. **Figure 7-29**

Wall pocket: An intentional void left in a concrete wall to support a large timber or I-beam.

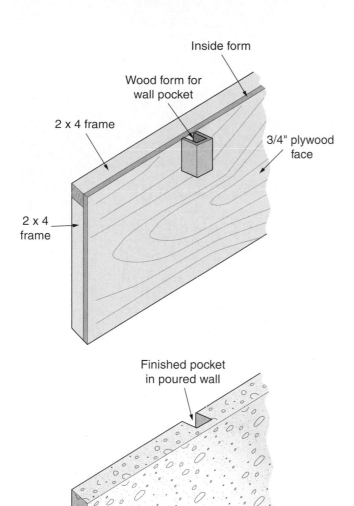

Inside form

Wood form for
wall pocket

2 x 4 frame

3/4" plywood
face

2 x 4
frame

Finished pocket
in poured wall

Figure 7-29. Forming for a wall pocket. Attach the pocket form to the face of the inside wall form. Drive duplex nails through from the exterior so they can be removed when the concrete has cured.

Pilaster: A part of a wall that projects not more than one-half its own width beyond the outside or inside face of a wall. Chief purpose is to add strength, but may also be decorative.

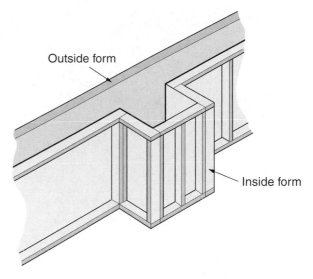

Outside form

Inside form

Figure 7-30. Inside section of a foundation form set up for a pilaster. Often, the studs at the sides are omitted.

shows installation of a buck on the face of the inside form. It should be held in place by double-headed nails driven through the wooden form.

7.5.6 Pilasters

Long walls may have pilasters. A *pilaster* is a thickened section of a concrete or masonry wall that strengthens the wall or provides extra support for beams. **Figure 7-30** shows a form setup for pouring a pilaster in a concrete foundation wall.

7.5.7 Column Forms

For standard square or rectangular columns, prefabricated forms save time. Those shown in **Figure 7-31A** and **7-31B** are quickly assembled and set up on a column footing. To prevent blowout, brace the bottom of the form.

Tubular fiber or paper forms for piers may also be used to save time. These are available in diameters from 8" to 48". For residential construction, sizes 12" and 16" are most often used. **Figure 7-31C** shows a pier formed using a fiber tube.

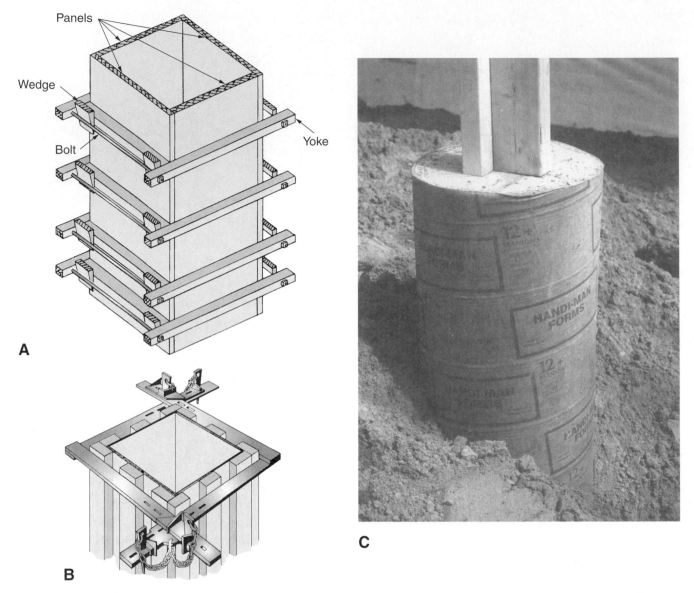

Figure 7-31. Prefabricated column forms save construction time. A—Yoke and wedge arrangement. B—Scissor clamps are adjustable. C—Fiber tube forms are easier to set up and are replacing older commercial forms. (The Burke Co.)

7.6 Concrete

Concrete is made by mixing carefully measured amounts of:
- Cement.
- Fine aggregate (sand).
- Coarse aggregate (gravel or crushed stone).
- Water.

First, the cement and aggregate are mixed dry. Then, water is added and a chemical change called *hydration* takes place. Hydration, not drying, causes hardening of the mixture.

The proper amount of water is important to the strength of the concrete. Too much water will cause weak concrete. Generally, the ratio should be no more than 1/2 lb. of water to 1 lb. of cement. The moisture in the sand must always be a consideration in fixing the ratio.

Hydration: Heat-producing chemical reaction occurring between cement and water to form concrete.

7.6.1 Effect of Improper Amount of Water in Mix

Excess water in the mix may rise to the surface and sit there. This is known as *bleeding*. Even though the water eventually evaporates, the concrete surface may craze, dust, and scale. Also, when the water evaporates it leaves pockets where water can later enter and freeze. With too little water in the mix, hydration will not take place. With too little mixing, water will not blend into the mix. Some ingredients will separate.

7.6.2 Proper Curing

Fresh concrete should be kept moist during the initial stage of hydration so it can properly cure. The curing process can go on for years after placement. However, most of the strength of the concrete develops during the first week or two. If allowed to dry out too soon, concrete may reach only 40% of its full strength.

Proper *curing* of concrete prevents loss of its moisture for a period of time. At seven days, with moisture, it is 50% stronger. After a month of curing, this strength is double. Another benefit of proper curing is a reduction in shrinkage, which also reduces cracking.

7.6.3 Curing Methods

Curing of the concrete can begin as soon as it is hard enough that the concrete does not mar. The easiest curing methods are flooding or keeping a sprinkler running. Other methods are used:
- Covering with burlap, straw, or any other material that will hold water and keep the material wet.
- In hot weather, covering the concrete with plastic and sealing the laps with tape or planking.

- Use of a liquid-membrane compound to seal in moisture, **Figure 7-32.** Carefully follow the manufacturer's directions.

Refer to the table in **Figure 7-33.** It compares several methods of curing.

7.6.4 Slump Testing

Sometimes a *slump test* is made to assess the consistency, stiffness, and workability of fresh concrete. The amount of slump is influenced by the amount of water in the mix. More water means more slump. However, there are other factors: type of aggregate, admixtures, air in the mix, ambient temperature, and proportions of the mix. Mixing time and standing time also affect slump. The test is made with a sheet metal cone 4″ in diameter at the top, 8″ in diameter at the bottom, and 12″ high.

Figure 7-32. A plastic-based material can be sprayed on concrete to form a continuous membrane that aids in proper curing.

Curing: Preventing loss of moisture in fresh concrete to strengthen it and reduce shrinkage, which also reduces cracking.

Comparing Concrete Curing Methods		
Curing method	Advantage	Disadvantage
Ponding	Inexpensive Needs little supervision Surface stays clean Good buffer for temperature Controls quality best	May leak Is not good in cold weather Traffic is restricted
Sprinkling/fogging	Surface stays clean	Involves more work/cost Is poor method when cold Traffic is restricted Hard to keep surface wet
Cover with wet burlap	Cheap; material reusable Surface remains clean	Traffic is restricted Needs frequent wetting
Black/grey mix	Absorbs heat Cheap Asphalt tile sticks	Gives surface a color
Cover with waterproof paper	Surface stays clean	Traffic deteriorates it
Cover with polyethylene sheet	Handles easily Cheap	Can be torn or punctured May leave surface blotchy
White pigmented	Heat is reflected	Needs to be stirred before use
Clear resin base	Can be painted	More expensive
Wax-resin base	Cheap	Leaves a gummy residue
Clear wax base	Cheapest	Leaves a gummy residue

Figure 7-33. Each of these methods of curing concrete for strength has both advantages and disadvantages.

PROCEDURE

Making a slump test

Figure 7-34 shows the steps involved in making a slump test:

1. Secure a slump cone, wet it, and place it upright on a solid level base. The large-diameter end should be down.
2. Fill the cone with wet concrete in three layers, each equal to about one-third of the volume of the cone. Rod (tamp) each layer 25 times before adding the next layer. See **Figure 7-34A.**
3. Strike off the top, then slowly and evenly remove the cone. See **Figure 7-34B.** This should take 5–12 seconds. Avoid jarring the mixture or tilting the cone during this step.
4. Turn the cone, small end down, and place it next to the concrete.
5. Lay the tamping rod across the top of the cone so it extends over the concrete. See **Figure 7-34C.**
6. Use a ruler to measure the amount of slump. The slump range recommended for different types of construction is shown in **Figure 7-35.**

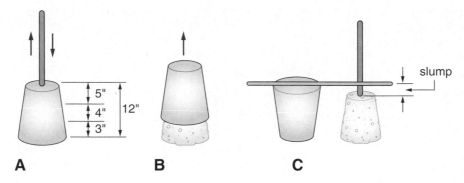

Figure 7-34. Method of performing a slump test. It should be done whenever the consistency of the concrete is critical to the job. Testing can be performed at the job site or at the ready mix plant.

Slump Range for Different Types of Construction		
Type of construction	Maximum slump	Minimum slump
Reinforced footings and foundation walls	3"	1"
Plain footings, caissons, and substructure walls	3"	1"
Beams and reinforced walls	4"	1"
Building columns	4"	1"
Pavement and slabs	3"	1"
Mass concrete	2"	1"

Figure 7-35. The recommended slump range for concrete depends on what is being constructed.

7.6.5 Giving Concrete Tensile Strength

The compressive strength of concrete is high, but its *tensile* strength (stretching, bending, or twisting) is relatively low. For this reason, concrete used for beams, columns, and girders must be reinforced with steel *rebar*. Normally, rebar is not used in footings or slabs unless the plans or codes call for it. No. 4 rebar may be used in a footing to prevent settling at one or more locations. When concrete must resist compression forces only, reinforcement is usually not necessary. Slabs for floors or similar applications are often reinforced with welded wire fabric. See **Figure 7-36.**

Safety Note

To avoid possible injury caused by a worker falling onto vertical rebar, plastic or wooden caps should be placed on the ends. An alternative is to bend the end at a 90° angle.

7.6.6 Cement

Most *cement* used today is portland cement. It is usually manufactured from limestone mixed with shale, clay, or marl. Each sack of portland cement holds 94 lb. This is equal to one cubic foot in volume. Cement should be a free-flowing powder. If it contains lumps that cannot be easily pulverized between thumb and fingers, it should not be used.

7.6.7 Aggregates

Aggregate may consist of sand, crushed stone, gravel, or lightweight materials such as expanded slag, clay, or shale. The large, coarse aggregate forms the basic structure of the concrete. This is seldom over 1 1/2″ in diameter. The voids

---

Cement: Binding material that, when combined with water and aggregate, forms concrete.

Aggregate: Materials such as sand, rock, and gravel used to make concrete.

Type of construction	Recommended style	Remarks
Barbecue foundation slab	6 x 6-8/8 to 4 x 4-6/6	Use heavier style fabric for heavy, massive fireplaces or barbecue pits.
Basement floors	6 x 6-10/10, 6 x 6-8/8 or 6 x 6-6/6	For small areas (15-foot maximum side dimension) use 6 x 6–10/10. As a rule of thumb, the larger the area or the poorer the subsoil, the heavier the gage.
Driveways	6 x 6-6/6	Continuous reinforcement between 25- to 30-foot contraction joints.
Foundation slabs (residential only)	6 x 6-10/10	Use heavier gage over poorly drained subsoil, or when maximum dimension is greater than 15 feet.
Garage floors	6 x 6-6/6	Position at midpoint of 5- or 6-inch thick slab.
Patios and terraces	6 x 6-10/10	Use 6 x 6-8/8 if subsoil is poorly drained.
Porch floor a. 6-inch thick slab up to 6-foot span b. 6-inch thick slab up to 8-foot span	6 x 6-6/6 4 x 4-4/4	Position 1 inch from bottom form to resist tensile stresses.
Sidewalks	6 x 6-10/10 6 x 6-8/8	Use heavier gage over poorly drained subsoil. Construct 25- to 30-foot slabs as for driveways.
Steps (free span)	6 x 6-6/6	Use heavier style if more than five risers. Position fabric 1 inch from bottom form.
Steps (on ground)	6 x 6-8/8	Use 6 x 6-6/6 for unstable subsoil.

Figure 7-36. Recommended styles of welded wire fabric reinforcement for concrete structures.

between these particles are filled with smaller particles. The voids between these smaller particles are filled with still smaller particles, and so on. Together, they form a dense mass.

7.7 Ordering Concrete

If there is a local ready mix supplier, it is generally more convenient to have concrete delivered rather than mixed on site. The mix is purchased by the cubic yard (27 cu. ft.) and a minimum order is usually 1 cu. yd. Ingredients are carefully measured, often in automated plants. Mixing takes place en route to the job site.

Another advantage of ready mix is that the plant personnel can help in selecting the right mix for the builder's needs. The mix proportions can be adjusted to give the strength needed for the type and location of the construction. Both climate and usage have a bearing on these proportions.

When placing the order, the carpenter should tell the plant how the concrete will be used (such as a footing, wall, slab-on-grade, or sidewalk). The amount needed should also be specified, as well as when it must be delivered. An order should be for about 5% more than the calculated amount to allow for consolidation, spillage, and form movement.

Since concrete must be placed as soon as possible after mixing, builders and suppliers must work together to avoid delays. Before the concrete truck arrives, forms should be checked for accuracy. The crew must be on site, equipped with the proper tools, and ready to work the concrete into the forms.

7.8 Placing Concrete

Usually, concrete can be placed directly into the forms from the ready mix truck. To move the concrete to other areas not accessible to the truck, a wheelbarrow or a pump is generally used, **Figure 7-37.** When placing concrete, follow these general guidelines:

- Place concrete as near as possible to where it will rest. Never allow it to run or be worked over long horizontal distances.

- Never allow concrete to drop further than 4′. To do so, or working over a long distance, could cause segregation. This is a condition in which large aggregates become separated from the cement paste and smaller aggregates.

- Promptly place concrete in forms after mixing.

- Concrete for walls should be placed in the forms in horizontal layers of uniform thickness not exceeding 1′ to 2′. As the concrete is placed, spade or vibrate it enough to thoroughly compact the concrete. This produces a dense mass.

- Working the concrete next to the form tends to produce a smooth surface. It prevents honeycombing along the form faces. A spade or thin board may be used for this purpose. Large aggregates are forced away from the forms and any air trapped along the form face is released. Mechanical vibrators are effective in consolidating concrete.

- Mechanical vibrators create added pressure on the forms. This factor must be considered in the form design. The vibrator should not be held in one location long enough to draw a pool of cement paste from the surrounding concrete.

- When pouring slabs, work in strips 4′ to 6′ wide. Avoid concentrating water at the ends, in corners, and along the face of the form.

A

B

C

Figure 7-37. Proper methods must be used to move concrete from the ready mix truck to where it is to be placed. A—The spout from the ready mix truck should be used wherever it will reach. B—Where height makes other methods impossible, concrete can be placed with a concrete pump. C—For horizontal transport, a wheelbarrow can be useful. (Kasten-Weiler Construction)

- Use a baffle on slopes to prevent segregation and collecting of aggregate at the bottom of the slope. Better still, start placing concrete at the bottom and work uphill.

7.8.1 Anchors

Wood plates are fastened to the top of foundation walls with 1/2" *anchor bolts* or *anchor straps.* They are spaced not more than 4' apart. See **Figure 7-38.** In concrete walls, anchors are set in place as soon as the pour is completed and leveled off.

7.8.2 Removing Forms

Wall forms help protect concrete from drying too fast. They should not be removed until the concrete is strong enough to carry the loads

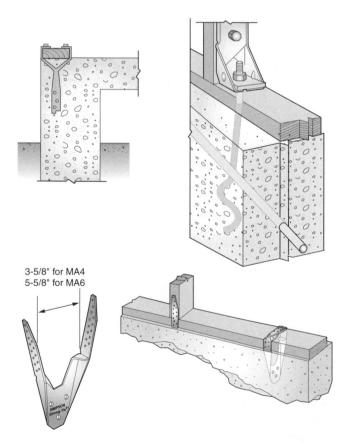

3-5/8" for MA4
5-5/8" for MA6

Figure 7-38. Patented anchor bolts and straps are embedded in the concrete. They secure the sills to the foundation. (The Panel Clip Co.; ©Simpson Strong-Tie Company, Inc.)

that will be placed on it. The material should be hard enough so the surface is not damaged by the stripping operation. Sufficient hardening of concrete will normally take a day or two.

7.9 Concrete Block Foundations

In some localities, *concrete blocks* are used for foundation walls and other masonry construction. The standard block is made from a mixture of portland cement and aggregates such as sand, fine gravel, or crushed stone. It weighs from 40 to 50 lbs.

Lightweight units are made from portland cement and natural or manufactured aggregates. Among these aggregates are volcanic cinders, pumice, and foundry slag. A lightweight unit weighs between 25 and 35 lb. It usually has a much lower U factor than a standard block. The *U factor* is a measurement of the heat flow or heat transmission through materials.

Blocks should comply with specifications provided by the American Society for Testing Materials. ASTM specifications for a Grade A, load-bearing unit requires that the compression strength equal 1000 lb. per sq. inch. Thus, an $8 \times 8 \times 16$ block must withstand about 128,000 lbs., or 64 tons.

7.9.1 Sizes and Shapes

Blocks are classified as solid or hollow. A solid unit is one in which the core or cell area (hollow portion) is 25% or less of the total cross-sectional area. Blocks are usually available in 4", 6", 8", 10", and 12" widths and 4" and 8" heights. **Figure 7-39** shows some of the sizes

---

Anchor bolts: Bolts embedded in concrete to hold structural members in place.

Anchor straps: Strap fasteners that are embedded in concrete or masonry walls to hold sills in place.

Concrete blocks: Blocks made from a mixture of portland cement and aggregates such as sand, fine gravel, or crushed stone. They weigh 40 lb.–50 lb.

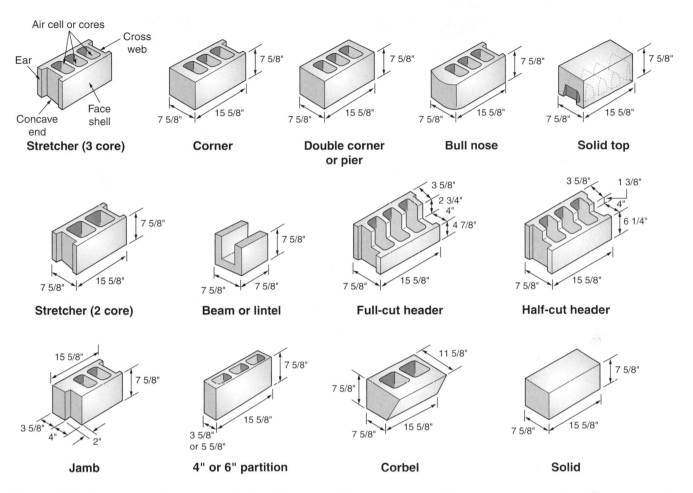

Figure 7-39. Concrete blocks are manufactured for many different purposes. These shapes are typical. Sizes are specified in this order: width, height, and length. Sizes are nominal, allowing for mortar joints. (Portland Cement Assn.)

and shapes. Sizes are actually 3/8″ shorter than their nominal (named) dimensions to allow for the mortar joint. For example: the 8 × 8 × 16 block is actually 7 5/8″ × 7 5/8″ × 15 5/8″. With a standard 3/8″ mortar joint, the laid-in-the-wall height is 8″ and the length 16″.

It is important to limit the moisture content of concrete block. Cover the blocks stockpiled at the job site to protect them from the elements. They should be secured on pallets or other supports to keep them off of the ground. Never wet blocks before or during laying a wall.

Mortar: A combination of cement, sand, lime, and water used to bind masonry blocks.

7.9.2 Mortar

To build a wall, blocks are bonded one to another with *mortar.* This is a mixture of cement, lime, and masonry sand. These materials give the mortar its strength, ability to retain water, durability, and bonding capabilities.

When water is added, just before use, the mortar retains its plastic properties long enough for the mason or carpenter to lay up the block. Mortar can be made to resist different pressures, depending on where it is to be used. ASTM has set standards for different uses. Mortar used in foundations must withstand pressures from 1800 psi–2500 psi when set. Mortar should be used within 1 1/2–2 hours after it has been mixed.

7.9.3 Laying Concrete Block

Foundation block should be laid on a good footing or concrete slab. The mason or carpenter first locates the corners on the footing and snaps a chalk line on all sides. This marks the building line and the outer edge of the blocks. The first course is then strung out dry (without mortar) to check the layout, as shown in **Figure 7-40.**

The next step is to lay down a mortar bed on the footing for the first course of blocks, **Figure 7-41.** Corner blocks are laid first. The *head joint* is the end of the block that will butt against the previously laid block. These are "buttered" with mortar before being laid.

The corners are usually built up first from four to six courses high, **Figure 7-42.** All corner blocks must be at the right height, in line, level, and plumb. The mason uses a level to check this

A

B

Figure 7-42. Checking for plumb and level. A—All corners are first laid up four to six courses high and carefully checked for plumb. B—The mason carefully checks for level as each course is laid. (Portland Cement Assn.)

Figure 7-40. After marking the building line on the footing with a chalk line, the mason always lays out the first course of block dry. This allows for adjustments that minimize cutting of block. (Portland Cement Assn.)

Figure 7-41. A full bed of mortar 1 1/2″ thick is laid down on the footing or slab for the first course of block. (Portland Cement Assn.)

Head joint: The end of the block that butts against the previously laid block.

Figure 7-43. A course pole can be created from a straight piece of lumber and used to check the height of each course. (Portland Cement Assn.)

A

B

Figure 7-44. Laying block. A—The head joints of the blocks are buttered before the blocks are laid. Blocks should be lightly bumped against the previously laid block for a good joint. B—A mason's line is used as a guide for laying block between corners. It maintains a straight line of block for each course. (Portland Cement Assn.)

after laying every three or four blocks. Sometimes, a *course pole* is used to check the height of each course, **Figure 7-43.**

Each course at the corners is stepped back one-half of a block from the previous course. Lay the level on an angle across the ends of each unit. All the edges of the last block in each course should fall along the straight line formed by the level. If not, the corner must be rebuilt.

With all four corners built up, laying the wall between corners can begin, **Figure 7-44.** Note how the blocks are buttered before being laid. A *mason's line* stretched at each course acts as a reference for proper vertical and horizontal alignment. The mason's line is attached to the corner by a line holder. This can be a line block, line pin, or line stretcher.

As each block is laid, be careful that it does not touch the line. The mason's level is now used only to check the face of each block to keep it lined up with the face of the wall. Use the edge of the trowel to cut off mortar squeezed out of the joint. Use a lifting motion to capture the excess with the trowel. Do not allow it to fall off of the trowel or smear the wall.

Some building codes require a bond beam between every fourth course of block. The beam is constructed on site of steel-reinforced poured concrete.

Course pole: A straightedge, often made from a piece of lumber, marked up to check the height of each course of concrete block.

Mason's line: Reference line (cord) used to keep masonry units aligned as they are laid up.

Closure block: The last concrete block to be placed in any course

Figure 7-45. The closure block is buttered at both ends and then carefully placed in the wall. (Portland Cement Assn.)

The last block to be placed in any course is called a *closure block.* Both ends of this block are buttered before the block is carefully placed, **Figure 7-45.**

Sometimes, solid block is used for the top course. Openings must then be cut in the top of the block for insertion of the anchor bolts.

Corbel blocks are concrete blocks that are dimensioned and shaped to form a shelf in a wall. They are used when brick or stone veneer will be used as the exterior wall finish. The corbeling provides a supporting base for such masonry materials. **Figure 7-46** shows another method of providing a ledge to support brick veneer siding. Here, the mason places standard 8″ block atop 12″ block. The ledge formed easily supports a veneer facing of 4″ masonry.

Block foundation walls should be topped with anchors to secure the sill, **Figure 7-47.**

These should be spaced 6′ on center and within 12″ of the end of any sill piece. Before the top two courses are laid, put down a strip of metal lath wide enough to cover the core spaces. After the top course has been laid, fill the cores with mortar and install the anchor bolts a minimum of 15″ deep. Check local codes. Anchor clips are installed in the same way. Anchors of either type should be about 18″ long. For areas with termite problems, top courses of stretcher block

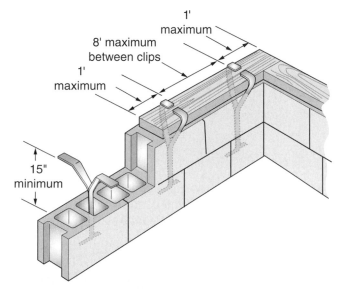

Figure 7-47. Install anchor straps or bolts in block walls to secure the sills. Always embed anchors in the second-from-top course. Place screening over voids beneath the second-from-top course to prevent mortar from falling away from the anchors.

Figure 7-46. Forming a ledge by laying standard 8 × 8 × 16 block on top of 12″ wide block. The ledge that is formed will support a stone or brick veneer facing.

Corbel block: Concrete block shaped and dimensioned to form a shelf in a wall.

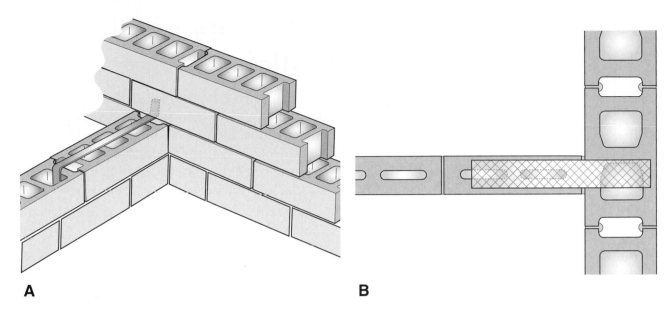

A   **B**

Figure 7-48. Connecting intersecting block walls. A—Intersecting walls are often tied together with an anchor strap. B—Another method of tying together intersecting walls involves the use of hardware cloth. (National Concrete Masonry Assn.)

should have all voids filled with concrete. This also helps distribute the load from floor joists.

7.9.4 Anchoring Intersecting Walls

Concrete block walls that intersect at 90° should use some type of connector to bind the intersecting walls together. Anchoring should be installed at every sixth course. There are several anchoring devices available, **Figure 7-48.** One is a metal strap known as a *fixed anchor.* The anchor ends are bent in opposite directions. It is placed in the wall between courses. One end projects upward into the hollow core of a block in the next course. The other end extends downward into a core from the intersecting wall. Both cores are then filled with mortar, concrete, or grout before laying the next course. Note: place metal lath or wire mesh across the corresponding cores in the course below to contain the mortar.

Hardware cloth or joint reinforcement is another device often used to connect two intersecting walls. It is also installed between courses and embedded in the mortar.

Another method uses a 1/4″ diameter tie with a rounded open loop on each end. The ends are hooked around embedded rebar.

7.9.5 Lintels

A structural unit called a *lintel* supports masonry located above openings for doors and windows, **Figure 7-49.** A lintel can be one of the following:

- A precast concrete unit that includes metal reinforcing bars (rebar).
- Standard blocks supported by steel angle iron.
- A course of lintel blocks across the opening, supported by a frame. Reinforcing bars are added and the blocks are filled with concrete.

Fixed anchor: Metal strap connector used to bind intersecting concrete block walls together.

Lintel: A horizontal structural member that supports the load over an opening, such as a door or window. Also called a *header.*

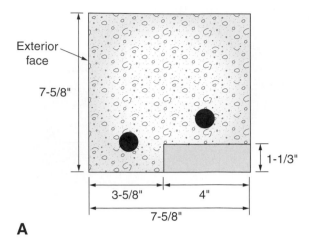

A

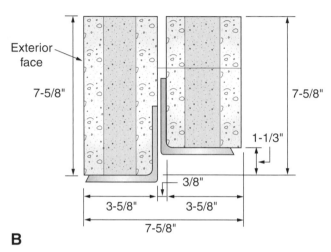

B

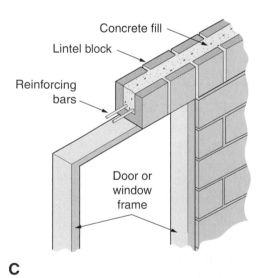

C

Figure 7-49. Lintel construction methods. A—This is a cutaway view of a precast solid lintel for a block wall. Note the reinforcing rebar (black dots). B—This is a cutaway view of lintel made up of standard block units supported by steel angle iron. Notches allow space for installing window or door frames. C—Open-sided lintel blocks allow adding rebar and concrete. Blocks are laid over a supporting frame.

7.10 Insulating Block Walls

In northern climates, insulating the foundation walls may be necessary when residential plans include a finished basement. There are several ways to reduce the heat flow.

- Lightweight masonry units that have a lower U factor may be used.
- Adhering rigid insulation sheets to the exterior surface.
- Cavity wall construction or the use of various forms of insulation applied to the interior or exterior surface. Building codes in some states do not permit the use of interior insulation on concrete or masonry foundation walls. Additional information about insulation is included in Chapter 15, **Thermal and Sound Insulation**.

7.10.1 Waterproofing

In most localities, the outside of poured concrete or concrete masonry basement walls should be waterproofed below the finished grade. Drain tile should also be installed at the footing to take away ground water.

Concrete masonry (block) walls may be waterproofed by an application of cement plaster, followed by several coats of an asphaltic material. The wall surface should be clean and dampened with a water spray just before the first coat of plaster is applied. The plaster can be made of cement and sand (1 to 2 1/2 mix by volume) or mortar may be used. When the first coat has partially hardened, it should be roughened with a scratcher to provide better bond with the second coat.

After the first coat has hardened at least 24 hours, apply a second coat. Again, dampen the surface just before applying the plaster. Both coats should extend from about 6″ above the finished grade to the footing.

A cove of plaster should be formed between the footing and wall. This precaution prevents water from collecting and seeping through the joint. The second coat should be kept damp for at least 48 hours.

Figure 7-50. Adhering a 6 mil membrane to a block wall. It is applied to a troweled-on layer of bituminous waterproofing before the waterproofing material cures.

Another waterproofing method involves troweling a bituminous product directly on the concrete blocks. Then, while the coating is still wet, a 6 mil water barrier is adhered to it, **Figure 7-50.**

In poorly drained soils or when it is important to secure added protection against moisture penetration, the plaster may require coating with asphalt waterproofing or hot bituminous material. See **Figure 7-51.**

To waterproof poured walls, bituminous waterproofing is generally used without cement plaster. Polyethylene sheeting is often used as a waterproofing material. It should cover the wall and footing in one piece.

Perimeter drains often are laid alongside a footing, **Figure 7-52.** This drainage system should carry water away from the foundation to an outlet that always remains open. In some localities, perimeter drains may be connected to a sump pump installed in the basement floor. The drain is usually installed at a slope of about 1″ in 20′. Concrete or clay drain tile once were common in this application, but have been largely replaced by perforated plastic drain piping. Perforated piping is lightweight and delivered in coils. It is widely used because it requires less installation labor than standard concrete tile. An alternate system called Form-a-Drain® builds drainage into a patented footing form. Refer to **Figure 7-18.** The system uses a plastic channel 4″ to 6″ high with perforations at regular intervals. The channel remains in place as part of the foundation drainage.

As an energy conservation measure, plastic foam board may be added to the outside face of the foundation wall. In such cases, no cement

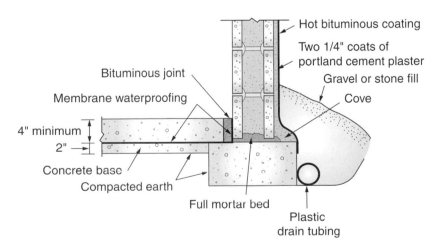

Figure 7-51. This section view shows a method of waterproofing basements that is especially effective in very wet soils. Plaster is omitted on the outside walls in poured concrete foundations.

A

B

Figure 7-52. Installing footing drains. A—Perforated plastic drain tubing is delivered to the site in large coils. Since it is flexible and already perforated, plastic drain tubing is easier to install than concrete or clay tile. B—Codes now call for socked flexible tile to screen out silt that would eventually fill the tile.

plastering or waterproofing should extend any higher than 2″ below the final grade. More information on this type of insulation is found in Chapter 15.

7.11 Backfilling

After the foundation has cured and water-proofing has been completed, the excavation

outside the walls needs to be refilled with earth. This is known as *backfilling.* If not done properly, it can cause serious foundation problems. There are four important elements to account for if the job is to be done right:

- Protecting the new foundation from damage.
- Using clean material (no debris).
- Compacting (tamping) the backfill.
- Sloping the final grading to keep water away from the foundation.

Since backfilling places considerable inward pressure on the walls of the foundation, the first floor framing and rough flooring are usually installed first. This is the best insurance against damage. However, many builders want to backfill as soon as the wall is cured. Doing so speeds up framing, because the foundation is more accessible. In such cases, bracing the walls becomes important. This is especially critical with concrete block foundations. If the foundation wall has windows that will be below grade, steel areaways must be installed to create window wells.

Backfill material ought to be the same that was taken out during excavation. However, if that soil is poor—such as heavy, moisture-laden clay—granular material should be substituted. Never bury construction lumber scraps or any other organic material against the foundation. It will only cause trouble later with uneven settling as the organic material decays. Such material may attract termites.

If a perimeter drain system has been installed, use care not to disturb or crush the lines. When heavy equipment is used to backfill, the operator must also be careful not to damage the walls. Compacting the backfill avoids problems caused by later settling. Final grading should slope the soil away from the foundation. A minimum slope should be 1/2″ per foot for the first 10′ away from the foundation.

Backfilling: The process of replacing soil around foundations after excavating.

7.12 ICF Foundation and Wall Systems

ICF is an acronym that stands for *insulated concrete form.* It is a relatively new system using forms made of light polystyrene foam or other types of rigid insulating material. ICFs are used both for foundation walls and exterior walls of a building. Once forms are erected and braced, concrete is poured into them. The forms are left in place, resulting in a structural wall that is insulated on both sides. Depending on the specific system, it also provides surfaces for the carpenter to install both exterior and interior wall coverings.

7.12.1 Code Restrictions for ICFs

Builders planning to use an ICF foundation should first check the local building code. Some codes prohibit the use of ICF systems under certain conditions. The ban concerns areas of the country where termite infestation is very heavy. The 1998 CABO One- and Two-Family Dwelling Code bans ICF use below grade in dwellings that will be built partially or completely of untreated wood framing. Several states have passed amendments that make the CABO code less restrictive. The International Residential Building Code has also placed a ban on below-grade use in heavily infested areas. Again, some exemptions apply. Carpenters should check local codes for details.

There are four basic types of ICF systems that take their names from the shape of the poured wall. In general, these forms take three basic shapes. One consists of insulating planks 4′–8′ long. These forms have interlocking ties and notches for stacking. A second shape has 4′ × 8′ panels held together and spaced with ties. Individual panels are glued or wired together. The third shape is a block hollowed out to provide a cavity for the concrete. **Figure 7-53** illustrates the four types:

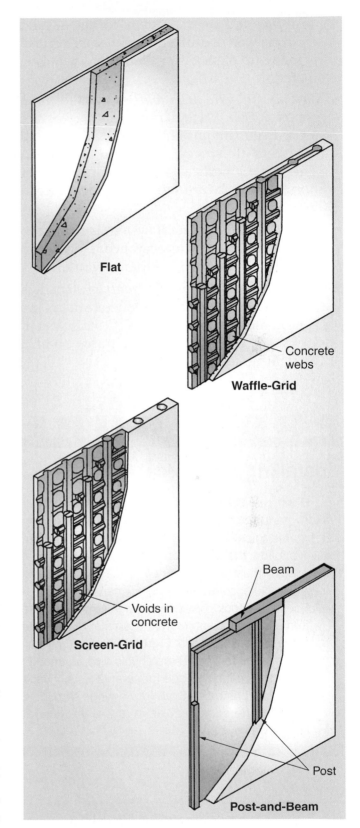

Figure 7-53. Insulated concrete forms (ICFs) are a lightweight alternative to conventional forms. They fall into the four general types shown.

- The *flat ICF wall* system forms a solid concrete wall of uniform thickness. The form is made up of sheets or planks of insulating material.

- The *waffle-grid ICF wall* system forms a concrete web. Vertical parts of this web are spaced a maximum of 12″ apart O.C. Horizontal members are spaced at most 16″ apart O.C. The thicker vertical members and thinner horizontal webs are similar in appearance to a breakfast waffle.

- A *screen-grid ICF wall* system is similar to the waffle-grid. Its appearance is nearer to that of a window screen than a waffle.

- The *post-and-beam ICF wall* system has vertical and/or horizontal concrete members spaced more than 12″ apart O.C. This system looks more like a concrete frame than a monolithic concrete wall.

Figure 7-54. IFCs may be easily transported to construction sites on pallets.

Block-type ICFs are usually transported on pallets, **Figure 7-54.** Either the masons or carpenters will install them after the footings are poured and cured.

PROCEDURE

Installing the typical flat ICF

Each manufacturer of ICFs supplies an installation manual for its specific system. The following steps are typical for setting up flat systems. Caution! *Always* follow the more detailed installation manual provided by the manufacturer of the ICF system being used. Check local codes for conformance, as well.

1. Footings should be wide enough to receive the forms with sufficient clearance to install 2 × 4 cleats along the outside edge, **Figure 7-55.** Securely nail cleats to the footing. There should also be enough room on the inside edge for scrap 2 × 4s. These should be placed 4′–6′ apart. Footings should be keyed in the same way as in any other foundation or wall system.

2. Insert spreaders, called **form ties,** in every course to keep the width of the wall cavity uniform. Continue stacking the forms and inserting the form ties to the proper height. If needed, install horizontal reinforcing rebar. Rest the rebar on top of the form ties. Insert vertical rebar as the concrete is being placed.

3. Bracing is extremely important. The ICF manufacturer may recommend that vertical braces be anchored to the form when the wall reaches a height of 4′. These should be 8′ apart along the outside of the form wall. **Figure 7-56** shows a typical system that also provides scaffolding where the carpenter can stand during pours or to continue adding courses to increase wall height.

Flat ICF wall: A solid concrete wall formed by sheets or planks of polystyrene rigid foam.

Waffle-grid ICF wall: Insulated concrete wall system that forms a grid similar to the appearance of a waffle.

Screen-grid ICF wall: Poured concrete wall using insulated forms. Resulting wall forms a grid similar to a window screen.

Post-and-beam ICF wall: Poured wall contained by insulated concrete forms. Pour is shaped into vertical and/or horizontal concrete members more than 12″ apart O.C.

Form ties: Wood or plastic devices that keep concrete wall forms a uniform distance apart.

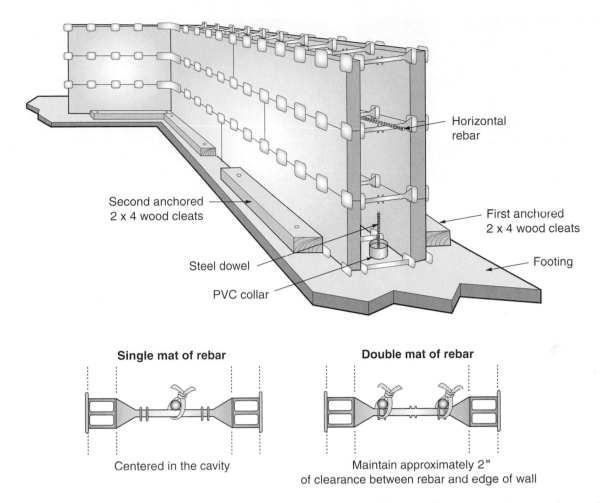

Horizontal rebar

Second anchored
2 x 4 wood cleats

First anchored
2 x 4 wood cleats

Steel dowel

Footing

PVC collar

Single mat of rebar

Centered in the cavity

Double mat of rebar

Maintain approximately 2"
of clearance between rebar and edge of wall

Figure 7-55. ICF planks are anchored by cleats fastened to footings. Note the placement of spacer ties and rebar. (Lite-Form International)

7.12.2 Placing the Concrete

Any method can be used to place concrete in the insulating form walls. For practical reasons, a concrete pump is often used. The height of the wall often makes other methods impossible. See **Figure 7-57.**

The concrete receives a "moist cure" because hydration continues over a longer period due to the insulating properties of the form. This increases the concrete's strength and reduces cracks and foundation leaks.

Normally, the delivery hose should be held horizontal over the form so the concrete falls naturally. If the wall is higher than 12', it is advisable to make pours in stages. Place reinforcing steel between the cold seams of the pours. Some manufacturers warn that making single pours higher than 4' may cause the insulating form to fail.

7.13 Slab-on-Grade Construction

Today, many commercial and residential structures are built without basements. The main floor is formed by placing concrete directly on the ground, referred to as *slab-on-grade* construction. Footings and foundations are similar to those for basements. However, they need to extend downward only to solid soil below the frost line. See **Figure 7-58.**

In slab-on-grade construction, insulation and moisture control are essential. The earth under the floor is called the subgrade and must be firm and completely free of sod, roots, and debris.

A coarse fill at least 4" thick is placed over the finished subgrade. The fill should be brought to

A

Safety handrails

16-gage steel
exterior vertical brace

Scaffold bracket

In-wall bracing

2 x 4 diagonal brace
anchored with stake
or turnbuckle

Vertical bracing is wire-tied to full
tie pads, every 2' up the wall

Vertical bracing can
rest on footing, pad,
or cleat

B

Figure 7-56. Bracing wall forms. A—A properly braced wall form. B—At 4' high, the assembled insulating forms must be braced as shown. Place vertical and angled braces 8' apart along one side of the form. (Arxx Building Products, Lite-Form International)

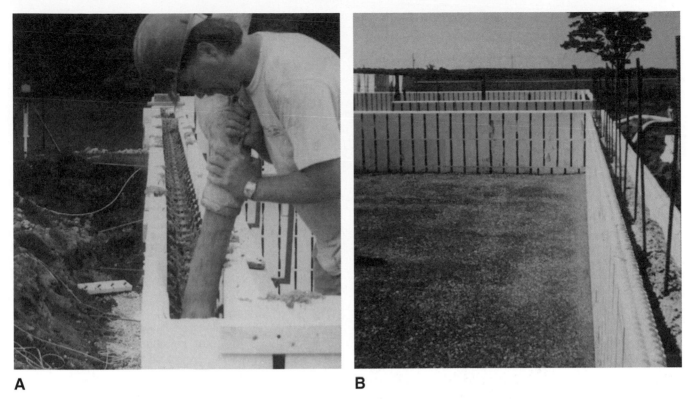

A   **B**

Figure 7-57. Placing concrete in insulating forms. A—A construction worker places concrete in a steel-reinforced insulated form. B—The wall has received the first pour. Vertical reinforcing rods have been placed. Horizontal reinforcing is also recommended before the second pour. (Arxx Building Products)

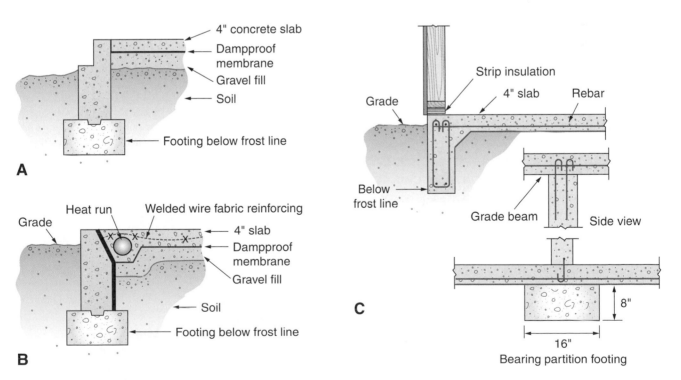

Figure 7-58. Types of slab-on-grade. A—Unreinforced slab with loads supported by footing and wall. Used where soil is coarse and well drained. B—Slab is reinforced with welded wire fabric. The inside of the foundation wall is insulated because of the perimeter heat duct included in the slab. C—Monolithic slab. This type is used over problem soils. Loads are carried over a large area of the slab. (Pierce Construction, Ltd.)

grade and thoroughly compacted. This granular fill may be slag, gravel, or crushed stone, preferably ranging from 1/2″ to 1″ in diameter. The material should be uniform, without fines, to ensure maximum air space in the fill. Air spaces add to the insulating qualities and reduce capillary attraction of subsoil moisture. Capillary action is the action by which moisture passes through fill. In areas where the subsoil is not well drained, a line of drain tile may be required around the outside edge of the exterior wall footings.

While preparing the subgrade and fill, make the various mechanical installations. Under-floor ducts are usually embedded in the granular fill. Water service supply lines, if placed under the floor slab, should be installed in trenches deep enough to avoid freezing. Connections to utilities should be brought above the finished floor level before placing the concrete.

After the fill has been compacted and brought to grade, a vapor barrier should be placed over the subbase. Its purpose is to stop the movement of water into the slab. Among materials widely used as vapor barriers are 55 lb. roll roofing, 4 mil polyethylene film, and asphalt-impregnated kraft papers. Strips should be lapped 6″ to form a complete seal. A vapor barrier is essential under every section of the floor. Instructions supplied by the manufacturer should be carefully followed.

Perimeter insulation is important. It reduces heat loss from the floor slab to the outside. The insulation material must be rigid and stable while in contact with wet concrete. **Figure 7-59** shows a foundation design with perimeter insulation typical of residential frame construction. Note how rigid insulation is applied to the exposed foundation exterior. The thickness of the insulation varies from 1″ to 2″, depending on the average outside temperature and type of heating. The insulation can be either horizontally or vertically placed along the foundation wall. Refer to Chapter 15 for additional information.

When the insulation, vapor barrier, and all mechanical aspects are complete, lay down a reinforcing mesh, if it is to be used. See **Figure 7-60.** Check local building code requirements. Usually a 6″ × 6″ × 10 gage mesh is sufficient for residential work. This should be carefully located from 1″–1 1/2″ below the surface of the concrete.

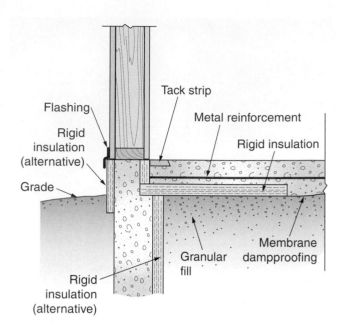

Figure 7-59. Perimeter insulation is important when using slab-on-grade construction in cold climates.

Terrazzo, ceramic tile, asphalt tile, wood flooring, linoleum, and wall-to-wall carpeting are coverings appropriate for use over concrete floors. When linoleum, asphalt tile, or similar resilient-type flooring materials are to be applied, the concrete surface is usually given a smooth, steel-troweled finish. Information on finished wood floors is provided in Chapter 17, **Finish Flooring**.

Figure 7-60. In construction of a slab foundation, reinforcing mesh is laid on top of a vapor barrier. As concrete is placed, the mesh is lifted up so that it is 1″ to 1 1/2″ below surface of finished slab. The perimeter of the slab is reinforced with rebar. (Witkowski Construction)

7.14 Basement Floors

Many of the considerations previously listed for slab-on-grade floors also apply to basement floors. Basement floors are placed later in the building sequence, sometime after the framing and roof are complete and after the plumber has installed drain/waste/vent (DWV) and water service lines. Essentials for a basement floor slab include:

- Subgrade beneath the slab should be well and uniformly compacted to avoid uneven settling.
- Sewer and water lines should be installed and covered with 4″–6″ of well-tamped crushed rock.
- Put down a vapor barrier of 6 mil polyethylene before placing the reinforcing and concrete. This not only keeps the slab dry, but also keeps the curing concrete from losing needed moisture to the crushed rock or gravel base.
- Screed the concrete immediately after it is placed. Float the concrete when the sheen has disappeared from the surface.
- If a smooth, dense surface is important for the application of finish flooring, give the slab a final smoothing with a steel trowel.

7.14.1 Super-Insulated Basement Floors

Most basement floors are placed over a base of gravel covered with a dampproof membrane. The typical super-insulated floor also has a 2″ layer of rigid foam insulation between the gravel base and the vapor barrier, **Figure 7-61.**

To prevent cracks, a grid of rebar is installed, as shown in **Figure 7-62.** This can be held above the surface with patented metal *chairs* or blocked up with scrap pieces of 2 × 4. If the owner has

Chair: Small metal fixture used to hold reinforcing bars away from the ground prior to the pouring of concrete slabs.

A

B

Figure 7-61. Installing insulation. A—Rigid insulation 2″ thick is sometimes used under a basement floor. Seams in the insulation are sealed with tape. B—A dampproofing film is rolled out over the rigid insulation. Overlaps in the film are sealed with tape. Note that rebar from a section already poured is being held up out of the way. (Kasten-Weiler Construction)

Figure 7-62. Making up a grid of rebar. Every intersection is wire-tied. High resistance wiring for in-floor heating is also placed at this stage. (Kasten-Weiler Construction)

specified radiant heat in the floor slab, special high-resistance cable can be installed in serpentine loops from wall to wall. This resistance wire is fastened at intervals to the metal reinforcing bars with plastic ties. This work must be completed before the ready-mix concrete arrives.

The concrete is delivered to the basement through a window, stairway, or an open section of the outside wall, **Figure 7-63.** The rough opening for a fireplace located on an outside wall may also provide easy access. Concrete should not be moved any great distance once placed in the form. A wheelbarrow is often used to carry it from the ready-mix truck's chute to the form.

Working from one side of the form to the other, workers screed the concrete as fast as the form fills. See **Figure 7-64.** Power equipment may also be used for screeding. The chief purpose of *screeding* is to level the surface by striking off (removing) the excess concrete. Striking off starts as soon as the concrete is placed in the form. The screed rides on the edges of the form or previously poured sections of the slab. Low spots found behind the screeded concrete are filled and struck off once more. Two people move the screed across the form. They use a sawing motion as they move the screed along.

Screeding leaves a level, but coarse, surface. *Floating* provides a smoother finish. It also fills up the hollows and compacts the concrete. The proper tool for floating is the wood float or the long-handled bull float, **Figure 7-65.** Floating is done when the concrete has begun to set. It must be firm enough so that a person's weight on it produces a footprint no more than 1/4″ deep. There should be no bleed water present on the surface. Floating too soon brings too many fines

A

Figure 7-64. Screeding levels the concrete as the form is filled. This operation also may be done with a power screed.

B

Figure 7-63. Delivering concrete for a basement slab. A—The chute from the ready mix truck enters the basement through the rough opening for the fireplace. B—Concrete is wheeled away from the chute and placed with a wheelbarrow. (Kasten-Weiler Construction)

Screeding: The process of leveling off concrete slabs or plastering on interior walls, using a screed.

Floating: Smoothing the surface of wet concrete using a large flat tool (float).

A

B

Figure 7-65. Finishing the surface. A—Floating the finished slab by hand with a bull float. This tool can be made of wood, steel, or magnesium. B—Troweling a smooth finish with a power trowel. (Kasten-Weiler Construction)

and more water to the surface. This produces fine cracks in the surface of the cured concrete.

Troweling produces a denser and smoother surface than floating. Trowels are made of metal and have a smooth finish on them. Troweling requires a firmer set than floating. It can proceed when working the surface does not bring up more water. Troweling can be done by hand for smaller jobs, but large slabs are usually finished with a power trowel, as shown in **Figure 7-65B.**

7.15 Entrance Platforms and Stairs

Entrance platform foundations can be constructed as a part of the main foundation. If not, they should be firmly attached to the main foundation. Reinforcing bars placed in the wall when it is constructed can provide a solid connection.

Steps may be formed up and poured as part of the platform, **Figure 7-66.** When the steps are more than 3′ wide, use 2″ stock to keep the risers from bowing. Detailed information on building

Troweling: Producing a dense, smooth surface on a concrete floor using a trowel (metal tool with a smooth finish).

2 x 8

Bevel 45°

10 1/2″

1 1/2″

2 x 8

Beveling form to provide toe room on treads (nosing)

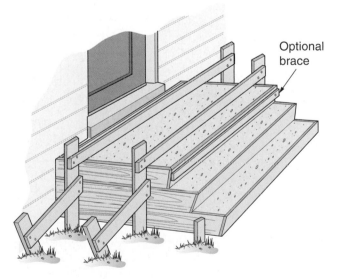

Optional brace

Figure 7-66. Constructing forms for steps. Nosing is formed by tilting the form boards inward at the bottom.

stairs is included in Chapter 18, **Stair Construction**. The 2 × 8 riser boards are set at an angle of about 15° to provide a slight overhang (nosing). Also, bevel the bottom edges of the boards so the mason can trowel the entire surface of the tread.

Some concrete steps must be placed against a wall or between two existing walls. A form can be constructed like the one shown in **Figure 7-67**.

7.16 Sidewalks, Steps, and Drives

Usually, sidewalks and drives are laid after the finished grading is completed. If there is extensive fill, wait until the grade has settled or tamp the soil.

Main walks leading to front entrances should be at least 4′ wide. Those to secondary entrances may be 3′ or slightly less. In most areas, sidewalks are 4″ thick and the formwork is constructed with 2 × 4 lumber. Walks and drives are usually laid directly on the soil. If there is a moisture problem and frost action, a coarse granular fill should be put down first.

When joining two levels of sidewalks with steps, it is usually best to pour the top level first. If retaining walls are used, these should be constructed next along with a segment of the lower sidewalk. Finally, form up and pour the steps.

When setting sidewalk forms, provide a slope to one side of about 1/4″ per foot, **Figure 7-68**. Also, increase the thickness to 6″ or add reinforcing where there will be heavy vehicular traffic. The concrete should not be permitted to bond against foundation walls or entrance platforms. It should be permitted to "float" on the ground. A 1/2″ thick strip of asphalt-impregnated composition board provides satisfactory separation.

Driveways should be 5″ or 6″ thick with reinforcing mesh included. A single driveway should be at least 10′ wide and a double driveway a minimum of 16′ wide. The minimum crosswise slope should be 1/4″ per foot.

Always place the concrete between the forms close to its final position. Do not overwork the concrete while it is still plastic. This tends to bring excess water and fine material to the surface. It also causes scaling and dusting after the concrete has cured.

7.16.1 Finishing Walks and Drives

After the concrete is roughly spread between the forms, immediately screed it. After

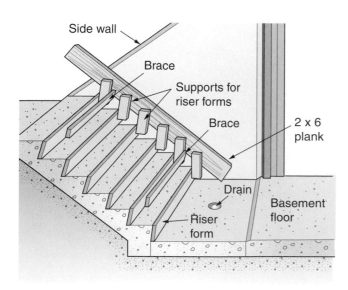

Figure 7-67. A method of building forms when concrete steps must be poured between two existing walls.

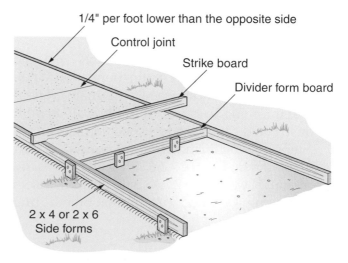

Figure 7-68. The basic setup for constructing a sidewalk. Control joints should be 1/5 of the thickness of the concrete.

screeding, move a float over the surface. When skillfully performed, this operation removes high spots, fills depressions, and smoothes irregularities. See **Figure 7-69.**

As the concrete stiffens and the water sheen disappears from the surface, round the edges of walks and driveways. This is done by working an edger tool along the forms. The edger tool can also be used to finish the control joints.

Start surface finishing operations after the concrete has hardened enough to become somewhat stiff. For a rough finish that will not

A

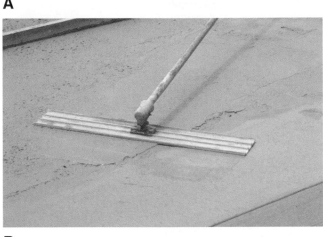

B

Figure 7-69. Finishing concrete flatwork. A—A manual screed is being used to level and compact a concrete driveway. B—Floating the surface smoothes irregularities such as high spots and depressions.

Permanent wood foundation (PWF): A special building system that saves time because it can be installed in almost any weather.

become slick during rainy weather, the surface can be stroked with a stiff bristle broom. When a finer texture is desired, the surface should be steel-troweled and then lightly brushed with a soft bristle broom.

7.16.2 Forming Control Joints

Control joints can be formed with a groover and straightedge. A power saw with a masonry blade can also be used. Start sawing 18 to 24 hours after the concrete is placed. These joints should extend to a depth of at least one-fifth the thickness of the concrete. For sidewalks and driveways, the distance between the control joints is usually about equal to the width of the slab.

For proper curing of concrete, protect it against too-rapid moisture loss during the early stages of hardening (hydration). Refer to **Figure 7-33,** which describes several different methods of controlling curing. A convenient method is spraying the concrete with a plastic-based curing compound, which forms a continuous membrane over the surface.

7.17 Wood Foundations

The *permanent wood foundation (PWF)* is a special building system that saves time because it can be installed in almost any weather, **Figure 7- 70.** Moreover, it provides comfortable living space in basement areas because the stud walls can be fully insulated. All wood parts are pressure-treated with a solution of chemicals that make the fibers useless as a food for the insects and fungus that cause damage and decay.

Foundation sections of 2″ lumber and exterior plywood can be panelized in fabricating plants or constructed on the building site. Pressure-treated wood foundations, have been approved by major code groups and accepted by FHA, HUD, and the Farmers Home Administration.

As for a conventional basement, the site is excavated to the required depth (below the frost line). Plumbing lines are installed and provisions are made for foundation drainage, following

Figure 7-70. Installing a wood foundation that has been prefabricated in a plant. All components are pressure-treated against rot and insect damage. Outside wall panels located below grade are waterproofed and protected by a 6 mil polyethylene film. (APA-The Engineered Wood Association)

local requirements. Some soils require a *sump* (a pit for water collection) that is connected to a storm sewer, pump, or other drainage system.

The subgrade is then covered with a 4"–6" layer of porous gravel or crushed stone and carefully leveled. Footing plates of 2 × 6 or 2 × 8 material are directly installed on this base. The wall sections of the foundation are then erected and securely fastened to the footing plate with noncorrosive fasteners. All fasteners should be made of stainless steel, silicon bronze, copper, or hot-dipped zinc-coated steel. Special caulking compounds must be used to seal all joints between sections and between footing plates and sole plates. Before pouring the basement floor, cover the porous gravel or crushed stone base with a 6 mil polyethylene film, **Figure 7-71.**

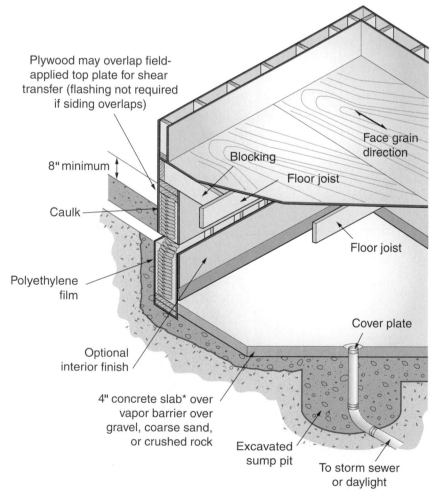

Plywood may overlap field-applied top plate for shear transfer (flashing not required if siding overlaps)

8" minimum

Caulk

Polyethylene film

Optional interior finish

4" concrete slab* over vapor barrier over gravel, coarse sand, or crushed rock

Blocking

Floor joist

Face grain direction

Floor joist

Cover plate

Excavated sump pit

To storm sewer or daylight

*A wood basement floor system is available; write APA for details

Figure 7-71. A cutaway of a typical wood foundation. Note the drainage sump. It keeps the gravel base and subsoil dry around the foundation. (APA-The Engineered Wood Association)

Use a screed board after the basement concrete is placed.

Install the first floor frame on the double top plate of the foundation wall. Pay special attention to method of attachment, so that inward forces will be transferred to the floor structure. Where joists run parallel to the wall, install blocking between the outside joist and the first interior joist.

Before backfilling, pour the basement floor and install first floor joists. This is necessary to avoid shifting and other damage to the foundation. Do not backfill until the concrete basement floor has cured. Also, before backfilling, apply a 6 mil polyethylene moisture barrier to sections of the outside wall below grade. Bond the top edge of the barrier to the wall at grade level with a special adhesive. Install a treated wood strip over this and caulk it. Later, the strip will serve as a guide for backfilling. Lap vertical joints in the polyethylene film at least 6". Seal joints with adhesive.

As with any foundation system, satisfactory performance demands full compliance with standards covering design, fabrication, and installation. Standards for the construction of wood foundations are contained in manuals prepared by several associations.

Carpenters installing wood foundations should make certain that each piece of treated lumber and plywood carries the mark "AWPB-FDN." This assures them that the materials meet requirements of code organizations and federal regulatory agencies.

7.18 Cold Weather Construction

Cold weather may call for some changes in the way concrete and masonry materials are handled and placed. It may be necessary to:
- Heat the materials.
- Cover freshly placed concrete or masonry.
- Erect an enclosure and keep the construction area heated.

When temperatures fall below 40°F (5°C), concrete should have a temperature of 50°F–70°F

(10°C –20°C) when it is placed. Since most concrete is ready mix, only the building site needs to be sheltered and heated. However, when blocks are being placed, the materials also need shelter and heat.

Temporary shelter can often be arranged with scaffolding sections, lumber, and canvas or plastic tarpaulins. If a shelter cannot be built, the bagged materials and masonry units should be wrapped with tarpaulins when the temperature is below 40°F (5°C). Be sure the material is stored so ground moisture cannot reach it. Follow recommendations listed in **Figure 7-72.**

7.18.1 Protecting Concrete

Protect freshly placed concrete with a covering. Hydration creates heat. The covering helps hold the heat in while the concrete cures.

Avoid placing concrete on frozen ground. When the ground thaws, uneven settling may cause cracks. Before pouring, make sure that reinforcing metal, embedded fixtures, and the inside surfaces of forms are free of ice.

When air temperature is below	Do this to materials	Do this to protect placed masonry
40°F	Heat mixing water. Keep mortar temperatures 40°F to 120°F.	Cover walls and masonry materials to protect from moisture and freezing. Use canvas or plastic.
32°F	Do all of above but also: Heat sand to thaw frozen clumps. Heat wet masonry units to thaw ice.	Provide windbreak for workers when wind speed is above 15 mph. Cover walls and materials after workday to protect against wetness and freezing. Keep masonry temperature above 32°F, using heaters or insulated blankets for 16 hours after placing of units.
20°F	Besides above: Heat dry masonry units to 20°F.	Enclose structure and heat the enclosure to keep temperature above 32°F for 24 hours after placing masonry units.

Figure 7-72. Follow these recommendations from the Portland Cement Association for cold-weather construction.

7.18.2 Mortar Temperatures

The temperature of the materials used in mortar is important. Water is the easiest to heat. It can also store more heat and help bring cement and aggregate up to temperature. Generally, it should be cooler than 180°F (80°C). There is danger that hotter water could cause the mortar to set instantly.

When the air temperature is lower than 32°F (0°C), sand should be heated to thaw frozen lumps. If desired, the sand temperature can be raised as high as 150°F (65°C). A 50 gallon drum open on one end or a metal pipe works well for containing flames. Heap the sand over and around the container.

7.18.3 Admixtures

Admixtures are materials added to concrete or mortar to change its properties. Cold weather admixtures used for concrete or mortar include:

- Antifreeze to lower the freezing point of the mixture.

- Accelerators to speed up curing. Accelerators include calcium chloride, soluble carbonates, silicates, fluosilicates, calcium aluminate, and organic compounds such as triethanolamine.

- Air-entraining agents that improve workability and freeze-thaw durability of mortar as it ages.

- Corrosion inhibitors. When reinforcement is placed in winter construction, these materials prevent rust formation.

The carpenter or contractor must obtain the admixtures that are to be mixed in the mortar. They are added on the job. Ready mix companies can include admixtures in the concrete mix when the carpenter, contractor, or architect specifies them.

7.19 Insect Protection

Wood-loving insects cause varying amounts of damage to structures throughout most areas of the country. Among these pests, the most destructive are termites. Other insects infesting and damaging wood structures include carpenter ants and several types of beetles. Control of these pests—especially termites—involves both following certain rules during construction and treating wood and soil with insecticides.

7.19.1 Termites

Two types of termites are found in temperate to tropical climates. *Subterranean termites* are the most destructive. They like warm, moist soil where wood or cellulose material is in good supply. In their search for wood, they build earthlike shelter tubes that give them passage over foundation walls, through cracks in walls, on pipes, or other supports leading into buildings.

Drywood termites are so called because they do not need moisture or contact with the ground. They are airborne and bore into wood structures as well as furniture, particularly when the material is not protected by some type of finish. Though these termites are more common to the tropics, infestations are found along coastal areas from Virginia to northern California.

7.19.2 Termite Control

In new construction, prevention of termite damage begins with the design and construction of the building. The following measures are recommended. Always check for the requirements of local codes.

Admixtures: Chemicals added to concrete to change the characteristics of the mix.

Subterranean termites: The most destructive type of termite. In their search for wood, they build earthlike shelter tubes that give them passage over foundation walls, through cracks in walls, on pipes, or other supports leading into buildings.

Drywood termites: Termites that do not need moisture or contact with the ground. They are airborne and bore into wood structures as well as furniture.

- Chemical treatment of the soil near the foundation walls.

- Chemical treatment (prior to construction) of soil under a slab-on-grade foundation.

- Removal of wood debris, stumps, and form boards from the building site during and after construction.

- Sealing foundation walls with material such as damp-proofing that prevents termites from entering through cracks that may develop.

- Maintaining a minimum of 8″ between grade and the wood sills and siding.

- Capping of foundation walls with termite shields and placing similar shielding around pipes that run from the soil into the building. Placement of shields is discussed in Chapter 8, **Floor Framing**. A shielding method developed by one manufacturer for its ICF foundation system is shown in **Figure 7-73**.

Control of termites in older buildings is also possible, **Figure 7-74**. Studies have shown that appropriate chemicals added to soil around

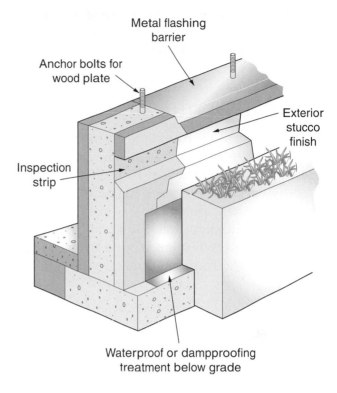

Figure 7-73. This detail shows a termite guard developed for use with an ICF foundation system. (Lite-Form International)

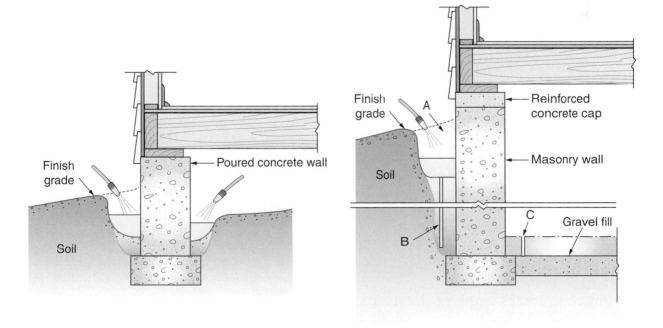

Figure 7-74. With existing buildings, treat the soil as shown to control termites. Left—Trench around both exterior and interior perimeters of a crawl-space foundation. Treat the soil as it is replaced. Right—Treating a full basement. A—Trench and treat as with the crawl space. B—Create rod holes down to the footings to aid distribution of the chemical. C—Drill holes in the concrete floor at 1′ intervals and fill with chemicals.

and under existing buildings prevent or control infestation for many years. Treatment is best left to a professional exterminator.

7.19.3 Carpenter Ants

Carpenter ants are found mostly in the Northeast, Midwest, and Northwest. They only nest in the wood, they do not eat it. Their tunneling in wooden structures causes the damage. Once in the wood, they are difficult to control, since the treatment must get into the nests. A 2′ wide band of chemical treatment around the building discourages ants from entering.

7.20 Estimating Materials

Concrete is measured and sold by the cubic yard. A cubic yard is 3′ square and 3′ high. It contains 27 cu. ft. ($3 \times 3 \times 3 = 27$). To determine the number of cubic yards needed for any square or rectangular area, use the following formula. All dimensions should be converted to feet and fractions of feet.

$$\text{Cubic yards} = \frac{\text{width x length x thickness}}{27}$$

For example, to find the concrete needed to pour a basement floor that is $30′ \times 42′ \times 4″$:

$$\text{Cubic yards} = \frac{30 \times 42 \times 1/3}{27}$$

$$= \frac{\overset{10}{\cancel{30}} \times \overset{14}{\cancel{42}} \times 1/3}{\underset{9}{\cancel{27}} \ \underset{1}{\cancel{3}}} = \frac{140}{9}$$

$$= 15.56 \text{ cubic yards}$$

Allow extra concrete for waste or slight variation in the cross sections of the form. Usually an additional 5%–10% is added.

The number of concrete masonry units needed can be estimated by determining the number of units required in each course and then multiplying by the number of courses between the footing and plate. For example: find the number of $8 \times 8 \times 16$ blocks required to construct a foundation wall with a total perimeter (distance around the outside) of 144′. The wall will be laid 11 courses high.

$$\text{Total number} = \frac{\text{perimeter}}{\text{unit length}} \text{ x number of courses}$$

$$= \frac{144'}{16''} \text{ x } 11$$

$$= \frac{144'}{(4/3)'} \text{ x } 11$$

$$= \frac{144 \times 3}{4} \text{ x } 11$$

$$= \frac{\overset{36}{\cancel{144}} \times 3}{\underset{1}{\cancel{4}}} \text{ x } 11$$

$$= 108 \qquad \text{ x } 11$$

$$= 1188 \text{ blocks}$$

Another method is to find the face area of the wall in square feet and divide by 100. This figure is then multiplied by 112.5. This gives you the number of $8 \times 8 \times 16$ blocks required to construct 100 sq. ft. of wall. Multiplying this figure by 2.6 yields the cubic feet of mortar required. See **Figure 7-75.**

Wall thickness	For 100 square feet of wall		For 100 concrete block
Inches	Number of block	Mortar* cu.ft.	Mortar* cu.ft.
8	112.5	2.6	2.3
12	112.5	3.9	3.5

Based on block having an exposed face of 7 5/8" x 15 5/8" and laid up with 3/8" mortar joints.

*With face shell mortar bedding – 10 percent waste included.

Figure 7-75. Needed quantities of concrete block and mortar can be calculated with this chart. (Portland Cement Assn.)

ON THE JOB

Cement Mason

Concrete is one of the most widely used building materials. In residential work, it is primarily used for foundations and basements, but in commercial and industrial construction, it often is a major above-ground structural material. Cement masons are the workers who assemble the forms and reinforcing material, place the concrete, and level its exposed surface. Concrete finishers are a related trade and usually work on larger projects. They work on the surface of the concrete after forms are stripped away, filling small voids and cutting away any high spots. As a final step, they usually coat the entire surface with a smooth, thin layer of portland cement. On smaller projects, cement masons usually handle finishing as well.

A cement mason must have a thorough knowledge of how concrete behaves when being placed in forms, how it cures, and what effects various tools, such as screeds and floats, have on the concrete surface. The mason must monitor temperature, wind, and moisture conditions and take appropriate actions to make sure that the concrete properly cures and attains maximum strength. The mason must also know the effects of various concrete additives and how to properly use them.

Working with concrete is physically demanding and fast paced, since concrete remains in a workable state for a limited time. Although concrete work is usually suspended in extremely wet or cold weather, the cement mason must be able to effectively work in a variety of weather conditions.

Most cement masons and concrete finishers work for general contractors or specialized concrete contractors. Only about 1 in 20 (the lowest percentage of any building trade) are self employed.

Many cement masons begin as construction laborers, working as helpers and gradually learning the trade from experienced workers. They may also enter the trade through an apprenticeship program lasting three to four years. In addition to on-the-job experience, apprenticeship programs require a minimum of 144 hours per year of classroom work in such areas as blueprint reading, safety, and applied mathematics.

Summary

Carpenters are sometimes responsible for laying out building lines and building the concrete forms for building foundations. Batter boards consist of horizontal boards and stakes. They are set up well away from where each corner of the new building will be. Cords stretched between the batter boards define the building outline. A control point is established on the site as a reference for the depth of excavation and height of the foundation. Footings must be placed around the building perimeter to support the foundation. There are a number of foundation types, including slab-on-grade, spread foundations (where a basement is required), piling and girder foundations, and grade beam foundations. Forms for concrete must be erected, reinforced, then removed after the concrete sets. In other cases, the foundation walls are constructed using concrete masonry units (blocks). Walls must be properly constructed and anchored to withstand stresses and insulated to prevent heating/cooling losses. Some buildings use insulated concrete forms that remain in place. In certain climates, permanent wood foundations are installed. Finally, the area around the foundation must be backfilled. Usually, sidewalks and drives are laid after the finished grading is completed.

Test Your Knowledge

Answer the following questions on a separate piece of paper. Do not write in this book.

1. *True or False?* Grading may be needed before building lines are laid out.
2. *True or False?* Carpenters always locate lot lines.
3. Batter boards should be located _____ feet or more away from the building lines.
4. *True or False?* Building sites on steep slopes or rugged terrain should be rough-graded before the building is laid out.
5. Both the depth of the excavation and the height of the foundation are established at the highest point on the perimeter of the excavation. This point is known as the:
   A. Grade level.
   B. Control point.
   C. Grade point.
   D. Building line.
6. What is a *slab-on-grade foundation?*
7. What is a *grade beam?*
8. In cold climates, foundations should be located below the frost line. However, CABO now allows frost-protected shallow footings that are only _____ inches below grade.
9. A(n) _____ footing is one that changes grade levels at intervals to accommodate a sloping lot.
10. In residential construction, a safe design is usually obtained by making the width of the footing _____ as wide as the foundation wall.
11. Foundation forms constructed of 1″ boards should be held in place with stakes placed _____ feet apart.
12. *True or False?* Loose dirt and debris should not be removed from the ground under a footing.
13. Concrete is made by mixing _____, sand, gravel, and water in proper proportions.
14. What is a *pilaster?*
15. Concrete hardens by a chemical action called _____.
16. List four methods for properly curing concrete.
17. A(n) _____ test is used to assess the consistency, stiffness, and workability of fresh concrete.
18. Each sack of portland cement contains _____ pounds.
19. When placing concrete in forms, working the concrete next to the forms tends to produce a _____ surface along the form faces.
20. Wood sill plates are fastened to the top of a foundation wall with anchor straps or _____.
21. A concrete block specified as 8 × 8 × 16 actually measures _____ × _____ × _____.
22. _____ are laid up several courses high before laying the rest of the block wall.
23. The last block laid in each course is called a _____ block.
24. Forms made of lightweight expanded polystyrene foam are known as _____ concrete forms.
25. Bracing for foam forms should be installed when the form reaches a height of four feet, with the braces spaced _____ feet apart.

26. In slab-on-grade construction, a(n) _____ should be laid over the subbase to stop the movement of water into the concrete slab.
27. What is the chief purpose of screeding a concrete slab?
28. Explain why a slab is floated after it has been screeded.
29. What may occur if a slab is floated too soon?
30. In most areas, sidewalks are _____ inches thick.
31. A sidewalk should have a slope to one side of about _____.
32. A wood foundation is installed on a layer of _____ that is 4″ to 6″ thick.
33. *True or False?* In cold weather, it is accepted practice to pour concrete over frozen ground.
34. *True or False?* Termite control is possible only by treatment of the soil before construction of the building.
35. How many 8 × 8 × 16 concrete blocks are required to lay 100 square feet of wall surface?

Curricular Connections

Science. Research the materials used to mix concrete. Determine what each material contributes to the finished product. Determine the proper ratio of the materials in the mix and the possible problems that incorrect proportions could cause. Describe what chemically happens as concrete cures. Write a short report or orally present your findings to the class.

Social Studies. Study the building code in your area to learn about such requirements as building setbacks from property lines, design of footings for residential structures, minimum depths for footings and foundations, and basic construction of concrete and masonry foundation walls. Summarize your findings in a written report for your class.

Outside Assignments

1. Visit a ready-mix concrete plant and study the operations. Obtain information about the following:
   - Source of aggregates.
   - Handling and storing cement and aggregates.
   - Equipment used to measure and proportion mixtures.
   - Size of truck-mounted mixers.
   - Distance trucks can travel without additional charges.
   - Cost of a cubic yard of concrete in various psi ratings.
   - Fractional parts of a cubic yard that can be ordered.

   Prepare a written report.
2. Find a set of house plans that includes a fireplace located on an outside wall. Prepare a scaled drawing (1 1/2″= 1′-0″) of the formwork needed for the footings under the fireplace wall. Show individual form boards and stakes. Include one or more section views to describe the shape of the footing.
3. Study reference materials and booklets prepared by such organizations as the Portland Cement Association. Obtain information about air-entrained concrete, slump tests, lightweight aggregates, ultra-lightweight concrete, thermal conductivity, compression tests, reinforcing, and prestressed concrete units. Prepare an outline and report to the class.

For safety, bracing is used to prevent collapse of the form.

Floor Framing

Learning Objectives

After studying this chapter, you will be able to:

- Explain the difference between platform and balloon framing.
- Identify the main parts of a platform frame.
- Calculate the load on girders and beams used in residential construction.
- Lay out and install sills on a foundation wall.
- Describe how layouts are made on a header joist.
- Explain the correct procedure to follow when assembling a floor frame.
- Identify the parts of a floor truss.
- Prepare a sketch that shows how overhangs and projections are framed.
- Describe materials used for subflooring.
- Estimate materials (sizes and amounts) required to construct a specific floor frame.

Technical Vocabulary

Backfilling	Cripple wall
Balloon framing	Cross bridging
Band joist	Crown
Blocking	Deflection
Bridging	Girders
Cantilevered	Glued floor system
Chord	Hanger
Header	Sill sealer
Herringbone	Solid bridging
Joists	Span
Ledger	Steel posts
Mudsill	Stirrup
Platform framing	Subflooring
Post anchor	Tail joist
Post-and-beam framing	Trimmer
Ribbon	Underpinning
Rim joist	Web
Rough flooring	Western framing
Sill	Wide flange
	Wood I-beams

When the building foundation is completed and the concrete or mortar has properly set up, floor framing can begin. Framing of the floor is often completed before the foundation is backfilled. *Backfilling* is the process of replacing soil that was removed during excavation for the building foundation. The soil is placed in the void between the foundation and the unexcavated

---

Backfilling: The process of replacing soil that was removed during excavation of the building foundation.

surrounding earth. Installing the floor frame first helps the foundation withstand the pressure placed on it by the backfill material.

On the other hand, backfilling may be done before the floor frame is installed. This offers two advantages: it makes delivery of lumber easier and allows convenient access for the start of framing. As described in Chapter 7, the carpenter must make sure the foundation wall is adequately braced against damage or cave-in from the pressure of the backfilling. This is critical with concrete block walls. In any case, the site should be brought to rough-grade level after backfilling.

The building's basic design—such as one-story, two-story, or split level—often determines the type of framing used. Other factors may also influence the decision. In some parts of the country, for example, buildings must be constructed with special resistance to wind and rain. In other areas, earthquakes may be the greatest hazard. In cold climates, heavy loads of wet snow may require sturdier and/or special roof framing. All structures should be built to reduce the effects of lumber shrinkage and warping. They must also resist the hazard of fire.

8.1 Types of Framing

There are three basic types of framing used in residential construction:

- *Platform framing,* also called *western framing,* is the most popular for residential construction. See **Figure 8-1.**

- *Balloon framing* is no longer used in new construction, but carpenters doing remodeling or repair of older homes should know about it. This type of framing is discussed in Chapter 26, **Remodeling, Renovating, and Repairing**.

- *Post-and-beam framing,* also called plank and beam, is different from the other two types. Its heavy structural members are 4" or more thick. This kind of construction is covered in Chapter 23, **Post-and-Beam Construction**.

In both platform and balloon framing, joists, studs, plates, and rafters are the common

structural members. Material with a nominal 2" thickness is used. While sawn 2" lumber is still used extensively for framing, engineered lumber often replaces the sawn materials. These new materials include wood I-beams, open truss joists, and laminated veneer lumber. They are lightweight, have consistent strength, and can be purchased in long lengths.

8.1.1 Platform Framing

In platform framing, the first floor is built on top of the foundation wall as a platform. It provides a work area for safely and accurately assembling and raising wall sections. Wall sections are one story high. Outside walls and interior partitions support rafters or upper stories. Each floor is separately framed. Platform framing is satisfactory for both one-story and multistory structures. Settlement due to lumber shrinkage occurs in an even and uniform manner throughout the structure.

Figure 8-2 shows the first-floor platform of a platform frame. Usually, the first-floor platform rests on a *sill.* This is a wood member that is attached by bolts or other metal fasteners embedded in the foundation. Sometimes, the platform rests on an *underpinning.* This is a wall made of short studs that extends above a

---

Platform framing: A system of framing a building where the floor joists of each story rest on the top plates of the story below and the bearing walls and partitions rest on the subfloor of each story.

Balloon framing: A type of building construction with upright studs that extend from the foundation sill to the rafter plate. A ribbon supports the second floor joists. This system is not used in new construction.

Post-and-beam framing: A framing method in which loads are carried by a frame of posts connected with beams, eliminating the need for load-bearing walls.

Sill: The lowest member of the frame of a structure that rests on the foundation and supports the uprights of the frame.

Underpinning: A short wall section between the foundation sill and first floor framing. Also called a *cripple wall.*

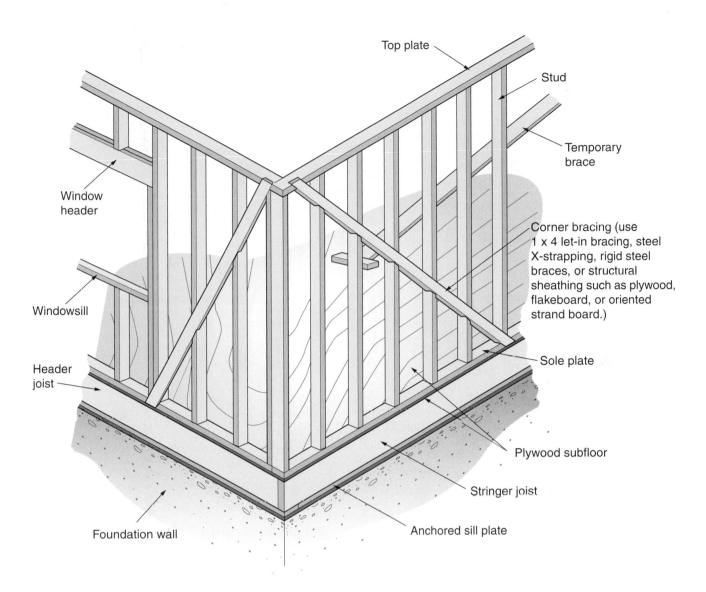

Top plate

Stud

Temporary brace

Corner bracing (use 1 x 4 let-in bracing, steel X-strapping, rigid steel braces, or structural sheathing such as plywood, flakeboard, or oriented strand board.)

Window header

Windowsill

Header joist

Sole plate

Plywood subfloor

Stringer joist

Foundation wall

Anchored sill plate

Figure 8- 1. In platform framing, a frame of joists provides support for the walls of the first level of the building. Ceiling joists or floor joists for the second level rest on the top plate.

low foundation. It is usually used on a house that is to have a full basement. Sometimes an underpinning is called a *cripple wall*.

Figure 8-3 illustrates typical construction methods used at the first- and second-floor levels. *Firestopping* is used to prevent the spread of a fire. The only firestopping needed is built into the floor frame at the second floor level. It prevents the spread of fire in a horizontal direction. It also serves as solid bridging, holding the joists in a plumb (vertical) position.

Referring to the illustrations in **Figure 8-3A,** note that the carpenter usually places a strip of

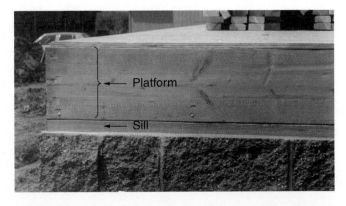

Platform

Sill

Figure 8-2. An example of platform framing. The first floor platform rests on a sill that is anchored to the foundation.

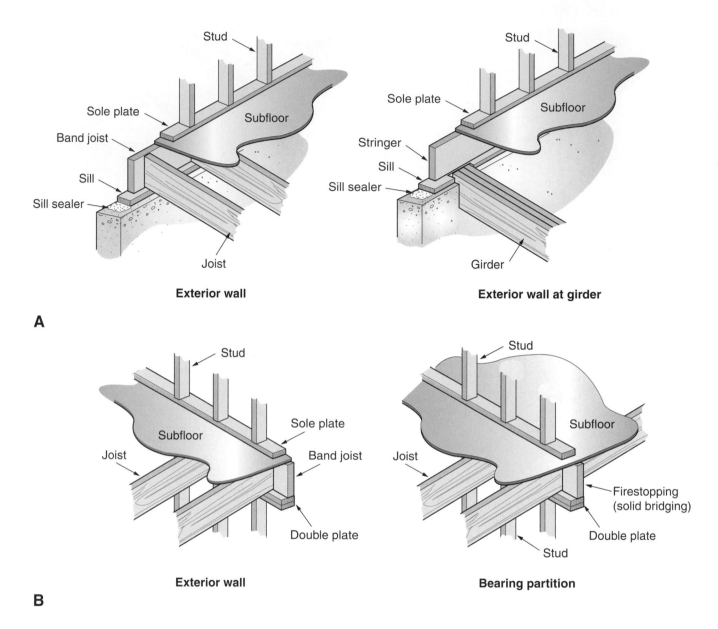

Figure 8-3. Platform framing details. A—First floor. The band joist also acts as a firestop. B—Second floor framing is similar to first. Sill sealer installed under the sill prevents air infiltration. Note the solid bridging, which acts as a horizontal firestop.

insulation called a *sill sealer* between the sill plate and the foundation. This prevents loss of heat at that point.

Architectural plans usually specify the type of framing. There will be section views of floors, walls, and ceilings. A typical detail drawing of first floor framing includes not only the type of construction, but also the size and spacing of the various members. See **Figure 8-4.**

Sill sealer: A strip of insulation between the sill plate and foundation.

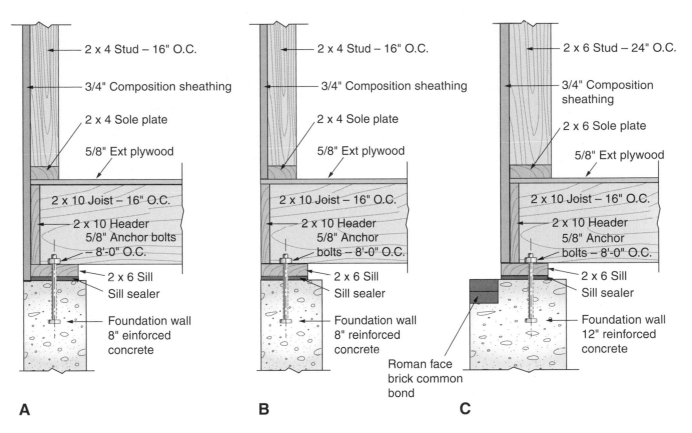

A   **B**   **C**

Figure 8-4. Architectural detail drawings show methods of construction as well as the materials to use. A—Sheathing brought to the foundation. B—Sheathing brought to the sole plate. C—Brick veneer construction.

8.1.2 Balloon Framing

In balloon framing, the studs are continuous from the sill to the rafter plate. A *ribbon* let into the studs supports the second floor joists. They are spiked to the stud as well, **Figure 8-5.** Fire-stopping must be added to the space between the studs. This stud space, which also occurs in load-bearing partitions, permits easy installation of mechanical systems.

In balloon framing, shrinkage is reduced because the amount of cross-sectional lumber is low. Wood shrinks across its width, but practically no shrinkage occurs lengthwise. Thus, the high vertical stability of the balloon frame makes it adaptable to two-story structures, especially where masonry veneer or stucco is used on the outside wall. Though once popular, this type of framing is rarely, if ever, used today. However, many balloon frames still exist and carpenters may encounter this design in remodeling or repair projects on older buildings.

8.2 Girders and Beams

Joists are the horizontal members of the floor frame. They rest on top of foundation walls, girders, and the top plates of second stories. The *span* is the distance between the walls. Usually, the span is so great that additional support must be provided. *Girders,* also called beams, rest on the foundation walls and on posts or

---

Ribbon: A narrow board let into a stud or other vertical member of a frame, thus adding support to joists or other horizontal members.

Joists: The horizontal members of the floor frame.

Span: The distance between structural supports.

Girder: A principal beam used to support other beams or the ends of joists.

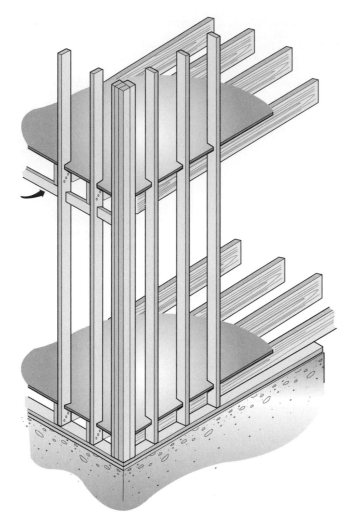

Figure 8-5. Balloon framing. The second floor joists rest on a ribbon (arrow) let into the studs.

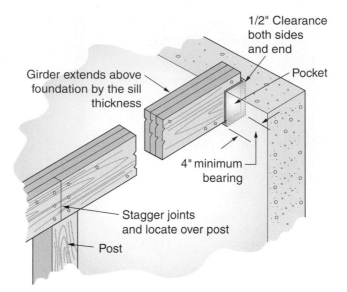

Figure 8-6. When built-up wood girders are fabricated on the job, nails should be spaced no farther apart than 32″ along the top and bottom edges. A metal bearing plate should be placed under the girder at the foundation wall. Clearance in the pocket allows air circulation and prevents rot from moisture.

columns. They support the inner ends of joists. Girders may be solid timbers, built-up lumber, engineered lumber, or steel beams. Sometimes, a load-bearing partition replaces a girder or beam.

Built-up girders can be made of three or four pieces of 2″ lumber nailed together with 20d nails. See **Figure 8-6.** Modern engineered lumber, such as laminated beams, may be used for girders. These are factory-made by gluing together thin strips of wood or veneer. Such beams can be manufactured in many different sizes and lengths. Joints should rest over columns or posts.

PROCEDURE

Sizing girders

1. Find the distance between girder supports.
2. Find the girder load width. A girder must carry the weight of the floors on each side to the midpoint of the joists that rest on it.
3. Find the total floor load per sq. ft. carried by the joists and bearing partitions to the girder. This is the sum of loads per sq. ft. listed in the diagram, **Figure 8-7.** The total floor load does not include roof loads. These are carried on the outside walls unless braces or partitions are placed under the rafters. Then a portion of the roof load is carried to the girder.
4. Find the total load on the girder. This is the product of girder span × girder load width × total floor load.
5. Select the proper size of girder according to the code in your area. The table in **Figure 8-8** is typical. It indicates safe loads on standard size girders for spans from 6′ to 10′. Shortening the span is usually the most economical way to increase the load a girder can carry.

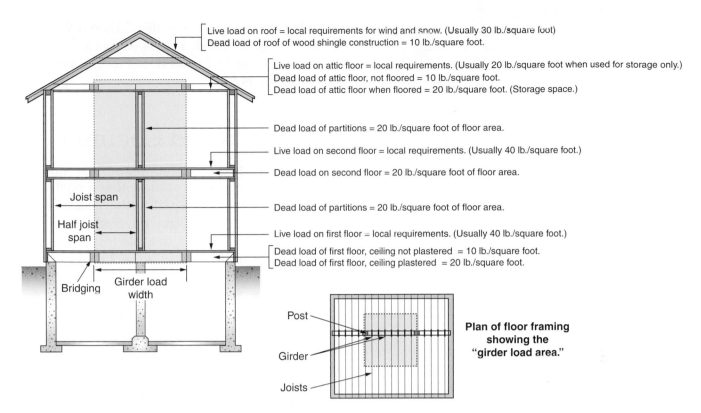

Figure 8-7. This diagram shows the method of figuring loads for the frame of a two-story home.

Girders	Safe Load in Pounds for Spans from 6 to 10 feet				
Size	6'	7'	8'	9'	10'
6 x 8 Solid	8,306	7,118	6,220	5,539	4,583
6 x 8 Built-up	7,359	6,306	5,511	4,908	4,062
6 x 10 Solid	11,357	10,804	9,980	8,887	7,997
6 x 10 Built-up	10,068	9,576	8,844	7,878	7,086
8 x 8 Solid	11,326	9,706	8,482	7,553	6,250
8 x 8 Built-up	9,812	8,408	7,348	6,544	5,416
8 x 10 Solid	15,487	14,782	13,608	12,116	10,902
8 x 10 Built-up	13,424	12,768	11,792	10,504	9,448

Figure 8-8. Typical safe loads for standard size wood girders.

8.2.1 Steel Beams

In many localities, steel beams are used instead of wood girders. Two types of steel beams are illustrated in **Figure 8-9.** The W-beam (*wide-flange*) is generally used in residential construction. Wood beams vary in depth, width, species, and grade. Steel beams vary in depth, width of flange, and weight.

Wide-flange: Type of steel beam generally used in residential construction.

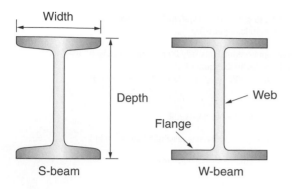

Figure 8-9. Steel beams are commonly used in residential construction. S means standard; W means wide flange.

Sizes depend on the load. The load is calculated in the same way as for wood girders. After the approximate load on a steel beam has been determined, the correct size can be selected from the table in **Figure 8-10.** This table lists a selected group of steel beams commonly used in residential structures. For example, if the total, evenly distributed load on the beam is 15,000 lb. and the

span between supports is 16'-0", then a W8 × 18 beam should be used. This specifies an 8" deep beam weighing 18 lb. per lineal foot. The width of the flange is 5 1/4".

8.2.2 Posts and Columns

For wood posts shorter than 9' or smaller than 6" × 6", it is safe to assume that a post with its greater dimension equal to the width of the girder it supports will carry the girder load. For example, a 6 × 6 post is suitable for a girder 6" wide. For a girder 8" wide, a 6 × 8 or 8 × 8 post should be used.

Adequate footings must be provided for girder posts and columns. Wood posts should be supported on footings that extend above the floor level, as shown in **Figure 8-11.** To make sure the posts will not move off of their footing, pieces of 1/2" diameter reinforcing rod or iron bolts should be embedded in the footing before the concrete sets. They should project about 3" into holes bored in the bottoms of the posts. A

Designation weight/foot	Nominal size depth x width	Span in feet									
		8'	10'	12'	14'	16'	18'	20'	22'	24'	26'
W8 x 10	8 x 4	15.6	12.5	10.4	8.9	7.8	6.9	—	—	—	—
W8 x 13	8 x 4	19.9	15.9	13.3	11.4	9.9	8.8	—	—	—	—
W8 x 15	8 x 4	23.6	18.9	15.8	13.5	11.8	10.5	—	—	—	—
W8 x 18	8 x 5 1/4	30.4	24.3	20.3	17.4	15.2	13.5	—	—	—	—
W8 x 21	8 x 5 1/4	36.4	29.1	24.3	20.8	18.2	16.2	—	—	—	—
W8 x 24	8 x 6 1/2	41.8	33.4	27.8	23.9	20.9	18.6	—	—	—	—
W8 x 28	8 x 6 1/2	48.6	38.9	32.4	27.8	24.3	21.6	—	—	—	—
W10 x 22	10 x 5 3/4	—	—	30.9	26.5	23.2	20.6	18.6	16.9	—	—
W10 x 26	10 x 5 3/4	—	—	37.2	31.9	27.9	24.8	22.3	20.3	—	—
W10 x 30	10 x 5 3/4	—	—	43.2	37.0	32.4	28.8	25.9	23.6	—	—
W12 x 26	12 x 6 1/2	—	—	—	33.4	29.7	26.7	24.3	22.3	20.5	
W12 x 30	12 x 6 1/2	—	—	—	—	38.6	34.3	30.9	28.1	25.8	23.8
W12 x 35	12 x 6 1/2	—	—	—	—	45.6	40.6	36.5	33.2	30.4	28.1

Figure 8-10. Allowable uniform loads for wide-flange steel beams. Loads are given in kips (1 kip = 1000 lb.). (Grosse Steel Co.)

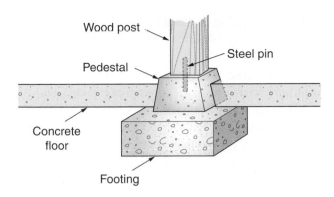

Figure 8-11. For moisture protection, footings for columns must extend above the floor level.

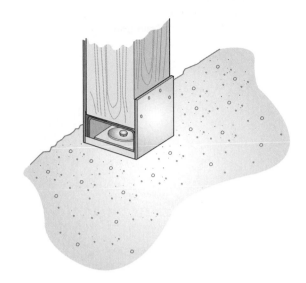

Figure 8-12. Steel post anchor permits lateral adjustment. Various designs are available. (Timber Engineering Co.)

post anchor can be used to securely hold the wood post, **Figure 8-12.** It supports the bottom of the post above the floor, protecting the wood from dampness. The bracket can be adjusted for plumb if the anchor bolt was improperly located.

When a wood column supports a steel girder, consider fitting the end of the column with a metal cap. If wood supports wood, install a metal cap to give an even bearing surface. The metal also prevents the end grain of the post from crushing the horizontal grain of the wood girder.

A built-up wood post may be made by spiking together three 2 × 6s. The pieces should be free from defects and securely nailed together. Otherwise, excessive loading may cause the members to buckle away from each other.

Steel posts are most popular for girder and beam support. The post should be capped with a steel plate to provide a good bearing area. A steel post designed especially for this purpose has a threaded hole at one end. A rod attached to a plate is threaded into this hole. This arrangement allows for adjustment as the wooden beam and other structural members shrink.

8.2.3 Framing over Girders and Beams

A common method of framing joists over girders and beams is shown in **Figure 8-13.** The steel beam is set level with the top of the foundation wall. The 2″ wood pad then brings the joists level with the sill. When a wooden girder is used, it is usually set so the top is level with the sill.

If the ceiling height under joists needs to be lowered, the joists can be notched and carried on a *ledger,* **Figure 8-14.** This arrangement is seldom used in new construction. When it is necessary for the underside of the girder to be flush with the joists to provide an unbroken ceiling surface, the joists should be supported with *hangers* or *stirrups.* See **Figure 8-15.**

Working Knowledge

Be sure the tops of posts and columns and the pockets in foundation walls are flat so the girder or beam is well supported with its sides plumb.

Post anchor: Supports the bottom of a post above the floor, protecting the wood from dampness.

Steel posts: The most popular posts for girder and beam support.

Ledger: A strip attached to vertical framing or structural members to support joists or other horizontal framing. Similar to a ribbon strip.

Hanger: Connecting hardware used to attach beam ends to other building frame members.

Stirrups: Hangers made of metal that support joists.

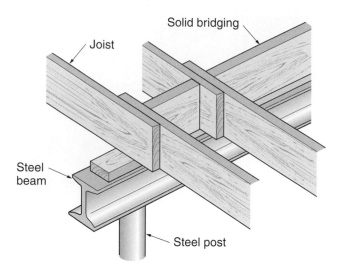

Figure 8-13. Joists supported on top of a steel beam. The top of the beam is set flush with the top of the foundation wall.

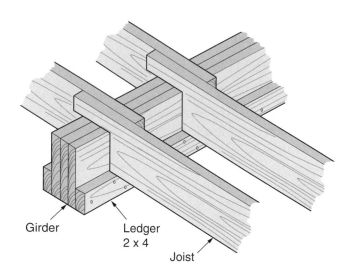

Figure 8-14. In this arrangement, the joists rest on the ledger strip, not on top of the girder.

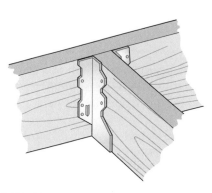

Figure 8-15. Hangers are often used to attach joists to headers.

Framing joists to steel beams at various levels can be accomplished with special hangers in somewhat the same manner as for wood girders. You must make allowance for the fact that joists will likely shrink while the steel beam will remain the same size. For average work with a 2 × 10 joist, an allowance of 3/8″ above the top flange of the steel girder or beam is usually sufficient.

Figure 8-16 shows a method of framing joists flush to the underside of a steel S-beam. Do not notch the joists so they rest on the lower flange of an S-beam. The flange surface alone does not provide a large-enough bearing surface. Use a welded extension as shown in the figure. Flanges on W-beams, however, *do* provide sufficient support surface for this method of construction. **Figure 8-17** shows two butted-joist methods of framing over girders.

8.3 Sill Construction

After girders and beams are in place, the next step is to attach the sill to the foundation wall. The sill is the part of the side walls or floor frame that horizontally rests on the foundation. It supports the frame of the building and ties the frame to the foundation. The sill is also called the sill plate or the *mudsill.* The latter term originated from the process of correcting

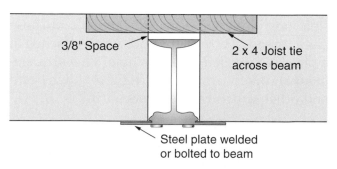

Figure 8-16. Edge view of joists supported on an S-beam. Allow 3/8″ space above the beam for shrinkage.

Mudsill: Part of the floor frame that rests on the foundation wall.

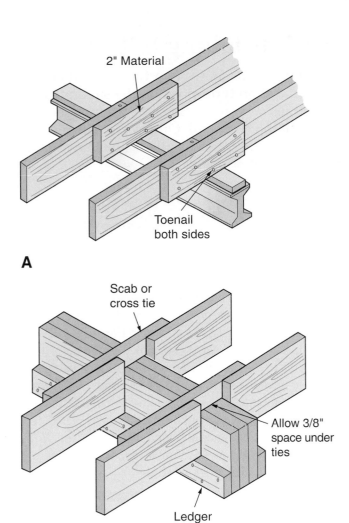

2" Material

Toenail both sides

A

Scab or cross tie

Allow 3/8" space under ties

Ledger

B

Figure 8-17. Some carpenters like to butt joists over the girder. A—Joists are butted over a steel I-beam and wood 2 × 4 and secured with toenailing and a short scab of 2" material. B—Joists are tied together with a scab or cross tie.

irregularities in masonry work by embedding the sill in a layer of fresh mortar or grout.

Codes specify that a sill must have a minimum level of protection against termite and decay damage. Also, it should be pressure-treated or made from a material with natural rot protection, such as heartwood of redwood, black locust, or cedar. The Southern Forest Products Association says that pressure-treated lumber with a 0.40-retention or ground-contact lumber is required wherever treated lumber contacts concrete that is in contact with the ground. To find the level of treatment, read the grade stamp or tag stapled to the lumber.

Location of the sill on the foundation depends on custom and the type of exterior covering the building will have. The sill may be set flush with the outside edge of the foundation or it may be set back to allow for the thickness of the sheathing. In the case of brick veneer, sills may be set back even farther.

To fill the crack between the bottom of the sill and the top of the foundation wall, a sill sealer is typically used. This seal stops passage of heated air through the space between the foundation and the sill. The sealer also keeps out dust and insects.

Sills are usually 2 × 6 lumber. However, the width may vary depending on the type of construction. They are attached to the foundation wall with anchor bolts or straps. The Council of American Building Officials (CABO) has a model code that says anchor bolts:

- May not be placed in less than 7" of concrete.
- Must be a minimum diameter of 1/2".
- Must be spaced no further apart than 6' O.C.
- Each sill section must have at least two anchor bolts and the bolts must be within 12" of the ends of the sill.
- The size and spacing of anchors, however, is a matter of concern to each community. Always check the specifications in local building codes.

8.3.1 Termite Shields

If termites are a problem in your locality, special shields must be installed. Some termites live underground and come to the surface to feed on wood. They may enter through cracks in masonry or build earthen tubes on the sides of masonry walls to reach the wood.

The wood sill should be at least 8" above the ground. Also, a protective metal shield should extend out over the foundation wall as shown in **Figure 8-18.** The shield should be not less than 26 gage.

Figure 8-19 is a map of the United States that locates various levels of termite infestation. Canada and Alaska are considered to be in region IV. Hawaii and Puerto Rico are in region I.

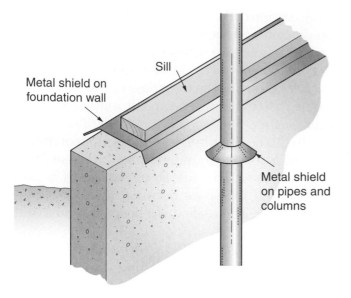

Figure 8-18. Termite shields. Use galvanized sheet steel or other suitable metal that resists rusting.

In areas where termite damage is great, additional measures should be taken. Sometimes it is necessary to use chemically-treated lumber for lower framing members. Also, the soil around the foundation and under the structure can be treated.

8.3.2 Installing Sills

Figure 8-20 shows two types of sill anchor. With the strap type, position the sill and attach the straps with nails. Some types must be bent over the top of the sill. Others are nailed on the sides. The other type of sill anchor shown in the figure is anchor bolts.

PROCEDURE

Using anchor bolts

1. Remove the washers and nuts.
2. Lay the sill along the foundation wall.
3. Using a square, draw lines across the sill on each side of the bolts, as shown in **Figure 8-20.**
4. Measure the distance from the center of the bolt to the outside of the foundation. If you plan to have the sheathing or brick veneer flush with the foundation, subtract the thickness of the sheathing or the brick veneer.
5. Use this distance to locate the bolt holes. You will probably need to make separate measurements for each anchor bolt.

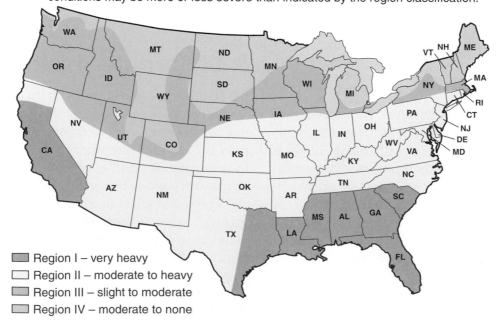

Note: Lines defining areas are approximate only. See local FHA Office for specific areas. Local conditions may be more or less severe than indicated by the region classification.

Region I – very heavy
Region II – moderate to heavy
Region III – slight to moderate
Region IV – moderate to none

Figure 8-19. Nearly all parts of the United States have some risk of termite infestation. Termites can damage buildings. (Forest Products Laboratory)

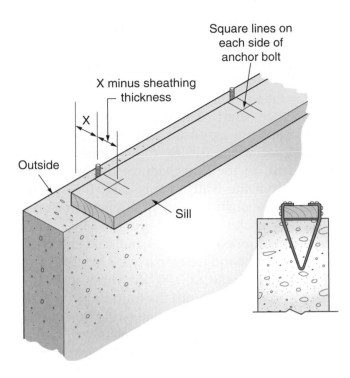

Figure 8-20. Laying out anchor bolt holes. An anchor strap, right, needs no layout. The sill is positioned and the strap is nailed to the sill. (TECO)

Since foundation walls are seldom perfectly straight, many carpenters snap a chalk line along the top where the outside edge of the sill should be located. This ensures an accurate floor frame, which is basic to all additional construction. Variations between the outside surface of the sheathing and foundation wall can be shimmed when siding is installed.

After marking all of the holes on the sill, place the sill on sawhorses and bore the holes. Most carpenters bore the hole about 1/4" larger than the diameter of the bolts to allow for some adjustment due to slight inaccuracies in the layout. As each section is laid out and holes bored, position the sill over the bolts to check for accuracy.

After fitting all sill sections, remove them from the anchor bolts. Install the sill sealer and

---

Deflection: Bending downward at the center of a joist.

Band joist: A joist set perpendicular to the floor joists and to which the ends of the floor joists are attached. Also called a *header.*

then replace the sill sections. Install washers and nuts. Before the nuts are tightened, ensure that the sills are properly aligned. If there is a setback for sheathing or brick veneer, check the distance from the edge of the foundation wall.

Now, check that the sill is level and straight. Low spots can be shimmed with wooden wedges. However, it is better to use grout or mortar.

8.4 Joists

Floor joists are framing members that carry the weight of the floor between the sills and girders. In residential construction, they may be:
- Nominal 2" lumber placed on edge.
- Wood I-beams.
- Open truss joists.

In heavier construction, steel bar joists and reinforced concrete joists are used.

The most common spacing of wood joists is 16" O.C. (on center). However, 12", 19.2", and 24" O.C. spacings are also used. A table in Appendix B **Technical Information** lists safe spans for joists under average loads. For floors, this is usually figured on a basis of 50 lb. per sq. ft. (10-lb. dead load and 40-lb. live load).

Joists must not only be strong enough to carry the load that rests on them, they must be stiff enough to prevent undue bending or vibration. *Deflection* is bending downward at the center. Building codes usually specify that the deflection of a joist must not exceed 1/360th of the span with a normal live load. This is 1/2" for a 15'-0" span.

8.4.1 Laying Out Joists

Carefully study the plans. Note the direction the joists are to run. Also, become familiar with the location of posts, columns, and supporting partitions. The plans may also show the center-lines of girders.

The position of the floor joists can be directly laid out on the sill, **Figure 8-21.** In platform construction, the joist spacing is usually laid out on the band joist rather than the sill. *Band joists*

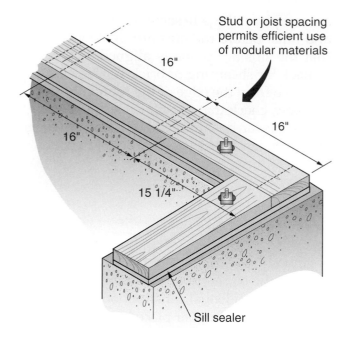

Figure 8-21. After the sill is attached to the foundation, locations for studs or joists may be marked using a square.

are those joists that sit on the sill and to which other floor joists are butted and attached. They are also called headers or *rim joists.* The position of an intersecting framing member may be laid out by marking a single line and then placing an X to indicate on which side of the line the joist is to be installed, **Figure 8-22.** The crown of the joist should be turned upward. The *crown* is a slight warping, also called a *crook.*

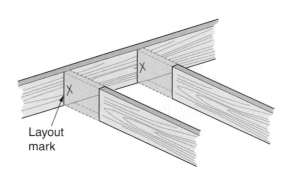

Figure 8-22. How to mark the actual location of framing members being attached to a band joist. Layout marks show where the edge of a joist should be. The X indicates to which side of the line the joist should be positioned.

Rather than having to measure each individual joist space with a steel tape, it is more accurate and efficient to make a master layout (called a *rod*) on a strip of wood. Use it to transfer the layout to band joists or the sill. The same rod is then used to make the joist layout on girders and the opposite wall. When the joists are lapped at the girder, the X (joist location) is marked on the other side of the layout line for the opposite wall. In this case, the spacing between the stringer and first joist is different from the regular spacing, **Figure 8-23.**

As mentioned earlier, some carpenters set back sills and band joists to leave a ledge around the outside of the sill. When they apply the sheathing, it is flush with the foundation wall. In **Figures 8-23B** and **8-23C,** the sill is flush to the edge of the wall, but the band joists are offset.

Joists are doubled where extra loads must be supported. When a partition runs parallel to the joists, a double joist is placed underneath. Doubled joists placed under partitions that are to carry plumbing or heating pipes are usually spaced far enough apart to permit easy access, **Figure 8-24.**

Joists must also be doubled around openings in the floor frame for stairways, chimneys, and fireplaces, **Figure 8-25.** These joists are called *trimmers.* They support the *headers* that carry the tail joists. The *tail joists* are short joists that run from the band joist to the header of the opening. The carpenter must become thoroughly familiar with the plans at each floor level so adequate support can be provided.

Select straight lumber for the band joist and lay out the standard spacing along its entire length, **Figure 8-26.** Add a line to mark the position for any doubled joists and trimmer joists that are required along openings. Where regular joists will become tail joists, change the X mark to a T as shown.

Trimmer: The beam or floor joist a header is framed into. Adds strength to the side of the opening.

Headers: A floor framing member that supports the weight around floor openings.

Tail joist: A short joist spanning from the band joist to the header of a floor opening.

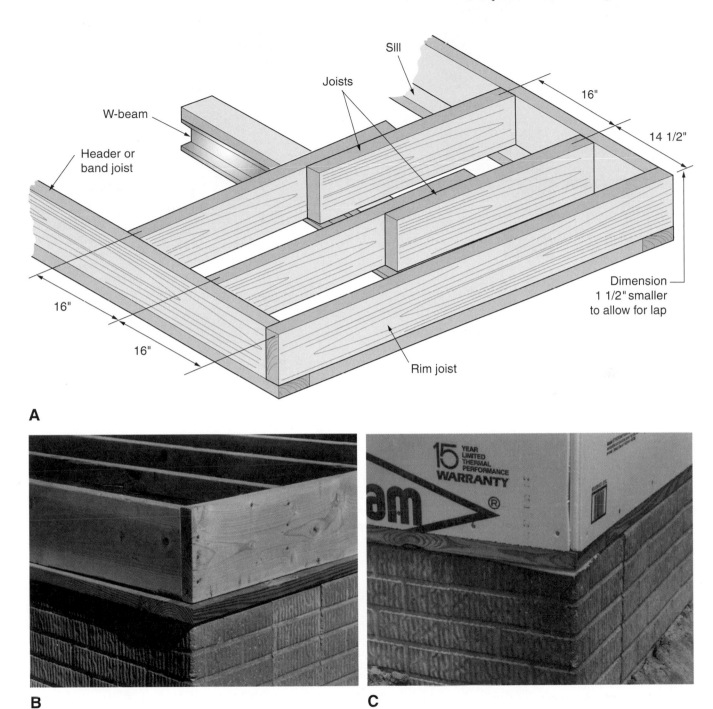

Figure 8-23. A—When joists lap over the center girder, spacing between the first joist and the rim joist is different. B—Some carpenters set back the stringer and headers, leaving a ledge on the sill. The ledge is the same width as the thickness of the sheathing. C—The sheathing is flush with the foundation wall.

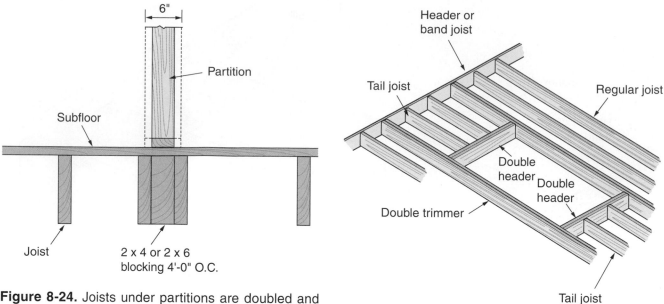

Figure 8-24. Joists under partitions are doubled and spaced to allow access for heating or plumbing runs. If the wall must hold a plumbing stack (vent to roof), the wall is framed with 2 × 6s.

Figure 8-25. Framing members are doubled around floor openings.

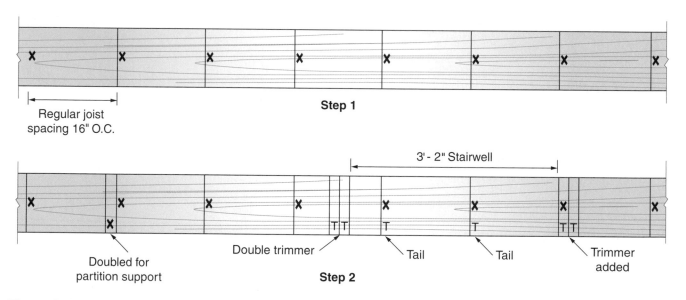

Figure 8-26. Laying out a band joist with joist positions. In step 1, a rod is used to mark regular spacing. In step 2, double and trimmer joist positions have been added.

8.4.2 Installing Joists

After the band joists are laid out, toenail them to the sill. Position all full-length joists with the crown turned up. Tightly hold the end against the header and along the layout line so the sides of the joist are plumb. Butt the joists against the header and fasten them with 16d nails.

Working Knowledge

The Uniform Building Code specifies that in standard framing nails should not be spaced closer than one-half of their length, nor closer to the edge of a framing member than one-fourth of their length.

Next, fasten the joists along the opposite wall. If the joists butt at the girder, they should be connected with a scarf or metal fastener. If they lap, they can be nailed together using 10d nails. Also use 10d nails to toenail the joists to the girder.

To improve the accuracy of the floor frame, some carpenters first create an assembly by nailing floor joists to the header (band) joists. Then, they carefully align the header joists with the sill or a chalk line on the sill or foundation. Finally, they toenail the assembly to the sill.

Nail doubled joists together using 12d or 16d nails spaced about 1′ along the top and bottom edges. First, check that the crowns of the joists are at the same height. If not, toenail through the higher one to bring them to the same level. Next, drive several nails straight through to tightly pull the two surfaces together. Clinch the protruding ends. Finish the nailing pattern, driving the nails at a slight angle.

Some carpenters lay a bead of caulk along the joint formed by the header and the sill to keep out air and dust. A sill sealer serves the same purpose.

8.4.3 Framing Openings

Place 1″ boards or sheets of plywood across the joists to provide a temporary working deck for installing header and tail joists for the opening. First, nail in the trimmer joists. A *trimmer* is a full-length joist or a stud that reinforces a rough opening. Sometimes, a regular joist is located where it can serve as the first trimmer. **Figure 8-27** is a plan view of the finished assembly.

The length of the headers can be determined from the layout on the main band joist. Cut headers and tail joists to length. Make the cuts square and true. Considerable strength will be lost in the finished assembly if the members do not tightly fit together. Lay out the position of the tail joists on the headers by transferring the marks made on the main header in the initial layout.

Working Knowledge

Be accurate in laying out and cutting floor framing members. The strength of the assembly depends on all of the parts tightly fitting together.

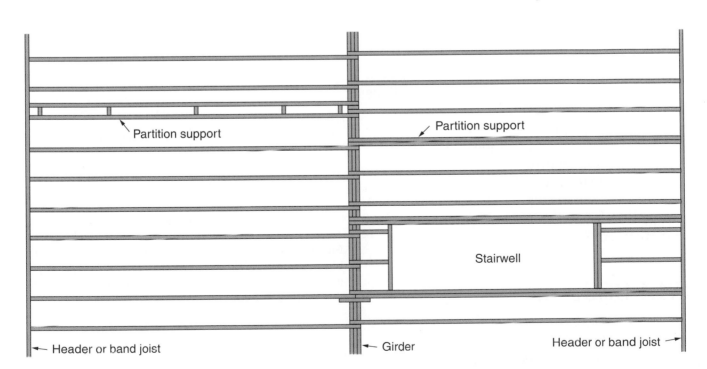

Figure 8-27. This is a partial plan view of a floor framed for a stair opening and partition supports.

Partition support

Partition support

Stairwell

Header or band joist

Girder

Header or band joist

When the assembly of tail joists and first headers is small, they are sometimes nailed together and then set in place. Usually, however, the headers are installed and then the tail joists are attached. One of the tail joists can be temporarily nailed to the trimmer on each side of the opening to accurately locate the header and hold it while it is being nailed.

Figure 8-28 illustrates the procedure for fastening tail joists, headers, and trimmers. After the first header and tail joists are in position between the first trimmers, nail the second or double header in place. Be sure to nail through the first trimmer into the second header. Use three 16d nails at each end. Finally, nail the second trimmer to the first trimmer.

A good nailing pattern for the entire assembly is shown in **Figure 8-29.** To allow more room for nailing, do not install joists adjacent to trimmers until the trimmers are doubled and nailed. This nailing pattern will support a concentrated load of 300 lb. at any point on the floor. It will also hold a uniformly distributed load of 50 lb. per sq. ft. with any spacing and span of tail joists ordinarily used in residential construction. This is true only if the long dimension of the floor opening is parallel to the joist. If the long way of the opening is at a right angle to the joists, excessive loading may be carried to the junction of headers with trimmers. Anticipated loads should be checked and more nails or additional supports should be provided at these junctions when needed.

Metal joist hangers are often used to assemble headers, trimmers, and tail joists, **Figure 8-30.** They are manufactured from 18 gage, zinc-coated sheet steel. The National Forest Products Association recommends the use of joist hangers or ledger strips to support tail joists that are over 12′ long.

8.4.4 Bridging

Studies have shown that *bridging* may be eliminated if these conditions are satisfied:
- Joists are properly secured at the ends.
- Subflooring is adequate and carefully nailed.

However, many local building codes include bridging requirements and general standards

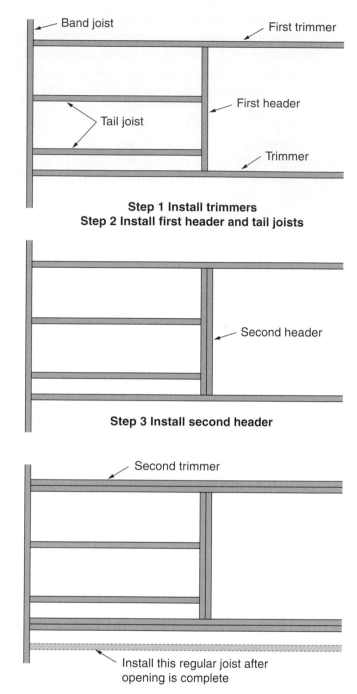

Step 1 Install trimmers
Step 2 Install first header and tail joists

Step 3 Install second header

Step 4 Install second trimmers

Figure 8-28. These steps can be followed in assembling frames for floor openings. In step 4, waiting to install the regular joist provides clearance for nailing the second trimmer.

Crown: A slight warping. Also called a *crook.*

Bridging: Wood or metal pieces fitted in pairs from the bottom of one floor joist to the top of adjacent joist. Used to distribute the floor load.

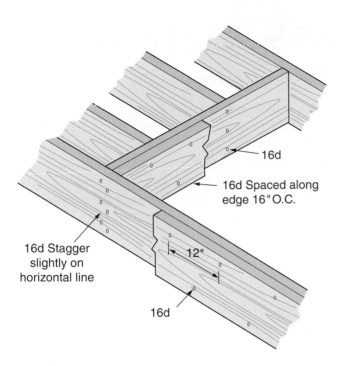

Figure 8-29. Nailing pattern for attaching floor opening members.

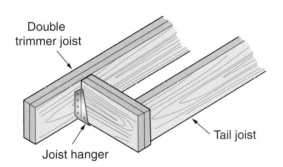

Figure 8-30. Joist hangers are often used for assembling floor framing members.

suggest that bridging be installed at intervals of no more than 8′.

Regular bridging, sometimes called *herringbone* or *cross bridging*, is composed of pieces of 1×3 or 2×2 lumber diagonally set between the joists to form an X. Its purpose is twofold:

- To keep the joists in a vertical position.
- To transfer the load from one joist to the next.

Subflooring: A covering of panels nailed to the joists over which a finish floor is laid.

Figure 8-31 shows how a carpenter's framing square can be used to lay out a pattern for bridging. Pieces can quickly be cut using a radial arm saw. As an alternative, a jig can be set up to use a portable electric saw or a handsaw.

As the name implies, *solid bridging* consists of solid pieces of 2″ lumber installed between the joists. This type of bridging is easier to cut and install when there are odd-sized spaces in a run of regular cross bridging. Solid bridging, also called *blocking*, is often installed above a supporting beam where its chief purpose is to keep the joist vertical. However, it also adds rigidity to the floor.

Several types of prefabricated steel bridging are available. The steel bridging can be installed very quickly. The type shown in **Figure 8-32** is manufactured from sheet steel. A V-shaped cross section makes it rigid. No nails are required and it is driven into place with a regular hammer.

After the bridging is installed, the floor frame should be carefully checked to see that nailing patterns have been completed in all members. After this is done, the frame is ready to receive the subflooring. *Subflooring* is a covering of boards or panels, nailed to the joists, over which a finish floor is laid.

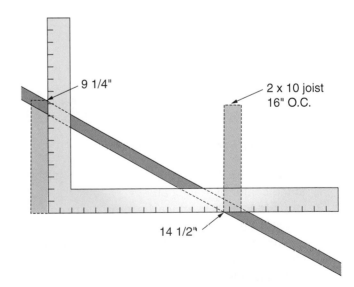

Figure 8-31. A carpenter's framing square can be used to lay out bridging. The line for the lower cut can be found by shifting the tongue of the square to the 14 1/2″ mark on the stock.

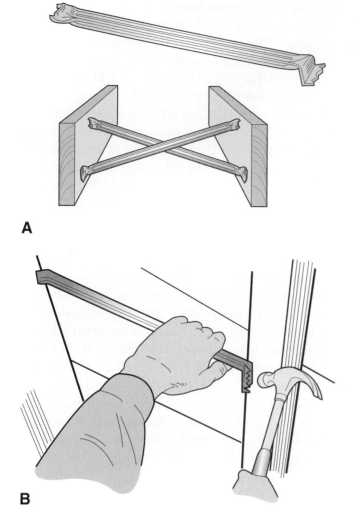

A

B

Figure 8-32. Using steel bridging. A—One type of steel bridging. B—Installation method. (Timber Engineering Co.)

PROCEDURE

Installing wood bridging

1. Snap a chalk line across the tops of the joists at the center of their span.
2. Use two 8d nails to attach the top of the bridging to each side of every joist. Alternate the positioning of the bridging, first on one side of the chalk line and then on the other.
3. After the subflooring is complete or before the under surfaces of the floor are enclosed, the lower ends of the bridging are nailed to the joist.

8.5 Special Framing Problems

A building's design may include a section of floor that overhangs a lower floor or basement level. When the floor joists run at right angles to the walls, use longer joists, **Figure 8-33.** If the run of joists is parallel to the supporting wall, extend *cantilevered* joists inward two to three times the length of the overhang, **Figure 8-34.** The exact spacing between joists and length of the members depends on the weight of the outside wall.

Entrance halls, bathrooms, and other areas are often finished with tile or stone that is installed on a concrete base. To provide room for this base, the floor frame must be lowered. When the area is not large, using doubled joists of a smaller dimension is acceptable, **Figure 8-35.** Reducing the distance between joists provides additional support. When the area is large, add steel or wood girders and posts.

Bathrooms must support unusually heavy loads—heavy fixtures and often the additional weight of a tile floor. The fixed dead load imposed by a tile floor averages around 30 lb. per sq. ft.

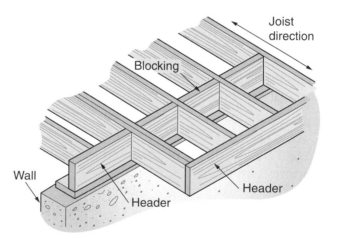

Figure 8-33. Framing a cantilevered section of a floor frame when the joists run perpendicular to the supporting wall. Blocking holds the joists vertical, adds rigidity, and closes up the space to provide a firestop.

Cantilevered: Extending horizontally beyond a supporting surface.

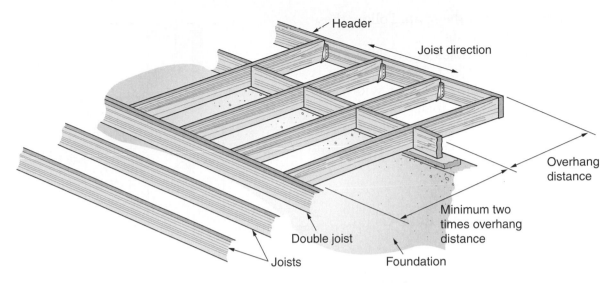

Figure 8-34. When cantilevered joists run parallel to the supporting wall, extend the overhanging joists inward two to three times the length of the overhang. Always check local codes for exact requirements.

The load from bathroom fixtures adds from 10 to 20 lb. per sq. ft., for a total of 40 to 50 lb. dead load. In addition, it is frequently necessary to cut joists to bring in water service and waste drains. Take special precautions, therefore, in framing bathroom floors to provide adequate support.

8.5.1 Cutting Openings in Floor Joists

Before cutting joists to install plumbing, it is useful to know how stress affects floor joists. This knowledge will help you determine where to make holes and cut notches. When the top of a joist is in compression and the bottom in tension, there is a point at which the stresses change from one to the other. At this point, there is neither tension nor compression. In the usual rectangular joist, this point is assumed to be midway between the top and bottom. Variations in the quality of lumber and other conditions may shift the point slightly. Still, this assumption is accurate enough. Since there is neither compression nor tension at the center, a hole has little effect on the joist's strength, provided it is not larger than one-fourth of the total depth of the joist. See **Figure 8-36.**

Weight produces the greatest bend if it is at the center of the span. Therefore, a hole is more likely to reduce the strength of a joist or beam

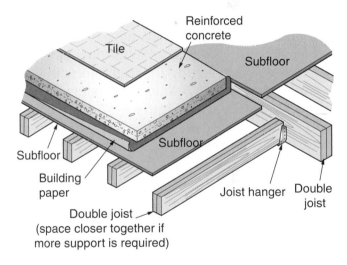

Figure 8-35. Smaller, doubled joists are used when a concrete base is needed for tile or stone surfaces.

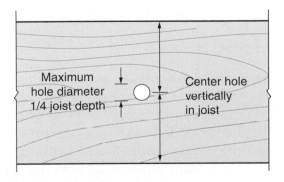

Figure 8-36. Holes for plumbing pipes should be vertically centered in the joist and have a diameter no more than 1/4 of the depth of the joist.

if it is near the center of the span. Follow these precautions in cutting joists:

- When possible, cut holes at or close to the vertical middle of a joist. If the opening is limited to 1/4 of the total joist width, the reduction in strength is insignificant.

- Wood I-beams require special attention when cutting openings for plumbing or electrical. These are discussed later.

8.5.2 Low-Profile Floor Frames

Some home buyers prefer a house with a low silhouette. Standard wood-floor construction, whether over a basement or a crawl space, requires adequate distance between the framing members and the ground. This requirement places the first floor level well above the finished grade.

Various framing systems have been devised to bring the floor level closer to the outside grade. These involve designing the foundation in such a way that the floor frame is surrounded and protected by the wall, **Figure 8-37.** In construction of this type, take special precautions to assure that the joists have an adequate bearing surface. Also, make allowance for shrinkage of the wood members.

Foundations for low-profile floor framing may be constructed of wood, poured concrete, or masonry units such as concrete block. Whichever type of building material is used, install proper barriers against moisture and heat or cold.

8.6 Open-Web Floor Trusses

Open-web floor trusses are widely used in new construction. They are designed with the aid of computers and factory-built to specifications for their intended use. These designs ensure that loading requirements are met through the use of a minimum amount of material. Engineered jigs are used in the assembly to build-in the proper camber (bend) in each unit. Open-web floor trusses offer several advantages:

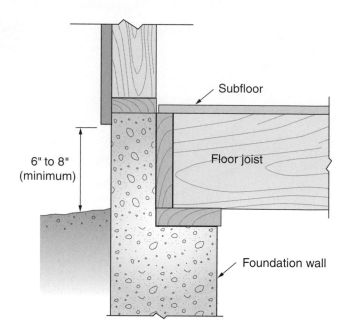

Figure 8-37. An offset in the foundation wall reduces the distance between the floor level and the finished grade.

- Since they are manufactured in a variety of lengths and depths, they are suited to many different loading conditions.

- The trusses provide long, clear spans with a minimum of depth (14″ and 16″ are most common).

- The open webs make them lighter and easier to handle.

- Installing plumbing pipes, heating ducts, and electrical systems through the web requires no cutting of material. Cutting is time-consuming and often weakens traditional lumber joists.

- Sound transmission is reduced.

Most trusses are fabricated with wooden *chords* (top and bottom members of a truss) connected by galvanized steel webs. The webs have metal teeth that are pressed into the sides of the chords. They also have a reinforcing rib that withstands both tension and compression forces. See **Figure 8-38.**

Chord: The top or bottom member of a floor or roof truss. Also called a *flange.*

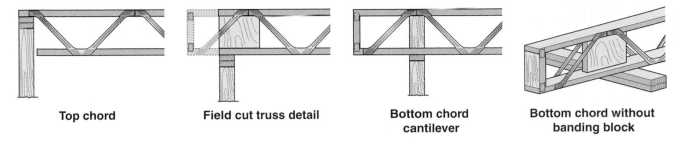

Top chord **Field cut truss detail** **Bottom chord cantilever** **Bottom chord without banding block**

Figure 8-38. Truss construction details. Chords are made of lumber. Webbing is galvanized steel. Trusses provide a wide nailing surface because the chord is laid flat. (TrusWal Systems Corp.)

8.7 Solid-Web Trusses

Solid-web trusses are generally called *wood I-beams.* The chords are made from Douglas fir in solid lumber or laminated veneer. The *web* is the material between the chords. It is made of 3/8″ plywood or 7/16″ nonveneered panels, usually oriented strand board.

The web is glued into grooves cut in the chords, **Figure 8-39.** No nails are used in the manufacture. These manufactured joists are not prone to shrinking or warping. This also reduces the occurrence of squeaking floors caused by drying, shrinking lumber. Special techniques must be used to fasten wood I-beams to other frame components. Use of metal connectors may be required, **Figure 8-40.**

Figure 8-39. Wood I-beams are constructed with solid wood chords and plywood or OSB webs. Wood I-beams must be placed upright during storage and transport to avoid damage.

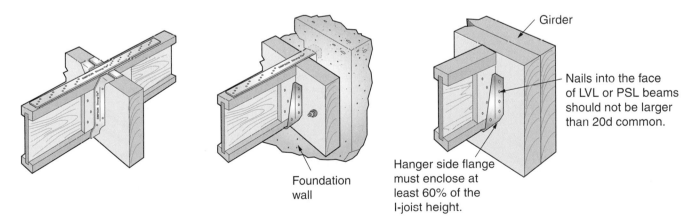

Figure 8-40. Strapping and hangers are used to fasten wood I-beams when butted to other frame parts. Note the special blocking on the I-beam joists. (©Simpson Strong-Tie Company, Inc.)

Wood I-beams: Beams consisting of flanges of structural composite lumber. Also called *solid-web trusses.*

Web: Material between the chords in trusses.

8.7.1 Nailing Solid-Web Trusses

The special construction techniques detailed in **Figure 8-41** are typical. Special nailing requirements might include:

- Nail joists at bearing points with two 8d nails, one on either side, no closer than 1 1/2" from the end to reduce splitting. If needed, 10d or 12d box nails can be substituted.
- Nail 1 3/4" or thinner rim joist to the wood I-beam using two 8d, one each at top and bottom flange.

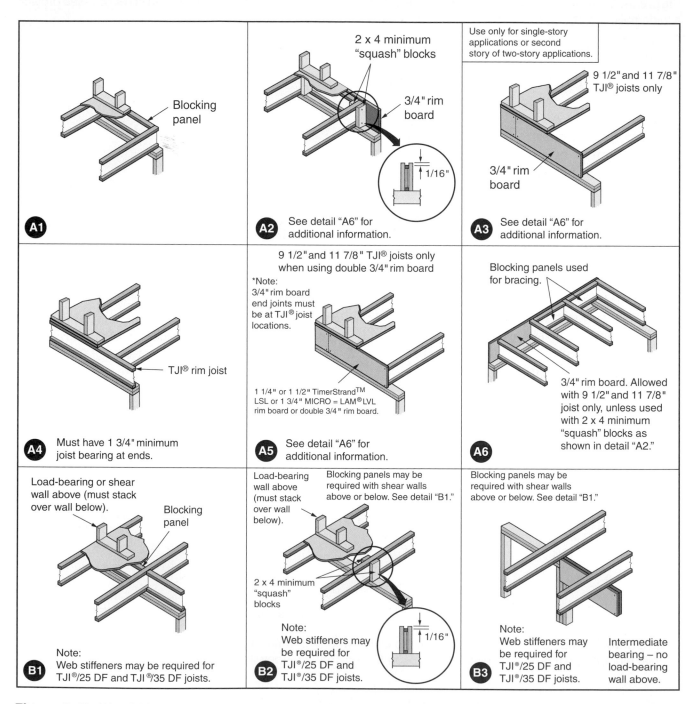

Figure 8-41. Wood I-joists are light, strong, and, if correctly installed, provide a rigid, quiet floor. (TrusJoist MacMillan)

- Attach 2 × 4 or wider "squash" plates to the top and bottom flanges to help support bearing walls. See details A2 and B2 in **Figure 8-41.**

Safety Note

Use great care in working with wood I-beams. They are unstable until properly braced. Failing to observe the following precautions and practices could cause accidents and damage:

- Never allow workers to walk on joists until braced.

- Do not stack building materials on unsheathed joists. Rather, place them only over beams or walls.

- Install and nail all blocking, hangers, rim boards and rim joists at the end supports of joists.

- To temporarily brace joists, provide a rigid structure such as a braced wall or permanent deck (sheathing) nailed to the first 4' of the joists.

- Attach 1 × 4 struts from the braced area across the tops of the rest of the joists. Unless this is done, sideways buckling or a rollover is likely.

- Subflooring must be completely attached to each joist before additional loading of the floor.

- When joists are cantilevered, struts must be attached to both the top and bottom flanges.

- Keep flanges straight with a tolerance of no more than 1/2" of true alignment.

8.7.2 Cutting Holes in Solid-Web Trusses

Special rules apply for cutting through wood I-beams to run pipes, ducts, and electrical conduit or cable.

Rough flooring: The subfloor.

- Leave 1/8" of the web on top and bottom of any hole. Do not cut flanges.

- A 1 1/2" hole can be made anywhere in the web.

- If more than one hole is to be cut in the web, the length of the uncut web section between the holes must be twice the length of the longest dimension of the largest adjacent hole. Holes may be vertically located anywhere in the web.

- If the span is simple (5' minimum), only one maximum size round hole may be cut in a uniformly loaded joist meeting the requirements of the manufacturer (as listed in its guide). The hole must be at the center of the span.

- Holes through a cantilever can be no more than 1/2" in diameter.

- Provide at least 1 1/2" between a hole and a bearing surface.

8.8 Subfloors

The laying of the subfloor or *rough flooring* is the final step in completing the floor frame. Panel materials are used for this purpose. Plywood and nonveneer panels are the materials of choice. These products are discussed in Chapter 1. The subfloor serves three purposes:

- Adds rigidity to the structure.

- Provides a base for finish flooring material.

- Furnishes a work surface where the carpenter can lay out and construct additional framing.

8.8.1 Plywood

In most construction, plywood is used for subflooring. It provides a smooth, even base and acts as a horizontal diaphragm that adds strength to the building. Plywood can be rapidly installed and usually ensures a squeak-free floor.

Although 1/2" plywood over joists spaced 16" O.C. meets the minimum FHA requirements, many builders use 5/8″ plywood. The long dimension of the sheet should run perpendicular to the joists. Joints should be staggered in successive courses, **Figure 8-42.** For 5/8″ or 3/4″ plywood, use 8d nails spaced 6″ along edges and 10″ along intermediate members.

Combined subfloor-underlayment systems make use of a special plywood panel with tongue and groove edges. This single layer provides adequate structural qualities and a satisfactory base for direct application of carpet, tile, and other floor finishes. Subfloor-underlayment panels are available for joists or beam spacing of 16″, 20″, 24″, or 48″. Maximum support spacing is stamped on each panel. A 3/4″ thickness is used for 24″ O.C. spacing. Be sure to follow the instructions supplied by the manufacturer.

8.8.2 Other Sheet Materials

Other sheet materials are also approved for use as subflooring. These include composite board, oriented strand board, and structural particleboard, **Figure 8-43.** These products have been rated by The Engineered Wood Association and meet all standards for subflooring. The specifications for application are the same as for plywood. For additional information on these materials, refer to Chapter 1.

8.8.3 Glued Floor System

In a *glued floor system*, the subfloor panels are glued and nailed to the joists. Structural tests have shown that stiffness is increased by about

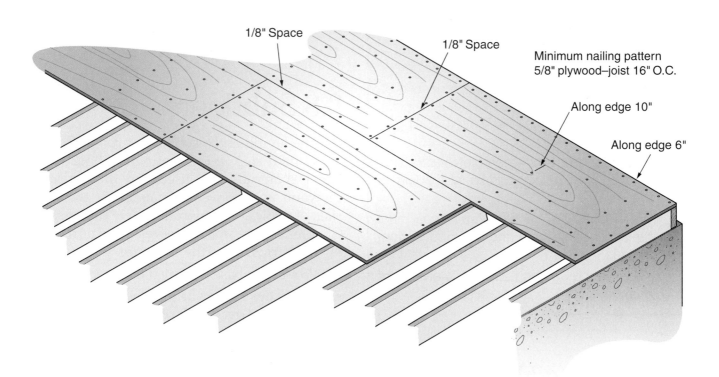

Figure 8-42. Properly installing plywood subfloor. Note the nailing pattern and how joints are staggered for added strength.

Glued floor system: The subfloor panels are glued and nailed to the joists.

Figure 8-43. Using an autofeed screw gun to install oriented strand board subflooring.

Figure 8-44. Applying construction adhesive for a glued floor system. A 1/4″ wide bead of adhesive applied to the top of the joist is sufficient. Two beads are applied where panel ends butt.

25% with 2 × 8 joists and 5/8″ plywood. In addition, this system ensures squeak-free construction, eliminates nail popping, and reduces labor costs.

Before each panel is placed, a 1/4″ bead of construction adhesive (glue) is applied to the joists, as shown in **Figure 8-44.** Spread only enough adhesive to lay one or two panels. Two beads of adhesive are applied on joists where panel ends butt together. All nailing must be completed before the adhesive sets.

When laying tongue-and-groove panels, apply adhesive along the groove. Use a 1/8″ bead so that excessive squeeze-out is avoided. The bead can be either continuous or spaced. Drive sheets into the groove of previously laid subflooring using a maul and a protective piece of scrap lumber. See **Figure 8-45.** Leave a 1/8″ space at all end and edge joints. Immediately fasten the subfloor with screws or nails. This operation is speeded up with the power tool shown in **Figure 8-46.**

Working Knowledge

Material specifications and application procedures for subflooring are provided in a booklet published by APA-The Engineered Wood Association.

Figure 8-45. Use a maul and protective strip of wood to close up the tongue-and-groove joints in subflooring. Leave a 1/8″ gap all around.

Figure 8-46. Fasten the glued subflooring with mechanical fasteners before the adhesive sets. A power screw gun avoids stooping and speeds the operation.

8.9 Estimating Materials

To estimate the number and size of floor joists on a job, first use a scale on the plan and determine the lengths that are needed. Be sure to allow sufficient length for full bearing on girders and partitions. Average residential structures require several different lengths. Multiply the length of the wall that carries the joists by 3/4 for spacing 16″ O.C. and add one more. Use 3/5 for 20″ O.C. and 1/2 for 24″ O.C. spacing. Also, add extra pieces for doubled joists under partitions and trimmer joists and headers at openings. For example:

No. of joists = length of wall × 3/4 + 1 + extras

Some carpenters estimate one joist for every foot of wall on which the joists rest (16″ O.C. spacing). The overrun allows for extra pieces needed for doubles, trimmers, and headers. Band joists are usually separately figured and then added to the above figures. Your estimate should include the cross-sectional size and the number of pieces of each length. For example, a complete estimate might be:

Required floor joists:
- 40 pcs.—2 × 10 × 16′-0″
- 36 pcs.—2 × 10 × 14′-0″

Required band joists:
- 4 pcs.—2 × 10 × 16′-0″
- 2 pcs.—2 × 10 × 12′-0″

Procedures for estimating the subflooring vary, depending on the type of material used. Usually the area is figured by multiplying the overall length and width and then subtracting major areas that are not to be covered. These include breaks in the wall line and openings for stairs, fireplaces, and other items. This will give the net area and the basic amount of material needed. Waste and other extras must be added to this.

When using sheet materials, there is practically no waste. The net area is divided by 32 (number of sq. ft. in a 4 × 8 sheet) and rounded out to the next whole number. This is the required number of pieces of plywood. Be sure to specify the type of sheet material and its span rating.

Working Knowledge

An alternative and more complete method of specifying plywood, as recommended by the APA, is provided in Chapter 1

Summary

There are three basic types of framing used in residential construction: platform (western) framing, balloon framing, and post-and-beam framing. Platform framing is by far the most common. Balloon framing is no longer used for new construction. In platform framing, the first floor is constructed on top of the foundation wall and used as a work area for constructing the remaining wall, floor, and roof framing components. Floor framing consists of joints that rest on the foundation walls or girders. The joist ends rest on a sill that is fastened to the foundation wall and are held in position by a joist header fastened around the perimeter. Bridging is installed between joints to hold joists vertical and help transfer loads from one joist to the next. Special framing must be used for floor areas cantilevered beyond the foundation walls. Open truss floor joists are increasingly being used. They are light, strong, and provide space for electrical, piping, and HVAC installations. Subfloors of plywood or other sheet materials are installed on top of the joists, making the whole floor structure more rigid. Proper fastening of the subfloor to the joists is important for strength, smoothness, and durability.

Test Your Knowledge

Answer the following questions on a separate piece of paper. Do not write in this book.

1. The type of framing used in most residential construction is _____.
2. The studs of a balloon-type frame run continuously from the _____ to the rafter plate.
3. Girders may be solid timbers, _____ lumber, _____ lumber, or _____ beams.
4. The sill may be flush with the outside of the foundation or spaced back a distance equal to the _____.
5. In residential construction, the deflection of first floor joists under normal live loads should not exceed _____ of the span.
6. A member of the floor frame that runs from the band joist to a header for an opening is called a _____ joist.

7. Joists must be _____ around openings in the floor for stairways.
8. Cantilevered joists should extend inward at least _____ times the distance that they overhang the supporting wall.
9. Open-web floor trusses are typically constructed with wooden _____ connected by a web made from galvanized metal.
10. When sheet material is used for the subflooring, the short dimension of the panel should run _____ to the joist.

Curricular Connections

Mathematics. Working from a set of architectural plans for a single-story house, develop a list of materials required to frame the floor. Select the type of subflooring (if not specified) and estimate the amount of material needed. Check material prices at several local sources, such as home improvement centers or lumberyards, and prepare a cost estimate. If the prices significantly differ among the sources, prepare a high estimate and a low estimate.

Language Arts. From the local building code in your area, find the requirements for floor framing. Prepare a list of the requirements along with sketches to clarify complicated written descriptions. Make an oral report to your class and pass around your sketches.

Outside Assignments

1. Obtain a set of architectural plans for a house with a conventional basement. Study the methods of construction specified in the section and detailed drawings. Then, prepare a first-floor framing plan. Start by tracing the foundation walls and supports shown in the basement or foundation plan and then add all joists, headers, and other framing members. Your drawing should be similar to the partial drawing shown in **Figure 8-27.**

2. Research the advantages and disadvantages of using different types of floor joists for residential construction. Compare solid dimension lumber joists, open web trusses, and solid web (wood I-beam) trusses. Consider ease of installation, material cost, strength-for-size, and other factors such as ease of installing plumbing and heating lines.

Wall and Ceiling Framing

Learning Objectives

After studying this chapter, you will be able to:

- Identify the main parts of a wall frame.
- Explain methods of forming the outside corners and partition intersections of wall frames.
- Show how rough openings are handled in wall construction.
- Explain plate and stud layout.
- Describe the construction and erection of wall sections and partitions.
- List the materials commonly used for sheathing.
- Demonstrate the process of ceiling frame construction.
- Estimate materials required for wall frames, ceiling frames, and sheathing.

Technical Vocabulary

Building paper	Master stud pattern
Ceiling frame	Metal strap bracing
Ceiling joists	Nailer
Cripple studs	Partitions
Flush beam	Racking
Headers	Rough opening
Housewrap	Sheathing
Jack studs	Soffit
Lintels	Sole plates
Story pole	System
Strongback	Top plates
Studs	Trimmers

Wall framing is assembling the vertical and horizontal members that support the outside and inside walls of a structure. This frame also supports upper floors, ceilings, and the roof. It serves as a nailing base for inside and outside wall-covering materials. Inside walls are called *partitions.*

The term *system* commonly means "methods and materials of construction." It is used in connection with floors, ceilings, roofs, and walls. Included are the design of the framework, the surface-covering materials, and the methods for applying those materials. For example, a floor system includes:

- Details of the sill construction.
- Size and spacing of joists.
- Type of subflooring.
- Application requirements.

Partition: A wall that subdivides space within any story of a building.

System: Methods and materials of construction. The term is used in connection with floors, ceilings, roofs, and walls.

9.1 Parts of the Wall Frame

Anyone planning to become a carpenter needs to know the correct names for the parts of a house frame. The wall-framing members used in conventional platform construction include *sole plates, top plates, studs,* headers or *lintels,* and *sheathing.* Studs are the vertical members of the wall frame. Except where interrupted by openings for windows and doors, they run full length from sole plate to top plate. Short studs are known as *cripple studs* or *jack studs,* depending on the area of the country. Full-length studs become cripple studs when they end due to an opening. See **Figures 9-1** and **9-2.** Note that extra studs are used at the corners, the sides of the rough openings for doors and windows, and where an interior wall meets an outside wall.

Trimmer studs are shortened studs that stiffen the sides of rough openings. They bear the direct weight of a header. Normally, carpenters install doors and windows so the tops are at the same height. This practice not only improves appearance, but allows all trimmers to be cut the same length. All of the trimmers can be cut at one time, speeding up production.

Studs and plates are made from 2 × 4 or 2 × 6 lumber. Headers usually require heavier material.

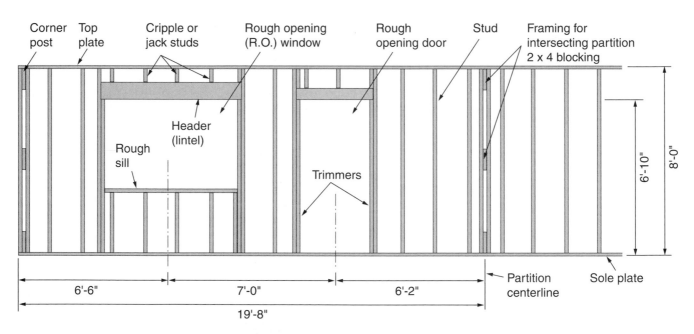

Figure 9-1. This is a drawing of a typical wall frame section with its members named. An extra stud and 2 × 4 blocking at an intersecting partition provide a nailing surface for an inside corner.

Sole plate: The lowest horizontal strip in wall and partition framing. The sole plate is supported by a wood subfloor, concrete slab, or other closed surface.

Top plates: Wall-framing members used in conventional platform construction.

Stud: One of a series of vertical structural members in walls and partitions.

Lintel: A horizontal structural member that supports the load over an opening, such as a door or window. Also called a *header.*

Sheathing: Boards or prefabricated panels that are attached to the exterior studding or rafters of a structure.

Cripple stud: A stud used above or below a wall opening. Extends from the header to the top plate or from the sole plate to the rough sill. Also called *jack stud.*

Trimmer: The stud a header is framed into. It adds strength to the side of the opening.

Figure 9-2. This typical framing example shows rough openings for windows and a corner post. The bracing is temporary. Cripple studs and trimmers are in place. Note the double top plate supporting open truss joists for a second-level floor.

A

B

Figure 9-3. Two alternatives to angle bracing. A—Plywood is only placed at the corners. B—The entire house is sheathed in nonveneered panels.

Bracing made of 1 × 4 stock or steel strips must be let into the wall frame at corners. This prevents the frame from *racking* (shifting out-of-square). An alternative method of bracing uses plywood or other sheet material at the corners, **Figure 9-3.**

In one-story structures, studs are sometimes placed 24″ O.C. (on center). However, 16″ O.C. spacing is more common. This spacing has evolved from years of established practice. It is based more on accommodating the wall-covering materials than on the actual calculation of imposed loads.

Stud intervals remain modular, regardless of interruptions by openings or intersecting walls. This is done so that modular sheet sizes can be installed with minimum cutting or waste. In conventional inch-foot measurement, this module is based on 48″ or multiples of 4″.

The height of walls in residential construction varies from one region to another. For example, 10′ high walls are not uncommon in warm climates. In colder climates, 8′ high walls may be more common to help minimize heating costs.

Wall-framing lumber must be strong and straight with good nail-holding power. Warped lumber is not acceptable, especially if the interior finish is drywall. Stud and No. 3 grades

are approved and used throughout the country. Species such as Douglas fir, larch, hemlock, yellow pine, and spruce are satisfactory. Where straightness is especially critical, such as kitchen walls where cabinets will be installed, engineered lumber studs can be used. See Chapter 1, **Building Materials**, for additional information.

9.1.1 Corners

Any of several methods can be used to form the outside corners of the wall frame. In platform construction, the wall frame is usually assembled in sections on the rough floor and then tilted up into place. Corners are formed when a sidewall and end wall are joined.

Usually, an additional stud is included in the sidewall frame. It should be spaced inward the width of a 2 × 4 and stiffened with three or

---

Racking: Twisting of a frame.

four blocks. Once the end wall is erected, the complete corner is formed, **Figure 9-4A.** An alternate method is to turn the extra stud 90°, as shown in **Figure 9-4B.**

Select only straight studs for corners. Assemble the wall with 10d nails spaced 12" apart. Stagger the nails from one edge to the other. Attach the filler blocks with nails as well.

Some carpenters prefer to separately build the corners for platform construction. These assemblies are set in place, carefully plumbed, and braced before the wall sections are raised. This makes it easier to plumb and straighten

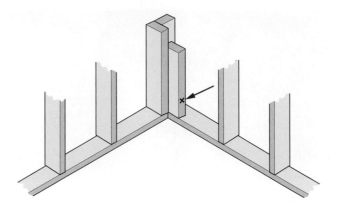

Figure 9-5. Typical corner construction in 2 × 6 framing. The inside corner is formed with a 2 × 4 (arrow).

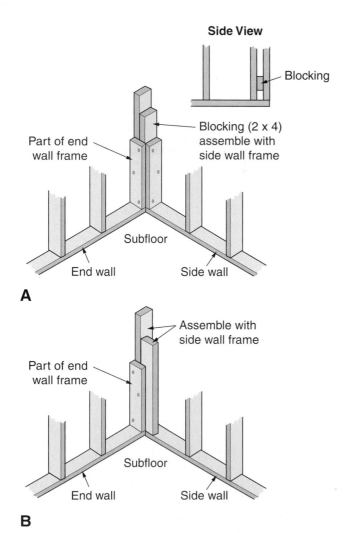

Figure 9-4. Placement of studs to form corners in platform construction. A—Corner built from three full studs and blocking. B—Corner built with three full studs and no blocking.

the wall sections. However, sheathing cannot be applied while the frame is still on the deck.

In climates that require the house to be well insulated, 2 × 6 studs are commonly used for exterior walls. This allows thicker insulation to be installed. **Figure 9-5** shows typical corner construction for 2 × 6 framing.

9.1.2 Partition Intersections

Partitions should be solidly fastened to the outside walls. This requires extra framing on the outside wall. The framing must also provide a nailing surface on inside corners for wall covering, such as drywall. Several methods can be used to accomplish these purposes:

- Install extra studs in the outside wall. Attach the partition to them.
- Insert blocking and nailers between the regular studs.
- Use blocking between the regular studs and attach nailers or backup clips to support inside wall coverings at all inside corners.

A **nailer** is lumber, such as 1 × 6 or 2 × 4, added as a backing at inside corners. **Figure 9-6** shows various methods.

Nailer: Lumber, such as 1 × 6 or 2 × 4, added as a backing for attaching other members or covering.

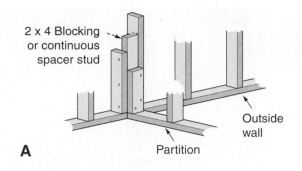

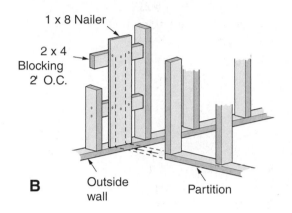

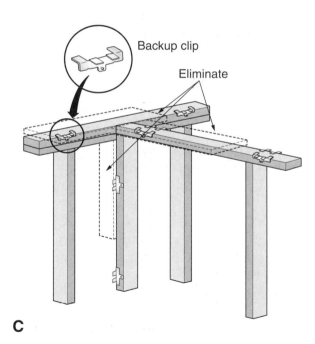

Figure 9-6. Framing details where partitions intersect. A—Using extra studs. B—Blocking installed between studs. C—Backup clips are sometimes used and take the place of some framing studs.

Rough opening: An opening formed by framing members to receive and support a window or door.

9.1.3 Rough Openings

Study the house plans to learn the size and location of the *rough openings.* A rough opening is often referred to as R.O. on drawings. Plan views have dimension lines. Usually, the measurement is taken from corners or intersecting partitions to the centerlines of the openings. Heights of rough openings are given in elevation and section views. Sizes are shown on the plan view or listed in a table called a *door and window schedule.*

Headers support the weight of the building across door and window openings. To make a header, cut and nail together two or three framing members. Insert 1/2″ plywood spacers between the pieces to make the header the same thickness as the wall, **Figure 9-7.** Use 12d or 16d nails and

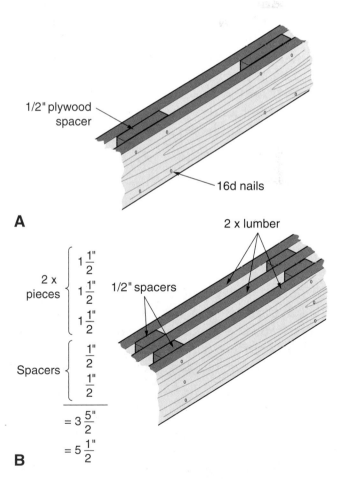

Figure 9-7. Normally, headers must be built up of several members. Place plywood spacers 16″ to 24″ apart on centers. A—A header for a 2 × 4 stud wall requires two members with 1/2″ spacers to equal the width of the studs. B—A 2 × 6 stud wall requires three members with 1/2″ plywood spacers between each member.

Material on edge	Supporting one floor, ceiling, roof (feet-inches)	Supporting only ceiling, and roof (feet-inches)
2 x 4	3-0	3-6
2 x 6	5-0	6-0
2 x 8	7-0	8-0
2 x 10	8-0	10-0
2 x 12	9-0	12-0

Figure 9-8. Recommended maximum header spans. Be sure to check local codes, which may vary.

stagger them 16″ on center. Fasten the header in the rough opening using 16d nails driven through the studs into the ends of the header.

The header length is equal to the rough opening plus the width of two trimmers (3″). The width of the lumber to be used in the header depends on the span of the opening. Local building codes or the building plan may include requirements for headers. The table in **Figure 9-8** gives the size of headers normally required for various rough opening widths for two different load conditions.

Headers are also required across openings in load-bearing partitions. If loads are very heavy or spans unusually wide, a *flush beam*, or strong-back, may be used. In such cases, hangers can be used to attach ceiling-for-floor joists to the flush beam. Flush beams are discussed later in this chapter. **Figure 9-9** shows a truss joist being attached to a doubled 2 × 12 header flush beam.

In platform construction, extra studs are included around rough openings, as shown in the standard assembly in **Figure 9-10**. The studs and trimmers support the header and provide a nailing surface for window and door casing. Rough sills may be doubled to provide a nailing base for trim.

Figure 9-9. When open-web joists are used for a second story floor frame, a doubled 2 × 12 header may be used to support a cripple joist. (Kasten-Weiler Construction)

is to increase the header size to completely fill the space to the plate. Most builders follow this practice and extend it to include all openings, regardless of the span. The cost of labor required to cut and fit the cripple studs is usually greater than the cost of the larger headers. A disadvantage of such construction is extra shrinkage. This may cause cracks above doors and windows unless special precautions are taken when applying the interior wall finish.

9.1.4 Alternate Header Construction

In large window openings, the size of the header may reduce the length of the upper cripple studs to a point where they cannot be easily assembled. In this case, the cripple studs should be replaced with flat blocking. Another solution

9.2 Plate Layout

Use only straight 2 × 4 stock for plates. Select two pieces of equal length and lay them side by side along the location of the outside wall. The length is determined by what can be easily lifted off of the floor and into a vertical position after it is assembled. Remember that the weight may include all of the framing for rough

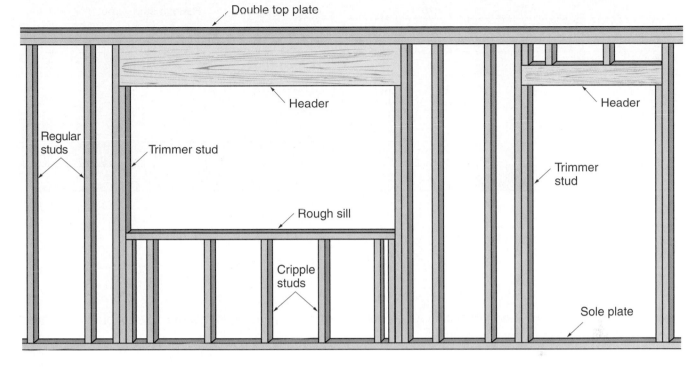

Figure 9-10. Framing door and window openings. Notice how the wide header above the window eliminates cripple studs.

openings, bracing, and sheathing. If wall jacks or a forklift are available for lifting, sections can be made larger. Where they must be lifted by hand, attach sheathing after the wall is up. Always locate joints over a stud. The centerlines of rough openings are marked first.

PROCEDURE

Laying out plates for the first outside wall

1. On the rough floor, mark the width of the plates on all sides from the outside in. Snap a chalk line as a guide for the inside edge of the wall frame. Place the bottom and top plates along the main sidewalls. Align the ends with the floor frame and then mark the regular stud spacing along both plates, **Figure 9-11.** Some carpenters tack the pieces to the floor with several nails so they will not move while the layout is being made.

2. Study the architectural plans and lay out the centerline for the rough opening of each door and window.

3. Measure and mark off one-half of the width of the opening on each side of the centerline.

4. Mark the plate for trimmer studs outside of these points. On each side of the trimmer stud, include marks for a full length stud. Identify the positions with the letter T for trimmer studs and X for full-length studs.

5. Mark all of the stud spaces located between the trimmers with the letter C. This designates them as cripple studs.

6. Lay out the centerlines where intersecting partitions butt. Add full length studs at these points, if required by the method of construction.

7. When blocking is used between regular studs, the centerline is needed as a guide for positioning the backing strip.

8. Carefully plan the layout of wall corners so they correctly fit together when the wall sections are erected.

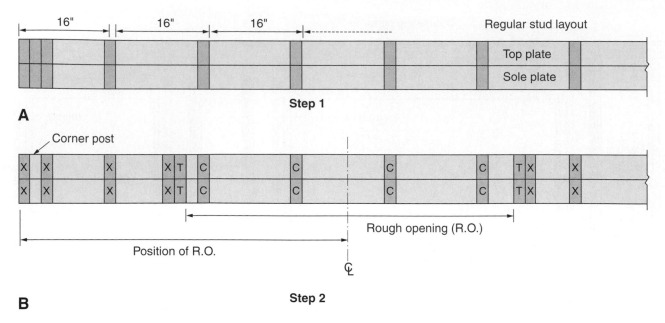

A

Step 1

B

Step 2

Figure 9-11. Layout of sole and top plates. A—Regular stud spacing has been marked. B—Layout is converted for a window opening. Stud type is marked.

Working Knowledge

Carefully check over your rough opening layouts for errors. Do the math before cutting and framing.

9.2.1 Laying Out the Second Exterior Stud Wall

Laying out the second exterior wall follows the same procedure, with one exception. If sheet material is used for rough siding, then the location of the first stud from the corner post must allow for the edge of the panel to be flush with the outside edge of the siding. Assuming the siding is 3/4″ thick and the studs are 16″ on center, then lay out the first stud 15 1/4″ from the end of the plate, as shown in **Figure 9-12.** Then, when the first siding panel is installed, the first edge will be even with the outside of the siding and the second edge is centered on a stud.

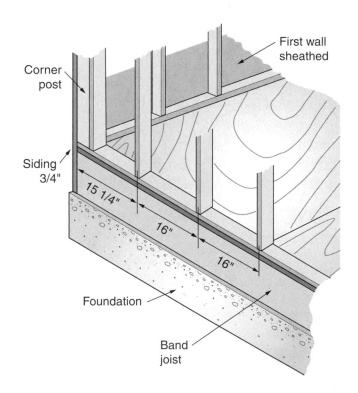

Figure 9-12. When laying out the second stud wall, the carpenter must adjust the spacing so the corner panel of the siding covers the edge of the siding from the adjoining wall.

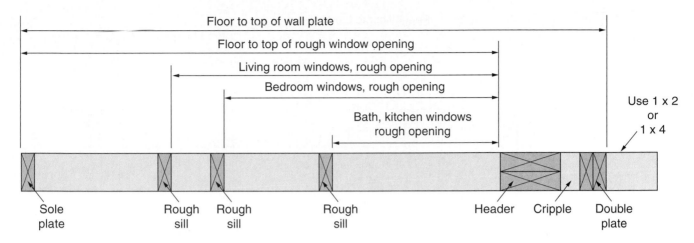

Figure 9-13. A story pole is a handy guide that marks height of every horizontal member and length of every vertical member of the wall frame. It usually extends only one story, but may include more.

9.2.2 Story Pole

A *story pole* is a long measuring stick made up by the carpenter on the job. It represents the actual wall frame with markings made at the proper height for every horizontal member of the wall frame—sole plate, rough windowsill, headers, and top plates. See **Figure 9-13.** Its use saves time that would be spent checking the drawings for these dimensions.

Since the pole must be light and easy for the carpenter to handle, it is usually a strip of 1×2 or 1×4 lumber. It must be long enough to reach from the rough floor to the underside of the ceiling or floor joists above.

When marking a story pole, transfer all of the heights for horizontal members from the drawings to the pole at one time. Measurements must be accurate and lines must be square across the pole. With this guide, there is no need to consult the plans time and again to find the lengths of studs, trimmers, and cripple studs. A story pole is particularly useful in split-level construction, multistory buildings, or where stub walls are needed.

A *master stud pattern* is like a story pole, but has information for only a portion of the wall, **Figure 9-14.** Do the layout on either a straight 1×2 or 1×4. First, lay out the distance from the rough floor to the ceiling. This dimension can be directly taken from the main story pole or from the plans. Mark the position of the sole plate and double top plate. Now, lay out the header. When several header sizes are used, they can be marked on top of each other. Lay out the height of the rough openings, measuring down from the bottom of the header. Then, draw in the rough sill. The length of the various studs (regular, trimmer, and cripple) can now be taken directly from this full-size layout.

When the header height of the doors is different from that of the windows, mark the height on the other side of the pattern. This keeps the two heights separate. In multistory or split-level structures, a master stud layout is probably required for each level.

Story pole: A strip of wood used to lay out and transfer measurements for door and window openings, siding and shingle courses, and stairways

Master stud pattern: Similar to a story pole, but has information marked on it for only a portion of the wall.

9.3 Wall Sections

Wall sections are assembled on their edges on the rough flooring and fastened together. Some carpenters first fasten the sole plate to the rough flooring and then nail together the studs

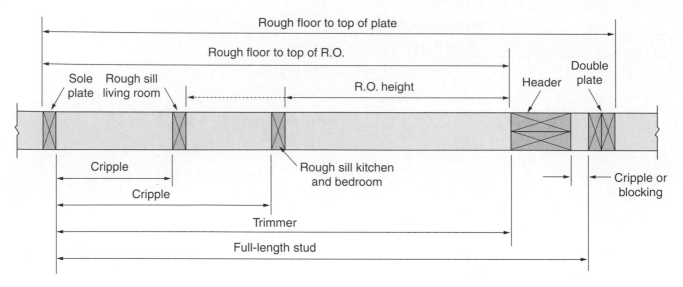

Figure 9-14. A master stud pattern is like a story pole, but covers only a portion of the wall. Several of them may be used on a single job.

and the top plate. After raising the wall, the studs are toenailed to the sole plate. The usual practice, however, is to fasten both top and sole plates to the studs while on the floor, then raise the wall into position.

Wall sheathing is often applied to the frame before it is raised. Make certain that the frame-work is square before starting the application. Check diagonal measurements across the corners; they must be equal. To keep the frame square while the sheathing is being applied, fasten a diagonal brace across one corner. If you prefer, temporarily nail two edges of the frame to the floor.

PROCEDURE

Constructing a wall section

1. Working from the master stud layout, cut the various stud lengths. It is seldom necessary to cut standard full-length studs. These are usually precision end trimmed (P.E.T.) at the mill and delivered to the construction site ready to assemble.
2. Cut the headers and rough sills. Take their lengths directly from the plate layout. Assemble the headers.
3. Move the top plate away from the sole plate about a stud length. Turn both plates on edge with the layout marks inward. Place a full-length stud, crown up, at each position marked on the top plate and sole plate. See **Figure 9-15.**
4. Nail the top plate and the sole plate to the full-length studs using two 16d nails.

5. Set the trimmer studs in place on the sole plate and nail them to the full-length studs.
6. Place the header so it is tight against the ends of the trimmers. Nail through the full-length stud into the header using 16d nails, **Figure 9-16.**
7. The upper cripples, if used, can be installed after the header is installed.
8. For window openings, transfer marks for the cripple studs from the sole plate to the rough sill and assemble the cripples with 16d nails, **Figure 9-17.** If the wall section is erected before installing the cripples, toenail the lower ends of the cripple studs to the sole plate. Install the rough windowsill.
9. Add studs or blocking at positions where parti-tions will intersect.
10. Install any wall bracing that may be required for special installations. Remember, the inside of the wall is face down.

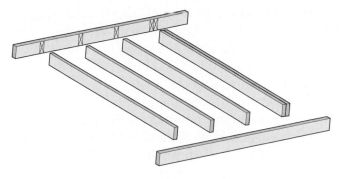

Figure 9-15. Assemble the full-length studs and plates as shown. Place a stud at every position marked on the plate. Turn crowns upward and nail through the plates into the ends of the studs.

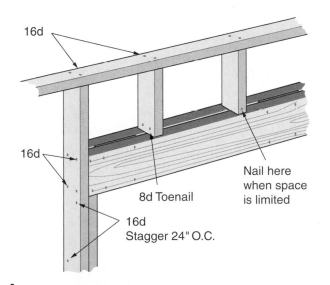

16d

16d

8d Toenail

Nail here when space is limited

16d
Stagger 24″ O.C.

A

B

Figure 9-16. A—How to assemble and fasten headers, cripple studs, and trimmers. B—Assembling a wall section with the header (arrow) filling the space to the underside of the top plate. This method is more common.

9.4 Erecting Wall Sections

Most one-story wall sections can be raised by hand, **Figure 9-18.** Larger sections or structures require the use of a crane or other equipment. When raising sections to which sheathing has not been applied, it is good practice to install temporary diagonal bracing if regular bracing is not included, as shown in **Figure 9-19.**

A

B

Figure 9-17. Installing cripple studs below a window opening. A—Cripple studs can be installed before the wall section is erected. B—If the wall sections are first erected, toenail the cripples to the sole plate.

A

B

Figure 9-18. Tilting up wall sections. A—Raising a section of wall constructed from 2 × 6s. B—The front wall of garage is being raised by student carpenters. Note the angle and cross bracing to keep the center from racking. (North Bennett Street School, Boston)

running to the subfloor at about a 45° angle. Next, make final adjustments in the position of the sole plate. Be sure it is straight. Then, nail it to the floor frame using 20d nails driven through the subfloor and into the joists.

Loosen the braces one at a time and plumb the corners and midpoint along the wall. On one-story construction, a carpenter's level is generally used for this. A plumb line can also be used. Braces temporarily attached to square a wall section can be removed when the permanent braces and sheathing are installed, **Figure 9-20.**

After one section of the wall is in place, proceed to other sections. No particular sequence needs to be followed. Most carpenters prefer to erect main sidewalls first and then tie in end walls and smaller projections. The exact procedures must be determined for each individual project. Design and construction methods help determine how to proceed.

Working Knowledge

When plumbing a wall with a carpenter's level, hold the level so you can look straight in at the bubble. If the wall framing member or surface is warped, you should hold the level against a long straightedge that has a spacer lug at the top and bottom.

Some carpenters attach temporary blocking to the edge of the floor frame before raising wall sections. The blocking keeps the wall section from sliding off the platform as it is raised.

Safety Note

Before raising a section, be sure it is in the correct location. Have bracing at hand and ready to be attached. If the section is large, have extra help available. Make sure each worker knows what to do.

Immediately after the wall section is up, secure it with braces attached near the top and

Extra bracing

Figure 9-19. The wall section is in place, nailed to the platform, plumbed, squared, and temporarily braced.

Figure 9-20. Temporary bracing used to square up wall sections can be removed when permanent braces and sheathing are in place.

9.5 Partitions

When the outside wall frame is complete, partitions are built and erected. At this stage, it is important to enclose the structure and make the roof watertight. Only bearing partitions are installed at this time. Bearing partitions are those that support the ceiling and/or roof. Roof and floor trusses require no other support than the outside walls, **Figure 9-21.** Erection of

Figure 9-21. The open-web trusses used here are supported by the outside walls alone. Partitions do not need to be installed until the roof is on and the building is closed in. (Kasten-Weiler Construction)

nonbearing partitions can be put off until after the building is enclosed.

Establish the centerlines of the partitions from a study of the plans. Mark the centerlines on the floor with a chalk line. Lay out the plates, studs, and headers. Cut the headers, then assemble and erect the partitions in the same way as outside walls. Erect long partitions first, then cross partitions. Finally build and install short partitions that form closets, wardrobes, and alcoves.

The corners and intersections are constructed as described for outside walls. Refer to **Figure 9-6.** The size and amount of blocking, however, can be reduced. The chief concern is to provide nailing surfaces at inside and outside corners for wall covering material.

9.5.1 Nonbearing Partitions

Nonbearing partitions do not require headers above doorways and other openings. Many rough openings can be framed with single pieces of 2×4 lumber since there is virtually no load on them. Trimmers are usually added for rigidity. They also provide added framework for attaching casing and trim. Door openings in partitions and outside walls are framed with the sole plate included at the bottom of the opening. After the framework is erected, the sole plate in the door opening is cut out with a handsaw. Rough door openings are generally made 2 1/2" wider than the finished door size.

The partitions between noisy areas and quiet areas are often soundproofed. This may require a special method of framing. See Chapter 15 for information on insulation.

Small alcoves, wardrobes, and partitions in closets are often framed with 2×2 material or by turning 2×4 stock sideways, thus saving space. This is satisfactory when the thinner constructions are short and intersect regular walls. Snap a chalk line across the rough floor to mark the position of these partitions, **Figure 9-22.**

During wall and partition framing, add various important details. Add basic provisions for recessed and surface-hung cabinets, tissue-roll holders, and similar items to the framing at this stage. Architectural plans usually provide information concerning their size and location.

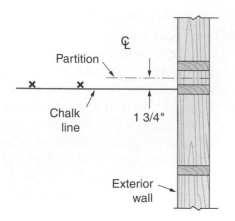

Figure 9-22. Snap a chalk line on the floor to mark the position of partitions.

Openings for the installation of heating ducts are easily cut and framed at this time, **Figure 9-23.** Bathtubs and wall-mounted toilets require extra support, **Figure 9-24.** Wall backing for drapery brackets, towel bars, shower curtains, and wall-mounted plumbing valves should also be added, **Figure 9-25.** Plumbing fixture rough-in drawings is helpful in locating the backing, **Figure 9-26.** For most items, 1″ thick backing material provides adequate support.

Small items that are not critical to the structure can often increase efficiency and quality of work during the finishing stages. For example, corner blocks make it possible to nail baseboards some distance back from their end, **Figure 9-27.** This eliminates the possibility of splitting the wood.

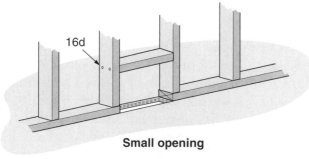

Small opening

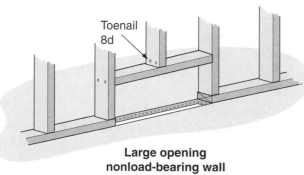

Large opening nonload-bearing wall

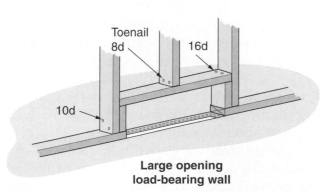

Large opening load-bearing wall

Figure 9-23. Special framing for heat ducts. Cripple studs should be added for large openings in load-bearing walls.

Working Knowledge

The carpenter must continually study and plan the sequence of the job so that neither the weather nor work of other tradespeople will cause slowdowns or bottlenecks.

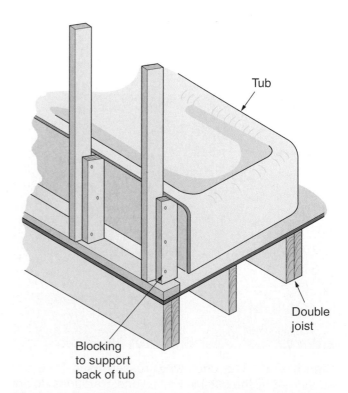

Figure 9-24. Extra joists and blocking are needed to support tubs.

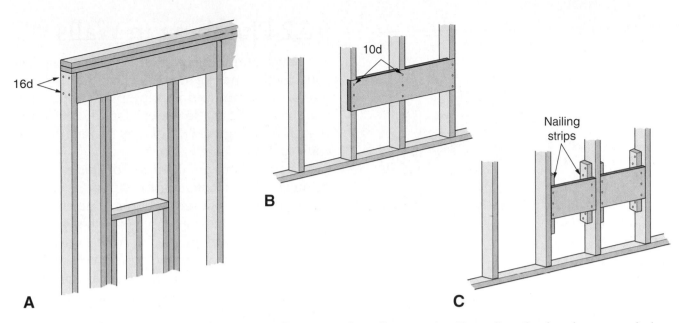

Figure 9-25. Backing for mounting various fixtures and appliances. A—Extending the header over windows provides a base for attaching drapery rod brackets. B—Backing let into studs. Never cut back more than 25% of the stud width on bearing walls or partitions. C—Backing attached to nailing strips.

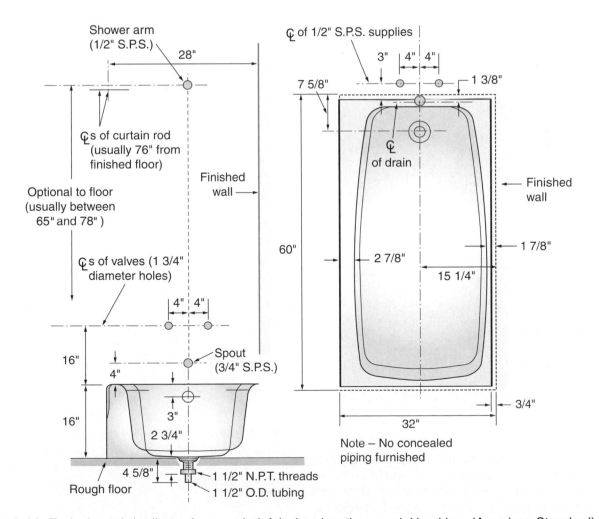

Figure 9-26. Typical rough-in dimensions are helpful when locating special backing. (American Standard)

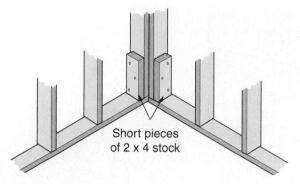

Figure 9-27. Adding blocking in corners provides a better nailing surface for attaching baseboards.

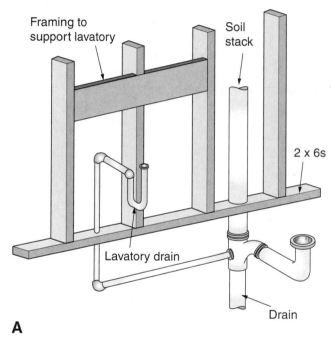

A

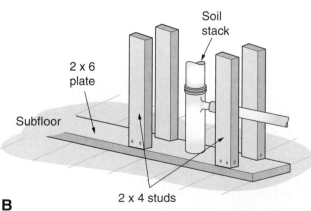

B

Figure 9-28. A partition containing plumbing pipes may need different construction to provide room for the plumbing. A—A wall constructed of 2 × 6 studs and plate. B—The use of 2 × 4 studs on a 2 × 6 plate eliminates the need to notch or bore studs for lateral runs.

9.5.2 Plumbing in Walls

Where plumbing is run through walls, special construction may be required. Depending on the size of the drain and venting pipes, a partition may have to be made wider. Usually a 6″ frame is sufficient. **Figure 9-28** shows two methods of construction.

Lateral (horizontal) runs of pipe require drilling holes or notching the studs. Sometimes, a wooden block is used to bridge a notch cut for plumbing. A metal strap can be attached to bridge the notch and strengthen the stud, **Figure 9-29**. The strap also protects the pipe from accidental damage should a nail be driven into the stud at this point. Similar protection should be provided for electrical wiring in walls.

9.5.3 Bracing

Exterior walls usually need some type of bracing to resist lateral (sideways) loads. Some types of material, such as plywood sheathing, provide sufficient rigidity. In these cases, additional bracing can be eliminated. Always check the exact requirements of the local building code.

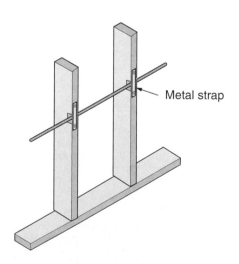

Figure 9-29. A metal strap can be used to reinforce studs that are notched or bored for plumbing. This method is also used to protect electrical wiring from fastener damage.

Metal strap bracing is widely used. It is made of 18 or 20 gage galvanized steel and is 2″ wide. One type includes a 3/8″ center rib,

Figure 9-30. The let-in corner bracing shown in **Figure 9-31** is no longer permitted by building codes in areas subject to earthquakes. In such

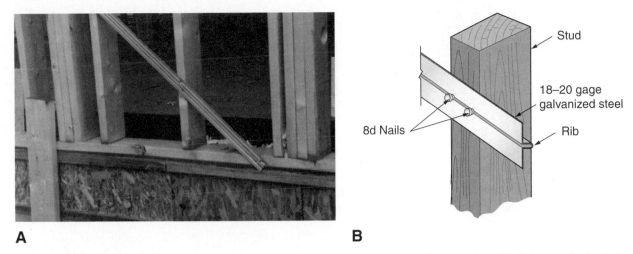

A   **B**

Figure 9-30. Metal strap bracing. A—The brace extends from the top plate to the sill. It is easily installed by cutting an angled saw kerf into the studs, top plate, and sole plate. B—Detail of metal strap bracing connection. The rib is designed to slip into a saw kerf at each member of the frame and is fastened with nails

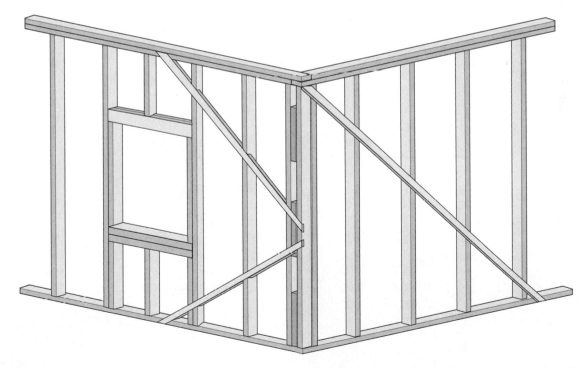

Figure 9-31. Bracing a wall with 1 × 4 lumber. Note how the bracing is applied when an opening interferes with the diagonal run.

Metal strap bracing: Used to help exterior walls resist lateral (sideway) loads. Typically made of 2″ wide, 18 or 20 gage galvanized steel.

areas, plywood shear panels must be used for greater strength and rigidity.

To install the ribbed strap bracing, snap a chalk line across the erected wall frame. Set a portable circular saw to a shallow cut. Make multiple passes to form a groove for the rib. Drive two 8d nails through the rib into each framing member. For 2 × 6 studs, use one 16d nail.

9.5.4 Double Plate

To add support under ceiling joists and rafters, double the top plate. See **Figure 9-32.** This also serves to further tie the wall frame together. Select long, straight lumber. Install the double plate with 16d nails. Place two nails near the ends of each piece. The others are staggered 16″ apart. Locate nails near or over the studs so subcontractors will not hit them with their drills while installing mechanical systems. Joints in the upper top plate should be located at least 4′ from those in the lower top plate. At corners and intersections, lap the joints as shown in **Figure 9-33.**

A structure that may be shaken by earthquakes or blown by winds of hurricane force requires a stronger frame. Metal ties are available that secure the building to its foundation. These ties strengthen the joints between walls and foundations and also between rafters and sidewalls. The ties help to reduce or eliminate the damage from natural disasters. See **Figure 9-34.** Many of these ties are shown in Chapter 1.

9.5.5 Straightening Walls

At this point, corner posts are plumb and top plates are doubled. Now, the wall needs to be straightened and braced between the corners. Obtain three blocks of the same thickness. Tack one at each end of the wall. Tightly stretch a line between the two blocks. Using the third block

A

B

Figure 9-32. A—Double top plates stiffen the wall. B—A 2 × 4 has been added as a nailer for the ceiling covering.

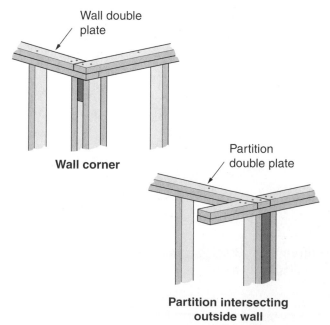

Figure 9-33. Double plates are lap-jointed for strength wherever they intersect.

A

B

Figure 9-34. Metal ties can be used to strengthen building frames against damaging winds and earthquakes. A—Metal fasteners secure studs to the foundation. B—Metal fasteners hold studs to the top plates. Fasteners can also be used to secure rafters to both top plates and studs.

as a gauge, align the wall and brace it at intervals. Be careful that the gauge comes up to, but does not touch, the line. If it touches the line, the line is no longer straight!

This procedure is a two-person job as described. It is difficult for one person to hold the wall while securing the brace. Some carpenters use a manufactured brace that can be adjusted

---

Soffit: The underside of the members of a building, such as staircases, cornices, beams, and arches. Relatively minor in size as compared with ceilings. Also called *drop ceiling* and *furred-down ceiling.*

to align the wall. Others fashion spring braces on the job.

9.6 Tri-Level and Split-Level Framing

Tri-level and split-level housing presents special challenges in wall framing. Generally, a platform type of construction is used. However, the floor joists for upper levels may be carried on ribbons let into the studs. The plans prescribe the type of construction. They should also include calculations of distances between floor levels. When working with split-level designs, a good carpenter prepares accurate story poles that show full-size layouts of these vertical distances for all levels of the building.

9.7 Special Framing

Framing carpenters are sometimes asked to build structures with special features. It helps if these features are carefully engineered and detailed by an architect. In such a case, construction details are included in the plans. When they are not, the carpenter must develop the plan.

Most bay windows are prefabricated by the window manufacturer. However, a carpenter may be asked to build a bay window. When asked to do so, the carpenter must visualize the details of construction, lay out and construct the floor frame to carry the project, and build the wall and roof frame. See **Figure 9-35.** Framing of the floor extension is best done by extending the floor joists, cantilevering them over the foundation. Double the joists forming each side of the bay to help carry the weight of the structure. A header over the opening in the wall carries the weight of the structure above. It also provides a nailing surface for the window's ceiling joists.

Another special framing project might be a cabinet *soffit* that closes in the space between the ceiling and the tops of cabinets. Soffits are most often found in kitchens, although bathrooms frequently have them, too.

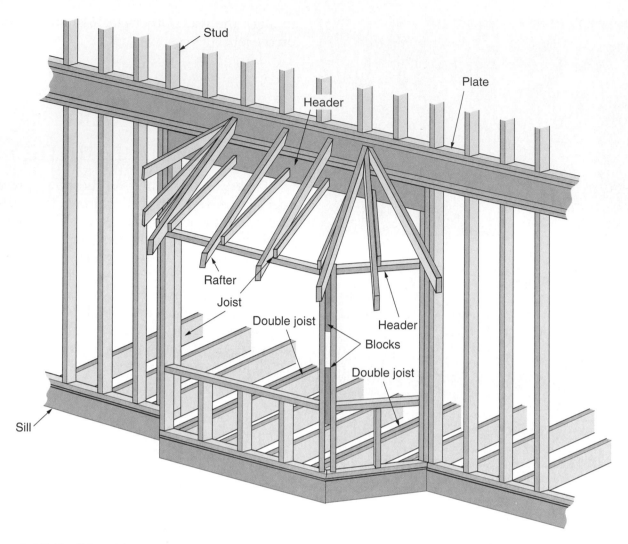

Figure 9-35. Traditional framing for a bay window. The subfloor is not shown to illustrate how joists are extended over the sill. Joists at either side of the extension should be doubled. (National Forest Products Assn.)

Construction of a soffit begins by snapping a chalk line on the studs at the bottom edge of the soffit. This is usually 84″ from the finished floor. Next, nail a 2 × 2 along this line using one 10d nail at each stud. Then, snap a second chalk line snapped along the ceiling at a distance that is at least the depth of the cabinets. Nail another 2 × 2 onto the ceiling along this line. If the ceiling joists run parallel to the soffit, 2 × 4 blocking is used to bridge between two joists and provide support for the 2 × 2. Next, cut a 2 × 4 or 2 × 2 for the lower front edge of the soffit and blocking to frame the front and bottom of the soffit. These pieces are attached with screws or 8d nails. See **Figure 9-36.**

Figure 9-36. A framed soffit before the firestops are installed. Note the 2 × 2 frame against studs and ceiling joists. Special blocking is used at the corner.

9.8 Wall Sheathing

Outside wall sections should be covered with sheathing before the roof framing is started. Sheathing adds rigidity, strength, and some insulating qualities to the wall. Plywood, oriented strand board, fiberboard, and rigid insulating panels are widely used for sheathing. These materials are available in large sheets that can be rapidly attached. Plywood and oriented strand board usually provide enough lateral strength so that diagonal bracing can be eliminated.

Fiberboard sheathing is made largely from wood fibers with weatherproofing ingredients added. It is commonly available in 4′ × 8′ sheets. However, sheets as large as 8′ × 14′ are manufactured. Regular fiberboard sheathing is also available in a 2′ × 8′ size with tongue and groove or shiplap joints. It is applied horizontally.

Regular fiberboard sheathing cannot be used as a nailing base for exterior wall finish materials. However, a special nail-base fiberboard is available. When fiberboard sheathing is used, corner bracing is required. Installing 1/2″ plywood sheets at each corner provides adequate bracing and is an alternative to angle bracing.

Standard thicknesses of fiberboard sheathing are 1/2″ and 25/32″. Use 1 1/2″ roofing nails for the 1/2″ thickness and 1 3/4″ for the 25/32″ thickness. Provide a 1/8″ space between all edge joints when applying fiberboard sheathing.

Gypsum sheathing is usually 1/2″ thick. When using gypsum sheathing, it is necessary to include corner bracing. This type of sheathing is usually installed with 1 3/4″ nails spaced 4″ around the edge and 8″ on intermediate supports.

Either interior or exterior plywood can be used for sheathing, provided it is structural grade. It should be at least 5/16″ thick for studs spaced 16″ on center and 3/8″ thick for studs spaced 24″ on center. A thickness of 1/2″ or more is often used so that the exterior finishing material can be nailed directly to the panel.

Most sheathing can be applied vertically or horizontally. Provide 1/8″ space on all edges. Space edge nails 6″ apart and field nails 12″ apart.

Plain wood boards can also be used for wall sheathing. They should be at least 3/4″ thick and not over 12″ wide. End joints should fall over the centers of the studs. Diagonal bracing is required when board sheathing is installed horizontally. When using tongue and groove boards for sheathing, the joints need not be over a stud. When applied diagonally, no corner braces are needed. Boards should be angled in opposite directions on adjoining walls for greatest strength and stiffness. Two 8d nails are needed on each stud for 6″ or 8″ boards. Three nails should be used for wider boards.

Foamed sheathing is made from polystyrene or polyurethane. Fiberglass sheathing is also used, **Figure 9-37.** The use of foamed and fiberglass sheathing has grown dramatically in recent years because of the high insulating properties of these materials. These panels are not very strong and must be handled with extra care. Corner bracing is required. Always follow the manufacturer's directions for application. Procedures may vary somewhat from one product to another.

9.9 Multistory Floor Framing

Framing the upper floor of a multistory dwelling is similar to framing the first floor, **Figure 9-38.** Joists are placed on top of the double plate along with headers (band joists). They

Figure 9-37. Both fiberglass and rigid polystyrene foam sheathing have high insulating properties. Panels are 1″ thick and 8′ long. (Owens-Corning)

Figure 9-38. In platform framing, the second story of a home is built just like the first story.

Figure 9-40. For wide spans, joists may be supported on a bearing wall, as well as the exterior walls. Note the large header needed for openings in bearing walls.

Header or band joist

A

B

Figure 9-39. Upper story framing. A—The floor frame is fastened to the doubled wall plate of the previous story. B—Joist are toenailed to the wall plate. (Kasten-Weiler Construction)

may be fastened to the plate with steel anchors or toenailed. See **Figure 9-39.** Some joists need the support of a bearing wall, **Figure 9-40.**

9.9.1 Ceiling Framing

A *ceiling frame,* as the name suggests, is the system of support for all components of the ceiling. This frame may be the underside of the floor joists for the next story or an assembly just below the roof. See **Figure 9-41.** The basic construction of an assembly just below the roof is similar to floor framing. The main difference is that lighter joists are used and headers are not included around the outside. When trusses are used to form the roof frame, a ceiling frame is not required. The bottom chords of the truss rafters form the ceiling frame. Refer to **Figure 9-41C.**

Main ceiling framing members are called *ceiling joists.* Their size is determined by the span and spacing. To coordinate with walls and permit the use of a wide range of surface materials, a spacing of 16″ O.C. is commonly used. Size and quality requirements must also be based on the type of ceiling finish (plaster or drywall) and how the attic space will be used, **Figure 9-42.** The architectural plans usually

Ceiling frame: The system of support for all components of the ceiling.

Ceiling joists: Main ceiling framing members.

A

B

C

Figure 9-41. Ceiling framing. A—Ceiling joists for a flat roof. B—Second floor trusses. Ceiling covering is fastened to the underside and subfloor sheathing to the top side. C—The underside of truss rafters form a frame for ceiling covering. (Kasten-Weiler Construction, Southern Forest Products Assn.)

include the specifications. These requirements should be checked with local building codes.

Ceiling joists usually run across the narrow dimension of a structure. However, second-floor joists can also be supported by bearing walls. This makes the span even shorter and may result in ceiling joists perpendicular to each other in the same frame, **Figure 9-43.**

In large rooms, the midpoint of the joists may need to be supported by a beam. This beam can be located below the joists or installed flush with the joists. In a flush installation, the joists may be carried on a ledger, or joist hangers can also be used. Sometimes a beam is installed above the joists in the attic area. It is tied to the joists with metal straps.

At their outer ends, the upper corners of ceiling joists may need to be cut at an angle to match the slope of the roof. To lay out the pattern for this cut, use the framing square as shown in **Figure 9-44.** When the amount of stock to be removed is small, the cuts can be made after the joists and rafters are in place.

When ceiling joists run parallel to the edge of the roof, the outside member is likely to interfere with the roof slope. This often occurs in low-pitched hip roofs. The ceiling frame in this area should be constructed with stub joists running perpendicular to the regular joists, **Figure 9-45.**

Lay out the position of the ceiling joists along the top plate. When a double plate is used, the joists do not need to align with the studs in the wall. The layout, however, should put the joists alongside the roof rafters so that the joists can be nailed to them. Ceiling joists are installed before the rafters unless the assembly is prefabricated. Toenail them to the plate using two 10d nails on each side.

Partitions or walls that run parallel to the joists must be fastened to the ceiling frame. A nailing strip or drywall clip to carry the ceiling material must also be installed. Various size materials can be installed in a number of ways. The chief requirement is that they provide adequate support for interior wall coverings. **Figure 9-46** shows a typical method of making such an installation. **Figure 9-6C** shows backup clip installation for a ceiling.

An access hole, also called a *scuttle*, must be included in the ceiling frame to provide entrance

Size	Spacing	Group A	Group B	Group C	Group D
2 x 4	12"	9'-5"	9'-0"	8'-7"	4'-1"
	16"	8'-7"	8'-2"	7'-9"	3'-6"
2 x 6	12"	14'-4"	13'-8"	13'-0"	9'-1"
	16"	13'-0"	12'-5"	11'-10"	7'-9"
2 x 8	12"	19'-6"	18'-8"	17'-9"	14'-3"
	16"	17'-9"	16'-11"	16'-1"	12'-4"
2 x 10	12"	24'-9"	23'-8"	22'-6"	19'-6"
	16"	22'-6"	21'-6"	20'-5"	16'-10"

Figure 9-42. Spans for ceiling joists are figured for a normal dead load of 10 psf and a live load of 20 psf. This permits the attic to be used for storage. Always check local codes. (National Building Code)

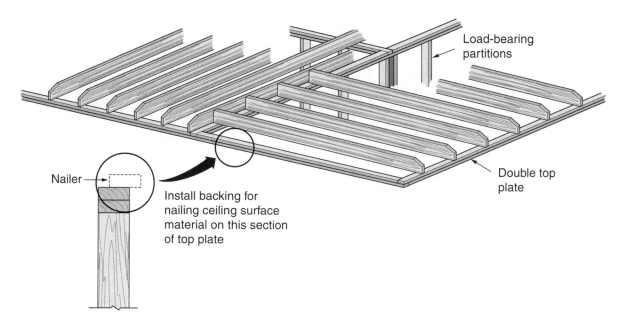

Figure 9-43. A ceiling frame. The joists in the foreground are turned at right angles to reduce the span.

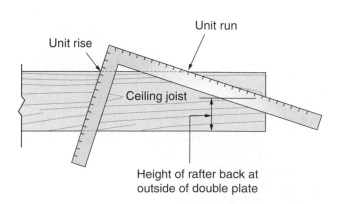

Figure 9-44. A framing square is used to lay out the trim cut on the end of a ceiling joist. Using the rise and run of the building, the cut matches the slope of the rafters.

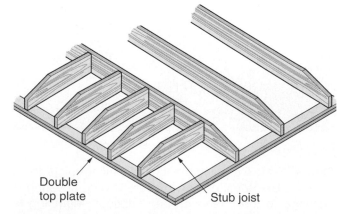

Figure 9-45. Stub ceiling joists butted to a full-length joist. The stub joists are required for a low-pitched hip roof due to the lack of space near the edge.

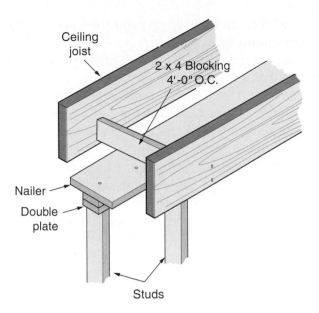

Figure 9-46. Use special blocking to anchor partitions to the ceiling frame when they run parallel to the joists.

to the attic area. Fire regulations and building codes usually list minimum size requirements. The building plans generally indicate the size and show where it should be located. The opening is framed following the procedure used for openings in the floor. If the size of the opening is small (2' to 3' square), doubling of joists and headers is not required.

9.10 Strongbacks

Long spans of ceiling joists may require a *strongback.* This is an L-shaped support constructed of 2-inch lumber. It is attached across the tops of joists to strengthen them and maintain the space between them. It also evens up the bottom edges of the joists so the ceiling is not wavy after the drywall is applied.

To construct a strongback, first mark the proper spacing (16" or 24" O.C.) on a 2 × 4. Position the 2 × 4 across the tops of the ceiling joists

Strongback: An L-shaped wooden support attached to tops of ceiling joists to strengthen them, maintain spacing, and bring them to the same level.

Housewrap: Plastic sheets used to seal exterior walls against air infiltration.

and fasten it with two 16d nails at each joist. Apply pressure against the joists as needed to maintain proper spacing.

Select a straight 2 × 6 or 2 × 8 for the second member. Place it on edge against either side of the 2 × 4 just attached to the joists. Attach one end to the 2 × 4 with a 16d nail. Work across the full-length of the strongback, aligning and nailing it. Stepping on either the 2 × 4 or the member on edge helps align each joist. Nail the vertical member to each joist and to the 2 × 4, **Figure 9-47.**

9.11 Housewrap and Building Paper

Housewrap is a thin, tough, plastic sheet material that is applied to side walls. It prevents movement of air into or out of a building. See **Figure 9-48.** Manufactured under such trade names as Tyvek® and Typar®, it comes in rolls of various widths up to 10'. It is designed to cover cracks at wall joints where air might enter or leave a building. It is efficient at preventing air

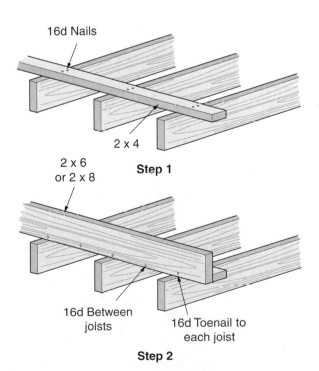

Figure 9-47. Building a strongback. Step 1—Nail a 2 × 4 to the joists. Step 2—Turn a 2 × 6 or 2 × 8 on edge and nail it to the 2 × 4 and the joists.

Figure 9-48. Builders often install housewrap over the sheathing on exterior walls. The wrap is a tough plastic fabric that prevents air infiltration.

infiltration. At the same time, it is waterproof. It prevents water from entering the structure, yet allows interior water vapor to pass through and escape. This property is known as *vapor permeability*.

The typical wrap is made from high-density, spun, polyethylene fibers. These fibers are virtually tearproof. The wrap is stapled to the sheathing before exterior doors and windows are installed. At openings, the wrap is trimmed, leaving enough to wrap around the inside of the rough openings.

Housewrap can also be applied to concrete block walls as an air- or weather-resistant barrier. Usually, block masonry is then faced with stucco or brick. Caulk is used at 6″ intervals as the preferred fastening method. Follow the manufacturer's instructions.

Building paper is a general term for papers, felts, and similar sheet materials that come in rolls. Like housewrap, these materials are designed to stop air from passing through the walls. At the same time, they allow passage of water vapor so that moisture is not trapped inside the wall. Builders prefer a 15-lb. paper that is paraffin-coated or saturated with asphalt.

Building paper should be applied horizontally. Begin installation at the bottom, lapping each layer about 4″. Smoothly apply the paper, avoiding bulges that allow air passage. Lap vertical joints. Place strips 6″ wide around openings for doors and windows. Do not leave building paper exposed to the weather for long periods. Install it just prior to siding application.

PROCEDURE

Installing housewrap

To install housewrap on erected walls:
1. Start from the bottom at a corner. Leave about a foot extra to wrap around the corner. Make sure the roll is perfectly vertical. Otherwise, the wrap will run either uphill or downhill on the wall. The bottom edge should run along the top edge of the foundation.
2. Secure the wrap with broad-headed or washer-headed nails. Start at a corner,
3. Continue to unroll a few feet at a time. Keep the sheet straight with no wrinkles. Check that you are following the foundation line.
4. Secure the wrap every 12″ to 18″ on a stud line.
5. Roll over window and door openings. Cut these openings later using an X pattern. Then, wrap the extra material through the openings and fasten it to the house frame.
6. Overlap every layer at least 3″. Upper layers should overlap lower layers. Continue all around the building. Use a 1 × 4 to push the wrap into inside corners.
7. Secure all laps with a special housewrap tape.

To install housewrap on walls that are not tilted up:
1. Unroll the housewrap over the wall section. Leave enough overhanging the sole plate to cover the band joist or wrap that was installed atop the foundation wall. Leave a flap at each side of the wall section to overlap adjacent wall sections.
2. Before raising the section, roll back the flaps so they will not become pinched in the joints.

(Continued)

Building paper: Paper and felt combination applied to the outside walls of buildings to prevent infiltration of air.

(Continued)

3. With the section raised, loosen the bottom flap and attach it to the band joist or the lower flap at the sill.
4. As other wall sections are raised, release the side flaps and fasten them to adjoining walls.

Where the building has more than one floor, wrap can be installed vertically:

1. Working from the top plate of the top story, staple a 2 × 2 to the end of the roll and lower down the wall to the band joist of the lowest floor.

2. Be sure to allow a flap at the beginning vertical corner. Then, secure the wrap at the band joist. Remove the 2 × 2.
3. Fold the flap around the corner and fasten.
4. Nail 12″ apart at every stud.
5. Overlap each succeeding strip at least 6″. Immediately tape the joints to keep wind from getting under the wrap.

9.12 Estimating Materials

To estimate wall and ceiling framing materials, first determine the total lineal feet. Add together the length of each wall and partition. The plans include the dimensions of outside walls. These can be added together. Partitions, especially those that are short, may not be dimensioned. In this case, use a scale on the drawing. It is a good idea to place a colored pencil check mark on each wall and partition as its length is added to the list.

For plates, multiply the total figure by three (one sole plate plus two top plates). Add about 10% for waste. Order this number of linear feet of lumber in random lengths or convert to the number of pieces of a specific length. Three pieces of 2 × 4 one foot long equals 2 bd. ft. Thus, the total lineal feet of walls and partitions can be quickly converted to board measure if required. For example:

Wall and partition length = 240
Total plate material   = length × 3 + 10%
                             240 × 3 + 10%
                             792 lineal ft.
                             or 57 pcs.,
                             2″ × 4″ × 14′-0″

9.12.1 Estimating Studs

The total length of all walls and partitions is also used to estimate the number of studs required. When studs are spaced 16″ O.C., multiply the total length (in feet) by 3/4 and then add two more studs for each corner, intersection, and opening. Using 240′ for the wall length, assume there are 12 corners, 10 intersections, and 20 openings. Find the number of studs needed, including outside walls and partitions:

$$\text{Total studs} = \text{total length} \times 3/4 + 2 \text{ (corners } + \text{ intersections } + \text{ openings)}$$

$$= \frac{240 \times 3}{4} + 2 \, (12 + 10 + 20)$$

$$= \frac{\overset{60}{\cancel{240}} \times 3}{\cancel{4}} + 2 \, (42)$$

$$= 60 \times 3 + 84 = 264$$

Many carpenters estimate the number of studs by simply counting one stud (spaced 16″ O.C.) for each lineal foot of wall space. About 10% is added for waste. The overrun on spacing provides the extras needed for corners and openings. This method is rapid and fairly accurate. However, it does not estimate enough material for a small house divided into many rooms. On the other hand, too many studs will likely be estimated for a large house with wide windows and open interiors.

Material for headers must be calculated by analyzing the requirements for each opening. Use the rough opening width plus the thickness of the trimmers.

Ceiling joists are estimated by the same method used for floor joists. Since ceilings are relatively free of openings, no extras or waste needs to be included. Because of this, the short method should not be applied to ceiling joists. Use the following formula and include the size of the joists required:

Number of ceiling joists = wall length × 3/4 + 1

9.12.2 Estimating Wall Sheathing

To estimate the amount of wall sheathing, first find the total perimeter of the structure. Multiply this figure by the wall height. When the sheathing extends over the sill construction, measure the height from the top of the foundation. The product is the gross square footage of the wall surface. Next, calculate the area of each major opening (windows and doors) and subtract this total from the original figure. Round the opening sizes downward to the nearest foot.

Net area = perimeter × height – wall openings

To the net area to be sheathed, add allowances for waste and other extras when common boards or shiplap are used. See Chapter 8 for more information. When using sheet materials for sheathing, there is only slight waste. Divide the net area by the square footage per sheet to secure the number of pieces required. For example, if the net area to be sheathed is 1060 sq. ft. and the fiberboard sheets selected are 4′ × 9′, calculate quantity needed as follows:

Net area = 1060
Fiberboard sheet size = 4 × 9 = 36
No. of sheets needed = 1060 ÷ 36 = 29 + (round up to 30)

Working Knowledge

Problems in estimating, although based on simple formulas, usually contain so many variables that good judgment must be applied to their solution. This judgment is acquired through experience.

ON THE JOB

Framing carpenter

Most of the workers employed in residential and light commercial construction are framing carpenters. They are sometimes referred to as *rough carpenters* to distinguish them from the finish carpenters who do trim and specialty work. Framing carpenters primarily work with wood as a structural material. They fabricate the floor structures, wall sections, and roof framing of the building. In recent years, framing carpenters have become increasingly involved with the cutting, fitting, and fastening of light steel structural members (joists and studs), especially in commercial construction.

Approximately 1/3 of all carpenters are self employed. Another 1/3 work for general contractors. The remaining 1/3 are spread among specialty contractors, manufacturing firms, government agencies, and other employers. The growth of manufactured housing has created jobs for framing carpenters in firms that prefabricate building sections or entire structures.

Basic skills for framing carpenters are the ability to make use of measuring devices and to read and accurately follow construction drawings. They must be able to efficiently and safely use a variety of hand and power tools. Framing carpentry is strenuous work, involving lifting and carrying, standing or kneeling for long periods, and working in situations where the danger of injury from falls or other accidents is present. Most of the work is done outdoors, often in rainy, dusty, cold, or hot weather conditions. Wearing appropriate clothing and proper personal protective gear is important.

Framing carpenters acquire their skills in a variety of ways. Many learn through informal on-the-job training working with more experienced carpenters. Others enter the field from vocational school programs or take part in formal apprenticeship training. A carpentry apprenticeship typically is 3–4 years in length and combines classroom training with practical experience and instruction on the job site. Because they are exposed to most aspects of the construction process, carpenters who work for general contracting firms have opportunities for advancement to positions such as foreman, carpentry supervisor, or general construction supervisor.

Summary

The wall framing consists of the vertical and horizontal members that form the outside of a structure. This frame also supports upper floors, ceilings, and the roof. Framing members include the vertical studs, horizontal members called top and sole plates, lintels or headers, and sheathing. Studs and plates are made from 2×4 or 2×6 lumber, usually placed 16″ on center. Wall sections are assembled on the rough flooring, then raised into place. Once the outside wall frame is completed, the inside walls (partitions) are built and erected. Many inside partitions are nonbearing, which means they do not carry the weight of the structure. Tri-level and split-level housing presents special challenges in wall framing. Ceiling framing may be the underside of the floor joists for the next story or an assembly just below the roof. After walls are sheathed, plastic housewrap is usually installed to prevent air infiltration.

Test Your Knowledge

Answer the following questions on a separate piece of paper. Do not write in this book.

1. What does the wall frame support?
2. Trimmer studs stiffen the sides of an opening and support the _____.
3. What function do headers in a wall serve?
4. The first layout to be marked on the plates is the _____ spacing.
5. What is a *story pole?*
6. Immediately after a wall section is up, secure it with _____.
7. What is used to bridge a notch cut in a stud for plumbing?
8. Joints in the upper top plate should be at least _____ from those in the lower top plate.
9. *True or False?* Regular fiberboard sheathing can be used as a nailing base for exterior wall finish materials.
10. The position of the ceiling joists along the double plate should be coordinated with the location of the _____.
11. The first step in estimating the number of studs required is to calculate the _____ of all walls and partitions.
12. A strongback is needed for _____.
    A. strengthening a long span of ceiling joists
    B. maintaining proper spacing between ceiling joists
    C. keeping joists even along their lower edges
    D. None of the above.
    E. All of the above.
13. What is *housewrap?*
14. _____ is a general term for papers, felts, and similar sheet materials that, like housewrap, provide a wind and moisture barrier for walls.
15. How many studs are required for a plain wall panel 8′-0″ long if the studs are spaced 16″ O.C.?

Curricular Connections

Social Studies. Conduct library or Internet research to learn about the history and development of the balloon framing method of residential construction. Identify the parts of the country where it was most extensively used and what role it played in the expansion of Midwestern and Western cities. Try to determine when and why it was replaced by the platform framing method. If facilities are available, develop your information into a PowerPoint presentation.

Language Arts. Construct a model wall section (preferably at scale of $1″ = 1′$) showing framing for a window opening. Make an oral presentation to the class on proper framing for windows using the model to describe each component and its function.

Outside Assignments

1. Obtain a set of architectural plans for a one-story house. Study the details of construction, especially typical wall sections. Prepare a scale drawing of the framing required for the front walls. Be sure the rough openings are the correct size and in the proper

location. Your drawing should look similar to **Figure 9-1.**

2. Working from the set of plans used in #1, develop an estimated list of materials for the wall frame and sheathing. Include the number and length of studs; number and size of lumber for headers; material for plates; and type and amount of sheathing. Obtain prices from a local supplier and figure the total cost of the materials.

3. Obtain literature about fiberboard, foamed plastic, and gypsum sheathing. Obtain this material from local building products dealers, download it from the Internet, or write directly to manufacturers. Also, study books and other reference materials. Prepare a report for the class, based on the information you obtain. Include grades, manufacturing processes, characteristics, and application requirements. Discuss current prices and purchasing information. Be prepared to discuss specifications. Relate these specs to your local code.

Roof Framing

10

Learning Objectives

After studying this chapter, you will be able to:

- List and describe the various types of roofs.
- Identify the parts of a common rafter.
- Define the terms *slope* and *pitch.*
- Use a framing square, speed square, and rafter tables.
- Lay out common rafters.
- Describe the layout and erection of a gable roof.
- Explain the design and erection of trusses.
- Describe the procedure for sheathing a roof.
- Estimate roofing materials.

Technical Vocabulary

Bird's mouth	Flat roof
Butterfly roof	Framing square
Camber	Gable roof
Collar ties	Gambrel roof
Common difference	Hip jack
Common rafters	Hip rafters
Cornice	Hip roof
Cripple jack	Hypotenuse
Dead load	King post truss
Dormer	Live load
Extended rake	Lookouts
Fascia	Mansard roof
Fink truss	Purlin

Rafters	Skip sheathing
Ridge	Slope
Rise	Speed square
Roof trusses	Truss
Run	Valley jack
Scissors truss	Valley rafters
Sheathing	W truss
Shed roof	

Roof framing provides a base to which the roofing materials will be attached. The frame must be strong and rigid. In addition, a carefully designed and well-proportioned roof can add a distinctive and decorative character to the structure.

Roofs must be strong enough to withstand the weight and stress of snow and high winds. This must be kept in mind in the design of the roof components. In the conventional rafter system, these components include not only the rafters, but ceiling joists, collar ties, and purlins. In roof trusses, roof components include the chords, webs, and connector plates.

In the design of roof framing systems, both live loads and dead loads must be considered. The ***live load*** on a roof includes the weight of

---

Live load: The total of all moving and variable loads that may be placed on a building.

snow and the pressure from wind. Obviously, this load varies from one locality to another. The *dead load* is the weight of the material used to construct the roof. The rafters, decking, and roof coverings are all part of the dead load.

10.1 Roof Types

Roofs types vary widely, as illustrated in **Figure 10-1.** Most of them can be grouped as follows.

- *Gable roof.* Two surfaces slope from the centerline (*ridge*) of the structure. This forms two triangular ends called *gables.* Because of their simple design and low cost, gable roofs are most often used for homes. They are the easiest to build.

- *Hip roof.* All four sides slope from a central point or ridge. The angles created where two sides meet are called hips. An advantage of this type is its strength and the protective overhangs formed over end walls and sidewalls.

- *Gambrel roof.* In this variation of the gable roof, each slope is broken at midspan. This style is used on two-story construction. It permits extra space for more efficient use of the second floor. Dormers are usually included. The gambrel roof is a traditional style typical of the American colonial period and the period immediately following.

- *Flat roof.* This roof is supported on joists that also carry ceiling material on the underside. It may have a slight pitch (slope) to provide drainage.

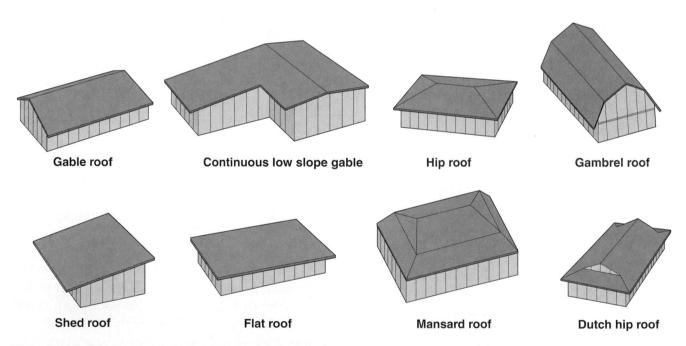

     Gable roof   Continuous low slope gable   Hip roof   Gambrel roof

     Shed roof   Flat roof   Mansard roof   Dutch hip roof

Figure 10-1. Common types of roofs used in residential construction. The gable is most often used.

Dead load: The weight of permanent, stationary construction and equipment included in a building.

Gable roof: A roof consisting of a single ridge with slopes in both directions. Made entirely of common rafters.

Hip roof: A roof that rises from all four sides of a building.

Ridge: The horizontal line at the junction of the top edges of two roof surfaces.

Gambrel roof: A roof slope formed as if the top of a gable roof is cut off and replaced with a less steeply sloped cap. This cap has a ridge in the center.

Flat roof: A roof that is either level or pitched only enough to provide for drainage.

- *Shed roof.* This simplest of pitched roofs is sometimes called a *lean-to roof.* The name comes from its frequent use on additions to a larger structure. It is often used in contemporary designs where the ceiling is attached directly to the roof frame.

- *Mansard roof.* Like the hip roof, the mansard has four sloping sides. However, each of the four sides has a double slope. The lower, outside slope is nearly vertical. The upper slope is slightly pitched. Like the gambrel roof, the main advantages are the additional space gained in the rooms on the upper level and making the house look lower than it is. The name comes from its originator, architect Francois Mansart (1598–1666).

- A lesser-used roof is the **butterfly roof.** This type looks like an inverted gable roof or like two shed roofs with the low ends attached to each other.

There are two basic systems used in framing the roof of platform structures: conventional, stick-built rafters and truss rafters. A carpenter builds conventional roof frames on-site. The ceiling joists and rafters are laid out, cut, and installed one at a time.

Rafters are sloped framing members that run downward from the peak of the roof to the plates of the outside walls. They are the supports for the roof load. As you learned in Chapter 9, the ceiling joists tie the outside walls together and support the ceiling materials for the rooms below. They also secure the bottom ends of the rafters, preventing them from pushing outward when loaded.

Shed roof: A single-slope roof, sometimes called a *lean-to roof.*

Mansard roof: A type of roof in which the pitch of the upper portion of a sloping side is slight and that of the lower portion is steep. The lower portion is usually interrupted by dormer windows.

Butterfly roof: An inverted gable roof with the low ends joined.

Rafter: One of a series of structural members of a roof designed to support roof loads. The rafters of a flat roof are sometimes called *roof joists.*

Roof truss: Engineered and prefabricated rafter assemblies. The lower chord also serves as a ceiling joist.

Roof trusses are engineered and prefabricated assemblies. They combine rafter and ceiling joist in one. Usually, trusses are factory-built and delivered to the building site, **Figure 10-2.** The lower chords also serve as ceiling joists and support the ceiling coverings.

10.2 Roof Supports

Roofs are supported by one or all of the following systems, depending on the type of rafter design:
- Outside walls.
- Ceiling joists (beams that hold the ceiling materials).
- Interior bearing walls (partitions that support structures above).

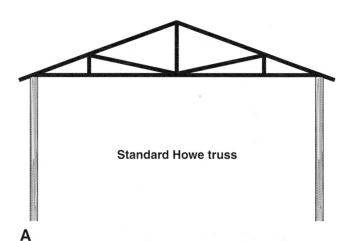

Standard Howe truss

A

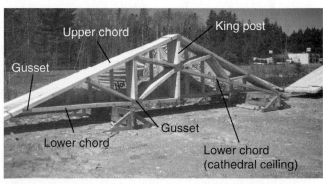

B

Figure 10-2. Roof trusses. A—The standard Howe truss efficiently supports heavy loads. B—Truss rafters come in many styles and sizes. One side of these trusses is designed for a cathedral ceiling. Note the names of the parts.

10.3 Parts of a Roof Frame

The plan view of the roof shown in **Figure 10-3** combines gable and hip roof types. The kinds of rafters are identified.

Common rafters are those that run at a right angle (in plan view) from the wall plate to the ridge. A gable roof has only rafters of this kind.

Hip rafters also run from the plate to the ridge, but only at a 45° angle. They form the support where two slopes of a hip roof meet.

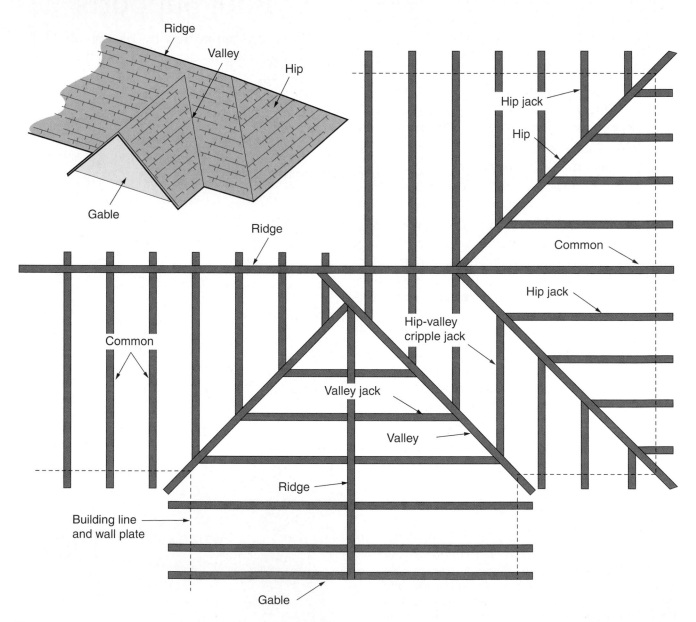

Figure 10-3. A roof frame in plan view. Note the types of rafters. The inset shows a view of this roof when finished.

Common rafter: A rafter connected to both the ridge and the wall plate.

Hip rafter: A rafter that runs from the wall plate to the ridge, always at a 45° angle.

Valley rafters extend diagonally from the plate to the ridge in the hollow formed by the intersection of two roof sections. These roof sections are usually at a right angle to each other. There are three kinds of jack rafters:

- *Hip jack.* This is the same as the lower part of a common rafter, but intersects a hip rafter instead of the ridge.

- *Valley jack.* This is the same as the upper end of a common rafter, but intersects a valley rafter instead of the plate.

- *Cripple jack.* Intersects neither the plate nor the ridge and is terminated at each end by hip and valley rafters. The cripple jack rafter is also called a *cripple rafter, hip-valley cripple jack,* or *valley cripple jack.*

Rafters are formed by laying out and making various cuts. **Figure 10-4** shows the cuts for a common rafter and the sections formed. The ridge cut allows the upper end to tightly fit against the ridge. The *bird's mouth* is formed by a seat cut and plumb (vertical) cut when the rafter extends beyond the plate. This extension is called the *overhang* or *tail.* When there is no overhang, the bottom of the rafter is ended with a seat cut and a plumb cut.

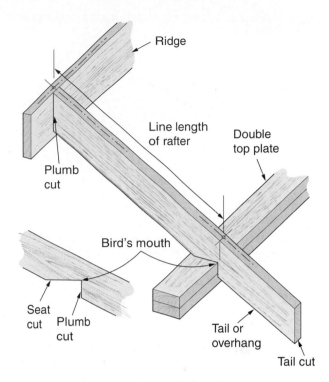

Figure 10-4. Rafter parts. Various cuts and surfaces are important because rafters must have a snug fit at all joints.

Valley rafter: A rafter that forms the intersection of an internal roof angle.

Hip jack rafter: A short rafter connecting to the wall plate and a hip rafter.

Valley jack: The same as the upper end of a common rafter, but intersects a valley rafter instead of the plate.

Cripple jack: A rafter that intersects neither the wall plate nor the ridge and is terminated at each end by hip and valley rafters. Also called a *cripple rafter, valley cripple jack,* or *hip-valley cripple jack.*

Bird's mouth: A notch cut on the underside of a rafter to fit the top plate. If the rafter end is flush with the top plate, the bird's mouth is not a full notch.

Hypotenuse: The side of a right triangle opposite of the right angle. In roof framing, the length of the rafter is the hypotenuse.

Run: The horizontal distance of rafters from the wall plate to the ridge.

10.4 Layout Terms and Principles

Roof framing is a practical application of geometry. This is an area of mathematics that deals with the relationships of points, lines, and surfaces. It is based largely on the properties of the right triangle:

- The horizontal distance (run) is the base.

- The vertical distance (rise) is the altitude.

- The length of the rafter is the *hypotenuse.*

As shown in **Figure 10-5,** if any two sides of a right triangle are known, the third side can be found mathematically. The formula used is $H^2 = A^2 + B^2$. *A* is the altitude or height, *B* is the base, and *H* is the line forming the third side of the triangle. To find the unknown length, extract the square root. Without a calculator that has a square root function, this is time-consuming. The same correct answer can be found using other methods.

In rafter layout, the base of the right triangle is called the *run.* It is the distance from the

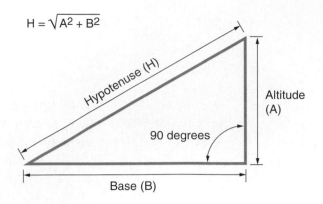

Figure 10-5. The rafter is like the hypotenuse of a right triangle. If you square the rise and run of a roof, then add the two figures, you have the square of the rafter's length.

outside edge of the plate to a point directly below the center of the ridge. The altitude, or *rise*, is the distance the rafter extends upward above the wall plate. Other layout terms and the relationship of the parts of the roof frame are illustrated in **Figure 10-6.**

Carpenters laying out conventional rafters use either the rafter tables on the framing square or a layout called the step-off method. Either way is fast, simple, and practical. The step-off method is very good, but must be double-checked against the calculation method found on the rafter tables on the square. Another option is to use a speed square. This is discussed later.

Small, hand-held construction calculators are sometimes used on the job. By entering the rise and the run of a roof, a carpenter can get a readout of the rafter length in feet, inches, and fractions. The same device also calculates lengths of hip and valley rafters.

10.4.1 Slope and Pitch

Slope refers to the incline of a roof. It is expressed as the relationship of the vertical rise to the horizontal run. Thus, slope is given as "x inches-in-12." In other words, a roof that rises at the rate of 4" for each 12" of run is said to have a 4-in-12 slope. A triangular symbol above the roofline in the architectural plans gives this information. The slope of a roof is sometimes called the *cut of the roof.*

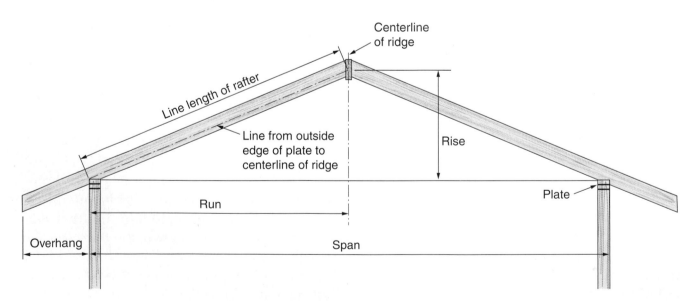

Figure 10-6. These are basic terms in rafter layout. Do you see how the rise, run, and rafter length form a right triangle?

Rise: The distance a rafter extends upward above the wall plate.

Slope: Incline of a roof.

The *pitch* is technically defined as the ratio of the rise to the span (twice the run). However, the word pitch is often used to mean the same as slope. It is simpler and easier to call out the rise and run, for example, 5-in-12 means the rise is 5″ for every unit (12″) of run. This system gives both slope and pitch.

10.5 Unit Measurements

The *framing square,* also called a *steel square* or *carpenter's square,* is the basic layout tool for roof framing. The side with the manufacturer's name is called the face and the opposite side the back. The longer, 24″ arm is called the blade or body, while the shorter arm (16″) is called the tongue.

The framing square is not large enough to make the rafter layout with one setting. It is necessary, therefore, to use smaller divisions called *units.* The foot (12″) is the standard unit for horizontal run. The unit for rise is always based on how many inches the roof rises in every foot of run. The unit run and rise is used to lay out plumb and level cuts required at the ridge, bird's mouth, and tail of the rafter. See **Figure 10-7.**

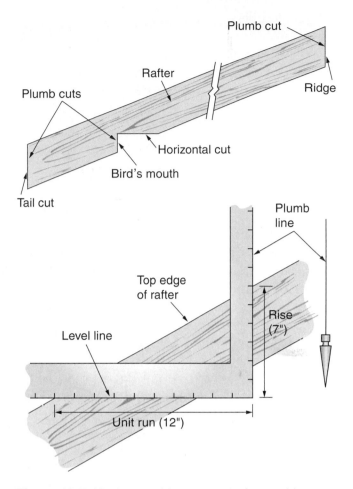

Figure 10-7. How to position a square for marking one unit of measure on a rafter. The run is measured along the blade. The rise (in inches per foot) is measured on the tongue. This same position is used to lay out and mark the bird's mouth and plumb cuts.

10.6 Framing Plans

When working with simple designs, the carpenter can easily visualize the roof framing. Since the wall framing is already erected, the only additional information needed is:
- Rise and run of the roof.
- Amount of required overhang.
- Subfascia and type of fascia.
- Heel height of the bird's mouth.

- Type of ridge.
- Any other information relating to roof construction.

These items are included in the house plans. When the structure has a complex roof design, the architect often includes a roof framing plan. If a framing plan is not included, the carpenter should prepare one. To do this, make a scaled drawing on tracing paper laid directly over the floor plans. Include ridges, overhang, and every rafter. Make the drawing similar to the one shown in **Figure 10-3.** However, you can use a single line to represent each framing member. In your drawing, maintain accurate spacing between rafters. Draw hips and valleys at a 45° angle and make jack rafters parallel to the common rafters.

Framing square: A square designed for carpenters with a short section (called the tongue) attached at a right angle to the longer section (called the blade). A number of tables are imprinted on its sides.

10.7 Rafter Sizes

As in floor and ceiling framing, the strength (size) of the rafter is determined by the spacing and span or length. Local building codes must be consulted for their specifications. **Figure 10-8** shows a table of rafter sizes for various loads and spacings. Stock for hips, valleys, and ridges is usually larger than for other roof framing members.

For purposes of estimating and ordering material, the rafter lengths can be determined with fair accuracy by making a scaled-down layout with the framing square. On the back of the square, the outside edge of the blade and tongue are divided into inches and twelfths. Assume the inches to be feet and each twelfth to be a full inch. First draw a triangle using the unit run and unit rise specified by the slope of the roof. If the total run is more than 12′, lengthen the base and hypotenuse. Now, lay the blade of the square along the base line until the point of total run on the square falls over the acute angle. Mark the point where the tongue crosses the sloping line, **Figure 10-9.** Use either the blade or tongue of the square to measure the hypotenuse. This gives the approximate length of the rafter. Add the overhang. Allow extra material for making cuts at the ridge and tail.

10.8 Laying Out Common Rafters

Rafters can be laid out by the step-off method, the rafter table on the framing square, or with a construction calculator. Any of these methods gives the correct rafter length. The step-off method is easy to use and the layout can be double-checked with the rafter tables.

Using any one of the methods, the carpenter first lays out, checks, and cuts a pattern rafter. The pattern is used to mark other rafters of the

Size of rafter (inches)	Spacing of rafter (inches)	Maximum allowable span (Feet and inches measured along the horizontal projection)			
		Group I	Group II	Group III	Group IV
2 × 4	12	10-0	9-0	7-0	4-0
	16	9-0	7-6	6-0	3-6
	24	7-6	6-6	5-0	3-0
	32	6-6	5-6	4-6	2-6
2 × 6	12	17-6	15-0	12-6	9-0
	16	15-6	13-0	11-0	8-0
	24	12-6	11-0	9-0	6-6
	32	11-0	9-6	8-0	5-6
2 × 8	12	23-0	20-0	17-0	13-0
	16	20-0	18-0	15-0	11-6
	24	17-0	15-0	12-6	9-6
	32	14-6	13-0	11-0	8-6
2 × 10	12	28-6	26-6	22-0	17-6
	16	25-6	23-6	19-6	15-6
	24	21-0	19-6	16-0	12-6
	32	18-6	17-0	14-0	11-0

Figure 10-8. This table shows the maximum runs allowed for rafters sloped 4-in-12 or greater. The group numbers refer to species of wood. Obtain this information from local building codes.

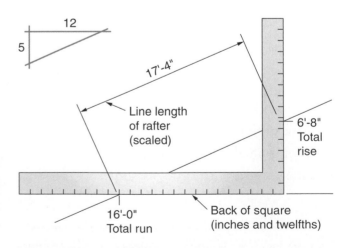

Figure 10-9. The framing square can be used to estimate the length of rafters. Use the square as a 1″ = 1′ scale.

same size and kind. For pattern layout, select a piece of lumber that is straight and true.

To cut the pattern, place the blank on a pair of sawhorses. Usually, the carpenter stands on the crowned side (top edge) of the rafter. It is easier to handle the framing square from that position. However, in order to describe rafter layouts that can be quickly understood, this position is reversed in **Figures 10-10** and **10-11**. The rafter in these illustrations is shown in its installed position.

As you lay out rafters, try to visualize how each will appear when it is set in the completed roof frame. Forming the habit of visualizing the rafter in its proper place helps to eliminate errors. An alternative to a full rafter pattern is shown in **Figure 10-12**. The length of the rafter is determined by one of the methods described. Then, the partial pattern pieces are used to mark ridge, bird's mouth, and tail cuts.

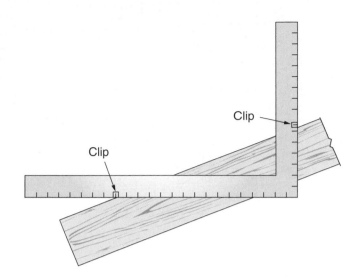

Figure 10-10. Some carpenters attach clips to the square at the rise and run positions. The clips automatically position the square against the rafter edge as each step is laid out.

PROCEDURE

Laying out rafters (step-off method)

In this example, the total run of the rafter (ridge line to building line) is 6′-8″ and the rise will be 5-in-12. There is a 1′-10″ overhang.

1. Place the framing square on the stock and align the numbers for the unit run (12″) and rise with the top edge of the rafter. The unit run is located on the blade and the unit rise on the tongue.

2. Set the rise and run on the square using clips or hand screws. To ensure accuracy, the correct marks on the square must be exactly positioned over the edge of the stock each time a line is marked. Be sure to use a sharp pencil to make the layout lines.

3. Starting at the top of the rafter, position the square and draw the ridge line along the edge of the tongue. Still holding the square, mark the length of the odd unit (8′ in this example).

4. Next, shift the square along the edge of the stock until the tongue setting is even with the 8″ mark. Draw a line along the tongue and mark the 12″ point on the blade for a full unit, **Figure 10-11.**

5. Move the square to the 12″ point just marked and repeat the marking procedure. Continue until the correct number of full units is laid out. This number is the same as the number of feet in the total run (6 in this example).

6. Form the bird's mouth by drawing a horizontal line (seat cut) to meet the building line so the surface is about equal to the width of the plate. The size of the bird's mouth may vary depending on the design of the overhang. In **Figure 10-11,** note that the square has been turned over to mark these cuts and also to lay out the overhang. This may or may not be necessary, depending on the length of the rafter blank.

7. To lay out the overhang, start with the plumb cut of the bird's mouth and mark full units first. Then, add any odd unit that remains. The tail cut may be plumb, square, or a combination of plumb and level. Check the cornice details shown in the architectural plans for exact requirements.

8. The final step in the layout consists of shortening the rafter at the ridge. With the square in position, draw a new plumb line back from the ridge line 1/2 of the thickness of the ridge board, **Figure 10-11.**

9. Make the ridge, bird's mouth, and tail cuts.

10. Label the rafter as a pattern, indicating the roof section to which it belongs.

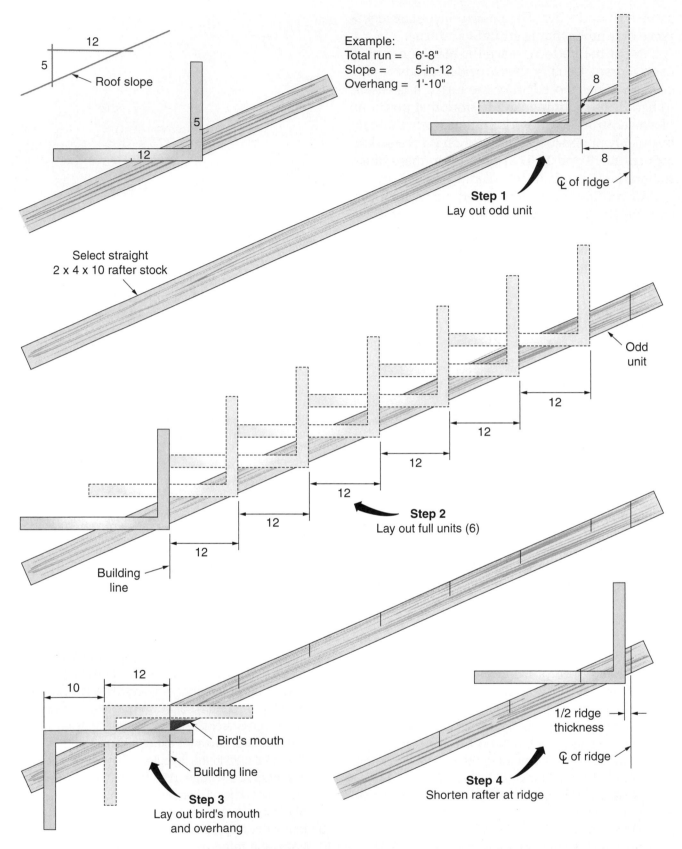

Figure 10-11. Carpenters can use this procedure for laying out a common rafter. It is known as the step-off method because each foot of run requires another step of the square.

Figure 10-12. Partial patterns such as these are sometimes used to make rafter cuts.

10.8.1 Using the Rafter Table

You can also calculate the length of a common rafter using the table on the framing square. This table is on the face side of the square. It is the one marked *Length Common Rafters Per Foot Run*. You need to know the slope and run before starting.

To see how this works, suppose the run of the building is 6'-8" and the rise per foot is 5". Find the inch mark for 5" on the blade of the square. See **Figure 10-13**. Under that number, in the top line of the table, look for the number nearest to the 5" mark. This is the length of the rafter in inches for one foot of run. To find the length of the rafter from the building line to the center of the ridge, multiply the units of run by the figure from the rafter table, as in the following examples.

Example 1: Run = 6'-8", Slope = 5-in-12
Run = 6'-8" = 6 2/3 units
Rise = 5", number from rafter table = 13
Rafter length = 6 2/3 × 13 = 86 2/3"
86 2/3" divided by 12 =
Rafter length: 7'-2 2/3"

Example 2: Run = 12', Slope = 4-in-12
Run = 12 units
Rise = 4", number from the rafter table = 12.65
Rafter length = 12 × 12.65 = 151.8"
151.8" divided by 12 = 12.65' =
Rafter length: 12'-7 4/5"

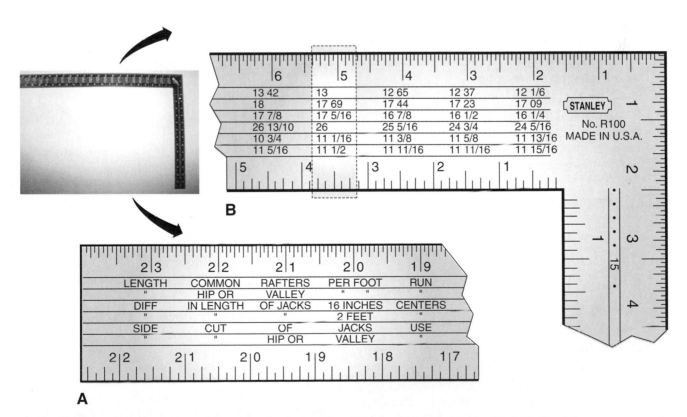

Figure 10-13. The rafter table is printed on the face side of the framing square. A—The blade of the square carries tables for figuring lengths of rafters. B—If the unit rise is 5", the rafter length for 12" of run is 13".

These calculations result in the line length of the rafter running from the center of the ridge to the outside of the plate. To make the pattern layout, add the overhang and subtract half the thickness of the ridge board from the length.

10.8.2 Using the Speed Square

Sometimes a carpenter uses a *speed square,* also called a *super square* or *quick square,* to find rafter lengths and determine the angle of cuts. This triangular measuring tool is smaller and easier to carry than a rafter square. To use a speed square, you need to know the pitch of the roof. This is given in the plans or it can be determined by some simple math. Rafter charts for every pitch are supplied by the square's manufacturer.

PROCEDURE

Laying out rafters (speed square method)

1. Once you know the rise per foot of run, refer to the chart for that rise. Read down the chart to the correct run to find the length of the rafter.
2. Mark the length on the rafter.
3. Place the pivot point of the speed square on the length mark while lining up the rise on another scale. See **Figure 10-14.** This step gives the angle for marking the plumb cut at the ridge.
4. Using the rafter pattern, cut the number required.

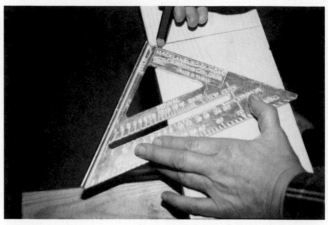

A

8-12 Pitch		
Ft. of run	Common rafter	Hip or valley rafters
1	1' 2-3/8"	1' 6-3/4"
2	2' 4-7/8"	3' 1-1/2"
3	3' 7-1/4"	4' 8-1/4"
4	4' 9-3/4"	6' 3"
5	6' 0-1/8"	7' 9-3/4"
6	7' 2-1/2"	9' 4-5/8"
7	8' 5"	10' 11-3/8"
8	9' 7-3/8"	12' 6-1/8"
9	10' 9-3/4"	14' 0-7/8"
10	12' 0-1/4"	15' 7-5/8"
11	13' 2-5/8"	17' 2-3/8"
12	14' 5-1/8"	18' 9-1/8"
14	16' 9-7/8"	21' 10-5/8"
16	19' 2-3/4"	25' 0-1/8"

B

Figure 10-14. A—Using a speed square to mark the plumb cut on a common rafter. B—A sample speed square chart for figuring common rafters and valley or hip rafters when pitch of roof is 8-in-12. (Macklanburg-Duncan)

10.9 Erecting a Gable Roof

Lay out the rafter spacing along the wall plate as the ceiling joists are laid out. When the rafter spacing is the same as ceiling joist spacing, every rafter is nailed to a joist. When rafters are spaced 24" O.C. and ceiling joists 16" O.C., the layout is arranged as in **Figure 10-15.** The plate layout is important, so carefully follow the roof framing plan. To allow more room over the wall plate for additional insulation, some carpenters attach rafters to a 2 × 4 nailed on top of the ceiling joists, **Figure 10-16.**

Speed square: Triangular measuring tool that can be used to mark any angle for rafter cuts by aligning the desired degree mark on the tool with the edge of the rafter being cut. Also called a *quick square* or *super square.*

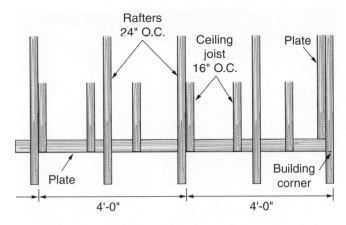

Figure 10-15. A plan view of ceiling joist and rafter layout for 24″ O.C. rafters and 16″ O.C. joists. A joist is nailed to every other rafter, acting as a tie beam to keep the walls from spreading.

Select straight pieces of ridge stock and lay out the rafter spacing by transferring the marking directly from the plate or a layout rod. Joints in the ridge should occur at the center of a rafter. Cut the lengths that will make up the ridge and lay them across the ceiling joists close to where they will be assembled with the rafters.

To make the initial assembly, some carpenters prefer to first attach the ridge to several

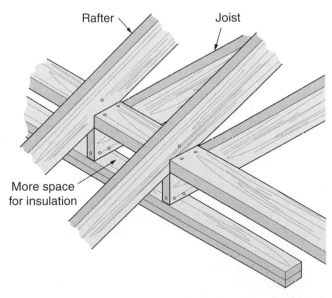

Figure 10-16. This alternate framing method has rafters resting on a 2 × 4 plate placed on top of the ceiling rafters. This allows room to place thicker insulation batts over the wall plate without restricting airflow to the soffit vents.

rafters on one side. This assembly is then raised and supported while the rafters are nailed to the plate. Then, several rafters are installed on the opposite side.

Another method of supporting a ridge board may be used. Mark two 2 × 4s for the height of the rise and attach one to either end of the ridge board. The edge of the ridge board should align with the mark. Lift up the assembly and tack the base of each 2 × 4 to the plate of the bearing wall. Plumb and secure the vertical 2 × 4s with braces. This method requires fewer carpenters.

PROCEDURE

Erecting a site-built roof

1. Select straight rafters for the gable end and nail one in place at the plate.
2. Install a rafter on the opposite side with a worker at the ridge supporting both rafters.
3. Place the ridge board between the two rafters and temporarily nail it in place.
4. Move about five rafter spaces from the end and install another pair of rafters.
5. Plumb and brace the assembly. Make any adjustments necessary in the nailing of the first rafters. **Figure 10-17** shows the assembly and nailing pattern at the plate. Special framing anchors are often used, especially in areas where hurricanes are possible, **Figure 10-18.**
6. After aligning the ridge, install the remaining rafters. First nail the rafter at the plate and then at the ridge, **Figure 10-19A**. On a small crew or when a carpenter works alone, temporary supports hold the ridge board until rafters are installed, **Figure 10-19B.**
7. Drive 16d nails through the ridge board into the rafter, **Figure 10-20**. The rafters on the opposite side of the ridge are toenailed. Install only a few rafters on one side before placing matching rafters on the opposite side. This practice makes it easier to keep the ridge straight.
8. Periodically, check the ridge to ensure it remains straight and level.
9. Continue to add sections of ridge and assemble the rafters. Add bracing when required. Always install rafters with the crown (curve) turned upward.

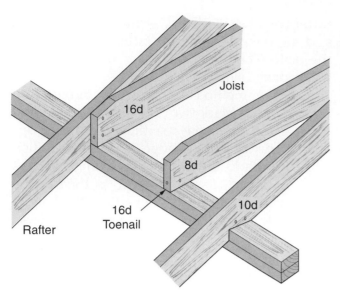

Figure 10-17. The nailing pattern for joists and rafters at the wall plate.

A

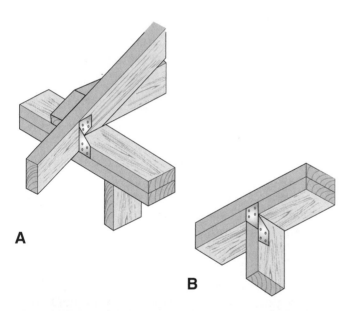

Figure 10-18. Special framing anchors. A—Anchors strengthen the joint between the rafter and wall. B—The same type of clip can be used to attach the plate to studs. (The Panel Clip Co.)

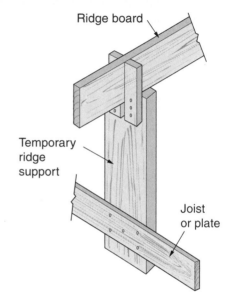

B

Figure 10-19. Methods of installing common rafters. A—The ridge board is supported by another member of the crew while the bird's mouth is tightly pushed against the plate and nailed. B—A temporary support, built on site, holds the ridge board level until a few rafters are installed. (Southern Forest Products Assn.)

Safety Note

When framing a roof, use extra care to prevent a fall. Erect solid scaffolding wherever it is helpful. Avoid working directly above another person. When working at heights over 25′, fall protection devices must be worn.

10.10 Gable End Frame

The end frames of a gable should be assembled after the rafters and ridge have been installed. The end frame consists of vertical studs running from the top plate of the bearing walls to the end rafters. The framing of the gable end is most easily done while it lies flat over the ceiling joists. Overhang, brick racks, frieze

Figure 10-20. Nailing rafters to the ridge board. Align the edge of the rafter with the layout lines, then toenail it. (Southern Forest Products Assn.)

trim, louvers, housewrap, and siding can all be installed at this time. After it is erected, the end gable should be well braced before installing the ridge board and rafters. The following procedure can be used to determine the length and location of studs.

PROCEDURE

Laying out a gable end frame

1. Drop a plumb line from the center of the ridge above the end wall plate. Place a stud directly below the ridge, if possible. Often, a ventilator is located in the center of the end frame. If this is the case, mark a distance 1/2 of the ventilator width to either side of the plumb line. This mark is the location of the first stud.
2. Position the stud on the top plate and use a level to check for plumb. Mark the location of the rafter on the side of the stud.
3. Lay out the next stud (16″ O.C. is common spacing), plumb it, and mark the rafter location on the stud's side. The difference between the two stud lengths is the *common difference,* **Figure 10-21.**
4. Using the common difference, continue marking, cutting, and placing studs until the edge of the frame is reached. If the rafters on both sides have the same rise and run, the end frame should be symmetrical (both sides identical).

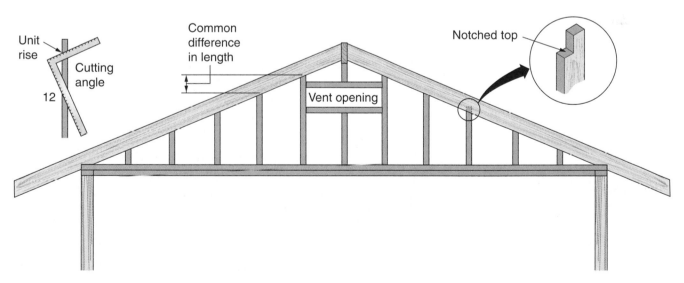

Figure 10-21. Lay out studs for a gable end as shown. The change in length from one stud to another is the common difference.

Common difference: The difference in length between any two adjacent studs or rafters when laying out a gable end frame or a hip roof.

PROCEDURE

Finding the common difference with the framing square

The common difference can also be obtained using the framing square. See **Figure 10-22.**

1. Set the square on a stud. Align the unit run on the blade with the edge of the stud. Then align the unit rise on the tongue to the same edge. This is the same as the step method used in determining the length of a rafter.
2. Mark a line on the stud along the outside edge of the blade.
3. Slide the square away from you. Keep the outer edge of the blade on this line. Watch the inch marks on the blade. Stop when the inch number for the stud spacing (16″ or 24″) aligns with the edge of the stud.
4. Read the inch mark on the tongue of the square that aligns with the edge of the stud. This is the common difference.

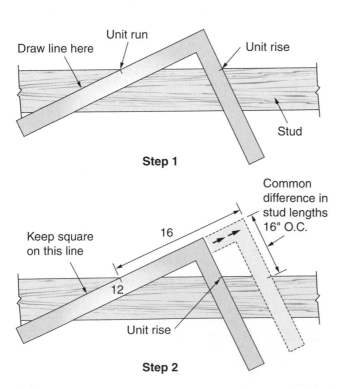

Step 1

Step 2

Figure 10-22. Using a framing square to find the common difference for gable-end studs. Step 1—Line up the run and rise on the edge of a 2 × 4. Step 2—Slide the square in the direction shown until the 16″ mark on the blade aligns with the edge of the stud.

Roof designs often include an *extended rake* (gable overhang). Typical framing is illustrated in **Figure 10-23.** It requires the gable end frame be constructed before the roof frame is completed.

When constructing the gable ends for a brick or stone veneer building, the frame must be moved outward to cover the finished wall. This projection can be formed by using *lookouts* and blocking attached to a ledger. When the top of the veneer is aligned on the sides and ends of the building, the ledger should be attached at the same level as the one used in the *cornice* construction. See **Figure 10-24.** Studs are mounted on this projection and attached to the roof frame in various ways. The architectural plans usually include details covering special construction features of this type.

10.11 Hip and Valley Rafters

Hip roofs or intersecting gable roofs have some or all of the following rafters: common, hip, valley, jack, and cripple. Two roof surfaces slanting upward from adjoining walls meet on a sloping line called a *hip*. The rafter supporting this intersection is known as a *hip rafter*.

First, cut and frame the common rafters and ridge boards. The ridge of a hip roof is cut to the length of the building minus twice the run plus the thickness of the rafter stock. It intersects the common rafters, **Figure 10-25.**

From each corner of the building, measure along the side wall a distance that is equal to

Extended rake: A gable overhang.

Lookout: Structural member running between the lower end of a rafter and the outside wall. It supports the soffit.

Cornice: Exterior trim of a structure at the meeting of the roof and wall. It usually consists of panels, boards, and mouldings. Also, an inclusive term for all of the parts that enclose or finish off the overhang of a roof at the eaves.

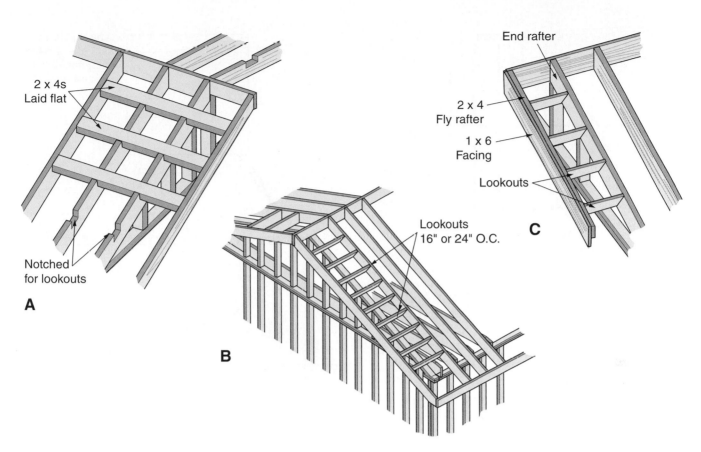

Figure 10-23. Methods of framing a gable overhang. A—The 2 × 4 lookouts are laid flat over notched rafters. B—The plate atop the gable end studs supports lookouts laid on edge. Make sure the top of the plate lines up with the bottoms of the rafters. C—A small overhang with short lookouts supports a 2 × 4 fly rafter and a fascia.

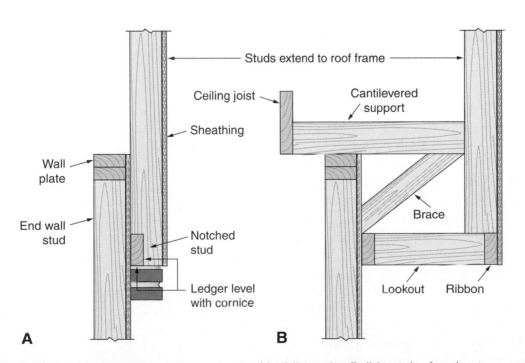

Figure 10-24. Gable-end framing for masonry veneered buildings. A—Build out the framing to cover the brick or stone veneer. B—An alternate framing that extends the gable end farther. Such framing might also be used over doors and for special features.

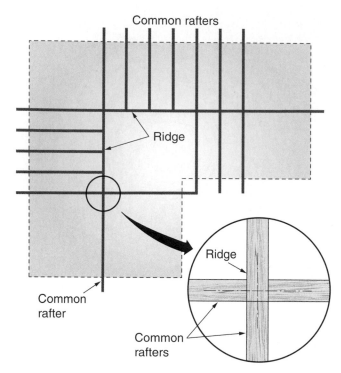

Figure 10-25. The first step in framing a hip or intersecting roof is to install the ridge boards and common rafters.

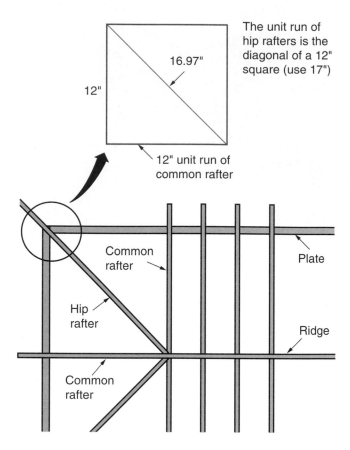

Figure 10-26. The length of a hip rafter is the diagonal of a square formed by the walls and two common rafters. Use the 17″ mark on the blade of the square for the run.

1/2 of the span. Mark the point, which is the centerline of the first common rafter. All other rafters are laid out from this position.

In a plan view, the hip rafter is seen as the diagonal of a square, **Figure 10-26.** The diagonal of this square is the total run of the hip rafter. Since the unit run of the common rafter is 12″, the unit run of the hip rafter will be the diagonal of a 12″ square. This distance is calculated as 16.97″. For actual application, round to 17″.

To lay out a hip rafter, follow the same procedure used for a common rafter. Use the 17″ mark on the blade of the square instead of the 12″ mark, **Figure 10-27.** The odd unit of run, if any, must also be adjusted. Its length is found by measuring the diagonal of a square whose sides are equal to the length of the odd unit.

Valley rafters are also the diagonal of a square. The sides are formed by ridges and common rafters. The layout is the same as described for hip rafters. Use 17″ on the framing rafter for the unit run. The length of hip and valley rafters can be determined from rafter tables just like common rafters. Using the same figures as used in the previous example, the calculations are:

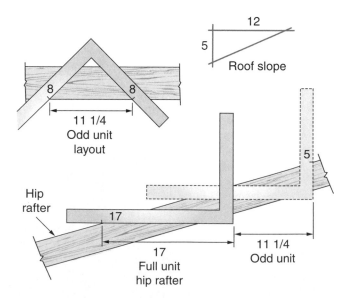

Figure 10-27. Starting the layout of a hip rafter. The slope and odd unit size are the same as those used in the common rafter layout.

Run = 6'-8", Slope = 5-in-12
Run = 6'-8" = 6 2/3 units
Rafter Table No. 2 = 17.69
Hip or valley length = 17.69 × 6 2/3
$$= 117.93$$
$$= 9'-9\ 15/16"$$
$$= 9'-10"$$

Hip and valley rafters must be shortened at the ridge by a distance equal to one-half of the 45° horizontal thickness of the ridge, **Figure 10-28.** The side cuts are then laid out as shown in the illustration. Another method is to use the numbers from the sixth line of the rafter table. For example, consider that the slope is 5-in-12. The number on the sixth line below the 5" mark is 11 1/2. All of

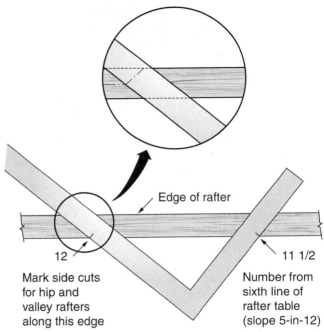

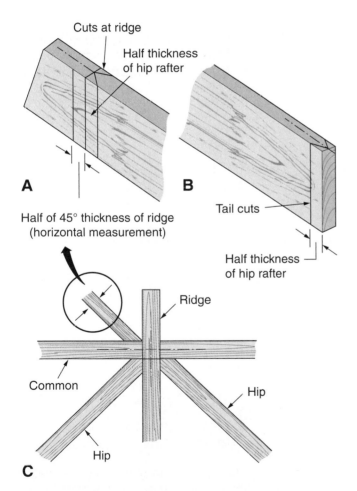

Figure 10-28. Shortening the hip rafter and making side cuts. A—Shorten the top of the hip rafter by 1/2 of the thickness of the ridge and mark. Then, measure back 1/2 of the thickness of the rafter. Mark both sides for angle cuts. B—Note how the tail is marked for cutting. C—This drawing explains why angled cuts must be 1/2 of the thickness of the ridge.

Figure 10-29. Using the framing square to lay out side cuts on hip and valley rafters.

the numbers in the table are based on or related to 12, so line up 12 on the blade and 11 1/2 on the tongue along the edge of the rafter, **Figure 10-29.** Draw the angle for each cut. Then, draw plumb lines. Use the 17" mark on the blade and the unit rise on the tongue. Tail cuts at the ends of the rafters are laid out using the same angle.

A centerline along the top edge of a hip rafter is where the roof surfaces actually meet. The corners of the rafter extend slightly above this line. Some adjustment must be made. The corners could be planed off. However, it is easier to make the seat cut of the bird's mouth slightly deeper, thus lowering the entire rafter, **Figure 10-30.** Using the framing square, align the 17" mark and the number for the unit rise on the bottom edge of the rafter. Mark the position with a short line drawn along the body of the square. Measure toward the tail 1/2 of the thickness of the rafter and mark. Shift the square toward the tail of the rafter. Be sure to maintain the alignment of 17" and the unit rise. When the square aligns with the previous mark, mark the seat of the bird's mouth. Do not mark the plumb cut until you have read the next paragraph.

The plumb (vertical) cut of the bird's mouth for valley rafters must be trimmed so it fits into

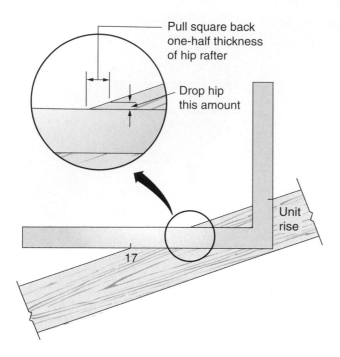

Figure 10-30. Using the square to find and mark the distance to drop a hip rafter. The bottom of the rafter is facing up.

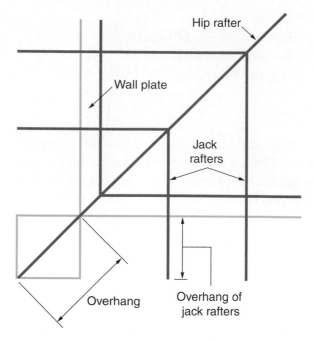

Figure 10-31. Finding the length of the hip rafter overhang. You can construct the square full size on paper, then measure the length of the diagonal. A rafter table on the square also gives this information.

the corner formed by the intersecting walls. Although side cuts of approximately 45° could be made, it is more practical to move the plumb cut toward the tail of the rafter by a horizontal distance equal to the 1/2 of the 45° thickness of the rafter.

The tail of the hip rafter is actually the diagonal of a square formed by extending the line of each of the walls the length of the jack rafters, **Figure 10-31.** You can find its length by constructing a full-size square on paper. Then, measure the diagonal. Mark the plumb cut using the run (17″) and the rise (in inches) on the square. The tail must form a nailing surface for intersecting fascia boards. Refer to **Figure 10-28** for directions on how to make the tail cut.

After hip and valley rafters are laid out and cut, they are installed on the roof. **Figure 10-32** is a rafter plan showing their positions.

To use the speed square in laying out hip and valley rafters, find the rafter length from the table corresponding to the rise per foot of run. Read down the table to the correct run. Then, read across to the column for hip and valley rafters. Mark the rafter for the ridge plumb cut, **Figure 10-33A.** Next, measure from the top of

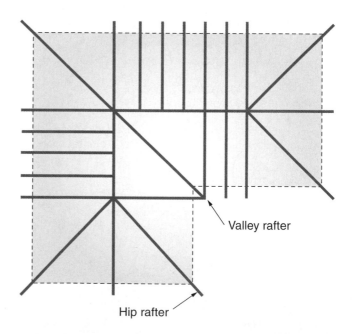

Figure 10-32. This rafter plan shows hip and valley rafters along with common rafters. In actual construction, jack rafters are cut and installed last.

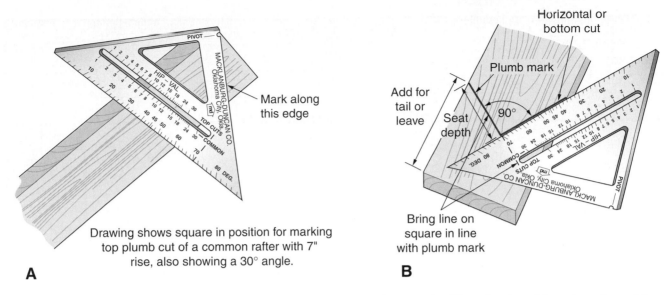

Mark along this edge

Drawing shows square in position for marking top plumb cut of a common rafter with 7" rise, also showing a 30° angle.

A

Horizontal or bottom cut

Plumb mark

Add for tail or leave

Seat depth

90°

Bring line on square in line with plumb mark

B

Figure 10-33. Using a speed square for plum and bird's mouth cuts. A—The plumb cut is set up for a hip or common rafter. Extra allowance must be marked for the miter cut. B—Marking the seat cut for the bird's mouth. (Macklanburg-Duncan)

the rafter down and mark the length obtained from the chart. Use the speed square to make the plumb seat mark. Next, rotate the speed square until the dashed line is aligned with the seat plumb mark just made. See **Figure 10-33B.** Mark the seat cut. Make the adjustments previously described and cut the rafter.

10.11.1 Jack Rafters

Hip jack rafters are short rafters that run between the wall plate and a hip rafter. They run parallel to common rafters and are the same in every respect except for their length. When equally spaced along the plate, the change in length from one to the next is always the same. This consistent change is the common difference, **Figure 10-34.** The common difference can be found in the third and fourth lines of the rafter table. For a roof slope of 5-in-12 with rafters 24" O.C., the figure from the table on the square is 26". See **Figure 10-35.**

The common difference can also be found with the layout method illustrated in **Figure 10-36.** Using the square, align the unit run and rise of the roof with the edge of a smooth piece of lumber. Draw a line along the blade. Then, slide the square along this line to a point on the blade

equal to the rafter spacing. Put a mark on the edge of the lumber against the outside edge of the tongue The distance between this mark and the first line is the common difference.

To lay out jack rafters, select a piece of straight lumber. Lay out the bird's mouth and overhang from the common rafter pattern. Next, lay out the line length of a common rafter. This is the distance from the plumb cut of the bird's mouth

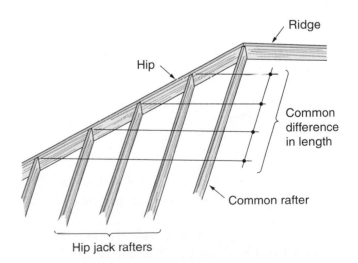

Ridge

Hip

Common difference in length

Common rafter

Hip jack rafters

Figure 10-34. When evenly spaced, hip jack rafters have the difference in length from one to the next. By using this information, all rafters can be cut without measuring each distance.

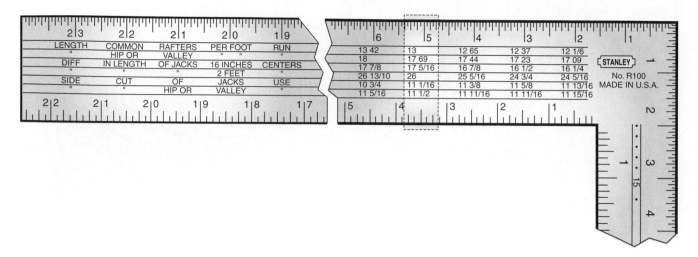

LENGTH	COMMON	RAFTERS	PER FOOT	RUN							
	HIP OR	VALLEY	"	"	13 42	13	12 65	12 37	12 1/6		
DIFF	IN LENGTH	OF JACKS	16 INCHES	CENTERS	18	17 69	17 44	17 23	17 09		
	"	"	2 FEET		17 7/8	17 5/16	16 7/8	16 1/2	16 1/4		
SIDE	CUT	OF	JACKS	USE	26 13/10	26	25 5/16	24 3/4	24 5/16		
"	"	HIP OR	VALLEY	"	10 3/4	11 1/16	11 3/8	11 5/8	11 13/16		
					11 5/16	11 1/2	11 11/16	11 11/16	11 15/16		

STANLEY

No. R100
MADE IN U.S.A.

Figure 10-35. The common difference of jack rafters spaced at 24″ O.C. can be found on the fourth line of the rafter table. Simply read down from the unit rise per foot.

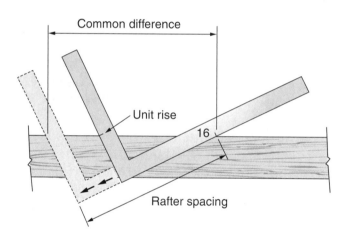

Figure 10-36. Using the framing square layout method of determining the common difference of jack rafters. Step 1—Place the square on the rafter as you would for the rise and run of the rafter. Draw a line along the edge of the blade. Step 2—Now, for a spacing of 16″ O.C., slide the square in the direction of the arrows until the 16″ mark on the blade aligns with the mark made on the rafter.

to the centerline of the ridge. For the first jack rafter down from the ridge, lay out the common difference in length.

Now, take off 1/2 of the 45° thickness of the hip, **Figure 10-37.** Square this line across the top of the rafter and mark the center point. Through this point, lay out the side cut as shown in the illustration. Another method is to use a square and the number from the fifth line of the rafter table, **Figure 10-38.** Mark the plumb lines to follow when the cut is made.

For the next hip jack, move down the rafter the common difference and mark the cutting line. A sliding T-bevel is a good tool to use. Continue until all jacks are laid out. Then, use the T-bevel to mark the jack rafters as required.

Each hip in the roof assembly requires one set of jack rafters made up of matching pairs. A pair consists of two rafters of the same length with the side cuts made in opposite directions.

10.11.2 Valley Jacks

Similar procedures are followed in laying out valley jack rafters. For these, however, it is usually best to start the layout at the building line and move toward the ridge, **Figure 10-39.** The longest valley jack is the same as a common rafter except for the side (angle) cut at the bottom.

Use the common rafter pattern and extend the plumb cut of the bird's mouth to the top edge. Lay out the side cut by marking (horizontally) 1/2 of the thickness of the rafter. If you prefer, use the framing square and apply the numbers located in the fifth line of the rafter table.

The common difference for valley jack rafters is obtained by the same procedure used for hip jacks. Lay out this distance from the longest valley jack to the next. Continue along the pattern until all lengths are marked. Now,

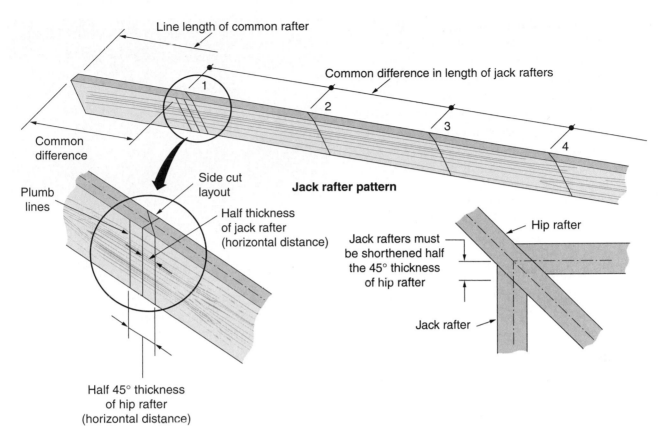

Figure 10-37. Laying out jack rafters. A master pattern may be made (top).

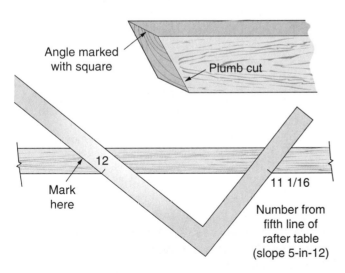

Figure 10-38. How to use a framing square to mark the angle cut on the top of the hip jack rafter. The mark is always made on the top of the rafter. Be sure to mark the plumb cut as well.

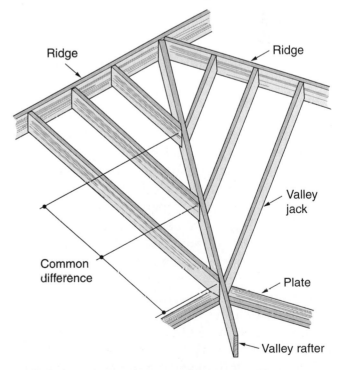

Figure 10-39. Valley jack rafters as assembled. Find the common difference in the same manner as for the hip jacks, except start at the plate.

use this pattern to cut all the valley rafters. They are cut in pairs in the same way as hip jacks.

When all jack rafters are laid out, they should be carefully cut. It requires a great deal of skill to make side cuts with a hand saw. Usually, it is better to use a portable electric saw with an adjustable base and guide. Radial arm saws are designed to do accurate cutting and are well suited for this kind of work, **Figure 10-40.**

Working Knowledge

The strength of a roof frame depends a great deal on the quality of the joints. Use special care on the side cuts of jack rafters so the joining surfaces tightly fit together.

Figure 10-40. Side cuts for jack rafters can be accurately made with a radial arm saw, if one is available on site. (Des Moines, Iowa, Public Schools)

10.12 Erecting Jack Rafters

When all jack rafters are cut, assemble them into the roof frame, **Figure 10-41.** The nailing patterns depend on the size of the various members. Use 10d nails. Space the nails so they are near the heel of the side cut as they go from the jack into the hip or valley rafter.

Jack rafters should be erected in opposing pairs. This prevents the hip and valley rafters from being pushed out of line. It is good practice to first place a pair about halfway between the plate and ridge. Carefully sight along the hip or valley to make sure that it is straight and true. Temporary bracing could also be used for this purpose. Be sure that the outside walls running parallel to the ceiling joists are securely tied into the ceiling frame before hip jack rafters are installed. These rafters tend to push outward.

After rafters have been erected and securely nailed, carefully check over the frame. If some rafters are bowed sideways, they can be held straight with a strip of lumber located across the center of the span. Sight each rafter and move as needed. Drive a nail through the strip into the rafter to hold it in place. When the roof has been sheathed to this point, remove the spacer strip.

The last material to be attached to the roof frame before the decking is the *fascia,*

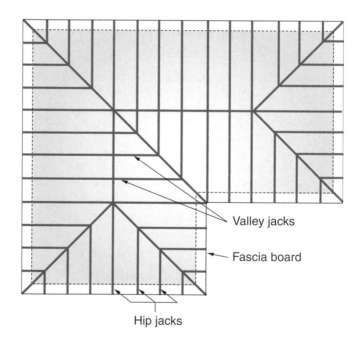

Figure 10-41. A plan view of a complete roof frame. All hip and valley jacks are in place and the fascia board has been installed.

Fascia: A wood member nailed to the ends of the rafters and lookouts and used for the outer face of a box cornice.

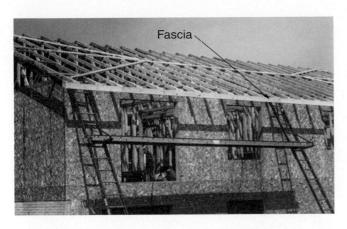

Figure 10-42. The fascia board is fastened to the rafter or truss tails before installing the sheathing.

Figure 10-42. This is the main trim member and is attached to the plumb-cut ends of the rafters. It conceals the rafter ends, provides a finished appearance, and furnishes a surface to which gutters can be attached.

Before attaching the fascia, however, stretch a chalk line along the rafter ends to assure the rafter ends are the same length. Trim any that are too long so the eave line is straight.

The fascia may be directly attached to the rafter ends. Some carpenters prefer to install 2 × 4s first to even up the ends of the rafters and provide a solid nailing surface for the 1" fascia board. Cut the upper edge of the fascia board at an angle to match the slope of the roof. Corners should be mitered and carefully fitted.

10.13 Special Problems

For intersecting roofs where the spans of the two sections are not equal, the ridges will not meet. To support the ridge of the narrow section, one of the valley rafters is continued to the main ridge. See **Figure 10-3.** The length of this extended or supporting valley is found by the same method used in the layout of a hip rafter. It is shortened at the ridge, just like the hip, but only a single side cut is required. The other valley rafter is fastened to the supporting valley with a square, plumb cut.

A rafter framed between the two valley rafters is called a *valley cripple jack*. The angle of the side cut at the top is the reverse of the side cut at the lower end. The run of the valley cripple is one side of a square, **Figure 10-43.** This run is equal to twice the distance from the centerline of the valley cripple jack to the intersection of the centerlines of the two valley rafters. Lay out the length of the cripple by the same method used for a jack rafter. Shorten each end 1/2 of the 45° thickness of the valley rafter stock. Make the side cuts in the same way as for regular jack rafters.

Rafters running between hips and valleys are called hip-valley cripple jacks. They require side cuts on each end. Since hip and valley rafters are parallel to each other, all cripple rafters running between them in a given roof section are the same length. The run of a hip-valley cripple jack rafter is equal to the side of a square, **Figure 10-44.** The size of the square is determined by the length of the plate between the hip and valley rafter. Use this distance and lay out the cripple in the same manner as a common rafter. Shorten each end by an amount equal to 1/2 of the 45° thickness of the hip and valley rafter stock. Now, lay out and mark the side cuts. Follow the same steps used for hip

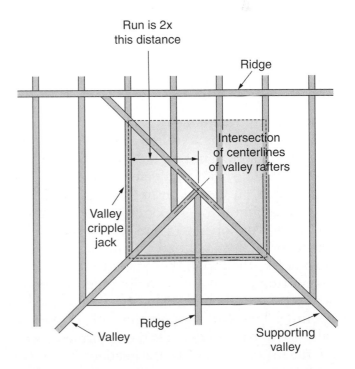

Figure 10-43. Calculating the run of a valley cripple jack.

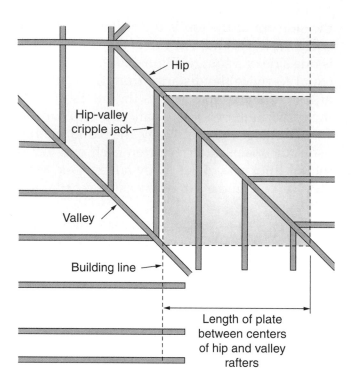

Figure 10-44. The run of a hip-valley cripple jack is equal to the length of the wall plate from the valley rafter to the hip rafter.

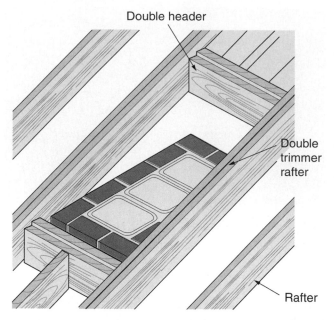

Figure 10-45. Framing an opening in the roof for a chimney. Allow 2″ clearance all around the opening. The headers above and below the chimney are plumb.

and valley jacks. The side cuts required on each end form parallel planes.

10.14 Roof Openings

Some openings may be required in the roof for chimneys, skylights, and other structures. To frame large openings, follow the same procedure used in floor framing. Frame a chimney opening as shown in **Figure 10-45.** To construct small openings, first complete the entire framework. Then, lay out and frame the opening.

Use a plumb line to locate the opening on the rafter from openings already formed in the ceiling or floor frame. Nail a temporary wooden strip across the top of the rafters to be cut. The supporting strip should be long enough to extend across two additional rafters on each side of the opening. This will support the ends of the cut rafters while the opening is being formed. Now, cut the rafters and nail in the headers. If the size of the opening is large, double the headers and add a trimmer rafter to each side.

10.15 Roof Anchorage

Rafters usually rest only on the outside walls of a structure. Rafters also lean against each other at the ridge, thus providing mutual support. This causes an outward thrust along the top plate that must be considered in the framing design.

Sidewalls are normally well secured by the ceiling joists, which are also tied to some of the rafters. End walls, however, are parallel to the joists. They need extra support, especially when located under a hip roof. Stub ceiling joists and metal straps are one method for reinforcing such walls. See **Figure 10-46.** Framing anchors can be substituted for the metal straps, especially when subflooring is part of the assembly.

10.16 Collar Ties

Collar ties tie together two rafters on opposite sides of a roof, **Figure 10-47.** They do not

Collar tie: A tie beam connecting rafters. It is located considerably above the wall plate. Also called *rafter tie.*

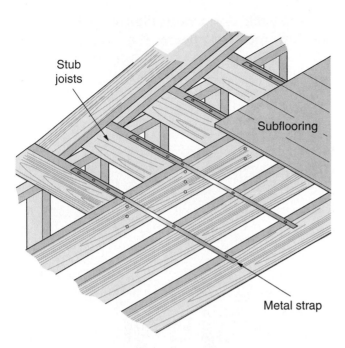

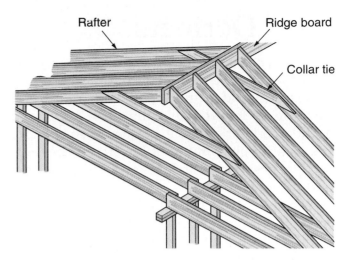

Figure 10-47. Collar beams tie the rafters and ridge together, thus reinforcing the roof frame. Together with the ceiling joists, they secure the frame against spreading outward.

Figure 10-46. Metal strapping and stub joists can be used to tie hip roofs to the end walls.

support the roof, but provide bracing and stiffening to hold the ridge and rafters together. In standard construction, 1 × 6 boards are installed at every third or fourth pair of rafters.

10.16.1 Purlins

Additional support must be provided when the rafter span exceeds the maximum allowed. A *purlin* provides this support. This is a structural member, usually a 2 × 4, that horizontally runs across the roof under the rafters. See **Figure 10-48.** The purlin is supported by bracing, also 2 × 4 stock, resting on a plate located over a supporting partition. Bracing under the purlin may be placed at an angle greater than 45° to transfer loads from the midpoint of the rafter to the support below. Purlins are also used over truss rafters when they are spaced further apart than 24″. In this case, roof sheathing is fastened to these purlins.

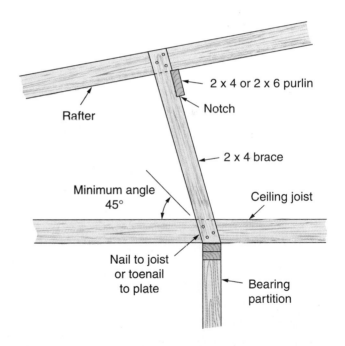

Figure 10-48. A purlin is a plate that supports long runs of rafters at their midpoint. Weight is transferred to a bearing partition through 2 × 4 braces.

Purlin: Horizontal roof members used to support rafters between the plate and ridge board.

10.17 Dormers

A *dormer* is a framed structure projecting above a sloping roof surface. It normally contains a vertical window unit. Although their chief purpose is to provide light, ventilation, and additional interior space, dormers also enhance the exterior appearance of the structure. There are two basic types of dormers, the *shed* dormer and the *gable* dormer.

The width of a shed dormer is not restricted by its roof design, **Figure 10-49**. It is used where a large amount of additional interior space is required. In the simplest construction, the front wall is extended straight up from the main wall plate. Double trimmer rafters carry the sidewall. The rise of the roof is sometimes figured from the top of the dormer plate to the main roof ridge. In such cases, the run is the same as the main roof. Be sure to provide sufficient slope for the dormer roof.

The gable dormers in **Figure 10-50** are tied into the roof well below the ridge. They are designed to provide openings for windows. Gable dormers can be located at various positions between the plate and ridge of the main roof.

A

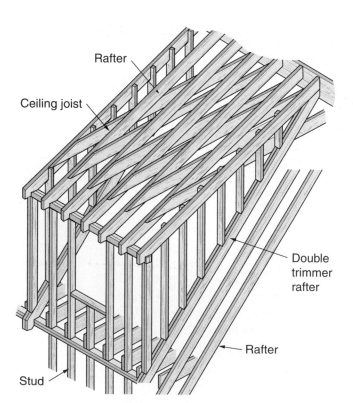

Figure 10-49. Typical framing for a shed dormer. A nailer strip (not shown) is added along the double trimmer to support roof sheathing.

B

Figure 10-50. A—Gable dormers add light and space to the rooms under the roof of this 1 1/2 story house. B—The studs in this gable dormer are 2 × 6 and rest on a sole plate. Note that they extend through the roof to their own wall plate.

Dormer: A framed structure extending from and fastened to a roof slope and having a roof of its own. It provides a wall surface for installing a window.

10.18 Framing Flat Roofs

Flat roofs provide a long, low appearance to the building. Sometimes, ceiling and roof members in a flat roof are combined as one system. The interior surfaces can be used as the finished décor or a place to fasten the finishing materials. In such cases, rigid insulation is installed between the decking and the built-up roofing materials.

Methods and procedures used to frame flat roofs are about the same as those followed in constructing a floor. Most designs require an overhang with the ends of the joists tied together by a header or band. Cantilevered lookout rafters are tied to doubled roof joists, **Figure 10-51**. Corners can be formed as shown in the figure or carried on a longer diagonal joist that intersects the double joist.

In mild climates, carpenters may use nominal 2″ planking as sheathing over widely spaced beams. Both can be exposed on the underside.

Builders in the southwestern United States sometimes build traditional homes in the Pueblo or Territorial style. This style of home includes a flat roof supported by round poles called *vigas*. The vigas are spaced anywhere from 16″ to 30″, depending on the span. The poles rest on the wall plates and extend through the exterior wall about 2′ to 3′. A through-bolt secures them to the plate. Usually, the outside wall extends 2′ to 3′ above the roof as an architectural feature called a *parapet wall*. **Figure 10-52** shows a traditional southwestern home under construction.

A

B

Figure 10-52. Flat roofs are a feature of some traditional home styles still built in southwestern U.S. A—Looking up at the ceiling of a Pueblo style house. Above the viga (pole) is a roof framed in 2 × 12s. Note the solid bridging. B—This exterior view shows the walls extending above the flat roof. (Pierce Construction, Ltd.)

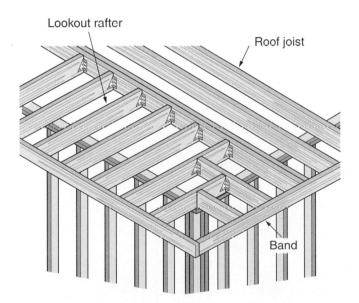

Figure 10-51. Lookout rafters are cantilevered over the outside wall to form an overhang for this flat roof.

10.19 Gambrel Roof

The gambrel roof is like a gable roof, but has four slopes. It is typical in an architectural style known as Dutch Colonial. Often used in two-story construction, it gives added living space with minimal outside wall framing, **Figure 10-53**. The upper roof surface usually forms a 20° angle with a horizontal plane, while the lower surface forms about a 70° angle.

In residential construction, this type of roof is usually framed with a purlin located where the two roof slopes meet. Rafters are notched to receive the purlin, which is supported on partitions and/or tied to another purlin on the opposite side of the building with collar ties.

The same procedures used to frame a gable roof can be applied to the gambrel roof. The rise and run of each surface is found on the architectural plans. The two sets of rafters are laid out in the same way as previously described for a common rafter.

It may help to make a full-size sectional drawing at the intersection of the two slopes. This makes it easier to visualize and proportion the end cuts of the rafters. **Figure 10-54** shows basic gambrel roof framing for a small building, such as a garden house or toolshed. Note the use of gussets to join the rafter segments. Also note the simple framing over the plate that is used to form the roof overhang.

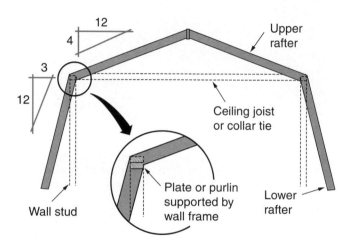

Figure 10-53. Basic framing for a gambrel roof. The gambrel roof provides extra living space on the second story with minimal wall framing.

10.20 Mansard Roof

Figure 10-55 shows a mansard-roofed dwelling. Like the gambrel roof, it has two slopes with the lower slope being steeper. The mansard roof, however, extends around all four sides of the building. Second floor joists extend beyond the first floor wall frame. This extension provides support for the lower rafters.

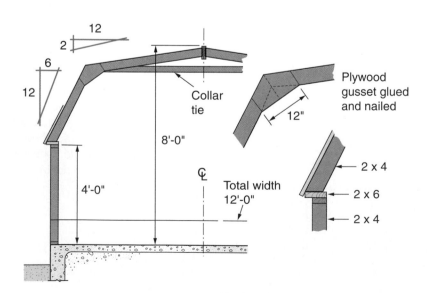

Figure 10-54. A gambrel roof for a small building is easy to design and build.

10.21 Special Framing

Figure 10-56 shows upper floor framing for a 1 1/2-story structure. Generally, this framing results in some saving of material, since walls and ceilings can be made a part of the roof frame. Knee walls are usually about 5' high. The ceiling height is typically 7'-6".

Low-sloping roofs like the one shown in **Figure 10-57** usually need support at several points. The strength derived from the triangular shape of regular-pitched roofs is greatly reduced in this design. Thus, carefully prepared architectural plans are essential for this type of roof structure.

Figure 10-55. An example of the mansard roof style. The main characteristics of the style are the two slopes and the overhang on all sides.

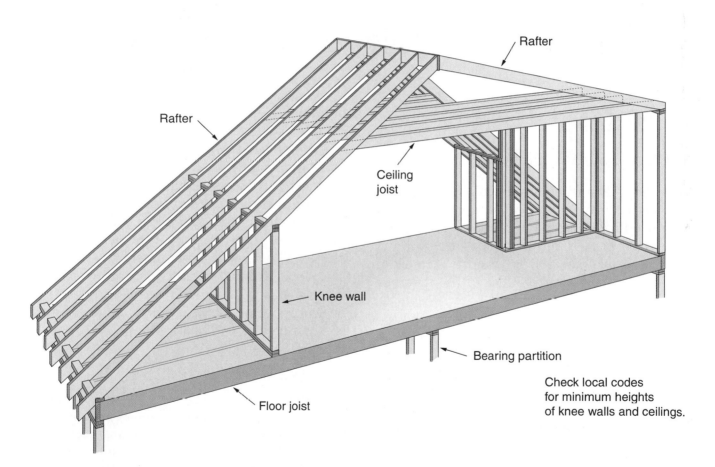

Figure 10-56. Wall, ceiling, and roof framing for a 1 1/2 story structure. (National Forest Products Assn.)

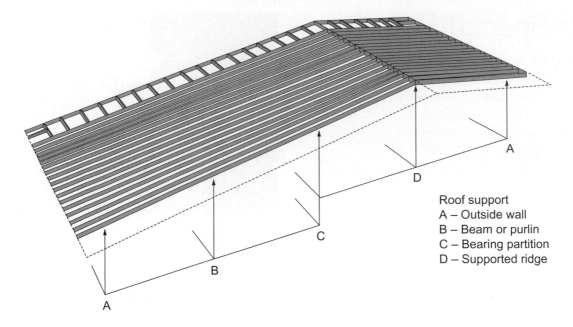

Figure 10-57. This low-slope roof is for a split-level design. The arrows indicate support points provided by the outside walls, bearing partitions, and purlins or beams.

Roof support
A – Outside wall
B – Beam or purlin
C – Bearing partition
D – Supported ridge

10.22 Roof Truss Construction

A *truss* is a framework designed to carry a load between two or more supports. The principle used in its design is based on the rigidity of the triangle. Triangular shapes are built into the frame in such a way that the stresses of the various parts are parallel to the members making up the structure.

Roof trusses are frames that carry the roof and ceiling surfaces. They rest on the exterior walls and span the entire width of the structure. Since no load-bearing partitions are required, more freedom in the planning and division of the interior space is possible. Roof trusses permit larger rooms without extra beams and supports. The use of roof trusses also allows surface materials to be applied to outside walls, ceilings, and floors before partitions are constructed.

There are many types and shapes of roof trusses, **Figure 10-58.** One commonly used in residential construction is the *W truss* or *Fink truss.* This type can be used for spans up to 50′. The *king post truss* has top and bottom chords and a vertical center post. It is used on shorter spans up to 22′. The *scissors truss* is used for buildings having a sloping ceiling. In general, the slope of the bottom chord is 1/2 of the slope of the top chord. Truss rafters can be designed for spans up to 50′.

Most carpenters prefer to use roof trusses that are factory built to engineered specifications. Architectural drawings should provide the manufacturer with all of the design specifications. The truss designer must know the span, roof slope, live and dead loads, and wind loads the roof must withstand. In most localities, engineering documents from a truss engineer must be submitted in order to obtain a certificate of occupancy.

Truss: A structural unit consisting of such members as beams, bars, and ties arranged to form triangles. Provides rigid support over wide spans with a minimal amount of material. Used to support roofs and joists.

W truss: Roof truss that can be used for spans up to 50′. Also called a *Fink truss.*

King post truss: Roof truss with top and bottom chords and a vertical center post. It is used on shorter spans up to 22′.

Scissors truss: Truss used for buildings having a sloping ceiling. In general, the slope of the bottom chord is 1/2 of the slope of the top chord.

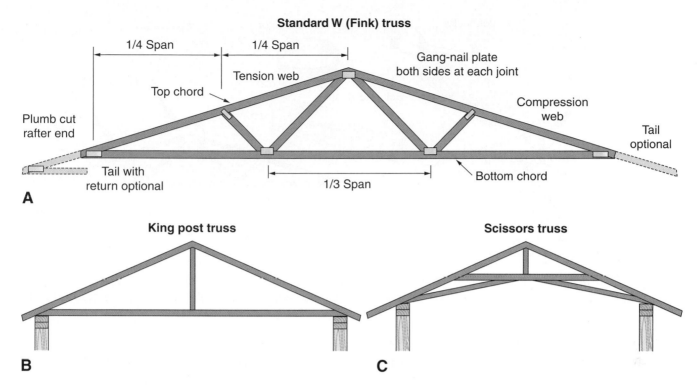

Figure 10-58. Three types of roof trusses. A—The standard W or Fink truss is commonly used in residential construction. B—The king post truss. C—The scissors truss permits cathedral ceilings.

If carpenters choose to build their own trusses, they must use carefully developed and engineered designs. Both APA-The Engineered Wood Association and the Truss Plate Institute have such designs and instructions. Refer to the different roof truss designs shown in Appendix B, **Technical Information**.

Roof trusses must be made of structurally sound 2 × 4 lumber. Joints must be carefully fitted and tightly fastened with metal connectors or gussets on both sides. The carpenter is seldom required to determine the sizes of truss members or the type of joints used. Site-built trusses are precut and assembled at ground level. Spacing of 24″ O.C. is common. However, 16″ O.C. or other spacing may be required in some designs.

When the truss is in position and loaded, there is a slight sag. To compensate for this, the

lower member (bottom chord) is slightly arched during fabrication of the unit. This adjustment is called *camber*. Camber is measured at the midpoint of the span. A standard 24′ long truss usually requires about 1/2″ of camber.

In truss construction, it is essential that joint slippage be held to a minimum. Regular nailing patterns are usually not satisfactory. Special connectors must be used. Various types are available, **Figure 10-59**. All of them securely hold the joint and are easy to apply. When plywood gussets are used, they must be applied with glue and nails to both sides of the joint.

Trusses can be laid out and constructed on any clear floor area. Make a full-size layout on the floor. Snap chalk lines for long line lengths. Use straightedges to draw shorter lines.

Trusses for residential structures can normally be erected without special equipment. Each truss is simply placed upside down on the walls at or near the point of installation. Then, the peak of the rafter is swung upward by hand. Use a pole as needed. On multistory dwellings or buildings with larger trusses, installation by hand may be impossible or impractical. In

---

Camber: A slight arch in a beam or other horizontal member so the bottom of the member is straight when under load.

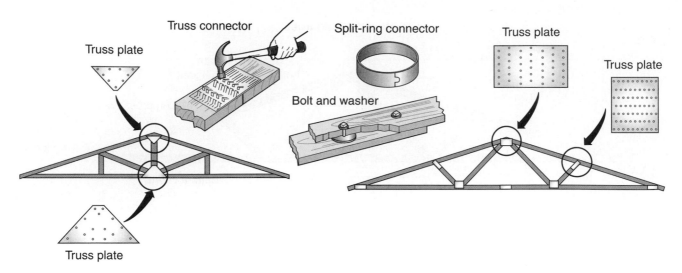

Figure 10-59. Plates and connectors for roof trusses. Truss plates are made in many sizes, shapes, and types. Some are perforated for nails, some require no nails. Split-ring connectors fit into recesses bored in mating joints.

such cases, a crane must be used. Always attach lifting chains at or near joints.

Working Knowledge

Use extra care when raising roof trusses. The first truss should be held with bracing or guy wires and all succeeding trusses carefully braced to prevent collapse.

Roofs framed with trusses need not be limited to gable types. Today, a wide range of configurations can be produced. Designs are based on carefully prepared data covering load and lumber specifications. Computers are used to apply these data to develop specific designs. Further efficiency results from the use of specialized methods, machines, and fasteners. **Figure 10-60** shows a variety of roof trusses used on a hip roof.

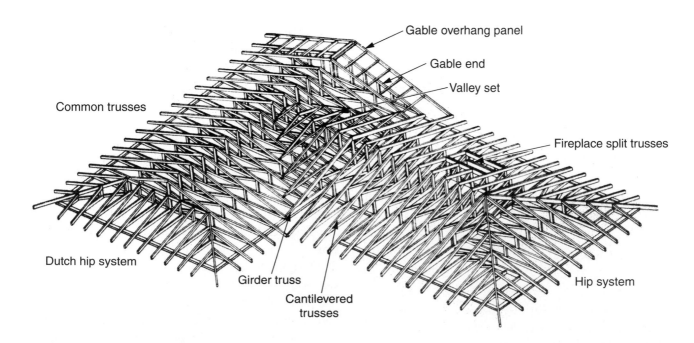

Figure 10-60. This roof frame is constructed with many different prefabricated truss units. (TrusWal Systems Corp.)

Always refer to the framing plans before installing roof trusses. If basic design requirements are to be met, the final erection and bracing of a roof system must be carried out according to plans and specifications. The installer is responsible for the proper storage, handling, and installation of truss rafters delivered to the job site, **Figure 10-61.** Trusses should not be stored on rough terrain or uneven surfaces that could cause damage to them.

10.23 Bracing of Truss Rafters

Proper ground bracing and temporary bracing of the truss rafter during erection is vital. Be prepared to ground brace the first rafter erected using either method shown in **Figure 10-62.** As additional truss rafters are placed, install lateral

 CAUTION: The builder, building contractor, licensed contractor, erector, or erection contractor is advised to obtain and read the entire booklet "Commentary and Recommendations for Handling, Installing & Bracing Metal Plate Connected Wood Trusses, HIB-91" from the Truss Plate Institute.

 CAUTION: All temporary bracing should be no less than 2 x 4 grade marked lumber. All connections should be made with minimum of two 16d nails. All trusses assumed 2' on-center or less. All multi-ply trusses should be connected together in accordance with design drawings prior to installation.

TRUSS STORAGE

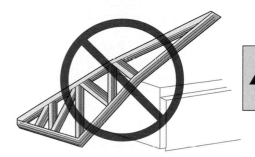

CAUTION: Trusses should not be unloaded on rough terrain or uneven surfaces that could cause damage to the truss.

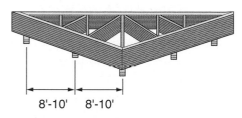

8'-10'   8'-10'

Trusses stored horizontally should be supported on blocking to prevent excessive lateral bending and lessen moisture gain.

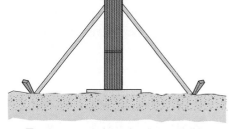

Trusses stored vertically should be braced to prevent toppling or tipping.

 WARNING: Do not break banding until installation begins or lift bundled trusses by the bands.

 WARNING: Do not use damaged trusses.

 DANGER: Do not store bundles upright unless properly braced.

 DANGER: Walking on trusses that are lying flat is extremely dangerous and should be strictly prohibited.

Figure 10-61. Observe these warnings for proper storage of trusses on the job site. (Reproduced from HIB-91, Courtesy of Truss Plate Institute)

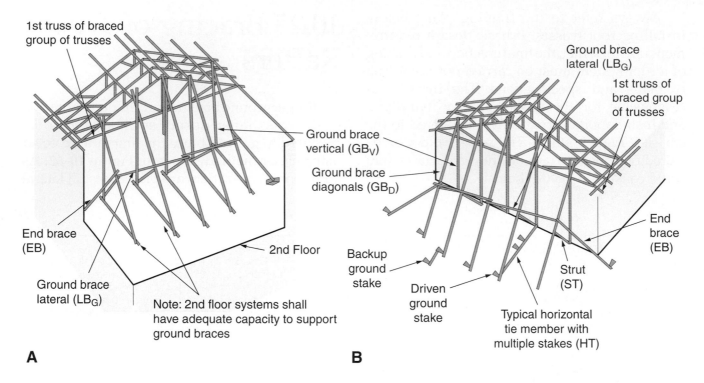

Figure 10-62. Temporary ground bracing of truss rafters. Consult a registered professional truss engineer for exact bracing. Bracing should be erected before raising the first truss. A—Ground bracing installed in the building's interior. B—Bracing installed outside the building. (Reproduced from HIB-91, Courtesy of Truss Plate Institute)

and diagonal bracing as shown in **Figure 10-63.** This bracing can be removed as roof decking is installed. Some carpenters prefer to install this bracing on the underside of the top chord. This avoids the work of removing them as sheathing is installed.

Additional inside permanent bracing is advisable. Remove the temporary ground bracing before sheathing the end of the roof. Observe the minimum pitch recommendations and bracing suggestions in the chart in **Figure 10-63.**

The theory of truss bracing is to apply sufficient support at right angles to the plane of the truss to permanently hold each member in its correct position. **Figure 10-64** shows a method of lateral and diagonal bracing applied once the trusses are secured to the wall plate.

Fasten truss rafters to the doubled wall plate as you would conventional rafters. Use toenailing or patented connectors. Sometimes both are used, **Figure 10-65.** Never toenail the lower chord of truss rafters to nonbearing walls. The trusses flex under snow loads and the bottom chord vertically moves. This tends to lift the wall,

causing cracks and gaps at the junction of the wall and the floor. **Figure 10-66** shows a clip that can be used to attach the truss so the truss can move up and down without disturbing the wall. When the roof frame is complete, check it carefully to see that all members are securely fastened and that nailing patterns are adequate.

10.24 Roof Sheathing

Sheathing (also called *decking*) provides a nailing base for the roof covering and adds strength and rigidity to the roof frame. Sheathing materials include plywood composites, oriented strand board, particleboard, and

---

Sheathing: Boards or panels attached to the exterior studding or rafters of a structure.

Span	Minimum pitch	Top chord lateral brace spacing (LB$_S$)	Top chord diagonal brace (DB$_S$) [# trusses]	
Up to 32'	4/12	8'	20	15
Over 32' - 48'	4/12	6'	10	7
Over 48' - 60'	4/12	5'	6	4
Over 60'	See a registered professional engineer			

DF = Douglas Fir-Larch   SP = Southern Pine
HF = Hem-Fir   SPF = Spruce-Pine-Fir

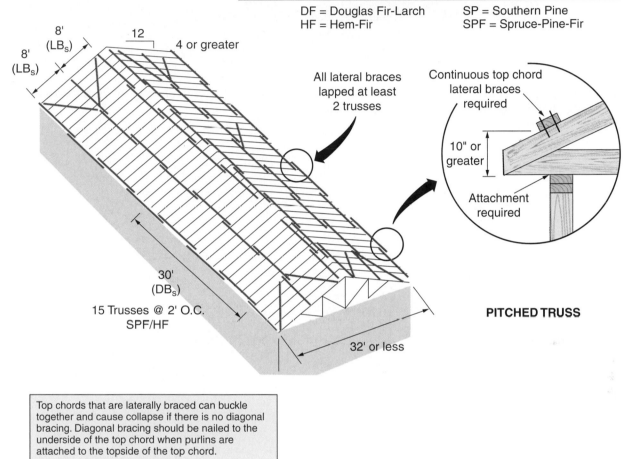

PITCHED TRUSS

Top chords that are laterally braced can buckle together and cause collapse if there is no diagonal bracing. Diagonal bracing should be nailed to the underside of the top chord when purlins are attached to the topside of the top chord.

Figure 10-63. Proper temporary bracing for roof trusses. Failure to follow these bracing recommendations can result in building damage as well as severe personal injury. (Reproduced from HIB-91, Courtesy of Truss Plate Institute)

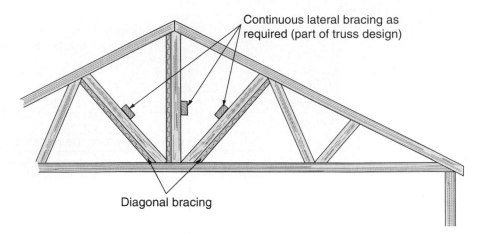

Figure 10-64. Truss rafters may require additional, permanent lateral and diagonal bracing. Always follow manufacturer's specifications when installing and bracing trusses.

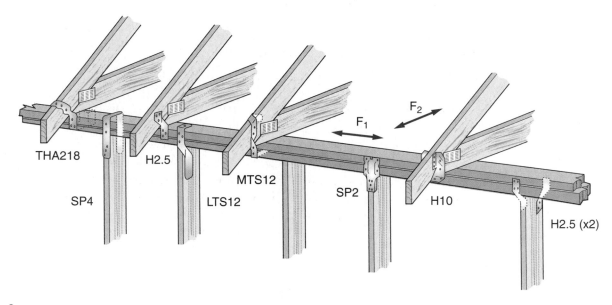

A

B

Figure 10-65. Attaching trusses to the wall plate. A—Trusses can be attached using metal tie-downs. B—Trusses can also be fastened with 10d nails, two on each side. (©Simpson Strong-Tie Company, Inc.)

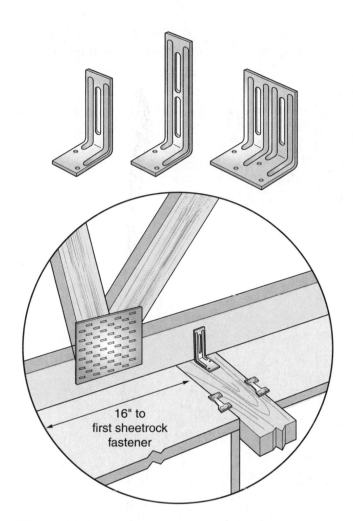

Figure 10-66. A roof truss clip keeps the truss and nonbearing wall in alignment, but allows for vertical movement of the lower truss chord. (©Simpson Strong-Tie Company, Inc.)

Figure 10-67. Oriented strand board has become popular as sheathing material. The textured side should be installed facing up to provide better footing.

up. To avoid bows in the sheathing, only stand where framing is supporting the panel as it is fastened.

Safety Note
Wear shoes or boots with skid-resistant soles when working on pitched roofs.

common boards, **Figure 10-67.** Other special materials are also available. For example, one product consists of panels formed with solid wood boards, bonded together with heavy paper.

Before applying sheathing to a roof, check for a level nailing surface. Use a long level or a straight piece of lumber 6–10′ long. Shim low trusses or rafters as needed. Blocking can be used to straighten bowed or warped top chords of trusses or rafters. Install the sheathing panels with screened surfaces or skid-resistant coatings

Before starting the sheathing work, erect the needed scaffolding. The scaffolding should make it easy and safe to install the boards or panels along the lower edge of the roof. If constructed of lumber on site, this scaffold can be later used to build the cornice after the roofing has been completed.

If asphalt shingles or other composition materials are used for the finished roof, common board sheathing must be applied solid (no gaps). For wood shingles, metal sheets, or metal tile, *skip sheathing* should be used. In this method, board sheathing is spaced so there are voids (gaps). This saves material and allows wood shingles to dry out, thus preventing rot. The spacing must be arranged according to the length of shingles. Nails must fall over wood, rather than voids. Boards should be attached with two 8d nails at each rafter. Joints must be located over the center of the rafter. For greatest rigidity, use long boards, particularly at the rake.

Skip sheathing: Board sheathing on a roof spaced to allow wood shingles to dry from the underside.

When end-matched (tongue-and-groove) sheathing is used, the joints may be made between rafters. Joints in the next board must not occur in the same rafter space. Boards should be long enough to extend over at least two rafters.

Carefully fit sheathing at valleys and hips and securely nail it. This ensures a solid, smooth base for the installation of flashing materials. Around chimney openings, allow a 1/2″ clearance from the masonry. Framing members must have a 2″ clearance. Always securely nail sheathing around roof openings.

10.24.1 Structural Panels (Sheets)

Structural panels are an ideal material for roof sheathing. They can be rapidly installed, hold nails well, and resist swelling and shrinkage. Because the panels are large, they add considerable rigidity to the roof frame. Always install plywood with the face grain perpendicular to the rafters. Locate end joints directly over the center of the rafter. Small pieces can be used, but they should always cover at least two rafter spaces.

The required thickness of the panels varies with the roofing material and rafter spacing. For wood or asphalt shingles with a rafter spacing of 16″, a thickness of 5/16″ is recommended. For a 24″ span, a 3/8″ thickness should be used. Slate and tile shingles require 1/2″ thicknesses for 16″ rafter spacing and 5/8″ for 24″ spacing. For a flat deck under built-up roofing, use 1/2″ thickness.

Panels should be nailed to rafters with 6d nails spaced 6″ apart on edges and 12″ apart elsewhere. If wood shingles are used and the sheathing is less than 1/2″ thick, 1 × 2 nailing strips spaced according to shingle exposure should be affixed to the sheathing.

If there are several carpenters on the job, sheets may be slid up a ladder, as shown in **Figure 10-68A.** A special rack can also be used, **Figure 10-68B.** This saves steps for the carpenter as it keeps sheathing panels close by. It is also a handy way to store panels until carpenters are ready to install them. On large construction jobs, a power panel elevator may be used to save time.

A

B

Figure 10-68. A—On small construction jobs, a ladder can be used as a slide to move roof sheathing to another worker. B—A rack secured to the fascia holds sheathing panels where carpenters working on the roof can easily reach them.

10.24.2 Installing Sheathing

Start installing the sheathing at the eaves and work up toward the ridge. Drive temporary fasteners at the corners, if necessary, to square panels on the rafters. Fasten one edge and then install intermediate fasteners, working inward from the edges. Lay down rows of panels from one edge of the roof to the other. Maintain the same fastening sequence for each panel. This procedure avoids internal stress in the panels. It is advisable to snap a chalk line on the sheathing to mark the center of rafters or trusses. Carpenters can do this on the ground. The factory usually places such lines on OSB sheathing, with different colors to indicate 16″ and 24″ O.C. spans.

When using power nailers or staplers, stand on the panel over the framing to ensure contact with the framing as the fastener is driven. Fastener heads must be driven flush with the panel surface. Maintain a 1/8″ space between edges and ends of panels. A 10d nail can be used to gauge the spacing. It is important to center each panel end on a rafter. Trim panel ends, if necessary.

After all sheathing is installed, trim the extra overhang at the ends of a gable roof. The cutting line should be carefully marked, preferably with a chalk line. It is important for the appearance of the finished roof that this edge is straight.

> **Safety Note**
>
> Use special care in handling sheet materials on a roof, especially if there is wind. You may be thrown off balance or the sheet may blow off of the roof and strike someone.

10.24.3 Panel Clips

Clips are manufactured to strengthen roof sheathing panels between rafters. See **Figure 10-69.** These clips are sometimes called *H clips.* They eliminate the need for blocking on long truss or rafter spans. The clips are slipped onto the panel edges midway between the rafter or truss spans. Two clips should be used where

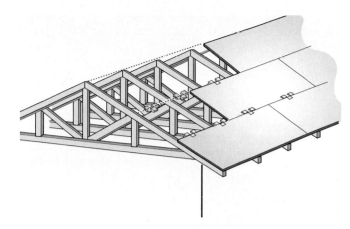

Figure 10-69. Panel clips eliminate blocking on long truss or rafter spans. (The Panel Clip Co.)

supports are 48″ on center. The clips are manufactured to fit five panel thicknesses: 3/8″, 7/16″, 1/2″, 5/8″, and 3/4″. An average house requires 250 panel clips.

> **Safety Note**
>
> Follow these safety rules when working on a roof.
> - When installing any roof covering, it is always good practice to work from some type of scaffolding.
> - On steeper pitches, beginning with 6-in-12 or 7-in-12, work from the scaffold to lay down the ice and water barrier, tar paper, and drip edge. Lay down material only as high as you can reach. At that height, install the first row of roof jacks. Use great care to attach them through the sheathing and into the center of a rafter or truss. Attach planking continuously end-to-end to the jacks through the provided holes. Leave no gaps that might lead to a slip or fall. Lower pitches, such as 4-in-12 and 5-in-12, are considered walkable. Then—though still important—safety measures can be simplified.
> - If shingles have not been delivered to the rooftop, they can be walked up the ladder to the scaffold platform. Use care not to overload the scaffolding. Move the material up the roof jack tiers as needed. Never
>
> *(Continued)*

(Continued)

overload the roof jacks or slam the materials onto the roof.

- Never place tools, materials, or debris where they can slide off of the roof. Never throw any discarded material off of the roof. Doing so can injure someone below.

- If debris of any kind must be cleared from a roof, assign a worker on the ground to secure and rope off the impact area. Maintain proper communication between workers on the roof and the ground during this entire procedure.

- In climates where extreme heat can be expected, keep hydrated to avoid heat stroke and dizziness. In such climates, roofers usually work split shifts (early morning and late afternoon) to avoid dangerous heat conditions.

Fall protection: Familiarize yourself with OSHA rules that may require the use of fall protection, guardrails on scaffolding, or hard hats in certain situations. The basic fall protection rules are:

- Any height above 6′ requires guardrails, safety nets, or a fall arrest system, such as ropes and harnesses.

- Use of body belts as a fall arrest is prohibited because the belts restrict breathing after a fall.

- Employers can choose among a variety of options, depending on circumstances on the job site. Often, a fall warning line is all that is needed.

10.25 Estimating Materials

The number of rafters required for a plain gable roof is easy to figure. Simply multiply the length of the building by 3/4 for spacing 16″ O.C., 3/5 for 20″, or 1/2 for 24″, round up, and add one more. Double this figure to include the other side of the roof. To determine the length of the rafter, use the scale on the framing square as previously described in this chapter.

For example, estimate rafters for the following building:

Building Size 28′ × 40′, roof slope 4-in-12, overhang 2′-0″, and rafter spacing 24″ O.C.

Total rafter run   = 16′-0″
Total rafter length   = 16′-11″
Nearest standard length = 18′-0″
Number of pieces = 2 (Length of wall × 1/2 + 1)
                       = 2 (40 × 1/2 + 1)
                       = 2 × 21
                       = 42
Rafter estimate: 42 pcs. 2″ × 8″ × 18′-0″

When estimating a hip roof, it is not necessary to figure each jack rafter. The number of jack rafters required for one side of a hip is counted to obtain the number of pieces of common rafter stock. This normally supplies sufficient rafter material for the other side of the hip. For a short method on a plain hip roof, proceed as if it is a gable roof. Add one extra common rafter for each hip. Then, figure and add the hip rafters required.

For complex roof frames, it is best to work from a complete framing plan. Apply the methods described for plain roofs to the various sections. Make check marks on the rafters as they are figured so you will not double up or skip members. In estimating material for the total roof frame, remember to include material for ridges, collar ties, and bracing.

To estimate the roof sheathing, first figure the total surface. Then apply the same procedures as used for subflooring and wall sheathing. Since the total area of the roof surface is also needed to estimate shingles, building paper, or other roof surface materials, it is worth the extra time to accurately figure the area.

To figure sheathing for a plain gable roof, multiply the length of the ridge by the length of a common rafter and double the amount. Figure a plain hip roof as though it is a gable roof. However, instead of multiplying the length of a common rafter by the ridge, multiply it by the length of the building plus twice the overhang.

When estimating sheathing for complex roof plans and intersecting roof lines, first determine the main roof areas. Multiply the common rafter length by the length of the ridge and that product by two. Now, add the triangles that make up the other sections located over jack rafters. The area of a triangle equals 1/2 of the base times the altitude. The altitude of most triangular roof areas is the length of a common rafter located in or near the perimeter of the triangle. A plan view of the roof lines is helpful since all horizontal lines (roof edges and ridges) are seen true length and can be scaled. Always add an extra percentage for waste when estimating sheathing requirements for roofs that are broken up by an unusually large number of valleys and hips.

10.26 Model and Small-Scale Construction

Students of carpentry can often gain worthwhile experiences through construction of small portable buildings or scaled-down models. Working with models requires much time. It is often best to construct only part of a building. Small buildings, such as a play house or storage shed, can be sold to recover the cost of materials.

A scale of 1 1/2″ = 1′-0″ usually makes it possible to apply regular framing procedures to the construction of a model that is not too large to handle and store. Cut framing members to their nominal size. For example, a 2 × 4 cut to this scale actually measures 1/4″ × 1/2″, while a 2 × 10 measures 1/4″ × 1 1/4″.

Make all framing materials from clear white pine or sugar pine. Both have sufficient strength and are easy to work. Use small brads and fast-setting glue for assembly.

Materials other than wood can often be simulated from a wide range of items. For example, foundation work can be built of rigid foamed plastic and then brushed with a creamy mixture of Portland cement and water.

Summary

Roof framing must be well designed and strong enough to withstand the stress of high winds and snow loads. There are various types of roofs, but gable and hip roofs are most common. Roof framing may be of the traditional stick-built rafter design or prefabricated roof trusses. Rafters used to construct roof frames are classified as common, hip, valley, or jack rafters, depending on their function and location. Roof framing is a practical application of geometry, requiring careful measurement during the layout process. Carpenters laying out common rafters may use the step-off method, a framing table on the framing square, or a construction calculator to determine the correct length. Gable roofs and gable end frames are the simplest to frame and erect. Hip roofs require more calculation and special cuts for the hip, valley, and jack rafters. Openings must be framed in roofs for chimneys, skylights, or other structures. Such openings are framed using methods similar to floor framing. Shed and gable dormers project from the roof. They require both wall and roof framing techniques. Specific framing techniques are required for flat, gambrel, and mansard roofs. Although roof trusses may be built on site, most are factory built. When being erected, they must be carefully braced to prevent collapse. The roof is structurally completed by applying sheathing to the rafters. Shingles or other weather protection are applied over the sheathing.

Test Your Knowledge

Answer the following questions on a separate piece of paper. Do not write in this book.

1. The _____ load is the external stresses placed on a roof, such as wind and snow.
2. The weight of the lumber and other materials used in the roof is called the _____ load.
3. *True or False?* A hip roof is easier to build because it has only two slopes from a central point or ridge.
4. A type of sloping roof that simplifies the construction of an overhang for all outside walls is called a _____ roof.
5. The pitch of a roof is indicated as the rise over the _____.
6. The tongue of a framing square is _____ inches long.
7. Referring to a chart used with a speed square shown in **Figure 10-14B,** give the length of a common rafter and a hip rafter when the run is 14′.
8. When assembling a roof frame, joints in the ridge should occur at the _____ of a rafter.
9. What is the *common difference?*
10. Figures used to make side cuts for hip and valley rafters are found in the _____ line of the rafter table on the framing square.
11. Hip jack rafters have the same bird's mouth and overhang as _____ rafters.
12. Jack rafters should be erected in _____ to keep the hip or valley rafters straight.
13. What is the *fascia?*
14. A _____ provides additional support when a rafter exceeds the maximum span.
15. The two general types of dormers are _____ and gable.
16. The adjustment made in the lower chord of a roof truss to compensate for sag is called _____.
17. The sheathing on a roof frame provides a nailing base for shingles and also adds _____.
18. The thickness of structural panels required for a sheathing application varies for different roofing materials and different _____.
19. To calculate the area of a plain gable roof, multiply the length of the ridge by the length of a(n)_____ and then double the product.
20. How many rafters are required for a gable roof for a building 42′ long with rafters spaced 16″ O.C., an overhang of 2′-0″, and a slope of 4-in-12?

Curricular Connections

Mathematics. Working from a set of architectural plans for a residential structure, make a layout for a common rafter in one of the roof sections. Use a good, straight piece of stock. A piece of 1″ material may be used. Make the layout by the step-off method and cover all operations, including the shortening at the ridge. When completed, put on a brief demonstration for the class, showing them the procedure you followed.

Science. Use basswood, balsa wood, or similar strip material to make small models of three types of trusses: W truss, Howe truss, and king post truss. All three models should be the same length. Devise a method to measure how large of a load each truss can support before breaking. Conduct your tests on the trusses, then write a report describing your methods and discussing your findings.

Outside Assignments

1. Obtain a set of house plans where the design includes a hip roof and/or inter-secting sections. It should not have a roof framing plan. Study the elevations and detail sections. Then, prepare a roof framing plan. Overlay the floor plan with a sheet of tracing paper. Trace the walls and draw all roof framing members to accurate scale. Be sure to include openings for chimneys and other items.

2. Prepare an estimate of the framing materials required for the roof used in #1. Include the dimensions for all needed lumber. Refer to the detail drawings or specifications to find lumber size requirements. If this information is not included in the plans, find it in the local building code.

Carpenter crews with several members are able to install even large roof trusses without using machinery.

Framing with Steel

Learning Objectives

After studying this chapter, you will be able to:
- List the advantages and disadvantages of steel framing.
- Describe the fastening methods used with steel framing components.
- Explain how wood and steel structural components are combined in floor framing.
- Demonstrate construction of walls using metal studs.
- Explain the use of a jig for fabricating steel roof trusses on the job site.
- Describe the safety precautions that must be used when working with steel framing components.

Technical Vocabulary

Metal studs	Structural sheathing
Self-tapping drywall screws	Track
	Web-type studs
Steel framing members	Weld bead
	X-bracing

Steel framing has been extensively used for years in light commercial construction. Today, mostly for reasons of cost, it is increasingly used in residential construction. Steel became more popular as lumber prices sharply increased in the 1990s. With some additional equipment, carpenters have been able to readily adapt to the different construction methods involved in steel framing.

11.1 Steel Framing

Steel framing members are manufactured in various widths and gages. They are used as studs, joists, and truss rafters, **Figure 11-1.** Most manufacturers use a color code to prevent the different gages from being mixed at the construction site. Prices are based on thousands of lineal feet.

Studs, joists, and **track** are manufactured by brake forming and punching galvanized coil and sheet stock. Strength of the steel varies from one manufacturer to another. Steel components used for structural framing are coated to resist rust and corrosion.

Steel framing members: Manufactured in various widths and gages and are used as studs, joists, and truss rafters.

Track: A steel framing member formed into a U-shaped channel that is used at the top and bottom of a wall. The steel studs fit into the track and are secured by screws or welds.

Sizes of Steel Framing
Studs Widths: 1 5/8"*, 2 1/2", 3 5/8", 4", 6" Gages: 25*, 20, 18, 16, 14, 12
Joists Widths: 4", 6", 8", 9 1/4", 10", 12", 14" Gages: 18, 16, 14, 12
*Non-load-bearing uses only Gage equivalents: 25 = .019", 20 = .0346", 18 = .0451", 16 = .0566", 14 = .0713", 12 = .1017"

Figure 11-1. Steel framing for studs and joists comes in several dimensions. (Journal of Light Construction)

Metal stud systems are most often used for non-load-bearing walls and partitions. Size and spacing is first taken from the architectural drawings. The studs are attached to base and ceiling channels by welding or with screws or clips. Often, wood is used for sole plates and wall plates, **Figure 11-2.** A typical stud consists of a metal channel with openings through which electrical and plumbing lines can be installed. See **Figure 11-3.**

Wall surface material, such as drywall or paneling, is attached to the metal studs with an adhesive or with *self-tapping drywall screws.* Some *web-type studs* have a special metal edge with a gap into which nails can be driven.

Framing with steel requires some special tools. These include a variable speed drill and screw gun, hearing protectors, clamping pliers, metal snips, metal punch, metal cutoff blade for a portable saw, magnetic level, metal cutoff saw, and a right-angle drill. In some cases, welding equipment is needed.

Safety Note

Be sure all portable electric power tools are properly grounded to avoid shock hazards when cutting or fastening steel framing members. Be careful to avoid cutting the insulation on power cords by contact with sharp metal edges.

A

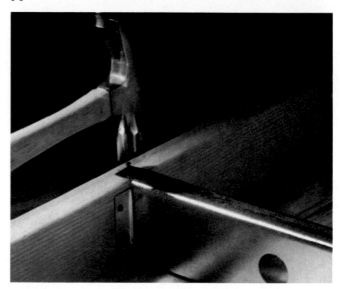

B

Figure 11-2. Some steel framing is designed to be fastened to wood plates using hammers, nailers, or screw guns. A—Using a nail gun. B—Using a hammer. (H.L. Stud Corp.)

Self-tapping drywall screws: Screws used to attach wall surface material, such as drywall or paneling, to metal studs.

Web-type studs: Studs with a special metal edge containing a gap into which nails can be driven.

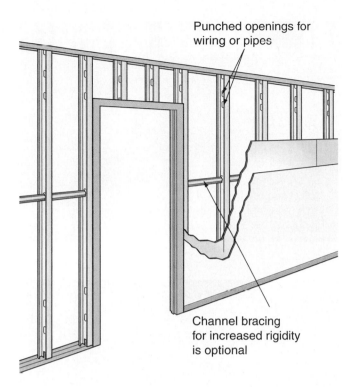

Punched openings for wiring or pipes

Channel bracing for increased rigidity is optional

Figure 11-3. Metal studs are stamped and prepunched to accommodate installation of mechanical systems, such as electrical wiring and plumbing. (National Gypsum Co.)

Figure 11-4. A wall section combining steel and wood. The wall section is so light one carpenter can carry it. (H.L. Stud Corp.)

11.1.1 Advantages of Steel Framing

Steel framing members, though slightly more expensive than wood, have remained relatively unchanged in price as the price of lumber has been rising. Steel members are lighter than their wood counterparts. This makes it easier for the carpenter to assemble and erect wall assemblies, **Figure 11-4.** Studs and joists are normally straight and consistent in size. They do not shrink, warp, swell, or have knots and other imperfections that affect the quality of the construction. There is also little limitation in the length of steel framing. Joists can be manufactured in lengths up to 40′.

Steel framing is noncombustible, does not absorb moisture, is impervious to insect destruction, and does not support mold or fungus. Steel framing can be designed to withstand the destructive forces of high winds, tornados, hurricanes, and earthquakes. Its strength and ductility properties allow it to easily meet the wind and seismic ratings established by the International Building Code.

Industry sources claim that scrap and waste amounts to only 2%, compared to 20% for lumber, **Figure 11-5.** Steel scrap is fully recyclable, while wood is not. On average, new steel manufactured today contains 25% scrap steel. This reduces the pressure on finite resources and the load on valuable landfill space.

When steel framing is used in commercial buildings, it has a distinct advantage over wood due to the different construction methods used. In commercial structures, the framework is typically made up of beams and columns of steel or reinforced concrete. This framework supports the floors, which are always installed before the walls. The floors always have a camber; they are not level. Further, mechanical systems are also installed before the walls. Taking into consideration all of these factors, it is necessary to install the walls piece by piece. First, the carpenter installs the plates and then the studs, fastening them around the mechanical systems in the wall. See **Figure 11-6.** If wood is used, each stud must be very accurately

Figure 11-5. The scrap resulting from cutting steel studs to size is recyclable. Note that this carpenter is wearing both safety glasses and hearing protection. (Jack Klasey)

A

measured and cut—a time-consuming, labor-intensive job. Steel studs, by contrast, can be cut in bundles of ten to within an inch or more of accuracy, nested in the channels of the plates, and fastened with a weld or a screw.

11.1.2 Disadvantages of Steel Framing

Steel has disadvantages that impact safety and cost. The standard stud for non-load-bearing walls has a metal thickness of 25 gage and is flimsy. Care must be taken when handling them. Steel edges are sharp and may cause injury, even when hemmed components are used, **Figure 11-7.**

Even though steel prices are competitive with wood, engineering and labor costs tend to be higher. Crews trained to frame in steel are hard to find. It also takes longer to drive screws into metal than to drive nails into wood. Installing blocking, sheathing, and siding takes longer with steel studs than with wooden studs.

In cold climates, there is an added cost for the thermal breaks needed to conserve energy. Steel also suffers a cost penalty in extra insulation for exterior walls. Some carpenters do not use steel studs on outside walls because they are poor

B

Figure 11-6. Installing steel framing in a commercial building. A—Plates are fastened in place where the walls will be located. The channels are fitted over or around mechanical systems, such as the electrical conduit in the foreground. B—Studs are installed into the floor and wall channels and fastened with screws or welding. These extra-long studs allow space for HVAC ducts and other systems above a suspended ceiling.

Track options

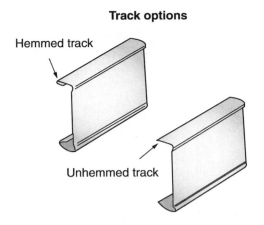

Hemmed track

Unhemmed track

Figure 11-7. Steel framing comes hemmed or unhemmed. A hemmed channel reduces the chance of injuries from sharp steel edges. (Journal of Light Construction)

insulators. The R-value of a wall can be reduced by as much as 50% when steel members are used.

11.2 Framing Floors

The methods for using steel in floor construction vary. Steel framing members can be exclusively used or combined with wood members. Depending on the materials, various fastening methods are used. **Figure 11-8** shows various methods of attaching floor joists.

Self-tapping screws are often used to fasten steel to steel. Special nails may be used to attach steel members to wood framing members. If steel

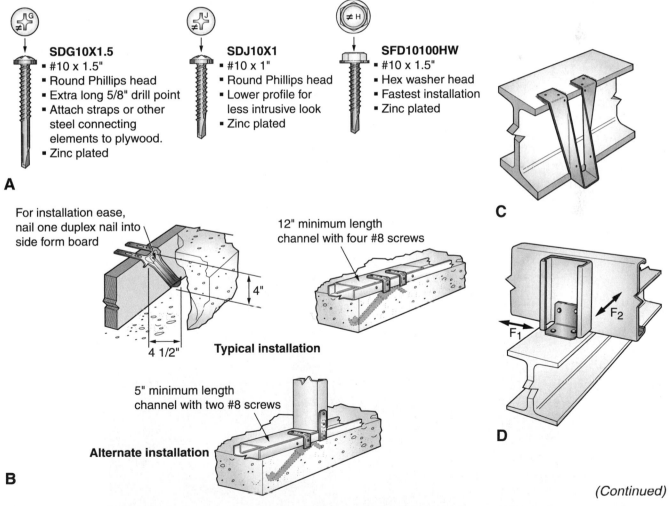

SDG10X1.5
- #10 x 1.5"
- Round Phillips head
- Extra long 5/8" drill point
- Attach straps or other steel connecting elements to plywood.
- Zinc plated

SDJ10X1
- #10 x 1"
- Round Phillips head
- Lower profile for less intrusive look
- Zinc plated

SFD10100HW
- #10 x 1.5"
- Hex washer head
- Fastest installation
- Zinc plated

A

For installation ease, nail one duplex nail into side form board

4"

4 1/2"

Typical installation

12" minimum length channel with four #8 screws

5" minimum length channel with two #8 screws

Alternate installation

B

C

F₁ F₂

D

(Continued)

Figure 11-8. Fasteners and joints for metal floor framing. A—Screw fasteners designed for light-gauge steel construction. Long drill points allow for installation through several layers of steel before the threads engage. Note the markings on the screw heads: G has an extra long point; J has rounded head for lower profile; H has a hex head for faster installation. B—A mudsill anchor embedded in concrete. C—A hanger that can be attached by welding or screws. D—This angle bracket attaches the joist to the I-beam. E—Tension-type bridging.

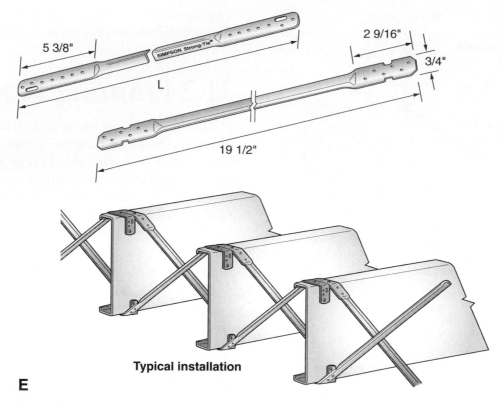

Typical installation

E

Figure 11-8. Continued.

frames are welded, electric arc welding equipment is required. Pneumatic or electric drills are used to drive screws. Some builders prefer powder-actuated tools to attach subflooring, such as oriented strand board, to steel joists.

Refer to Chapter 5 for additional information on power tools. Subflooring also can be attached to joists with adhesives and self-tapping screws.

Figure 11-9 shows a floor framed entirely in steel. The headers (band joists) are fastened to

Figure 11-9. A floor frame done in steel. The headers are attached to the foundation wall with steel fasteners driven in by a powder-actuated tool.

the foundation with steel pins. A powder-actuated tool drives the pins through a flange on the header into the concrete wall. Joists spaced 16″ on center are first fastened to the headers with screws, then welded. Another method involves attaching a metal pan across the joists and pouring a 2 1/2″–3″ thick, fiber-reinforced concrete floor on top of the joists, **Figure 11-10.**

A

B

Figure 11-10. A steel-framed concrete floor. A—The view from the underside. Joists are one piece from wall to wall. The steel I-beam provides support. The floor pan has been attached to the joists with screws in preparation for pouring a concrete floor. B—A thin concrete base has been poured over the floor joists. The concrete is reinforced with fibers to prevent cracks.

11.3 Framing Walls and Ceilings

As noted earlier, steel-framed residential construction is increasing in popularity, **Figure 11-11.** *Metal studs* can be used with

Figure 11-11. This building is being constructed entirely with metal framing members. Because steel is light, studs must have bracing, such as the x-braces shown, to maintain spacing and rigidity.

either metal plates or wood plates. Joints are fastened with self-tapping screws or welds, **Figure 11-12.** Some carpenters first use screws to hold the joint and then secure it with a weld.

Although several types of welding are used on the light gage steel, shielded metal arc welding (SMAW) is the most popular. SMAW equipment is portable and the welds are strong. A 200A (ampere) "hot box" electric welder or a 200A gasoline generator is adequate as an electrical power source. Wherever possible, welds are made before walls are erected, **Figure 11-13.**

As with wood construction, headers must be installed over openings in load-bearing walls. **Figure 11-14** shows a window roughed out in steel and a first-floor frame completely erected. As with wood wall framing, proper fastening and bracing is important. Wall sections should have horizontal bridging to space studs, **Figure 11-15.** Temporary diagonal bracing of sections may be required to keep the walls square and plumb until permanent bracing can be installed. Permanent bracing of load-bearing walls is needed to withstand shear forces that could cause racking. *Structural sheathing* is normally applied to both sides of the wall frame and sufficient for this purpose. Type 2 plywood or oriented strand board are

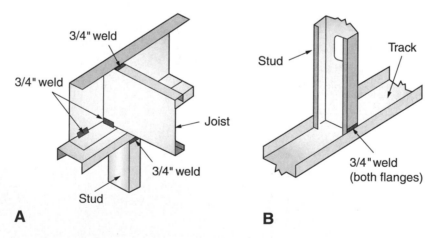

A   **B**

Figure 11-12. Joint assembly details. A—Assembling and welding a ceiling joist and a header in a load-bearing wall. B—A welded sole plate and stud assembly. (CEMCO)

Metal studs: Studs used with either metal or wood plates.

Structural sheathing: Type 2 plywood or oriented strand board, normally applied to both sides of the wall frame. Used for permanent bracing of load-bearing walls.

Figure 11-13. A carpenter welding cripple studs in a rough opening for a window. Note the use of heavy gloves and a welding helmet.

rated for wind speeds of 100 mph. Sheets should be applied with the long side parallel to the studs. If plywood is used, it should extend from the top track to the bottom track. Shear strength can also be assured by using *x-bracing*. The x-braces are steel straps that diagonally extend from the top to the bottom of the wall. They are attached to the tracks and studs with screws. Unlike diagonal bracing used in wood framing, they rest entirely on the surface of the wall frame (not let in).

Construction of steel-framed partitions is similar to wood construction. Sills, plates, and studs are used. Plumbing and electrical systems make use of prepunched or specially cut openings in the studs. Drywall or other surfacing material is attached to the studs with self-tapping screws. See **Figure 11-16.** A thermal break or insulation should be applied to the outside face of exterior walls, **Figure 11-17.** This improves the R-value of the outside wall by preventing the transfer of heat through the metal studs and joists.

Installation of drywall, often called gypboard, is similar to installation over wood framing. Construction adhesive can be used to attach it, but if the steel is oil treated, screws must be used. Install drywall with its length parallel to the studs and joists. If a second layer is to be used, install it horizontally. Space fasteners 24″ O.C. for the base layer and 16″ O.C. for the face layer.

A

B

Figure 11-14. Wall framing. A—Squaring up wall sections and preparing to secure a corner. Note the header over the window opening and the temporary angle brace. B—The first-floor wall is erected and preparations are under way for placing ceiling joists.

X-bracing: Steel straps that diagonally extend from the top to the bottom of the wall. They are attached to the tracks and studs with screws.

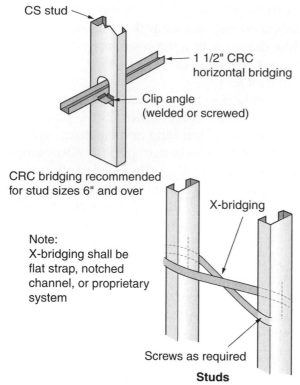

CS stud

1 1/2" CRC
horizontal bridging

Clip angle
(welded or screwed)

CRC bridging recommended
for stud sizes 6" and over

X-bridging

Note:
X-bridging shall be
flat strap, notched
channel, or proprietary
system

Screws as required

Studs

Solid blocking:
• Locate blocking at each end of wall,
  adjacent to openings, and as required
• For track:
  Where blocking material thickness
  allows, notch and bend flanges 90°
  or anchor to verticals w/clip angles

Screw as
required at
each stud

Flat straps, notched channel,
X-bridging, or proprietary
bridging (each side) lap
splice straps minimum 4" (10 cm)

Screws as required
(each side)

Note:
Number of rows of bridging as required by design

Figure 11-15. Details for spacing and bracing steel wall frames. (CEMCO and Dale Industries Inc.)

A

B

C

Figure 11-16. Steel-framed partitions. A—Water and waste piping is installed through openings cut in the steel studs. Protective bushings are used where flexible piping passes through a stud. B—Rigid electrical conduit is routed through openings in studs and connected to junction boxes that are attached to the studs with self-tapping screws. C—Screws are also used to mount drywall on the steel studs.

Figure 11-17. Thermal insulation is installed between the steel studs and a concrete exterior wall to control heat transfer. This electrician is installing conduit in the wall. (Jack Klasey)

11.4 Framing Roofs

Like their wooden counterparts, steel roof trusses are normally engineered and constructed by companies specializing in this work. However, they can be fabricated on-site using the same methods and calculations as for wooden members. In such cases, be sure to follow engineering specifications.

Truss members are cut with a portable electric saw, **Figure 11-18.** Next, the chords and webs are placed in a jig set up on the ground, **Figure 11-19.** The lapped and butt joints are fastened with self-tapping screws, a *weld bead*, or both. Welds are made with arc welding or submerged arc welding equipment, **Figure 11-20.**

Erection of the steel trusses follows the same procedure as for wood trusses. After they are

Weld bead: The thickened area of metal forming a joint between two pieces that have been melted together by the welding arc.

fastened to the wall plate, permanent bracing is welded to the webs. **Figure 11-21** shows prefabricated trusses in place, fastened to the plate, and permanently braced. Plywood or OSB sheathing is then fastened to the rafters with sheet metal screws, **Figure 11-22.**

Figure 11-18. Cutting steel rafters to length using a portable power saw fitted with a carbide blade.

Figure 11-19. A wooden jig holds steel truss chords and webs as carpenters secure joints with self-tapping screws.

Figure 11-20. Welding the metal trusses. This may be done while the truss is in the jig or after it is stacked and ready for installation.

Figure 11-22. A screw gun is used to attach sheathing to steel rafters. Self-tapping screws are used. It is a good idea to mark the center of the rafter with a chalk line.

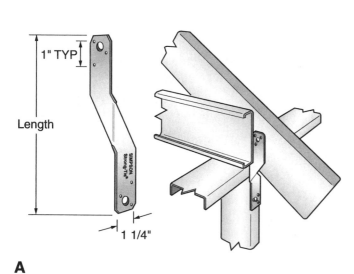

A

B

C

Figure 11-21. Steel truss installation. A—This all-purpose tie is suitable for securing steel trusses to the top plate, as shown. B—Fastening a gable-end stud to the top chord. C—Rafter installation is complete when bracing (lighter-colored pieces) has been installed. (©Simpson Strong Tie Company, Inc.)

Summary

Steel framing has long been used in commercial construction, but is becoming more widespread in residential work. Framing members are formed out of light-gage steel and fastened in place with self-tapping screws or by welding. Some structures combine wood and steel framing. Steel framing has a number of advantages and disadvantages when compared to conventional wood construction. Construction and erection methods used for steel framing are similar to those used for wood. Extra caution must be used when handling steel framing materials to avoid injury caused by sharp edges.

Curricular Connections

Social Studies. Steel framing is often used in light commercial construction, but has been slow to gain acceptance in residential building. Interview a small general contractor in your area that does primarily residential construction. With the contractor's permission, record your conversation as you explore his or her views on steel framing materials and whether they will become more widely used in home building. Unless the interview is very long, play it for the class.

Science. Steel framing members are sometimes joined by the use of arc welding. Research the subject of arc welding to find out how the process permanently joins two pieces of metal. Make sketches showing the welding process and how a weld joint holds two metal pieces together. Use them to illustrate a short written report.

Test Your Knowledge

Answer the following questions on a separate piece of paper. Do not write in this book.

1. Pricing for steel construction materials is based on _____.
2. Steel studs, joists, and track are manufactured by _____ forming and _____.
3. Which of the following fastening methods is *not* used to join steel components to other steel components?
   A. screws
   B. nails
   C. weld beads
4. *True or False?* Framing in steel requires the same tools and methods as wood framing.
5. *True or False?* Steel and wood framing members can be combined in the same structure.
6. On average, new steel manufactured today contains _____% scrap steel.
7. Steel band joists are often fastened to foundations using _____.
8. What is the most popular type of welding used on steel construction?
9. If steel studs are oil treated, _____ must be used to fasten drywall to the studs.
10. *True or False?* Steel roof trusses may be built on site.

Outside Assignments

1. Visit a lumberyard or building supply company in your area and secure literature on steel framing materials. Report to your class information on available sizes, description of units, and costs. Compare these costs with similar members in wood.
2. Contact local building contractors to find one that uses steel framing. If you find such a contractor, try to arrange a class field trip to observe steel framing methods.
3. With your instructor's permission, secure a video on steel framing to show to the class.

Section 3

Closing In

Simpson Door Co.

Roofing Materials and Methods

Learning Objectives

After studying this chapter, you will be able to:

- List the covering materials commonly used for sloping roofs.
- Define roofing terms.
- Describe how to prepare the roof deck.
- Describe reroofing procedures for both asphalt and wood shingles.
- Demonstrate correct nailing patterns.
- Select appropriate roofing materials for various slopes and conditions.
- Describe the application procedure for a built-up roof.
- Explain how various roofing products are applied.
- Demonstrate the proper positioning of gutters.
- Estimate materials needed for a specific roofing job.

Technical Vocabulary

Asphalt shingles
Base flashing
Battens
Built-up roof
Cant strip
Cap flashing
Closed-cut valley
Counterflashing
Coverage
Drip edge
Eaves flashing strip
Eaves trough
Exposure
Flashing
Four-inch method
Gravel stop
Gutter
Head lap
Ice-and-water barrier
Nailer boards
Open valley
Plies
Roll roofing
Roofing tile
Saddle
Saturated felts
Selvage
Shingle butt
Side lap
Six-inch method
Square
Starter strip
Sweat sheet
Terne metal roofing
Underlayment
Wood shakes
Woven valley

Roofing materials protect the structure and its contents from the sun, rain, snow, wind, and dust. In addition to weather protection, a roof should offer some measure of fire resistance and be extremely durable. Since a large amount of surface is usually visible, especially on sloping roofs, the roofing materials can contribute to the attractiveness of the building. Roofing materials can add color, texture, and pattern, **Figure 12-1.**

Preparing a roof for its protective covering involves a number of operations. Most of these operations must follow a definite sequence. All items that project through the roof should be built or installed before roofing begins. These include chimneys, vent pipes, and special facilities for electrical and communications service. Performing any of this work after the finished roof is applied could damage the roof covering.

Figure 12-1. The roofing material not only protects the structure from weather, it adds to the overall appearance.

12.1 Types of Materials

Materials used for pitched (sloping) roofs include:

- Asphalt, wood, metal, and mineral fiber shingles.

- Slate.

- Tile made from clay or cement.

- Sheet materials, such as rubberized single-ply membrane, roll roofing, galvanized iron, aluminum, tin, and copper.

For flat or low-slope roofs, a membrane system is used. It consists of a continuous, watertight surface usually obtained by using a built-up roof or seamed-metal sheets.

Built-up roofs are fabricated on the job. Roofing felts are *laminated* (stuck together) with asphalt or coal tar pitch. Then, hot asphalt is mopped over the felt layers. Finally, this surface is coated with crushed stone or gravel.

Metal roofs are assembled from flat sheets. Seams are soldered or sealed with special compounds to ensure that the roof is watertight.

When selecting roofing materials, it is important to consider such factors as:

- Initial cost

- Maintenance costs

- Durability

- Appearance

Built-up roof: A roof covering composed of several layers of felt or jute saturated with coal tar, pitch, or asphalt. The top is finished with crushed slag or gravel. Generally used on flat or low-pitched roofs.

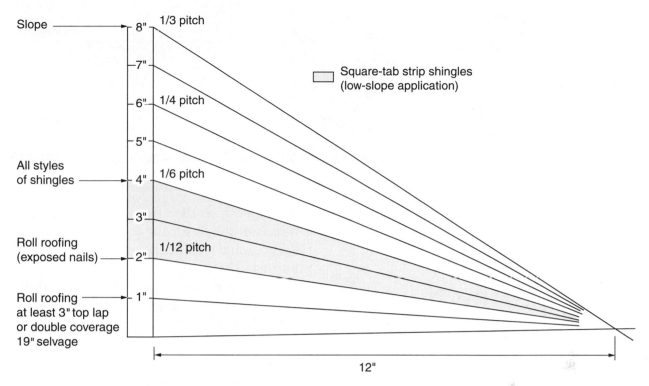

Figure 12-2. Roofing manufacturers list minimum slope requirements for various asphalt roofing products. (Asphalt Roofing Manufacturers Assn.)

The pitch of the roof also determines the selection. Low-sloped roofs require a more watertight system than steep roofs, **Figure 12-2.** Materials such as tile and slate require heavier roof frames.

Local building codes may prohibit the use of certain materials. In some cases, the materials may be a fire hazard. Some materials cannot resist high winds or other elements found in a certain locality.

Square: A unit of measure equal to 100 square feet. It is applied to roofing material and to some types of siding.

Coverage: The amount of weather protection provided by the overlapping of shingles.

Exposure: The amount of material exposed to the weather in siding or shingles.

Head lap: The distance in inches from the butt of an overlapping shingle or roll roofing to the top edge of the shingle or roll roofing beneath.

12.2 Roofing Terminology

There are a number of specialized terms used in the roofing trade. Refer to **Figure 12-3.** Two of these, slope and pitch, are defined in Chapter 10, **Roof Framing.** Another widely used term is the *square.* This is the unit of measure for estimating and purchasing shingles. It is the amount of roofing material needed to provide 100 sq. ft. of finished (shingled) roof surface.

Coverage is the amount of weather protection provided by the overlapping of shingles. Depending on the material and method of application, shingles may furnish one, two, or even three thicknesses of material on the roof. This is referred to as *single coverage, double coverage,* or *triple coverage.*

Exposure is the distance between the edge of one course of shingles and the edge of the next higher course. The distance from the lower edge of an overlapping shingle or sheet to the top edge of the shingle or sheet beneath is called **head lap.** The overlap length for side-by-side

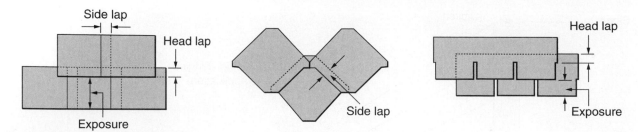

Figure 12-3. Exposure, head lap, and side lap are important terms used in the application of roofing materials.

elements of roofing is *side lap*. A *shingle butt* is the lower, exposed edge of a shingle.

12.3 Preparing the Roof Deck

The roof sheathing should be smooth and securely attached to the frame. It must provide an adequate base to receive and hold the roofing nails and fasteners. In areas subject to hurricane-force winds, ring-shank nails should be used to install sheathing because of their greater holding power. The state of Florida requires sheathing on all residential pitched roofs to be fastened with 8d ring-shank nails spaced 6″ on center.

Inspect the roof deck to see that nailing patterns are complete and that there are no nails sticking up. Joints should be smooth and free of sharp edges that might cut roofing materials. Repair large knot holes over 1″ diameter by covering with a piece of sheet metal. Clean the roof surface of chips or other scrap material.

All types of shingles can be applied over solid sheathing. However, skip sheathing or application of shingle breather underlayment are better choices for wood shingles or shakes. Wood absorbs a certain amount of moisture. Either application method provides ventilation and promotes faster drying after a rain. Shingle breather underlayment is discussed later in this chapter. Solid boards have often been used for skip sheathing. Boards over 6″ wide should not be used as sheathing, however.

Attics should be properly ventilated to remove moisture. Moisture vapor originating in lower stories may sometimes enter the attic. If the vapor becomes chilled below the dew point, it will condense on the underside of the roof deck. This causes the sheathing (especially sheet materials) to warp and buckle. To avoid this condition, install openings in locations that provide adequate ventilation, such as in the soffits or gable end. Vents or louvers should provide 1/2 sq. in. of opening per square foot of attic space.

12.4 Asphalt Roofing Products

Asphalt roofing products are widely used. These products fall into three broad groups: saturated felts, roll roofing, and shingles.

Saturated felts are used under shingles as sheathing paper. They are also used as laminations for built-up roofs. Saturated felts are made of dry felt soaked with asphalt or coal tar. They are available in different weights, with 15 lb. the most common. This number indicates the weight of the felt needed to cover 100 sq. ft. of the roof deck with a single layer.

Roll roofing and shingles are outer roof coverings. They must be weather-resistant. Their base material is organic felt or fiberglass. This base is saturated and then coated with a

---

Side lap: The amount of overlap for side-by-side elements of roofing.

Shingle butt: The lower, exposed edge of a shingle.

Saturated felts: Dry felt soaked with asphalt or coal tar. Used under shingles for sheathing paper and as laminations for built-up roofs.

Roll roofing: Mineral granules on asphalt saturated felts or fiberglass. Roll roofing is the uncut form of mineral-surfaced shingle material.

special asphalt that resists weathering. A surface of ceramic-coated, opaque mineral granules is then applied. The mineral granules shield the asphalt coating from the sun's rays, add color, and provide protection against fire. Asphalt shingles are available in varying degrees of durability. Shingles come in durability ratings of 15 to 40 years.

Figure 12-4 displays data on shingles and other asphalt products. Additional products within each group differ in weight and size. For example, the three-tab, square-butt shingle is

| Product | Configuration | Per square | | | Size | | Exposure | Underwriters laboratories listing* |
		Approximate shipping weight	Shingles	Bundles	Width	Length		
Self-sealing random-tab strip shingle 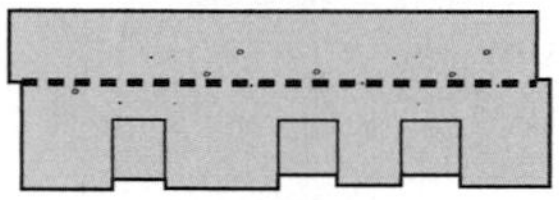 Laminates	Various edge, surface texture and application treatments	285# to 390#	66 to 90	3 to 5	11 1/2" to 14"	36"	4" to 6"	A or C Many wind resistant
Self-sealing random-tab strip shingle 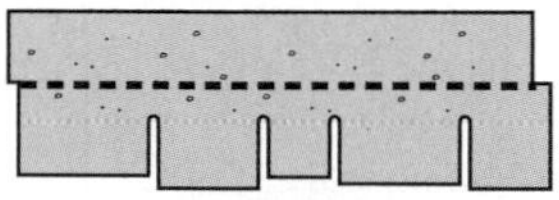 Single-thickness	Various edge, surface texture and application treatments	250# to 300#	66 to 80	3 or 4	12" to 13 1/4"	39.37	5" to 5 5/8"	A or C Many wind resistant
Self-sealing square-tab strip shingle 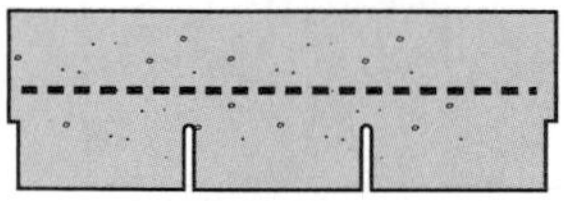 Three-tab	Two-tab or Four-tab	215# to 325#	66 to 80	3 or 4	12" to 13 1/4"	39.37	5" to 5 5/8"	A or C All wind resistant
	Three-tab	215# to 300#	66 to 80	3 or 4	12" to 13 1/4"	39.37	5" to 5 5/8"	
Self-sealing square-tab strip shingle 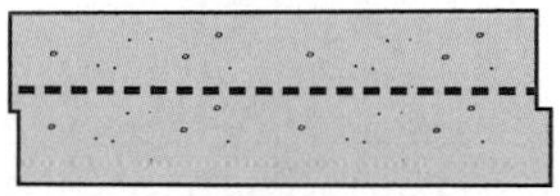 No-cutout	Various edge and surface texture treatments	215# to 290#	66 to 81	3 or 4	12" to 13 1/4"	39.37	5" to 5 5/8"	A or C All wind resistant
Individual interlocking shingle 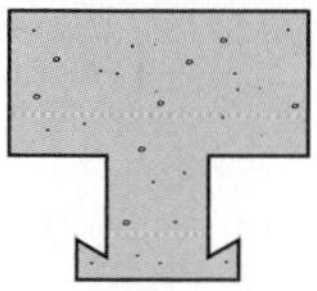 Basic design	Several design variations	180# to 250#	72 to 120	3 or 4	18" to 22 1/4"	20" to 22 1/2"	—	C Many wind resistant

(Continued)

Figure 12-4. This chart provides installation data for common asphalt roofing products. *UL rating at time of publication. Refer to the manufacturer's product literature at the time of purchase. (Asphalt Roofing Manufacturers Assn.)

Product	Approximate shipping weight		Square per package	Length	Width	Side or end lap	Top lap	Exposure	Underwriters laboratories listing*
	Per roll	Per square							
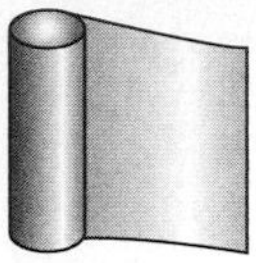 Mineral surface roll	75# to 90#	75# to 90#	1	36' to 38'	36"	6"	2" to 4"	32" to 34"	C
	Available in some areas in 9/10 or 3/4 square rolls.								
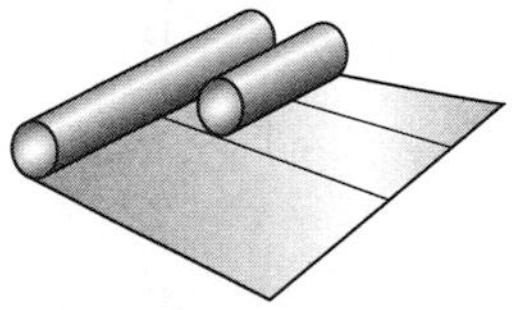 Mineral surface roll (double coverage)	55# to 70#	110# to 140#	1/2	36'	36"	6"	19"	17"	C
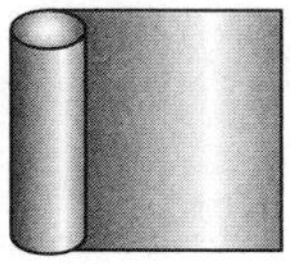 Smooth surface roll	40# to 65#	40# to 65#	1	36'	36"	6"	2"	34"	None
 Saturated felt (nonperforated)	60#	15# to 30#	2 to 4	72' to 144'	36"	4" to 6"	2" to 19"	17" to 34"	None

Figure 12-4. Continued.

available in many quality levels and colors and in weights from 215 to 300 pounds per square.

Asphalt shingles are the most common type of roofing material used today. They are manufactured as strip shingles, interlocking shingles, and large individual shingles. Dimensions of a standard three-tab strip shingle are shown in **Figure 12-5.** Most shingles have a strip of factory-applied, self-sealing adhesive. Heat from the sun softens the adhesive. This results in a bond between each shingle tab and the shingle below. This bond prevents tabs from being raised by wind. The self-sealing action usually takes place within a few days during warm weather. In winter, the sealing time can be considerably longer, depending on the climate.

Asphalt shingles: The most common type of roofing material used today. Manufactured as strip shingles, interlocking shingles, and large individual shingles.

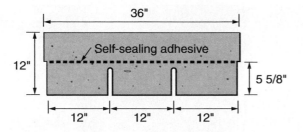

Figure 12-5. These are the most common dimensions for a standard three-tab asphalt strip shingle. Metric sizes measure 1000 mm by 333 mm.

Safety Note

Safety considerations are very important in roofing work. Be sure to erect a secure scaffold that will support workers at a waist-high level with the eaves. Study Chapter 31, Scaffolds and Ladders, for more information.

12.4.1 Underlayment

A roof *underlayment* is a thin cover of asphalt-saturated felt or other material. It has a low vapor resistance. This underlayment serves these purposes:

- Protects the sheathing from moisture until the shingles are laid.
- Provides additional weather protection by preventing the entrance of wind-driven rain and snow.
- Prevents direct contact between shingles and resinous areas in the sheathing.

Underlayment: A thin cover of asphalt-saturated felts or other material that protects sheathing from moisture until shingles are laid, provides weather protection, and prevents direct contact between shingles and resinous areas in the sheathing.

Drip edge: A metal edging placed along the rake and eaves before installing shingles.

Ice-and-water barrier: A self-sealing waterproof covering installed at eaves and in valleys before installing roofing.

Materials such as coated sheets or heavy felts should not be used. They may act as a vapor barrier and allow moisture and frost to gather between the covering and the roof deck. Although 15 lb. roofer's felt is most commonly used as an underlayment, requirements vary depending on the kind of shingles and the roof slope.

General application standards for underlayment suggest a 2″ head lap at all horizontal joints and a 4″ side lap at all end joints. Lap at least 6″ on each side of the centerline of hips and valleys.

Working Knowledge

Do not put down underlayment on a damp roof. It may trap moisture and damage the roof.

12.4.2 Drip Edge

The roof edges along the eaves and rake (the inclined edge of a gable roof) should have a metal *drip edge*, **Figure 12-6.** Various shapes formed from 26 gage galvanized steel are available. They extend back about 3″ from the roof edge and are bent at the factory to go downward over the edge. This causes the water to drip free of the underlying cornice construction. It also keeps the shingles from drooping over the edge when heated by the sun.

Install the drip edge with shingle nails 8″ to 10″ apart. Keep the nails near the upper or inner edge. Lap joints in the drip edge about 2″. At the eaves, the underlayment should be laid over the drip edge. At the rake, place the underlayment under the drip edge.

12.4.3 Ice-and-Water Barrier

It is recommended that an *ice-and-water barrier* be installed at the eaves for buildings in cold climates. Installed on new construction or during reroofing, it prevents leak-through from ice dams or wind blown rain. It is sold under various trade names, such as Winter-Guard, Water Shield, and Weather Watch. These

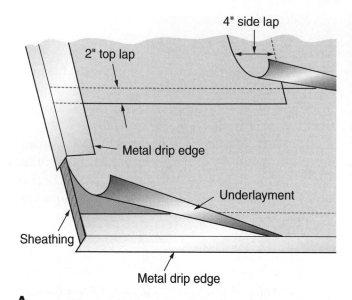

A

B

Figure 12-6. Drip edge. A—Proper application of underlayment and metal drip edge. B—This carpenter is installing a drip edge on the rake with a pneumatic nailer during a reroofing job.

materials are self-sealing around nails and deck joints.

Available in 36″ wide rolls, some types have an adhesive backing and some are reinforced with fiberglass. Nonadhering types may require a coat of hot tar on the sheathing to hold the fabric in place.

The lower edge of the ice-and-water barrier should be placed even with the lip of the drip edge. When using the adhesive-backed type, try to start the strip parallel with the drip edge. The barrier should cover the sheathing and drip edge from the roof's edge to at least 24″ inside the building's outside wall line. See **Figure 12-7.**

Figure 12-7. In cold climates, put down an ice-and-water barrier and extend it up the roof slope beyond the building line.

12.5 Flashing

Intersections with other roofs, adjoining walls, and such projections as chimneys and soil stacks complicate the installation of roof coverings. Making these areas watertight requires a special building material called flashing. *Flashing* is water-resistant sheet material designed to keep joints in the roof watertight. Materials used for flashing include: tin-coated metal, zinc-coated (galvanized) metal, copper, lead, aluminum, asphalt shingles, and roll roofing.

12.5.1 Valley Flashing

A *valley* is the junction of two sloping roofs. The slopes direct water into this area. Thus, water drainage is heavier at this point, so it is important that the area is protected against leakage. Flashing is a method of sealing valleys against leakage. Proper installation of flashing

Flashing: Sheet metal or other material used in roof and wall construction (especially around chimneys and vents) to prevent rain or other water from entering.

for the type of valley shingle application used is critical to producing a roof that does not leak.

The following procedure is recommended by the Asphalt Roofing Manufacturers Association. First, center a 36″ wide strip of asphalt-saturated felt in the valley. Sparingly secure it, using only enough nails to hold it down. Then, lay down horizontal courses of felt underlayment. Lap each strip over the valley strip at least 6″.

Further preparation depends on how you plan to shingle the valley:

- Open valley.
- Woven valley.
- Closed-cut valley.

Whatever the style of flashing, make certain that it is smooth, has no obstructions, has the capacity to quickly move water away, and can handle occasional water backup.

12.5.2 Installing Open Valley Flashing

For valley flashing under asphalt shingles, manufacturers recommend 26 gage galvanized sheet metal or a similar material that will not corrode or leave stains. Install *open valley* flashing as shown in **Figure 12-8**. In areas where rains are often heavy, it is best to lay down ice-and-water barrier under the flashing. Fasten the sheet metal flashing without puncturing the surface. All underlying joints are lapped at least 12″.

Before applying the shingles, snap a chalk centerline in the valley. Also snap a line on each side of the center to mark the width of the waterway. This should be 6″ wide at the ridge and gradually widened. The lines should move away from the valley at the rate of 1/8″ for every foot as they approach the eave. A valley 8′ long would be 7″ wide at the eave. When a course

Open valley: Method of roofing in which the roof covering is set back from the middle of the flashed valley.
Woven valley: Method of shingling valleys in which shingles run across the valleys.

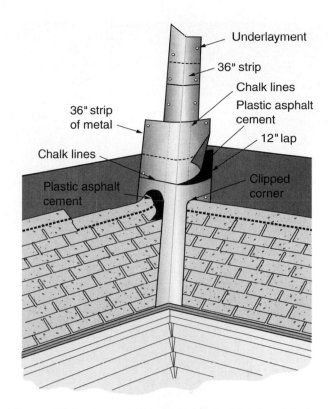

Figure 12-8. Open valley flashing is laid down before shingling begins. The bottom is trimmed to match the eave line. Shingles should be sealed with asphalt cement where they lap over the flashing.

of shingles meets the valley, the outside chalk lines serve as guides in trimming the last unit. After the shingle is trimmed, cut off the upper corner at about a 45° angle with the valley line. Cement the end of the shingle over the flashing with asphalt cement.

12.5.3 Installing Woven Valley Flashing

Some roofers prefer to run shingles across valleys to create a *woven valley*. This method is often used when reroofing. See **Figure 12-9**. Only asphalt strip shingles may be applied this way.

Flashing must be wide enough to straddle the valley with a minimum of 12″ of material on either side. In order to provide this margin, some of the preceding shingle strips must be cut away. Fasteners must be kept at least 6″ away from the valley centerline.

A

36" roll roofing
50 lb. or heavier

Nails 6" minimum
from valley

Extend each strip
at least 12" beyond
center of valley

Two extra nails
in end of strip

B

Another method uses ice-and-water barrier for flashing. It may be installed either parallel to the valley or across it. If the barrier material is run across the valley, allow a 6" lap.

When laying shingles across the valley, firmly press them into place. Keep fasteners at least 6" away from either side of the centerline. Use two nails at the end of each terminal strip.

PROCEDURE

Installing a woven valley

1. Apply a 36" wide strip of 50 lb. (minimum) roll roofing or a 36" wide ice-and-water barrier across the valley.
2. Lay the first course of shingles along the eave of one roof surface.
3. Extend one strip at least 12" across the valley.
4. Lay the first course on the intersecting roof and extend it across the valley over the previously applied shingle.
5. Alternate succeeding courses, first along one roof surface and then the other. Refer again to **Figure 12-9.**

12.5.4 Installing Closed-Cut Valley Flashing

Another method of flashing a valley is the *closed-cut valley* method. In this method, the two intersecting roof surfaces (except the first course) are individually shingled. Refer to **Figure 12-10.**

C

Figure 12-9. Woven valley. A—Ice-and-water barrier is recommended in addition to valley flashing. B—Method of laying woven valley shingles. C—An example of woven valley shingling.

Closed-cut valley: Method of shingling valleys by carrying each course across the valley onto the adjoining slope. When the adjoining slope is shingled, the overlapping shingles are cut at the valley.

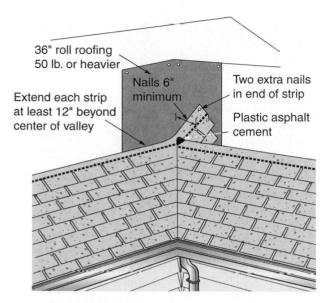

Figure 12-10. Details of construction for a closed-cut valley. Shingles on the right are cut along a line 2″ back from the valley centerline.

PROCEDURE

Installing a closed-cut valley

1. Install the valley flashing.
2. Apply the first course on both surfaces. At the valley, interweave this course only.
3. Apply all of the shingles on one roof surface. Carry each course across the valley and onto the adjoining roof at least 12″. If a shingle should fall short, insert a 1- or 2-tab section well away from the valley.
4. Apply remaining shingles in the same way, extending each course across the valley. Press shingles firmly into the valley.
5. When the first roof surface is complete, snap a chalk line 2″ from the centerline of the valley on the second, unshingled surface.
6. Apply the second course of shingles along the eaves of the intersecting roof.
7. Where each course meets the valley, trim the shingle where it falls on the chalk line.
8. Trim off 1″ of the upper corner of the shingle at a 45° angle to prevent water from running back along the top edge.
9. Embed the end of the shingle in a 3″ wide strip of plastic asphalt cement.
10. Apply and complete succeeding courses in the same way.

Working Knowledge

When applying a new roof over the top of old shingles, build up the trough in an open valley to the average level of the roof surface. This can usually be done with strips of beveled wood. Then, flash the valley as described in the text.

12.5.6 Flashing at a Wall

Where the sloping part of a roof abuts a vertical wall, step flashing should always be used. This flashing should be 10″ wide and 2″ longer than the exposed face of the regular shingles. Bend the 10″ width at a right angle so that it extends 5″ over the roof and 5″ up the wall, as shown in **Figure 12-11.**

Install metal flashing as each course of shingles is laid. Nail it to the roof at the top edge as shown. Do not nail flashing to the wall, since settling of the roof frame could damage the seal.

Install wall siding after the roof is completed. The siding then serves as cap flashing, **Figure 12-12.** Position the siding just above the roof surface. Allow enough clearance to paint the lower edges.

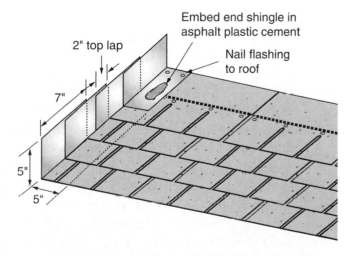

Figure 12-11. Apply metal step flashing against vertical siding with each course of shingles. The vertical portion goes under the siding.

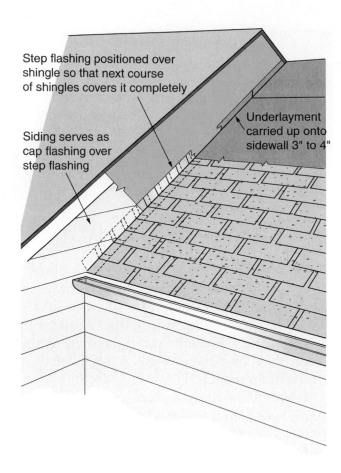

Step flashing positioned over shingle so that next course of shingles covers it completely

Siding serves as cap flashing over step flashing

Underlayment carried up onto sidewall 3" to 4"

Figure 12-12. Metal step flashing is applied to waterproof joints between the sloping roof and the walls. Note that the siding serves as cap flashing.

12.5.7 Chimney Flashing

Flashing around a masonry chimney must allow for some movement caused by settling or shrinkage of the building framework. This must be done in such a way that movement causes no damage to the water seal. The flashing has two parts that move independently of each other:

- **Base flashing,** which is attached to the roof.
- **Cap flashing** (also called *counterflashing*), which is attached to the chimney.

Sheet metal is usually preferred for base flashing. It should be applied by the step method previously described in the section on wall flashing. However, **Figure 12-13** shows how mineral-surfaced roll roofing can be used to form the base flashing. First, cement the front unit into place and then attach the sidepieces. The flashing on the high (back) side is installed

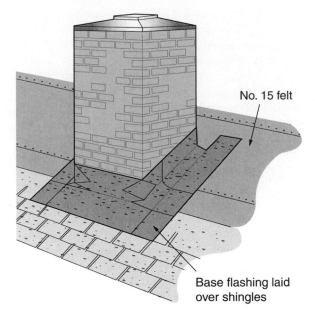

No. 15 felt

Base flashing laid over shingles

Figure 12-13. Base flashing seals the joints between the chimney and the roof. Mineral-surfaced roll roofing is cut and cemented into place. The installation is ready for cap flashing, which covers the base flashing.

last. All sections are cemented together as they are applied.

Housings for prefabricated chimneys require flashing similar to masonry chimneys. Some prefabricated chimney units have flashing flanges that simplify their installation.

Cap flashing is sheet metal shaped to cover the top of the base flashing. When used around a masonry chimney, it is set into the mortar joints and bent down over the base flashing. In new construction, the cap flashing is mortared into the joints when the brick chimney is built. If the chimney is laid up without flashing, the mortar joints must be chiseled out and the flashing forced in. It may also be held in place by nails driven into the mortar. Finally, the joints must be filled or pointed with good mortar. The metal is set into the joints 1 1/2". Cap flashing on the front of the chimney may be one continuous

Base flashing: The part of chimney flashing that is attached to the roof.

Cap flashing: Flashing used on chimneys at the roofline to cover base flashing and prevent leaks. Also called *counterflashing.*

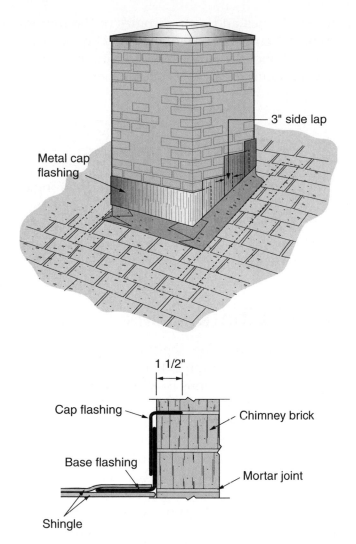

Figure 12-14. Metal cap flashing is set into the mortar joints as the chimney is built. Cap flashing must go over the top of the base flashing as shown in the cross section.

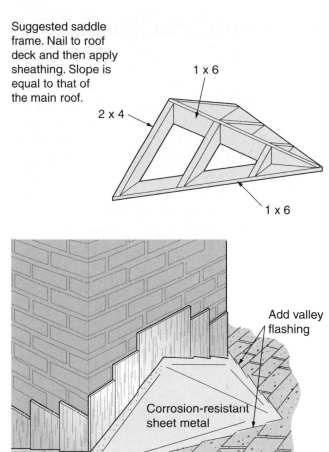

Figure 12-15. A chimney saddle, also called a cricket, diverts water away from the upper side of a chimney. Small saddles need not be framed. They can be formed from triangular pieces of 3/4″ exterior plywood.

piece. Sides must be stepped in sections because of the roof slope, **Figure 12-14.**

12.5.8 Chimney Saddle

Large chimneys on sloping roofs generally require an auxiliary roof deck on the high side. This structure is called a *saddle* or a *cricket*. It

Saddle: A small gable roof placed behind a chimney on a sloping roof to shed water and debris. Also know as a *cricket*.

diverts water from behind the chimney, preventing a buildup of ice and snow. If water, ice, or snow is allowed to collect behind the chimney, roof leaks could result.

Figure 12-15 shows a chimney saddle framing design. The frame is nailed to the roof deck and then sheathed. A small saddle could be constructed without framing from triangular pieces of 3/4″ exterior plywood.

Saddles are usually covered with corrosion-resistant sheet metal. However, mineral-surfaced roll roofing can be used. Valleys formed by the saddle and main roof should be carefully flashed in the same way as for regular roof valleys.

12.5.9 Vent Stack and Skylight Flashing

Pipes and skylights in the roof must also be carefully flashed, **Figure 12-16.** The roofing must be laid up to the protrusion. Cut and fit the shingles around the protrusion, then carefully cement a flange in place and lay shingles over the top. The flange must be large enough to extend along the roof surface at least 4″ below, 8″ above, and 6″ on each side of the stack. Most vent stack flashing is fitted with a neoprene boot that seals out moisture, as shown in **Figure 12-16.**

One-piece plastic skylights have a one-piece flashing that must be fitted under shingles at the top and sides and over shingles at the bottom. A wood curb is sometimes used to raise the skylight above the roof. The flashing procedure is very similar to flashing a chimney.

12.6 Strip Shingles

On roofs less than 30′ long, strip shingles may be laid starting at either end. On longer roofs, it is usually best to start at the center and work both ways. In this case, snap a chalk line at a right angle to ridge from the eaves to ridge.

12.6.1 Chalk Lines

Asphalt shingles slightly vary in length, usually plus or minus 1/4″ in a 36″ strip. There may be some variations in width, as well. Therefore, chalk lines should be used to achieve the proper placement so shingles are accurately aligned both horizontally and vertically. See **Figure 12-17.** Snap a number of chalk lines between the eaves and ridge. They serve as reference marks for starting each course. Space them according to the type of shingle and layout pattern. Full or cut shingles are aligned with the vertical chalk lines to form the desired pattern.

Chalk lines parallel to the ridge are used to maintain straight horizontal lines. Usually, a line is snapped for every fifth or sixth course. However, if the roofers are inexperienced, chalk lines may need to be snapped for every course, or at least every second or third course. To do this, mark the spacing of courses (using the shingle exposure) all the way to the ridge and snap chalk lines as needed. Some carpenters snap a horizontal chalk line every 10″ all the way to the ridge. Assuming a 5″ exposure, the top of every other course then falls on a chalk line.

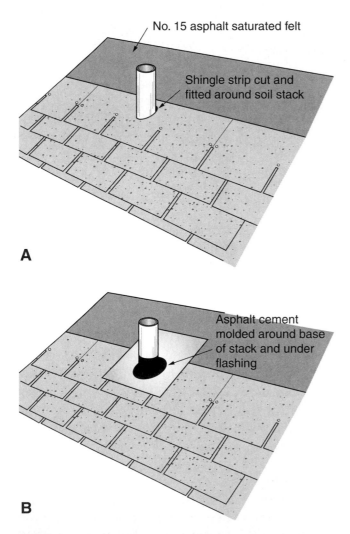

A

B

Figure 12-16. Flashing stacks. A—Lay shingles up to the stack and fit the last course around it. B—Install a flange and apply shingles over the upper side of the flange.

Working Knowledge

To get the best performance from any roofing material, always study and follow the manufacturer's directions. They are usually included on the packaging material.

Figure 12-17. Snap chalk lines to keep shingles in alignment. These should be laid down wherever the carpenter will start a course.

Figure 12-18. A hoist saves the labor of carrying roof coverings to the roof. Some roofers use an elevator or conveyor to bring shingles to the point of application.

12.6.2 Fastening Shingles

When roofing materials are delivered to the building site, handle them with care and protect them from damage. If a lift is available, use it to hoist the materials to the roof, **Figure 12-18.** Try to avoid handling asphalt shingles in extreme heat or cold.

Nails used to apply asphalt roofing must have large heads (3/8″–7/16″ diameter) and sharp points. **Figure 12-19** shows standard nail designs and suggests lengths. Most manufacturers recommend 12 gage, galvanized-steel nails with barbed shanks. Aluminum nails are also used. The length should be sufficient to penetrate nearly the full thickness of the sheathing (or 3/4″ into wood boards).

The number of nails and correct placement are both vital factors in proper application of shingles. For three-tab, square-butt shingles, use a minimum of four nails per strip, as shown

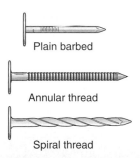

Plain barbed

Annular thread

Spiral thread

Nailing application	1" sheathing	3/8" plywood or wafer board
On sheathing (new construction or tear off)	1 1/4"	7/8"
Over old asphalt layer	1 1/2"	1"

Figure 12-19. Nails suited for installing asphalt shingles must be long enough to penetrate the roofing materials and sheathing without going entirely through the sheet material.

in **Figure 12-20.** Generally speaking, fasteners should be placed 5/8″ below the adhesive band found on the shingle strip. Carefully align each shingle and start the nailing from the end next to the one previously laid. Proceed across the shingle. This will avoid buckling the shingle. Drive nails straight so the edge of the head does not cut into the shingle. The nail head should be driven flush with, not sunk into, the surface. If for some reason the nail fails to hit solid sheathing, drive another nail in a slightly different location.

Pneumatic-powered staplers and nailers are often used to install asphalt shingles, **Figure 12-21.** Always follow the manufacturer's recommendations for staples and special power nailing equipment. Nailers are loaded with coils of shingle nails. Special staples with an extra wide crown should be used with staplers. In general, 16 gage staples with a minimum length of 3/4″ should be used to attach asphalt strip shingles to new construction. See **Figure 12-22** for examples of well-set and poorly set staples. Power nailers and staplers must be adjusted for proper staple application. Follow the staple recommendations in **Figure 12-23.**

Working Knowledge

If a fastener must be removed from a shingle, repair the hole with asphalt cement applied according to the manufacturer's directions.

A

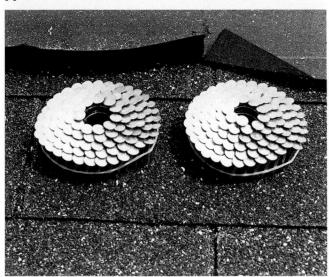

B

Figure 12-21. Pneumatic tools are widely used to install asphalt shingles. A—A pneumatic nailer that feeds roofing nails from the round magazine located below the handle. B—Coils of nails ready to be inserted in the nailer's magazine.

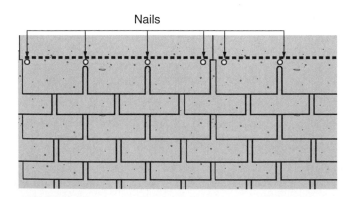

Nails

Figure 12-20. An approved nailing pattern for three-tab, square-butt shingles. This placement catches the tops of the preceding course, providing additional holding power. (Asphalt Roofing Manufacturers Assn.)

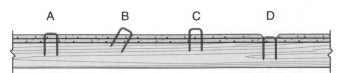

A   B   C   D

Figure 12-22. Stapling of shingles. A—This is correct. The crown should be parallel to and tight against shingle surface without cutting into it. B—Avoid driving staples at an angle. C—Set the staple deeper if you can see daylight under crown. D—If set too deep, the staple will cut the shingle. (Senco Products, Inc.)

Thickness of wood deck	Minimum staple leg length
3/8"	7/8"
1/2"	1"
5/8" and thicker	1 1/4" 1 1/2"

Figure 12-23. Recommendations for staple length in new construction. These specifications apply in most parts of the country. However, always check local building codes.

12.6.3 Starter Strip

A *starter strip* backs up the first course of shingles and covers the gap between the tabs. Failure to install it will result in damage to the exposed underlayment and sheathing.

An inverted (upside down) row of shingles is often used as the starter strip, **Figure 12-24.** A strip of mineral-surfaced roll roofing, 9" or wider, of a weight and color to match the shingles, can also be used. Let the strip slightly overhang the drip edge. Secure it with nails spaced 3"–4" above the edge. Space the nails so they are not exposed at the cutouts between the tabs of the first course of shingles.

12.6.4 First and Succeeding Courses

Some roofers start the first course with a full shingle. Succeeding courses are then started with either full or cut strips, depending on the type of shingle and the pattern.

Three-tab, square-butt shingle strips are commonly laid so the cutouts are centered on the middle of the tab in the course directly

Inverted row of shingles

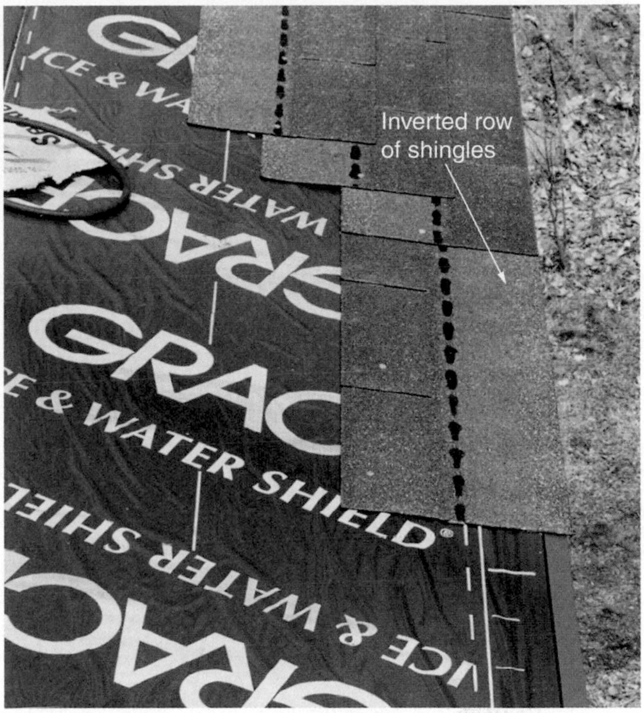

Figure 12-24. Often, standard asphalt shingle strips are inverted and used as a starter strip.

below. Thus, the cutouts in every other course are exactly aligned. This is called the *six-inch method.*

For this pattern, cut 6" from a strip to start the second course. A pair of tin snips or a utility knife can be used to cut the shingles. The third course is started with one full tab (12") removed from the strip. The fourth course is started with half of the strip (18") removed. Continue as shown in **Figure 12-25.** Reduce the strip lengths in the same sequence for subsequent courses.

Figure 12-26 shows a pattern where the cutouts break joints on thirds. Called the *four-inch method,* this pattern starts the second course with a strip shortened by 4" and the third by 8". The fourth course starts with a full strip.

Using an approved nailing pattern for three-tab shingles is very important in securing both the best appearance and full weather protection. Manufacturers recommend that four nails be used, as shown in **Figure 12-20.** When shingles are applied with an exposure of 5", nails should be placed 5/8" above tops of cutouts. Locate one nail above each cutout and one nail in 1" from each end. Nails should not be placed in or above the factory applied adhesive strip.

Starter strip: Strip of mineral-surfaced material placed beneath the first course of shingles to cover the gaps between shingle tabs.

Six-inch method: A shingling pattern where the cutouts on every other course align.

Four-inch method: A shingling pattern where the cutouts break joints on thirds.

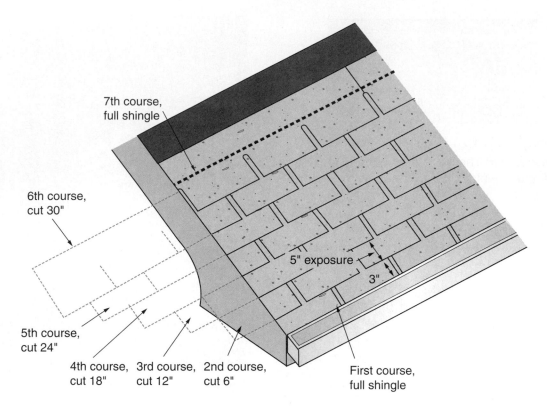

Figure 12-25. These three-tab, square-butt shingles are laid so the cutouts are centered over the tabs in the course directly below. This is called the six-inch method. (Johns Manville)

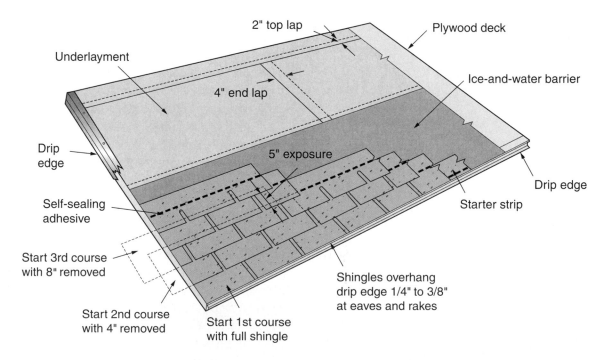

Figure 12-26. In this application, cutouts break joints on thirds. It is called the four-inch method. (Bird Division, CertainTeed Corp.)

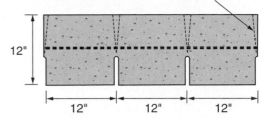

Cut along dotted line, tapering top portion slightly

12"

12"   12"   12"

A

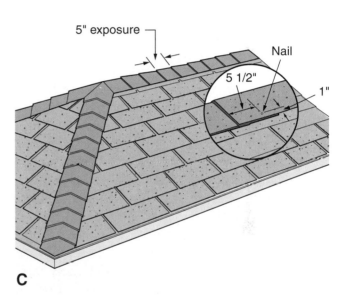

Nail

5"

1"

5 1/2"

B

5" exposure

Nail

5 1/2"

1"

C

Figure 12-27. Hip and ridge shingles. A—Making cap shingles for hips and ridges. B—Properly placed nails catch the lap end of the previous cap. C—Nail hip and ridge shingles 5 1/2″ back from their edges. Use one nail on each side. (Asphalt Roofing Manufacturers Assn.)

12.7 Hips and Ridges

Special hip and ridge shingles (also called *Boston ridge*) are sometimes available from the manufacturer. These shingles also can be easily made on site. Metal ridge roll is not recommended for asphalt shingles. Corrosion may discolor the roof.

PROCEDURE

Making and installing ridge shingles

1. Cut three-tab shingles into three equal pieces.
2. Taper the lap (covered) portion of each ridge shingle with a utility knife. See **Figure 12-27A.**
3. After the ridge shingles are cut, bend them lengthwise in the centerline, **Figure 12-27B.** In cold weather, the shingle should be warmed before bending to prevent cracks and breaks.
4. Beginning at the bottom of the hips or at one end of the ridge, lap the units to provide a 5″ exposure, as illustrated in **Figure 12-27C.** Secure with one nail on each side, 5 1/2″ back from the exposed end, and 1″ from the edge.

12.8 Wind Protection

The factory-applied adhesive above the tabs on shingles prevents wind from lifting and damaging tabs. Only a few warm days are needed to seal tabs to the course underneath. However, in windy localities, a different nailing pattern is recommended, **Figure 12-28.** This precaution is most important on roofs with low slopes. On these roofs, it is easy for wind to get under the shingles.

Self-sealing shingles are satisfactory for roofs with slopes up to about 60°. For very steep slopes, like those used on mansard roofs, special application steps must be followed. You may

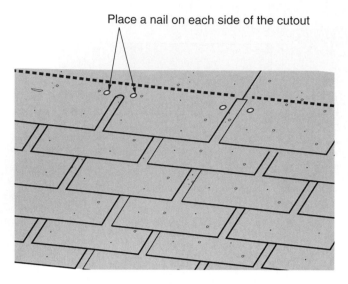
Place a nail on each side of the cutout

Figure 12-28. Nailing pattern for fastening asphalt shingles in windy localities. (Shakertown Corp.)

have to seal them in place with quick-setting asphalt cement. Follow the recommendations provided by the roofing manufacturer.

If shingles without a sealing strip are used, the tabs can be cemented. Apply a spot of special tab cement, about 1" square, with a putty knife or caulking gun and then press the tab down. Avoid lifting the tab any more than necessary while applying the cement.

Interlocking shingles are designed to resist strong winds. Details of the interlocking devices and methods of application considerably vary by manufacturer. Always study and follow the manufacturer's directions when installing all types of shingles.

12.9 Individual Asphalt Shingles

Roof surfaces may be laid with individual asphalt shingles. There are several sizes and designs available. One commonly used design is 12" wide and 16" long. Several patterns can be used. See **Figure 12-29.** Follow the same procedure described for strip shingles. Use horizontal and vertical chalk lines to ensure accurate alignment.

12.10 Low-Slope Roofs

Roofs with slopes as low as 2-in-12 can be made watertight and wind resistant with asphalt shingles. However, for slopes less than 4-in-12, certain additional procedures should be followed. First, use two layers of felt underlayment. Lap each course of felt over the preceding one by 19". In areas where the daily average temperature in January is 25°F (−5°C) or colder, cement the two felt layers together from the eaves up the roof to 24" inside of the interior wall line of the building. As an alternative, an ice-and-water barrier can be used. **Figure 12-30** demonstrates how the barrier should be installed.

Shingles provided with factory-applied adhesive and manufactured to conform to the Underwriters Laboratories *Standard for Class "C" Wind-Resistant Shingles* should then be installed. Free tab, square-butt strips can be used *if* you cement all of the tabs. See **Figure 12-31** for special application methods.

12.11 Roll Roofing

Asphalt roll roofing is manufactured in a variety of weights, surfaces, and colors. It is used as a main roof covering and sometimes as a flashing material. For best results, install it at temperatures of 45°F (10°C) or above.

In residential construction, a double-coverage roll roofing provides good protection. It can be used on slopes as low as 1" rise per foot. The 36" width includes a granular-surfaced area that is 17" wide and a smooth surface, called a *selvage,* that is 19" wide.

Although double-coverage roll roofing can be applied parallel to the rake, it is usually applied parallel to the eaves, as shown in **Figure 12-32.** The starter strip can be made by cutting off the granular-surfaced portion. Use two rows of nails to install the starter strip—one 4 3/4" below the upper edge and the other 1" above the lower edge.

Selvage: The part of the width of roll roofing that is smooth.

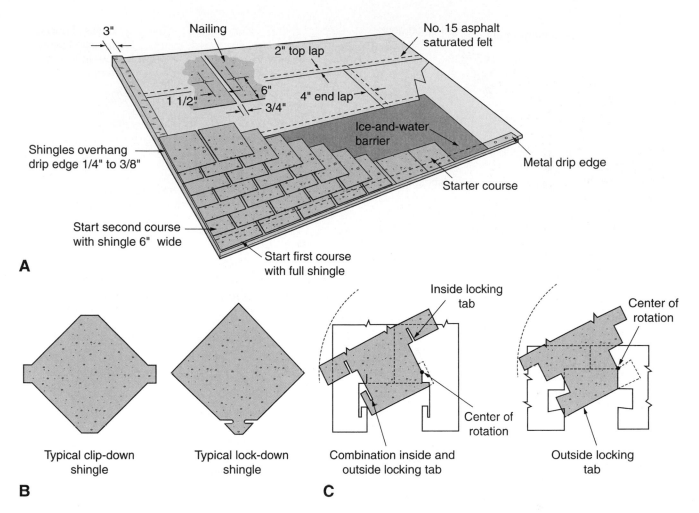

Figure 12-29. Individual shingles. A—A method of installing giant individual shingles. This is called the American method. B—Two types of hex shingles primarily intended for application over old roofing. The slope must be 4″ per foot (4-in-12) or greater. C—Interlocking devices on individual shingles provide increased wind resistance. (Asphalt Roofing Manufacturers Assn.)

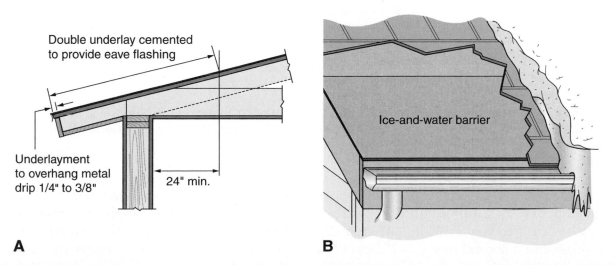

Figure 12-30. Pay special attention to underlayment materials for low-slope roofs. A—Two plies are cemented together for a watertight eave flashing. B—This eave flashing is a single layer of a thick, polymer-modified asphalt reinforced with a fiberglass mat. This membrane must extend at least 24″ upward from the side walls. (CertainTeed Corp.)

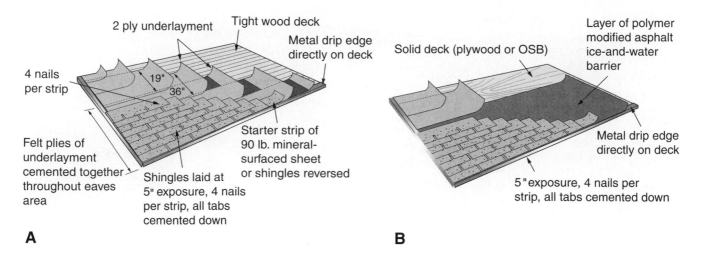

Figure 12-31. Use special application methods for shingling low-slope roofs. A—Double plies of tar paper cemented together. B—Ice-and-water barrier used as eaves flashing. The rest of the underlayment is cemented two-ply tar paper.

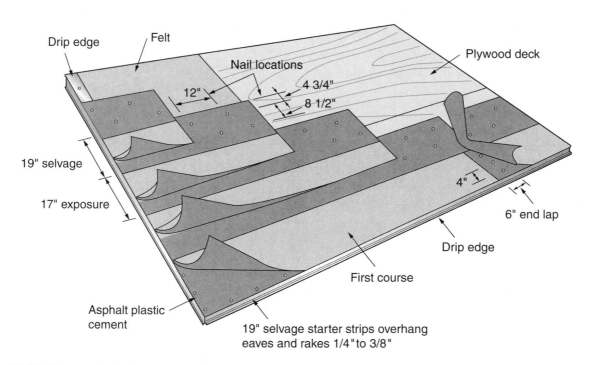

Figure 12-32. One method of applying double-coverage roll roofing parallel to the eaves. The roll can also be laid parallel to the rake. Ice-and-water barrier can be used along the eaves. (Bird Division, CertainTeed Corp.)

Coat the entire starter strip with asphalt cement and overlay a full-width sheet. Attach the sheet with a row of nails 4 3/4″ from the upper edge and a second row 8 1/2″ below the first row. The nail interval should be about 12″.

Position each succeeding course so that it overlaps the full 19″ selvage area. Nail the sheet in place at the upper edge and then carefully turn the sheet back to apply the cement coating. Spread the cement to within about 1/4″ of the granular surface. Firmly press the overlaying sheet into the cement using a stiff broom or roller. Avoid excessive use of cement. Be sure to follow the manufacturer's recommendations.

12.12 Reroofing

When reroofing, choose between removing the old roofing or leaving it in place. It is usually not necessary to remove wood shingles, asphalt shingles, or roll roofing before putting on a new asphalt roof provided these conditions are met:

• The strength of the existing deck and framing is adequate to support the weight of workers and the additional roofing, snow loads, and wind loads.

• The existing deck is sound and will provide good anchorage for the nails used in applying new roofing.

• The additional layer of shingles is allowed by the governing building code.

When putting on new roofing over old wood shingles, remove all loose or protruding nails. Renail the shingles in new nail locations. Renail loose, warped, and split shingles. Replace missing shingles. At the eaves and rakes, cut back the shingles far enough to allow the application of 4″–6″ wide, nominal 1″ thick strips. These strips should be nailed in place with their outside edges projecting beyond the roof deck the same distance as the old wood shingles.

When the old roof consists of square-butt asphalt shingles with a 5″ exposure, new self-sealing strip shingles can be applied as shown in **Figure 12-33.** This application pattern ensures a smooth, even appearance. In addition, it establishes a new nailing pattern about 2″ below the old one.

The joint between a vertical wall and roof surface should be sealed when reroofing. First, apply a strip of smooth roll roofing about 8″ wide. Firmly nail each edge, spacing the nails about 4″ on center. As the shingles are applied, spread asphalt cement on the strip. Thoroughly bed the shingles. To ensure a tight joint, use a caulking gun to apply a final bead of cement between the edges of the shingles and the siding.

Sometimes the old shingles are to be removed before applying a new roof. A roofing shovel, pitchfork, or standard shovel is used, **Figure 12-34.** Work from the bottom up, sliding the shovel under the shingles and prying upward. Clear away all old materials down to the sheathing. Check the surface for any protrusions, such as old fasteners,

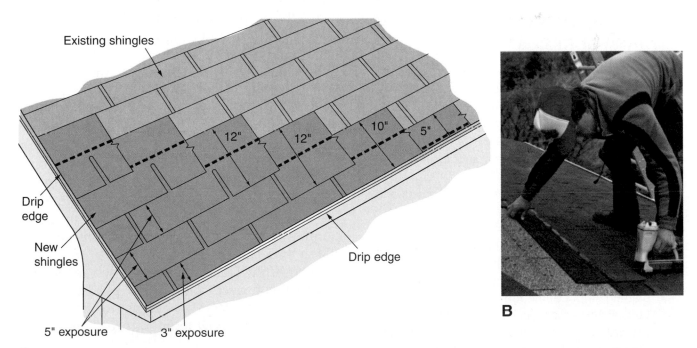

Figure 12-33. Reroofing over old asphalt shingles. A—Lay down a starter strip the same width as the exposure. Then, trim the first course to butt against the third course of old shingles. B—Butt full width strip shingles against the next old course. Offset cutouts so they do not fall over cutouts in the lower old course. (Asphalt Roofing Manufacturers Assn., Johns Manville)

Figure 12-34. A roofing shovel, pitchfork, or regular shovel can be used to quickly remove old shingles. Push the tool under the tabs and lift or pry. (Rick Pomerlou)

that might puncture the new coverings. Inspect the sheathing and replace any areas that are damaged. Sweep the roof to remove any remaining debris. Then proceed as you would for installing a new roof.

PROCEDURE

Installing asphalt shingles over asphalt shingles

1. For the starter course, remove the tabs from the new shingles and also 2″ from the tops. The remaining portion should now be equal to the exposure of the old shingles.
2. Cut off 3″ from the rake end of the first starter strip. This assures that all cutouts of the old shingles will be covered.
3. Overhang the starter course at the eaves and rake edges about 1/4″. Nail it in place.
4. Cut 2″ from the butts for the first course. All other courses are full depth. Align the cut edge with the butts of the old shingles.
5. Start by removing 6″ from the width on the starting strip. Let the top edge butt against the butts of the old shingles in the next course. While this will reduce the exposure of the preceding course, the guttering will conceal the difference.
6. Increase the cutoff of the starting strip by an additional 6″ through the sixth course. Start the seventh with a full-width shingle.

12.13 Built-Up Roofing

A flat roof or a roof with very little slope must be covered with a watertight system. Many flat roofs are covered with *built-up roofing.* Such a roof is very durable. Companies that manufacture the components provide detailed specifications and instructions for making the installation. The components are mainly saturated felt and asphalt.

On a clean wood deck, first lay down a heavy layer of saturated felt. Nail this down with galvanized nails, **Figure 12-35.** Nails must have a large head or be driven through tin caps. Mop each succeeding layer in place with hot asphalt. Built-up roofs for residential structures normally have three or four *plies* (layers of asphalt-saturated felts). When the felts are all in place, they are coated with hot asphalt and covered with slag, gravel, crushed stone, or marble chips. See **Figure 12-36.** These materials provide a weathering surface and improve the roof's appearance. Three to four hundred pounds of the mineral covering are used on a 100 sq. ft. section of roof.

The asphalt used between each layer and to bed the surface coating is a product of the petroleum industry. It begins to flow, very slowly, at a fairly low temperature. This results

Plies: Layers of asphalt-saturated felts used on built-up roofs.

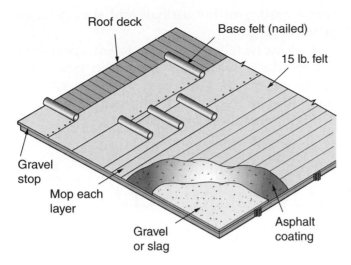

Figure 12-35. Constructing a built-up roof.

in a self-healing property that is essential for flat roofs, since water is likely to stand on the roof. A special low-temperature asphalt known as *dead-flat* asphalt is used for flat roofs. For sloping roofs, *steep* asphalt (an asphalt with a high melting point) is used. In hot climates, use only steep asphalt.

A *gravel stop* is attached to the roof deck to serve as a trim member. This is usually fabricated from galvanized sheet metal. It helps keeps the mineral surface and asphalt in place, **Figure 12-37.** Gravel stops are installed after the base felt has been laid. Joints between sections of gravel stop are bedded in a special mastic that permits expansion and contraction in the metal.

Flashings around chimneys and vents or where the roof joins a wall must be constructed with special care. Leaks are most likely to occur at these locations. The best flashing materials include lead jackets, sheet copper, and special flashing cement.

Basic flashing construction is shown in **Figure 12-38.** The *cant strip* provides support for the felt layers as they curve from a horizontal to a vertical attitude.

Bare spots on a built-up roof should be repaired. First, clean the area. Then, apply a

A   **B**

Figure 12-36. Built-up roofing. A—After hot asphalt and saturated felt layers are laid down, more hot asphalt is poured over the felt underlayment to bind the gravel. The gravel stop is installed after the base felt is laid. B—Applying washed gravel, 400 lb. per square.

Gravel stop: Galvanized sheet metal trim member attached to the roof deck to help keep the mineral surface and asphalt in place.

Cant strip: A strip of wood, triangular in cross section, used under shingles at gable ends or under the edges of roofing on flat decks.

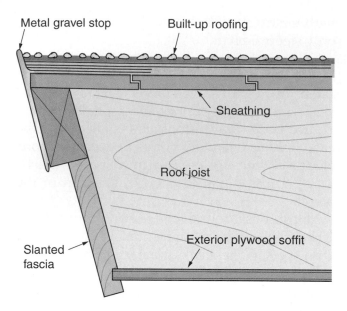

Figure 12-37. This section view through the edge of a flat roof overhang shows the metal gravel stop installation.

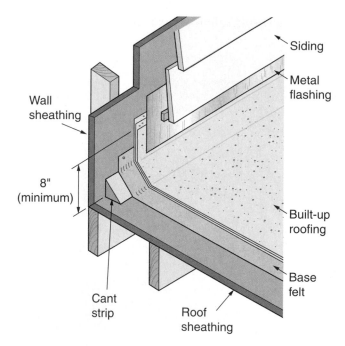

Figure 12-38. Basic flashing construction details where a flat roof meets a wall. The cant strip is a triangular wooden strip that provides a gently curving base for layers of roofing felt or rubberized roof membrane.

heavy coating of hot asphalt and spread more gravel or slag.

Cut away felts that have fallen apart and replace them with new felts. The new felts should be mopped in place, allowing at least one additional layer of felt to extend not less than 15" beyond the other layers.

12.14 Ridge Vents for Asphalt Roofing

Moisture buildup and high temperatures can be problems in the attic space during certain seasons. Venting at the eaves, gable ends, and at or near the ridge helps dissipate heat and moisture. **Figure 12-39** shows two types of ridge vent. One type is rolled over the ridge opening.

A

B

Figure 12-39. Ridge vents. A—A roll vent is installed over the opening at the ridge that allows free passage of air. B—This rigid plastic vent is attached to the ridge with large-headed nails.

This opening is a gap of about 2" or 3" left in the sheathing at the ridge to allow air to pass through. Shingles are secured over the vent to keep out weather. The second type is rigid plastic. It is fastened over the vent opening with large-headed asphalt shingle nails.

12.15 Wood Shingles

Wood shingles are a traditional material used in residential construction. They are available with or without a polymer fire-retardant treatment. Building codes often prohibit untreated wood roofing materials. Since wood weathers to a mellow color after exposure, wood shingles provide an appearance that is desired by many homeowners. When properly installed, they also provide a very durable roof.

Wood shingles are made from western red cedar, redwood, or cypress. All of these woods are highly resistant to decay. The shingles are taper sawed and graded No. 1, No. 2, No. 3, and No. 4 (a utility grade). The best grade is cut in such a way that the annular rings are perpendicular to the surface. Butt ends vary in thickness, as shown in **Figure 12-40**. Wood shingles

are manufactured in random widths and in lengths of 16", 18", and 24". They are packaged in bundles. Four bundles contain enough shingles to cover 100 square feet of roof using a standard application.

The exposure of wood shingles depends on the slope of the roof. When the slope is 5-in-12 or greater, standard exposures of 5", 5 1/2", and 7 1/2" are used for 16", 18", and 24" sizes respectively. On roofs with lower slopes, the exposure should be reduced to 3 3/4", 4 1/4", and 5 3/4" respectively. This provides a minimum of four layers of shingles over the entire roof area. In any type of construction, there should be a minimum of three layers at any given point to ensure complete protection against heavy, wind-driven rain.

12.15.1 Sheathing

Solid sheathing for wood shingles may consist of matched or unmatched 1" boards, but usually it is plywood or oriented strand board. Open, skip, or spaced sheathing is sometimes used because it costs less and permits shingles to quickly dry out, **Figure 12-41A**. One reason for using solid sheathing is to gain the added

Grade	Length	Thickness (at butt)	No. of courses per bundle	Bdls/cartons per square	Description
No.1 Blue label	16" (Fivex) 18" (Perfections) 24" (Royals)	.40" .45" .50"	20/20 18/18 13/14	4 bdls. 4 bdls. 4 bdls.	The premium grade of shingles for roofs and sidewalls. These top-grade shingles are 100% heartwood. 100% clear and 100% edge-grain.
No. 2 Red label	16" (Fivex) 18" (Perfections) 24" (Royals)	.40" .45" .50"	20/20 18/18 13/14	4 bdls. 4 bdls. 4 bdls.	A good grade for many applications. Not less than 10" clear on 16" shingles, 11" clear on 18" shingles and 16" clear on 24" shingles. Flat grain and limited sapwood are permitted in this grade.
No. 3 Black label	16" (Fivex) 18" (Perfections) 24" (Royals)	.40" .45" .50"	20/20 18/18 13/14	4 bdls. 4 bdls. 4 bdls.	A utility grade for economy applications and secondary buildings. Not less than 6" clear on 16" and 18" shingles, 10" clear on 24" shingles.
No. 4 Under-coursing	16" (Fivex) 18" (Perfections)	.40" .45"	14/14 or 20/20 14/14 or 18/18	2 bdls. 2 bdls. 2 bdls. 2 bdls.	A utility grade for undercoursing on double-coursed sidewall applications or for interior accent walls.
No. 1 or No. 2 Rebutted-rejointed	16" (Fivex) 18" (Perfections) 24" (Royals)	.40" .45" .50"	33/33 28/28 13/14	1 carton 1 carton 4 bdls.	Same specifications as above for No. 1 and No. 2 grades but machine trimmed for parallel edges with butts sawn at right angles. For sidewall application where tightly fitting joints are desired. Also available with smooth sanded face.

Figure 12-40. Wood shingles are made in several grades and to certain specifications for various applications. (Cedar Shake and Shingle Bureau)

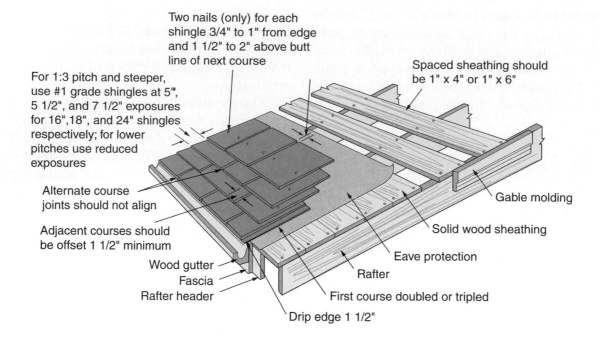

Two nails (only) for each
shingle 3/4" to 1" from edge
and 1 1/2" to 2" above butt
line of next course

For 1:3 pitch and steeper,
use #1 grade shingles at 5",
5 1/2", and 7 1/2" exposures
for 16",18", and 24" shingles
respectively; for lower
pitches use reduced
exposures

Spaced sheathing should
be 1" x 4" or 1" x 6"

Alternate course
joints should not align

Gable molding

Solid wood sheathing

Adjacent courses should
be offset 1 1/2" minimum

Eave protection

Wood gutter

Rafter

Fascia

Rafter header

First course doubled or tripled

Drip edge 1 1/2"

A

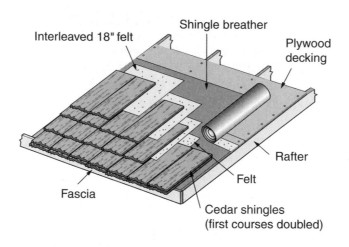

Interleaved 18" felt

Shingle breather

Plywood
decking

Rafter

Felt

Fascia

Cedar shingles
(first courses doubled)

B

Figure 12-41. General application details for wood shingles. A—Install double or triple layers at eaves and allow the butts of the first course to project beyond fascia by 1 1/2″. Note the use of both tight and spaced solid wood decking. B—To allow shingles to dry over solid decking, a layer of shingle breather is first installed. It provides an air space between the deck and the shingles. (Cedar Shake and Shingle Bureau, Benjamin Obdyke, Inc.)

insulation and resistance to air infiltration that such a deck offers. If applying wood shingles over solid sheathing, it is advisable to use an underlayment that allows the shingles to breathe. One such product on the market is known as *cedar breather.* Made up in rolls of stiff fiber, it maintains a thin air space between the decking and the shingles. The breather material allows drying of shingles after a rain. Refer to **Figure 12-41B.**

One method of applying skip roof sheathing is to space 1 × 3, 1 × 4, or 1 × 6 boards the same distance apart as the anticipated shingle exposure. Each course of shingles is nailed to a separate board.

12.15.2 Underlayment

Normally, an underlayment is not used for wood shingles, except when applied over solid sheathing. If roofing felt is used to prevent air infiltration, rosin-sized building paper or dry unsaturated felts are suitable. Saturated paper is usually not recommended. It may cause condensation problems.

To prepare a solid roof deck for application of wood shingles, first install 30 lb. roofing felt. Allow 1/4" overhang at the eaves. Allow 4" of overlap as you work toward the ridge.

Next, tack on the breather, placing a nail or tack every three square feet. Butt each course against the previous course.

Begin applying shingles after the first course of breather. This is to avoid walking on the breather. Follow the manufacturer's directions when applying shingles. Use a nail length that gives a 3/4" penetration into the deck.

12.15.3 Fire Resistance

Recent tests have shown that flame-spread and burn-through rates for wood shingles and shakes can be reduced. This is achieved by pressure-treating the shingles with fire retardants. Flame-penetration time can also be increased by using 1/2", Type X gypsum board under solid or spaced sheathing. For more information, see *Uniform Building Code Standard No. 32-14.*

12.15.4 Flashing

In areas where outside temperatures drop to 0°F (–15°C) or colder, there is a possibility of ice forming along the eaves. An *eaves flashing strip* or ice-and-water barrier is recommended. The installation procedure is identical to that used for asphalt shingles.

It is important to use good-quality materials for valleys and eave flashing. Materials used for this purpose include tinplate, lead-clad iron, galvanized iron, lead, copper, aluminum sheets, and ice-and-water barrier. Galvanized iron is mild steel coated with a layer of zinc. If it

is selected, use 24 or 26 gage metal. Tin or galvanized sheets with less than 2 oz. of zinc per sq. ft. should be painted on both sides with a rust-proof primer. Allow the primer to dry before installing the flashing.

When making bends in the flashing, use care not to crack the zinc coating. On roofs of 6-in-12 pitch or steeper, the valley sheets should extend up on both sides of the center of the valley for at least 7". On roofs of less pitch, wider valley sheets should be used. The minimum extension should be at least 10" on both sides, **Figure 12-42.** The open portion of the valley is usually about 4" wide and should gradually increase in width toward the low end. The low end is where drainage is heaviest.

Tight flashing around chimneys is also essential. **Figure 12-43** shows two methods of installation. In either method, the base flashing goes on first. Bend the flashing to 90°, allowing upward projection of about 10". Allow at least as much to lay over the sheathing.

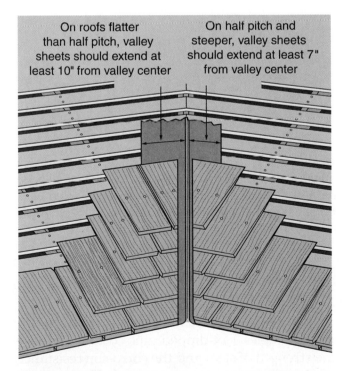

Figure 12-42. Valley flashing for wood shingles is similar to flashing for asphalt shingles. (Cedar Shake and Shingle Bureau)

Eaves flashing strip: Used under wood shingles in areas where the outside temperatures drop to 0°F (–15°C) or colder and there is a possibility of ice forming along the eaves.

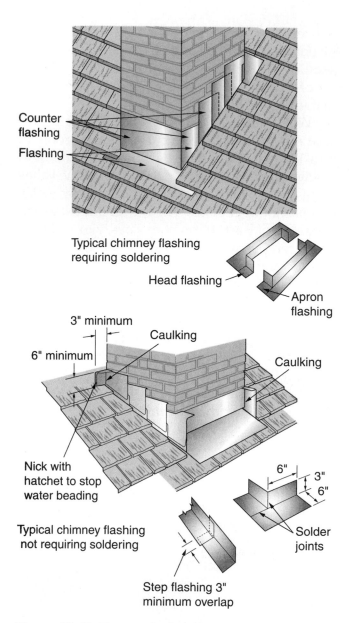

Typical chimney flashing requiring soldering

Head flashing

Apron flashing

3" minimum

Caulking

6" minimum

Caulking

Nick with hatchet to stop water beading

Typical chimney flashing not requiring soldering

6"

3"

6"

Solder joints

Step flashing 3" minimum overlap

Figure 12-43. Two methods of flashing around a brick chimney. In either method, cap flashing should be mortared into the joints as the chimney is being built. For an existing chimney, it is necessary to carefully remove old mortar to install the flashing.

12.15.5 Nails

Only rust-resistant nails should be used with wood shingles. Hot-dipped, zinc-coated nails have the strength of steel and the corrosion resistance of zinc. These are recommended. **Figure 12-44** shows sizes of nails for various jobs.

Most carpenters prefer to use a shingler's or lather's hatchet to lay wood shingles, **Figure 12-45.** This tool has a blade for splitting and trimming.

Some have a gauge for spacing the weather exposure.

12.15.6 Applying Shingles

The first course of shingles at the eaves should be doubled or tripled. Horizontally space all shingles 1/4″–3/8″ apart. Because wood shingles absorb moisture, this spacing allows expansion when they become rain soaked.

Use only two nails to attach each shingle. The proper placing of these two nails is of considerable importance. The nails should be near the butt line of the shingles in the next course. Under no circumstances should nails be driven below this line. This would expose the nails to the weather. Driving the nails 1″ to 1 1/2″ above the butt line is good practice. Two inches above is an allowable maximum. Place nails not more than 3/4″ from the edge of the shingle at each side. When so nailed, the shingles lie flat and give good service.

The second layer of shingles in the first course should be nailed over the first layer so the joints in each course are at least 1 1/2″ apart. A good shingler uses care in breaking the joints in successive courses so they do not line up in three successive courses. Joints in adjacent courses should be at least 1 1/2″ apart.

For shingles containing both flat and vertical grain, joints should not be aligned with centerline of the heart grain. Split flat grain shingles in two before nailing. Treat knots and other defects as the edge of the shingle.

It is good practice to use a board as a straightedge to line up courses of shingles, **Figure 12-46.** Temporarily tack the board in place to hold the shingles until they are nailed. Two shinglers often work together. One distributes and lays the shingles along the straightedge while another nails them in place. As shingling progresses, check the alignment every five or six courses with a chalk line. Measure down from the ridge occasionally to be sure shingle courses are parallel to the ridge.

On a roof section where one end terminates at a valley, carefully cut valley shingles to the proper angle at the butts. Use wide shingles. Nail the shingles in place along the valley first.

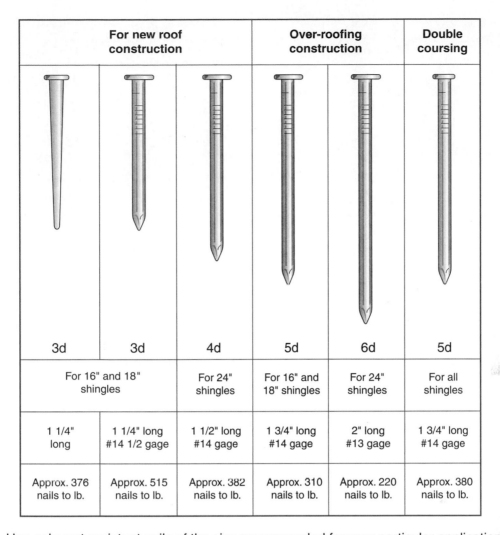

For new roof construction			Over-roofing construction		Double coursing
3d	3d	4d	5d	6d	5d
For 16" and 18" shingles		For 24" shingles	For 16" and 18" shingles	For 24" shingles	For all shingles
1 1/4" long	1 1/4" long #14 1/2 gage	1 1/2" long #14 gage	1 3/4" long #14 gage	2" long #13 gage	1 3/4" long #14 gage
Approx. 376 nails to lb.	Approx. 515 nails to lb.	Approx. 382 nails to lb.	Approx. 310 nails to lb.	Approx. 220 nails to lb.	Approx. 380 nails to lb.

Figure 12-44. Use only rust-resistant nails of the size recommended for your particular application.

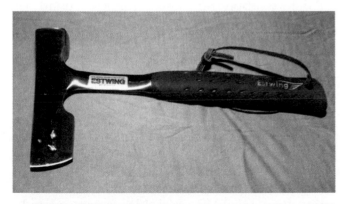

Figure 12-45. A shingler's hatchet is especially suited for laying wood shingles and shakes. Some have adjustable gages for measuring weather exposure. (True Value Hardware, Ashland, WI)

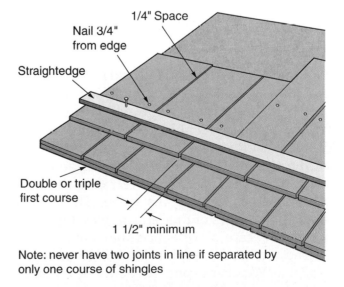

Note: never have two joints in line if separated by only one course of shingles

Figure 12-46. Use a wooden straightedge as a guide in laying wood shingles. It keeps the courses straight and maintains proper shingle exposure.

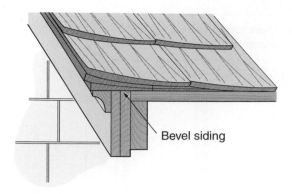

Figure 12-47. Beveled siding installed along the rake tilts shingles inward to prevent water dripping off of the edge.

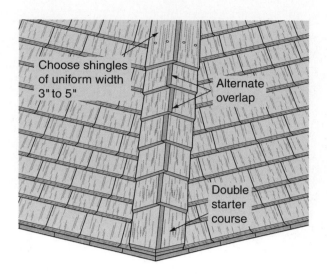

Figure 12-48. Installing wood cap shingles at hips and ridges.

Dripping from gables can be prevented by installing a wedge. Use a piece of 6″ bevel siding along the edge and parallel to the end rafter. Place the wedge under the shingles as shown in **Figure 12-47.**

Working Knowledge

Use care when nailing wood shingles. The wood is soft. It can be easily crushed and damaged under the nail heads. Drive the nail just flush with the surface.

12.15.7 Shingled Hips and Ridges

Carefully and tightly cover ridges and hips to avoid roof leaks. In the best type of hip construction nails should not be exposed to the weather, **Figure 12-48.** Select shingles of approximately the same width as the roof exposure. Snap lines on the shingled roof, one on each side of the ridge. Mark the lines the correct distance back from the centerline of the ridge. These lines indicate the edges of the cap shingles. On small houses, hip caps may be made narrower.

Factory assembled hip and ridge units are available. Weather exposure should be the same as that used for the regular shingles. Be sure to use longer nails that will penetrate well into the sheathing.

12.15.8 Reroofing with Wood Shingles

Wood shingles may be applied to old as well as new roofs, **Figure 12-49.** If the old wood-shingle roofing is in reasonably good shape, it need not be removed. In reroofing houses covered with composition material, whether in the form of roll roofing or asphalt shingles, it is usually best to strip off the old material. Otherwise, moisture may condense on the roof deck below. Decay of sheathing could follow.

Before applying new shingles, renail or replace all warped, split, and decayed shingles. To finish the edges of the roof, cut off the exposed portion of the first two rows of old shingles along the eaves. Use a sharp hatchet or portable circular saw. Nail a 1″ wood strip in this space. Place the outer edge flush with the eave line. Prepare shingle edges along the gable ends in a similar manner.

Raise the level of the valleys by applying wood strips. Install new flashings over the strips. Remove old hip and ridge caps to provide a more even base for new shingles. As new shingles are applied, space them 1/4″ apart to allow for their expansion in wet weather. Let them project 1/2″–3/4″ beyond the edge of the eaves.

Now, follow the same procedure described for new roofs. However, longer nails are needed. For 16″ and 18″ shingles, rust-resistant or

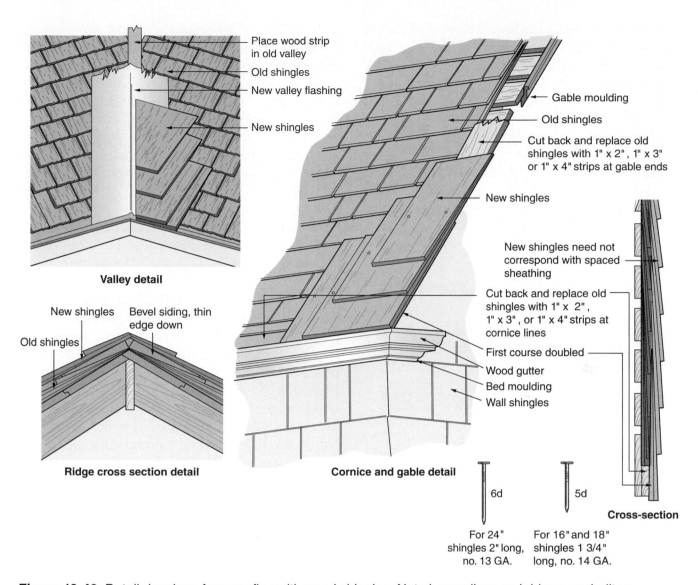

Place wood strip in old valley
Old shingles
New valley flashing
New shingles

Valley detail

New shingles
Bevel siding, thin edge down
Old shingles

Ridge cross section detail

Gable moulding
Old shingles
Cut back and replace old shingles with 1" x 2", 1" x 3" or 1" x 4" strips at gable ends
New shingles
New shingles need not correspond with spaced sheathing
Cut back and replace old shingles with 1" x 2", 1" x 3", or 1" x 4" strips at cornice lines
First course doubled
Wood gutter
Bed moulding
Wall shingles

Cornice and gable detail

6d
5d

For 24" shingles 2" long, no. 13 GA.
For 16" and 18" shingles 1 3/4" long, no. 14 GA.

Cross-section

Figure 12-49. Detail drawings for reroofing with wood shingles. Note how valleys and ridges are built up.

zinc-clad 5d box nails or special 1 3/4" long, 14 gage *over-roofing* nails should be used. A 6d, 13 gage rust-resistant nail is used for 24" shingles.

Usually, no particular attention needs to be given to how the nails penetrate the old roof. It does not matter whether they strike the sheathing strips or not because complete penetration is obtained through the old shingles with the longer nails. An adequate number of nails needed to anchor all of the shingles of the new roof will strike sheathing or nailing strips.

Wood shakes: Durable, hand-split, wood roofing material.

Place new flashings around chimneys, but do not remove the old flashing. Liberally use high-grade, nondrying mastics to get a watertight seal between the brick and metal.

12.16 Wood Shakes

Hand-split *wood shakes* provide a very pleasing surface texture. They are often called the "aristocrat" of roofing materials. If properly installed, shakes are durable and may last as long as the structure itself.

Generally, wood shakes are available as straight split, hand-split and resawn, or taper split, **Figure 12-50.** Like regular wood shingles, they are available in random widths. Various lengths and thicknesses are standardized, **Figure 12-51.**

Shakes should not be used on roofs with not enough slope for good drainage. The recommended minimum slope is 4-in-12. The maximum weather exposure is 13″ for 32″ shakes, 10″ for 24″ shakes, and 8 1/2″ for 18″ shakes. Shakes can be applied over solid sheathing or over skip sheathing on roofs with a 4-in-12 or steeper slope. They can also be installed on mansard roofs where the pitch is not greater than 20° from vertical. Underlayment is the same as for wood shingles. Use a shingle breather over solid sheathing, as previously described for wood shingles.

Start the application by laying down an ice-and-water barrier or a 36″ strip of 30 lb. felt along the eaves. Double the beginning course of shakes. A narrow strip of 30 lb. felt must be laid between each course. This must be wide enough to cover the top portion of the shakes and extend onto the sheathing. For example, if 24″ shakes are being laid at a 10″ exposure, place the roofing felt 20″ above the butts of the shake, **Figure 12-52.** Space individual shakes from 1/4″–3/8″ apart to allow for expansion. Offset joints at least 1 1/2″ from course to course.

Proper nailing is important. Use rust-resistant nails, preferably the hot-dipped zinc-coated type. The 6d size, which is 2″ long, normally is adequate. Use longer nails if needed due to unusual shake thickness or weather exposure. Nails should be long enough for adequate penetration into the sheathing. Use two nails for each shake. Drive them at least 1″ from each

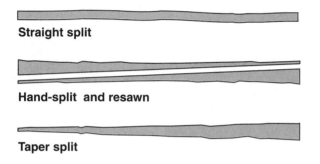

Straight split

Hand-split and resawn

Taper split

Figure 12-50. These are the three basic types of wood shakes. The hand-split and resawn shake type has one side sawn flat.

Grade	Length and thickness	18″ Pack**		Description
		Courses per bdl.	Bdls. per sq.	
No. 1 Hand-split & resawn	15″ Starter-finish	9/9	5	These shakes have split faces and sawn backs. Cedar logs are first cut into desired lengths. Blanks or boards of proper thickness are split and then run diagonally through a bandsaw to produce two tapered shakes from each blank.
	18″ x 1/2″ Mediums	9/9	5	
	18″ x 3/4″ Heavies	9/9	5	
	24″ x 3/8″	9/9	5	
	24″ x 1/2″ Mediums	9/9	5	
	24″ x 3/4″ Heavies	9/9	5	
No. 1 Taper-sawn	24″ x 5/8″	9/9	5	These shakes are sawn both sides.
	18″ x 5/8″	9/9	5	
No. 1 Taper-split	24″ x 1/2″	9/9	5	Produced largely by hand using a sharp-bladed steel froe and a wooden mallet. The natural shingle-like taper is achieved by reversing the block, end-for-end, with each split.
		20″ Pack		
No. 1 Straight split	18″ x 3/8″ True-Edge*			Produced in the same manner as taper-split shakes except that by splitting from the same end of the block, the shakes acquire the same thickness throughout.
	18″ x 3/8″	14 Straight	4	
	24″ x 3/8″	19 Straight	5	
		16 Straight	5	

Note: *Exclusively sidewall product, with parallel edges.
 **Pack used for majority of shakes.

Figure 12-51. Wood shakes are manufactured in four styles with various specifications and sizes. (Cedar Shake and Shingle Bureau)

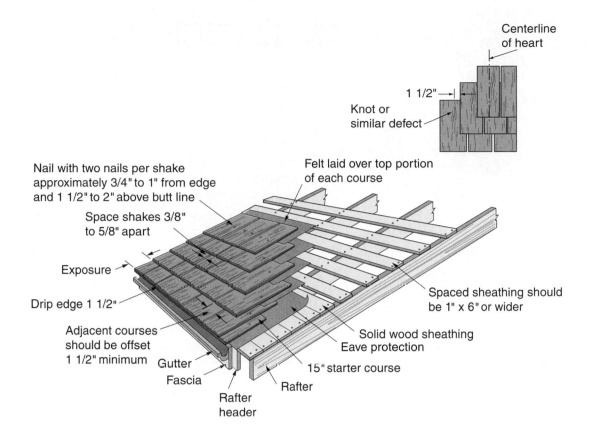

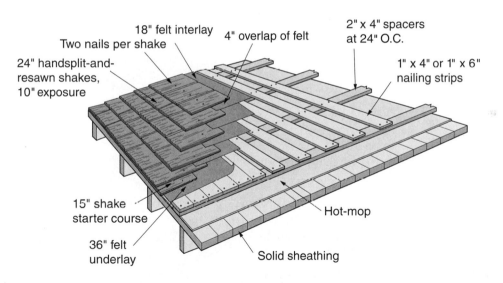

A

B

Figure 12-52. Shake application. A—Protect eaves with ice-and-water barrier. Lay an 18″ wide strip of No. 30 (30 lb.) asphalt-saturated felt between each course. Straight-split shakes should be laid with the froe-end (end from which the shake has been split) toward the ridge. B—Recommended method for applying shakes to low-slope roofs. The lattice framework is embedded in a bituminous surface coating. (Cedar Shake and Shingle Bureau)

edge and about 1″ or 2″ above the butt line of the following course. Do not drive nail heads into the shakes.

Valleys are laid as recommended for regular wood shingles. Underlay all valleys with 30 lb. roofing felt. Metal valley flashing must be at least 20″ wide.

For the final course at the ridge line, try to select shakes of uniform size of about 6″. Trim off the ends so they meet evenly. Carefully

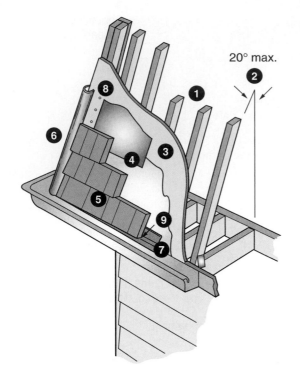

20° max.

1) 16" O.C. stud spacing
2) Pitch 20° from vertical maximum
3) 3/8" plywood sheathing
4) 15 lb. asphalt impregnated non-vapor-barrier-type building paper
5) Shake shingle siding
6) Metal corner post
7) 6" drip edge flashing
8) Corner flashing
9) 3/8" x 1 1/2" starter strip

Figure 12-53. Applying shakes to a mansard roof requires special flashing. (ABTco, Inc.)

apply a strip of 30 lb. felt lengthwise along all ridges and hips. Nail them in place following the procedure described for regular wood shingles. Prefabricated hip-and-ridge units are available. Their use will save time and provide uniformity.

Chimneys or other structures that project through the roof must be flashed and counter-flashed on all sides. Flashing should extend at least 6" under the shakes. Flashing should be applied at top and bottom of the roof, at inside and outside corners, and around any openings. Inside and outside corners of wood, vinyl, or aluminum are required. This material must be properly flashed and caulked. See **Figure 12-53.**

12.17 Tile Roofing

The most commonly used types of *roofing tile* are manufactured from concrete or clay. The clay is made from hard-burned shale or mixtures of shale and clay. A few types of roofing tile are made from metal.

When well made, concrete and clay tiles are hard, dense, and very durable. Colors, textures, and shapes come in a great variety, **Figure 12-54.** Most tile is made of concrete; some is glazed. Typical applications are shown in **Figure 12-55.** While most applications are on new construction, tile may be applied over old roofs provided:

- The old covering is in reasonably good condition.

- The roof framing and sheathing is strong enough to support the added weight. Typically, tile weighs from 5.8–10.25 pounds per square foot.

On new construction, roof trusses are engineered to support the tile. Additional roof framing or

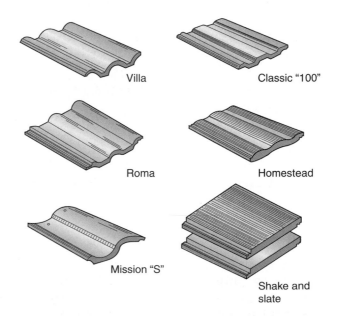

Villa

Classic "100"

Roma

Homestead

Mission "S"

Shake and slate

Figure 12-54. Roof tiles come in curved, barrel shapes and flat. Special tile shapes are manufactured for ridges, hips and rakes. (Monier Roof Tile)

Roofing tile: Tile manufactured from concrete or clay.

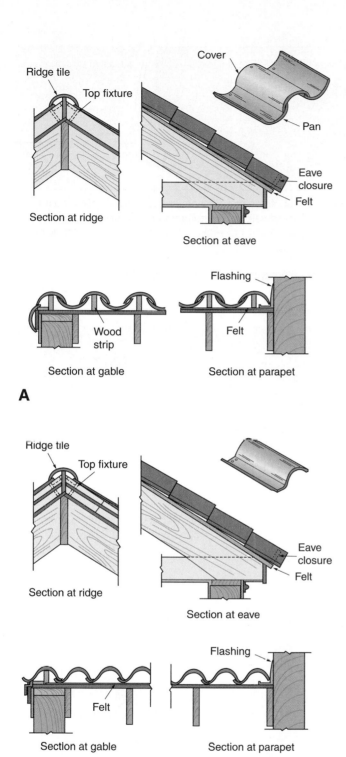

Figure 12-55. Typical application methods for tile roofs. A—Two-piece pan and cover commonly known as Mission tile. B—Spanish tile.

Batten: A horizontal or vertical strip placed on the roof to provide spacing or to hold tile.

bracing may be required on existing construction. Sheathing should be at least a nominal 1″ thick and bridge no more than 24″.

12.17.1 Installing Tile Roofing Units

Requirements for underlayment depend on the roof pitch and local climate. Below a 3-in-12 pitch, the tile roofing is considered decorative. Therefore, a minimum of two plies of Type 15 felt, hot-mopped between layers, is recommended. This is topped with vertical battens spaced 24″ apart from the eave to the ridge that are also mopped with asphalt. Then, horizontal battens are laid and fastened at intersections with the vertical battens. *Battens* are strips of wood installed on the roof to hold the tiles in place. Normally, 1×2 battens are used over solid sheathing, spaced 24″ apart. The first row of battens should be placed so the tiles overhang the eaves by 1 1/2″.

Prior to laying tile, valleys must be prepared. **Figure 12-56** shows proper installation of flashing and underlayment. Then, tile roofing is laid from right to left, beginning at the right rake, **Figure 12-57.** Lugs located on the underside of each tile hook over the battens. Flat tiles are usually laid with each course overlapping the previous one. **Figure 12-58** shows proper method of laying tile up to the valley. A circular saw fitted with a masonry cutting wheel is used to make angle cuts on tile.

Working Knowledge
Be careful while moving about on the roof—step only on the lower one-third of the tiles on the overlapped area.

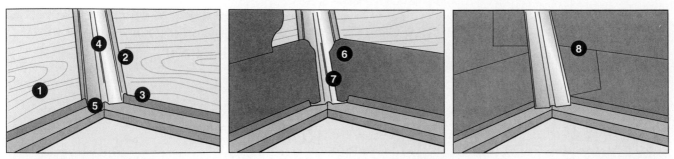

1) Decking. 2) 36" sweat sheet under valley flashing is recommended. 3) Cut off top of eave riser strip to permit valley drainage. 4) Standard galvanized iron (G.I.) valley flashing with crimped edges. 5) Extend valley flashing beyond eave riser strip. 6) Cut top corner of underlayment to ensure proper diversion of water into valley flashing. 7) Overlap valley flashing with underlayment. 8) Optional weaved underlayment treatment of valley.

Figure 12-56. Flashing a valley for a tile roof. Preformed standard flashing is available. (Lifetile)

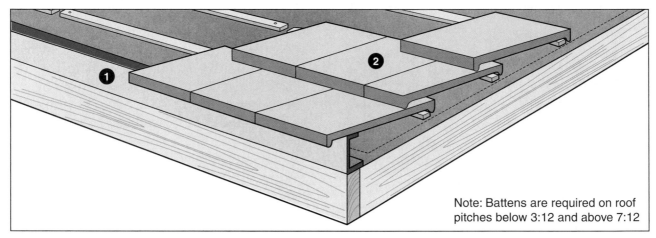

Note: Battens are required on roof pitches below 3:12 and above 7:12

1) Eave riser strip. 2) Shingle-lap vertical courses.

A

Profile Tiles

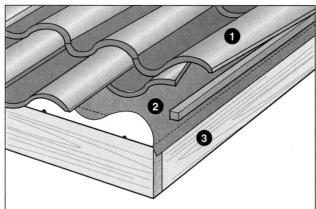

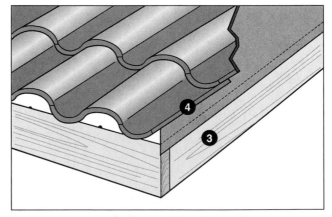

Right rake 2" x 2" wood strip starter   **Right rake pan section starter**

1) When full tile is used, edge of tile must not extend beyond center of nominal 2" x 2" starter. 2) Recess nominal 2" x 2" wood starter 4" to 6" up from eave line and 3/4" in from edge of barge rafter. 3) Barge rafter. 4) Start half-tiles at least 1" from edge of barge rafter.

B

Figure 12-57. Starting to lay tile roofing. A—Flat tiles have joints staggered between courses like shingles. B—Profiled (curved) tiles have joints aligned from the eaves to the ridge. (Lifetile)

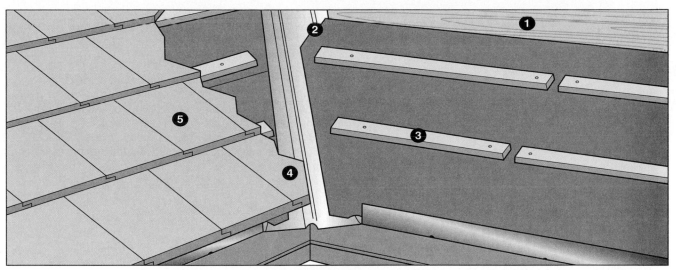

1) Decking. 2) Cut top corner of underlayment to ensure proper diversion of water into valley flashing. 3) Battens are optional on roof pitches between 3:12 and 7:12. 4) Remove lugs under tiles that rest on valley flashing. 5) Shingle-lap tile courses on flat tile installation.

Figure 12-58. Proper tile installation at a valley. Lugs must be removed when the top of the tile rests on the flashing. (Lifetile)

PROCEDURE

Preparing a tile valley

1. Lay down a 36″ wide sweat sheet, dividing the width across the valley. A **sweat sheet** is a strip of felt or ice-and-water barrier.
2. Install the flashing. Valley flashing should be 28 gage corrosion-resistant metal extending at least 11″ each way. In the center of the flashing there should be a diverter rib not less than 1″ high. Ends should be lapped 6″.
3. Cut off the top of the eave riser strip and extend the metal valley flashing slightly over the eave and upward the length of the valley. Edges along the length of the flashing should be turned up 1/2″ by 30°. Standard flashing usually comes preformed, ready for installation.
4. Put down underlayment, lapping it over the flashing. Another method involves putting down the underlayment and weaving it across the valley before installing the flashing.

12.17.2 Hips, Ridges, and Rakes

Nailer boards are wood strips installed on edge at hips and ridges to support trim tiles. The height of the ridge board is 2″ to 6″. The board must be high enough to maintain an even plane of trim tile. Trim tiles are attached with one corrosion-resistant 10d nail. Nose ends should be set in a bead of roofer's mastic that also covers the nail head. See **Figure 12-59.**

Fasten rake tile with two nails. The joints between field tile and trim tile should be weatherproofed with a bed of mortar or an approved dry ridge-hip system.

Tile is extremely heavy. Use a mechanical method—conveyor or lift truck—to deliver it to the roof. This should be done after the deck has been prepared with flashing, underlayment, and battens. Be sure to include flashing around chimneys, pipes, and vents. **Figure 12-60** shows proper installation of vent flashing for tile roofing.

Sweat sheet: A strip of felt or ice-and-water barrier in tile roofing.

Nailer boards: Wood strips installed on edge at hips and ridges to support trim tiles.

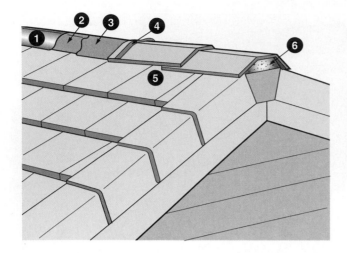

1) Ridge nailer to be of sufficient height to maintain even plane of ridge tiles. 2) Underlayment carried over or under ridge nailer. 3) Optional second layer of felt over nailer board as weatherblock. 4) Apply continuous bead of approved roofers' mastic at overlapping areas and over nail holes. 5) Provide minimum 3" headlap. 6) Optional mortar fill end treatment.

Note: Use one 10-penny nail per ridge tile.

A

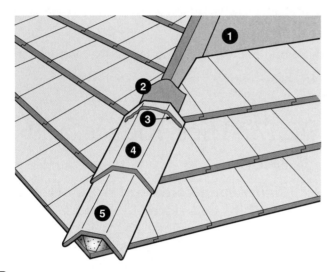

1) Underlayment carried over or under hip nailer. 2) Optional second layer of felt over nailer board as weatherblock. 3) Apply continuous bead of approved roofers' mastic at overlapping areas and over nail holes. 4) Provide minimum 3" headlap. 5) Hold back hip nailer 6" from eave edge.

B

Figure 12-59. Details of ridge and hip trim tile installation. A—Typical ridge design. B—Typical trim tile installation at a hip. (Lifetile)

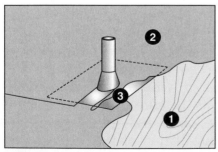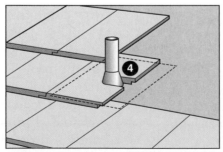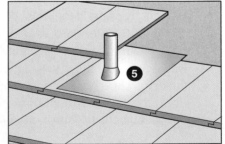

1) Decking. 2) Underlayment. 3) Standard galvanized iron (G.I.) base flashing. 4) Notch tile to accept flashing. 5) Standard galvanized iron (G.I.) top flashing. Seal with approved roofers' mastic.

Figure 12-60. Vent flashing for tile roofing is similar to the method used for asphalt or wood shingles. With some high-profile tiles, lead or other flexible flashing is used. (Lifetile)

Severe weather conditions and taller structures require extra construction steps when installing roofing tile. Where wind velocities may exceed 80 mph or where the roof is more than 40′ above the ground, observe the following guidelines. Be sure to consult building codes in these cases.

- Nail the head of every tile.
- Fasten the noses of eave courses with special clips.
- Secure rake tiles with two nails.
- Set noses of ridge, hip, and rake tiles in a bead of appropriate, approved roofer's mastic.
- Tiles cut too small for nailing should be set in an approved mastic or secured to the roof with wire.

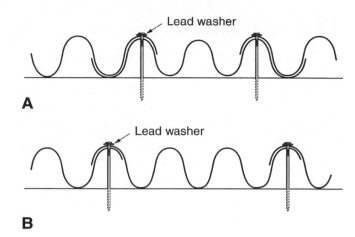

Figure 12-61. Applying corrugated sheet metal roofing. A—Sheets properly laid with one-and-one-half corrugation lap. B—Single corrugation lap is not recommended and may leak in a driving rain.

12.18 Metal Roofing

Structures in areas where heavy snow occurs may be given a roof covering of metal sheeting. This is because snow tends to slide off of the metal roof before building to a deep accumulation. Sheets heavily coated with zinc galvanizing (2 oz. per sq. ft.) or aluminum-zinc alloy are approved for permanent structures. Sheets with lighter coatings of zinc are less durable and are likely to require painting every few years.

Sheets are generally 26 or 28 gage. The heavier gage has no particular advantage except its added strength. A zinc coating for durability is more important than strength for this type of roofing. On temporary buildings and in cases where the most economical construction is required, lighter metal can be used. It will give satisfactory results if protected by paint.

12.18.1 Slope and Laps

Metal roofing sheets may be laid on slopes as low as 3-in-12 (1/8 pitch). If more than one sheet is required to reach the ridge, ends should lap no less than 8″. When the roof pitch is 1/4 or more, 4″ of end lap is usually satisfactory.

To make a tight roof, lap sides of corrugated sheets by 1 1/2 corrugations, **Figure 12-61.** If

only a single-corrugation lap joint is used, wind-driven rain will likely be forced through the joint. When using roofing 27 1/2″ wide with 2 1/2″ corrugations and 1 1/2 corrugation lap, each sheet covers a net width of 24″ on the roof.

Sheet metal roofing is also available in a form that looks like clay tile. See **Figure 12-62.** Computer-controlled roll-forming machines produce a continual stepped-panel with tile forms about 7″ wide and 12″ long. Standard sheets are 36″ wide and measure from 2′–20′ long. The product is available in a variety of colors.

Figure 12-62. This sheet metal roofing is roll-formed to resemble Mission tile. It is available in a variety of colors. (Met-Tile Inc.)

12.18.2 Fasteners

If 26 gage sheets are used, supports may be 24" apart. If 28 gage sheets are used, supports should not be more than 12" apart. For best results, galvanized sheets should be fastened with lead-headed nails or galvanized nails and lead washers. Drive nails only into the tops of the corrugations. To avoid corrosion, use the nails specified by the sheet manufacturer.

Typically, No. 10×1" hex washer-head screws are installed spaced 18"–24" on center. In areas of high winds, closer spacing may be required. Consult local codes. Use a driving tool with a variable speed of 200–2500 rpm and a depth-sensing nosepiece.

Various accessories are available for closures and fittings. Their application is shown in **Figure 12-63**.

12.18.3 Aluminum Roofing

Corrugated aluminum sheets usually make a long-lasting roof, if properly applied. Exposure tests reported by the Bureau of Standards indicate this material is capable of resisting corrosion in coastal areas unless subjected to direct contact with saltwater spray. Where this is likely to happen, aluminum roofing is not recommended.

Aluminum alloy sheets available for roofing usually have a corrugation spacing of 1 1/4" or 2 1/2". Recommendations for the installation of sheet metal regarding side lap and end lap are applicable to the laying of aluminum sheets.

An important precaution to observe in laying aluminum roofing is to make sure that contact with other kinds of metal is avoided. Contact between dissimilar metals in the presence of moisture can result in *galvanic corrosion* that eats away one of the metals. Where it is not possible to avoid such contact, both metals should be given a heavy coating of asphalt paint wherever the surfaces touch.

Aluminum is soft and the sheets used for roofing are relatively thin. Thus, they should be laid on tight sheathing or on decks with openings no more than 6" wide. Aluminum roofing should be nailed with no less than 90 nails to a square, or about one nail for each square foot. Use aluminum alloy nails and place nonmetallic washers between nail heads and the roofing.

If desired, the sheathing may be covered with water-resistant building paper or asphalt impregnated felt. Paper that absorbs and holds water should never be used.

To avoid corrosion, aluminum sheets should be stored so that air will have free access to all sides. Otherwise, a white deposit will form. This deposit can very quickly create pinholes.

12.18.4 Aluminum Shakes

Aluminum shakes are manufactured of an aluminum-magnesium alloy with a nominal .019" thickness. They are available in brown, red, dark gray, white, and natural aluminum. For installation, see manufacturer's instructions.

12.18.5 Terne Metal Roofing

Terne metal roofing, also called *terneplate,* is made of copper-bearing steel. It is heat treated to provide the best balance between malleability (ease of forming) and toughness. The metal is also hot-dip coated with terne metal. This is an alloy of 80% lead and 20% tin. The high weather resistance of terne metal roofing is primarily due to the lead. Tin is included because the alloy makes a better bond with steel.

Grades are expressed as the total weight of the coating on a given area. This area is the total area contained in a box of 112 sheets that are $20" \times 28"$. It is equal to 436 square feet. The best grade of terne coating is 40 lb. coating. It provides a roof surface that will last for many years.

Terne metal roofing is available in a wide variety of sheet sizes, as well as in 50' seamless rolls of various widths. This permits its use for many different types of roofs and methods of application. It is extensively used for flashing around both roof and wall openings.

Terne metal roofing: Copper-bearing steel sheets dip coated with an alloy of lead and tin.

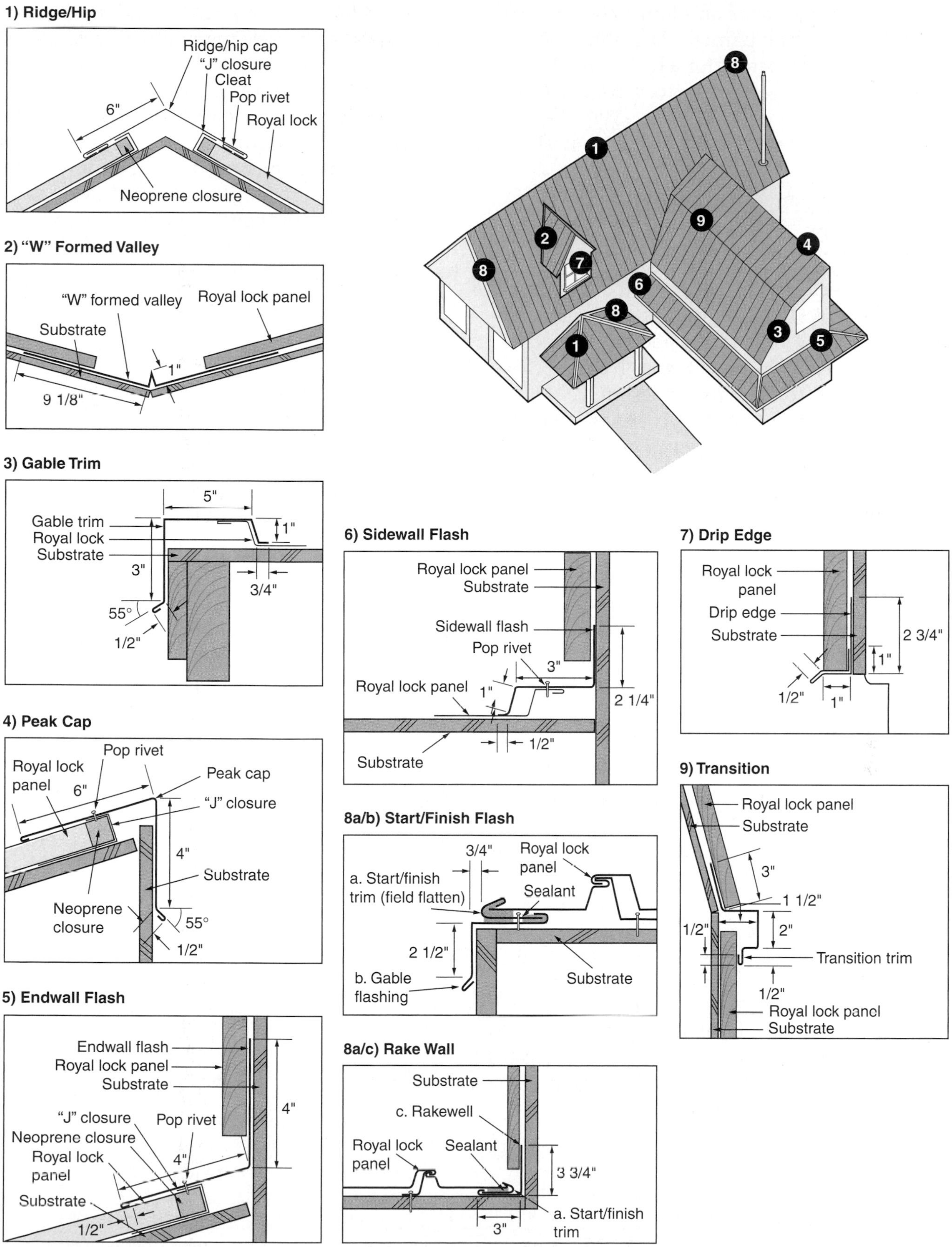

1) Ridge/Hip

Ridge/hip cap
"J" closure
Cleat
Pop rivet
Royal lock
6"
Neoprene closure

2) "W" Formed Valley

"W" formed valley   Royal lock panel
Substrate
1"
9 1/8"

3) Gable Trim

5"
Gable trim
Royal lock
Substrate
1"
3"
3/4"
55°
1/2"

4) Peak Cap

Pop rivet
Royal lock panel
6"
Peak cap
"J" closure
4"
Substrate
Neoprene closure
55°
1/2"

5) Endwall Flash

Endwall flash
Royal lock panel
Substrate
4"
"J" closure   Pop rivet
Neoprene closure
Royal lock panel
4"
Substrate
1/2"

6) Sidewall Flash

Royal lock panel
Substrate
Sidewall flash
Pop rivet
Royal lock panel   1"   3"   2 1/4"
Substrate   1/2"

8a/b) Start/Finish Flash

3/4"   Royal lock panel
a. Start/finish trim (field flatten)   Sealant
2 1/2"
b. Gable flashing   Substrate

8a/c) Rake Wall

Substrate
c. Rakewell
Royal lock panel   Sealant
3 3/4"
a. Start/finish trim
3"

7) Drip Edge

Royal lock panel
Drip edge
Substrate
2 3/4"
1"
1/2"   1"

9) Transition

Royal lock panel
Substrate
3"
1 1/2"
1/2"   2"
Transition trim
1/2"
Royal lock panel
Substrate

Figure 12-63. Various accessories are available for proper installation of metal roofing. (American Building Components)

For best appearance and longest wear, terne metal roofs must be painted. Use a linseed oil–based, iron oxide primer for a base coat. Almost any exterior paint and color can be used over this base.

12.18.6 Zinc-Aluminum Coated Steel Roofing

Some steel roofing has a zinc-aluminum coating. A typical zinc-aluminum coating is 55% aluminum, 43% zinc, and 2% silicon. This coating is applied by a hot-dip process. Next, a chromate treatment adds corrosion resistance. This is followed by a primer and top coatings of polyesters, silicone, fluorocarbons, or plastisols. The top coats are available in a large selection of colors.

Before any coatings are applied, the flat, 29 gage sheets are passed through forming rolls to add ribs to the sheets. These ribs stiffen and strengthen the sheets. See **Figure 12-64.**

Apply zinc-aluminum-coated roofing sheets over solid decking. The recommended under-layment is 30 lb. asphalt-saturated roofing felt. Other suitable barrier materials may be substituted. An ice-and-water barrier should be applied at the eaves in cold climates.

If it is being used for reroofing, remove the old roof covering. If this is not practical, hot mop a layer of underlayment and then put down 2 × 2 vertical battens at regular intervals. Mop more asphalt over this construction. If additional roof insulation is needed, sheet insulation can be applied between the battens. Cross battens are horizontally laid across the roof and roofing sheets are fastened to the cross battens. Use only approved, self-penetrating, self-tapping screws. Do not overdrive screws, which can cause panel distortion. In some panel designs, screws are concealed by attaching the next panel.

12.18.7 Valley Flashing

Valley flashing for metal roofs consists of a layer of ice-and-water barrier and a preformed metal valley flashing. The metal flashing should extend at least 9″ on either side of the valley.

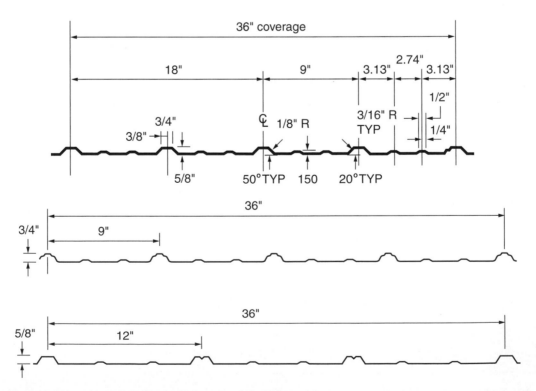

Figure 12-64. The ribs in metal roofing help stiffen the sheets and provide for leakproof overlapping. (American Building Components)

Ends of the flashing are lapped 12″. If other types of flashing are used, they should usually extend up to 20″ on either side of the valley.

12.18.8 Cutting Metal Panels

Metal roofing panels must be cut at valleys and hips. Use a circular saw with an appropriate blade. Cut with the exterior surface turned down. This prevents damage to the finish. Remove steel chips, which can rust and spoil the finish.

12.19 Gutters

The term *gutter* refers to a separate unit that is attached to the eave. The term *eaves trough* usually applies to a waterway built into the roof surface over the cornice. Gutters or eaves troughs collect rainwater from the edge of the roof and carry it to downspouts. For proper drainage, a gutter should slope 1/4″ for every 4′ in the direction of the downspout. Vertical pipes called *downspouts* direct water away from the foundation or into a drainage system.

For best results, gutters must be sized to suit the roof areas from which they receive water. For roof areas up to 750 sq. ft., a 4″ wide trough is suitable. For areas between 750 and 1400 sq. ft., 5″ troughs should be used. For larger areas, a 6″ trough is recommended. Quality of gutters, like that of flashing, should correspond to the durability of the roof covering. If galvanized steel guttering is used, it should have a heavy zinc coating.

The size of downspouts or conductor pipes also depends on the roof area. For roofs up to 1000 sq. ft., downspouts of 3″ diameter have sufficient capacity, if properly spaced. For larger roofs, 4″ downspouts should be used.

Gutter: A metal, plastic, or wood trough attached to the edge of a roof to collect and conduct water.

Eaves trough: A waterway built into the roof surface over the cornice.

Working Knowledge

An eaves trough must be carefully designed and built since any leakage will penetrate the structure. Because of this and the extra cost of construction, they are seldom used in new construction or refurbishing work.

PROCEDURE

Installing gutters

1. At the high end of the eaves, attach a chalk line 3/4″ below the shingles.
2. Establish the proper slope at the other end and snap a line. The upper edge of the gutter will be located on this line.
3. Attach an end cap to the end of the gutter. The design of this assembly varies, depending on the gutter material and the manufacturer. In some cases, mastic is used where parts join. In others, gaskets are provided.
4. Begin attaching the gutter at the end opposite the downspout. Fasten the gutter at 2′ or 3′ intervals using the fastening devices provided. In some cases, spike and ferrule fasteners are provided. In others, brackets or hangers are used.
5. Carefully check that the assembly is maintaining the proper slope.
6. If the gutter must wrap around corners, install miters, making sure that joints are made watertight.
7. Attach downspout elbows and downspouts. Each section has a large end and a crimped end. These are installed with the large end up so that debris does not catch on the crimped end.
8. Attach downspouts to the wall using downspout bands. Use two on each section.
9. If leaf guards are specified, install them.

12.19.1 Metal and Plastic Gutters

A wide variety of metal gutters systems is available to control roof drainage. Manufacturers have perfected gutter and downspout systems that include various component parts. Whole systems can be quickly assembled and installed on the building site. Materials consist of galvanized iron and aluminum. Many systems are available either primed or prefinished to match a wide range of colors. Plastic guttering systems are also extensively used.

Gutter systems include inside and outside mitered corners, joint connectors, pipes, brackets, and other items. All are carefully engineered and fabricated. Parts easily slip together and are generally held with soft pop-rivets or sheet metal screws. **Figure 12-65** shows standard parts of a typical gutter and downspout system and how they are assembled.

12.19.2 Wood Gutters

Wood gutters are made of fir or red cedar. When properly installed and maintained, the life of wood gutters is usually equal to that of the main structure. When wood gutters are specified, the architectural plans usually include installation details.

Wood gutters are usually installed before laying down the shingles. One type is attached to the fascia board after the roof is sheathed. Some designs are coordinated with the fascia board and are installed with this trim unit before the roof sheathing is complete. In this case, the sheathing overhangs the fascia and gutter, thus reducing the possibility of leakage.

Cutting, fitting, and drilling is done on the ground before the various units are set in place. Most wood guttering is primed or prepainted. Always use galvanized or other types of weatherproof nails and/or brass wood screws.

Gutter ends may be sealed with blocks, returned and mitered, or butted against an extended rake frieze board. In general, the gutter is treated like cornice molding and should present a smooth, trim appearance.

The roof surface should extend over the inside edge of the gutter with the front top edge at approximately the height of a line extended from the top of the sheathing. For correct appearance, wood gutters are set nearly level. They will satisfactorily drain if kept clean and adequate downspouts are provided.

12.20 Estimating Material

To estimate roofing materials, first calculate the total surface area to be covered. In new construction, the figures used to estimate the sheathing can also be used to estimate the underlayment and finished roofing materials. When these figures are not available, they can be calculated by the same methods used for roof sheathing, as described in Chapter 10. Another method used to estimate roof area is to determine the total footprint (ground area) of the structure. Include all eave and cornice overhang. Convert the ground area to roof area by adding a percentage determined by the roof slope, as follows:

- Slope of 3-in-12, add 3% of the area.
- Slope of 4-in-12, add 5 1/2% of the area.
- Slope of 5-in-12, add 8 1/2% of the area.
- Slope of 6-in-12, add 12% of the area.
- Slope of 8-in-12, add 20% of the area.

There is a simple method for determining the roof pitch when it is not known. You can estimate it from the ground with the help of a folding carpenter's rule. Stand some distance away from the building and fold the rule into a triangle. Hold the folded rule at arm's length and frame the roof inside of the triangle. Adjust the triangle until the slope of its sides line up with the roof, as in **Figure 12-66.** Be sure the base of the triangle is level. Read off the dimensions on the base of the rule that is marked in **Figure 12-66** as the "reading point". Then, refer to the chart in **Figure 12-67** to locate the proper pitch and slope.

To find the number of squares to be covered, divide the total square feet of roof surface by 100. For example, if the total ground area, including overhang, is found to be 1560 and the slope of the roof is 4-in-12:

$$\text{Roof Area} = 1560 + (1560 \times 5\,1/2\%)$$
$$= 1560 + (1560 \times .055)$$
$$= 1560 + 85.80 \text{ (or 86)} = 1646$$

Number of squares   = 16.46 or 16 1/2

After the number of squares is established, additional amounts must be added. For asphalt

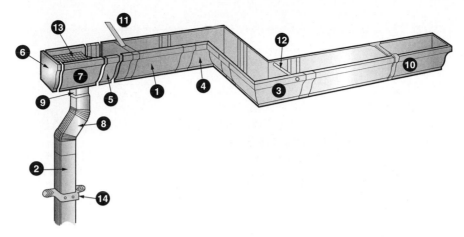

Key	Description	Key	Description
1	5" K gutter	**9**	K Outlet tube (with flange)
2	3" Square corrugated downspout	**10**	5" K Fascia hanger
3	5" K Miter (outside)	**11**	5" K Strap hanger
4	5" K Miter (inside)	**12**	7" Spike (aluminum) 5" Ferrule (aluminum)
5	5" K Slip joint connector	**13**	5" K Strainer
6	5" K End cap left or right	**14**	3" Pipe band (ornamental)
7	5" x 3" K End section with outlet tube		Touch-up paint
8	3" Square corrugated 75° elbow or 60° elbow style a and b		Gutter seal (tube or cartridge)

Figure 12-65. A metal gutter system. The parts can be assembled on the job site. (Crown Aluminum Industries)

Figure 12-66. Determining the roof slope. Stand a sufficient distance from the gable end and frame the roof inside of a folded carpenter's rule held at arm's length. Adjust the end of rule to get a proper reading at the reading point. (Asphalt Roofing Manufacturers Assn.)

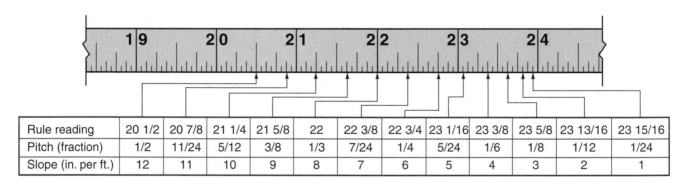

Rule reading	20 1/2	20 7/8	21 1/4	21 5/8	22	22 3/8	22 3/4	23 1/16	23 3/8	23 5/8	23 13/16	23 15/16
Pitch (fraction)	1/2	11/24	5/12	3/8	1/3	7/24	1/4	5/24	1/6	1/8	1/12	1/24
Slope (in. per ft.)	12	11	10	9	8	7	6	5	4	3	2	1

Figure 12-67. Reading point conversions. Locate the reading point on the chart and read downward to find the pitch and slope. (Asphalt Roofing Manufacturers Assn.)

shingles, it is generally recommended that 10% be added for waste. This, however, may be too much for a plain gable roof and too little for a complex intersecting roof. Certain allowances must also be added for reduced exposure on low-sloping roofs. Usually the 10% waste figure can be reduced if allowance is made for hips, valleys, and other extras. For asphalt or wood shingles, one square is usually added for each 100 lineal feet of hips and valleys.

Quantities of starter strips, eaves flashing, valley flashing, and ridge shingles must be added to the total shingle requirements. All of these are figured on lineal measurements of the eaves, ridge, hips, and valleys. For a complex structure, a plan view of the roof is helpful in adding together these materials. All ridges and eave lines are seen as true length and can simply be measured with a scale to find their length. Hips and valleys are not seen as true length and a small amount must be added. Using the percentage listed for converting ground area to roof area usually provides sufficient accuracy.

Safety Note

Worker safety is important everywhere on a construction site. Working on a pitched roof can be made safer by observing certain precautions. These include:

- Wear boots or shoes with rubber or crepe soles. Soles and heels must be in good condition. If the soles are worn, discard the shoes or have them repaired.

- If the roof is steep, workers should wear safety harnesses with lines tied off to a solid anchoring point. This same precaution applies for shallower pitches if there is danger of serious injury or death from a fall.

- Rain, frost, and snow make a roof slippery. Wait until the surface is dry before working. If work must go on under these conditions, workers must tie off or wear special roof shoes with skid-resistant cleats.

- Keep a broom or brush handy to sweep the roof clear of sawdust, loose debris, and dirt.

- Install shingle underlayment as soon as possible. Such material reduces the danger of slipping.

- Install temporary 2 × 4 cleats as toe holds. They can be removed as shingles are installed.

- Remove unused tools, cords, and other loose items from the roof; they can be serious hazards.

In addition to these precautions, check local, state, and OSHA requirements. Be alert to other potential hazards and practice common sense. Taking chances often leads to injuries.

Unlike foundations and framing that are normally hidden from view, exterior finish materials must be attractive. This doorway allows entrance to the home and is pleasing to the eye.

ON THE JOB

Roofer

All types of buildings—residential, commercial, or industrial—must have a weather-tight roof or they will rapidly become unusable. The roofer is the tradesperson who installs roofing material on new buildings and repairs or replaces roofs on existing structures.

The type of roofing installed on a building depends on the roof structure. Flat or slightly sloped roofs are primarily used on industrial, commercial, and multiple unit residential (apartment) buildings. These roofs have a "built-up" roofing system. This type of roofing consists of several layers of felt impregnated with a waterproofing material laid down and sealed with a molten, tar-like material called bitumen. The combination provides a waterproof and quite durable covering for the roof. Usually, bitumen and gravel are applied as a top layer to provide wear-resistance. A newer technology used for flat roofs is a thin sheet of rubber or thermoplastic material that is rolled out onto the roof surface in long strips. The seams between adjoining strips are sealed so the entire roof is covered with a single waterproof layer. Various fastening and covering methods are used to hold the sheet in place.

Residential buildings typically have pitched (sloping) roofs and are covered with shingles. The most common form of shingles is made of asphalt-impregnated paper or felt covered with a layer of fine stone granules for durability. Fiberglass, slate, rot-resistant wood, metal, tile, and other materials are also used for shingles. The shingles are laid in overlapping rows, starting at the lowest edge of the roof and proceeding up the ridge or peak, to form a weatherproof covering.

Two out of three roofers are employed by contractors that perform both types of roofing work, either as new construction or repair/replacement. The remaining one-third own small roofing companies, typically specializing in residential installations.

The work is hot, dirty, and strenuous. Roofers must constantly be aware of their surroundings because of the danger of injuries from hot bitumen or falls. Roofers must be able to do heavy lifting and work under adverse weather conditions.

People often enter the field as helpers to experienced roofers, gradually acquiring skills on the job. Some participate in a three-year apprenticeship program that requires 144 hours of classroom time and 2000 hours of on-the-job experience each year.

Summary

The roofing materials applied to a structure protect it from sun, wind, rain, and snow. They should also provide some degree of fire protection and add to the attractiveness of the building. The most common roof surfacing material is asphalt shingles, but mineral fiber shingles, wood shingles and shakes, and clay tiles are also used. Sheet materials include roll roofing, rubberized membranes, and metals such as galvanized iron, steel, aluminum, tin, and copper. For most roofs, a continuous layer of wood sheathing is applied over the rafters to form a base for the roofing materials. Spaced or skip sheathing may be used for wood shingles and is typical for tile or slate roofs. In cold climates, an ice-and-water barrier is applied along roof edges and valleys before the surfacing material is put in place. Metal flashing is used in roof valleys, at wall-roof intersections, around chimneys, around vent stacks, and around skylights. Shingle application begins at the eaves and proceeds upward, course by course, to the ridge. Low-slope or flat roofs normally require a seamless, membrane-type covering or the use of "built-up" roofing constructed from layers of felt, asphalt, and crushed stone. Since all roofing work is done at a height, safety practices must be carefully observed to prevent injury.

Test Your Knowledge

Answer the following questions on a separate piece of paper. Do not write in this book.

1. The selection of roofing materials is influenced by such factors as initial costs, maintenance costs, appearance, and _____.
2. In shingle application, the distance between the bottom edge of one course and the bottom edge of the next higher course is called _____.
3. Saturated felts are available in different weights. What does the weight reflect?
4. The most commonly used underlayment for asphalt shingles is _____.
5. _____ is a self-sealing barrier applied to the eaves in cold climates.
6. From the ridge to the eaves, the waterway of a valley should diverge (grow wider) at a rate of _____ inch per foot.
7. Chimney flashing consists of two parts: the cap or counterflashing and the _____ flashing.
8. Where is a chimney saddle used and why?
9. The minimum number of nails recommended for the application of each three-tab square-butt shingle unit is _____.
10. Indicate the correct answer(s). When using a powered nailer or stapler for attaching asphalt shingles, _____.
    A. install the fasteners 5/8″ below adhesive strips
    B. adjust the gun for proper staple application
    C. be sure the fasteners are sunk well into the shingle surface
    D. if a fastener must be removed, repair hole with asphalt cement
    E. the fastener should be driven flush with the top of the shingle
11. Asphalt shingles can be used on roofs with a slope as low as _____, if special application procedures are used to make the roof watertight and wind-resistant.
12. List three tools commonly used by roofers to remove old asphalt shingles when reroofing.
13. List four mineral materials that may be used in the top coat of a built-up roof.
14. A formed metal strip called a _____ is attached to the edge of built-up roofs.
15. What is the purpose of a ridge vent?
16. In the best grade of wood shingles, the annular rings run _____ to the surface.
17. When laying wood shingles, they should be horizontally spaced _____″–_____″ apart horizontally to provide for expansion when they become rain soaked.
18. A narrow strip of 30 lb. roofing felt is placed between each course when applying _____.
19. Roofing tile installed on a roof with less than a _____ pitch is considered decorative. What are the recommendations to prevent failure of the roofing materials?
20. *True or False?* When installing corrugated steel roofing, the joints should be lapped one corrugation.
21. What is the main advantage of a metal roof in cold climates?
22. _____ roofing consists of sheets of copper-bearing steel coated with an alloy of lead and tin.

23. The vertical pipes of a gutter system are called _____.

24. For proper drainage, a gutter should slope _____" for every _____' in the direction of the downspout.

25. To determine the number of squares of asphalt shingles needed for a roof, divide the total square feet of roof area by _____.

Curricular Connections

Social Studies. Before the development of asphalt roofing shingles, many different kinds of materials were used for roofing. Many of these were natural: slate shingles, wood shingles, dried grasses (thatch), and even prairie sod. Manufactured materials, such as clay tiles, thin iron plates, and copper sheets, were also used. Research several of these materials and their application methods. Try to determine why a given material (such as thatch) was used in some areas and not in others.

Mathematics. Obtain a building plan for a small, one-story residence. From the plan, determine the dimensions of the roof—length, width, and height (top wall plate to ridge). Use this information to calculate the square footage of the roof. Then, determine the quantity of shingles you would have to order for the roof. Allow for a 5" exposure on each course.

Outside Assignments

1. Study the types and qualities of asphalt shingles used in your locality. Obtain manufacturer's literature from a local building supply center. Prepare a report including information about types, grades, and costs. Also include information about materials for underlayment, valley flashing, hip and ridge finish, and fasteners.

2. Report on application procedures used to install a roof on a residence in your community. Visit the building site and observe the methods used. Do not enter the building site without permission. Note the type of sheathing, special preparation of the roof deck, how valley flashing is applied, use of drip edges, type of starter courses, procedures used to align shingle courses, and how ridges are finished.

3. Construct a full-size visual aid showing the application of wood or asphalt shingles. Use a piece of 3/4" plywood about 4' square to represent a lower corner of a roof deck. Apply underlayment (if required), drip edges, and starter strips. Then, carefully lay the shingles according to an approved pattern. By making only a partial coverage of the various layers, all of the application steps and materials used can be easily studied and observed.

Windows and Exterior Doors

Learning Objectives

After studying this chapter, you will be able to:
- Discuss standards for window and door fabrication.
- Identify the various types of windows.
- Calculate required rough openings.
- Interpret a window schedule.
- Explain how window frames are adjusted for wall thickness.
- Summarize procedures for installing a standard window.
- Describe procedures for installing a replacement window.
- Prepare a rough opening for installation of a door frame.
- Describe the procedure for sliding glass door installation.
- Explain the correct construction of garage door frames.
- Select appropriate garage door hardware.
- Describe the procedure for installing a bow or box bay window unit.

Technical Vocabulary

Awning window
Casement window
Double-hung window
Emissivity
Fixed window
Float glass process
Flush door
Glass block
Hopper window
Horizontal sliding window
Jalousie window
Jamb extension
Light tube
Mullions
Multiple-use window
Muntin
Panel door
R-value
Sash
Storm sash
Window and door schedule

Windows and doors are an important part of a structure. **Figure 13-1** shows some of the many designs available for exterior doors. The carpenter should:
- Have a basic understanding of the various types, sizes, and standards of construction.
- Be able to recognize good quality in materials, fittings, weatherstripping, and finish.
- Understand the importance of careful installation of the various units.
- Be an expert in installation.

Placement of the outside doors and the windows starts after the sheathing and building paper or house wrap has been installed. Careful installation assures a close fit so that air infiltration is kept at a minimum.

Figure 13-1. Exterior doors are manufactured from many different materials and in many different designs. Standard widths are 2'-6", 2'-8", 3'-0", and 3'-6". Standard heights are 6'-8" and 7'-0". (Simpson Door Company)

13.1 Manufacture

Windows and doors are built in large mill-work plants. They arrive at the building site as completed units ready to be installed in the openings of the structure.

13.1.1 Materials

Some windows used in residences are made from aluminum, steel, or vinyl. Most, however, are made from wood or a combination of these materials. Since wood does not transmit heat as readily as metal, there is less tendency for wooden window frames to condense humid air inside the residence. Ponderosa pine is commonly used in the fabrication of windows. It is carefully selected and kiln dried to a moisture content of 6%–12%.

Wood will decay under certain conditions and must be treated with preservatives. Exposed surfaces must be painted or clad with metal or plastic. Metal is stronger than wood and thus permits the use of smaller frame members around the glass. Aluminum has the added advantage of a protective film of oxide, which eliminates the need for paint. To eliminate the problem of heat transfer, better quality metal frames have thermal stops built into the frames.

13.1.2 Manufacturing Standards

The control of quality in the manufacture of windows is based on standards established by the National Woodwork Manufacturers Association (NWMA). These standards cover every aspect of material and fabrication. Included are such details as the projection of the drip cap and the slope of the sill.

For example, NWMA industry *Standard I.S. 2-74* states that the weatherstripping for a double-hung window "shall be effective to the point that it will prevent air leakage in excess of .5 cfm (cu. ft. per minute) per linear foot of sash crack when tested at a static air pressure of 1.56 psf (lb. per sq. ft.)." In simple terms, the weatherstripping cannot allow leakage greater than .5 cfm when subjected to pressure equivalent to a 25 mph wind. Standards also cover the methods and procedures used in preservative treatments.

13.2 Parts of Windows

As shipped from the factory, windows are complete and assembled except for the interior trim. The carpenter must know the names of all the window parts, **Figure 13-2,** their functions, and where in the assembly they are located.

Sash: The framework that holds the glass in a window.

Mullions: Slender bars forming divisions between units of windows, screens, or similar generally nonstructural frames.

Knowing all this, he or she will understand or be able to give instructions about them.

A *sash* is a frame that holds many window parts. Essentially, it contains and holds the glass in place. The sides of the sash are called stiles. The other parts of the frame are the top rail and the bottom rail.

The sash is encased in another frame that is firmly secured to the wall frame. The sides of this frame are called jambs. The top is known as the head and the bottom is the sill.

Because much of the actual construction of a window is hidden, sectional views are used on architectural drawings to show the parts and how they fit together. It is standard practice to use sections through the top, side, and bottom of a window. These drawings also include wall framing members and surface materials, as shown in **Figure 13-3.** A drip cap, shown in the head section of **Figure 13-3,** is designed to carry rainwater out over the window casing. When the window is protected by a wide cornice, this element is seldom included.

A *mullion* is formed by the window jambs when two units are joined together. The section

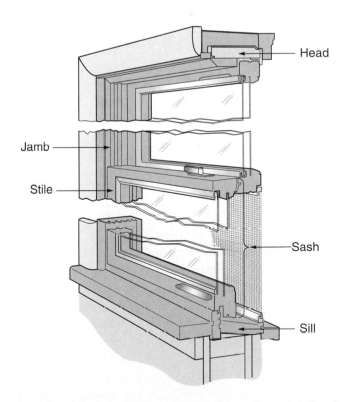

Figure 13-2. Standard sections used to show details of all types of windows. Note the part names. (Pella Corp.)

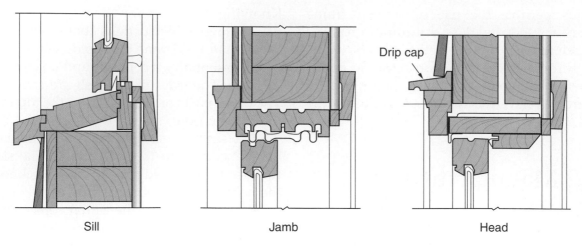

Sill   Jamb   Head

Figure 13-3. Typical detail drawing shows a section through a window. Such details are included with the house plans. They help the carpenter to see how windows should be installed.

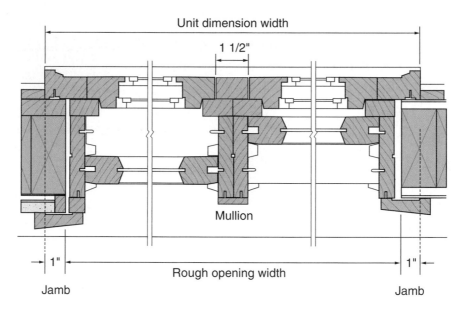

Figure 13-4. Section drawing of a mullion. This is where two window jambs are joined. (Rock Island Millwork)

view of a typical mullion in **Figure 13-4** shows how window units fit together.

Years ago, window glass was available only in small sheets. By using rabbeted strips called *muntins,* a number of small panes of glass could be used to fill large sashes. See **Figure 13-5.** Today, muntins are applied as an overlay on a large sheet of glass. They do not actually separate or support small panes of glass. Made of wood or plastic and available in various patterns, they snap in and out of the sash for easier painting and cleaning.

13.3 Types of Windows

In general, windows can be grouped under one or a combination of three basic types:
- Sliding.
- Swinging.
- Fixed.

Muntins: The vertical and horizontal sash bars separating the different panes of glass in a window.

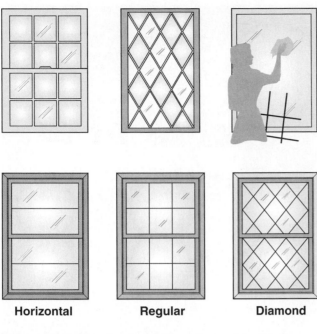

Horizontal   **Regular**   **Diamond**

Figure 13-5. Standard muntin patterns. Muntin assemblies in modern windows easily detach for cleaning and painting.

Each of these includes a variety of designs or methods of operation. Sliding types include the double-hung window and horizontal sliding window. Swinging units that are hinged on a vertical plane are called casement windows, while those hinged on a horizontal plane can be either awning windows or hopper windows.

13.3.1 Double Hung

A *double-hung window* consists of two sashes that slide up and down past each other in the window frame. These are held in any vertical position by a friction fit against the frame or by springs and various balancing devices. Double-hung windows are widely used because of their economy, simplicity of operation, and adaptability to many architectural designs.

Double-hung window: Window that consists of two sashes that slide up and down past each other in the window frame.
Horizontal sliding window: Window with two or more sashes, at least one of which horizontally moves within the window frame.

Figure 13-6 shows a window unit consisting of three double-hung windows. Screen and storm sashes are installed on the outside of the windows.

13.3.2 Horizontal Sliding

Horizontal sliding windows have two or more sashes. At least one of them can horizontally move within the window frame, **Figure 13-7.** The most common design consists of two sashes, both of which are movable. In units with three sashes, the center one is usually fixed.

Figure 13-6. Three double-hung units with storm sashes installed.

Figure 13-7. A horizontal sliding window is sometimes called glide-by unit. (Marvin Windows and Doors)

13.3.3 Casement

A *casement window* has a sash that is hinged on the side and swings outward. Units usually consist of two or more windows separated by mullions. Sashes are operated by a cranking mechanism or a push-bar mounted on the frame, **Figure 13-8.** Latches are used to close, lock, and tightly hold the sash against the weatherstripping. Crank operators make it easy to open and close windows located above kitchen cabinets or other built-in fixtures.

The swing sash of a casement window permits full opening of the window. This provides good ventilation. Frequently, where a row of windows is desirable, fixed units are combined with operating units. Screen and storm sashes are attached to the inside of standard casement windows.

13.3.4 Awning

Awning windows have one or more sashes that are hinged at the top and swing out at the bottom, **Figure 13-9.** They are often combined with fixed units. Several operating sashes can be vertically stacked in such a way that they close on themselves or on rails that separate the units.

Most awning windows have a "projected action" in which the top rail moves down as the bottom of the sash swings out. Crank and push bar operators are similar to those on casement windows. Screens and storm sash are mounted on the inside.

Awning windows are often installed side by side to form a "ribbon" effect at the top of a wall. Such an installation provides privacy for bedroom areas and also permits greater flexibility in furniture arrangements along outside walls. Often, the "ribbon" is placed above fixed windows to provide ventilation.

Figure 13-8. Casement windows are hinged on the side and are operated by a crank or push bar. (Andersen Corp.)

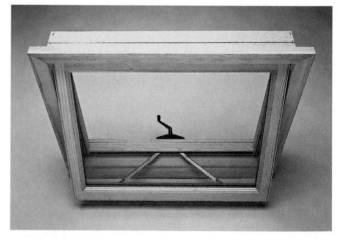

Figure 13-9. Awning windows have one or more sashes that are hinged at the top. (Marvin Windows and Doors)

Working Knowledge

Consideration of outside clearance must be given to both casement and awning windows. When open, they may interfere with movement on porches, patios, or walkways that are located adjacent to outside walls.

Casement window: A window in which the sash pivots on its vertical edge, so it may be swung in or out.

Awning window: A window type with a sash that is hinged at the top, allowing the bottom to swing outward.

13.3.5 Hopper

The *hopper window* has a sash that is hinged along the bottom and swings inward, similar to an upside-down awning window. It is operated by a locking handle located in the top rail of the sash. Hopper windows are easy to wash and maintain. However, their operation may interfere with drapes, curtains, and the use of inside space near the window.

13.3.6 Multiple Use

The *multiple-use window* has a single, outswinging sash designed so it can be installed in either a horizontal or vertical position. These windows are simple in design and do not require complex hardware.

13.3.7 Jalousies

A *jalousie window* is a series of horizontal glass slats held and controlled at each end by a movable metal frame. The metal frames are attached to each other by levers. The slats tilt together in about the same manner as a mini or venetian blind. Jalousie windows provide excellent ventilation. However, they are not very weather tight. Their use in northern climates is usually limited to three-season rooms and breezeways.

Hopper window: A type of window with a sash hinged at the bottom. It swings inward.

Multiple-use window: A unit with a single outswinging sash designed so it can be installed in either a horizontal or vertical position.

Jalousie window: A window with a series of small horizontal overlapping glass slats that simultaneously rotate when operated.

Fixed window: A window that does not open.

Float glass process: A process used to create glass in which melted glass flows onto a flat surface of molten tin.

13.3.8 Fixed

The *fixed window* can be used in combination with any of the other units. Its main purpose is to provide daylight and a view of the outdoors. When used in this type of installation, the glass is set in a fixed sash and mounted in a frame that matches the ventilating windows. Large sheets of plate glass are often separated from other windows to form window walls. They are usually set in a special frame formed in the wall opening.

13.4 Window Heights

One of the important functions of a window is to provide a view of the outdoors. The architect must be aware of the dimensions shown in **Figure 13-10**. These dimensions may need to be adjusted when designing for persons with disabilities.

In residential construction, the standard height from the bottom of the window head to the finished floor is 6'-8". When this dimension is used, the heights of window and door openings are the same. If inside and outside trim must align, 1/2"–3/4" must be added to this height for thresholds and door clearances. Window manufacturers usually provide exact dimensions for their standard units. See **Figure 13-11.**

13.5 Window Glass

Sheet glass used in regular windows is produced by the *float glass process.* In this process, melted glass flows onto a flat surface of molten tin in a vat more than 150' long. As it flows over the tin, a ribbon of glass is formed that has smooth, parallel surfaces. The glass cools, becoming a rigid sheet, and is then carried through an annealing oven on smooth rollers. Finally, the continuous sheet of glass is inspected and cut to usable sizes. **Figure 13-12** lists standard thicknesses.

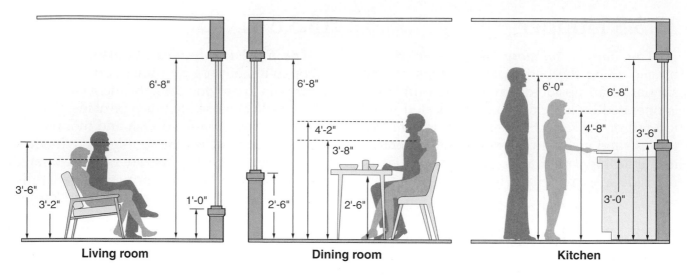

Figure 13-10. Architectural drawings should take into account the standards for window heights. Avoid placing horizontal framework at the eye levels shown. Designs may need to be adjusted for people with disabilities.

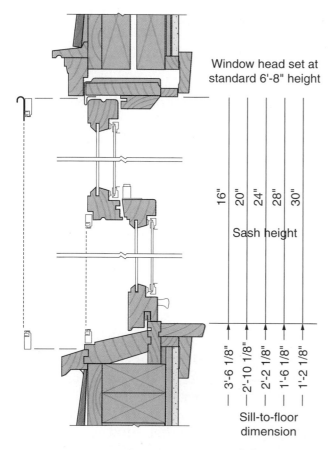

Figure 13-11. Consult manufacturer's product literature for window head and sill heights. (Pella Corp.)

Float glass		Glazing quality	Nominal thickness	
			inches	mm
	SS*	B	3/32	2.5
	DS°		1/8	3
Clear		B, Select	5/32	4
			3/16	5
		Mirror	1/4	6
		Select		
			5/16	8
			3/8	10
Heavy duty clear		Select	1/2	12
			5/8	15
			3/4	19
			7/8	22

SS*= Single strength

DS°= Double strength

Figure 13-12. Standard thicknesses of glass used in residential construction are shown on this chart. (Libby-Owens-Ford Co., Glass Div.)

13.6 Energy-Efficient Windows

Glass areas of a dwelling account for much of the heat loss in winter and heat gain in summer. Glass more readily conducts heat than most other building materials. The resistance of a material to the passage of heat is measured as an *R-value.* When a building material has a low R-value, it means that the material has little resistance to the passage of heat.

A single pane of glass has an R-value of about .88. Adding a second pane of glass with a 1/2" dead air space in between increases the R-value to about 2.00. Storm sashes have long been used to improve the R-value of window space. The *storm sash* is attached to the outside of the window frame. They are double- or triple-track units with movable screens and glass panes. In warm months, the screen is moved into position. In cold months, the glass is moved into position.

Another variation of the storm sash is the storm panel, **Figure 13-13.** Panes of glass mounted in metal frames can be attached to either the inside or outside of the window sash. It is generally used on horizontal sliding and casement or other hinged sashes. When a higher R-value is required, the storm panel can be equipped with sealed double glazing.

13.6.1 Double- and Triple-Sealed Glazing

For movable sashes, two or three layers of 1/8" glass are fused together with a 3/16" air space between layers. Double or triple layers of plate glass are used in large fixed units. Special seals are used to trap the air between panes. Air spaces are generally from 1/4"–1/2" wide. The air is dehydrated (has the moisture removed) before the space is sealed. **Figure 13-14** shows

Figure 13-13. A removable interior insulating panel is easy to install. The inside surface of the panel has a low-emissivity coating that reflects radiant heat. (Pella Corp.)

R-value: A number related to the efficiency of an insulating material.

Storm sash A sash unit attached to the outside of the window frame. Used to increase the R-value of the window unit.

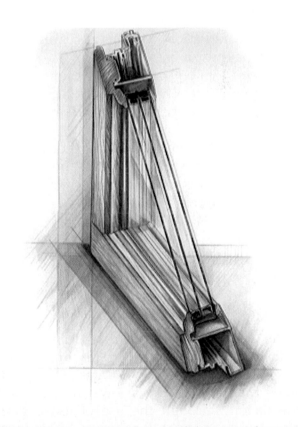

Figure 13-14. A triple-glazed unit has two air spaces. (Pella Corp.)

sealed triple glazing in a standard casement unit. Double and triple glazing offer advantages:
- Lower heat loss in cold weather.
- Reduced downdrafts along the window surface.
- Reduced heat penetration in summer months.
- Decreased or eliminated sweating and fogging of windows in cold weather.
- Less outside noise is transmitted through the window.

13.6.2 Low-Emissivity Glazing

Emissivity is the relative ability of a material to absorb or re-radiate heat. Research in the area of glazing technology has developed a method of raising the R-value of double-glazed windows. Commonly referred to as *low-e* or *high performance* windows, these units have a clear outer pane, an air space, and a special coating on the air-gap side of the inner pane. The special, factory-applied coating consists of an extremely thin layer of metal oxide. It reflects infrared (heat wave) radiation, but allows regular light waves to pass through.

During winter months, warm surfaces within a room (wall, floor, furniture) radiate heat waves. When these waves strike the low-emissivity surface of the window, they are reflected back into the room. This action reduces heat loss. During summer months, heat waves from walks, drives, and other outside surfaces are prevented from entering. This lowers air conditioner loads.

13.6.3 Argon-Filled Insulating Glass

The R-value of sealed, double-glazed windows can be raised by replacing the air between the glass with argon gas. This gas is heavier than air and has a lower heat conductance factor. Through the use of low-e glass surfaces and argon gas, R-values of 4.50 or more can be attained.

13.7 Screens

Windows opened for ventilation require screens to keep out insects. The mesh, usually made from fiberglass or aluminum, should have a minimum of 252 openings per square inch. Manufacturers have perfected many methods for mounting and storing screen panels. Most screens have a light metal frame.

13.8 Windows in Plan and Elevation Drawings

Carpenters study the plans and elevations of the working drawings to determine the types of windows called for and where they are to be located. It is common practice in frame construction to locate the horizontal position of windows and exterior doors by dimensioning to the center of the opening, as shown in **Figure 13-15**. In masonry construction, the dimension is to the edge of the opening. This method is sometimes used in frame construction.

Elevations show the type of window and may include glass size and heights, **Figure 13-16**. The position of the hinge line (point of dotted line) indicates the type of swinging window. Sliding windows require a note to indicate they are not fixed units. Supporting mullions are indicated in the plans and elevations.

13.9 Window Sizes

Besides the type and position of the window, the carpenter should know the size of each unit or combination of units. Window sizes may include:
- Glass size.
- Sash size.

Emissivity: The ability of a material to absorb heat or emit heat by radiation.

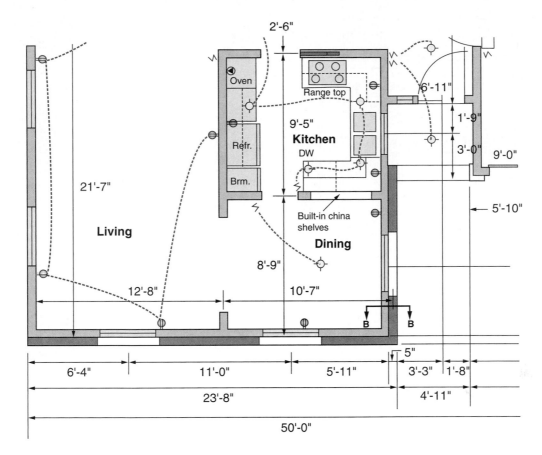

Figure 13-15. Floor plans show locations of windows and doors. Can you locate the door shown on this partial floor plan?

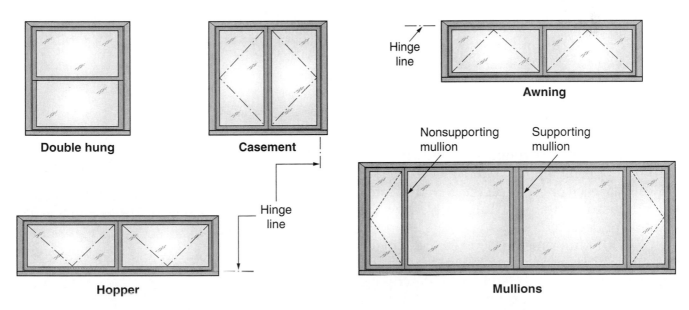

Figure 13-16. Windows shown in elevation views. Horizontal sliding windows are noted to distinguish them from fixed units. Elevations show supporting and nonsupporting mullions.

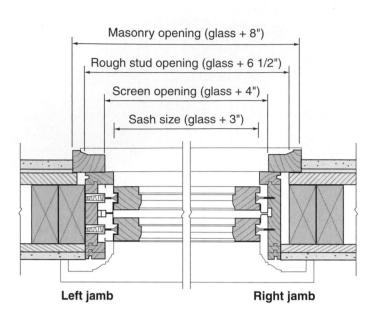

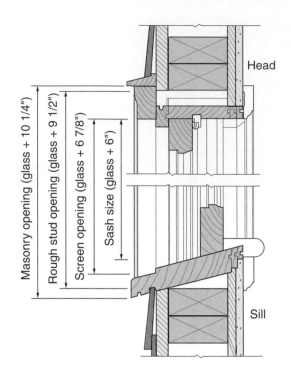

Figure 13-17. Window sizes and location of measurements. The rough opening is larger than the overall size of the frame to permit alignment and leveling when the unit is installed.

- Rough frame opening.
- Masonry or unit opening.

Figure 13-17 shows the position of these measurements and approximately how they are figured from the glass size. Measurements slightly vary from one manufacturer to another.

A complete set of architectural drawings should provide detailed information about window sizes. This information is usually listed in a table called a *window and door schedule.* It includes, among other things, the manufacturer's numbers and rough opening sizes for each unit or combination. An identifying letter is located at each opening on the plan. The same letter is used in the schedule to identify the window or door.

When this information is not included in the architectural plans, the carpenter must study manufacturer's catalogs and other descriptive literature. Sizes for basic units are often given in diagrams as shown in **Figure 13-18.** The size of the rough opening is of major importance to the carpenter. However, the other dimensions are also needed—for example, the height of the rough opening above the floor.

13.10 Detail Drawings

Section drawings are helpful to the carpenter. They show each part of the window and how the unit is placed into the wall structure. **Figure 13-19** shows detail drawings for a typical double-hung window. Similar drawings are available for other

Window and door schedule: A list of the doors and windows for the structure.

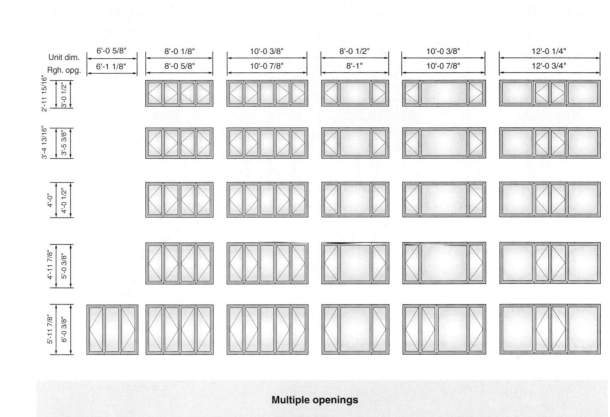

Multiple openings

A number of suggested combinations are shown above using the narrow (no support) mullion.
Additional combinations using support mullion and transom joining can be arrived at by using the dimensions in the formulas listed below.

Casement narrow mullion

Joining basic casement units to form multiple units or picture window combinations without vertical support between units.

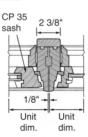

Overall unit dimension width – The sum of individual unit dimensions, plus 1/8" for each unit joining.
Overall rough opening width – Add 1/2" to overall unit dimension width.

Casement support mullion

Joining basic casement units using a 2 x 4 vertical support between units.

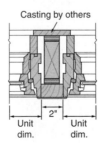

Overall unit dimension width – The sum of individual unit dimensions, plus 2" for each unit joining.
Overall rough opening width – Add 1/2" to overall unit dimension width.

Casement transom

Joining basic casement units by stacking units to form combinations.

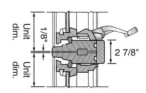

Overall unit dimension height – The sum of individual unit dimension heights, plus 1/8" for each unit joining.
Overall rough opening height – Add 1/2" to overall unit dimension height.

Figure 13-18. Catalogs and brochures picture the variety of window units a manufacturer offers. This page shows the sizes of available casement window units. (Andersen Corp.)

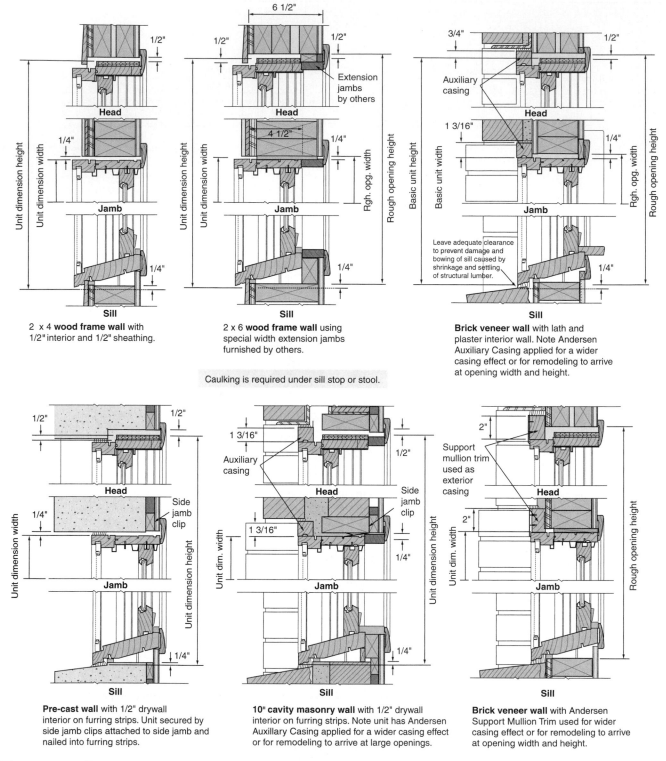

2 x 4 wood frame wall with 1/2" interior and 1/2" sheathing.

2 x 6 wood frame wall using special width extension jambs furnished by others.

Brick veneer wall with lath and plaster interior wall. Note Andersen Auxiliary Casing applied for a wider casing effect or for remodeling to arrive at opening width and height.

Caulking is required under sill stop or stool.

Pre-cast wall with 1/2" drywall interior on furring strips. Unit secured by side jamb clips attached to side jamb and nailed into furring strips.

10" cavity masonry wall with 1/2" drywall interior on furring strips. Note unit has Andersen Auxillary Casing applied for a wider casing effect or for remodeling to arrive at large openings.

Brick veneer wall with Andersen Support Mullion Trim used for wider casing effect or for remodeling to arrive at opening width and height.

Figure 13-19. These detail drawings show a standard double-hung window installed in wall structures of different materials and thicknesses. (Andersen Corp.)

types of windows. The architect often includes selected detail views in the set of drawings. Whether included in the architectural drawings or made available through manufacturer's catalogs, detail drawings such as these are essential in building the rough frame of the wall structure and in installing the window units. See **Figure 13-20.**

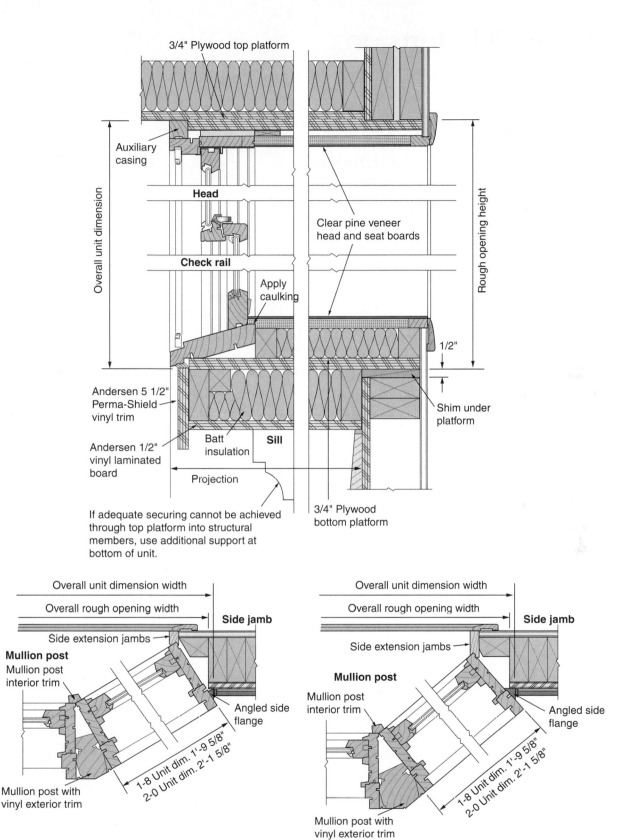

Figure 13-20. Bay window construction is more complicated than standard window construction. Details for installation are extremely important for the carpenter. (Andersen Corp.)

13.11 Jamb Extensions

The thickness of window units may be adjusted to various walls. For example, a frame wall with 3/4″ sheathing and a standard interior surface of lath and plaster may be 5 1/8″ thick, compared with a single-layer drywall construction of about 4 3/4″ thick. A *jamb extension* applied to the window frame, as in **Figure 13-21,** provides the additional thickness needed. Some manufacturers build their frames to a basic size, such as 4 5/8″, and equip the unit with an extension as specified by the builder or architect.

13.12 Aligning Tops of Windows

A door and window story pole is helpful when installing doors and windows, **Figure 13-22.** It helps ensure alignment of the tops of doors and window heads. To use the pole, hold it against the trimmers and transfer the marks.

Another method some carpenters use is to snap a chalk line all around the building at the height of the header trim piece. They then install all windows with the trim piece on the line.

The construction details of the window head are normally the same throughout a building. The sill height may vary, however. Additional positions can be superimposed over other layouts if each one is carefully labeled or indexed so the correct distance is always applied to the proper opening.

Residential doors are normally 6′-8″ high. The tops of windows are usually held at this

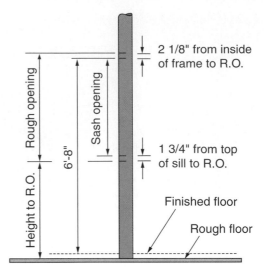

Figure 13-22. Use a story pole to check height of certain parts of the window frame.

same height. An extra 1/2″ is added to allow for thresholds under entrance doors and clearance under interior doors.

13.13 Installing Windows

If rough openings are plumb, level, and correctly sized, it is easy to install windows. Manufacturers furnish directions that specifically apply to their various products. Regardless of the manufacturer, installation of windows is always similar. However, the carpenter should carefully follow the instructions supplied by the window manufacturer.

When windows are received on the job, store them in a clean, dry area. If they are not fully protected by packaging, some type of cover should be used to protect them from dust and dirt. Allow wood windows to adjust to the humidity of the location before they are installed.

Based on the carton labels, move the still-packaged window units into the various rooms and areas where they will be installed. Unpack the window and check for shipping damage.

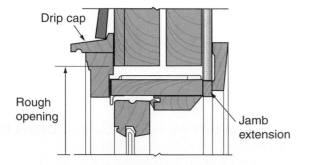

Figure 13-21. Jamb extensions are attached to standard window frames to adjust for various wall thicknesses.

Jamb extension: Narrow strips added to a window or door jamb to increase its width.

Do not remove any diagonal braces or spacer strips until after the installation is complete. Make certain each unit is the correct size for the intended rough opening. Most windows require at least 1/2″ clearance on each side and 3/4″ above the head for plumbing and leveling.

If wood windows have not been primed at the factory, this should be done before installation. Weatherstripping and special channels should not be painted, however. Follow the manufacturer's recommendations.

Most window units are installed from the outside of the structure. From inside the structure, the unit can be turned at an angle and moved to the outside through the rough opening. Be sure to have plenty of help when handling the window entirely from the inside.

Figure 13-23. Rest the window sill on the bottom of the rough opening and then swing the top into place.

PROCEDURE

Installing a window with a flange

1. Place the unit into the rough opening. The sash should be closed and locked. The flange should overlap the exterior sheathing.
2. While resting the window sill on the rough opening, swing the top into place. See **Figure 13-23.**
3. Shift the window as needed to horizontally center it in the rough opening.
4. Use shims (wedge blocks or shingles) under the sill to raise the frame to the correct height as marked with the story pole or the chalk line.
5. Place a level on the sill and adjust the shims until the sill is perfectly level. There may be a tendency for the sill to sag on multiple units.
6. Temporarily secure one corner by tacking a roofing nail (1 1/2″ long) through the lower flange.
7. Double check for level once more. Adjust as needed.
8. Secure the opposite corner, tacking a nail through the flange.
9. Plumb the side jambs with whims, **Figure 13-24.**
10. Check the corners with a framing square or measure the diagonals. If the unit is not square, adjust as needed.
11. Check for front-to-back plumb. Place the level on the outside face of the frame and make sure the window is not tilted outward or inward.

12. Temporarily drive several nails into the top of the side flange.
13. Check over the entire window again to see that it is square and level. Use additional shims if necessary.
14. Check the sash for ease of operation. Make sure the spacing is even between the sash and frame.
15. Permanently nail the flanged window in place with 1 3/4″ or longer galvanized roofing nails. Space the nails 12″–16″ O.C., or as specified by the window label. See **Figure 13-25.**
16. Cover the flange on all sides with adhesive-backed, bituminous tape.
17. To seal the top flange against water penetration, cut a slit in the house wrap and slip a 6″ wide strip of building paper under the slit and over the flange. Use tape to cover and seal any overcut in the house wrap.
18. At this point, many builders cover the inside of the window with a sheet of polyethylene film to protect it during the application of inside wall surface materials.

Installing window units without flanges requires the same general procedure. The only difference is that tacking and nailing is done through the exterior casing into the sheathing. Use galvanized casing nails and nail every 16″. Casing thickness vary, but nails should be long enough to go through the sheathing and into the rough frame.

Figure 13-24. Use a carpenter's level to plumb the jambs. Also check the front-to-back plumb.

Figure 13-25. A double-hung window with a flange is held in place by nails driven through the flange and sheathing into the building frame.

Working Knowledge

When installing window units in masonry walls and walls with brick veneer, attach them to the wood bucks using the same procedure as explained for frame construction. When brick veneer is the exterior finish, allow enough clearance for caulking between the masonry and the window sill. In case of shrinkage and settling of the underlying wood framing, the window will be protected from damage.

13.14 Installing Fixed Units

When the fixed glass panel is of medium size, it is usually mounted in a sash and frame. **Figure 13-26** is a detail drawing of a fixed unit. Often such units are combined with matching ventilating units. The installation of fixed units is essentially the same as for regular windows. They are, of course, larger and heavier. Take extra precautions in handling the units and making the installation. Have extra help.

Large insulating units have 1/4″ thick glass and are heavy. They are seldom installed in window frames at the factory, although they are sometimes mounted in a sash. Usually, large glass units are glazed as a separate operation after the frame and/or sash have been installed. *Glazing* is setting the glass in the opening using a glazing sealant.

Openings must be square, free of twists, and rugged enough to bear the weight of the heavy glass unit. Use only high-grade wood materials

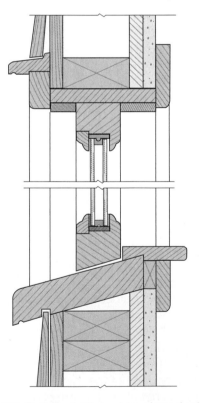

Figure 13-26. Detail of 1″ double-glazed window in a conventional sash and frame.

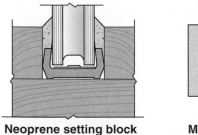 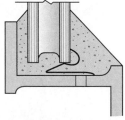

Neoprene setting block in wood sash or frame   **Metal glazing clip in metal sash or frame**

Figure 13-27. Setting blocks and clips. The use of metal clips is limited to welded glass units.

that are dry and free from warp. Special setting blocks and clips may be used to position the glass with clearance on all edges, **Figure 13-27.** For wood frames or sashes, use two neoprene setting blocks at least 4″ long located at quarter points along the lower edge. The width of the blocks should be equal to or greater than the thickness of the glass unit.

When the glass is glazed, its edge should be completely surrounded with a high-quality, nonhardening glazing sealant. There must be no direct contact between the glass unit and frame. **Figure 13-28** shows clearances recommended by one manufacturer.

All glazing systems should be designed and installed to ensure that the seal (organic type) between the layers of glass is not exposed to water for long periods of time. Weep holes as recommended by the Sealed Insulating Glass Manufacturers Association (SIGMA) should be included.

Insulating glass units cannot be altered in any way on the building site. It is essential, then, that sash and frames be carefully designed and that specified dimensions are followed. Wooden window frames should be treated with a wood preservative.

Step-by-step procedures for building a simple wood frame and installing insulating glass are given in Appendix B, **Technical Information.** Also provided are standard thicknesses of insulating glass for fixed window units. Details for the construction of a window wall are given as well.

13.15 Glass Blocks

A *glass block* is made of two formed pieces of glass that are fused together to leave an insulating air space between. They come in different patterns and are usually available in three nominal sizes: 6″ × 6″, 8″ × 8″, and 12″ × 12″.

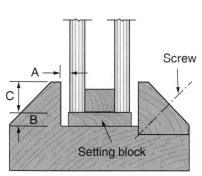

Glass Thickness		Dimensional Tolerance				Minimum Clearance					
		Up to 48″ (1220 mm)		Over 48″ (1220 mm)		Face (A)		Edge (B)		Bite (C)	
inches	mm	inches	mm	inches	mm	inches	mm	inches	mm	inches	mm
1/2	12	±1/16	±1.6	+1/8 – 1/16	+3.2 – 1.6	1/8	3.2	1/8	3.2	1/2	12.7
5/8	15	+1/8	+3.2	+3/16	+4.8						
23/32	18	–1/16	–1.6	–1/16	–1.6						
3/4[1]	19										
3/4[2]	19	±1/16	±1.6	+1/8 – 1/16	+3.2 – 1.6	3/16	4.8	1/4	6.4	1/2	12.7
7/8	22	+1/8	+3.2	+3/16	+4.8						
31/32	24	–1/16	–1.6	–1/16	–1.6						
1	25										

[1] 1/4 ″(6 mm) Air space
[2] 1/2″ (12 mm) Air space

Figure 13-28. Recommended clearances for sealed insulating glass.

Glass block: Two formed pieces of glass fused together to leave an insulating air space between.

Square blocks

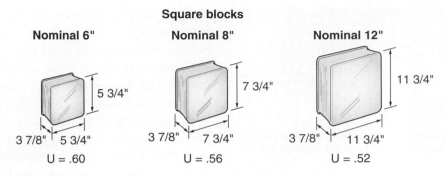

Figure 13-29. Nominal and actual sizes of commonly used glass block.

See **Figure 13-29.** All are 3 7/8″ thick. Special shapes are available for turning corners and for building curved panels. The blocks come in both light-diffusing and light-directing types.

Glass blocks have good insulating properties. Used in outside walls, they provide light, help prevent drafts, muffle disturbing noises, cut off unpleasant views, and ensure privacy, **Figure 13-30.** Inside the home, partitions and screens of glass blocks add a pleasant touch to rooms they divide.

Installing glass blocks into a panel is not difficult. Regular masonry tools are used. Even though carpenters seldom make the actual installation, they are required to build the framework. Therefore, knowledge of the design requirements is helpful.

Figure 13-30. This glass block window provides a high level of security in a bedroom area. It also reduces noise transmission so that acoustical privacy is maintained. (Pittsburgh Corning Corp.)

13.15.1 Installing Small Glass Block Panels

Details for installing glass block panels of 25 sq. ft. or less are given in **Figure 13-31.** In such panels, the height should not exceed 7′, nor the width 5′. Panels may be supported by a mortar key at jambs in masonry construction or by wood members in frame construction. Wall anchors or wall ties are not required in the joints. Expansion space is only required at the head.

13.15.2 Installing Large Glass Block Panels

When the panel area exceeds 25 sq. ft., additional requirements must be met. First,

install expansion strips to partially fill expansion spaces at jambs and heads of larger panel openings. *Expansion strips* are strips of resilient material.

Also, provide support at jambs with wall anchors. In masonry walls, a portion of each anchor is embedded in masonry and in the mortar joint for the glass block. Anchors should be crimped within expansion spaces and spaced on 24″ centers to rest in the same joints as wall ties. Anchors are corrosion resistant and 2′ long by 1 3/4″ wide.

Wall ties should be installed on 24″ centers in horizontal mortar joints of larger panels. Lap ties not less than 6″ whenever it is necessary to use more than one length. Do not bridge expansion

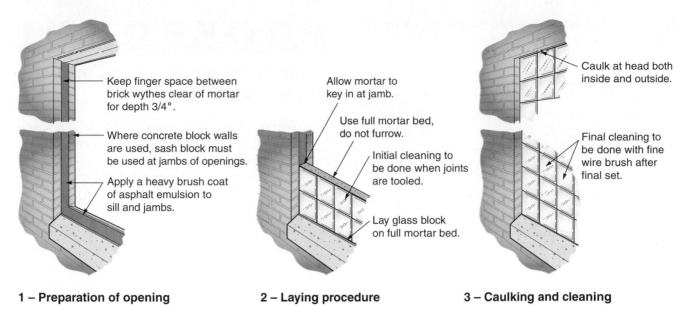

Keep finger space between brick wythes clear of mortar for depth 3/4".

Where concrete block walls are used, sash block must be used at jambs of openings.

Apply a heavy brush coat of asphalt emulsion to sill and jambs.

Allow mortar to key in at jamb.

Use full mortar bed, do not furrow.

Initial cleaning to be done when joints are tooled.

Lay glass block on full mortar bed.

Caulk at head both inside and outside.

Final cleaning to be done with fine wire brush after final set.

1 – Preparation of opening   **2 – Laying procedure**   **3 – Caulking and cleaning**

For panels 25 sq. ft. and less

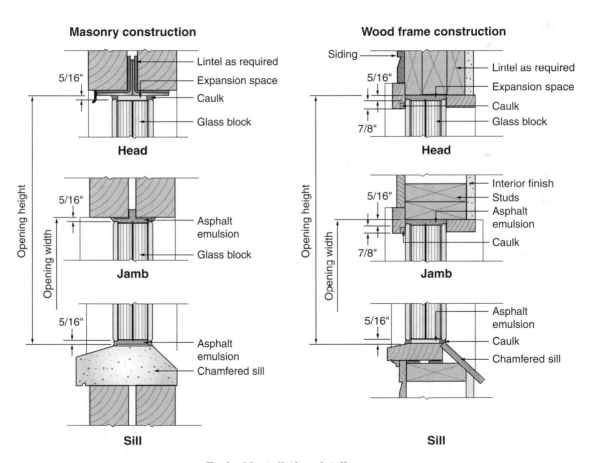

Masonry construction

Head
- 5/16"
- Lintel as required
- Expansion space
- Caulk
- Glass block

Jamb
- 5/16"
- Asphalt emulsion
- Glass block

Sill
- 5/16"
- Asphalt emulsion
- Chamfered sill

Opening height
Opening width

Wood frame construction

Head
- Siding
- 5/16"
- 7/8"
- Lintel as required
- Expansion space
- Caulk
- Glass block

Jamb
- 5/16"
- 7/8"
- Interior finish
- Studs
- Asphalt emulsion
- Caulk

Sill
- 5/16"
- Asphalt emulsion
- Caulk
- Chamfered sill

Opening height
Opening width

Typical installation details

Figure 13-31. Construction details for installing glass block.

spaces. Wall ties are also corrosion resistant. They are 8′ long and 2″ wide.

13.14.3 Openings for Glass Blocks

To determine the height or width of the opening required for a glass block panel, multiply the number of units by the nominal block size and then add 3/8″. For example, a panel that is four 8″ blocks wide by five 8″ blocks high requires an opening 32 3/8″ wide and 40 3/8″ high.

13.16 Replacing Windows

Older windows waste a tremendous amount of heating and cooling energy. Homeowners often replace these windows with new units that have excellent weatherstripping and double or triple glazing. Manufacturers provide a wide range of sizes and also a variety of special trim members that are helpful in making the replacement. They also provide detailed instructions for installing their products.

First consideration in window replacement is the type and size of the new units. To determine the size, it is best to remove the inside trim and measure the rough opening. If the trim members are to be reused, pry them off carefully. Remove nails by pulling them through the trim from the back side.

PROCEDURE

Removing old windows and installing new units

1. Remove the inside stops and lift out the lower sash. If there are weights or counterbalances, disconnect them.
2. Remove the parting stops and lift out the upper sash.
3. With the window sash removed, pry off the outside casing. Make saw cuts through the sill and frame, collapse the members, and remove them from the opening as shown in **Figure 13-32A.**
4. Clean the rough opening to remove dirt, plaster, putty, and nails.
5. Carefully measure the opening to determine the thickness of material needed to bring the rough opening to size required for the new window.
6. Install furring strips as needed, **Figure 13-32B.** Constantly check with a square and level. Use shims where necessary.
7. Check the rough opening for plumb and level.
8. Insert the new unit, **Figure 13-32C.**
9. With the window resting on the rough sill, make adjustments with shims.
10. Attach the unit to the rough opening by nailing through special flanges, metal clips, or the exterior casing. Check the manufacturer's directions for their recommendations on specific units.
11. With the window unit firmly secured in the opening, install outside and inside trim, **Figure 13-32D.** Manufacturers furnish a variety of trim pieces especially made for this purpose. Since the new window unit may not be exactly the same size as the one removed, standard casing may require various widths of filler strips.
12. Bed all of exterior trim members in a high-quality construction mastic to seal against air infiltration.
13. If the top of the window is not well protected by the roof overhang, install metal flashing.
14. Completely seal the window by caulking the joints between the house siding and side trim, head trim, and sill.
15. On the inside, carefully pack all openings between the wall and frame with insulation.
16. Apply the interior trim members.

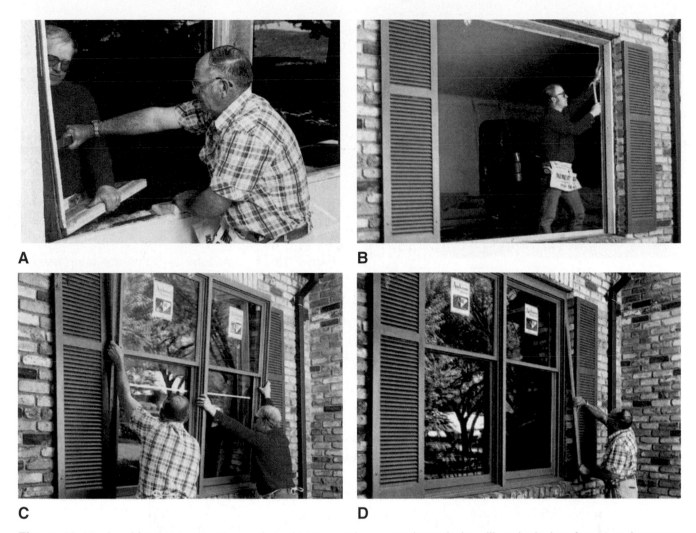

Figure 13-32. A—After removing the sash and trim members, cut through the sill and window frame and remove them. B—Install wood furring to bring the rough opening to the correct size. C—Place the new window unit in the rough opening. D—Install the outside trim furnished by manufacturer or fashion your own. (Andersen Corp.)

13.17 Skylights

Skylights can make a dramatic difference in the appearance and feeling of a room, **Figure 13-33.** Skylights in bathrooms provide good light and privacy. They may be the best way to secure natural light in stairwells and hallways. Skylights can convert attic space into living space at a fraction of the cost of dormers with regular windows.

Light tube: Type of skylight, typically about 12″ in diameter, that consists of a clear dome and a tube with a reflective lining.

In recent years, the *light tube* skylight has come into use. This type of skylight, typically about 12″ in diameter, consists of a clear dome and a tube with a reflective lining that concentrates the light. As shown in **Figure 13-34,** the tube can direct the light to areas not directly beneath the dome. Since only a small opening is made in the roof, light tube units are easier and faster to install than a traditional skylight.

Rough openings for skylights are framed in about the same way as described for chimneys. A detail drawing of a typical installation is shown in **Figure 13-35.** Flashing flanges included in the skylight frame should be installed according to the manufacturer's directions. Skylights of this type should not be installed in roofs with less than a 3-in-12 slope.

Figure 13-33. Skylights provide natural overhead lighting for this bedroom. (Andersen Corp.)

Figure 13-34. Light tube skylights are often used to supply light to windowless bathrooms or interior hallways. (Solatube North America, Ltd.)

In standard frame construction, a shaft is required to connect the ceiling opening with the roof. **Figure 13-36** shows several types of shafts. It is very important to construct the shaft as airtight as possible. Attach to the sides of the shaft insulation that has the same thickness as the ceiling insulation.

13.18 Installing Bow and Box Bay Windows

Many homes have a box bay or bow window, **Figure 13-37.** Since these units are larger than standard windows, as well as projecting a

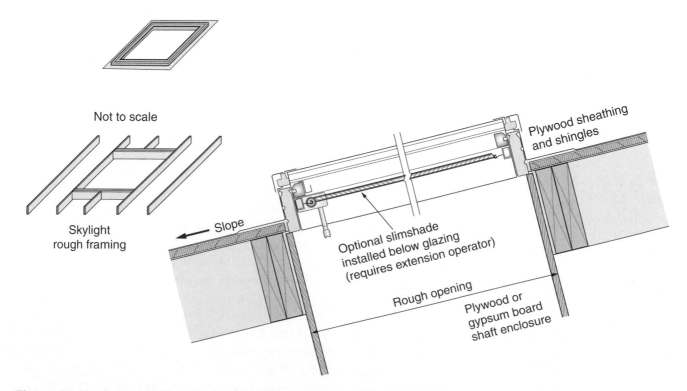

Not to scale

Skylight rough framing

Slope

Optional slimshade installed below glazing (requires extension operator)

Plywood sheathing and shingles

Rough opening

Plywood or gypsum board shaft enclosure

Figure 13-35. Cross-section detail for installation of a skylight. (Pella Corp.)

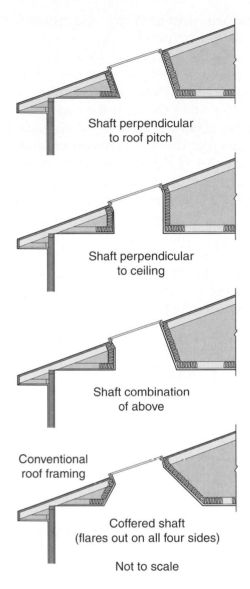

Figure 13-36. A skylight shaft may take any of several shapes.

A

B

Figure 13-37. A bow window being installed in new construction. A—Two workers slide the unit into the rough opening. B—When shimmed and plumbed, the unit is secured by driving nails through the exterior trim.

considerable distance outward from the building wall, they require extra support. This support is important to prevent sagging of the window and reduce stress on the wall framing.

Cable supports are one way of providing bracing. Cables extend from the outer edge of the window's base to framing members above the window. The cables are concealed with wall covering materials.

The carpenter may choose any of several methods of attaching the cables to the building frame. In one method used for single-story houses, cables are attached to rafter tails in the soffit, **Figure 13-38.** It may be necessary to install

bracing between two rafter tails. The cables are secured to cleats at the top and to T-nuts at the bottom. In another variation, the support might be provided by a horizontal or vertical member of the building's wall frame. In a gable installation, the cleats may be anchored to a lintel, studs, second floor band joist, or sill plate.

Extra support along the base is necessary for larger units. If box or angle bay units have a center sash of 5′ × 5′ or larger, use a steel angle or channel or a properly sized wood member across the bottom, spanning the center sash. Drill holes at either end of the support member for the steel cables. See **Figure 13-39.**

Figure 13-38. Supporting a bay window. Cables are attached either to rafter tails or to 2″ lumber spanning the distance between two rafters. (Andersen Corp.)

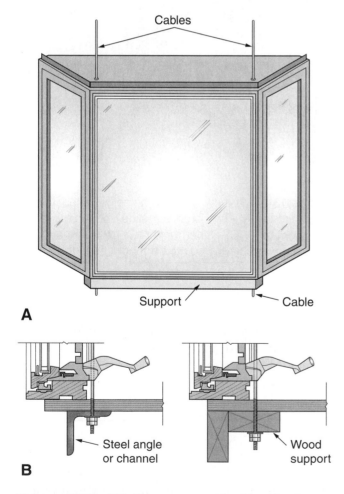

Figure 13-39. Window support. A—Heavier box or angle bay units must have extra support at their bases. B—Channel steel, angle steel, or wood is recommended. (Andersen Corp.)

Kits are provided by the manufacturer along with instructions and suggestions for cable locations. See **Figure 13-40.** When preparations are completed, support the window unit with jacks, lever arms, or some other method. Thread the cable through the top and bottom platforms, allowing an inch of threads below the bottom platform. Attach washers and nuts to each cable. Anchor the cables to the cleats previously attached overhead and draw the cables taut. Remove the temporary support and check unit for plumb, level, and sash reveal. Readjust as necessary. Finally, tighten the upper nuts to prevent movement and then tighten the lock nut. Notch the head and seat boards and slide them into place. Always follow specific instructions provided by the manufacturer.

13.19 Exterior Doors and Frames

Exterior doors are offered in different types and styles. Wood, metal, and fiberglass are the materials of choice for exterior doors. Metal and fiberglass doors may be more common because wood doors can be expensive.

Flush doors have a flat face applied to each side of a light framework. A core of the same thickness as the frame is inside of the frame. Solid core construction consists of wood blocking or particleboard. Hollow core doors use spacing materials at intervals between the facing materials. Metal flush doors usually have a core of rigid insulation that reduces heat transfer through the entryway.

Panel doors have a solid wood framework of stiles (vertical members) and rails (horizontal members) that hold panels. Panels are thinner, often decorative, parts that fill in spaces between the rails and stiles. The panels may be glass, solid wood, or wood louvers.

Flush door: A door that fits into the opening and does not project outward beyond the frame.

Panel door: Door that has a frame and separate panels of plywood, hardboard, or steel set into the frame.

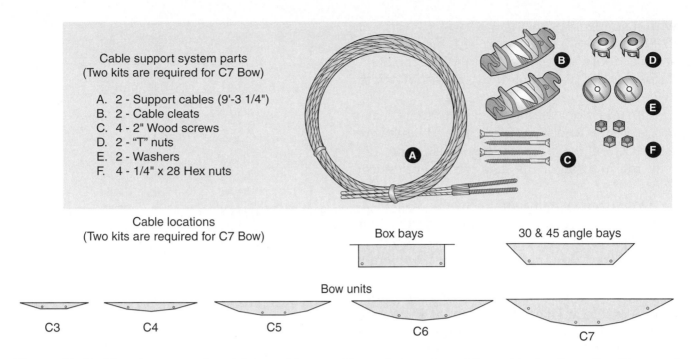

Cable support system parts
(Two kits are required for C7 Bow)

A. 2 - Support cables (9'-3 1/4")
B. 2 - Cable cleats
C. 4 - 2" Wood screws
D. 2 - "T" nuts
E. 2 - Washers
F. 4 - 1/4" x 28 Hex nuts

Cable locations
(Two kits are required for C7 Bow)

Box bays

30 & 45 angle bays

Bow units

C3   C4   C5   C6   C7

Figure 13-40. Manufacturers of prefabricated bow and bay windows have kits and instructions for all types of installations. (Andersen Corp.)

Outside door frames and doors are installed at the same time as the windows following similar procedures. Secondary and service entrances usually have frames and trim members to match the windows. Main entrances, however, often contain additional elements that add an important decorative feature, **Figure 13-41.**

Exterior doors in residential construction are nearly always 6'-8" high, although 7'-0" sizes are available. Main entrances usually are equipped

A

B

Figure 13-41. A—Main entry doors. B—Frames and doors for main entrances as they appear on elevation drawings.

with a single door that is 3'-0" wide. Narrower (2'-8" and 2'-6") sizes are used for rear and service doors. FHA Minimum Property Standards specify a minimum exterior door width of 2'-6".

Outside door frames, like windows, have heads, jambs, and sills. The head and jambs are made of 5/4" stock since they must carry not only the main door but also screen and storm doors.

Door frames are manufactured at a millwork plant and arrive at the building site either assembled and ready to install (prehung) or disassembled (knocked down, or K.D.). Sometimes K.D. units are assembled by the dealer or distributor. It is relatively easy to assemble door frames on the job when the joints are accurately machined and the parts are carefully packaged and marked. While details of a door frame may vary by manufacturer, the general construction is the same, **Figure 13-42.** The head and jambs are rabbeted, usually 1/2" deep, to receive the door. In residential construction, outside doors swing inward and the rabbet must be located on the inside.

Stock door frames are designed for standard wall framing. However, they can also be adapted to stone or brick veneer construction, as shown in **Figure 13-43A.** Stock frames can also be fitted with extension strips that convert the frames to fit a greater wall thickness, **Figure 13-43B.** Extension strips are also used on window frames.

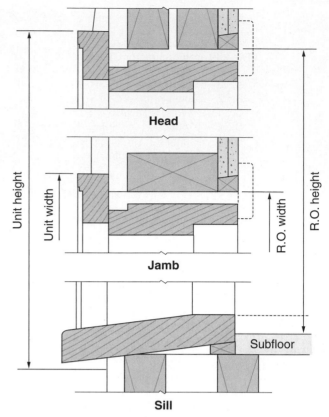

Calculations for determining frame and rough opening sizes:

Unit height = Door + 4 1/2"
Unit width = Door + 4"
R.O. height = Door + 2 1/2"
R.O. width = Door + 2 1/2"

Figure 13-42. Always check drawings for exterior door frame details and sizes. Make sure rough openings are the correct sizes.

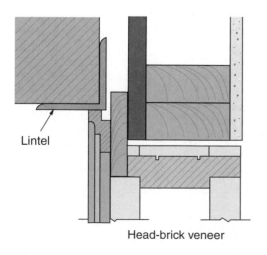

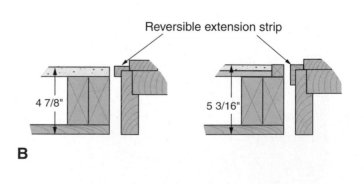

Figure 13-43. A—Detail of how to install the head of a door frame in brick veneer construction. B—Stock frames can be made wider. These reversible extension strips convert 4 5/8" jambs to either 4 7/8" or 5 3/16". (C-E Morgan)

Doorsill design varies considerably. However, the top of the sill is always level with the finished floor. Sills may be made of wood, metal, stone, or concrete. The outside stoop at the entrance is placed just below the doorsill or it may be lowered by a standard riser height (7 1/2″), **Figure 13-44.**

Setting a doorsill so it will be level with the finished floor requires cutting away a section of the rough floor. Part of the top edge of the floor joist must also be cut away. This is done at the time the door frame is installed. The framing of the rough opening, however, comes earlier—before the door frame is delivered to the job. The carpenter must carefully check the working drawings. The size of the rough opening is usually included in the door and window schedule. The height of the opening is shown in detail sections. For standard construction, the rough opening can be calculated by adding about 2 1/2″ to the door width and height. When this information is not included in the working drawings, consult the manufacturer's literature. This contains not only rough opening requirements, but also detail drawings, **Figure 13-45.** The latter are especially important for front entrances that consist of more complex structures or have fixed sidelight window units

Setting the threshold and hanging the door are a part of the interior finishing operation. These operations are covered in Chapter 19, **Doors and Interior Trim.** After the installation is complete, lightly tack a piece of 1/4″ or 3/8″ plywood over the sill to protect it during further construction work. At this time, many builders prefer to hang a temporary combination door so that the interior of the structure can be secured, providing a place to store tools and materials.

Figure 13-44. This door is elevated a height equivalent to two standard risers above the walk. Two concrete steps are added. (Pease Industries, Inc.)

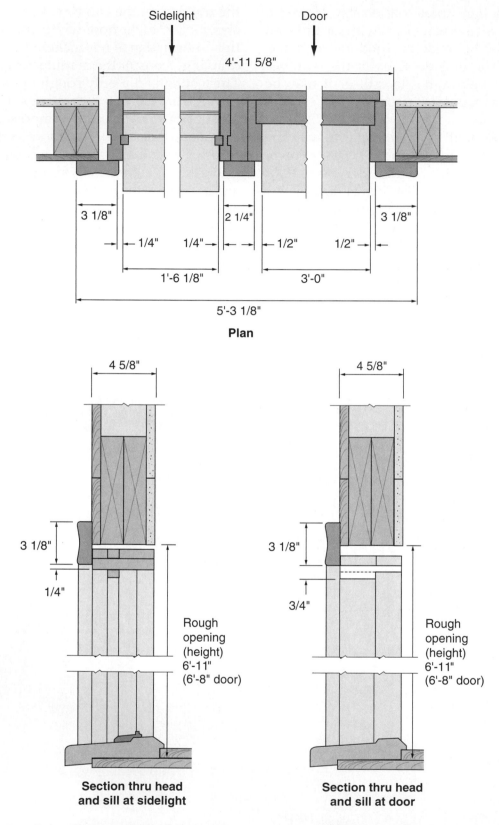

Figure 13-45. Detail drawings of a front entrance door frame with sidelight. The dimensions given by the manufacturer's literature serve as a guide when framing the rough opening. (C-E Morgan)

PROCEDURE

Installing a door frame

1. Check the size of the rough opening to make sure that proper clearances have been provided.
2. Cut out the sill area, if necessary, so the top of the sill will be the correct height above the rough floor.
3. If necessary, install flashing over the bottom of the opening.
4. Place the frame in the opening, center it horizontally, and secure it with a temporary brace.
5. Using blocking and wedges, level the sill and bring it to the correct height. Be sure the sill is well supported. For masonry walls and slab floors, the sill is usually placed on a bed of mortar.
6. With the sill level, drive a nail through the casing into the wall frame at the bottom of each side.
7. Insert blocking or wedges between the studs and at top of the jambs. Adjust until the frame is plumb. Use a level and straightedge as shown in **Figure 13-46.**
8. Place additional wedges between the jambs and rough opening in the approximate location of the lock strike plate and hinges. Adjust the wedges until the side jambs are well supported and straight. Then, secure the wedges by driving a nail through the jamb and wedge into the stud.
9. Nail the exterior casing in place with casing nails spaced 16″ on center. Follow the same precautions given for window frame installation.

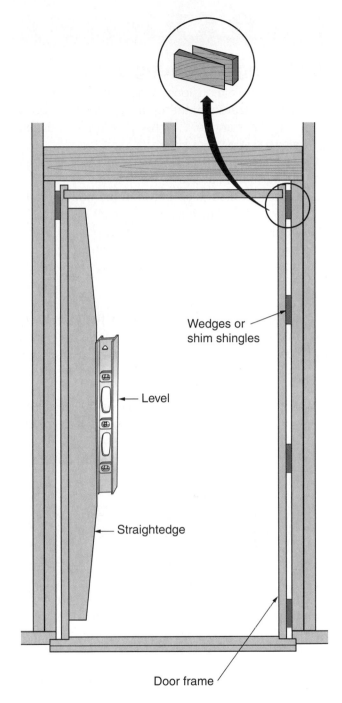

Wedges or shim shingles

Level

Straightedge

Door frame

Figure 13-46. Use a long level or a straightedge and shorter level to plumb doorjambs.

PROCEDURE

Installing a prehung entry door

1. Check the dimensions of the rough opening. For proper clearance, the opening should be about 3/4″ wider and 1/2″ higher than the unit dimensions to allow for shims.
2. Check that the opening is plumb in both directions, square, and level.
3. Check for clearance between the bottom of the door and the finish floor.
4. Apply a double bead of caulking compound to the subfloor to seal under the door sill.
5. Lay the door unit flat and apply caulk to the underside of the sill.
6. Set the unit in place.
7. Check the hinge side jamb in both directions for plumb. If there are thin spacer shims located between the frame and door, do not remove them until the frame is firmly attached to the rough opening.
8. Slip shims into the space between the hinge side jamb and the rough opening and tack to prevent movement.
9. Insert shims on the hinge side jamb behind the hinges between the side jambs and rough opening. Install them at the top, bottom, and midpoint.
10. Place shims behind the hinges to maintain the spacing when the longer hinge screws are installed that secure the hinges through the jamb and into the rough opening. Manufacturers usually recommend that at least two of the screws in the top hinge be replaced with 2 1/4″ screws.
11. Shim the lock-side jamb top and bottom, being sure to maintain spacing between the jamb and door. Tack it temporarily and check for plumb and square.
12. Drive 16d finish nails through the jambs and shims, into the structural frame members.
13. If the unit has an adjustable threshold, adjust it for a smooth contact with the bottom edge of the door.

Working Knowledge

After a prehung door unit is installed, some builders remove the door from its hinges and carefully store it to avoid damage during continuing construction. However, for security reasons, it may be necessary to install a lock and leave the door in place.

13.20 Sliding Glass Doors

To accommodate outdoor living, patio doors are often included in residential designs. The French or casement type door was once used for this purpose. However, the sliding glass door is now more common.

A sliding glass door rides on nylon or stainless steel rollers. This makes it easy to operate. When equipped with quality weatherstripping and insulating glass, heat loss and condensation are limited to satisfactory levels, even in cold climates. A sliding glass door is available as a factory assembled frame and door unit. Parts and installation details are similar to those for sliding windows.

Sliding glass door units contain at least one fixed and one sliding panel. Some may have three or four panels. The type of unit is commonly designated by the number and arrangement of the panels as viewed from the outside. See **Figure 13-47.** The letters L or R indicate which way the operating panel opens. Some manufacturers use the letter X to indicate operating panels and O for fixed panels. Sliding glass door units are glazed with tempered (safety) glass.

Manufacturers of sliding glass doors provide detailed instructions for installing their particular product. Study this material carefully before proceeding with the work. Construction details vary from one manufacturer to another, as do unit sizes and rough opening requirements. See **Figure 13-48.** If detailed drawings

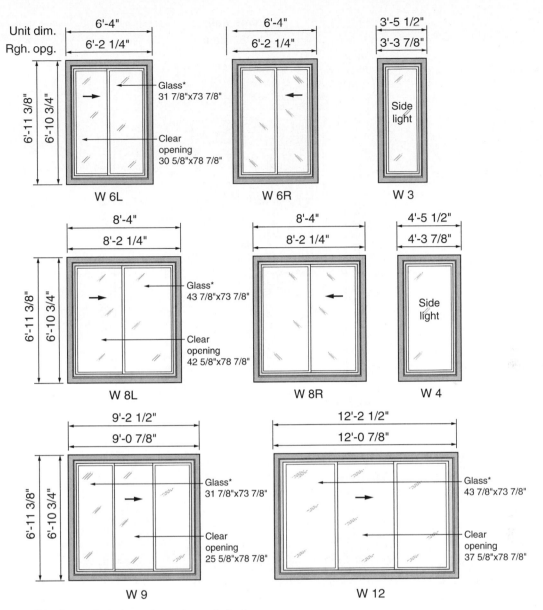

*Unobstructed glass sizes shown in inches.

Numbering system
6L and 8L left-hand operating panel
6R and 8R right-hand operating panel
9 and 12 center panel operates to right only.
    (All handing as viewed from outside)
Door operation may be specified either left- or
right-hand as viewed from outside.

Figure 13-47. Sliding glass door units come in standard sizes and shapes. (Andersen Corp.)

or rough opening sizes are not included in the architectural plans, obtain the information from the manufacturer's literature.

The installation of sliding door frames is similar to the procedure described for regular outside doors. Before setting the frame in place, lay a bead of sealing compound across the opening to ensure a weathertight joint. If the heavy, glass doors are to properly slide, the sill must be level and straight.

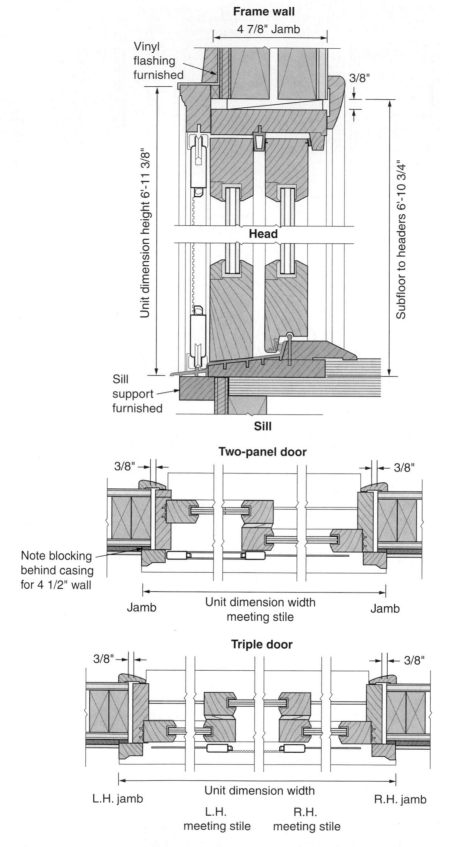

Frame wall

4 7/8" Jamb

Vinyl flashing furnished

3/8"

Unit dimension height 6'-11 3/8"

Subfloor to headers 6'-10 3/4"

Head

Sill support furnished

Sill

Two-panel door

3/8"

3/8"

Note blocking behind casing for 4 1/2" wall

Jamb

Unit dimension width meeting stile

Jamb

Triple door

3/8"

3/8"

L.H. jamb

Unit dimension width

R.H. jamb

L.H. meeting stile

R.H. meeting stile

Figure 13-48. Construction details like these differ from one manufacturer to the next. Be sure to study the manufacturer's information before installation.

Plumb side jambs and install wedges in the same way as you would install regular door frames. After careful checking, complete the installation of the frame by driving galvanized casing nails through the side and head casings into structural frame members.

At this point, many carpenters prefer to check the fit between the metal sill cover and doors and then, rather than installing them, carefully store these items. The opening is temporarily enclosed with plywood or polyethylene film attached to a frame. Then, during the finishing stages of construction, after inside and outside wall surfaces are completed, the sill cover, threshold, doors, and hardware are installed, **Figure 13-49.**

Working Knowledge

After installing large glass units, it is considered good practice to place a large X on the glass while the building remains under construction. Use masking tape or washable paint. This alerts workers so they will not walk into the glass or damage it with tools and materials.

13.21 Garage Doors

Basically, there are three types of garage doors—hinged or swinging, swing-up, and roll-up. The roll-up door has almost completely replaced the hinged and swing-up door types. See **Figure 13-50.** Wooden garage doors are constructed much like exterior entry doors.

A

B

Figure 13-50. A—Roll-up garage doors are hinged in sections. Note how the sections move upward and tilt to horizontal position. B—Rollers attached to each section run in a special steel track to support the door. (Steve Olewinski)

Figure 13-49. Sliding glass doors are installed during the finishing stages of construction. (Andersen Corp.)

Flush-type doors with foamed plastic cores are available for installations that require high levels of sound and thermal insulation. Garage doors are also manufactured from steel, aluminum, and fiberglass. Many designs are available to match contemporary or traditional architecture. Stock sizes are listed in **Figure 13-51.**

For use in areas subject to severe storms with high winds, hurricane-proof garage doors have been developed. These doors feature heavier construction than conventional doors and are equipped with steel bracing posts and locking hardware to withstand hurricane-force winds. See **Figure 13-52.**

13.21.1 Garage Door Frames

Frames for garage doors include side jambs and a head similar to exterior entry doors. No

rabbet is required. The frame is usually included in the millwork order along with windows and doors so all outside trim matches. The size of the frame opening is usually the same size as the door. However, the manufacturer's specifications and details should be checked before placing the order.

Figure 13-53 shows typical garage door jamb sections for both wood frame and masonry construction. Note the thickness (2″ nominal) of the heavy inside frame to which the track and hardware will be mounted. The width of this member should be at least 4″ with no projecting bolt or lag screw heads.

The rough opening width for frame construction is normally about 3″ greater than the door size. The height of the rough opening should be the door height plus about 1 1/2″, as measured from the finished floor.

Give careful consideration to the inside height. A minimum clearance between the top of the door and the ceiling must be provided for hardware, counterbalancing mechanisms, and the door itself when open. On some special, low-headroom designs, this distance may be as little as 6″. Always be sure to check the manufacturer's requirements for each door.

To replace an existing overhead garage door, take exact measurements of the opening before placing the order. Dimensions should include both width and height in feet and inches. Check the old frame for damage. Replace it if necessary.

Width	Height
8'-0"	7'-0" or 6'-6"
9'-0"	7'-0" or 6'-6"
10'-0"	7'-0" or 6'-6"
16'-0"	7'-0" or 6'-6"
18'-0"	7'-0" or 6'-6"

Figure 13-51. Stock garage door sizes.

A   B

Figure 13-52. Hurricane-proof garage door. A—Exterior view. B—The inside view shows steel posts and heavy-duty locking hardware. (Stanley Door Systems)

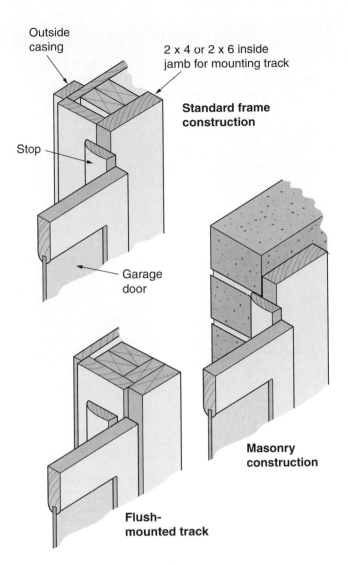

Figure 13-53. Typical jamb construction for garage door frames when a roll-up door is used.

Also, check that there is sufficient side room for attaching the vertical track. Normally, about 3 1/2″ is needed for standard extension or torsion bar springs. Clearance requirements should be checked with the supplier.

When installing garage door frames, follow the general procedure used for regular door frames. Make certain the jambs are plumb and the head is level. Before beginning the installation, set up several sawhorses to support the door sections as they are readied for installation. Manufacturers always supply detailed directions and procedures for the installation of their products. Carefully follow these printed materials. See **Figure 13-54.**

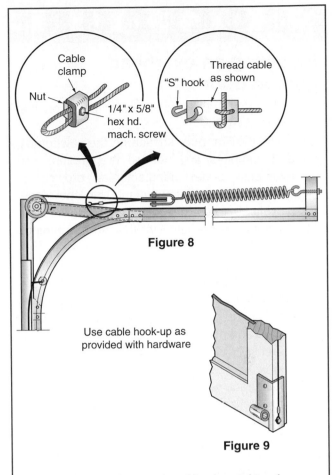

14. Install eye bolts at rear of horizontal tracks.
15. Assemble springs and sheave wheels. Hook springs to eye bolts as shown above.
16. Hook clip ends of cables to bottom bracket pins. See Figure 9.
17. Thread cables over sheaves. Put "S" hooks in about center of row of holes in top of horizontal channel. Pull cables thru "S" hooks until springs are stretched enough to hold door open. Install cable clamps as shown above.
18. Close door. Lengthen or shorten springs to correctly balance door by moving "S" hooks in channel holes or by moving cable clamps. This should always be done with door in open position.
19. Apply pull rope handles and lock assembly.
20. Final spring adjustment should be made after door has been glazed and finish painted.
21. Keep all hinges, rollers, and bearings well oiled.

Figure 13-54. Partial sample of a manufacturer's installation instructions. These are generally well illustrated and easy to understand.

PROCEDURE

Installing an overhead garage door

1. Temporarily tack door stops in place on the door jambs.
2. Place the first door section on the sawhorses and complete any needed preparation. For example, some manufacturers provide a channel for weatherstripping. This should be installed before proceeding.
3. Position the section in the door opening, centering it and leveling it using shims.
4. Temporarily tack the section to the stops to maintain the position.
5. Assemble the door sections, one on top of the other, in the opening. As you assemble, attach hinges and other hardware.
6. Attach rollers to door.
7. Slip vertical track sections onto the rollers and attach track sections to the jambs.
8. Mount horizontal track sections. This may involve nailing wood supports to bridge across joists.
9. Raise and prop door in open position.
10. Attach the counterbalancing mechanism.
11. Open and close door and make necessary adjustments to the track.
12. Adjust stops for a smooth, tight fit.

13.21.2 Hardware and Counterbalances

Garage door hardware must be well designed so the door will easily operate. Track, hinges, and bolts should be made of galvanized steel. The metal must be heavy enough to last the life of the door. To offset the weight of the door, various counterbalancing devices are used. Two of the most common types are the extension spring and the torsion spring. The torsion spring and its mechanism is more expensive, but provides a smoother and more consistent action. It is especially recommended for wide doors and those that are heavier in construction. See **Figure 13-55.** Most garage doors in new construction are equipped with an electric-powered door opener, **Figure 13-56.**

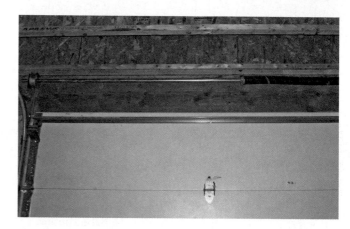

Figure 13-55. The torsion spring counterbalance is widely used for garage doors.

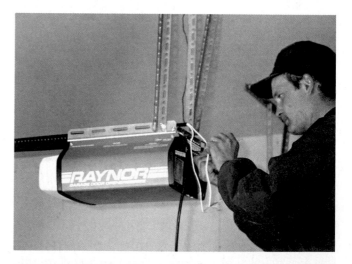

Figure 13-56. Electric-powered garage door openers are common. They allow remote-control opening and closing of the door without leaving the car. (Wayne-Dalton Corp.)

Summary

To make a house weathertight and minimize air infiltration, windows and exterior doors must be properly installed. Large millwork plants manufacture doors and windows in a wide variety of types and sizes. Doors are fabricated from wood, steel, or fiberglass. Windows are made in wood, aluminum, steel, or vinyl. There are three basic types of windows: sliding (double hung and horizontal), swinging (casement, hopper, or awning), and fixed (usually used in combination with swinging or sliding units). To control heat gain and loss, energy-efficient windows are double or triple glazed and many have a low-e coating on the glass to reflect infrared rays (heat waves). Building drawings usually include tables called window and door schedules that list all of the sizes and types of doors and windows used in the project. This information is important to the carpenter, since it shows the rough opening size needed for each door or window. In some structures, glass blocks are installed as a type of fixed window. They admit light but protect privacy. Skylights, bow windows, and bay windows have special installation requirements. Doors are made in flush and panel styles and are installed in much the same way as windows. Prehung doors simplify and speed up installation. Sliding glass doors leading to patios or decks are widely used. Sills must be level and straight for such doors to operate properly. Another specialized type of exterior door is the garage door. Most widely used today are sectional overhead doors that roll upward on a track to provide a clear opening.

Test Your Knowledge

Answer the following questions on a separate piece of paper. Do not write in this book.

1. Windows and doors are built in large _____ plants.
2. The wood used in the manufacture of windows is typically kiln dried to a moisture content of _____%– _____%.
3. The side section of a window frame that holds the sash is called the _____.
4. When there is little or no roof (cornice) overhang to protect the window, a(n) _____ should be installed above the head casing.
5. A type of window that is hinged on the side and swings outward is called a(n) _____ window.
6. The standard height from the finished floor to the bottom of the window head is _____, which is the standard height of a door.
7. A single pane of glass in a window has an R-value of about _____. Adding another pane of glass with 1/2″ of air space in-between increases the R-value to about _____.
8. If the rough opening for a window is listed as 3′-6″ × 3′-5″, the height of the rough opening is _____.
9. To adjust for various wall thicknesses, a(n) _____ is applied to standard window frames.
10. After a window unit has been temporarily set in the rough opening, the next step is to:
    A. Fasten it securely in place.
    B. Align the top to the proper height.
    C. Level the sill.
    D. Plumb the side jamb.
11. Large, insulating, fixed window units are made from glass that is _____ thick.
12. The thickness of standard glass block is _____ inches.
13. When figuring the size of the opening for glass block panels, multiply the number of units by the nominal block size and add _____.
14. List one method of supporting bow and bay windows.
15. The top of the doorsill is always level with the _____.
16. The outside wood casing for windows and doors is attached with nails spaced _____ inches on center.
17. A sliding glass door unit contains at least one _____ panel and one sliding panel.
18. The most popular type of garage doors is the _____ type.
19. *True or False?* Hardware should be attached to overhead garage door sections before the section is set in the opening.
20. The two common types of spring counterbalances for garage doors are extension and _____ types.

Curricular Connections

Language Arts. Prepare a written report on the manufacture of glass. Include such headings as historical development, early production methods, modern processes, float method, drawing method, and grinding and polishing plate glass. Study encyclopedias, reference books, Internet sources, and booklets from glass manufacturers.

Science. Some substances conduct heat more readily than others. Research the three common materials used for exterior residential doors—wood, fiberglass, and steel—to determine which would be best at preventing heat loss from the house during cold months. Rank them from most efficient (least heat loss) to least efficient (greatest heat loss). What criteria did you use to make this determination?

Outside Assignments

1. The rising cost of energy has resulted in special emphasis being placed on window design and construction. Some manufacturers produce triple-glazed window units for homes. Gather information about these units, including R-values, prices, special installation directions, and predicted fuel savings.

2. Collect information from lumberyards or home improvement centers on window and door units. Compare costs and R-values in a written or oral report.

Exterior Wall Finish

14

Learning Objectives

After studying this chapter, you will be able to:
- Identify the parts of a cornice and rake.
- Describe cornice and rake construction.
- Illustrate approved methods of flashing installation.
- Describe how wood siding and shingles are applied.
- Estimate the amount of siding or shingles required for a specific structure.
- Discuss the proper application of bevel siding.
- List the most common siding choices and their characteristics.
- Discuss exterior insulation and finish systems (EIFS) and their application.
- Demonstrate installation techniques for various siding materials.

Technical Vocabulary

Aluminum siding
Bevel siding
Board-and-batten siding
Board-on-board siding
Box cornice
Channel rustic siding
Cornice
Double coursing
Drop siding
Exterior insulation and finish systems (EIFS)
Exterior finish
Exterior trim
Fascia
Fascia backer
Fiber-cement siding
Frieze
Hardboard siding
Hard-coat systems
Heel
Ledger strip
Lookouts
Mortar
Open cornice
Preacher
Rake
Single coursing
Snub cornice
Soft-coat systems
Soffit
Starter strip
Starter track
Toe
Veneer wall
Vinyl siding
Wall hanger strips
Wood shingles

The term *exterior finish* includes all exterior materials of a structure. It generally refers to the roofing materials, cornice trim boards, wall coverings (such as siding), and trim members around doors and windows. The installation of special architectural woodwork at entrances or the application of a ceiling to a porch or breezeway area is also included under this broad heading.

Previous chapters describe the application of the finished roof and the installation of the trim around windows and outside doors. This chapter

---

Exterior finish: All exterior materials of a structure. Generally refers to siding, roofing, cornice, and trim members.

covers the construction and finish of cornice work and the materials and methods used to provide a suitable outside wall covering.

14.1 Cornice Designs and Terms

Cornice is an inclusive term for all of the parts that enclose or finish off the overhang of a roof at the eaves. It usually includes the fascia board, lookouts and a soffit for a closed cornice, and any mouldings used to conceal joints. Cornices provide a finished connection between the wall and roof. The style of the house determines the cornice design. Basically, there are three different cornice styles: box, open, and snub. A fourth, which might be considered a variation of the box style, is used with truss rafters.

The *box cornice* is the most common. It completely encloses the rafter and other parts of the overhang. The overhang protects sidewalls from the elements while providing shade for windows.

An *open cornice* may have no enclosing parts. There is no soffit covering the underside and often no fascia board concealing the rafter tails. The underside of the roof sheathing, since it is exposed at this point, must be attractive. Thus, tongue-and-groove, beaded, or V-grooved lumber is used. Open cornices are used when the overhang is very large or when rafters are large, laminated or solid beams.

A *snub cornice* has no overhang. The rafters end at the sidewall. This style does not provide walls or windows with any protection from the weather. It is sometimes used to save material and labor. However, most houses have overhangs. These provide shade for large window areas, protect the walls from the weather, and add to attractiveness of the structure.

Roof trusses are more common than site-built rafters. They accommodate a closed cornice very well. The bottom chord of the truss extends to the tail of the rafter. This extension supports the soffit and takes the place of the lookouts used on stick-built rafters. Some trusses, however, are designed for use with a snub cornice.

Figure 14-1 shows details of different cornice constructions. Several closed or box cornice designs are illustrated in **Figure 14-2.**

14.1.1 Parts of Cornice and Rake Section

Figure 14-3 shows structural and trim parts of a box cornice. The *fascia* is the main trim member. It is horizontally attached along the ends of the rafters. A *ledger strip* is a 1" or 2" ribbon horizontally nailed along the wall to support the lookouts. *Lookouts* are usually 2 × 4 pieces attached between the rafters and the wall. Their purpose is to provide support for soffit materials. With light soffit material, lookouts are not needed for support. A metal channel at the wall provides enough support.

The *soffit* is usually sheet material applied between the wall and the fascia board. Soffits can be plywood, hardboard, solid lumber, vinyl, or metal. Vents are provided along its width or at intervals to ventilate the attic space. Sometimes, a 1" or 2" wood nailing strip is attached to the rafter ends in a box cornice to support the fascia. This strip is called a *fascia backer* or

---

Cornice: Exterior trim of a structure at the meeting of the roof and wall; usually consists of panels, boards, and mouldings. Also, an inclusive term for all of the parts that enclose or finish off the overhang of a roof at the eaves.

Box cornice: Cornice that encloses the rafters at the eaves.

Open cornice: Cornice with no enclosing parts.

Snub cornice: Cornice with no overhang. Rafters end flush with the sidewalls.

Fascia: A wood member nailed to the ends of the rafters and lookouts and used for the outer face of a box cornice.

Ledger strip: A 1" or 2" ribbon horizontally nailed along the wall to support the lookouts.

Lookout: Structural member running between the lower end of a rafter and the outside wall to support the soffit.

Soffit: The underside of the members of a building, such as staircases, cornices, beams, and arches.

Fascia backer: A 1" or 2" wood nailing strip attached to the rafter ends in a box cornice to support the fascia.

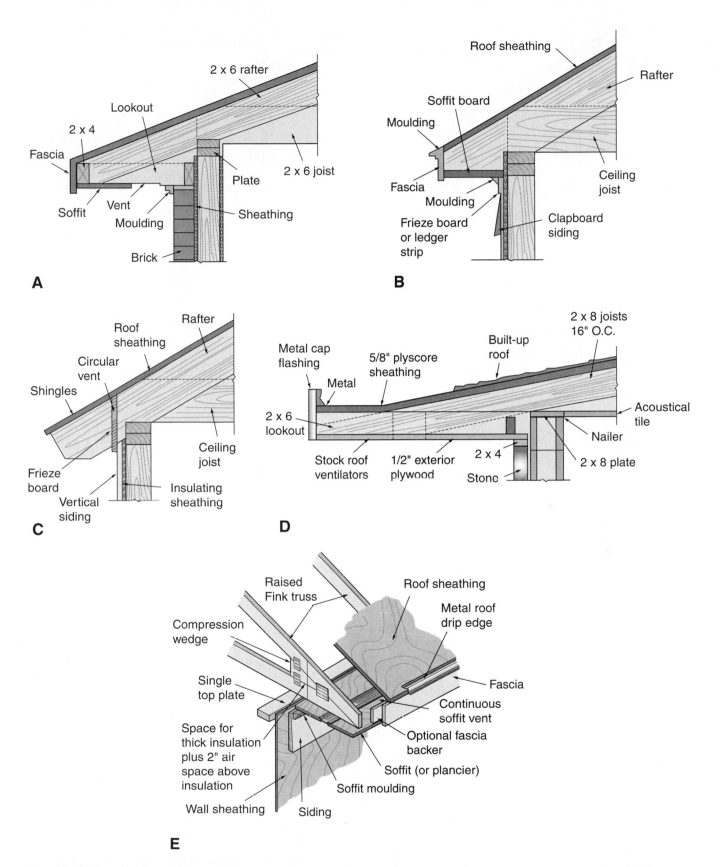

Figure 14-1. Typical cornice details for different architectural styles. A—Normal box cornice. B—Narrow box cornice. C—Open cornice. D—Wide box cornice. E—Box cornice applied to truss rafters. (Wood Frame House Construction, U.S. Dept. of Agriculture)

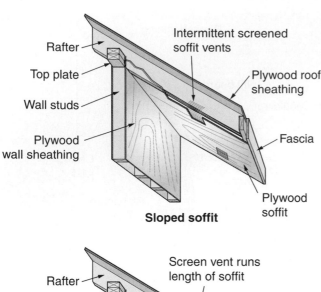

Sloped soffit

Rafter
Top plate
Wall studs
Plywood wall sheathing
Intermittent screened soffit vents
Plywood roof sheathing
Fascia
Plywood soffit

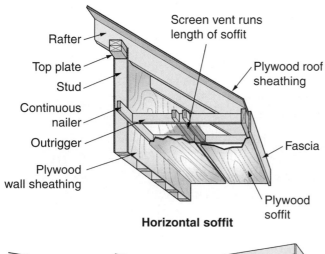

Horizontal soffit

Rafter
Top plate
Stud
Continuous nailer
Outrigger
Plywood wall sheathing
Screen vent runs length of soffit
Plywood roof sheathing
Fascia
Plywood soffit

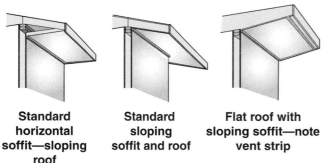

| **Standard horizontal soffit—sloping roof** | **Standard sloping soffit and roof** | **Flat roof with sloping soffit—note vent strip** |

Figure 14-2. Three different box cornice designs. (Council of Forest Industries of British Columbia)

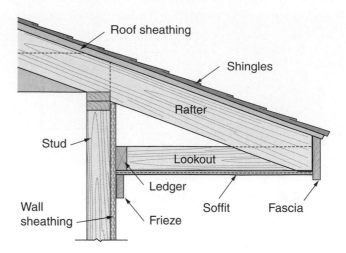

Roof sheathing
Shingles
Rafter
Stud
Lookout
Ledger
Wall sheathing
Frieze
Soffit
Fascia

Figure 14-3. Parts of a typical box cornice. Soffit materials are often prefabricated, usually from hardboard, plywood, or metal.

rake section is supported by the projecting roof boards, **Figure 14-4**. In addition, lookouts or nailers are fastened to the end rafter and the roof sheathing. As in cornice construction, these serve as a nailing base for the rake soffit and fascia.

When the rake projects a considerable distance, the sheathing does not provide adequate support. In such cases, the roof framing should be extended.

The parts of the cornice and rake structure that are exposed to view are generally called *exterior trim.* Typically, these parts are cut on the job. The properties needed in material used for exterior trim include good painting and weathering characteristics, easy working qualities, and maximum freedom from warping.

Where materials might absorb moisture, decay resistance is also desirable. Cedar, cypress, and redwood have high decay resistance. Less-durable species may be treated to make them decay-resistant. End joints or miters of members

false fascia. Along with the lookouts and ledger, it provides a frame to which the soffit material can be applied.

The *frieze* is the horizontal member that is placed flat against the wall below the soffit. Its lower edge is often rabbeted or furred to receive the siding material. The frieze is often decorative.

The *rake* is the part of a roof that overhangs a gable end. It is usually enclosed with carefully fitted trim members. The trim used for a boxed

Frieze: A boxed cornice wood trim member attached to the structure. It is used where the soffit and wall meet.

Rake: The trim members that run parallel to the roof slope and form the finish between the roof and wall at a gable end.

Exterior trim: Parts of the cornice and rake structure that are exposed to view.

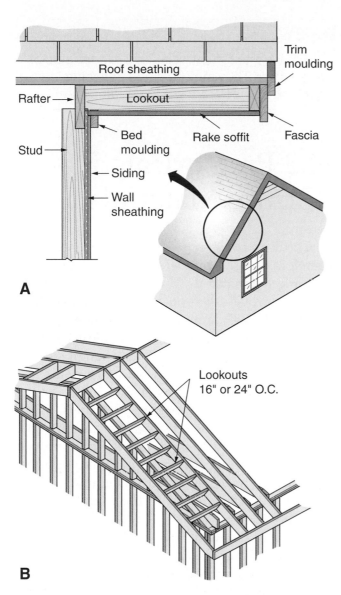

A

B

Figure 14-4. Lookouts. A—Typical boxed rake section. Lookouts provide a nailing surface and support for the soffit. B—Sometimes, the lookouts are cantilevered to support a wide overhang on the gable.

exposed to heavy moisture must be coated. Special caulking compounds are commonly used for this purpose.

14.1.2 Cornice and Rake Construction

In most construction, the fascia boards are installed on the rafter ends at the time the roof is sheathed. It is important that they be straight, true, and level with well-fitted joints.

Before attaching the fascia, some builders prefer to nail on a 2 × 4 ribbon to align the tails of the rafters. The fascia may be attached to this ribbon. Corners of fascia boards should be mitered. End joints should meet at a 45° angle, as illustrated in **Figure 14-5.** A rough fascia is often nailed to the rafters and then covered with a prefinished, maintenance-free vinyl or metal strip material. See **Figure 14-6.**

The rake should be constructed to match the cornice. After the main trim members are installed, mouldings can be set in corners to cover irregularities. However, the use of moulding is minimal to maintain a smooth trim appearance.

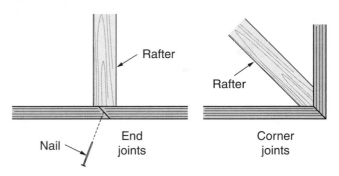

Figure 14-5. Plan view of joints for fascia boards. Miter corners and make matching angled cuts on ends. Note how end joints are nailed.

Figure 14-6. Installing a prefinished metal fascia cover. Follow the manufacturer's recommendations.

Most construction makes use of truss rafters. The raised Fink truss is designed to allow thick ceiling insulation over the exterior wall without blocking the airway between the soffit and roof venting. Refer to **Figure 14-1E**. It also allows steeply sloped roofs with wide overhangs that do not interfere with windows and doors. Soffits remain at the same height as interior ceilings regardless of roof slope or projection of the cornice. Soffit material is attached to the bottom chord of the truss, which extends to the very end of the rafter. A compression wedge carries the weight of the roof where the top and bottom chords extend over the outside wall.

PROCEDURE

Framing a box cornice

1. If lookouts are to be used, install a ledger strip along the wall.
2. With a level, locate and mark points on the wall that are level with the bottom edges of the rafter tails. Snap a chalk line through these points.
3. Nail on the ledger strip (or a metal channel, depending on soffit material to be used).
4. Cut the lookouts. Lookouts are usually made from 2 × 4 stock. Locate them at each rafter or every other rafter, depending on the kind of soffit material to be used.
5. Toenail one end of the lookout to the ledger. Check that other end is flush with the bottom of the rafter. Then, nail the other end to the rafter.
6. Rip cut the top edge of the fascia to match the slope of the roof.
7. Install the fascia to the rafter ends.
8. If using thin material for the soffit, attach a nailing strip along the inside of the fascia to provide a nailing surface. Sometimes the back of the fascia is grooved to receive soffit material. If the fascia is to be clad with vinyl or aluminum, an F channel is installed that covers the fascia and provides a slot for supporting the soffit.
9. Cut the soffit material to size.
10. Secure the soffit with rust-resistant nails or screws. When regular casing or finish nails are used, they must be countersunk and the holes filled with putty. Do this after the prime coat of paint. Installation of metal soffits is explained later.

Working Knowledge

Always use rust-resistant nails for outside finish work. They may be made from aluminum, galvanized steel, or cadmium-plated steel.

14.2 Prefabricated Cornice Materials

Cornice construction is time-consuming. Therefore, many builders prefer to purchase prefabricated materials. Various systems are available that provide a neat, trim appearance. One consists of 3/8″ laminated, wood-fiber panels. These are factory primed and available in a variety of standard widths (12″–48″) and lengths up to 12′. Panels can be equipped with factory-applied screened vents. See **Figures 14-7** and **14-8.**

When installing large wood panels, allow some clearance at the edges for expansion of the wood. Fasten with 4d rust-resistant nails spaced about 6″ along edges and intermediate supports. Start nailing at the edge butted against a previously placed panel. Nail to the main supports first and then along the edges, **Figure 14-9.** Drive nails flush with the panel surface.

Lookouts may be left out of a soffit system where special supports are attached to the upper surface of the panels. In one system, these supports are made of 20 gage steel channels with prongs that make it easy to attach them to the back of the panels. The supports provide rigidity, so the panels only need to be attached at the front and back edges.

Porch and carport ceilings can be covered with factory-primed panels similar to those used for cornice soffits. Standard 4′ × 8′ and larger units are designed for either 16″ or 24″ O.C. framing. Always leave a 1/8″ space along all edges for expansion.

Start nailing in the center of the panel and move toward the outside. Space nails about 6″ apart. Edges of panels can be secured and joints covered with special H-strips.

Metal soffit material is provided in rolls and sheets in several widths. Aluminum systems require little maintenance, incorporate venting,

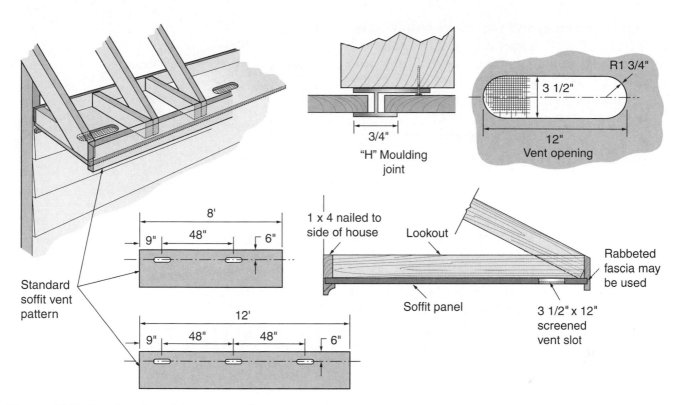

Figure 14-7. Details of prefabricated soffit system. Soffits are a manufactured material, usually hardboard or metal.

Figure 14-8. This soffit arrangement has a line of narrow ventilation holes near the outer edge.

Figure 14-9. Soffit can be more efficiently installed with the use of power tools.

are self-supporting, and will never rust. Widths vary from 12″–48″. Edges are held in U-shaped channels called *wall hanger strips.* These are attached to fascia and walls. See **Figure 14-10.** Panels may be vented or unvented. Metal soffit systems consist of three basic units:

- Wall hanger strips (frieze strips or runner guides).
- Soffit panels.
- Fascia covers.

A continuous soffit ventilation system is shown in **Figure 14-11.** Designed to be used along with a continuous ridge vent, it comes in 8′ lengths that are attached at the junction of the wall and the soffit. It can be used with either aluminum or vinyl soffit cover material. A channel in the upper edge receives the soffit material. It will accommodate soffit thicknesses of 1/4″ to 1/2″.

Some aluminum soffit coverings are sold in 50′ coils with widths from 12″–4′. To install, first attach runners at the wall and fascia board. Feed the coil into the runners. On a hip roof with soffits on all sides, leave one end open so that the last section can be fed into the runners. **Figure 14-12** shows material and accessories used for soffits.

PROCEDURE

Hanging metal soffit

1. Snap a chalk line along the wall that is level with the bottom edge of the fascia board.
2. Either attach a hanger strip flush with the bottom of the fascia or plan to nail the metal soffit panels to the bottom of the fascia board.
3. Attach a metal U-shaped wall hanger strip above or below the chalk line established in Step 1.
4. Insert the panels, one at a time into the strip or strips.
5. After all soffit panels are installed, cut the metal or vinyl fascia cover to fit.
6. Hook the bottom edge of the fascia cover over the end of the soffit panels.
7. Nail the fascia cover into place through prepunched slots located along the top edge. Always study and follow the manufacturer's directions when making an installation of this type.

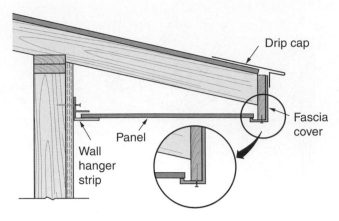

Figure 14-10. Section view of a prefinished metal soffit system. Fascia cover must be slid under drip cap to prevent moisture from damaging the fascia.

14.3 Wall Finish

After covering the cornice and rake sections of the roof, you are ready to apply siding to the walls. When the structure has a gable roof, the wall surface material is usually applied to the gable end before the lower section is covered. This permits scaffolding to be directly attached to the wall while siding the gable end. All exterior trim members should be given a primer coat of paint as soon as possible after installation, if not pre-primed.

14.4 Wall Sheathing and Flashing

Siding can be applied over various sheathing materials. When the sheathing is solid wood, plywood, oriented strand board, or nail-base fiberboard, the siding is directly nailed to the material at about 24″ intervals. End joints in the siding may occur between framing members. Gypsum board and regular fiberboard sheathing cannot be used as a nailing base. In these cases, the siding should be attached by nailing through

Wall hanger strips: U-shaped channels that hold the edges of metal soffit material; attached to fascia and walls.

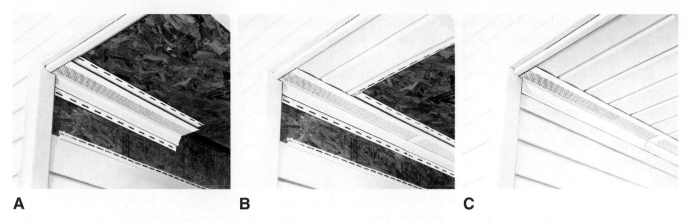

A   **B**   **C**

Figure 14-11. This patented, cove-like moulding provides ventilation and has a built-in channel to receive various types of soffit covering. A—When applying over a covering, cut a 2″ wide slot close to the wall and nail on the continuous vent. B—Install the soffit and siding. C—The finished installation. (Tapco)

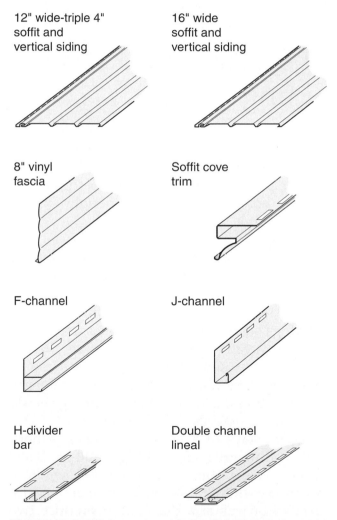

12" wide-triple 4" soffit and vertical siding

16" wide soffit and vertical siding

8" vinyl fascia

Soffit cove trim

F-channel

J-channel

H-divider bar

Double channel lineal

Figure 14-12. Various types of panels and trim pieces are available for cladding fascia and soffits. Panels usually have 10″ or 12″ exposure and are available in lengths of 12′ or 12′-6″ packaged 12, 16, or 20 to a bundle. Trim pieces are generally 12′-6″ long and packaged in bundles of 10, 20, 24, 36, or 40. (CertainTeed Corp.)

the sheathing and into the frame. Additional information about wall sheathing can be found in Chapter 9.

An insulation board of rigid polystyrene may be used in place of sheathing. It adds substantial R-value to walls. The foamed sheets have shiplap edges for tight fit. They are usually available in 1/2″, 3/4″, and 1″ thicknesses. **Figure 14-13** shows

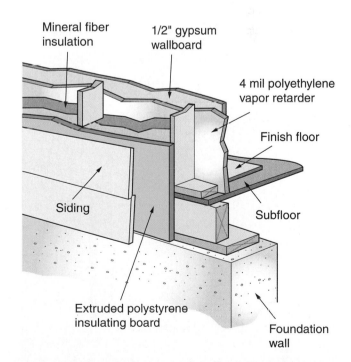

Mineral fiber insulation

1/2" gypsum wallboard

4 mil polyethylene vapor retarder

Finish floor

Siding

Subfloor

Extruded polystyrene insulating board

Foundation wall

Figure 14-13. Using rigid polystyrene insulation board on an outside wall. Fasten with galvanized roofing nails (3/8″ head) or 16 gage staples (3/4″ crown). (Amoco Foam Products Co.)

a typical application. Housewrap is typically used instead of traditional sheathing paper. It seals the walls against air infiltration.

Certain special application methods may require wood furring strips to form a nailing base. It is considered good building practice to use furring to create a vented rain screen between building paper or housewrap and wood siding. It prevents contact between the siding and building paper or housewrap. Although results of current research are not conclusive, there is some evidence that this contact could affect the ability of the paper or housewrap to shed moisture.

Before the application of siding, install flashing where it is required around openings. Metal flashing is usually installed over the drip caps of doors and windows, **Figure 14-14.** In areas not subjected to wind-driven rain, the head flashing may be omitted when the vertical distance between soffit and the top of the finished trim of the opening is equal to or less than 1/4 of the overhang width. For structures with unsheathed walls, flash jambs of doors and windows with a 6″ wide strip consisting of either metal, 3 oz. copper-coated paper, or a 6 mil polyethylene film.

14.5 Horizontal Wood Siding

One of the most common materials used for the exterior finish of American homes is wood siding. The shadow cast on the wall by the butt edge emphasizes the horizontal lines preferred by many homeowners. This is especially evident in bevel siding.

Siding is usually applied over a base consisting of sheathing and building paper or housewrap. The sheathing may be oriented strand board, plywood, boards, or rigid insulation board. However, in mild climates or on buildings such as summer cottages, the siding may be directly applied to the studs. Where sheathing is omitted or where the type of sheathing does not provide sufficient strength to resist a racking load, the wall framing should be braced as described in Chapter 9.

Edge views of a number of types of horizontal siding are shown in **Figure 14-15.** *Bevel siding* is most commonly used and available in various widths. It is made by sawing plain-surfaced boards at a diagonal to produce two

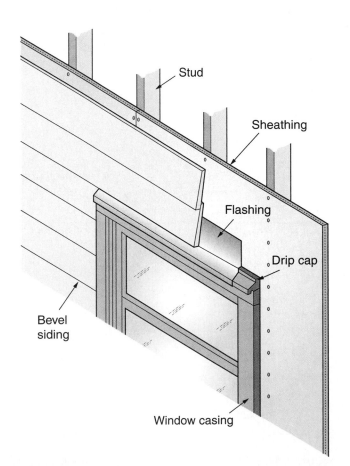

Figure 14-14. Apply metal flashing over the drip caps above windows and doors that are not protected by the roof overhang.

Bevel siding: Wedge-shaped siding used as finish covering on the exterior of a structure.

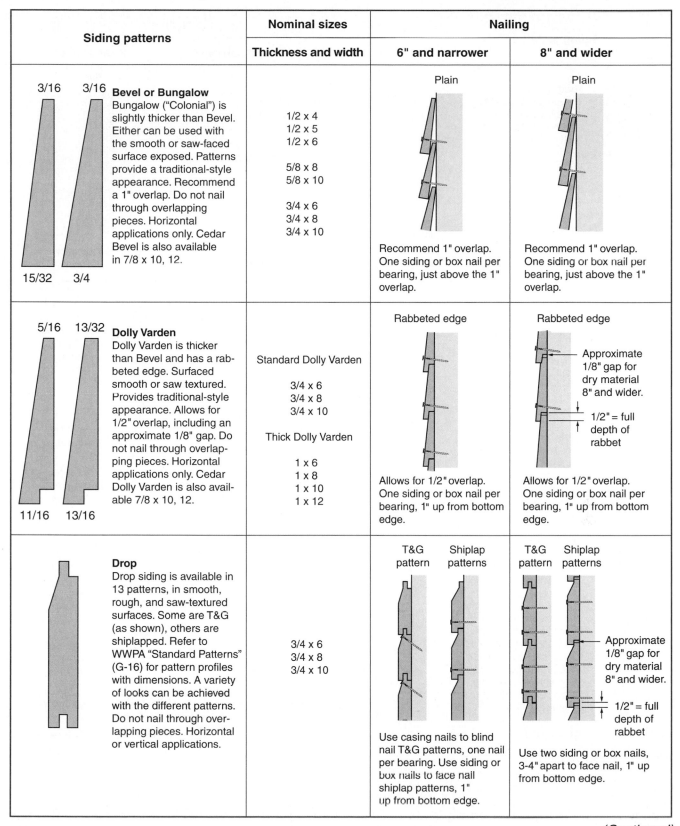

Siding patterns	Nominal sizes	Nailing	
	Thickness and width	6" and narrower	8" and wider
Bevel or Bungalow 3/16 3/16 15/32 3/4 Bungalow ("Colonial") is slightly thicker than Bevel. Either can be used with the smooth or saw-faced surface exposed. Patterns provide a traditional-style appearance. Recommend a 1" overlap. Do not nail through overlapping pieces. Horizontal applications only. Cedar Bevel is also available in 7/8 x 10, 12.	1/2 x 4 1/2 x 5 1/2 x 6 5/8 x 8 5/8 x 10 3/4 x 6 3/4 x 8 3/4 x 10	Plain Recommend 1" overlap. One siding or box nail per bearing, just above the 1" overlap.	Plain Recommend 1" overlap. One siding or box nail per bearing, just above the 1" overlap.
Dolly Varden 5/16 13/32 11/16 13/16 Dolly Varden is thicker than Bevel and has a rabbeted edge. Surfaced smooth or saw textured. Provides traditional-style appearance. Allows for 1/2" overlap, including an approximate 1/8" gap. Do not nail through overlapping pieces. Horizontal applications only. Cedar Dolly Varden is also available 7/8 x 10, 12.	Standard Dolly Varden 3/4 x 6 3/4 x 8 3/4 x 10 Thick Dolly Varden 1 x 6 1 x 8 1 x 10 1 x 12	Rabbeted edge Allows for 1/2" overlap. One siding or box nail per bearing, 1" up from bottom edge.	Rabbeted edge Approximate 1/8" gap for dry material 8" and wider. 1/2" = full depth of rabbet Allows for 1/2" overlap. One siding or box nail per bearing, 1" up from bottom edge.
Drop Drop siding is available in 13 patterns, in smooth, rough, and saw-textured surfaces. Some are T&G (as shown), others are shiplapped. Refer to WWPA "Standard Patterns" (G-16) for pattern profiles with dimensions. A variety of looks can be achieved with the different patterns. Do not nail through overlapping pieces. Horizontal or vertical applications.	3/4 x 6 3/4 x 8 3/4 x 10	T&G pattern Shiplap patterns Use casing nails to blind nail T&G patterns, one nail per bearing. Use siding or box nails to face nail shiplap patterns, 1" up from bottom edge.	T&G pattern Shiplap patterns Approximate 1/8" gap for dry material 8" and wider. 1/2" = full depth of rabbet Use two siding or box nails, 3-4" apart to face nail, 1" up from bottom edge.

(Continued)

Figure 14-15. Edge views of six different types of horizontal siding. Nominal sizes are used in figuring footage of siding. Note nailing suggestions. (Western Wood Products Assn.)

Siding patterns	Nominal sizes	Nailing	
	Thickness and width	6" and narrower	8" and wider
Tongue and Groove Tongue and Groove siding is available in a variety of patterns. T&G lends itself to different effects aesthetically. Refer to WWPA "Standard Patterns" (G-16) for pattern profiles. Sizes given here are for Plain Tongue and Groove. Do not nail through overlapping pieces. Vertical or horizontal applications.	1 x 4 1 x 6 1 x 8 1 x 10 Note: T&G patterns may be ordered with 1/4", 3/8", or 7/16" tongues. For wider widths, specify the longer tongue and pattern.	Plain Use one casing nail per bearing to blind nail.	Plain Use two siding or box nails 3-4" apart to face nail.
Channel Rustic Channel Rustic has 1/2" overlap (including an approximate 1/8" gap) and a 1" to 1-1/4" channel when installed. The profile allows for maximum dimensional change without adversely affecting appearance in climates of highly variable moisture levels between seasons. Available smooth, rough, or saw textured. Do not nail through overlapping pieces. Horizontal or vertical applications.	3/4 x 6 3/4 x 8 3/4 x 10	Use siding or box nail to face nail once per bearing, 1" up from bottom edge.	Approximate 1/8" gap for dry material 8" and wider. 1/2" = full depth of rabbet Use two siding or box nails 3-4" apart per bearing.
Log Cabin Log Cabin siding is 1-1/2" thick at the thickest point. Ideally suited to informal buildings in rustic settings. The pattern may be milled from appearance grades (Commons) or dimension grades (2x material). Allows for 1/2" overlap, including an approximately 1/8" gap. Do not nail through overlapping pieces. Horizontal or vertical.	1 1/2 x 6 1 1/2 x 8 1 1/2 x 10 1 1/2 x 12	Use one siding or box nail face nail once per bearing, 1 1/2" up from bottom edge.	Approximate 1/8" gap for dry material 8" and wider. 1/2" = full depth of rabbet Use two siding or box nails, 3-4" apart per bearing to face nail.

Figure 14-15. Continued.

wedge-shaped pieces. The siding is about 3/16″ thick at the thin edge and 1/2″–3/4″ thick on the other edge, depending on the width of the piece.

Wide bevel siding often has shiplapped or rabbeted joints. The siding lies flat against the studding instead of touching it only near the joints, as ordinary bevel siding does. This reduces the apparent thickness of the siding by 1/4″, but permits the use of extra nails in wide siding and reduces the chance of warping. It is also economical, since the rabbeted joint requires less lumber than the lap joint used with plain bevel siding. The rabbet, however, must be deep enough so that—when the siding is applied—the width of the boards can be adjusted upward or downward to meet windowsill, head casing, and eave lines.

Channel rustic siding and *drop siding* are usually 3/4″ thick and 6″, 8″, or 10″ wide. Channel rustic or rustic has shiplap-type joints. Drop siding usually has tongue-and-groove joints. Drop siding is heavier, has more structural strength, and makes tighter joints than bevel siding. Because of this, it is often used on garages and other buildings that are not sheathed.

Wood used for exterior siding should be a select grade, free of knots, pitch pockets, and other defects. The best grade has edge grain because it is less likely to warp than a flat grain. The moisture content at the time of application should be what it will reach in service. This is about 12%, except for the southwestern United States, where the moisture content should average about 9%.

Siding should be carefully handled when it is delivered to the building site. The wood from which it is made is usually quite soft. The surface can be easily damaged. Try to store siding inside the structure or keep it covered with a weatherproof material until it is applied.

14.5.1 Installation Procedures

Wood siding is precision-manufactured to standard sizes. It is easily cut and fitted. Plain beveled siding is lapped so it will shed water and provide a windproof and dustproof covering. A minimum lap of 1″ is used for 6″ widths, while 8″ and 10″ siding should lap 1 1/2″. With lap

siding, it is an advantage to be able to vary the exposure. This allows you to have a full course below and above windows and over the top of doors.

PROCEDURE

Preparing a story pole and the layout

1. Measure the distance from the soffit to about 1″ below the top of the foundation, **Figure 14-16.**
2. Divide this distance into spaces equal to the width of the siding minus the lap.
3. Adjust the lap allowance (maintain minimum requirements) so the spaces are equal.
4. When possible, adjust the spacing so courses of siding are continuous above and below windows or other wall openings without notching.
5. Mark the position of the top of each siding board on the story pole.
6. Hold the story pole against the wall at each inside and outside corner of the structure. Transfer the layout marks to the wall as in **Figure 14-17.** Also transfer the story pole layout marks along both sides of window and door casings.
7. Some carpenters prefer to tack nails at these layout points, since lines can be quickly attached to them and used to align the siding stock. Nails can also be useful for holding a chalk line if guidelines are laid out by snapping.
8. After making the layout, carefully check it.

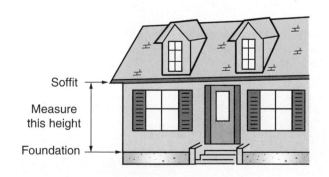

Figure 14-16. Begin layout of the story pole by taking the total measurement from beneath the soffit to about 1″ below the top of the foundation.

Channel rustic siding: Horizontal wood siding that is usually 3/4″ thick and 6″, 8″, or 10″ wide and has shiplap-type joints.

Drop siding: Siding usually 3/4″ thick and 6″ wide; machined into various patterns with tongue-and-groove or shiplap joints.

Figure 14-17. Transfer the story pole layout to all inside and outside corners, as well as to door and window casings.

Working Knowledge

When snapping a chalk line stretched over a long distance, it is a good idea to hold it against the surface at the midpoint, then snap it on each side. This results in a more accurate line.

14.5.2 Wood Corner Boards

A square piece of solid lumber can be used for inside corners, **Figure 14-18.** The thickness of the piece depends on the thickness of the siding. The siding should never project above the corner board.

Figure 14-18. Inside corners can be formed from solid lumber. When necessary, use two thicknesses to secure the required size.

Inside or outside corners can be formed with formed metal pieces, as shown in **Figure 14-19.** Although outside corners could be lapped or mitered, metal corners are almost universally used. They can be quickly installed and provide a neat, trim appearance.

Outside wood corners can be formed with two pieces of trim lumber. The thickness, again, depends on the siding. One of the two corner pieces should be narrower by the thickness of

A

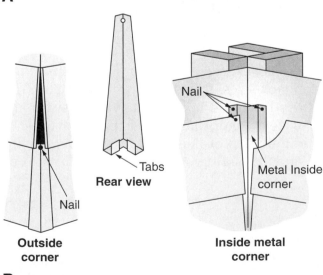

B

Figure 14-19. Metal corners are almost always used for horizontal siding. A—A finished installation. B—Attachment details. All nails are covered.

the other trim piece. When butted together, they will appear to have the same width. Attach these to the structure before installing the siding.

Corner boards may be plain or moulded, depending on the architectural treatment required. After the corner boards are in place, installation of siding can begin.

Some carpenters mark cuts at corner boards and window and door casings using a *preacher,* sometimes called a siding gauge. As shown in **Figure 14-20,** a preacher is a small block made from 3/8" or 1/2" hardwood. It is notched to fit over the siding. To use the preacher, make certain the siding piece is properly located where it is to be installed. Position the notched piece over the siding and firmly hold it against the edge of the corner board or trim. Mark the siding where it is to be cut off.

For quality work, the carpenter first makes the cuts with a fine-tooth saw and then smoothes the ends with a few strokes of a block plane. Square butt joints are used between adjacent pieces of siding. Stagger the joints as widely as possible from one course to the next.

Consider giving wood siding a coat of water-repellent preservative before it is installed.

However, you can also do this after the installation. Preservatives are sold by lumber dealers and paint stores. They contain waxes, resins, and oils that protect the wood from weather and rot. In addition to this treatment, joints in siding may be bedded in a special caulking compound to make them watertight.

PROCEDURE

Installing bevel siding

1. Nail a spacer strip along the foundation line. Its thickness should be the same as the thin edge of the siding, **Figure 14-21.** This will give the first course the proper tilt.
2. Apply the first piece of siding. Allow the butt edge to extend below the spacer strip to form a drip edge.
3. Cut and fit horizontal wood siding tightly against corner boards, window and door casings, and adjoining boards. Cut the siding long enough so it has a hairline overlap. If then bowed slightly, it can be snapped in for a tight fit.
4. Apply the remaining rows, staggering joints.

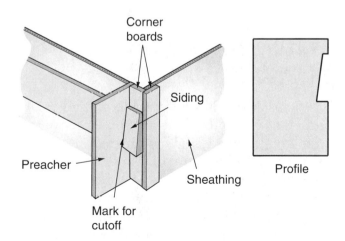

Figure 14-20. Some carpenters use a guide called a preacher to accurately mark siding cuts.

Preacher: A small block made from 3/8" or 1/2" hardwood and notched to fit over the siding. It is used to mark the siding where it is to be cut off. Also called a *siding gauge.*

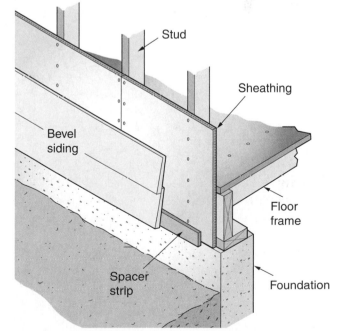

Figure 14-21. Install a spacer strip under the first course of bevel siding to give it the proper tilt to match succeeding courses.

14.5.3 Nailing

To fasten siding, stainless steel, high-tensile-strength aluminum, hot-dipped galvanized, or other noncorrosive nails are recommended. Avoid fasteners that might produce unsightly rust stains. Even small-headed plain steel nails, countersunk and puttied, are likely to rust. If there is a possibility the material will split when nailing end joints, drill holes for the nails. **Figure 14-22** show wood siding nailing patterns for different types of siding. Also refer to **Figure 14-15.**

Face nail horizontal siding to each stud. Nails should penetrate into the wood at least 1 1/2″. For 1/2″ siding over wood or plywood sheathing, use 6d nails. Over fiberboard or gypsum sheathing, use 8d nails. For 3/4″ siding over wood or plywood sheathing, use 7d nails and over fiberboard or gypsum sheathing, use 9d nails. For 1″ thick siding, it is best to use 8d or 10d nails. Avoid using nails that are longer than necessary. They may interfere with electrical wiring or plumbing in the wall.

For narrow siding, the nail is generally placed about 1/2″ above the butt edge. In this location, the fastener passes through the upper edge of the lower course.

When applying wide (4″–6″) bevel siding, drive the nail through the butt edge 1″ above the lap so that it misses the thin edge of the piece of siding underneath. This permits expansion and contraction of the siding boards with seasonal changes in moisture content. It eliminates the tendency of the siding to cup or split when both edges are nailed. Since the amount of swelling and shrinking is proportional to the width of the material, vary the nail distance proportionally.

Although it is usual to install horizontal siding from the bottom up, some carpenters prefer to start at the top and work down. This is useful for multistory or split level structures where scaffolds are attached to the wall or where freestanding scaffolds are already erected. Also, the siding is less likely to be damaged after it is applied. When following such procedure, set chalk lines at the butt edge of the siding instead of the top edge.

14.5.4 Painting and Maintenance

Wood siding is subject to decay and weathering. Neither will occur if simple precautions are taken. Decay is the disintegration of wood caused by the growth of fungi. These fungi grow in wood when the moisture content is too high. If the structure is built on a foundation that has been carried well above the ground and the construction is such that water runs off instead of into the walls, decay should not be a problem.

A wide range of finishing materials is on the market. Most fall into one of four general categories: clear water repellents, bleaching oils, stains, and paints. More will be said about each of these finishes in a later chapter. Certain factors need to be considered when selecting a finish. These may include: desired appearance,

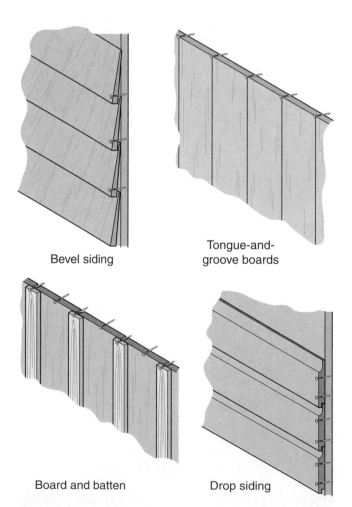

Bevel siding

Tongue-and-groove boards

Board and batten

Drop siding

Figure 14-22. Different types of wood siding require different nailing patterns.

preparation and maintenance requirements of the finish, location of the structure, climate, and current condition of the siding.

If the siding is to be painted, a priming coat should be put on as soon as possible. If an unexpected rain should wet unprimed wood siding, the first coat of paint should not be applied until the wood has dried.

To avoid future separation between primer and topcoat, apply the first topcoat of paint within two weeks of the primer. The second topcoat should be applied within two weeks of the first.

14.5.5 Estimating Siding

To determine how much siding to order, it is necessary to increase the footage to make up for the difference between nominal and finished sizes. More must also be added for waste resulting from the cutting of joints and the overlap in beveled siding. The table in **Figure 14-23** provides a factor. Multiply the net square footage of the wall surface to be covered by this factor. The following example shows the steps.

Type	Size (inches)	Lap (inches)	Multiply net wall surface by
Bevel siding	1 x 4	3/4	1.45
	*1 x 5	7/8	1.38
	1 x 6	1	1.33
	1 x 8	1 1/4	1.33
	1 x 10	1 1/2	1.29
	1 x 12	1 1/2	1.23
Rustic and drop siding (shiplapped)	1 x 4		1.28
	*1 x 5		1.21
	1 x 6		1.19
	1 x 8		1.16
Rustic and drop siding (dressed and matched)	1 x 4		1.23
	*1 x 5		1.18
	1 x 6		1.16
	1 x 8		1.14

*Unusual sizes.

Figure 14-23. When estimating horizontal wood siding, multiply the net wall surface to be covered by the factor in the right-hand column.

1 × 10 bevel siding with 1 1/2″ lap
Wall height = 8′
Wall perimeter = 160′
Door and window area = 240 sq. ft.
Total area to be covered = (8 × 160) − 240
= 1280 − 240
= 1040 sq. ft.
Siding needed = 1040 × 1.29
= 1342 sq. or bd. ft.

Calculate the area of gable ends by multiplying the height above the eaves by the width, then divide by two. Since considerable waste occurs in covering triangular areas, add at least 10% to this calculation. When the structure includes many corners due to projections and recesses in the wall line, add an additional .05 to the factors shown in **Figure 14-23.**

14.6 Vertical Siding

Vertical siding is commonly used to set off entrances or gable ends. It is also often used for the main wall areas, **Figure 14-24.** Vertical siding may be plain-surfaced, matched boards; pattern-matched boards; tongue-and-groove boards; or square-edge boards covered at the joint with a batten strip.

Figure 14-24. Vertical wood boards with battens make a durable and beautiful siding material. (APA-The Engineered Wood Association)

If the siding is directly applied to the building frame, backing blocks (called *bearings*) should be horizontally installed at 16″ to 24″ intervals between studs. These provide a good nailing surface. If structural panels, such as oriented strand board, are used as sheathing, the panels provide a satisfactory nailing surface.

Matched vertical siding made from solid lumber should be no more than 8″ wide. It should be installed with two 8d nails not more than 4′ apart. Backing blocks should be used between studs to provide a good nailing base. The bottoms of the boards are usually undercut to form a drip edge.

If the siding material is wide, such as 4′ wide panels, there is no need to install blocking between studs. Simply nail the sheets onto the studs.

Board-and-batten siding applications are designed around wide, square-edged boards spaced about 1/2″ apart. They are fastened at each bearing with one or two 8d nails. Use 10d nails to attach the battens, **Figure 14-25.** Locate nails in the center of the batten so the shanks will pass between the boards and into the bearing.

Board-and-batten effects are also possible using large vertical sheets of exterior plywood or composition material. Simply attach vertical solid wood strips over the joints and at intervals between the joints.

A variation of the board-and-batten siding is the *board-on-board siding.* Refer to **Figure 14-25.** Note the nailing pattern.

Normally, corner boards are not used on vertical siding. If installing tongue-and-groove siding, the groove must be ripped from the corner piece. To mark the edge of the corner piece for ripping, place the board on the starting edge, using a level to assure it is vertical. The bottom end should be about an inch below the sheathing. The top end should butt against or be covered by the trim. Mark the groove edge for ripping and make the cut, slightly undercutting the edge.

Install the corner board using face nailing at the corner and blind nailing at the tongue. Next, temporarily tack another board at the opposite edge and stretch a string between them as a guide for the rest of the siding.

When nearing an opening in the wall, such as a window or door, cut and fit the siding piece just before the piece that must be trimmed. Set it

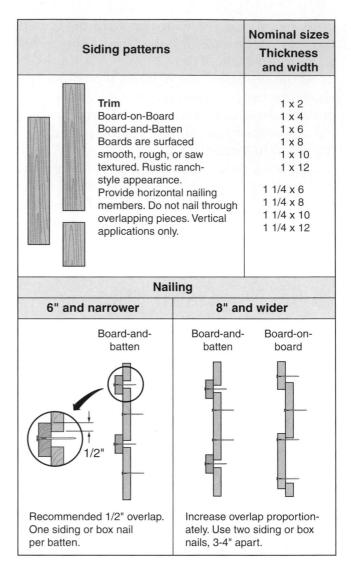

Siding patterns	Nominal sizes
	Thickness and width
Trim Board-on-Board Board-and-Batten Boards are surfaced smooth, rough, or saw textured. Rustic ranch-style appearance. Provide horizontal nailing members. Do not nail through overlapping pieces. Vertical applications only.	1 x 2 1 x 4 1 x 6 1 x 8 1 x 10 1 x 12 1 1/4 x 6 1 1/4 x 8 1 1/4 x 10 1 1/4 x 12

Nailing	
6″ and narrower	**8″ and wider**
Board-and-batten	Board-and-batten Board-on-board
1/2″	
Recommended 1/2″ overlap. One siding or box nail per batten.	Increase overlap proportionately. Use two siding or box nails, 3-4″ apart.

Figure 14-25. Board-on-board and board-and-batten siding give a building a rustic appearance. They also provide for expansion and contraction. (Western Wood Products Assn.)

aside for the moment. Cut and fit the next piece around the opening. Tack it in place of the one just cut and set aside. Use a level to transfer a mark from the top and bottom of the opening to the siding. Now, take a short piece of siding the same width to use as a marking gauge. If the

Board-and-batten siding: Siding applications designed around wide, square-edged boards spaced about 1/2″ apart.

Board-on-board siding: A variation of the board-and-batten siding with boards replacing battens.

siding is tongue-and-groove, you will need to cut away the groove to avoid a false marking. Hold this gauge against the trim and draw a cutting line on the siding as you slide the piece from the top to the bottom of the trim. Remove the piece of siding and cut away the waste material along the marks. Finally, install both pieces.

14.7 Wood Shingles

Wood shingles are sometimes used for wall covering, **Figure 14-26.** A large selection of types is available. Some are especially designed for sidewall application with a grooved surface and factory applied paint or stain.

Figure 14-26. Wood shingles or shakes are an attractive and durable siding material. (Shakertown Corp.)

Wood shingles: Individual wooden pieces with a wedge-shaped cross section used for roofing and sidewall applications. Wood shingles are made from cedar or other rot-resistant woods.

Single coursing: Applying shingle siding in single layers.

Double coursing: Applying shingle siding with a second layer over the first course.

Shingles are very durable and can be applied in various ways to provide a variety of architectural effects. Handsplit shingles are occasionally used, but they are expensive and difficult to install.

Most shingles are made in random widths. No. 1 grade shingles vary from 3″–14″ wide. Only a small number of the narrow width is permitted. Shingles of a uniform width, known as *dimension shingles,* are also available. For sidewall application, follow these recommendations for maximum exposure:

- For 16″ shingles, 7 1/2″ exposure
- For 18″ shingles, 8 1/2″ exposure
- For 24″ shingles, 11 1/2″ exposure

14.7.1 Application of Shingles to Sidewalls

Shingles can be applied to walls using one of two basic methods: single course or double course. **Single coursing** is similar to applying wood shingles to a roof. However, greater weather exposure of the shingle is allowed. In **double coursing,** a second layer is applied over the first. This method permits use of a lower grade shingle under the shingle exposed to the weather. The exposed shingle butt should extend about 1/2″ below the butt of the under course.

When double coursing shingle siding, secure the outer course with two 5d, small-head, rust-resistant nails at a point 2″ above the butt. This makes a greater weather exposure possible. Frequently, exposures of as much as 12″ for 16″ shingles, 14″ for 18″ shingles, and 16″ for 24″ shingles are satisfactory. **Figure 14-27** lists the sizes of standard sidewall shingles with a grooved surface. Approximate coverage for various lengths and exposures is included.

In mild climates, solid sheathing on sidewalls is not always needed where shingles are installed. Skip sheathing can be substituted. Sheathing boards may be spaced the same distance as the shingle exposure. This provides a nailing base for each shingle course. A high-grade shingle provides a satisfactory wall covering. Roofing felt should be used with such construction. Place it either between the shingles and sheathing or between the sheathing and the studs.

Grade	Length	Thickness (at butt)	No. of Courses per bdl/carton	Bdls/Cartons per square	Shipping weight	Description
No.1	16" (Fivex) 18" (Perfections) 24" (Royals)	.40" .45" .50"	33/33 28/28 13/14	1 carton 1 carton 4 bdls.	60* lbs. 60* lbs. 192 lbs.	Same specifications as rebutted-rejointed shingles, except that shingle face has been given grain-like grooves. Natural color or variety of factory-applied colors. Also in 4' and 8' panels.

Note: *70 lbs. when factory finished.

Length and thickness	Approximate coverage of one square (4 bundles) of shingles based on following weather exposures																									
	3 1/2"	4"	4 1/2"	5"	5 1/2"	6"	6 1/2"	7"	7 1/2"	8"	8 1/2"	9"	9 1/2"	10"	10 1/2"	11"	11 1/2"	12"	12 1/2"	13"	13 1/2"	14"	14 1/2"	15"	15 1/2"	16"
16" x 5/2"	70	80	90	100*	110	120	130	140	150**	160	170	180	190	200	210	220	230	240†	—	—	—	—	—	—	—	—
18" x 5/2 1/4"	—	72 1/2	81 1/2	90 1/2	100*	109	118	127	136	145 1/2	154 1/2**	163 1/2	172 1/2	181 1/2	191	200	209	218	227	236	245 1/2	254 1/2†	—	—	—	—
24" x 4/2"	—	—	—	—	—	80	86 1/2	93	100*	106 1/2	113	120	126 1/2	133	140	146 1/2	153**	160	166 1/2	173	180	186 1/2	193	200	206 1/2	213†

Note: *Maximum exposure recommended for roofs. **Maximum exposure recomended for single coursing on sidewalls. †Maximum exposure recommended for double coursing on sidewalls.

Figure 14-27. Sizes and coverage for sidewall shingles. Shingle thickness is based on the number of butts required to equal a given measurement.

Spaced sheathing is also satisfactory on implement sheds, garages, and other uninhabited structures. Here, protection from the elements is the principal consideration, not prevention of air infiltration.

To obtain the best effect and avoid unnecessary cutting of shingles, butt lines should be even with the upper edges of window openings. Likewise, they should line up with the lower edges of such openings. Steps for adjusting course spacing to accomplish this are explained later in this section.

> **Working Knowledge**
>
> It is better to tack a straight edge to the wall to use as a guide for maintaining straight and level courses of shingles, rather than attempt to shingle to a chalk line.

14.7.2 Single Coursing of Sidewalls

The single-coursing method for sidewall shingle application is much like roof application. The major difference is in the exposures employed. In roof construction, maximum permissible exposures are slightly less than 1/3 of the shingle length. This produces a three-ply covering. Vertical sidewall surfaces present fewer weather-resistance problems than do roofs. Accordingly, a two-ply covering of shingles is usually adequate.

In single-coursed sidewalls, weather exposure of shingles should never be greater than 1/2 of the length of the shingle minus 1/2". Thus, two layers of wood are found at every point in the wall. For example, when 16" shingles are used, the maximum exposure should be 1/2" less than 8", or 7 1/2".

Single-coursed shingle sidewalls should have concealed nailing, **Figure 14-28.** This means that the nails must be driven about 1" above the butt line of the next course. The shingles of this course then adequately covers the nails. Two nails should be driven in each shingle up to 8" wide. Place each nail about 3/4" from the edge of the shingle. On shingles wider than 8", drive a third nail in the center of the shingle at the same distance above the butt line as the other nails. Use rust-resistant, 3d nails that are 1 1/2" long.

For obvious reasons, shingles are applied from the bottom up. Attach a shingle to each end of the wall. Butts should extend about 1" below the top of the foundation. Stretch a line between them at the butts. Since even tightly stretched lines may sag, it may be necessary to nail on other shingles at intervals. Attach the line to them, as well. Install the starter course. Bring the butts close to the line without touching it. Apply a top course over the starter course, offsetting joints at least 1 1/2".

To keep succeeding courses straight and level, use a straightedge. Space untreated shingles at least 1/8" to allow for swelling when wet. Treated shingles can be laid much closer together.

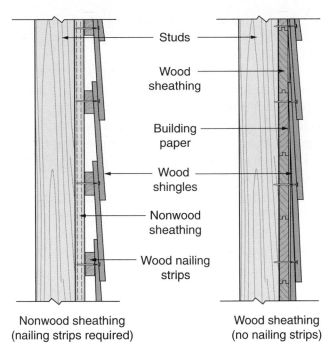

Figure 14-28. Single-course method of applying shingles to sidewalls. Solid backing and nailing base is provided by wood sheathing or nailing strips.

Depending on the customer's preference, it may be necessary to adjust exposure of courses so they evenly break at the top of the window trim. This is practical only if all windows are the same size and height along the wall. Refer to **Figure 14-29** as you study the following steps.

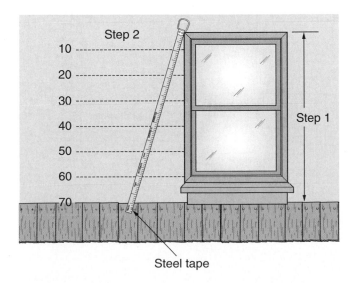

Figure 14-29. Method for laying out shingle siding courses so they evenly break above and below windows.

PROCEDURE

Adjusting course spacing

1. Measure the height of the window from the bottom trim to the top trim.
2. Divide this measurement by the shingle exposure you are using and round to the nearest whole number. For example, suppose the height is 50″ and the shingle exposure 7 1/2″. Dividing the height by the exposure gives you 6.67″. Go to the closest whole number, which is 7″. This is the number of even courses that will bring the break to the top of the window trim.
3. To mark the wall for these courses, extend a steel tape to some large multiple of 7″ (for example 70″).
4. Run the tape at a slant from the top of the window trim to a level line even with the bottom of the stool (if one is to be installed later). The 70″ mark on the tape should rest on this line.
5. Mark the sheathing at 10″, 20″, 30″, 40″, 50″, 60″ and 70″. These are the course lines that will break at the top of the window.
6. Mark these lines on a scrap of lumber so you can transfer them to the other side of the window and to all the windows as needed. If there is a second story, mark those windows, too.

14.7.3 Estimating Quantities

In estimating the quantity of shingles required for sidewalls, areas to be shingled should be calculated in square feet. Deduct window and door areas. Consult the table in **Figure 14-27**. The coverage of one square (4 bundles) at the exposure to be used should be divided into the wall area to be covered. The resulting figure is the number of squares needed. Add 5% to allow for waste in cutting and fitting around openings and for the double starter course.

14.7.4 Double Coursing of Sidewalls

In double coursing, a low-cost shingle is generally used for the bottom layer. This is covered with a No. 1 grade shingle or a processed shingle or shake. Many types of shingles are available for the outer course. Prestained shingles are available in attractive colors and particularly suitable. Although wide exposures usually require the use of long shingles, this effect is obtained in double coursing by the application of doubled layers of regular 16″ or 18″ shingles. The maximum exposure to the weather of double-coursed 16″ shingles is 12″. For 18″ shingles it is 14″.

The proper application of shingles on a double-coursed sidewall is illustrated in **Figure 14-30**. Most procedures used for regular siding can be followed. When the application is made over composition or spaced sheathing, mark the position of the furring strips on the story pole when it is laid out.

There are variations to straight-line coursing. In staggered coursing, butts of alternating shingles are offset below the line. Offsets are never more than 1″ for 16″ and 18″ shingles or 1 1/2″

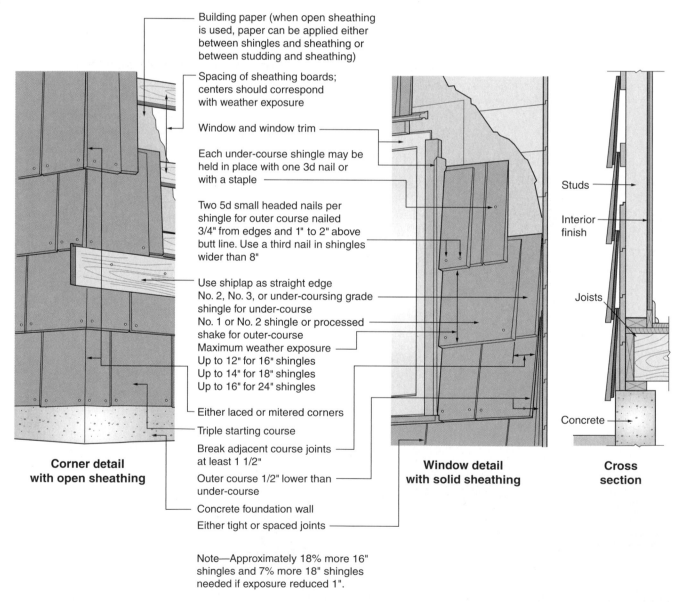

Building paper (when open sheathing is used, paper can be applied either between shingles and sheathing or between studding and sheathing)

Spacing of sheathing boards; centers should correspond with weather exposure

Window and window trim

Each under-course shingle may be held in place with one 3d nail or with a staple

Two 5d small headed nails per shingle for outer course nailed 3/4″ from edges and 1″ to 2″ above butt line. Use a third nail in shingles wider than 8″

Use shiplap as straight edge
No. 2, No. 3, or under-coursing grade shingle for under-course
No. 1 or No. 2 shingle or processed shake for outer-course
Maximum weather exposure
Up to 12″ for 16″ shingles
Up to 14″ for 18″ shingles
Up to 16″ for 24″ shingles

Either laced or mitered corners

Triple starting course

Break adjacent course joints at least 1 1/2″

Outer course 1/2″ lower than under-course

Concrete foundation wall

Either tight or spaced joints

Studs

Interior finish

Joists

Concrete

Corner detail with open sheathing

Window detail with solid sheathing

Cross section

Note—Approximately 18% more 16″ shingles and 7% more 18″ shingles needed if exposure reduced 1″.

Figure 14-30. Double-coursed shingle siding. Sheathing may be solid or spaced on-center to the nailing line.

for 24" shingles. In ribbon coursing, both layers are oriented in a straight line. The top course is raised roughly an inch above the under course. Both courses are normally butt nailed.

14.7.5 Shingle and Shake Panels

Shingles and shakes for sidewall application are available in panel form. The panels consist of two courses of individual shingles (usually western red cedar) permanently bonded to a backing, **Figure 14-31**. Standard length panels are 8'. End shingles of each course are offset for staggered joints that match up with adjacent panels. Panels are available in various textures, either unstained or factory finished in a variety of colors. Special metal or mitered wood corners are also manufactured.

Shingle panels are applied following the same basic precautions and procedures described for regular shingles, **Figure 14-32**. Installation time, however, is greatly reduced. Additional on-site labor is also saved when factory-primed or factory-finished units are

Figure 14-31. Applying panelized shakes. Panels are 8' long and consist of two 7" courses bonded to a plywood or veneer core base. (Shakertown Corp.)

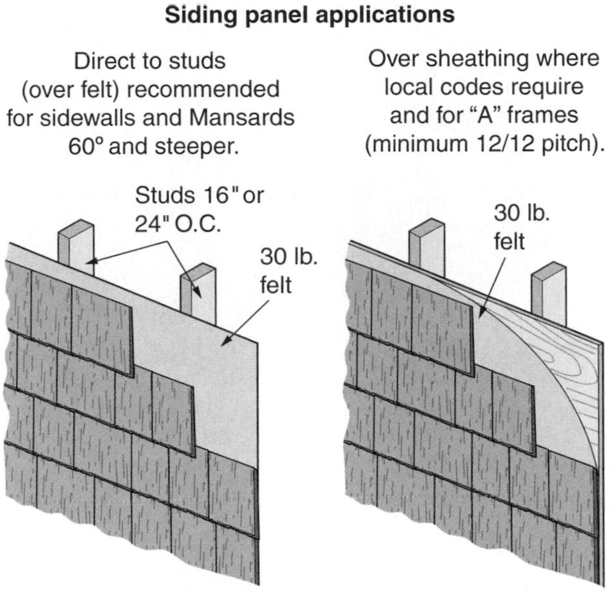

Siding panel applications

Direct to studs (over felt) recommended for sidewalls and Mansards 60° and steeper.

Over sheathing where local codes require and for "A" frames (minimum 12/12 pitch).

Studs 16" or 24" O.C.

30 lb. felt

30 lb. felt

Figure 14-32. Proper application of panelized siding. Panels are self aligning. (Shakertown Corp.)

used. When applying the latter, the installation should be made with nails of matching color supplied by the manufacturer.

Working Knowledge
When making an application of a specialized or prefabricated product such as shingle panels, be sure to follow the recommendations furnished by the manufacturer.

14.7.6 Applying Wood Shingles over Old Siding

Shingles or shakes can be applied over old siding or other wall coverings that are sound and will hold furring strips. First, apply building paper or housewrap over the old wall. Next, attach furring. Furring spacing should correspond to the exposure spacing of the shingles.

Usually it is necessary to add new moulding strips around the edge of window and door casings. This provides a new trim edge for the shingles. **Figure 14-33** illustrates how nailing strips are applied over an old stucco surface.

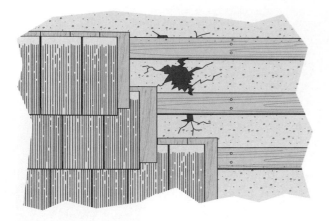

Figure 14-33. Re-siding a damaged stucco wall with double-coursed wood shingles. Furring strips are spaced to correspond to the nailing lines.

14.8 Plywood Siding

The use of plywood as an exterior wall covering permits a wide range of application methods and decorative treatments, **Figure 14-34.** Plywood can also be used alongside other building materials. A few of the ways it can be used include:

• Vertical treatment for gable ends.

• Fill-in panels above and below windows.

• Continuous, decorative band at various levels along an entire wall.

All plywood siding must be made from exterior-type plywood. Douglas fir is the most commonly used species. However, cedar and redwood are also available. Panels come in either a sanded condition or with factory applied sealer or stain. For information on grading standards, see Chapter 1.

Panel sizes are 48″ wide by 8′, 9′, or 10′ long. A 3/8″ thickness is normally used for direct-to-stud applications. A 5/16″ thickness may be used over approved sheathing. Thicker panels are required when the texture treatment consists of deep cuts. For unsheathed walls, plywood thickness should not be less than 3/8″ on 16″ stud spacing, 1/2″ for 20″ stud spacing, and 5/8″ for 24″ stud spacing.

Application of large sheets is generally made with the long dimension vertical. This eliminates the need for blocking to support horizontal joints. **Figure 14-35** shows a home sided with grooved vertical panels having shiplap edges.

In vertical installation, center the joints over studs. In horizontal installations, place solid blocking behind joints. Standard application requirements are given in **Figure 14-36.** Battens are an option, but should not be used with plywood siding that is textured.

Figure 14-37 shows several ways to handle joints between plywood panels. All edges of plywood siding—whether butted, V-shaped, lapped, covered, or exposed—should be sealed with a heavy application of high grade exterior

A

B

Figure 14-34. Examples of two plywood siding styles. A—Rough-sawn surface. B—Reverse board-and-batten. (APA-The Engineered Wood Association.)

Figure 14-35. This grooved plywood is applied over a sheathed wall and has a rough-sawn surface. (APA-The Engineered Wood Association)

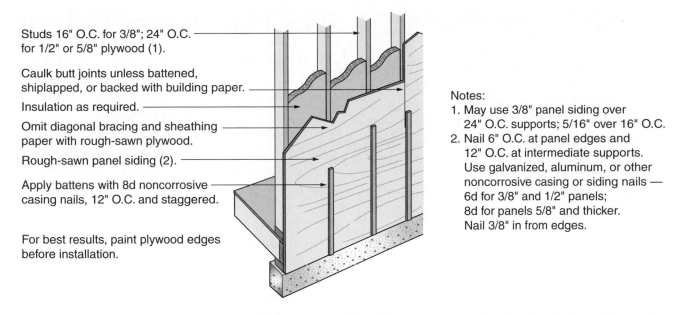

Studs 16" O.C. for 3/8"; 24" O.C. for 1/2" or 5/8" plywood (1).

Caulk butt joints unless battened, shiplapped, or backed with building paper.

Insulation as required.

Omit diagonal bracing and sheathing paper with rough-sawn plywood.

Rough-sawn panel siding (2).

Apply battens with 8d noncorrosive casing nails, 12" O.C. and staggered.

For best results, paint plywood edges before installation.

Notes:
1. May use 3/8" panel siding over 24" O.C. supports; 5/16" over 16" O.C.
2. Nail 6" O.C. at panel edges and 12" O.C. at intermediate supports. Use galvanized, aluminum, or other noncorrosive casing or siding nails — 6d for 3/8" and 1/2" panels; 8d for panels 5/8" and thicker. Nail 3/8" in from edges.

Figure 14-36. These recommendations are the standard for thickness of panel and method of installing vertical siding. Always check local codes to assure compliance. (APA-The Engineered Wood Association)

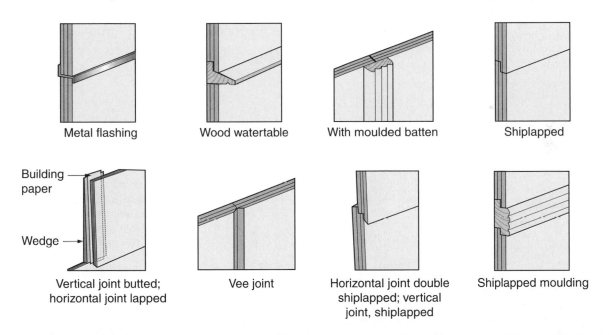

Metal flashing   Wood watertable   With moulded batten   Shiplapped

Building paper

Wedge

Vertical joint butted; horizontal joint lapped   Vee joint   Horizontal joint double shiplapped; vertical joint, shiplapped   Shiplapped moulding

Figure 14-37. Joint details for plywood siding. All edges should be sealed with paint before installing or with a special caulking compound at the time of installation. (APA-The Engineered Wood Association)

primer, aluminum paint, or oil paint. Special caulking compounds are also recommended.

Installing vertical panel siding is not difficult, although handling the panels may be clumsy. Start from a corner with the edge flush, square, and plumbed to the corner. Make sure the trailing edge is centered on a stud. Use 6d

siding nails for 1/2″ panels and 8d nails for thicker panels. Space nails 6″ around edges and 12″ when field nailing. Apply successive panels, keeping a straight line.

Plywood lapped siding may look the same as regular, beveled siding. Heavy shadow lines are secured by using spacer strips at the lapped edges. Application requirements are given in **Figure 14-38.** A bevel of at least 30° is recommended. The lap should be at least 1 1/2″.

Vertical joints of lapped siding should be butted over a shingle wedge and centered over a stud, unless 3/4″ wood sheathing is used. Nail siding to each of the studs along the bottom edge and not more than 4″ O.C. at vertical joints. Nails should penetrate studs or wood sheathing at least 1″.

If plywood lap siding is wider than 12″, a wooden tapered strip, such as a shingle, should be used at all studs with nailing at alternate studs. Outside corners should butt against corner boards or be covered.

Working Knowledge

Large sheets of plywood and hardboard siding provide tight, draft-free wall construction. It is important to have an effective vapor barrier. This should be placed between the insulation and the warm surface of the wall.

14.9 Hardboard Siding

Hardboard siding materials are durable, easy to apply, and adaptable to various architectural effects. Installation methods are similar to those described for plywood sidings.

Hardboard siding may expand more than plywood. Special precautions should be observed in the application. Manufacturers usually recommend leaving a 1/8″ space where hardboard siding butts against adjacent pieces or trim members.

Hardboard siding panels for vertical application are available in standard widths of 4′ and lengths of 8′, 9′, and 10′. Lap siding boards are usually 12″ wide by 16′ long. However, narrower widths can be purchased. The most common thickness is 7/16″.

Like plywood, hardboard siding is available in a wide range of textures and surface treatments. Most panels are given a primer coat at the factory. Prefinished units with matching battens and trim members are also available.

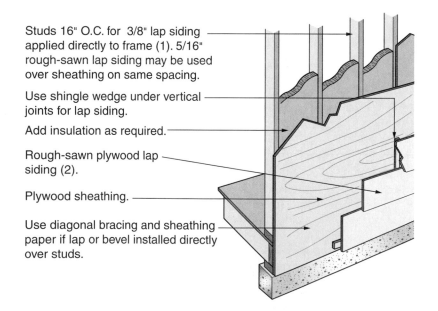

Studs 16″ O.C. for 3/8″ lap siding applied directly to frame (1). 5/16″ rough-sawn lap siding may be used over sheathing on same spacing.

Use shingle wedge under vertical joints for lap siding.

Add insulation as required.

Rough-sawn plywood lap siding (2).

Plywood sheathing.

Use diagonal bracing and sheathing paper if lap or bevel installed directly over studs.

Notes:
1. Use same nail schedule to install lap siding over studs or over sheathing. With sheathing, siding need not join over studs.
2. Use one nail per stud along bottom panel edges and 4″ O.C. at vertical joints. Nail 8″ O.C. at intermediate studs where siding is wider than 12″. Use galvanized, aluminum, or noncorrosive casing or siding nails. 6d for 3/8″ lap and 8d for thicker siding.

Figure 14-38. Application requirements for lapped plywood siding. If siding is directly applied to the frame, install corner bracing. Always check local codes for variance from these standards.

Hardboard siding: A siding material manufactured by treating ground-up wood with steam and pressure to form it into strips or sheets.

Figure 14-39 shows prefinished panels that have been installed around windows and doors.

When applying hardboard siding, follow standard installation procedures described for regular siding materials. Studs should not be spaced greater than 16″ on center. A firm and adequate nailing base is essential. Use a fine-tooth handsaw or power saw equipped with a combination blade to cut the panels. Nails must be galvanized or otherwise rustproofed. Wood trim and corner boards should be at least 1 1/8″ thick. Space nails at least 1/2″ in from edges and ends. Use an approved caulking compound at joints.

PROCEDURE

Applying hardboard lap siding

1. Apply building paper or housewrap to sheathing or studs (if there is no sheathing).
2. Measure and mark off equal distances from the top to the bottom on both ends of the wall, **Figure 14-40.** Siding should be at least 8″ above the ground.
3. Measure equal distances down from the eaves and/or windows and snap a chalk line.
4. Install inside and outside corner boards as described for installation of wood siding. Use metal, wood, or vinyl. Wood trim should be a minimum of 1″ thick.
5. Measure off and snap parallel chalk lines as guides for locating the top edge of each course of siding. Each course should overlap the previous course at least 1″. Increase the lap to 1 3/8″ when installing beaded lap siding. See **Figure 14-41.**
6. Nail on a lath starter strip 1/4″ above the bottom of the sheathing. Substitute a wood strip, 3/8″ × 1 1/2″, if lath is not available.
7. Install the first course. Nail from the center toward the ends or from one end to the other. Nail at all stud locations 3/4″ above bottom edge to penetrate both courses. Stay 3/8″ away from butt ends. Nails should penetrate studs at least 1 1/2″. For best results use hot-dipped, galvanized nails with minimum of 1/4″ diameter heads.
8. Install flashing over drip caps.
9. Caulk joints with a high-grade acrylic latex or similar product.
10. Apply finish coat as soon as possible. Re-prime if left unfinished for more than 180 days.

Hardboard siding is also available for hidden-nail application, **Figure 14-42.** Application is similar to the method just described.

Figure 14-39. A—Window trim with panels. A drip cap is not needed over windows because of the flange used on the metal-clad window. B—Drip cap is required over this wood-trim door.

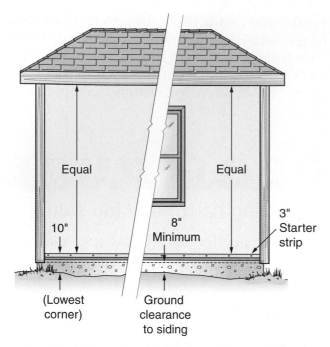

Figure 14-40. To establish a starting point for the top edge of horizontal siding, first find the lowest corner of building. Then, measure and mark equal distances between eaves and starting point. Snap a level chalk line to mark top edge of first course. (ABTco, Inc.)

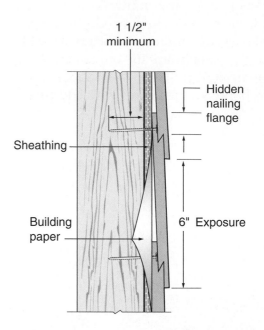

Figure 14-42. Application of a concealed-nail siding product. Always be careful that the succeeding courses are fully seated in the locking channel. (ABTco, Inc.)

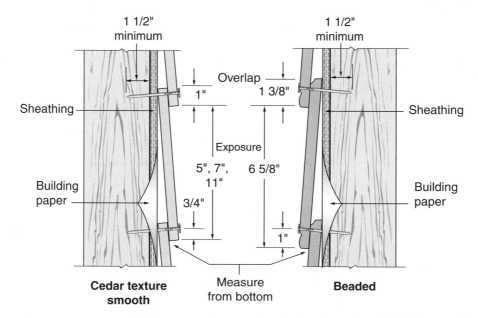

Figure 14-41. One manufacturer's recommendations for application of horizontal siding. Note nail penetration, exposure, and use of building paper. Always follow the manufacturer's recommendations. (ABTco, Inc.)

14.10 Fiber-Cement Siding

In recent years, siding manufactured of fiber-reinforced cement has seen considerable growth in the residential market. *Fiber-cement siding* is made in various forms, including traditional lap siding, vertical panel siding, shingle panels and individual shingles, trim planks, and perforated soffit panels. Surfaces are wood-grain textured, stucco textured, or smooth and may be painted. Factory-finished siding is also available in a number of colors.

The material is strong, fire- and impact-resistant, and will not deteriorate from weather exposure. It is installed by nailing, like wood or vinyl siding, and can be cut with saws or shears. The use of shears is preferred because it does not generate dust, **Figure 14-43.**

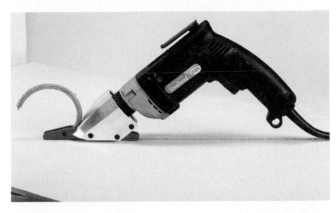

Figure 14-43. Shearing is the preferred method for cutting fiber-cement siding since it does not create dangerous dust. (Pacific International Tool and Shear)

Safety Note

When cutting fiber-cement siding materials with a saw, wear an approved dust mask or respirator to prevent breathing crystalline silica dust. Excessive exposure to the dust can cause silicosis, a serious lung disease. Whenever possible, sawing should be done outdoors so that air movement can carry the dust away from the person doing the cutting. Workers in the immediate vicinity should wear dust masks while cutting is being done.

14.11 Aluminum and Vinyl Siding

Aluminum and vinyl siding, though very different materials, are similar in many respects. Both closely resemble painted wood siding and offer low maintenance costs to the homeowner. Both are made with interlocking edges and offer designs for both vertical and horizontal application. Many of the installation steps applying to aluminum also apply to vinyl.

14.11.1 Aluminum

Aluminum siding is factory finished with baked-on enamel. It is designed for use on new or existing construction. Aluminum can be applied over wood, stucco, concrete block, and other structurally sound surfaces. Basic specifications (alloy and gage) for aluminum siding are established by the FHA and the Aluminum Siding Association.

A variety of horizontal and vertical panel styles in both smooth and textured designs are produced with varying shadow lines and size of face exposed to the weather. A panel that has an insulating fiberboard material laminated to the back surface is also produced.

Manufacturers supply directions and suggest needed tools and equipment for the installation of their products. See **Figure 14-44.** These instructions and suggestions should be carefully noted before starting the job.

Panels have prepunched nail holes. Special interlocking designs allow fastening of parts with the nails concealed.

Aluminum panels can be cut with a utility knife, tin snips, or a portable electric circular saw. The electric saw is the tool preferred

---

Fiber-cement siding: Strong, damage resistant siding material made from cement reinforced with cellulose fiber.

Aluminum siding: Factory-finished metal siding with baked-on enamel.

Tools	Materials
Steel tape	Aluminum trim sheet
Folding rule	Aluminum or galvanized
Level (2' minimum)	nails
Steel square	
Hand saw (crosscut)	
Hacksaw (fine-tooth,	1 1/2", general use
metal cutting)	2" for re-siding
Chalk line	2 1/2" through siding
Claw hammer (or power	with backer board
hammer/stapler)	1–2" trim nails
Portable power saw	(color-matched)
(fine-tooth blade)	
Screw driver	**Equipment**
Pliers	
Aviation snips	Ladders/scaffolds
Shears	Cutting table
Snap-lock punch	Portable brake
Steel awl	
Line level	
Utility knife	
Nail slot punch	

Figure 14-44. Certain tools, equipment, and materials are best suited for shaping and installing aluminum siding. The brake (bender) is needed to shape custom trim around window casings, door casings, and window sills.

since it is faster and more accurate. It is best to set up a sawing guide (jig) such as that shown in **Figure 14-45.** The jig is easy to make and holds the saw above the workpiece where it cannot cause damage to the finish. Use a 10 point aluminum cutting blade in the saw. If using tin snips, the duckbill type makes better straight cuts.

Special corners and trim members are often formed in the shop or on the job site, **Figure 14-46.** Manufacturers also offer pre-formed accessories such as starter strips, corner boards, flashing, stiffeners, and trim pieces used on windows and door frames.

PROCEDURE

Using a power nailer

Manufacturers provide instructions for the proper use of power nailers with their products. In general, follow these instructions:

1. Hold the power nailer perpendicular to the siding. Center the tool guide in the nailing slot. Maintain 1/32″ clearance between the nail crown and the siding panel so panel can move during expansion and contraction.

2. Use corrosion-resistant fasteners. They should be aluminum, galvanized steel, or cadmium coated.

3. Power nailers operate at specific air pressure settings. Follow manufacturer's instructions for initial settings. Adjust as necessary as the first course is installed.

Figure 14-45. A simple jig saves time cutting siding units to length.

Figure 14-46. Forming aluminum pieces for window and fascia trim requires careful measuring and accurate bends.

Safety Note

As a precaution against faulty electrical wiring or appliances that might energize aluminum siding and create an electrical hazard, grounding should be included. The Aluminum Siding Association recommends that a No. 8 or larger copper wire be connected to any convenient point on the siding and to the cold water service or the electrical service ground. Connectors should be UL (Underwriters Laboratories) approved.

14.11.2 Vinyl Siding

Vinyl siding is made from a rigid polyvinyl chloride compound. It is tough and durable and very common. The vinyl material is extruded into either horizontal or vertical siding units and accessories. The panel thickness is about 1/20″ and panels are available in various widths up to 8″.

Horizontal strips are manufactured to resemble wood drop siding and shiplap, **Figure 14-47.** Each double strip hooks into the course below and is nailed through slots in the top edge. Special inside and outside corner posts cover the ends. They permit expansion and contraction of the siding resulting from temperature changes.

As with aluminum, vinyl siding components can be easily cut with a portable circular saw. Use a fine-toothed blade and mount it in reverse. A jig makes cutting faster and easier.

Working Knowledge

When installing vinyl siding, be sure to read and follow the directions furnished by the manufacturer.

Vinyl siding: Siding manufactured from rigid polyvinyl chloride.

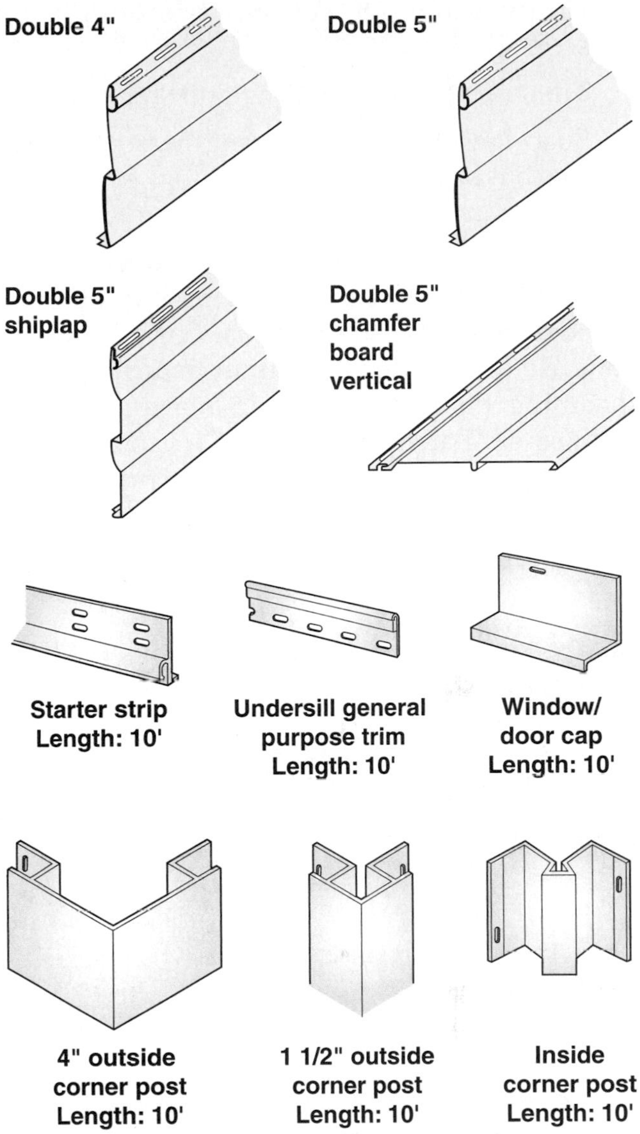
Double 4"
Double 5"
Double 5" shiplap
Double 5" chamfer board vertical
Starter strip Length: 10'
Undersill general purpose trim Length: 10'
Window/ door cap Length: 10'
4" outside corner post Length: 10'
1 1/2" outside corner post Length: 10'
Inside corner post Length: 10'

Figure 14-47. Various accessories are made for use with vinyl siding. (Bird Division, CertainTeed Corp.)

14.11.3 Installing Vinyl Siding and Soffits

Vinyl siding can be installed with essentially the tools already in the carpenter's toolbox. The following tools are recommended by the Vinyl Siding Institute:

- Power saw, table or radial, with fine-tooth blade (12–16 teeth per inch). Some installers prefer a fine-tooth hand saw.
- Other standard tools—hammer, square, chalk line, level, steel tape, and safety glasses.

- Utility knife for cutting, trimming, and scoring.
- Tin snips to speed cutting and shaping.
- Snap-lock punch for cutting lugs on siding that has been trimmed for the top or finishing course under the eaves.

No special preparation of the exterior walls is needed when using vinyl siding on new construction. In some situations, installers may use an exterior insulation board over the sheathing. Installation of horizontal and vertical siding is similar. However, there are some differences, as described in the following procedures. **Figure 14-48** shows before and after photos of a re-siding project. Various elements of vinyl siding are shown in **Figure 14-49.**

The top course under the eave requires special trim and preparation of the panel. Every manufacturer provides a system for concealing and securing the top siding panel. Generally speaking, this requires a trim piece designed to grip the trimmed top edge of the panel.

Gable ends are covered in the same manner as the walls. Use two scrap pieces of panel to make a pattern for cutting the proper angle. Lock one piece into the panel below. Hold the other piece against the gable. Mark a line across the bottom piece and cut. Follow the same procedure for the other side of the gable.

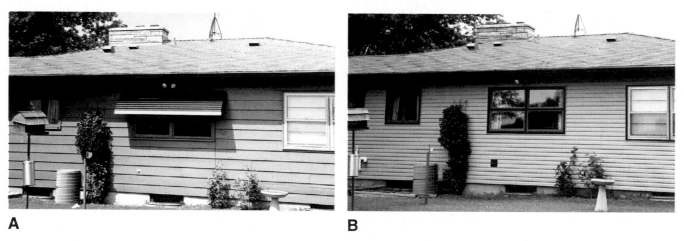

A   **B**

Figure 14-48. Done properly, re-siding enhances the appearance of a building. A—This siding is old and needs to be replaced. B—A vinyl recladding updates the look of the home and offers low maintenance.

Aluminum gutter

Aluminum clad fascia

Soffit

Soffit vent

Inside corner

Outside corner

Vinyl siding

Aluminum downspout

Figure 14-49. The different elements of vinyl exterior finish. (Benjamin Obdyke, Inc.)

PROCEDURE

Installing horizontal siding

1. Snap a chalk line at the bottom of the wall. Refer to steps given for installation of hardboard siding. Carpenters usually start at the back of the building and work toward the front. Finish each side before starting the next. Always cover largest areas first. Short panels can be used up on the smaller surfaces such as dormers.

2. Using the chalk line as a guide, nail on a ***starter strip,*** **Figure 14-50.** Allow space for inside and outside corner trim and J-channel.

3. Position and fasten all inside and outside corner posts allowing 1/4″ space at the top, **Figure 14-51.** Place the first nail at the top of the top slot. Nail in the center of all other slots, spacing nails 6″–12″ apart.

4. If splicing of corner posts is necessary, cut away 1″ from all but the outer face of the top of the lower corner post. Lap 3/4″ of the upper post over the lower. This allows 1/4″ for expansion.

5. Install J-channel around windows and doors, **Figure 14-52.** Cut side J-channels slightly longer than the trim. Notch the top ends. Cut top J-channel longer to provide a tab. Miter the faces of the channel pieces. Bend down the tabs on the top J-channel to act as a flashing over the side pieces.

6. Place the first panel in the starter strip. Lock it in place.

7. Nail the first panel using the same spacing as for corner posts. Never pull siding panels taut before nailing. This deforms the panel and causes a bad lap with the panel beneath.

8. Overlap each panel 1″, facing the butt edge away from the main traffic areas. Stagger laps. None should be directly above another unless separated by three courses. Always allow a 1/4″ expansion gap where siding meets accessories. In cold climates, allow additional room for expansion.

9. Check every five or six courses for alignment with eaves.

10. Install the top course and special trim according to manufacturer directions.

Starter strip: A strip placed at the bottom of a wall to which vinyl siding is attached.

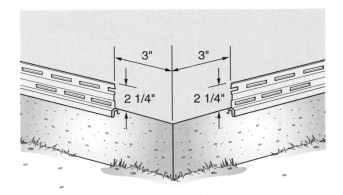

Figure 14-50. Installing the starter strips. A 3″ setback at corners allows room for installing corner posts at inside and outside corners. Allow 1/4″ clearance for expansion between the siding and the corner post.

A

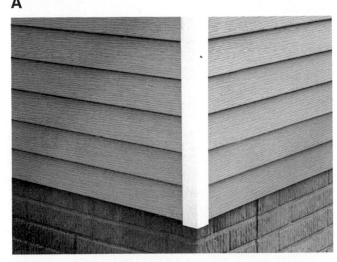

B

Figure 14-51. Corner post installation. A—Install corner posts after starter strips. Start the corner post 3/4″ below the bottom of the starter strip. B—The same view after siding is installed.

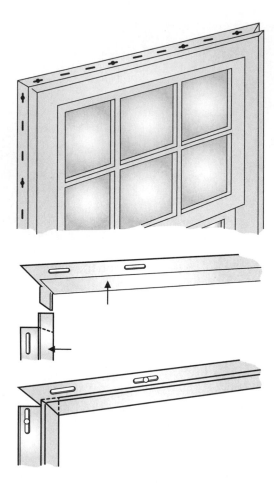

Figure 14-52. Install J-channel around windows and doors to provide a finished appearance to siding and to seal joints against water. Cut and shape the flange and miter the corners as shown. (CertainTeed Corp.)

PROCEDURE

Typical installation of top course

1. Install trim with a double J-channel against the overhang.
2. To determine how much of the panel to cut off, measure the distance between the top, inside slot of the dual under-sill trim and the top lock of the last installed siding panel. Subtract 1/4″ for expansion. Cut the top panel to this dimension.
3. Punch the top panel with a snap-lock punch to 1/4″ below the cut edge. This creates raised points that snap into the double J-channel.
4. Install the top panel.
5. Install the trim.

PROCEDURE

Installing vertical siding

1. Install J-channel at the top and bottom of the wall. Locate these using the same method as for horizontal installations.
2. Apply horizontal furring strips 12″ O.C. to provide a nailing surface. Normally, 1 × 4s are used since they are wide enough to accommodate nailing slots and thick enough to give 3/4″ penetration of the fasteners.
3. Carefully measure the distance from the inside of the top J-channel to the inside of the bottom J-channel. Subtract 1/2″ from the total length to allow for panel expansion. Cut the panels to this dimension.
4. Install corner posts as described for horizontal applications. Be careful not to distort the post in any way. Make sure posts are plumb, since this affects installation of the panels.
5. Install a starter strip into the corner post channel, **Figure 14-53.** Check that there is sufficient clearance inside of the post channel for the panel to lock onto the starter strip. Drive a nail at the top of the first slot, then check for plumb and finish nailing.
6. Cut, shape, and fit J-channel around windows and doors as in horizontal installations.
7. Trim and fit vertical panels around doors and windows, **Figure 14-54.** Use furring strips as needed.
8. Ending panels at corners are trimmed and fitted into a J-channel inserted under the corner post. Unless the trim cut is in a vee, use vertical shimming to maintain the surface plane and keep the panel pressed against the inside of the corner post, as shown in **Figure 14-54.**

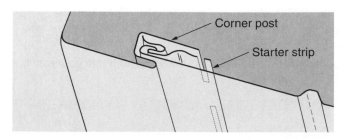

Figure 14-53. In a vertical application, use a starter strip to lock in the first panel at the starting edge of each corner. (Rollex Corp.)

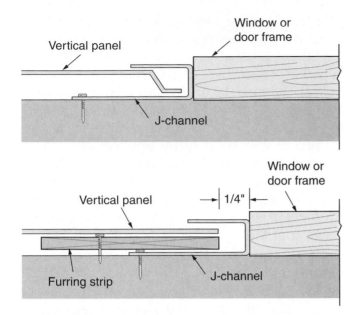

Figure 14-54. Fitting trimmed panels into J-channels. If the trim cut is not made in the vee of the panel, install a furring strip to produce a snug fit in the J-channel. (Rollex Corp.)

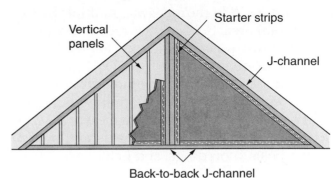

Figure 14-55. Finishing a gable end with vertical siding. Install J-channel on all sides along with starter strips. (Rollex Corp.)

PROCEDURE

Finishing gable ends for vertical siding

1. Use a level to establish and draw a vertical line in the center of the gable.
2. Install starter strips along the rake and base of the gable, as shown in **Figure 14-55.**
3. Attach two J-channels back to back on this line.
4. Install starter strips inside each of these J-channels.
5. Make a pattern for angle cuts. Measure and cut panels.
6. Install panels, starting at the center and working out to each side.

PROCEDURE

Preparing old siding

1. Nail loose boards and trim. Replace any rotted boards.
2. Remove old caulk from joints around windows, doors, and corner boards. It will interfere with placement of J-channel trim.
3. Remove downspouts, lighting fixtures, and mouldings where they will interfere with siding installation.
4. Tie back shrubbery and trees that are close enough to be damaged or interfere with work.
5. To ensure a smooth surface for the new siding, install furring strips. Shim low areas as needed. As an alternative, sheath the wall with foam panels. Either method will enhance the appearance of the finished job.

14.11.4 Re-Siding over Old Siding

New aluminum or vinyl siding is often installed without removing the old siding. In some re-siding jobs, window sills, heads, and casings are clad with aluminum to reduce maintenance. Cut aluminum from prefinished aluminum sheet and form it on a metal brake (bender). This trim should be installed before installing the vinyl J-channel. When installing new siding over existing siding, preparation requires several steps.

14.11.5 Estimating Aluminum and Vinyl Siding

Panels are sold by the square (100 sq. ft.). Find the area of one side and one end wall of the building. Add the areas and multiply by two to get total area. Subtract the area of openings. Add 10% for waste. Divide by 100 to find the number of squares needed.

Order accessories after taking careful linear measurements. Carefully determine where each accessory will be needed as you measure. Accessories include starter strips, inside and outside corner posts, mouldings, and J-channels.

14.12 Stucco

When properly applied, stucco provides an exterior wall finish that is virtually maintenance-free. The finish coat may be tinted by adding coloring or the surface may be painted with a suitable material. Where stucco is used on houses more than one story high, the use of balloon framing for the outside walls is desirable. If stucco is applied to walls built with platform framing, shrinkage at band joists may cause distortion or cracks in the stucco surface.

Stucco is usually applied in three coats over concrete block, wood sheathing, or cement board. The three layers include a base or scratch coat 3/8" to 1/2" thick, brown coat 1/4" to 3/8" thick, and 3/8" color or finish coat.

Stucco can be directly applied to a clean concrete block wall. On wood structures, "D" or better sheathing paper or housewrap and metal lath are necessary as a substrate, **Figure 14-56.** The metal lath should be heavily galvanized. Space it at least 1/4" away from the sheathing so the scratch coat can be easily forced through, becoming thoroughly keyed into the metal lath, **Figure 14-57.** Metal or wood moulding with a groove that also keys the stucco is applied at edges and around openings.

Galvanized furring nails, metal furring strips, and self-furring wire mesh are available. Nails should penetrate the sheathing at least 3/4". When fiberboard or gypsum sheathing is

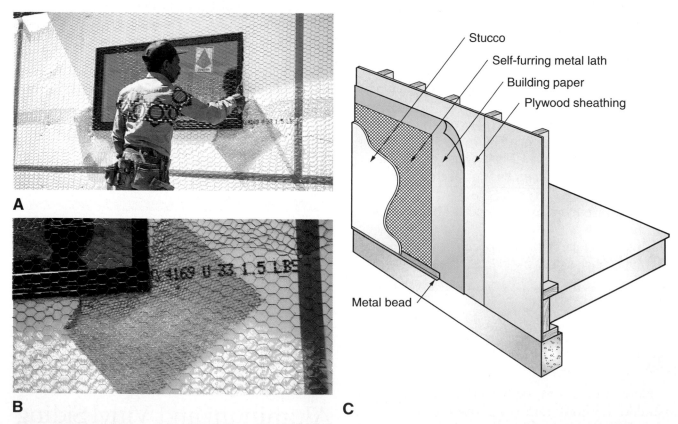

Figure 14-56. Preparation for stucco. A—A worker applies metal lath over insulating board. B—Reinforcing is added at window corners. C—Construction details for stucco finish. Sheathing paper or housewrap is recommended over both interior and exterior plywood sheathing. (Pierce Construction, Ltd.; APA-The Engineered Wood Association)

Figure 14-57. The base coat is going on over metal lath. Sufficient pressure is applied to push some of the stucco through the lath.

Figure 14-58. The top coat is being applied. When dry, it will be painted.

used, nailing with adequate penetration should occur over studs.

A metal rake is employed to create scratch lines in the still-wet first coat. The brown coat is applied when the first has dried. It is usually given a smoother surface using straightedges and wooden floats. To produce an adobe-like appearance, such as found in the Southwest, workers use a steel trowel when the brown coat is the final coat.

The first two coats should cure for one or two weeks before the final coat goes on. Usually, the final coat is given a smoother, pebbled finish, **Figure 14-58.** If color is preferred, iron oxide pigments are added to the finish material. Typically, the finished stucco wall is 7/8″ thick.

Working Knowledge

Stucco plaster is usually one part portland cement, three parts sand, and hydrated lime that is 10% less by volume than the cement. Once applied, keep it wet for three days. Never apply stucco in temperatures below 40°F (5°C).

Exterior insulation and finish systems (EIFS): Multi-layered wall systems that consist of a layer of insulation board attached to the wall sheathing, a water-resistant base coat with a reinforcing mesh, and a finish coat that resembles stucco or stone.

Soft-coat systems: Polymer-based EIFS that is typically thin (1/8″), flexible, and attached with adhesives.

Hard-coat systems: Polymer-modified EIFS that is about 1/4″ thick and mechanically attached.

14.13 Exterior Insulation and Finish Systems

Exterior insulation and finish systems (EIFS) are similar in appearance to stucco and are often referred to as *synthetic stucco.* Several manufacturers produce these systems, which offer high R-values, low maintenance, design flexibility, and a wide variety of colors.

EIFS are available either as polymer based (PB) or polymer modified (PM). The PB type, or *soft-coat systems,* are typically thin (1/8″), flexible, and attached with adhesives. The PM type, or *hard-coat systems,* are thicker (about 1/4″) and are mechanically attached. EIFS are offered in different textures.

Earlier systems have been redesigned to handle problems with water penetration and retention. The systems currently being installed drain away any water penetration before it can cause structural damage. **Figure 14-59** shows two typical EIFS applications in cutaway. Water management is accomplished in several ways:

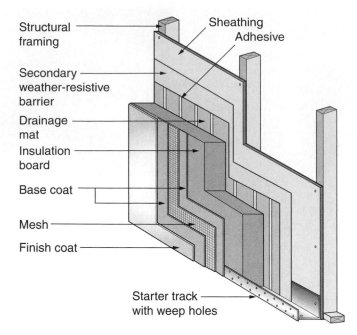

Structural framing

Sheathing

Adhesive

Secondary weather-resistive barrier

Drainage mat

Insulation board

Base coat

Mesh

Finish coat

Starter track with weep holes

Note: Secondary weather resistive barrier as required by local code, (e.g., Southern Building Code, minimum one layer Type 15 felt, waterproof building paper or equivalent).

A

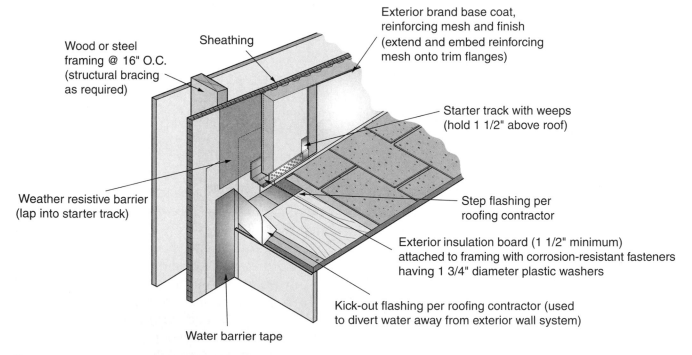

Wood or steel framing @ 16" O.C. (structural bracing as required)

Sheathing

Exterior brand base coat, reinforcing mesh and finish (extend and embed reinforcing mesh onto trim flanges)

Starter track with weeps (hold 1 1/2" above roof)

Step flashing per roofing contractor

Exterior insulation board (1 1/2" minimum) attached to framing with corrosion-resistant fasteners having 1 3/4" diameter plastic washers

Weather resistive barrier (lap into starter track)

Kick-out flashing per roofing contractor (used to divert water away from exterior wall system)

Water barrier tape

B

Figure 14-59. Typical EIFS installations. A—This cutaway shows the components of an EIFS-sided structure that incorporates a drainage cavity and water-resistive barrier. B—This cutaway illustrates a water-managed EIFS that incorporates grooves in the insulation board to channel any incidental moisture to weep holes. (Sto Finish Systems Division, U.S. Gypsum Co.)

- Application of housewrap or other water barrier materials to the sheathing.

- Placing grooves on one surface of the foam panel to channel incidental water that has entered the wall to the outside through weep holes.

- Installation of a drainage mat between the exterior insulation board and the water barrier.

- Careful application of flashing at points where leakage is likely to occur.

Water-managed systems require accessories, many designed expressly for preventing water damage. See **Figure 14-60.**

Installation should be attempted only after carefully studying manufacturer's instructions. Final details must be approved by a licensed professional.

An exterior insulation finish system can be installed over gypsum sheathing, glass mat–reinforced gypsum, plywood, oriented strand board, concrete, and concrete block. The surface of the substrate must be flat

---

Sloped sill wedge

Description: Exterior grade rigid polyvinyl chloride (PVC) trim component used to provide a positive slope over the sill to direct intruding water to the outside face of weather-resistive barrier for drainage out of system. May require increased rough opening size.
Size: 1/4" high, 3 7/8" and 5 7/8" deep.
Length: 10'

Starter track

Description: Exterior grade rigid polyvinyl chloride (PVC) trim component with weep holes used at all horizontal terminations (foundations, bottom of wall/roof intersections, bottom of chimney/roof intersections, etc.).
Size: Sized to match system thickness specified.
Length: 10'

Water barrier tape

Description: A polyethylene-film coated, self-sealing tape that is primarily used to seal window sills to the building paper prior to the installation of the windows. The tape is self-adhesive with a disposable silicone-treated release sheet.
Thickness: 30 mils.
Width: 9"
Color: Black

Drip flashing

Description: Exterior grade rigid polyvinyl chloride (PVC) trim component used to provide flashing and drainage of the system above windows, doors, and other penetrations.
Size: Sized to match system thickness specified.
Length: 10'

Weather-resistant barrier

Description: Spunbonded olefin material is engineered to improve the effectiveness of wall systems that require a weather-resistive barrier. Features include: engineered surface design, superior water resistance, high permeance value, excellent tear resistance, pliability, light weight.
Size: 5" wide, 200 long rolls, 0.22 oz./sq. ft.

Casing bead

Description: Exterior grade rigid polyvinyl chloride (PVC) trim component used to terminate system at edges of all windows and doors to create a space for sealant application when foam or panel bands are used for insulation board.
Size: Sized to match system thickness specified.

Wood and steel screws

Description: Wafer head design provides greater holding power. Special coating provides corrosion resistance. For 14 to 20 ga.
Steel Framing: 1 1/4" and 1 5/8" steel screws.
For Wood Framing: 1 1/4", 1 5/8", or 2 1/4" wood screws.

45° bead

Description: Exterior grade rigid polyvinyl chloride (PVC) trim component used to finish around window and door penetrations as a backing for perimeter sealant application.
Size: 1/2" long, 45° angled return, 10' length.

Figure 14-60. A typical list of accessories supplied by an EIFS manufacturer. (U.S. Gypsum Co.)

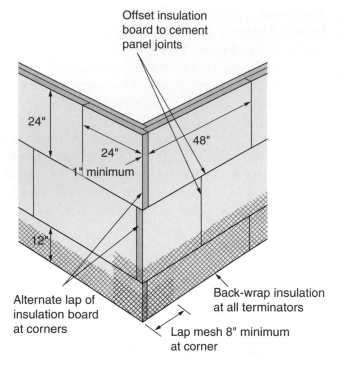

Offset insulation
board to cement
panel joints

24"

24"

48"

1" minimum

12"

Alternate lap of
insulation board
at corners

Back-wrap insulation
at all terminators

Lap mesh 8" minimum
at corner

A

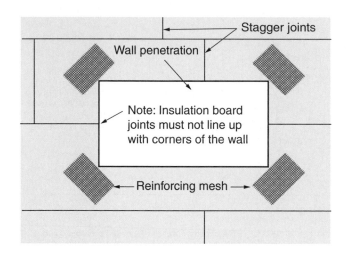

Stagger joints

Wall penetration

Note: Insulation board
joints must not line up
with corners of the wall

Reinforcing mesh

B

Figure 14-61. Insulation board alignment. A—Alternate the lap of insulation board at outside corners and offset the vertical joints. B—Avoid lining up joints, either horizontal or vertical, with window or door openings. Apply reinforcing at the corners of wall penetrations. (U. S. Gypsum Co.)

before beginning installation. The following is a general method of installing EIFS using grooved insulation board. Manufacturers supply detailed instructions. Always carefully read and follow them.

PROCEDURE

Installing the insulation board

1. Install **starter track** at the bottom of the walls, over the tops of windows, and at roof/wall intersections.
2. Horizontally staple a weather-resistive barrier over the sheathing, working from the bottom up. Lap the first course over the flange of the starter track. Lap succeeding courses 4″–6″.
3. Prepare window and door openings by cutting the barrier and securing it at these openings as when applying housewrap.
4. Install flashing at top of openings to channel water to the outside in case of penetration. This should be done at the time you install windows and doors. Flashing should be installed through a slit made above the window head. Secure it with nails. Apply water barrier tape to provide a seal. Since windows vary from one manufacturer to the next, consultation between window manufacturer, EIFS manufacturer, and architect is critical for proper installation of both products.
5. Install vertical trim around windows and doors.
6. Install control joints following the architect's drawings.
7. Install insulation board so grooves are vertical. Allow 1/16″–3/16″ clearance between the insulation board and the starter strip. Attach to every stud with approved fasteners. Some manufacturers use a drainage mat to direct incidental water to weep holes. Follow their instructions for installing this mat and the insulation board.
8. Precut L-shaped pieces to fit around openings. Avoid having joints at these locations.
9. Joints in the insulation board should be offset between sheathing and window openings. See **Figure 14-61.** Attach insulation board with fasteners no more than 8″ apart in the field, remaining 3″–5″ away from all edges. Fasteners should be long enough to penetrate 3/4″ into studs. Gaps 1/8″ or larger should be filled with slivers of insulation board cut to fit. Do not use adhesive.
10. Sand or rasp the insulation board to improve the bond with the base coat or added architectural features.

Starter track: Channels used at the top and bottom of walls and around openings in an EIFS application.

PROCEDURE

Applying the EIFS

1. Apply the base coat (an adhesive) to any foam pieces used as architectural features and firmly press them in place, **Figure 14-62.**
2. Coat fastener heads and allow them to dry four hours.
3. Using a steel trowel, apply the base coat to the sanded surface. Uniformly apply the base coat (adhesive).
4. Embed the reinforcing mesh immediately into the wet base coat. Smooth it with a trowel. At this time the base coat should be no thicker than the mesh. Lap the mesh 2″ on all sides.
5. When the base coat has firmed, trowel on a second coat and smooth to 3/32″ thickness.
6. When the base coat has dried a minimum of 24 hours, it is ready for primer or a finish coat.

PROCEDURE

Applying primer and finish material to EIFS

1. Inspect the base coat for defects and repair before continuing.
2. Mix the primer following the manufacturer's instructions. If desired, tint or shade it to the finish color.
3. Apply the primer with a napped roller or quality latex paint brush. Allow to completely dry before applying finish coats. Work only in temperatures above 40°F (5°C). Protect from rain.
4. Mix the finish material and apply with a trowel to minimum 1/16″ thickness, **Figure 14-63.** Work rapidly, maintaining a wet edge.
5. Allow the finish to dry for 24 hours. During drying, protect it from adverse weather, dust, or physical contact.

Figure 14-62. Insulation board being applied over oriented strand board sheathing.

A

B

Figure 14-63. Applying the final finish to an EIFS. A—A worker mixes colored stucco finishing materials. B—Other construction workers apply the final finish.

14.14 Brick or Stone Veneer

A *veneer wall* is usually not referred to as a "masonry wall," although it is. It is a wall framed in wood or metal to which stone, brick, or concrete block is attached rather than siding. However, the weight of the veneer is supported directly by the foundation, **Figure 14-64.** The foundation must be wide enough to provide a base for the masonry units.

A base flashing of noncorroding metal should extend from the outside face of the wall, over the top of the ledge, and at least 6" up behind the sheathing. When sheathing is plywood and an air space of 1" is included, sheathing paper is not required.

A

B

Figure 14-64. A—General construction details for brick veneer siding with plywood sheathing. One tie should be anchored to a stud for every 2 sq. ft. of brick area. B—A brick veneer being installed. (APA-The Engineered Wood Association)

14.14.1 Brick Tools

A trowel is the most-used tool in the mason's tool kit. See **Figure 14-65.** The pointed end of the trowel is the *toe* or point; the wide end is the *heel.* Another important tool is the mason's level. A mason's level should have both vertical and horizontal vials that can be read from either side. Some levels are aluminum, while others are wood with metal edges designed to withstand the rough environment of the construction site.

The mason usually has two kinds of rules. One rule is a 6' folding rule, often with a 6" sliding scale on the first section for taking inside measurements. The folding rule usually has markings on its back side giving course heights for various masonry unit sizes and joint thicknesses. The other rule is a 10' steel tape.

Jointers, or jointing tools, are used to compress, smooth, and shape the surface of the mortar joints. Several jointers are commonly used. Their shape determines the shape and style of the mortar joint. See **Figure 14-66.** Masons also sometimes use joint rakers to remove a portion of the mortar when a raked mortar joint is desired.

A brick hammer is used to drive nails, strike chisels, and break or chip masonry units. Chisels are used for cutting brick and block. Different sizes and shapes are available.

A mason's line is strung across a wall as a guide to keep courses level and walls straight.

Veneer wall: A masonry facing, such as a single thickness of brick, applied to a frame building wall.

Toe: The pointed end of a mason's trowel.

Heel: The wide end of a mason's trowel.

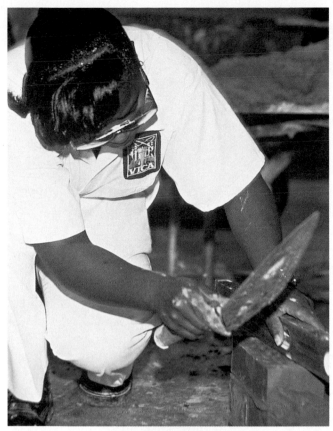

A

B

Figure 14-65. Masonry tools. A—Mason's trowel is used to handle mortar and strike off mortar joints. B—A mason's level is used to check masonry walls for plumb and level. (SkillsUSA)

The line is secured at each end by line holders. For more accurate and faster cutting of brick or block, masons may use a power-driven masonry saw.

Figure 14-66. A jointing tool shapes mortar joints before the mortar sets up. (SkillsUSA)

14.14.2 Masonry Materials

Bricks are structural units made to several sizes from clay or shale. This material is mixed with water and then dried in large kilns.

Brick comes in various sizes, some of them modular and some nonmodular. Modular sizes are based on 4″. This includes dimensions of one-half of 4″ as well as two or three times 4″. See **Figure 14-67.**

Sizes include allowance for the thickness of the mortar joint. Thus, the nominal size is smaller than the actual space the brick will occupy in a wall.

Bricks are engineered for various uses. The two general types are building or common and facing. Each of these types has different grades.

Building or common brick is a strong, general purpose brick intended for use where strength is more important than appearance. There are three grades:

- SW grade resists freezing and is used often for foundation courses or retaining walls.

- MW grade is used where there may be exposure to below-freezing temperatures, but in dry locations.

- NW grade is used to back up interior masonry.

Facing brick is used where appearance is important. In addition, the brick must be strong and durable. There are three grades:

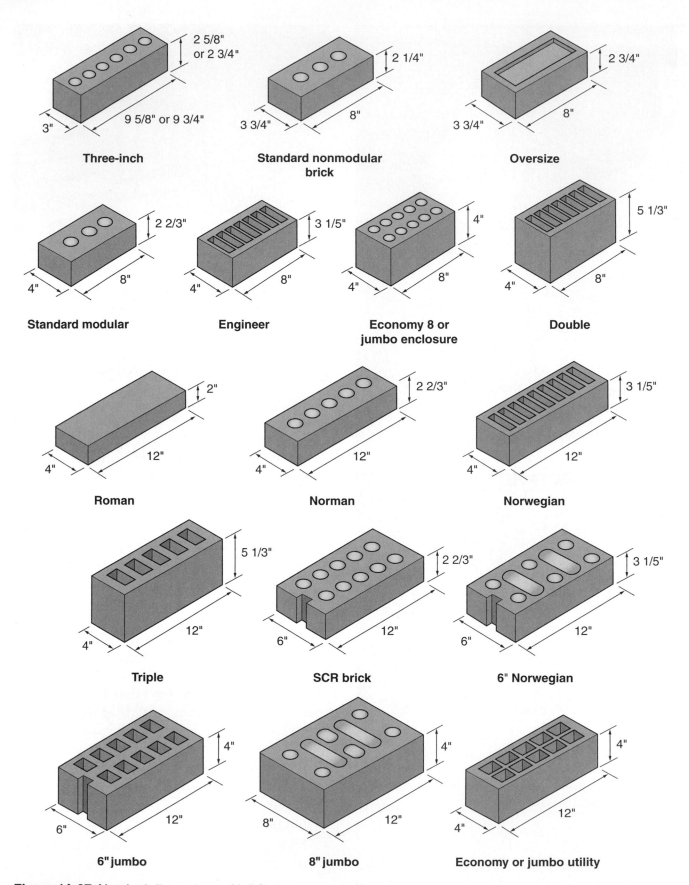

Figure 14-67. Nominal dimensions of brick.

- FBX is for general use in exposed interior or exterior walls or partitions. Color and size are uniform.

- FBS is for general use in exposed exterior and interior walls and partitions where wider color variations and sizes are permitted.

- FBA is used to produce architectural effects from the lack of uniformity in size, color, and texture.

Mortar is mostly portland cement with the addition of hydrated lime and sand. It is designed to bond bricks and blocks into a strong, waterproof wall. Various types have been developed to meet the needs of different situations. See **Figure 14-68.**

Corrosion-resistant metal ties are used to secure the veneer to the framework. These are usually horizontally spaced 32″ apart and vertically spaced 15″ apart. Where other than wood

Mortar type	Parts by volume of portland cement or portland blast furnace slag cement	Parts by volume of masonry cement	Parts by volume of hydrated lime or lime putty	Aggregate, measured in a damp, loose condition
M	1 1	1 (Type II) —	— 1/4	Not less than 2 1/2 and not more than 3 times the sum of the volumes of the cements and lime used.
S	1/2 1	1 (Type II) —	— Over 1/4 to 1/2	
N	— 1	1 (Type II) —	— Over 1/2 to 1 1/4	
O	— 1	1 (Type I or II) —	— Over 1 1/4 to 1 1/2	
K	1	—	Over 2 1/2 to 4	

A

ASTM mortar type designation	Construction suitability
M	Masonry subjected to high compressive loads, severe frost action, or high lateral loads from earth pressures, hurricane winds, or earthquakes. Structures below grade, manholes, and catch basins.
S	Structures requiring high flexural bond strength, but subject only to normal compressive loads.
N	General use in above grade masonry. Residential basement construction, interior walls and partitions. Concrete masonry veneers applied to frame construction.
O	Non-load-bearing walls and partitions. Solid load bearing masonry of allowable compressive strength not exceeding 100 psi.
K	Interior non-load-bearing partitions where low compressive and bond strengths are permitted by building codes.

B

Figure 14-68. A—This chart shows the materials in mortar and their proportions. B—Uses of different types of mortar.

Mortar: A combination of cement, sand, lime, and water used to bind masonry blocks.

sheathing is used, the ties should be secured to the studs.

Weep holes are small openings in the bottom course. They permit the escape of any water or moisture that may penetrate the wall. Space weep holes about 4′ apart.

Select a type of brick suitable for exposure to the weather. Such brick is hard and low in water absorption. Sandstone and limestone are most commonly used for stone veneer. These materials widely vary in quality. Be sure to select materials locally known to be durable. Procedure for laying brick and stone are essentially the same as laying concrete block. **Figure 14-69** shows typical patterns for both brick and stone veneer.

14.15 Shutters

Some architectural designs may require the installation of shutters at the sides of window units. These consist of frame assemblies with solid panels or louvers. In the early days of our country, shutters served an important practical function. They could be closed over the window to protect the glass. Closing and locking the shutters also provided some security to the inhabitants.

Today, except where hurricanes are likely to occur, shutters and blinds are decorative. They tend to extend the width of windows, stressing horizontal lines of the structure. Hinges are seldom used. Shutters are usually attached to the exterior wall with screws or other fasteners so they can be easily removed for painting and maintenance. Stock sizes include various heights to fit standard window units. Widths range from 14″ to 20″ in 2″ increments.

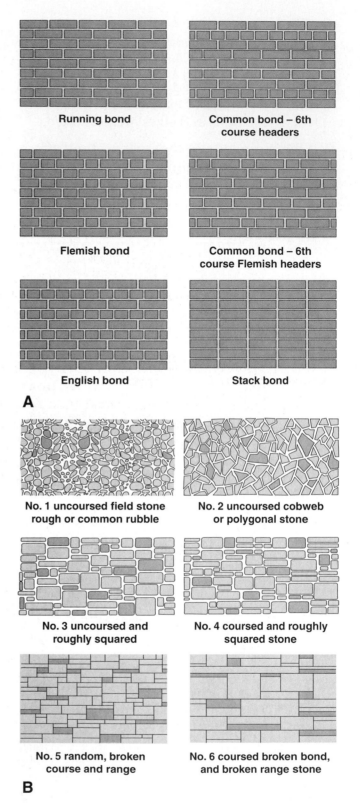

Running bond

Common bond – 6th course headers

Flemish bond

Common bond – 6th course Flemish headers

English bond

Stack bond

A

No. 1 uncoursed field stone rough or common rubble

No. 2 uncoursed cobweb or polygonal stone

No. 3 uncoursed and roughly squared

No. 4 coursed and roughly squared stone

No. 5 random, broken course and range

No. 6 coursed broken bond, and broken range stone

B

Figure 14-69. These are common patterns for brick and stone veneer siding. A—Basic structure bonds for brick. B—Common stone patterns.

Summary

The exterior finish of a structure includes installation of siding or other exterior wall finish materials, cornices, and the trim around windows and doors. A cornice is the finished connection between the exterior wall and the roof. It includes the fascia board, lookouts, soffit, and trim moulding. A box cornice conceals the rafter ends with a fascia board and soffits. An open cornice does not use soffits and may not have a fascia board covering the rafter tails. If the rafters do not extend past the sidewall, the cornice is called a snub cornice. Prefabricated, metal soffits are extensively used. Sheathing is nailed to wall studs to provide structural stiffening and a solid base for the application of siding. Siding types include horizontal, vertical, and shingles. Horizontal siding—wood, hardboard, aluminum, or vinyl—is overlapped for weathertightness and has an attractive three-dimensional appearance due to its beveled cross section. Shingles are traditionally made of rot-resistant wood, but are also available in a fiber-cement material. These are applied with methods similar to those for shingling roofs. Vertical siding may be individual boards or large sheets of material such as plywood. Where large sheets are used, narrow wooden strips called battens are often applied over the vertical joints to seal them. Exterior walls may be covered with a layer of cement-like material called stucco or with an applied exterior insulation and finish system (EIFS). An EIFS application can mimic stucco, stone, or other materials. Some buildings use a thin layer of brick or stone, called a veneer, as the exterior finish.

Test Your Knowledge

Answer the following questions on a separate piece of paper. Do not write in this book.

1. Which one of the following is *not* considered part of the exterior finish of a house?
   A. An entry door.
   B. Ceiling of a porch or breezeway.
   C. Cornice.
   D. Exterior wall coverings.
2. In cornice construction, the strip nailed to the wall to support the lookouts and soffit is called a(n) _____.
3. Some prefabricated soffit systems use steel channels to provide rigidity to the soffit material, thus eliminating the need for _____.
4. Name the three parts of a metal soffit system.
5. In areas not subject to wind-driven rain, head flashing can usually be omitted over an opening if the vertical distance to the soffit is less than _____ the width of the overhang.
6. Of the various types of horizontal wood siding, _____ siding usually has the most strength and tightest joints.
7. The best grade of solid wood siding is made from _____-grain material.
8. Plain beveled siding in a 10″ width should be lapped about _____ inches.
9. Wood corner boards are installed _____ (before, after) the horizontal siding units are fastened in place.
10. When plain square-edged boards, spaced about 1/2″ apart, are used for vertical siding, the joint between boards is usually covered with a(n) _____.
11. Standard wood shingle lengths for sidewall coverage are 16″, 18″, and _____ inches.
12. Shingle siding is often *double coursed*. Explain this term.
13. When using plywood siding over an unsheathed wall, the thickness should not be less than _____ inches when studs are spaced 16″ O.C.
14. All edges of plywood siding should be sealed with high grade _____, _____, or _____.
15. For plywood siding, it is important that a vapor barrier be included on the (warm, cold) side of the wall.
16. The most common thickness for hardboard siding material is _____″.
17. Nails used to secure hardboard siding must be _____ or otherwise rustproof.
18. Aluminum siding can be applied over _____, _____, _____, and other structurally sound surfaces.
19. Vinyl siding is made from rigid _____.
20. *True or False?* When re-siding with aluminum or vinyl siding, the old siding must be removed.
21. Aluminum and vinyl siding panels are sold by the _____.

22. Stucco is usually applied in _____ coats or layers. Name them.
23. List the four ways in which water is diverted from an EIFS installation.
24. The weight of the veneer in a veneer wall is supported by the _____.
25. Small openings located in the bottom course of brick or stone veneer construction that permit moisture to escape are called _____.

Curricular Connections

Mathematics. Using the pricing information gathered from a lumber yard or home improvement center, estimate the cost of materials to cover the roof and walls of a single story, 40′ × 20′ house with a 4-in-12 gable roof. Select a mid-priced vinyl siding and asphalt roof shingles.

Science. Do research to compare the properties of five major types of siding: wood, wood composite (fiberboard), aluminum, vinyl, and cement fiber. Compare impact resistance, fire resistance, useful life, maintenance needed, and cost per square foot. Create a table showing the results of your research.

Outside Assignments

1. Study the cornice work on your home or some other residence in your neighborhood. Prepare a detail drawing (use a scale of 1″ = 1′-0″) showing a typical cross section. Since most of the construction will likely be hidden, you will need to develop your own structural design and select sizes for some of the parts. Study the details shown in various architectural plans and those in **Figure 14-1**. Be sure to include the sizes of all materials.

2. Visit a building site in your neighborhood where the exterior wall finish is being applied. Be sure to get permission from the builder or head carpenter before entering the site. Observe the methods and materials being used. Make notes about: type of framing and stud spacing; type of sheathing; type of flashing; type, size, and quality of siding material; layout methods; how units are cut and fitted; nails and nailing patterns; and special joint treatment. Carefully organize your notes and make an oral report to your class.

An early adobe home. Walls are built of thick adobe brick, then plastered over with adobe.

484

Section **4**

Finishing

PhotoDisc

Thermal and Sound Insulation

Learning Objectives

After studying this chapter, you will be able to:

- Summarize the principles of conduction, convection, and radiation.
- Define technical terms relating to thermal and acoustical properties of construction materials.
- Interpret thermal ratings charts.
- Describe the types of insulation.
- Select appropriate areas for insulation in a given structure.
- Explain the principle of condensation.
- Describe methods of controlling moisture problems.
- List general procedures for installing batt and blanket, fill, and rigid insulation.
- Define acoustical terms.
- Describe methods of construction that raise STC ratings in desired areas.

Technical Vocabulary

Acoustical plaster
Batt insulation
British thermal unit (Btu)
Coefficient of thermal conductivity (k)
Conductance (C)
Conduction
Convection
Convection currents
Crawl space
Decibels (dB)
Degree day
Dewpoint
Draft excluder
Expanded polystyrene foam (EPS)
Extruded polystyrene foam (XPS)
Flexible insulation
Foamed-in-place insulation
Frequency
Impact sounds
Infiltration
Insulation
Insulation board
Loose-fill insulation
Masking sounds
Noise
Noise reduction coefficient (NRC)
Permissible exposure limit (PEL)
Polyisocyanurate foam (PIR)
Radiation
Reflective insulation
Resistance (R)
Rigid insulation
Sound transmission class (STC)
Sound transmission loss (STL)
Threshold of pain
Total heat transmission (U)
Vapor barrier
Water vapor

In construction, *insulation* involves materials that do not readily transmit energy in the form of heat, electricity, or sound. This chapter primarily deals with its use to prevent the transfer of heat either into or out of a building. To a lesser extent, the use of insulation to prevent sound transmission is also discussed. Acoustical

---

Insulation: Any material high in resistance to heat, sound, or electrical transmission.

treatments and sound control use many of the same materials as for thermal insulation. They are, therefore, both described in this chapter.

Insulation requirements of homes and other buildings have radically changed since the energy crisis of 1973, which first made most people aware of the need for energy conservation. As energy costs have increased, so has the amount of insulation being placed in walls, floors, and ceilings, **Figure 15-1.** At the same time, the insulating qualities of doors and windows have been upgraded to hold down energy costs. Today, ductwork is likely to be insulated if it is not placed in a heated space.

Insulation is necessary in buildings where the temperature of the interior space must be controlled. In cold climates, the major concern is to retain heat in the building. In warmer regions, insulation keeps external heat from entering the building.

A wide range of insulation materials is available to fill energy-efficiency requirements. The materials are engineered for efficient installation and come in convenient packages that are easy to handle and store.

15.1 Building Sequence

While carpenters are completing the exterior finish of a structure, other tradespeople are installing mechanical systems on the inside. These systems must be completed before interior walls are insulated and closed in with drywall or plaster. Ductwork for heating and air conditioning is installed in the floors, walls, and ceilings. The electrician installs wiring for electrical circuits, attaching conduit or cable to the building's framework. The boxes that enclose connections for convenience outlets, switches, and lighting fixtures are also located and attached, **Figure 15-2.**

The plumber installs the water supply piping and drainage/waste/venting (DWV). These pipes supply fresh water and drain away wastewater. Built-in plumbing fixtures are also installed at this time. They must be carefully covered with building paper or polyethylene

Figure 15-2. Before thermal and sound insulation materials are installed, other construction trades must make their installations. In this commercial structure, the electrical conduit and outlet boxes for a series of offices have been installed.

Figure 15-1. Thick batts of fiberglass insulation are being placed between the ceiling joists to reduce a home's heat loss through the attic. (Owens-Corning)

film to protect them during the interior finish operations.

During the heating and plumbing rough-in, tradespeople often need to cut through structural framework. The carpenter should check this work. Framing members may have been seriously weakened and must be reinforced before interior walls are covered.

15.2 How Heat Is Transmitted

Although carpenters are seldom required to design buildings or figure heat losses, they should have some knowledge and understanding of the theory of heat transfer and the factors involved.

Heat seeks a balance with surrounding areas. When the inside temperature is controlled within a given comfort range, there is some flow of heat. Heat moves from the inside to the outside in winter and from the outside to the inside during hot summer weather.

Heat is transferred through walls, floors, ceilings, windows, and doors at a rate directly related to the difference in temperature and the resistance to heat flow provided by intervening

materials. Heat moves from one place to another by one or more of three methods: conduction, convection, and radiation. See **Figure 15-3.** Actually, heat transmission through walls, ceilings, and floors is a result of all three transmission methods. In addition to this, some heat is moved by convection as air flows through cracks around doors, windows, and other openings in the structure.

15.2.1 Conduction

Conduction is transmission of heat from one molecule to another within a material or from one material to another when they are held in direct contact. Dense materials, such as metal or stone, more rapidly conduct heat than porous materials, such as wood and fiber products. Any material conducts some heat when a temperature difference exists between its surfaces.

15.2.2 Convection

Convection is the transfer of heat by another agent, such as air or water. In large spaces, molecules of air can carry heat from warm surfaces to cold surfaces. When air is heated, it becomes lighter and rises. Thus, airflow (called

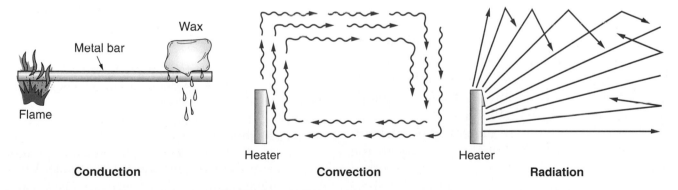

Figure 15-3. How heat is transferred. Conduction takes place only in solids. Convection uses air or water to move heat to cooler space. Radiation makes use of wave motion.

Conduction: The movement of heat through substances from one end to the other.

Convection: Movement of heat through liquids and gases as a result of heated material rising and being replaced by cooler material.

convection currents) is created within the space. Air is a good insulator when it is confined to a small space (cavity) in which convection flow is limited or absent. In walls and ceilings, trapped air restricts convection currents and will reduce the flow of heat.

15.2.3 Radiation

Heat can be transmitted by wave motion in about the same manner as light. This process is called *radiation* because it represents radiant energy.

Heat received from the sun is a good example of radiant heat. The waves do not heat the space through which they move. However, when they come in contact with a conducting surface, a part of the energy is absorbed, while some may be reflected.

Effective resistance to radiation heating comes about through reflection. Shiny surfaces, such as aluminum foil, are often used to provide this type of insulation.

15.3 Thermal Insulation

All building materials resist the flow of heat (mainly via conduction) to some degree. The amount of resistance depends on their porosity or density. Air is an excellent insulator when confined to the tiny spaces or cells inside of a porous material. Dense materials such as masonry or glass contain few, if any, air spaces and are poor thermal insulators.

Fibrous materials are generally good insulators, not only because of the porosity in the fibers themselves, but also because of the thin film of air that surrounds each individual fiber.

Commercial insulation materials are made of glass fibers, glass foam, mineral fibers, organic fibers, and foamed plastic. A good insulation material should be fireproof, vermin proof, and moistureproof. In addition, it should be resistant to any physical change that would reduce its effectiveness in preventing heat flow. Selection is based on initial cost, effectiveness, durability, and the adaptation of the insulating material's form to that of the construction and installation methods.

15.4 Terminology

A *coefficient* is a number that serves as a measure of a property—in this case, the ability to transfer heat. The thermal properties of common building materials and insulation materials are known or can be accurately measured. Heat transmission (the amount of heat flow) through any combination of these materials can be calculated. First, it is necessary to understand certain terms.

A *British thermal unit (Btu)* is a measure of heat. One Btu is the amount of heat needed to raise the temperature of 1 lb. of water 1°F. Btu = weight × temp. diff.

The *coefficient of thermal conductivity (k)* is the amount of heat (Btu) transferred in one hour through 1 sq. ft. of a given material that is 1″ thick and has a temperature difference between its surfaces of 1°F.

The *conductance (C)* of a material is the amount of heat (Btu) that will flow through the material in one hour per sq. ft. of surface with 1°F of temperature difference. The thickness of the material is not a factor. For example, the C-value for an average hollow concrete block is .53.

The *resistance (R)* is the reciprocal (opposite) of conductivity or conductance. Most insulation

---

Convection currents: Airflow produced when heated air becomes lighter and rises.

Radiation: Transmission of heat by wave motion, much like light movement.

British thermal unit (Btu): The amount of heat needed to raise the temperature of 1 lb. of water 1°F.

Coefficient of thermal conductivity (k): The amount of heat (Btu) transferred in one hour through 1 sq. ft. of a given material that is 1″ thick and has a temperature difference between its surfaces of 1°F.

Conductance (C): The amount of heat (Btu) that will flow through a material in one hour per sq. ft. of surface with 1°F of temperature difference.

Resistance (R): The reciprocal (opposite) of conductivity or conductance.

has an R-value listed. A good insulation material has a high R-value.

$$R = \frac{1}{k} \ or \ \frac{1}{C}$$

The *total heat transmission (U)* is measured in Btu per sq. ft. per hour with 1°F temperature difference for a structure (wall, ceiling, and floor). The structure may consist of several materials or spaces. A standard frame wall with composition sheathing, gypsum lath and plaster, and a 1″ blanket insulation has a U-value of about .11. To calculate the U-value where the R-values are known, apply this formula:

$$U = \frac{1}{R_1 + R_2 + R_3 + \dots R_n}$$

The harshness of climate is measured in degree days. The higher the number of degree days, the colder the climate and the more insulation needed. A *degree day* is the product of one day and the number of °F that the mean temperature is below 65°F. Figures are usually quoted for a full year. Degree days are used by a heating engineer to determine the design and size of the heating system. For example, if the high is 60°F and the low is 30°F:

$$\text{Degree day} = 65 - \frac{60 + 30}{2}$$
$$= 65 - \frac{90}{2}$$
$$= 65 - 45$$
$$= 20$$

Figure 15-4 shows how insulation reduces the U-value for a conventional frame wall. Note that a 5 1/2″ thick blanket reduces the U-value from .29 to .053. This is about an 82% reduction.

Total heat transmission (U): Represented in Btu per sq. ft. per hour with 1°F temperature difference for a structure (wall, ceiling, and floor) that may consist of several materials or spaces.

Degree day: Method of measuring the harshness of climate for insulation and heating purposes. A degree day is the product of one day and the number of degrees the mean temperature is below 65°F.

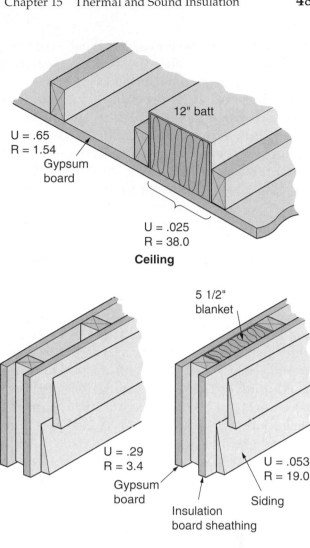

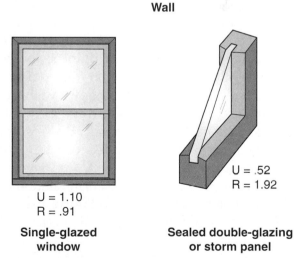

Figure 15-4. Insulation reduces flow of heat. Double-glazed windows enclose trapped air, which is a good insulator.

Actually, the U-value for the total wall structure is slightly higher because the wood studs have a lower R-value than the blanket insulation.

R-values of wall structures with various types and amounts of insulation are shown in **Figure 15-5.** The R-value provides a convenient measure for comparing heat loss in materials and structural designs. However, to determine the total heat loss (or gain) through a wall, ceiling, or floor, you must convert R-values to U-values. The total of these U-values is needed to calculate the size of heating and cooling equipment. Refer to Appendix B, **Technical Information.**

It is important to understand that heat transmission decreases as the insulation thickness is increased, but not in a direct relationship. This can be noted through a study of U- and R-values for various materials and structures. For example, in a frame wall with a U-value of .24, the addition of 1″ of insulation reduces the heat loss by 46%, to U-.13. A second inch of insulation reduces the loss by 16% to U-.09. A third inch reduces the loss by 10% to U-.065. Additional thicknesses continue to lower the U-value, but at a still-lower percentage. At some point, it becomes useless to add more insulation.

Windows provide another example of how U-values decrease. The heat loss through a sash with a single pane of glass is reduced from U-1.10 to about U-.52 when a second pane is added. This provides a reduction of about 50%. A third pane (triple glazing) reduces the heat loss to U-.35 and a fourth pane (quad glazing) results in a U-value of .27. Windows with three and four layers of glass are very expensive. However, in northern climates, the resulting lower heating costs and added comfort often justify this extra expense.

R-values for commonly used insulation and building materials are listed in **Figure 15-6.** The original source of most data on this subject is the American Society of Heating, Refrigerating, and Air Conditioning Engineers (ASHRAE). R-values can be converted to U-values by calculating the reciprocal (dividing the value into 1). Values of R-18 or higher in residential wall construction can be obtained by:

- Using 2 × 6 studs, which provide a thicker wall cavity for insulation, **Figure 15-7.**

Uninsulated 2 x 4 stud wall

Air films	R = .9
3/4″ wood exterior siding	1.0
1/2″ insulation board	1.2
Air space	1.2
Vapor barrier	0
1/2″ gypsum board	.5
	4.8 total R

2 x 4 stud wall with batt insulation

Air films	R = .9
3/4″ wood exterior siding	1.0
1/2″ insulation board	1.2
3 1/2″ batt or blanket insulation	11.0
Vapor barrier	0
1/2″ gypsum board	.5
	14.6 total R

2 x 4 stud wall with rigid board

Air films	R = .9
3/4″ wood exterior siding	1.0
1″ polystyrene rigid board	5.0
3 1/2″ batt or blanket insulation	11.0
Vapor barrier	0
1/2″ gypsum board	.5
	18.4 total R

Improved insulated 2 x 4 stud wall

Air films	R = .9
3/4″ wood exterior siding	1.0
3/4″ insulation board	2.0
3 5/8″ batt insulation	13.0
Vapor barrier	0
5/8″ urethane insulation board	5.0
1/2″ gypsum board	.5
	22.4 total R

2 x 6 insulated stud wall

Air films	R = .9
3/4″ wood exterior siding	1.0
3/4″ insulation board	2.0
5 1/2″ insulating blanket	19.0
Vapor barrier	0
1/2″ gypsum board	.5
	23.4 total R

Improved 2 x 6 insulated stud wall

Air films	R = .9
3/4″ wood exterior siding	1.0
3/4″ insulation board	2.0
5 1/2″ batt or blanket insulation	19.0
Vapor barrier	0
5/8″ urethane insulation	5.0
5/8″ gypsum board	.6
	28.5 total R

Figure 15-5. Types of wall construction and their R-values. Materials are listed in order from the outside in. "Air films" refer to the inside and outside film of stagnant air that forms on any surface. (Iowa Energy Policy Council)

Material	Kind	Insulation value
Masonry	Concrete, sand, and gravel, 1"	R-.08
	Concrete blocks (three core)	
	Sand and gravel aggregate, 4"	R-.71
	Sand and gravel aggregate, 8"	R-1.11
	Lightweight aggregate, 4"	R-1.50
	Lightweight aggregate, 8"	R-2.00
	Brick	
	Face, 4"	R-.44
	Common, 4"	R-.80
	Stone, lime, sand, 1"	R-.08
	Stucco, 1"	R-.20
Wood	Fir, pine, other softwoods, 3/4"	R-.94
	Fir, pine, other softwoods, 1 1/2"	R-1.89
	Fir, pine, other softwoods, 3 1/2"	R-4.35
	Maple, oak, other hardwoods, 1"	R-.91
Manufactured wood products	Plywood, softwood, 1/4"	R-.31
	Plywood, softwood, 1/2"	R-.62
	Plywood, softwood, 5/8"	R-.78
	Plywood, softwood, 3/4"	R-.93
	Hardboard, tempered, 1/4"	R-.25
	Hardboard, underlayment, 1/4"	R-.31
	Particleboard, underlayment, 5/8"	R-.82
	Mineral fiber, 1/4"	R-.21
	Gypsum board, 1/2"	R-.45
	Gypsum board, 5/8"	R-.56
	Insulation board sheathing, 1/2"	R-1.32
	Insulation board sheathing, 25/32"	R-2.06
Siding and roofing	Building paper, permeable felt, 15 lb.	R-.06
	Wood bevel siding, 1/2"	R-.81
	Wood bevel siding, 3/4"	R-1.05
	Aluminum, hollow-back siding	R-.61
	Wood siding shingles, 7 1/2" exp.	R-.87
	Wood roofing shingles, standard	R-.94
	Asphalt roofing shingles	R-.44
Insulation	Cellular or foam glass, 1"	R-2.50
	Glass fiber, batt, 1"	R-3.13
	Expanded perlite, 1"	R-2.78
	Expanded polystyrene bead board, 1"	R-3.85
	Expanded polystyrene extruded smooth, 1"	R-5.00
	Expanded polyurethane, 1"	R-7.00
	Mineral fiber with binder, 1"	R-3.45
Inside finish	Cement plaster, sand aggregate, 1"	R-.20
	Gypsum plaster, light wt. aggregate, 1/2"	R-.32
	Hardwood finished floor, 3/4"	R-.68
	Vinyl floor, 1/8"	R-.05
	Carpet and fibrous pad	R-2.08
Glass	(see Chapter 13 Windows and Exterior Doors)	

Figure 15-6. R-values for commonly used construction and insulation materials.

Figure 15-7. One of the easiest ways to secure R-18+ values in outside walls is to use 2 × 6 studs and install insulation in the stud space.

- Using 2 × 4 studs sheathed with thick, rigid insulation panels made from foamed polystyrene plastic.
- Using a double 2 × 4 frame to provide a thicker wall cavity.

15.5 How Much Insulation?

Insulation is required in any building where a temperature above or below the outdoor temperature must be maintained inside of the building. Comfort, health, and economy are the three considerations for thermal insulation.

Comfortable and healthy indoor temperatures depend not only on the temperature of the air, but also on the temperature of the wall, ceiling, and floor surfaces. It is possible to heat the air in a room to the correct comfort level and still have cold room surfaces. The human body will lose heat by radiation (or conduction, if there is direct contact) to these colder surfaces.

The amount of insulation for a given structure must be based not only on comfort standards, but also on factors such as insulation costs (material and labor), probable fuel costs in the future, and local climatic conditions. **Figure 15-8** shows a map of the continental United States and a chart of recommended insulation for each area according to winter/summer temperatures. In warmer climates, smaller amounts of insulation are needed—not for protection against cold, but to keep out heat during the summer.

15.6 Types of Insulation

Thermal insulation is made in many forms, **Figure 15-9.** It may be grouped into four broad classifications:
- Rigid
- Flexible
- Loose fill
- Reflective

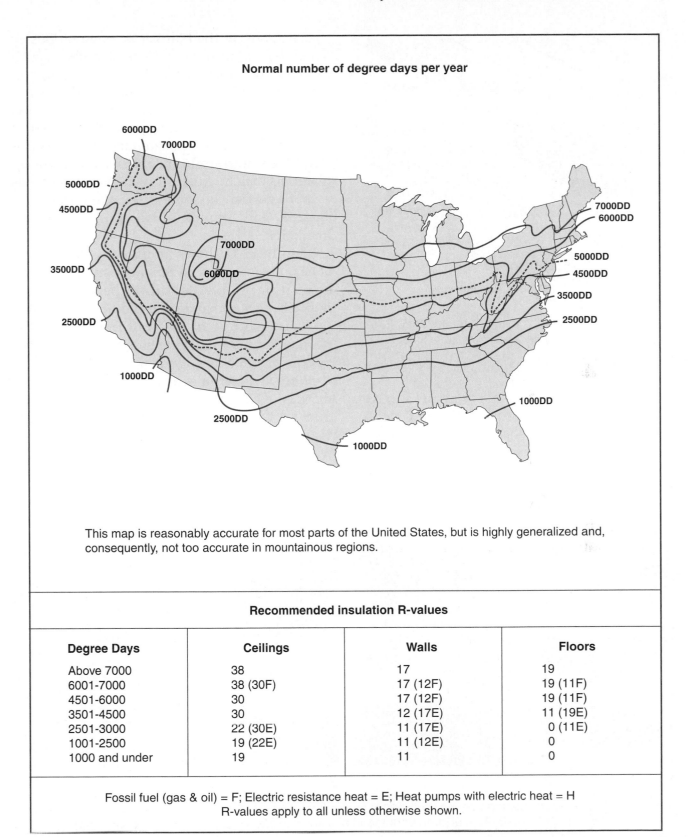

Normal number of degree days per year

This map is reasonably accurate for most parts of the United States, but is highly generalized and, consequently, not too accurate in mountainous regions.

Recommended insulation R-values

Degree Days	Ceilings	Walls	Floors
Above 7000	38	17	19
6001-7000	38 (30F)	17 (12F)	19 (11F)
4501-6000	30	17 (12F)	19 (11F)
3501-4500	30	12 (17E)	11 (19E)
2501-3000	22 (30E)	11 (17E)	0 (11E)
1001-2500	19 (22E)	11 (12E)	0
1000 and under	19	11	0

Fossil fuel (gas & oil) = F; Electric resistance heat = E; Heat pumps with electric heat = H
R-values apply to all unless otherwise shown.

Figure 15-8. Temperature zones for different parts of the United States, as expressed in degree days per year. Use 9000 degree days for Alaska and 0 degree days for Hawaii. The R-values listed below the map are recommended by the U.S. Department of Energy (DOE). (Owens-Corning)

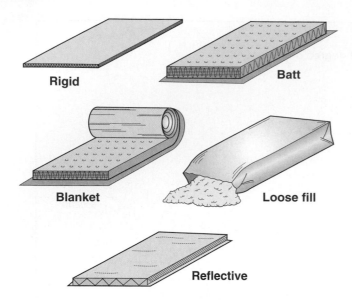

Figure 15-9. Insulation is made in these basic forms. Batt and blanket are both types of flexible insulation.

Flexible insulation is manufactured in two types: batt and blanket (quilt). Blanket insulation is generally furnished in rolls or strips of convenient length and in various widths suited to standard stud and joist spacing, **Figure 15-10.** It comes in thicknesses of 3/4″–12″.

The body of the blanket is made of loosely felted mats of mineral or vegetable fibers, such as glass fiber, rock wool, slag, wood fiber, and cotton. Organic fiber mats are usually chemically treated to make them resistant to fire, decay, insects, and vermin. Blanket insulation is often enclosed in paper with tabs on the side for attachment to framing members. The covering materials may be treated on one side to serve as a vapor barrier. In some cases, the covering is surfaced with aluminum foil or other reflective insulation.

Figure 15-11 shows an unfaced blanket that is easy to install between wall studs. Friction holds it in place. After all of the insulation is in place, a vapor barrier should be applied over the entire surface of the wall frame, since the blanket is unfaced. It prevents vapor inside the building from entering the insulation.

Batt insulation is made of the same fibrous material as blankets. Thickness can be greater in this form and may range from 3 1/2″–12″, **Figure 15-12.** Batts are generally available in widths of 15″ and 23″ and in 24″ and 48″ lengths. They are available with a single flanged cover or with both sides uncovered.

A   B

Figure 15-10. Blanket insulation may be kraft-paper faced, foil faced, or unfaced. A—This R-25 unfaced blanket insulation is wrapped in film. B—Installation of unfaced blanket insulation in an attic. (Owens-Corning)

Flexible insulation: Insulation made of loosely felted mats of mineral or vegetable fibers.

Batt insulation: Insulation made of the same fibrous material as blankets, but cut into shorter segments.

Figure 15-11. Unfaced blanket insulation must snugly fit between studs as it is held there by friction. (Owens-Corning)

Figure 15-12. Batts are sold in various thicknesses. Thicker batts are designed to give high insulation values for ceilings and attics. (Owens-Corning)

Loose-fill insulation is made from such materials as rock wool, glass, slag wool, wood fibers, shredded redwood bark, shredded paper, granulated cork, ground or *macerated* (softened by soaking in a liquid) wood pulp products, vermiculite, perlite, powdered gypsum, sawdust, and wood shavings. It is used in bulk form and supplied in bags or bales, **Figure 15-13.** Loose insulation may be poured or blown. It is commonly used to insulate spaces between studs or to build up any desired thickness on a flat surface, such as above ceilings. One of the chief advantages of this type is that when insulating an older structure, only a few boards need to be removed in order to blow the material into the walls. **Figure 15-14** shows typical coverage

of loose fiberglass insulation blown into walls and attics.

Rigid insulation is manufactured in sheets and other rigid forms. These foam panels have higher R-values per inch than other forms of insulation. They are also more expensive. Their durability and rigidity allow them to be used in conditions where other types of insulation would

Figure 15-13. Loose insulation is often used over ceilings in attic areas. Coverage and R-values are listed on the bag.

Loose-fill insulation: Insulation used in bulk form and supplied in bags or bales.
Rigid insulation: Insulation manufactured in rigid panels of expanded or extruded polystyrene foam.

Sidewall coverage information				
Nominal R-value	**Thickness (framing timber)**	**Density**	**Minimum weight per sq. ft.**	**Maximum coverage per bag****
To obtain a thermal resistance (R) of:	Installed insulation should not be less than: (inches)	Pounds per cubic foot:	Weight per square foot of installed insulation should not be less than: (lbs.)	Contents of bag should not cover more than: (square feet)
R-14	3.5" (2 x 4)	1.8	.525	55
R-22	5.5" (2 x 6)		.825	35
R-29	7.25" (2 x 8)		1.088	26
R-37	9.25" (2 x 10)		1.387	21
R-15	3.5" (2 x 4)	2.3	.670	43
R-23	5.5" (2 x 6)		1.054	27
R-31	7.25" (2 x 8)		1.389	21
R-39	9.25" (2 x 10)		1.773	16

Attic/open blow coverage information				
R-value	**Bags per 1000 sq. ft.***	**Maximum sq. ft. per bag***	**Minimum weight per sq. ft.**	**Minimum thickness**
To obtain a thermal resistance (R) of:	Bags per 1000 square feet of net area:	Contents of bag should not cover more than: (square feet)	Weight per square foot of installed insulation should not be less than: (lbs.)	Installed insulation should not be less than: (inches)
R-60	43.5	23.0	1.307	20.50"
R-50	35.2	28.4	1.057	17.50"
R-44	31.0	32.3	.928	15.75"
R-38	26.7	37.4	.803	14.00"
R-30	20.2	49.5	.606	11.00"
R-26	18.2	54.8	.547	10.00"
R-22	15.0	66.5	.451	8.50"
R-19	13.1	76.1	.394	7.50"
R-13	9.3	107.5	.279	5.50"
R-11	7.9	126.6	.237	4.75"

Coverages per bag do not include framing members, which will increase coverage, depending on 16″ or 24″ O.C.
** **For sidewall applications...** *To compensate for wall framing, the net coverage per bag should be increased. To calculate for conventional wood stud framing, when framing is 16″ O.C. multiply coverage value shown by 1.14. When framing is 24″ O.C. multiply coverage value shown by 1.11.*

Figure 15-14. These R-values and coverage charts are for a blown-in type of insulation. (Ark Seal, Inc.)

not be suitable. The uses and properties of rigid foam insulation are covered in **Figure 15-15.**

Expanded polystyrene foam (EPS) is similar to the material used for disposable coffee cups. It is made from a petroleum by-product and recycled polystyrene and is impermeable to moisture. Pentane is the blowing agent used to expand the foam. It does not deplete the ozone layer.

Extruded polystyrene foam (XPS) is the same plastic as EPS, but a different blowing

---

Expanded polystyrene foam (EPS): Material made from a petroleum by-product and recycled polystyrene; impermeable to moisture. Pentane, the blowing agent used to expand the foam, does not deplete the ozone layer.

Extruded polystyrene foam (XPS): The same plastic as EPS, but uses a different blowing agent that slightly depletes ozone. It is also impervious to moisture and has a higher R-value than EPS.

Properties/uses of rigid insulation			
	Extruded polystyrene (XPS)	**Expanded polystyrene (EPS)**	**Polyisocyanurate (PIR)**
R-value	5 per inch	3.6 to 4.2 per inch	6.6 to 7 per inch
Permeability	Almost impervious to moisture	Voids between beads absorb moisture	Least resistant to moisture penetration
Insect infestation	Very low resistance (borate treatment deters nesting; not infestation)	Very low resistance (borate treatment with same results as extruded polystyrene)	Same as for XPS and EPS
UV exposure	Loses thickness and R-value	Loses thickness and R-value	Loses thickness and R-value
When faced with foil, polyethylene or kraft paper	Reduces handling damage; improves surface cohesion for nailing; increases "perm" rating; gives bonding surface for adhesives; protects against UV degradation	The same as for XPS	Reduces handling damage
Fiberglass facing	NA	NA	Suitable for EIFS or stucco underlayment
Organic asphalt	NA	NA	Roofing applications
Adhesive compatibility	Petroleum-based solvents dissolve XPS	Petroleum-based solvents dissolve EPS	Petroleum-based solvents cause no harm; polystyrene adhesives may not bond to PIR
Fire resistance	Will soften at 165°F Melt at 200°F Most codes require 1/2" drywall covering	Same as XPS	Thermoset will not melt, but will burn giving off asphyxiants; requires drywall covering
Applications			
Under stucco	Yes	Yes	Yes
Foundations	Yes	Yes	No
Roof insulation	Yes, under plywood or oriented strand board	Yes, under plywood or oriented strand board	Yes
Wall sheathing	Yes	Yes	Yes
Underlayment for residing	Yes	Yes	Yes
Interior wall insulation	Yes	Yes	Yes

Figure 15-15. Rigid insulation has higher insulation values and is more rugged than other types. It has many uses in residential construction.

agent is used that depletes ozone. It is impervious to moisture and also has a higher R-value than EPS.

Polyisocyanurate foam (PIR) uses the same blowing agent as XPS and is moisture resistant. Its R-value is higher than other rigid foam products.

15.6.1 Reflective Insulation

Reflective insulation is usually a metal foil or foil-surfaced material. It differs from other insulating materials in that the number of reflecting surfaces, not the thickness of the material, determines its insulating value. In order to be effective, the metal foil must be exposed to an air space, preferably 3/4" or more in depth.

Aluminum foil is available in sheets or corrugations supported on paper. It is often mounted on the back of gypsum lath. One effective form of reflective insulation has multiple spaced sheets.

15.6.2 Other Types of Insulation

There are insulation materials that do not fit the other classifications. Some examples: a confetti-like material mixed with adhesive and sprayed on the surface to be insulated; foams that expand; multiple layers of corrugated paper; and lightweight aggregates like vermiculite and perlite used in plaster to reduce heat transmission.

Lightweight aggregates made from blast furnace slag, burned clay products, and cinders are commonly used in concrete and concrete blocks. They improve the insulating qualities of these materials.

15.7 Where to Insulate

Heated areas should be surrounded with insulation by placing it in the walls, ceiling, and floors, **Figure 15-16.** It is best to have the insulation as close to the heated space as possible. For example, if an attic is unused, the insulation

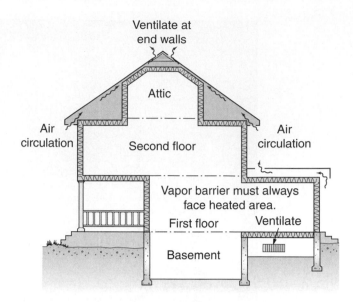

Figure 15-16. Where to insulate residential structures. In many cases, basement walls and crawl space walls are also insulated.

should be placed in the attic floor rather than in the roof structure.

If the attic space or certain portions of it must be heated, insulate the walls and overhead. If the insulation is placed between the rafters, be sure to a allow space between the insulation and sheathing for free air circulation. The floors of rooms above unheated garages or porches also require insulation for maximum comfort.

Where a basement is used as a living or recreation area, insulate the walls as shown in **Figure 15-17.** Walls may be framed with studs or furred out and then rigid insulation installed. Be sure to check local codes; some place restrictions on materials used in basement living spaces. Insulating is highly recommended in basements as an energy conservation measure. Not only does it save heat and provide comfort,

---

Polyisocyanurate foam (PIR): Moisture-resistant insulation material. It uses a blowing agent that slightly depletes the ozone. Its R-value is higher than other rigid foam products.

Reflective insulation: A metal foil or foil-surfaced material. The number of reflecting surfaces, not the thickness of the material, determines its insulating value.

Figure 15-17. Insulating a basement wall with R-11 unfaced fiberglass. A framework of 2 × 4 studs has been installed against the masonry wall and provides support for insulation and drywall or paneling. (Owens-Corning)

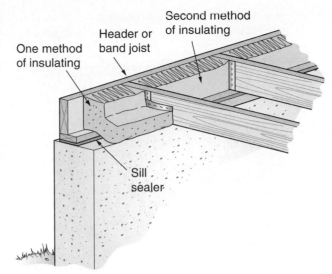

Figure 15-18. Insulate band joists atop basement walls. As you can see, there is little protection from loss of heat otherwise.

it has the added benefit of giving the basement better acoustical qualities.

Even if basement walls are not insulated, insulating material should be installed over the band joists and headers, **Figure 15-18.** This area has little protection against heat loss.

15.7.1 Insulating Crawl Spaces

Floors over unheated space directly above the ground require the same degree of insulation as walls in the same climate zone. This area is called a *crawl space.* It is usually enclosed by foundation walls. Such areas require ventilation and, therefore, have about the same temperature as the outside. The International Building Code sets requirements for crawl space ventilation:

"Crawl spaces must be cross-ventilated either through the foundation wall or through the rim joist. For each 150 sq. ft. of area, allow 1 sq. ft. of venting. Vent covering requirements:

- Perforated sheet metal plates not less than .070" thick.
- Expanded sheet metal plates not less than .047" thick.
- Cast iron grills or gratings.
- Extruded load-bearing vents.
- Hardware cloth of .035" wire or heavier.
- Corrosion-resistant wire mesh with the least dimension not greater than 1/8"."

Figure 15-19 shows crawl space insulation installed between the floor joists. Install a vapor barrier either on top of the insulation or between the rough and finish floor.

Covering the soil surface in the crawl space with 6 mil polyethylene plastic film or roll roofing weighing at least 55 lb. per square can control moisture coming up through the ground. Lay the material on the soil with edges overlapping at least 4". If the ground is rough, it is best to put down a layer of sand or fine gravel before placing the vapor barrier. Covers of this kind greatly restrict the evaporation of ground water, so less ventilation is needed.

The grade level beneath the building should be higher than the outside grade if water is likely to migrate inside the foundation wall. A ground cover in the crawl space is especially valuable where the water table is near the surface or if the soil has high capillarity (easily absorbs water).

Crawl space: The space below the first floor of a building directly above the ground.

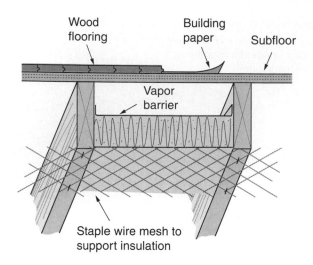

Figure 15-19. Basic construction and insulation requirements in a crawl space. Insulation can be held in place with bowed wire, chicken wire, or fishing line. (Manville Building Products, Inc.)

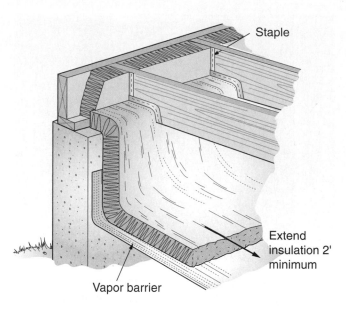

Figure 15-20. Proper insulation of crawl space walls and the ground when the crawl space is heated. Staple fiber blankets to wood members. Beads of mastic adhesive can be used to secure blankets to masonry walls.

Be sure the covering extends well up along the foundation wall.

Some crawl spaces may be heated along with other spaces in the structure. For example, closed crawl spaces are often used as a plenum (large duct) to distribute heated or cooled air throughout the structure. No venting is needed in such spaces. Insulation is installed along the inside of the foundation wall and the band joist. **Figure 15-20** shows the general method of installing insulation blankets. Extruded polystyrene is an alternative to batts or blanket insulation. It can provide a high R-value in this type of installation. Sections should be carefully cut and fitted to band joists and walls. Use a compatible mastic adhesive to secure the insulation. Lay loose, 2' wide sheets on top of the vapor barrier.

15.7.2 Insulating Slab Foundations

Many homes, as well as other structures, are built on concrete slab floors. Such floors should be insulated and a vapor barrier laid down to prevent heat loss and vapor penetration. Very little heat is lost into the ground under the central area of the floor. Only the perimeter needs to be insulated. The vapor barrier, however, must

be continuous under the entire floor. The insulation can be horizontally installed extending about 2' under the floor. It can also be vertically installed along the foundation walls, as shown in **Figure 15-21.**

15.7.3 Insulating Existing Foundations

Insulating walls and ceilings of existing structures is fairly simple. However, insulating floors or foundations of basementless structures can be a difficult task. Crawl spaces may not be easy to enter or may not provide room enough to work. Attaching a rigid *insulation board* to the outside surface of the foundation is usually the best solution.

Figure 15-22 shows extruded polystyrene applied to the exterior wall of a crawl space. Dig a trench along the wall. Clean the wall surface and repair cracks.

Insulation board: Rigid insulation attached to the outside surface of the foundation to insulate floors or foundations of basementless structures.

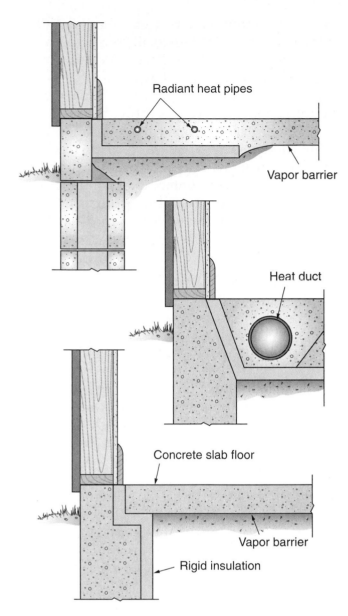

Figure 15-21. Slab floors need insulation along their perimeters. The vapor barrier should be continuous under the entire floor.

The insulation board and its protective cover will extend outward beyond the wall siding. To waterproof this joint, install a metal flashing. Extend the flashing at least 1″ upward behind the siding.

Cut the polystyrene panels to size and then attach them to the wall with 3/8″ beads of mastic. Follow the manufacturer's directions. Only a few minutes are required for the mastic to make an initial set.

Polystyrene panels must be protected against ultraviolet light, wear, and impact forces. An approved method is to cover the surface with special plaster made from cement, lime, glass fibers, and a water-resistant agent. This mixture is troweled onto the surface 1/8″–1/4″ thick. It should extend downward to about 4″ below finished grade. The buried portion of the panel does not require protection.

Cementitious panels, exterior plywood, or other durable weatherproof panels can also be used to

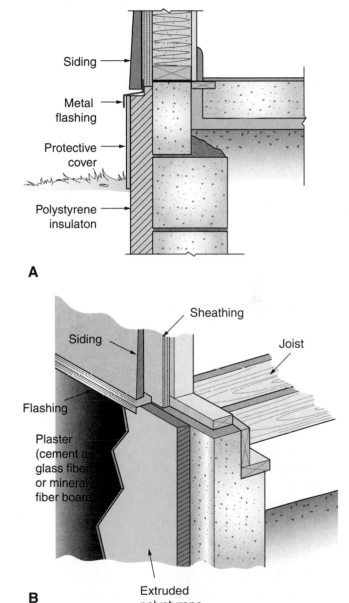

Figure 15-22. Extruded polystyrene can be installed on the outside surface to provide insulation for all types of existing foundations. A—Slab foundation. B—Crawl space or basement foundation.

protect the insulation. These panels can usually be attached with an approved mastic. Their use is practical when only a small area of the insulation board projects above the finished grade.

Polystyrene insulation is available in panels that have a weatherproof coating of fiber-reinforced cement on one side. They are especially designed to insulate the outside surface of foundations. These panels must be cut with a circular saw equipped with a masonry blade.

Existing concrete slab floors that require additional insulation can be handled in about the same manner. Insulation panels should extend downward at least 2′ or to the top of the footing.

15.8 Condensation

Water vapor is always present in the air. It is able to penetrate wood, stone, concrete, and most other building materials. Warm, moisture-laden air inside of a heated building creates a vapor pressure that tries to escape and mix with the colder, drier outside air. This is how water vapor penetrates building materials.

Water vapor has many sources within a living space. It is generated by activities such as cooking, bathing, clothes washing and drying, and even breathing. During the heating season, many people use humidifiers, which add moisture to the air.

When warm air is cooled, some of its moisture is released as condensation. The temperature at which this occurs (when the air is completely saturated with moisture) is called the *dewpoint.* If you live in a cold climate—any region where the average January temperature is 25°F (–5°C) or colder—the dewpoint can occur inside the wall structure or even in the insulation itself. The resulting condensation reduces the efficiency of the insulation and may eventually damage structural members.

Moisture that collects in a wall during the winter months usually finds its way to the exterior finish in the spring and summer. It causes deterioration of siding and/or paint peeling. The siding is usually a porous material and allows a considerable amount of moisture to pass through. The paint, however, is nonporous

and the moisture gathers under the paint film. This causes the paint film to blister and separate from the wood surface. Recent developments in paint manufacturing have provided products that are somewhat porous. They permit migrating moisture to pass through.

During warm weather, condensation may occur in basement areas or on concrete slab floors in contact with the ground. When warm, humid air comes in contact with cool masonry walls and floors, some of the moisture condenses causing a wet surface. Covering these surfaces with insulation reduces or stops condensation. Operating a dehumidifier in areas surrounded by cool surfaces also helps.

15.9 Vapor Barriers

A *vapor barrier* or *vapor diffusion retarder* is a membrane through which water vapor cannot readily pass. When properly installed, it protects ceilings, walls, and floors from moisture originating within a heated space. If you could check the temperature inside an insulated wall in the winter, you would find that it is warm on the room side and cool on the outside. The vapor barrier must be located on the warm side to prevent moisture from moving through the insulation to the cool side where it could condense. See **Figure 15-23.**

Many insulation materials have a vapor barrier already applied to the inside surface. The vapor barrier should be on the warm (in winter) side of the wall. In very hot and humid areas, such as along the Gulf Coast, the vapor barrier may be omitted or installed to the outside of the wall. Also, many interior wall surface materials

---

Water vapor: Moisture in air.

Dewpoint: Temperature at which air is sufficiently cooled for water vapor to condense out.

Vapor barrier: A watertight material used to prevent the passage of moisture or water vapor. Also called a *vapor diffusion retarder.*

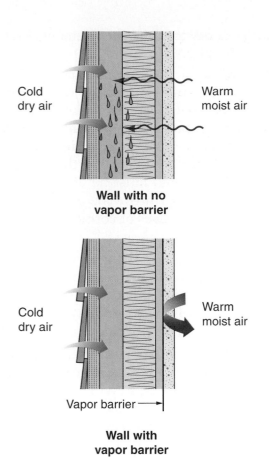

Figure 15-23. Always install a vapor barrier on the warm side of an insulated wall to prevent moisture from condensing inside of the wall.

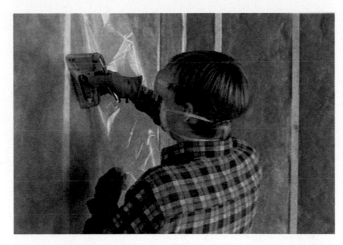

Figure 15-24. Installing a 6 mil polyethylene film over unfaced insulation.

are backed with vapor barriers. When these materials are properly applied, they usually provide satisfactory resistance to moisture penetration.

If the insulating materials do not include a satisfactory vapor barrier, then one should be installed, **Figure 15-24.** Vapor barriers in wide, continuous rolls include:

• Asphalt-coated paper.

• Aluminum foil.

• Polyethylene films.

Some architects specify the use of 4 mil or 6 mil polyethylene film.

To prevent accidental puncturing, vapor barriers should be installed after heat ducts, plumbing, and electrical wiring are in place. Cut and carefully fit the barrier around openings such as outlet boxes. Apply the vapor barrier just before the plaster base or drywall is installed. It should form a continuous covering over walls,

ceilings, and windows. The covering protects the window unit during plastering operations and permits light to enter. It is trimmed from window openings just before the finished wood trim is installed.

Use a staple gun and securely attach the film to the top plate and then to the sole plate. Smooth out the wrinkles as you go, stapling at 6″–12″ intervals on every stud. Staple at all window and door openings. Lap the joints and repair any tears with tape.

Vapor barriers are not generally recommended for ceilings. It is considered better to allow vapor to pass into attics where good ventilation will carry the moisture away. There are exceptions, however. Where roof slope is less than 3-in-12 and average temperatures drop below –20°F (–30°C), use of a ceiling vapor barrier may be advisable.

15.10 Ventilation

Proper placement of vapor barriers alone will not protect the structure against moisture. Steps must be taken to provide good attic ventilation. The cold side (outside) of walls should be weathertight, but still permit the wall to breathe. Building paper or housewrap can be used over wood sheathing to reduce infiltration while still permitting moisture in the wall to escape, **Figure 15-25.**

Figure 15-25. Installing a special infiltration barrier on the outside surface before siding is applied. It prevents air infiltration while still allowing moisture to escape. (Simplex Product Division)

It is especially important to ventilate an unheated attic or the space directly under a low-pitched or flat roof. **Figure 15-26** shows the most common systems for ventilation and the recommended size of the vent area. The International Building Code sets these requirements for ventilating attic spaces:

"There shall be a minimum of 1″ space between insulation and the roof sheathing. Net free ventilation shall not be less than 1/150th of the area of the space. Half of the requirement must be provided by ventilators in the upper portion of the space being ventilated. Vents must be at least 3′ above the eaves or cornices. The balance must be located in the eaves or cornices.

Openings to the outside of the attic space must be covered with corrosion-resistant wire cloth screening, hardware cloth, perforated vinyl, or similar material to prevent entry of birds or insects. Minimum mesh opening shall be 1/8″ and maximum 1/4″."

Insufficient insulation or ventilation directly under low-pitched roofs may cause a special problem, **Figure 15-27.** In winter, heat escaping from rooms below may melt the snow on the roof and cause water to run down the roof. At the overhang, where the roof surface is colder, the water may freeze again, causing a ledge or dam of ice to build up. Water may back up under the shingles and leak into the building.

It is current practice to insulate attics to R-38 or more in cold climates. This requires an

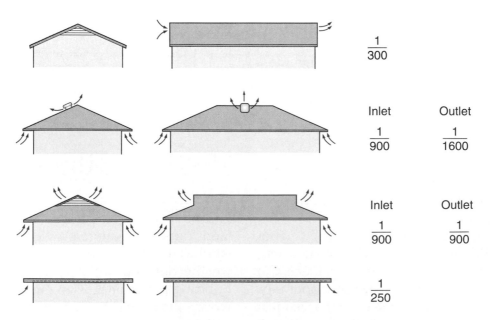

Figure 15-26. Ventilation requirements for various types of roofs, with recommendations on location. Numbers indicate the ratio of vent area to total ceiling area.

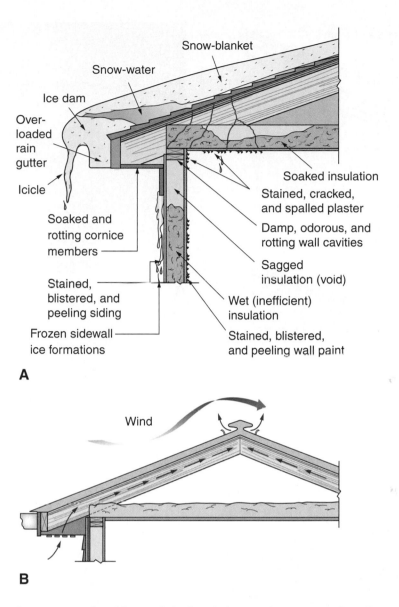

Snow-blanket

Snow-water

Ice dam

Over-
loaded
rain
gutter

Icicle

Soaked and
rotting cornice
members

Stained,
blistered, and
peeling siding

Frozen sidewall
ice formations

Soaked insulation

Stained, cracked,
and spalled plaster

Damp, odorous, and
rotting wall cavities

Sagged
insulation (void)

Wet (inefficient)
insulation

Stained, blistered,
and peeling wall paint

A

Wind

B

Figure 15-27. Ice dams form on roofs with too little insulation and not enough attic ventilation. A—Escaping heat melts the snow, causing runoff. Water freezes in the overloaded gutter, damming up water. B—Insulation over the outside wall and good ventilation should solve the problem of ice dams. (Agricultural Extension Service, University of Minnesota)

insulation thickness of at least 12″. At the same time, it is suggested that the insulation be extended over the outside wall plate. Roof trusses allow extra height at the eaves for this much insulation over the plate. However, older construction could present a problem. In any case, when installing thicker insulation at the eaves, use care not to block the airway from the soffit vent into the attic. Installation of baffles will keep airways open. **Figure 15-28** shows how baffles are used to assure an open airway for adequate ventilation.

Gable roofs usually have louvered ventilators in the gable ends. **Figure 15-29** shows two models. There are many sizes available.

Ventilators located on the roof slope may leak if not properly installed. Whenever possible, these ventilators should be installed on a section of the roof that slopes to the rear. This hides them from view on the front of the house.

Sometimes, it may be possible to use a false flue or a section of a chimney for attic ventilation, **Figure 15-30.** In most cases, it is best to install continuous ridge venting. An example

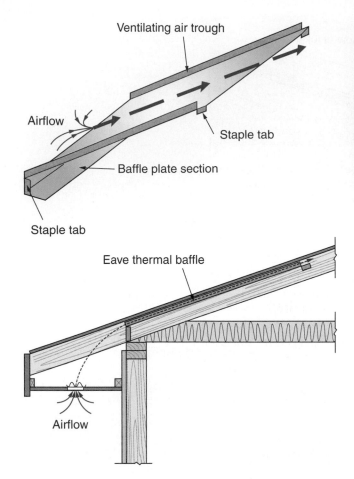

Ventilating air trough

Airflow

Staple tab

Baffle plate section

Staple tab

Eave thermal baffle

Airflow

A

B

Figure 15-28. Maintaining an airway under the eaves for proper attic ventilation. A—These baffles are to be attached to the roof between rafters. They prevent insulation from shutting off airflow. B—This view of the attic shows the baffles installed at the eaves. (Pease Industries, Inc.; Owens-Corning)

A

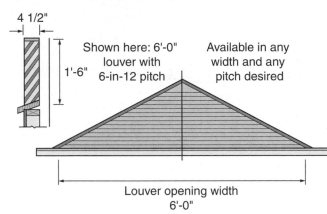

4 1/2"

1'-6"

Shown here: 6'-0" louver with 6-in-12 pitch

Available in any width and any pitch desired

Louver opening width 6'-0"

B

Figure 15-29. Gable-end ventilation. A—Prefabricated metal gable-end vent. B—Vents can also be constructed from wood and are available in a variety of shapes. (Ideal Co.)

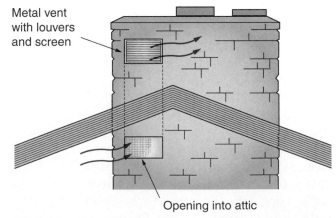

Metal vent with louvers and screen

Opening into attic

Figure 15-30. A section of a chimney can sometimes be used for attic ventilation.

of this method is discussed and illustrated in Chapter 12, **Roofing Materials and Methods.**

Walls of existing frame buildings that have inadequate "cold side" ventilation can be vented by installing small ventilators, **Figure 15-31.** Most are simply a metal tube with a louvered cover. To install, bore a hole through the siding and sheathing with a hole saw and simply press the vent into place. For maximum ventilation, install one at the bottom and top of every stud space.

15.11 Safety with Insulation

Installing insulation is not particularly hazardous. The Occupational Safety and Health Administration (OSHA) does not have specific recommendations for working with fiberglass. However, it has proposed a *permissible exposure limit (PEL).* The American Conference of Governmental Hygienists has suggested limits.

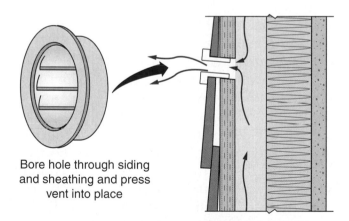

Bore hole through siding and sheathing and press vent into place

Figure 15-31. Small vents can be installed in the siding of existing structures to provide needed side wall ventilation. Sizes range from 1"–4". The 1" size is usually big enough.

Permissible exposure limit (PEL): The maximum amount or concentration of a chemical or other material that a worker may be exposed to without causing harm.

Safety Note

When working with insulation:
- Wear loose clothing. Long-sleeved shirts or blouses that are loose at the neck and wrists, caps, and long trousers will prevent most fibers from coming into contact with the skin. Loose clothing will prevent chafing where fibers do contact the skin. Gloves may be recommended in some circumstances.
- Protect eyes. Use goggles or safety glasses with side shields when applying fiberglass materials overhead or where loose particles or fibers may get into your eyes.
- Wear a mask covering your nose and mouth. This will prevent or reduce the inhaling of airborne fibers.
- Do not rub or scratch your skin. Instead, thoroughly wash, but gently, with warm water and soap. Barrier cream applied before working with fiberglass will minimize the effect of skin contact.
- Separately wash work clothes. This practice will remove all possibility of fibers being transferred to other clothing. Rinse the washing machine thoroughly before reuse.
- Dispose of scrap materials. Fiberglass scraps allowed to accumulate remain troublesome. Use a vacuum or wet sweeping to pick up dust.

15.12 Installing Batts and Blankets

To efficiently perform, insulation materials must be properly installed. Even the best insulation will not provide its rated resistance to heat flow if the manufacturer's instructions are not followed or if the material is damaged.

Blankets or batts can be cut with shears or a utility knife, **Figure 15-32.** Compress the insulation with a straight edge and use it as a guide for the utility knife. Measure the space and then cut the insulation 2"–3" longer. It is usually best to cut with the kraft paper facing up. On

Figure 15-32. Use a utility knife to cut fiberglass insulation to the required length. To save time, lay out and cut several pieces at once.

kraft-paper-faced batts, remove a portion of the insulation from each end so that you will have a flange of the backing or vapor barrier to staple to the framing. When working with blanket insulation, it is usually best to mark the required length on the floor, unroll the blanket, align it with the marks, and cut the pieces.

For wall installation of blankets, first staple the top end to the plate and then staple through the flanges down along the studs, aligning the blanket carefully, **Figure 15-33.** Finally, secure the bottom edge to the sole plate.

To install batts in a wall section, place the batt at the bottom of the stud space and press it into place. Start the second batt at the top, pressing it tight against the plate. Sections can be joined at the midpoint by butting them together. The vapor barrier should be overlapped at least 1″, unless a separate one is installed. Some batts are designed without covers or flanges and are held in place by friction between the batt and the studs.

Figure 15-33. Install blanket insulation by stapling it to the studs, working from top to bottom. (Owens-Corning)

Flanges are common to most blankets or batts. These are stapled to the face or side of the framing members, **Figure 15-34.** Pull the flange smooth and space the staples no more than 12″ apart. The interior finish serves to further seal the flange in place when it is fastened to the stud face. Some blankets have special folded flanges, which enable them to be fastened to the face of the framing. These flanges also form an air space, as shown in **Figure 15-34C.**

In drywall construction, specifications may require that the faces of the studs be left uncovered. Fit the flanges of the insulation smoothly along the sides of the framing and space the staples 6″ or closer. Be sure there are no gaps or fish mouths (wrinkles). To secure maximum vapor protection, apply a separate vapor barrier over the entire wall or ceiling area. Fully lap joints and avoid making perforations.

New construction usually requires that plumbing be installed in interior walls. In cold climates, water supply pipes should never be located in outside walls. In older dwellings, you must carefully thread the insulation behind any pipes that might be located in the outer wall. If water supply pipes are present, add a separate vapor barrier between the pipe and the interior surface to prevent condensation on the cold pipe. Insulation should also be carefully fitted around and behind electrical boxes, **Figure 15-35.**

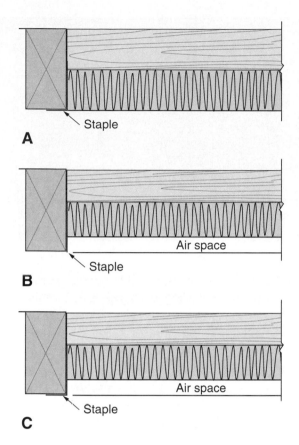

A

Staple

B

Air space

Staple

C

Air space

Staple

Figure 15-34. Three methods of installing blankets and batts. A—Flush with the inside surface of the stud. B—With the flange stapled along the side of the stud to form an air space. C—With a special flange providing an air space

Ceiling insulation can be installed from below or above, if attic space is accessible. When batts are used, they are usually installed from below. Follow the same general procedure recommended for walls. Snugly butt pieces together at their ends and carry insulation over the outside wall, as shown in **Figure 15-36.** In cold climates, extra thicknesses are recommended up to R-49. When constructing stick-built rafters, it may be necessary to place a 2 × 4 on top of the ceiling joists above the outside wall. Fasten the rafters to it, rather than to the wall plate below. See **Figure 15-37.** A raised roof truss design also provides extra space for insulation. Illustrations of both designs can be found in Chapter 10, **Roof Framing.**

In multistory construction, the floor frame should be insulated at the band joist, **Figure 15-38.** Insulation should also be installed in the perimeter of the first floor. Cut and fit pieces so they snugly fit between the joists and against the header.

Insulate large wall and ceiling areas first. Then, insulate the odd-sized, smaller spaces above and below windows. Small cuttings remaining from the main areas can be used.

Figure 15-35. Carefully fit insulation around plumbing located in outside walls, as well as around and behind electrical boxes, to eliminate insulation voids.

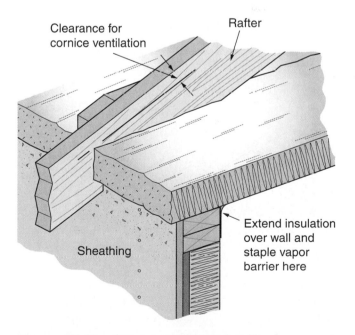

Clearance for cornice ventilation

Rafter

Extend insulation over wall and staple vapor barrier here

Sheathing

Figure 15-36. Ceiling insulation should extend over the top of the wall plate to help avoid ice dams on the roof. Be sure to leave an airway between the cornice and attic for adequate ventilation.

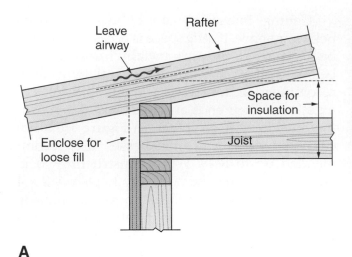

A

B

Figure 15-37. Allowing for thicker insulation. A—One method of framing a low-pitched roof to get the extra space needed for ceiling insulation. Rafters rest on a 2 × 4 added over the top of the ceiling joists. B—An alternate truss rafter design for insulation/airway clearance.

Take the time to carefully apply the insulation and vapor barrier around electrical outlets and other wall openings. Be careful that you do not cover outlet boxes or they may be missed when the wall coverings are applied.

For a thorough insulation job, all voids must be filled. Be sure to include the voids between window and door frames and the rough framing, as shown in **Figure 15-39.** Use scraps left over from larger spaces. Loosely push the insulation into small voids and cracks with a

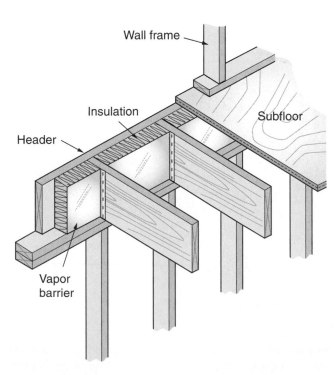

Figure 15-38. Always insulate the perimeter of the floor frame, as shown, for each floor of the building.

Figure 15-39. Insulate window and door frames by carefully stuffing pieces of batt insulation into the cavities around the frames. Try not to compress the material too much. (Bullard-Haven Technical School)

stick or screwdriver. Be careful not to compress the insulation, since this will reduce its insulating qualities. Cover the area with a vapor barrier, **Figure 15-40.**

Insulate cantilevered floor projections that carry a chimney chase, bay window unit, or extend a room over an outside wall. See **Figure 15-41.** To seal against air infiltration,

the sheathing must be tight and the insulation flange and/or vapor barrier must be carefully stapled to the sides of the joist as shown. If weather conditions permit, this segment of insulation can be installed before the subfloor is laid. This simplifies installation since the work can be done from above.

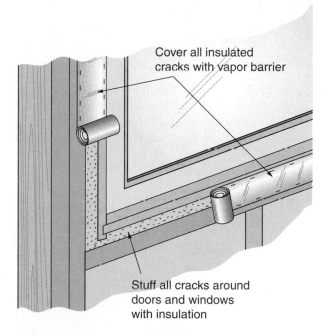

Figure 15-40. Cover insulated cracks around wall openings with vapor barrier to seal against air infiltration.

15.13 Installing Loose and Foamed Insulation

Loose insulation is most used above ceilings, where it can be placed by pouring or blowing, **Figure 15-42.** It can be directly poured from bags into the spaces between joists. Mineral fibers made from rock, slag, or glass are widely used. Blown-in fill insulation is also made from cellulose fibers.

Begin by laying down a vapor barrier between the ceiling covering and the loose insulation. This barrier also prevents the movement of moisture into the insulation and stops the fine particles present in some forms of fill insulation from sifting through any cracks that may develop in the ceiling.

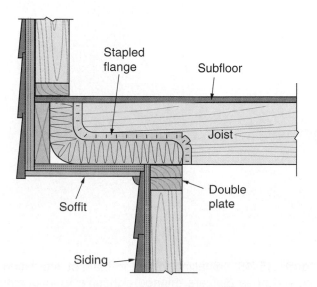

Figure 15-41. Floors that are cantilevered over an outside wall should be insulated to at least R-19.

Figure 15-42. Blowing loose-fill insulation into an attic area. Wear a cap, face mask, goggles, and gloves. (Owens-Corning)

When pouring or blowing loose insulation in an attic, contain the fill around the soffit area with batt insulation. This is required since breezes tend to blow the loose materials away from the eave area. Either install thick batts next to outside walls or install baffles that will direct the incoming air upward away from the insulation.

Be careful when insulating around lights or exhaust fans recessed into the ceiling. As a safeguard against fire, the National Electrical Code requires a 3″ space around any heat-producing device. See **Figure 15-43.**

It is easy to blow loose insulation into ceilings or walls, whether in new or remodeled structures. Special blowers are needed for this method. Usually, arrangements for their rental or free use can be made through a supplier.

One blown-in insulation system mixes a thin coating of binder adhesive into fiberglass fibers and then sprays the fiberglass into cavities behind netting. The adhesive eliminates settling problems that account for voids in the insulating blanket. The "blown-in-blanket" material is mixed on the site. The system uses a fine-mesh nylon netting that is glued or stapled to the building studs, **Figure 15-44.** The netting restrains the bonded fibers injected into each wall cavity, **Figure 15-45.** A properly filled cavity has a slight bulge. A wide roller is used to

Figure 15-44. Fine-mesh nylon netting has been attached to studs prior to insulating with adhesive-coated loose insulation. Drywall will be installed on the opposite side before the coated insulation is blown in.

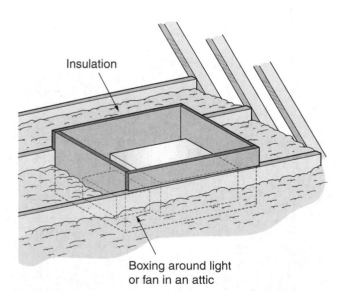

Figure 15-43. Before adding loose-fill insulation, box in fans and light fixtures that project into ceilings. The National Electrical Code (NEC) requires a 3″ air space on all sides.

Figure 15-45. Installing blown-in-blanket insulation. The red hose delivers thinned adhesive to insulating fibers. Netting holds the insulation in the cavity. (Ark Seal, Inc.)

bring it flush with the studs. Small voids around windows and doors can be filled by hand or sealed with a foamed insulation. Two methods are suggested for filling the cavities: two-three hole method and one hole method.

For the *two-three hole method,* insert the tip of the nozzle through the netting 2'–3' from the bottom. Fill the cavity to within a foot above the point of insertion. Then, pointing the nozzle tip upward, continue filling until the cavity is about half full. Next, reinsert the nozzle tip 2' from the top. With the tip aimed downward, fill the lower portion. Then, point the tip upward to complete the fill. If necessary, insert the tip at remaining voids and underfilled areas to fill the entire cavity.

The *one hole method* is only used for cavities 3 1/2" thick by 8' high and 16" on center. Attach a 5'–8' length of flexible hose to the end of the nozzle. Insert the hose into the center of the cavity and push it in until it is about 2' from the bottom. As the cavity fills, pull the hose upward. When the cavity is filled halfway, reinsert the hose upward to within about 2' of the top, slowly retracting the hose as the remaining cavity fills.

Blown-in insulation can also be used to fill cavities in concrete block walls. Thermal resistance is greatly increased. For example, the R-value of a standard concrete block (R-1.9) is increased to R-2.8 when the cores are filled with insulation. A lightweight 8" block will be increased from R-3.0 to R-5.9. Attach a 5' length of 2 1/2" PVC pipe to a flexible hose. Lower the pipe into each cavity and slowly remove it as the cavity fills. For retrofit of existing walls, 1 1/2" holes are drilled at intervals for access with the nozzle tip.

The same method can also be used to retrofit walls in old buildings. Working from the outside, drill holes in the sheathing after removing sections of siding. Be careful not to damage the siding, since it must be re-installed later.

Foamed-in-place insulation: Insulation materials that can be installed only in open cavities since they expand in volume 100 times beyond the applied thickness.

R-value	Minimum thickness	Maximum net coverage area (square feet)	Minimum weight per square foot (lbs.)
R-38	11 1/4"	8.5	2.344
R-33	9 3/4"	9.8	2.031
R-30	8 7/8"	10.8	1.849
R-26	7 3/4"	12.4	1.615
R-22	6 1/2"	14.8	1.354
R-19	5 5/8"	17.1	1.172
R-11	3 1/4"	29.5	.677

Figure 15-46. R-values for mineral wool pouring insulation.

The installed R-value for fill insulation varies depending on the installation method (pouring or blowing). Manufacturers include these figures on the bag or bale labels. R-values for a 20 lb. bag of poured mineral wool insulation are listed in **Figure 15-46.**

Foamed-in-place insulation is sometimes used, especially in new construction. Some of these materials can be installed only in open cavities, since they expand in volume 100 times beyond the applied thickness. In such cases, they cannot be used in existing walls unless the drywall or plaster is first removed. Foamed-in-place products seal up gaps and cracks and are nearly impermeable to moisture. Their value also lies in the superior insulating qualities they provide. Special training may be required for persons applying the foam.

15.14 Installing Rigid Insulation

Insulating board is widely used for exterior walls. Its application is covered in Chapters 9 and 12. Sometimes it is used:

- As the sheathing material for roofs.
- As an insulating material installed over the roof deck.
- As exterior wall insulation applied over sheathing.
- As a base for application of "synthetic stucco" exterior coverings.

Plastic foam (polystyrene) insulation is widely accepted as a rigid insulating material and has been successfully applied to a wide variety of construction types. The installation methods for slab or block insulation vary with the type of product. Always study the manufacturer's specifications.

A number of products are especially designed to insulate concrete slab floors, such as sheets of rigid polystyrene laid down inside the perimeter before pouring a concrete slab foundation. Sometimes a builder lays down sheets of the same material outside of the slab perimeter. The insulating value of these extra sheets makes deep, below–frost line footings unnecessary.

Figure 15-47 shows a section view of plastic foam insulation board applied to the interior of a masonry wall. It is bonded to the wall surface with a special mastic. This provides a permanent insulation and vapor barrier. After the boards are installed, conventional plaster coats can be applied to the surface.

15.15 Insulating Basement Walls

Where basements are to be used as living space, exterior walls should be insulated. Check local codes for any restrictions on insulating basement walls. The outside surface of concrete or masonry walls should be waterproofed below grade and should include a footing perimeter drain. See Chapter 7, **Footings and Foundations.**

The inside surface may be finished by using studs or furring strips to form a cavity for the insulation and provide a nailing base for surface coverings. In cold climates, a framework of 2×4 studs spaced 16″ O.C. is best. Use concrete nails, screws, or mastic to secure the sole plate and fasten the top plate to the joists. Unfaced insulation requires a separate vapor barrier. **Figure 15-48** shows one method of insulating the band joist in a basement.

15.16 Insulating Existing Structures

Use special care when insulating an existing structure where no vapor barrier can be installed. The humidity inside of the building should be controlled. Cold-side ventilation is essential. In the attic space, make sure that there

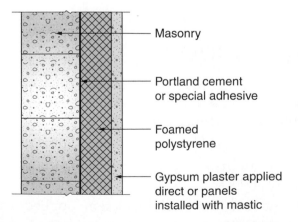

Figure 15-47. Cross section of a masonry wall. The warm side has been insulated with rigid foamed polystyrene.

- Masonry
- Portland cement or special adhesive
- Foamed polystyrene
- Gypsum plaster applied direct or panels installed with mastic

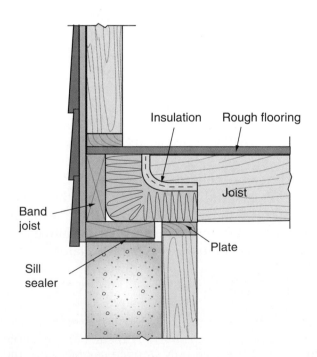

Insulation   Rough flooring

Joist

Band joist

Plate

Sill sealer

Figure 15-48. Approved method for insulating band joist between the basement wall and first floor.

is an unrestricted path for air to travel from the soffit vents to vents in the roof. Venting of walls through siding might also be required. A satisfactory vapor barrier can be secured by applying to room surfaces one of:

- A vapor-barrier wallpaper. The application should be carefully made on all surfaces of outside walls and ceiling.
- Two coats of oil-base paint, rubber emulsion paint, or aluminum paint.

15.17 Stopping Air Infiltration

Infiltration refers to air that leaks into buildings through cracks. It occurs around windows and doors and through other small openings in the structure. Air also leaks out of the building through these cracks. In the construction process, infiltration can be reduced by properly assembling materials and by sealing joints. Caulking and sealing is usually required at these locations:

- Joints between the sill and foundation.
- Joints around door and window frames.
- Intersections of sheathing with the chimney and other masonry work.
- Cracks between drip caps and siding.
- Openings between masonry work and siding.

Be sure to caulk around the electrical service entrance and hose bibbs. The preferred types of caulking compound include polysulfide, polyurethane, and silicone materials.

Inside of the structure, give special attention to recessed light fixtures and any built-in units located in outside walls. Also, seal electrical conduit and plumbing pipes that run from the attic into walls and partitions located in the living space. Seal conduit where it enters elec-

trical boxes and seal the boxes to the inside wall surface.

Modern windows are built and assembled in a factory. Appropriate weatherstripping is applied during fabrication. Outside doors, however, are often fitted to the door frame on the job and the weatherstripping is installed by the carpenter. Each type of weatherstripping requires a different method of installation. Follow the manufacturer's instructions. **Figure 15-49** shows a standard type of metal weatherstripping.

Working Knowledge

Sometimes, you may think that a window is leaking air when you are near it and feel a slight draft. This is usually due to the air in contact with the cold glass becoming colder and heavier than the rest of the room air. It moves downward to the floor and across the room. Heating registers and convectors are usually positioned to offset or minimize these downdrafts.

15.18 Estimating Thermal Insulation Materials

The amounts of insulating materials are figured on the basis of area (square feet). The thickness is then specified as separate data. The size of packages varies, depending on the type and thickness. For example, one manufacturer packages 1 1/2″ blankets in rolls of 140 square feet. A 3″ blanket in the same width comes in rolls of 70 square feet. Batts are furnished in packages (sometimes called tubes) that contain as much as 100 square feet. A 6″ batt usually contains 50 square feet.

To determine the amount of insulation for exterior walls, first calculate the total perimeter of the structure. Then, multiply that number by the ceiling height. Deduct from the total the area of doors and windows. Many carpenters deduct only large windows or window-walls

---

Infiltration: Movement of air into an enclosed space through cracks and other openings.

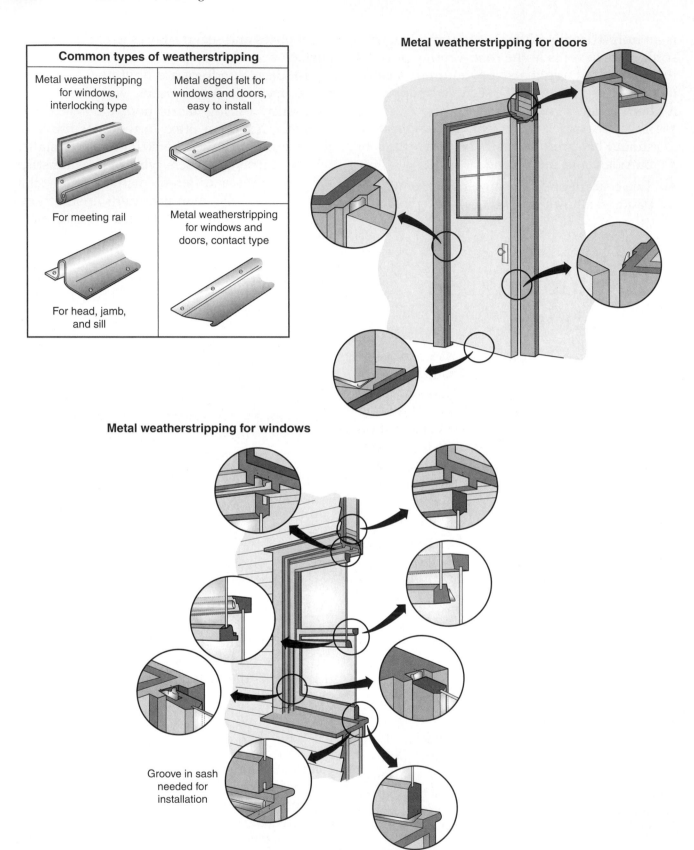

Common types of weatherstripping

Metal weatherstripping for windows, interlocking type

Metal edged felt for windows and doors, easy to install

For meeting rail

Metal weatherstripping for windows and doors, contact type

For head, jamb, and sill

Metal weatherstripping for doors

Metal weatherstripping for windows

Groove in sash needed for installation

Figure 15-49. The manufacturer's instructions for installing door and window weatherstripping are usually well illustrated.

and disregard doors and smaller openings. This extra allowance makes up for loss in cutting and fitting and also provides for additional material needed around plumbing pipes, recessed lighting boxes, and other items. Here is an example:

Perimeter	$= 30' + 40' + 36' + 20' + 6' + 20'$	
	$= 152'$	
Area	$= 152' \times 8'$	$= 1216$ sq. ft.
Window Wall	$= 12' \times 8'$	$= 96$ sq. ft.
Net Area	$= 1216 - 96$	$= 1120$ sq. ft.

The same procedure can be used to estimate reflective insulation. Make a greater allowance for cutting and waste, especially when using the accordion type. This type cannot be effectively spliced. Rolls of reflective insulation hold from 250 to 500 square feet.

Rigid insulation is also estimated on the basis of area. It can be calculated from dimensions shown on the working drawings.

When estimating the amount for floors and ceilings, use the same figures calculated for the subfloor area. Stairwell openings and openings for large fireplaces can be deducted. However, not deducting these amounts may provide the extra needed for waste and special packing around fixtures.

For a perimeter insulation strip in a concrete slab floor, multiply the perimeter of the building by the width of the strip. There is little waste on such an installation, since even small pieces can be used.

Fill insulation comes in bags that usually contain 3 or 4 cubic feet. The cubic feet required can be calculated as:

Area	$= 1200$ sq. ft.
Thickness	$= 4''$
	$= 1/3'$
Cu. ft. required	$= 1200 \times 1/3 = 400$
Less 10% allowance	
for joists 16" O.C.	$= 400 - 40$
Net Amount	$= 360$ cu. ft.
Number of bags	
(4 cu. ft. per bag)	$= 90$

Manufacturer's directions and specifications usually have tables that provide a direct reading of the number of bags required for a certain thickness of application. These are especially helpful in estimating the amount needed for such items as filling the cores of concrete blocks. See **Figure 15-50**.

15.19 Acoustics and Sound Control

Noise is unwanted sound. Noise is unpleasant. It reduces human efficiency and can cause undue fatigue. A by-product of our modern world, it has reached a magnitude that usually requires some measure of sound control in every home and building.

The wall and roof structures of houses, especially those built in cold climates, are heavy enough to block average outside noises. Therefore, noise or sound control mainly applies to interior partitions, floors, and surface finishes.

Sounds in an average home are generated by conversation, television, radios, stereos, computers, and musical instruments. Vacuum cleaners, washing machines, food mixers, and garbage disposals are examples of mechanical equipment that create considerable noise. Plumbing, heating, and air conditioning systems may be a source of excessive noise if poorly designed and installed. The activities in playrooms and workshops may create sounds

Approximate Coverage							
Number of Bags (4 cubic feet) Required							
Wall area (square feet)	Core fill block size			Cavity fill cavity width			
	6"	8"	12"	1"	2"	2 1/2"	3"
100	5	7	12	2	4	5	6
500	23	33	58	10	21	26	31
1000	46	65	118	21	42	52	62
2000	91	130	236	42	84	104	125
3000	137	195	354	62	124	155	187

Figure 15-50. This table provides estimates of fill insulation needed for masonry walls. (Perlite Institute, Inc.)

Noise: Unwanted sound, which is a vibration or wave motion that can be heard.

that are undesirable if transmitted to relaxing and sleeping areas.

The solution to problems of sound and noise control can be separated into three parts:

- Reducing the source.
- Controlling sound within a given area or room.
- Controlling sound transmission to other rooms, **Figure 15-51.**

The carpenter should have some understanding of how the latter two can be accomplished and be familiar with sound conditioning materials and construction techniques. As in the case of thermal insulation, the carpenter must appreciate the importance of careful work and proper installation methods.

15.19.1 Sound Intensity

The number of *decibels (dB)* indicates the loudness or intensity of the sound. Roughly

speaking, the decibel unit is about the smallest change in sound that is audible to the human ear. Refer to **Figure 15-52.** Notice that the rustle of leaves or a low whisper is on the threshold of audibility. That is, the sound is barely heard by the human ear. At the top of the scale are painfully loud sounds of over 130 decibels, which is often referred to as the *threshold of pain.*

Common description	Decibels	Threshold of feeling
Threshold of pain	130	Space shuttle (180+) (lift-off) Boeing 747
	120	
Deafening	110	Thunder, artillery Nearby riveter Elevated train Boiler factory
	100	
Very loud	90	Loud street noise Noisy factory Truck unmuffled Police whistle
	80	
Loud	70	Noisy office Average street noise Average radio Average factory
	60	
Moderate	50	Noisy home Average office Average conversation Quiet radio
	40	
Faint	30	Quiet home or private office Average auditorium Quiet conversation
	20	
Very faint	10	Rustle of leaves Whisper Soundproof room Threshold of audibility
	0	

Figure 15-52. Decibel levels for a wide range of sounds. Those above 90 decibels are disagreeable and can even be painful or damaging to the hearing.

Figure 15-51. Installing batts of acoustical insulation between partition studs reduces room-to-room sound transmission. (Owens-Corning Fiberglas Corp.)

Decibel (dB): Unit of sound intensity.

Threshold of pain: Painfully loud sounds of over 130 decibels.

There is a logarithmic relation, on the decibel scale, to the amount of sound energy involved. If a given sound level is ten decibels greater than another, its intensity is 10^1, or ten times, greater than the first level. If the sounds differ by 20 decibels, the ratio of their intensities is 10^2 or 100 times greater; if by 30 decibels, the ratio is 10^3 or 1000 times greater, and so on up the scale.

15.19.2 Sound Transmission

When sound is generated within a room, the sound waves strike the walls, floor, and ceiling. Much of this sound energy is reflected back into the room. The rest is absorbed by the surfaces. If there are cracks or holes through the wall (no matter how tiny), part of these sound waves travel through as airborne sounds.

The sound waves striking the wall also cause it to vibrate as a diaphragm, reproducing these waves on the other side of the wall. Sound transmission through theoretically airtight partitions is the result of such a diaphragm action. These are the sounds carried through a building by the vibrations of the structural materials themselves. Footsteps heard through the floors of a structure are an example of impact sounds. The sound insulation value of such substances is almost entirely a matter of their relative weight, thickness, and area. In partitions of normal dimensions, this value depends mostly on weight.

As the sound moves through any type of wall or barrier, its intensity is reduced. This reduction is called *Sound Transmission Loss (STL)* and is expressed in decibels. A wall with a STL of 30 dB will reduce the loudness level of sound passing through it from, for example, 65 dB to 35 dB. See **Figure 15-53.** The transmission loss of any floor or wall is determined by the materials, design, and quality of the construction techniques.

Although transmission loss rated in decibels is still used, a system that rates the sound-blocking efficiency is widely accepted. It is called the *Sound Transmission Class (STC)* system. Standards have been established through extensive research by such associations as the Insulation Board Institute and the National Bureau of Standards.

STC numbers have been adopted by acoustical engineers as a measure of the resistance to sound transmission of a building element. Like the resistance in thermal insulation (R), the higher the number, the better the sound barrier. **Figure 15-54** shows how a composite STC rating is applied to a given wall construction through a specified range of frequencies. **Figure 15-55** further describes these STC ratings by making a simple application to a wall separating two apartments.

Actually, there is another factor present that should be considered when designing any sound-insulating panel. These are usually referred to as *masking sounds.* In theory, an inaudible sound rating of zero on the decibel scale is for a perfectly quiet room. Since there are noises in every habitable room that tend to mask the sound entering, it is only necessary to reduce sound below the ambient (existing) sound level within the space to be insulated. Assume that there is a radio playing soft music in

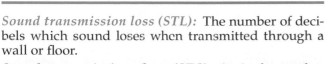

Sound transmission loss (STL): The number of decibels which sound loses when transmitted through a wall or floor.

Sound transmission class (STC): A single number that represents the minimum sound-deadening performance of a wall or floor at all frequencies.

Masking sounds: The normal sounds within habitable rooms, which tend to hide or mask some of the external sounds entering the room.

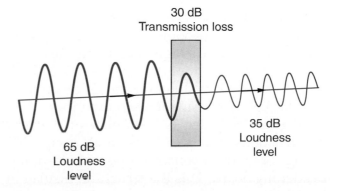

30 dB
Transmission loss

65 dB
Loudness
level

35 dB
Loudness
level

Figure 15-53. Sound transmission losses occur as sound waves travel through a wall. Values vary, depending on the frequency of the sound waves.

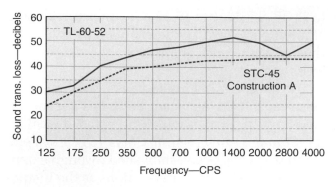

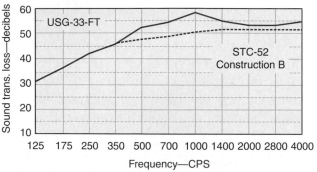

Figure 15-54. These graphs show the transmission loss values in decibels at various frequencies for two STC-rated constructions. The dotted lines represent these losses.

the listening room (about 30 dB). Thus, sound of less than 30 dB entering the room is completely masked by the ambient sound in the room.

15.19.3 Wall Construction

How high must an STC rating be for a given wall? This largely depends on the types of areas it separates. For example, partitions between bedrooms in an average home usually do not require special soundproofing, while those between bedrooms and activity or living rooms should have a high STC rating. Partitions surrounding a bathroom should also have a high STC number. Extra attention should be given to the placement of insulation around pipes.

Figure 15-56 shows a number of practical constructions for partitions and their STC ratings. Some of the constructions include sound-deadening board. This is a structural insulation board product designed especially for use in sound control systems. It is principally made from wood and cane fibers in a nominal 1/2″

Figure 15-55. How different STC ratings of partitions affect transmission of noise between two apartments.

30 STC — Loud speech can be understood fairly well

35 STC — Loud speech audible but not intelligible

42 STC — Loud speech audible as a murmur

50 STC — Loud speech not audible

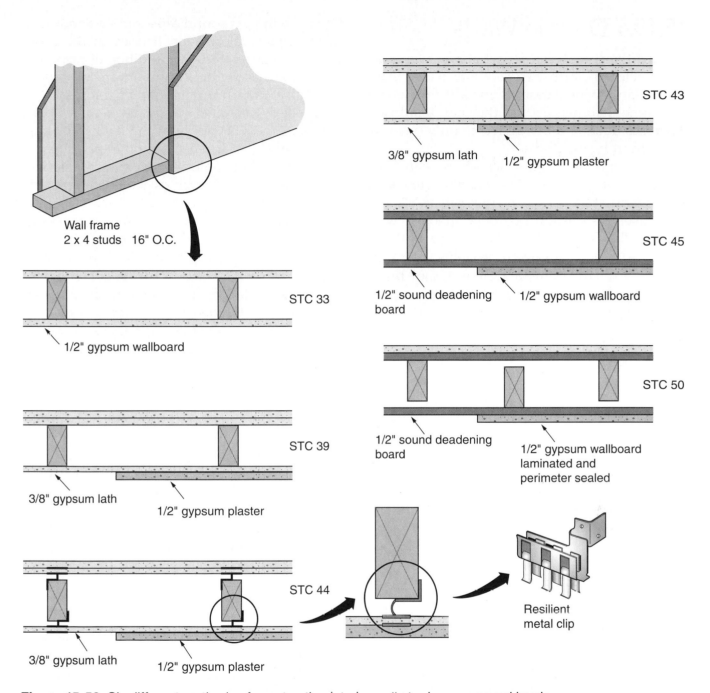

Figure 15-56. Six different methods of constructing interior walls to dampen sound levels.

thickness. Standard sizes of sound deadening board are the same as regular insulation board materials. This type of board is usually identified with the words "IBI–rated sound deadening board" on each sheet or package to distinguish it from other insulation board products.

To secure high STC ratings in a wall structure when using a sound deadening board, application details supplied by the manufacturer should be carefully followed. Where nails are used, the size, type, and application patterns are critical. In drywall construction, the joints should be taped and finished and the entire perimeter sealed. Openings in the wall, such as for convenience outlets and medicine cabinets, require special consideration. For example, electrical outlets on opposite faces of the partition should not be located in the same stud space.

15.19.4 Double Walls

Partitions between apartments are often constructed to form two separate walls. Standard blanket insulation is installed in about the same manner as for thermal insulation purposes. It should be stapled to only one row of the framing members.

For economical and space-saving construction, strips of special, resilient channel are nailed to standard stud frames. See **Figure 15-57.** The base layer of gypsum board is attached with screws. The surface layer is bonded with an adhesive. Since laminated systems like this minimize the use of metal fasteners, they result in a finer appearance along with better sound and fire resistance.

15.19.5 Floors and Ceilings

In general, the considerations for sound-proofing that were used for walls apply to floors and ceilings, as well. Floors are subjected to *impact sounds.* These are the noises from activities such as walking, moving furniture, or operating vacuum cleaners and other equipment. Sound control is somewhat more difficult.

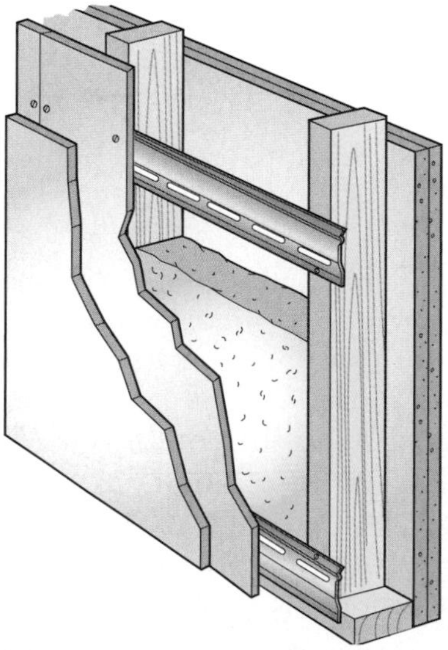

Figure 15-57. Resilient channels can be added to wood studs on one side with a double drywall layer and 3" insulation batt for an STC rating of 50. (U. S. Gypsum Co.)

Often, an impact sound causes more annoyance in the room below than it does in the room where it is generated. The addition of carpeting or similar material to a regular hardwood floor is effective in minimizing impact sounds.

Properly installed sound deadening board increases the STC rating of the floor or ceiling. See **Figure 15-58.** The use of metal clips to attach the ceiling material is a practical solution. Various suspended ceiling systems also provide high levels of sound control.

Figure 15-59 shows the installation of a floor/ceiling system with an STC rating over 52. It consists of 2 × 10 joists placed 16" O.C. with a standard wood subfloor and finished floor. The floor is covered with carpet and pad. Resilient metal channels are attached to the joists with 1 1/4" screws. Nails must not be used. Gypsum panels are attached to the channels with screws. The system includes a 3" insulation blanket.

An existing floor can be soundproofed using the method shown in **Figure 15-60.** Sleepers of 2 × 3 wood are laid over a glass wool blanket, but *not* nailed to the old floor. When the new floor is laid, be sure the nails do not go all the way through the sleepers. The only contact between the new and old floor is the glass wool blanket. It compresses to about 1/4" under the sleepers. The system makes the floor resilient, in addition to reducing sound transmission.

15.19.6 Doors and Windows

Sound tends to spread out after passing through an opening. Thus, cracks and holes should be avoided in every type of construction where sound insulation is important. Doors between rooms are probably the greatest transmitters of sound. A 1/4" crack around a 1 3/4" thick wood door admits four times as much sound of medium intensity as the door itself. Felt, rubber, or metal strips around the jambs and head help deaden sound. Conditions can

Impact sounds: The sounds that are carried through a building by the vibrations of the structural materials themselves.

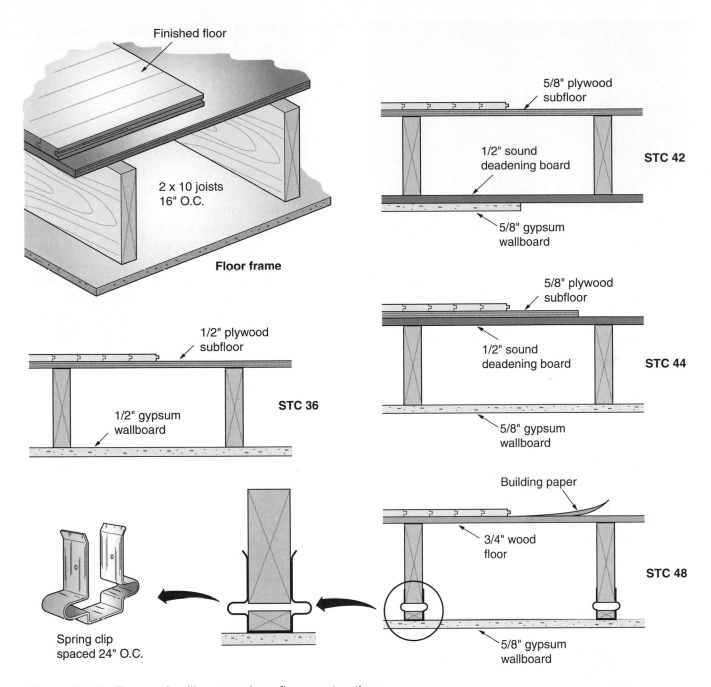

Figure 15-58. Floor and ceiling soundproofing constructions.

be further improved by some form of ***draft excluder*** at the sill, such as a threshold or felt weatherstop.

Similar precautions should be observed with glazed openings. Cracks around these openings may cancel out other efforts to cut down sound transmissions. Windows or glazed openings should be as airtight as is practical. Double or triple glazing greatly reduces the amount of sound passing through the opening. Glass blocks have a sound reduction factor of about 40 dB. They are effective where transparent glazing is not needed.

Hollow-core interior doors that are well fitted have a sound reduction value from 20 to 25 dB. Similar double doors hung with at least 6″ air

Draft excluder: A threshold or felt weatherstop.

Figure 15-59. Soundproofing the ceiling above a basement room. The resilient channels will carry the ceiling panels. When the floor above is covered with a carpet and pad, an STC of 52 can be attained. (Owens-Corning)

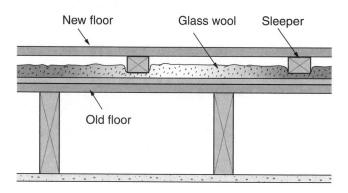

Figure 15-60. Sleepers that are floated over insulation provide significant soundproofing to an existing floor.

space between have a sound reduction factor as high as 40 dB. Using felt or rubber strips around the stops increases this factor.

For special installations, soundproof doors are available. They can be built to suit almost any condition. Special hardware is used on this type of door to prevent sound transmission through the doorknobs.

15.20 Noise Reduction within a Space

While it is important to design walls and floors that reduce sound transmission between spaces, it is also advisable to treat the enclosure so that sound is trapped or reduced at its source.

Reducing the noise level within the room not only cuts sound transmission to other rooms, but improves living conditions within the room. Areas in the home where noise reduction is most important include the kitchen, utility room, family room, and hallways.

There are a number of different types of acoustical material available to the builder. These come in a wide range of sizes, from 12" × 12" tiles to 4' × 16' boards. Those with the best acoustical properties absorb up to 70% of the sound that strikes them. The most common types are perforated or porous fiberboard units, perforated metal pan units, cork acoustical material, and acoustical plaster. All of these materials provide high absorption qualities and (except for acoustical plaster) have a factory-applied finish. Since the sound-absorbing properties of any of these materials depends on its sponge-like quality, the materials are relatively light-weight. No building reinforcement or structural changes are required.

15.21 How Acoustical Materials Work

The sound-absorbing value of most materials depends on a porous surface. Sound waves entering these pores, or holes, get "lost" and are dissipated (scattered) as heat energy. See **Figure 15-61.**

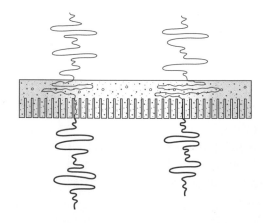

Figure 15-61. Porous materials absorb sound waves because the waves are captured in the material's many voids.

Other materials depend on a similar absorption action to reduce sound. The material used has a vibration point approaching zero. Heavy draperies or hangings, hair felt, and other soft flexible materials function in this manner.

The efficiency of an acoustical unit or product is measured by its ability to absorb sound waves. Since noise is a mixture of sounds each with a different *frequency*, this efficiency is measured by the *noise-reduction coefficient (NRC)* of a material for the average middle range of sounds. For most installations, the NRC can be used to compare the values of one material over another. However, some materials are designed to do a better job for either high or low frequency sounds. For special cases, such as music studios, auditoriums, and theaters, an acoustical engineer should be consulted. Many manufacturers of acoustical materials furnish this service.

Perforated fiberboard acoustical materials have a low-density, fibrous composition. They are formed into tile shapes. Holes of various sizes are drilled almost through the tile. The sounds that strike these units are trapped in the holes. Since the walls of the holes are relatively soft and fibrous in nature, they form tiny pockets that absorb the sound.

There are also some fibrous products that are not drilled. These depend on the surface porosity and the low vibration point of the material to absorb the sound. Most have a relatively smooth surface, providing a high degree of light reflection without glare. This type of product is usually installed on the ceiling or upper wall surface by gluing or stapling. Where the surface is in poor condition, the material can be nailed or glued to furring strips. Each manufacturer has specifications for installation.

The perforated-metal, pan-type material performs in somewhat the same manner as the other materials. The sound enters through the

Figure 15-62. Removable 2′ × 2′ panels rest on a metal grid in this suspended ceiling. They can be removed easily to provide access to pipes, electrical wiring, or HVAC ducts above the ceiling. Lighting fixtures also rest on the grid.

holes in the surface of the pans and is trapped by the backing material. The backing material is soft and resilient. The perforated-metal, pan-type material is commonly used in institutional and commercial buildings where ease of maintenance and fire resistance are important factors.

15.21.1 Suspended Ceilings

The panels of a suspended ceiling are installed on metal runners that form a grid hung below the actual ceiling, **Figure 15-62.** This system allows the large panels to simply be dropped into place. Suspended ceiling systems are used to conceal pipes, electrical wiring, and structural beams. The entire ceiling, or any part of it, may be removed and relocated without damage to the material.

The panels are made from ground cork, glass fibers, or other types of porous materials. The material is pressed into acoustical panels of various sizes and thicknesses. Because of the resistance of both cork and glass fiber to moisture, panels made from them are ideal for use as ceiling material in indoor swimming pools, commercial kitchens, or any place where humidity is a problem. Additional information on suspended ceilings is included in Chapter 16, **Interior Wall and Ceiling Finish.**

Noise reduction coefficient (NRC): The average percentage of sound absorption at 250, 500, 1000, and 2000 hertz (cycles per second).

Frequency: Rate at which sound-energized air molecules vibrate; the higher the rate, the more cycles per second (cps).

15.21.2 Acoustical Plaster

An *acoustical plaster* is generally used where an unlined or plain-surface wall or ceiling is desired. It is also used where curved or intricate planes make the use of sheet materials impractical. A number of lightweight or fibrous materials are used in the preparation of acoustical plaster. For best appearance and highest sound-absorption qualities, the plaster should be sprayed onto the surface. If acoustical plaster is to be installed, omit the finished plaster coat. In its place, apply two coats of the acoustical plaster.

15.21.3 Installation of Acoustical Materials

Carefully follow the manufacturer's recommendations when applying acoustical materials. If they are not properly installed, the sound-deadening materials may not do the job. In many cases, the amount of air space in back of the material is a factor in its sound absorption qualities.

Because most acoustical materials are soft, they are usually installed on the ceiling or upper portion of sidewalls. For sounds originating in the average room, the ceiling offers a sufficient area for sound absorption. Directions concerning the methods and procedures for installing ceiling tile are included in Chapter 16.

15.21.4 Maintenance

If properly done, painting of perforated insulation material usually does not lower the material's efficiency. However, if improperly applied, paint will soon fill the pores of the material and destroy its efficiency. Spray painting is usually preferable to brush painting for these materials. The thinner mixture is less likely to clog the pores of the material.

Any dirt clogging the pores of the material should be first removed. This may be done with a vacuum cleaner or by brushing with a soft-bristle brush. Some acoustical material can be cleaned by washing.

Working Knowledge

Manufacturers of acoustical tiles and other products have prepared detailed instructions for installation and maintenance. Be sure to carefully follow them.

Acoustical plaster: Sound-deadening plaster generally used where an unlined or plain surface wall or ceiling is desired or where curved or intricate planes make the use of sheet materials impractical.

ON THE JOB

Insulation Installer

Increased emphasis on energy efficiency in homes and commercial structures has led to increased job opportunities for insulation installers. Installers work with many kinds of insulation, including fiberglass batts, loose-fill materials, rigid sheets of foam plastic, and expanding foam sprays. More complex installation techniques are typically needed for industrial plants and some commercial applications than for residential buildings.

Most insulation workers are employed by building finishing contractors and typically work on larger residential developments and commercial/industrial buildings in urban areas. Smaller building contractors (primarily residential) usually do their own insulation work. Carpenters, drywallers, or HVAC installers usually handle the insulating tasks.

Insulation installers typically work indoors, but working conditions often are dirty and dusty. The conditions can also be uncomfortable during hot, humid weather. Fine particles from fiberglass and other insulating materials can cause skin, eye, or respiratory system irritation. For this reason, following safety guidelines, ensuring adequate ventilation, and wearing proper personal protective gear are very important.

Installation skills are most often learned on the job by working with an experienced installer. High school or vocational center classes in construction, woodworking, and blueprint reading are useful preparation for this field. A formal apprenticeship program combining classroom work and four years of on-the-job training is available in some localities. Apprenticeship programs help prepare workers for more complex types of installation work in industrial plants and similar settings and can lead to advancement into supervisory roles.

Summary

Insulation materials do not readily transmit energy in the form of heat, electricity, or sound. In construction work, much of the applied insulation is intended to prevent the transmission of heat energy—keeping heat out in the summer and holding it in during the winter. Insulation is used inside walls, in attics, under floors, and around foundations. The efficiency of insulation is determined by its R-value. The higher the R-value, the more the material resists transmitting heat. Types of insulation are rigid, flexible, loose fill, and reflective. Batt-type, flexible insulation often has a vapor barrier applied to the side that faces the living spaces of the house. This prevents water vapor from moving through the insulation to the cooler side where it could condense and cause damage. Some safety precautions must be taken when working with fiberglass insulation to avoid possible skin or eye irritation. Insulation is also used to a more limited extent to control sound transmission.

Test Your Knowledge

Answer the following questions on a separate piece of paper. Do not write in this book.

1. *True or False?* Requirements of sound control and insulating against heat loss are so different that the same materials cannot be used for both.
2. When heat moves from one molecule to another within a given material, the method of heat transmission is called _____.
3. Heat can be transmitted by wave motion. This method is referred to as _____.
4. A material with a(n) _____ U-factor would not be suitable as insulation.
5. Define *degree day*.
6. What are the four broad classifications of thermal insulation?
7. The temperature at which condensation occurs for a given sample of air is called the _____.
8. The vapor barrier in a wall structure should be located on the _____ side of the insulation.
9. Blanket or batt insulation with flanges should be stapled to stud faces at intervals of no more than _____ inches.
10. Loose insulation can be poured or _____ into place.
11. _____ insulation can be installed either on the outside or inside of exterior walls and is often used to insulate concrete slab floors.
12. Air leakage around windows and doors is called _____.
13. The unit of measure used to indicate the loudness or intensity of a sound is called the _____.
14. A wall with an STL of 30 dB reduces the loudness of a given sound traveling through it from 65 dB to _____ dB.
15. A method of rating the sound-blocking efficiency of a wall, floor, or ceiling structure is called the _____ system.
16. Sound-deadening board is made largely from wood and _____ fibers.
17. To ensure a high STC rating, electrical outlets on opposite faces of a partition should not be located in the same _____.
18. What are *impact sounds*?
19. The efficiency of an acoustical ceiling unit is expressed in an NRC rating. NRC is an abbreviation for _____.
20. Why is spray painting preferred to brushing when refinishing acoustical ceiling tile?

Curricular Connections

Language Arts. From a study of reference books and trade magazines, prepare a report on radiant heating systems. Include information about panels that are heated with electricity as well as those heated with hot water. Place special emphasis on the methods of installation, since a traditional structural design may need to be modified for these systems. Also, collect information concerning types and amounts of insulation materials. Discuss special application procedures that may be required.

Science. Using a sound recorder, carefully record the various sounds produced by equipment and devices found in your home. Include laundry equipment, dishwashers, food mixers, garbage disposals, plumbing fixtures, and vacuum cleaners. Also include such noises as walking on hard-surfaced floors and the opening or closing of passage or cabinet doors. Play the recording for your class. Then, lead a discussion on how to control each sound through proper design and the use of special materials and construction.

Outside Assignments

1. Obtain samples of various thermal insulating materials, such as glass, mineral, organic fibers, and foamed plastic. Include loose-fill insulation made from such material as vermiculite. Enclose fibrous and granular materials in small envelopes made of polyethylene plastic film. Mount the samples on a display board with descriptive titles, including k and R factors.

2. Prepare a brief study of the most common types of heating and cooling systems used in your region. Make several drawings showing how they operate and how they are controlled. Study local building codes and outline the basic requirements that must be followed in the installation. Visit a heating contractor and obtain basic information concerning the heating and cooling load calculations for residential structures. Get information about manufacturers, approximate costs, and installation procedures. Give a report in your class.

3. Residential construction emphasizes the use of vapor barriers in walls and floors and weatherstripping on windows and doors. This nearly airtight situation can result in high levels of humidity and indoor air pollution. These conditions are potential health hazards and can cause damage to interior surfaces and fixtures. Air-to-air heat exchangers can provide a desirable standard of ventilation. Study trade magazines and visit local heating/cooling contractors to learn about this kind of equipment. Prepare a written report.

This worker is installing gypsum wallboard in a steel-framed building. (National Gypsum)

Interior Wall and Ceiling Finish

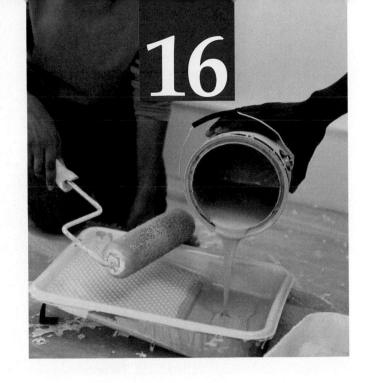

16

Learning Objectives

After studying this chapter, you will be able to:

- Explain wall and ceiling covering materials.
- Describe wallboard cutting, nailing, and adhesive techniques.
- Describe the characteristics of gypsum plaster.
- Explain how gypsum and metal lath are applied.
- Illustrate the use of plaster grounds.
- Describe plastering methods.
- Illustrate how double layer and predecorated wallboard are applied.
- List the procedures for installing wood paneling.
- Lay out ceiling tile and install furring strips.
- Describe methods for leveling and installing a suspended ceiling.
- Estimate quantities of lath, wallboard, and ceiling tiles for a specific interior.

Technical Vocabulary

Backing board	Darby
Banjo	Diagonal paneling
Brown coat	Double layer
Cement board	Expanded metal lath
Chevron paneling	Field

Finish coat	Moisture-resistant
Grounds	wallboard
Gypsum wallboard	Plaster
Gypsum lath	Plastic laminates
Hardboard	Scratch coat
Herringbone paneling	Single-layer
Insulating fiberboard	Suspended ceiling
lath	Wood lath
Interior finishing	Working time

The installation of cover materials to walls and ceilings is referred to as *interior finishing*. This stage of construction can start after the mechanical systems (plumbing, heating, and electrical wiring) and insulation are installed. Exterior doors must also have been hung and windows installed. They will protect the finishing materials from the weather damage. Interior walls can be covered with any of a number of materials, **Figure 16-1:**

- **Gypsum wallboard.** Commonly called drywall or sheetrock, *gypsum wallboard* is a laminated material with a gypsum core and paper covering on either side. It usually

---

Interior finishing: The installation of cover materials to walls and ceilings.

Gypsum wallboard: Wall covering panels consisting of a gypsum core with facing and backing of paper. Also called *drywall* or *sheetrock*.

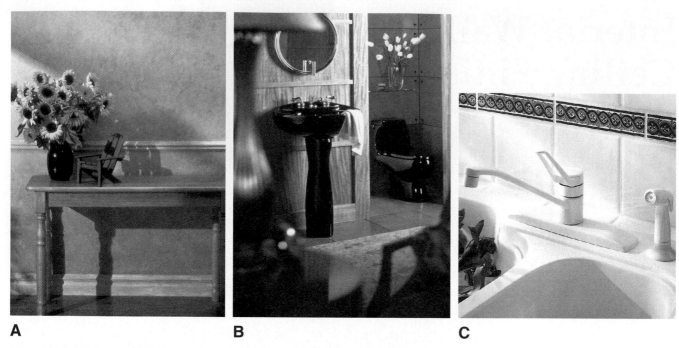

A   **B**   **C**

Figure 16-1. Various materials are used as wall coverings. A—Gypsum wallboard is the most widely used material. It accepts many different surfacing materials, such as wallpaper and paint. B—Wood is used in solid form, as shown here, or in various engineered panels. C—Ceramic tile is used extensively for walls in kitchens and bathrooms. (Wagner, American Standard, Moen)

comes in 4′ × 8′ sheets. However, it is also available in 7′, 9′, 10′, 12′, and 14′ lengths. It is sold in several thicknesses: 1/4″, 5/16″, 3/8″, 1/2″, and 5/8″. Gypsum wallboard is used on both walls and ceilings. *Backing board* is a gypsum board with a gray liner paper on both sides. It is used as the base sheet on multilayer applications.

- **Gypsum lath for plaster veneering.** This is a base of 16″ × 48″ gypsum board usually 3/8″ or 1/2″ thick. It is applied as a backing for a thin finish coat of plaster.

- **Predecorated gypsum paneling.** This is the same as gypsum wallboard. However, decorative vinyl finishes have been applied and edges have received special treatment so that no other finishing work need be done after the panels have been installed. The finishes are tough and easily cleaned.

- **Plywood and particleboard.** Plywood is fabricated in 4′ widths. Lengths include 7′, 8′, 9′, and 10′. Panels are manufactured in thicknesses of 1/4″, 3/8″, 7/16″, 1/2″, 5/8″, and 3/4″. Usually, the sheets are prefinished in a variety of colors and patterns. The

surface material may be either a hardwood or a softwood. Surfaces can be embossed, stained, or color toned. Some have veneers of paper with a wood grain printed on them. Flakeboard is sometimes used for rustic finishes.

- **Hardboard and fiberboard.** These are produced from wood fibers in sizes and thicknesses similar to plywood. The face finish is simulated to look like wood. Other decorative patterns are also applied. Sheets may be embossed and grooved to take on the look of random planking, leather, or wallpaper. Surfaces may also be coated with plastic. Variations of fiberboard are used as ceiling coverings.

- **Solid wood paneling.** These are boards or pieces of solid wood. Widths of boards vary

---

Backing board: In a two-layer drywall system, the base panel of gypsum drywall. It has gray liner paper as a facing and is not suitable as a top surface. Also referred to as *backer board*.

from 2″–12″ and thicknesses are a nominal 1″ or 2″. Faces may be rough-sawed, plain, or molded in a variety of patterns. Lengths vary from 4′–10′. Shingles, usually considered a siding or roofing material, are occasionally used on interior walls.

- **Plaster.** For many years, *plaster* was the most popular wall covering. It is made of powdered gypsum to which other materials are added to improve drying time. A plastered wall system includes a base support, such as metal or gypsum lath, over which coats of wet plaster are applied.

- **Cement board.** Available under several different brand names, *cement board* is a versatile fiber-reinforced cement panel material. It is not considered a finishing material itself, but serves as a base (underlayment) for finishing materials on floors, countertops, and exterior or interior walls. It is fireproof, not damaged by water, and resists impact. Some cement board products are lightweight with fiberglass-reinforced matting and silicone-treated cores. **Figure 16-2** shows a sample of a typical cement board product.

- **Special finishes.** These include a variety of products and materials: brick, stone, glazed tile, plastic tile, and plastic laminates. They are used either as accent materials or to provide a wear-resistant surface. They are often found in kitchens and bathrooms.

Ceilings can be covered with many of the same materials used for walls. Composition tiles are especially suitable because they are easy to install.

Before beginning the wall and ceiling application, check the framework. Be certain that sufficient backing has been installed for fixtures and appliances. Refer to Chapter 9. Nailers provide a surface for fastening wall coverings

---

Plaster: Mixture of gypsum and water that can be troweled wet onto interior walls and ceilings.

Cement board: Fireproof, moisture-resistant, fiber-reinforced cement panels used as a base for finishing materials on walls, floors, and countertops.

Figure 16-2. A sample of a fiber-reinforced cement underlayment. It can be used as backing for ceramic tile, marble, and plastic laminates. The dot at the edge is a guide for nailing. (James Hardie Building Products)

at all intersections of wall and ceiling surfaces. They must be included at all vertical corners and at all intersections between walls and ceilings. Special or built-in equipment, such as a prefabricated fireplace, must be installed before the interior wall surface is applied.

16.1 Drywall Construction

Drywall materials are the most common wall coverings used in construction. Most builders prefer to use drywall because it saves time. Regular plaster requires considerable drying time. Either type of finish presents advantages and disadvantages. Drywall construction, for example, requires that studs and ceiling joists be perfectly straight and true; otherwise, the wall surface will be uneven. This can sometimes be corrected in ceiling joists by installing a strongback. Where steel or engineered wood joists and studs are used in framing, this is not a problem. The wood framing material must also have a moisture content very near to what it will eventually reach in service. This helps prevent nail pops and joint cracks. Using screws to fasten drywall also solves this problem.

Figure 16-3. Transporting drywall into a new residence using power equipment.

16.1.1 Handling and Storage

Drywall should be delivered only a few days before it is to be installed. If stored on the job site for too long, it may be damaged. Likewise, joint compound and veneer plaster finishes have a short shelf life and should not be stored for long periods.

Handle drywall as carefully as you would millwork. It can be manually transported to the point of use or by machine, **Figure 16-3.** In multistory construction, use of a machine may be the most practical and timesaving way to deliver the material to the point where it will be installed.

Stack drywall flat on a clean floor in the center of the largest rooms. Place those sheets to be used on the ceiling on the top of the pile. This is because the ceilings will be drywalled before the walls. Never stack longer sheets on top of shorter ones—the overhanging portion could crack.

Store metal corner beads, casing beads, and trim where they will not be bent. Bags of veneer plaster should never rest on damp concrete. They are best stored on planks, skids, or shelves.

16.1.2 Types, Thickness, and Styles

Gypsum wallboard has a fireproof core. **Figure 16-4** lists a variety of thicknesses, edge joint designs, and types. **Figure 16-5** illustrates

Type	Thickness (inches)	Edges
Regular (ASTM C36, FS SSL30d)	1/4 5/16 3/8 1/2 5/8	Tapered Square Square Tapered Bevel
Fire resistant type "X" wallboard	1/2 5/8	Square Tapered Bevel
Insulating wallboard (aluminum foil on back surface)	3/8 1/2 5/8	Square T & G Tapered Round
Regular backing board (ASTM C442, FS SSL30d)	1/4 3/8 1/2	Square Square T & G
Foil-backed backing board	3/8	Square
Fire resistant type "X" backing board	1/2 5/8	T & G
Coreboard (Homogeneous or laminated)	3/4 1	Square T & G Ship lap
Pre-decorated	3/8 1/2 5/8	Bevel Round Square Bevel Square

Figure 16-4. Main types of gypsum wallboard.

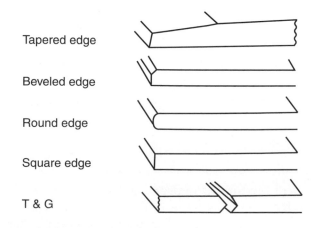

Tapered edge

Beveled edge

Round edge

Square edge

T & G

Figure 16-5. Gypsum wallboard is manufactured with several different edge styles.

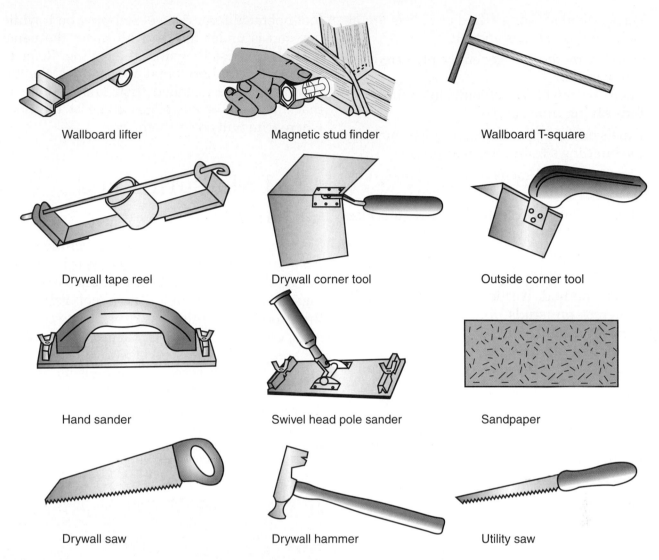

Wallboard lifter

Magnetic stud finder

Wallboard T-square

Drywall tape reel

Drywall corner tool

Outside corner tool

Hand sander

Swivel head pole sander

Sandpaper

Drywall saw

Drywall hammer

Utility saw

Figure 16-6. These tools are used for preparing and installing drywall. Note the names. (Harrington)

several standard edge designs for gypsum wallboard. The tapered edges form a shallow depression between adjacent sheets. This depression is brought level with drywall tape and filler. The result is a smooth, uninterrupted surface. Drywalling requires some special tools for marking, cutting, installing, and finishing. See **Figure 16-6.**

Single-layer: Gypsum board construction used where economy, fast installation, and fire resistance are important.

16.2 Single-Layer Construction

Single-layer construction is used where economy, fast installation, and fire resistance are important. It is well suited to remodeling jobs, and for resurfacing damaged or cracked plaster walls. For both new construction and remodeling, use 1/2″ or 5/8″ gypsum wallboard. The 5/8″ sheet is preferred for high quality construction. Cover ceilings first, then the walls.

There are two methods of arranging the drywall sheets:
- Parallel—long edges of panels run in the same direction as studs and joists.

- Perpendicular—long edges of panels are at right angles to studs and joists.

The second method is generally preferred for several reasons:

- There are fewer feet of joints to be finished, thus saving time and costs.
- Panels bridge more studs and joists, making the building's frame stronger.
- The strongest dimension of the panel runs across the frame.
- There are fewer problems with irregularities in alignment and spacing of the frame.
- Horizontal joints are easier to treat because they are lower on the walls.

In either method, vertical wall joints must be centered on studs for proper fastening. **Figure 16-7** shows both parallel (vertical) and perpendicular (horizontal) applications. As a general rule, use whichever method results in the fewest joints. Stagger end joints and locate them as far away from the center of walls and ceilings as possible.

Loosely butt wall panels against the ceiling panels. In a parallel application, use a wallboard lifter to raise the panel to the ceiling. A lifter is a foot-operated lever device. Stepping on it while one end is under the drywall raises the panel enough to press it against the ceiling. Refer to **Figure 16-6.** In horizontal applications, the top wall panels are installed first. This is done so that any gaps or cut edges come at the floor where trim will cover them.

16.2.1 Measuring and Cutting

All measurements should be carefully taken from the spot where the wallboard will be installed. Usually, it is best to make two readings, one for each side of the panel. Following this procedure eliminates errors. It also corrects for openings and framing that are not plumb or square. Use a 12′–25′ steel tape to take measurements. Transfer these measurements to the drywall panel. Where the cut will not be entirely across the panel, draw a line along the straightedge, as in **Figure 16-8.** Then, use a drywall saw to make the cut or cuts.

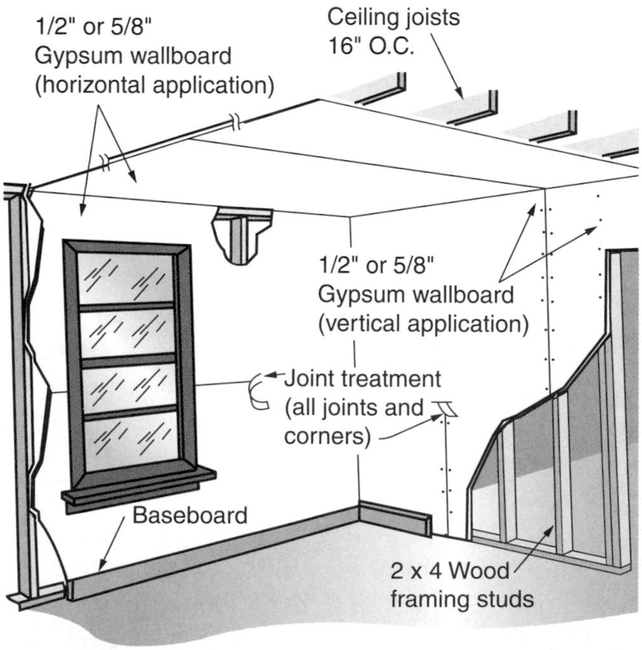

Figure 16-7. Single-layer drywall application. The left-hand wall and ceiling have a horizontal application. The right-hand wall has a vertical application.

1/2" or 5/8"
Gypsum wallboard
(horizontal application)

Ceiling joists
16" O.C.

1/2" or 5/8"
Gypsum wallboard
(vertical application)

Joint treatment
(all joints and
corners)

Baseboard

2 x 4 Wood
framing studs

Figure 16-8. A drywall square can be used to mark a square cut line on the panel. Where only a portion of the width will be cut away, a pencil is used to first mark the cut line. (Construction Training School, St. Louis, MO)

If the cut is straight and across the width or length of a board, first score the face with a sharp utility knife. Pull the knife along a metal straightedge as a guide. The scoring cut should be deep enough to penetrate the paper facing and enter the gypsum core. Support the main section of the sheet close to the scored line. With one hand firmly holding the sheet on the table or bench, snap the core by sharply pressing downward on the overhang. Support the cutoff with the other hand. Cut the backing paper and remove the cutoff piece. When necessary, the cut can be smoothed with a file or with coarse sandpaper mounted on a block of wood.

Another common method used for cutting does not require laying the panel flat. With the panel resting on edge, make the cut using a drywall T-square or straightedge. Hold the T-square with one hand and make the cut. Lift the board slightly and snap the cutoff portion backward. Cut the backing paper to separate the parts.

Irregular shapes and curves can be cut in drywall with either hand or power tools. Use a coping saw, compass saw, electric saber saw, or rotary power tool. See **Figure 16-9.**

Figure 16-9. A portable power tool makes cutting holes in drywall for electrical boxes easier and faster. The tool can also be used to cut circles or irregular shapes.

> ### Working Knowledge
>
> When scoring wallboard, always use a sharp knife. This will make a "clean" cut through the paper face without tearing it or rolling it up in front of the blade.

16.2.2 Nail Fastening

Nail spacing varies depending on the materials being used. For single-layer construction, space nails no farther apart than 7″ on ceilings and 8″ on walls. Keep nails at least 3/8″ from ends and edges. Annular ring nails that are 1 1/4″ long with a 1/4″ diameter head are generally recommended. **Figure 16-10** shows examples of drywall fasteners.

Field: The middle area of a sheet of wallboard.

Tightly draw the drywall against the framing so there can be no movement of the board on the nail shank as it is being driven. Press the board tightly against the stud or joist to avoid breaking through the wallboard face.

Braces can be used to support large panels while they are fastened to a ceiling. Start nailing ceiling panels at the abutting edge. Next, nail the *field* (area between edges), working away from the abutted edge. Then, finish with the opposite

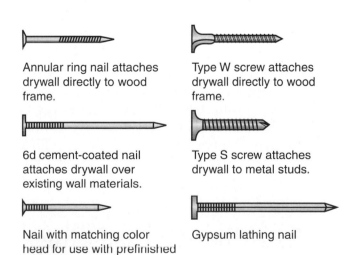

Annular ring nail attaches drywall directly to wood frame.

Type W screw attaches drywall directly to wood frame.

6d cement-coated nail attaches drywall over existing wall materials.

Type S screw attaches drywall to metal studs.

Nail with matching color head for use with prefinished gypsum wallboard.

Gypsum lathing nail

Figure 16-10. Gypsum wallboard and gypsum lath fasteners. Others are available.

side. If the perimeter is nailed first, panels will sag and not draw up tight to joists. After the ceiling is completed, drywall can be applied to the walls, **Figure 16-11.**

Drive nails straight and true. An angled nail will tear the face paper as it is countersunk. Use extra care during the final strokes so the nail head rests in a slight dimple formed by the crowned head of the hammer or stapler. Be careful not to break the paper face of the board.

The *double nailing method* of attachment ensures firm contact with framing. Panels are applied as required for conventional nailing, except that nails in the field of the board should be spaced 12″ on center. After the panel is secured, another nail is driven approximately 2″ away from the first. If necessary, the first nail should receive another blow to assure snug contact.

16.2.3 Screw Fastening

Screw application requires a screw gun. This is a positive-clutch electric power tool designed for attaching drywall. It uses a Phillips bit and has an adjustable head that controls screw depth. To adjust for proper screw depth, the control head is rotated. The screw gun is designed so that it does not operate until the screw and gun tip are pressed against the drywall.

Gypsum board screws are an alternative to nails. They provide a firm, tight attachment to wood or metal framing, **Figure 16-12.** Special, self-tapping screws are used for metal-framed wallboard systems. Fasteners of this type must be driven so the screw head rests in a slight dimple formed by the driving tool. The paper face of the drywall should not be cut, nor should the gypsum core be fractured. Since screws hold the wallboard more securely than nails, ceiling spacing can be extended to 12″ and side walls to 16″.

16.2.4 Adhesive Fastening

Drywall may also be fastened with a special adhesive that is sold in cartridges. Apply it in a continuous bead over the frame surface using

Figure 16-11. After ceilings are drywalled, apply drywall to walls. Try to get a close fit in the corners.

a hand or powered gun. One main advantage of adhesive is there are only a few depressions from mechanical fasteners that need to be filled later. Adhesives also produce a sturdier wall that is more resistant to impact sounds.

Figure 16-12. An electric screw gun with a special depth-adjusting clutch is used to drive drywall screws into steel studs. The clutch disengages when the nose strikes the panel surface. (St. Paul Technical College)

PROCEDURE

Applying adhesives

1. Select the proper adhesive and read the manufacturer's directions for use.
2. Apply adhesives only when temperatures are between 50°F and 100°F (10°C–40°C). Keep adhesive containers closed. Evaporation of the solvent can affect the adhesive's performance and ability to bond.
3. Check all surfaces. They must be free of dirt, oil, or other contaminants.
4. Observe manufacturer's open time for the adhesive. If exceeded, a poor bond is sure to result. As a general rule, apply no more adhesive than can be covered within 15 minutes.
5. Apply a continuous bead of adhesive to the center of all studs, joists, or furring.
6. Where two pieces of wallboard join on a framing member, use a zigzag bead pattern. The bead should be 1/4″–3/8″ wide. Then, when the board is in place, it will be held by a band at least 1″ wide and 1/16″ thick.
7. Use temporary nailing or bracing to ensure full contact between the adhesive and the drywall. Go over each surface applying hand pressure to force the panel into the adhesive. All of these precautions help the adhesive to develop proper bonding strength.

16.2.5 Concealing Joints and Fastener Heads

Mechanical taping tools are often used to apply the compound and tape. **Figure 16-13**

Banjo: Taping tool used to apply joint compound and tape at the same time. This tool has a reservoir to hold the compound and a reel to hold the tape.

A

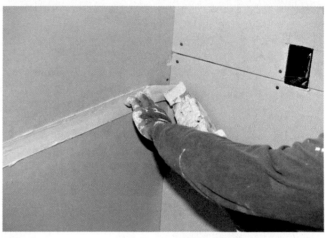

B

Figure 16-13. Using a banjo for applying tape to drywall joints. A—The reservoir is being filled with compound. The roll of tape is visible under the reservoir cover. B—The operator squeezes out compound as the tape is pulled across the joint. (Loren LaPointe Drywalling)

illustrates the use of a ***banjo*** to apply compound and tape at the same time. This tool has a reservoir to hold the compound and a reel to hold the tape. Manual pressure is applied to spread the compound and apply the tape as the tool is moved along the joint. After the tape and compound are applied, the joints are smoothed with a broad knife.

The introduction of pressure-sensitive, glass-fiber tape has reduced the time required to conceal and reinforce joints and interior angles. It has an open weave of 100 meshes per square inch. This mesh provides excellent reinforcing and keying of plaster or compound coats. Simply use hand pressure to attach it to the wall, then bond it by pulling a finishing knife or trowel along its length. The tape is also easily applied to inside corners.

Next, apply two coats of fast-setting joint compound over the tape and sand smooth. If the surface is to be covered with a texture paint, the joints can be finished with a single coat of compound. When applying joint compound to ceilings and the upper walls, it will be necessary to stand on a platform to reach the work. Often, stilts are used to reach high areas, **Figure 16-14.**

A wide variety of finish coatings is available for drywall construction. Some are in a powder form ready to be mixed with water. Others are ready to use from the containers.

Figure 16-14. Stilts are very helpful when working on ceilings and upper walls. The pair shown here is adjustable in height, making it possible to work on high ceilings and tall walls.

Always read and follow the manufacturer's recommendations.

PROCEDURE

Concealing joints and fastener heads

1. Apply a bedding coat of joint compound into the depression formed at all butt joints by the tapered edges of the board. Use a 5″ or 6″ joint knife, **Figure 16-15.**
2. Center the reinforcing tape over the joint. Press tape into compound by drawing the knife along the joint with enough pressure to remove excess compound. Smooth it out to avoid wrinkling or buckling. Apply a skim coat over the tape.
3. After the embedding coat is completely dry, apply a second coat over the tape. Feather the edges approximately 1/2″–3/4″ beyond the edges of the first coating.
4. When this coat is completely dry, apply a third coat with the edges feathered out about 2″ beyond the second coat.
5. After the last coat is dry, sand lightly, if necessary. Fasteners are also concealed with compound, each coat being applied at the same time the joints are covered.

16.2.6 Corners

Outside corners are reinforced with a metal corner bead, which is made in various styles. The bead is installed after horizontal and vertical joints have been taped. Fasten the bead by nailing through the wallboard and into the framing, **Figure 16-16.** After installation, the bead is concealed with joint compound in about the same manner as regular joints. To finish and reinforce edges around doors, windows, and other openings, metal-channel trim is available.

At inside corners, reinforcing tape is used. First, apply a bedding coat of joint compound to both sides of the corner. Then, fold the tape along the centerline and smooth it into place. Remove excess compound and finish surfaces along with the other joints.

A

B

C

Figure 16-15. Taping wallboard joints with taping knives. A—First apply compound to the channel at the joint. B—Then, embed the tape. Be sure it is centered over the joint. C—Immediately apply a skim coat over the tape and smooth the edges. Use a broader knife for this step. (National Gypsum Co.)

Figure 16-16. Attaching metal corner bead to an outside corner with drywall nails.

16.2.7 Attaching Drywall to Steel Framing

Single-layer application to steel framing members is similar to wood frame application. Arrange panels either parallel or perpendicular to framing and attach them with 1″ long Type S screws, **Figure 16-17.** The leading ends or edges of panels must first be attached to the open edge of the stud or joist flange.

Figure 16-17. Screws are used to attach drywall to steel framing in this office building project.

16.3 Double-Layer Construction

Double-layer (also called *two-ply*) gypsum board applications over wood framing ensure a strong wall surface, **Figure 16-18.** Fire protection and sound insulation qualities are also improved. This method is adaptable to the use of either predecorated panels or standard beveled drywall with treated joints.

The base layer may be a regular gypsum wallboard or a specially designed base called *backing board.* Backing board is the same as regular gypsum board but covered with a gray liner paper. It is not suitable for decorating and should not be used as a top surface.

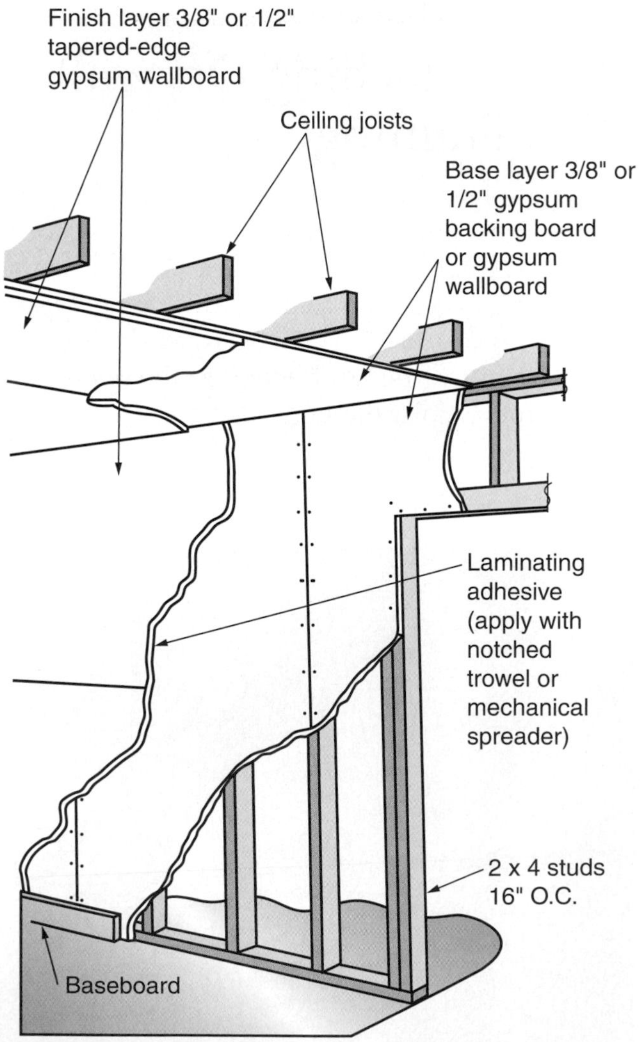

For areas where there is likely to be moisture coming into direct contact with the wall, there are highly water-resistant backing boards, including some that are impervious to water. Their use is especially recommended in shower areas as a base for tile and other protective coverings.

Sound-deadening backing board is sometimes used for the base of double layered walls. It is specified where high sound transmission control (STC) ratings are needed.

16.3.1 Attaching the Layers

Apply backing board to framing with staples, nails, or screws. The finish layer is laminated to the base layer with an adhesive or joint compound. Joints of the finish layer should be offset at least 10″ from the joints of the base layer. Finish layers can be applied parallel to the base layer or at right angles to it.

16.3.2 Mixing Compound

Place a quantity of compound in a clean container. Mix in only clean water according to instructions on the bag. Use care that all compound becomes uniformly damp. Avoid contaminating the mixture with dirty water or previously mixed compound. To do so will affect setting time. Mix only what can be used up within the *working time* indicated on the bag. Adding water to retemper compound once it begins to set is useless—it will not prevent setting or increase working time.

Ready-mixed compound should be used as it comes from the container. Add cool water in half-pint increments for a thinner mix. Use a

Finish layer 3/8" or 1/2" tapered-edge gypsum wallboard

Ceiling joists

Base layer 3/8" or 1/2" gypsum backing board or gypsum wallboard

Laminating adhesive (apply with notched trowel or mechanical spreader)

2 x 4 studs 16" O.C.

Baseboard

Figure 16-18. A cutaway of double-layer gypsum wallboard construction. The finish layer may be applied at a right angle to the base (as shown) or running parallel.

Double-layer: Gypsum board construction used where strength, insulation qualities, and improved fire resistance are important.

Working time: The period during which joint compound can be used before it sets.

potato-masher type mixer and lightly remix after each addition of water. Test the compound after each addition to avoid too thin of a mixture.

16.3.3 Applying Adhesive

First, cut and fit drywall sheets. Adhesive is usually applied to the entire surface. However, strip lamination is used in some applications. This method uses strips with ribbons of adhesive spaced at regular intervals.

Many methods of applying adhesive are acceptable. Trowels and powered devices are available. Whatever method is used, the spacing and size of the bead of adhesive must provide the required spread when panels are pressed into position. A notched spreader is often used to apply adhesive to the entire back surface of a finish-layer panel. Strip lamination is frequently used for sidewall panels. The application can be made either on the base surface or on the face panel. Temporary bracing may be used to hold panels in position until bonding has taken place. Recommended fasteners are:

- On-ceiling applications when using a laminating adhesive—Space fasteners 16″ O.C. along ends and edges. At mid-width, use one fastener for every framing member.

- When laminating with compound—Provide permanent supplementary fastening or temporary fastening (usually overnight) until the compound has dried.

When nails are used, they should provide a minimum penetration of 3/4″ into the wood framing members. Consult a nail chart.

16.3.4 Finishing Double Layer Wallboard

If the wallboard is to receive other covering material, joints should be taped, nails concealed, and corners finished in the same way as for single-layer construction. If a veneer of plaster is to be applied, use reinforcing tape and a single bedding coat of compound over joints. Tape inside corners and apply a special bead to outside corners to form grounds for the veneer coat. Always use a single length of corner bead to extend from the floor to the ceiling.

16.4 Special Backing

Special backing is available as a base for tile in areas where walls are frequently wet, for example, bathrooms and showers. One type is known as cement board. It is manufactured from a slurry of portland cement reinforced with polymer-coated, fiberglass mesh embedded in both sides. Some backing board products are rigid, while others are somewhat flexible. They are manufactured under such trade names as Hardibacker, Durock, and DensShield.

These materials, depending on the brand, can be used in many different applications on both interior and exterior surfaces. Applications include:

- Floor underlayment for tile, resilient coverings, carpeting, or thin brick.

- Floor or wall heat shields for stoves.

- Backing for tiles on walls in shower stalls or tub surrounds.

- Base for countertops, **Figure 16-19.**

- Base for exterior finishes such as ceramic tile, thin brick, or synthetic stucco.

Figure 16-19. Cement board is used as backing for ceramic tile on walls, countertops, and other locations that are frequently exposed to water. (U.S. Gypsum Co.)

16.4.1 Working with Cement Board

Cement board can be worked with ordinary carpenter's tools. Panels can be fastened with nails, screws, or staples. Common panel thicknesses are 1/4″, 7/16″, and 1/2″. Panel dimensions are 3′ × 5′ and 4′ × 8′.

To cut cement board, score it several times with a tungsten-tipped knife. Use a straightedge or drywall square as a guide, **Figure 16-20.** Use the straightedge to apply topside pressure at the score and snap the panel upward. Rough edges can be smoothed with a rasp or coarse sandpaper. A circular saw or a handsaw can also be used to make cuts, **Figure 16-21.** Small holes should be outlined with a series of drilled holes. Use a tungsten carbide-tipped masonry bit. On larger holes, score all sides and then make a diagonal score across the opening. In either case, break out the hole from the face side with a hammer. See **Figure 16-22.**

Cement board is installed in about the same way as conventional wallboard. **Figure 16-23** shows a detail of the construction at the edge of a tub. Stud spacing should not be greater than 16″ O.C. Note the furring strip. It ensures alignment between the tub lip and the wallboard. Also note the 1/4″ space that should be maintained along the tub edge.

Figure 16-21. Sawing with a hand or power saw is an alternate method of cutting cement board. (U.S. Gypsum Co.)

Figure 16-22. Cutting holes in cement board. For small openings, make a series of small holes and break out the waste with a hammer. Score larger openings and saw where possible. (James Hardie Building Products)

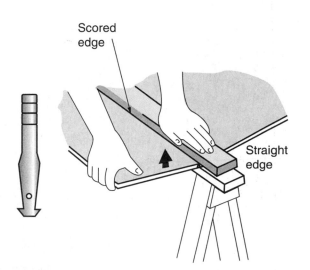

Figure 16-20. Cutting cement board. (James Hardie Building Products)

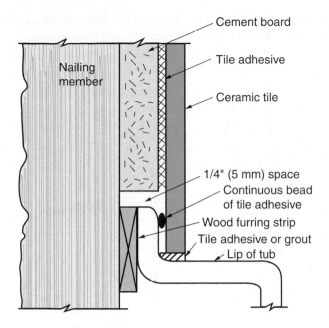

Figure 16-23. Installation of cement board around a tub. (Gold Bond Building Products)

16.5 Moisture-Resistant (MR) Wallboard

Moisture-resistant wallboard is a type of gypsum wallboard processed to withstand the effects of moisture and high humidity. Its facing paper is light green so it can be easily identified. Because of this coloring, it is often referred to as *greenboard*. Moisture-resistant wallboard is not as water-resistant as cement board, but is still extensively used. Standard thicknesses include 1/2″ and 5/8″. Standard width is 4′ and lengths range from 8′ through 12′. This product is not used as a base under ceramic tile or other nonabsorbent finishing materials in showers and tub alcoves. However, it can be used in other areas of the bathroom.

Moisture-resistant (MR) wallboard: A type of gypsum wallboard processed to resist the effects of moisture and high humidity. It is not used as a base under ceramic tile and other nonabsorbent finishes used in showers and tub alcoves.

16.6 Veneer Plaster

Veneer plaster is a high-strength material applied as a coat less than 1/8″ thick. Because of the composition and thinness of the coat, it very rapidly dries. Trim and decoration work may proceed after a minimum drying time of 24 hours.

A special gypsum board is used for the base. Its face surface consists of several layers of paper. The outer layer rapidly absorbs moisture and makes it easier to apply the plaster coat. The inner layer keeps the gypsum core dry and rigid. To identify the face surface, note that the outer layer is rolled over the long edge. Other than this, the materials and methods are nearly the same as those for regular drywall construction.

Veneer plaster can be applied as a one- or two-coat system. Either system can be given a smooth or a textured surface. Corner bead, trim, and grounds must be carefully set for a 1/16″ thickness in one-coat applications and 3/32″ for two-coat applications. **Figure 16-24** shows a two-coat application of veneer plaster.

Figure 16-24. Application of a two-coat veneer plaster system to a shower stall. The first coat is being applied with a standard trowel. The palette held in the right hand of the plasterer is called a hawk.

The manufacturer's directions should be carefully followed for best performance and workability of veneer plaster. Proper mixing is especially important. This is usually accomplished with a cage-type paddle mounted in an electric drill.

16.7 Predecorated Wallboard

A variety of predecorated gypsum wallboard is available. This type of wallboard is usually vertically applied because of the difficulty involved in successfully matching and finishing butt joints.

Wall surfaces must be dry before installation can begin. Panels should be unpacked and stood on their long edges, exposing both sides to room air, for 24–48 hours before being attached.

The panels can be attached to furring strips, studs, or other solid surfaces. On remodeling jobs, first remove wallpaper and loose paint. Also, repair any damaged plaster. The use of an adhesive to bond the panels to a base layer is common practice. However, color-matched nails are available, as well. For best results, manufacturers recommend both gluing and nailing for 1/4″ panels. To avoid damaging the finish, be sure to drive colored nails with a plastic-headed hammer, rawhide mallet, or a special cover placed over the face of a regular hammer. Nails should be spaced 8″ apart and should never be closer than 3/8″ to the ends or edges of the wallboard. Avoid a tight fit at floors and ceilings. An expansion space of 1/16″ should be allowed to avoid buckling.

When adhesives alone are used for fastening, check each panel after about 15–30 minutes to ensure that the adhesive is set. Firmly press along the edges and framing members. Use a rubber mallet or cover a block of wood with a soft cloth and tap along all areas where adhesive was applied. Adhesive on the decorated surface must be immediately removed with a soft cloth and mineral spirits.

To trim edges and joints of predecorated panels, you can use aluminum trim made to match the finished surface of the wallboard, **Figure 16-25.** Cut the trim with a hacksaw. Attach it with flat-head wire nails spaced 8″–10″ apart. When attaching divider strips, first place the trim on one panel that is carefully aligned, nail the exposed flange in place, then insert the next panel. **Figure 16-26** shows a completed panel installation.

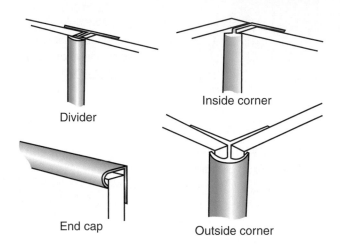

Divider   Inside corner

End cap   Outside corner

Figure 16-25. Trim can be used to cover raw edges of predecorated gypsum wallboard.

Figure 16-26. This bathroom has walls with prefinished paneling. (ABTco, Inc.)

16.8 Wallboard on Masonry Walls

Gypsum wallboard can be installed over metal or wood furring strips attached to a masonry wall. Where the structure consists of an interior wall that is straight and true, the panels can be directly laminated to the masonry surface with a special adhesive.

Exterior walls must be thoroughly waterproofed. Insulation should be included if the structure is located in a cold climate. **Figure 16-27** shows a masonry wall application made with furring strips. The insulation may be rigid foam or batts. It is best to use a powder-actuated nailer and special fasteners to attach wood furring strips to concrete or masonry surfaces. Such strips should be a nominal 2″ wide and 1/32″ thicker than the insulation. Rigid foam insulation is usually bonded to the masonry surface with adhesive. Wallboard joints and nail holes are concealed following the finishing steps previously described.

Installations of wallboard on interior masonry walls, especially those below grade, must be carefully done. Be sure to follow recommendations furnished by manufacturers.

16.9 Installing Plywood Paneling

To provide greater fire resistance, some building codes require a drywall base under sheet paneling, such as plywood. This practice is recommended even where it is not required. In addition to its fire resistance, the wall will be stronger and provide some insulation against sound transmission. Further, it provides a more rigid, smooth surface for the paneling.

Most of the plywood paneling used for interior walls has a factory-applied finish that is tough and durable, **Figure 16-28**. Manufacturers can furnish matching trim and moulding that is also prefinished and easy to apply. Color-coordinated putty sticks are used to conceal nail holes.

Joints between plywood sheets can be treated in a number of ways. Some panels are fabricated

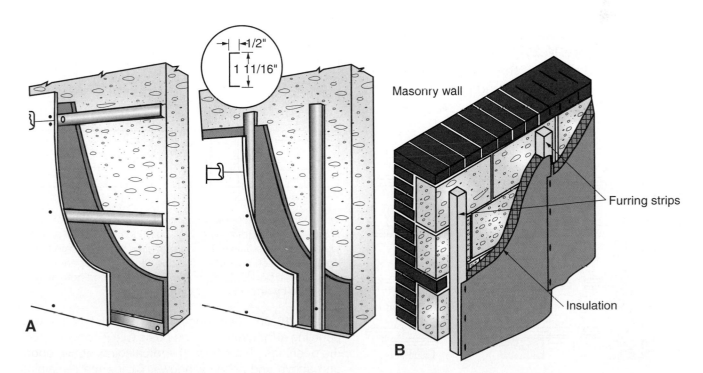

Figure 16-27. Two methods of preparing masonry walls for interior finish. A—Wallboard can be attached to metal furring channels. Rigid insulation is used. B—Wood furring strips and blanket insulation.

Figure 16-28. Prefinished plywood for interior walls is available in a variety of finishes. A base consisting of drywall is recommended. (ABTco, Inc.)

with machine-shaped edges that permit almost perfect joint concealment. Usually, it is easier to accentuate the joints with grooves or battens and strips, **Figure 16-29.**

Before installation, the panels should be allowed to become conditioned (adjusted) to the temperature and humidity of the room. Remove

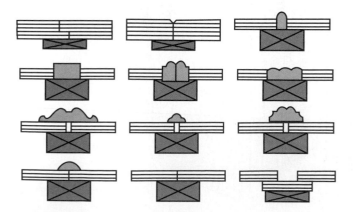

Figure 16-29. Different styles of battens can be used to conceal joints in plywood paneling.

prefinished plywood from cartons and carefully stack it flat. Place 1″ spacer strips between each pair of face-to-face panels. This allows air to circulate between the panels. After at least 48 hours, the panels can be installed.

Plan the layout to reduce the amount of cutting and number of joints. **Figure 16-30** shows two application designs. It is important to align panels with openings whenever possible. If the finished panels have a grain, stand the panels around the walls and shift them until

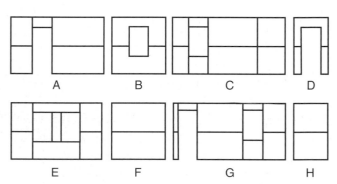

In design 1, a simple two-panel horizontal arrangement has the single continuous horizontal joint placed midway between the floor and ceiling. The panel design is defined by the vertical joints at openings, such as those in elevations A, C, E, and G. However, key panels may be omitted where the panel and length exceed the wall element width, as in elevations B and D.

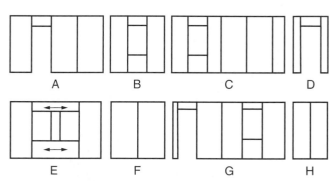

In design 2, a vertical-arrangement panel design involves lining up vertical joints with wall openings. Then, the plain wall space is vertically divided in widths proportionate to the openings. When the width of a door or window opening exceeds the panel width, panels may be placed horizontally, as shown by arrows in elevation E.

Figure 16-30. Two commonly used plywood wall paneling arrangements. The basic rule is to "work from the openings." First, line up vertical joints above doors and above and below windows. Divide the remaining plain wall space into an orderly pattern as stud location allows.

you have the most pleasing effect in color and grain patterns. To avoid confusion, number the panels in sequence after their position has been established.

When cutting plywood panels with a portable saw, mark the layout on the back of the panel. The cutting action of the saw blade will be upward against the panel face. Splintering will be minimal. This is even more important when working with prefinished panels. When using a handsaw, mark on the face and cut from that side. Carefully support the panel and check for clearance below.

When studs are poorly aligned, it is advisable to install a backing of drywall, or at least furring strips, to provide an even surface. If installation is made over an existing surface in poor condition, it is usually advisable to use furring.

16.9.1 Installing Furring

Horizontally nail 1 × 3 or 1 × 4 furring strips across the studs. Start at the floor line and continue up the wall. Spacing depends on the panel thickness. Thin panels need more support. Install vertical strips every 4′ along the wall to support panel edges. Level low areas by shimming behind the furring strips. Use a line stretched from corner to corner to locate low spots. **Figure 16-31** shows furring strips being attached to a concrete block wall. Prefinished plywood panels will be installed over the furring.

16.9.2 Installing Panels

It is important to mark stud locations before putting up paneling, plywood or otherwise. Use a pencil to mark their location both on the floor and ceiling. Often the panels have surface grooves in them at modular increments. If the first panel is installed so the trailing edge centers on a stud, some of the grooves will fall on studs. Nailing in the groove helps conceal the fasteners.

Plywood can be directly attached to the wall studs with nails or special adhesives. For 1/4″

Figure 16-31. Paneling can be directly attached to masonry walls, but furring strips produce a better-looking wall.

paneling, space nails about 6″ apart along the edges and about 12″ apart over studs for field nailing. When using 3/8″ or thicker plywood, space nails further apart. When a drywall base is used, the plywood panel can be bonded to it with adhesive. In general, follow the same procedures described in double-layer drywall construction.

Working Knowledge

Before installing any panels, apply narrow strips of black paint floor-to-ceiling. These strips should be centered wherever there will be a joint between panels. If shrinkage during the heating season opens a joint between panels, it will not be noticeable because of the paint.

PROCEDURE

Marking the first panel

1. Begin installing panels at a corner. Plumb the panel and temporarily tack it. The leading edge of the plumbed panel, at some point, should touch the intersecting wall.
2. Check that the trailing edge of the panel is centered over a stud. If not, measure the amount of adjustment needed. The distance between points of the dividers should be set to include the widest point of the gap between the wall and the distance the panel overlaps the center of a stud. For wide distances, using a block of wood the correct width will produce a more accurate scribe.
3. Scribe a trimming line with one end of the dividers against the wall and the other end marking a line on the panel. Be sure to keep the dividers at right angles to the abutting wall as you scribe from top to bottom.
4. Remove the panel and place it on a pair of sawhorses. Support the panel with 2 × 4 lumber or several sheets of paneling.
5. If a portable power saw is used, the scribed line must be transferred to the back side of the panel. If the scribed line is irregular, using a power saw is not practical. Use a handsaw to make the cut following the scribed line.
6. Offset the panel before starting the cut to avoid cutting into the supporting panels or 2 × 4s. Undercut the edge slightly (about 5°) so the facing edge will snugly fit.
7. Check again that the trailing edge is centered on a stud. Fasten it in place before fitting the next panel. Allow about 1/4″ clearance at the top and bottom. Fasten all panels with 5d or 6d casing or finishing nails.

PROCEDURE

Making cutouts for electrical boxes

1. Carefully measure the location of switch and outlet boxes on the wall. Transfer these to the panel. For accuracy, vertical measurements must be plumb and horizontal measurements must be level.

(Continued)

(Continued)

2. If the opening is near the floor, plumb both sides of the box and make marks on the floor. If the box is near the ceiling, take the vertical measurements to the ceiling. Measure the distance from the box to both marks.
3. Take level measurements from the top and bottom of the box and extend the measurements to the edge of the previous panel. Make careful measurement of these distances.
4. Transfer the measurements to the panel and make the cuts.

PROCEDURE

Trimming the last panel

The last panel on the wall needs to be trimmed to properly fit in the corner.

1. Take careful level measurements at several points. Usually, top, middle, and bottom are sufficient.
2. Transfer these measurements to the panel, draw a cut line from mark to mark. Then, make the cut. The fit does not need to be snug if the intersecting wall is to be paneled or a corner moulding will be used.
3. Repeat these steps for all walls to be paneled.
4. After all panels are in place, use mouldings to cover the gap along the ceiling. Baseboards will conceal the gaps at the floor line.

PROCEDURE

Fitting snug corners

If no corner moulding is used and the intersecting wall is not paneled, you may use the following procedure to obtain a snug fit at the corner.

1. Cut the last sheet about 1/2″ wider than the width of the widest gap.
2. Hold or temporarily tack the sheet in place with the trailing edge overlapping the next-to-last sheet. Make sure the sheet is plumb.
3. Set the dividers to the width of the overlap.
4. Scribe a line that width on the edge of the sheet next to the corner.
5. Cut the sheet to the scribed line and install.

16.10 Hardboard

Through special processing, *hardboard* (also called *fiberboard*) can be manufactured with a low moisture-absorption rate. The face is often scored to form a tile pattern for use in bathrooms and kitchens. Panels for wall application are usually 1/4″ thick.

Since hardboard is made from wood fibers, the panels slightly expand and contract with changes in humidity. Panels should be installed when they are at their maximum size. There will be a tendency for them to buckle between the studs or attachment points if installed when moisture content is low. Manufacturers of prefinished hardboard panels recommend the panels be unwrapped and then separately placed around the room for at least 48 hours before installation.

Installation methods are similar to those previously described for plywood. When applying factory-finished wallboard, plywood, or hardboard materials, always follow the recommendations furnished by the manufacturer. Special adhesives are available, as are metal or plastic trim in matching colors. Drill nail holes for the harder types. **Figure 16-32** shows an attractive room paneled in hardboard.

16.11 Plastic Laminates

Plastic laminates are sheets of a synthetic material that is hard, smooth, and highly resistant to scratching and wear. Although basically designed for use on tables and countertops, plastic laminates are also installed as wainscoting and wall paneling. Since the material is thin (1/32″–1/16″), it must be bonded to other

Figure 16-32. This prefinished hardboard wall paneling is made to look like pine. (Masonite Corp.)

supporting panels. Contact cement is commonly used as the bonding agent.

Manufacturers have developed prefabricated panels with the plastic laminate already bonded to a base or backing material. Edges are tongue-and-groove, so that units can be blindnailed. When blindnailed, the nail heads do not appear on the finished surface. Use a 6d finish nail and drive it at a 45° angle into the base of the tongue and on into the bearing point. See **Figure 16-33**. Matching corners and trim are available.

16.12 Solid Lumber Paneling

Solid wood paneling makes a durable and attractive interior wall surface. It may be used in nearly any type of room. A number of

---

Hardboard: A board material manufactured of wood fiber formed into a panel with a density of approximately 50–80 pounds per cubic foot. Also called *fiberboard*.

Plastic laminates: Sheets of synthetic material. Hard, smooth, and highly scratch resistant.

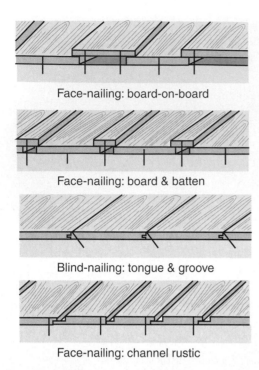

Face-nailing: board-on-board

Face-nailing: board & batten

Blind-nailing: tongue & groove

Face-nailing: channel rustic

Figure 16-33. Nailing techniques for a variety of paneling styles. (Western Wood Products Assn.)

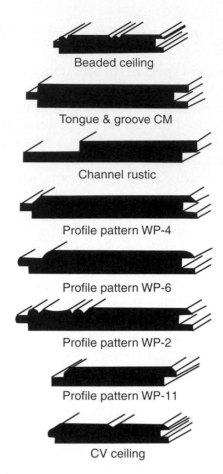

Beaded ceiling

Tongue & groove CM

Channel rustic

Profile pattern WP-4

Profile pattern WP-6

Profile pattern WP-2

Profile pattern WP-11

CV ceiling

Figure 16-34. Smooth-surfaced solid wood paneling is available in several pattern profiles. (Western Wood Products Assn.)

different species of hardwood and softwood are available. Sometimes, grades that contain numerous knots are used for a special appearance. Defects, such as the deep fissures in pecky cypress, can provide a dramatic effect.

Softwood species most commonly used for solid lumber paneling include pine, spruce, hemlock, and western red cedar. Boards range in widths from 4"–12" (nominal size) and are dressed to 3/4" thickness. Board-and-batten or shiplap joints are sometimes used, but tongue-and-groove (T&G) joints combined with shaped edges and surfaces are more popular.

Paneling patterns are often reversible, offering two choices in a single panel. Some patterns are smooth on one side and saw-textured on the other. **Figure 16-34** shows several patterns offered by one manufacturer. Dozens of variations are possible, varying by species of wood, texture, finish, and how the paneling is applied to a wall.

As with plywood paneling, allow the boards to adjust to the temperature and humidity of the room before installation. Stand them around the room. At the same time, match the boards for color and grain. If tongue-and-groove boards

are to be stained or finished later, apply the same finish to the tongues. Unless this is done, later shrinkage of the wood will expose unfinished surface of the tongue.

When solid wood paneling is horizontally applied, furring strips or blocking is not required. The boards are directly nailed to the studs. Inside corners are formed by butting the paneling units flush with the other walls.

Vertical installations require furring strips at the top and bottom of the wall and at various intermediate spaces. Sometimes, 2 × 4 blocking is installed between the studs to serve as a nailing base. Even when heavy tongue-and-groove boards are used, these nailing members should not be spaced more than 48" apart.

Narrow widths (4"–6") of tongue-and-groove paneling are blindnailed. This eliminates the need for countersinking and filling nail holes. It also provides a smooth, blemish-free surface.

This is especially important when clear finishes are used.

Exterior wall constructions where the interior surface consists of solid wood paneling should include a tight application of building paper or housewrap. This will prevent the infiltration of wind and dust through the joints. In cold climates, insulation and vapor barriers are important.

Working Knowledge

If random widths are used, boards on adjacent walls must match and be accurately aligned.

PROCEDURE

Installing vertical solid paneling

1. Start installation at a corner. Choose a straight board and cut it slightly shorter than the height of the room—normally about 1/4″ shorter. Divide the space between the top and the bottom, unless you do not plan installing moulding at the top.

2. With tongue-and-groove boards, plumb the board with the groove edge to the wall. Tack it in place while you scribe the edge to match the adjoining wall.

3. Rip the board along the scribed line and face nail the edge next to the wall. Use finishing nails 16″ apart. Then, edge nail the trailing edge.

4. Continue installing boards by slipping the groove into the tongue of the previously installed board and edge nail only. Slightly warped boards can be pried into place with a wood chisel. Badly warped pieces should not be used.

5. To mark the last board so it tightly fits the adjoining wall, temporarily install it in place of the next-to-last board. Then, with a block or dividers at the same width as the finished face of the next-to-last board, scribe a line. Remove the board and cut to the scribed line.

6. Install and fasten the next-to-last board and the last board.

16.12.1 Installing Solid Paneling at an Angle

Though more difficult and time-consuming to install, angled paneling is handsome and goes well with informal designs. *Diagonal paneling* angles in only one direction. Angles of 22 1/2°, 30°, and 45° are most popular. *Chevron paneling* is installed in a V or inverted V pattern. *Herringbone paneling* alternates the direction of the angle at regular intervals. All three types are shown in **Figure 16-35** along with tables on coverage.

To determine the proper angle for diagonal paneling, first establish a vertical line in the middle of the wall using a plumb line. Draw an intersecting horizontal line about one foot off of the floor. Mark 3′ from the intersection in each direction (horizontal and vertical). Draw a diagonal line from each point on the horizontal line through the point marked on the vertical line. The result is two 45° angles. For other angles, use a protractor to establish the diagonal lines.

For chevron and herringbone panel application, it is important to have plumb nailing bases wherever the diagonals change direction. This can be a stud or furring strip. For chevron paneling, the nailing base is located midway across the wall. For herringbone paneling, locate a plumbed vertical nailer every 36″, **Figure 16-36.** Center an additional nailing surface at 18″.

To start chevron and herringbone patterns, install triangles of paneling at the centerline(s). These should be glued or blindnailed. For tight, well-matched joints, mark the cutting angle for each board as it is installed. Use a level or straightedge as a guide. Manufacturers, or their trade associations, provide detailed instructions for such installations.

Diagonal paneling: Solid paneling installed at a 45° angle.

Chevron paneling: Solid paneling installed diagonally in a chevron pattern.

Herringbone paneling: Angled solid paneling that alternates the direction of the angle at regular intervals.

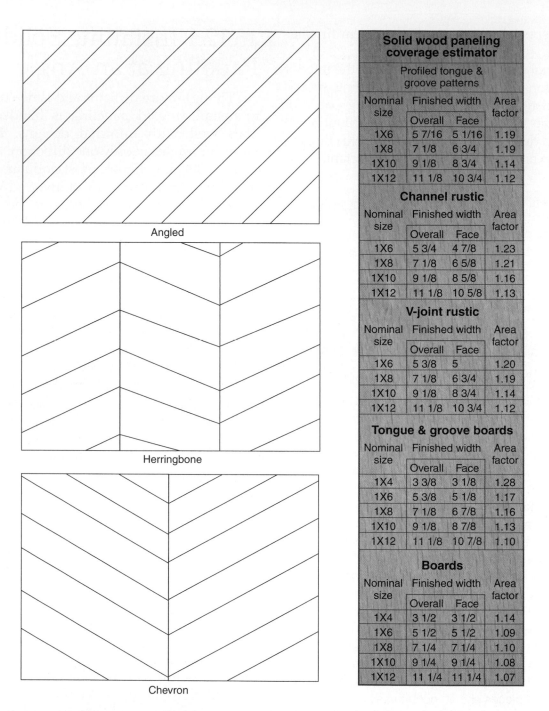

Angled

Herringbone

Chevron

Solid wood paneling coverage estimator

Profiled tongue & groove patterns

Nominal size	Finished width		Area factor
	Overall	Face	
1X6	5 7/16	5 1/16	1.19
1X8	7 1/8	6 3/4	1.19
1X10	9 1/8	8 3/4	1.14
1X12	11 1/8	10 3/4	1.12

Channel rustic

Nominal size	Finished width		Area factor
	Overall	Face	
1X6	5 3/4	4 7/8	1.23
1X8	7 1/8	6 5/8	1.21
1X10	9 1/8	8 5/8	1.16
1X12	11 1/8	10 5/8	1.13

V-joint rustic

Nominal size	Finished width		Area factor
	Overall	Face	
1X6	5 3/8	5	1.20
1X8	7 1/8	6 3/4	1.19
1X10	9 1/8	8 3/4	1.14
1X12	11 1/8	10 3/4	1.12

Tongue & groove boards

Nominal size	Finished width		Area factor
	Overall	Face	
1X4	3 3/8	3 1/8	1.28
1X6	5 3/8	5 1/8	1.17
1X8	7 1/8	6 7/8	1.16
1X10	9 1/8	8 7/8	1.13
1X12	11 1/8	10 7/8	1.10

Boards

Nominal size	Finished width		Area factor
	Overall	Face	
1X4	3 1/2	3 1/2	1.14
1X6	5 1/2	5 1/2	1.09
1X8	7 1/4	7 1/4	1.10
1X10	9 1/4	9 1/4	1.08
1X12	11 1/4	11 1/4	1.07

Figure 16-35. Three patterns for angled solid paneling. Tables such as those at the right are used for estimating coverage of various patterns of solid paneling. (Western Wood Products Assn.)

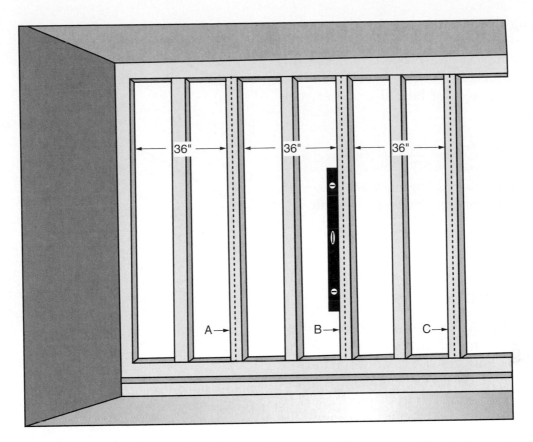

Figure 16-36. Herringbone paneling requires furring strips every 18″ on center. Plumb lines should be drawn every 36″ on center. (Western Wood Products Assn.)

Grounds: Strips of wood installed as guides at the floor line and at openings in a wall used to strike off plaster.

16.13 Plaster

Through the years, gypsum plaster has provided desirable qualities in a wall and ceiling finish: beauty, durability, economy, fire protection, structural rigidity, and resistance to sound transmission. It is also highly adaptable, since it can be readily applied to curved or irregular surfaces.

Fire protection engineers have developed accurate fire-protection ratings for plaster. These are listed by the American Insurance Association (formerly known as the National Bureau of Standards and the National Board of Fire Underwriters). These ratings are used as a basis for establishing requirements in various building codes.

When plaster is used for the interior wall and ceiling surfaces, the carpenter usually installs grounds that serve as guides for the plasterer. *Grounds* are strips of wood or metal placed along floors and around wall openings as a thickness guide. The carpenter also applies the lath that forms the plaster base. For this reason, it is important for the carpenter to have a general knowledge of how the plaster coats are applied.

16.13.1 Plaster Base

A plaster finish requires some type of base. For many years, *wood lath* was used for this purpose. The lath consisted of thin, narrow strips of soft wood that were nailed to studs. The strips were spaced a short distance apart, so that the wet plaster could be forced through the gaps. As the wet plaster was pushed through, it slumped downward to form a key. When dry, this key firmly held the plaster to the lath. Other materials have generally replaced wood lath, but carpenters may encounter it in remodeling jobs on older buildings. Today, commonly used plaster bases include *gypsum lath* and *expanded metal lath*, **Figure 16-37.** Plaster is also sometimes applied directly to masonry surfaces or special gypsum block units.

Gypsum lath consists of a rigid gypsum filler with a special paper cover. A standard panel measures 16″ × 48″ and is horizontally applied to the framing members of the structure. For a stud or ceiling joist spacing of 16″ O.C., 3/8″ gypsum lath is used. For a 24″ spacing, 1/2″ is required. Gypsum lath is also made with a backing of aluminum foil vapor barrier. Lath with perforations improves the plaster bond and extends the time the wall surface remains intact when exposed to fire. Some building codes specify this type.

Expanded metal lath consists of a copper alloy steel sheet that is slit and expanded to form openings for keying the plaster. The two most common types are diamond mesh and flat rib. Standard-size pieces are 27″ wide by

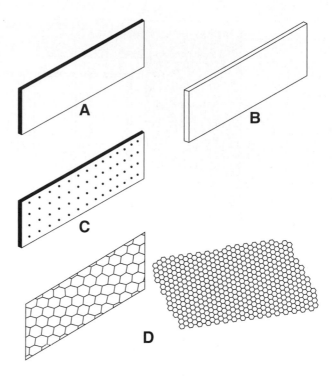

Figure 16-37. Plaster base materials. A—Gypsum lath. B—Insulating fiberboard lath. C—Perforated gypsum lath. D—Expanded metal lath.

96″ long. To improve its rust-resistance, metal lath is usually galvanized or dipped in black asphaltum paint.

Insulating fiberboard lath is also used as a plaster base. It comes in a 3/8″ and 1/2″ thickness with a shiplap edge. Fiberboard lath has considerable insulation value and is often used on ceilings or walls adjoining exterior or unheated areas. **Figure 16-38** shows the available sizes.

Wood lath: Plaster base consisting of thin, narrow strips of soft wood nailed to studs.

Gypsum lath: A common plaster base consisting of a rigid gypsum filler with a special paper cover.

Expanded metal lath: Plaster base consisting of a copper alloy steel sheet that is slit and expanded to form openings for keying the plaster.

Insulating fiberboard lath: Plaster base that comes in a 3/8″ and 1/2″ thickness with a shiplap edge. It has considerable insulation value and is often used on ceilings or walls adjoining exterior or unheated areas.

PROCEDURE

Installing gypsum lath

Before lath is applied, inspect the framing for proper spacing and alignment. Check to see that corners, ceiling lines, and openings have nailers to support the ends of the lath. Insulating lath is installed in the same manner as gypsum lath, except that 13 gage 1 1/4″ blued nails should be used.

1. Install the ceiling and then the walls. Work from the top down on walls. Apply with the long dimension at right angles to the framing.
2. Stagger the end joints between adjacent courses, **Figure 16-39.**
3. Gypsum lath is easily cut to size by scoring one or both sides with a pointed or edge tool. Break the lath along the line. Be sure to make neatly fitted cutouts for plumbing pipes and electrical outlets.
4. Turn the folded or lapped paper edges of gypsum lath toward the framing. Edges and ends of lath should be in moderate contact.
5. Apply 3/8″ thick gypsum lath with 13 gage gypsum lathing nails 1 1/8″ long. Nail sizes for other installations are given in **Figure 16-40.** The lath must be nailed at each stud or joist. Begin nailing from the center and work toward outer edges. As you nail, firmly press the lath against the frame. Drive nails flush without cutting the face paper. Keep nails 3/8″ away from ends and edges. Other types of fasteners may be used, as well. This includes screws, staples, and special clips. Staples should be flattened wire. Drive them so the crown lies in the same direction as (parallel to) the wood framing. Screws must always be used when attaching to steel frame members.

Type	Thickness (inches)	Width (inches)	Length (inches)
Plain	3/8 1/2	16 16	48 or 96 48
Perforated	3/8 1/2	16 16	48 or 96 48
Insulating	3/8 3/8 or 1/2	16 24	48 or 96 as requested to 12′
Long length	1/2	24	as requested to 12′

Figure 16-38. Gypsum and insulating lath are available in these sizes.

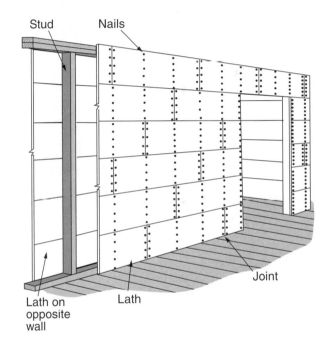

Figure 16-39. Gypsum lath over wood studs. Stagger joints.

16.13.2 Metal Lath

In commercial construction, metal lath is common. In residential work, it is often used only in shower stalls or tub alcoves, **Figure 16-41.** When the construction is properly designed, the metal lath provides a rigid wall surface.

Metal lath contains many small voids. These openings allow proper keying of the plaster.

Produced in different weights, each bundle has some identifying mark for weight. For example, one manufacturer uses a color code. Ends of bundles are spray painted, 3.4 lb. in red, 2.5 and 2.75 lb. in white. Diamond mesh 1.75 lb. is not painted.

Metal lath must be of the proper type and weight for the support spacing. Sides and ends are lapped and corners are returned (overlapped). Studs are usually first covered with

Type framing	Base thickness		Fastener	Max. frame spacing		Max. fastener spacing	
	inch	mm		inch	cm	inch	cm
Wood	3/8	10	Nails 13 gage, 1 1/8" long, 19/64" flat head, blued	16	40	5	13
			Staples— 16 gage galvanized, flattened wire flat crown 7/16" wide, 7/8" divergent legs				
	1/2	13	Nails— 13 gage, 1 1/4" long, 19/64" flat head, blued	24	60	4	10
			Staples— 16 gage galvanized, flattened wire flat crown 7/16" wide, 1" divergent legs				
USG steel stud	3/8	10	1" Type S screws	16 24	40 60	12	30
TRUSTEEL stud	3/8	10	Clips	16	40	16	40

Figure 16-40. Follow these fastening requirements for gypsum lath.

a 15 lb. asphalt-saturated felt. Portland cement plaster is often used as the first coat when the surface will be finished with ceramic tile. Gypsum plaster is used for top coats.

16.13.3 Reinforcing

Since some drying always occurs in wood frame structures, shrinkage can be expected. This is likely to cause plaster to crack around openings, in corners, or wherever there is a concentration of cross-grained wood.

To minimize cracking, expanded metal lath is often used in key positions over the plaster base. Strips 8″ wide are applied at an angle over the corners of doors and windows as shown in **Figure 16-42.** Lightly tack or staple these into place so they become a part of the plaster base only. If nailed securely to the framing, warping, shrinking, and twisting of the frame will be transmitted into the plaster and cause cracks.

Metal lath should also be used under and around wood beams that will be covered with plaster. Be sure to extend the edges of the reinforcing well beyond the structural element being covered.

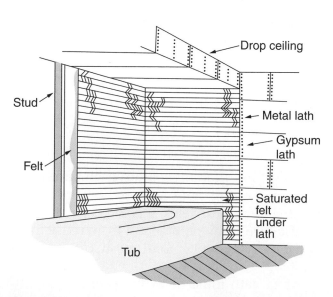

Figure 16-41. Metal lath is installed over a layer of waterproof felt paper in areas subject to moisture. The paper can be eliminated if cement board is used under the lath.

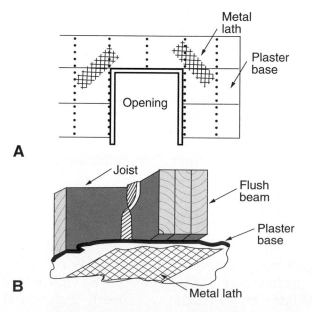

Figure 16-42. Reinforcing the plaster base. A—Place metal lath around the jamb area of large openings, especially where large headers are used. B—Provide reinforcing under flush beams.

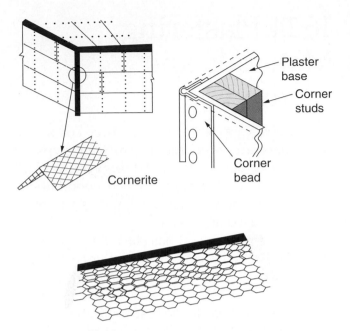

Figure 16-43. Cornerite and corner beads. Prefabricated reinforcing is made for both inside and outside corners. (Cemco)

Inside corners may be reinforced with a specially formed metal lath or wire fabric sometimes called Cornerite™, **Figure 16-43.** Minimum widths should be 5″ so that there is 2 1/2″ on each surface of the internal angle. In some plaster base systems that employ a special attachment clip, Cornerite is not recommended.

Outside corners where no wood trim will be applied are reinforced with metal corner beads. They must be carefully applied and plumbed or leveled. This is because they serve not only as a reinforcement for the plaster, but as a ground (guide) for its application. When applying corner bead, use a straightedge and level or plumb line. Also, use a spacer block to check the distance between the corner of the bead and the surface of the plaster base. This distance must be equal to the thickness of the plaster coats.

16.13.4 Plaster Grounds

Plaster grounds are usually wood strips as thick as the plaster base and plaster. For average residential construction, this is 3/8″ plus 1/2″ for a total thickness of 7/8″. Grounds are installed before the plaster is applied. The plasterer uses

them as a gauge for thickness of the plaster. They also help keep the plaster surface level and even. Later, the grounds may become a nailing base for attaching trim members. They are used around doors, windows, and other openings. Sometimes, they are included at the bottom of walls along the floor line, **Figure 16-44.** In some wall systems—especially large commercial and institutional buildings—metal edges and strips serve as grounds.

Windows are usually equipped with jamb extensions that are adjusted for the various thicknesses of materials used in the wall structure. These jamb extensions serve as plaster grounds. The carpenter seldom needs to make any changes.

Grounds around some openings are removed after the plastering is complete. Those used at door openings must be carefully set (plumbed) and conform to the width of the doorjamb to ensure a good fit of the casing. The width of standard interior doorjambs is usually 5 1/4″.

Carpenters often construct a jig or frame that is temporarily attached to the door opening. Grounds can then be quickly nailed in place along the straight edges of the jig. Instead of using two strips, some carpenters prefer to use a single piece of 3/8″ exterior plywood ripped to the same width as the doorjamb, **Figure 16-45.** Such grounds, if carefully removed, can be reused on future jobs.

When the plaster base is complete, mark lines on the subfloor at the centerline of each

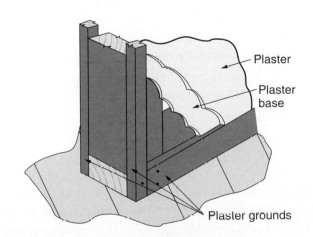

Figure 16-44. Plaster grounds are installed around openings and sometimes along the floor.

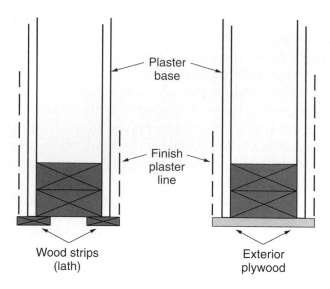

Figure 16-45. Two methods of installing removable grounds at door openings. These are top views.

stud. After the plaster has been applied and is dry, these marks are transferred to the wall. They help when installing baseboards, cabinets, and fixtures.

16.13.5 Plaster Base on Masonry Walls

Gypsum plaster can be directly applied to most masonry surfaces. However, furring strips are usually installed on outside walls to prevent excessive heat loss through the walls. These carry the plaster base materials and provide an air space. Strips are attached in one of the following ways:

- Driving case-hardened nails into the mortar joints or using a powder-actuated nailer.

- Nailing into metal or wood plugs placed in the wall during construction.

- Application of adhesive.

When heat loss needs to be further reduced, blanket insulation can be installed. Some carpenters use rigid polystyrene panels. These are directly attached to the wall with adhesives. If furring strips are not used, the plastic foam can serve as a plaster base. Follow manufacturer's recommendations when making an installation of this kind.

16.14 Plastering Materials and Methods

Plaster is applied in two or three coats. The base coats are prepared by mixing gypsum with an aggregate, either at the gypsum plant or on the job. The aggregate may be wood fibers, sand, perlite, or vermiculite. Sand is the most commonly used aggregate.

Discard plaster that has started to set. Wash out the mixer with clean water after each batch has been prepared. Keep tools and equipment clean.

In three-coat work, the first application, called the *scratch coat,* is directly applied to the plaster base. It is cross-raked, or scratched, after having "taken up" (stiffened), **Figure 16-46.** The scratch coat is then allowed to set and partially dry. The second application, or *brown coat,* is then applied and leveled with the grounds and screeds. A long flat tool called a *darby* and a rod (straightedge) are used. When the brown coat has set and is somewhat dry, it is time to apply the third or *finish coat.* This coat is about 1/16″ thick.

In two-coat work, the scratch coat and brown coat are applied at almost the same time. The cross raking of the scratch coat is omitted. The brown coat of plaster is usually applied (doubled-back) within a few minutes. This application method is the one most frequently used over gypsum or insulating lath plaster bases in residential construction.

The minimum plaster thickness for all coats is 1/2″ when applied to regular gypsum or insulating lath bases. A 5/8″ thickness is usually required over brick, tile, or masonry.

Scratch coat: First layer of plaster that has its surface roughened to provide tooth for succeeding layers.

Brown coat: Second layer of plaster that is applied on top of the scratch coat to form a base for the finish coat.

Darby: A long flat tool used to level the brown coat.

Finish coat: Third layer of plaster that is applied when the brown coat has set and is somewhat dry.

When plaster is applied to metal lath, it should measure 3/4″ in thickness from the backside of the lath.

A plaster job should be constantly inspected for base coat thickness. Unless grounds and screeds are used on ceilings or large wall areas, it is extremely difficult to keep the thickness uniform. Should the thickness be reduced, the possibility of checks and cracks is much greater. A 1/2″ thickness of plaster possesses almost twice the resistance to bending and breaking as a 3/8″ thickness.

The final or finish coat consists of two general types: sand-float (or textured) finish and putty (or smooth) finish. In the sand-float finish, special sand is mixed with gypsum or lime and cement. After the plaster is applied to the surface, it is smoothed with a float to produce various effects. The final finish depends on the floating method and the coarseness of the sand. A smooth finish is produced by applying a putty-like material consisting of lime and gypsum or cement. It is troweled perfectly smooth, like concrete, **Figure 16-47.**

Working Knowledge
Sometimes, a plastering machine is used instead of hand troweling. It not only saves a great deal of labor but also improves the quality of the plaster application. The lapsed time between mixing and application is shortened. The machine also makes possible the control of the plaster coat's density.

16.15 Ceiling Tile

Ceiling tiles are found in both old and new construction. They can be installed over engineered metal strips, wood furring strips, solid plaster, drywall, or any smooth, continuous surface. There is a considerable range in the types of material used to make ceiling tile. Some types are fiberboard tile, mineral tile, perforated metal tile, and glass-fiber tile. When selecting a product, consider its appearance, light reflection, fire resistance, sound absorption, maintenance, cost, and ease of installation.

Figure 16-46. Cross-raking a scratch coat of plaster. The raking helps the brown coat adhere to the scratch coat. (National Gypsum Co.)

Figure 16-47. A putty-like mixture of lime and gypsum or cement is used for the finish coat of plaster. It is troweled to an even thickness with a smooth finish. (National Gypsum Co.)

A standard size tile is 12″ × 12″. However, the tiles are available in larger sizes; for example, 24″ × 24″ and 16″ × 32″. A wide range of surface patterns and textures is manufactured.

Figure 16-48 illustrates an overlapping, tongue-and-groove edge that provides a wide flange to receive staples. This type of joint also permits efficient installation when an adhesive is used to hold the tile in place.

PROCEDURE

Layout procedure

1. Measure the two short walls and locate the midpoint of each one.
2. Snap a chalk line to establish the centerline. In the middle of this line, establish a chalk line that is at right angles to the long centerline. All tiles are installed with their edges parallel to these lines.
3. For an even appearance, the border courses along opposite walls should be the same width. For example, in a room 10′-8″ wide, use nine full tiles and two border tiles trimmed to 10″ each.
4. After determining the width of the border tile, snap chalk lines parallel to the centerlines to provide a guide for the installation of the border tiles or the furring strips that will support them.

16.15.1 Furring

Furring should be directly nailed to the ceiling joists on new construction. If applied over an old ceiling surface, be sure to locate and mark the joists before attaching the strips. Place the first furring strip flush against the wall at a right angle to the joists. Nail it with two 8d nails at each joist. Nail the second strip in place so that it will be centered over the edge of the border tile. Use the line previously laid out. All other strips are then installed on center. For 12″ × 12″ tile, locate strips every 12″. **Figure 16-49** shows completed furring.

Some carpenters prefer to start from the center and work each side toward the walls. If the furring strips are uniform in width, a spacer jig may speed up the job. Double-check the position of the furring strips from time to time. Make sure they will be centered over the tile joints.

It is essential that the faces of the furring strips be level with each other. Check alignment with a carpenter's level, straightedge, or line stretched across the strips. To align strips, drive tapered shims between the strips and the joist. If only one or two joists extend below the plane

Figure 16-49. A completed furring installation. Note that the two strips along the wall are closer together. This is to allow for the narrower width of the border tiles.

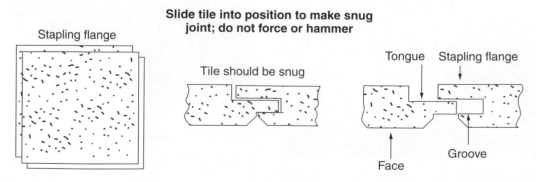

Figure 16-48. The tongue-and-groove joint used on standard ceiling tile. Flanges receive staples, while the groove holds one edge of the next tile.

of the others, it may be best to notch these joists before installing the furring strips.

If pipes or electrical conduit are located below the ceiling joists, it may be necessary to double the furring strips, **Figure 16-50.** The first course is spaced 24″–32″ on center and attached directly to the joists. The second course is then applied at a right angle to the first. Space strips according to the width of the tiles, as previously described. Large pipes or ducts that project below the ceiling joists should be boxed in with furring strips before the tile is installed. Wood or metal trim can be used to finish corners and edges.

There is an alternate to furring. It consists of installing gypsum wallboard to the joists or an existing ceiling surface. The tile is then attached to the wallboard either with adhesive or special staples.

Working Knowledge

Most manufacturers recommend that fiberboard tile be unpacked in the area where it will be installed at least 24 hours before application. This will allow the tile to adjust to room temperature and humidity.

16.15.2 Adhering Ceiling Tile

Figure 16-51 shows tile being attached with a stapler. Be sure the tile is correctly aligned.

PROCEDURE

Installing ceiling tile

1. Carefully check over the furring.
2. Snap chalk lines on the strips to provide a guide for setting the border tile. Do this along each wall. Be certain the lines are correct. Double-check to see that the chalk lines form 90° angles at the corners.
3. To cut the border tile, first deeply score the face with a knife drawn along a straightedge. Break it along this line by placing it over a sharp edge. Make any irregular cuts around light fixtures or other projections with a coping or compass saw. Power tools can also be used. Some tiles are made of mineral fiber, which rapidly dulls regular cutting edges.
4. Start the installation with a corner tile and then set border tile out in each direction. Fill in full-size tile.
5. When you reach the opposite wall, trim the border tile to size. If border tiles are full-width, remove the stapling flanges and face-nail the tile. Locate these nails close to the wall so they will be covered by the trim.

Firmly hold it while setting the staples. For 12″ × 12″ tile, use three staples along each flanged edge. Use four staples for 16″ × 16″ tile. Staples should be at least 9/16″ long.

There are several types of adhesive designed especially for installing ceiling tile. The thick-putty

Figure 16-50. Doubling the furring at right angles provides space for the conduit located below the ceiling joists.

Figure 16-51. Staples can be used to attach ceiling tile to furring strips. (Duo-Fast Corp.)

type is applied in daubs about the size of a walnut. Apply adhesive on the back at each corner of 12″ × 12″ tile, about 1 1/2″ away from each edge. Position the tile, then slide it back against the adjoining tile. This motion, along with firm pressure, spreads the adhesive so the daubs are about 1/8″ thick. Some adhesives are thinner and are applied with a brush.

In remodeling work, be sure the old ceiling is clean and that any paint or wallpaper is adhering well. Always follow the recommendations and directions provided by the manufacturer of the products being used.

> **Working Knowledge**
>
> Be sure to keep your hands clean while handling ceiling tile.

16.15.3 Metal Track System

Another ceiling tile system uses 4′ long metal tracks to replace the wood furring strip. The track is nailed or screwed to the old ceiling or to joists at 12″ O.C. intervals, **Figure 16-52.** Tongue-and-groove panels are slipped into place. A clip snapped into the track slides over the tile lip. No other fasteners are used, **Figure 16-53.**

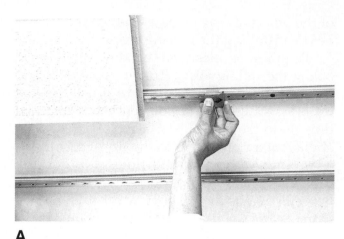

A

B

Figure 16-53. This track system uses clips to hold the ceiling tiles. Use one clip for each tile. Border tiles must have two clips. A—Snap the clip into the track. B—Snugly slide the clip against the tile. (Armstrong World Industries, Inc.)

Figure 16-52. Attach the predrilled track to the ceiling at 12″ intervals. (Armstrong World Industries, Inc.)

16.16 Suspended Ceilings

When heating ducts and plumbing lines interfere with the application of a finished surface, a *suspended ceiling* is practical. In other instances, it provides a simple way of lowering high ceilings. The installation consists of a metal framework designed to support tile or panels, **Figure 16-54.** The framework is mainly

Suspended ceiling: A metal framework designed to support tile or panels.

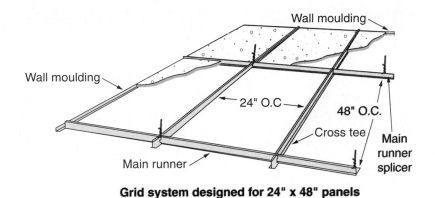

Grid system designed for 24" x 48" panels

Figure 16-54. This sketch is typical of the framework used for suspended ceilings. (U.S. Gypsum Co.)

supported by wires tied to the building structure directly above.

The height of the suspended ceiling must first be determined. Then, trim must be attached to the perimeter of the room. Use a level chalk line or a laser level as a guide, **Figure 16-55.**

Carefully calculate room dimensions, lay out positions of the main runners, and then install screw eyes 4′ O.C. in the existing ceiling structure. The panels next to the wall (border panels) may need to be reduced in width to provide an arrangement that is symmetrical (the same on both sides of room). Plan the layout as previously described for regular ceiling tile.

Install the main runners by resting them on the wall trim and attaching them to wires

tied to the screw eyes. Use chalk lines or string stretched between the wall trim to ensure a level assembly. Sections of runners are easily spliced. Odd lengths can be cut with a fine-tooth hacksaw or aviation snips.

After all main runners are in place, recheck the level and adjust the wires as necessary. Next, check the required spacing for tiles or panels and install the cross tees. Simply insert the end tab of the cross tee into the runner slot , **Figure 16-56.**

When the suspension framework is complete, panels can be installed. Each panel is tilted upward and turned slightly on edge so it will "thread" through the opening. After the entire panel is above the framework, turn it flat and lower it onto the grid flanges.

Figure 16-55. This installer is using a laser level to plumb the suspended ceiling framework. (Spectra-Physics Laserplane, Inc.)

Figure 16-56. Cross tees are installed between the main runners. (Armstrong World Industries, Inc.)

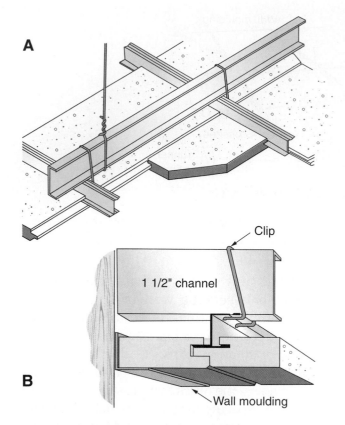

Figure 16-57. A concealed suspension system for ceiling tile. A—Channels of the suspension system are concealed by the ceiling panels. B—Tongue-and-groove joint detail for the panels.

A concealed suspension system is shown in **Figure 16-57.** The tongue-and-groove joint is similar to the joint on regular ceiling tile. Joints are interlocked with the flange of the special runner when the installation is made.

16.17 Estimating Materials

The amount of material required to cover walls or ceilings is determined by first calculating the total number of square feet of wall or ceiling area. Regular-size door and window openings are disregarded. This provides an allowance for waste. However, larger openings, such as window walls or picture windows, should be subtracted.

16.17.1 Determining Area of Rooms

The ceiling area is usually the same as the floor area. It is much easier to take floor dimensions and multiply the length times the width. To find wall area, add all of the wall lengths and multiply by the wall height.

16.17.2 Sheet Materials

When ordering sheet materials such as wallboard or paneling, be sure to specify the length. Always plan to use the longest practical sheet. This will keep the number of butt joints to a minimum or may even eliminate them. Divide the total length of the walls by the width of the sheets to find the number of sheets needed for a vertical application.

Separately estimate each room. Directly take dimensions from the walls or carefully scale the plan. Consult tables and charts of suppliers for estimates of joint compound, adhesives, and nails.

When working with expensive hardwood panels, it is usually advisable to make a scaled layout. Refer to **Figure 16-30** for typical horizontal and vertical arrangements.

16.17.3 Estimating Solid Paneling

Estimates for solid paneling are based on its nominal (unfinished) size. Seasoning and planing bring it to its dressed size. Forming of joints may further reduce the actual face widths. For example, a 1 × 4 tongue-and-groove board has a face width of 3 1/8″.

First, calculate the square footage of the wall to be covered. Then, multiply by factors taken from lumber tables like the one in **Figure 16-35:**

- For 1 × 4 tongue-and-groove boards, use 1.28.
- For 1 × 6 tongue-and-groove boards, use 1.17.

On standard vertical applications, add 5% for waste when required lengths can be selected or when the lumber is end-matched.

16.17.4 Estimating Gypsum Lath

Since gypsum lath is produced in sections smaller than full sheets, you will need to use a different method to figure the needed quantity. First, determine the area of the ceiling and add to this the area of the walls. For example:

Ceiling area (same as floor area)
= 1250 sq. ft.
Total length of walls = 14 + 12 + 14 + 12 + 10
= 62
Wall area = 62 × 8
= 496 sq. ft.
Total area = 496 + 1250
= 1746 sq. ft.

A standard lath bundle contains 64 sq. ft. of lath. Therefore, the bundle estimate is equal to 28 bundles.

Plasterers usually base their prices and estimates on the number of square yards to be covered. To convert square feet to square yards, divide by 9 (1 sq. yd. = 9 sq. ft.).

16.17.5 Estimating Ceiling Tile

Quantities of ceiling tile are estimated by figuring the area (square footage) to be covered. Round off any fractional parts of a foot in width and length to the next larger full unit when making this calculation.

Add extra units when it is necessary to balance the installation pattern with border tile along each wall. When using 12″ × 12″ tile, the number required equals the square footage plus the extra allowance described. Standard 12″ × 12″ ceiling tile are packaged 64 sq. ft. to a carton.

ON THE JOB

Plasterer

Plastering is one of the oldest building trades, dating back at least to the ancient Romans. The plasterer applies interior wall finish material in several coats to form a smooth, hard finish. The finish material is a mixture of water and lime- or gypsum-based plaster. The plaster is applied over a solid surface, such as concrete block, or a perforated metal material called lath attached to wall studs. Plasterers also may use gypsum wallboard (drywall) as a base, applying a thin "skim coat" of lime-based plaster as a smooth, abrasion-resistant finish.

Stucco masons are members of a related trade. They apply exterior finish coats to buildings in much the same way that plasterers cover interior walls. Their material, however, is a weather-resistant mixture of portland cement, lime, and water. Plasterers and stucco masons also apply another type of exterior finish, known as an exterior insulated finishing system (EIFS), that has achieved wide acceptance in recent years. The EIFS consists of rigid foam insulation board and a reinforcing mesh attached to building walls, then covered with polymer-based finish similar to stucco. The polymer coat can be finished to mimic various types of stone or other traditional wall materials.

Plasterers and stucco masons must be in good physical condition, since their work involves standing for long periods and engaging in fairly strenuous physical activity, especially involving arm and shoulder muscles. Working conditions may be dusty and the materials used can cause skin, eye, or respiratory irritation. Personal protective equipment should be worn where appropriate.

Most plasterers and stucco masons work for small-to-medium-sized specialty contractors. Fewer than 10% are self employed.

While some apprenticeship programs are available, most plasterers and stucco masons learn their skills on the job. They start as helpers, carrying and mixing materials, setting up scaffolding, and cleaning tools and the work site. They learn their job skills from experienced tradespeople, gradually assuming more responsibility. Apprentice training, where available, typically lasts 2–3 years and involves at least 144 hours of classroom work in addition to on-the-job training.

ON THE JOB

Drywall Installer

Gypsum wallboard, generally referred to as drywall, has replaced plaster as a wall and ceiling finish in many applications. The wide acceptance of drywall, with the resulting employment opportunities for drywall installers, is due to its lower cost and easier, faster installation process.

Drywall installers are responsible for measuring and cutting to size the standard-size drywall sheets, then fastening them to walls and ceilings. The installation involves fitting material around doors, windows, and openings such as electrical boxes, plumbing pipes, and HVAC ducts. While nailing was once extensively used for fastening to wood studs and ceiling joists, screws are now more likely to be used. Screws are used with both wood and metal studs.

On large construction projects, finishers called *tapers* may handle the next installation step. These workers apply joint compound and paper or mesh tape to cover the joints between drywall sheets. Compound is also used to cover screw heads and other imperfections. Several coats of compound are applied and sanded. A skilled taper makes joints virtually invisible. On smaller jobs, the same person often does both installation and taping.

A related occupation to drywall installation is ceiling tile installation. A ceiling tile installer may mount acoustical tiles directly on drywall surfaces or may install a grid of metal channels for a suspended ceiling. Acoustical panels fit into the channels.

Another related occupation is the *lather*. These workers fasten metal mesh or rockboard (a more strong and rigid variation of drywall) to studs and ceiling joists as a base for traditional plaster application. As the popularity of drywall has increased, the number of lathers has declined.

Drywall installers work in a protected environment, since the building must be closed in before their work can begin. Sheets of drywall are heavy and often handled with aid from a helper or mechanical devices. Large amounts of fine dust are generated during finishing work. Respiratory protection is required.

Although apprenticeship programs do exist, most drywall installers learn their skills on the job, starting as a helper. Because of the large amount of measuring involved, drywall installers must have good math skills. Approximately 20% of drywall installers are self employed. Business skills are also useful for those installers.

Summary

Interior finish work—installing cover materials for walls and ceilings—cannot begin until all exterior windows and doors have been installed. Many different covering materials are used to finish interior walls and ceilings, but the most popular is gypsum wallboard. It is a laminated material with a gypsum core and paper covering on either side. The wallboard is directly attached to wooden or steel studs, usually with specially designed screws. Nails and adhesives are also used. Application may be in a single layer, which is most common. Double-layer applications provide greater sound insulation and fire resistance. Special water-resistant gypsum boards are used in potentially wet areas, such as bathrooms. Cement board is used as a base for tile application in showers and tub areas. Other sheet materials applied to walls are usually referred to as paneling. They include hardboard with a printed surface pattern, veneered plywood, and plastic-laminate-faced sheets. Paneling is also available in the form of individual boards or planks. A traditional wall covering that is not widely used today is plaster and lath. When used, the lath is usually a metal mesh fastened to the wall studs. Three layers of plaster are normally applied. Ceilings are often finished in the same way as walls, but may have acoustical tiles installed to aid in sound control. Some tile is attached to furring strips; other types are placed in a suspended metal grid.

Test Your Knowledge

Answer the following questions on a separate piece of paper. Do not write in this book.

1. Not counting special finishes, there are eight common materials used as interior wall and ceiling coverings. Name them.
2. When storing drywall on the job site, (select all true statements):
   A. Store it inside and flat on a clean floor in the center of the largest room.
   B. Separate all sheets and line them against the wall to reach room humidity.
   C. Lay it flat with the ceiling drywall on top.
   D. Store it flat with the ceiling panels on the bottom to keep them from warping.
3. In single-layer drywall construction, the gypsum wallboard thickness should be _____" or 5/8". For higher quality construction, 5/8" is preferred.
4. Name and describe two methods of installing drywall in the single-layer method.
5. When nailing drywall in place, the nails are spaced closer together on the _____ (walls, ceilings).
6. Since drywall screws have greater holding power than nails, ceiling spacing can be _____" and wall spacing _____".
7. When making a double-layer wallboard application, joints of the finishing layer should be offset at least _____ inches from joints on the base layer.
8. *True or False?* Adding more water to retemper joint compound will help extend its working time.
9. Cement board is available in three thicknesses: _____, _____, and _____.
10. Before installing paneling, it is important for the carpenter to mark stud locations on the _____ and _____.
11. *True or False?* Some solid lumber paneling can be reversed for a different surface pattern.
12. Name the three patterns used for installing solid wood paneling at an angle.
13. Solid wood wall paneling should have a moisture content of about _____%–_____% for most areas of the United States.
14. Strips of wood or metal installed at the edge of openings to provide a guide for the plasterer are called _____.
15. A standard-size panel of gypsum lath measures 16" × _____".
16. To reinforce the plaster base at the corners of window and door openings, a strip of _____ is used.
17. When applied to regular gypsum lath, the total thickness of the plaster coats should not be less than _____".
18. A standard ceiling tile is _____ × _____.
19. For a symmetrical ceiling tile pattern in a room 11'-6" wide, use ten full tiles (standard size) and a _____ inch-wide border tile on each edge.
20. The metal framework of a suspended ceiling is mainly supported by _____ tied to the building structure directly above.

Curricular Connections

Mathematics. Take the dimensions of a room and calculate the number of sheets of drywall needed to cover it. Be sure to include the ceiling and allow for door and window openings. From your calculations, determine the cost of the drywall. Check prices from several sources, including local lumberyards, builders' supply firms, and home improvement centers, to determine the lowest cost.

Language Arts. Obtain and study literature from companies that manufacture gypsum products. Learn about the history and development of plastering, including the methods used today to apply plaster in new home construction or remodeling. Prepare a written report or carefully outline the information and make an oral presentation to your class.

Outside Assignments

1. Visit a local drywall contractor and learn about some of the special problems and remedies typical to this type of wall finish. Obtain information on the following topics: care and handling of materials; repairing and adjusting warped studs and framework; repairing damaged boards and surfaces; causes and remedies of tape blisters; and definition and prevention of nail pops. Organize the information under appropriate headings and make a presentation to your class.

2. Determine the coverage of one bag of plaster. Then, determine the number of bags needed for a single-coat coverage, 3/8″ thick, of a 12′ × 14′ ceiling. Have your instructor check your computations.

3. From various reference books, including manuals on architectural standards, learn about methods and construction details for ceiling systems that include radiant heating. Cover both hot water and electrical systems. Prepare scaled drawings larger than actual size of sections through various ceiling constructions, showing the heating pipes and elements. Include notes concerning material specifications and critical temperatures for the various constructions.

4. Suspended ceilings in institutional and commercial buildings often include a ventilation system that provides both heating and cooling. The space above the ceiling serves as a plenum. Air enters the room through small slots or holes either in the tile or at special joints between the tiles. Study this method of air distribution and prepare a written report with drawings and other illustrations.

Finish Flooring

17

Learning Objectives

After studying this chapter, you will be able to:

- Describe strip, plank, and unit block wood flooring.
- Lay out and install strip flooring on concrete or plywood subfloors.
- Describe the procedure for applying hardboard, particleboard, waferboard, and plywood underlayment.
- Outline the basic steps for installing resilient flooring.

Technical Vocabulary

Bisque	Permeability
Blindnailed	Plank flooring
Block flooring	Pull bar
Cementitious	Quarry tile
Cove base	Racking
Face nail	Sanded grout
Finish flooring	Self-adhering tile
Full-adhesive bonded	Setup line
Laminated wood strip	Side-and-end matched
flooring	Sleepers
Lugged tile	Spline
Mosaic tile	Strip flooring
Paver tile	Undercut
Perimeter bonded	Underlayment

Finish flooring is any material used as the final surface of a floor. A wide selection of materials is manufactured for this purpose. Many new and improved materials are being used in the production of composition (resilient) flooring. Notable among these are the vinyl plastics, which have largely replaced asphalt tile, rubber tile, and linoleum.

Flagstone, slate, brick, and ceramic tile are frequently selected for special areas. These areas include entrances, bathrooms, kitchens, and multipurpose rooms. Floor structures usually need to be designed to carry the greater weight of this type of flooring material.

The finish flooring is usually laid after wall and ceiling surfaces are completed, but before interior door frames and other trim are added. The floor surface should be covered to protect it during other inside finish work. Sanding and finishing of a wood floor surface is the last major operation as the interior is completed.

Finish flooring: The final floor covering.

17.1 Wood Flooring

Wood is popular as flooring in residential structures. Wood flooring, especially hardwood, has the strength and durability to withstand wear, while providing an attractive appearance, **Figure 17-1.** Hardwood and softwood flooring is available in a variety of widths and thicknesses, and as random-width planks or unit blocks.

Oak is a widely used species. However, maple, birch, beech, and other hardwoods also have desirable qualities. Softwood flooring includes such species as pine and fir. These are fairly durable when produced with an edge-grain surface.

17.1.1 Types of Wood Flooring

The general types of wood flooring used in residential structures are:

- Strip
- Laminated strip
- Plank
- Block

As the name implies, *strip flooring* consists of pieces of solid wood cut into narrow strips. It is the most widely used wood flooring and laid in a random pattern of end joints. Most strip flooring is tongue-and-groove on the sides and ends. This design is also referred to as *side-and-end matched.* Another feature of strip flooring is the *undercut,* as shown in **Figure 17-2.** This is a wide groove on the bottom of each piece that enables it to lie flat and stable even when the subfloor surface is slightly uneven.

Unfinished strip flooring is manufactured with sharp, square corners on the face edges. After the floor is laid, the surface is sanded to remove unevenness. Prefinished strip flooring is sanded, finished, and waxed before it leaves the factory. Both edges of the face are chamfered. When the strips are laid, the chamfered edges form a vee joint. This makes any unevenness unnoticeable.

Plank flooring is also solid wood, **Figure 17-3A.** It gives an informal appearance

Figure 17-1. Oak strip flooring is very durable and has a natural beauty that is brought out by a clear finish. A wide variety of hardwood flooring with a factory applied finish is available. (Weather Shield)

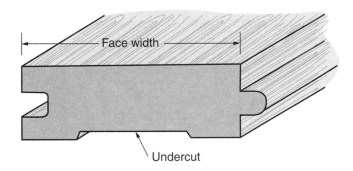

Figure 17-2. Typical section of strip flooring. The ends are also tongue-and-groove. The undercut helps compensate for slightly uneven subfloors.

Strip flooring: Usually hardwood cut into narrow strips, tongued and grooved, and laid in random pattern of end joints.

Side-and-end matched: Strip flooring that is tongue-and-groove on the sides and ends.

Undercut: Cutting a wood member so that the back of the member is slightly shorter than the front surface.

Plank flooring: Like strip flooring, but bored and plugged at the factory to simulate the wooden pegs used to fasten the strips in colonial times.

A

B

Figure 17-3. Two types of wood floors. A—Random planks with walnut plugs. B—Unit block parquet flooring. (Robbins/Sykes, Tibbias Flooring Inc.)

to the floor. This is particularly appropriate for colonial and ranch style homes. Plank floors are usually laid in random widths. The pieces are bored and plugged in the millwork shop to simulate the wooden pegs traditionally used to fasten the flooring in place. Today, this type of flooring has tongue-and-groove edges. It is laid in about the same manner as regular strip floors. Widths of the strips vary in different combinations.

Block flooring looks like conventional parquetry. Often, this type is referred to as *parquet flooring.* Each block has an elaborate design formed by smaller wood blocks. The block unit consists of short lengths of flooring held together with glue, metal splines, or other fasteners. Blocks are laminated units produced

by gluing together several layers of wood. Square and rectangular units are produced. Generally, each block is laid with its grain at a right angle to the surrounding units, as shown in **Figure 17-3B.**

Laminated wood strip flooring is designed to float on the subfloor, **Figure 17-4.** It is manufactured by laminating several plies and the top ply is prefinished. Strips are made up of 9/16″ thick, 7 1/2″ wide and 7′-11 1/2″ long. The face veneer is made up of three hardwood strips tightly edge joined to resemble solid strip flooring. The thickness of the veneer and the joining are so precise that chamfering of edges is not necessary.

17.1.2 Sizes and Grades

Hardwood strip flooring is generally available in widths ranging from 1 1/2″–3 1/4″. Standard thicknesses include 3/8″, 1/2″, and 3/4″. Solid planks are usually 3/4″ thick. However, greater thicknesses are available. Widths range

---

Block flooring: Wood flooring cut in square blocks.
Laminated wood strip flooring: Flooring manufactured in three layers of hardwood veneer with the face layer prefinished. Edges are tongue-and-groove.

Figure 17-4. Laminated strip flooring has become very popular with homeowners, and is available in many wood grain patterns. (Warmly Yours)

from 3″–9″ in multiples of 1″. Unit blocks are also commonly produced in a 3/4″ thickness. Dimensions (width and length) are in multiples of the width of the strips from which they are made. For example, squares assembled from 2 1/4″ strips will be 6 3/4″ × 6 3/4″, 9″ × 9″, or 11 1/4″ × 11 1/4″.

Uniform grading rules are established by manufacturers working with the U.S. Bureau of Standards and such organizations as the National Oak Flooring Manufacturers Association and the Maple Flooring Manufacturers Association. Grading is largely based on appearance. Consideration is given to knots, streaks, color, pinworm holes, and sapwood. The percentage of long and short pieces is also a consideration.

Oak, for example, is separated into two grades of quarter-sawed stock and five grades of plain-sawed stock. In descending order, plain-sawed grades are: clear, select and better, select, No. 1 common, and No. 2 common. White and red oak species are ordinarily separated in the highest grades. A chart on grades for hardwood flooring is included in Appendix B, **Technical Information.**

17.1.3 Delivery

Manufacturers recommend that wood flooring be delivered four or five days before installation. This period of *conditioning* permits the wood to match its moisture content with that in the building. Avoid transporting flooring in rain or snow. In damp or foggy conditions, protect the lumber with a tarp. Loosely pile the flooring throughout the structure. Do not store where floors are less than 18″ above the ground or where there is poor air circulation under the floor. A minimum inside temperature of 70°F (20°C) should be maintained. Avoid damp buildings. Make sure concrete work and plaster are thoroughly dry before bringing in flooring.

17.1.4 Subfloors

In conventional joist construction, most building codes specify a sound and rigid subfloor. It adds considerable strength to the structure and serves as a base for attaching the finish flooring.

For regular strip or plank flooring, the subfloor should consist of good-quality plywood or boards 1″ thick and not more than 6″ wide. In most construction, plywood is commonly used for the subfloor. It must be installed according to recognized standards. Refer to Chapter 8, **Floor Framing,** for sizes and installation methods. If using boards, space them about 1/4″ apart and solidly face nail them at every bearing point. Inadequate or improper nailing of subfloors usually results in squeaky floors.

17.2 Installing Wood Strip Flooring

Normally, laying the flooring should be the last building operation in the house, except for interior trim. Mechanical systems and wall and ceiling coverings should have been completed.

Before installing wood strip floors, carefully check the subfloor to make certain it is clean and that nailing patterns are complete. Put down a good quality (usually 15 lb.) building paper at a right angle to the long dimension of the finish flooring. Extend it from wall to wall with a 4″ lap, as shown in **Figure 17-5.** Joist locations should be marked with a chalk line on the paper. Flooring should be nailed into the joists. This is especially important when plywood is used as the subfloor.

Over the heating plant or hot air ducts, it is advisable to use a double-weight building paper. Insulation may also be directly attached to the underside of the subfloor. This extra precaution will prevent excessive heat from reaching the finish flooring, where it can cause cracks and open joints.

When the subfloor is plywood, lay strip flooring at a right angle to the floor joists. Where boards are used for the subfloor, lay strips at a right angle to the boards. Strip flooring looks best when its direction aligns with the longest dimension of a rectangular room. Since floor joists normally span the shortest dimension of the living room, this will establish the direction the flooring runs in other rooms.

Stagger end joints so that several are not grouped in one small area. Shorter pieces should be used up in closets and in areas where there is the least traffic. Long strips should be used at entrances and for starting and finishing off a room.

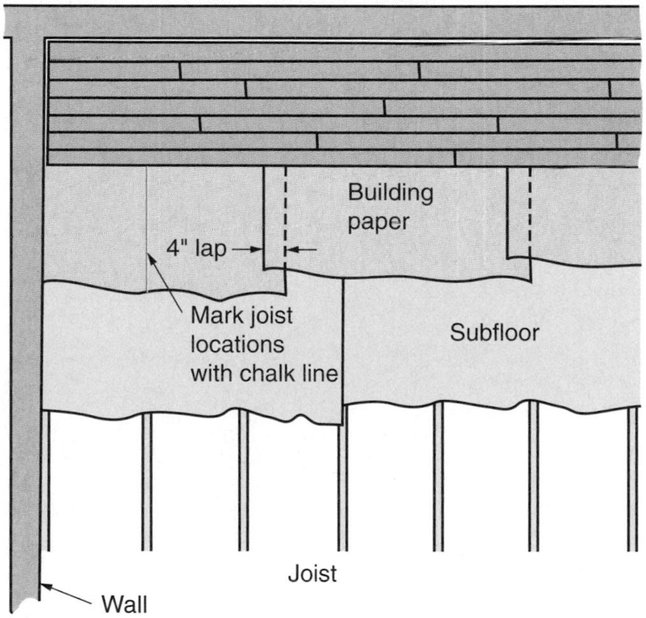

Figure 17-5. This cutaway view shows the various layers of material under wood strip flooring. The subfloor is usually plywood with 15 lb. building paper overlaid.

Face nailing: A fastening method in which the nail is driven perpendicular to and through the surface of a piece.

Blindnailed: Nailing concealed by installation of another strip of wood; used in tongue-and-groove flooring.

Working Knowledge

The moisture content usually recommended for flooring at the time of installation is 6% for the dry, southwestern states, 10% for the more-humid, southern states, and about 7% or 8% for the remainder of the country.

17.2.1 Nailing

Floor squeaks or creaks are caused by the movement of one board against the other. The problem may be in either the subfloor or the finished floor. Using enough nails of the proper size will reduce these undesirable noises. When possible, the nails used for the finish floor should go through the subfloor and into the joist. When plywood is used for the subfloor, place a nail at each joist and one in between. Nail sizes are specified for different types of floors in **Figure 17-6.**

Start the installation by laying the first strip along either sidewall of the room. Select long pieces. If the wall is not perfectly straight and true, set the first strip along a chalk line. Place the groove edge next to the wall, staying at least 1/2″ away from the wall to allow for expansion of the flooring. This space will be covered later by the baseboard and base shoe.

Make sure the first strip is perfectly aligned. Then, face nail it as illustrated in **Figure 17-7.** In *face nailing,* the nail head must be *set* (sunk below face) and the hole filled. Predrilling holes will prevent splitting.

Succeeding strips are *blindnailed.* The nail penetrates the flooring where the tongue joins the shoulder. Drive the nail at an angle of about

Flooring nominal size, inches	Size of fasteners	Spacing of fasteners
3/4 x 1 1/2 3/4 x 2 1/4 3/4 x 3 1/4	2" machine-driven fasteners; 7d or 8d screw or cut nail.	10" - 12" apart
Following flooring must be laid on a subfloor.		
1/2 x 1 1/2 1/2 x 2	1 1/2" machine-driven fastener; 5d screw, cut steel or wire casing nail.	10" apart
3/8 x 1 1/2 3/8 x 2	1 1/4" machine-driven fastener, or 4d bright wire casing nail.	8" apart
Square-edge flooring as follows, face-nailed — through top face.		
5/16 x 1 1/2 5/16 x 2	1", 15 gage fully barbed flooring brad. 2 nails every 7 inches.	
5/16 x 1 1/3	1", 15 gage fully barbed flooring brad. 1 nail every 5 inches on alternate sides of strip.	
Plank flooring		
3/4 x 3" to 8" plank	2" serrated-edge barbed nails. 7d or 8d screw or cut nail. Use 1 1/2" length with 3/4" plywood subfloor or slab.	8" apart into and between joists.

Figure 17-6. This chart suggests fastener size and spacing for application of strip and plank flooring. Always follow the flooring manufacturer's instructions, which may vary from brand to brand.

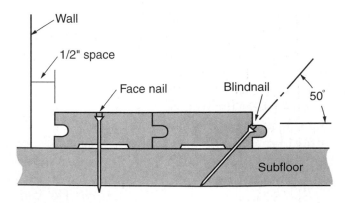

Figure 17-7. Nail the starter strip through the face. Succeeding strips are blindnailed through the tongue.

50°. Countersink all nails. Use care so that the edge of the flooring strips are not damaged. See **Figure 17-8.**

Each strip should fit tightly against the preceding one. When it is necessary to drive strips into position, use a piece of scrap flooring as a driving block.

Lay out seven or eight rows of flooring at a time. Stagger the pattern, cutting and fitting as you go. This step is known in the trade as *racking* the floor. Use different strip lengths. Match the strips so they will extend wall-to-wall with about 1/2″ clearance at each end.

Joints in successive courses should be 6″ or more from each other. Try to arrange the pieces

Figure 17-8. Strip floor installation using a portable nailer. Never attempt to set nails with a hammer. You will damage the flooring. (Bullard-Haven Technical School)

Racking: In flooring, laying out seven or eight rows of strip flooring prior to nailing them in place.

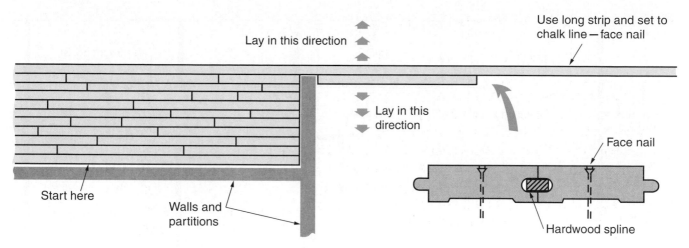

Figure 17-9. Follow this procedure when laying strip floors around a wall or partition.

so the joints are well distributed. Blend color and grain for a pleasing pattern. Lengths cut from the end of a course should be carried back to the opposite wall to start the next course. This is the only place it can be used since its cut end has no tongue or groove.

Flooring strips should run uninterrupted through doorways and into adjoining rooms. When there is a projection into the room, such as a wall or partition, follow the procedure shown in **Figure 17-9.**

When strip or plank flooring a large area, it sometimes helps to set up a starter strip at or near the center. Use a *spline* in the groove of the starter strip to create a tongue. Be sure to accurately measure and align the starter strip with the walls on each side of the room.

When the flooring has been brought to within 2′–3′ of the far wall, check the distance to the wall at each end to see if the strips are parallel to the wall. If not, slightly dress the grooved edges at one end until the strips have been adjusted to run parallel. This is necessary so that the last piece may be aligned with the baseboard.

Spline: A thin strip of hardwood that fits into mortises or grooves machined into boards that are to be joined.

Setup line: A chalk line the width of two flooring strips plus 1/2″ away from a partition.

PROCEDURE

Laying flooring around projections

1. Lay the main floor area to a point even with the projection.
2. Extend the next course all of the way across the room. Set the extended strip to a chalk line and face nail it in place.
3. Form a tongue on the grooved edge by inserting a spline.
4. Install the flooring in the second room in both directions from this strip.

17.2.2 Multiroom Layout

Sometimes, a strip flooring installation is carried through all or a major part of a building. Before beginning, study the floor plan to determine the most efficient procedure to follow.

Figure 17-10 shows a typical residential floor plan with a method for installing strip flooring. Following this plan, the carpenter first snaps a chalk line the width of two flooring strips (plus 1/2″) away from the partition as shown. Carpenters may call this a *setup line.* Align the starter courses with this setup line before face nailing. Continue installation across the living room, across the hall, and into and through bedrooms No. 1 and No. 2.

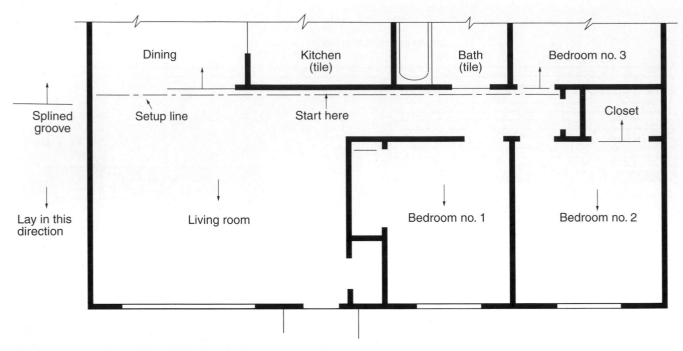

Figure 17-10. This floor plan has been marked with arrows to show the common procedure for laying strip flooring through a number of rooms.

Next, set a spline in the groove of the starter strip and lay the floor from this point into the dining room and bedroom No. 3. In bedroom No. 2, use a spline to lay the floor back into the closet. In small closet areas, such as shown in bedroom No. 1, it is usually impractical to reverse the direction of laying with a splined groove. Simply face nail the pieces in place.

The last strip laid along the wall of a room usually needs to be ripped so it has the required clearance (1/2″ minimum). Due to lack of clearance for blindnailing, the last several strips must be face nailed. Do not place ripped strips where they might detract from the floor's appearance. It is recommended that full-width strips always be used around entrances and across doorways.

Working Knowledge

In general, plank flooring is installed following the same procedures used for strip flooring. In addition to the regular blindnailing, screws are set and concealed in the face of wide boards.

17.2.3 Wood Flooring over Concrete

Finished wood flooring systems can be successfully installed over a concrete slab. When the concrete floor is suspended with an air space below, a moisture barrier is usually not required. Below-grade installation over concrete is not recommended.

New concrete is heavy with moisture. On-grade slabs must be properly installed with a vapor retarder, such as 6 mil polyethylene film, between the 6″ gravel fill and the slab. This keeps ground moisture from entering the slab, but it also slows down curing of the slab. Always test for dryness—even though the slab may have been in place for two years or more.

For slab-on-grade installation, put down *sleepers* (also called *screeds*). These are wood strips attached to or embedded in the concrete

---

Sleepers: Wood strips laid over or embedded in a concrete floor to which finish flooring is attached. Also called *screeds*.

surface. The sleepers serve as a nailing base for the flooring material.

Beware of damp slab-on-grade concrete floors. If a vapor barrier was not laid down under the concrete during construction, moisture can work up to the surface. This dampness will damage wood flooring.

To test for moisture presence, place several smooth-backed (noncorrugated) rubber mats on the slab. Weight them down so the moisture cannot escape. Allow the mats to remain 12 to 24 hours. If there are damp spots or even darkened areas when the mats are removed, an excess of moisture is coming up through the slab.

An alternative to the mat test is to duct tape 2′ squares of clear polyethylene to the floor. Inspect after 12 to 24 hours for evidence of moisture.

Whenever concrete is directly placed on grade, either at or below ground level, an approved, membrane moisture barrier must be installed between the flooring and the concrete. Evenly apply a coat of asphalt mastic thinned with an appropriate solvent over the concrete surface. Cover this with polyethylene film. Next, apply another coat of a special asphalt mastic. Then set wood sleepers into the mastic. The sleepers should be 2 × 4 lumber about 30″ long, laid flat and running at a right angle to the flooring. Lap them at least 3″ for 2 1/4″ flooring and 4″ for 3 1/4″ flooring. Spacing between centers is usually 12″–16″.

A newer method of laying strip floors over a concrete slab-on-grade offers savings over older methods. A double layer of 1 × 2 wood sleepers is nailed together with a moisture barrier of 4 mil polyethylene film placed between them, **Figure 17-11.** This is accepted by FHA. An alternate system described by the Wood Flooring Manufacturers Association (NOFMA) uses 3/4″ or thicker sheathing-grade plywood in place of sleepers.

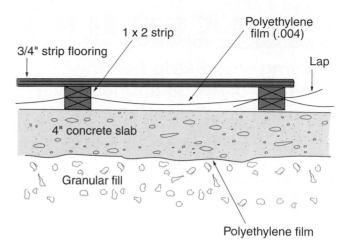

Figure 17-11. Strip flooring over a concrete slab. The sleepers are doubled 1 × 2 strips.

PROCEDURE

Laying wood strip flooring over concrete

1. Clean and prime the floor.
2. Snap chalk lines 16″ apart.
3. Put down bands (rivers) of adhesive along the layout lines.
4. Embed treated wood sleepers in the adhesive.
5. Secure the sleepers with 1 1/2″ concrete nails or special concrete screws about 24″ apart. If screws are used, drill holes down through the sleepers into the concrete. Use a carbide-tipped masonry drill and a hole shooter. Strips can also be attached with powder-actuated equipment.
6. Place a layer of 4 mil–6 mil polyethylene film over the strips.
7. Join the polyethylene sheets by forming a lap over a sleeper.
8. Nail a second layer of untreated strips to the bottom sleepers, sandwiching the barrier. Use 4d nails about 16″ apart.
9. Install strip or plank flooring as previously described. No two adjoining flooring strips should break joints between the same two sleepers.

Working Knowledge

Installation of wood flooring over concrete must be done carefully. Secure detailed specifications from manufacturers of the products used. Organizations such as the Wood Flooring Manufacturers Association can also provide instructional materials and advice.

PROCEDURE

Laying wood strip flooring using a plywood subfloor over concrete slab

1. Put down an appropriate vapor retarder, such as 6 mil polyethylene.
2. Lay 3/4″ plywood loosely over the entire floor. Place the plywood at a diagonal to the direction of the strip flooring to help prevent cracks associated with panel edges. Stagger plywood joints every 4′. Allow 3/4″ spacing at walls and 1/4″ to 1/2″ between panels.
3. Cut to fit around door jambs and other obstructions allowing 1/8″ gap.
4. Fasten the plywood to the slab. Use a powder-actuated concrete nailer or a hammer and concrete nails. Start at the center of the panel and work outward. Use at least nine nails on each panel.
5. You also may glue the plywood to the vapor retarder. When using this method, cut 3/4″ plywood into 4′ squares and score the back 3/8″ deep on a 12″ grid pattern. Apply cut-back (reduced) mastic to the vapor barrier with a 1/4″ × 1/4″ notched trowel and lay down the plywood squares, pressing them into the mastic.

17.3 Estimating Strip Flooring

To determine the number of board feet of strip flooring needed to cover a given area, first calculate the area in square feet. Then, add the percentage listed in **Figure 17-12** for the particular size being used. The figures listed are based on laying flooring straight across the room. They provide an allowance for side matching plus 5% for end matching. For example:

Total area	= 900 sq. ft.
Flooring size	= 3/4″ × 2 1/4″
Bd. ft. of flooring	= 900 + 38% of 900
	= 900 + 342
	= 1242
Number of bundles	= 52 (24 bd. ft. to bundle)

Flooring board size	Additional percentage
3/4″ x 1 1/2″	55%
3/4″ x 2″	43%
3/4″ x 2 1/4″	38%
3/4″ x 3 1/4″	29%
3/8″ x 1 1/2″	38%
3/8″ x 2″	30%
1/2″ x 1 1/2″	38%
1/2″ x 2″	30%

Figure 17-12. When estimating strip flooring, add the listed percentage to total area to allow for waste.

There are 24 board feet in a standard bundle of wood strip flooring. Where there are many breaks and projections in the wall line, allow additional amounts.

17.4 Wood Block (Parquet) Flooring

Block or parquet flooring is made up of squares of wood in various patterns. It requires application methods different from strip flooring. Flooring blocks are usually squares 6″, 8″, or 12″ in size. They are produced in two different ways:

- Unit blocks are glued up from several short lengths of flooring.
- Laminated blocks are made by bonding three plies of hardwood with moisture-resistant glue.

Parquet flooring may use intricate patterns of small pieces in producing the squares. Each block is tongue-and-groove, as shown in **Figure 17-13.**

Styles and types of block, as well as installation methods, somewhat vary among manufacturers. Detailed instructions usually come with the product. If not, they can be secured from the manufacturer or distributor.

The Wood Flooring Manufacturers Association (NOFMA) recommends that installation be made over a double layer of subflooring. Block flooring is nearly always laid in a mastic. The mastic is evenly spread over the subfloor to a thickness of 3/32″. Some installations may require a layer of 30 lb. asphalt-saturated felt.

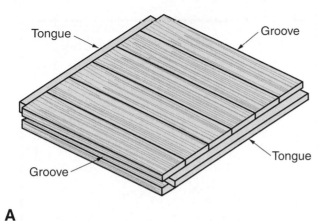

A

B

C

Figure 17-13. Examples of typical block or parquet flooring. Edges are always tongue-and-groove. A—One style of strip flooring. B—In this style, squares run perpendicular to each other. C—The wood strips in B are held together with metal strips mortised into the reverse side. (Bayfield Lumber Company)

Lay the felt in a mastic coat and then coat the top surface with mastic to receive the flooring. Lay blocks in this coat. If installation is over a concrete slab, use a moisture barrier as for installation of strip flooring.

Allow some clearance on unit blocks for expansion. Rubber strips are sometimes used for this purpose. Generally, it is recommended that unit blocks be installed with a 1″ space at the wall.

Working Knowledge

If blocks are to be nailed rather than set in mastic, blindnail them through the tongued edges as with strip or plank floors.

17.4.1 Installation Patterns

Installation can be made either on a square or diagonal pattern. When laying on a square pattern, never use a wall as a starting line. Walls are not always truly straight. Snap chalk lines equal distances from the sidewalls at a 90° angle to each other so that they cross in the center of the room. See **Figure 17-14.** Begin laying blocks from the center point where the chalk lines intersect. When using a diagonal pattern, start in a corner and measure equal distances along each of the walls. Snap a chalk line between these two points. This is your base line. Next, run a test line to intersect the base line at 90°. Begin layout at the intersection of the base line and the test line.

Layout procedures are similar to those described for resilient flooring materials later in this chapter. Block and parquet flooring and adhesive materials vary from one manufacturer to another. Secure detailed information concerning installation procedures from the manufacturer of the product.

17.5 Prefinished Wood Flooring

Most types of wood flooring materials are available with a factory-applied finish,

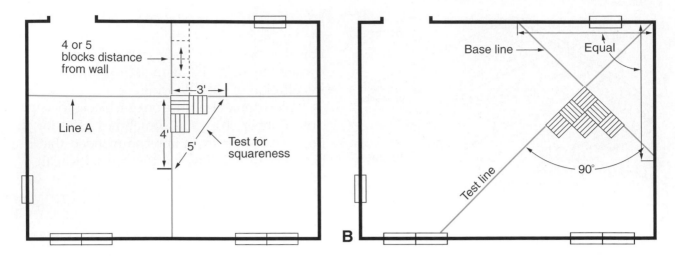

Figure 17-14. Laying out chalk lines for parquet flooring patterns. A—Square pattern. B—Diagonal pattern. (Oak Flooring Institute)

Figure 17-15. These finishes are generally considered to be superior to one applied on the job. Another advantage of prefinished flooring is that it speeds up construction. Floors are immediately ready for service.

Figure 17-15. Prefinished flooring. The edges of each unit have a slight bevel. (Memphis Hardwood Flooring Co.)

A disadvantage is that the installation must be made with greater care to avoid damaging the finish. Face nailing must be held to a minimum when working with prefinished materials. However, the holes can be covered with a special filler material. Hammer marks caused by careless nailing are extremely difficult to repair.

A prefinished floor must be the last step in the interior finish sequence. Other trim that abuts the finished floor must be set beforehand with proper clearances. For example, interior door jambs and casing require spacer blocks equal to the floor thickness. Place these under the units while they are being installed. They are removed before installing the finish flooring. Unless prefinished baseboards are installed after the floor is laid, they must also be placed with the necessary clearance. Other special provisions need to be made around built-in cabinets and at stairways and entrances.

17.6 Laminated Wood Strip Flooring

Laminated strip flooring was first introduced to the American market in 1994. It consists of a 1/4″ core of medium- or high-density fiberboard. This is sandwiched between layers of melamine paper and a protective topcoat. The top layer carries the wood-grained pattern. Sizes vary, but most planks are 46″–50″ long and 8″ wide.

The laminate is not fastened to a subfloor. Instead, it floats over the subfloor. An underlayment of foam takes care of cushioning and absorbs sound. Tongue-and-groove construction allows the planks to be snapped together as they are installed. Although the first types introduced required gluing of the joints, most manufacturers now offer flooring in both glueless and glued-joint types. Glued joints are used for installations where moisture could be present, such as bathrooms and kitchens.

Follow the manufacturer's recommendations on allowing the laminated flooring to acclimate to the building's humidity level. Some types require storing the unopened cartons in the room for two days before installation, but others can be installed without this waiting period.

Before installation, inspect each plank for defects. Since they are very thin, the strips can be installed over almost any kind of subfloor—even old floors. However, carpeting usually must be removed. Each manufacturer supplies instructions for installation of its product. Read and follow these instructions carefully.

Working Knowledge

Normally, radiant heating creates no problem for laminate flooring and requires no precautions. But, if in doubt, consult the distributor.

PROCEDURE

Installing laminated strip flooring

1. Clean the floor.
2. If the subfloor is concrete, lay down a vapor barrier—usually an 8 mil polyethylene film. Overlap edges 8″. This step is important even when the concrete is covered with other types of floor coverings.

(Continued)

(Continued)

3. Put down a foam underlayment as recommended by the manufacturer.
4. Determine which direction to lay the planks. As a rule, run them in the same direction as the incoming light. There are other considerations. Small or narrow rooms look best with planks running parallel to the longest dimension of the room.
5. On existing construction, remove existing baseboards and/or quarter-round moulding. When replaced, the quarter-round will conceal the expansion gap between the flooring and the baseboards.
6. Measure the room width at a right angle to the direction planks are to be installed. Most likely the last row of planks will have to be ripped to fit. If the last row will be less than 2″ wide, reduce the width of the first row, as well.
7. Assemble the first two rows. If using glued-joint flooring, do a dry assembly of the two rows.
8. Start at the left-hand corner of the longest, straightest wall. Lay down the first row as shown in **Figure 17-16.** The groove should be toward the wall. Be sure to use spacers between the wall and the planking. This should maintain a gap for expansion as directed by the manufacturer.
9. Mark and cut the end plank. If using a handsaw, mark and cut from the top side. With a power saw, work from the underside.
10. Use the leftover piece to start the next row. If the cut piece is less than 8″ long, start with a new plank cut in half. Check that joints are offset at least 8″ from adjacent rows.
11. The double row will allow you to see whether the wall is straight. If it is uneven, use a spacer to trace a cut line. Number the planks consecutively and make the cuts.
12. Allow the first two rows of glued-joint planks to set for an hour before installing the rest of the floor.
13. Assemble planks row by row, snapping or gluing them together as instructed by the manufacturer. For glued-joint flooring, apply a ribbon of glue to grooves, completely filling them. A groove against the wall or other fixed object does not require glue.
14. Tightly fit all joints. Use a tapping block and hammer or mallet to tighten planks, following

(Continued)

(Continued)

the manufacturer's instructions. On glued-joint installations, some glue should ooze to the top of the joint for a good seal. Wipe off excess with a damp cloth.

15. Cut the last row to fit. Place a full row of planks to be marked directly on top of the second-to-last row that has already been installed. Then, using a piece of full-width scrap plank, place the tongue edge against the wall. Hold a marking pencil against the opposite edge and mark a line as the scrap is pulled along. Placing the tongue edge against the wall allows for the expansion room needed next to the wall. Trim the planks along the pencil line, then install the final row.

16. Use a tapping block to tap planks into place. See **Figure 17-17.** Where tight quarters do not permit use of the tapping block, use a ***pull bar.***

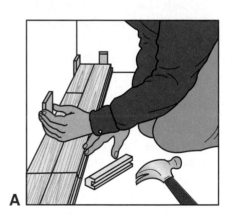

A

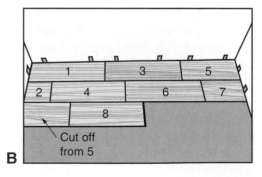

B

Figure 17-16. Begin installing wood laminate flooring at the left-hand wall with the groove to the wall. A—Use spacers to keep the flooring 1/2″ away from wall. B—Lay the planks in the order shown here. If the cutoff from the first row (#5) is longer than 8″, it can be used to start the third row. (Boen Hardwood Flooring, Inc.; Pergo, Inc.)

17.7 Underlayment for Nonwood Floors

If resilient flooring materials are laid over rough or irregular surfaces, such as existing flooring or wood subflooring, they will eventually telegraph (show on their own surface) even the slightest irregularities. An ***underlayment*** is required to provide a satisfactory surface for resilient materials. This is thin sheet material laid down to provide a smooth, even surface for the finish flooring. The material can be plywood, hardboard, cement board, or other

A

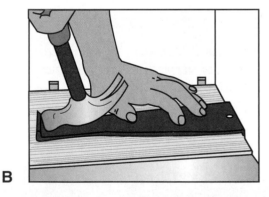

B

Figure 17-17. Achieving tight joints. A—Use a tapping block to drive tongues into grooves. B—When driving planks together in tight quarters, such as near a wall, use a pull bar and hammer. (Boen Hardwood Flooring, Inc.; Pergo, Inc.)

Pull bar: Tool used with a hammer when driving planks together in tight quarters, such as near a wall.

Underlayment: A thin sheet material laid down to provide a smooth, even surface for the finish flooring.

manufactured product. On concrete floors, a special mastic material is sometimes used when the surface condition does not meet the requirements of the finish flooring. An underlayment also prevents the finish flooring materials from checking or cracking when slight movement takes place in a wood subfloor. When used under carpeting and resilient materials, the underlayment is usually installed as soon as wall and ceiling surfaces are complete.

17.7.1 Cementitious Underlayment

Cementitious (cement board) underlayment has the advantage of being dimensionally stable. It is not necessary to leave expansion room between panels or at walls. Further, it is impervious to water and particularly suited for use under ceramic tile floors. Cement board is sold under various trade names. It comes in 4′ × 4′ panels that are 5/16″ thick.

17.7.2 Hardboard and Particleboard

Both hardboard and particleboard meet the requirements of an underlayment board. They will bridge small cups, gaps, and cracks. The standard thickness for hardboard is 1/4″. Particleboard thicknesses range from 1/4″ to 3/4″.

Large irregularities in the subfloor should be repaired before underlayment application. High spots should be sanded down. Low areas should be filled.

Panels should be unwrapped and separately placed around the room for at least 24 hours before they are installed. This allows them to adjust to humidity conditions in the room.

To apply the panels, start in one corner and securely fasten each panel before laying the next. Some manufacturers print a nailing pattern on the face of the panel. Allow at least a 1/8″ to 3/8″

space along an edge next to a wall or any other vertical surface.

Stagger the joints of the underlayment panels. Their direction should be at a right angle to those in the subfloor. Be especially careful to avoid alignment of any joints in the underlayment with those in the subfloor. Leave a 1/32″ space between the joints of hardboard panels. Particleboard panels can be lightly butted.

Underlayment panels should be attached to the subfloor with approved fasteners. These fasteners include ring-grooved and screw-shank nails. Spacing of nails for particleboard varies with the material's thickness. Be sure to drive nail heads flush with the surface. When fastening underlayment with staples, use a type that is etched or galvanized and at least 7/8″ long. Space staples not over 4″ apart along panel edges.

Special adhesives can also be used to bond underlayment to subfloors. An advantage of using adhesive is that it eliminates the possibility of nail popping under resilient floors. A nail pop will eventually damage the resilient flooring.

17.7.3 Plywood

Install plywood underlayment with the smooth side up. Since a range of plywood thicknesses is available, vertical alignment of the surfaces of various finish flooring materials is easy, **Figure 17-18.**

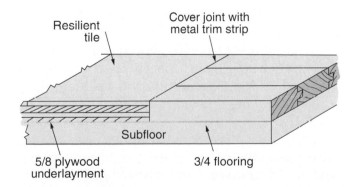

Figure 17-18. Alignment of resilient floor tile with strip flooring is easy since plywood is available in different thicknesses.

Cementitious: Made of cement.

Follow the installation procedures described for hardboard. Turn the grain of the face ply to run at a right angle to the joists. Stagger end joints. When applying 1/4" plywood, use 3d ring-grooved or screw-shank nails. Nails should not be more than 3" from the edges. Field spacing of nails should be 6" each way.

Working Knowledge

When applying an underlayment over a solid board subfloor, never drive long nails into the joists. Shrinkage of the subfloor and framing will likely cause nail pops in the resilient flooring.

17.8 Resilient Floor Tile

Surfaces of the underlayment must be smooth and joints level. Double-check that no fasteners are sticking above the surface. Fill cracks and open areas with a floor-leveling compound. Use a sander or block plane to remove rough edges and high spots. With the underlayment securely fastened and leveled, carefully sweep and vacuum the surface.

Smoothness is extremely important, especially under the more pliable materials, such as vinyl, rubber, and linoleum. Over a period of time, these materials will telegraph even the slightest irregularities such as the texture of wood grain or rough surfaces. Linoleum, either in tile or sheet form, is especially susceptible to telegraphing. For this reason, a base layer of felt is often applied over the underlayment when this material is to be installed.

Underlayment panels should resist denting and punctures from concentrated loads. In addition, look out for any substances that might stain vinyl. These include patching compounds, marking inks, paints, solvents, adhesives, asphalt, and dyes. Some fasteners are coated with resin, rosin, or cement containing a dye that could discolor vinyl. They should only be used if they are known to be safe.

Working Knowledge

There are many resilient flooring materials on the market. Make the application according to the recommendations and instructions furnished by the manufacturer of the product.

PROCEDURE

Preparing the floor for resilient tile

1. Locate the center of the end walls of the room. Disregard any breaks or irregularities in the contour.
2. Establish a main centerline on the floor by snapping a chalk line between the two center points. When snapping long lines, hold the line at various intervals and snap only short sections.
3. Lay out a centerline at a right angle to the main one. Use a carpenter's square or set up a right triangle (base 4', altitude 3', hypotenuse 5'). A chalk line can be used or you can draw the line along a straightedge.
4. With the centerlines established, make a trial layout of tile along the centerlines, as shown in **Figure 17-19.**
5. Measure the distance between the wall and last tile, **Figure 17-20.** If the distance is less than 2" or more than 8", move the centerline closer to the wall by half of the tile's dimension. This adjustment eliminates the need to install border tiles that are too narrow. Since the original centerline is moved exactly half the tile size, the border tile width will remain uniform on opposite sides of the room.
6. Check the layout along the other centerline in the same way.
7. Remove the loose tile put down in the trial layout.
8. Clean the floor surface again.
9. Using the type of spreader (trowel or brush) recommended by the manufacturer, spread the adhesive over one-quarter of the total area. Bring the adhesive right up to the chalk line, but do not cover it.
10. The next step is to lay the tile.

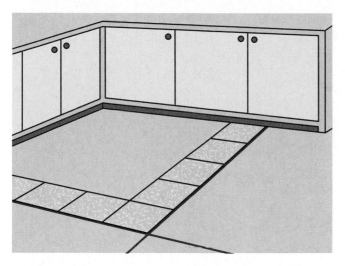

Figure 17-19. Lay down tiles along both centerlines to the walls.

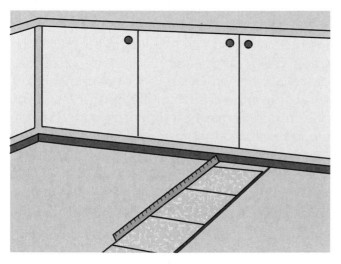

Figure 17-20. Measure the width of the border tile and adjust the centerline, if necessary.

17.8.1 Laying Tile

Allow the adhesive to take an initial set before a single tile is laid. The set time will vary upward from a minimum of about 15 minutes, depending on the type of adhesive used. Test the surface with your thumb. It should feel slightly tacky but should not stick to your skin.

Start laying the tile at the center of the room. Make sure the edges of the tile align with the chalk line. If the chalk line is partially obscured by the adhesive, snap a new line or tack down a thin, straight strip of wood to act as a guide in placing the tile.

Carefully lay each tile in place. Do not slide the tile—sliding can cause the adhesive to work up between the joints and prevent a tight fit. Squarely butt each tile to the adjoining one with the corners in line, **Figure 17-21.** Take the time to see that each tile is correctly positioned. There is usually no hurry, since most adhesives can be "worked" over a period of several hours.

Rubber, vinyl, and linoleum are usually rolled after a section of the floor is laid. Asphalt tile does not need to be rolled. Be sure to follow the manufacturer's recommendations.

After the main area is complete, set the border tile as a separate operation. To lay out a border tile, place a loose tile that will be cut and used over the last tile in the outside row. Then, take another tile and place it against the wall. Mark a line on the first tile with a sharp pencil, **Figure 17-22.** Cut the tile along the marked line using heavy duty household shears or tin snips. Some types of tile require a special cutter, while others may be scribed and snapped. Asphalt tile can be readily cut with snips, if the tile is heated.

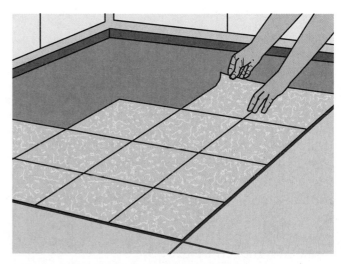

Figure 17-21. When laying tile, align joining edges first, then lower the rest of the tile.

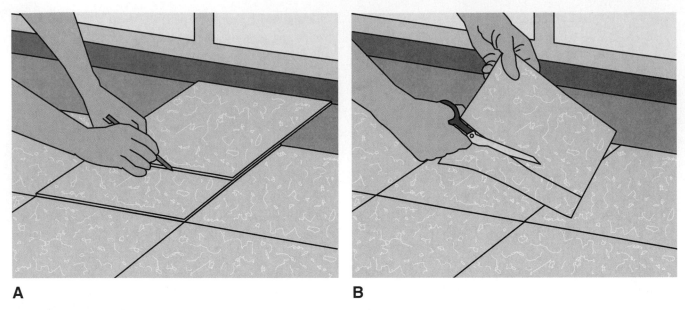

A

B

Figure 17-22. Finishing off the borders. A—Mark the border tile. B—Cut the tile to fit using shears.

Various trim and feature strips are available to customize a tile installation. They are laid by following the general procedures previously described for regular tile.

After all four sections of the floor have been completed, *cove base* can be installed along the wall and around fixtures. A special adhesive is made for this operation. Cut the proper lengths and make a trial fit. Apply the adhesive to the cove base and press the base into place.

Carefully check over the completed installation. Remove any spots of adhesive on the floor or cove base. Carefully work using cleaners and procedures approved by the manufacturer.

17.8.2 Installing Resilient Tile over Concrete

Before installing resilient tile over concrete, make sure the surface is properly prepared. It must be dry, smooth, and structurally sound. There must be no depressions, pits, scale, or foreign deposits. Remove any paint, varnish, oil, or wax. Trisodium phosphate mixed with hot water will remove most paints, except those with a chlorinated rubber or resin base. These must be removed by grinding.

Fill joints and cracks with a latex underlayment or crack filler. Follow the manufacturer's

instructions for on-grade or subgrade applications. Cover dusty or chalked surfaces on suspended concrete floors with a single coat of primer before spreading adhesive. Chalking or dusty floors on or below grade may be a sign of leaching. Perform a bond and moisture test before going ahead. To make a bond and moisture test, install 3' squares of the flooring material at different places. If they are securely bonded after 72 hours the floor is dry enough for application of the tile.

17.9 Self-Adhering Tiles

Self-adhering tiles are similar to standard floor tile, except the adhesive is applied to the tile back at the factory. These tiles are easy to install. Remove the paper from the back of the tile, position the tile on the floor, and press it down. Follow the same layout procedure as for

---

Cove base: A flexible moulding glued in place to hide the joint between a tile floor and a wall.

Self-adhering tile: Floor tile with factory-applied adhesive.

standard tiles, except adhesive is not spread on the underlayment.

It is very important that floors be dry, smooth, and completely free of wax, grease, and dirt. Generally, self-adhering tiles can be laid over smooth-faced resilient floors. However, embossed tile, urethane finish, or cushioned floors should be removed.

Tiles should be kept at a room temperature of at least 65°F (20°C) for 24 hours before installation. The room should be kept warm for one week after installation. This ensures a firm bond to the subfloor surface. Always study and follow the manufacturer's directions.

Percentage of tile waste for various areas	
Area (square feet)	Percentage
Up to 50	14
50–100	10
100–200	8
200–300	7
300–1000	5
1000–5000	3
5000–10,000	2–3
Over 10,000	1–2

Figure 17-23. Tile flooring waste percentages for 12″ × 12″ floor tile.

17.10 Estimating Tile Flooring and Adhesive

To estimate the number of tiles needed, first find the area of the room. Multiply the length in feet and inches by the width in feet and inches. If the room has an irregular shape, section the room into squares or rectangles. Then, figure each area and add the results.

There must be an allowance for waste. Multiply the square footage by the factor shown in **Figure 17-23.** For example, suppose that the area of a room to be floored is 88 square feet. Looking at the chart, 10% should be added for waste. Thus, 88 × 1.10 = 96.8 or 97 square feet. If one tile is 12″ × 12″, you will need about 97 tiles to complete the job. Add 1/3 more when laying down 9″ × 9″ tiles.

Estimate adhesive for laying resilient tile according to square yardage or square footage. Do not include the extra added for the waste. Divide the square yardage or footage by the per-gallon spreading capacity of the adhesive. This is shown on the container.

17.11 Sheet Vinyl Flooring

Sheet vinyl flooring is extremely flexible. This property makes installation much easier. Since sheets are available in 12′ widths, many installations can be made free of seams. This material usually can be installed over concrete, plywood underlayment, or noncushioned vinyl flooring

There are two types of sheet vinyl flooring, full-adhesive bonded and perimeter bonded. *Full-adhesive-bonded* flooring must be fastened to the substrate with a coat of adhesive covering the entire surface. *Perimeter-bonded* flooring is fastened down only around the edges and at seams.

Full-adhesive-bonded flooring is normally cut to fit the space using a template before installation. It is then laid down to test the fit. If a seam is needed, the two pieces are pattern matched and slightly overlapped. One-half of the material is then folded back and adhesive applied to the substrate. The folded portion is then laid in the adhesive. The procedure is repeated with the other half. If there is a seam, the process may be done in quarters instead of halves. The overlapped seam is cut through both pieces of vinyl and the scraps removed, **Figure 17-24.** As the final step, a heavy roller is used to tightly bond the seam and the rest of the flooring.

To install perimeter-bonded flooring, smoothly spread the sheet over the floor. Let excess material turn up around the edges of the

---

Full-adhesive bonded: Type of sheet vinyl flooring that must be fastened to the substrate with a coat of adhesive covering the entire back surface.

Perimeter bonded: Flooring that is fastened down only around the edges and at seams.

room. When there are seams, carefully match the pattern and fasten the two sections to the floor with adhesive.

Trim edges to size. After all edges are trimmed and fitted, secure them with a staple gun or use a band of double-faced tape Always carefully study the manufacturer's directions before starting the work.

17.12 Ceramic Floor Tile

Ceramic tiles are made from a mixture of clay and other materials that may include ground shale, gypsum, talc, vermiculite, and sand. Sometimes, ceramic tiles are made from pure clay. During manufacture, the dry ingredients are mixed with water and other materials and shaped into a *bisque.* This is the body of the tile. The bisque may be formed by extrusion or die pressing, cut from a sheet, or formed by hand. The bisques are allowed to air dry for a short time before being placed in a kiln and fired at temperatures ranging from 1900°F–2200°F (1000°C–1200°C). Ceramic tiles are available in a variety of colors and shapes.

Figure 17-24. Where seams overlap, trim flexible vinyl sheet linoleum with a utility knife drawn along a straightedge. Be sure the straightedge is parallel to the wall.

17.12.1 Ceramic Tile Types

The types of tile are grouped according to their permeability. *Permeability* is the tendency of the tile to absorb water. Permeable and impermeable tile are used on floors, walls, and countertops, **Figure 17-25.** The following list is organized from most to least permeable:

- Nonvitreous.
- Semivitreous.
- Vitreous.
- Impervious.

Tile can also be organized by their characteristics. These are the most common:

- Paver tile
- Quarry tile

Figure 17-25. Ceramic tile comes in a variety of colors and textures. This versatile product provides durable and attractive surfaces not only for floors, but for walls and countertops. (Riviera Marmitec)

Bisque: The body of a ceramic tile.
Permeability: Tendency to absorb water.

- Mosaic tile
- Lugged tile

Paver tile is either glazed or unglazed. They are at least 1/2″ thick. The smaller ones are 4″ × 6″ and larger ones are 12″ square. *Quarry tile* is very dense and always unglazed. They are ideal for floors. Generally, quarry tile is semivitreous or vitreous and ranges in thickness from 1/2″ to 3/4″. *Mosaic tile* is any tile 2″ square or smaller. They are usually vitreous and range from 3/32″ to 1/4″ square. See **Figure 17-26.** *Lugged tile* is any kind of tile that has protrusions around it to ensure exact spacing. The lugs are eventually concealed by grout. **Figure 17-27** shows several examples of tile designed and manufactured specifically for floors, walls, and counters.

A

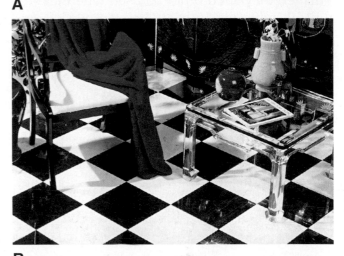

B

Figure 17-26. The bottom of this fountain is laid with mosaic tile. (American Olean Tile Co.)

C

Figure 17-27. Various flooring applications with tile. A—A vitreous tile with a Mexican saltillo look suitable for walls, counters, and floors in residences and light commercial buildings. B—Quarry stone in granite and marble. These materials vary in hardness, veining, color, and finish. C—Matte-finished, white body glazed tile for use in floors, walls, and counters in residential and light commercial buildings. Not recommended for use where subjected to standing water or accumulations of grease. (American Olean Tile Co.)

Paver tile: Concrete-based masonry units used for finish flooring and walkways.

Quarry tile: Ceramic tile used in construction as finish flooring and counter tops.

Mosaic tile: Small pieces of ceramic tile of different colors laid to form a pattern or picture.

Lugged tile: Tile with projections at all edges to maintain proper spacing between individual tiles.

17.12.2 Ceramic Tile Adhesives

Tiles are attached with various types of adhesives. Cement mortar was the preferred adhesive 40 years ago and is still used. However, the most-used adhesives are mastics, dry-set, and latex-portland cement mortars.

Mastics are the least expensive tile adhesive and come ready to use. However, they do not have the strength or flexibility of mortar. Further, they should not be used where the tile will be subjected to heat (for example, around a fireplace). It is also thought that mortars are a better choice of adhesive where tile will be exposed to water.

Mortar mixes are basically mixtures of sand and portland cement, but also contain additives that improve bonding. Dry-set can be mixed with plain water or a liquid modified with latex, acrylic, or epoxy resins. While expensive, epoxy dry-set mortar has high bonding strength and resistance to impact. Furthermore, it can be applied to nearly any surface—plastic laminate, plywood, and steel.

17.12.3 Estimating Ceramic Tile

To determine the quantity of tile needed, first calculate the square footage of the area to be covered. Measure the length and width and multiply the two figures. Add an additional 10% to compensate for breakage, cuts, and extras for future repairs. Manufacturer's coverage charts are available where tile is sold. A sample is shown in **Figure 17-28.**

17.12.4 Installing Ceramic Tile

Surfaces to be tiled should be tight and even. It may be necessary to check the nailing pattern of the substrate or to add an underlayment such as cement board or another type of backing board. Cement backing board sheets should be

Square feet of coverage		
	Tile size	
Number of cartons	4 1/4" x 4 1/4" 6" x 6"	4" x 4" 8" x 8" 10" x 10" 12" x 12"
	Carton size	
	12.5 square feet per carton	11.1 square feet per carton
1	12	11
2	25	22
3	37	33
4	50	44
5	62	55
6	75	66
7	87	77
8	100	88
9	112	99
10	125	111

Figure 17-28. A coverage chart aids in determining the amount of tile needed. (American Olean Tile Co.)

installed in a "brick" pattern. Do not allow four corners to meet at one point. Keep panels 1/8" away from walls. Wherever possible, place cut edges to the outside.

Before installing the underlayment, lay down a bed of mastic to a minimum thickness of 3/32". This will give support to the backing board. Panels should be installed before the mastic films over. Allow edges of sheets to touch, but do not force them together. Fasten the sheets with 1 1/4" corrosion-resistant roofing nails or screws 1"–1 1/4" long. Space fasteners 6" on center around the perimeter, staying 3/8"–3/4" in from the edges. Drive fastener heads flush with the surface. Expansion joints should be used with ceramic tile in certain situations, **Figure 17-29.**

- Over existing structural joints in the floor.

- Where a floor dimension is greater than 15' in any direction.

- At doorways where tile is carried through to the next room.

Sanded grout: Cement to which sand is added.

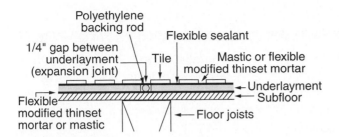

Figure 17-29. Expansion joints are recommended in certain applications of ceramic tile. (James Hardie Building Products.)

Figure 17-30. Applying grout. Allow ceramic or clay tile installations to dry for 24 to 36 hours before grouting.

• Where directional changes occur, such as in L-shaped rooms.

It is important for good adhesion of the tiles that the mortar or mastic completely cover the back of the tile. A notched trowel should be used to apply a full coat to the underlayment. Use the notched edge to evenly spread the material.

Two methods of mortar installation are used: thick bed and thin set. In thick bed, the mortar is 3/4″–1 3/4″ thick. The mortar is allowed to set and cure before tiles are adhered with a dry-set adhesive. This method produces a surface that is strong and unaffected by frequent and prolonged wetting. In thin set, the adhesive (mortar) is applied in a thin layer to the substrate. Thickness may vary from 1/32″ to 1/8″. Thin-setting mortar is less expensive, easier to install, and results in a quicker job.

The spaces between individual tiles must be filled with a form of mortar called grout. This step prevents moisture and dirt from getting between the tiles. Grout comes in dry form and can be mixed with water or latex- or acrylic-modified liquids.

Plain grout is cement with additives that give it a smooth consistency. It is used when spaces between tiles are less than 1/16″ wide. *Sanded grout* consists of cement to which sand is added. It is used where joints are wider than 1/16″.

Allow ceramic or clay tile installations to dry for 24–36 hours before grouting. Grout is applied with a grouting trowel. It has a cushioned surface that allows spreading of grout over the tiles without damaging them. First, the grout is spread over the surface, **Figure 17-30.** Then, it is worked into the joints. The final operation is to use a rough cloth or sponge to clean off grout from the tiles before it sets.

ON THE JOB

Flooring Installer

Residential and commercial buildings, and some areas of industrial structures, make use of decorative or protective floor coverings. Depending on the application, these range from resilient tiles and "sheet goods" to wood or laminate flooring, ceramic tile products, or various types of carpeting.

The general term *flooring installer* covers several specialized trades, all dealing with some form of floor covering. Floor installers or layers work with wood, vinyl, and other types of resilient tile. They also install sheet goods. Sheet goods are wide lengths of resilient flooring material that can often cover an entire room without the need for seams. Floor sanders and finishers specialize in finishing or refinishing wood floors. Carpet installers or layers work with different types of carpeting, using various installation methods. Some of these methods are stretched, glue down, or loose laid. Tile installers or setters cover floors with ceramic tiles or natural stone materials, such as marble.

Flooring installers generally work in fairly comfortable conditions, since their work is done after buildings are closed in and heated and construction debris has been removed. They must be in good physical condition, since the work requires a considerable amount of bending, kneeling, and carrying. Flooring installers encounter few safety hazards, but should wear personal protective gear as needed. For example, floor sanding generates considerable dust and adhesives and other chemicals can produce irritating fumes.

Carpet installers make up more than half of the flooring installer workforce. While many flooring installers work for flooring contractors or retail stores, more than 40% are self employed. This is more than double the rate of self employment for the construction trades as a whole.

While some apprenticeship programs exist, most flooring installers learn their trades through informal on-the-job training. Trainees start as helpers and are gradually given additional responsibility as their skills develop. Since flooring installation involves a great deal of measuring and calculation, good mathematical ability is an asset.

Summary

Finish flooring is available in many materials, ranging from stone, brick, and ceramic tile through resilient tiles and sheet vinyl to various wood products. Wood, resilient tile, and sheet vinyl floors are usually installed after all wall and ceiling surfaces are completed. Wood strip or plank (wider strip) flooring is installed in an unfinished form, then sanded and top coated after all other interior work is done. It must be installed over a solid subfloor. If the subfloor is concrete, sleepers and a vapor barrier must usually be installed before wood strip flooring can be laid. Strip flooring is fitted together with tongue-and-groove joints and is blindnailed through the groove. Prefinished wood flooring materials are available in parquet (block) form, traditional strip or plank, and laminated strips. Parquet flooring is usually laid in a mastic, while prefinished strip flooring is installed with the same nailing method as unfinished strip flooring. Prefinished laminated flooring floats on a sheet foam underlayment. Tongue-and-groove joints between strips snap together to firmly hold the pieces together. In potentially wet situations—kitchen or bath—the joints are glued to prevent water penetration. Resilient flooring, such as tile or sheet vinyl, requires a smooth underlayment that will not telegraph irregularities through to the flooring. Plywood, hardboard, or cement board are used as underlayments. Resilient tile and sheet vinyl are usually adhered with mastic. Some resilient tiles are supplied with adhesive already applied and can merely be pressed in place on the underlayment. Ceramic tiles must be installed over a rigid underlayment, such as cement board. Although sometimes laid in a mortar bed, ceramic tiles are more commonly set in mastic. Spaces between tiles are filled with a mortar-like material called grout.

Test Your Knowledge

Answer the following questions on a separate piece of paper. Do not write in this book.

1. Define *finish flooring*.
2. List the four types of wood flooring used in residential structures.
3. Standard thicknesses of hardwood strip flooring include 3/8", _____, and 3/4".
4. *True or False?* Wood flooring should be delivered on the day it is to be installed.
5. When the starter strip is located in the center of an area, install a(n) _____ in the groove to permit laying of flooring in both directions.
6. How many board feet are there in a standard bundle of hardwood flooring?
7. The two types of block flooring most commonly used in residential installations are unit blocks and _____ blocks.
8. *True or False?* Laminated wood strip flooring is attached to a plywood subfloor.
9. Laminated wood strip flooring may have _____ or _____ type joints.
10. Why does cementitious underlayment require no expansion gap between panels?
11. A resilient flooring material that is especially susceptible to telegraphing small irregularities in the base is _____.
12. When laying resilient tiles, the room is divided into quarters and tiling is started at the _____ of the room.
13. How is installation of self-adhering resilient tiles different from that of standard resilient tiles?
14. Which type of ceramic tile would you likely place on a bathroom floor?
15. Which type of grout is used to fill joints between ceramic tiles when the gap is wider than 1/16"?

Curricular Connections

Science. Obtain samples of each of these types of finish flooring: unfinished oak strip, prefinished oak strip, laminate strip flooring, ceramic tile, resilient tile, vinyl sheet flooring. Devise tests that will allow you to compare various properties of the different materials, such as scuff (wear) resistance, impact resistance, water resistance, heat resistance. Be sure to follow all safety rules for such testing and wear proper personal protective gear (safety glasses, face shield, gloves, etc.). Carry out your tests and record the results. Make a chart listing each material and each type of test. Grade the materials on an A-F scale. Make a presentation to the class using an overhead projector to display a chart of the results. Also, display the test materials.

Outside Assignments

1. Study reference books in the library or pamphlets obtained from a local building supply firm. Prepare a written report describing the procedures and processes used in the manufacturing of hardwood flooring. Include kiln-drying requirements and moisture content standards. Also, include grading rules that are applied to the species and qualities commonly used in your geographical area.

2. Prepare an oral report on resilient flooring materials. Obtain samples from a floor covering contractor or home furnishing store. Include information about thicknesses, colors, tile sizes, and approximate costs. Include a list of the advantages and disadvantages of the various types. If time permits, include information about the composition (basic materials used in manufacture) of each type and the general requirements for installation.

Stair Construction

Learning Objectives

After studying this chapter, you will be able to:
- Identify the various types of stairs.
- Define basic stair parts and terms.
- Calculate the rise-run ratio, number and size of risers, and stairwell length.
- Prepare sketches of the types of stringers.
- Lay out stringers for a given stair rise and run.
- List prefabricated stair parts that are commonly available.

Technical Vocabulary

Balusters	Run of stairs
Balustrade	Stairwell
Built-up stringer	Straight run
Carriages	Total rise
Cut-out stringers	Total run
Handrail	Tread
Headroom	Unit rise
Housed stringers	Unit run
Newel	Wall rail
Nosing	Winders
Platform	Winding stairs
Riser	

A *stair* is a series of steps, each elevated a measured distance, leading from one level of a structure to another. When the series is a continuous section without breaks formed by landings or other constructions, the terms *flight of stairs* or **run of stairs** are often used. Other terms that can be properly used include *stairway* and *staircase*.

For a period of time, the popularity of the one-story structure in residential construction minimized the frequency of stair construction. Framing carpenters could usually handle the relatively simple task of constructing the service stairs leading from the first floor to the basement level. However, revival of traditional two-story styles along with split-level and multilevel designs has again made fine stair construction an important skill.

Because of European influence, main stairs have often been the chief architectural feature in an entrance hallway or other area, **Figure 18-1.** However, in new construction, public rooms are usually on the first floor. Due to this, there is a trend to move the stairs to a less conspicuous location.

Stair construction requires a high degree of skill. The quality of the work should compare with that found in fine cabinetwork. The parts for main stairways are usually made in millwork

Run of stairs: A series of steps that is a continuous section without breaks formed by landings or other constructions. Also called a *flight of stairs.*

597

Figure 18-1. An attractive main stairs is a desirable architectural feature in a residence. Stair design and construction has long been considered one of the highest forms of joinery.

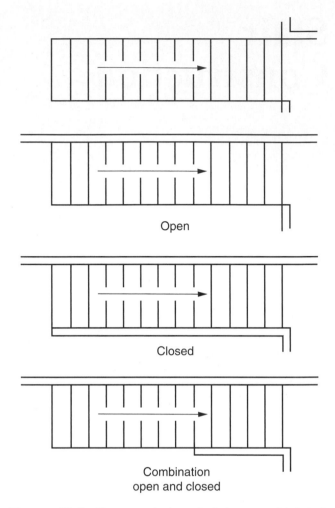

Open

Closed

Combination
open and closed

Figure 18-2. Open and closed stairways. A stair is called open even if one side is enclosed by a wall.

plants and then assembled on the job. Even so, the assembly work must be performed by a skillful carpenter who understands the basic principles of stair design and knows layout and construction procedures.

Main stairways are usually not built or installed until after interior wall surfaces are complete and finish flooring or underlayment has been laid. Basement stairs should not be installed until the concrete floor has been placed.

Carpenters build temporary stairs from framing lumber to provide access until the permanent stairs are installed. These are often designed as a detachable unit so they can be moved from one project to another. Sometimes, permanent carriages are installed during the rough framing and temporary treads are attached. *Carriages,* or stringers, are the inclined

supports that carry the treads and risers. In this case, after the interior is finished, the temporary treads are replaced with finished parts.

18.1 Types of Stairs

Basically, there are two stair categories: service stairs and main stairs. Either of these may be closed, open, or a combination of open and closed. See **Figure 18-2**. In addition, the type of stairs may be straight run, platform, or winding.

Carriage: A sloping member that supports the risers and treads of stairs. Also called a *stringer.*

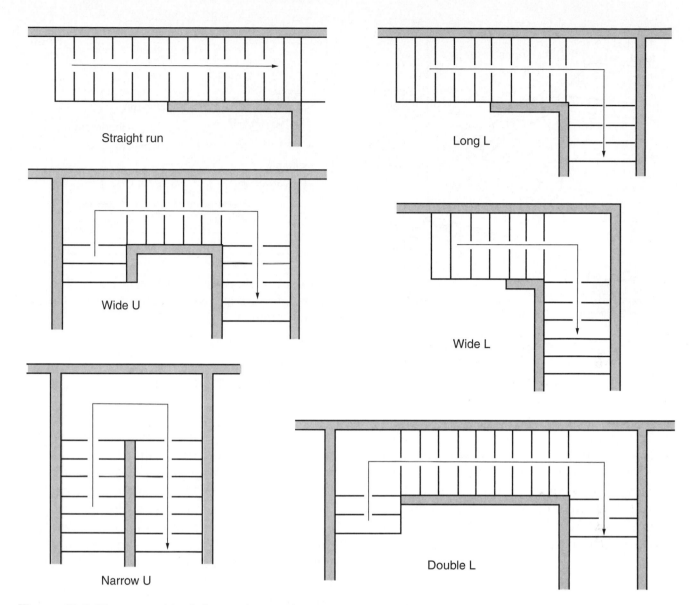

Straight run

Long L

Wide U

Wide L

Narrow U

Double L

Figure 18-3. Terms used to define various stair types.

The *platform* type includes landings where the direction of the stair run is usually changed. Such descriptive terms as L-type (long L and wide L), double L-type, and U-type (wide U and narrow U) are commonly used. See **Figure 18-3.**

In split-level houses, platform stairs with short and generally straight runs are used. Usually, stair runs of this type are located so that the stair run directly above automatically provides headroom, **Figure 18-4.**

The *straight run* stairway is continuous from one floor level to another without landings or turns. It is the easiest to build. Standard multistory designs require a long *stairwell* in the floor above to provide headroom. This often presents a problem in smaller structures. A long run of 12 to 16 steps also has the disadvantage of being tiring. It offers no chance for a rest during ascent.

Platform: A horizontal section between two flights of stairs. Also called a *landing*.

Straight run: A stairway that does not change direction.

Stairwell: The rough opening in the floor above to provide headroom for stairs.

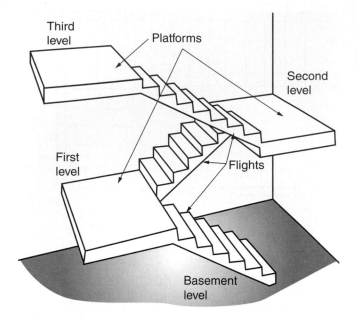

Figure 18-4. Stair runs are often made one above the other to gain headroom. This one is designed for a split-level home.

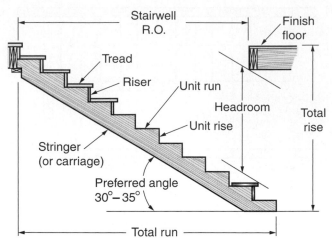

Figure 18-5. Basic stair parts and terms. The total number of risers is always one greater than the total number of treads.

Winding stairs, also called *geometrical,* are circular or elliptical. They gradually change directions as they ascend from one level to another. These often require curved wall surfaces that are difficult to build. Because of their expense, winding stairs are usually only found in high-end homes.

18.2 Stair Parts

Stairs are basically sets of *risers* and *treads* supported by stringers, **Figure 18-5.** The relationship between the riser height (*unit rise*) and the tread width (*unit run*) determines how easily the stairs may be negotiated. Research has indicated that the ideal riser height is 7″, while the ideal tread width is 11″.

Headroom is measured from a line along the front edges of the treads to the enclosed surface or header above. This distance is usually specified in local building codes. Refer to **Figure 18-5.** The Federal Housing Administration (FHA) requires a minimum headroom of 6′-8″ for main stairs and 6′-4″ for basement or service stairs. Local codes may have other requirements.

18.3 Stairwell Framing

Methods of stair building differ from one locality to another. One carpenter may cut and install a carriage (stringer) during the wall and floor framing. Another may put off all stairwork until the interior finishing stages. **Figure 18-6** shows several stages of stair building.

Regardless of procedures followed, the rough openings for the stairwell must be carefully laid out and constructed. If the architectural drawings do not include dimensions and details of the stair installation, then the carpenter must calculate the sizes. Follow recognized standards and local code requirements.

Winding stairs: A curving stairway that gradually changes direction; usually circular or elliptical in shape. Also called *geometrical.*

Riser: The vertical stair member between two consecutive stair treads.

Tread: Horizontal walking surface of a stair.

Unit rise: The height of the stair riser; the vertical distance between two treads.

Unit run: The width of a stair tread minus the nosing.

Headroom: The clear space between the floor line and ceiling.

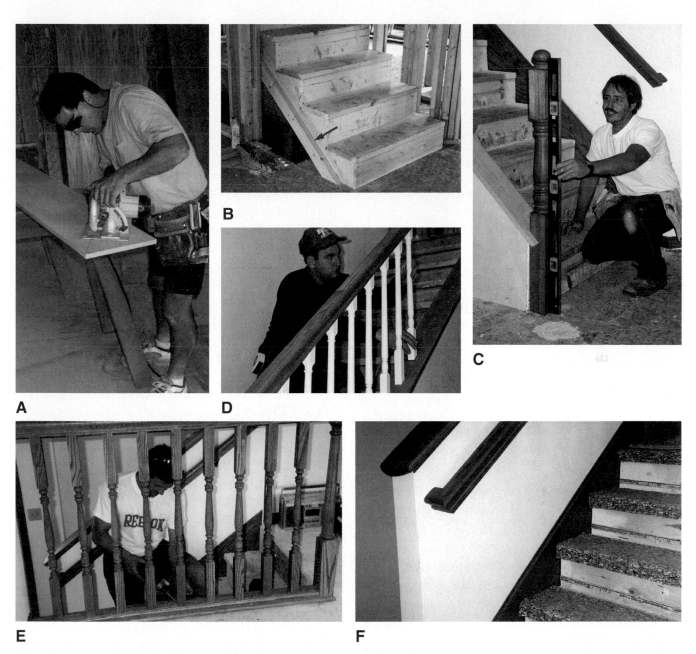

Figure 18-6. This series of photos shows various stages in building stairs. A—This carpenter is making a plumb cut on a housed (closed) stringer. Grooves are cut in the stringer to receive the treads and risers. B—This stringer is a cut-out type. The 2 × 4 spacer (arrow) gives clearance for installation of the wall finish. C—Newel post is being installed. The carpenter is checking for plumb. D—The handrail has been installed and the carpenter is cutting and placing the prefinished balusters. E—This carpenter is fastening the lower rail of a banister to the floor. Since the banister is made of oak, it is necessary to drill pilot holes for the nails. Glue is also applied to each joint. F—These stairs are completed except for the installation of carpeting.

Trimmers and headers in the rough framing should be doubled, especially when the span is greater than 4′. Headers more than 6′ long should be installed with framing anchors, unless supported by a beam, post, or partition. Tail joists over 12′ long should also be supported

by framing anchors or a ledger strip. Refer to Chapter 8 for additional information on framing rough openings.

Providing adequate headroom is often a problem, especially in smaller structures. Installing an auxiliary header close to the main

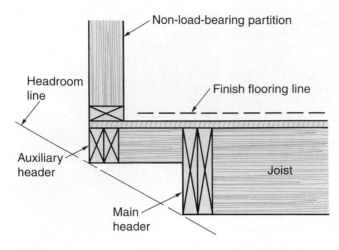

Figure 18-7. Extending the upper floor area with a shallow auxiliary header to provide headroom. The partition over the auxiliary header must not be load bearing.

header permits a slight extension in the floor area above a stairway, **Figure 18-7.** When a closet is located directly above the stairway, the closet floor is sometimes raised for additional headroom.

18.4 Stair Design

Most important in stair design is the mathematical relationship between the riser and tread. There are three generally accepted rules for calculating the rise-run or riser-tread ratio. It is wise to observe them:

- The sum of two risers and one tread should equal 24"–25".
- The sum of one riser and one tread should equal 17"–18".
- The height of the riser times the width of the tread should equal 70"–75".

According to the first rule, a riser 7 1/2" high requires a tread of 10". A 6 1/2" riser requires a 12" tread.

The current edition of the International Residential Code, developed by the International Code Council, includes a change in the allowable tread width and height of risers. The previous standard for residences was an 8 1/4" rise and a 9" tread. Under the new code, a 7 3/4" rise and a 10" tread (assuming a nosing of at least 3/4") are required. The International

Residential Code, which is updated every three years, has no legal force until it is adopted by state or local governments.

In residential structures, treads (excluding nosing) are seldom less than 9" or more than 12". Nosing is a small extension of the tread. In a given run of stairs, it is extremely important that all of the treads be the same size. The same is true of the risers. A person tends to subconsciously measure the first few steps and will probably trip if subsequent risers are not the same.

When the rise-run combination is wrong, climbing the stairs will be tiring and cause extra strain on leg muscles. Further, the toe may kick the riser if the tread is too narrow. A unit rise of 7"–7 5/8" with an appropriate tread width provides both comfort and safety. Main or principal stairs are usually planned to have a rise in this range. Service stairs are often steeper, but risers should be no higher than 8". As stair rise is increased, the run must be decreased. See **Figure 18-8.**

A main stair should be wide enough to allow two people to pass without contact. Further, it should provide space so furniture can be moved up or down, **Figure 18-9.** A minimum width of 3' is generally recommended, **Figure 18-10.** FHA permits a minimum width of 2'-8", measured clear of the handrail. On service stairs, the requirement is reduced to 2'-6". Furniture moving is an important consideration and extra clearance should be provided in closed stairs of the L- and U-type, especially those that include wedge-shaped treads, or *winders.*

Stairs should have a continuous rail along the side for safety and convenience. A *handrail* (also called a *stair rail*) is used on open stairways that are constructed with a low partition or banister. In closed stairs, the support rail is called a *wall rail.* It is attached to the wall with special metal brackets. Except for very wide stairs, a rail on

---

Winders: Wedge-shaped treads installed where stairs turn.

Handrail: A pole installed above and parallel to stair steps to act as a support for persons using the stairs. Also called a *stair rail.*

Wall rail: In closed stairs, the support rail that is attached to the wall.

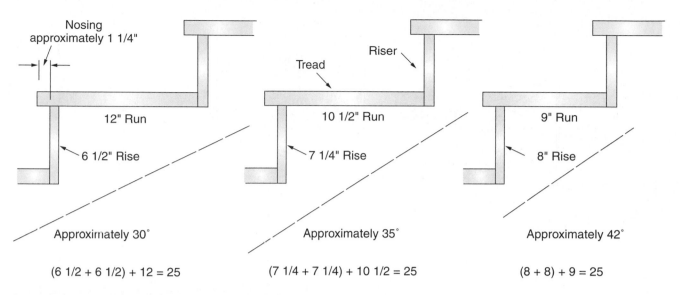

Figure 18-8. Be careful about rise-run relationships in stair design.

Figure 18-9. This L-shaped stairs is wide enough for moving furniture up and down. (C-E Morgan)

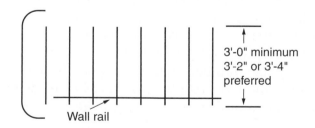

Figure 18-10. A main stair should be at least 3′ wide for easy movement of people and furniture.

only one side is sufficient. **Figure 18-11** illustrates the correct height for a rail.

A complete set of architectural plans should include detail drawings of main stairs, especially when the design includes any unusual features. For example, the stair layout in **Figure 18-12** shows a split-level entrance with open-riser stairs leading to upper and lower floors. An exact description of tread mountings, overlap, nosing requirements, and height of the handrail is not included. These items of construction are the responsibility of

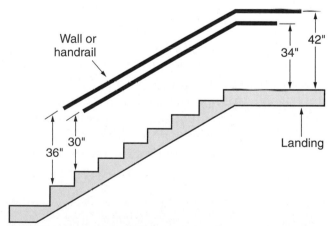

Figure 18-11. A handrail height of 30″ at the rake (slope) and 34″ at the landings has been an accepted standard. Recently, building codes in some places have been adopting heights of 36″ at the rake and 42″ at landings. (C-E Morgan)

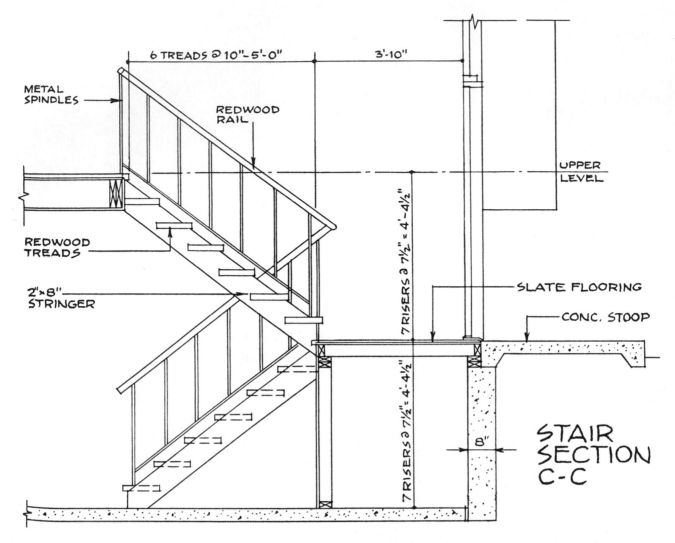

Figure 18-12. Architectural drawings show stair layouts. Note the information given for riser-tread ratios.

the carpenter, who must have a thorough understanding of basic stair design and how to lay out and make the installation.

All stairs, whether main or service, are shown on the floor plans. When details of the stair design are not included in the complete set of plans, the architect usually specifies on the plan view the number and width of the treads for each stair run. Sometimes, the number of risers and the riser height are also included.

18.5 Stair Calculations

To calculate the number and size of risers and treads (less nosing) for a given stair run, first divide the *total rise* by 7 to determine the number of risers. Some carpenters divide by 8. Either number is accurate enough. For example, if the total rise for a basement stairway is 7'-10" (94"), dividing by 7 yields 13.43. Since there must be a whole number of risers, round 13.43 to 13. Divide the total rise by that number to determine the unit rise:

Total rise: Vertical distance from one floor to another.

$94'' \div 13$ $= 7.23$ or $7\ 1/4''$
Number of risers $= 13$
Riser height $= 7\ 1/4''$

In any stair run, the number of treads is one less than the number of risers. A 10 1/2" tread is correct for this example. The *total run* is calculated as follows.

Number of treads $= 12$
Total run $= 10\ 1/2'' \times 12$
                 $= 126''$
                 $= 10'\text{-}6''$

The stairs in this example will have 13 risers 7 1/4" high, 12 treads 10 1/2" wide, and a total run of 10'-6".

Since this example is for basement stairs, the total run can be shortened by using a steeper angle. To do so, decrease the number of risers and shorten the treads. The calculations are:

$94'' \div 12$ $= 7.83$ or $7\ 5/6''$
Number of risers $= 12$
Height of risers $= 7\ 5/6''$
Number of treads $= 11$
Tread width selected $= 9''$
Total run $= 9'' \times 11''$
                     $= 99''$
                     $= 8'\text{-}3''$

Some manufacturers supply tables for determining rise and run, riser, and tread ratios. See **Figure 18-13.**

18.6 Stairwell Length

The length of the stairwell opening must be known during the rough framing operations. If not included in the architectural drawings, it can be calculated from the size of the risers and treads.

It is also necessary to know the headroom required. Add to this the thickness of the floor structure and divide this total vertical distance by the riser height. This gives the number of risers in the opening.

When counting down from the top to the tread from which the headroom is measured, there is the same number of treads as risers. Therefore, to find the total length of the rough opening, multiply the tread width by the number of risers previously determined. Some carpenters prefer to make a scaled drawing (elevation) of the stairs and floor section to check the calculations.

Well Openings Based on Minimum Head Height of 6'-8" Dimensions Based on 2 x 10 Floor Joist									
Total rise floor to floor H	Number of risers	Height of riser R	Number of treads	Width of run T	Total run L	Well opening U	Length of carriage	Use stock tread width	Dimension of nosing projection
8'-0"	12	8"	11	9 1/2"	8'-8 1/2"	9'-1"	11'-4 5/8"	10 1/2"	1"
	14	6 7/8"	13	10 5/8"	11'-6 1/8"	10'-10"	13'-8 1/2"	11 1/2"	7/8"
8'-4"	13	7 11/16"	12	9 13/16"	9'-9 3/4"	10'-0"	12'-5 1/2"	10 1/2"	11/16"
	14	7 1/8"	13	10 3/8"	11'-2 7/8"	11'-1"	13'-7 5/8"	11 1/2"	1 1/8"
8'-6"	13	7 7/8"	12	9 5/8"	9'-7 1/2"	9'-2"	12'-5 1/4"	10 1/2"	7/8"
	14	7 5/16"	13	10 3/16"	11'-0 1/2"	10'-8"	13'-7"	11 1/2"	1 5/16"
8'-9"	14	7 1/2"	13	9 1/4"	10'-0 1/4"	9'-5"	12'-10 3/4"	10 1/2"	1 1/4"
	14	7 1/2"	13	10"	10'-10"	10'-1"	13'-6 1/2"	11 1/2"	1 1/2"
8'-11"	14	7 5/8"	13	9 3/8"	10'-1 7/8"	9'-5"	13'-1 1/4"	10 1/2"	1 1/8"
	14	7 5/8"	13	9 1/16"	9'-9 7/8"	9'-0"	12'-10"	10 1/2"	1 7/16"
	14	7 5/8"	13	10 1/4"	11'-1 1/4"	10'-2"	13'-10 1/4"	11 1/2"	1 1/4"
9'-1"	14	7 13/16"	13	9 11/16"	10'-6"	9'-5"	13'-5 3/4"	10 1/2"	13/16"
	15	7 1/4"	14	10 1/4"	11'-11 1/2"	10'-8"	14'-7 3/4"	11 1/2"	1 1/4"

Figure 18-13. A table can be used to determine the number of risers and treads and their dimensions. (C-E Morgan)

18.7 Stringer Layout

To lay out the stair stringer, first determine the riser height. Place a story pole in a plumb position from the finished floor below through the rough stair opening above. On the pole, mark the height of the top of the finished floor above.

Set a pair of dividers to the calculated riser height and step off the distances on the story pole. There will likely be a slight error in the first layout, so adjust the setting and try again. Continue adjusting the dividers and stepping off the distance on the story pole until the last space is equal to all of the others. Measure the setting of the dividers. This length is the exact riser height to use in laying out the stringers.

To create a cut-out stringer for a simple basement stair, select a straight piece of 2 × 10 or 2 × 12 stock of sufficient length. Place it on sawhorses to make the layout. Begin at the end that will be the top and hold the framing square in the position shown in **Figure 18-14.** Let the blade represent the treads and the tongue represent the risers. For example, if the risers are 7 3/4", align that mark on the tongue with the edge of the stringer. If the treads are 10", align that mark on the blade with the edge of the stringer.

Draw a line along the outside edge of the blade and tongue. Now move the square to the next position and repeat. The procedure is similar to that described for rafter layout in Chapter 10. Continue stepping off with the square until the required number of risers and treads have been drawn, **Figure 18-15.**

The stair begins with a riser at the bottom, so extend the last tread line to the back edge of the stringer. At the top, extend the last tread and riser line to the back edge.

One other adjustment must be made before the stringer is cut. Earlier calculations that gave the height of the riser did not take into account the thickness of the tread. Therefore the total rise of the stringer must be shortened by one tread thickness. Otherwise the top tread will be too high. The bottom of the stringer must be trimmed, as shown in **Figure 18-16.**

Working Knowledge

Extreme accuracy is required in laying out the stringer. Be sure to use a sharp pencil or knife and make the lines meet on the edge of the stock. Accuracy can be assured in this layout by using framing square clips or by clamping a strip of wood to the blade and tongue.

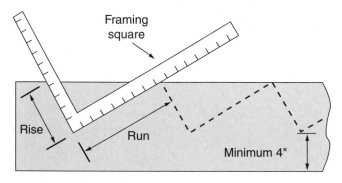

Figure 18-14. Using a framing square to lay out a stringer.

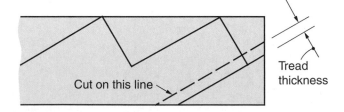

Figure 18-16. Trim the bottom end of the stringer to adjust the riser height for the tread thickness.

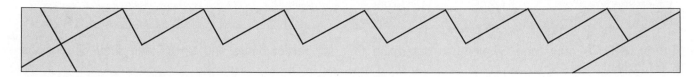

Figure 18-15. A completed stringer layout will look something like this.

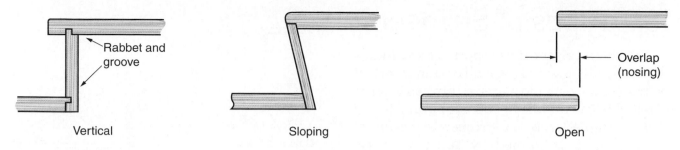

Figure 18-17. Basic stair riser shapes. For the open riser, the tread should overlap the riser at least 2″.

18.8 Treads and Risers

The thickness of a main stair tread is generally 1 1/6″ or 1 1/8″. Hardwood or softwoods may be used. FHA requires that stair treads be hardwood, vertical-grain softwoods, or flat-grain softwoods covered with a suitable finish flooring material.

Lumber for risers is usually 3/4″ thick and should match the tread material. This is especially important when the stairs are not covered. In most construction, the riser drops behind the tread, making it possible to reinforce the joint with nails or screws driven from the back side of the stairs. **Figure 18-17** shows basic types of riser designs. A sloping riser is sometimes used in concrete steps since it provides an easy way to form a nosing.

Where the top edge of the riser meets the tread, glue blocks are sometimes used. A rabbeted edge of the riser may fit into a groove in the tread. A rabbet and groove joint may also be used where the back edge of the tread meets the riser.

Stair treads must have a *nosing.* This is the part of the tread that overhangs the riser. Nosings serve the same purpose as toe space along the floor line of kitchen cabinets. They provide toe room. The width of the tread nosing may vary from about 1″–1 1/2″. It should seldom be greater than 1 3/4″. In general, as the tread width is increased, the nosing can be decreased. **Figure 18-18** illustrates a number of nosing forms. Cove molding may be used under the nosing to cover the joint between riser and tread and conceal nails used to attach the riser to the stringer or carriage.

Basement stairs may be constructed with an open riser (no riser board installed). Sometimes an open riser design is built into a main stair to provide a special effect. Various methods of support or suspension may be used. Often custom-made metal brackets or other devices are needed.

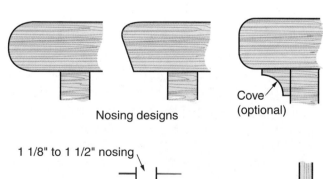

Nosing designs

Cove (optional)

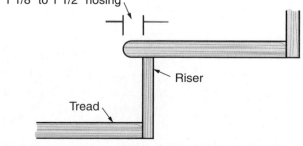

1 1/8″ to 1 1/2″ nosing

Riser

Tread

Figure 18-18. Common tread nosings.

Total run: The horizontal distance occupied by the stairs; measured from the foot of the stairs to a point directly beneath where the stairs rest on a floor or landing above.

Nosing: The part of a stair tread that projects beyond the riser.

18.9 Types of Stringers

Treads and risers are supported by stringers that are solidly fixed to the wall or framework of the building. For wide stairs, a third stringer is installed in the middle to add support.

The simplest type of stringer is the ***built-up stringer.*** It is formed by attaching cleats on which the tread can rest. Another method consists of cutting dados into which the tread will fit, **Figure 18-19.** This type is often used for basement stairs where no riser enclosure is called for.

Standard ***cut-out stringers*** are commonly constructed for either main or service stairs. This is the type created in the earlier layout description. Prefabricated treads and risers are often used with this type of stringer. An adaptation of the cut-out stringer, called *semihoused* construction, is shown in **Figure 18-20.** The cut-out stringer and backing stringer may be assembled and then installed as a unit or each part may be separately installed.

A popular type of stair construction has a stringer with tapered grooves into which the treads and risers fit. It is commonly called *housed construction.* ***Housed stringers*** can be purchased completely cut and ready to install. They can be cut on the job, using an electric router and template. Wedges with glue applied are driven into the grooves under the tread and behind the riser, **Figure 18-21.** The treads and risers are joined with rabbeted edges and

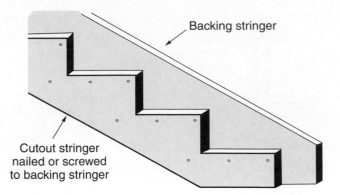

Figure 18-20. This is a semihoused stringer, an adaptation of the cut-out stringer.

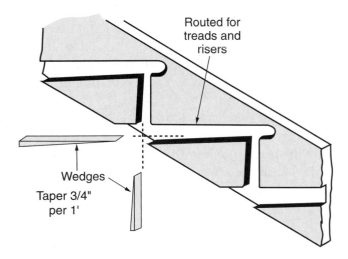

Figure 18-21. In a housed stringer, risers and treads are let into the stringer. This type of housing is difficult to make.

grooves or glue blocks. To assemble the stairs, the housed stringer is spiked to the wall surface and into the wall frame. The treads and risers are then set into place. Work is done from the top downward.

Housed construction produces a stair that is strong and dust tight. It seldom develops

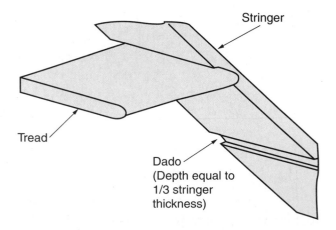

Figure 18-19. Open riser stairs. Treads are set into dados cut in the stringer.

Built-up stringer: A stringer to which blocking has been added to form a base for adding treads and risers.

Cut-out stringer: A stair stringer into which the rise and run are cut.

Housed stringer: A stair stringer where the edges of the steps are covered with a board.

squeaks. Housed stringers show above the profiles of the treads and risers and provide a finish strip along the wall. The design should permit a smooth joint where it meets the baseboard of the upper and lower levels.

18.10 Winder Stairs

Winder stairs present stair conditions that are frequently regarded as undesirable, **Figure 18-22.** In fact, some localities do not allow them. Check local building codes to see if this type of stairs is allowed. The use of winder stairs may sometimes be necessary, however, where space is limited. When used, it is important to maintain a winder-tread width along the line of travel that is equal to the tread width in the straight run. When winders are used, it is best if they are at the bottom of a straight run.

An adaptation of the standard winder layout is illustrated in **Figure 18-23.** Here, if you extend the lines of the risers, they meet outside the stairs. This provides some tread width at the inside corner. Before starting the construction of this type of stairs, the carpenter should make a full-size or carefully scaled layout in plan view. The best radius for the line of travel can then be determined.

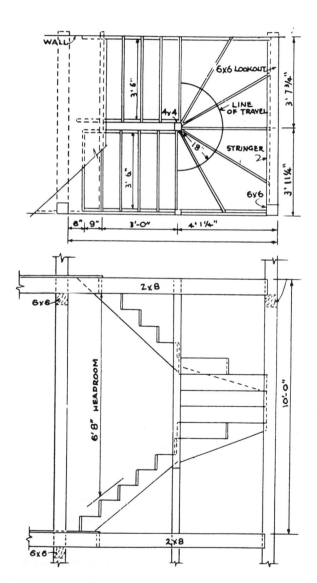

Figure 18-22. Typical drawing of a winder stairs. The tread width on the winding section should be the same at line of travel (near middle of stairs) as the tread in the straight run.

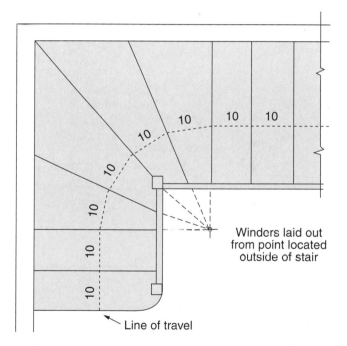

Figure 18-23. Laying out a winder stairs with lines representing the tread nosings converging outside of the construction.

PROCEDURE

Splitting angles for miter cuts

The carpenter sometimes faces odd angles that must be accurately split to make a miter cut. This can be mathematically calculated, but the following method avoids the math and assures great accuracy.

1. Select a plywood scrap about 6″ wide and 1′ long with a factory edge to use as a storyboard.
2. Draw a line near to and parallel with the factory edge.
3. Use a T-bevel to find the angle to be mitered and transfer the angle to the storyboard.
4. Draw a line along the blade of the T-bevel, as shown in **Figure 18-24A.**

5. Open a pencil compass or scribe about 3″ to 4″. Place the point of the instrument at the intersection of lines AB and AC and draw arcs of equal length across both lines, **Figure 18-24B.**
6. Swing arcs of equal distance from points B and C to create point D, **Figure 18-24C.** You may need to open up the compass or dividers somewhat more to create this point.
7. Draw a line to connect points A and D, **Figure 18-24D.** This is the miter angle.
8. Adjust the T-bevel to this angle and use it to set the miter saw.
9. Make a test cut on scraps to verify the accuracy of the angle.

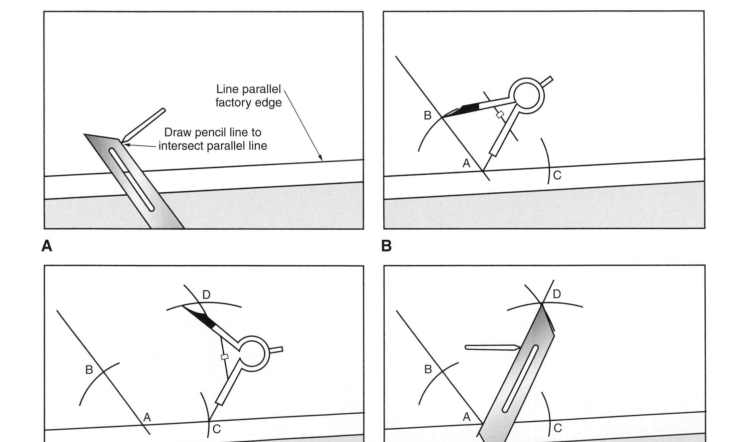

A

B

C

D

Figure 18-24. Splitting an angle for a miter cut. A—After drawing a line parallel to the factory edge, transfer the miter angle to the board. B—Draw arcs of equal lengths from the intersection of lines AB and AC. C—Swing arcs to create point D. D—Draw line AD. This is the miter angle.

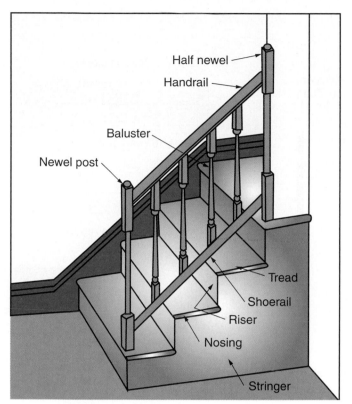
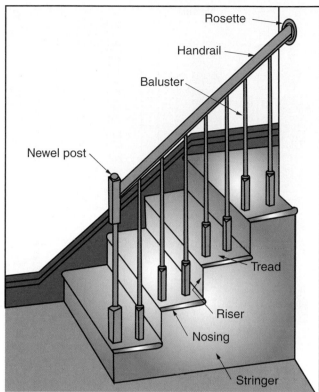

Figure 18-25. Parts of an open stair. An assembly including a newel, balusters, and rail is called a balustrade.

18.11 Open Stairs

Stairs that are open on one or both sides require some type of decorative enclosure and support for a handrail. Typical designs consist of an assembly of parts called a *balustrade,* **Figure 18-25.** The principal members of a balustrade are newels, balusters, and rails. They are usually made in a factory and assembled on the job by the carpenter.

The starting *newel* must be securely anchored either to the starter step or carried down through the floor and attached to a floor joist. *Balusters* are joined to the stair treads using either a round or square mortise. Two or three may be mounted on each tread.

Working Knowledge

The main purpose of balusters is to prevent anyone, children especially, from slipping under the railings and falling to the floor below. Codes usually require baluster spacing of no more than 6″, although 4″ is required in some localities.

Balustrade: An assembly with a railing resting on a series of balusters that, in turn, rest on a base, usually the treads.

Newel: The main post at the start of a stair and the stiffening post at the landing.

Baluster: The vertical member (spindle) supporting the handrail on open stairs.

18.12 Using Stock Stair Parts

While many parts of a main staircase could be cut and shaped on the job, the usual practice is to use factory-made parts. These are available in a wide range of stock sizes and can be selected to fill requirements for most standard stair designs. See **Figures 18-26** and **18-27.** Stair parts are ordered through lumber and millwork

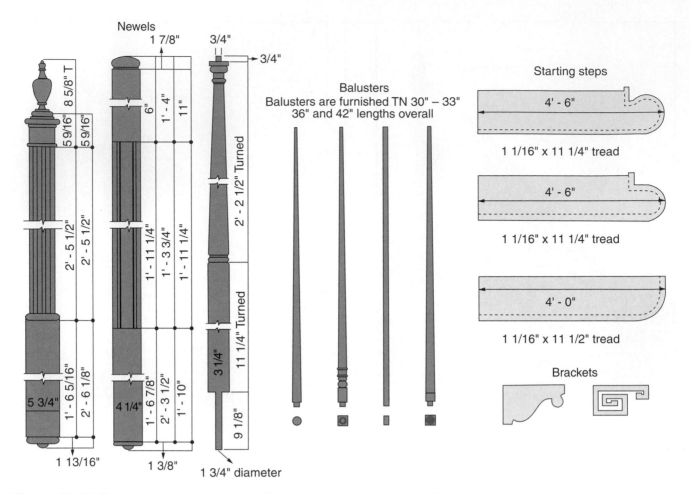

Figure 18-26. Typical stock parts commonly available for stair construction.

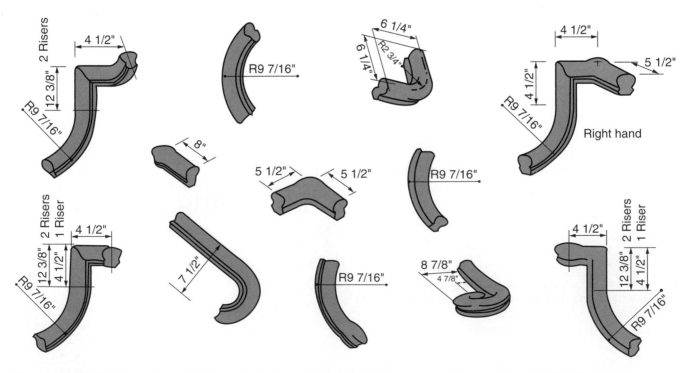

Figure 18-27. Preformed handrails and stock parts for special shapes can be purchased. (C-E Morgan)

dealers. They are shipped to the building site in heavy, protective cartons along with directions for fitting and assembly.

A completely prefabricated stairway and a factory assembly are shown in **Figure 18-28.**

Stringers are made in two sections for easier shipping. The system is available in lengths up to 18 steps and widths of 36″ and 48″.

Figure 18-29 shows some suggested assemblies of balustrades using stock parts. Hardware

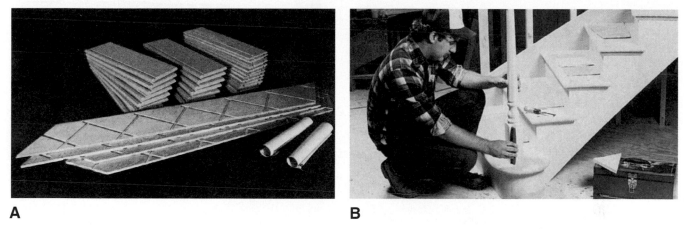

A   B

Figure 18-28. Prefabricated stairway system. A—Parts for the system. Stringer sections lock together with a common tread. B—This installed mock-up shows the assembled section mounted on substringers. (Visador Co.)

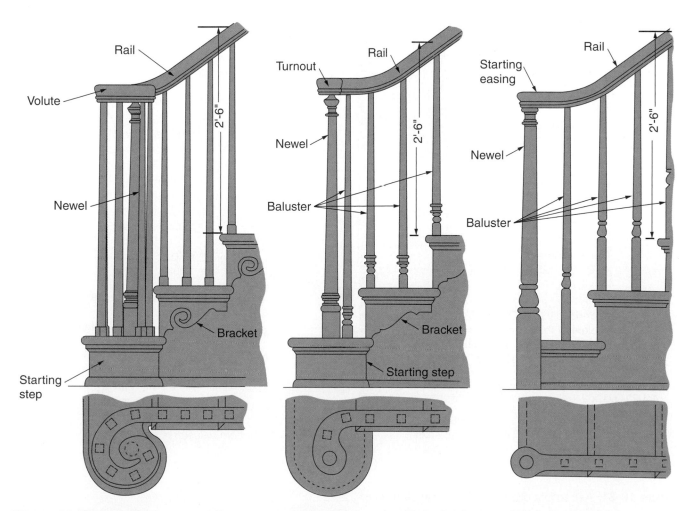

Figure 18-29. Balustrade assemblies produced from stock parts. (Colonial Stair and Woodwork Co.)

especially designed for stair work is illustrated in **Figure 18-30.**

18.13 Spiral Stairways

Metal spiral stairways eliminate framing and save space. See **Figure 18-31.** Units are available in aluminum or steel in a variety of designs to fit requirements up to 30 steps and heights up to 22'-6". Use of spiral stairs is often restricted by building codes. Some codes permit use of a spiral stairway for exits in private dwellings or

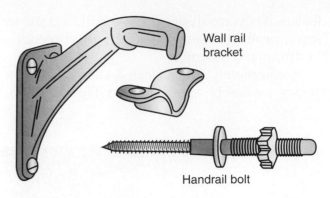

Wall rail bracket

Handrail bolt

Figure 18-30. Hardware for stair rails. The handrail bolt is concealed in the center of a joint. The nut is accessible from below and can be adjusted with a screwdriver or a hammer and nail set.

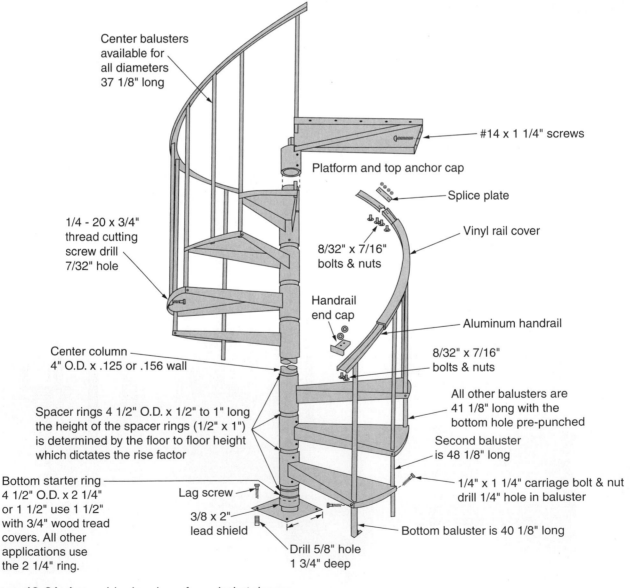

Center balusters available for all diameters 37 1/8" long

#14 x 1 1/4" screws

Platform and top anchor cap

Splice plate

1/4 - 20 x 3/4" thread cutting screw drill 7/32" hole

Vinyl rail cover

8/32" x 7/16" bolts & nuts

Handrail end cap

Aluminum handrail

Center column 4" O.D. x .125 or .156 wall

8/32" x 7/16" bolts & nuts

All other balusters are 41 1/8" long with the bottom hole pre-punched

Spacer rings 4 1/2" O.D. x 1/2" to 1" long the height of the spacer rings (1/2" x 1") is determined by the floor to floor height which dictates the rise factor

Second baluster is 48 1/8" long

Bottom starter ring 4 1/2" O.D. x 2 1/4" or 1 1/2" use 1 1/2" with 3/4" wood tread covers. All other applications use the 2 1/4" ring.

Lag screw

3/8 x 2" lead shield

Drill 5/8" hole 1 3/4" deep

1/4" x 1 1/4" carriage bolt & nut drill 1/4" hole in baluster

Bottom baluster is 40 1/8" long

Figure 18-31. Assembly drawing of a spiral stairway.

in some other situations when the area served is not more than 400 square feet.

18.14 Disappearing Stair Units

Where attics are used primarily for storage and where space for a fixed stairway is not available, hinged or disappearing stairs are often used. Such stairways may be purchased ready to install. They operate through an opening in the ceiling and swing up into the attic space when not in use, **Figure 18-32.** Where such stairs are to be provided, the attic floor should be designed for regular floor loading and the rough opening should be constructed at the time the ceiling is framed.

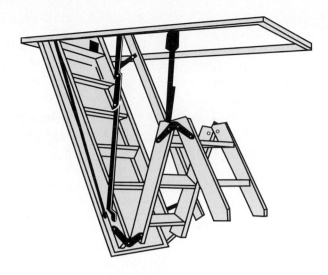

Figure 18-32. This disappearing stair unit is designed to fold into the ceiling. The ceiling opening should be framed as the ceiling joists are installed. (Rock Island Millwork)

Summary

A stairway is a series of steps, each elevated a measured distance, leading from one level of a structure to another. Main stairways are often constructed in a factory, then assembled on-site by the carpenter. Stairways are either straight run or winding (circular or elliptical). Stairway components include the stringers that support risers and treads, vertical risers between treads, and the horizontal treads (steps). In a given run of stairs, all treads must be the same size to assure safe use. Risers must also be the same size. A main stair should be wide enough for two people to pass each other without contact. The number of steps and risers is calculated using the total rise (vertical height) and total run (horizontal distance) occupied by the stairs. Stairways that are open on one side require use of a decorative enclosure that also supports a handrail. This enclosure is called a balustrade and consists of newels, horizontal rails, and balusters. Balustrades are typically factory made and shipped to the job site for installation by the carpenter. Metal spiral stairways are available. They are usually installed to save space. Disappearing stair units fold into the ceiling. They provide access to attic areas without consuming space in the room below.

Test Your Knowledge

Answer the following questions on a separate piece of paper. Do not write in this book.

1. The platform type of stairway includes _____ where the direction of the stair run is usually changed.
2. The minimum headroom for a main stairway, as specified by FHA, is _____.
3. One of the rules used to calculate riser-tread relationship states that the sum of two risers and one tread should be _____.
4. The front edge of the tread that overhangs the riser is called the _____.
5. A stairs in a split-level home has six risers with a tread width of 11". The total run of the stairs is _____.
6. A semihoused stair stringer is formed by attaching a(n) _____ stringer to a backing stringer.
7. *True or False?* Winder stairs are allowed by all building codes.
8. _____ are glued and driven into the stringer to assemble risers and treads in housed stringers.
9. The three principal members of a balustrade are the newels, rails, and _____.
10. When a disappearing stair unit is used to provide attic access, the attic floor should be designed for _____ floor loading.

Curricular Connections

Social Studies. The Dutch artist M.C. Escher created a number of drawings featuring stairways that were optical illusions—they led nowhere but back to themselves. Use the library or Internet to view several Escher stairway prints. Try to determine how he achieved the optical illusion. Also, search for ways that other artists have featured stairways in their paintings or photographs. See how many different kinds of stairways you can find depicted. If possible, determine why the artist made the stairway the focal point of the work.

Outside Assignments

1. Obtain a set of architectural plans where the main or service stairway is not drawn in detail. Carefully study the stair requirements and then prepare a detail drawing similar to **Figure 18-12.** Use a scale of 1/2" equals 1'. Carefully select and calculate the riser-tread ratio. Be sure the number and size of risers is correct for the distance between the two levels. Check the headroom requirements against your local building codes and determine the stairwell sizes. Submit the completed drawing and size specifications to your instructor.

2. Study a millwork catalog and become familiar with the stock parts shown for a main stairway. Working from a set of architectural plans or a stair detail that you may have drawn, prepare a list of all of the stair parts needed to construct the stairway. Include the number of each part and its size, quality, material, and catalog number. Take the list to a building supply dealer and obtain a cost estimate for the materials. Be prepared to discuss the materials and costs with your instructor and class.

This freestanding curved stair was completely fabricated in a manufacturing plant and then disassembled and shipped to the building site. (L.J. Smith, Inc.)

Doors and Interior Trim

19

Learning Objectives

After studying this chapter, you will be able to:

- Describe how door frames and casings are installed.
- Explain the difference between panel- and flush-type doors.
- List the steps for hanging a door.
- Name lock parts and describe typical installation procedures.
- Compare the pocket and bypass types of sliding doors.
- Outline the order in which window trim members should be applied.
- Cut, fit, and nail baseboard trim.

Technical Vocabulary

Baseboard	Panel door
Base shoe	Plinth block
Core	Pocket door
Dead bolts	Prehung door unit
Door casing	Rail
Door frame	Reveal
Door stop	Scarf joint
Flush door	Side jambs
Gains	Stile
Head jamb	Stool
Lug	Threshold
Mouldings	

This chapter deals with the methods and materials of an important part of interior finish. It includes:

- Installing door frames.
- Hanging doors.
- Fitting trim around openings.
- Fitting trim at intersections of walls, floors, and ceilings.

This aspect of carpentry requires great skill and accuracy. Well-fitted trim greatly improves the appearance and desirability of a home.

19.1 Mouldings

Mouldings are decorative wood or plastic strips. They are designed to be functional as well as decorative. For example, window and door mouldings cover the space between the jamb and the wall covering. They also make the installation more rigid.

A wide range of types, patterns, and sizes of mouldings is used in residential construction. Common shapes and typical uses are shown

---

Moulding: A relatively narrow strip of wood, usually shaped to a curved profile throughout its length, used to accent and emphasize the ornamentation of a structure and to conceal surface or angle joints.

in **Figure 19-1.** Other standard mouldings include stools, aprons, door and window stops, mullion casings, quarter rounds, bed moulds, cove mouldings, and base shoe mouldings. See **Figure 19-2.** In addition to those shown, a complete list includes cove moulding, brick moulding, battens, glass beads, drip caps, picture moulding, and screen mould. Information on moulding patterns along with a numbering system and grading rules are included in a manual that is available from Western Wood Products Association, Portland, Oregon.

19.1.1 Classifying Mouldings

Some mouldings take their names from their shape. Bed mouldings, coves, and crowns are named for their shapes. Full rounds, half

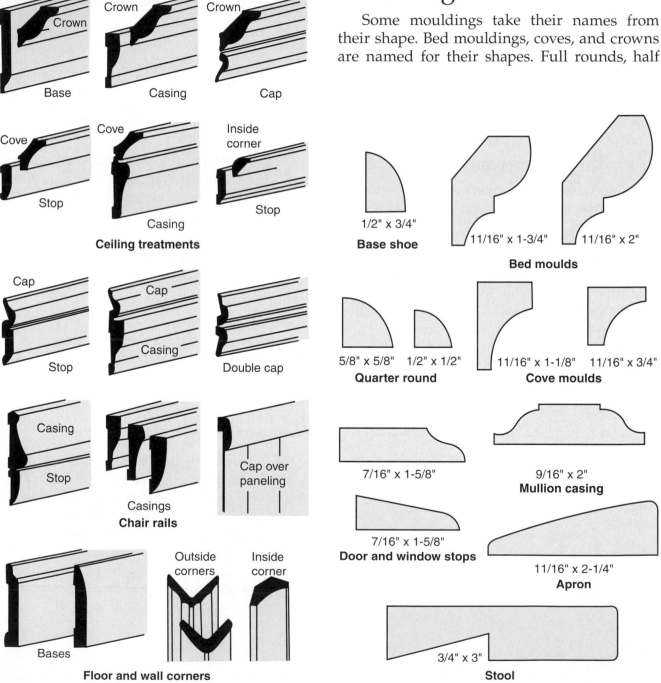

Figure 19-1. Mouldings perform various functions, depending on where they are used. (ABTco, Inc.)

Figure 19-2. Profiles of some typical moulding patterns.

rounds, and quarter rounds are other examples. They may be used in many different locations to finish interior space. For example:

- Full rounds are often installed as closet poles.

- Half rounds may serve to cover the raw edges of shelving.

- Baseboard (or mop boards), base shoe, and base cap are installed where the wall and floor meet. Quarter round used as a base cap dresses off the top of the base when the base's top edge is squared off. Quarter round is used as the base shoe, which covers the joint between the base and the flooring.

Names of other mouldings come from their use. They are generally used to cover intersections of walls and ceilings.

19.2 Interior Door Frames

The interior *door frame* covers the unfinished edges of the door opening. It provides support for the door and its hardware. It also supports the trim pieces that are attached after installation.

The frame consists of two *side jambs* and a *head jamb*, **Figure 19-3.** Interior door frames are simpler than those for exterior doors. The jambs are not rabbeted and there is no door sill. Refer to Chapter 13, **Windows and Exterior Doors.**

Standard jambs for regular 2 × 4 stud partitions are made from nominal 1″ material. For plastered walls, the jambs are 5 1/4″ wide. For drywall, the jambs are 4 1/2″ wide. The back side of the jamb is usually kerfed to reduce the tendency of the material to cup (warp). The edge of the jamb is slightly beveled so the casing snugly fits against it with no visible crack.

Door frame: An assembly of wood parts that form an enclosure and support for a door.

Side jambs: Parts of a door frame that are dadoed to receive the head jamb.

Head jamb: Part of a door frame that fits between the side jambs, forming the top of the frame.

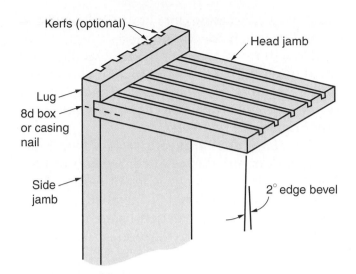

A

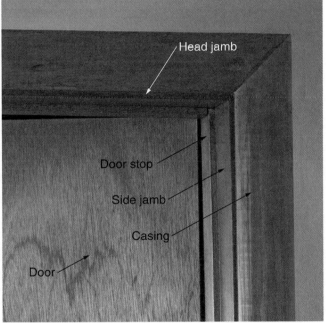

B

Figure 19-3. A—Section view of an interior door frame. Parts are listed. Note that edges of all jambs are beveled slightly so trim will snugly fit. B—Parts of a standard inside passage door frame.

Side jambs are dadoed to receive the head jamb. The side jambs for residential doorways are 6′-9″ long (measured to the head jamb). This provides clearance at the bottom of the door for flooring materials, while allowing a 6′-8″ opening.

Interior doorjambs are sometimes adjustable. These are designed to fit walls of different thicknesses. The three-piece type depends on a rabbet joint and a concealing door stop, **Figure 19-4.** A second type is made in two pieces.

Door frames are usually cut, sanded, and fitted (prehung) in millwork plants. This allows quick assembly on the job. Door frames should receive the same care in storage and handling as other finished woodwork.

A *prehung door unit* consists of a door frame with the door already installed. See **Figure 19-5.** The frame includes both sides of casing. Lock hardware may or may not be installed, although machining for its installation has usually been completed. Quite often, the door is prefinished.

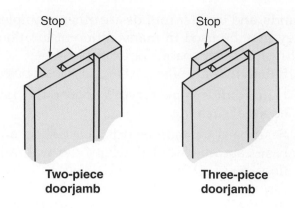

Two-piece doorjamb   **Three-piece doorjamb**

Figure 19-4. Adjustable doorjambs fit any thickness of wall. The stop conceals the joints.

1. Remove door unit from carton and check for damage. Separate the two sections. Place tongue side (not attached to door) outside of the room.

2. Slide frame section that includes the door into the opening. Side jambs should rest on finished floor or spacer blocks.

3. Carefully plumb door frame and nail casing to wall structure. Be sure all spacer blocks are in place between the jambs and door.

4. Move to the other side of the wall and install shims between side jambs and rough opening. Shims should be located where spacer blocks make contact with door edges. Nail through jambs and shims.

5. Install remaining half of door frame. Insert the tongue edge into the grooved section already in place. Nail casing to wall structure.

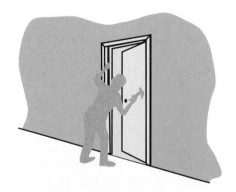

6. Nail through stops into jambs. Remove spacer blocks and check door operation. Make any adjustments required. Drive extra nails where shims are located. Install lock set.

Figure 19-5. General procedure for installing a prehung interior door. (Frank Paxton Lumber Co.)

Prehung door unit: A door frame with the door already installed.

PROCEDURE

Installing a prehung door

1. Check the length of the head and side jamb against the rough opening to determine if the door is correct for the opening.
2. Place the unit into the opening, **Figure 19-5.** Let the side jambs rest on the finish flooring or on spacer blocks of the right thickness. Spacers are needed only if the final flooring surface has not been laid.
3. Level the head jamb, shimming a side jamb as required.
4. Plumb the jambs side-to-side in the frame with a straightedge and level or a long carpenter's level. Make adjustments using doubled shims between the frame and the stud.
5. Plumb the frame front-to-back in the opening.
6. Temporarily fasten the top and bottom of each side jamb with an 8d casing nail.
7. Add more doubled shims in back of each jamb as illustrated in **Figure 19-6.** On the hinge jamb, locate one block 11″ up from the bottom and one 7″ down from the top. Set a third block halfway between these two.
8. Reverify plumb.
9. Nail through the shims into the studs. Generally, it is best to use two 8d nails staggered front to back.
10. If recommended by the manufacturer, install longer screws in the hinges.

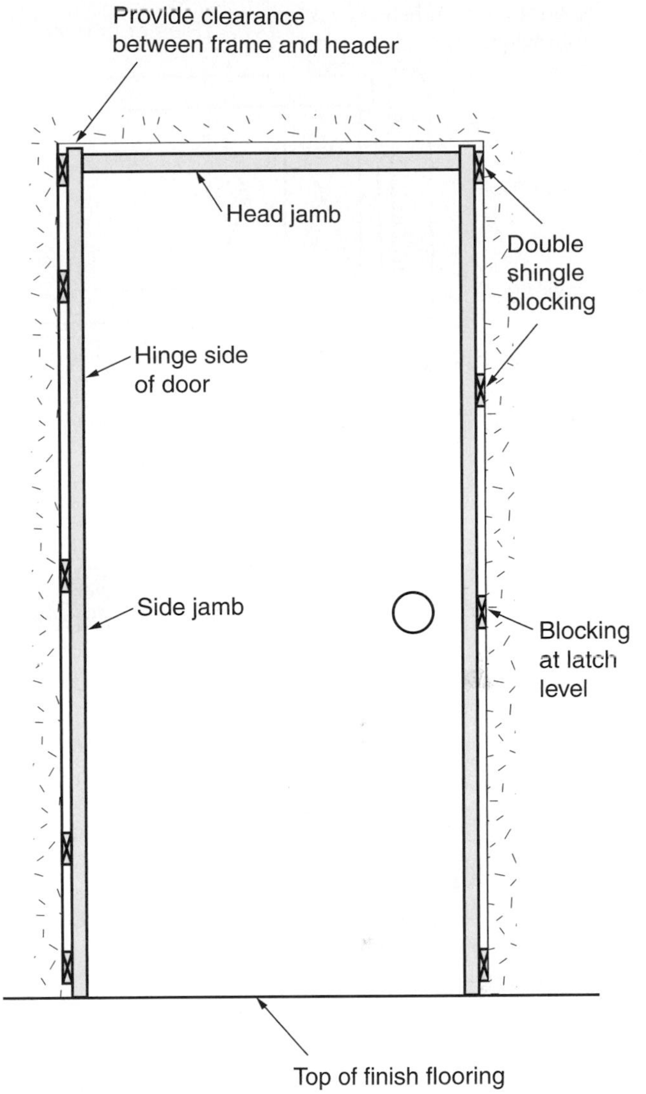

Working Knowledge

When setting a door frame, do not fully drive any of the nails until all shims have been adjusted and the jambs are straight and plumb.

Door casing: Trim applied to each side of the door frame to cover the space between the jambs and the wall surface.

Reveal: Amount of setback of a window or door casing from the inside edge of the jamb.

Figure 19-6. Set the door into the opening and add blocking (double shims).

19.2.1 Door Casing

Trim, known also as *door casing*, is applied to each side of the door frame to cover the space between the jambs and the wall surface. This secures the frame to the wall structure and stiffens the jambs so they can carry the door. **Figure 19-7** shows a section view through a doorjamb. The casing covers the blocking and is attached to both the jamb and wall surface. It is usually installed with a *reveal*. This is a setback from the edge of the frame of about 1/8″ or 3/16″. The setback allows room for hinges and striker plate, while improving the appearance of the trim.

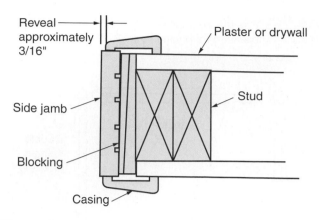

Figure 19-7. This section view shows the position of casing, which is attached to the doorjamb.

Figure 19-8. A power miter saw can be used to make compound cuts.

PROCEDURE

Installing casing

1. Mark the reveal. Some carpenters prefer to draw a light pencil line on the edge of the side and head jambs. Use a combination square with the blade extended to the reveal as a marking guide.
2. Select the casing material.
3. Check the bottom end of each side casing to see that it is square on what will be the bottom end. This is necessary for a tight fit against the finished floor.
4. Install either the side casings or the head casing. Many carpenters prefer to do the head casing first.
5. Determine the length of the head casing, being sure to include the width of the side casings.
6. Make miter cuts. Use a miter box, a wood trimmer, or a power miter saw to make accurate cuts, **Figure 19-8.**

7. Align the head casing with the pencil marks for reveal and install it.
8. Mark the length of the side casings. Carefully place each piece against the jamb, aligned with the reveal setback. Mark the location of the miter joint at the top.
9. Make the miter cuts.
10. If the miters do not properly fit, trim them with a block plane.
11. Temporarily nail casing with casing or finish nails.
12. When the fit is satisfactory, fully drive the nails and complete the nailing pattern. Use 4d or 6d nails along the jamb edge and drive 8d nails through the outer edge into the studs making up the rough opening. Each pair of nails should be spaced about 16″ on center.
13. Use a nail set to sink the nails below the surface, **Figure 19-9.**

Working Knowledge

To avoid making hammer dents in finish lumber that must have a fine appearance, cut a saw kerf in the butt end of a shim shingle. Start the finish nail, and then slip the kerfed shim around the nail. Drive the nail flush with the shim shingle and finish by using a nail set. Any missed hammer blows will be absorbed by the shim, leaving the face of the finish trim unblemished.

19.2.2 Using Plinth Blocks

A *plinth block* is a decorative corner trim mainly used on door frames and window frames. It allows the finish carpenter to make butt joints instead of the more difficult mitered corners. Installation of the casing for plinth blocks is slightly different from that used with mitered corners. Usually, the head casing and plinths are installed first. Sometimes plinth blocks are also installed at the floor level on door frames.

Figure 19-9. When installing casing, use a nail set to drive casing nails below surface. The resulting holes will be filled before finishing.

The side casings are typically fitted, marked, cut, and installed last. See **Figure 19-10.**

19.3 Panel Doors

There are two general types of doors: panel and flush. The panel door is also referred to as a stile and rail door. This type of construction is used in sash, louver, storm, screen, and combination doors. Sash doors are similar to panel doors in appearance and construction, but have one or more glass lights in place of the wood panels.

A *panel door* consists of stiles and rails with panels of plywood, hardboard, or solid stock, **Figure 19-11.** A *stile* is an outside vertical member and a *rail* is a horizontal member. The rails and stiles are usually made of solid material. However, some are veneer applied over a lumber core. In some cases, the doors are molded

---

Plinth block: Decorative, carved block set at left and right top corners of window and door trim.

Panel door: Style of door that has a frame with separate panels of plywood, hardboard, or steel set into the frame.

Stile: The upright or vertical outside pieces of a sash, door, blind, screen, or face frame.

Rail: Cross or horizontal members of the framework of a sash, door, blind, or other assembly.

A

B

Figure 19-10. Plinth blocks are decorative corners. A—Most carpenters install the plinth blocks before the head casing. However, this carpenter attaches the side casing after the plinth blocks. B—Close-up of an installed plinth block.

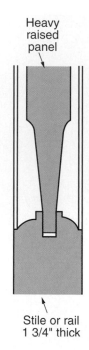

Heavy
raised
panel

Stile or rail
1 3/4" thick

A

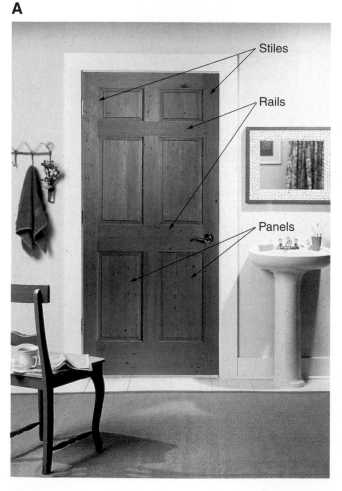

Stiles

Rails

Panels

B

Figure 19-11. Panel door construction. A—This section through the stile and panel shows the joint detail. B—A panel door installed in a bathroom. (Jeld-Wen, Inc.)

from a wood fiber, **Figure 19-12.** A variety of designs is formed by changing the number, size, and shape of the panels.

Special effects are secured by installing raised panels, which add line and texture. This panel is formed of thick material that is reduced around the edges where it fits into the grooves in the stiles and rails.

A

B

Figure 19-12. A—Molded doors are made to simulate panel doors. A section does not have the steel skin to show the hardboard base. B—A molded door appears like a standard panel door when installed. (Masonite Corp.; Jeld-Wen, Inc.)

19.4 Flush Doors

A *flush door* consists of a wood frame with thin, flat sheets of material applied to both faces. These doors are strong and durable. Flush doors account for a high percentage of wood doors used in homes and commercial structures.

The face panels are also called *skins*. These are commonly made of 1/8″ plywood. However, hardboard, plastic laminates, fiberglass, and metal are also used.

Flush doors are made with either solid or hollow cores. The *core* is the inside of the door between the face panels. Cores of wood or various composition materials are used in solid (or slab) construction, **Figure 19-13.** The frame inside of the door is usually made of softwood that matches the color of the face veneers. The most common type of solid core construction uses wood blocks bonded together with the end joints staggered.

Flush doors with hollow cores are widely used for interior doors. They may also be used for exterior doors, if made with waterproof adhesives. Hollow core flush doors do not ordinarily provide as much thermal (heat) and sound insulation as solid core doors. Usually their fire resistance rating is lower, as well. Some flush exterior doors have compression-molded fiberglass face panels. They are attached to a wooden frame and can be formed to reproduce various traditional and contemporary designs. The core is a high-density polyurethane foam that has a high R-value. Unlike steel-faced doors, the fiberglass door can be trimmed for a precision fit. The surface is textured like wood and can be stained or painted.

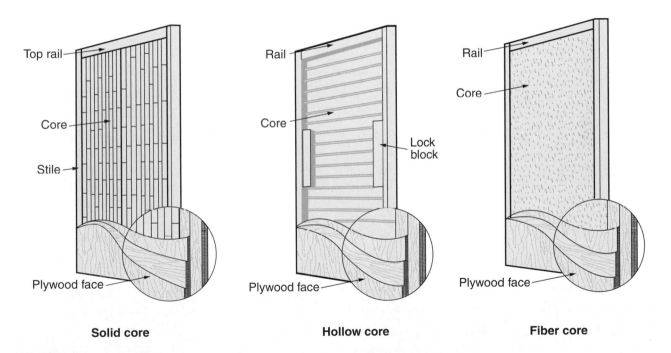

Figure 19-13. Basic types of cores used in wood flush door construction. The basic hollow core types include lattice (also called mesh or grid), ladder, and implanted blanks. Frames are usually made of softwood.

Flush door: A door that fits into the opening and does not project outward beyond the frame.

Core: Interior area between the face veneers of a flush door; may be solid or hollow.

19.5 Sizes and Grades

Interior door widths vary with the installation. Widths of 2'-8" and 3'-0" are most commonly used for interior doors. The FHA specifies a minimum width of 2'-6" for bedrooms and 2' for bathrooms. Closet doors may also be 2' wide. Standard thickness for interior passage doors is 1 3/8". Interior door sizes and patterns are illustrated in **Figure 19-14**.

Interior residential doors have a standard height of 6'-8". A 7'-0" door is sometimes used for entrances or special interior installations. A 7'-0" height is usually considered standard for commercial buildings.

Grades and manufacturing requirements for doors are listed in *Industry Standards* developed by the National Woodwork Manufacturing Association (NWMA) and the Fir and Hemlock Door Association (FHDA). The purpose of these standards is to establish nationally recognized

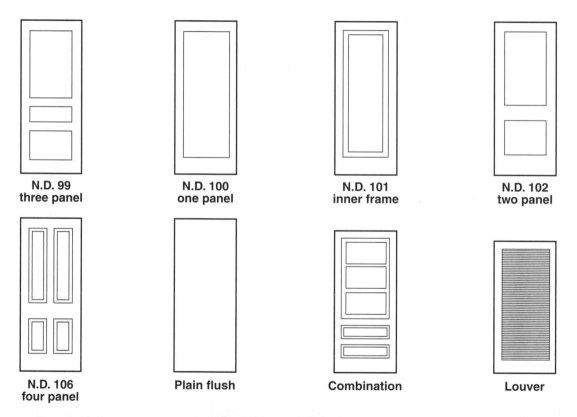

Constructions Details

Design No.	Stiles	Top rail	Cross rail	Lock rail	Intermediate rails	Mullions or muntins	Bottom rail	Panels
N.D. 99	4 3/4"	4 3/4"	4 5/8"	4 5/8"	—	—	9 5/8"	Flat
N.D. 100	4 3/4"	4 3/4"	—	—	—	—	9 5/8"	Flat
N.D. 101	4 1/4" Face	4 1/4" Face	—	—	—	—	9 1/4" or 9 1/2" Face	Flat
N.D. 102	4 3/4"	4 3/4"	—	8"	—	—	9 5/8"	Flat
N.D. 106	4 3/4"	4 3/4"	—	8"	—	4 5/8"	9 5/8"	Raised
N.D. 107	4 3/4"	4 3/4"	—	—	4 5/8"	—	9 5/8"	Raised
N.D. 108	4 3/4"	4 3/4"	—	8"	3 7/8" or 4 5/8"	3 7/8" or 4 5/8"	9 5/8"	Raised
N.D. 111	4 3/4"	4 3/4"	—	8"	3 7/8" or 4 5/8"	3 7/8" or 4 5/8"	9 5/8"	Raised

Figure 19-14. Sizes and patterns commonly used for interior doors. The table lists construction details of the various parts. (National Woodwork Manufacturers Association)

dimensions, designs, and quality specifications for materials and work.

19.6 Door Installation

Check the architectural drawings and door schedule to determine the hand of the door and the correct type of door for the opening. The correct type is determined by checking the door schedule in the plans.

The hand of the door is determined by the location of the hinges when viewed from the outside, **Figure 19-15.** In the case of interior doors, it is viewed from the side where the hinges are not visible. For example, if the hinges (hidden) are on the right as the door is viewed, the door is a right-hand door. Conversely, if the hidden hinges are on the left of the viewer, it is a left-hand door.

Doors may be trimmed somewhat so they fit with enough clearance on each side to prevent the door from binding, even with some swelling during humid weather. *Never* attempt to cut down a door to fit a smaller opening. If a large amount of material is removed from its edges, the structural balance of the door may be disturbed. Warping may result and the door is weakened.

Cutouts for glass inserts in flush doors should never be more than 40% of the face area. The opening should not be within 5″ of the edge.

Carefully handle doors. Avoid soiling unfinished doors. If they are to be stored for more than a few days, horizontally stack the doors on a clean level surface. Cover them to keep them clean. Before installing doors, place them in the room for several days so they will reach the average prevailing moisture content before being hung or finished.

19.6.1 Fitting the Door

If the door is not prehung, it must be fitted to the frame. Mark the doorjamb that will receive the hinges. Also mark the edge of the side jamb where they will be mounted. This is opposite of the edge viewed to determine hand. Next, using a rule or tape measure, check measurements of both the door and the door opening.

Trim the door to fit the opening. Most doors are carefully sized at the millwork plant, leaving only a slight amount of on-the-job fitting and adjustment. The small amount of material can be removed by planing. Door trimming can be done with a hand plane, but carpenters today generally use a power plane, **Figure 19-16.** While the door is being planed, it may be securely held on edge either by clamping it to sawhorses or using a special door holder.

Clearances should be 3/32″ on the lock side and 1/16″ on the hinge side, **Figure 19-17.** A clearance of 1/16″ at the top and 5/8″ at the bottom is generally satisfactory. If the door is to swing across heavy carpeting, increase the bottom clearance. Thresholds are generally used only under exterior doors. The threshold may be installed before or after the door is hung. Where weatherstripping is used around exterior door openings to reduce air infiltration, additional clearance is needed. About 1/8″ on each side and on the top edge is enough.

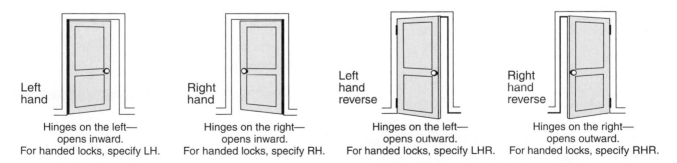

Left hand
Hinges on the left— opens inward.
For handed locks, specify LH.

Right hand
Hinges on the right— opens inward.
For handed locks, specify RH.

Left hand reverse
Hinges on the left— opens outward.
For handed locks, specify LHR.

Right hand reverse
Hinges on the right— opens outward.
For handed locks, specify RHR.

Figure 19-15. To determine the "hand" of a door, imagine viewing it with hinges concealed. When ordering, use the abbreviations suggested.

Figure 19-16. A power plane speeds up the task of planing a door. It has an edge guide fence that provides consistently accurate planing across the door. (Bosch Power Tool Corp.)

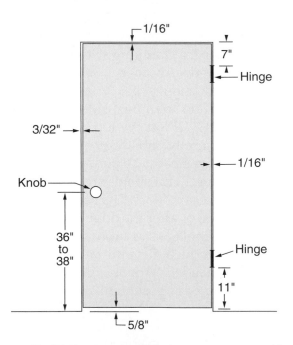

Figure 19-17. Recommended clearances around interior doors. Some carpenters use a quarter (25-cent) coin to check the clearance at the top and lock side.

After the door is brought to the correct size, plane a bevel on the lock side to provide clearance for the edge when it swings open. This bevel should be about 1/8" in 2" (approximately 3 1/2°). Narrow doors require a greater bevel than wide doors since the arc of swing is

smaller. The type of hinge and the position of the pins should be considered in determining the exact bevel required. After the bevel is cut and the fit of the door is checked, use a block plane to soften (round) corners on all edges of the door. Smooth with sandpaper.

> **Working Knowledge**
>
> Millwork plants can furnish prefitted doors that are machined to the size specified, with the lock edge beveled and corners slightly rounded. Doors can also be furnished with gains cut for hinges and holes bored for lock installation.

19.6.2 Installing Hinges

Gains are the recesses cut into the edge of a door to receive the hinges. If the door is not prehung, these must be cut. Gains are best cut with a router. Used with a door-and-jamb template, the router saves time and ensures accuracy.

Adjustments can be made for various door thicknesses and heights, as well as for different sizes of butt hinges. See **Figure 19-18.** The design of most templates makes it nearly impossible to mount them on the wrong side of the door or jamb. For this type of equipment, hinges with rounded corners may be used. It will save the time required to square the corner with a wood chisel. Once the gains have been cut, hang the door.

PROCEDURE

Cutting gains

1. Position the template on the door.
2. Make the cuts with the router.
3. Attach the template to the doorjamb.
4. Cut the matching gains.

Gain: Recess or mortise cut to receive a hinge leaf.

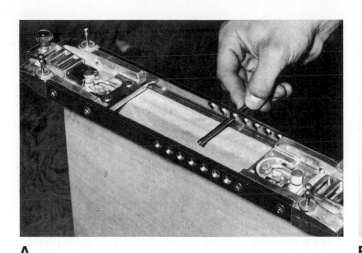

A

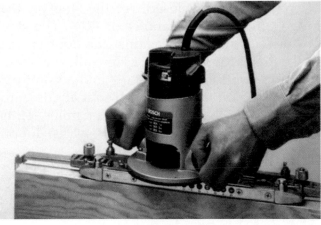

B

Figure 19-18. A—Hinge templates can be adjusted for different door sizes as well as for different hinge sizes. B—A router and template are being used to cut a gain for a round-cornered hinge.

PROCEDURE

Hanging a door

1. Place the hinge in the gain so the head of the removable pin will be on top when the door is hung.
2. Drive the first screw in slightly toward the back edge to tightly draw the leaf of the hinge into the gain.
3. Repeat steps 1 and 2 to attach the free leaf of the hinge to the jamb. Set only one or two screws in each hinge leaf.
4. Check the fit and then install the remaining screws.
5. After all hinges are installed, hang the door and check clearance on all edges.
6. Make required corrections by planing the door edges or adjusting the depth of the hinge gains.
7. Minor adjustments can be made by applying cardboard or metal shims behind the hinge leaf, **Figure 19-19.**

Door stop: A moulding nailed to the faces of the door frame jambs to prevent the door from swinging through.

19.6.3 Door Stops

A *door stop* is a narrow strip of wood attached to the side and top jambs. Its purpose is to stop the door as the door is closed. Door stops are usually the last trim members to be installed. If the door is prehung, the door stop is installed at the factory. For doors that are not prehung, many carpenters cut the stops and tack them in place before installing the lock. Permanent nailing comes after the lock installation has been completed.

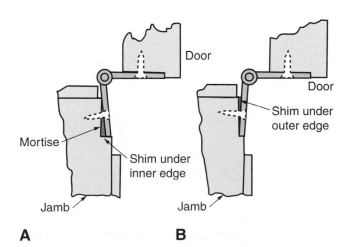

A   **B**

Figure 19-19. Cardboard or metal shims can sometimes be used to make minor adjustments in door clearance. A—Providing more clearance along the lock jamb. B—Closing the space along the lock jamb.

With the door closed, set the stop on the hinge jamb with a clearance of 1/16″. The stop on the lock side is set against the door except in the area around the lock. Here allow a slight clearance for humidity changes and decorating. Set the stop on the head jamb so it aligns with the stops on the side jambs. Cut miter joints and attach the stop with 4d nails spaced 16″ on center.

19.7 Door Locks

Four types of passage door locks are illustrated in **Figure 19-20.** Cylindrical and tubular locks are used most often because they can be easily and quickly installed. Unit locks are installed in an open cutout in the edge of the door and need not be disassembled when installed. Such locks are commonly used on entrance doors for apartments and some commercial buildings where locks must be changed from time to time.

Cylindrical locks have a sturdy, heavy-duty mechanism that provides security for exterior doors, **Figure 19-21.** They require boring a large hole in the door face, a smaller hole in the edge, and a shallow mortise for the front plate. Often, doors come with these holes already bored at the factory. The tubular lock is similar, but requires a smaller hole in the door face.

When ordering door locks, it is sometimes necessary to describe the way in which the door swings. This is referred to as the "hand" of the door. Refer to **Figure 19-15** for the procedure used to determine this specification.

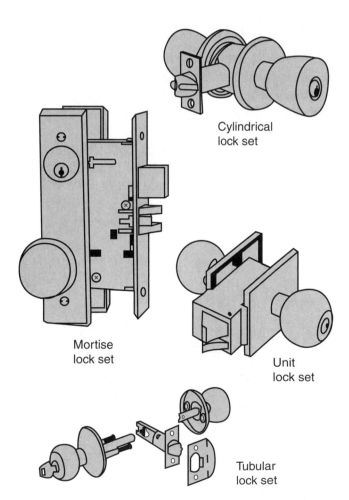

Figure 19-20. Four basic types of door lock sets. Mortise lock sets provide high security and are often found in apartment buildings. Cylindrical and tubular locks are most often chosen for residential, with cylindrical being more secure for entrance doors. Unit lock sets are best for apartments where locks are frequently changed.

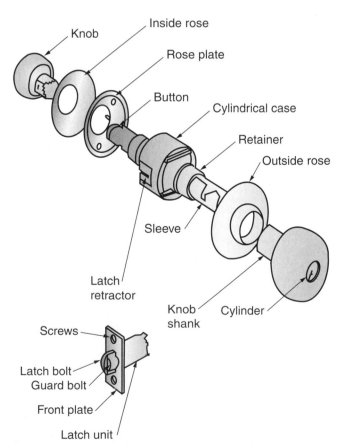

Figure 19-21. Parts of a cylinder lock set. (Eaton Yale & Towne, Inc.)

Deadbolts (also called *deadlocks* or *throwbolts*) are extra locks that provide more security against unauthorized entry. Units are made with single-cylinder or double-cylinder action. Double-cylinder deadbolts require key use on both sides of the door. They should be used if the door contains a window. However, there is a safety concern related to an emergency exit. **Figure 19-22** shows two types of deadbolts.

A

B

Figure 19-22. Two types of dead bolts. A—Tubular dead bolt. B—A keyless deadbolt uses a numeric code. (Weiser Lock)

Deadbolt: Special door security consisting of a hardened steel bolt and a lock. Lock is operated by a key on the outside and either a key or handle on the inside. Also called a *deadlock* or *throwbolt.*

Threshold: A member beveled or tapered on each side and used to close the space between the bottom of a door and the sill or floor underneath. Also called a *saddle.*

PROCEDURE

Installing a lock in a non-pre-drilled door

1. Open the door to a convenient working position and block it with wedges placed underneath.
2. Measure up from the floor a distance of 38" (35", 36", or 40" are sometimes used) and mark a light horizontal line. This is the center of the lock.
3. Position the template furnished with the lock set on the face and edge of the door.
4. Lay out the centers of the holes, **Figure 19-23.**
5. Bore the holes according to the instructions provided with the lock set. Use of a boring jig assures accurate work, **Figure 19-24.**
6. Lay out and cut the shallow mortise on the door edge. A faceplate mortise marker, also called a marking chisel, is faster and more accurate than standard wood chisels.

19.8 Thresholds and Door Sweeps

Entry doors require a **threshold,** or *saddle,* to seal the space between the bottom of the door and the sill. Normally, the carpenter will not install the threshold until most of the interior work is done. This prevents accidental damage as materials are moved in. However, the threshold is part of a prehung exterior door unit.

At one time, oak or other hardwoods were milled to a special shape for thresholds. Modern wood units have a rubber or vinyl strip that provides a tight seal. However, aluminum has become the material of choice. These thresholds are available in plain anodized or gold finish. Vinyl seals are inserted in special slots. See **Figure 19-25.** Sometimes a matching seal is attached to the underside of the door. Always follow the provided installation instructions.

Interior doors do not require a threshold, but may be equipped with various sealing strips, such as the automatic door sweep shown

1 Mark door

Mark height line across edge of door. 38" is the usual height above the floor. Fold template over edge of door, centering on height line. Mark centers of 7/8" and 2 1/8" holes.

2 Install latch unit

Bore 2 1/8" hole through door, and 7/8" hole into edge of door at points marked on template.

Cut out for latch front and install latch unit.

3 Install strike

Mark height line for strike on jamb. Mark vertical centerline on jamb. This centerline must be same distance from stop molding as latch case centerline is from edge of door that will hit stop molding.

Cut mortise in jamb for strike and box.

Insert box and strike and tighten screws securely.

4 Adjust lock

5/16"
For 1 3/8" door

Case cutout

Outside rose plate

1/2"
For 1 3/4" door

Case

Case cutout

To adjust this lock for a 1 3/8" door, unscrew outside rose plate 5/16" from case cutout. To adjust for a 1 3/4" door, unscrew outside rose plate to provide 1/2" between rose plate and case cutout. To adjust for any thickness between 1 3/8" and 1 3/4", set the rose plate at a suitable intermediate position.

5 Install lock

With latch case in place, insert lock assembly into 2 1/8" hole, making sure that lock case hooks retainer legs and retractor hooks bolt tails. Do not force.

Retractor
Retainer leg

Bolt tail

Lock case

6 Install rose plate

Slip on rose plate and locate screw holes on vertical centerline with "Top" up. Insert machine screws and tighten alternately to obtain secure attachment.

See instructions above

Vertical

7 Install inside rose

Place inside rose over rose plate with notch in rose over spring retainer and snap rose down so rose is flush with door.

Spring

Notch

8 Install inside knob

Align lug on knob with narrow slot on side of spindle and push knob all the way in until retainer clicks into slot on knob.

9 Important

This lock is set for a right-hand door.

Right way   **Wrong way**

To change lock hand

Knob must be in unlocked position. Turn outside knob counterclockwise approximately 45° and insert small nail in hole of trim cap. Depress retainer and slide knob off. Turn knob 180° and replace knob.

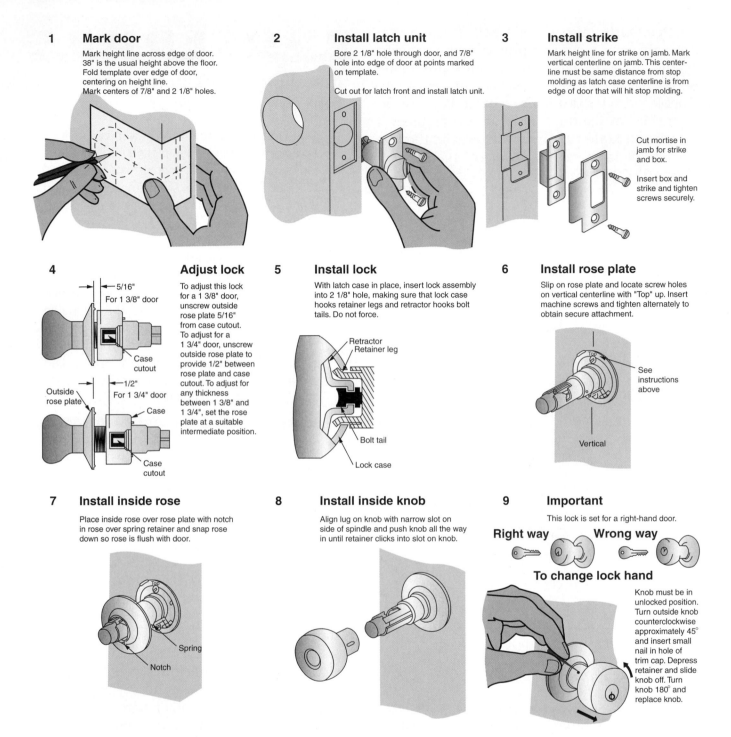

Figure 19-23. Manufacturers furnish detailed instructions for installation of their lock sets. Generally, cylindrical locks can be installed following similar instructions.

Figure 19-24. When installing locks, a boring jig for door hardware saves time and ensures accuracy. (Porter-Cable Corp.)

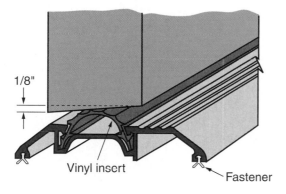

1/8"

Vinyl insert

Fastener

Figure 19-25. An aluminum threshold with a vinyl seal strip. These are frequently found in residential construction. (Pemko Manufacturing, Inc.)

in **Figure 19-26.** This prevents unwanted drafts and reduces sound transmission. The unit pictured has a movable strip that drops to the floor when the door is closed and lifts when the door is opened.

Pocket door: A sliding door suspended on a track that slides into a wall cavity when opened. Also called a *recessed door.*

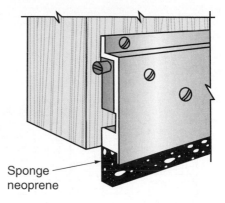

Sponge neoprene

Figure 19-26. An automatic door sweep. The sealing strip moves upward when the door is opened.

19.9 Sliding Pocket Doors

A **pocket door,** or *recessed door,* slides into a recess or pocket in a partition wall. This is a space-saving feature used where there is no clearance for a swinging door. The pocket door frame consists of a split side jamb attached to a framework built into the wall. The rough opening in the structural frame must be large enough to include the finished door opening and the pocket. The usual height of the rough opening is 6'-11 1/2", as shown in **Figure 19-27.** The preassembled pocket framework and track is installed during the rough framing stage. Follow the manufacturer's instructions when framing up the interior partition and installing the pocket door.

Pocket-frame units are available from millwork plants in a number of standard sizes. When doing the rough-in, the carpenter should have the manufacturer's specifications for the door selected.

Manufacturers that specialize in builder's hardware have developed steel pocket door frames. They are easy to install and provide a firm base for wall surface materials.

Figure 19-28 shows a typical track and roller assembly for a pocket door. The hanger (wheel assembly) snaps into the plate attached to the top of the door. It can be easily adjusted up or down to plumb the door in the opening.

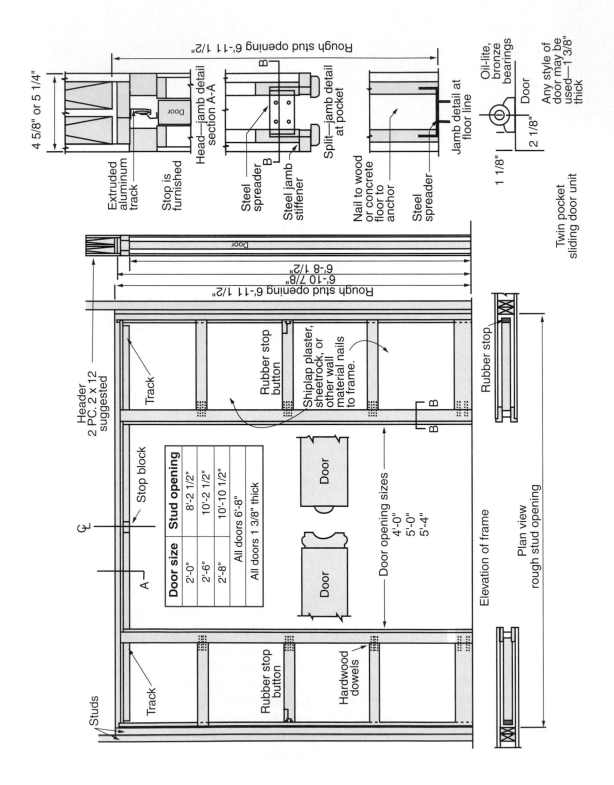

Door size	Stud opening
2'-0"	8'-2 1/2"
2'-6"	10'-2 1/2"
2'-8"	10'-10 1/2"
All doors 6'-8"	
All doors 1 3/8" thick	

Rough stud opening 6'-11 1/2"

4 5/8" or 5 1/4"

Door

Extruded aluminum track

Stop is furnished

Head—jamb detail section A-A

B
B

Steel spreader

Steel jamb stiffener

Split—jamb detail at pocket

Nail to wood or concrete floor to anchor

Steel spreader

Jamb detail at floor line

Oil-lite, bronze bearings

Door

Any style of door may be used—1 3/8" thick

1 1/8"

2 1/8"

Twin pocket sliding door unit

Door

6'-8 1/2"

6'-10 7/8"

Rough stud opening 6'-11 1/2"

Header 2 PC. 2 x 12 suggested

Track

Rubber stop button

Shiplap plaster, sheetrock, or other wall material nails to frame.

Rubber stop

B B

Stop block

℄

A

Door

Door

Door opening sizes
4'-0"
5'-0"
5'-4"

Elevation of frame

Plan view rough stud opening

Studs

Track

Rubber stop button

Hardwood dowels

Rubber stop

Figure 19-27. Structural details of a pocket door. The ladder-like framing on both sides of the pocket supports drywall. Nails should not be so long as to penetrate this framework. (Ideal Co.)

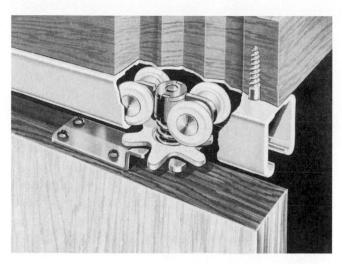

Figure 19-28. A track and roller assembly for a pocket door. The star wheel adjusts the door up and down.

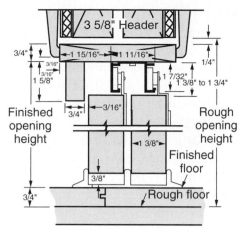

**Track mounted
under head jamb
1 3/8" doors**

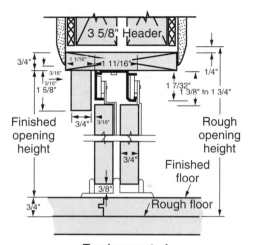

**Track mounted
under head jamb
3/4" doors**

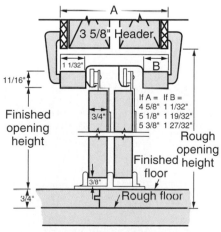

**Track recessed
in head jamb
3/4" doors**

Figure 19-29. Drawings of typical bypass sliding door installations. Regular interior doors may be used.

PROCEDURE

Installing a pocket door

1. Install the header that will support the door, its tracking mechanisms, and the 1″ framing.
2. If provided, fasten a steel channel to the floor. The channel keeps the proper clearance needed in the pocket framework.
3. Attach the steel track to the header.
4. Install the hardware onto the door. Attach rollers to the top of the door where specified by the manufacturer. Pocket doors also require an edge pull and a recessed side pull for opening and closing.
5. Tilt the door outward to engage the rollers in the overhead track.
6. Slide the door into the floor channel. Slide it back and forth. It should easily move while evenly butting against side jambs.
7. Adjust the rollers as needed.
8. Apply wall coverings and trim to conceal the framing and track.

19.10 Sliding Bypass Doors

Standard interior door frames can be used for a bypass sliding door installation. When the track is mounted below the head jamb, the height of a standard door must be reduced and a trim strip installed to conceal the hardware, **Figure 19-39.** Head jamb units are available with a recessed track that permits the doors to ride flush with the underside of the jamb. Hardware for sliding bypass doors is packaged complete with track, hangers (rollers), floor guide, screws, and instructions for making the installation.

A disadvantage of bypass sliding doors is that access to the total opening at one time is not possible. They are, however, easy to install and practical for wardrobes and many other interior wall openings.

19.11 Bifold Doors

Bifold doors consist of two doors that are hinged together. The folding action is guided by an overhead track. A complete unit may consist of a single pair or two or more pairs of doors or panels. See **Figure 19-30.** These door units are well suited to wardrobes, closets, pantries, and certain openings between rooms.

The opening for bifold units is trimmed with standard jambs and casing. **Figure 19-31** illustrates an installation of hardware. The pivot brackets and center guides have self-lubricating, nylon

bushings. The weight of the doors is supported by the pivot brackets and hinges between the doors, not by the overhead track and guide. Two-door units generally range from 2′ to 3′ wide, while four-door units are available in widths from 3′ to 6′.

Folding door hardware comes in a package that includes hinges, pivots, guides, bumpers, aligners, nails, and screws. Instructions for the installation are also included. Millwork plants can supply matching doors with prefitted hardware.

When the total opening for a four-door unit is greater than 6′ or wider than 3′ for a two-door unit, it is usually necessary to install heavier hardware. Also, a supporting roller-hanger is used instead of a regular center guide.

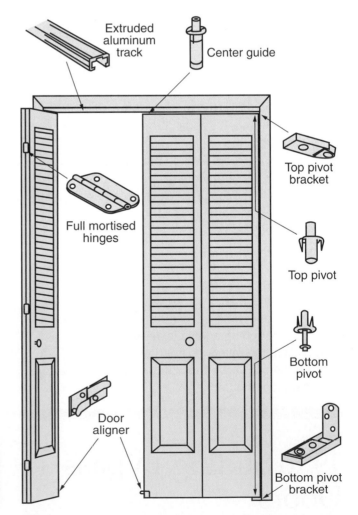

Figure 19-30. Bifold doors are often used to close off wardrobe space. (LTL Home Products, Inc.)

Figure 19-31. Typical hardware used for folding door installation. The hinged units are supported by brackets. No weight is carried by the center guide. (Ideal Co.)

19.12 Multifold Doors

Multifold doors are built from narrow panels with some type of hinge along the edges. See **Figure 19-32.** One manufacturer produces a design where the hinge action is provided by steel springs threaded through the panels. The entire door assembly is supported by nylon rollers located in an overhead metal track. The track is wood-trimmed to match the door. **Figure 19-33** shows how the track is installed in either a wood-framed or plastered opening. The track for bifold doors is installed in the same way. Manufacturers furnish door units in a complete package that includes track, hardware, latches, and instructions for making the installation.

Door panel surfaces are available in a variety of materials and finishes. The best panel grades are made from wood veneers bonded to wood cores. Panels are also made of stabilized particleboard wrapped with woodgrain-embossed vinyl film.

An important advantage of this type of folding door is its space-saving feature. As the door is opened, the panels fold together forming a "stack" that requires little room space. For example, the stack dimension for an 8′ opening is only 11 1/2″. When it is desirable to clear the entire opening, the stack can often be housed in a special wall cavity. Bifold doors and regular passage doors must have swinging clearance in the room as they are opened and closed. **Figure 19-34** shows general details of construction with the door in both opened and closed positions.

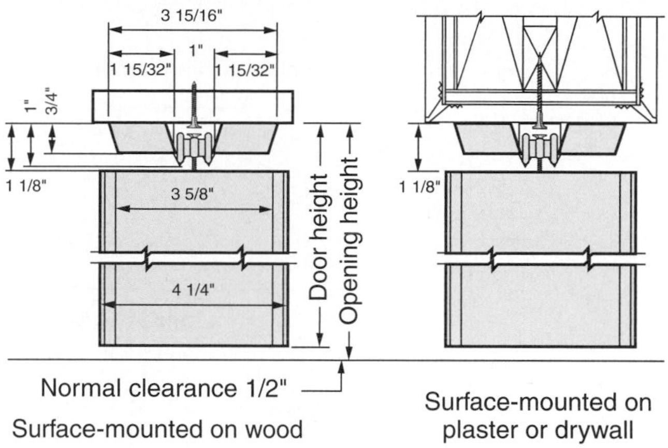

Figure 19-33. Typical head sections show the installation of an overhead track for folding doors. (Pella Corp.)

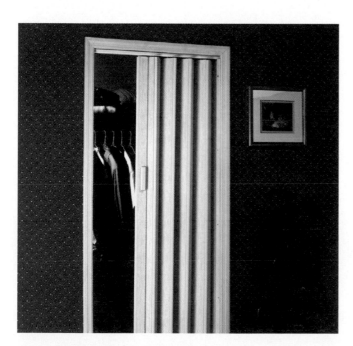

Figure 19-32. A typical multifold door.

Working Knowledge

Folding doors are available in a variety of sizes. They can be used to separate room areas, laundry alcoves, and general storage space.

19.13 Window Trim

Interior window trim consists of the casing, ***stool,*** apron, and stops, **Figure 19-35.** Millwork companies select and package the proper length trim members to finish a given unit or combination of units.

When the face of the apron is curved, the ends should be returned or coped so that the shape of the ends are the same. The returned end is commonly used and formed with miter cuts as illustrated in **Figure 19-36.**

Sometimes, the stool and apron are eliminated. Instead, a piece of beveled sill liner is

Stool: Trim forming the interior sill cap of a window.

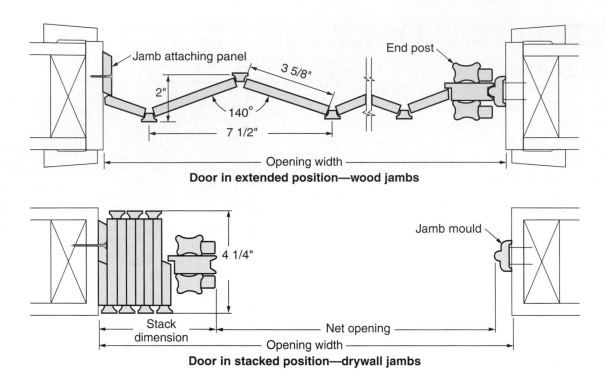

Door in extended position—wood jambs

Door in stacked position—drywall jambs

Figure 19-34. General details showing the operation of a folding door. (Pella Corp.)

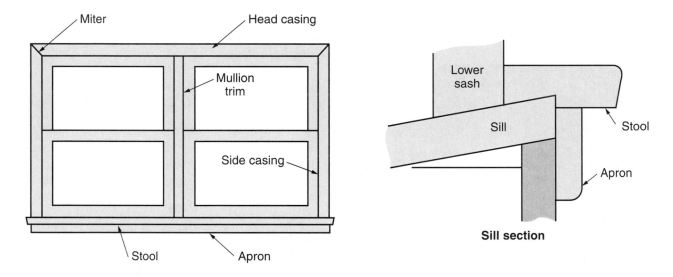

Sill section

Figure 19-35. Trim members used for a standard double-hung window.

installed to match the window jamb. Regular casing is then applied around the entire window. See **Figure 19-37.** This is known as *picture-frame trimming.*

PROCEDURE

Installing trim for double-hung windows

1. To find the length of the stool, position a piece of side casing and measure beyond it by about 3/4″. Do this on both sides of the window and mark light lines on the wall.
2. Position the stool and mark the cutoff lines for the ends of the stool using the light lines on the wall.
3. Hold the stool level with the sill. Mark the inside edges of the side jambs. Also mark a line on the face of the stool where it will fit against the wall surface. For a standard double-hung window this line is usually directly above the square edge of the stool rabbet. This is the notch where it fits over the top of the sill.
4. Carefully cut out the marked notches on the ends of the stool and check the fit. You will have to open the lower sash to slide the stool into position.
5. Carefully lower the sash on top of the stool and draw the cutoff line so the sash will clear the stool when closed. Allow about 1/16″ clearance between the front edge of the stool and the window sash.
6. Cut the ends, sand the surface, and nail the stool into place. Some installations require that the stool be bedded in caulking compound.
7. Set a length of side casing in position on the stool.
8. Mark the position of the miter on the inside edge. Usually, the casing is installed with an 1/8″ reveal.
9. Cut the miter.
10. Nail side casing in place. Drill nail holes if the casing is made of hardwood.
11. Cut and install the head casing.
12. Measure and cut the apron. Its length should be the distance to the outer edges of the side casing.
13. Sand any visible saw cuts on the apron and nail it in place.

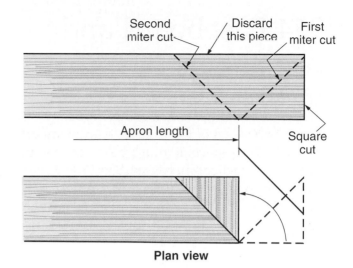

Figure 19-36. How to mark and cut a returned end on an apron. Use glue to attach the end piece.

A

B

Figure 19-37. A—Notice how casing can be used to replace the stool and apron. B—Using a cordless drill to make holes for attaching hardwood window casing.

19.14 Baseboard and Base Shoe

The *baseboard* covers the joint between the wall surface and the finish flooring, **Figure 19-38.** It is among the last of the interior trim members to be installed, since it must be fitted to the door casings and cabinetwork. Baseboards run around the room between door openings, cabinets, and built-ins. The joints at internal corners should be coped. Those at outside corners are mitered, **Figure 19-39.**

Base shoe is used to seal the joint between the baseboard and the finished floor. It is usually fitted at the time the baseboard is installed, but not nailed in place until after any surface finishes have been applied. Base shoe is often used to cover the edge of resilient tile or carpet.

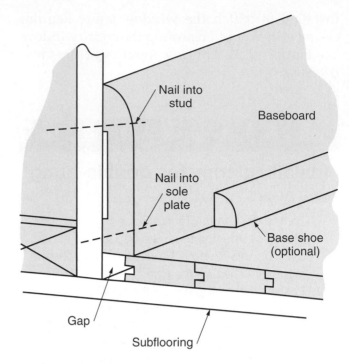

Figure 19-38. Cutaway of baseboard and base shoe. They conceal the gap between the flooring and finished wall.

19.15 Installing Baseboard and Base Shoe

Baseboard and base shoe are installed in the same way. Select and place the baseboard material around the sides of the room. Sort the pieces so there will be the least amount of cutting and waste. Where a straight run of baseboard must be joined, use a mitered-lap joint (also called a *scarf joint*). Be sure to locate the joint so it can be nailed over a stud. See **Figure 19-40.**

Baseboard installation is much easier if stud locations have been marked. First, mark on the rough floor ahead of the plaster or drywall installation. Then, mark on the wall surface before installing the finish flooring or underlayment.

If the stud positions have not been marked, locate them with a stud finder. If a stud finder is not available, tap along the wall with a hammer until a solid sound is heard. Drive nails into the wall to locate the exact position and edges of the stud. Then mark the location of others by measuring the stud spacing (usually 16" O.C.).

Start installing the baseboard at an inside corner. Make a square cut on the end of the first

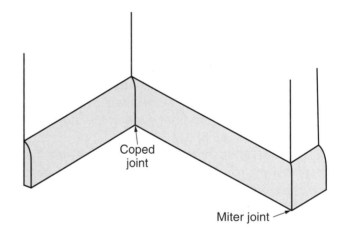

Figure 19-39. Coped joints are suitable for inside corners. Use a mitered joint for outside corners.

Baseboard: A finishing board that covers the joint where a wall intersects the floor.

Base shoe: Narrow moulding used around the perimeter of a room where the baseboard meets the finish floor.

Scarf joint: A mitered-lap joint used where a straight run of baseboard must be joined.

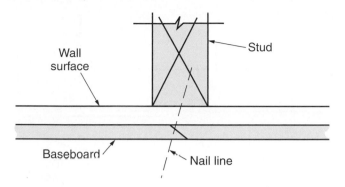

Figure 19-40. Cut a scarf joint when joining a straight run of baseboard.

piece of baseboard. The butt joint should be tight with the intersecting wall surface. Make scarf joints as needed to complete the run to the opposite inside corner. The next piece butts against the base just installed. This joint is coped, as shown in **Figure 19-41.**

To form a coped joint, first cut an inside miter on the end of the baseboard trim piece. Then, using a coping saw, cut along the line where the sawed surface of the miter joins the curved surface of the baseboard, **Figure 19-42.** Make the cut perpendicular to the back side of the baseboard. This forms an end profile that matches the face of the baseboard. Slightly

Figure 19-41. A coped joint follows the contour of the piece it will butt against.

undercutting the coped end ensures a tight fit at the face.

The coped joint takes longer to make than a plain miter, but it makes a better joint at an inside corner. When nailed into place, it will not open up. Also, if the wood shrinks after installation, a noticeable crack will not appear.

All outside corners are joined with a miter joint. Hold the baseboard in position and mark at the back edge in line with the intersecting surface. Make the 45° cut using a miter saw, **Figure 19-43.**

Before installing each section of baseboard, check both ends to make sure the cut and fit is

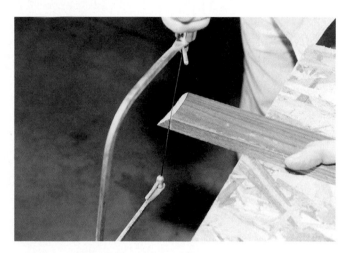

Figure 19-42. Form a coped joint by first making an inside miter cut, then making a perpendicular cut following the curve of the front edge of the miter cut.

Figure 19-43. Using a power miter box to cut a miter joint for outside corners.

correct. To install, tightly hold the board against the floor. Fasten it with finishing nails long enough to penetrate well into the studs. The lower nail is slightly angled down so it is easier to drive and will enter the sole plate. Set the nail heads.

Baseboards are normally butted against the door casing, as illustrated in **Figure 19-44.** The casing should be thick enough to accommodate the slightly thinner profile of the baseboard. Base shoe, if used, is ended at the casing with a miter cut as shown.

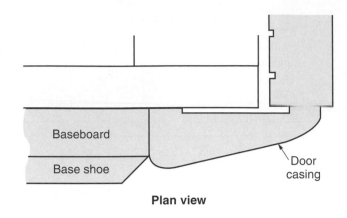

Plan view

Figure 19-44. The baseboard is butt-jointed against door casing. The base shoe gets an outside miter cut.

ON THE JOB

Finish carpenter

The finish carpenter normally has all of the knowledge and skills of the framing carpenter plus specialized knowledge and skills in such areas as interior and exterior trim work, stairway planning and building, door and window installation, and the installation of kitchen cabinets and countertops. The finish carpenter's skill set falls between that of the framing carpenter and the cabinetmaker.

Larger general contracting firms or housing development companies usually employ both framing and finish carpenters. In large residential developments, the carpenters often are organized into crews that move from one building to another at the appropriate construction stage. A finish carpenter crew, for example, normally completes all of the trim work and other finishing tasks in one structure, then moves onto the next structure and repeats the same tasks. Carpenters who work for smaller general contractors and remodeling companies or operate their own business usually perform both framing and finish carpentry work.

Finish carpenters must be skilled in measuring, cutting, and joining of wood components. The work calls for a higher degree of precision, especially in making joints, than framing carpentry. Practical math skills and the ability to read prints are important to success in finish carpentry. Although their work may be somewhat less strenuous than that of the framing carpenter, the finish carpenter must still be in good physical condition. The work requires standing for long periods and can involve some heavy lifting. While most work is indoors, finish carpenters may be exposed to inclement weather. Some of their responsibilities often involve door and window installation and the application of the building's exterior trim.

Finish carpenters typically enter the trade by the same routes as the framing carpenter—vocational school programs, on-the-job training, or formal apprenticeship. Many began as framing carpenters and gradually acquired the skills needed to do finish carpentry work. Apprenticeship programs combine classroom instruction and 3–4 years of on the job experience. These programs prepare apprentices for both framing and finish carpentry work.

Like framing carpenters, finish carpenters are exposed to almost all aspects of construction work. This suits them for advancement to supervisory positions within larger contracting firms and can also provide preparation for self employment.

Summary

Installing interior doors and trim requires great skill and accuracy. Mouldings are decorative strips, available in either wood or plastic and in either finished or unfinished form. There are many different shapes of moulding available. Moulding is typically installed where the floor and walls meet in a room and around doors and windows. Interior door frames consist of two side jambs and a head jamb. They cover the unfinished edges of the door opening and provide support for the door and its hardware. Mouldings, known as door casing, are installed to trim the edges of the doorway where they meet the wall. Plinth blocks are decorative items used for the corners of window and door trim, allowing use of butt joints for moulding, rather than the more-difficult mitered corners. There are two types of interior doors, panel and flush. Panel doors are built of stiles and rails that frame solid wood or plywood inserts. Flush doors have a smooth, thin sheet of material applied to both faces. Interior doors are available in two standard heights, but in a variety of widths for different applications. Installation steps for traditional doors include cutting gains to install hinges, and boring the door to receive the lock mechanism. Prehung door units ease installation, since they are already fully assembled. They are positioned in the door opening, adjusted, then nailed in place. Specialized types of doors include sliding pocket doors, bypass-type sliding doors, and folding doors. Window trim is similar to door casing, but also involves installing the stool (window sill). Moulding used to trim the joint between the wall and floor is known as baseboard. A narrow moulding called a base shoe is often used to cover the baseboard-to-floor junction.

Test Your Knowledge

Answer the following questions on a separate piece of paper. Do not write in this book.

1. The name of a moulding may come from its _____ or use.
2. Interior doorjambs for a standard framed wall with a plaster finish should be _____ inches wide.
3. A(n) _____ consists of a door frame with the door already installed.
4. To cover the space between the wall surface and the doorjambs, door _____ is applied to each side of the frame.
5. What is a *reveal?*
6. Where is a *plinth block* used?
7. *True or False?* Installing a plinth block involves cutting compound miter joints.
8. A panel door consists of rails, panels, and _____.
9. The face panels of flush doors, also called _____, are usually made of 1/8″ plywood.
10. Flush doors are made with either _____ or _____ cores.
11. The minimum width for a bedroom door is _____ , as specified by the FHA.
12. When setting the door stop on the hinge jamb, provide a clearance of _____ with the door.
13. The two most commonly used types of residential door locks are cylindrical and _____.
14. _____-type deadbolts should be installed if the door contains a window.
15. What is the purpose of the *threshold?*
16. The type of door that slides into an opening in the wall or partition is a(n) _____.
17. The type of door that consists of two doors that are hinged together is a(n) _____.
18. Sometimes, the _____ and _____ are eliminated from window trim to create picture-frame trim.
19. When a straight run of baseboard must be joined, use a(n) _____ joint.
20. When installing baseboards, a(n) _____ joint should be used at internal corners to mate against another baseboard.

Curricular Connections

Language Arts. Use a camcorder to record a demonstration of how to make a coped joint when installing baseboard. Before recording, take time to plan each step of the demonstration, then practice the steps several times. Write a script to accompany the video—it can be recorded as the video is being made.

or done separately after recording (a "voice-over"). If possible, add titles to give your program a finished appearance. Show the completed video to the class.

Science. Pneumatic (air-powered) nailers have become very widely accepted on the job site, especially by carpenters installing trim work. Do research to find out how air-powered tools work. Make a series of sketches to show the operating sequence of the nailer. List the advantages and disadvantages of air-powered tools compared to conventional electric tools or manually operated tools (such as a hammer).

Outside Assignments

1. Visit a residential building site in your community where the inside finish work is in progress. Obtain permission from the site supervisor or head carpenter before making the visit. Note the types of trim being applied and the procedures being followed. See if one of the carpenters will discuss with you the advantages and disadvantages of prehung door units. Prepare carefully organized notes of your observations and make an oral report to the class.

2. Develop a door schedule for a preliminary or presentation drawing of a residence. Door sizes of doors are not usually shown on these drawings. You must make the selection. A door schedule should include the location, width, height, thickness, type and design, kind of material, and quality requirements. Obtain cost estimates by referring to a builder's supply catalog or visiting a local lumber dealer.

3. Build a full-size, sectional mock-up of a partition with a doorway. Include all parts—studs, wall finish, doorjamb, casing, baseboard, and door. Carry the section up about 16" above the floor and extend the partition out from the door only about the same amount. A 3/4" thickness of particleboard can serve as the base and represent the finish flooring. It will be best to glue most of the parts together since the structure may not withstand much nailing.

Cabinetry

Learning Objectives

After studying this chapter, you will be able to:

- Select prefabricated cabinets for a specific floor plan.
- Install prefabricated base and wall cabinets.
- Compare the common alternative procedures for building cabinets on the job.
- Lay out and frame a cabinet from drawings.
- Describe the three types of drawer guides.
- List the steps in cutting and assembling drawers.
- Describe material choices for cabinet shelves and doors.
- Explain how to install a plastic laminate surface.

Technical Vocabulary

32 mm construction system	Frameless construction
Backing sheet	Frame-with-cover construction
Base level line	Gain
Build-up strips	Kicker
Cabinetwork	Lipped drawers
Drawer guides	Lipped door
Facing strips	Master layout
Flush door	Overlay door
Flush drawers	Plastic laminate
Frame construction	

Pulls	Sliding doors
Rails	Slip sheet
Shelves	Stiles

Cabinetwork, as used in the interior finish of a residence, refers to built-in kitchen and bathroom storage. In a general way, it also refers to such work as closet shelving, wardrobe fittings, desks, bookcases, and dressing tables. The term *built-in* emphasizes that the cabinet or unit is attached to the structure. The use of built-in cabinets and storage units was an important development in architecture and design.

Kitchen cabinets present an attractive appearance and help to increase kitchen efficiency, **Figure 20-1.** In the kitchen, and in other areas of the home, storage units should be designed for the items that they will contain. Space must be carefully allocated. Drawers, shelves, and other elements should be proportioned to satisfy specific needs. Three types of cabinetwork used in homes are:

- Units that are built on the job by the carpenter.

Cabinetwork: A term that refers to built-in kitchen and bathroom storage and, in a general way, to such work as closet shelving, wardrobe fittings, desks, bookcases, and dressing tables.

Figure 20-1. These kitchen cabinets are attractive and provide efficient storage space in shelving and drawers. They also provide countertop working space for food preparation around sinks and cooktops. (Merillat Industries, Inc.)

- Custom-built units constructed in local cabinet shops or millwork plants.
- Mass-produced cabinets from factories that specialize in this area of manufacturing.

Except in large housing projects, combinations of these types of cabinetwork are found on most jobs. Even when most of the cabinets are factory produced, the carpenter is responsible for the installation. This task requires skill and careful attention to detail.

20.1 Drawings for Cabinetwork

Architectural plans usually include details of built-in cabinetwork. The floor plan shows cabinetwork location. Elevations provide detailed dimen-sions. A typical drawing of a base cabinet, desk, and room divider is illustrated in **Figure 20-2.**

Drawings of this type are measured with a scale. Thus, the need for extensive dimensioning is eliminated. The drawings serve as a construction guide to the carpenter. They are also followed in the selection of factory-built components. When cabinets are built on the job, the detail of joints and structure becomes the responsibility of the carpenter, who should be skilled in cabinetmaking.

Architectural drawings may include more specific details concerning the cabinetwork. This is justified on large commercial contracts, where a number of individuals will be involved in the project. It also becomes important when the cabinetwork contains special materials and constructions. **Figure 20-3** shows kitchen cabinets designed for an apartment complex. Written specifications define the type of joinery and quality of materials.

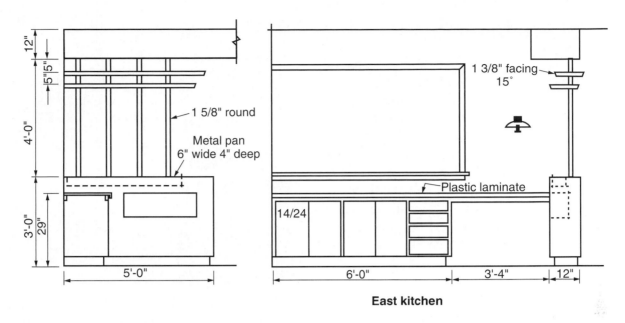

Figure 20-2. Typical built-in cabinetwork detail. The drawings are carefully scaled, so many dimensions have been eliminated.

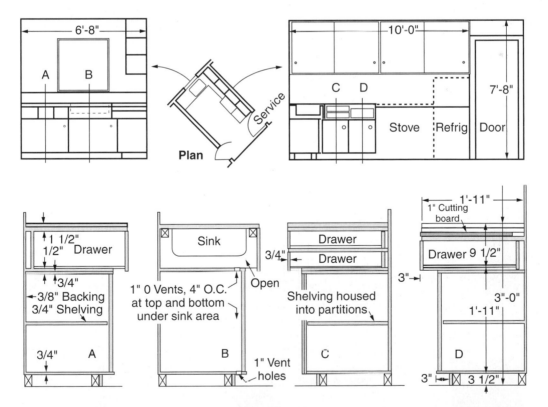

Figure 20-3. Drawings for cabinetwork to be installed in a large apartment building. (Architectural Woodwork Institute)

20.2 Standard Sizes

Overall heights and other dimensions of built-in units are usually included in the architectural plans. However, the carpenter should be familiar with basic design requirements.

Base cabinets for kitchens are typically 36" high and 24" deep. The countertop extends about 1" beyond the base cabinets, **Figure 20-4.** The vertical distance between the top of the base unit and the bottom of the wall unit may vary from 15" to 18". The FHA specifies a minimum distance of 24" when the wall cabinet is located over a cooking unit or sink.

In bathrooms or dressing rooms, cabinets with built-in lavatories are normally 31" high, **Figure 20-5.** The depth depends on the type of fixture. The counter surface may be plastic laminate, stone, synthetic stone, or tile. Some lavatories provide knee room; others do not.

Standards for closets and wardrobes widely vary. The determining factors are the items to be stored. The minimum clear depth for clothing on hangers is 24". When hooks are mounted on doors or the rear wall, this distance must be extended. Refer to architectural standards manuals for additional information.

Figure 20-5. Bathroom lavatories are often built into a cabinet called a vanity. Face-frame construction was used on this unit. (Dal-Tile Corp.)

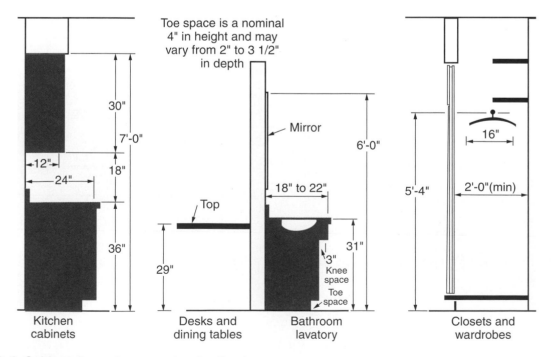

Figure 20-4. Cabinet dimensions are standardized.

20.3 Types of Construction

Cabinets may have a frame, frame with cover, or may be frameless. In *frame construction,* a skeleton of solid lumber supports shelves or a work surface. This type of construction is often used in building furniture, but may be used for bookshelves or other built-ins.

Frame-with-cover construction is also called *face-frame construction.* It is similar to frame construction, except that panels of plywood or fiberboard cover the frame. Since support is provided by the frame, the panels are lightweight. The front edges of the cabinet are faced with solid lumber, **Figure 20-6.**

A popular type of cabinet construction is *frameless construction.* As the name implies, this cabinet type has no internal supporting frame. The panels are heavy enough to carry the weight of the assembly and materials that will be stored in the cabinet. Front edges of these cabinets are not framed. See **Figure 20-7.**

Frameless construction often uses the standardized *32 mm construction system.* This system has a set of standard hole sizes and spacings on panels designed to be used on various case types and sizes. Central to this system are two vertical rows of 5 mm holes drilled in side panels on 32 mm centers. The vertical centerlines are 37 mm from the front and rear edges of the panel. Precise spacing assures proper fit of installed hardware. While the system was initially developed for mass production, it can be adapted to custom building as well. Modular cabinet systems are often based on the 32 mm construction system. The term *modular* refers to units assembled from parts having standard sizes. Thus, a variety of finished assemblies can be made from the same parts.

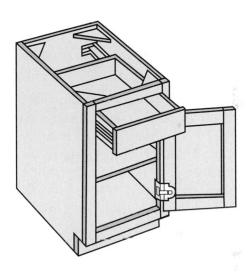

Figure 20-6. This drawing shows the face-frame type of cabinet construction. Edges of the cabinet front are often covered with solid wood. Framed panel doors are common. (Merillat Industries, Inc.)

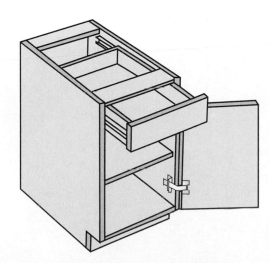

Figure 20-7. Frameless construction is shown in this drawing. In this unit, the front edges of the cabinet are not faced. Frameless cabinets are often called European since the style originated in Germany. (Merillat Industries, Inc.)

Frame construction: Cabinet construction method in which solid lumber provides a frame that is not covered with panels.

Frame-with-cover construction: Similar to frame construction, except that panels of plywood or fiberboard cover the frame. The front edges of the cabinet are faced with solid lumber. Also called *face-frame construction.*

Frameless construction: Cabinet construction method in which heavier panels provide support, rather than the usual framework of narrow pieces of solid wood.

32 mm construction system: Frameless construction system that has a set of standard hole sizes and spacings on panels. Central to this system are two vertical rows of 5 mm holes drilled in side panels on 32 mm centers.

20.4 Factory-Built Cabinets

A major part of the cabinetwork for residential and commercial buildings is constructed in factories that specialize in this work. Modern production machines and tools can save time and produce high-quality work. Mass-produced parts are assembled with the aid of jigs and fixtures. Factory-built cabinets may be purchased in one of three forms:

- Disassembled.
- Assembled but not finished (*in-the-white*).
- Assembled and finished.

Disassembled cabinets, often referred to as *knocked down* or *KD*, consist of parts cut to size and ready to be assembled on the job. Assembled, but not finished cabinets are ready to set in place. Hardware is included, but not installed. All surfaces are sanded and ready for finish to be applied. After installation, finishing materials and procedures can be coordinated with doors and inside trim—thus ensuring an exact match.

Assembled and finished cabinets are widely used because they save time during finishing stages of construction. Manufacturers offer a variety of shades and colors that are factory-applied by experts, **Figure 20-8.** Because of the

controlled conditions and special equipment, finishing materials can be applied that have high resistance to moisture, acids, and abrasion. Once the finishing process is complete and hardware has been installed, the units are carefully packaged and shipped to a distributor or directly to a construction site.

Manufacturers of cabinetwork offer a variety of standard units, especially in kitchen cabinets. Most kitchen layouts can be entirely made from these units. When necessary, matching custom-built units can be ordered, **Figure 20-9.** Sometimes, factory-built cabinets are combined with units constructed in custom cabinet shops or with various units built on the job. **Figure 20-10** shows standard base and wall units offered by one cabinet manufacturer. Other standard units offered by various manufacturers include oven and utility cabinets, peninsular base units, and bathroom vanity and sink cabinets.

Figure 20-9. Custom-built cabinets are often offered by manufacturers to match their product line. (Bullard-Haven Technical School)

Figure 20-8. Premium-quality cabinet doors receive a coat of hand-rubbed stain from highly skilled wood finishers. (Riviera Kitchens, Evans Products Co.)

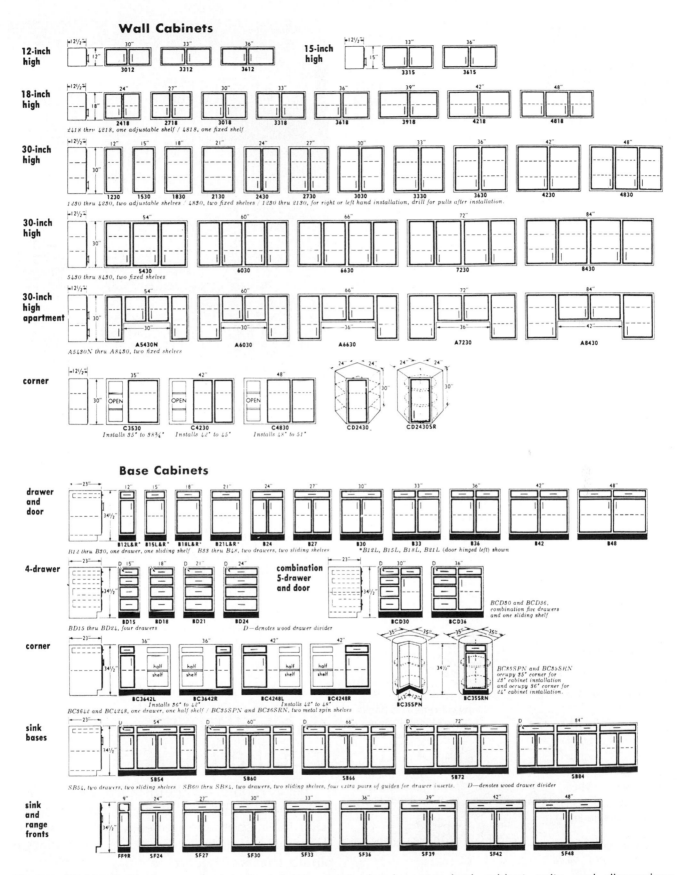

Figure 20-10. Cabinet manufacturer's catalog sheet showing standard cabinet units and dimensions. (I-XL Furniture Co.)

20.5 Cabinet Materials

A wide range of materials is used in factory-built cabinets. Low-priced cabinets are usually made from panels of particleboard with a vinyl film applied to exposed surfaces. The vinyl is printed with a wood grain pattern.

High-quality cabinets are made from veneers and solid hardwoods such as oak, birch, ash, and hickory. Hardboard, particleboard, and waferboard may be used for certain interior panels, drawer bottoms, and as the base for plastic laminate countertops. Frames are assembled with accurately made joints. Dovetail joints are generally used in drawer assemblies. See **Figure 20-11.** A completed kitchen installation is shown in **Figure 20-12.**

A variety of storage features can be added to standard cabinet units. Examples include revolving shelves (lazy Susan), special compartments and dividers in drawers, slide-out breadboards, and slide-out shelves. Storage units for canned goods provide extra convenience with

Figure 20-12. This kitchen installation of factory-built cabinets includes a built-in sink and wall oven. (Merillat Industries, Inc.)

swing-out wood shelving, **Figure 20-13.** Other available storage features include slide-out or tilt-out wire racks for under-sink storage; wire racks for special lid storage; swing-out storage trays; file cabinets; wall tambour storage; pull-out ironing boards; and swing-out, multi-storage wire shelves for base cabinets.

20.6 Cabinet Installation

Before cabinets can be installed, the carpenter must check walls and floors for uneven spots. After these spots are located, cabinets must be shimmed or scribed to make the installation plumb, true, and square.

Floors can be checked with a long straightedge or straight 2 × 4 and a level. The floor should be checked within 22″ of the walls where base

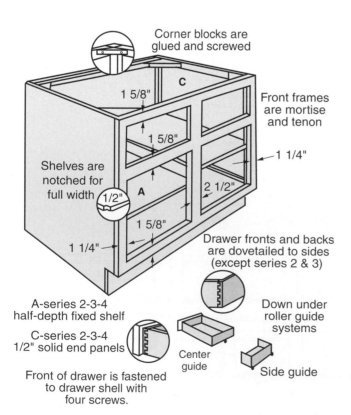

Figure 20-11. Construction details of a factory-built cabinet. (Brammer Mfg. Co.)

A   B

Figure 20-13. Efficient kitchen storage. A—Swing-out pantry units have large amounts of storage for food cans and boxes in a relatively small space. B—Tilt out wire bins and racks provide handy storage under sinks and in other base cabinets. (KraftMaid Cabinetry, Inc.; Merillat Industries, Inc.)

cabinets will be installed. A level line should be snapped on the wall from the high point around the wall as far as the cabinets will extend. This line is called the ***base level line.***

To check the wall, first mark the outlines of all cabinets on it. Use a straightedge to check for low and high points. High spots must be removed by scraping or sanding. Low spots can be shimmed with thin pieces of wood or wood shingles. See **Figure 20-14.**

There are two basic procedures for installing factory-built cabinets:

Base level line: A level line that is snapped on the wall from the high point around the wall as far as the cabinets will extend.

- Some manufacturers recommend that the wall cabinets be installed first. Layouts are made and wall studs located with a stud finder or by tapping the wall with a hammer. Then, the wall units are lifted into position. They are held with a padded T-brace (sometimes called a *story stick*) that allows the worker to stand close to the wall while making the installation. See **Figure 20-15.** After the wall cabinets are securely attached and checked, the base cabinets are moved into place, leveled, and secured.

- In the alternate procedure, base cabinets are installed first. The tops of the base cabinets can then be used to support braces (story sticks) that hold the wall units in place. This procedure is illustrated in **Figure 20-16.**

Floors and walls are seldom exactly level and plumb. Shims and blocking must be used as necessary so the cabinets are not racked or twisted. Doors and drawers will not properly operate if the cabinet is distorted by improper installation.

Screws should go through the hanging strips (if used) and into the stud framing. Never use

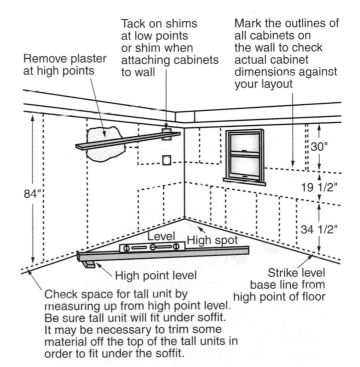

Figure 20-14. Before installing cabinets, mark the outline of every base and wall unit on the wall. Check for high and low spots, walls for plumb, and floors for level. (KraftMaid Cabinetry, Inc.)

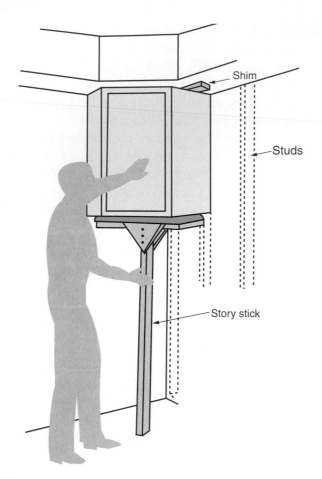

Figure 20-15. Use a story stick to support a wall cabinet. Mark the stud locations on the inside of the cabinet for mounting. (KraftMaid Cabinetry, Inc.)

nails. Use toggle bolts when studs are inaccessible. Join units by first clamping them together and aligning them then installing fasteners, **Figure 20-17.**

Countertops are usually prefabricated in a millwork shop, but a carpenter may choose to build them to the proper measurements and specifications. Before beginning installation, double check the base cabinet tops. They must be level and even across the top. If necessary, shim them at the floor or base. *Build-up strips* are narrow wood pieces that are used to attach the countertop to the base cabinets. They are usually provided by the millwork shop. If not provided, the carpenter must cut them.

Figure 20-18 shows the installation of a mitered countertop. Miters must be accurately cut and fitted, a difficult procedure requiring great skill. The joint is secured and drawn tight from the bottom with bolts that are let into the substrate.

PROCEDURE

Installing prefabricated countertops

Normally, the prefabricated top will come with instructions for installation. In general, follow these steps:

1. If build-up strips are not provided, cut 3/4″ lumber into strips 2″ wide and 24″ long.
2. Space build-up strips 2′ apart, but be sure to place beginning and end strips 2″ in from the cabinet ends. This space allows room for end caps on the countertop. See **Figure 20-19.**
3. To mount the build-up strips, drill screw holes at each end. Then, using 1 1/4″ drywall screws, attach one end to the top of the face frame and the opposite end to the back of the cabinet.
4. Position the countertop on top of the cabinets and tight against the wall.
5. Use a marking pen or a carpenter's pencil to scribe a line transferring any irregularities in the wall to the backsplash.
6. Use a belt sander and medium-grit sandpaper to remove material from the backsplash up to the marked line. Always have the belt moving into the laminate to avoid chipping.
7. Check your work and repeat steps 5 and 6 until the fit is satisfactory.
8. If a sink or drop-in range is to be installed, check the manufacturer's instructions for size and location. Apply masking tape to the countertop and trace the cut lines on it. Again, this avoids chipping.
9. Drill a relief hole at each corner.
10. Install a sharp saw blade in a jigsaw and make the cut.
11. Position the countertop on the cabinet base and check the fit.
12. Working from underneath, drill pilot holes in the build-up strips.
13. Drive 1 1/4″ black drywall screws up through the build-up strips to anchor the countertop.
14. Place caulk at the joint between the backplash and the wall.

Build-up strips: Narrow wood pieces that are used to attach the countertop to the base cabinets.

1.
Locate the position of all wall studs where cabinets are to hang by tapping with a hammer. Mark their position where the marks can easily be seen when the cabinets are in position.

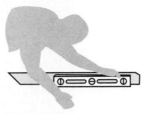

2.
Find the highest point on the floor with a level. This is important for both base and wall cabinet installation later. Remove the baseboard from all walls where cabinets are to be installed. This will allow them to go flush against the walls.

3.
Start the installation with the corner or end unit. Slide it into place then continue to slide the other base cabinets into the proper position.

4.
When all base cabinets are in position, fasten the cabinets together. This is done by drilling a 1/4" diameter hole through the face frames, and using the 3" screws and T-nuts provided. To get maximum holding power from the screw, one hole should be close to the top of the end stile and one should be close to the bottom.

5.
Check the position of each cabinet with a spirit level, going from the front of the cabinet to the back of the cabinet. Next shim between the cabinet and the wall for a perfect base cabinet installation.

6.
Starting at the high point in the floor, level the leading edges of the cabinets. Continue to shim between the cabinets and the floor until all the base cabinets have been brought to level.

7.
After the cabinets have been leveled, both front to back and across the front, fasten the cabinets to the wall at the stud locations. This is done by drilling a 3/32" diameter hole 2 1/4" deep through both the hanging strips for the 2 1/2" x 8 screws that are provided.

8.
Fit the counter top into position and attach it to the base cabinets by predrilling and screwing through the front corner blocks into the top. Use caution not to drill through the top. Cover the counter top for protection while all the cabinets are being installed.

9.
Position the bottom of the 30" wall cabinets 19" from the top of the base cabinet, unless the cabinets are to be installed against a soffit. A brace can be made to help hold the wall cabinets in place while they are being fastened. Start the wall cabinet installation with a corner end cabinet. Use care in getting this cabinet installed plumb and level.

10.
Temporarily secure the adjoining wall cabinets so that the leveling may be done without removing them. Drill through the end stiles of the cabinets and fasten them together as was done with the base cabinets.

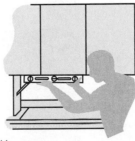

11.
Use a spirit level to check the horizontal surfaces. Shim between the cabinet and the wall until the cabinet is level. This is necessary if doors are to fit properly.

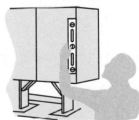

12.
Check the perpendicular surface of each frame at the front. When the cabinets are level, both front to back and across the front, permanently attach the cabinets to the wall. This is done by predrilling a 3/32" diameter hole 2 1/4" deep through the hanging strip inside the top and below the bottom of the cabinets at the stud location. Enough number 8 screws should be used to fasten the cabinets securely to the wall.

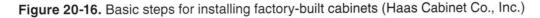

Figure 20-16. Basic steps for installing factory-built cabinets (Haas Cabinet Co., Inc.)

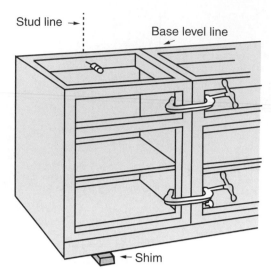

Figure 20-17. To install base cabinets, shim as needed at the floor to level the units and bring them up to the base level line. Clamp adjoining units, as shown, to hold them in alignment, then fasten cabinets to each other and to the wall studs. (KraftMaid Cabinetry, Inc.)

Figure 20-19. Build-up strips provide clearance for cabinet doors and drawers. The strips may run parallel or perpendicular to the cabinet run. In this case, the carpenter is using both methods to accommodate a new sink. Be sure to allow a 2″ setback for the strips at each end of the cabinet run to provide clearance for the end reinforcement attached to the countertop.

Figure 20-18. Properly fitting a mitered countertop is a job for a skilled carpenter or woodworker.

20.7 Cabinets for Other Rooms

Some manufacturers of kitchen cabinets build storage units for other areas of the home. Built-in units provide drawers and cabinets that save space. They improve efficiency and also offer greater convenience. For example, many homes are being equipped with closet organizer systems for more efficient use of storage space, **Figure 20-20.** These systems consist of modular

Figure 20-20. A combination of wall- and floor-mounted cabinetry is used in this walk-in closet to provide various types of storage. (Schulte Distinctive Storage)

cabinet and shelving units that can be installed in various combinations, depending on the space available and specific storage needs.

In other rooms, carefully planned built-in cabinets provide efficient storage and display for items used in work, recreation, and hobby activities. Their custom-built appearance usually adds to room decor. They may even eliminate the need for some movable pieces of furniture.

20.8 Building Cabinets

Two different procedures are commonly used when building cabinets on the job. The first consists of cutting the parts and assembling them "in place" one piece at a time. The carpenter or cabinetmaker attaches each piece to the floor, wall, or other member. After the basic structure is assembled, facing strips, doors, and other fittings are marked, cut to size, and attached.

In the second procedure, the entire unit is first assembled. Then, it is moved into place and fixed to the floor or attached to the wall.

The cabinet structure is formed with end panels, partitions, and backs. Horizontal frames join the parts. See **Figure 20-21.** Basically, this is the method used in custom cabinet shops or for mass-production of cabinets in factories. These must be of a size easily moved to the building site. When units are built on the job, they can be much larger (longer sections) than is practical in a shop or factory.

20.8.1 Master Layouts

Before cutting out the parts for a cabinet, it will be helpful to prepare a *master layout* on plywood or cardboard, **Figure 20-22.** This is especially true when fully assembling the unit and then installing it. Several layouts may be needed.

Master layout: Drawing of each cabinet member full size on cardboard or plywood showing all clearances that may be required.

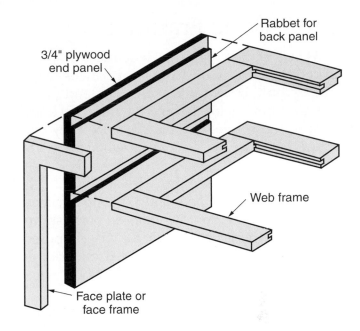

Figure 20-21. Typical frame construction for a cabinet built on-site.

Follow the overall dimensions provided in the architectural plans and details. Draw each member full size. Show all clearances that may be required. Where the structure is complex with drawers, pull-out boards, and special shelving, these should be drawn as section views.

The master layout will be valuable when cutting side or end panels to size and locating joints. It can also be referred to for exact sizes and locations of drawer parts and other detailed dimensions not included in the regular drawings.

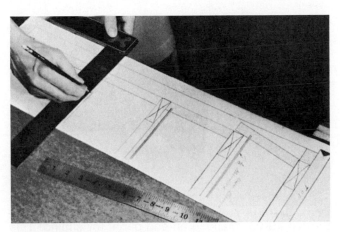

Figure 20-22. Drawing a full-size master layout for a base cabinet.

20.8.2 Basic Framing

When following the assembled-in-place procedure, some of the basic layout can be made directly on the floor and wall. Each end panel and partition is represented with two lines. Be sure these lines are plumb, since they can be used to line up the panels when they are installed.

Construct the base first. Use straight 2 × 4s and nail them to the floor and to a strip attached to the wall. If the floor is not level, place shims under the various members of the base. Exposed parts can be faced with a finished material or the front edge can be made of a finished piece, such as base moulding.

After the base is completed, cut and install the end panels, **Figure 20-23.** Attach a strip along the wall between the end panels and level with the top edge. Be sure the strip is level throughout its length. Nail it to the wall studs.

Next, cut the bottom panels and nail them in place on the base. Follow this with the installation of the partitions. These must be notched at the back corner of the top edge so they will fit over the wall strip.

Plumb the front edge of the partitions and end panels. Secure them with temporary strips nailed along the top, as shown on the left-hand cabinet in **Figure 20-24.**

Figure 20-24. Partly completed base cabinets. The temporary strip at the top, left is used to hold the partition in alignment until the face frames are installed.

Wall units are constructed in basically the same way as the base units. Make layout lines directly on the wall and ceiling. Attach mounting strips by nailing through the wall surface into the studs. At inside corners, end panels can be directly attached to the wall.

During the basic framing operations, use care when constructing the openings for built-in appliances. They must be the correct size. Obtain rough-in drawings and specifications, **Figure 20-25.** These are furnished by the manufacturer of the appliances; carefully follow them.

Figure 20-23. Marking an end panel for cutting. This procedure is typical of assembled-in-place cabinet construction.

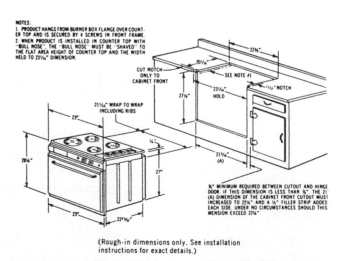

(Rough-in dimensions only. See installation instructions for exact details.)

Figure 20-25. This rough-in drawing shows the required cabinet dimensions for a built-in range unit. (Whirlpool Corp.)

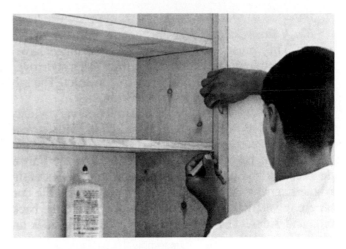

Figure 20-26. Marking a face frame stile for positioning and cutting.

20.8.3 Facing

Finished *facing strips* are applied to the front of the cabinet frame. In factories and cabinet shops, these strips are often assembled into a framework (called a *face plate* or *face frame*) before they are attached to the basic cabinet structure. The vertical members are called **stiles** and the horizontal members are called **rails.**

For assembled-in-place cabinets, each piece is separately cut and installed. Establish the size by positioning the facing stock on the cabinet and marking it as shown in **Figure 20-26.** Then, make the finished cuts. A marked part can be used as a pattern to lay out duplicate pieces.

Generally, stiles are installed first and then the rails, **Figure 20-27.** Sometimes, an end stile is attached plumb and rails are installed to determine the position of the next stile.

The parts are glued and nailed with finishing nails. When nailing hardwoods, drill pilot holes where splitting is likely to occur.

Many kinds of joints can be used to join the stiles and rails. A practical design is illustrated in **Figure 20-28.** The depth of the **gain,** or dado, is about 3/8" so it will be covered by the lip of the door or drawer. Lap joints and dowel joints are also commonly used to join stiles and rails.

20.8.4 Drawer Guides

Drawer guides are devices that keep drawers in place and allow them to slide in and out. Three common types of drawer guides are used:
- Corner guides.
- Center guides.
- Side guides.

Figure 20-29 shows all three types. The *corner guide* may be formed in the cabinet by the side panel and frame. It may be necessary to add a spacer strip to hold the drawer in alignment with the front facing.

A *center guide* is often used in cabinetwork and consists of a strip or runner fastened between

Figure 20-27. Face frame stiles and rails are both glued and nailed in place on the cabinets.

Facing strips: Finish material applied (as stiles and rails) to the front of the cabinet frame.

Stiles: The upright or vertical outside pieces of a sash, door, blind, screen, or face frame.

Rails: Cross or horizontal members of the framework of a sash, door, blind, or other assembly.

Gain: Recess or mortise cut to receive the end of another structural member, hinge, or other hardware.

Drawer guides: Wood strips or metal devices that support drawers at the sides or lower corners.

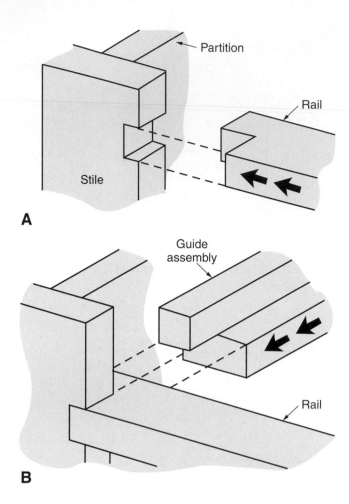

Figure 20-28. Practical joint designs for assembled-in-place face frames. A—Stile-and-rail joint. B—Drawer guide assembly joint. It fits behind the facing and rests on the rail.

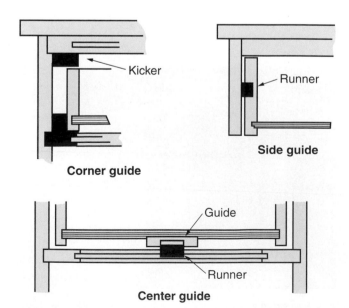

Figure 20-29. Three basic types of drawer guides.

the front and back rails. A guide that is attached to the underside and back of the drawer rides on this runner. The runner is attached to the frame or rails with screws. It can be adjusted so the clearance on each side is equal and the face of the drawer aligns with the front of the structure. **Figure 20-30** shows how to attach a commercially made back bearing for a center slide.

In drawer openings where there is no lower frame, a *side guide* may be used. Grooves are cut in the drawer side before it is assembled and matching strips are fastened to the side panels. This type of guide can be used for shallow drawers or trays in wardrobe units where framing between each drawer opening is unnecessary and would waste space.

The drawer carrier arrangement may require a *kicker.* It keeps the drawer from tilting

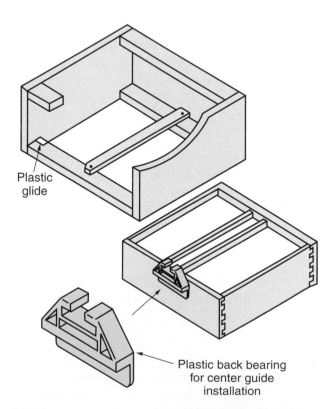

Figure 20-30. This center guide installation uses a patented back bearing. (Ronthor Plastics Div., U.S. Mfg. Corp.)

Kicker: A strip of wood centered between the rails above a drawer to keep it level as it is opened.

downward when it is opened. The kicker may be located over the drawer side or centered to hold down the back edge of the drawer. See **Figure 20-29.**

Wooden drawer guides should be carefully fitted. The parts should be given a coat of sealer, then lightly sanded and waxed. The drawer will then smoothly slide on the guides.

For large drawers or those that will carry considerable weight, special drawer slides may be used. Some of these will support weights up to 50 lb. Units are available in several sizes and types. See **Figure 20-31.**

20.8.5 Drawers

There are two general types of drawers:
* Flush.
* Lipped.

Flush drawers are those that fit into the drawer opening and are *flush* (even) with the face frame. They must be carefully fitted. *Lipped drawers* have a rabbet along the top and sides of the front. This style overlaps the opening and is much easier to construct. The rabbet is used mainly in built-in cabinetwork.

Sizes and designs in drawer construction widely vary. **Figure 20-32** shows several types of joints commonly used in standard drawer construction.

The joint between the drawer front and the drawer side receives the greatest strain. It should be carefully designed and fitted. Often, in high-quality work, the corners are dovetailed and the bottom grooved into the back, front, and sides. The top edges of the drawer sides are rounded.

Flush drawer fronts may be constructed with the sides set about 1/16"–1/8" deeper into the fronts so that a slight lug (projection) is formed. After the drawer is assembled, this lug is trimmed to form a good fit.

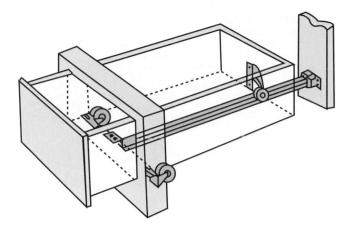

Tri-roller system

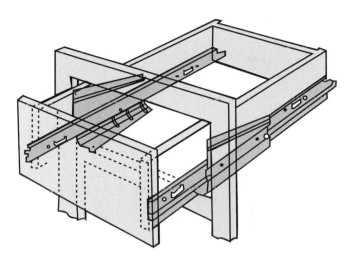

Side rail system

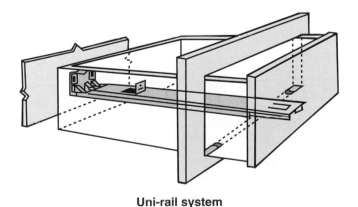

Uni-rail system

Figure 20-31. Several available designs of drawer slide assemblies. (Amerock Corp.)

Flush drawer: Drawer whose front fits flush in its opening.

Lipped drawer: Drawer with a front that has a lip covering the joint between the drawer and the cabinet's faceplate.

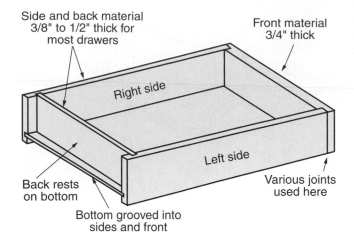

Side and back material 3/8" to 1/2" thick for most drawers

Front material 3/4" thick

Right side

Left side

Back rests on bottom

Various joints used here

Bottom grooved into sides and front

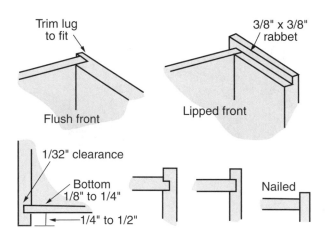

Trim lug to fit

3/8" x 3/8" rabbet

Flush front

Lipped front

1/32" clearance

Bottom 1/8" to 1/4"

1/4" to 1/2"

Nailed

Back and side joints

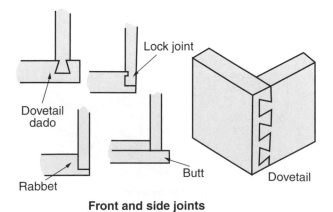

Lock joint

Dovetail dado

Rabbet

Butt

Dovetail

Front and side joints

Figure 20-32. Standard drawer construction. Various types of suitable joints are shown.

PROCEDURE

Constructing a drawer

1. Select the material for the drawer fronts. Solid stock or plywood may be used. Grain patterns should match or blend with each other. Prefabricated drawer fronts may be used. They are offered in many standard sizes and several styles.

2. If flush type, cut the drawer fronts to the size of the opening, allowing a 1/16" minimum clearance on each side and on the top. This clearance will vary, depending on the depth of the drawer and the kind of material. Deep drawers with solid fronts require greater vertical clearance. For lipped drawers, add the depth of the rabbets to the dimensions.

3. Select and prepare the stock for the sides and back. A less-expensive hardwood or a softwood can be used. Plywood is usually not satisfactory for these parts. The surfaces should be sanded either before or after cutting to length.

4. Select material for the drawer bottom. Hardboard or plywood should be used. Trim to final size after the joints for the other drawer members are cut.

5. Cut a groove in the front and sides for the bottom.

6. Cut joints in the drawer fronts to hold the drawer sides. **Figure 20-33** shows a sequence for cutting a lock joint. For drawer fronts with a lip, cut the rabbet first, then the joint.

7. Cut the matching joint in the drawer sides. Be sure to cut a left and right side for each drawer.

8. Cut the required joints for the drawer sides and backs.

9. Trim the bottom to the correct size. Make a trial assembly and perform any adjustments in the fit that may be needed. The parts should smoothly fit together. If they are too tight, they will be difficult to assemble after glue is applied.

10. Disassemble and sand all parts. Rounding of the top edge of the sides can be done at this point. Stop the rounding about 1" from each end.

11. Make the final assembly. Any of a number of procedures can be used. One method is to first

(Continued)

(Continued)

glue the bottom onto the front and then glue one of the front corners. Be sure the bottom is centered. The side grooves are usually not glued. Turn the drawer on its side and glue the back into the assembled side. Glue and attach the remaining side. Clamps or a few nails can be used to hold the joints together until the glue has completely hardened.

12. Carefully check the drawer for squareness. Then, drive one or two nails through the bottom into the back. If the bottom was carefully squared, you should have little trouble with this operation. Wipe off the excess glue.

13. After the glue has cured, fit each drawer to a particular opening. Trim and adjust drawer guides. Place an identifying number or letter on the underside of the bottom. This label will make it easy to return the drawer to its proper opening after sanding and finishing operations. The inside surfaces of quality drawers should always be sealed and waxed.

be necessary to fit and glue the shelves into dados cut into the sides of the cabinet. This adds strength to the joint.

If possible, make the shelves adjustable. This allows the storage space to be used for various purposes.

Figure 20-34 presents several methods of installing adjustable supports. Carefully and accurately lay out the shelf support system so the shelves will be level. Usually, it is best to do the cutting and drilling before the cabinet is assembled. If shelf standards are of the type set in a groove, it is absolutely necessary to cut the groove before assembly. Some patented adjustable shelf supports are designed to mount on the surface.

Standard 3/4″ shelving should be supported every 42″ or at shorter intervals. This applies especially to shelves that will carry heavy loads. The front edge of plywood shelving should be overlaid with a strip of wood material that matches the cabinet wood. This may be solid stock or thin strips that are glued to the plywood edge.

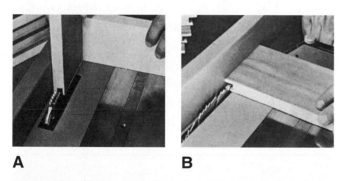

A   **B**

Figure 20-33. Cutting a lock joint in a drawer front. A—Cut the groove first. B—Trim the tongue to length in the second cut.

20.8.6 Shelves

Shelves are widely used in cabinetwork, especially in wall units. In some designs, it may

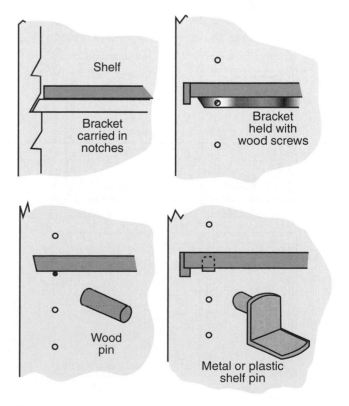

Figure 20-34. Various methods of supporting removable shelves.

Shelves: Fixed or adjustable horizontal dividers in cabinets designed to provide storage.

20.9 Doors

Either swinging (hinged) or sliding doors may be used with built-in cabinets. Swinging doors are commonly used for kitchen cabinets. **Figure 20-35** shows the six basic styles of cabinet doors.

Doors may be made of plywood, preferably with a lumber core, or they may consist of a frame with a panel insert. Fine cabinet doors often have a frame covered on each side with thin plywood. This construction is similar to a hollow core door. Because of its tendency to warp, solid stock is seldom used, except on very small doors.

Many carpenters use prefabricated cabinet doors. These are available in various styles and

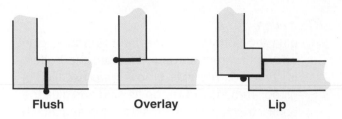

Flush   **Overlay**   **Lip**

Figure 20-36. Three basic types of cabinet doors.

in a wide range of stock sizes. Doors are delivered to the job cut to the exact dimensions with lipped edges (if specified) already machined.

Some manufacturers offer panel doors with special "see-through" inserts. These are useful for displaying items on cabinet shelves. Simulated leaded glass and perforated metal inserts are of this type.

Hinged doors for cabinets are of three basic types, depending on how they fit into or over the door opening. **Figure 20-36** shows the basic shapes of the three types:

- *Flush door.* The door fits into the opening and does not project outward beyond the frame.

- *Overlay door.* Though its edges are square like the flush door, it is mounted on the outside of the frame, wholly or partly concealing it.

- *Lipped door.* The door is rabbeted along all edges, so that part of the door is inside the door frame. A lip extends over the frame on all sides to conceal the opening.

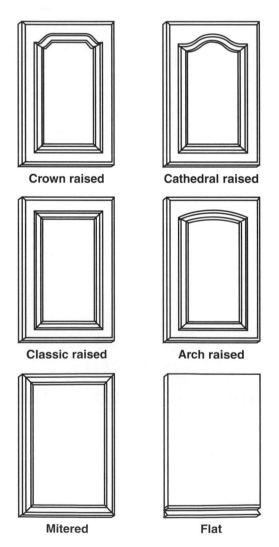

Crown raised   **Cathedral raised**

Classic raised   **Arch raised**

Mitered   **Flat**

Figure 20-35. Six basic styles used for cabinet doors. (Merillat Industries, Inc.)

> **Working Knowledge**
>
> When specifying the size of the opening or the size of a cabinet door, always list width first and height second. This is standard practice among carpenters and is commonly followed by lumber suppliers and millwork plants.

Flush door: A door that fits into the opening and does not project outward beyond the frame.

Overlay door: A slab door in cabinetry that covers the cabinet frame.

Lipped door: A cabinet door rabbeted (partially recessed) along its edges. The resulting lip conceals the joint.

20.9.1 Installing Flush Doors

The flush door is usually installed with butt hinges. However, surface hinges, wraparound hinges, knife hinges, or various semi-concealed hinges can be used. **Figure 20-37** shows styles that are often used for cabinetwork.

The size of a hinge is determined by its length and width when open. Select the size to suit the size of the door. Large wardrobe or storage cabinet doors should have three hinges; smaller doors typically have two.

Flush doors must be carefully fitted to the opening with about 1/16″ clearance on each edge. The total gain (space) required for the hinge can be entirely cut in the door. However, for fine work, it is best to equally cut into the door and stile. On large doors, it is practical to use a portable router for this operation.

Install the hinges on the door, then mount the door in the opening. Initially, use only one screw in each hinge leaf. Adjust the fit and then set the remaining screws. It may be necessary to plane a slight bevel on the door edge opposite of the hinges for better clearance.

Stops are set on the door frame so the door will be held flush with the surface of the opening when closed. The stops may be placed all around the opening or only on the side where the lock or catch is located.

Overlay doors provide an attractive appearance in some contemporary styles of cabinetwork. Butt hinges can be used on overlay doors. However, it is usually best to make the installation with a hinge type that is designed for the purpose, **Figure 20-38.**

20.9.2 Cutting and Fitting Lipped Doors

Lipped doors are easier to cut and fit than flush doors, because the clearance is covered. They must be installed with a special offset hinge. See several styles in **Figure 20-39.**

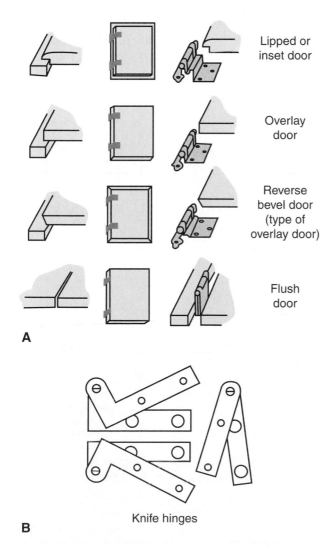

Lipped or inset door

Overlay door

Reverse bevel door (type of overlay door)

Flush door

A

Knife hinges

B

Figure 20-37. Hinge types used for cabinet doors. A—Common cabinet-door hinge types. B—Knife hinges. (Baldwin Hardware Corp., Amerock Corp.)

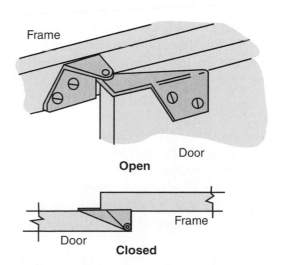

Frame

Door

Open

Frame

Door

Closed

Figure 20-38. This pivot hinge is designed for use with overlay-style doors.

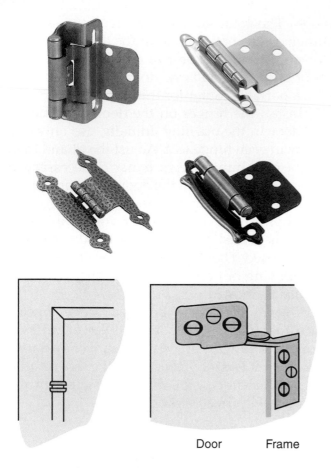

Semi-concealed pivot hinge

Figure 20-39. Hinges for lipped doors. Note how semi-concealed pivot hinges show only the pin. (Amerock Corp.)

Select the door blanks for a given series of openings. Try to match grain and color when a natural finish will be applied. Cut the doors to the size of the openings plus the amount required for the lip all around. Allow clearance of about 1/16″ on each edge. Make an additional allowance for the hinges, which are not gained into the door or frame.

The lip is formed by cutting a rabbet along the edge. This can be accomplished on the table saw or with a shaper.

Check the fit of each door in its proper opening and mark its position in small letters on the back. Remove all machine marks with abrasive paper.

Soften edges and corners. Hardware may be prefitted and then removed before the finishing operations.

20.9.3 Sliding Doors

Sliding doors are often used where the swing of regular doors would be awkward or cause interference. They are adaptable to various styles and structural designs.

Construction details of a sliding door arrangement are shown in **Figure 20-40.** Grooves are cut in the top and bottom of the cabinet before assembly. The doors are sometimes rabbeted so the edge formed will match the groove with about 1/16″ clearance. Cut the top rabbet and groove deep enough that the door can be inserted or removed by simply raising it into the extra space. The doors will smoothly slide if the grooves are carefully cut, sanded, sealed, and waxed. Avoid too much finish.

Sliding glass doors are sometimes specified in cabinets. They are heavy and require a special plastic or roller track. Follow the manufacturer's recommendations for installation. Sliding glass

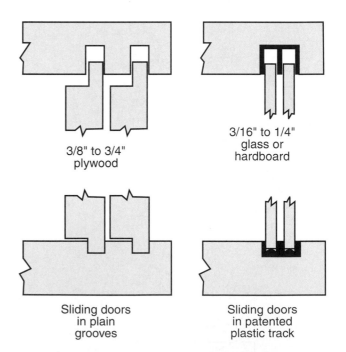

Figure 20-40. Sliding doors. Thicker stock must be rabbeted, as shown at the left, to fit the track.

Sliding doors: Cabinet doors often used where the swing of regular doors would be awkward or cause interference.

doors are adaptable to wall units and add variety to an installation. They are usually made of 1/4″ plate glass with polished edges.

Many variations of sliding door track and rollers are available. **Figure 20-41** illustrates a self-lubricating plastic track. It is easy to install and provides smooth operation for furniture and cabinet doors. An overhead track and roller system is normally used for large wardrobe doors and passage doors.

20.10 Counters and Tops

A high-pressure type of *plastic laminate* is commonly used as a surface for counters and tops. This material, usually 1/16″ thick, offers high resistance to wear and is unharmed by boiling water, alcohol, oil, grease, and ordinary household chemicals, **Figure 20-42.**

Although the laminate is very hard, it does not possess great strength and is serviceable only when bonded to plywood, particleboard, waferwood, or hardboard. This base or core material must be smooth and dimensionally stable. Hardwood plywood (usually 3/4″ thick) makes a satisfactory base. However, some plywoods, especially fir, have a coarse grain texture that may telegraph (show through). Particleboard, which is less expensive than plywood, provides a smooth surface and adequate strength.

When the core or base is free to move and is not supported by other parts of the structure, the laminated surface may warp. Bonding a backing sheet of the laminate to the second face can counteract this. It will minimize moisture gain or loss and provides a balanced unit with identical materials on either side of the core. For a premium grade of cabinetwork, Architectural Woodwork Institute standards specify that a *backing sheet* be used on any unsupported area exceeding 4 square feet. Backing sheets are like the regular laminate without the decorative finish and are usually thinner. A standard thickness for use opposite of a 0.060″ (1/16″) face laminate is 0.020″.

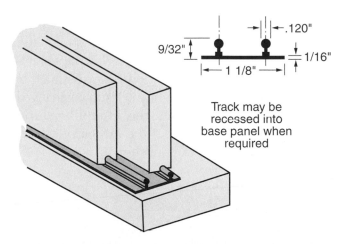

Track may be recessed into base panel when required

Figure 20-41. Plastic track for sliding doors. Similar track is available in metal. (Kentron Div., North American Reiss)

Plastic laminate: Material used as a surface for counters and tops that offers high resistance to wear and is unharmed by boiling water, alcohol, oil, grease, and ordinary household chemicals.

Backing sheet: Sheet that is like regular laminate without the decorative finish and is usually thinner.

Figure 20-42. Plastic laminate countertops are hard and durable, resisting stains and wear. (Formica Corp.)

20.10.1 Working with Laminates

Plastic laminates can be cut to rough size with a handsaw, table saw, portable saw, or portable router, **Figure 20-43.** Use fine-tooth blades and support the material close to the cut. Laminates used on vertical surfaces, which are 1/32″ thick, can be cut with tin snips.

It is best to make the roughing cuts 1/8″–1/4″ oversize. Trim the edges after the laminate has been mounted. Handle large sheets carefully, because they can be easily cracked or broken. Be careful not to scratch the decorative side.

Contact bond cement is used to apply the plastic laminate because no sustained pressure is required. It is applied with a spreader, roller, or brush to both surfaces being joined. On large horizontal surfaces, it is best to use a spreader, **Figure 20-44.**

For soft plywoods, particleboard, or other porous surfaces, the spreader is held with the serrated edge perpendicular to the surface. On hard, nonporous surfaces and on plastic laminates, hold the edge at a 45° angle. A single coat should be sufficient.

An animal hair or fiber brush may be used to apply the adhesive to small surfaces or those in a vertical position. Apply one coat, let it thoroughly dry, and then apply a second.

All of the surface should be completely covered with a glossy film. Dull spots, after drying, indicate that the application was too thin and that another coat should be applied.

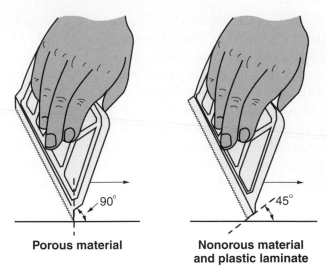

Porous material   **Nonorous material and plastic laminate**

Figure 20-44. To spread contact bond cement on large horizontal surfaces, use a toothed metal spreader. Note the different angles used, depending on the material.

Thoroughly stir the adhesive before using it and follow the manufacturer's recommendations. Usually, brushes and applicators must be cleaned in a special solvent.

After the adhesive has been applied to both surfaces, let it dry (usually for at least 15 minutes). You can test the dryness by lightly pressing a piece of paper against the coated surface. If no adhesive sticks to the paper, it is ready to be bonded. This bond can usually be made any time within an hour. Time varies with different manufacturers. If the assembly cannot be made within this time, the adhesive can be reactivated by applying another thin coat.

> **Safety Note**
>
> Some types of contact cement are extremely flammable. Nonflammable types may produce harmful vapors. Be sure the work area is well ventilated. Follow the manufacturer's directions and observe precautions.

20.10.2 Adhering Laminates

Once the adhesive is ready, bring the two surfaces together in the exact position required.

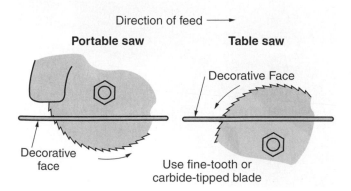

Direction of feed ⟶

Portable saw   **Table saw**

Decorative Face

Decorative face

Use fine-tooth or carbide-tipped blade

Figure 20-43. Saws used to cut plastic laminate must have fine-toothed blades.

The pieces cannot be shifted once contact is made. When joining large surfaces, place a sheet of heavy wrapping paper, called a *slip sheet,* over the base surface. Then, slide the laminate into position. Slightly withdraw the paper so one edge can be bonded and then remove the entire sheet and apply pressure.

Total bond is secured by the application of momentary pressure. Hand rolling provides satisfactory results if the roller is small (3″ or less in length). Long rollers apply less pressure per square inch. Work from the center to the outside edges and be certain to roll the entire surface. In corners and areas that are hard to roll, hold a block of soft wood on the surface and tap it with a rubber mallet.

Trimming and smoothing the edges is an important step in the application of a plastic laminate. A plane or file may be used, but many carpenters prefer an electric router equipped with an adjustable guide. When making cutouts for sinks or other openings, a saber saw is practical.

The corner and edges of a plastic laminate application should be beveled, **Figure 20-45.**

This makes it smooth to touch and less likely to chip. The bevel angle can be formed with a smooth mill file, **Figure 20-46.** Stroke the file downward and use care not to damage the surfaces of the laminate. Some routers can be equipped with an adjustable guide and a special bit that will make this cut. Final smoothing and a slight rounding of the bevel should be done with a 400 wet- or dry-abrasive paper.

When working with plastic laminates, be especially careful that files, edge tools, or abrasive papers do not damage the finished surfaces. Such damage is difficult to repair.

Figure 20-46. Laminate edges and corners can be beveled by carefully working with a mill file. (Bullard-Haven Technical School)

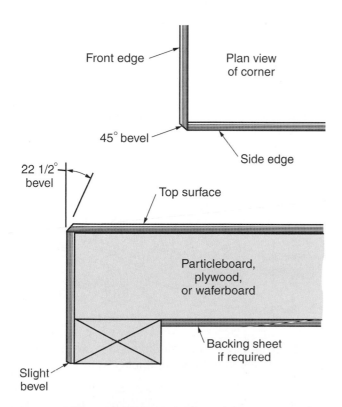

Figure 20-45. Properly beveled corners and edges on plastic laminate countertops improve the appearance and durability of the installation. Note the angles shown in the two views.

Slip sheet: A sheet of heavy wrapping paper placed over the base surface when adhering laminates.

20.11 Cabinet Hardware

After counters and tops have been covered and surface finishes applied to cabinets, hardware is installed. This consists of knobs, pulls, and various other metal fittings. Use care to select an appropriate size, style, material, and finish. **Figure 20-47** shows several designs of hardware for drawers and doors.

Some hardware (mainly hinges) should be prefitted before the final surface finish is applied. Pulls and catches can be installed after the finish. Use care in the layout. Drill holes with sharp bits that will not splinter the surface.

Drilling jigs aid in the fast and accurate installation of hardware.

Drawer *pulls* usually look best when they are located slightly above the centerline of the drawer front. They should be horizontally centered and level. Door pulls are convenient when located on the bottom third of wall cabinet doors and the top third of base cabinet doors.

Hinged doors require some type of catch to keep them closed. Several common types are illustrated in **Figure 20-48**. An important consideration in their selection is the noise they produce when the door is opened or closed. In general, the catch should be placed as near as practical to the door pull. The package usually contains instructions for installation. Carefully follow these.

20.12 Other Built-In Units

The information to this point in the chapter has been largely directed toward kitchen cabinets. The same general procedures can be applied

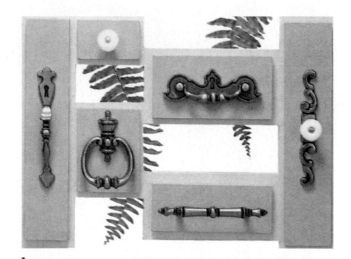

A

B

Figure 20-47. Some examples of cabinet door and drawer hardware. (Amerock Corp.)

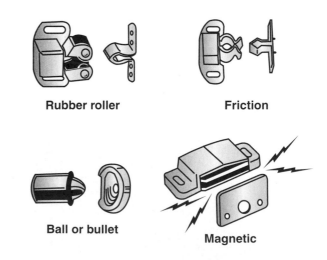

Rubber roller   Friction

Ball or bullet

Magnetic

Figure 20-48. Various mechanisms are used by cabinet catches to hold doors shut.

Pulls: Hardware that provides a means of grasping and opening a door or drawer.

to wardrobes, room dividers, and various built-in units for living rooms and family rooms. Built-in features are popular in other areas of the house for these reasons:

- Increased storage space.

- Efficient organization and use of space.

- Attractive, customized appearance of well-designed and carefully constructed units.

Study the details of the construction provided in the architectural plans or develop a carefully prepared working drawing using the actual dimensions secured from the wall, floor, and ceiling surfaces. When drawings are provided, check the space available to make sure that it agrees with the drawings.

Wall and floor surfaces are seldom perfectly level or plumb. Slight adjustments will be needed as the cabinetwork is constructed. Room corners or the corners of alcoves designed for built-in units will likely not be square. Here, the cabinetwork will need to be trimmed or strips and wedges used to bring the work into proper alignment.

Keep a square and level constantly at hand, especially during the rough-in stages. Do not carry inaccuracies from the wall or floor into the cabinet framework. Doing so will make it difficult to keep the cabinet facing plumb and level. Errors will also be a source of annoyance during the hanging of doors and fitting of drawers.

Built-in units vary in so many ways that no particular procedure can be recommended for their construction. For example, a bookcase that also serves as a room divider could be constructed during the regular interior finishing stages. It could also be built as a detachable unit and installed just before interior painting and decorating.

ON THE JOB

Cabinetmaker

Built-in shelving units, ornate wood paneling, furniture built to order, and custom-made cabinets for kitchens or other rooms are all products produced by the cabinetmaker. While some cabinetmakers are employed by high-end contractors and remodelers, most of those who work in the construction field are self employed. Outside of the construction industry, cabinetmakers are employed by custom woodworking firms building furniture and cabinets to order.

Although a number of large companies produce stock and semicustom kitchen cabinets, their workers are generally considered skilled machine operators and finishers, rather than cabinetmakers. Cabinetmakers are usually responsible for the design and building of custom wood products to meet customer specifications. In some cases, they may execute a design produced by an architect or other person. Skills required of a cabinetmaker, in addition to design ability, include knowledge of wood species and their suitability for cabinet use; mastery of layout, cutting, and joining techniques; the ability to use all types of power and hand tools; familiarity with surface preparation methods and materials; and thorough knowledge of surface finish materials and application techniques.

Working conditions for cabinetmakers are seldom harsh, since they are in a shop or closed-in building site. Job hazards are primarily the chance of injury from using hand or power tools, dust from sanding, and fumes from finishing materials. Carefully using tools and wearing proper personal protective gear minimizes the hazards.

In addition to vocational training in woodworking, persons wishing to enter the cabinetmaking field should take classes in math, design and drafting, and printreading. Most cabinetmakers acquire their skills over a period of years by working under the direction of an experienced cabinetmaker.

Summary

Cabinetwork involves the building of kitchen and bathroom storage units that are located within the building. Built-in units are attached to the structure. Cabinetwork may be built in, custom built, or mass produced. Even when custom-built or mass-produced cabinets are used, the carpenter is usually responsible for their installation. Building plans may include details of built-in cabinets. Cabinet construction may be framed and built of solid lumber; face framed, with a solid frame and sheet material covering; or frameless, with thicker panels that provide strength and rigidity. The 32 mm construction system is a popular standard for frameless construction. Factory-built, or mass-produced, cabinets may be ordered disassembled; assembled, but unfinished; or assembled and finished. Factory-built cabinet materials range from particleboard covered with a printed vinyl film to a combination of solid hardwoods and veneers. Cabinet installation calls for precision to be sure the cabinets are plumb, true, and square. Shims and blocking often must be used. When building cabinets on the job, some carpenters build them in place, piece by piece. Others build the entire cabinet, then move it into place. A master layout is used to ensure uniformity. Doors for built-in cabinets may either be flush or lipped. The lipped door style overlaps the door opening in the cabinet and conceals it. Cabinet doors are also built in either a flush or a lipped style. Sliding doors are sometimes used for cabinets, but are much less common than swinging doors. Countertops are made from a variety of natural and synthetic materials, but plastic laminate over particleboard or waferboard is extensively used. Some carpenters build countertops on site, but most order them from a custom countertop fabricator.

Test Your Knowledge

Answer the following questions on a separate piece of paper. Do not write in this book.

1. List the three types of cabinetwork used in homes.
2. Dimensions of cabinets are usually found in the _____ for the house.
3. The base unit of a standard kitchen cabinet is _____ inches high.
4. In cabinetry using _____ construction, front edges are not face framed.
5. *True or False?* Central to the 32 mm construction system for cabinetry are two vertical rows of 6 mm holes drilled in the side panels.
6. A padded T-brace that supports wall cabinets during installation is known as a _____.
7. A(n) _____ is a full-size sectional view of a cabinet drawn on plywood or cardboard.
8. When building assembled-in-place base cabinets, when are the partitions installed?
9. The vertical members used to face a cabinet are called _____.
10. Three common types of drawer guides include: side guides, corner guides, and _____.
11. Standard shelving that is 3/4″ thick should be carried on supports spaced no greater than _____ inches apart.
12. Three types of hinged doors that may be used in cabinetwork include: lipped, flush, and _____.
13. When are sliding doors typically used?
14. When making a plastic laminate installation, it is recommended that a(n) _____ be used on any unsupported area greater than 4 square feet.
15. Where are drawer pulls usually located in relation to the centerline?

Curricular Connections

Social Studies. Kitchens in American homes have become larger in recent years, and have seen many changes in finishes and materials. One major change has been in the materials used for countertops. While plastic laminate was once almost universal, many kitchens in new construction have countertops fabricated from solid polymers, metals, ceramic materials, and various types of stone. Survey a number of homeowners and determine why they selected the material used for their countertops. If they are planning a change to a different material, ask them to explain why they wish to do so.

Mathematics. Secure a set of architectural plans that include detail drawings of kitchen cabinets. After a study of the details, determine the number of pieces of hardware needed to complete the installation. List the number of hinge pairs, cabinet pulls, and door catches that will be needed. Take your list to a home improvement center and examine the different styles and price ranges of hardware available. Compute the cost for the kitchen using budget-priced hardware, mid-priced hardware, and premium-priced hardware.

Outside Assignments

1. Using a straightedge or straight 2 × 4, check a wall and floor for high spots, plumb, and level. Report the findings to your instructor.

2. If you have access to shop equipment, construct a typical cabinet drawer. Select appropriate joints that can be used for on-the-job built cabinets. Carefully cut and fit the parts. You may prefer not to glue the drawer together so that when you make a presentation to your class it will be easier to show the joints. Another alternative is to glue the drawer, then cut it apart in several places so joint sections could be easily viewed.

3. Study the various methods and devices used to install sinks in countertops. From manufacturer's literature, prepare a section drawing about four times the actual size that you can use in making a presentation to your class.

4. Take careful measurements of an existing kitchen countertop, then make a scaled drawing that a fabricator could follow to make a replacement. Have your instructor check your work.

This bathtub is being given a tile surround. Notice the decorative pattern being used and the cement board under the tile. (National Gypsum)

Painting, Finishing, and Decorating

Learning Objectives

After studying this chapter, you will be able to:

- Cite safety rules that apply to painting and finishing.
- List tools and equipment and demonstrate their use.
- Select proper materials for various painting, finishing, and decorating jobs.
- Prepare exterior and interior surfaces for painting.
- Explain proper procedures for painting, finishing, and wallpaper hanging.

Technical Vocabulary

Alligatoring	Intermediate colors
Binder	Mildew
Blistering	Peeling
Blooming	Pigment
Booking	Plugs
Checking	Pressure-feed spray
Coatings	gun
Complementary colors	Primary colors
Cracking	Putty
Crawling	Related colors
Curtains	Runs
Cutting in	Sags
Excessive chalking	Scaling
Fillers	Sealers
Hue	Secondary colors
Shade	Wallpapering
Specks in surface	Wet edge
Suction-feed spray gun	Wood bleaching
Tint	Wood putty
Value	

Paint is more than an attractive covering—it is also a protective coating. Wood and some other covering materials used inside and outside a building require protection against soiling, rot, and other types of deterioration caused by the environment.

Exterior wood siding and trim, if left unprotected, will deteriorate from exposure to ultraviolet light, moisture, and microscopic organisms. Frequent wetting of wood, even for short periods, breaks down the lignins that bind the cellulose fibers. Then, rain washes away the loose fibers. Unfinished woods also can discolor, shrink, swell, check, and warp.

The term *coatings* is used to describe all types of finishes, whether they are designed for wood or for other materials, such as metals and drywall. Coatings can be clear or colored. They include paints, stains, varnishes, and various

---

Coatings: All types of finishes, including paints, stains, varnishes, and various synthetic materials.

synthetic materials. Anodizing is a process used to color and preserve the surface of aluminum. It is also considered a coating.

Figure 21-1. Respiratory protection is important when operating a paint sprayer. Depending on the material being sprayed, wear a mask or approved respirator.

21.1 Painting and Finishing Tools

A variety of tools may be needed to tackle painting or finishing tasks. These include ladders, scaffolding, compressor and paint sprayer, sanders, brushes, rags, paint roller and pan, paint pads, tack rags, putty knives, and paint guards.

21.1.1 Brushes

Brushes are sold in many sizes and grades and are designed for various purposes. The most common brushes are shown in **Figure 21-2.**

A brush has several parts, as shown in **Figure 21-3.** The bristles used in brushes are either synthetic fibers or natural fibers (animal hair). Natural-fiber brushes are traditionally made from the hair of hogs, especially Chinese hogs, which grow long hair. Hog bristles are

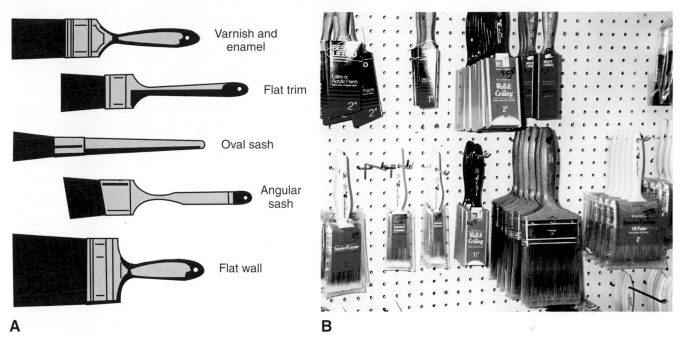

Figure 21-2. Brushes. A—Most painting or finishing tasks can be accomplished with these five kinds of brushes. B—Good painting and decorating stores carry a wide variety of brushes for all purposes. (Dutcher Glass and Paint)

oval in cross section and have naturally tapered and flagged (split) ends. These flagged tips provide paint-holding ability through capillary action. Other natural fibers used for brushes include hair from horses, oxen, and fitch (a type of weasel) and tampico, which is derived from cactus.

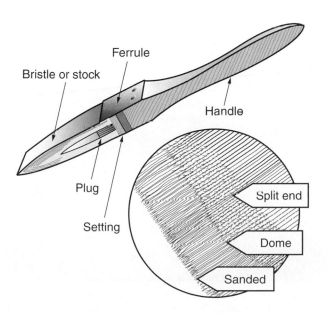

Figure 21-3. Parts of a brush. The inset shows the tip of a quality brush, pointing out its features.

The synthetic fibers of nylon or polyester are manufactured from petroleum products. They are widely used for certain types of brushes. Nylon and polyester bristles are flagged during manufacture. These bristles will outlast natural bristles. They are especially suited for water-based finishes.

Quality brushes have a smooth taper from the ferrule to the tip. These are formed by varying the length of the bristles. Inside the brush, completely surrounded by the bristles, is a plug made from wood, metal, fiber, or plastic. It creates a reservoir to hold the paint.

Wire brushes are useful for cleaning plaster off of subflooring and removing rust from metal, **Figure 21-4.** They are also used to remove old paint that is peeling or flaking.

21.1.2 Rollers, Pans, and Pads

Paint rollers are sold in various widths. They are designed to hold tube-shaped pads that are available in many different types, naps, and sizes. **Figure 21-5** shows various rollers, roller covers, and pads. Rollers are ideal for painting

Figure 21-4. Wire brushes are useful for removing rust and other unwanted foreign materials from a surface to be painted.

large surfaces and can be used to apply many types of interior and exterior finishes. These include oil-base paint, water-base paints, floor and deck paints, masonry paint, and aluminum coatings. For smooth surfaces, short-nap roller covers are available. The rougher the surface to be painted, the longer the nap should be.

Painting mitts, with a thick, towel-like surface, are made for applying paint to unusually rough or odd-shaped surfaces. Pad painters consist of a rectangular frame with a hinged handle for flexibility. Disposable pads with naps similar to paint rollers snap into place on the frame. The coating is wiped onto the surface with a motion similar to that used with a large paintbrush. Pad painters are often used for the wider types of exterior siding. Also available are various trim-painting devices that are variations on the roller and pad painter.

A

B

Figure 21-5. Paint rollers. A—Rollers and roller covers are made in different widths. Covers are designed in different naps for smooth to rough surfaces. B—Pads and mitts are also used for paint application.

Pans for roller painting have a sloped bottom with the lowest part designed to hold a supply of paint, **Figure 21-6.** The pan has a textured sloping surface for rolling off excess paint from the roller.

21.1.3 Mechanical Spraying Equipment

Nearly all types of paint, stain, and varnish can be applied with a compressor and spray gun, **Figure 21-7.** It is the quickest and most practical method for painting very large surfaces.

Paint spray guns are either suction type or pressure type. The suction type works like an atomizer, **Figure 21-8.** It depends on air passing over a tube extending into the paint. This action draws the paint up into the air stream, where it is broken up into fine particles. A *suction-feed spray gun* is designed for applying light-bodied materials such as shellacs, stains, varnishes,

A

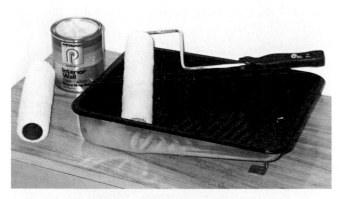

Figure 21-6. A roller pan is designed for even spreading of paint on a roller cover. The deepest section holds a supply of paint, while the sloping, upper part provides a surface for removing excess paint. This pan has a disposable plastic insert for easy cleanup.

B

Figure 21-7. Equipment needed for paint spraying. A—The spray gun uses compressed air to apply paint to large surfaces. B—A compressor may be powered by a gasoline engine, like this unit, or an electric motor.

Suction-feed spray gun: Spray gun designed for applying light-bodied materials such as shellacs, stains, varnishes, lacquers, and synthetic enamels.

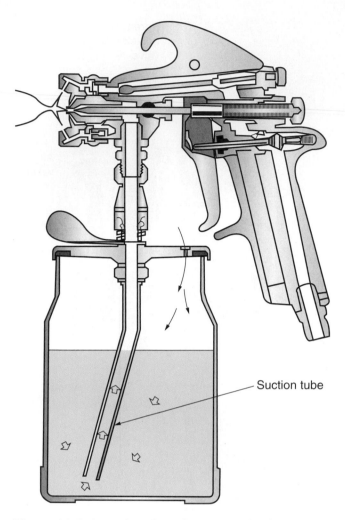

Figure 21-8. A section view of an external-mix, suction-feed spray gun. (DeVilbiss Co.)

lacquers, and synthetic enamels. A ***pressure-feed spray gun*** diverts some air into the paint container. The pressure created forces finishing material into the air stream. Pressure-feed guns are intended for spraying heavy-bodied paints.

Spray guns may have either an internal-mix or external-mix spraying head, **Figure 21-9.** The external-mix head is used to apply fast-drying materials and provides greater control of the spray pattern. The internal-mix head is designed for slow-drying materials and is best used with low air pressure.

> **Working Knowledge**
>
> Great care must be taken to clean painting tools after each use. Paint that is allowed to dry in a brush, roller, or spray gun may be difficult or even impossible to remove. Of course, such a tool is useless unless the paint can be removed.

21.1.4 Ladders and Scaffolds

Both ladders and scaffolds are widely used in painting and decorating. They provide a safe method of reaching high places with painting

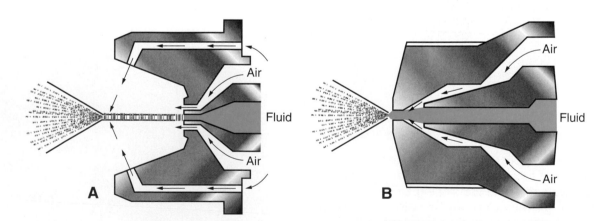

Figure 21-9. The spray head, or nozzle, consists of an air cap and a fluid tip. There are two basic types of heads. A—External-mix spray head. B—Internal-mix spray head.

Pressure-feed spray gun: Spray gun that diverts some air pressure into the paint container to force

the finishing material into the air stream. This gun is intended for spraying heavy-bodied paints.

tools. OSHA rules regarding safety in high places may apply to painting. Review the information in Chapter 31, **Scaffolds and Ladders.**

21.1.5 Paint Removal Tools

An electric sander speeds the work of removing old paint, foreign materials, or roughness from a surface. A belt sander or a disc sander are useful for removing material rapidly, **Figure 21-10A.** Orbital and vibrating sanders are best for fine smoothing, **Figure 21-10B.** New sander shapes and sizes make most sanding locations accessible. Hand sanders are available, as well.

A

B

Figure 21-10. Electric sanders. A—This disc sander is designed to remove paint from flat surfaces. B—Orbital sanders are designed for fine sanding to produce a very smooth surface for finishing. (Porter-Cable)

A heat gun, like the one shown in **Figure 21-11,** can be used to soften paint for removal with a hand scraper or putty knife. It is a much safer method for paint removal than the use of an open flame from a hand-held propane torch. The heat gun can also be used to soften old putty before reglazing windows.

21.2 Painting, Finishing, and Decorating Materials

There are various materials for painting, varnishing, and paperhanging. These include paints, vehicles or binders, solvents and thinners, varnishes, stains, fillers, sealers, bleaches, pumice, sandpaper, steel wool, wallpaper, and wallpaper paste.

Figure 21-11. A heat gun may be used to remove old paint. An open flame should *never* be used for paint removal. (Bosch)

21.2.1 Paints

Paints, varnishes, and stains are adhesive coatings, in a manner of speaking, **Figure 21-12.** They are spread in a thin, unbroken film that is designed to protect and beautify the surface. The term *paint* was once reserved for oil-base paints. However, it now includes coatings based on various resins such as acrylic latex. These are often referred to as water-base paints because they can be thinned or cleaned up with water.

The principal ingredients of paint are the *pigment,* which provides color and opacity, and the vehicle. *Vehicles* are oils or resins that make the paint a fluid and allow it to be applied to, and spread over, a surface. The vehicle also includes a **binder** that remains behind after the oils or resins have dried. The binder then holds the particles of pigment together.

21.2.2 Varnishes

Varnishes, lacquers, and sealers are clear coatings applied over wood. They not only protect the wood surface, but enhance the beauty of the grain and the wood color. Sealers are also used as an undercoat to keep stains and other clear coatings from being absorbed too deeply into porous wood. They also "tie down" stains and fillers that have been applied. Shellac is often used as a sealer to improve adhesion of topcoats and prevent bleeding of certain stains through topcoats.

Fillers are heavy-bodied liquids used to fill the pores of open-grain woods before application of stains and topcoats. *Putty* is a plastic, dough-like material used to fill holes and large depressions in wood surfaces. It is also used to fill nail holes before painting.

21.2.3 Stains

Many stains are designed for interior wood finishing, but stains for exterior uses on siding, decks, and fences are also popular. Interior

A

B

Figure 21-12. Paint stores are usually well stocked with paints and aids that help buyers in selection of colors. A—Overall view of paint stock. B—Close-up view showing labeling that helps the user make a product selection.

Pigment: Ingredient of paint that provides color and opacity.

Binder: Ingredient of paint that remains behind after the oils or resins have dried.

Fillers: Heavy-bodied liquids used to fill open grain in wood before the application of stains or topcoats.

Putty: A plastic, dough-like material used to fill holes and large depressions in wood surfaces. Also used to fill nail holes before painting.

stains are grouped as spirit stains, oil stains, and water stains. Spirit stains rapidly set up and dry because their base material is either alcohol or acetone. Their use is generally limited to spray applications. Oil stains are either penetrating or pigmented. Penetrating stains are brushed on and the excess is removed with a dry cloth. Pigmented stains are designed to be applied with either a brush or a cloth pad, **Figure 21-13.** For deeper tones, the stain is brushed on and allowed to dry without wiping. Water stains are a mixture of powders and water.

Exterior stains are oil base and classified by the amount of pigment they contain. Semi-transparent stains are lightly pigmented. They color the wood, but allow the grain to show through. Solid stains are much like paint, with a heavy pigmentation that provides an opaque color finish. Exterior stains designed for use on decks and porches have a high abrasion resistance.

Figure 21-13. Stains can be applied with a cloth, pad, or brush. (Deft, Inc.)

21.3 Color Selection

Color selection is important when working with paints. A color wheel is useful in selecting colors that go well together, **Figure 21-14.** There are a few terms that are important to understand when working with color schemes.

- *Hue*—The class of a particular color, such as red or blue.
- *Primary colors*—Red, blue, and yellow.
- *Secondary colors*—Violet, green, and orange.
- *Intermediate colors*—Red-violet, blue-violet, blue-green, yellow-orange, red-orange, and yellow-green.
- *Related colors*—Hues that are side-by-side on the color wheel.
- *Complementary colors*—Hues that are opposite each other on the color wheel.
- *Value*—The lightness or darkness of a color.
- *Tint*—Color nearer white in value (pure color with white added).
- *Shade*—Color nearer black in value (pure color with black added).

Paints are available in factory-mixed colors or as bases to which color may be added at the paint store. For a custom-mixed color, the customer can select a color from an array of paint chips, **Figure 21-15.** Paint manufacturers supply instructions and proportions of coloring agents needed so the store can exactly match the chip. Many stores today also have color analyzers that allow them to mix paint to perfectly match a color sample from a fabric, picture, or other source.

Hue: The class of a color.

Primary colors: The three colors (red, blue, and yellow) that can be combined to form all other colors.

Secondary colors: A color (violet, green, and orange) that is produced when two primary colors are combined.

Intermediate colors: A color combination of a primary color and a secondary color already containing that primary color. Intermediate colors are red-violet, blue-violet, blue-green, yellow-orange, red-orange.

Related colors: Hues that are side-by-side on the color wheel.

Complementary colors: Hues opposite of each other on the color wheel.

Value: The lightness or darkness of a color.

Tint: Color nearer white in value (pure color with white added).

Shade: Color nearer black in value (pure color with black added).

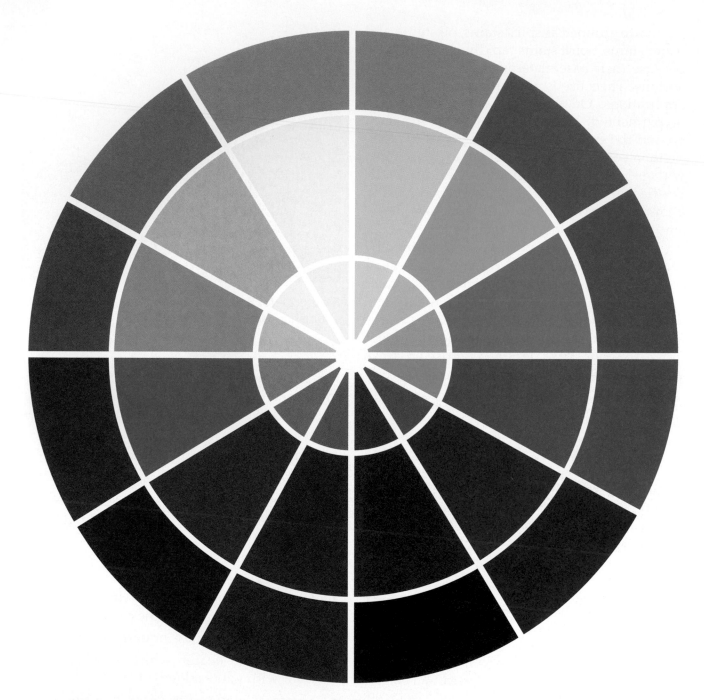

Figure 21-14. A color wheel may be used to assist in picking color combinations that are pleasing and complementary.

21.4 Preparing Surfaces for Coating

A good paint or varnishing job is not possible without proper preparation of the surface. On new work, all that is necessary may be removal of any foreign matter, such as plaster, grease, dirt, and pitch, followed by a light sanding to get a smooth surface.

Sandpaper is used to smooth surfaces prior to painting or varnishing. Steel wool and pumice are used to smooth and enhance the surface of clear coatings once they have dried. Sandpaper may be used between coats to provide a "tooth" for better adhesion.

Figure 21-15. Chips representing thousands of different colors assist the buyer in selecting a paint color. Information is included on the chip to aid in correctly mixing the color. (Dutcher Glass and Paint)

Figure 21-16. Removing dry, brittle glazing compound from windows can be difficult and cause damage to the muntins. Using a heat gun will soften the compound so it can be easily lifted off.

apply a compatible drywall compound. Sand and prime repaired areas before repainting.

Renewal or recoating of old exterior finishes requires much more work. Remove loose paint by scraping, wire brushing, and sanding. Edges around bare spots must be feathered. Work with the grain while sanding. Sometimes, complete removal of the paint may be necessary because of extensive blistering, cracking, and alligatoring of the film. Fill cracks and nail holes with putty and then sand smooth. Before painting window frames, check the old finish and glazing. A heat gun will soften dried putty, making it easy to remove, **Figure 21-16.**

Coat all bare spots with primer before applying the finish coating. On extremely weathered surfaces, sand the surface down to "bright" wood. Dry, weathered wood will quickly draw the vehicle out of the paint, leaving a painted surface that quickly fails. To remedy this problem, many painters apply a 50/50 mixture of turpentine (or mineral spirits) and linseed oil. This is very important on heavily weathered surfaces if the new paint job is to last even one season. Allow this coating 48 hours of drying time before priming.

Before painting interior walls, repair any cracks in plaster or drywall that persist or reopen. Cover the crack with fiberglass tape and

21.5 Choosing Paint Applicators

Selecting an applicator for painting is important. The applicator must be well adapted to the surface and the coating being used. You should be familiar with the applicator and how to use it. Brushes have long been the primary applicator for painting, for a number of reasons:

- They work well on small, detailed surfaces. Control is accurate—you can put the paint exactly where needed. This reduces the need for masking.

- Since brushes do not produce overspray or spattering, there is less need to protect surrounding areas.

- Brushes produce better results on textured surfaces. This is important on open-grain, rough-sawn, or brushed siding. When rollers or sprayers are used on such surfaces, they are apt to leave voids.

- Brushes work well in conjunction with rollers. The brush is used on the detail work; the roller coats the larger surfaces.

21.6 Using Proper Brushing Technique

Gripping the brush the right way helps you do more accurate work. At the same time, a proper grip is less tiring. See **Figure 21-17.**

The pencil grip is usually best for small brushes used on detail work. This grip would be used, for example, to paint muntins in a window. The fingers wrap around the metal ferrule, giving a full range of motion.

The pinch grip gives good control of the brush. It is most comfortable for painting walls, siding, and trim. All of the fingers will fit on the ferrule of a 3″ or 5″ brush. Do not grip the brush too tightly, or your hand may cramp from fatigue.

The wrap grip is preferred for overhead painting. In this grasp, the fingers and thumb wrap around the handle.

A brush is handy for cutting in the edges of boards. Several edges can be cut in at once, so long as the edges do not dry before the face of boards is painted.

21.7 Roller and Pad Application

Rollers are useful for coating large areas. They rapidly coat interior walls and exterior siding. Other advantages include:

- The roller's nap provides a uniform texture in the coating.
- Rollers are less expensive than good-quality brushes.
- By using an extension handle, the painter often can paint ceilings and walls without using ladders or scaffolds.
- Preparation and cleanup of rollers takes less time than brushes.

When painting siding, topcoats of a water-base exterior finish can be applied over an oil-base primer using a roller or paint pad. If using a roller, paint across one board at a time, working from top to bottom of the wall.

Paint pads have a napped fabric over a foam base. The pad is secured in a plastic holder. Special pans used with the pads hold about

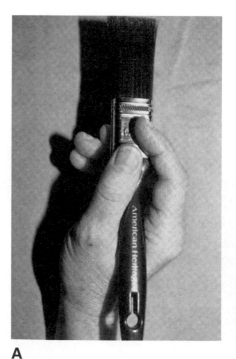

A

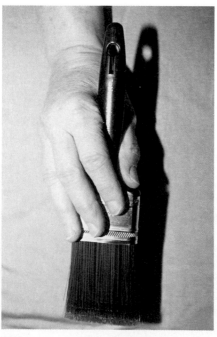

B

C

Figure 21-17. Proper methods of holding a brush. A—Pencil grip is used with small brushes for control of paint on small areas. B—Painters prefer the pinch grip on vertical surfaces. C—Wrap grip is less tiring when used on overhead surfaces.

1/2 gallon of paint. To fill the pad, the pan is tipped to flood a grid area with paint. The pad is pressed on the grid to load it. Three or four boards can be completed to a stopping point before moving to a new area.

21.8 Spray Painting

Exterior spray painting is sometimes preferred to painting with brushes, rollers, or pads. It is most economical where there are large surfaces to be painted in one color. Areas that are to receive no paint must be carefully masked, as shown in **Figure 21-18.**

Proper mixing of paints is important for all types of paint application. Some painters *box* the paint by intermixing different containers of paint, **Figure 21-19.** This procedure assures that the paint will be uniform in color and texture.

> ### Safety Note
>
> Safety precautions must be observed when spray painting to avoid the hazards of explosions, fire, and personal injury. Spraying should be done in a well-ventilated area. Flammable liquids should be kept in safety containers. Fumes and fine spray are toxic—always wear a respirator to avoid inhaling them. Wear protective clothing to keep paint spray off your skin. Always have a fire extinguisher handy.

21.9 Painting Wood Exteriors

Before painting new wood, seal all knots and pitch-containing areas with a shellac or knot sealer. Previously painted surfaces may have to be scraped or sanded to remove loose paint. Repair any imperfections and apply putty to holes.

The first coat of finish should be a pigmented primer or a sealer. Allow this coat to thoroughly dry before applying finish coats. Top coatings

may be exterior oil paints, oleoresinous finishes, alkyd coatings, or water-base paints. The thickness and smoothness of a paint coat is controlled, to an extent, by the amount of paint carried in the brush. Do not dip the brush more than half of the length of the bristles.

Oil paints are applied in a way slightly different from other coatings. The paint is daubed on in spots, then spread out with full-arm level strokes. Each brush load should be

Figure 21-18. When using a sprayer to apply paint, it is important to mask off surfaces not being painted, such as the foundation.

Figure 21-19. When painting large surfaces, such as the outside of a house, it is important to box (intermix) the paint so that color will be uniform throughout the paint job. (Greco Painting)

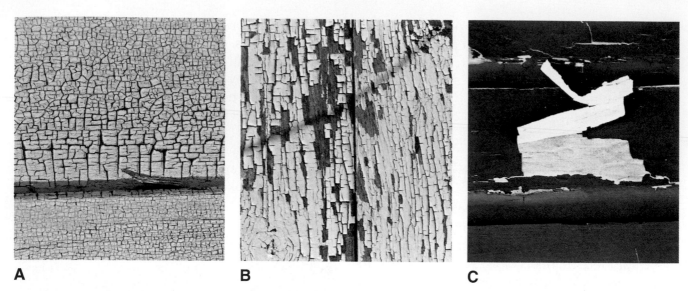

A   **B**   **C**

Figure 21-20. Various types of paint failure. A great deal of prep work will be required before these surfaces can be repainted. A—Checking. B—Advanced alligatoring. C—Blistering and peeling.

brushed out with sweeping strokes. Use the tip for brushing and gradually lift the brush at the end of each stroke. This will give a thin feather edge at the end of each stroke.

Alkyds and water-base paints require less brushing and rapidly dry. Thus, there is less time for brushing out than with oil paints. If too-rapid drying becomes a problem, inhibitors may be added; different types are used for latex paint, oil, and alkyd paints.

It is important to always paint to a *wet edge.* The paint band should be narrow enough so that the next band can be completed before the first has dried. This eliminates lap marks. Start painting at the highest point and move downward in horizontal bands. Never stop painting partway across a band. Always end a day's work at a window, door, edge of a board, or a corner.

Working Knowledge

The lower edges of clapboard siding should be painted first. Rough surfaces require additional brushing in all directions. This will assure that pores and crevices are well covered with paint. Poking at the surface with the brush tip will eventually destroy the flags on the bristle ends. If poking is necessary, use an old brush.

21.10 Problems with Coatings

Painters often experience problems with previously painted exteriors. When repainting previously painted exteriors, it is helpful to recognize certain defects in the old coatings. **Figure 21-20** illustrates some of these defects:

- *Alligatoring* and *checking*—Alligatoring is a severe case of checking. It is the result of applying relatively hard finishing coats over primers that are softer. To avoid this situation, apply progressively more flexible coats from primer to finishing coats. Allow more drying time between coats and follow the manufacturer's recommendations for thinning.

- *Cracking* and *scaling*—These conditions are usually the result of using paint lacking

Wet edge: A paint band narrow enough so that the next band can be completed before the first has dried, eliminating lap marks.

Alligatoring and checking: Paint defects resulting from application of relatively hard coatings over primers that are softer.

Cracking and scaling: Paint defects resulting from application of paint lacking elasticity.

in elasticity. The coating becomes hard and brittle and unable to contract and expand with the wood. The conditions are aggravated by too-thick of an application. To avoid this condition, use high quality paint and avoid building up thick layers.

- *Excessive chalking*—Oils in the finish coat render a glossy finish. Ultraviolet rays cause the oils to break down, leaving loosely bound pigment on the surface. Some chalking is desirable to keep the paint surface clean. Heavy chalking, however, wears away the paint and leaves the surface unprotected. Using poor-quality paint is the major cause of excessive chalking. Before repainting, remove the chalked paint by scrubbing or wire brushing.

- *Blistering* and *peeling*—Blistering always precedes peeling. This condition is due to the pressure of water vapor under the coating pushing the paint film from the substrate. The only cure is to eliminate the source of the moisture. Moisture barriers, such as aluminum paint or Mylar film, will stop the migration of moisture-laden air from the interior to the exterior. Repair of external leaks also may be necessary. Install proper flashing and caulk joints and cracks. Make sure grading is correct to eliminate ground

moisture coming in contact with siding. Proper ventilation of attics and crawl spaces will also reduce moisture buildup. Finally, allow surfaces to thoroughly dry before painting.

- *Mildew*—This is a form of plant life. It flourishes on soft paints. Eliminate mildew by washing the affected surface with a mild alkali such as washing soda, trisodium phosphate (TSP), or sodium metasilicate.

Certain defects in varnishes and other clear finishes are the result of improper application methods or application under unfavorable conditions.

- *Crawling*—The finish gathers up into small globules and does not cover the surface. This may be caused by applying the finish over a wet, cold, waxy, or greasy surface. Other causes are mixing of different kinds of varnish, using too much drier, or thinning with naphtha instead of turpentine or other approved thinner.

- *Blistering*—As with paints, blistering of clear finishes is caused by moisture vapor from below the surface.

- *Blooming* (clouding)—This condition may be caused by high humidity in the room where finishing is done. Another cause is the use of water and pumice instead of oil and pumice for rubbing the finish.

- *Runs, curtains,* and *sags*—Conditions caused by too-heavy of an application.

- *Specks in surface*—Results from hardened particles in the container of finish, dirty brushes, improperly prepared surfaces, or airborne dust in the room.

Excessive chalking: Loosely-bound pigment on the surface caused by ultraviolet rays breaking down oils in the finish coat.

Blistering and peeling: Paint defect caused by water collecting under the paint film.

Mildew: A fungus that attacks materials, especially in the presence of moisture. It flourishes on soft paints.

Crawling: Coating problem in which the finish gathers up into small globules and does not cover the surface.

Blooming: Coating defect that may be caused by high humidity in the room where finishing is done. Another cause is the use of water and pumice instead of oil and pumice for rubbing the finish. Also called *clouding.*

Runs, curtains, sags: Coating problems caused by too-heavy of an application.

Specks in surface: Coating problem resulting from hardened particles in the container of finish, dirty brushes, improperly prepared surfaces, or airborne dust in the room.

21.11 Interior Painting

Surfaces must be clean. Normally, a thorough dusting is enough. However, glossy surfaces or surfaces soiled with oily dirt will need to be washed with a strong washing compound such as trisodium phosphate to degrease and dull the surface. Rinse with clean, hot water. Pay particular attention to surfaces surrounding sinks and stoves, as well as frequently touched surfaces around switches, doorknobs, handrails, and the edges of doors. There are also special chemical deglossers that will prepare glossy painted surfaces for recoating. These are especially effective over latex paint.

Cover surfaces not being painted to protect them from paint drips or spatters. Spread drop cloths over floors and furniture. Mask off windows, trim, and edges that are not to be painted. Standard masking tape is available in different widths. Some tape is designed to cleanly peel off after being attached a week or more.

Remove all items attached to walls and trim that are not to be painted. This includes wall hangings, pictures, switch and receptacle plates, and hardware.

Assemble your tools. Thoroughly mix paints, even if they have recently been agitated on a paint store vibrator. Using a paddle or a power mixer, rapidly stir to suspend settled pigments.

When painting an entire room:
1. Paint the ceiling.
2. Paint the walls.
3. Paint the wood trim and doors.

Either rollers or brushes may be used; most painters prefer the roller because it allows the painting to progress more quickly.

Paint large surfaces in narrow strips so that you are always painting against a wet edge. Dry edges that are overlapped will show up as lap marks or have a different sheen when the paint is dry.

To paint into the corners formed by the ceiling and wall, a process known as *cutting in*, a small brush is necessary. Cutting in provides a narrow painted strip of paint at the ceiling line, around doors and windows, and along baseboards. These edges cannot be reached with the roller without depositing paint where it is not wanted.

If painting with a brush, dip the bristles into the paint to one-third of their length. Tap the brush gently against the side or edge of the paint container to release paint that might otherwise drip. Some painters make several small nail holes in the rim of the paint container. This allows paint collected there to drain back into the container. Paint stores have available a grid of expanded metal that hangs inside a five-gallon container for striking off excess paint from brush or roller. See **Figure 21-21**.

When loading a roller, do it carefully to avoid dripping paint. If using a tray, pour a small amount of paint into the deep end. Dip the roller into the paint and work it into the nap by moving the roller back and forth on the slanted area of the tray while applying light pressure. This removes excess paint.

Start rolling in one corner of the wall or ceiling, painting a narrow strip. Overlap each stroke, working first in one direction and then in another, until the surface is covered. Gently work the roller avoiding quick, choppy strokes that cause paint spatters. Continue rolling until the newly painted section is covered and the roller needs reloading. **Figure 21-22** shows proper rolling and brushing techniques. Unless you are repainting with the same color, applying a second coat makes the end result look more professional than applying one coat.

Trim painting should be done only after walls are thoroughly dry. A 1 1/2" brush is suitable for windows; use a 2" brush for other trim.

Figure 21-21. A grid of expanded aluminum can be used in a five-gallon can to remove excess paint from a roller or brush.

Cutting in: Painting into the inside corners using a brush.

A

B

C

Figure 21-22. Roller and brush technique. A—Cutting in the edge between ceiling and wall. B—Painting with a roller. C—Rolling paint onto a wardrobe shelf.

To avoid drips and runs, be careful not to overload the brush.

On multipaned windows, start painting on center mullions first, working outward in either direction. Trim should be painted last. Use masking tape to avoid spreading paint onto the window glass. Accidental spatters on the glass, however, can be removed with a razor blade when the paint has dried.

Paint doorjambs and casing before painting the door. To avoid laps and brush marks when painting paneled doors, paint the molded edges of the panels first, starting with the top panel. Then, fill in the panels using brush strokes with the grain. Next, paint the rails (horizontal cross boards). Finish by painting the stiles (vertical side boards). Refer to **Figure 21-23.** If the door swings away from the side being painted, paint the hinge edge the same color. If it swings toward the side being painted, paint the lock edge the same color. If both sides of the door are to have the same color, all edges should be painted.

Some painters prefer to finish multipaneled doors with a roller. First the recessed panels and decorative edges are brushed. Then stiles and rails are rolled.

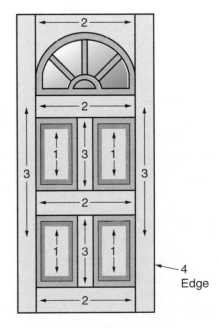

Figure 21-23. Efficient method of painting a paneled door. Paint the panels first, then rails, then stiles, and finally, edges. Arrows indicate the direction of the brushing.

Paint the baseboards last. A plastic, metal, or cardboard guard should be used to protect the finished floor or carpeting when painting the lower edge of the baseboard. As an alternative, painter's gummed masking paper can be used to protect hard-surfaced finish flooring.

Working Knowledge

When painting a ceiling, an extension handle attached to the roller will be useful and easier than climbing up and down a stepladder or a platform.

21.12 Working with Stains and Clear Finishes

Items to be finished should be smooth, free of dust, dirt, glue, and grease. Before applying stains or finishes, remove all hardware. Before staining and finishing a wood surface, remove or repair all surface dents. Minor defects such as machine marks, small scratches, and excess glue can be removed with light sanding. More serious conditions, such as dents and holes, require repairs.

Small dents can generally be removed by applying a drop of warm water to the spot. This causes the grain to swell and rise to or above the surface. Let it dry and then sand smooth.

Lift deeper dents with steam, using a damp cloth and a hot iron, **Figure 21-24.** Place the damp cloth over the dent and apply the hot iron to the cloth. Move the iron around to avoid burning the wood. Sand the dried surface smooth.

Holes include open joints, cracks, splits, and gouges. Fillers are needed to repair such defects. Basic fillers include plugs, Plastic wood®, wood putty, stick shellac, and wax sticks. *Plugs* are wooden pieces shaped to fit and forced into the defect. Plastic wood® is a combination of wood powder and a plastic hardener. *Wood putty* is a mixture of wood and adhesive in powder form. Stick shellac is hardened, colored shellac. It

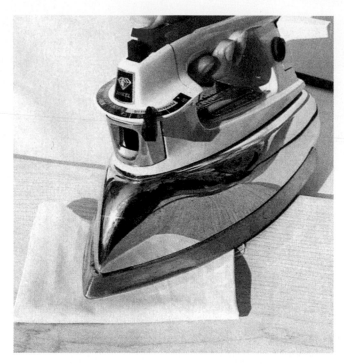

Figure 21-24. Dents may be steamed out using a damp cloth and heat from an iron.

must be applied with a heated knife. Stick wax is a colored wax that closely resembles crayons. Apply the wax by rubbing the stick over the defect. This forces the wax into it.

Wood bleaching removes some of the natural color from the wood. Bleaches can be purchased ready to mix and usually consist of two solutions. Mixing is done in a glass or porcelain container. The solution is applied to the bare wood with a rubber sponge or a cotton rag. Wear gloves and a face shield for this operation. Most bleaches are caustic and, therefore, dangerous.

Surfaces with all defects repaired are ready for staining and final finishing. While not necessary for every surface, stains are used to change the color of natural wood. When used on

---

Plugs: Wooden pieces shaped to fit and forced into a defect.

Wood putty: A mixture of wood and adhesive in powder form.

Wood bleaching: Applying bleach to remove some of the natural color from the wood.

exterior surfaces, stains have a protective quality. They can be applied by wiping, brushing, or spraying, **Figure 21-25.** Wipe off excess, if any, and allow the stain to dry for 24 hours. Then, coat it with a sealer. The sealer "ties down" the stain and filler. Some fast-drying sealers are ready for final finishes in about 30 minutes. Follow directions on the container.

Clear finishes are brushed or sprayed, **Figure 21-26.** Brushes should be the finest available. Flow the finish on rather than brushing it out as with paint. Be sure not to leave *holidays* (bare spots). Do not apply too heavy of a coat or sags and drips will mar the finish.

Smooth the dried finish coats with fine steel wool and treat them to a rubbing with wax. Allow at least 48 hours of drying time before attempting to rub out the final coat.

Figure 21-25. Stain, varnish, and clear finishes can be applied to trim before it is installed.

PROCEDURE

Estimating paint needed

1. Measure the height and width of the surface to be covered. Convert inches to feet, dividing the inches by 12. Carrying decimals two places to the right of the decimal point provides enough accuracy.
2. Multiply the two measurements to find total area in square feet.
3. Find the dimensions of windows, doorways, and any other spaces that do not require paint. Multiply height by width.
4. Deduct the resulting area from the area calculated in Step 2.
5. Check the label on the paint container to see how many square feet each container will cover.
6. Divide that number into the area to be covered. The resulting number indicates the number of containers required. If two coats are needed, double the number of containers.

Figure 21-26. With proper masking, stains and finishes can be applied after trim has been installed.

Wallpapering: The process of hanging wallcoverings.

21.13 Wallcoverings

Hanging wallcoverings, or *wallpapering,* is an important phase of decorating. Doing professional-quality work demands a thorough knowledge of materials and procedures.

Most new wallcoverings are made from vinyl resins. The vinyl is made in a continuous, flexible film and applied to a paper or fabric backing. Such materials have the advantage of being long lasting and washable. Moreover, they can be made highly decorative. Traditional wallcoverings are made from cotton, linen, hemp, wood, and waste paper. More expensive coverings include the grasscloths, flocks, and foil-based materials.

A more exotic type of wall hanging is embossed wallpaper simulating styles popular in the Victorian era. One of the embossed products, Lincrusta, is a rigid product made from a linseed oil mixture with a deep relief. Anaglypta is a similar, lightweight embossed material made from cotton pulp. These materials are used by restorers or by interior decorators seeking a period look.

Wallcoverings are measured by the roll. There are two standards for dimensions. In the American standard, a single roll covers about 36 square feet. Single rolls in the European system contain from 27 to 29 square feet. With trimming, about 21 to 23 sq. ft. are available in actual coverage.

For convenience sake, the coverings usually come in bolts containing two or three single rolls. Most are 20", 21 1/2", or 22" wide. English-made papers are generally 22" wide and rolls are 36' long. French papers are generally 18" wide and 27' long in single rolls. Other foreign coverings vary in length and width. The amount in a single roll is only an approximation.

Working Knowledge

Always keep a record of wallcovering purchases or retain a label with the collection, pattern number, and dye lot or printing run number. This information is important if you need to reorder to finish a job.

21.13.1 Wallcovering Estimate Considerations

Estimating wallcovering needed for a particular job requires careful calculation. It is wise to order more material than you think you will need. Consider these points:

- A great deal of waste occurs in matching patterns and in cutting around openings.
- Stopping work to order more rolls from the supplier is frustrating and wastes time.
- The new rolls may not be from the same color run.
- Most suppliers will take back unopened rolls. There might be a restocking charge, but it will be cheaper than the probable cost of running short.

There are two methods of estimating the quantity of wallpaper needed for a room. The *quick method* is less accurate and will work only for rectangular rooms with 8' ceilings. The *area method* of estimating is more accurate.

Paperhangers require a number of hand tools, a paste table, and an expandable plank. Tools include shears for cutting paper, several sizes of rollers, smoothing and paste brushes, several knives, a razor blade and holder, a broad knife, a 6' or 7' straightedge, plumb bob, and a chalk line and chalk. See **Figure 21-27.**

PROCEDURE

Estimating wallcovering—quick method

1. Add the width of two adjacent walls.
2. Multiply by .5 for the number of American standard rolls.
3. Multiply by .6 for European standard rolls.
4. Since the method takes no account for waste, deduct nothing for one door and one window unless they are larger than normal. For more openings or very large openings, subtract 1/2 roll.

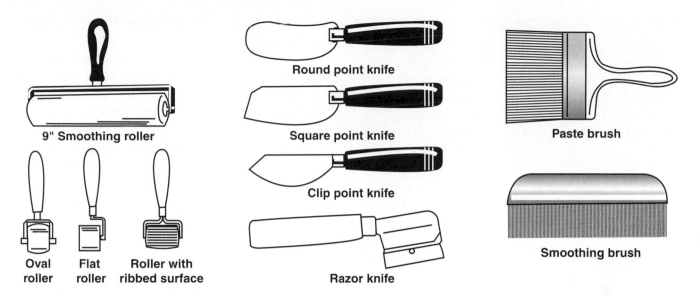

Figure 21-27. These tools are recommended for wallpapering.

PROCEDURE

Estimating wallcovering—area method

1. Take measurements of the four walls. Round off inches to the next larger foot or half-foot.
2. Make a sketch of the room showing doors, windows, fireplace, built-in shelving, and so on. Write down the measurements of wall lengths and heights, excluding baseboards and moldings. Subtract areas that will not be covered.
3. Take this information to your supplier or use the chart in Figure 21-28.

21.13.2 Preparing Walls for Covering

Allow newly erected walls, whether plastered or drywalled, to thoroughly dry before covering. Treat plastered walls with a solution that neutralizes the lime and alkali in the plaster. *Sizing* is a glue solution that is brushed or rolled onto a newly plastered or drywalled surface to seal the pores and prevent the paste from being absorbed into the wall.

Old wallpaper is not a good foundation for new paper. Take the time to remove it. Some old coverings are "strippable" and can be easily removed. Cloth-backed vinyl coverings are usually strippable. However, much depends on whether the right adhesive was used and whether proper surface preparation was done. Use a putty knife to lift a bottom corner of the strip. Lift upward at an angle nearly parallel to the wall. Other methods of stripping old coverings include:

- **Steaming**—This requires a machine to produce a heated vapor that penetrates the covering to soften the adhesive. Scoring the old covering helps the steam get to the adhesive quicker. Special tools are available for scoring.
- **Soaking**—In this method, use a sponge or a garden sprayer to apply water. As in the steaming method, scoring makes it easier for the water to penetrate. On nonbreathable coverings like foil, vinyl, vinyl backed, or cloth-backed vinyl, scoring before soaking is necessary for penetration to the adhesive.
- **Peeling**—Some coverings have a vinyl outer layer that can be pulled away. This exposes a paper backing that is easily removed by soaking or another method.
- **Chemical strippers**—There are three types: wetting agents, solvent removers, and enzymes. These degrade the starch in the adhesive.

Wall, Room & Border Estimating Chart For English/Metric Single Rolls					
Single rolls for wall area(s)					
Distance of wall in meters / feet	Ceiling Height				
	8 feet or 2.4 meters	9 feet or 2.7 meters	10 feet or 3.0 meters	11 feet or 3.4 meters	12 feet or 3.7 meters
1.8 6	3	3	3	4	5
2.4 8	3	4	4	5	6
3.0 10	4 to 5	5	5	6	7
3.7 12	5	5	5	7	8
4.3 14	5	5	7	7	8
4.9 16	5 to 7	7	8	8	9
5.5 18	7	8	8	8	10
6.1 20	8	8	9	10	11
6.7 22	8 to 9	9	10	11	12
7.3 24	9	9	10	12	13
7.9 26	9 to 10	10	12	13	14
8.5 28	10	12	13	14	16
9.1 30	10	12	13	15	16
9.8 32	12	13	14	16	18
10.4 34	13	14	15	16	18
11.0 36	14	14	16	18	20
Room size in meters / feet					
2.4 x 2.4 8 x 8	12	13	14	16	18
2.4 x 3.0 8 x 10	14	14	16	18	20
3.0 x 3.0 10 x 10	14	16	18	20	22
3.0 x 3.7 10 x 12	16	18	20	22	24
3.0 x 4.3 10 x 14	18	20	22	24	26
3.7 x 3.7 12 x 12	18	20	22	24	26
3.7 x 4.3 12 x 14	20	22	24	26	28
3.7 x 4.9 12 x 16	20	22	26	28	30
3.7 x 5.5 12 x 18	22	24	28	30	32
3.7 x 6.1 12 x 20	24	26	30	32	34
4.3 x 4.3 14 x 14	20	22	26	28	30
4.3 x 4.9 14 x 16	22	24	28	30	32
4.3 x 5.5 14 x 18	24	26	30	32	34
4.3 x 6.1 14 x 20	24	28	32	34	38
4.3 x 6.7 14 x 22	26	30	32	36	40
4.9 x 4.9 16 x 16	24	26	30	32	34

After finding the number of rolls needed for the wall(s) or room you are hanging, deduct one half-roll for each standard size door or window. Design repeat of the pattern and length of roll may increase your rollage needs. Your dealer can advise you if the wall covering selected will signifcantly affect your rollage needs.

Figure 21-28. This chart may be used to determine how much wallpaper is needed for a room. (National Decorating Products Assn.)

tion run. Coverings from other runs may vary in color and ruin the job.

Wallcoverings are brittle. They must be softened before being applied to walls. On prepasted papers, place the prepared strip in a water tray for a time specified by the manufacturer. Then, place it on the wall.

There are also products that will activate the paste on prepasted papers. These watery products are rolled onto the back of the covering. One gallon will normally cover 10 single rolls.

On papers that are not prepasted, apply the paste. The water it contains will soften the paper. The paste was once made from cooked flour or starch and water, but today it is usually purchased from paint or home improvement stores.

Booking is a method of pasting and folding wallcoverings. This is an important step as it allows the paste to soak into the wallcovering. Allow the booked strip to "relax" for three to five minutes before hanging. Check the manufacturer's recommendations for the exact time.

When using coverings with a design, you must match the pattern of the strips on the paste table. Do this before cutting them. Do not cut the strips too short—allow about 2" excess at both top and bottom. Trim strips to exact length after they are on the wall. Replace the razor blade after trimming each strip.

After hanging every three or four strips, go over the seams with a roller. Keep the roller clean and wipe away adhesive squeezed out of the seams. Do not press too hard—it will cause stretching or shrinking problems later. Never use seam rollers on flocked or embossed wallcoverings.

21.13.3 Preparing the Wallcovering

Before wallcoverings are cut to length, check to make sure that there is enough material to do the job. More will have to be purchased if there is too little to finish the job. Care must be taken that the new covering is from the same produc-

Booking: A method of pasting and folding wallcoverings that allows the paste to soak into the wallcovering.

PROCEDURE

Booking wallcovering strips

1. Turn all cut, unpasted strips face down on the paste table.
2. Paste only one strip at a time. Paste two-thirds of the strip.
3. With clean hands, book the strip. To book, pick up the corners of the pasted end and fold it over only the pasted portion (pasted side in). Do not go beyond the pasted portion and do not crease the fold.
4. Paste and fold the other one-third. Edges must be even. If there is selvage on the covering, trim it at this time.
5. Loosely roll the booked strip and let it sit.

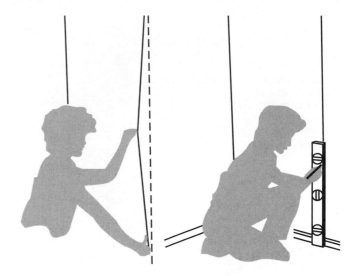

Figure 21-29. A plumb line is being set up using a level and a straightedge. This is usually done near a corner so the first roll of paper wraps around the corner at least 1/2″. A chalk line can be snapped for the plumb line. (National Decorating Products Assn.)

PROCEDURE

Hanging wallcovering

1. Using a plumb bob or a level, make a plumb starting line on the wall, as shown in **Figure 21-29.** This is used as a guide for hanging the first strip. It is better to start the first strip at a corner with a portion of the strip wrapping around the corner. The corner lap should be at least 1/2″ wide.
2. Overlap the top edge of the first strip 2″ onto the ceiling and align the edge with the plumb line.
3. Unfold the top one-third of the strip. Smooth the opposite side against the wall.
4. When the entire top portion is smoothed against the wall, as in **Figure 21-30,** unfold the second portion and align it with the plumb line.
5. Smooth out the strip from the center to the edges, working from the top down. Use brushing strokes in a sequence like that shown in **Figure 21-31.** Use care not to stretch the wet covering. When satisfied that the strip is properly aligned, smooth the entire surface again to remove remaining air bubbles.
6. Trim the ceiling and baseboard ends with a razor knife. Use a wall scraper or a broad knife as a guide. Tightly hold the scraper or broad knife against and parallel to the wall as you trim with the razor knife, **Figure 21-32.** Move the scraper or broad knife along the strip, but keep the razor knife in contact with the wall.
7. Rinse the surface of the strip to remove excess adhesive. Wipe away adhesive from the ceiling and baseboard at the same time. Frequently change the rinse water and use a good-quality, natural or synthetic sponge.
8. Match each succeeding strip to the previous strip. Butt the edge against the previous strip for a tight seam. Brush and smooth in the same way. Be sure to wipe off excess adhesive as before.

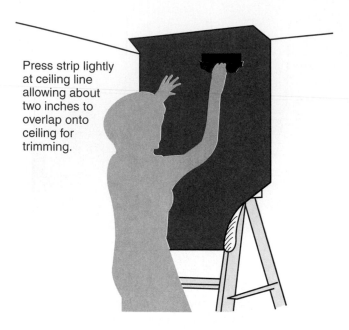

Press strip lightly at ceiling line allowing about two inches to overlap onto ceiling for trimming.

Figure 21-30. Attach the first strip of wallpaper as shown, aligning or centering it on the plumb line. (National Decorating Products Assn.)

21.13.4 Hanging Wallcovering around Openings

Never attempt to precut a strip to fit around a window or door. Apply the paper over the opening and use a razor knife to trim the excess, **Figure 21-33.** Use this method when papering around light switches and receptacles, as well.

Safety Note

Make sure the circuit breaker has been turned off before working around any electrical device.

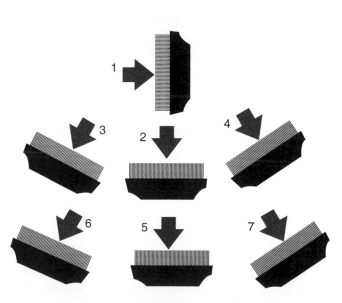

Figure 21-31. Wallcovering should be brushed out in this sequence to remove air pockets and firmly adhere the strips to the wall.

Trim the excess wallcoverings at the ceiling and baseboard.

Figure 21-32. Trim off each strip at the ceiling and baseboard using a broad knife or scraper and a razor knife. (National Decorating Products Assn.)

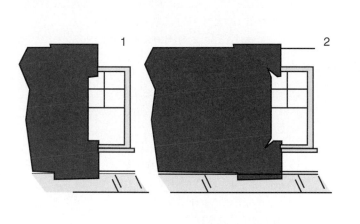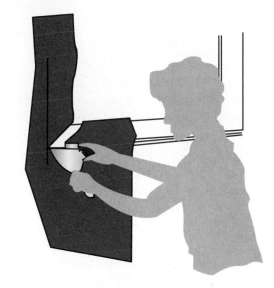

Figure 21-33. Proper procedure for papering around windows and doors. Before making final trim cuts, cut into all corners at a 45° angle. (National Decorating Products Assn.)

ON THE JOB

Painter/Paperhanger

Two out of five (40%) of all painters and paperhangers are self employed—a percentage double that of the construction trades as a group. The nature of the trade lends itself to setting up a contracting business operated by the owner with possibly one or two employees. An approximately equal number of painters and paperhangers are employed by contractors who work on new construction, remodeling, and restoration. The remainder of workers in the trade are employed by building management firms, industrial plants, or government agencies and perform primarily maintenance work.

Painters are responsible for preparing the surface to be painted, selecting the right finishing material and applicator for the application, and applying the finishes. Paperhangers follow a similar set of procedures, although the wall covering material is typically selected by the customer. Both painters and paperhangers often use ladders, scaffolds, or similar devices to work in areas that are sometimes high above the ground. In addition to possible falling hazards, painters and paperhangers may be exposed to toxic or dangerous materials when preparing surfaces or applying wall coverings. Paying close attention to safety rules and wearing proper protective gear is important.

While many painters and paperhangers gain their skills on the job, apprenticeship programs exist and are recommended for continuing success in the field. Apprenticeship programs generally include 2–4 years of on-the-job training under the direction of experienced workers, and 144 hours of classroom training each year. Advancement to supervisory or estimating jobs within an organization is possible. As noted earlier, many painters and paperhangers start their own contracting businesses.

Summary

Paint is not only decorative, but protective. It keeps interior and exterior surfaces from being damaged by moisture and other environmental factors. Safety is important. Finishing materials and supplies are chemicals that can cause skin, eye, and lung problems. They also are flammable. Painting and finish tools include those used for surface preparation, coating application, and access (ladders and scaffolds). Proper tool selection and use are important. So is choice of the proper coating material for the task. Paints are made up of a pigment and a vehicle that includes a binder; varnishes and lacquers are clear coatings. Stains are usually transparent, but may include pigments. A color wheel is useful when selecting paint colors. Proper surface preparation is vital to a good painting or varnishing job. This may involve cleaning, scraping and sanding, or even complete removal of the old finish, then priming the surface. Depending on the task, most coatings can be applied with brush, roller, pad, or spray gun. Wallpaper is another category of surface finish. Most wallpapers are long lasting, washable, and made from vinyl resins. Before applying new wallpaper, the existing wallpaper must be removed by peeling, soaking, or steaming. The new covering is then adhered to the wall with an adhesive. Careful attention must be paid to getting the first strip plumb and to edge-butting all succeeding strips. Paper is applied over windows, doors, and electrical switches and receptacles. The excess is then trimmed away.

Test Your Knowledge

Answer the following questions on a separate piece of paper. Do not write in this book.

1. List three factors in the environment that degrade unpainted wood siding.
2. *True or False?* Sanding of lead-base paint poses no hazard.
3. Name the parts of a typical paintbrush.
4. The rougher the surface, the _____ (longer; shorter) the nap of the roller should be.
5. A spray gun with an external-mix head is intended for spraying _____ coating materials.
6. Which spraying head provides greater control of the spray pattern: internal mix or external mix?
7. _____ is the ingredient that gives color and hiding qualities to a paint.
8. Which product is used to fill the pores of open-grain wood surfaces?
9. Red, blue, and yellow are called the _____ colors.
10. Explain why weathered wood siding should be coated with a turpentine/linseed oil mixture before applying paint.
11. How would you *box* paint? Why do it?
12. Cracking and scaling is usually the result of using paint lacking in _____.
13. List the three general steps used to paint an entire room.
14. Painting inside corners first is called _____.
15. Explain how to use warm water or steam to remove a dent from wood.
16. The walls in a 12′ × 10′ room are to be painted. The ceiling is 8′ high. How many containers of paint are needed for two coats if each container covers about 300 square feet? Assume one 36″ door that is 6′-6″ high.
17. Most wallpaper is made of what material(s)?
18. *True or False?* The *quick method* of estimating wallpaper is less accurate than the area method, but it works on any rectangular room regardless of height.
19. *True or False?* Booking wallpaper is the process of ordering, or "booking", it from the supplier.
20. How often should you replace the blade in a razor knife?

Curricular Connections

Science. Research how humans see a color as a result of different wavelengths of light reflected from a surface. Make sketches to show which wavelengths are reflected and which ones absorbed by surfaces that are yellow, purple, and white. Find out whether the type of light (daylight, fluorescent, incandescent) affects how color is seen. Make an oral or written report.

Social Studies. Colors have certain associations, such as blue for boy babies, pink for girl babies, and red for danger. Research the emotional associations of color (such as green conveying freshness and a peaceful atmosphere). Make a list of 10 colors and take a survey of at least 15 people. Ask each person to respond with one word that describes their association to each of the colors. Collate your results to see if you can find any patterns. For example, did males and females respond differently to certain colors? Write a short report describing your results and any conclusions you were able to make.

Mathematics. Measure a room and determine the number of gallons of paint required to paint it. Assume that one gallon covers 400 sq. ft. and that two coats will be needed.

Outside Assignments

1. Research a wood-finishing system. Prepare a report on the proper procedure for finishing an unpainted pine cabinet to a desired wood tone. Indicate both equipment and materials.
2. Take photos of paint defects and bring them to a class discussion on probable causes of paint failure.
3. Measure a room and estimate the number of rolls required to paper it. Use the quick method described in this chapter. Assume that the paper does not require matching a pattern.

Section 5

Special Construction

Chimneys and Fireplaces

22

Learning Objectives

After studying this chapter, you will be able to:

- Explain how masonry chimneys are constructed around flue linings.

- Name the parts of a typical masonry fireplace.

- Describe procedures for the construction of the chimney, hearth, walls, and throat.

- Define the functions of the damper and smoke shelf.

- Calculate the flue area for a given fireplace.

- Describe the common types of factory-built fireplaces.

- List special considerations for installing factory-built fireplace units.

- Install a prefabricated flue.

Technical Vocabulary

Ash dump	Flue lining
Chase	Hearth
Chimney	Smoke chamber
Circulator	Smoke shelf
Corbelled	Splay
Damper	Termination tops
Downdraft	Throat
Fireclay	Zero clearance
Flues	

A *chimney* is a vertical shaft that exhausts the smoke and gases from heating units, fireplaces, and incinerators. When properly designed and built, a chimney may also improve the outside appearance of a home.

The fireplace once was the only source of heat for dwellings. The fireplace remains popular, even though its efficiency as a source of heat is low compared with modern heating systems. The desire for a fireplace results from the cheerful, homelike atmosphere it creates and the value it adds to the home.

The rising cost of energy has led to many improvements in fireplace design. Masonry fireplaces are often equipped with glass doors to save heat. They may also include ductwork to circulate room air around the firebox and heat it. More important, they may have provisions to draw combustion air from outside of the house. There is a wide variety of prefabricated fireplaces. Prefabricated fireplaces are easy to install and usually less expensive than built-in-place masonry units.

Chimney: A vertical hollow shaft that exhausts smoke and gases from heating units and incinerators.

22.1 Masonry Chimneys

A masonry chimney usually has its own footing and is independent of the building structure. This means it is built in such a way that it provides no support to, nor receives support from, the building frame. Footings should extend below the frost line and project at least 6″ beyond the sides of the chimney. The walls of a chimney with a clay *flue lining* should be at least 4″ thick. Foundations for a chimney or fireplace, especially when located on an outside wall, may be combined with those used for the building structure.

The size of the chimney depends on the number, arrangement, and size of the *flues* (vertical openings in the chimney). The flue for a heating plant should have appropriate cross-sectional area and height to create a good draft. This permits the heating equipment to develop its rated output. When deciding on flue sizes, always follow the recommendations of the heating equipment manufacturer.

Building codes require that chimneys be constructed high enough to avoid downdrafts caused by the turbulence of wind as it sweeps past nearby obstructions or over sloping roofs.

Minimum heights generally required are illustrated in **Figure 22-1**. The chimney should always extend at least 2′ above any roof ridge that is within a 10′ horizontal distance.

Combustible materials, such as wood framing members, must be located at least 2″ away from the chimney wall, **Figure 22-2**. The open spaces between the framework and chimney can be filled with mineral wool or other noncombustible material.

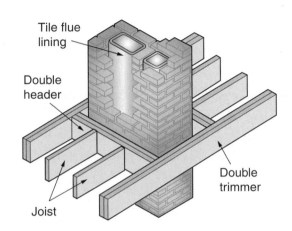

Figure 22-2. Floor framing around a chimney. Combustible materials, such as wood framing, must be spaced at least 2″ from the chimney wall. When multiple flues are used in a chimney, stagger the joints in adjacent flue columns by at least 7″.

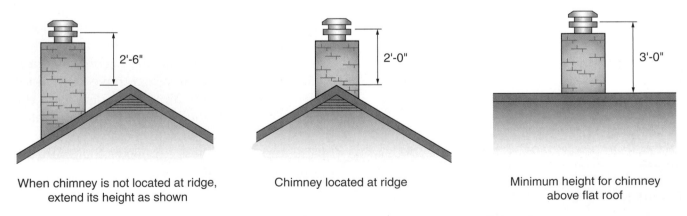

When chimney is not located at ridge, extend its height as shown

Chimney located at ridge

Minimum height for chimney above flat roof

Figure 22-1. Minimum chimney heights above the roof. Always check local building codes.

Flue lining: A noncombustible sleeve, usually made of fireclay, that is used to provide a smooth surface inside of a masonry chimney.

Flue: The passage in a chimney through which smoke, gas, and fumes rise.

22.1.1 Flue Linings

The National Board of Fire Underwriters recommends *fireclay* flue linings for masonry chimneys. Most local building codes also require them. Linings are available in square, rectangular, and round shapes. Each shape is available in several sizes, as listed in **Figure 22-3.** Note that different methods of measurement are used for the three types:

- Outside dimensions for the old standard.
- Nominal outside dimensions for the modular.
- Inside diameter for the round linings.

Small and medium flue linings are usually available in 2' lengths. Flue rings are placed during the assembly of the chimney to provide a form for the masonry.

Although unlined fireplace chimneys with masonry walls (minimum thickness 8″) will operate well, a glazed flue lining is usually recommended. Pitch and tar are less likely to accumulate on the smooth inside surface of a glazed lining. If these are allowed to accumulate, the passage will become restricted.

As a rule, connecting of a single gas-fired water heater to a furnace flue is often permitted. Do not combine larger flues.

22.1.2 Construction

Before constructing the chimney, allow its footing to cure for several days to give it proper strength. Use care in starting the masonry work. Make all lines level and plumb.

Each section of flue lining is set in place before the surrounding part of the chimney wall is built. The lining is a guide for the brick work. Joints in the flue lining are bedded in mortar or fireclay. Use care in placing the lining units so they are square and plumb. Brick work is carried up along the lining. Then, the next lining unit is placed.

Where offsets or bends are necessary in the chimney, miter the ends of the abutting sections of lining. This prevents reducing the flue area. The angle of offset should be limited to 60° from horizontal. The center of gravity of the upper

Standard liners		Modular liners		Round liners	
Outside dimensions (inches)	Areas of passage (square inches)	Nominal outside dimensions (inches)	Areas of passage (square inches)	Inside diameter (inches)	Areas of passage (square inches)
8 1/2 x 8 1/2	52.56	8 x 12	57	8	50.26
		8 x 16	74		
8 1/2 x 13	80.50			10	78.54
		12 x 12	87		
8 1/2 x 18	109.69	12 x 16	120	12	113.00
13 x 13	126.56	16 x 16	162	15	176.70
13 x 18	182.84	16 x 20	208		
18 x 18	248.06	20 x 20	262	18	254.40
		20 x 24	320	20	314.10
		24 x 24	385	22	380.13
				24	452.30

Figure 22-3. Flue liner dimensions and clear cross-sectional areas (areas of passage). The National Building Code requires flue liner walls to be a minimum of 5/8″ thick.

Fireclay: Heat-resistant clay material used for flue linings in masonry chimneys.

section must not fall beyond the centerline of the lower wall.

Chimneys are often *corbelled* to increase the size of the upper section. This means that successive courses extend outward for several courses. Corbelling is done just before the chimney projects through the roof, **Figure 22-4.** The larger exterior appearance is often more attractive. Also, breakage due to wind is less likely for thicker masonry. Corbelling should not exceed a 1″ projection in each course. The final size should be reached at least 6″ below the roof framing.

Openings in the roof frame should be formed before constructing the chimney. Refer to Chapter 10, **Roof Framing.** Water leakage around the chimney can be prevented by proper flashing. Corrosion-resistant metal, such as sheet copper, is often used for this purpose. The flashing is attached to the roof surface and extends up along the masonry. Cap flashing (counterflashing) is bonded into the mortar joints and is then lapped down over the base flashing. Refer to Chapter 12, **Roofing Materials and Methods,** for details concerning the installation of flashing.

At the top of the chimney, the flue lining should project at least 4″ beyond the top brick course or cap. Surround the lining with cement mortar at least 2″ thick. Slope the cap so wind currents are directed upward and water drains

away. When several flues are located in the same chimney, extend each flue to a different height. Horizontally space them no closer than 4″ apart.

22.2 Masonry Fireplaces

Figure 22-5 shows a cutaway view of a masonry fireplace. The chimney and fireplace are combined in a single unit. This unit often includes a flue for the regular heating equipment.

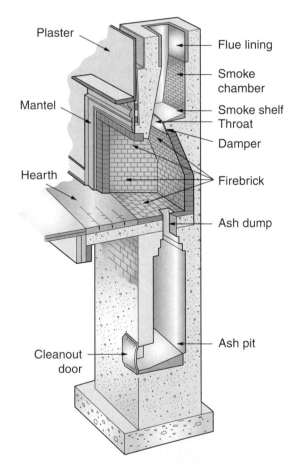

Figure 22-5. Major components of a masonry fireplace. Carpenters who do remodeling work must know about many different types of structures.

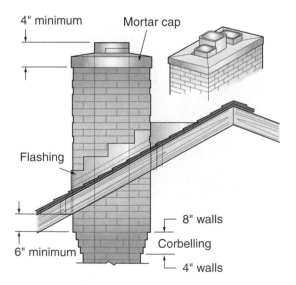

Figure 22-4. Corbelling is used to enlarge the top section of the chimney. This helps it resist wind pressure and results in a more attractive appearance.

Corbel: To extend outward from the surface of a masonry wall one or more courses to form a supporting ledge.

The *hearth* consists of two parts: one is located in front (front hearth) and the other is below the fire area (back hearth). The back hearth, as side walls, and back wall are lined with firebrick that can withstand direct contact with flame. The side and back walls are sloped to reflect heat into the room.

A *damper* is located above the fire to control combustion. The damper also prevents loss of heat from the room when the fireplace is not being used.

The throat, smoke shelf, and smoke chamber are also important parts of the fireplace. They must be carefully designed to ensure good operation.

Architectural plans often include details of the fireplace construction, **Figure 22-6.** The drawings include overall dimensions. They can be scaled (measured) to find other lengths.

22.2.1 Design Details

The size of the fireplace opening should be based on the size of the room and matched to the style of architecture. Some authorities recommend the fireplace be able to accommodate firewood 2' long. This is the length of standard cordwood cut in half. **Figure 22-7** shows common fireplace dimensions. Masonry structure and wood framing details for a fireplace are shown in **Figure 22-8.** This design includes a flue for the furnace.

22.2.2 Hearth

The hearth, including the front section, must be completely supported by the chimney. The support is constructed by first building temporary supports and forms. Then, a minimum of 3 1/2" thick concrete with reinforcing is poured. In the best construction, a cantilevered design is

---

Hearth: The part of a fireplace that holds the fuel and contains the fire.

Damper: A venting device in fireplaces used to control combustion, prevent heat loss, and redirect downdrafts.

Ash dump: A metal frame with pivoted cover provided in the rear hearth for clearing ashes.

secured by recessing the back edge of the rough hearth into the rear wall of the chimney.

An *ash dump* may be provided in the rear hearth for clearing ashes, if there is space below for an ash pit. The dump consists of a metal frame with pivoted cover. The basement ash pit should be of tight masonry and include a clean-out door.

When the floor structure consists of a slab-on-grade, an ash pit may be formed by a raised hearth, as illustrated in **Figure 22-9.** This design may be used when the fireplace is located on an outside wall. However, not where snow

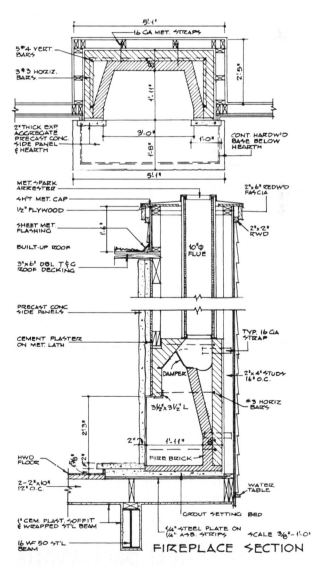

Figure 22-6. Detail drawings of fireplaces are included in architectural plans. Note the 2 × 4 framing on the outer edges in both views. A wooden chimney frame that juts out from a wall is called a chase.

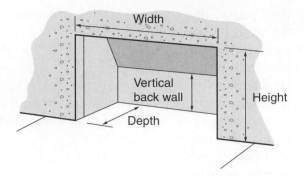

A

Fireplace opening		Depth (inches)	Minimum back (hori-zontal) (inches)	Vertical back wall (inches)	Inclined back wall (inches)	Outside dimensions of standard rectangular flue lining (inches)	Inside diameter of standard round flue lining (inches)
Width (inches)	Height (inches)						
24	24	16–18	14	14	16	8 1/2 x 8 1/2	10
28	24	16–18	14	14	16	8 1/2 x 8 1/2	10
24	28	16–18	14	14	20	8 1/2 x 8 1/2	10
30	28	16–18	16	14	20	8 1/2 x 13	10
36	28	16–18	22	14	20	8 1/2 x 13	12
42	28	16–18	28	14	20	8 1/2 x 18	12
36	32	18–20	20	14	24	8 1/2 x 18	12
42	32	18–20	26	14	24	13 x 13	12
48	32	18–20	32	14	24	13 x 13	15
42	36	18–20	26	14	28	13 x 13	15
48	36	18–20	32	14	28	13 x 18	15
54	36	18–20	38	14	28	13 x 18	15
60	36	18–20	44	14	28	13 x 18	15
42	40	20–22	24	17	29	13 x 13	15
48	40	20–22	30	17	29	13 x 18	15
54	40	20–22	36	17	29	13 x 18	15
60	40	20–22	42	17	29	18 x 18	18
66	40	20–22	48	17	29	18 x 18	18
72	40	22–28	51	17	29	18 x 18	18

B

Figure 22-7. Recommended dimensions for a wide range of fireplace sizes.

may block the outside access. In some designs, especially when no ash pit is included, the rear hearth is lowered several inches so ashes are contained in this area.

22.2.3 Side and Back Walls

The side and back walls of the combustion chamber extend upward to the level of the damper. The wall must be lined with firebrick at least 2″ thick. The firebrick is set in a special clay mortar that can withstand the heat of a fire. The total thickness of the walls, including the firebrick, should not be less than 8″.

Side walls are angled to reflect heat into the room. This angle, called *splay,* is usually laid

Splay: The rear face of the hearth that slopes toward the front.

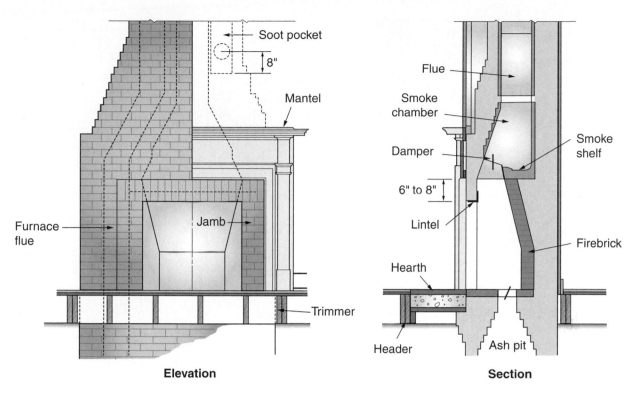

Figure 22-8. Elevation and section views show construction details for a typical masonry fireplace.

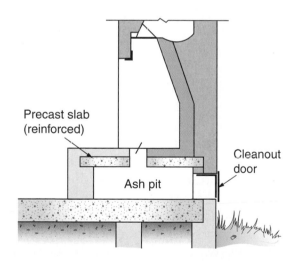

Figure 22-9. In slab-on-grade construction, a raised hearth provides space for an ash pit. This type of construction cannot be used in areas where snow may cover the cleanout door.

Downdraft: A flow of air down a chimney.

out at 5″ per foot. The back wall vertically rises from the base for a distance slightly less than one-half of the opening height. Then, it slopes forward. This slope directs the smoke into the throat of the fireplace. The slope also keeps an area clear for the smoke shelf located above.

22.2.4 Damper and Throat

Part of the fireplace throat is formed by the damper, **Figure 22-10.** A stationary front flange is angled a small amount away from the masonry of the front wall. The back flange is movable.

The damper affects a flow called the *downdraft.* In **Figure 22-10,** notice the upward flow of hot air and smoke from the throat into the front side of the smoke chamber. The rapid upward passage of hot gases creates a downward current (downdraft) on the opposite side of the flue. One purpose of the damper is to change the direction of the downdraft so smoke is not forced into the room.

The damper is installed so that the masonry work above it does not interfere with full

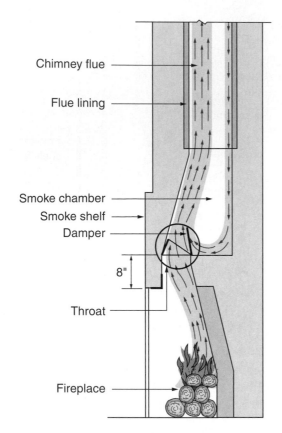

Figure 22-10. Operating details of a fireplace. Arrows indicate smoke and air currents. Note the role of the damper in keeping the downward current from entering the room.

operation. Also, the ends have a slight clearance in the masonry to permit expansion. Manufacturers provide data and recommendations for installing each model and size.

The ***throat*** size controls the efficiency of a fireplace. The highest efficiency occurs when the cross-sectional area of the throat is equal to that of the flue. The length of the throat along the face of the fireplace is the same as the opening width. The horizontal space is much less than the flue depth to get the proper cross section.

22.2.5 Smoke Shelf and Chamber

The ***smoke shelf*** helps the damper change the direction of the downdraft. The deeper the shelf behind the damper, the better the fireplace works. The depth may vary from 4" to 12",

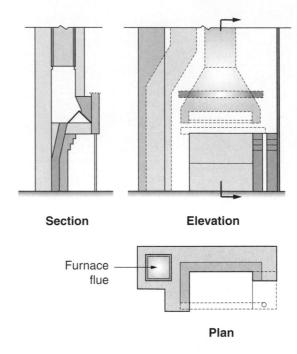

Figure 22-11. Design for a projecting corner fireplace.

depending on the depth of the fireplace. Some smoke shelves are curved to reduce turbulence in the airflow. The length along the face for all types is equal to the full width of the throat.

The ***smoke chamber*** is the space extending from the top of the throat up to the bottom of the flue. The area at the bottom of the chamber is quite large, since its depth includes that of the throat plus the depth of the smoke shelf. This space temporarily holds accumulated smoke if a gust of wind across the top of the chimney cuts off the updraft. Without this chamber, smoke would likely be forced out into the room. A smoke chamber also lessens the force of the downdraft by increasing the area through which it passes. Side walls are generally drawn inward one foot for each 18" of rise. The surfaces of most smoke chambers are plastered with about a 1/2" thickness of cement mortar.

Throat: In a fireplace, the narrowed passage above the hearth and below the damper.

Smoke shelf: The horizontal shelf located adjacent to the damper in a fireplace.

Smoke chamber: Fireplace chamber located between the smoke shelf and the entrance to the flue.

22.2.6 Flue Size

The cross-sectional area of the flue is based on the area of the fireplace opening. As a general rule, the flue area should be at least 1/10 of the total opening when a flue lining is used. A flue lining is highly recommended. This applies to chimneys that are at least 20' high. A somewhat larger flue may be needed to compensate for the lower chimney heights normally used in single-story construction. The upward movement of smoke in a low chimney does not reach a high velocity. Thus, a greater cross-sectional area is required for the flue.

One recommended method of calculating the area for a flue is to allow 13 sq. in. of area for the chimney flue to every square foot of fireplace opening. For example: if the fireplace opening equaled 8.25 sq. ft., then a flue area of at least 107 sq. in. is needed. If the flue is to be built of brick and unlined, it would probably be made 8" x 16", or 128 sq. in., because brickwork can be laid to better advantage if the dimensions of the flue are multiples of 4".

22.2.7 Construction Sequence

Masonry fireplaces are nearly always built in two stages. The first begins during the rough framing of the structure. Masonry work is carried up from the foundation and the main walls of the fireplace are formed. After a steel lintel is set above the opening, the damper is installed and the smoke chamber built. Then, the chimney is carried upward through the roof and the exterior masonry is completed. These steps usually occur before the roof deck is laid.

The second, finishing stage of the fireplace takes place during the application of interior trim

after the finish wall surface is applied. Decorative brick or stone may be set over the exposed front face. The surface of the front hearth can be finished at this time, since the reinforced concrete base was placed during the rough masonry construction. The wood trim (mantel) is installed when masonry work is complete.

22.2.8 Special Designs

Some fireplace designs have openings on two or more sides. For these, follow the same principles in planning as previously described for conventional designs. When calculating the flue area, the sum of the area of all faces must be used. **Figure 22-11** shows a corner fireplace design in which the flue area is based on the total of the front face opening plus the end face opening. In this particular construction, the side walls are not splayed. However, the rear wall is sloped in the usual manner.

Multiface fireplaces must incorporate a throat and damper with requirements similar to standard designs. Special dampers with square ends and sides are available for two-way fireplaces that serve adjoining rooms.

22.2.9 Built-In Circulators

The heating efficiency of a fireplace can be increased by using a factory-built metal unit called a *circulator*, **Figure 22-12**. The sides and back are double walled, providing a space where air is heated. Cool air enters this chamber near the floor level. When the air is heated, it rises and returns to the room through registers at a higher level.

Built-in circulators, also called *modified fireplaces*, include not only the firebox and heating chamber, but also the throat, damper, smoke shelf, and smoke chamber. Since all of these parts are carefully engineered, proper flue draw is assured when the flue size is adequate and installation is made according to the manufacturer's directions.

To install a circulator unit, first position it on the hearth, then build the brick and masonry work around the outside. Steel lintels are required across the top of the opening and

---

Circulator: A factory-built metal unit that increases the heating efficiency of a masonry fireplace. The sides and back are double walled providing a space where air is heated. Cool air enters this chamber near the floor level. When the air is heated, it rises and returns to the room through registers at a higher level.

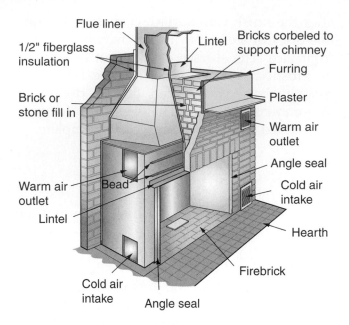

Figure 22-12. Cutaway view of a circulator installation. Masonry is erected around the metal unit. (Majestic Co.)

may be required in other locations to provide support. The unit itself should not be used for support of any masonry work.

When installing a unit, follow specifications furnished by the manufacturer. Some type of fireproof insulating material is usually placed around the metal form not only to prevent the movement of heat but also to provide some expansion space between the metal and masonry.

Working Knowledge

Correct operation of a fireplace depends on an adequate flow of air into the building to replace the air exhausted through the flue. Air infiltration around doors and windows may be sufficient. However, to improve the energy efficiency of the house, combustion air should be piped to the fireplace from the outside.

22.3 Prefabricated Chimneys

Lightweight chimney units are available that require no masonry work. They provide flues

for heating equipment or fireplaces and can be installed in single-story or multistory structures, **Figure 22-13.** Prefabricated chimneys usually consist of double- or triple-walled sections of pipe that are assembled to form the flue. Special flanges and fittings are used to fasten the flue to the building frame and provide proper clearance from wood members. **Figure 22-14** shows details of a typical installation.

The rooftop section of a prefabricated chimney unit must be sealed to prevent roof leaks. **Figure 22-15** shows the basic parts of a simple pipe projection. Bed the flashing unit in mastic with the roofing material overlapping the top and side edges. The storm collar diverts rainwater from the pipe to a conical section of the flashing. Some type of cap should be installed on the top of the flue to keep out rain or snow.

A *termination top* is the part of the chimney that extends above the roof. Many types of prefabricated units are available. For best appearance, they should blend with the architectural style of the building, **Figure 22-16.** Be sure to follow the manufacturer's directions for assembly and installation.

Figure 22-13. Prefabricated chimneys serve both fireplaces and furnaces in this multiple-unit housing complex. The carpenter may be required to install the chimney sections as well as building the framework for the exterior chase. (Preway, Inc.)

Termination top: The part of the chimney that extends above the roof.

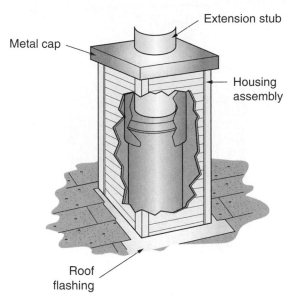

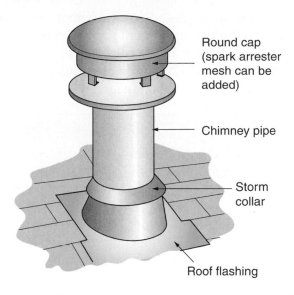

Figure 22-15. The basic parts of a rooftop projection for a standard flue pipe.

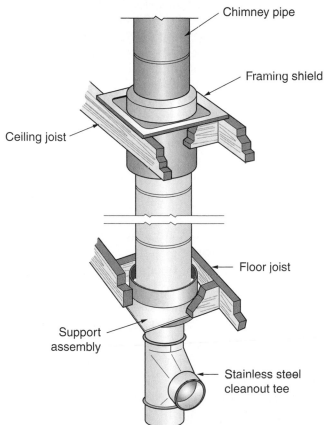

Figure 22-14. Triple-wall metal pipe sections are used to erect this prefabricated chimney. Depending on the application, inside diameter of the pipes can range from 6″ to 14″. A rain cap is installed on the extension stub.

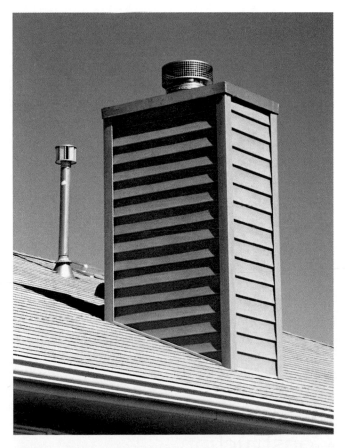

Figure 22-16. Termination tops (rain caps) and chimney housings are used with prefabricated chimneys. This example has a wood framework covered with vinyl siding to match house walls.

Smoke and gases from incinerators and solid-fuel boilers may contain corrosive acids. When a prefabricated chimney is used, the inside pipe should be made of stainless steel or have a porcelain-coated surface.

22.4 Prefabricated Fireplaces

Many of the fireplaces in new homes are factory-built units. These units are generally less expensive than masonry units. Manufacturers of prefabricated fireplaces provide a wide range of designs that are easy to install with ordinary tools, **Figure 22-17.** In some cases, multiple-steel-wall construction and special firebox linings permit *zero clearance.* This means that the outside of the housing can rest directly on wood floors and touch wood framing members.

Manufacturers provide detailed drawings and instructions for the installation of their fireplace units. These should be carefully studied and followed throughout installation. Follow the manufacturer's recommendations when installing. Other precautions should be taken if gas or electric starters are used or if gas logs are installed. Check local fire codes before starting work.

22.4.1 Operation and Construction

Prefabricated fireplaces operate like a metal circulator built into masonry fireplaces. Room air enters intakes at floor level and flows through chambers around the firebox. As the air is heated, it rises. This motion carries the air through the

Figure 22-17. This cutaway view shows the operation of a prefabricated fireplace. Room air (blue arrows) is drawn from the floor level into the heating chamber. The air is heated, then returned to the room (red arrows). Combustion air can also be drawn from the outside using a pipe connected to the inlet shown at lower left. Note the triple walls of the flue. (Preway, Inc.)

grillwork at the top of the unit and back into the room. Units are sometimes equipped with circulating fans that are controlled by switches.

Some units can be attached to vertical ducts that carry the heated air to various locations within the room or to an adjoining room. Units may also have blowers that increase the flow of air and thus improve heating efficiency.

Figure 22-18 shows framing and masonry built around a prefabricated unit. In new construction, the basic frame is usually built at the same time as exterior walls and partitions. The facing side of the frame should remain open until the fireplace is installed. If support is required above this opening, it should be framed with a header like that for a door. After the fireplace unit is installed, any front framing below the header can be added.

Many materials, including both wood and masonry, can be used to finish the wall and trim

---

Zero clearance: A quality of a prefabricated fireplace that allows the fireplace to be placed in direct contact with combustible material such as wood framing.

A

B

C

Figure 22-18. Installation of a prefabricated unit in new construction. A—A partially completed fireplace installation with the prefabricated insert surrounded by concrete block. B—Laying brickwork for the hearth and mantel. C—The completed installation. (Superior Fireplace Co.)

the fireplace opening. According to FHA specifications, wooden parts should not be placed closer than 3 1/2″ to the edge of the opening. Greater clearance is required when the parts project more than 1 1/2″. For example, a wooden mantel shelf must be 12″ above the opening.

Prefabricated, free-standing fireplaces are usually constructed with double walls somewhat like built-in units. Many rooms in the home can support these units and the units are easy to install. A single-walled pipe runs from the unit to the ceiling where it is fitted to a triple-walled section to permit zero-clearance installation.

Figure 22-19 shows a wall-mounted unit that is partially supported by the floor. Triple-wall construction is used around the firebox so the fireplace unit can be placed against any combustible material.

Free-standing fireplaces are available with many of the same features as built-in models. They can be equipped with a blower to force the circulation of air around the firebox. Air for combustion can be drawn from the outside.

Before selecting and installing any type of factory-built fireplace, be sure to check local building codes. Failure to comply with these requirements can result in the installation being rejected by the local building inspector.

Figure 22-19. A wall-mounted, prefabricated fireplace with a cantilevered hearth. It is equipped with a high-volume blower and a glass door. (Malm Fireplaces, Inc.)

22.4.2 Chimneys for Prefabricated Fireplaces

Chimney (flue pipe) systems for prefabricated fireplaces are designed for specific units and usually sold as a package along with the fireplace. **Figure 22-20** shows the typical assembly of standard components that run through the ceiling and roof structure.

Special attention must be given to clearance and support. Openings through the ceiling and roof should be carefully framed to provide the correct size openings for firestop spacers and support boxes. **Figure 22-21** shows the installation of a firestop spacer in an attic area.

The chimney pipe must be supported at the roof. If not, the entire weight of the pipe will rest on the fireplace. By supporting the chimney pipe at the roof, its weight is better distributed and the pipe is more stable.

When installed on an exterior wall, the chimney and fireplace unit can be located in a special projection called a *chase*. This is a box-like structure that is built as a part of the

Figure 22-21. A firestop spacer is used to close off the opening around a triple-walled fireplace flue. To prevent the spread of fire, all vertical channels in a building frame should be closed off at one or more points. (Majestic Co.)

floor, wall, or roof frame, **Figure 22-22**. When a chase is not used, the fireplace and chimney are located within the room. This reduces the amount of available living space. By using a chase, the room area is not reduced by the presence of the fireplace.

The outer walls of the chase are usually finished to match the outer walls of the house, as shown in **Figure 22-23**. Insulate the walls of the chase and seal them against infiltration. Be sure insulation does not touch the flue pipe.

22.5 Glass Enclosures

Glass enclosures improve the efficiency of fireplaces. They can be installed on either masonry structures or prefabricated units. They reduce the amount of heated room air that escapes up the chimney, even when there is no fire. Energy-efficiency is improved because the rate of burning can be controlled by adjusting draft vents and dampers.

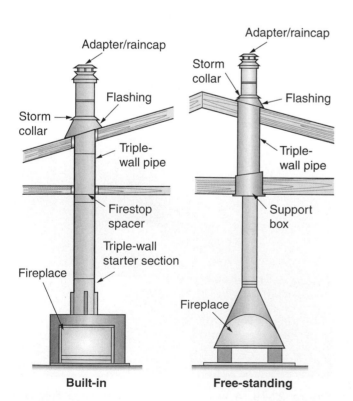

Adapter/raincap

Adapter/raincap

Storm collar

Flashing

Storm collar

Flashing

Triple-wall pipe

Triple-wall pipe

Firestop spacer

Support box

Triple-wall starter section

Fireplace

Fireplace

Built-in

Free-standing

Figure 22-20. Chimney system used for prefabricated fireplaces. (Preway, Inc.)

Chase: A wood frame jutting from an outside wall that supports a prefabricated chimney and fireplace.

Figure 22-22. Typical wood framing for a chimney chase. The roof structure helps hold the chase steady on three sides. (APA–The Engineered Wood Association)

Figure 22-23. A completed chase for a prefabricated fireplace and chimney. The base of the framework supports relatively little weight, since the fireplace is a metal shell.

ON THE JOB

Mason

Workers who build walls, floors, walkways, or other structures from brick, concrete block, or stone are called masons. They often are referred to by the type of material they work with: brickmasons, blockmasons, or stonemasons. A special category is formed by the small number of masons who lay firebrick and heat-resistant refractory tile. These refractory masons work mostly in steel mills and iron foundries.

While brickmasons and blockmasons work on both residential and commercial structures, the majority of work for stonemasons is in such structures as office buildings, hotels, and churches. Stone installed in residential settings is primarily used for fireplaces, floors, and outdoor walkways or patios. The material used by brickmasons and blockmasons is supplied in standard sizes and shapes, while stonemasons often must cut and fit pieces of irregular shape and size.

Approximately 1/4 of all masons are self employed. The rest work for general contracting firms or specialty businesses. A small number work for manufacturing firms, government agencies, and other employers. Masons should be able to accurately read and follow construction drawings. They make extensive use of measuring skills. Masons must be able to perform strenuous work that includes heavy lifting, often in unfavorable weather conditions. Another required skill is the ability to efficiently and safely use a variety of hand and power tools.

Most masons acquire their skills through informal on-the-job training learning from experienced workers. Some graduate from vocational school programs. A small number—usually those working for large nonresidential contractors—complete a formal apprenticeship that is typically three years in length. The apprenticeship combines approximately 144 hours of classroom training with practical experience and instruction on the job site.

Masons who work for larger contracting firms have opportunities for advancement to supervisory positions. Some leave wage employment to open their own businesses. Opportunities also exist in related fields such as building inspection and construction management.

Summary

A chimney is a vertical shaft that exhausts the smoke and gases from heating units, fireplaces, and incinerators. Masonry chimneys are placed on their own footings independent of the house structure. Chimney size depends on the number of flues needed for heating equipment and fireplaces. Fireplaces, once the primary source of household heating, are now popular for the atmosphere they provide. Increased energy costs have led to improved, more-efficient fireplace designs. The fireplace consists of a hearth, damper and throat, smoke shelf, smoke chamber, and flue. Built-in circulators increase the heating efficiency of a fireplace. Prefabricated chimneys and fireplaces are available to eliminate the need for masonry work. Some prefabricated fireplaces have a zero-clearance design, which means they can directly rest on floors and touch wood framing members. Glass enclosures on fireplaces improve efficiency by preventing heated air loss from the room, even when the fireplace is not being used.

Test Your Knowledge

Answer the following questions on a separate piece of paper. Do not write in this book.

1. The overall size of a chimney depends on the materials from which it is constructed and the number and size of the _____.
2. Wood framing should be located at least _____ inches away from a masonry chimney.
3. The offset angle in a chimney should be limited to _____ degrees.
4. To prevent heat loss from a room when the fireplace is not in operation, the fireplace must be equipped with a _____.
5. The length of the throat along the face should be equal to the _____ of the fireplace opening.
6. As a general rule, the cross-sectional flue area should equal about _____ of the total area of the fireplace opening.
7. Prefabricated chimneys consist of pipe sections with double or _____ walls.
8. When a prefabricated metal chimney is used for an incinerator, the inside pipe should be coated with porcelain or made of _____.
9. Zero clearance means that the outside housing of a prefabricated fireplace can _____.
10. A _____ is a special projection on the outside of a house that contains a prefabricated chimney and fireplace.

Curricular Connections

Language Arts. Write a report on the historical development of the fireplace. Start with the simple designs used by primitive humans and highlight developments through the years. Place special emphasis on the materials used for building fireplaces and chimneys. Is it possible to determine:

A.   When metal parts became popular?
B.   When safe designs for use in wood structures became common?

Ask your school librarian or media specialist to suggest reference materials that would be helpful.

Outside Assignments

1. Obtain a copy of the local building code and study the sections that deal with chimneys, fireplaces, and venting systems. Make a list of the specific requirements concerning residential structures. Include information about masonry and prefabricated metal chimneys for heating plants, incinerators, and fireplaces. Also include requirements for and restrictions on the use and installation of prefabricated fireplaces. Make a report to your class.
2. Interview a firefighter on proper installation and use of fireplaces. Prepare a list of safety rules from the interview.

Post-and-Beam Construction

Learning Objectives

After studying this chapter, you will be able to:
- List the advantages and disadvantages of post-and-beam construction.
- Describe general specifications for supporting posts.
- Compare transverse and longitudinal beams.
- Describe how roof and floor planks should be selected and installed.
- Sketch basic construction details of stressed skin panels and box beams.

Technical Vocabulary

Box beams
Longitudinal beams
Mortise-and-tenon joints
Plank-and-beam
Planks

Post-and-beam construction
Sandwich panels
Stressed-skin panels
Transverse beams

Post-and-beam construction, or *plank-and-beam construction,* consists of large framing members—posts, beams, and planks. Because of their great strength, these members may be spaced farther apart than conventional framing members. Frames of this type are similar to the "mill construction" once used for barns and heavy-timbered buildings. It is often used today for upscale residential building, since it permits greater flexibility than conventional framing methods for contemporary and traditional designs. See **Figure 23-1** for a comparison of construction methods.

Post-and-beam construction is often combined with conventional framing, **Figure 23-2.** For example, the walls might be conventionally built and the roof framed with beams and planks. In such a case, the term plank-and-beam could only be applied to the roof structure. Similarly, it would be correct to refer to a heavy-timbered floor structure as a plank-and-beam system.

23.1 Advantages of Post-and-Beam Construction

The most obvious advantages of post-and-beam construction are the distinctive architectural effect created by the exposed beams in

---

Post-and-beam construction: Construction consisting of large framing members—posts, beams, and planks. Because of their great strength, these members may be spaced farther apart than conventional framing members.

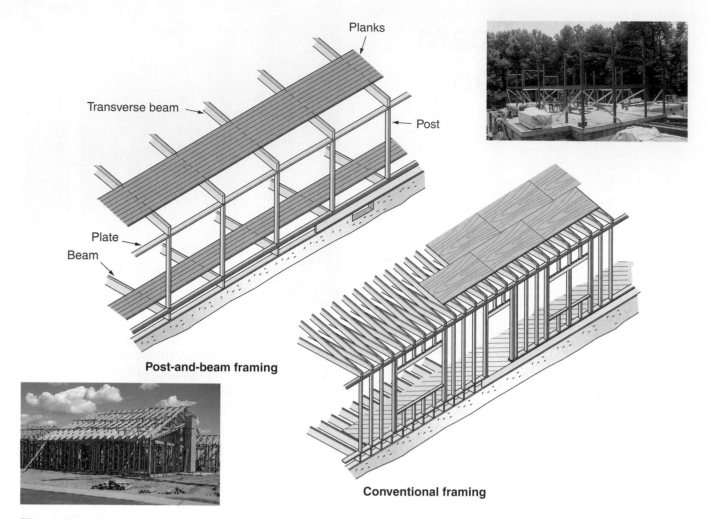

Figure 23-1. Post-and-beam framing compared to conventional framing. In post-and-beam construction, headers can be eliminated, which simplifies framing around windows and doors. (National Forest Products Assn.)

the ceiling and the added ceiling height. See **Figure 23-3.** The underside of the roof planks may serve as the ceiling surface, thus providing savings in material.

Post-and-beam framing may also provide some savings in labor. The pieces are larger and fewer in number. They can usually be more rapidly assembled than conventional framing.

One of the chief structural advantages of this method is the simplicity of framing around door and window openings. Loads are carried by posts spaced at wide intervals in the walls. Large openings can be framed without the need for headers, **Figure 23-4.** Window walls can be formed by merely inserting window frames between the posts. Another advantage is that wide overhangs can be built by simply extending the heavy roof beams.

In addition to its flexibility in design, post-and-beam construction also provides high resistance to fire. Wood beams do not transmit heat, nor do they collapse in the manner of unprotected metal beams. Exposure of wood beams to flame results in a slow loss of strength.

Most limitations of post-and-beam construction can be resolved through careful planning. The absence of concealed spaces in outside walls and ceilings makes installation of electrical wiring, plumbing, and heating somewhat more difficult. Also, the plank floors, for example, are designed to carry moderate, uniform loads. Therefore, extra framing must be provided under load-bearing partitions, bathtubs, refrigerators, and other places where heavy loads are likely. Extra members may be needed to provide lateral stability to the frame and walls.

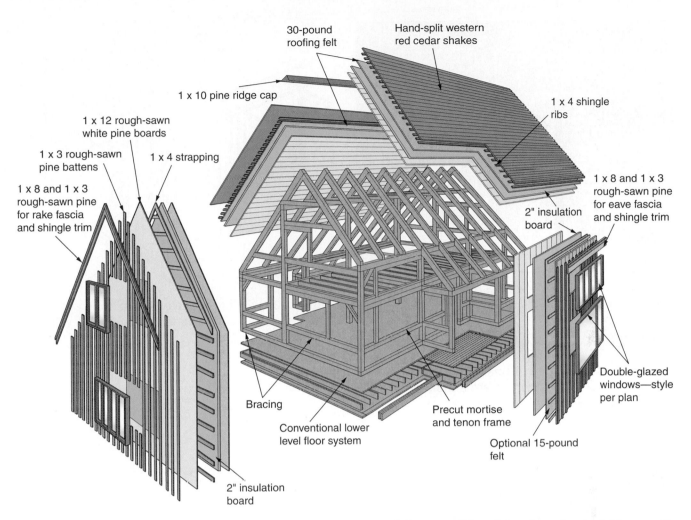

Figure 23-2. A rustic variation of a post-and-beam house that follows traditional framing of colonial barns. On the first level, a conventional flooring system is used, while heavy beams are used on the second level. The pine board-and-batten siding helps stiffen the walls against racking. (Timberpeg)

This might be achieved with various types of bracing. It is more common, however, to enclose some of the wall area with large panels and use conventional stud construction as shown in **Figure 23-5.**

23.2 Foundations and Posts

Foundations for post-and-beam framing may consist of continuous walls or simple piers located under each post. Refer to Chapter 7, **Footings and Foundations.** Either continuous or pier foundations must rest on footings that meet the requirements of local building codes.

Posts must be strong enough to support the load and large enough to provide full bearing surfaces for the ends of the beams. In general, posts should not be less than 4 × 4 nominal size. They may be made of solid stock or built up from 2″ pieces. Where the ends of beams are joined over a post, the bearing surface should be increased with bearing blocks as shown in **Figure 23-6.**

When posts extend to any great height without lateral bracing, a greater cross-sectional area is required to prevent buckling. Requirements are usually listed in local building codes through an l/d ratio, also called the *slenderness*

Figure 23-3. The high ceilings and exposed beams of post-and-beam construction are attractive architectural features. (Georgia-Pacific Corp.)

Figure 23-4. Post-and-beam construction makes possible wide expanses of windows, since headers are not required. (Yankee Barn Homes, Inc.)

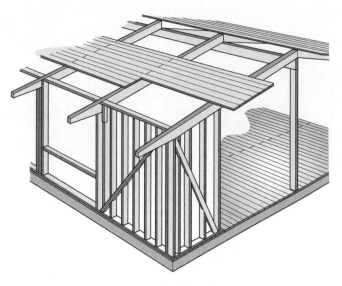

Figure 23-5. To provide lateral stability in a post-and-beam structure, a conventional stud wall with let-in bracing can be used.

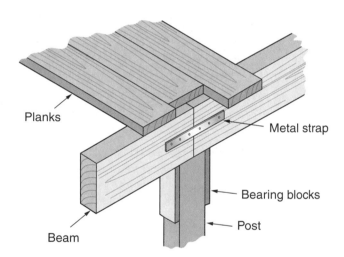

Planks

Metal strap

Bearing blocks

Post

Beam

Figure 23-6. Where beams are joined over a post, bearing blocks are used to ensure an adequate bearing surface. A heavy steel plate can be used in place of the bearing blocks.

ratio. The "l" represents the length in inches and the "d" stands for the smallest cross-sectional dimension (actual size). For example, a 4 × 4 piece 8′ long has a ratio of about 27. This is within the limits usually prescribed.

The distance between posts is determined by the basic design of the structure. This spacing must be carefully engineered. Usually, posts are evenly spaced along the length of the building and within the allowable free span of the floor or roof planks. Due to modular dimensions, construction costs will be lower if post-and-beam positions occur at standard increments (intervals) of 16″, 24″, and 48″.

In single-story construction, a plate is attached to the top of the posts in about the same way as conventional framing. The roof beams are then positioned directly over the posts as shown in **Figure 23-7.**

23.3 Floor Beams

Beams for floor structures may be solid, glue-laminated, or built-up, **Figure 23-8.** Sometimes, the built-up beams are formed with spacer blocks between the main members. Box beams may also be used. These are discussed later in this chapter. For single-story structures, where under-the-floor appearance is not critical, standard dimension lumber can be nailed together to form a beam of any desired size.

Design of sills for a plank-and-beam floor system can be similar to regular platform construction, **Figure 23-9.** When it is desirable to keep the silhouette of the structure low with the floor near grade level, the beams can be supported in pockets in the foundation wall.

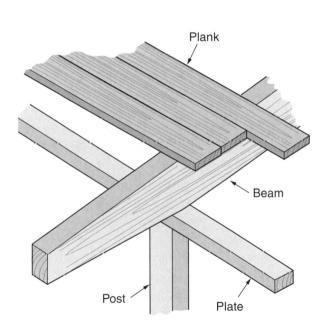

Figure 23-7. To prevent sagging, roof beams must rest directly over supporting posts.

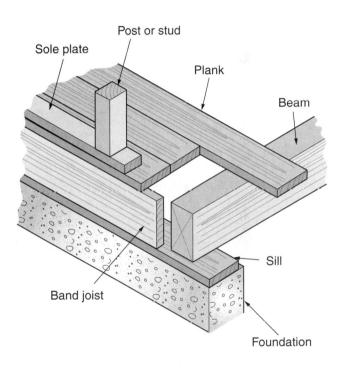

Figure 23-9. Typical sill construction for a post-and-beam frame. If a post is not located over a beam, blocking must be used to provide proper support.

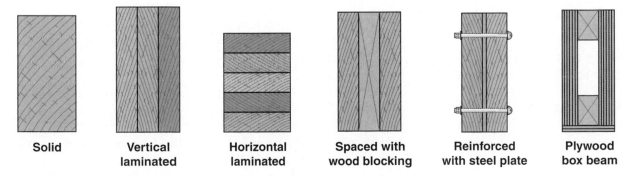

| Solid | Vertical laminated | Horizontal laminated | Spaced with wood blocking | Reinforced with steel plate | Plywood box beam |

Figure 23-8. Beams can be solid or built-up from smaller dimension lumber.

23.4 Beam Descriptions

In general, it is best to use solid timbers when beam sizes are small or when a rustic architectural appearance is desired. Where high stress factors demand large sizes or a finished appearance is required, it is usually more economical to use laminated beams. They are manufactured in a wide range of sizes and finishes. Solid timbers are available in:

- A range of standard cross sections.
- Lengths of 6′ and longer.
- Longer lengths in multiples of 1′ or 2′.

The surface finish may be either rough sawn or planed. When beams are exposed, appearance becomes an important consideration. **Figure 23-10** illustrates *casing-in,* an on-the-job treatment that may be applied to exposed beams.

Beam sizes must be based on the span (spacing between supports), the deflection permitted, and the load they must carry. Design tables available from lumber manufacturers may be used to determine sizes for simple buildings. See **Figure 23-11.** Refer to Chapter 8, **Floor Framing,** for additional information on calculating beam loads.

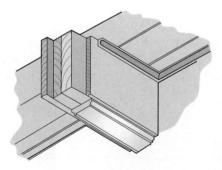

Figure 23-10. Built-up beams can be covered (cased) for better appearance when they will be left exposed.

23.5 Roof Beams

Beam-supported roof systems are one of two basic types. See **Figure 23-12.**

- *Transverse beams,* which are similar to exposed rafters on wide spacings.
- *Longitudinal beams,* which run parallel to the supporting sidewalls and ridge beam.

In either case, the purpose of the beams is to support roof planking or panels.

Longitudinal roof beams, also called *purlin beams,* are usually larger in cross-sectional area than transverse beams. This is due to the fact that they have greater spans and carry heavier loads. The use of longitudinal beams permits many variations in end-wall design. Extensive use of glass and extended roof overhangs are special features.

Either type of beam must be adequately supported either on posts or stud walls that incorporate a heavy top plate. When supported on posts, the connection can be reinforced with a wide panel frame that extends to the top of the beam, **Figure 23-13.** A similar method used to support a ridge beam is shown in **Figure 23-14.**

Transverse beams are joined to the sides of the ridge beam or supported on top of the beam, as illustrated in **Figure 23-15.** Metal tie plates, hangers, and straps are required to absorb the horizontal thrust.

Flat roof designs often consist of a plank-and-beam system. Details of construction are similar to those illustrated for low, sloping roofs.

23.6 Fasteners

A post-and-beam frame consists of a limited number of joints. Therefore, the loads and forces exerted on the structure are concentrated at these

---

Transverse beams: Roof beams that are similar to exposed rafters on wide spacings.

Longitudinal beams: Roof beams that run parallel to the supporting sidewalls and ridge beam.

Table 1

Size (actual)	Lbs. per lineal foot	Simple Span in Feet													
		10	12	14	16	18	20	22	24	26	28	30	32	34	36
		Load bearing capacity – lbs. per lineal foot													
3" x 5 1/4"	3.7	151	85	—	—	—	—	—	—	—	—	—	—	—	—
3" x 7 1/4"	4.9	362	206	128	84	—	—	—	—	—	—	—	—	—	—
3" x 9 1/4"	6.7	566	448	300	199	137	99	—	—	—	—	—	—	—	—
3" x 11 1/4"	8.0	680	566	483	363	252	182	135	102	—	—	—	—	—	—
4 1/2" x 9 1/4"	9.8	850	673	451	299	207	148	109	—	—	—	—	—	—	—
4 1/2" x 11 1/4"	12.0	1036	860	731	544	378	273	202	153	—	—	—	—	—	—
3 1/4" x 13 1/2"*	10.4	1100	916	784	685	479	347	258	197	152	120	—	—	—	—
3 1/4" x 15"*	11.5	1145	1015	870	759	650	473	352	267	206	163	128	104	—	—
5 1/4" x 13 1/2"*	16.7	1778	1478	1266	1105	773	559	415	316	245	193	154	124	101	—
5 1/4" x 15"*	18.6	1976	1647	1406	1229	1064	771	574	438	342	269	215	174	142	116
5 1/4" x 16 1/2"*	20.5	2180	1810	1550	1352	1155	933	768	586	457	362	290	236	183	160
5 1/4" x 18"*	22.3	2378	1978	1688	1478	1308	1113	918	766	598	474	382	311	254	204

*Horizontally laminated beams

Table 1 Roof Beam
Example: Clear span   = 18'-0"
            Beam spacing = 8'-0"
            Dead load   = 8 lbs./sq. ft. (decking + roofing)
            Live load   = 20 lbs./sq. ft. (snow)
            Total load   = (20 + 8) (8) = 224 lbs./lineal ft.

   From Table 1 – Select 3" x 11 1/4" beam with capacity of 252 lbs./lin.ft.

Table 2

Size (actual)	Lbs. per lineal foot	Simple Span in Feet													
		10	12	14	16	18	20	22	24	26	28	30	32	34	36
		Load bearing capacity – lbs. per lineal foot													
3" x 5 1/4"	3.7	114	64	—	—	—	—	—	—	—	—	—	—	—	—
3" x 7 1/4"	4.9	275	156	84	55	—	—	—	—	—	—	—	—	—	—
3" x 9 1/4"	6.7	492	319	198	130	89	—	—	—	—	—	—	—	—	—
3" x 11 1/4"	8.0	590	491	361	239	165	119	—	—	—	—	—	—	—	—
4 1/2" x 9 1/4"	9.8	738	479	298	196	134	96	—	—	—	—	—	—	—	—
4 1/2" x 11 1/4"	12.0	900	748	541	359	248	178	131	92	—	—	—	—	—	—
3 1/4" x 13 1/2"*	10.4	956	795	683	454	316	228	169	128	98	—	—	—	—	—
3 1/4" x 15"*	11.5	997	884	756	626	436	315	234	178	137	108	—	—	—	—
5 1/4" x 13 1/2"*	16.7	1541	1283	1095	732	509	367	271	205	158	123	96	—	—	—
5 1/4" x 15"*	18.6	1713	1423	1219	1009	703	508	376	286	221	173	137	109	—	—
5 1/4" x 16 1/2"*	20.5	1885	1568	1340	1170	939	678	505	384	298	235	187	151	—	—
5 1/4" x 18"*	22.3	2058	1710	1464	1278	1133	886	660	503	391	309	247	200	—	—

*Horizontally laminated beams

Table 2 Roof Beam
Example: Clear span   = 18'-0"
            Beam spacing = 5'-0"
            Dead load   = 7 lbs./sq. ft. (decking + covering)
            Live load   = 40 lbs./sq. ft. (furniture, occupants, etc.)
            Total load   = (40 + 7) (5) = 235 lbs./lineal ft.

   From Table 2 – Select 4 1/2" x 11 1/4" beam with capacity of 248 lbs./lin.ft.

Figure 23-11. Tables can be used to determine the allowable spans for glue-laminated beams. Table 1 is for roof beams where a deflection of 1/240 of the span is permitted. Table 2 is used for floor beams and is based on an allowable deflection of 1/360 of the span or less. (Weyerhaeuser Co.)

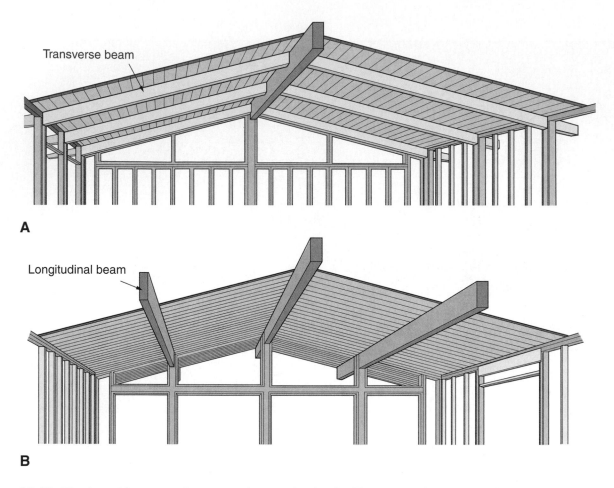

Figure 23-12. Plank-and-beam roof construction methods. A—Transverse beam, with beams running perpendicular to the ridge. B—Longitudinal beam, with beams running parallel to the ridge.

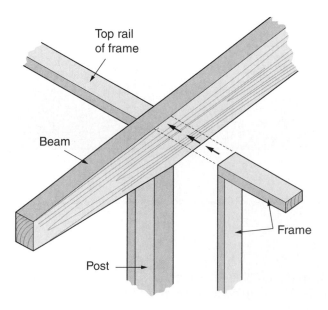

Figure 23-13. A transverse beam that bears on a post needs support to prevent lateral (side-to-side) movement. The reinforcement is provided by the filler frame panel.

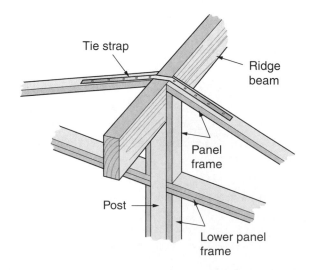

Figure 23-14. A metal tie strip and panel frames prevent lateral movement of the ridge beam.

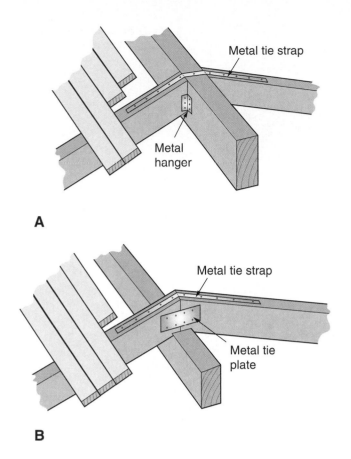

A

B

Figure 23-15. Attaching transverse beams to a ridge beam. A—Beams attached to the side of the ridge. B—Beams supported on top of the ridge.

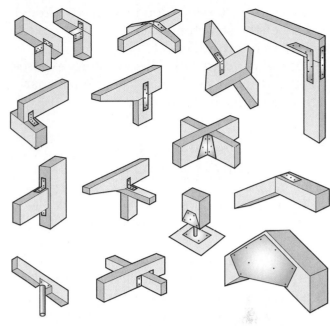

Figure 23-16. Many different kinds of metal fasteners are made for post-and-beam construction. (Western Wood Products Assn. and Timber Engineering Co.)

points. Traditionally, *mortise-and-tenon joints* have been used with wood pegs as fasteners. If butt joints are used instead of mortise-and-tenon joints, metal fasteners are needed. Regular nailing patterns used in conventional framing usually do not provide a satisfactory connection. The joints need to be reinforced with metal connectors. See **Figure 23-16.** To increase the holding power of metal connectors, they should be attached with lag screws or bolts.

Since beam structures are usually exposed, some connectors will likely detract from the appearance. Concealed devices need to be used.

Mortise-and-tenon joint: A type of joint in which a "tongue" (tenon) on the end of one wood member is inserted into a pocket or slot (mortise) cut into another member. Joints may be glued or fastened with wood pegs or mechanical fasteners.

Steel or wood dowel pins of appropriate size can be used.

Notches and gains cut in the members may provide an interlocking effect or a recess for metal connectors. **Figure 23-17** shows a heavy beam-and-truss system. Note how the metal

Figure 23-17. A heavy beam-and-truss system is used here to support roof beams at the midpoint and ridge. The rough-sawn surface finish helps mask joints in the laminated beams. (Boise-Cascade Corp.)

plates and fasteners blend with the rough surface of the structural members to provide a special architectural effect.

Use extra care when assembling exposed posts and beams. Tool and hammer marks detract from the final appearance.

23.7 Partitions

Interior partitions are more difficult to construct under an exposed beam ceiling. Except for a load-bearing partition under a main ridge beam, it is usually best to make the installation after beams and planks are in place. Partitions running perpendicular to a sloping ceiling should have regular top plates with filler sections installed between the beams.

Partitions parallel to transverse beams have a sloping top plate. Sometimes it is best to construct these partitions in two sections. First, build a conventional lower section the same height as the sidewalls. Then, add a triangular section above the lower section.

When nonbearing partitions run at a right angle to a plank floor, no special framing is necessary. However, when nonbearing partitions run parallel, additional support must be provided, **Figure 23-18.** Replace the sole plate with a small beam or add the beam below the plank flooring.

23.8 Planks

Planks for floor and roof decking can be anywhere from 2″ to 4″ thick, depending on the span. Edges may be tongue-and-groove or grooved for a spline joint that can be assembled into a tight, strong surface. **Figure 23-19** illustrates standard tongue-and-groove designs.

Planks: Floor and roof decking material that can be anywhere from 2″ to 4″ thick, depending on the span. Edges may be tongue-and-groove or they may be grooved for a spline joint that can be assembled into a tight, strong surface.

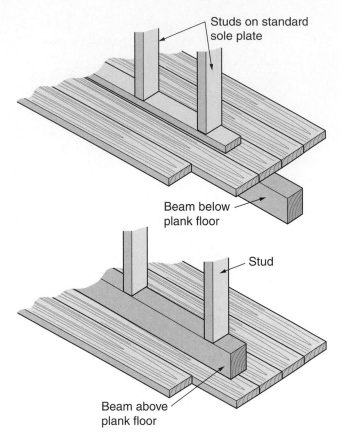

Figure 23-18. Two methods used to support nonbearing partitions that run parallel to flooring planks.

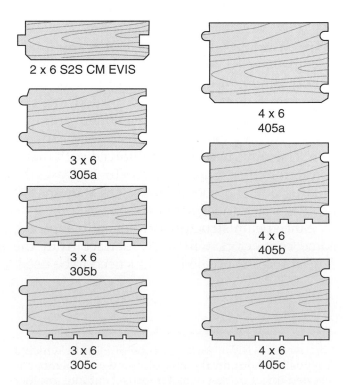

Figure 23-19. Standard plank patterns in end view. Faces are machined, while edges are tongue-and-groove.

The identification numbers given in the figure are those listed by the Western Wood Products Association. When planks are end matched (tongue-and-groove), the joints do not need to meet over beams.

Planks can support greater loads if they continue over more than one span. This rule can also be applied to beams, plywood, and other support material. See **Figure 23-20.**

Roof planks should be carefully selected, especially when the faces will be exposed. Solid materials should have a moisture content that closely corresponds to the Equilibrium Moisture Content (E.M.C.) of the interior structure when placed in service. Because of the large cross-sectional size of posts, beams, and planks, special precautions should be observed in selecting material with proper moisture content levels. Otherwise, difficulties due to excessive swelling or shrinkage may be encountered.

In cold climates, plank roof structures that are directly over heated areas must have insulation and a vapor barrier. The thickness of the insulation depends on the climate. Refer to Chapter 15, **Thermal and Sound Insulation.** The insulation should be a rigid type that will support the finished roof surface and workers who must walk on the roof. An approved vapor barrier should be installed between the planks and the insulation, as shown in **Figure 23-21.**

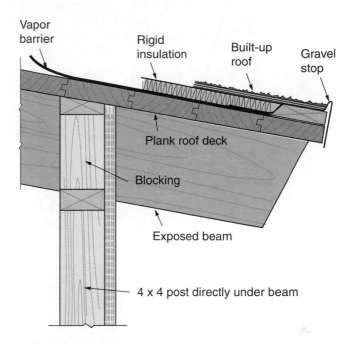

Figure 23-21. When a plank roof deck is directly over a heated space, a vapor barrier and insulation must be applied. This cross section shows the roof deck, vapor barrier, insulation, and built-up roofing.

Several types of heavy structural composition board are available for roof decks. These come in thicknesses of 2″ to 4″. The panel sizes are large and the material is lightweight. Edges usually have some type of interlocking joint that provides a tight, smooth deck. When the underside (ceiling side) is prefinished, no further decoration is usually necessary. Always follow the manufacturer's recommendations when selecting and installing these materials.

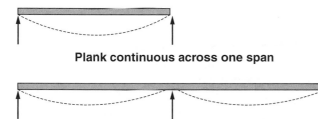

Figure 23-20. The stiffness of a plank is increased if it continuously extends over two or more spans.

23.9 Stressed-Skin Panels

Roof and floor decking and wall sections can be formed with plywood stressed-skin panels. These can be designed to carry structural loads over wide spans. **Figure 23-22** shows paneled roof deck sections being installed.

Stressed-skin panels are made by gluing sheets of plywood (skins) to longitudinal framing members or other core materials. They form a structural unit with a supporting action similar

Stressed-skin panels: Two facings are glued to opposite sides of an inner structural framework to form a panel. Facings may be of plywood or other suitable material.

Figure 23-22. These panels are being laid to form a roof deck. Note how the panel joints rest on the purlins. (Yankee Barn Homes, Inc.)

to a series of built-up wooden I-beams. See **Figure 23-23.** Panels are usually produced in factories where rigid specifications in design and construction can be maintained. **Figure 23-24** provides constructional details. Note that insulation can easily be installed as a part of the manufacturing process.

Sandwich panels with plywood skins and cores of such material as rigid polystyrene or paper honeycomb are similar to the stressed-skin panel. They do not provide as much rigidity and strength, however. Such panels are used for curtain walls and various installations where the major support is carried by other components. Skins can be made from a wide variety of sheet materials, including plywood, hardboard, plastic laminates, and aluminum.

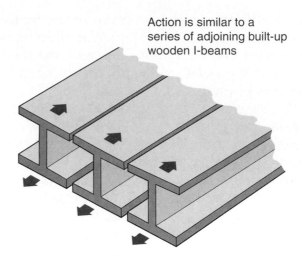

Action is similar to a series of adjoining built-up wooden I-beams

Figure 23-23. This is how a stressed-skin panel provides support.

23.10 Box Beams

Box beams made of plywood webs offer a structural unit that can span distances up to 120′. The high strength-to-weight ratio offers a tremendous advantage in commercial structures where wide, unobstructed areas are required.

The basic plywood box beam consists of one or more vertical plywood webs that are laminated to seasoned lumber flanges, **Figure 23-25.** The flanges are separated at regular intervals by vertical spacers (stiffeners) that help distribute the load between the upper and lower flange. Spacers also prevent buckling of the plywood webs. The strength of the unit depends to a large extent on the quality of the glue bond between the various members. Plywood box beams must be carefully designed and fabricated under controlled conditions.

Sandwich panels: Panels with cores of such material as rigid polystyrene or paper honeycomb. Skins can be made from a wide variety of sheet materials, including plywood, hardboard, plastic laminates, and aluminum. They are used for curtain walls and various installations where the major support is carried by other components.

Box beam: A beam made of one or more plywood webs connected to lumber flanges.

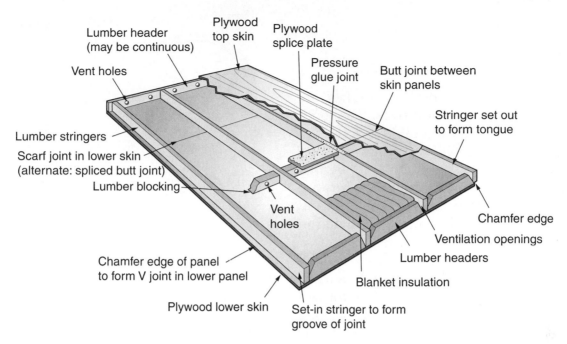

Figure 23-24. Stressed-skin panels are strong and conserve lumber.

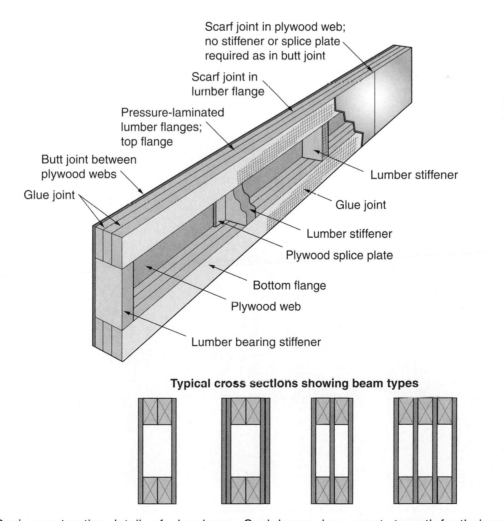

Typical cross sections showing beam types

Figure 23-25. Basic construction details of a box beam. Such beams have great strength for their weight.

23.11 Laminated Beams and Arches

Laminated wood beams and arches are available in many shapes and sizes. They offer a great deal of flexibility in building design. In addition to its natural beauty, laminated wood offers strength, safety, economy, and permanence. Most laminated structural members are made of softwoods. They are manufactured and prefinished in industrial plants specializing in such production.

In residential work, beams are usually straight or tapered. In institutional and commercial buildings, however, they are often formed into curves, arches, and other complex shapes, **Figure 23-26.** Some of the basic curved forms are illustrated in **Figure 23-27.**

In the fabrication of beams and arches, lumber is carefully selected and machined to size. To get the required length, pieces must often be end joined. Since end grain is hard to join, a special finger joint may be used, as shown in **Figure 23-28.** A number of these joints may be

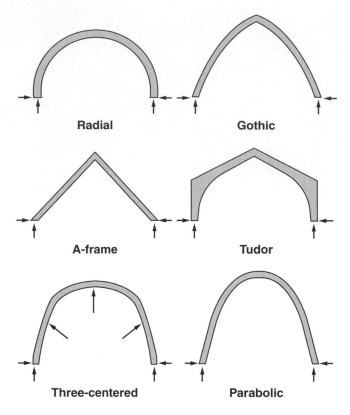

Figure 23-27. Different styles of laminated wood arches. The arrows indicate the support and lateral-thrust-reinforcement points.

Figure 23-26. Gracefully curved beams of laminated wood support the roof of this hockey rink. (APA–The Engineered Wood Association)

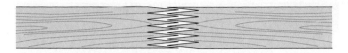

Figure 23-28. Ends of laminations are fastened together with glued finger joints. A special cutter head is used to make the joints. (American Institute of Timber Construction)

required in each ply. The joints are staggered at least 2′ from a similar joint in an adjacent layer.

23.12 Prefabricating Post-and-Beam Structures

Prefabrication of post-and-beam homes is a major industry in certain regions of North America. As with other factory-built systems, the design process consists of customizing homes by combining any of several standard modules or structures to suit the wishes of the homeowner.

The client reviews standardized structures working with architects and engineers. They work with models, combining them in ways that produce a plan to match their wishes and lifestyle, **Figure 23-29.** They also select the type of beams—planed, rough sawn, stained, and so on. See **Figure 23-30.**

Timbers are selected from the factory stock. These are then measured, marked, and cut. Large production machines are used to cut mortises and tenons, **Figure 23-31.**

Wall and roof sections are fabricated into large panels on the factory floor. Rigid insulation is cut and installed in the sections. Windows, exterior siding, and interior wall coverings are also installed. See **Figure 23-32.** Completed panels are moved with overhead cranes, wrapped to protect them from weather, and loaded onto trailers for transport to the building site, as shown in **Figure 23-33.** On site, builders erect the frame, **Figure 23-34.** Then, wall and roof panels are installed, **Figure 23-35.**

A

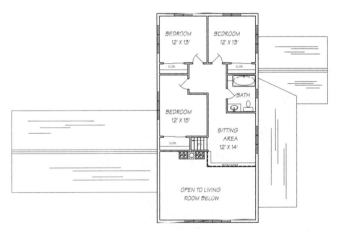

B

Figure 23-29. Planning a prefabricated post-and-beam house. A—The process begins by arranging simple block components on a house model. B—The floor plan derived from the model. (Yankee Barn Homes, Inc.)

A

B

Figure 23-30. Finishing timbers for prefabricated post-and-beam buildings. A—If client wants a smooth finish, beams can be planed. B—Stains can be applied to meet the client's wishes. (Yankee Barn Homes, Inc.)

Figure 23-31. Cutting a mortise in a beam with a large saw. (Yankee Barn Homes, Inc.)

A

B

Figure 23-32. Factory assembly. A—Trimming and placing rigid insulation in a wall panel. B—Applying vertical wood siding to the exterior of a panelized gable end. (Yankee Barn Homes, Inc.)

A

B

C

Figure 23-33. Delivering components for a building. A—The finished panel is moved from the production floor to the shipping area. B—Loading roof panels on a flatbed trailer truck. C—Weatherproof wrapping protects the panels in transit. (Yankee Barn Homes, Inc.)

Figure 23-34. The first construction step is to erect the frame on the building's foundation. (Yankee Barn Homes, Inc.)

A

B

Figure 23-35. Closing in. A—Once the frame is complete, wall panels are put in place. B—Roof panels complete the assembly. (Yankee Barn Homes, Inc.)

Summary

Because of their larger size and strength, post-and-beam framing members can be spaced farther apart than conventional framing members, increasing design flexibility. The post-and-beam method eliminates the need for window and door headers and makes large window walls possible. Foundations may be continuous or may consist of piers that carry the load of each post. Beams may be solid, glue laminated, or built-up. Roof beam systems may be transverse (similar to conventional rafters) or longitudinal (running parallel with the sidewalls and ridge beam). Pegged mortise-and-tenon joints are traditional in post-and-beam construction, but metal fasteners are sometimes used where appearance is less important. Most interior partitions are not load bearing, but are somewhat difficult to construct because of the exposed beam ceilings used. Floors and roof decking are made from 2″ to 4″ thick planks, usually with tongue-and-groove joints. Plywood stressed-skin panels are also used. In cold climates, plank roof structures that are directly over heated areas must have insulation and a vapor barrier. Plywood box beams have a high strength-to-weight ratio and can span distances up to 120′. Laminated beams and arches offer many design possibilities. Prefabricated post-and-beam buildings are available in a variety of sizes and styles. Structures are fabricated in a factory, then shipped to the building site for erection.

Test Your Knowledge

Answer the following questions on a separate piece of paper. Do not write in this book.

1. Post-and-beam construction consists of large framing members—_____, _____, and _____.
2. List three advantages of post-and-beam construction.
3. In general, posts should not be less than _____ × _____ nominal size.
4. The slenderness ratio of a post compares the total height in inches with the _____ dimension of its cross section.
5. The two types of roof beams are longitudinal beams and _____ beams.
6. Traditionally, _____ joints have been used in post-and-beam framing.
7. Planks for floor and roof decking are available in thicknesses of 2″ to _____.
8. Stress-skin panels are made by gluing _____ to longitudinal framing.
9. Box beams can span up to _____ feet.
10. *True or False?* Factory-built post-and-beam buildings offer the owner little choice in customizing their home.

Curricular Connections

Social Studies. Wooden barns on many American and Canadian farms were built using post-and-beam construction. As they age and deteriorate, however, they are being replaced with other types of structures. If you live in or near a rural area, use a camera to record some of the barns built using the post-and-beam method. Include both those in good condition and those that have deteriorated with age. Find a farm with a barn that is in good condition. Ask permission to go inside and photograph some of the construction details. Also, look for farms where pole barns have been built. Pole barns are wood-framed buildings with metal roofs and siding. Ask for permission to photograph interior construction details. Compare the post-and-beam barn photos with the pole barn photos. If possible, ask several farmers why they replaced traditional wooden barns with pole barns. Collect your photographs in an album. Write captions describing what each photo shows. Report the results of your conversations with farmers who erected pole barns.

Language Arts. Prepare a scale model of a small, single story post-and-beam house. Consult your instructor regarding types of fasteners and dimensions of posts and beams. Use the model as a visual aid while making a class presentation on the features of post-and-beam construction.

Outside Assignments

1. Visit a building supply center and obtain descriptive literature about floor and roof decking that can be used in post-and-beam construction. Include both solid and laminated planks and composition panels. Be sure to obtain prices. Thoroughly study these materials for qualities, characteristics, and installation procedures. Prepare a written or oral report.

2. Build a mockup of a stressed-skin panel in which you can experiment with the design and size of the various parts. Use a scale of about 3″ = 1′-0″. Skins can be made of 1/8″ plywood or 1/16″ veneer.

Often, systems-built housing cannot be distinguished from traditional stick-built housing by appearance. All of the homes shown here are manufactured homes. (Manufactured Housing Institute)

Systems-Built Housing

Learning Objectives

After studying this chapter, you will be able to:
- Describe the changes that have taken place in the technology of systems-built housing.
- Identify the variety of factory-built components that are utilized in a systems-built home.
- List and differentiate between the basic types of systems-built structures.
- Explain the erection sequence of a panelized home.
- Define terms used in the systems-built housing industry.

Technical Vocabulary

Closed panels
Manufactured home
Mechanical cores
Modules

Modular homes
Panelized homes
Precut homes
Systems-built housing

Systems-built housing refers to houses built from components assembled in a factory following precise design specifications. At one time, the term *prefabrication* was used to describe such construction. Either term indicates a process in which parts are cut and assembled into sections, modules, or entire homes in factories. Then, the units are shipped to the building site for final assembly and erection. Once erected, a systems-built house cannot easily be distinguished from a conventionally constructed (or stick-built) home, **Figure 24-1.**

Systems- or factory-built housing also includes *modules.* These are three-dimensional units that are fully assembled before they leave the manufacturing plant. Manufacturers of these

Figure 24-1. A systems-built home cannot be distinguished from a stick-built home. (Manufactured Housing Institute)

Systems-built housing: Housing built of components designed, fabricated, and assembled in a factory.

Modules: Three-dimensional assembled housing units built in a factory and transported to a building site to be combined with other modules.

modular structures also build commercial buildings such as banks, schools, office buildings, motels, and hotels. **Figure 24-2** shows a retail shopping center assembled from factory-built components.

At one time, prefabricated or factory-built housing limited the buyer to a few styles and plans offered by the manufacturer. All that has changed, partly because computer-assisted design can quickly adapt architectural drawings to the buyer's needs and lifestyle. Thus, architects and engineers can quickly produce any style of home and easily incorporate custom designs. Systems-built homes include geodesic domes and log homes, as well as traditional and classic styles.

To make quality components, many tasks must be carried out under controlled assembly-floor conditions using special tools. Overhead cranes lift heavy assemblies and power tools are used for fabrication. Stressed-skin panels and plywood box beams require accurate glue applications, special presses, and handling equipment. Often, the simplest prefabrication process in modern plants is done on large production equipment. The machine reduces the amount of human energy required and increases production. Efficiency is increased both in the plant and on the building site.

Designs are completed with the assistance of computers. During the manufacturing process, the components move from one workstation to another where all building trade activities are represented. A quality control process assures that the work is in conformity with state and local building codes.

Figure 24-2. This shopping center structure was constructed of super-insulated components built in a factory. (Enercept, Inc.)

24.1 Factory-Built Components

In home construction today, many factory-built components are used in place of job site finish work and stick-built framing work. Some of these components are windows, door units, soffit systems, stairs, and built-in cabinetwork.

One major framing component is the roof truss, **Figure 24-3.** These are used in all types of residential and light commercial construction. Manufacturing plants can usually furnish either simple Fink trusses (described in Chapter 10) or matched units for a complex roof. Roof trusses must be carefully designed by structural engineers and built to exact specifications. High-production saws are used to cut truss members to length. Special presses and jig tables are used to fasten the assembly with gang-nail connectors.

The floor truss is another example of a prefabricated component, **Figure 24-4.** It permits a wide, unsupported span and uses a minimum of material. It also has openings for heating and air conditioning ducts, electrical circuitry, and plumbing lines.

Wall and roof sections or panels are made in various shapes and sizes. Panels provide structural strength in addition to forming inside and outside surfaces. Some panel systems are constructed of conventional framing with 2″ lumber, standard insulation, and coverings. Two other types of panels are flat stressed-skin panels and flat sandwich panels. These are described in Chapter 23, **Post-and-Beam Construction.** They can be made to various heights and widths, depending on the design of the building. Voids, called *channels* or chases, can also be designed into the panels for running wires and pipes for mechanical systems. A roof panel may have channels simply to provide air circulation to prevent heat buildup. **Figure 24-5** shows a wall panel

A

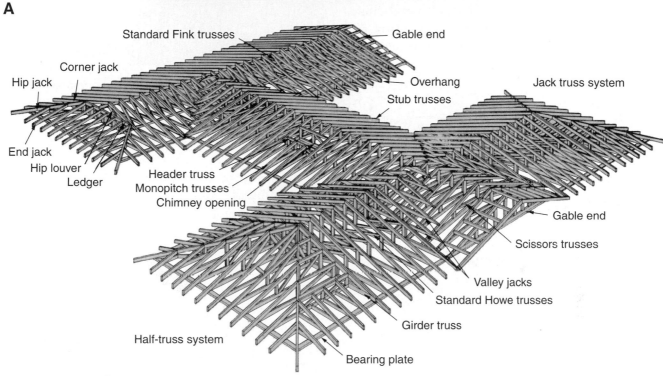

Standard Fink trusses

Gable end

Corner jack

Hip jack

Overhang

Jack truss system

Stub trusses

End jack

Hip louver

Ledger

Header truss

Monopitch trusses

Chimney opening

Gable end

Scissors trusses

Valley jacks

Standard Howe trusses

Half-truss system

Girder truss

Bearing plate

B

Figure 24-3. Factory-built truss rafters are used in all types of residential and light commercial structures. A—Trusses have been lifted into place and are held in alignment with temporary bracing until the sheathing can be applied. B—Prefabricated roof framing includes all the types of roof trusses used to form hips, valleys, overhangs, and gable ends. (Gang-Nail Systems, Inc.)

Figure 24-4. This floor truss system provides ample space for installation of plumbing lines. (Gang-Nail Systems, Inc.)

Figure 24-5. Workers attach a large half-round window to a tall wall panel. (Wausau Homes, Inc.)

assembly line where workers are fabricating a section that includes a large window with a half-round top. **Figure 24-6** shows various types of panels designed for different applications in walls and roofs.

Sandwich panels made with a rigid insulation core provide enough strength for walls and partitions in some structures. See **Figure 24-7.** These lightweight panels may have an aluminum or oriented strand board outer skin and a hardboard or plywood interior surface. Special metal channels and angles are available for connecting the panels, thus reducing labor at the building site.

Figure 24-7. This building is being constructed with super-insulated floor and wall panels. Such components are basic to all prefabricated construction. (Enercept, Inc.)

A building in various stages of completion can be prepared in a factory, then shipped as a "package" to the building site. Most such buildings are single-family homes. However, small commercial buildings, farm structures, and multi-family homes are also produced and moved this way.

24.2 Types of Systems-Built Homes

The Building Systems Council includes manufacturers in several categories of systems-built buildings. The Council groups systems-built housing in one of three categories:
- Modular homes.
- Panelized homes.
- Log homes.

Other organizations have additional categories. The North Carolina Manufactured Housing Institute adds two other categories:
- Precut homes.
- Manufactured homes (formerly called mobile homes).

Regardless of the type of housing chosen by the owner, all offer rapid construction, quality materials, lower building costs, and customizing.

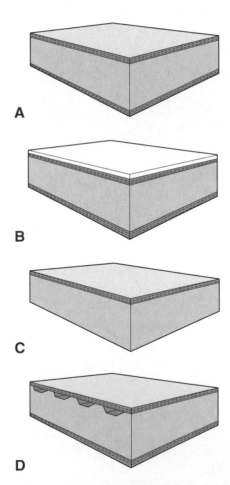

Figure 24-6. These four sandwich panels are typical of closed panels. A—A structural panel with facings of oriented strand board (OSB) and a polystyrene rigid insulation center that may be up to 11 3/8" thick. B—Drywall panel has a drywall interior facing over the OSB. C—Nail-base panels are designed to add a thick layer of insulation over a completed roof deck or wall. Sometimes they have foil on the interior face. D—A vented panel is designed to provide roof ventilation where desired. (Insulspan)

24.3 Modular Homes

According to the Modular Buildings Systems Association (MBSA), approximately 38,000 modular homes were sold in the United States during 2003. This represented an increase of more than ten percent from sales in 2002. Based on sales, modular homes are currently most popular in the states of North Carolina, New York, Michigan, Virginia, and Pennsylvania. Industry experts predict that in the near future modular homes will account for 7%–8% of all housing starts nationwide.

Modular homes are made up of two or more three-dimensional units. These units are called *modules*, sometimes referred to as *mods*. The modules are produced on a factory assembly line. Many modular homes are single-story. By stacking modules on each other, however, two- and three-story homes are possible. Modules also may be smaller units, such as a bathroom or a kitchen, **Figure 24-8.**

Each module is 12′ or 14′ wide and may be up to 60′ long. Such units are nearly completed at the factory. Some include cabinets and plumbing fixtures. At the building site, crews set the module on a foundation with a crane. Many manufacturers of modules have their own setup crews. When the module is in position, it is fastened with bolts, **Figure 24-9.**

In this type of prefabrication, entire sections (modules) of the structure are built and finished in manufacturing plants. The sections or modules are then hauled to the site where they are assembled. The width of a module seldom exceeds 14′. Trucks and roads cannot handle wider units.

An advantage of modular construction is that nearly all of the detailed finish work can be done at the factory. Kitchen cabinets can be attached to the walls and other built-in features can be installed. Also, wall, floor, and ceiling surfaces can be applied and finished. Electrical wiring chases, heating and air conditioning ducts, plumbing lines, and even plumbing fixtures can be installed under close control in manufacturing plants, using labor-saving tools. A section that has a group of plumbing and heating facilities and includes most of the utility hookups is called a *mechanical core.* It is often included with other sections that consist mainly of panels.

Mechanical cores group the kitchen, bath, and utilities in one unit that requires only three connections at the site. A core with a bathroom and kitchen on opposite sides of the same wall is typical. Some units are designed to include heating and air conditioning equipment, as well as electrical and plumbing equipment.

Figure 24-8. Some modules are small, consisting of only one room. Such modules usually include all plumbing, wiring, and fixtures. (Wausau Homes, Inc.)

Figure 24-9. A crane is being used to place a module on its masonry foundation. Units are 95% complete when they arrive on site. (Cardinal Industries, Inc.)

Modular homes: Homes consisting of two or more three-dimensional, factory-built units that are assembled on the building site.

There are disadvantages to sectionalized prefabrication. These include the problems involved with storage and transportation and the need for large cranes to handle the units at the construction site, **Figure 24-10.**

24.4 Panelized Homes

While most homes today have some prebuilt parts, such as roof trusses and engineered lumber, *panelized homes* leave the factory as a series of wall panels. These panels are 8′ high and may be from 4′ to 40′ long. Exterior walls are often sandwich panels of plywood, oriented strand board, or waferboard on either side of a core of rigid insulation. In other cases, the panels may have 2 × 6 studs and an exterior sheath of plywood. When prefabricated panels are finished on both the inside and outside surface, they are called *closed panels.* "Closed" means that no additional material must be installed once the panels are erected. However, drywall and mechanical systems may be added later. Some panelizers cut door and window openings, hang windows, and predrill studs for electrical wiring and plumbing runs. Channels may be formed in rigid insulation for running electrical wiring. A few producers make panelized ceilings and floors in the same way.

Panels arrive on the building site numbered according to their location in the house. Erection of the building can begin as the panels are unloaded from the truck.

In panelized prefabrication, flat sections of the structure are built on assembly lines. Large woodworking machines cut framing members to the desired length and angle, **Figure 24-11.** Parts are stored and delivered to the assembly stations as needed. Wall and floor frames are formed by placing the various members in positioning jigs on the production line. The parts are then fastened with pneumatic nailers. Electrical wiring or other mechanical facilities may be installed while the frame is being built.

Figure 24-11. A double-end sawing machine cuts bottom chord members for roof trusses. Each saw unit (right) has three blades that can be set at various angles. (Speed Cut, Inc. Corvallis, OR)

Mechanical cores: Prefabricated building modules that contain one or more of the following utilities: electrical, plumbing, heating, ventilating, and air conditioning. Floor, ceiling, and wall framing are fully formed at the factory. Modules are joined at the building site.

Panelized homes: Homes built in sections in a factory. Components such as wall sections and roof sections are then shipped to the building site and erected by carpentry crews.

Closed panels: Factory-built housing wall panels that are finished on both sides.

Figure 24-10. A modular home is being placed on an all-weather foundation. Sections are factory-built and completely finished inside and out. They will be bolted together. (APA-The Engineered Wood Association)

As the completed frames move along conveyor lines, wall surface materials are placed in position and nailed. Nailing is done with powered gang nailers or staplers, **Figure 24-12.** Farther along the line, insulation is put in, **Figure 24-13.** Some of the wall sheathing steps may be repeated to close the panel. Then, siding is attached.

In other areas of the plant, roof units are prepared. For post-and-beam structures, panels that form both the roof and ceiling surfaces are common. To avoid painting in high places after erection, the ceiling side is painted at the factory.

Although most panelized prefabricated houses use a first floor deck built by conventional methods, some manufacturers design and build floor panels. Full-length joists are assembled with headers. Long plywood sheets are formed using scarf joints. These are then glued in place. The resulting stressed-skin construction is rigid and strong. Use of these panels prevents nail pops and squeaks.

Figure 24-13. Installing blanket insulation in a wall panel. Insulation is arranged around any electrical equipment. (Wausau Homes, Inc.)

As the panels near the end of the production line, they receive a final inspection. Each panel is marked with a number for easy assembly, then is moved to storage or loaded onto a truck. When all materials and millwork are added to complete the "package", it goes to the building site.

24.5 Precut Homes

For *precut homes,* lumber is cut, shaped, and labeled. Then, it is shipped to the job site. This reduces labor and saves time on the building site. Manufacturers of this type of house include materials needed to form the outside and inside surfaces. Also shipped are such millwork items as windows, doors, stairs, and cabinets. Optional items include electrical, plumbing, and heating equipment. Kitchen cabinets and other built-in units are usually made in plants specializing in these items. The units may be shipped either to the home fabricator or directly to the building site.

Figure 24-12. A gang of pneumatic-powered nailers attaches fiberboard sheathing to a wall frame in less than two minutes. (Duo-Fast Corp.)

Precut homes: Structures for which lumber is cut, shaped, and labeled to reduce labor and save time on the building site. Manufacturers of this type of house include materials needed to form the outside and inside surfaces.

Working Knowledge

At one time, the terms *manufactured* and *industrialized* were used when referring to prefabricated housing.

24.6 Log Homes

Log homes are essentially a type of precut home. Walls of log homes are built by stacking precut and machined logs on top of each other, rather than by framing with studs. This is the only difference between a log home and conventional stick-built homes or homes that are of modular or panelized construction.

Logs are combined in the factory with other modern building materials to produce any style of home for the buyer. They can be customized as the buyers wish. High-speed machines mill the logs to uniform shapes and lengths. At the same time, the logs receive the tongues, grooves, notches, and splines that hold them together.

24.7 On-Site Building Erection

Preparation for on-site building erection begins at the factory. Workers load the prefabricated units onto semitrailer trucks. Each unit is carefully marked and placed on the truck so it can be removed in the order it is to be assembled on site, **Figure 24-14.**

At the building site, the owner's contractor prepares a foundation for the house. It is built according to the dimensions given on the foundation plan supplied by the building's manufacturer.

Many systems-built manufacturers send along their own setup crews as an efficiency measure. Usually the crews and trucks arrive at the site early in the morning. Since prefabricated units are heavy, a crane is used to lift the units from the trailers and set them in place on the foundation.

The floor deck is built in the standard way or is assembled from panels. Then, mechanical core units are set in place, walls and partitions are joined, ceiling-roof units are installed, and roof panels close the structure. **Figure 24-15** shows the construction of a systems-built home. The shell of a home can be finished in one day.

One disadvantage of prefabrication is soon discovered at the building site: weather prevents work much of the time. The builder

A

B

Figure 24-14. On-site building erection. A—On the home site, a crane lifts prefabricated panels in the order they are assembled. This speeds up the erection process. B—For larger homes, some factories send their own setup crews. After set-up, the customer's contractor takes over. This crew completed closing in of the structure in a single day.

Figure 24-15. This sequence shows on-site erection of a prefabricated home. A—Floor panels have been laid onto the foundation and a mechanical core for the kitchen is being lowered into place. Note the bath module already in place. B—The front wall panel for the living room is being installed. An outside surface has not been applied since this section of the house will be finished with brick veneer. C—A crane lowers ceiling-roof panels into place over the bedroom area. The combined unit is hinged and closed for shipment. The unit is opened for fitting at the building site. D—Installing the final roof panel on the garage. No ceiling panels were required. Note that carpenters have started shingle work over the bedroom area. This factory-built home was enclosed in less than a day. (Wausau Homes, Inc.)

must choose a day without wind or rain. Rain will damage inside wall materials applied at the factory. Wind makes it hard to unload and position large panels. There is also the problem of soft ground. It limits the size of the crane that can be used. Unless the reach of the crane is great, there is no way to safely bring large units into wet areas.

In many ways, however, the weather is less important for systems-built construction than for conventional construction. A skilled crew can erect and enclose a systems-built single-family home in 5–20 hours. A few weeks are usually required to complete the wiring, heating, plumbing, and decorating. With systems-built structures, the inside work begins on dry structures. In conventional construction, the inside often gets wet before the building is closed in. Therefore, the finish work goes faster on a systems-built home. For this and other reasons, prefabrication competes well with other building methods.

24.8 Assembling a Panelized Home

Typically, a panelized structure is assembled on site. Sometimes a block and tackle is used to lift and move the panels. **Figure 24-16** shows details of how to attach ceiling joists to the top plate and end wall when the wall panels extend to the roof. Adhesive and 12d nails are used to attach the edge joist. In multistory buildings, the construction method varies somewhat, **Figure 24-17.** The second-floor joists can be hung on joist hangers or they may rest on top of the first-floor wall.

Ceiling panels can be attached to walls and ridge beams in two ways. **Figure 24-18** shows metal clips or ties being used as a means of

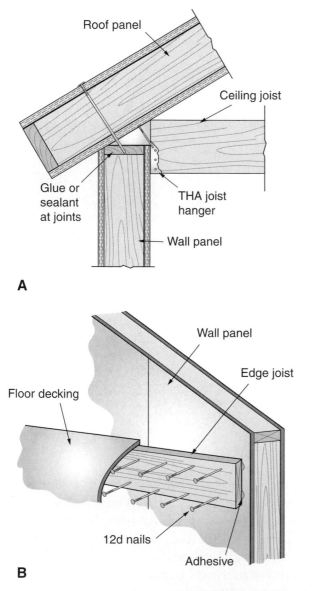

A

B

Figure 24-16. Panel installation. A—Attaching a roof panel to the top of the wall plate. Note that the top plate must be angled to match the roof angle. Fasteners are threaded thin-shank nails 1 1/2″ to 2″ longer than the panel thickness. B—When gable-end wall panels extend past the ceiling level, edge joists are glued and nailed to the panel to hold ceiling joists. (Insulspan)

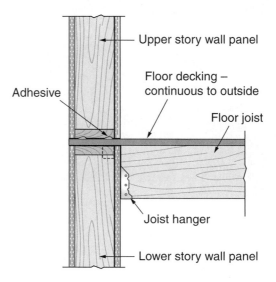

A

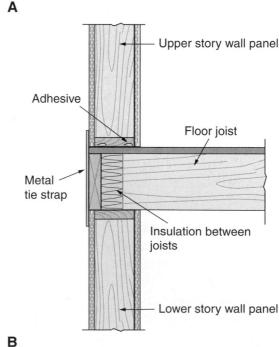

B

Figure 24-17. Methods vary for securing and supporting second-floor joists. A—Using joist hangers secured to top of plate. B—Resting second-floor joists on top of the wall plate. (Insulspan)

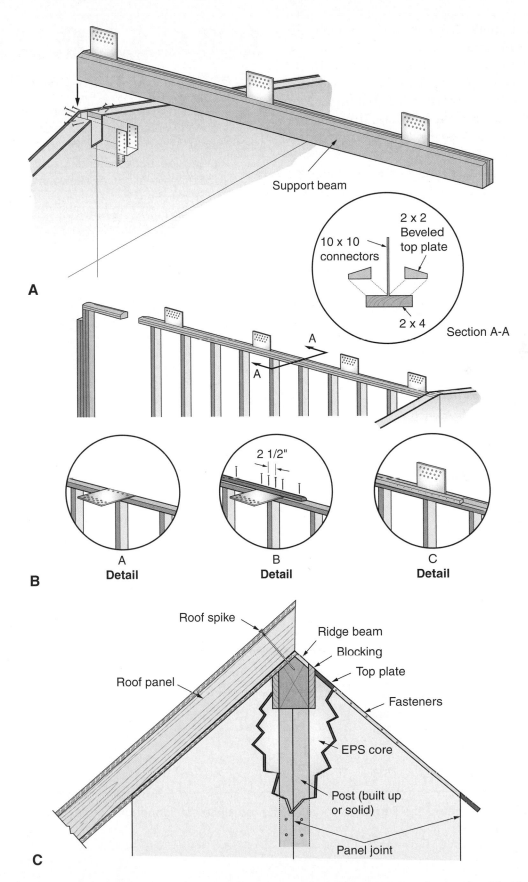

Figure 24-18. Alternate methods for securing roof panels. A—Steel connectors attached to the ridge beam secure the tops of roof panels. B—Clips attached to the wall plate secure the bottoms of roof panels. C—End view of a roof panel, ridge beam, and built-up king post. The beam must be solidly attached to the king post. (Enercept, Inc.; Insulspan)

securing the panels. Also shown is an attachment with roof spikes. These are thin-shank, threaded nails 1 1/2" to 2" longer than the thickness of the panels. Asphalt or fiberglass shingles can be applied directly over structural roof panels. When wood, tile, or slate shingles are used, it is customary to first apply horizontal wood battens.

The following assembly procedure is for sandwich panels. These steps are typical of this type of construction.

PROCEDURE

Erecting sandwich panels

1. Lay down a strip of polypropylene sill sealer along the outer edge of the subfloor.
2. Install 2 × 6 plates all around the perimeter with the outside edge extending 1/2" beyond the edge of the subfloor, as shown in **Figure 24-19.**

(Continued)

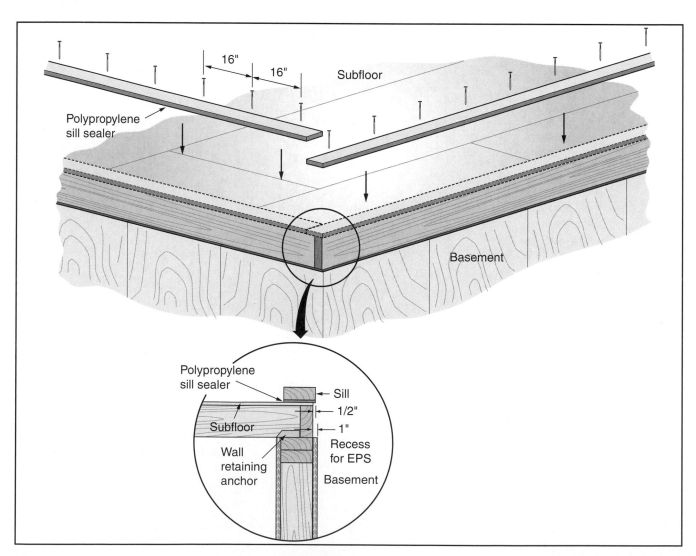

Figure 24-19. This view shows the sequence for attaching the sill to subflooring prior to setting up wall panels. (Enercept, Inc.)

(Continued)

3. Exterior wall panels are numbered in the order of their placement. Install them over the plate, **Figure 24-20.** Check each panel for vertical electrical wiring channels and drill a hole through the plate and subflooring in these locations.
4. Lay down a wavy line of caulk on the plate and vertical edges of each panel.
5. Carefully tip each panel into place. Avoid disturbing the caulk.
6. Plumb each panel and nail panels to the plate and to each other (on each side). Use 8d nails spaced 6″ apart.
7. Slide lintel panels into position, but do not immediately nail. Door and window panels are shown in **Figure 24-21.** Window rough openings have a lower panel and a lintel (upper) panel. Door openings have only a lintel panel.
8. Make height adjustments to lintels and then nail them in place.
9. Mark and trim the inner facing on mating corner panels so the panels can be plumbed and joined, **Figure 24-22.**
10. Toenail the king post on gable ends to the plate at the centerline of the building, **Figure 24-23.**
11. Mark the panels that have vertical wiring cores. Place the mark on the inside panel.
12. Double up the wall plate. Stagger-nail the plates every 6″.
13. Drill through both plates into the wiring cores.

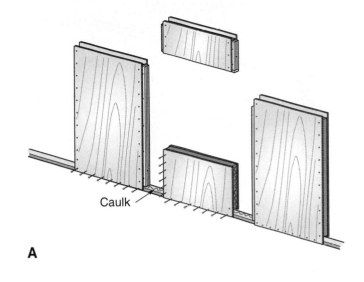

A

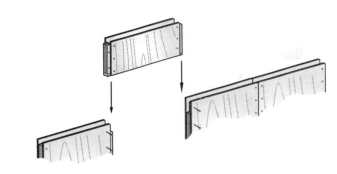

B

Figure 24-21. Door and window openings. A—Sequence for installing a window panel. B—The door header panel is dropped into place after wall panels are positioned. (Enercept, Inc.)

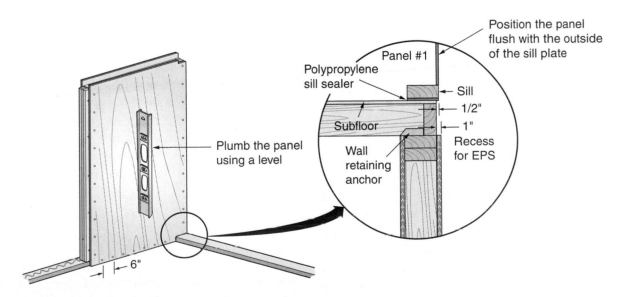

Figure 24-20. Setting the first panel in place. A bead of sill sealer is laid down and the panel is plumbed before nailing it to the plate.

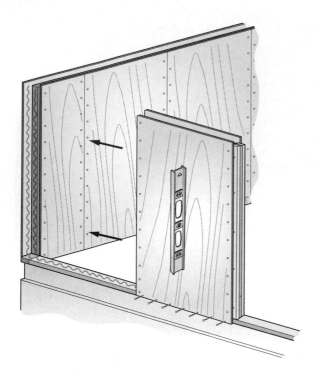

Figure 24-22. The inside face of a corner panel may need to be marked and cut shorter. Note the nailing pattern. Tilt the panel backward as it is moved into the corner. (Enercept, Inc.)

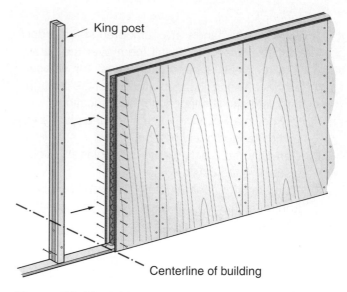

Figure 24-23. A king post is located at the center of the wall directly under the gable. It must align with the center of the building. The post must be toenailed to the plate with 16d nails. (Enercept, Inc.)

24.9 Manufactured Homes

Manufactured homes are defined by manufacturers as a trailer longer than 28′ and heavier than 4500 pounds. The most common width between 12′ and 14′. Lengths may be greater than 68′. HUD regulations require that units retain the chassis on which they were delivered.

Nearly 100 different companies build manufactured homes in factories located throughout North America. During the 1990s, this segment of the residential building industry grew an average of 17% a year.

The variety of choices available to the buyer also has grown. What used to be a market of metal-clad, single-wide units now includes multisectional units with brick siding and marble countertops. The average size of units has grown to over 1400 square feet.

There have also been significant changes in terms of public image. At one time, homes that were trucked to a site on their own wheels were called "mobile" homes or "trailer" homes. However, a survey done in the 1990s reported that 98% of such homes were never moved from their original site. To emphasize this and improve public image, the industry has replaced the term *mobile home* with **manufactured home.**

Production methods used in building manufactured homes are like those used for standard factory-built housing. Some of the structures, however, are different. For example, the floor must be one rigid unit. It usually consists of a wood frame attached to a welded steel chassis.

Designs sometimes include an expandable feature that permits a wider living room. A slide-out section is carried inside while the home is being transported. It is then expanded to the side after the home is on the site.

A manufactured home is finished and fully equipped at the factory. It is then pulled to the dealer or directly to the homesite. Since

---

Manufactured home: A home built on a chassis and finished and fully equipped at the factory, then pulled to the dealer or directly to the homesite. Manufacturers define it as a trailer longer than 28′ and heavier than 4500 pounds.

manufactured homes are mounted on a chassis, they do not need a permanent foundation, although some are placed on foundations. Utility hook-ups are ready to be made at the site. Interior furnish-ings include all major appliances, carpeting, drapes, furniture, and lamps. **Figure 24-24** shows a new double-wide manufactured home being installed in a retirement community.

A

B

Figure 24-24. A double-wide manufactured home is being moved onto a site in a retirement subdivision. A—The two sections are separately towed in and then positioned on the lot ready to be joined. B—A carpenter prepares to move the units together. The tongue jack rests on a roller that rides in a U-channel. A come-along winch operated by the carpenter or a helper slides the unit up against the mating unit.

ON THE JOB

Equipment Operator/Operating Engineer

Operators of heavy equipment on construction sites are responsible for such tasks as excavation and the movement of building materials around the site. They operate equipment such as bulldozers, graders, lifts, loaders, excavators, and cranes. Persons who are qualified to operate a number of different kinds of heavy equipment are classified as *operating engineers*.

Heavy equipment is typically controlled by using such devices as levers, switches, and dials. More-complex equipment makes use of computerized controls. In addition to operating equipment, operators and operating engineers are often responsible for setting up and inspecting equipment, maintaining and adjusting machines, and sometimes performing minor repairs.

Most operators learn through on-the-job training, but formal training is available through apprentice-ship programs and private vocational schools. Apprenticeship programs for operating engineers are three years in length and involve 1600 hours of jobsite training and more than 400 hours of class-room work. Employers generally prefer to hire high school graduate. Experience in large vehicle opera-tion, such as trucks, tractors, or military vehicles, is helpful in obtaining a job in this field.

The demand for operators and operating engi-neers is strong and expected to remain so well into the future. Continuing demand for housing and other structures, as well as highway and bridge construc-tion and repair, should generate a large number of jobs in this field.

Almost one-half of a million workers are employed as equipment operators or operating engineers. About 3/4 of them work in building construction and related fields. Most of the remaining jobs are in highway and street maintenance (paving and surfacing).

Working conditions can be harsh, since construc-tion equipment operators are exposed to all types of weather conditions—heat, cold, rain, wind, snow, and sleet. Working conditions are also noisy and many types of equipment subject their operators to vibration and jarring impacts as they run over rough ground.

Pay for operating engineers is comparable to other skilled members of the building trades. Annual earnings, however, may be reduced by severe weather conditions that prevent operating equipment on the jobsite.

Summary

Houses built from components assembled in a factory are known as systems-built housing. Some manufacturers also assemble complete modules that are then attached to each other to form the finished structure. Prefabricated components include roof and floor trusses and wall and roof panels. Complete "packages" are shipped to the building site for erection. Modular houses are built as two or more large finished sections that are then fastened together on site. Panelized homes consist of prefabricated wall sections, often complete with windows, doors, and electrical wiring. Precut homes are delivered with all lumber cut to size and labeled for proper assembly sequence. Both conventional stick-built homes and log homes are offered in precut form. Manufactured homes are complete structures built on a chassis equipped with wheels.

Test Your Knowledge

Answer the following questions on a separate piece of paper. Do not write in this book.

1. _____ refers to houses built from components assembled in a factory.
2. Which major factory-built framing component is used in all types of residential and light commercial construction?
3. _____ are systems-built structures made up of two or more three-dimensional structural units.
4. A structural unit that contains plumbing, heating, and electrical systems is called a _____ core.
5. In _____ prefabrication, flat sections of the structure are built on assembly lines.
6. A log home is a factory-built structure of the _____ type.
7. Describe the building site at the point where the systems-built structure is delivered.
8. *True or False?* Prefabricated construction is not affected by weather conditions during erection.
9. The time it takes a skilled crew to erect and enclose a systems-built single-family home is about _____.
10. Manufactured homes are constructed on a metal chassis and do not require a permanent _____.

Curricular Connections

Social Studies. Research the development of factory-built housing in this country. Highlight early experiments, but devote most of your study to progress since World War II. Find information in reference books, encyclopedias, trade magazines, or on the Internet. Prepare a written report that includes discussion of the factors that have led to the growth of this type of housing.

Outside Assignments

1. Visit a local builder or the manager of a building supply center that represents a "prefab" home manufacturer in your region. Get descriptive literature and information concerning sizes, design, fabrication features, material quality, and erection procedures. Also, check prices for standard models and the time required for delivery. Carefully organize the information and make an oral presentation to your class. Supplement your oral descriptions with appropriate visual aids.
2. Visit a manufactured home sales center and get information and descriptive literature concerning size, features, and prices. Inspect models that are on display and give special attention to the quality of materials, work, and finish. Also note the quality of equipment and furnishings. Compare prices with costs of other systems-built and conventionally built houses. Prepare carefully organized notes and make an oral report to your class on your findings.

Passive Solar Construction

Learning Objectives

After studying this chapter, you will be able to:
- Define conduction, convection, radiation, and thermosiphoning.
- Explain the difference between passive and active solar construction.
- List and describe the three types of passive solar construction.
- Calculate the amount of glazing and storage needed for a passive solar system.
- Locate a dwelling on a lot for maximum solar gain.
- Design and install various passive solar systems.

Technical Vocabulary

Active solar construction	Orientation
Conduction	Passive solar construction
Convection	Radiant energy
Direct gain	Solar collectors
Greenhouse effect	Solar radiation
Indirect gain	Sun space
Internal heat	Thermal mass
Isolated gain	Thermosiphoning
Movable insulation	Trombe wall

Because sunshine is free, it makes sense to capture it for heating buildings. Constructing buildings to take advantage of the sun requires some planning and an understanding of how to make the sun provide heat when it is most needed.

25.1 How Radiation and Heat Act

Solar radiation is transformed into heat energy by striking a surface. This radiation can travel through glass almost as well as it travels through the atmosphere. However, heat energy cannot readily pass back through the glazing. This is known as the *greenhouse effect*. This effect is important to all solar construction. See **Figure 25-1.**

The very basis of solar construction is found in the way that heat moves from one place to another. Heat always travels from an area of

---

Solar radiation: Radiation from the sun.

Greenhouse effect: Heating effect of solar energy that can pass through glass easily, creating heat that cannot easily pass back through the glass.

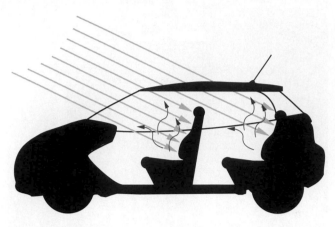

Figure 25-1. Solar radiation can pass through glass much more easily than can heat energy. It is the reason the interior of a vehicle becomes so hot when it is parked in direct sunlight with the windows closed.

higher temperature to one of lower temperature. It can travel by three different methods:
• Conduction.
• Convection.
• Radiation.

25.1.1 Conduction

In *conduction*, heat travels point-by-point through solid matter. One part of the solid body must be in contact with a heat source. Gradually, the whole body becomes heated. For example, if the sun heats a block of concrete, the entire block will eventually become hot, **Figure 25-2.** This is explained by the kinetic theory of matter. The molecules making up the concrete become

more active as they become warmer. As they move farther and faster, they bump into other molecules. This contact moves the heat from one molecule to another.

25.1.2 Convection

Convection occurs only in fluids or gases. Heat causes a fluid or gas to expand and become lighter (less dense). Lighter (warmer) elements always rise and cooler elements descend. Through this constant motion, heat moves through a liquid or a gas, such as air. See **Figure 25-3.** A fireplace heats a room through convection. Warm air exiting the fireplace rises to the ceiling as cold air from the floor is drawn in.

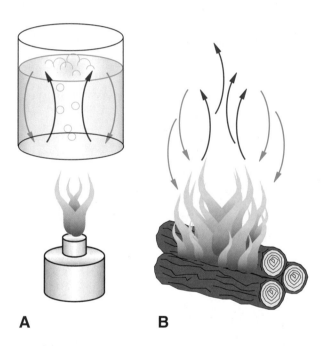

A   **B**

Figure 25-3. Convection is the heat transfer method in liquids and gases. Heated liquids and gases expand, become lighter, and rise. Cooler liquids and gases are heavier and sink. A—Convection in a liquid. B—Convection in a gas.

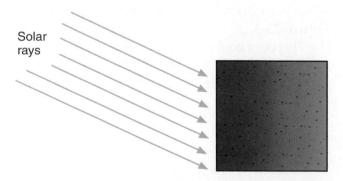

Figure 25-2. In a solid substance, heat travels by conduction. Heated molecules bump into cooler ones, transferring the heat.

Conduction: Method of heat transfer in solid material. The movement of heat through substances from one end to the other.

Convection: Movement of heat through liquids and gases as a result of heated material rising and being replaced by cooler material.

25.1.3 Radiation

Radiant energy moves through space in waves. When the waves strike a solid object, the heat energy is absorbed by the solid matter, **Figure 25-4.** The internal energy of the molecules in the matter increases and the temperature of the matter rises. An electric heater works in this way, as do the sun's rays.

25.1.4 Thermosiphoning

Thermosiphoning is simply the result of a liquid or gas expanding and rising. This principle is put to work in both active and passive solar heating. Sunlight is captured in a closed space where a liquid or gas, such as water or air, is heated. The action of the expanded water or air rising makes the system operate. Cooler air or water comes into the lower space to take its place, **Figure 25-5.**

25.2 Types of Solar Energy Systems

There are two commonly used methods of incorporating solar energy into a building: active solar construction and passive solar construction. *Active solar construction* is actually a system of collecting and transporting solar energy apart from the structure of the house. It is a separate system that may be added to a dwelling during or after it is constructed. It has little, if any, effect on the way the building is built. These systems are called active because blowers or pumps are needed to move the air or fluid. **Figure 25-6** shows an active solar system.

Radiant energy: Energy that moves through space in waves.

Thermosiphoning: Movement of heat through a fluid by the action of heated material rising in a narrowed space.

Active solar construction: A system that uses mechanical means to transfer heat via a medium heated by solar energy.

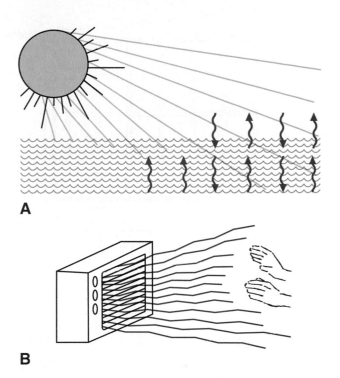

Figure 25-4. Examples of radiant heating. A—Solar radiation strikes the surface of a lake, heating the water. B—The glowing element of an electric heater radiates energy through the air.

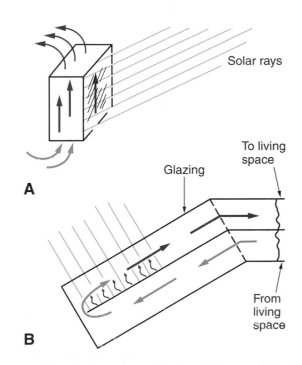

Figure 25-5. A simple example of thermosiphoning. A—Sun shines into an open-ended box. Heated air rises from the open top, pulling cooler air in at the bottom. B—When a baffle is added, cool air moves in through the lower side of the box, and heated air moves out of the top.

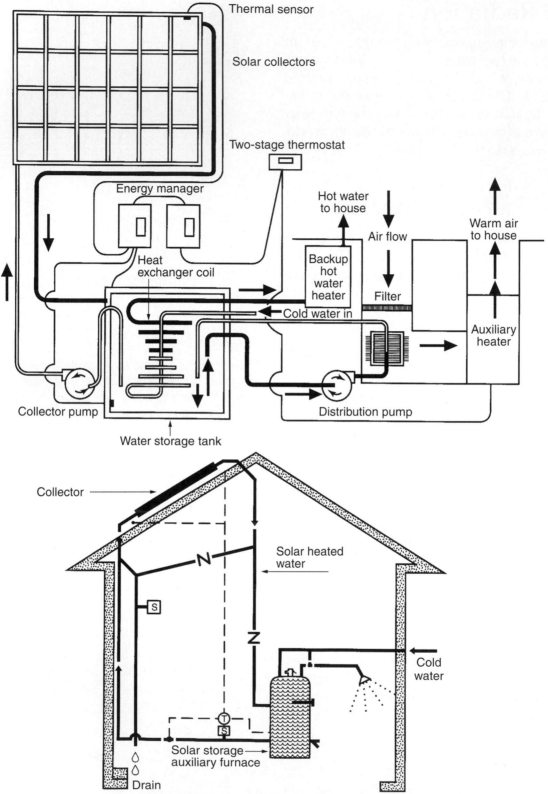

In a pumped draindown unit, solar-heated water flows to the storage tank for direct use by the household. When the pump shuts off, whatever water remains in the collector drains away by gravity flow.

Figure 25-6. Schematics of active solar heating systems that use flat-plate solar collectors like the one illustrated in Figure 25-7.

Passive solar construction, on the other hand, is designed so that energy is absorbed by the mass of the structure. Heat is not distributed by any mechanical means. These systems are called passive because they have few, if any, parts that must move or require power. By its very nature, this system requires a change in the structural makeup of the building. Collection, storage, and transporting of heat is naturally done by the materials used in construction.

Solar collectors are, in simplest terms, boxes that trap solar energy, **Figure 25-7.** In some cases, the solar collectors are made a part of the roof. In other cases, the collectors are attached to the existing roof by a metal framework.

Depending on the individual house design, more than one approach may be used, such as combining active and passive solar systems. Different systems may be used in the same house for greater efficiency, affordability, and beauty. No single floor plan is required—passive solar systems may be incorporated into many architectural styles. Passive designs work well with Cape Cod, colonial, contemporary, or ranch styles.

25.3 Passive Solar Construction

Passive solar energy systems are named for the way in which they operate. There are three basic types of passive solar construction:
- Direct gain.
- Indirect gain.
- Isolated gain (sun space).

The word *gain* refers to how the heat is picked up from the solar radiation.

25.3.1 Direct-Gain System

Direct gain means that the sun directly shines into the living space and heats it up. **Figure 25-8** shows a direct-gain structure. In the simplest of direct-gain systems, there is no massive structure to store the heat. The furniture and air in the room soak up the heat. However,

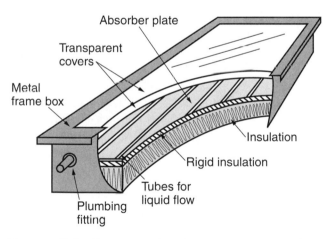

Figure 25-7. Cutaway of a flat-plate solar collector designed to heat liquid. It is used in an active solar system. (National Solar Heating and Cooling Center)

Passive solar construction: A building design that uses solar energy as a heat source with no mechanical means to transport the heat.

Solar collectors: Box-like structures with clear glazing to trap solar energy.

Direct-gain system: Passive solar construction in which the sun directly shines into the living space to heat it.

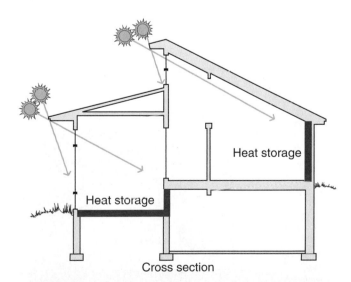

Figure 25-8. Basic design for a direct-gain, passive solar heating system. Heat energy from sunlight is stored by the walls and floor. (District 1 Technical Institute, Eau Claire, WI)

there is usually some means of storing the heat for use at night or on cloudy days. Thick masonry walls or floors, a masonry fireplace, a heavy stone planter, or containers of water are examples of storage systems that may be used. Suitable materials include stone, adobe, brick, or concrete.

At night, stored heat is distributed to the living space by radiation and convection. If the storage is adequate, it should give off heat throughout the night.

One major disadvantage of direct-gain systems is the wide range of heat fluctuation (changes). During the hours of strongest sunlight, heat can build up to uncomfortable levels.

25.3.2 Indirect-Gain System

In an *indirect-gain* system, the sun's rays enter through glazing and heat up a thermal mass rather than the room itself. Often the *thermal mass* is a thick masonry or concrete wall located directly behind large windows or sliding glass doors. Sometimes the wall is vented at the top and bottom. This allows hot air to flow upward and quickly enter the room by way of convection currents, **Figure 25-9.**

The thermal mass wall is the most popular storage structure for indirect solar heating. It is often called a *Trombe wall* for the French scientist who designed it, Felix Trombe. The outer face

is dark-colored for greater absorption of heat. By conduction, the heat travels through the wall to its inner face. Then, radiation and convection work together to distribute the heat throughout the interior space.

Thermal walls still perform their function even if they have windows cut in them. The inner masonry surfaces can be made to fit the interior design by covering them with stucco, gypsum plaster, ceramic tile, or other heat conducting materials. Pictures may be hung on the room side of the wall without greatly affecting thermal performance.

25.3.3 Water Storage Wall

Water is one of the most effective and inexpensive materials for storage of solar heat. In fact, it can absorb about twice as much heat as can stone. The materials for holding the water are not usually cheap, but barrels and other empty containers rescued from salvage yards can be used.

The water wall is placed directly behind a south-facing window, **Figure 25-10.** The container holding the water absorbs the solar heat by radiation. The water absorbs the heat from the container by conduction. The reverse process transfers heat to the interior room.

A more-sophisticated water wall may use phase-change materials. These are materials that change from a solid to a liquid state as they absorb heat. They are able to absorb much more heat than materials that do not change their form. Usually, some type of salt solution is used. A solution of Glauber's salt is common.

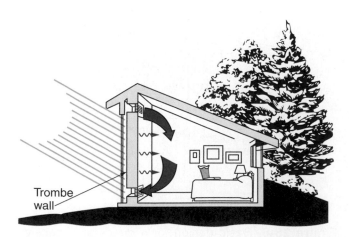

Figure 25-9. The Trombe wall is the most popular method of storing and distributing solar heat by the indirect-gain method. (Iowa Energy Policy Council)

Indirect-gain system: Passive solar construction in which solar heat is stored in masonry, water, or another medium. It is then passed along to living space by radiation, conduction, or convection.

Thermal mass: A thick wall, often masonry or concrete, directly located behind large windows or sliding glass doors.

Trombe wall: Thick wall of masonry placed next to exterior glazing to store solar energy in passive solar construction. Named for French physicist Felix Trombe.

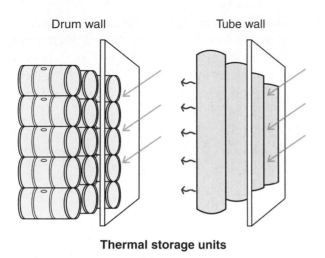

Figure 25-10. Thermal storage units, such as drums or tubes, can be used to hold water or phase-change materials as a storage medium for solar heat. (PPG Industries)

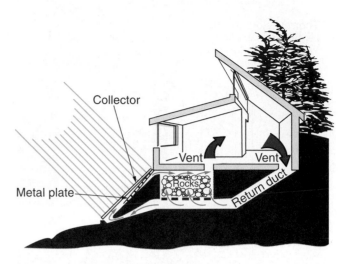

Figure 25-11. A thermosiphoning, isolated, passive solar heating system. The metal plate collector at ground level is the thermosiphon.

Working Knowledge

It is generally easiest and more troublefree to use masonry with conventional structural features, such as walls, floors, and interior fireplaces. Such materials and structures rarely need attention once they are in place.

Figure 25-11 shows a simple sketch of a thermosiphon, isolated-gain system. This is one type. Another is called a sun space. The *sun space* is a common method of providing solar heat, **Figure 25-12.** It is a separate room with large areas of its south wall glazed. Sometimes, the room has thermal mass built into the inside wall or the floor. Additional thermal storage can be provided in water containers or bins of rock.

25.3.4 Isolated-Gain System

In an *isolated-gain* system, solar heat is collected and stored in an area remote (apart) from the living or working space. Its advantages are:
* The living/working area is not directly exposed to the sun.
* The heat it collects is more easily controlled.

Isolated-gain system: Passive solar construction in which generated heat is stored in a separate sun space. Heat is transported to the living space by mechanical means.

Sun space: A separate room with large areas of its south wall glazed to provide solar heat. Sometimes the room has thermal mass built into the inside wall or the floor.

25.4 Passive Solar Advantages

Passive designs have several advantages over active solar heating and conventional heating. Some of these advantages include:
* Common building materials—glass, concrete, and masonry—can be used. Other manufactured materials are not needed, although they are available.
* Conventional carpentry and masonry skills are sufficient for installation. No additional skills must be learned.
* Passive components do not wear out and need little maintenance. Life expectancy is greater than active systems.
* Properly designed passive features can supply nearly 50% of heat required.

① Thermal storage wall
② Nighttime insulation
③ Vertical south glazing
④ Additional thermal
   storage
⑤ Circulation vents

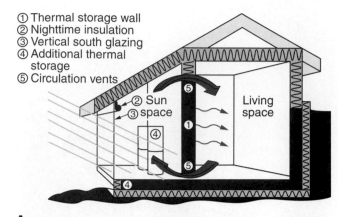

A

B

Figure 25-12. A sun space is a separate room with large, south-facing areas of glass. A—A large storage mass absorbs solar energy. Note the insulation to protect the roof, north wall, and floor from colder surroundings. B—An add-on sun room/conservatory with glass walls and roof functions as a sun space. (Iowa Energy Policy Council, Gorrell Windows & Doors)

- Passive solar is nonpolluting.
- Since heat is generated close to or in the space being heated, little heat is lost during transfer.

25.5 Passive Solar Disadvantages

Although there are many advantages of passive solar systems, there are also disadvantages. Disadvantages include:

- Control of the heat is not as responsive as either conventional heat systems or active solar systems. Most people are used to temperature swings of 3°F–5°F (2°C–3°C), but many passive systems have swings of 10°F–15°F (6°C–8°C).
- Controlling heat and heat distribution in a passive system is not a simple matter. Careful planning is required.

25.6 Solar Heat Control

Adequate control methods must be provided when passive solar heating is being planned. There are times when solar heat is not wanted, such as the summer months. During that time, the same structural features that allow sunlight to enter must be sheltered from the sun, **Figure 25-13.**

Shade can be either naturally or artificially provided. Deciduous trees lose their leaves in winter. They should be planted in locations to protect south-facing windows in summer. It may be possible to locate a new building to take advantage of existing trees on the property. Fast-growing trees can also be planted to eventually provide this shade. They should be located to provide shade for west-facing walls as well. Artificial shade includes shutters and roof structures, such as overhangs.

25.6.1 Overhangs

The correct width of a protective overhang is determined by three factors. See **Figure 25-14.**
- Height of the window or collector.
- Height of the header above the window or collector.
- Latitude of the building.

As a rule of thumb, the overhang in southern states (roughly 36° north latitude) should be 25% of the combined height of header and window. In northern states (around 48° latitude), the percentage should be 50%.

A simple mathematical formula provides a quick and accurate determination of how far the overhang should project:

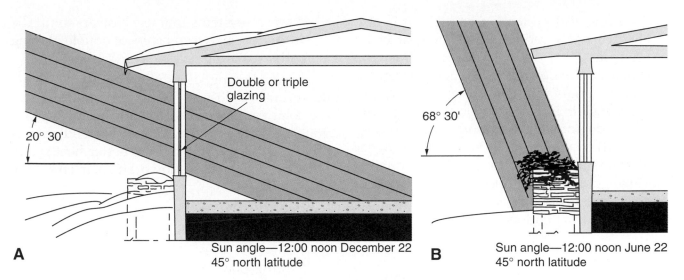

Figure 25-13. Roof extensions or overhangs can be used to control solar rays in a passive heating system. A—At the latitude of Minneapolis, MN, in December, the low sun angle allows rays to pass through windows to heat the thermal mass of the house. B—At the same latitude in June, the higher sun angle permits the overhang to shade the window, preventing heat gain inside of the building.

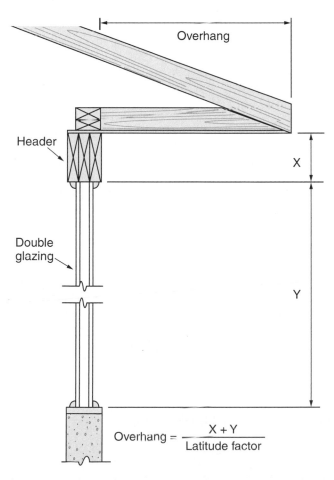

Figure 25-14. The proper width for an overhang can be computed by adding the height of the header (X) and the window (Y), then dividing by a factor based on latitude.

Overhang = (window height + header height) ÷ latitude factor

Use the chart in **Figure 25-15** to find the latitude factor.

25.6.2 Movable Insulation

While outdoor structures and natural shade are more effective protection against the hot summer sun, movable insulation is sometimes

Latitude (in degrees)	Approximate location	Latitude factor
28	Central Florida	5.6 to 11.1
32	Central Texas	4.0 to 6.3
36	Northern Oklahoma	3.0 to 4.5
40	Northern Missouri	2.5 to 3.4
44	Iowa-Minnesota Border	2.0 to 2.7
48	Northern Minnesota	1.7 to 2.2
52	Southern Canada	1.5 to 1.8
56	Central Canada	1.3 to 1.5

Figure 25-15. Factors used to calculate the width of an overhang for latitudes ranging from central Florida to central Canada.

the only practical protection possible. It is also effectively used to reduce nighttime heat losses through glass.

Movable insulation is usually sheet or blanket materials that can be temporarily put over glazing. It is placed over the glass only when you do not want radiation or heat passing through. Movable insulation solves a major problem of glass—its poor insulating quality. The insulation cuts nighttime heat losses and helps keep the building cool during summer days. Movable insulation is available in a variety of types:

- Sheets of rigid insulation.
- Framed-and-hinged or track-mounted insulation panels, **Figure 25-16.**
- Exterior-mounted plastic or metal roller shades, **Figure 25-17.**
- Padded, roller-mounted, flexible cloth panels (insulated shades), **Figure 25-18.**

- Powered systems that use blowers to fill the air cavities between layers of double glazing with insulating beads.

Triple glazing of window areas also cuts down on heat losses. However, it also reduces the solar radiation transmission from 74% (for double glazing) to 64%. Removable insulation is preferred because it provides better heat retention and does not cut down on solar efficiency of the window.

25.6.3 Venting

Venting plays an important role in controlling solar heat in both indirect and isolated-gain systems. Trombe walls sometimes have vents to provide daytime heating for the living space. Opening the vents allows convection currents to carry heated air into the room through the vents. Meanwhile, cool air is drawn into the

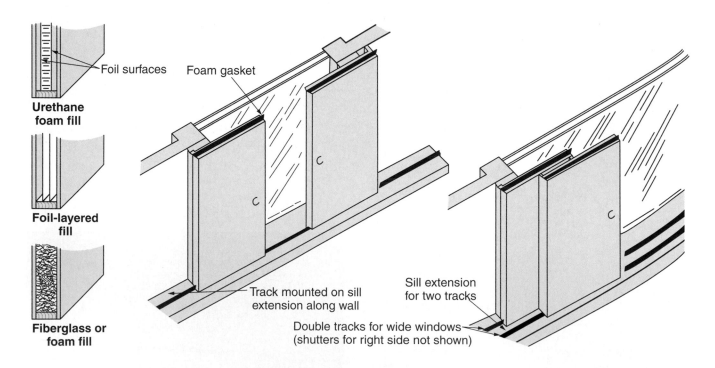

Figure 25-16. Insulated sliding shutters are used to prevent heat loss through glazing at night.

Movable insulation: Sheet or blanket insulation that can be temporarily placed over windows to insulate against solar energy.

A   **B**

Figure 25-17. Power-operated metal or plastic shutters for solar-heated buildings. A—Shutters are opened to permit heating from the sun. B—Shutters can be closed to block unwanted sunlight during the day or to prevent heat loss at night. (Pease Industries, Inc.)

Figure 25-18. An insulated-fabric, rollup shade can be raised or lowered to control heat gain or loss through windows.

Orientation: Locating a building so one wall catches the sun's rays all day long.

space between the glazing and the Trombe wall. Closing the top and bottom vents prevents heat from entering the living space via convection.

Isolated sun spaces are often vented to the outdoors to exhaust unwanted solar heat during hot weather. The vent acts like a chimney. The hot air rises through the roof. Cooler air is pulled into the sun space through vents at the floor level. This provides additional cooling.

Natural venting of buildings is also possible using the chimney effect. This is accomplished with a little sunlight and vents high in the building. A window or skylight lets in sunlight to heat a pocket of inside air. The air rises and passes through the vent. Cooler air is then brought into the building from another area of the living space. **Figure 25-19** shows two methods of using venting for cooling.

25.6.4 Orientation

The first step in solar construction is to locate the building so one wall catches the sun's rays all day long. This is called *orientation*, **Figure 25-20.** Most of the windows should be placed in the south-facing wall, so that the solar radiation can be collected. Few windows should

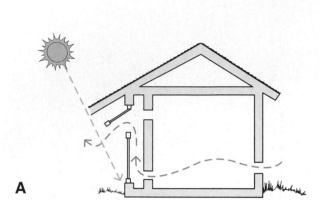

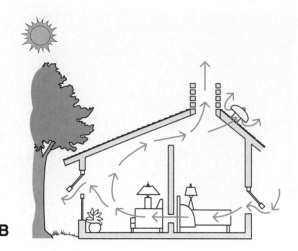

Figure 25-19. Two venting methods used for warm-weather cooling. A—The vent in the glazing allows warm air to escape, drawing cooler air from the shaded side of the building through openings in the Trombe wall. B—Thermal-chimney-effect venting permits warm air to escape through vents at the top of the house, drawing in cooler air through windows on the shaded side. (PPG Industries)

be placed on the north side. Garages should be placed to the northwest, if possible, to block cold winter winds.

25.6.5 Energy Balance

On a typical winter day, homes experience heat gains from three sources:

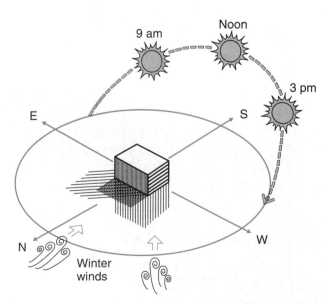

Figure 25-20. Proper orientation to the sun is important for a solar-heated home. The south-facing wall will capture solar energy. A deviation of 10°–20° from true south will not greatly reduce efficiency. (PPG Industries)

- Conventional heat source, such as a stove or furnace.
- Solar radiation.
- Appliances and occupants. This is called *internal heat.*

Energy use is reduced when the heating system can operate at the lowest level that will provide comfort. Solar and internal heat gains can permit a cutback in operation of the heating plant if the house design can balance internal heat gain against solar gain. This means locating heat-producing appliances on the opposite side of the house from space receiving solar gain. **Figure 25-21** shows the best use of internal and solar gains in the northern hemisphere.

25.7 Building Passive Solar Structures

Thermal storage walls used in passive solar construction need to store enough heat to even out wild fluctuations of temperature, which

Internal heat: Heat gain in a home from appliances and occupants.

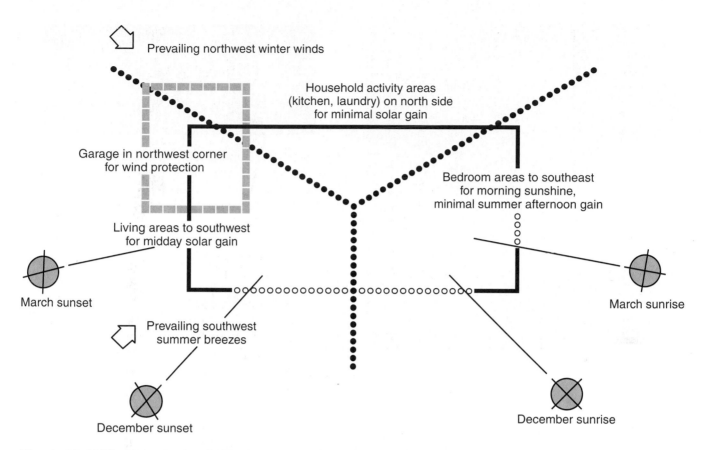

Prevailing northwest winter winds

Household activity areas
(kitchen, laundry) on north side
for minimal solar gain

Garage in northwest corner
for wind protection

Living areas to southwest
for midday solar gain

Bedroom areas to southeast
for morning sunshine,
minimal summer afternoon gain

March sunset

March sunrise

Prevailing southwest
summer breezes

December sunset

December sunrise

Figure 25-21. Solar and internal heat gains can be properly balanced by locating living areas of the house as shown. Appliances generate internal heat gains. (Forest Products Laboratory)

could make a building uncomfortable. All materials are capable of storing heat. However, some materials more effectively store heat than other materials. Some of the most commonly used storage materials include:

- **Water.** Though water is cheap and has greater heat capacity than solid materials, it might be expensive to construct or purchase containers to store it.

- **Concrete and concrete block.** Both have good heat storage capacity and provide structural support for the building, as well. Cores of hollow block can be used as warm air ducts. However, the wall will hold more heat if they are filled with sand-mix concrete. Walls can be plastered or painted to conceal the mass.

- **Brick.** Brick's properties are similar to those of concrete or block. Attractive in appearance, bricks are easier to blend with the decor. Though more expensive than concrete, their appearance often makes them the preferred material.

These materials are good choices for thermal storage, because of their cost, appearance, and suitability. **Figure 25-22** compares the heat-carrying capacities of these materials with other building materials.

25.8 Sizing Thermal Storage Systems

The size of a thermal storage system is determined by the square footage of the living space. For example, in climates where the average winter temperature is 20°F–30°F (–5°C to 0°C), between .43 and 1 sq. ft. of masonry per square foot of living area is adequate. **Figure 25-23** lists the adequate mass for different areas by degree-days. Consult the chart, "Normal Number of Degree-Days per Year" in Appendix B, **Technical Information,** for degree-day information on your locality.

Substance	Specific heat (BTU/lb-°F)	Density (lbs/cu ft)	Heat capacity (BTU/cu ft-°F)
Water*	1.00	62.4	62.4
Wood, oak	0.57	47.0	26.8
Fir, pine, and similar softwoods	0.33	32.0	10.6
Expanded polyurethane	0.38	1.5	0.57
Wool, fabric	0.32	6.9	2.2
Air	0.24	0.075	0.018
Brick*	0.20	123.0	25.0
Concrete*	0.156	144.0	22.0
Steel	0.12	489.0	59.0
Gypsum or plaster board	0.26	50.0	13.0
Plywood (Douglas fir)	0.29	34.0	9.9

*Preferred materials.

Figure 25-22. The high heat-carrying capacities of water, brick, and concrete make them good choices for heat storage. (Iowa Energy Policy Council)

Climate	Winter temperature for coldest months in degree-days/month	Square footage for each square foot area	
		Masonry wall	Water wall
Cold	1500	0.72 to 1	0.55 to 1
	1350	0.62 to 1	0.45 to 0.85
	1200	0.51 to 0.93	0.38 to 0.70
	1050	0.43 to 0.78	0.31 to 0.55
Temperate	900	0.35 to 0.60	0.25 to 0.43
	750	0.28 to 0.46	0.20 to 0.34
	600	0.22 to 0.35	0.16 to 0.25

Figure 25-23. The solar storage mass must be calculated based on the average winter temperature and the size of the area to be heated.

25.8.1 Wall Thickness

Storage wall thickness is important. A space will overheat if more energy is transmitted through the wall than is needed. This will happen if the wall is too thin. Further, a wall that is too thin will not be able to transmit

Material	Thickness (inches)
Water	6 or more
Dense concrete and block	12 to 18
Brick	10 to 14
Adobe	8 to 12

Figure 25-24. General recommendations for the thickness of storage walls made from different materials. Thicker walls will store more heat and release it into the living space over a longer period of time.

heat throughout the night. It will cool before morning.

Figure 25-24 suggests thicknesses for storage walls of various materials. The more rapidly the material conducts heat, the thicker it must be. Otherwise, it may deliver its heat load too soon. This makes the living space uncomfortably warm.

25.8.2 Sizing Direct-Gain Storage

A rule of thumb for walls and other structures used for storage in direct-gain systems is to provide 150 lb. of concrete or masonry for every square foot of glazing. This assumes that the storage is subjected to direct sunlight. At the same time, there should be at least 150 lb. of concrete or masonry for every 2 sq. ft. of area to be heated. It should be about 4″–6″ thick. If the structure is thinner than 4″, it will not be able to store enough heat per square foot. Floors used for thermal storage should not be carpeted wall to wall. However, scatter rugs, furniture, and other obstructions have little effect on the performance of the floor.

25.8.3 Effect of Color on Collecting Surface

Since dark colors more readily absorb heat than light colors, the wall surface facing the glazing should be dark. Although black is most efficient as an absorber, other dark colors may be used with nearly the same efficiency. Dark blue, for example, works almost as well. Inside wall surfaces may be any color.

25.8.4 Wall Construction

Solid Trombe walls store more heat than vented ones. The outside surface of the wall may reach 150°F (65°C) or more on a sunny day. The wall facing the room, however, maintains a fairly uniform temperature throughout a 24-hour period. Vents are necessary if the living space requires extra warmth during the day. Then, the air space between the wall and the glazing sets up a thermosiphon (natural convection).

Since the Trombe wall is a bearing wall, construct it based on the principles given in Chapter 7, **Footings and Foundations.** Footings should be twice as wide as the wall's cross section and equal in thickness. **Figure 25-25** shows a vertical cross section through a typical Trombe wall.

Some provision for venting of unwanted heat in warm weather is advisable. **Figure 25-26** is a detail drawing for venting the Trombe wall at the header. For aesthetic purposes, windows can be included in a Trombe wall. Adjustments should be made to the dimensions to assure sufficient area for proper heat collection and storage. **Figure 25-27** shows several designs that incorporate windows.

25.8.5 Special Concerns

It is desirable to construct thermal glazing so that the glazing can be occasionally removed for cleaning the glass or repainting the Trombe wall. Thus, it is helpful if the glass panels are supported in their own sash. Panels should not be so large that the windows are difficult to handle.

Single panes of glazing are not as efficient as double glazing, but triple glazing is not generally recommended. Triple glazing cuts down on the effectiveness of the solar radiation and does not appreciably reduce the heat loss over double glazing.

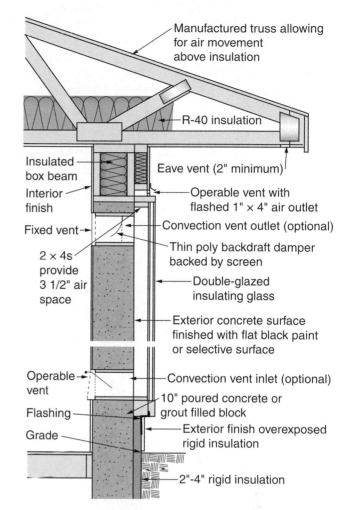

Figure 25-25. Typical Trombe wall section. Note the use of insulation in the hollow header. The air space between the glazing and the wall can vary from 2"–5". (Iowa Energy Policy Council)

25.9 Designing an Isolated-Gain System

Basically, the isolated-gain system is a small room attached to the main structure of the building. **Figure 25-28** shows a variety of designs for bringing the heat from the sun space into the living space. A greenhouse attached to a south exposure of a house is a good example of this type of structure.

The sun space is sometimes "embedded" in the house. That is, it is within the walls of the house and is separated from the house only by its interior partitions. One of these partitions can be a storage wall of concrete, masonry, or water.

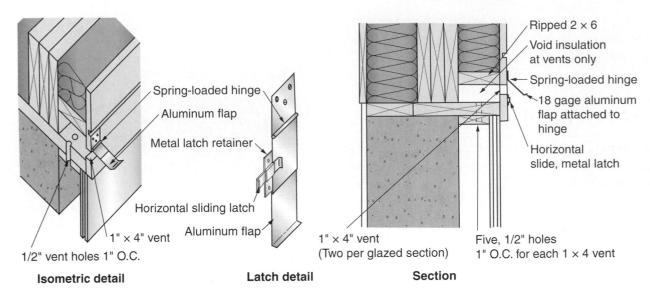

Figure 25-26. A method used to vent unwanted heat from a Trombe wall during warm weather.

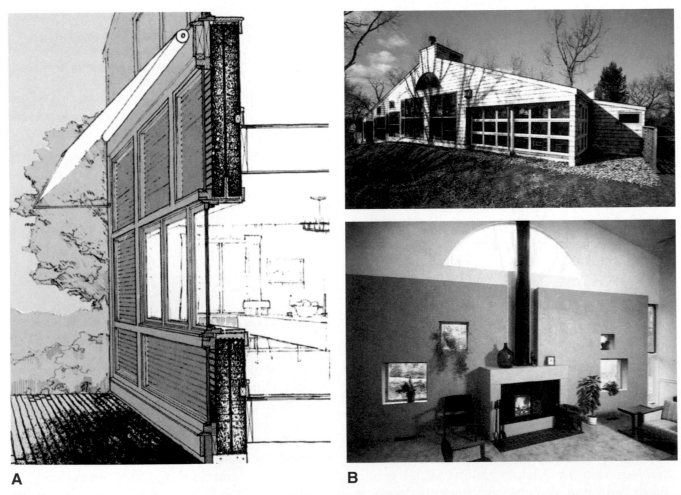

A

B

Figure 25-27. Incorporating windows in a Trombe wall. A—A section of a brick wall with casement window installed. The roll-up awning takes the place of an overhang to shade the window in warm weather. B—Outside and inside views of a design that uses small windows in a thick storage wall. The clerestory window above the fireplace provides additional light to the room. (PPG Industries)

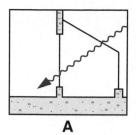
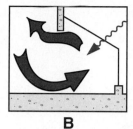
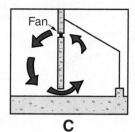
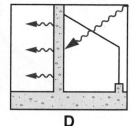
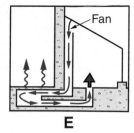

| A | B | C | D | E |

Figure 25-28. Various methods can be used to bring heat from a sun space into living areas. A—Solar rays transmitted through the sun space into the living space. B—Natural air exchange. C—Mechanical movement using a fan. D—A storage wall providing heat by radiation. E—Heat radiation from gravel in a storage bin to the building above.

25.10 Passive Thermosiphon System

Thermosiphoning space heating systems are not well researched or understood. Passive system designers suggest that the rock bed needed for storage have a cross-sectional area that is about 50%–75% of the area of the collectors. For example, if the area of the collector is 200 sq. ft., the cross section of the bed should be 100 sq. ft.–150 sq. ft. Four-inch-diameter rock is recommended for easy circulation of the air around it. The depth of the bed should be 20 times the rock diameter, about 6'-6" for 4" rock. To find the proper volume of storage, multiply the collector area by a factor of 3.25–5.0. The 200 sq. ft. collector requires 650 cu. ft.–1000 cu. ft. of storage space.

Rock bins need to be insulated against heat loss on their sides and bottoms. Dampers should be located in the ducts leading to the collector.

These are closed every evening to prevent the loss of heat to the outside at night.

25.11 Insulating Passive Solar Buildings

To take advantage of solar heating, buildings should be well insulated. Refer to Chapter 15, **Thermal and Sound Insulation.** Good insulating practice for passive solar construction is shown in **Figure 25-29.**

When insulating, it pays to provide thermal barriers anywhere that heat is likely to leak out. There should be no place that air can flow between the inside and outside without going through insulation. Seal up all areas where air could possibly leak through cracks.

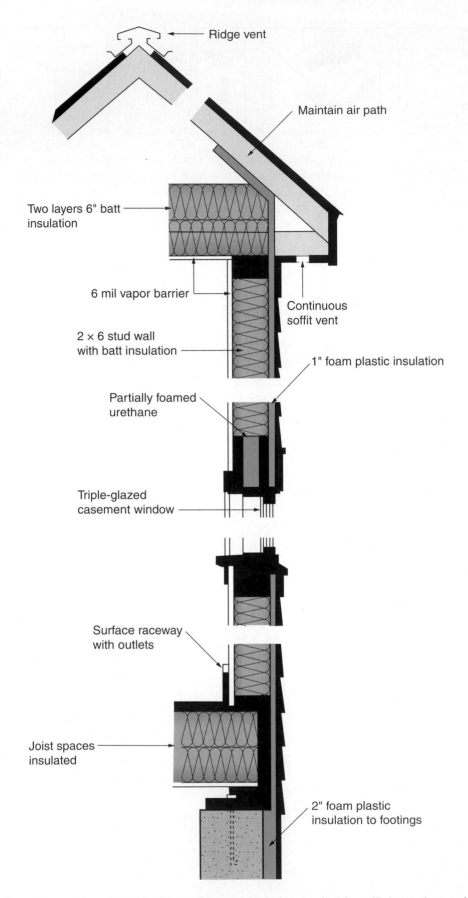

Ridge vent

Maintain air path

Two layers 6" batt insulation

6 mil vapor barrier

Continuous soffit vent

2 × 6 stud wall with batt insulation

1" foam plastic insulation

Partially foamed urethane

Triple-glazed casement window

Surface raceway with outlets

Joist spaces insulated

2" foam plastic insulation to footings

Figure 25-29. Cross section of a well-insulated building. Proper insulation is vital for efficiency in a solar-heated home. (Iowa Energy Policy Council)

Summary

The greenhouse effect occurs when solar radiation is trapped as heat energy after passing through glass. It is important to all types of solar construction. Heat always moves from areas of higher temperature to those of lower temperature by means of conduction, convection, or radiation. The two types of solar construction are active and passive. Active solar systems use collectors to trap solar energy and heat air or a liquid. Blowers or pumps move the heated air or fluid to the point where it will be used or stored. An active system can be added to an existing structure. A passive solar system uses the mass of the structure itself to absorb the heat energy, then gradually release it for heating. The building must be specifically designed for passive solar. There are three different types of gain used in passive solar systems. In direct gain, the sun directly shines into living spaces. Heat is absorbed by furniture and such structural elements as stone fireplaces or concrete floors. Indirect gain systems use the sun's energy to heat a thermal mass such as a large concrete wall located directly behind glass panels. Isolated gain uses a separate space, such as a greenhouse, to collect and absorb heat energy, which is then moved by convection to living spaces. Passive solar systems have environmental and cost advantages, but also results in less even and controllable heating than non-solar methods. Control of solar heating through structural and other methods must be planned before the building is built. Sizing of passive solar installations is based on the square footage of the living space. Charts are available to determine the glass area and amount of storage mass needed.

Test Your Knowledge

Answer the following questions on a separate piece of paper. Do not write in this book.

1. Solar radiation passes through glass readily, but heat cannot pass back through the glass as readily. This phenomenon is called the _____ effect.
2. List the three different ways heat is transferred.
3. _____ is simply the result of a liquid or gas expanding and rising.
4. The main characteristics of passive solar construction are (select all correct statements):
   A. It has few, if any, moving parts.
   B. It is usually a stationary, structural part of the house.
   C. Maintenance of the system is more expensive.
   D. Collection, storage, and transportation of heat is naturally done by the materials used in construction.
   E. There is always ductwork to carry the heat to living space.
5. List two disadvantages of passive solar energy.
6. What factors govern the size of the overhang above solar glazing?
7. Internal heat is the heat gained from _____.
8. As a general rule, how much heat storage (in pounds of concrete) should be provided for every square foot of glazing?
9. *True or False?* Triple glazing is more effective than double glazing in passive solar construction.
10. A storage bin must be constructed to house the rock needed for a thermosiphon system. The glazed area of the collector is 8' by 20'. What amount of rock must the bin hold?
    A. 160 cu. ft.–180 cu. ft.
    B. 520 cu. ft.–800 cu. ft.
    C. 1600 cu. ft.–1800 cu. ft.

Curricular Connections

Science. Construct a small thermosiphoning solar heat unit. Devise a simple test to determine whether it is operating. Take photographs as you assemble and operate the unit. Use them as visual aids when you report to your class on the thermosiphon project.

(Continued)

(Continued)

Mathematics. Research the cost for installing a solar collection system to provide for the hot water needs of a family of four in your area. Visit a store that sells water heaters and check the energy efficiency labels on several 50-gallon, gas-fired heaters to obtain an average annual operating cost. Using this figure and the cost of the solar collection system, determine the payback period. This is the time, usually in years, needed to repay the investment made in the system from the savings on gas use.

Outside Assignments

1. From your local library, select books on designing passive solar systems for housing. Report to the class on your findings.
2. Write to companies offering passive solar structures. Study the literature and report to the class on items that seem to have practical advantages.
3. Using the charts and information in this chapter, design an indirect-gain system using a Trombe wall. Size the glazing, storage wall, and overhang according to the conventional principles.

Remodeling, Renovating, and Repairing

26

Learning Objectives

After studying this chapter, you will be able to:

- Identify different types of residential construction by visual inspection.
- Set up a proper sequence of renovation or repair for the interior and exterior of a house.
- Repair and replace deteriorated components and systems.
- Remove parts of the structure without damaging the total structure.
- List steps for removal of wall sections prior to remodeling.
- Identify bearing walls.
- Determine loads and calculate the correct header size for the span.
- Install and support headers, concealed headers, and saddle beams.
- Follow accepted methods in replacing all types of doors.
- Make repairs to wood and asphalt shingles.
- Make a solar retrofit on an older home.

Technical Vocabulary

Concealed header
Firestops
Flitch
Jack posts
Jamb clip
King studs
Ribbon
Saddle beam
Shoring
Sistering
Thermosiphon

In some ways, remodeling and renovation work are more painstaking than new construction. It is more difficult to work with structures that may have defects such as sagging floors, rotting framework, and walls that are no longer plumb. Sections or components of the old building often must be dismantled without destroying what is to be retained or reused. Frequently, the carpenter must be able to visualize how a structure was built so that no damage is done to it or any of its systems during demolition and removal of old walls.

Remodeling and renovating are well worth the effort, **Figure 26-1.** Remodeling brings greater comfort and usability for the owner. Further, remodeling and repairing have other benefits, such as forestalling the structure's deterioration or enhancing its value.

26.1 What Comes First?

In a long-neglected home, halting further deterioration with temporary repairs may be the

Figure 26-1. This solid old house stood in the way of a commercial development. It was moved to a new location, placed on a new foundation, and renovated.

first order of business. Leaks should be remedied before they cause further damage to the structure. Replace missing shingles and seal leaks with roofing tar. If the shingles and roof sheathing are beyond repair, replace the rotted sheathing and reshingle the roof. See **Figure 26-2.** However, if structural changes are planned that would change rooflines, the roof should be repaired after those changes.

Structural repairs may be necessary for personal safety. Before attempting to repair or replace wall coverings, doors, and windows, examine the house to see if there are structural repairs that need to be made. Such items could be seriously affected by jacking or replacement of foundations, sills, studs, or other frame members. Start repairs with the foundation and sills and work your way up. Do not fix a structural problem involving the roof, then jack up the frame. Everything will have shifted.

26.1.1 Sequence of Exterior Renovation

Renovating the exterior of the building should be the first order of business. Not all of the following steps will need to be taken. However, whatever steps are needed should be taken in order.

A

B

C

Figure 26-2. Roofing repairs are the most common renovation project. A—Old shingles have been removed, revealing the need for replacing some of the sheathing. B—After rotted sheathing is replaced, 1/4″ waferboard is fastened in place over the entire roof area. C—Roofers apply a new drip edge prior to shingling.

PROCEDURE

Renovating an exterior

1. List what needs to be done.
2. Demolish whatever must be eliminated and clear away the debris.
3. Perform all needed structural work. Work from the bottom up. This step should include chimneys and masonry or concrete work. Take measures to protect open areas from the weather.
4. Complete the site work. Regrade and provide drainage as needed.
5. Repair or replace the roof, including flashing, gutters, and roof vents.
6. Scrape or strip paint.
7. Repair masonry and tuckpoint.
8. Repair or replace windows and outside doors, **Figure 26-3.**
9. Repair or replace siding, **Figure 26-4.**
10. Stain or prime wood siding and trim.
11. Caulk, glaze, and putty.
12. Paint.

A

B

C

Figure 26-3. Replacing exterior windows and doors. A—The new, energy-efficient windows installed in this house are smaller than the units they replaced. One is located where a door was removed. B—Carpentry students check the operation of a replacement window after installing it. C—A new doorway is being framed in place of a window in this remodeling job. The siding at the left was removed to provide access to an electrical receptacle that had to be relocated. (Des Moines, Iowa, Public Schools)

Figure 26-4. Replacing siding and soffits on a house located in a beach community (note the pilings) along the Gulf of Mexico. (Monday Construction)

26.1.2 Interior Renovation

Interior work can proceed while exterior work is underway. If the building is unoccupied, it is usually best to bring all phases of the work along at the same rate. All of the demolition, mechanical system repair, drywalling, and stripping and painting can be done in sequence.

If the house is occupied, consideration must be given to the comfort of the occupants. In this situation, there are two choices:

- Completing a single room or area to the point where the space is livable, but without finishes or decoration.

- Zone-by-zone completion. In this approach, a floor or wing is completely finished before going on to the next floor or area.

PROCEDURE

Renovating an interior

1. Demolish what is to be eliminated and remove the debris, **Figure 26-5.** This includes removal of structural elements such as studs. Provide temporary shoring as needed.
2. Perform structural work, such as alterations to bearing walls, **Figure 26-6.**
3. Insulate exterior walls and ceiling.

(Continued)

(Continued)
4. Make changes to or remove nonbearing partitions; install or remove soffits or pipe chases; repair subfloors; install nailers for built-ins, plumbing fixtures, and light fixtures.
5. Rough-in plumbing and electrical wiring.
6. Install drywall, repair lath and plaster, tape drywall joints, apply a skim coat of plaster. See **Figure 26-7.**
7. Put down underlayment for new flooring or tile.
8. Repair or install ceramic tile.
9. Install fixtures for plumbing and heating. Set fixtures before any finishing to avoid damage to floors, walls, or trim.
10. Install light switches and receptacles.
11. Install unfinished floor coverings.
12. Repair or install woodwork or trim. Refinish old woodwork and cabinetry, if in good repair, **Figure 26-8.**
13. Prime and paint or wallpaper walls.
14. Apply finish to floors, if unfinished.
15. Install prefinished flooring materials.
16. Touch up painted and clear-finished surfaces.
17. Install light fixtures and hardware such as cover plates and handrails.
18. Clean finished areas and wash windows.

Safety Note

When refinishing floors or woodwork that has been varnished, be sure to wear a dust mask or respirator and use a sander with a HEPA filter attachment to capture as much dust as possible. Old coats of varnish often contain significant levels of lead. After sanding, vacuum the walls and other areas where dust may have collected.

26.2 Design of Old Structures

Many older homes were built using balloon framing. This method was popular in the United States from about 1850 through the 1930s. The main feature of this construction is long studs that run from the sill all the way to the plate on

A   B

Figure 26-5. Demolition work. A—Using a hammer drill to remove a masonry wall. B—A dumpster is usually located on a construction site to collect debris and scrap materials. When material is being discarded from upper stories, a chute is used to direct it into the dumpster. (Des Moines, Iowa, Public Schools)

Figure 26-7. Drywall is installed after plumbing and electrical rough-in work is completed. A screw shooter is being used to fasten drywall to steel studs. (Des Moines, Iowa, Public Schools)

Figure 26-6. This carpenter is fastening one end of a large laminated beam that will serve as a header for an enlarged opening between an entryway and family room. The post in the foreground is part of temporary shoring used to support the structure during changes to the bearing wall.

Figure 26-8. Renovation work often involves refinishing woodwork or cabinetry. (T.W. Lewis Construction Company) which the rafters rest. These are called *building height studs.* They may be spaced anywhere from 12″ to 24″ on center.

In balloon framing, the second floor joists rest on a horizontal member called a **ribbon.** This member is let into the studs. It is usually a 1 × 6 board. *Firestops* are short pieces of 2″ thick

blocking installed between studs and joists at each floor level. Their purpose is to prevent spread of fire from one part of the dwelling to another.

Figure 26-9 shows details of balloon frame construction. Additional information on housing designs is located in Chapters 8, 9, 10, and 23.

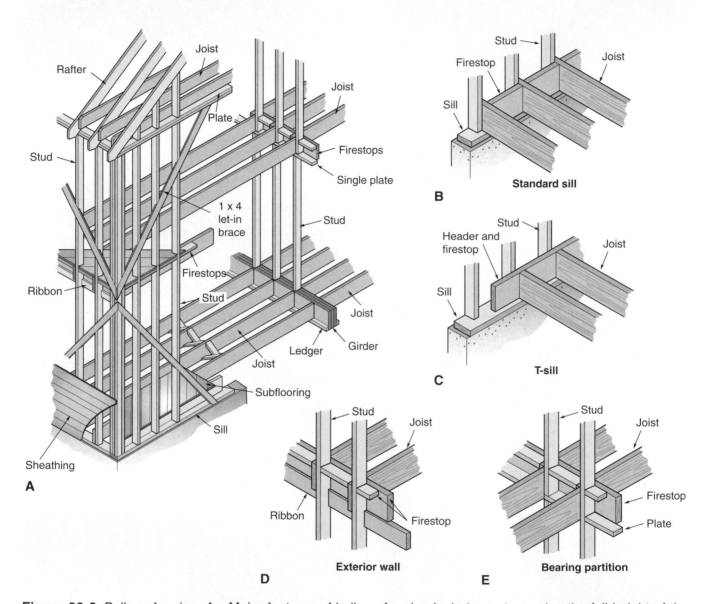

Figure 26-9. Balloon framing. A—Major features of balloon framing include studs running the full height of the building and ribbons used to support floor joists for the second story. B—Detail of a standard sill. C—Detail of a T-sill. D—Detail showing how joists are attached to the ribbon for a second story. E—Detail of framing used for a bearing partition.

Ribbon: A narrow board let into a stud or other vertical member of a frame, thus adding support to joists or other horizontal members.

Firestops: Blocks used in a building wall between studs to prevent the spread of fire and smoke through the air space.

Figure 26-10 shows details of post and beam construction. This type of construction was popular in colonial times. Some new homes are constructed with this method.

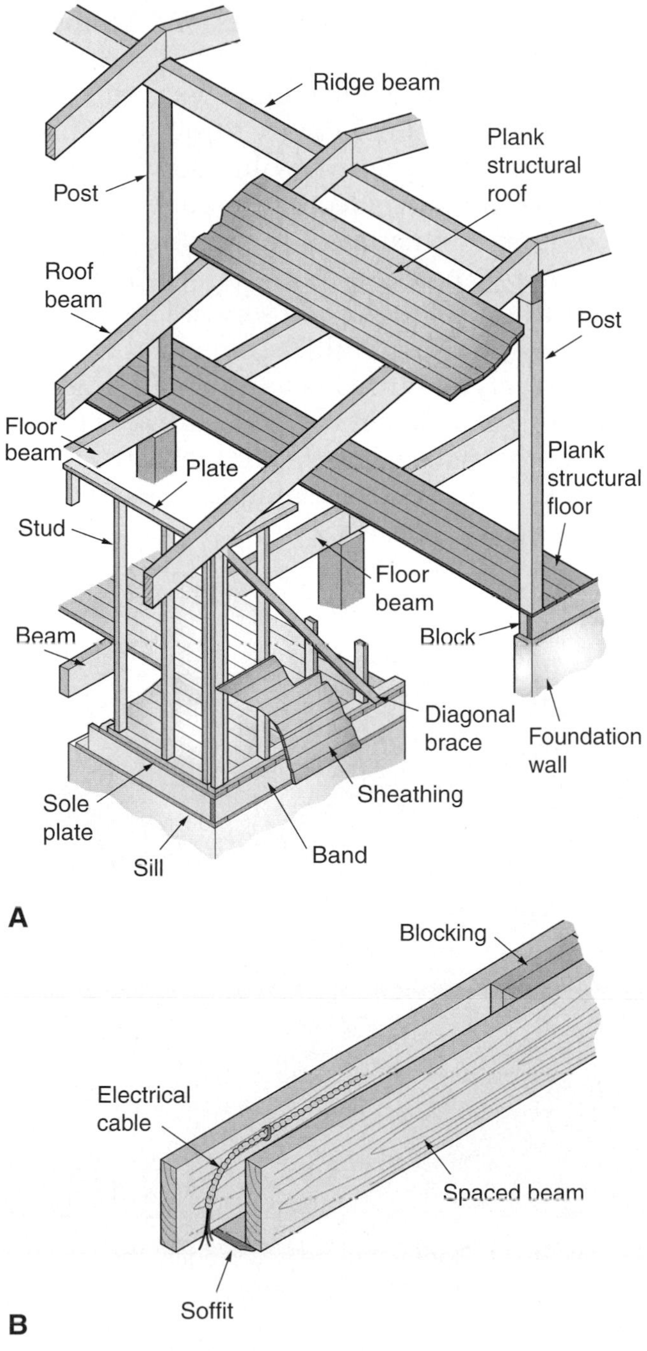

26.3 Replacing Rotted Sills

Rotted sills and other structural members are not uncommon in older homes. Often, leaks and other problems have not been corrected in these homes. Replacing a sill in platform construction requires lifting and supporting the weight of the building while the old sill is removed and replaced with a new, pressure-treated sill. See **Figure 26-11**.

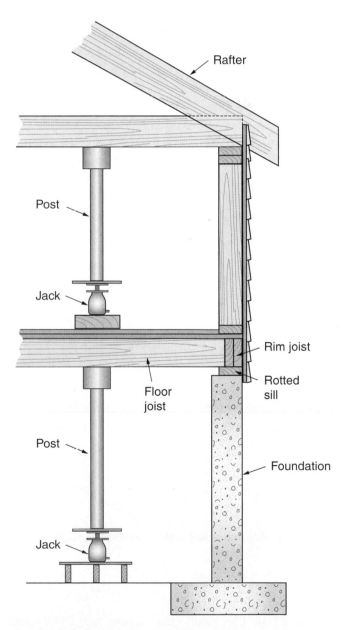

Figure 26-11. Using shoring at two levels to raise and support the structure when replacing a rotted sill.

Figure 26-10. Post and beam construction. A—General construction detail. B—Spaced beams used in this construction method allow electrical wires and plumbing lines to be hidden.

Reinforcing rotted studs usually does not require temporary support, unless the rot has caused the wall to settle and sag. Reinforcing is best done by *sistering*. This is accomplished by nailing another stud alongside the rotted member. Installing the sister requires removal of the interior wall covering or exterior siding.

26.4 Hidden Structural Details

Before demolition of any kind is attempted for inside changes, study the building to determine what type of construction was used. This will make removal of structural elements much easier and safer.

If the house has a basement or crawl space, examine the areas beneath first-floor walls. Try to determine if there are gas or fuel lines, electric conductors, or plumbing running through the walls or floors where you will be working. See **Figure 26-12.** Locate and close valves to stop gas or water flow in service lines. As a precaution, cut power to any electrical circuit that might be disturbed during demolition. Pull the fuse or trip the circuit breaker. If there are other services on the same circuit that you do not want interrupted, an electrician should be called to do the work.

Go outside and study the roof. Note where vents and chimneys are located. See **Figure 26-13.** Greater care must be exercised in removing wall coverings where exhaust vents and plumbing stacks are located.

Building additions require excavating for piers or foundations, **Figure 26-14.** Be sure to locate any underground utility services. Pipes and cables could accidentally be cut or damaged by digging. Sometimes, inspecting sills and joists in a basement will reveal markings indicating where underground services are located. Checking with local utility companies and city departments for buried lines is usually required by law. In some states or local areas, a joint utility locating agency exists. Such agencies offer a "one-call" service for locating all underground utilities prior to excavation.

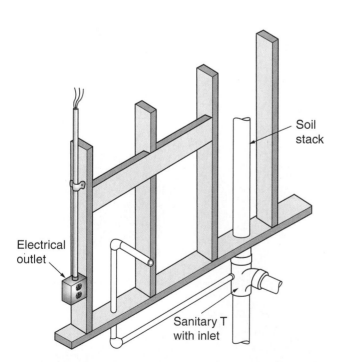

Figure 26-12. Electrical, water supply, and DWV lines are usually concealed within walls. Use care in removing wall coverings to avoid damaging these lines.

Figure 26-13. Roof vents help to identify the location of DWV piping. (CertainTeed Corp.)

Sistering: Reinforcing framing members by nailing another member alongside the first member.

Figure 26-14. Excavating for a building addition must be done with care to avoid damaging utility lines. Water, gas, and sewer lines must be located before beginning to dig. (Jack Klasey)

26.5 Removing Old Walls

The first step in opening up or tearing out a wall is to remove trim such as baseboards, cove strips at the ceiling, and window and door trim. If these materials are to be reused, carefully remove them.

Next, remove the wall covering to expose the framing, **Figure 26-15.** Be sure to wear a mask to avoid breathing in dust and particles of construction materials. Most construction done during the past 40 years used drywall. In homes older than that, wood lath and plaster were used.

Tools useful for demolition include a sledge hammer, ripping or wrecking bar, hammer or hatchet, and rip chisel, **Figure 26-16.** The rip chisel is essential if any of the trim is to be saved and reused. A reciprocating saw is also useful for cutting framing.

Figure 26-15. Removing wall coverings exposes the framing and utilities running inside the walls. In this older home being renovated, the large pipes at left supply the radiators of the hot-water heating system. Also visible are electrical conduit and water supply pipes.

Figure 26-16. A sledge hammer is sometimes used in demolition work. In this case, a partition made from hollow masonry is being removed. (Des Moines, Iowa, Public Schools)

Many carpenters use a flat-bladed garden spade to remove lath and drywall. This tool will also remove drywall and lathing nails as it is run over the studs. If the wall is a partition, removal of the covering is all that is necessary to strip the framing. In outside walls, insulation and siding also need to be removed to strip the framing.

For good housekeeping, it is often useful to rent a large trash container like the one shown in **Figure 26-17.** Then, debris can be immediately

Figure 26-17. A portable trash container, or dumpster, is rented and positioned at the job site to help keep it clean. When full, the dumpster is loaded on a truck and transported to a disposal site.

cleared away so that it does not clutter up the job site. Loose debris is an eyesore as well as a safety hazard.

26.5.1 Recognizing Bearing Walls

Before removing a wall or a portion of it, determine whether or not it is a bearing wall. Bearing walls need shoring (temporary support) while the wall is being removed. Permanent supports of some kind must be in place before the remodeling is completed.

Determining which walls are bearing walls is quite simple with outside walls; interior partitions are another matter. All outside walls support some weight of the structure. End walls of gabled, single story houses are the sole exception. They carry little weight except that of the wall section from the peak of the roof to the ceiling level. Outside walls running parallel to the ridge of the roof carry the most load since they provide almost all of the support for the roof.

Identifying which partitions (interior walls) support weight from structures above is not always simple. The following conditions are usually an indication of a bearing wall. See **Figure 26-18.**

• The wall runs down the middle of the length of the house.

• Ceiling joists run perpendicular to the wall.

• Overhead joists are spliced over the wall, indicating they depend on the wall for support.

• The wall runs at right angles to overhead joists and breaks up a long span. The joists may not be spliced over the wall. Check span tables for load-bearing ability of the joist.

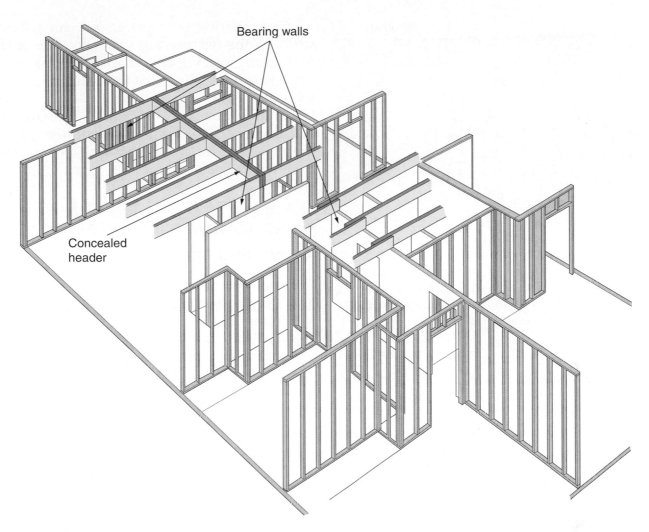

Figure 26-18. Bearing walls can often be most easily identified by examining an unfloored attic space. Walls running the length of the building center usually support the joists.

- The wall is directly below a parallel wall on the upper level. If the walls are parallel to the run of the joists, the joists may have been doubled or tripled to carry the load of the upper wall. The only way to be certain is to break through the ceiling for a visual inspection.

Familiarize yourself with the framing principles used for the type of structure you are remodeling. Study the framing chapters in this book for details.

Shoring: Bracing used to provide temporary support.

26.6 Providing Shoring

Shoring is temporary support of a building's frame. It must be installed to support ceilings or upper floors before all or a portion of a bearing wall is removed. **Figure 26-19** shows one type. It is simply a short wall of 2 × 4 plates and studs. The shoring wall can be assembled on the floor and lifted into place. To strengthen it, use double plates top and bottom.

In some cases, it may be necessary to place a second shoring on the other side of the bearing wall. Usually, the joists are lapped and spiked well enough to provide stiffness to the joint so a second shoring is not necessary.

A second shoring method is used by many carpenters. Larger dimension lumber—4 × 6s,

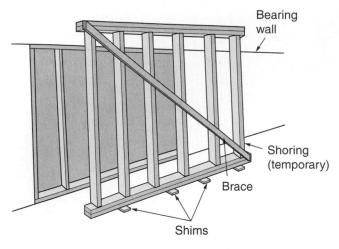

Figure 26-19. Shoring must be used to support the weight of the structure when a bearing wall is being removed or altered. The shoring is positioned next to and parallel with the bearing wall.

for example—are used for plates. Adjustable steel posts, or *jack posts,* are used in place of the studs, **Figure 26-20.**

Be sure that the shoring is resting on adequate support. That is, it should rest across several joists. If shoring is running parallel to floor joists, lay down planking to distribute weight across several joists. Do the same at the ceiling.

Leave the shoring in place until a permanent support has been framed in to take the place of the bearing wall. This may be a lintel or a concealed header.

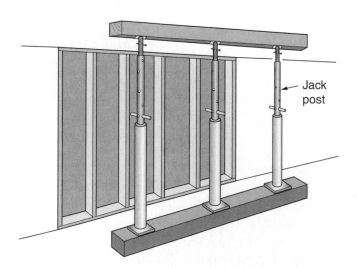

Figure 26-20. An alternate shoring method uses heavy timbers and adjustable steel pillars called jack posts.

Once shoring is in place, you can remove studs in the old bearing wall. Remove studs first, being careful not to disturb plaster or drywall at the ceiling. Try to salvage the studs for future use. Use a reciprocating saw to cut the studs away from the sole plate if the nails are not accessible. If the studs are not being recycled, use a sledge at the bottom to loosen them from the sole plate.

PROCEDURE

Installing shoring

1. Measure the distance from floor to ceiling.
2. Cut the studs shorter than the floor-to-ceiling distance to allow 1/4″ to 1/2″ clearance.
3. Nail the studs to one top and one bottom plate.
4. Attach the second top plate. Leave the second bottom (or sole) plate off to allow clearance when raising the shoring.
5. Square up the shoring wall and attach a 1 × 4 diagonal brace.
6. Lift the shoring in place. If necessary, get a helper to do some of the lifting. Keep the shoring far enough from the partition so you will have room to work.
7. Slide the second plate under the bottom plate. Use wood shingle shims at regular intervals along the bottom of the wall to bring the temporary wall to full height.

26.7 Framing Openings in a Bearing Wall

When studs are removed to open up a wall for a passageway or room addition, a header (lintel) must be installed to provide support. The header can be built up from 2″ lumber. Shim between the lumber with thin plywood to build out the width to that of the framed wall, **Figure 26-21.** On very long spans, the carpenter

Jack posts: Adjustable steel posts used in place of studs for shoring.

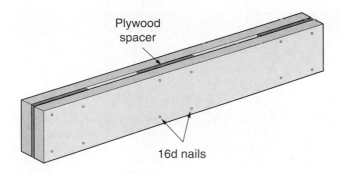

Figure 26-21. A header is built from two lengths of 2″ lumber with a plywood spacer between them to match finished width to the wall studs.

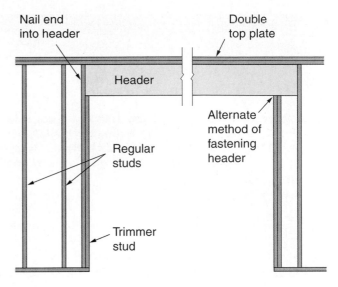

Figure 26-22. Standard and alternate methods of supporting a header. The doubled studs of the alternate method provide greater support for headers spanning longer distances.

may prefer to construct or purchase a box beam. Refer to Chapter 23, **Post-and-Beam Construction,** for detailed information.

26.7.1 Supporting Headers

Headers are supported by resting them on the top of short studs called trimmers, or trimmer studs. Additional support is provided by nailing a full length stud alongside the trimmer at each end of the header. These are sometimes called *king studs.* Stagger nail the trimmers to the king studs. Toenail the header to the trimmer and end nail it to the king studs, **Figure 26-22.** On long spans, the header may be extended beyond the rough opening to the next full-length stud. This helps make the framed opening more rigid.

26.7.2 Sizing Headers

Headers that replace sections of bearing walls should be sized according to recommended standards for load-bearing ability. The size is specified in the architectural drawings for the remodeling job. The carpenter should

also be aware of what the standards are and how the sizes are determined. Essential information is given in **Figures 26-23** and **26-24.** Also check local building codes before fabricating the header.

To use the table in **Figure 26-23,** find the column that describes the load. Move down the column until you find the span corresponding with the opening you need. Then, read to the left for the depth needed for the header.

The tables in **Figure 26-24** require some arithmetic. First, compute the load using the table of average weights for different areas of the building. Once you know the total load, locate the load in the column for the correct span. The size of the header needed is at the top of the column.

If headroom is a concern, a steel-reinforced header provides the needed strength with less depth than a built-up wood header. A steel plate is sandwiched between lengths of 2″ lumber. The plate is called a *flitch* and can be made by a metal fabricator.

26.7.3 Computing the Load

To compute the load on a header, refer to **Figure 26-25.** First, find the number of square

---

King studs: Full length studs nailed alongside the trimmer at each end of the header.

Flitch: A steel plate placed between two pieces of 2″ lumber to form a reinforced header.

Nominal depth of header (in inches)	Outside walls			Inside walls			
	Roof, with or without attic storage	Roof, with or without attic storage + one floor	Roof, with or without attic storage + two floors	Little attic storage	Full attic storage, or roof load, or little attic storage + one floor	Full attic storage + one floor, or roof load + one floor, or little attic storage + two floors	Full attic storage + two floors, or roof load + two floors
4	4'	2'	2'	4'	2'	—	—
6	6'	5'	4'	6'	3'	2'- 6"	2'
8	8'	7'	6'	8'	4'	3'	3'
10	10'	8'	7'	10'	5'	4'	3'-6"
12	12'	9'	8'	12'- 6"	6'	5'	4'

Figure 26-23. This table lists allowable spans for headers under different load conditions.

Average weight of house by area

Unit	Load/square foot
Roof	40 lb.
Attic (low)	20 lb.
Attic (full)	30 lb.
Second floor	30 lb.
First floor	40 lb.
Wall	12 lb.

Built-up wood header (double or triple on two 4" x 4" posts)

Span (in feet)	Weight (in pounds) safely supported by:							
	Two 2 x 6	Two 2 x 8	Two 2 x 10	Two 2 x 12	Three 2 x 6	Three 2 x 8	Three 2 x 10	Three 2 x 12
4	2250	4688	5000	5980	3780	5850	7410	8970
6	1680	3126	5000	5980	2520	4689	7410	8970
8	—	2657	3761	5511	—	3985	5641	8266
10	—	2125	3008	4409	—	3187	4512	6613
12	—	—	2507	3674	—	—	3760	5511
14	—	—	—	3149	—	—	—	4723

Steel plate header (on two 4" x 4" posts)

Span (in feet) / Plate	Weight (in pounds) safely supported by wood sides and plate								
	Two 2 x 8 + 7 1/2" by			Two 2 x 10 + 9 1/2" by			Two 2 x 12 + 11 1/2" by		
	3/8"	7/16"	1/2"	3/8"	7/16"	1/2"	3/8"	7/16"	1/2"
10	6754	7538	8242	10,973	12,199	13,418	15,933	17,729	19,604
12	5585	6216	6827	9095	10,131	11,106	13,224	14,517	16,265
14	4756	5293	5811	7751	8623	9463	11,295	12,561	13,876
16	—	4481	5036	6746	7494	8221	9815	10,953	12,086
18	—	—	—	5942	6606	7158	8675	9652	10,647
20	—	—	—	—	—	6466	7746	8618	9408

Figure 26-24. These tables are used to properly size headers. These tables require calculating the load that will be supported by the header.

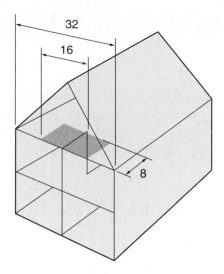

Attic: 20 lb./ft² x 16' x 8' = 2560 lb. = total load

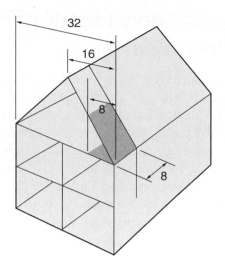

Roof: 40 lb./ft² x 16' x 8'   = 5120 lb.
Attic: 20 lb./ft² x 8' x 8'   = 1280 lb.
Roof + Attic   = 6400 lb. = total load

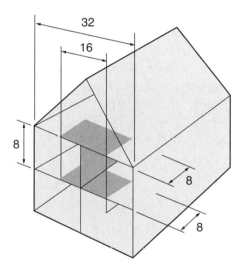

Attic: 20 lb./ft² x 16' x 8'   = 2560 lb.
Wall: 12 lb./ft² x 8' x 8'   = 768 lb.
Second floor: 30 lb./ft² x 16' x 8' = 3840 lb.
Attic + Wall + Second floor   = 7168 lb. = total load

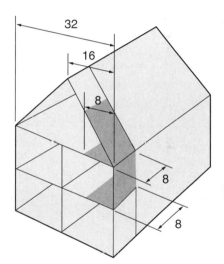

Roof: 40 lb./ft² x 16' x 8'   = 5120 lb.
Attic: 20 lb./ft² x 8' x 8'   = 1280 lb.
Wall: 12 lb./ft² x 8' x 8'   = 768 lb.
Second floor: 30 lb./ft² x 8' x 8'   = 1920 lb.
Roof + Attic + Wall + Second floor = 9088 lb. = total load

Figure 26-25. Load calculations for different configurations.

feet. Multiply the length of the span by half of the distance between load-bearing walls. You can always assume that the header must support the load halfway to the next bearing wall. Thus, if the distance to the next bearing wall is 12', the

square footage is 6' × the span. Add any other loads that the header must support to find total load. On roof sections, the load is figured in the same as for the floor—use the horizontal distance covered by the roof.

26.7.4 Concealed Headers and Saddle Beams

There are times when it is not desirable to have a header extending below the ceiling level. In such cases, the header must be at the same level or above the joists it supports. A *concealed header* is butted against the ends of the joists it is supporting. A *saddle beam* is positioned above the joists it supports. See **Figure 26-26.**

A concealed header is simple to add to an exterior wall in platform construction, **Figure 26-27.** Simply attach a second piece of 2″ lumber to the band joist. Add studs at each end to provide support for the header.

Because they are placed above the joists, saddle beams can only be used in the attic space. They should rest on top of joists supported by the remaining portions of the bearing wall.

Working Knowledge

An architect or engineer should check the loads and specify the header sizes.

26.8 Small Remodeling Jobs

Most of the chapters in this book deal with new construction. In general, these same construction steps can be used with some modification for small remodeling jobs.

26.8.1 Replacing an Outside Door

An old, worn, ill-fitting exterior door, in addition to being an eyesore, wastes energy. It permits excessive air infiltration and heat loss. Prehung units make replacement easy. Standard replacement units have an insulated steel, steel-clad, fiberglass, or wood door mounted in a steel or wood frame. **Figure 26-28** shows a unit with a steel frame.

To install a new door, remove the old door, hinges, threshold, trim, and frame. Then, slide the new unit into the opening. Replacement doors are generally available in standard sizes:

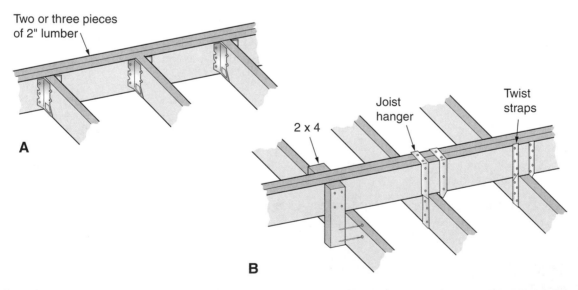

Figure 26-26. Hidden support methods using joist hangers. A—Concealed header. B—Saddle beam showing three attachment methods.

Concealed header: A header that is butted against the ends of the joists it is supporting.

Saddle beam: A beam positioned above the joists it supports.

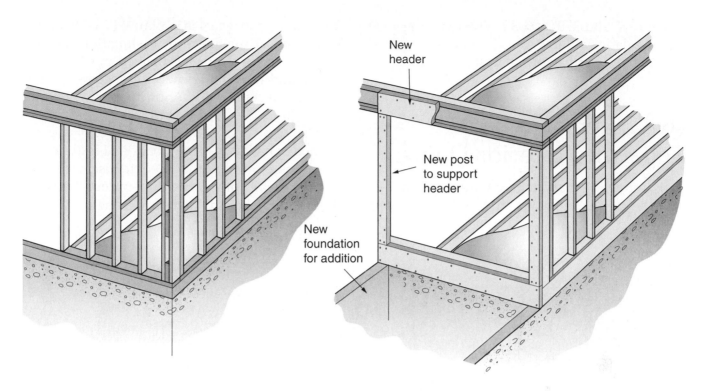

Figure 26-27. When remodeling a wall in platform construction, a header can often be constructed by adding to a band joist on a second floor.

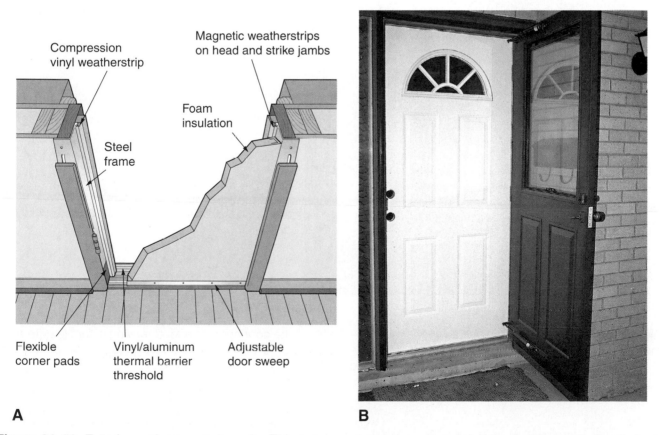

A

B

Figure 26-28. Exterior replacement door. A—This cutaway shows the general details of a door unit. B—A steel-clad entrance door and frame are set in place. (General Products, Inc.)

2'-6" × 6'-8" and 3'-0" × 6'-8". When making such an installation, follow the directions provided by the manufacturer of the unit.

26.8.2 Replacing or Repairing Interior Doors

Remodeling may include replacing, moving, or adding interior doors and doorjambs. The framing of doorways is explained in Chapter 9, **Wall and Ceiling Framing.** Installation of jambs and hanging of doors is explained in Chapter 19, **Doors and Interior Trim.** Several steps used in replacing a standard interior door are shown in **Figure 26-29.**

Use of a special *jamb clip* presents another method of hanging an interior door, **Figure 26-30.** The clips eliminate the need to use wedging. They are made of 19 gage, cold-rolled steel. One size (4 1/2") fits the standard rough opening.

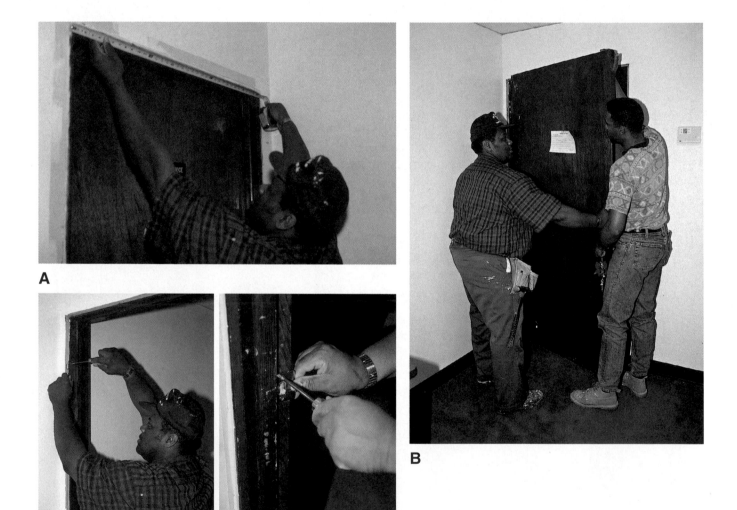

Figure 26-29. Preparing to replace an interior door. A—Measure the door opening after removing of trim to determine the replacement door size needed. B—Removing the old door. C—Removing the old hinge leaves from the jamb. D—Removing the striker plate. (McDaniels Construction Co., Inc.)

Jamb clips: Steel clips used to hang an interior door, eliminating the need to use wedging.

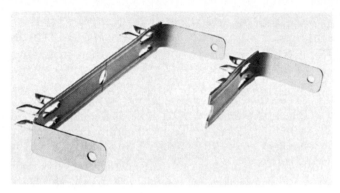

Figure 26-30. Jamb clips allow door installation without the use of shims. For jambs wider than standard, the clip can be snapped in half, as shown at the right. (Panel Clip Co.)

Figure 26-31 shows a cutaway with a jamb clip on a 2 × 4 plus 1/2″ of drywall on each side. For other jamb sizes, the clip snaps in two at a breakaway point. Eight clips should be used with hollow core doors, ten for solid core doors. Follow the installation instructions given in **Figure 26-32.**

As a building ages, day-to-day usage, settling, and other factors cause problems with doors. Hinges wear and latches no longer work. Also, raising the floor level with new solid flooring material or carpeting makes it necessary to trim the door to clear the new flooring material. If the door itself is still in good condition, repairs are in order. Consider the following conditions and suggested repairs.

- If the door binds or will not tightly close, check for loose hinges. Reinstall the hinges. You may need to use longer screws or fill oversized screw holes and reinstall screws. If the door still binds, note where it is binding on the frame and mark the spot with a pencil, **Figure 26-33.** Then, use a hand plane to remove a small amount of material. Be careful not to remove too much wood.

- Installation of new flooring may make it necessary to remove 1/2″ or more from the bottom of the door. Remove the door from its hinges. Have a helper hold the door in the door opening in its normal position with about 1/16″ space at the top or with hinge leaves aligned. Mark the door bottom. If the door has veneered facings, make a shallow cut on the line with a utility knife. Use a steel straight edge as a guide. Transfer the cut line to the opposite side and make a shallow cut there, as well. This will prevent splintering of the veneer when the excess is sawed or planed away.

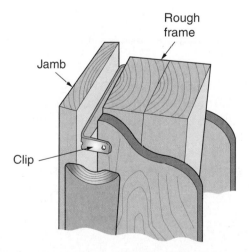

Figure 26-31. This cutaway view of the jamb and rough frame shows how a jamb clip is applied. (Panel Clip Co.)

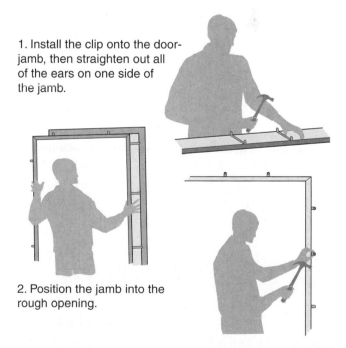

1. Install the clip onto the doorjamb, then straighten out all of the ears on one side of the jamb.

2. Position the jamb into the rough opening.

3. Bend the ears back and nail.

Figure 26-32. Steps for attaching jamb clips and installing a door. (Panel Clip Co.)

Figure 26-33. Marking the point where a door is sticking provides a guide for removing the excess material with a plane.

- If the door binds on the threshold or floor, check for a loose upper hinge or worn hinges. Replace the hinges if worn. Check for sagging of the door by measuring diagonals. Sagging is confirmed if the diagonal from the top of the hinge side of the door is longer. Replace the door or plane the bottom where it is binding.

- Before replacing a seemingly defective door latch, check to see if it aligns with the striker. Reset the striker if alignment is the problem.

26.8.3 Installing New Windows

Replacement windows are available in a variety of sizes and styles to fit rough openings of many old windows. Before ordering replacements, take careful measurements. Measure again before installing to ensure proper fit.

Check around the opening for signs of rot or leakage that a new window will not solve. Repair any problem before installing the new window.

Specific instructions for installation are provided by the window manufacturer. Carefully read them before beginning installation. Also, refer to Chapter 13, **Windows and Exterior Doors.**

26.8.4 Repairing Wood Shingles

Wood shingles are used for both roofing and exterior wall coverings. Over time, individual shingles may sustain damage and require repair or replacement.

Split shingles can be repaired. If the crack is small, fill it with asphalt roofing cement. If the shingle appears loose, secure it with small, galvanized nails. Cover the nail heads with asphalt roofing cement. If the crack is large or a piece of the shingle is missing, install a piece of flashing under the shingle. Make sure the flashing extends well above the top of the crack or that it extends under the next course. Secure the patch with galvanized nails. Use asphalt roofing cement over the patch.

Badly damaged wood shingles can be replaced without disturbing the surrounding shingles. Use a wood chisel or screwdriver to split the damaged shingle into narrow pieces. These can be pulled out with your fingers. Use an old hacksaw blade to cut off the nails. Slide a new shingle of the same size into place, but leave it about 1/4″ below adjoining shingles. Secure it with two galvanized nails driven at an angle against the upper course. Then, drive the new shingle upward until the nail heads are concealed.

26.8.5 Repairing Asphalt Shingles

Wind sometimes causes damage to asphalt shingles by flipping them up and even tearing them. If the shingle is not too badly damaged, shingle. Cover the nail heads with asphalt cement.

26.9 Building Additions onto Homes

When building additions to older homes, check all dimensions of the existing construction carefully. In particular, check the following.

- Foundations—New masonry units may not match the size of older units. You may need to adjust the level of footings or alter the mortar joint thickness.

- Ceiling heights—In balloon framing, ceiling heights may vary from house to house. Study local codes to see if your changes are acceptable.

- Dimensions of framing members—The standards for dimension lumber (2 × 4s, 2 × 6s, etc.) have changed with time. Check the actual dimension of older lumber.

In addition, in recent years regions that experience serious damage from earthquakes, high winds, tornadoes, and hurricanes have revised building codes to reduce future damage. **Figure 26-34** shows how and where to install ties that anchor a wood frame building from the foundation to the rafters. Check local codes to see if such ties must be installed in an addition.

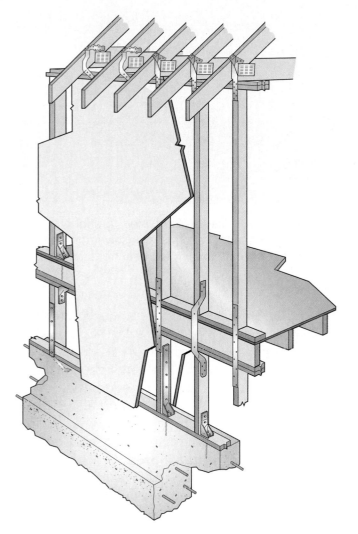

Figure 26-34. To help buildings withstand earthquakes and high winds, special metal ties are used to tie the various components of the building together and anchor the structure to its foundation. (©Simpson Strong-Tie Co., Inc.)

apply a generous dab of asphalt cement to the underside. Press it down and secure it with broad-headed roofing nails. Apply asphalt cement to the nail heads.

Asphalt shingles are usually manufactured in two- or three-tab sections or strips. If a strip is badly damaged, it can be replaced without destroying the adjacent shingles. Lift the upper shingle and remove the nails from the damaged strip using a small, flat ripping claw or cut them off with a hacksaw blade. Remove the damaged strip, slip the new one into place, and secure it by blind nailing. As an alternate method, secure the new shingle by nailing through the upper

26.10 Solar Retrofitting

Older homes with a long wall to the south where the sunlight is not blocked by other structures are likely candidates for a passive solar retrofit. *Solar retrofitting* means adding some form of solar heating (space or water) or electricity generation to a structure. A first consideration, however, is to investigate the insulation levels in the house. For insulation standards and methods, refer to Chapter 15, **Thermal and Sound Insulation.**

26.10.1 Basic Solar Designs

The three basic passive solar designs are:
- Direct gain
- Indirect gain
- Isolated gain (sun space)

In a direct gain system, south-facing double-glazed windows allow the sun direct access to living space. A masonry wall or floor is needed to act as a storage mass for the solar heat. In an indirect gain system, a storage wall is located a few inches from the glazing. The wall soaks up the solar heat and radiates it to the living space. An isolated gain system is a separate space, such as a greenhouse, that has solar storage systems for collecting and distributing the heat to other parts of the house. See **Figure 26-35.** Additional information on these systems and how to construct them can be found in Chapter 25, **Passive Solar Construction.**

Direct gain and isolated gain systems are most often used as part of a remodeling program. Adding a direct gain system involves removing a section of a south-facing wall and replacing it with windows. If the glazed area is small, no storage is required. However, for large areas of glass, a storage wall or floor may be needed. One way of getting storage is to install a masonry fireplace on the opposite wall where sunlight can strike it. A brick planter can also serve as heat storage. These are not always practical because of weight.

The amount of thermal mass is important. For a large expanse of glass, about 150 lb. of masonry is needed for every square foot of south-facing glass. This could present serious structural problems for many homes when you consider that 200 sq. ft. of glazing requires 15 tons of masonry. In most situations, solar retrofit through adding sun space or a thermosiphon seems more practical.

26.10.2 Thermosiphon

Sometimes called a solar furnace, a ***thermosiphon*** is basically a glazed box that captures heat from sunlight. Natural convection (rising of heated air) moves air into the living space. See

Figure 26-35. This prefabricated greenhouse will act as a sun space, which is a passive solar heating system that collects and stores heat from the sun.

Figure 26-36. Thermosiphons can be attached to window sills to provide solar heat for individual rooms.

One experimental thermosiphon design involved glazing the entire south wall of a two-story house. At regular intervals, 6″ wide boards were vertically run edgewise up the wall. Blocking was added to top and bottom enclosing the individual boxes. At the first floor ceiling level, more blocking was added. Then, vents were cut through the wall at floor and ceiling levels on both floors. Finally, glazing was attached to the outer edge of the vertical boards. As sunlight falls on the glass, the air in the enclosed spaces heats up, setting up a convection. Warm air rises, entering the house at the ceiling level, while cooler room air moves into the space from the bottom. **Figure 26-37** is a simple diagram of this system.

26.11 Responsible Renovation

A major remodeling of a building may seem to generate a great deal of waste. Even so, it is

---

Thermosiphon: A solar collector that draws in cool air, heats it, and returns the air to the living space or storage without mechanical assistance.

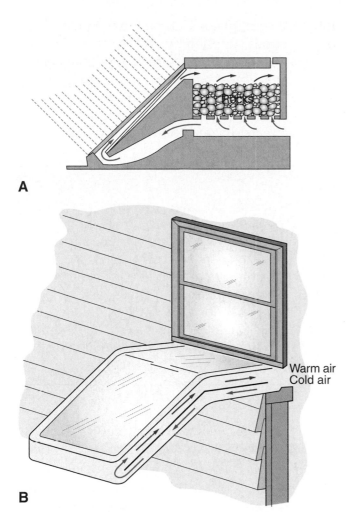

A

B

Warm air
Cold air

Figure 26-36. Two thermosiphon units. A—A large unit designed to store heat in rock. B—A small window unit serving a single room.

Figure 26-37. This thermosiphon installation occupies a portion of a south-facing wall. A series of boxes is constructed from lumber, then covered with a glazing material. Vents at the top and bottom of each box allow circulation of the heated air.

usually far less damaging to the environment than new construction. Inevitably, some waste ends up in a landfill. This waste can be kept to a minimum by careful salvage and attention to what can be recycled.

Window trim, doors, wood paneling, hardwood flooring, tile, brick, and framing lumber can be set aside and reused. If not used in the remodeling, someone else may use it. Careful removal of recyclable materials is more time-consuming than demolition, however. Many carpenters and contractors find salvage too costly. This approach is most feasible for do-it-yourself projects.

Salvaged material should be carefully cleaned and stored out of the weather until it can be reused or sold. Some material will need

to be given away. A "Free Material" sign on the lawn works well for giveaways. The savings in disposal costs may even pay for the extra time spent in careful removal.

Inefficient fixtures, such as old flush toilets, are water wasters and should not be recycled. Old windows are equally inefficient and waste energy. They should be used only for unheated garages or sheds.

Finally, in selecting building materials for remodeling, choose those made from recycled material wherever possible. By doing so, you save natural resources and help to create markets for recycled products. This is a key requirement to every recycling program.

For information on recycled products, check a local library or bookstore. Another source of information is *Environmental Building News*, 122 Birge Street, Suite 30, Brattleboro, VT 05301. On their Web site (www.buildinggreen.com), they offer a bibliography of useful publications.

26.11.1 Fall Protection

OSHA fall protection rules allow residential carpenters and contractors to write their own fall protection plans. The plan has to meet a set of guidelines supplied by OSHA.

The residential plan, which applies at heights of 6′ or more, must be specific to the site, properly supervised, and only applied in situations where conventional fall protection procedures would cause a greater hazard than the work itself.

ON THE JOB

Asbestos Removal Technician

For many years, the mineral fiber asbestos was used for fireproofing, pipe insulation, and other applications in buildings. It is no longer used in construction since airborne particles of the material have been shown to cause cancer and other diseases of the lungs. Many older buildings have large amounts of asbestos in them, which poses danger to workers involved in demolition or remodeling work. This material must be removed for health and safety of workers and occupants. Asbestos contamination has been a particular concern in school buildings where young children can be exposed to asbestos particles.

These concerns have created a new building trades specialty, the asbestos removal technician. A related field is lead removal, since old lead-based paints are a health hazard as well. Many people working in this field are trained in both lead and asbestos removal. Older structures frequently have both types of material.

Since buildings usually cannot be occupied while asbestos is being removed, asbestos removal technicians typically work in teams to complete the work in as short a time as possible. The building or portion of the structure where the work is being done is sealed to prevent the escape of airborne asbestos particles. To protect their health, asbestos removal technicians (and, often, lead removal workers) wear full body suits with an air supply. They physically remove the asbestos materials from the structure, enclosing all debris in sealed containers for safe disposal. Specialized vacuum cleaners with highly efficient filters are used to clean up dust resulting from the removal activities.

Both asbestos and lead removal work can be dangerous, although strict preventive measures are taken to prevent the breathing of toxic dust. As in most construction occupations, working with hand and power tools presents dangers, as do the necessary climbing, lifting, and other physical activities. Working inside full body suits is hot and uncomfortable and may cause claustrophobia in some people. Work hours may often be long, since there is usually time pressure to complete the work. These specialty technicians are generally well-paid, however, ranking with the more well-compensated members of the traditional building trades.

Workers in both asbestos and lead removal must be licensed. Although formal education beyond high school is not needed, licensing requires successful completion of a formal 32-to-40 hour training program. Employment in these fields is expected to continue to grow as more and more aging buildings containing asbestos and lead are demolished or renovated.

Summary

Remodeling and renovation is often more difficult than new construction. Existing structures have defects that must be addressed and the carpenter must be able to visualize how a structure was built so that no structural damage is done during demolition and removal of old walls. Work should be done in the proper sequence, beginning with structural repairs and changes working from the bottom up. Roofing repairs and window/door replacement should be done early to ensure weathertightness. Interior renovation work can proceed at the same time as exterior work. A major project in renovation is often replacing rotted building sills. If walls are to be removed, care must be taken to find hidden structural details—especially whether or not the wall is a bearing wall. When a bearing wall is removed or changed, proper shoring must be provided to support the weight of the structure. New openings in bearing walls require calculating header size to properly support loads above the openings. There are many small remodeling jobs that may be individually done or a part of a larger project. These include door and window replacement and repairs to roofing materials. Building additions onto existing structures requires checking dimensions of the current structure. Solar retrofitting of structures for either active or passive systems is possible, but has special requirements. Disposal of waste from remodeling projects should be done in an environmentally responsible manner. As in all construction activity, safe working practices and use of proper protective gear are vital.

Test Your Knowledge

Answer the following questions on a separate piece of paper. Do not write in this book.

1. The first concern when planning the renovation of a long-neglected house is: *(select the best answer)*
   A. Condition of the siding.
   B. Condition of the foundation.
   C. Assuring that mechanical systems are working.
   D. Repairing any leaks found.

2. What is the first step in a sequence of tasks for renovating the exterior?
3. The last step in renewing the exterior of a house is to _____.
4. Before rotted sills can be repaired:
   A. The foundation should be repaired.
   B. The building must be supported in some way to remove the load from the sills.
   C. Measurements should be taken of building's height.
   D. The carpenter must obtain a license to perform this complicated task.
5. What does the term *sistering* mean?
6. Before beginning a remodeling job, why is it wise to find out the type of construction used?
7. Before starting excavation for an addition, determine the location of _____ utility services.
8. List five indications that a partition is a bearing wall.
9. _____ is a temporary wall installed to prevent the collapse or sagging of a structure when part of a bearing wall is removed.
10. A built-up wooden header must span 8′ in a bearing partition and carry the load of a shallow attic above it. How deep should the header be?
11. When a header must be flush with the ceiling, one of two types of beam or header must be used. Name the two types.
12. What does *solar retrofitting* mean?
13. *True or False?* A home using direct gain solar heating does not need a method of heat storage.
14. A(n) _____ is a passive solar method that can provide heat for individual rooms.
15. _____ is usually far less damaging to the environment than new construction.

Curricular Connections

Language Arts. Make sketches and write the copy for a set of storyboards depicting a new Web site advertising your remodeling business. Think about how to convey the needed information in a few pictures and carefully chosen words. Remember to give potential customers several ways to contact you.

(Continued)

(Continued)

Mathematics. Compute the depth of the header needed to span an 8' opening in a bearing partition of a one-story house. The partition is in the center of a 28' span. Use the load tables in **Figures 26-23** and **26-24**. As necessary, consult **Figure 26-25**.

Outside Assignments

1. Write a report on the steps for constructing a 12'× 16' addition on the side of a ranch home built by the platform method. Include sketches for shoring and for final framing.

2. Discuss with your instructor the possibility of a field trip to a vacant house in your community. Purpose of the visit: a visual inspection of the exterior to determine what renovation, if any, might be necessary. Each student should then prepare a written report on his or her recommendations.

Building Decks and Porches

Learning Objectives

After studying this chapter, you will be able to:

- Identify the different types of decks and porches.
- Describe the advantages and disadvantages of various structural and decking materials.
- Select and install the appropriate types of fasteners for deck construction.
- Prepare the site, then lay out and construct a deck.
- Describe the differences between deck and porch construction.

Technical Vocabulary

Ammoniacal copper quaternary ammonium compound (ACQ)
Attached deck
Balusters
Beams
Chromated copper arsenate (CCA)
Closed porch
Composite decking
Copper azole (CA)
Dadoes
Disodium octaborate tetrahydrate (DOT)
Engineered wood
Flashing
Freestanding deck
Galvanic corrosion
Glue laminated timber (glulam)
Joists
Ledger
Multi-level deck
Open porch
Parallel strand lumber
Pergola
Piers
Plumbed
Ring-shank nails
Semi-enclosed porch
Site preparation
Stringers

Lifestyles that value the outdoors have given rise to the construction of various types of decks. These decks become an extension of the main dwelling, functioning as outdoor rooms. There are two basic types of decks: attached and freestanding. Either type may be a single-level or multi-level deck.

An ***attached deck,*** or *elevated deck,* is shown in **Figure 27-1.** The principles of construction for an attached deck are very similar to those used in building the main structure. It is supported by posts and at least one side is attached to the house. Height of the supporting posts vary, depending on the slope of the lot and the deck design. If soil conditions permit excavation for footings, they can be dug by hand or with a power post-hole auger. See **Figure 27-2.** Some augers are powered by a small gasoline engine mounted on the device itself or the auger may be attached to and controlled by a tractor or other mobile power source.

Attached deck: A deck that has a supporting ledger physically attached to the building wall. Also called an *elevated deck.*

Figure 27-1. An example of an attached deck, which is always connected to a building.

Figure 27-3. Freestanding decks are ground-level decks that do not depend on a building for support. (Southern Pine Council)

A *freestanding deck,* or *grade-level deck,* is not attached to the house, **Figure 27-3.** However, it may be placed next to the house. It is low to the ground and does not require a railing for safety, although one may be used for appearance or to define the edges of the deck. A grade-level deck does not require a foundation.

A *multi-level deck* is simply a deck with more than one level, **Figure 27-4.** It may be attached or freestanding. On a sloping site, the upper level of the deck is typically attached to the building, while one or more lower levels might be freestanding.

Figure 27-2. Power augurs can be used to quickly excavate holes for foundation piers. This large power augur is mounted on a tractor.

Figure 27-4. An example of a multi-level deck. Some multi-level decks have three or more levels. (Southern Pine Council)

Freestanding deck: An independent structure that is not attached to the building. Also called a *grade-level deck.*

Multi-level deck: A deck structure with connected sections built to different vertical elevations.

27.1 Structural Materials

Before construction begins, the carpenter must choose materials suited to the construction plan. Among the choices in lumber are the grade, species, and size. Engineered lumber is another choice for structural members. If a composite or plastic is the chosen material, then there are various materials and designs available. There are also choices to make in fasteners, connectors, anchors, and other hardware.

27.1.1 Lumber

Resistance to decay is a major concern when choosing lumber for a deck. Deck materials are always subject to weather conditions and may not be protected by paint or other coatings. Some woods naturally resist rot, but are more expensive than commonly used treated lumber. The chart in **Figure 27-5** describes various species that are suitable for decks. As shown in the chart, certain woods are better suited than others for deck framing. Most prized for its resistance to rot and termite damage is redwood cut from the heartwood of the tree. As a rule of thumb, all heartwood grades have the word "heart" in the grade name.

27.1.2 Treated Lumber

Because of material cost, treated lumber is more often used than redwood, cedar, or tropical hardwoods. At one time, the most common preservative used for lumber was *chromated copper arsenate (CCA)*. Lumber treated with this preservative can be identified by its greenish

Softwoods Suited for Decks	
Western Red Cedar	Highly resistant to warping and weathering; like redwood, though coarser; the heartwood resists decay but not termite attacks; easy to work though weak and brittle; moderate ability to hold nails.
Douglas Fir Western Larch	Heavy, strong, stiff, holds nails well; not easily worked with hand tools; when pressure treated, resists decay and termites.
True Fir Eastern and Western Hemlock	While firs are mostly lightweight and moderately soft, hemlocks are moderately strong and stiff; firs are easy to work, but hemlocks more difficult; shrinkage could be a problem; resistant to termites and decay only when pressure treated.
Ponderosa Pine	Somewhat strong and stiff, but not as strong as southern pine; very resistant to warping; holds nails well; when pressure treated, resists termites and decay.
Red Pine	Strong and stiff, but not to degree of southern pine; easier to work and holds nails well; resistant to termites and decay when pressure treated.
Southern Yellow Pine	Hard, very stiff, and strong with good nail-holding quality; moderately easy-working; decay and termite resistant when pressure treated.
Redwood	Heartwood famous for durability and resistance to decay and termites; moderately light, but limited in structural strength; workable, but brittle; moderate ability to hold nails.

Figure 27-5. A basis for choosing woods for deck building. Refer to samples of woods shown in Chapter 1 of this text.

Chromated copper arsenate (CCA): Wood preservative treatment once widely used for lumber. No longer available for residential use due to toxic emissions.

tint. CCA treated lumber is no longer produced for residential use. Studies show that arsenic, a toxic substance, could leach out of the wood and contaminate the soil and groundwater. The Environmental Protection Agency (EPA) ordered production to end January 1, 2004. Lumber treated with CCA may be found in some areas, since remaining stocks of the material can still be sold. One special use of CCA treatment is still allowed—it can still be applied to lumber intended for below-ground use in permanent wood foundations. For information on permanent wood foundations, refer to Chapter 7, **Footings and Foundations.**

Other preservatives have replaced arsenic as a wood treatment. One that is widely used is *ammoniacal copper quaternary ammonium compound (ACQ).* Also available is lumber treated with *copper azole (CA).* Treatment with ACQ or CA usually leaves the lumber with a brownish color. Both of these treatments are high in copper content, which makes them more expensive than CCA lumber. The high copper content also creates corrosion problems with some flashing and fastener materials. See the Fasteners and Connectors section of this chapter for more detailed information.

Lumber treated with a borate preservative is also available, although harder to find. Lumber treated with *disodium octaborate tetrahydrate (DOT)* is insect, mold, and rot resistant. It is basically nontoxic to humans and animals. Also, it does not corrode fasteners. DOT has been primarily used in locations protected from weather. This was due to concerns about borate leaching from the material and reducing its effectiveness as a preservative when the wood gets wet. Recent studies indicate that leaching is less of a problem than originally thought and new techniques have better bonded the borates to the wood.

27.1.3 Engineered Lumber

Engineered wood construction materials are derived from smaller pieces of wood that are either laminated or modified in some other way to make them stronger and eliminate warping and shrinking. Given the same dimensions,

engineered structural units are stronger than sawn lumber. In deck applications, engineered lumber is used for beams and joists. Some engineered wood products are not suited to exterior use where they will be exposed to the weather.

Glue laminated timber (glulam) is manufactured by stacking finger-jointed layers of standard lumber and adhering the layers with adhesives. Performance is improved by placing different grades of wood in the laminations. Glulam is a traditional choice for exposed applications. Pressure treating glulams makes them resistant to rot or termite damage. The material can be curved during manufacture for use in special applications.

Parallel strand lumber is made up of long strips of wood running parallel to each other and saturated with adhesives. It is sold under the trade name Parallam®. For exterior use, pressure treatment renders Parallam® resistant to both decay and termites.

27.2 Decking Materials

Although treated lumber, redwood, cedar, or tropical hardwoods such as mahogany or teak have traditionally been the material used

---

Ammoniacal copper quaternary ammonium compound (ACQ): Nontoxic preservative used for treating lumber; requires the use of stainless steel or heavily galvanized fasteners.

Copper azole (CA): Copper-based, nontoxic wood preservative; requires the use of stainless steel or heavily galvanized fasteners.

Disodium octaborate tetrahydrate (DOT): Borax-based wood preservative that is nontoxic and does not cause galvanic corrosion to fasteners.

Engineered wood: Lumber that has been altered by manufacturing processes. Included are wood I-beams, glue laminated beams, laminated strand lumber, laminated veneer lumber, and parallel strand lumber.

Glue laminated timber (glulam): Large beams made by gluing and then applying heavy pressure to a stack of four or more layers of 1 1/2" thick stock.

Parallel-strand lumber: Engineered lumber made up of strands of wood up to 8' long that are bonded together using adhesives, heat, and pressure.

for the decking, both composites and vinyl have captured part of the market in recent years.

27.2.1 Composite Decking

Composite decking is a blend of 30%–50% plastic with wood fibers or sawdust, **Figure 27-6.** The plastic is usually recycled. The combination of wood and plastic produces a skidproof and paintable surface. Other advantages include low maintenance, no splintering, and easy workability. Screws sink into the surface and disappear. Some brands offer 10 year warranties; others as long as 20 years.

There are a number of firms manufacturing composite decking boards in standard sizes. Standard thicknesses are 1 1/4″ (5/4″) and 2″. Standard widths are 4″, 6″, and 8″. Boards are either plain or tongue-and-groove and may be solid or hollow. Some have channels that allow concealing electrical conductors for lights and electrical outlets. Composites weather to a light gray and can be painted or stained for appearance, if desired.

Figure 27-6. Composite decking materials offer low maintenance and easily worked with the same tools as conventional lumber. Surface finishes are available in wood grain or other textures. This installation makes use of a hidden fastener system. (Trex Company, Inc.)

Composite decking: Material made with a blend of wood fibers and plastic resins.

Galvanic corrosion: Eating away of metal fasteners due to a chemical reaction between dissimilar metals.

Some composites may have an artificial (rather than wood-like) appearance. Over time, color may fade from exposure to sunlight. During construction, sawdust and shavings must be collected on a drop cloth, since the debris is not biodegradable.

27.2.2 Vinyl Decking

Deck systems made from extruded or molded vinyl plastic offer decking boards, railings, spindles, and fascia. The boards are usually hollow with internal reinforcement for stiffness. If factory-treated, vinyl resists the effects of ultraviolet rays.

27.3 Fasteners and Connectors

Various types of fasteners and connectors are available for installing decks, **Figure 27-7.** For use with ACQ or CA pressure-treated lumber, all connectors, related hardware, and fasteners must be corrosion resistant. The copper-rich lumber treatment will set up galvanic corrosion with fasteners made from common steel or steel with a light galvanized coating. *Galvanic corrosion* occurs when dissimilar metals are in contact in a damp environment. A weak electrical current is set up. Over time, the corrosion eats away the fasteners and hardware, weakening the deck structure to the point of possible collapse. Stainless steel fasteners and hardware are preferred, but the best grades of galvanized steel can be used. Hot-dipped galvanized coatings should have a G-185 rating. Electrogalvanized items should have a class rating of 40 or higher. Manufacturers of decking screws have developed ceramic coatings for their products that provide corrosion protection by serving as a barrier between the steel of the fastener and the copper in the preservative.

In addition to fasteners and other hardware, any flashing used at points where the decking adjoins the structure must be corrosion resistant. Stainless steel or copper flashing is preferred. Aluminum should never be used, since it is particularly susceptible to galvanic corrosion.

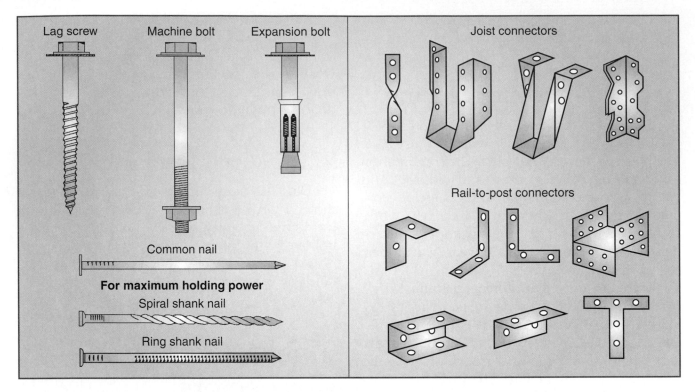

Figure 27-7. Fasteners typically used in deck construction. Joist connectors and rail-to-post connectors are formed out of sheet metal and designed to make deck frames stronger. (Wolmanized® Wood)

Special, concealed fasteners are available that install on the underside of all types of deck boards. This method produces a deck surface unmarred by screw holes and eliminates a condition that causes rot in wood deck materials. Follow the manufacturers instructions for ordering the quantity needed. Usually, the fasteners can be installed working from the top side of the deck.

PROCEDURE

Installing a deck with concealed fasteners

1. Nail the fasteners to the top of all joists, alternating from side to side, **Figure 27-8.**
2. Begin attaching the deck boards, working from the top. Use two screws for each board. Joints should meet over a joist and a fastener must be used on each side, **Figure 27-9.**
3. Continue attaching deck boards one at a time until the job is completed. **Figure 27-10** shows the underside of a completed deck.

27.4 Deck Planning and Layout

Before layout begins, make sure that plans conform to building codes and zoning restrictions. Be sure to check local requirements. Your design must meet these requirements. Draw up a plan and have it approved by appropriate community officials before beginning to build. Refer to Chapter 3, **Plans, Specifications, and Codes,** for additional details. In most cases, a building permit is required. This permit must be displayed at the site during construction.

Before the deck is laid out, site preparation should be done. *Site preparation* involves any modification of the location where the deck will be built. This could involve several steps:

- Leveling the ground.
- Providing necessary drainage.
- Removing sod or laying down sheet material (black polyethylene or landscape fabric) to eliminate weed growth.
- Excavating and installing piers.

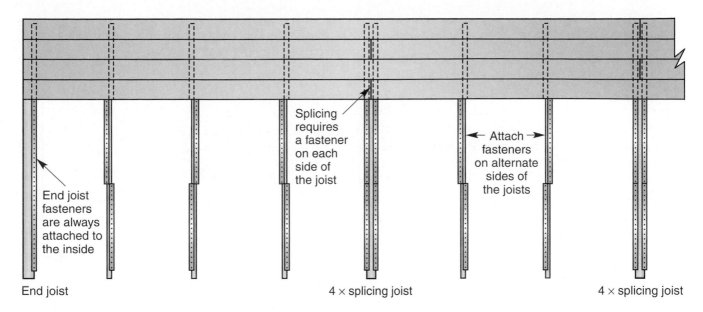

Figure 27-8. This layout is a guide for one method for installing concealed deck fasteners. (Deckmaster)

Layout is done most easily using an optical instrument such as a builder's level or a transit. If such an instrument is not available, it is possible to lay out the deck with measuring tapes.

After initial layout, you can check for square corners once more by installing a double rim joist hanger on one end of the ledger. Slip a side rim joist in place and apply the 3-4-5 rule (a shorter-distance form of the 6-8-10 rule). Mark 3′ on the ledger and 4′ on the rim joist. Then, shift the rim joist back and forth until the diagonal measurement is 5′. Refer to **Figure 27-11.**

For a detached deck, the location will be determined by the carpenter after consulting with the homeowner. Layout is similar to the procedure for an attached deck, except that batter boards are needed at all four corners and at intermediate locations to locate support beams.

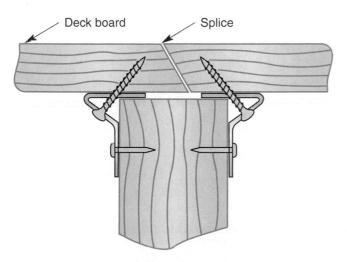

Figure 27-9. This cutaway view shows the proper method for installing spliced deck boards using a concealed fastening system. (Deckmaster)

Figure 27-10. A properly installed concealed deck fastening system, as viewed from beneath the deck.

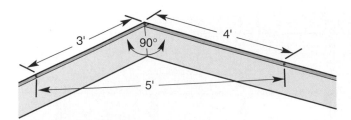

Figure 27-11. Checking a corner of a deck frame to assure it is indeed square.

PROCEDURE

Laying out an attached deck

1. Establish one corner of the deck. This is along one wall of the house. Usually, this point will be at one end of the ledger, which should already be installed. Installing a ledger is described later in this chapter.
2. With a tape, mark the length of one side of the deck.
3. Drive a 2″ × 2″ stake at that point.
4. Drive a nail in the top of the stake to mark the exact location of a pier footing.
5. Using the 6-8-10 rule for checking that corners are square, lay out all sides and drive stakes to mark each corner. Refer to Chapter 6, **Building Layout,** for information on the 6-8-10 rule.
6. Install batter boards at each corner, **Figure 27-12.**
7. Stretch a mason's line from the nails driven at each end of the ledger to the batter boards. Refer to the section Laying out Building Lines in Chapter 7, **Footings and Foundations.**
8. Shift the lines at the batter boards as necessary until the diagonal measurements are equal.
9. Locate the positions of other piers needed to support beams.

27.5 Constructing the Deck

Like a house, a deck is built upward from a solid foundation. Its beams or girders must rest on firm ground, concrete piers, posts resting piers, or a ledger solidly attached to the house

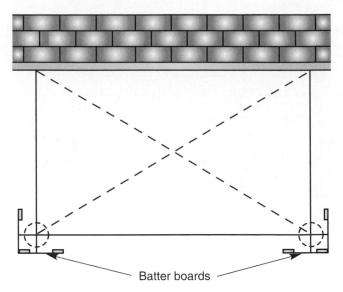

Figure 27-12. Use batter boards and attach string to indicate the layout of the deck.

structure. Joists are installed to form the base for the decking surface. Railings, stairs, and other finishing touches are added to complete the deck.

27.5.1 Installing Piers

In cold climates or unstable soils, *piers* (concrete bases) supporting decks should rest on a concrete footing located below the region's frost line, **Figure 27-13.** Location of the piers is important, since they must support the entire load of the deck itself as well as persons using it.

Piers support *beams* or girders that, in turn, support *joists.* Before locating the piers, be sure to check span tables for joists. The distance between beams is governed by the recommended span of

Site preparation: Excavation and other activities needed to ready the ground for foundation work and building activities.

Piers: Concrete bases, often cylindrical in shape, used to support a deck structure.

Beams: Used to support joists or other beams; may be attached to the building walls or rest on posts or columns. Also called *girders.*

Joists: The horizontal members of the deck frame that support the decking surface.

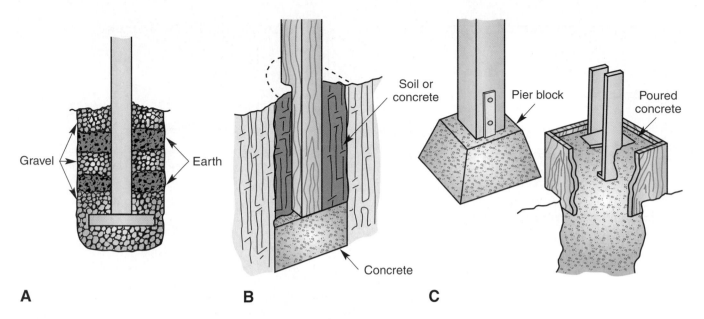

Figure 27-13. Three methods of setting posts/piers for the deck. It is important to get below the frost line in cold regions. A—Gravel and earth setting. B—Concrete setting. C—Concrete footings. (Wolmanized® Wood)

the joist to be used. **Figure 27-14** shows design details for an attached deck.

Piers are constructed of concrete. The underground portion should include or rest on a footing that extends several inches below the frost line. The above-ground portion is usually 8″ high. If poured, a supporting form such as a Sonotube® is required, **Figure 27-15**. This is a round form made of either plastic or fiber. The tube is placed in the hole after the hole has been partly filled with concrete to form the footing. The tube is then filled with concrete. A metal fastener is embedded in the concrete to secure either a post or beam to the pier, depending on the deck's height. **Figure 27-16** shows a post secured to a pier with a metal fastener.

27.5.2 Installing Posts

Deck posts can be installed on top of piers or sunk below ground on a footing. Usually

above-ground installation is preferred, since it protects the posts against rot and insect damage. Regardless of the installation method used, posts must be *plumb* (exactly vertical). Use a torpedo level or other short level on two adjacent sides to plumb each post. As shown in **Figure 27-17,** nail two temporary braces at right angles to each other to hold the post in position. After bracing, double-check to assure that the post is still plumb.

Working Knowledge

Whenever pressure-treated wood is cut or drilled, the exposed, untreated area is vulnerable to rot and deterioration. Thoroughly treat all cut edges and bore holes in posts, beams, ledgers, joists, and deck boards with a generous coating of preservative solution.

27.5.3 Installing the Ledger

When one side of the deck is attached to a building, a *ledger* must be attached to the wall. It must be at least the same size as the deck joists (for example, 2″ × 6″ or 2″ × 8″) and run the full length of the deck.

Plumb: Exactly perpendicular or vertical; at a right angle to the horizon or ground.

Ledger: A structural member attached to the building wall to support one end of the deck joists.

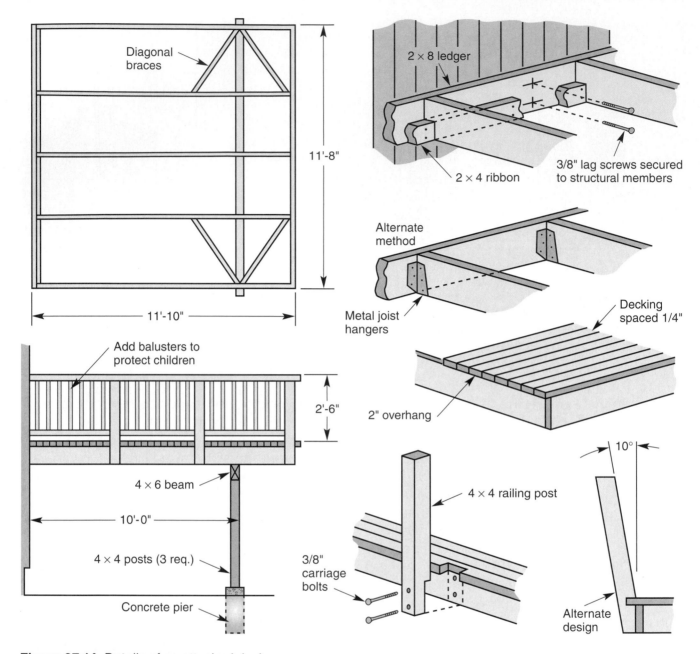

Figure 27-14. Details of an attached deck.

Figure 27-15. Sonotubes®. A—Tubes are nested for shipping to save space. B—The tubes are installed and ready for pouring concrete.

Figure 27-16. A metal fastener is often used to secure a deck post to a pier. The air space below the fastener allows water to drain away, preventing rotting of the post.

Prepare the building wall by removing siding material so the ledger can be attached to the house band joist. The top of the ledger should be a comfortable distance below a doorway to the deck. In cases where the ledger must be attached to a poured foundation or masonry wall, use expansion anchors in drilled holes.

To protect both the house structure and the ledger from water that could cause rot, install metal *flashing* and spacers, as shown in **Figure 27-18.** Use pieces of treated wood or metal washers to space the ledger 1/4"–1/2" away from the band joist to permit drainage. Bolt holes through the flashing should be caulked for watertightness. As noted earlier, only stainless steel or copper flashing should be used with copper-rich (ACQ or CA) treated lumber. Aluminum flashing will rapidly corrode.

Figure 27-17. This carpenter is checking for plumb on this post. Bracing on two directions holds the post exactly vertical.

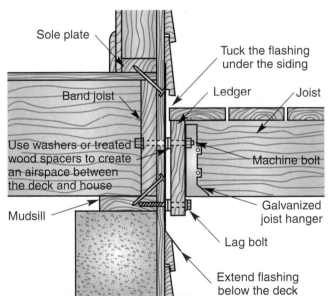

Figure 27-18. Proper flashing and the use of spacers will help protect the house and deck ledger from water damage.

Flashing: Sheet metal or other material used at the point where a deck attaches to the building wall to divert rain or other water from the joint.

27.5.4 Placing Beams and Joists

With the ledger in place and posts plumbed and braced, the post tops must be cut down to the proper level. Measure down from the top of the ledger the depth of the beam to establish the cutoff line for the posts. It is general practice to slope the deck away from the building 1/8″ per foot. Stretch a string from the mark on the ledger to the farthest post, **Figure 27-19.** Use a line level to level the string, then measure down and mark the calculated slope. For example, if the deck extends 20′ from the building, the mark on the farthest post should be 2 1/2″ lower than the mark on the ledger. Drive a small nail into the post at the mark and attach the string. Mark all intermediate posts at the point where the string crosses them. Follow the same procedure with the remaining posts. Cut off the excess height with a portable circular saw.

Beams should be installed with the crown up. For maximum load-bearing ability, mount the beam on top of the posts. Attach the beam to the post with a metal post-and-beam connector. If splicing of the beam is necessary, locate the splice over a post. Install joists, crown up, resting them on top of the beam or hanging them between beams, flush with the top. Use hangers to attach the joists to the beam and ledger. If joists rest on top of the beam, fasten them down with framing anchors. For added strength, double the rim joists.

Safety Note

Since the ledger acts as a beam to support joists and other beams, it must be firmly attached to the building's structure with threaded fasteners (bolts, lag screws, or masonry anchors). Do not use nails.

27.5.5 Installing Deck Boards

As a general rule, attach decking perpendicular to the joists. Because sawed lumber is seldom exactly straight, it is best to snap a chalk line as a guide for laying the beginning row. Straighten boards as you go, **Figure 27-20.** For redwood, cedar, or tropical hardwood decking, maintain a 1/4″ space between rows to allow drainage. If you are using treated lumber, no spacing is required. As the boards dry, the shrinkage takes care of the spacing. Spans for composites are essentially the same as for wood—5/4 × 6 deck boards for residential construction can span joists 16″ O.C., while regular 2 × 6 boards can span 20″ O.C. joists.

Since sawed lumber is prone to cupping, it is good practice to turn the board so the bark side is up. You can determine this by checking the annular rings visible at the end of the board. The convex (upward-arching) side is the bark side. See **Figure 27-21.**

Use either nails or screws as fasteners. Usually, it is best to drill pilot holes. This avoids splitting of the boards when nailing and makes driving of screws easier. For greater holding power, always drive nails at an angle. *Ring-shank nails* have greater holding power and are less subject to popping that exposes the

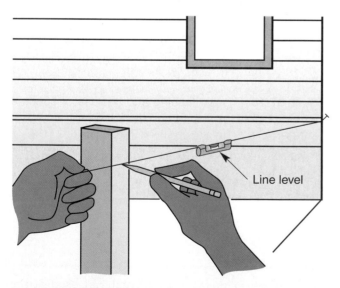

Figure 27-19. Using string and a line level to determine the cutoff line for posts. (Wolmanized® Wood)

Ring-shank nails: Fasteners with annular rings formed around the shaft to provide increased holding power.

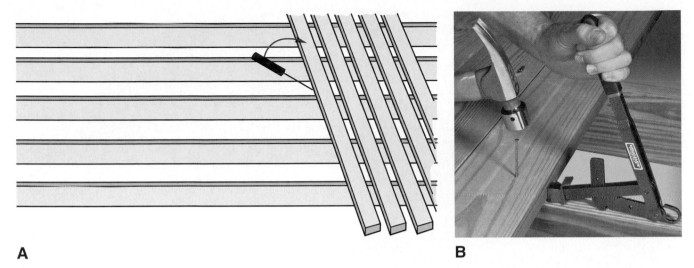

A

B

Figure 27-20. Straightening a deck board. A—Insert the sharp point of tool into a joist, then pry against the board to close up spacing. To open space between boards, insert the tool between them and pry outward. B—A lever-type tool specifically designed for straightening deck boards. (Vaughan and Bushnell Manufacturing Co.)

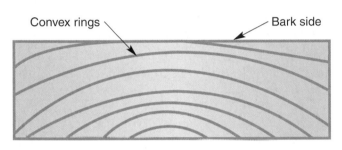

Figure 27-21. To minimize cupping of deck boards, install them bark-side-up. You can identify the bark side by observing the annular rings at the cut end of the board. Upward-arching (convex) rings point to the bark side.

nail heads over time. Composite decking is not subject to nail pops.

Working Knowledge

On an attached deck, start installing the deck boards from the outer edge and work toward the building. A narrower board at the building is less noticeable than one at the outer edge of the deck.

Baluster: Vertical stair member that supports the stair rail.

27.5.6 Installing Railings

Design of railings must follow local code. As a rule, railings must be at least 36″ high whenever the deck is 2 1/2′ or more above ground. Codes may also govern spacing of *balusters.* These are narrow, vertical strips of wood fastened between the top and bottom rails. The ability of the railing to withstand lateral loads may also be specified by code. Always check code requirements before building.

Railings require posts, top rails, bottom rails, and balusters. Some railings may also include an intermediate rail. Latticework replaces the balusters in some designs. Posts for railings are usually attached to the rim joist with through-bolts or lag screws. Posts are normally spaced about 4′ apart. Complete rails in many shapes for decks and steps may be purchased through home improvement centers and lumber yards, **Figure 27-22.**

27.5.7 Installing Stairs

Unless the deck is built at ground level, it will have stairs to provide access from the ground. Stairs may have as few as two treads. If the deck is at a second-floor level, the stairs may have a dozen or more treads. Stairs have three basic components:

Figure 27-22. Factory-built stair railings are available from manufacturers. (Trex Company, Inc.)

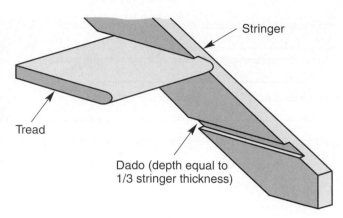

Figure 27-23. Dados may be cut in stringers to support treads.

27.6 Pergolas

A *pergola* is a structure that is often built over all or a part of a deck to provide partial shade, Figure 27-24. Posts supporting the structure may be extensions of the deck posts. On an attached deck, one side may be fastened to the

- Carriage
- Treads
- Cleats or dadoes

The carriage supports all other parts of the stairs. The carriage consists of two sloping boards (*stringers*) supported at the top by the deck's rim joist or fascia. The bottom of the stringers should rest on concrete. The concrete base can be a slab built to support the stairs or it may be one end of a sidewalk. Do not rest the stringers on the ground where they would be subject to deterioration.

The treads are the horizontal parts that form the steps. Treads are what people step on to climb onto the deck.

Cleats are ridges attached to the stringers to support the treads. Cleats may be special metal parts or they may be cut from 1″ lumber. In either case, they must be attached to the stringers and treads with screws or nails. *Dadoes* are slots cut into the stringers to hold the treads, Figure 27-23. Treads must be secured with waterproof glue and screws.

Figure 27-24. This pergola is supported by posts fastened to horizontal frame members.

Stringers: Sloping members that support the risers and treads of stairs.

Dadoes: Rectangular horizontal grooves cut in stair stringers to support the treads.

building. A pergola may also be built as a free-standing structure and is often used in gardens in this manner, **Figure 27-25.** Unlike a roof, the open canopy of the pergola is not intended to provide protection from the elements.

27.7 Porches

A porch deck is structurally very much like an open deck, but its supporting members must be heavier to provide adequate support for the roof, **Figure 27-26.** As a rule, porches are narrower than decks (8' is the norm). The porch width is often dictated by the extension of the house roof to cover the porch. There are three types of porches:

- An *open porch* has no railing—just a deck, roof, and supporting columns. The height of the porch deck above the ground may be governed by a local code.

- A *semi-enclosed porch* has either an ornamental balustrade or a low wall enclosure. The wall may have a wide sill. Some designs allow the addition of removable screens and/or storm panels to accommodate the change of seasons.

Figure 27-25. A freestanding pergola is frequently used in gardens. Many different designs are used for such structures. (Southern Pine Council)

Pergola: A structure with an open roof that provides partial shade from the sun's rays.

Open porch: An attached structure that has no railing—just a deck, roof, and supporting columns.

Semi-enclosed porch: An attached structure with either an ornamental balustrade or a low wall enclosure.

Closed porch: An attached structure that is framed with studs and sided like the rest of the house. It is more like an extension of the house than a transitional area.

Figure 27-26. A spacious, semi-enclosed porch with interesting architectural details. Semi-enclosed and open porches are especially popular in warmer climates. (Southern Pine Council)

- A *closed porch* typically is framed with studs and sided like the rest of the house. It is more like an extension of the house than a transitional area.

Framing of the porch deck should allow a slope of 1″ in 8′ for proper drainage away from the building. Unlike deck boards, the floorboards of the porch are laid tight to one another. Often, the boards are tongue-and-groove. If not treated by the manufacturer, the boards can be soaked in preservative. Flooring should be painted on all faces and edges before installing, if possible. Each joint should be caulked before the board is nailed down. Since they have a number of exposed joints and faces that can collect water, porches are more susceptible to rot than other parts of the house. Be careful to design joints that shed water.

Summary

Basic types of decks are the attached (elevated) form, which is supported by posts and a ledger beam fastened to the house, and the freestanding (grade-level) structure, which is not attached to the house. Either type of deck may have a single level or several levels. The supporting structure of the deck is typically wood—either conventional pressure-treated lumber or engineered lumber. The deck surface itself is traditionally wood, although vinyl materials and composites have rapidly gained ground in recent years. CCA preservatives are no longer allowed. Copper-based preservative treatments are nontoxic, but must be used with stainless steel or specially treated fasteners that will not be weakened by galvanic corrosion. Concealed fastener systems are gaining popularity. Layout of the deck must be done using the same techniques as a house foundation. For attached decks in cold climates or where soil is unstable, concrete piers must be installed for support. A ledger beam is solidly attached to the house structure to support one end of the deck. Posts are located on the piers to support the rest of the structure. Beams or girders rest on the posts and ledger and, in turn, support the deck joists. Decking material, railings, and steps complete the structure. A porch is structurally like a deck with the addition of a roof. It may be open, semi-enclosed, or closed.

Test Your Knowledge

Answer the following questions on a separate piece of paper. Do not write in this book.

1. The two basic deck types are _____ and _____.
2. Which type of deck is placed on grade?
3. *True or False?* An attached deck and a freestanding deck are built the same way.
4. Redwood cut from the _____ is most resistant to rot and insect damage.
5. What materials go into the manufacture of composite deck boards?
6. Fasteners used with ACQ or CA pressure-treated lumber must be _____.
7. *True or False?* It is not necessary to check local codes before laying out a deck.
8. When must piers rest on a concrete footing below the frost line?
9. A(n) _____ is a structure designed to provide partial shade.
10. List the three basic porch types.

Curricular Connections

Mathematics. Make a materials list for a 20′ × 20′ attached deck with the deck surface 24″ above grade. Include quantities of all materials such as concrete, lumber, fasteners, and stain. Check the cost of materials from a catalog or visit to a lumber yard or home improvement center. Prepare two different cost estimates: one using ACQ or CA treated lumber decking and the other with composite decking material. Present your plan, materials list, and cost estimates to your instructor for evaluation.

Outside Activities

1. Design a deck based on the information presented in this chapter.
2. Using the plan in Activity 1, research span tables and specify sizes of beams and joists.
3. Research different products for treating lumber against rot and insect damage. Correspond with various manufacturers. Compare the various claims made for each product.

Section 6

Mechanical Systems

Electrical Wiring

28

Learning Objectives

After studying this chapter, you will be able to:
- Define basic electrical terms.
- Explain what is included in an electrical wiring system.
- List the tools, devices, and materials required to do electrical wiring in a residential building.
- Demonstrate the proper use of tools and handling of materials.
- Demonstrate an understanding of basic circuit theory.
- Use approved methods for simple wiring installation tasks.
- Perform simple electrical troubleshooting.
- Describe the components of a home security alarm system.

Technical Vocabulary

Alternating current
Branch circuit
Circuit
Circuit breakers
Conductors
Conduit
Current
Electric current
Entrance panel

Fish tape
Ground fault circuit
  interrupter (GFCI)
Induction
Mechanical systems
National Electrical
  Code (NEC)
Receptacles
Service

Short circuit
Structured wiring

Switches
Transformer

Once a building is constructed and closed in (roof, windows, and exterior doors installed), it is ready for the installation of electrical wiring. This is one of three installations, along with plumbing and heating/ventilating/air conditioning, that are known as *mechanical systems.* Normally, these systems are the responsibility of persons skilled in building trades other than carpentry. These workers are generally hired by or work for subcontractors. Mechanical systems are installed before insulation and wall coverings, because they must be placed in the frame of the building.

Electrical wiring consists of conductors (wires), boxes, and various devices that control the distribution and use of electricity in the building. The system provides the *current* (flow of electrons) that powers lights, heating units, and appliances.

Mechanical systems: Installations in a building including plumbing; electrical wiring; and heating, ventilating, and air conditioning.

Current: Flow of electrons through electrical wiring.

Electricity is dangerous if it accidentally comes in contact with people, combustible materials, or metal that is not part of the electrical path. Current accidentally contacting a person can cause serious injury and even death from electrical shock. An electrical short may cause a fire. For safety and proper operation of the building's electrical system, a qualified electrician must install electrical wiring according to national and local electrical codes.

The *National Electrical Code (NEC)* is a collection of rules for proper installation of electrical conductors and related devices. It has been developed by experts. This code is revised on a three-year cycle. It becomes law—and is thereby enforceable—only when adopted by a local unit of government (such as a city, village, or county). Communities hire inspectors to examine new installations and enforce the local code when violations are found. When doing any type of electrical work, it is necessary to be familiar with local building code requirements. The National Electrical Code (NEC) is the model code used by most municipalities.

28.1 Tools and Equipment

Certain tools, equipment, and materials are essential when installing electrical wiring. **Figure 28-1** shows a basic tool list. Electricians must be comfortable using all carpentry tools, especially hammers and drills.

Hammers are used for driving fasteners and staples, attaching hangers and electrical boxes, and striking chisels. An electrician's hammer is preferred for attaching boxes because it has a longer neck than a carpenter's hammer, **Figure 28-2**.

Saws are useful for making cuts in a building's frame. A handsaw can be used for notching or cutting studs. For heavier cuts, a power circular saw is better. A keyhole saw or jab saw is useful for sawing in tight quarters and for cutting openings in wall coverings for electrical boxes, **Figure 28-3**. A reciprocating saw also works wherever a keyhole saw is used. It is faster and will handle thicker materials.

Basic Tool List	
Striking tools	**Cutting and sawing tools**
Claw hammer	Files
Lineman's or electrician's hammer	Crosscut saw
	Keyhole saw
Drilling tools	Hacksaw
Cordless drill, 1/2″ chuck	Circular saw, 7″
Cordless drill, 3/8″ chuck	Compound miter saw
Cordless drill, 1/4″ chuck	Reciprocating saw
Drill bits, various sizes	Pocketknife or electrician's knife
wood twist	Cable cutters
metal	Wood chisel
masonry	Wire strippers
spade	Cable strippers
bit extenders	
	Pliers
Soldering and wire-joining tools	Slip joint pliers
Soldering iron	Lineman's pliers
Soldering gun	Side cutting pliers
Propane torch	Diagonal pliers
Solder, rosin core	Long nose pliers
Soldering paste	End cutting pliers
Blow torch	Curved jaw pliers
Crimping tool	
	Special and miscellaneous tools
Fastening tools	Fish tape wire puller
Standard screwdriver	Wire pulling lubricant
Phillips screwdriver	Conduit or pipe cutter
Offset screwdriver	Reamer
Torque head screwdriver	Conduit bender (hickey)
Adjustable wrench	Fuse puller
Allen wrenches	Tap and die set
Socket/ratchet wrenches	Flashlight
Box end wrenches	Plumb bob
	Test light, continuity tester
Measuring tools	Level
Folding ruler	Conduit threader
Carpenter's extension ruler	Trouble light
Steel tape	Gas generator, about 1500 W
12″ ruler or meter stick	Portable space heater
Wire gage	Assorted wood or fiberglass ladders
VOM	Wire grips
	Chalk line
	Tool pouch

Figure 28-1. An electrician's toolbox should be stocked with tools from this basic list.

National Electrical Code (NEC): Collection of rules developed to control and recommend methods of installing electrical systems.

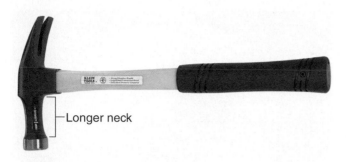

Figure 28-2. An electrician's hammer has a longer neck than the framing hammer used by carpenters. The longer neck and narrow striking face are useful when installing outlet boxes or reaching into tight places. (Klein Tools)

Longer neck

Figure 28-3. A jab saw is useful when cutting openings in drywall for electrical boxes. The large triangular teeth cut fast and do not easily clog. (Klein Tools)

A hacksaw cuts conduit and cable and may be used in place of a handsaw on wood.

Wood chisels are handy for trimming away small amounts of wood on framing members for mounting boxes and fixtures. They are preferred for notching studs, joists, plaster, flooring, and old-style lath.

A number of different tools can be used for wire and cable cutting. Multipurpose tools cut, strip, and crimp, **Figure 28-4.** A cable stripper is used on larger conductors.

Pliers are good for holding, shaping, and cutting. The basic types used are slip joint pliers and lineman's pliers or side cutters, **Figure 28-5.** A versatile tool is the locking pliers. It serves as pliers, lock wrench, open-end wrench, or a pipe wrench.

The standard straight-blade screwdriver and the Phillips screwdriver are the most-used fastening tools. They are primarily needed to tighten terminal screws, attach switches and receptacles, and fasten wires to devices.

Drilling tools are used to bore holes when running conductors or conduit through studs and joists. For tight quarters, use a right-angle drill.

The electrician's tool kit may also contain soldering tools—either a propane torch or an electric soldering iron. Even though the use of wire connectors has all but eliminated soldering, there is still some use for soldered connections.

A folding 6′ rule and a steel tape are important for measuring. Typical rules are 6′, 8′, and 10′.

A fish tape is necessary for pulling wires through conduit, **Figure 28-6.** A *fish tape* is a long strip of metal or other strong, fairly stiff material wound on a reel for ease of handling. The tape is pushed through installed electrical conduit, then used to pull wires into place.

Various other tools round out the electrician's tool kit. These might include a trouble light, pipe cutter, continuity tester, neon tester, conduit bender, conductor bender, and a set of socket wrenches. For electrical work, use only wood or fiberglass ladders. Neither is a good conductor of electricity.

Figure 28-4. This multipurpose tool is used to cut electrical wire, strip insulation from either solid or stranded wire, and crimp various types of connectors onto conductors. (Klein Tools)

Figure 28-5. A side cutting, or lineman's, pliers is an essential tool for everyone doing electrical installations. (Klein Tools)

Fish tape: A long, flat, flexible steel tape used to pull electrical wires through conduit or walls.

Figure 28-6. One type of fish tape. (Klein Tools)

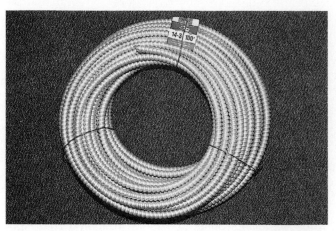

A

B

Figure 28-7. Electrical conductors are the copper or aluminum wires that provide a path for current in the building's circuits. The conductors are made up in different ways. Codes vary on which types can be used. A—Armored cable, also called BX, has a flexible metal cladding protecting several insulated conductors. B—Nonmetallic (NM) cable is similar to armored cable, but has a plastic protective covering.

> **Safety Note**
>
> Never use a metal ladder when doing electrical work. If the ladder accidentally contacts a current-carrying conductor, serious injury or even death could result.

28.2 Electrical System Components

An electrical system in a building is made up of:

- Conductors
- Boxes and box covers
- Circuit breakers or fuses
- Ground fault circuit interrupters (GFCIs)
- Switches and receptacles
- Conduit or raceways
- Connectors
- Circuit

Conductors are the wires that carry current. See **Figure 28-7.** The wires may be enclosed in conduit, a metal sheath, or a plastic sheath.

Various kinds of boxes and box covers are available, **Figure 28-8.** Their purpose is to enclose devices that control electric current or protect connections between two or more conductors. Covers conceal and protect devices, such as switches and receptacles that are placed in boxes.

Circuit breakers or fuses are devices that protect conductors from overloads, **Figure 28-9.** They shut off power to the circuit in case of a fault in the circuit or a too-heavy current draw.

Conductor: A material, such as a copper wire, that carries an electric current.

Circuit breaker: An over-current protection device that opens a circuit, preventing the flow of electricity, when an overload or short is detected so that the circuit will not be damaged.

A

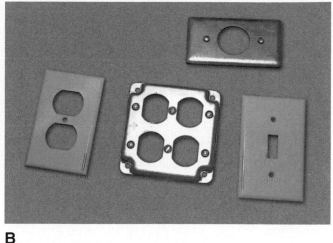

B

Figure 28-8. A—Several types of plastic and metal boxes used for either new construction or remodeling work. B—Covers protect devices and conductor connections in the boxes.

Figure 28-9. Circuit breakers (left) and fuses protect circuits from overloads that could cause damage to the circuit or to the building.

Ground fault circuit interrupters (GFCIs): An overcurrent protection device that is able to detect a short circuit more rapidly than a fuse or circuit breaker; used where water is present.

Switches: Devices that control current to lights and appliances.

Receptacle: Devices that allow for current transfer between conductors and appliances. Also called an *outlet* or *convenience outlet.*

Conduit: A tube of metal or plastic in which wiring is installed.

If this is not done, the wire will heat up and fail or cause a fire.

Ground fault circuit interrupters (GFCIs) are safety devices installed either in an electrical circuit or at an outlet as a protection against electrical shock. GFCIs are required by electrical code in kitchens, bathrooms, laundry rooms, and other locations where water is present. They are able to detect a short circuit more rapidly than a fuse or circuit breaker.

Various types of switches and receptacles are used in residential and light commercial applications, **Figure 28-10.** *Switches* control the current to lights and appliances. *Receptacles* allow convenient connection of appliances and lights to a circuit. They are sometimes called *outlets* or *convenience outlets.*

Conduit or raceways are pipes or enclosed channels through which conductors are run, **Figure 28-11.** *Conduit* is rigid tubing. Lengths of the tubing can be cut and bent as needed to fit the structure. Conduit lengths are connected with special couplings.

Connectors are fastening devices that ensure tight connections between two or more conductors or lengths of conduit, **Figure 28-12.** Wire nuts form a sound mechanical and electrical connection between conductors. Color coding is used to identify the appropriate conductor sizes. Couplings are used to connect conduit, while other fittings are used to securely clamp armored or nonmetallic cable to electrical boxes.

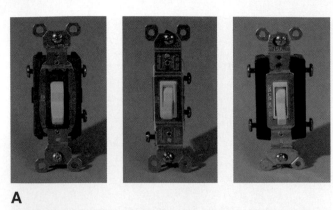

A

B

Figure 28-10. Devices most commonly placed in electrical boxes are switches and receptacles. A—Switches interrupt the path of the current. They control the flow of electricity, just as a faucet controls the flow of water. B—Receptacles allow quick temporary connection of appliances and lights to a circuit.

The *circuit* is the path through which current flows from the source back to the source. It includes the conductors, switches, receptacles, and any appliances or lights.

28.3 Basic Electrical Wiring Theory

Electricity is generated when a conductor made from a metal, such as copper, silver, or

Circuit: A path for electrical power provided by wires (conductors) between a power source and the power-using device (load).

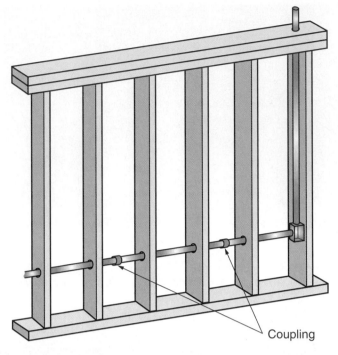

Figure 28-11. Conduit is used to protect electric wires.

A

B

Figure 28-12. Fittings connect and hold cables and conduit. A—Wire nuts. B—Couplings are needed for conduit.

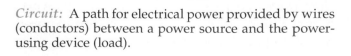

aluminum, is passed through a magnetic field. This causes electrons to move through the conductor in one direction. This electron flow is known as ***electric current.*** When the conductor (wire) passes through the magnetic field in the opposite direction, the current flows in the opposite direction. This changing of direction is known as ***alternating current.*** Alternating current (ac) is the form of electric current found in homes. Another form of current, used for various purposes, is known as direct current (dc). In dc, electrons flow in only one direction. This is the type of current supplied by automotive or flashlight batteries, for example.

Alternating current is generated at an electric power station and sent out through conductors, **Figure 28-13.** Transformers along the way step up (increase) or step down (reduce) the voltage as required by power customers. A ***transformer*** is a device that can make the voltage it receives either higher or lower. It does so by way of a magnetic process called ***induction.***

A transformer is made up of two coils. A coil is a core of iron around which are wrapped a number of turns of conducting wire. One coil in the transformer has more turns of wire than the other. There is no electrical connection between the two coils. When current enters one coil, it sets up a magnetic field. This field reaches the other coil and ***induces*** a current. This induced current does not have the same voltage because of the difference in number of turns of wire on the two coils.

Step-up transformers at the power station increase the voltage generated, because it is easier to send high-voltage current a long distance over conductors. However, before current is delivered to a house, it must pass through a step-down

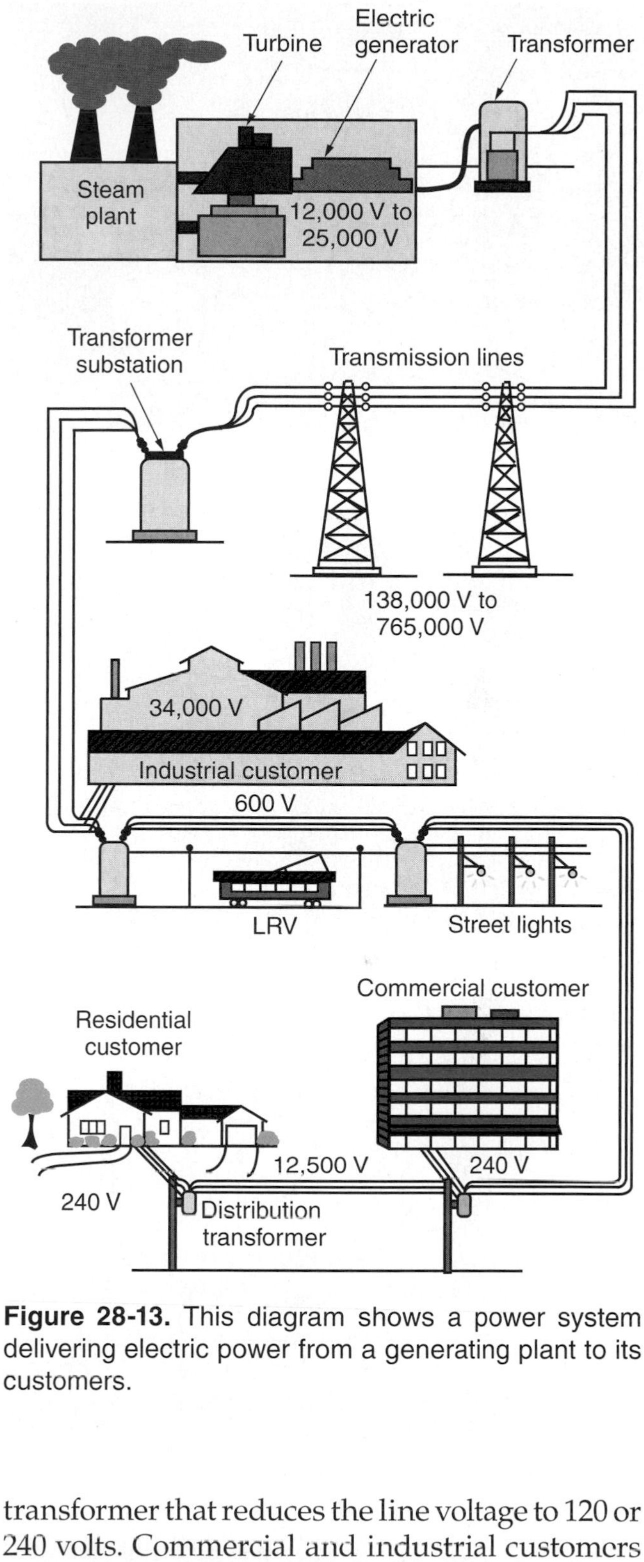

Figure 28-13. This diagram shows a power system delivering electric power from a generating plant to its customers.

Electric current: Flow of electrons through a conductor.

Alternating current: An electric current that regularly reverses direction; used in home wiring.

Transformer: A device for changing the voltage of an electric current.

Induction: A process by which an electric current is produced when a conductor cuts through a magnetic field.

transformer that reduces the line voltage to 120 or 240 volts. Commercial and industrial customers may use much higher voltages.

Figure 28-14 shows three kinds of transformers for changing voltage. One is called a *substation.* It serves a whole area, such as a town. The second, smaller "pole" transformer serves one residence or a group or residences. The third transformer is used within the house to

A

B

C

Figure 28-14. In an electrical distribution system, voltage is changed several times before being delivered to customers. A—A substation is a series of transformers that step down voltage before it is delivered to a large number of customers in a neighborhood or town. B—Step-down transformer further reduces voltage. C—Transformers can be very small, such as this 14 volt unit used to power doorbells and some other low-voltage energy users. (Ship and Shore General Store, Dauphin Island)

reduce current to low voltage for devices such as a doorbell.

From the residential transformer, conductors called a *service* bring electricity to the residence. The conductors may be strung from poles or buried in the ground. See **Figure 28-15.** Two 120 V conductors and a neutral are supplied to the house, which allows for 240 V for certain applications, such as an air conditioner or electric hot water heater.

Safety Note

Electrical current is dangerous if a person accidentally touches bare conductors or other energized parts of a circuit. Always follow these safety rules when you work around electricity:

• Make certain that a circuit is de-energized before working on it; trip the circuit breaker or pull the fuse.

• If you must work on a live circuit, use only tools with insulated handles.

• Remove jewelry, watches, rings, and other metal objects before beginning work.

• Keep work areas dry and clean. Cover wet floors with wood planks.

• Wear rubber boots or rubber-soled shoes.

• Use only wood or fiberglass ladders, never metal.

• Wear clothing that is neither too loose nor too tight.

• Wear safety glasses when using striking tools.

28.4 Installing the Service

The *service* is the conductors that bring power from a transformer to a building. It extends only from the transformer, through a metering device, to a box in the dwelling called a distribution panel

Service: The conductors that bring power from a transformer to a building.

A

B

Figure 28-15. Residential electrical service. A—A trench must be dug to lay down underground service. The electric code specifies how deeply cables or conduit must be buried. B—Here, an electric power company worker completes a hook-up.

or entrance panel. The *entrance panel* houses a main breaker, circuit breakers, or fuses for all of the home's circuits, **Figures 28-16** and **28-17**. **Figure 28-18** is a drawing of all the components of the service entrance at the dwelling. Installation of these components is the responsibility of the electrician hired by the homeowner.

Entrance panel: Where the service ends inside of the house; contains a main breaker and circuit breakers or fuses for all of the home's circuits.

Short circuit: Accidental grounding of a current-carrying electrical conductor. Also called a *ground fault*.

28.5 Grounding and Ground Faults

An electrical system is designed so that electrical current safely travels through insulated conductors. When current leaves the conductors, it enters a device that converts electrical energy to provide light, heat, or motion. Should a conductor come loose or a device break, a *short circuit*—also known as a ground fault—can occur. When this happens, electrical current energizes something that should not be energized. A person touching energized metal will get a dangerous shock. Grounding minimizes the possibility that a short circuit will cause a shock.

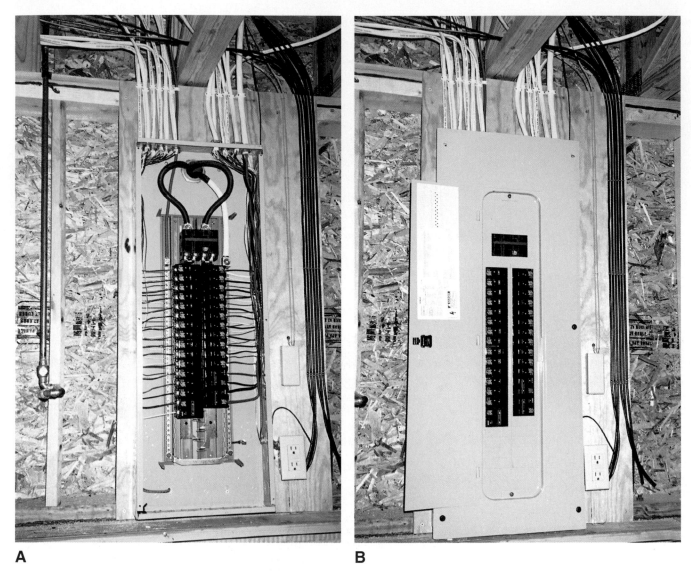

A   **B**

Figure 28-16. The service entrance or distribution panel is located inside the building. A—Each circuit is connected to a circuit breaker (black objects in the box). The heavy cables at top are the conductors that bring power from the transformer. The large black block just below the cables is the main breaker. B—When connections have been completed, a protective cover is placed over the installation.

In a properly grounded circuit, every device is connected to ground. The ways of providing this connection are:

- With a ground wire, either bare (uninsulated) or green, that leads from the device to a neutral bar in the service panel.

- Using metal boxes and conduit that house the conductors provide a path for current along the conduit to the neutral bar.

The ground is completed by a conductor that runs from the neutral bus bar to a cold-water pipe or to grounding rods driven deep into the soil. See **Figure 28-19.**

When a ground fault happens, the grounding path carries the current back to the panel. The extra path lowers resistance, which, in turn, causes a surge in power. This causes the circuit breaker to trip or fuse to blow. At the same time, current is directed into the earth where no harm results.

A   **B**

Figure 28-17. The exterior box for electrical service is called a meter socket. A—Meter socket is shown before completion of service. B—When the wiring system of a building is completed, the meter is installed and the system is ready to be used.

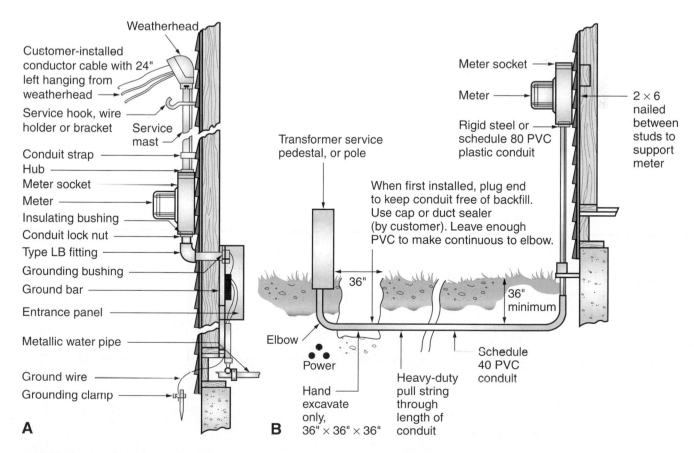

Weatherhead

Customer-installed
conductor cable with 24"
left hanging from
weatherhead

Service hook, wire
holder or bracket   Service
                       mast

Conduit strap
Hub
Meter socket
Meter
Insulating bushing
Conduit lock nut
Type LB fitting
Grounding bushing
Ground bar
Entrance panel

Metallic water pipe

Ground wire
Grounding clamp

A

Transformer service
pedestal, or pole

When first installed, plug end
to keep conduit free of backfill.
Use cap or duct sealer
(by customer). Leave enough
PVC to make continuous to elbow.

36"

Elbow

Power

Hand
excavate
only,
36" × 36" × 36"

Heavy-duty
pull string
through
length of
conduit

B

Meter socket

Meter

Rigid steel or
schedule 80 PVC
plastic conduit

2 × 6
nailed
between
studs to
support
meter

36"
minimum

Schedule
40 PVC
conduit

Figure 28-18. Diagrams of two types of three-wire service to a building. A—An above-ground installation. B—An underground installation. (KCPL)

28.6 Reading Prints

The electrical plan for the house is usually a part of the drawings done by an architect. If there is no electrical print, the electrician prepares one. Symbols on the drawing show the types of conductors and devices to install at each location, **Figure 28-20.** These symbols are standardized so that there can be no misunderstanding, **Figure 28-21.**

After studying the electrical plan, the electrician may draw a cable layout for each room. This layout shows where in the room electrical cable is to be run, the size of the conductors, and how many conductors are in each run of cable. This layout may also be used in other ways. For example, the electrician may use it as a billing reference or as a way of checking off work that has been completed. **Figure 28-22** shows an example of a cable layout.

Figure 28-19. Electrical circuits must have a grounding system for safety. Copper rods driven 8′ deep into the ground to damp soil provide a satisfactory ground. Another grounding method involves the use of the building's cold water piping.

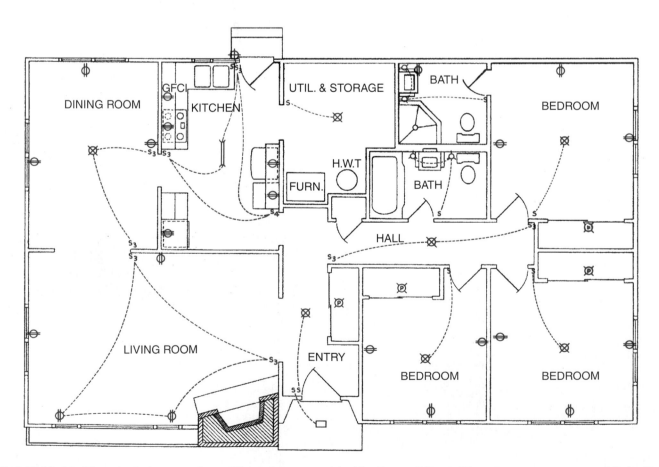

Figure 28-20. Electrical plans such as this one are provided by the architect. The plan serves as a guide to the electrician when installing wiring, devices, and fixtures.

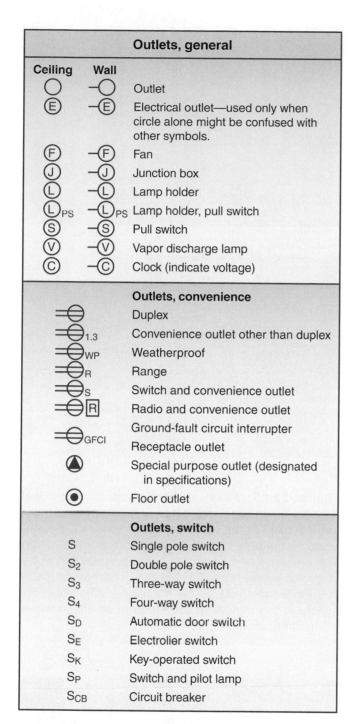

Figure 28-21. Electrical symbols are a type of short-hand that tells the electrician which type of device or fixture is to be installed. This is a sample of some symbols for electrical outlets.

Branch circuit: A single circuit running from a distribution panel.

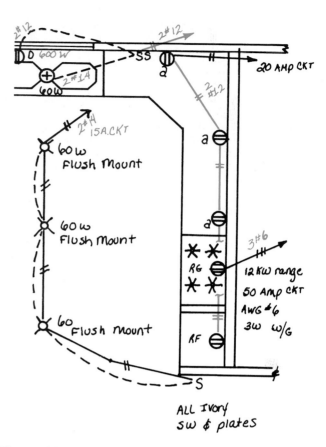

Figure 28-22. A cable layout. Electricians sometimes produce such drawings as a reference while wiring a room.

28.7 Running Branch Circuits

Think of a circuit as a loop or continuous path from the source of electric power, through various current-using devices (such as lights or electric motors), and then back to source. The source in this case is the service panel containing the circuit breakers or fuses. **Figure 28-23** shows a sketch of a simple circuit.

Branch circuits are circuits that include all of the outlets on one run of conductors. The circuit begins at the distribution panel, where it is connected to a fuse or circuit breaker. The electrician first installs junction, switch, and receptacle boxes. Then, conductors are run (installed) through the building frame into these boxes. In metal-framed structures, holes are already made in the studs and joists. In wood-framed buildings, the electrician drills holes through the wood

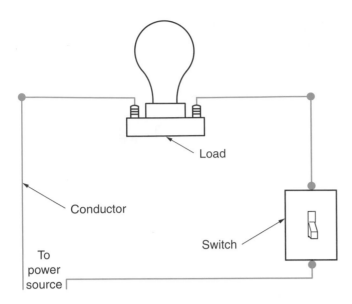

Figure 28-23. A circuit is simply a path from a source of power, through energy-using devices, back to the power source.

Figure 28-24. An electrician often must drill holes through the building frame to run electrical cable or conduit.

framing members so cable or conduit can pass through, **Figure 28-24.** The cable or conduit must be run before insulation and drywall or plaster are installed, **Figure 28-25.** The conductors are run to each box and securely connected to them. See **Figure 28-26.**

A

B

Figure 28-25. Running conductors through wall framing. A—When running nonmetallic electrical cable through steel studs, electrical codes require the use of plastic grommets to protect the cable from abrasion. The grommets are placed in holes prepunched in the studs for wiring runs. B—Running nonmetallic cable through wood studs. Holes should be drilled approximately in the middle of the studs to minimize weakening of the structure.

Figure 28-26. Cable or conduit must be securely held to boxes with the proper connectors.

Electrical cables must also be securely fastened to framing members, as shown in **Figure 28-27.** The National Electrical Code requires that fasteners must be installed at intervals not greater than 3′ and within 12″ of their entrance into an electrical box.

There is always a danger of accidentally piercing the conductors with nails driven into the studs while installing drywall. Metal plates are available to protect the conductors. See **Figure 28-28.**

28.8 Device Wiring

To complete a circuit, it is necessary to connect wires to one another. It is also necessary to connect wires to devices such as switches, receptacles, and large appliances. The NEC tells how these connections are to be made. Before attempting to connect wires, several inches of insulation must be stripped away. To make such a connection, wrap the bare wires together, then install a wire nut, **Figure 28-29.**

The terminals on switches and receptacles are usually screws. The screws are tightened to grip the conductors. **Figure 28-30** shows the correct method of bending a conductor for attaching to a terminal.

Switches control current through a circuit. The hot (current carrying) wire in the circuit is interrupted by the switch as shown in **Figure 28-31.**

A single pole switch is simple to wire, but three-way and four-way switches can be complicated unless the electrician fully understands what happens when a switch is activated.

Figure 28-27. Nonmetallic cable must be properly fastened to framing members. Conductors must be secured with a cable staple (arrow) within 12″ of entering a receptacle box.

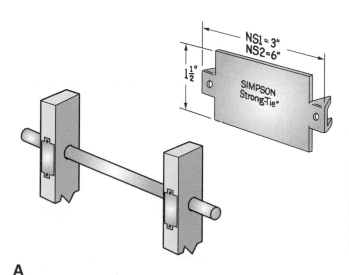

A

B

Figure 28-28. Protecting conduit and cable. A—Special metal plates are attached to wood studs so that nails cannot accidentally pierce the conductors and cause a dangerous short. B—Installing a plate. (Simpson Strong-Tie Co., Inc.; Hadley-Hobley Construction)

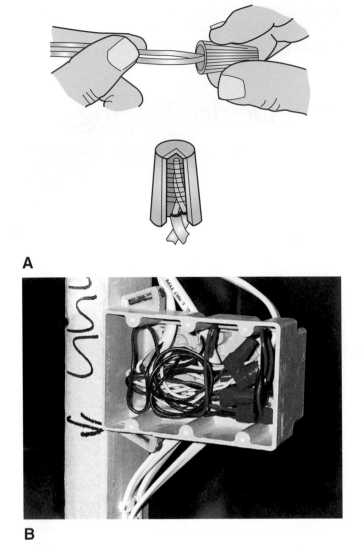

A

B

Figure 28-29. The proper use of wire nuts to connect bared ends of conductors. A—Twist the conductors together, then insert the bare conductors into the wire nut and twist the nut clockwise until it is tight. B—Connected wires are insulated against accidental shorting by the wire nut.

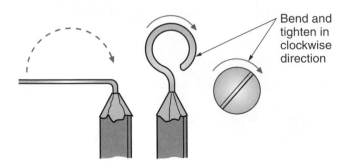

Figure 28-30. The correct method of preparing a conductor for attaching to a screw-type terminal. Use a needle-nosed pliers to form the loop.

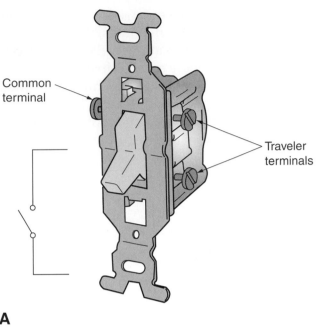

A

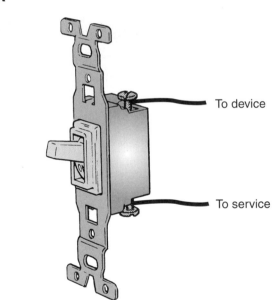

B

Figure 28-31. A switch interrupts a current-carrying (hot) wire. It is always placed on a black or red wire A—Three-way switch. B— Single-pole switch.

Figure 28-32 illustrates how current flows when three-way switches are operated.

Receptacles allow the transfer of electrical energy from the circuit conductors to lamps, toasters, and a variety of appliances. They have terminals on both sides. The hot wire is attached to the copper-colored terminal. The neutral (white) wire is attached to the lighter, silver-colored terminal, **Figure 28-33.**

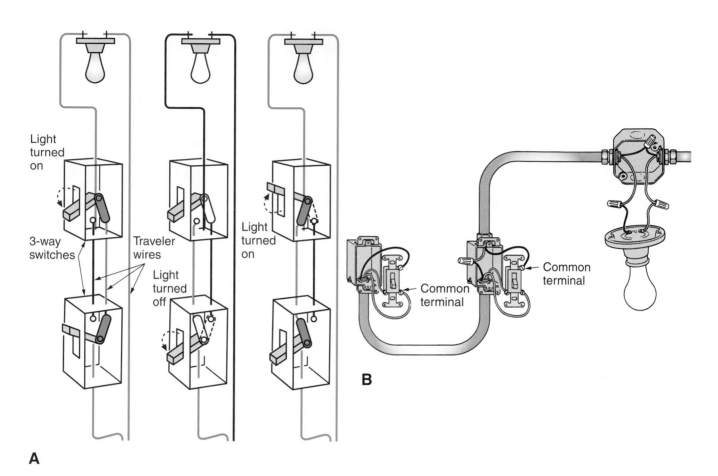

A

Figure 28-32. Three-way switch installation. A—How a three-way switch arrangement actually works. B—How an electrician might wire it up.

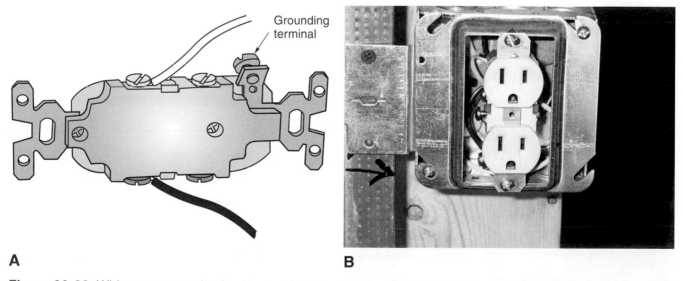

A   **B**

Figure 28-33. Wiring a receptacle. A—Neutral (white) wire is connected to light colored terminals; black or red to darker terminal. B—A receptacle properly wired and installed in a box.

28.9 Electrical Troubleshooting

Electrical troubles are indicated by a wide range of symptoms. Usually, the problems involve faulty receptacles, switches, lighting fixtures, fuses, or circuit breakers. Complex problems include open circuits, broken conductors, voltage fluctuation, and ground faults or current leakage.

PROCEDURE

Testing receptacles

When a receptacle is not functioning, either the receptacle or the circuit is at fault.

1. Insert the leads of a neon tester into the receptacle. If the tester does not light, check the line conductors.
2. Remove the wall plate and check for voltage at the terminals of the receptacle, **Figure 28-34.**
3. If the tester lights up, the receptacle is defective. If the tester does not light, the problem is in the circuit. Sometimes an open neutral is a problem.
4. To test for an open neutral, place one test probe in the ground slot and the other in the hot slot of the receptacle. See **Figure 28-35.**
5. If the tester lights, indicating voltage, there is an open in the neutral conductor.
6. Check other receptacles between the problem receptacle and the distribution panel. Check at the distribution panel, as well.

Figure 28-34. To check for voltage, place the probes on the two terminals that have wires attached to them. If the tester lights up, voltage is reaching the device.

Figure 28-35. Place probes in the hot side slot and the ground. If the tester lights, there is an open in the neutral conductor.

PROCEDURE

Testing switches

To test switches, remove the cover plate and determine if there is power to the switch.

1. Touch one probe of a neon tester to the metal box or the ground wire if the box is plastic while touching the other probe to the hot side terminal.
2. If the tester lights, the circuit is live.
3. With the switch on, touch the probe to the load terminal of the switch and the other probe to ground.
4. If the tester lights, the switch is good and the fixture or the wiring to it is defective.

PROCEDURE

Testing fixtures

Nonfunctioning fixtures can be checked by testing whether power reaches the supply conductors at the fixture outlet, **Figure 28-36.**

1. Place the probes of a neon tester on the supply conductors.
2. If the tester lights, voltage is present.
3. Repair or replace the fixture.

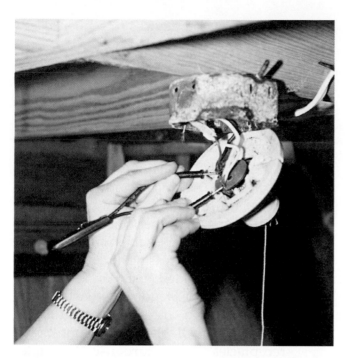

Figure 28-36. Test a fixture by checking if line voltage is reaching the terminals. Place one probe on each terminal. If the test causes the tester to glow, but the fixture is dead, replace or repair the fixture.

28.10 Home Security and Automation Wiring

More and more new homes are being prewired with low-voltage wiring systems to handle everything from a basic home security system to full-scale home automation. The most sophisticated systems include control and distribution of audio and video signals, remote heating/cooling controls, telephone and computer system wiring, wiring for Internet access, security and fire alarm systems, and other features. Installing such systems is most efficient and least expensive when the walls are still open. Thus, most installations are in new homes or in major remodeling projects where

Structured wiring: An integrated system of low-voltage wiring and centralized control units used to provide a home with a variety of communication, automation, and security services.

walls have been stripped down to the studs. Retrofitting of systems in existing structures is time- and labor-intensive, greatly increasing costs.

The idea of the "smart house" with advanced communication and automation features has been discussed in magazines and books for a number of years. It is no longer seen as a future development, but as a growing trend among homebuilders. Within a few years, structured wiring may be considered as a basic household system, like plumbing and electrical.

28.10.1 Installation

The installation of home security system wiring is relatively simple, involving such components as entry/exit keypads used to set or disable the system, door and window switches, motion detectors, smoke and heat detectors, security cameras, external lights and audible alarms, the system controller, and associated low-voltage wiring. The system typically includes a telephone connection to a monitoring agency or the local police and fire stations.

Installation of home automation systems, referred to as *structured wiring,* can be considerably more complex. It typically involves one or more distribution panels and runs of multiple cables from the distribution box to wall outlets in each room in the house. The distribution panels are similar to electrical distribution panels, but contain various electronic modules instead of circuit breakers. In some installations, wiring may be run through conduit. In other installations, bundled cables alone are used. An advantage of conduit is the ability to later install additional cables.

Working Knowledge

To simplify the task of pulling additional cables through a conduit at a later date, pull a length of strong nylon fishing line through each conduit run at the time of original installation. Make each nylon line twice as long as the conduit run and loosely coil the excess at one end. This will allow you to pull additional cable and still have the line in place for later use.

28.10.2 Wiring

Several types of wiring are used in home automation systems: twisted-pair cables with a number of wire pairs in each cable, shielded coaxial cable, and fiber optic cables. The coaxial cable handles video signals and Internet access (with a cable modem). Twisted pair cables are often referred to as Category 5 cables. These are multi-purpose wires capable of carrying audio, high speed computer data transmissions, telephone, and security system signals. Fiber optic cables are included for future use.

Both security system wiring and structured wiring is often installed by specialized subcontractors, but an increasing number of electricians and electrical subcontractors are performing the work. In most cases, they install the conduit, distribution boxes, cables, and wall outlets following the drawings developed by a structured wiring specialist.

28.10.3 General Considerations

There are some general considerations to consider when installing structured wiring. These include:

- Whenever possible, install structured wiring after the electrical and plumbing work are done. This avoids potential damage and electrical interference problems.
- To provide the best-quality service, all wiring should be set up as home runs. These are direct runs from the panel to the outlet. Daisy chains are connec-

tions of several devices in sequence and should be avoided.

- Provide vertical chases or raceways (usually, 2" conduit) between floors to simplify later wiring additions or changes.
- Install wiring runs with slack and avoid 90° bends. Tightly stretched cables and tight bends can degrade system performance, especially in computer networks.
- To minimize electrical interference, run all systems wiring at least 12" away from electrical wiring (in the next joist or stud space, for example). If cables must cross electrical wires, make the crossing at a 90° angle.
- The recommended installation for most rooms is two Category 5 (twisted pair) cables and two coaxial cables, all terminated in a wall plate with appropriate receptacles. For future use, two fiber optic cables are often bundled with the twisted pair and coaxial cables.
- Place the distribution panel in a central, accessible location such as a closet or mechanical/utility room. Place it 24" or more from the electrical distribution panel.
- In a large home, two or more distribution panels may be used with functions divided between them. For example, security and HVAC systems may be in a panel located in the garage or basement. Entertainment, communications, and related functions may be in a panel centrally located in the living area.
- To aid in troubleshooting or system changes, carefully label every cable in the distribution panel with its function and the room where it terminates.

ON THE JOB

Electrician

Electricians install electrical systems in new or existing residential, commercial, and industrial structures. They also are often responsible for installing low-voltage wiring and devices for alarm and security systems, computer networking, communication systems, and similar applications. Approximately 1/4 of all electricians work in the construction industry, while the remaining 3/4 are employed in maintenance roles. Most large institutions, from hospitals to industrial plants, employ maintenance electricians. An estimated one in ten electricians is self employed.

Electricians must be familiar with all aspects of the National Electrical Code, as well as state and local building codes, and must be able to read construction drawings. Those who work with low-voltage systems need a good working knowledge of electronics.

Working conditions, especially in construction, are often strenuous—workers stand for long periods and may have to repeatedly climb ladders or scaffolding. Although most jobs are indoors, electricians may have to work in difficult weather conditions. Safety is a major concern, since the work has the potential for electrical shock and injury from falls. Wearing proper personal protective gear and following good safety practices is very important.

Most electricians learn the trade by completing an apprenticeship of three to five years that involves at least 144 hours of classroom instruction and a minimum of 2000 hours of on-the-job training. Apprenticeship programs are typically operated by a joint effort involving the electrician's union (International Brotherhood of Electrical Workers) and one of several electrical contractors' associations. Those who successfully complete the program are qualified to work in either construction or maintenance areas. Smaller numbers of electricians learn their trade through shorter-term training programs or by learning on the job as helpers for experienced electricians.

Electricians must be licensed in most states and local jurisdictions. Licensing is obtained through an examination that typically covers electrical theory and the National Electrical Code. Electricians who work in larger organizations can advance to supervisory positions. Experienced electricians often obtain positions with the government as electrical inspectors. Some choose to open their own electrical contracting business.

Summary

While electrical wiring is normally installed by an electrical contractor, carpenters need to be aware of the basics of electrical installation. Basic tools for electrical work include an electrician's hammer, saws, various types of pliers and screwdrivers, drills, measuring and fish tapes, and a wooden or fiberglass ladder. Metal ladders are unsafe around electricity. The components of an electrical system include conductors (wires), conduits or raceways, device and junction boxes and covers, switches and receptacles, and connectors for joining wire. Electricity is transmitted by the power utility over wires to a transformer that steps down the voltage to a level suitable for household use. The wires from the transformer to the building are called the service. They pass through a metering device and terminate in an entrance panel. This panel contains circuit breakers or fuses for each of the household electrical branch circuits. These circuits are shown on the electrical plan for the house, which is part of the construction drawing set. Once an electrical system is installed, it may require troubleshooting, which involves testing of receptacles, switches, and fixtures. Low-voltage systems for home security and home automation are becoming increasingly important in residential building. More electricians are becoming involved with their installation.

Test Your Knowledge

Answer the following questions on a separate piece of paper. Do not write in this book.

1. What is included in a building's electrical wiring?
2. Define *current*.
3. The _____ is a collection of rules for the proper installation of electrical conductors and related devices.
4. _____ control the current to lights and certain appliances.
5. Electrical current that goes one direction and then another is known as _____ current.
6. *True or False?* Transformers are devices that change voltage by means of deduction.
7. Electrical service enters the house through a meter and into a(n) _____.
8. A(n) _____, also called a ground fault, occurs when electric current energizes something that should not be energized.
9. *True or False?* An electrical plan is the same as a cable layout.
10. What is a *branch circuit?*
11. A(n) _____ is used to connect the bared ends of two wires.
12. The neon tester does not light when the probes are placed on the terminals of the receptacle. Where does the problem lie?
13. *True or False?* Home security and automation systems generally cost less to install before the wall covering has been applied.
14. Define *structured wiring*.
15. _____ cable is used in structured wiring systems to carry video signals.

Curricular Connections

Science. Conduct research to determine how series, parallel, and series-parallel circuits are wired. Using a low-voltage power source (such as a 6-volt lantern battery), bell wire, and appropriate lamps, switches, etc., design and build examples of each type of circuit. Study the circuits to determine how they are alike and how they differ.

Outside Assignments

1. Wire a sample circuit that includes a plug, convenience outlet, and box-to-house outlet. Have your instructor inspect your circuit before testing it.
2. Using a junction box and two lengths of No. 14 nonmetallic or armored cable, show the proper method for making connections between two conductors.
3. Prepare a drawing that shows the operation of two three-way switches wired to control a light from two locations. Discuss the drawing with your instructor and then wire the circuit on a wall mockup.

Plumbing Systems

29

Learning Objectives

After studying this chapter, you will be able to:

- Cite codes that govern the installation of plumbing systems.
- List necessary plumbing tools and explain how to use them.
- Describe the different types of materials used in plumbing systems.
- Name and recognize devices and fixtures that are part of the plumbing system.
- Explain the proper design and installation of plumbing systems.
- Read plumbing prints.
- Demonstrate a basic understanding of well and pump systems.
- Unplug drains using chemicals and the special tools designed to clear out clogs.
- Cite safety measures that plumbers must observe.

Technical Vocabulary

Bored wells
Casing
Compression fittings
Compression valves
Drainage, waste, and venting (DWV) system
Drilled wells
Faucets
Fixtures
Flare fittings
Gate valves
Hub
Oakum
O-rings
Pipe compound
Plumber's putty
Potable
Rough-in
Slip-joint washers
Stub-in
Submersible pump
Sweat soldering
Teflon® tape
Venting
Vent stack
Washerless
Water supply system

Plumbing includes all of the piping and fixtures that supply water for drinking, cooking, bathing, and laundry. Also included are drainage pipes that provide a means of disposing of wastewater. See **Figure 29-1.**

To install rough plumbing, plumbers must cut holes and notches in the building frame. It is important for the plumber to closely work with the carpenter to be sure that the holes will not weaken joists and studs. The work should be neatly done. Holes should be smooth, cleanly made, and only large enough to receive the pipe. Notches should be square or rectangular.

Once plumbing is installed, notched framing members should be reinforced with metal strapping. In certain situations, a joist may need to be partially removed. In such cases, a header should be installed to reinforce the floor frame at that point. **Figure 29-2** illustrates how the plumber should handle these situations. Note also the ease with which plumbing runs are made through truss joists.

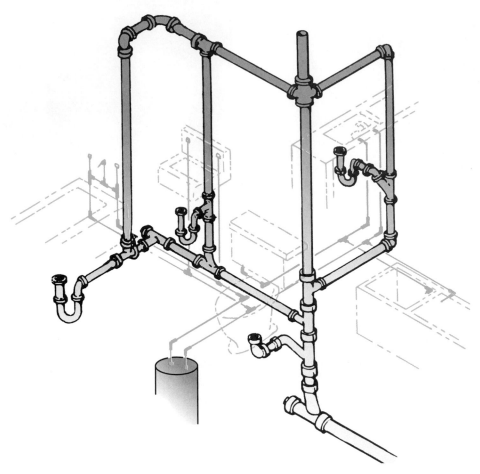

Figure 29-1. A pictorial view of a plumbing system. The large pipes are parts of the drainage, waste, and venting (DWV) system. The upper portions of the DWV comprise the venting. The small pipes are the water supply system.

Accidents are usually the result of unsafe acts or failure to wear protective clothing or safety equipment. Keep these safety precautions in mind as you work:

- Eye protection, such as safety goggles, is necessary when doing hazardous tasks. These tasks include sawing, drilling, chipping, spraying, sand blasting, welding, operating powder-actuated tools, and using compressed air for cleaning.

- OSHA regulations require filter lenses of the proper shade and number for soldering, brazing, and welding operations. Laser safety goggles should be worn when using laser equipment for leveling and plumbing operations.

- Synthetic clothing should be carefully checked. Some of these fabrics are highly flammable. Welding sparks or other high-temperature heat sources could set them afire.

- Danger of fire is always present on a construction site. Promptly clear away combustible construction debris. Protect compressed gas cylinders from damage. Highly flammable materials such as liquid fuel, certain adhesives, and solvents should never be stored in large quantities on the job site. Keep fire extinguishers on hand. Study the charts in **Figures 29-3** and **29-4** for classes of fire and the type of fire extinguisher required for each class.

- Wear an approved hard hat when there is danger from falling objects. Wear gloves to protect hands when handling plumbing materials. They should *not* be worn when

(Continued)

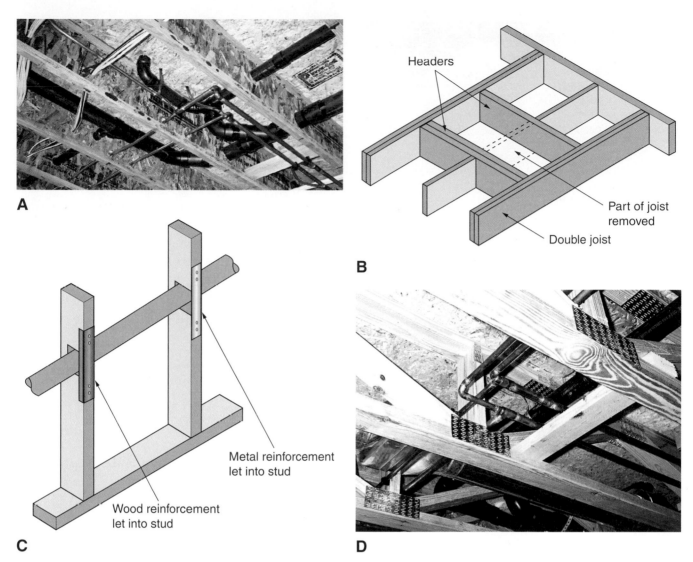

Figure 29-2. A plumber must be careful not to weaken framing when altering it to run pipes. A—Approved method of cutting holes in wood I-beams. B—Approved method of supporting joists to get large openings for plumbing. Joists have been doubled and headers installed. C—Where studs must be notched, wood or metal reinforcing must be installed across the notch. D—When open-truss joists are used, no cutting is necessary. All mechanical systems can be run through the voids in the truss.

(Continued)

operating tools. Safety shoes protect the feet from falling objects as well as from puncture wounds.

- Always remove jewelry before starting a job. Rings are especially dangerous since they may catch on moving parts.

- Use only grounded electrical tools. For added safety, use an extension cord with a ground fault circuit interrupter (GFCI). A GFCI can also be installed in a distribution panel.

29.1 Plumbing Codes

Installation of the plumbing system is regulated by codes. It is important to comply with plumbing codes for these reasons:

- An improperly installed system will not pass inspection. It must be torn out and redone—a costly remedy.

- A poorly designed system will not perform well. The owner will be dissatisfied and may insist on changes.

Class	Description
A	Fires in common combustible materials such as paper, wood, cloth, rubber, and many plastics. The cooling effects of water or solutions of water and chemicals will extinguish the fire. Also, these fires can be controlled by the coating effects of selected dry chemicals that retard combustion.
B	Fires in flammable liquids such as gasoline, grease, and oil. Preventing oxygen (air) from mixing with the vapors from the flammable liquid results in smothering the fire.
C	Fires in "hot" electrical equipment are especially dangerous because only nonconductive extinguishing agents can be safely used. Note that once the electricity is turned off, Class A or Class B extinguishers may be safe to use.
D	Fires in combustible metals such as sodium, magnesium, and titanium. These fires require the use of extinguishing agents that will not react with the burning metals. Also, the extinguishing agent needs to be heat-absorbing.

Figure 29-3. Fires are classified as type A, B, C, or D according to the combustible materials involved.

- A poorly designed system could be unhealthy because it could allow wastewater to contaminate *potable* (drinking) water.

Regulations for installing plumbing systems are covered by several agencies. One is the International Code Council (ICC), which sponsors the International Building Code. This agency is made up of three code-writing agencies: Building Officials and Code Administrators International, Inc. (BOCA); International Conference of Building Officials (ICBO); and Southern Building Code Congress International, Inc. (SBCCI).

Some states adopt codes other than the International Building Code for the plumbing trade. These include:

- The Uniform Plumbing Code (UPC) by the International Association of Plumbing and Mechanical Officials.
- National Plumbing Code by the American Standards Association.

Local governments may adopt all of one code or parts of several plumbing codes. Once adopted, codes can be enforced as law. Since codes may vary in their requirements from one community to another, a plumber must always be familiar with the local code.

Plumbing requires a variety of skills. The plumber must use many different kinds of tools and equipment. Further, he or she must be skilled in woodworking, metalworking, pipe threading, welding, soldering, brazing, and caulking.

Plumbers must make accurate measurements and calculations. They must add and subtract dimensions, figure pipe offsets, and determine the volumes of tanks. As a precaution against costly errors, a good plumber checks each measurement and each calculation twice before marking or cutting.

29.2 Two Separate Systems

Plumbing includes two subsystems that have important differences:

- Water supply system
- Drainage, waste, and venting system

The *water supply system* distributes water under pressure throughout the structure for drinking, cooking, bathing, and laundry. This is a two-pipe system. One pipe carries cold water and the other hot water.

Drainage piping, commonly referred to as *drainage, waste, and venting (DWV) system* carries away wastewater and solid waste from bathrooms, kitchens, and laundries. This subsystem is not under pressure.

While both subsystems are watertight to prevent leakage, it is doubly important for the

Potable: Fit to drink.

Water supply system: System of pipes that distribute water under pressure to kitchens, bathrooms, and laundry areas.

Drainage, waste, and venting (DWV) system: The unpressurized part of a plumbing system that carries away wastewater and solid waste.

Fire Extinguishers and Fire Classifications

Fires	Type	Use		Operation
Class A Fires Ordinary Combustibles (Materials such as wood, paper, textiles.) *Requires... cooling-quenching.* Old / New	**Soda-acid** Bicarbonate of soda solution and sulfuric acid	Okay for use on: A Not for use on: B C D		Direct stream at base of flame.
Class B Fires Flammable Liquids (Liquids such as grease, gasoline, oils, and paints.) *Requires...blanketing or smothering.* Old / New	**Pressurized Water** Water under pressure	Okay for use on: A Not for use on: B C D		Direct stream at base of flame.
Class C Fires Electrical Equipment (Motors, switches, and so forth.) *Requires... a nonconducting agent.* Old / New	**Carbon Dioxide (CO_2)** Carbon dioxide (CO_2) gas under pressure	Okay for use on: B C Not for use on: A D		Direct discharge as close to fire as possible, first at edge of flames and gradually forward and upward.
Class D Fires Combustible Metals (Flammable metals such as magnesium and lithium.) *Requires...blanketing or smothering.* D	**Foam** Solution of aluminum sulfate and bicarbonate of soda	Okay for use on: A B Not for use on: C D		Direct stream into the burning material or liquid. Allow foam to fall lightly on fire.

Dry Chemical row:

	Dry Chemical	Multi-purpose type	Ordinary BC type	Direct stream at base of flames. Use rapid left-to-right motion toward flames.
		Okay for A B C	Okay for B C	
		Not okay for D	Not okay for A D	

| | **Dry Chemical** Granular type material | Okay for use on: D Not for use on: A B C | | Smother flames by scooping granular material from bucket onto burning metal. |

Figure 29-4. There are four classifications of fire extinguisher. Each class has its own distinguishing symbol.

supply subsystem. Supply piping is under pressure, so the smallest pinhole or fault in a connection will cause leaking.

It is also extremely important that the DWV be properly vented. *Venting* permits air to circulate in the pipes. This prevents backpressure and siphoning of water from traps. Under certain conditions, it also prevents the introduction of wastewater into the potable water supply.

29.3 Tools

A plumber's tool kit includes many of the same tools used in carpentry and electrical work. To perform simple plumbing tasks, only a few tools are needed. One of the plumber's most used tools is the wrench, **Figure 29-5.** Some tools designed to perform tasks specific to plumbing are shown in **Figure 29-6.**

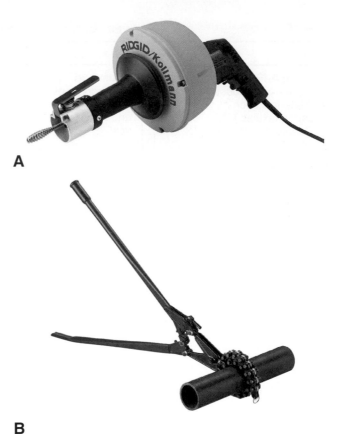

A

B

A

B

Figure 29-5. Wrenches are the plumber's most-used tools. A—Pipe wrenches are used to tighten threaded fittings on metal pipe. B—Adjustable wrenches work on fittings with flats. (Reed Manufacturing Co.)

C

D

Figure 29-6. Certain tools are used only for plumbing. A—A power auger is used to clear away clogs in drains. B—A chain cutter is used for cutting large-diameter soil pipe. C—This reamer removes burrs from any size of pipe. D—Pipe wrench pliers grip and turn pipe. (Ridge Tool Company, Reed Manufacturing Co.)

Among the more important tools are those used for measuring and to produce accurate lines, circles, and other markings. Plumbing measurements must be accurate to fractions of an inch and the tools must be capable of this accuracy. Measuring tools include rules, tapes, squares, levels, transits, plumb bobs, chalk lines, compasses, and dividers.

Tools are also necessary for determining when pipes are level and plumb (exactly vertical). A level is needed to check both conditions. A good all-purpose tool, the level has at least three indicator vials, **Figure 29-7.** The outer set of lines on a level indicator vial is used to determine the proper slope of a drainpipe. When the edge of the bubble touches one of the outer lines, the slope is at a 2° grade (1/4″ per foot of run).

A plumb bob vertically transfers locations of a pipe between floors. For accurate measurement, the string on which the bob is suspended must come out of the center of the bob. Marking tools such as pencils, chalk line, pencil compasses, and dividers should also be part of the plumber's toolbox.

Necessary cutting tools include saws, files, chisels, snips, and pipe cutters, **Figure 29-8.** Boring tools are used for making holes in wood framing members. Some cutting and boring tools should be suitable for metal-cutting, since steel framing is often found in residential construction.

Reaming and threading tools are used on metal pipe. Reaming the end of a pipe removes the burrs left inside by cutting. Dies are used for cutting threads on steel pipe.

A

B

Figure 29-7. A level is used to find level and plumb. Spirit levels, like this one, have an air bubble in a liquid-filled tube. When the bubble is centered between the lines on the tube, the pipe or construction member being checked is level. (Ridge Tool Company)

C

Figure 29-8. Pipe cutters. A—Standard steel pipe cutter. B—Tubing cutter. C—Plastic pipe shears. (Reed Manufacturing Co., Ridge Tool Company.)

Various types of wrenches are designed for holding and turning pipe and pipe fittings. Pipe wrenches, pliers, locking pliers, chain wrenches, strap wrenches, monkey wrenches, and adjustable wrenches should be included in the toolbox.

> **Working Knowledge**
>
> Power tools save time and are less tiring to use than hand tools.

29.4 Plumbing Supplies

There are several types of plumbing supplies. These include:

- Pipe and tubing for supply lines carrying pressurized water to faucets and fixtures.
- Large-diameter pipe to drain away wastewater.
- Valves and faucets for controlling water supply.
- Washers and O-rings for both supply and drainage systems.
- Tape, compounds, and putties to help seal connections against leaks.

29.4.1 Pipe

Several different types of material are used for pipes and pipe fittings. Depending on codes, supply system pipe may be made of copper, galvanized steel, or plastic. DWV pipe may be of malleable iron, cast iron, copper, galvanized steel, or plastic.

29.4.2 Copper Tubing

Copper tubing is available as rigid, straight lengths or in flexible coils, **Figure 29-9.** It is sold in a variety of diameters and seven different weights. Only three weights—types K, L, and M—are typically used for water supply lines.

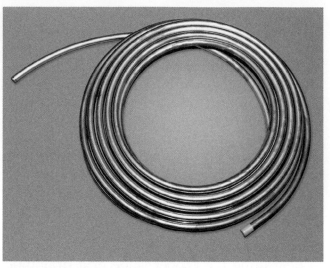

A

B

Figure 29-9. Typical copper materials. A—Copper tubing. B—Forged copper fittings. These must be sweat soldered to connect to the pipe.

Copper tubing can be cut with a pipe cutter or hacksaw. The flexible tubing is often joined using *compression fittings* that slide over the tubing. Another method is to sweat solder the fittings. Only lead-free solder is permitted, avoiding the hazard of lead getting into the water.

Compression fittings: Connectors that seal joints by compressing a ring of soft material, such as brass.

29.4.3 Plastic Pipe

Plastic pipe can be used for both supply and drainage systems. It is available in lengths that can be cut with a handsaw or tubing cutters. It is joined with special fittings or by using a special adhesive.

- Chlorinated polyvinyl chloride (CPVC) is a buff-colored thermoplastic. It is rigid, light, easy to handle, and resistant to cracking in freezing conditions. Connections may be threaded or solvent welded. CPVC is suitable for hot water piping and has a rating of 180°F and 100 psi.

- Polybutylene (PB) is flexible, which allows easier installation. Connections are made mechanically with compression fittings.

- Cross-linked polyethylene (PEX) is flexible tubing. It can be used for hot and cold water supply lines. Connections are made with crimped connectors. A centralized manifold is usually installed to distribute the water.

Plastic piping has become popular for drainage systems, as well. It is used in most new construction. There are two major types:

- Acrylonitrile butadiene styrene (ABS) resists chemical attack and is inexpensive. It is usually joined with a one-step solvent.

- Polyvinyl chloride (PVC) has lower thermal expansion that makes long runs easier to control. It is joined with a two-step primer/ solvent and suitable fittings. See **Figure 29-10**.

A third type, polypropylene (PP), is often used for sink traps and other short connectors.

29.4.4 Cast Iron

Cast iron was once extensively used in plumbing, and still is to some extent. Both pipe and fittings are cast from gray iron. Its strength and corrosion resistance stems from the formation of large graphite flakes during casting. Cast iron pipe will not leak or absorb water. It is often chosen because it is a quiet drainage system. The Cast Iron Soil Pipe Institute specifies two

grades of soil pipe: Service (SV) for above-grade use and Extra Heavy (XH) for below-grade use.

Joining methods depend on which of two styles of cast iron soil pipe is used. The difference is in the shape of the ends. Both styles are shown in **Figure 29-11**.

There are two methods of sealing joints in hub and spigot soil pipe. In the first, molten lead is poured into the *hub* (bell-shaped end of the

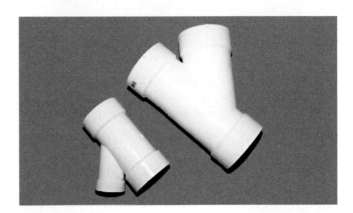

A

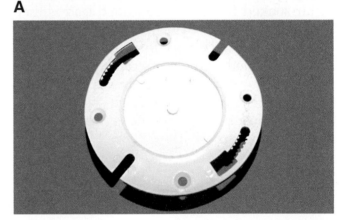

B

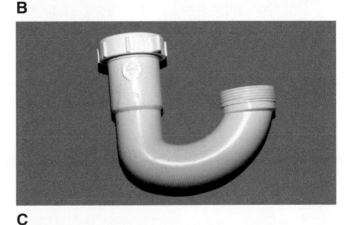

C

Figure 29-10. Plastic fittings like these are used in drainage systems. A—Wye. B—Closet flange. C—P-trap.

Hub: The bell-shaped end of a cast-iron soil pipe.

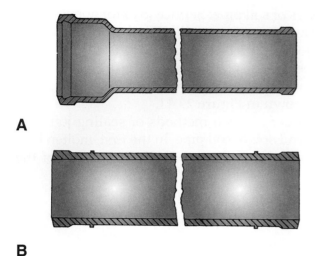

A

B

Figure 29-11. Two types of cast iron soil pipe. A—Hub and spigot type. B—The no-hub type does not have a bell and must be connected with neoprene seals and stainless steel clamps.

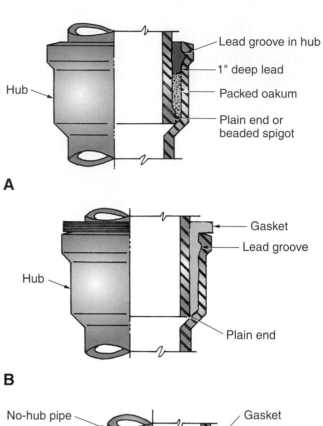

A

Lead groove in hub

1" deep lead

Packed oakum

Plain end or beaded spigot

Hub

Gasket

Lead groove

Hub

Plain end

B

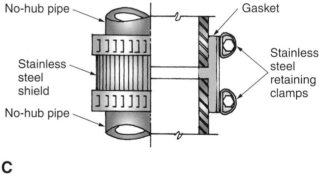

No-hub pipe

Gasket

Stainless steel shield

Stainless steel retaining clamps

No-hub pipe

C

D

Figure 29-12. Methods of joining cast iron soil pipe. A—Lead and oakum are packed into the joint. B—Hub and spigot soil pipe sealed with a neoprene gasket. C—No-hub soil pipe joint sealed with a neoprene gasket and secured by a clamp. D—A rubber coupling with stainless steel clamps. (E.I. duPont de Nemours & Co.)

joint) after it has been sealed with an oakum packing. *Oakum* is a rope-like material made up of tar-soaked fibers. The solidified lead is then tamped to form a seal against the bell. In the second method, the joint formed by a hub and spigot is sealed with a neoprene gasket. Joints of the no-hub type of soil pipe are also sealed with neoprene held in place by a stainless steel clamp. **Figure 29-12** shows the various methods of sealing joints.

29.4.5 Steel Pipe

The steel pipe used in plumbing is galvanized with a coating of zinc to retard rusting. It is manufactured in standard lengths of 21′ and in diameters from 1/8″ up to 2 1/2″. The ends are given a tapered thread so that the joint will seal. Fittings are made of malleable iron to make them more resistant to stress. To help seal the joints, either a putty-like pipe compound or Teflon® tape is applied to threads before assembly.

Oakum: Asphalt-saturated, rope-like material used as packing in cast-iron soil pipe joints prior to pouring in molten lead.

A

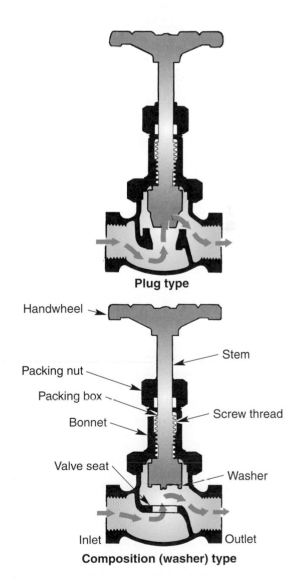

Figure 29-13. Valves interrupt and control water flow. A—A gate valve must always be fully open or fully closed to properly work. B—This angle compression valve is used where plumbing needs to change direction.

29.4.6 Valves

Valves are devices that control the flow of water in the water supply system, **Figure 29-13.** They are installed at certain places in the lines

Gate valves: Valves that control water with a flat disk that slides up and down perpendicular to the supply pipe; never used to restrict flow.

Compression valves: Valves with a tapered plug or a washer that can be fully seated to stop water flow or opened by different amounts to control the flow.

so that water can be shut off or its flow restricted. *Gate valves* control water with a flat disk that slides up and down perpendicular to the supply pipe. It is never used to restrict flow; it is either fully open or fully closed. *Compression valves* can stop or restrict water flow. They make use of a tapered plug or a washer that can be fully seated to stop water flow, or opened by different amounts to control the flow. Two types of compression valves are shown in cutaway in **Figure 29-14.** Valves are made from a variety of materials including bronze, brass, malleable iron, cast iron, thermoset plastic, and thermoplastic. Threads may be on the inside or the outside. Some have sweat connections.

Plug type

Handwheel
Stem
Packing nut
Packing box
Bonnet
Screw thread
Valve seat
Washer
Inlet
Outlet

Composition (washer) type

Figure 29-14. Two types of compression valves. (William Powell Co.)

29.4.7 Faucets

Faucets are compression or washerless valves that permit the controlled flow of water as needed. They usually deliver the water to fixtures such as sinks, lavatories, showers, and bathtubs, but may deliver it to an appliance, hose, or bucket. See **Figure 29-15.**

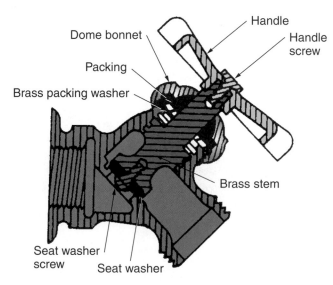

A

B

Figure 29-15. Faucets control water at a fixture. A—A hose bibb (faucet) is used for outside water supply and is threaded to receive a hose. The flange allows it to fit flush to an outside wall. B—A single-control lavatory faucet. (Kunkle Valve Co., Inc.; American Standard)

29.5 Fixtures

Fixtures are water-using devices, such as sinks, lavatories, bathtubs, bidets, urinals, stools, or showers. They receive water from the supply system and have a means for delivering wastewater to the DWV system. Certain water-using appliances also require attachment to plumbing systems. These include clothes washers, dishwashers, and automatic icemakers in refrigerators.

29.6 Sealing Plumbing Systems

Some supplies and materials seal the plumbing system against leaks, **Figure 29-16.** *Slip-joint washers* are used to seal P-trap connections beneath sinks. *O-rings* are used

Faucets: Devices that deliver water, usually to a fixture, hose, appliance, or bucket.

Fixtures: Devices that receive water from the water supply system and have a means for discharge into the DWV system.

Figure 29-16. Sealing plumbing pipes requires different materials and devices. A—Slip-joint washers and a slip-joint nut secure a watertight connection between the tailpiece and the P-trap. B—O-rings. C—Teflon® tape, pipe joint compound, and plumber's putty.

on faucets to produce leakproof seals between the faucet body and internal parts. *Teflon® tape* and *pipe compound* are used to seal threaded connections between pipe and other parts. *Plumber's putty* is used to produce a seal on sink rims, strainers, faucets, and toilet bowls.

29.7 Reading Prints

To save time, certain drawing symbols have been standardized for the plumbing trade. These symbols are a type of shorthand. **Figures 29-17** and **29-18** show generally-accepted symbols for fixtures, appliances, mechanical equipment, pipes, and fittings.

Since architectural plans may not include pipe drawings, the plumber needs to develop some sketches as a guide to installation. This is especially important if others are working on the same plumbing job. There are three types of drawings or sketches that will be made: riser diagrams, plan view sketches, and isometric sketches. These are shown in **Figure 29-19.**

29.8 Installing Plumbing

Installation of plumbing is usually done by a subcontractor. It can begin as soon as the building is closed in. Measurements are made to determine exact location of pipes and fixtures. First, measure runs of pipe from face to face. This is the distance from the end of one

---

Slip-joint washers: Washers used to seal P-trap connections beneath sinks.

O-rings: Rings used on faucets to produce leakproof seals between the faucet body and internal parts.

Teflon® tape: Tape used to seal threaded connections between pipe and other parts.

Pipe compound: Used to seal threaded connections between pipe and other parts.

Plumber's putty: Putty used on sink rims, strainers, faucets, and toilet bowls to produce a seal.

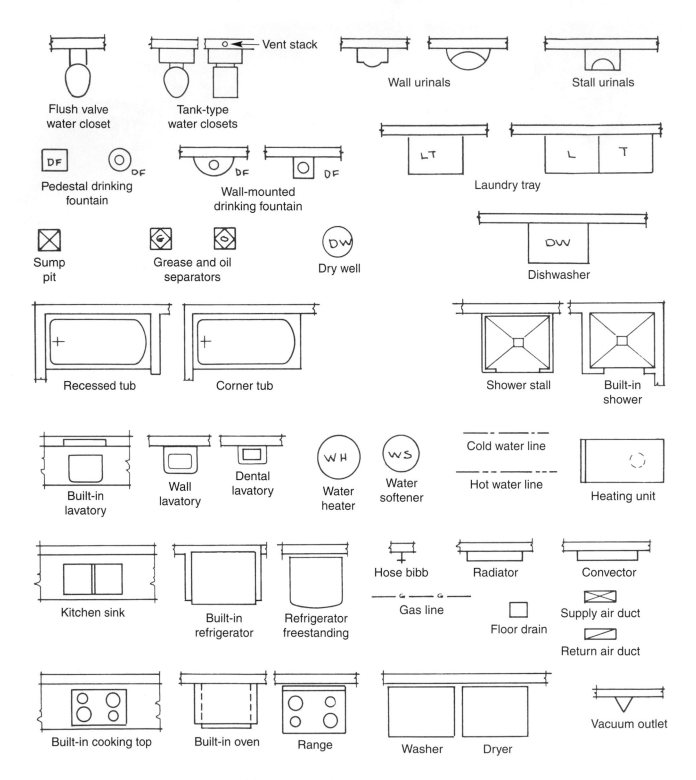

Figure 29-17. These symbols are used for plumbing fixtures, appliances, and mechanical equipment.

Fitting or valve	Type of connection			Fitting or valve	Type of connection		
	Screwed	Bell and spigot	Soldered or cemented		Screwed	Bell and spigot	Soldered or cemented
Elbow–90°				T–outlet down			
Elbow–45°				Cross			
Elbow–turned up				Reducer–concentric			
Elbow–turned down				Reducer–offset			
Elbow–long radius				Connector			
Elbow with side inlet–outlet down				Y or wye			
Elbow with side inlet–outlet up				Valve–gate			
Reducing elbow				Valve–globe			
Sanitary T				Union			
T				Bushing			
T–outlet up				Increaser			

Figure 29-18. Pipe and fitting symbols that appear in drawings.

fitting to the end of the other fitting. Then, add the distance the pipe will extend into the fitting. Use a steel tape or plumber's rule to take careful measurements.

Cutters or saws are used to cut lengths of pipe or tubing. Be certain to ream pipes to remove interior ridges or ragged edges caused by sawing or cutting.

Plumbing is concealed within the building frame. The single exception to this rule is plumbing below first-floor joists. Here, runs can made below, rather than through joists, unless the basement will be finished as living space. However, use of engineered lumber—specifically open-truss joists—allows mechanical systems to be run through the open spaces of the truss braces, **Figure 29-20.**

Some bathroom fixtures, such as bathtubs and showers, are too large to fit through doorways of framed-up partitions. They must be moved

in prior to installation of studs for partitions, **Figure 29-21.**

The first phase of plumbing installation is known as a *rough-in.* All runs are placed and tested with pressurized air or water to make sure that no leaks exist. The vent pipes are connected to a large riser, or *vent stack,* that extends through the roof. It serves as an exhaust for any odors from the waste system. It also allows the system to take in air and thus avoid siphoning problems with traps. The pipe ends that extend out from the walls and above the floors are known as *stub-ins.* **Figure 29-22** shows stub-ins where fixtures will be connected.

Rough-in: The first phase of plumbing, which includes installing all pipes to the point where connections are made with the plumbing fixtures.

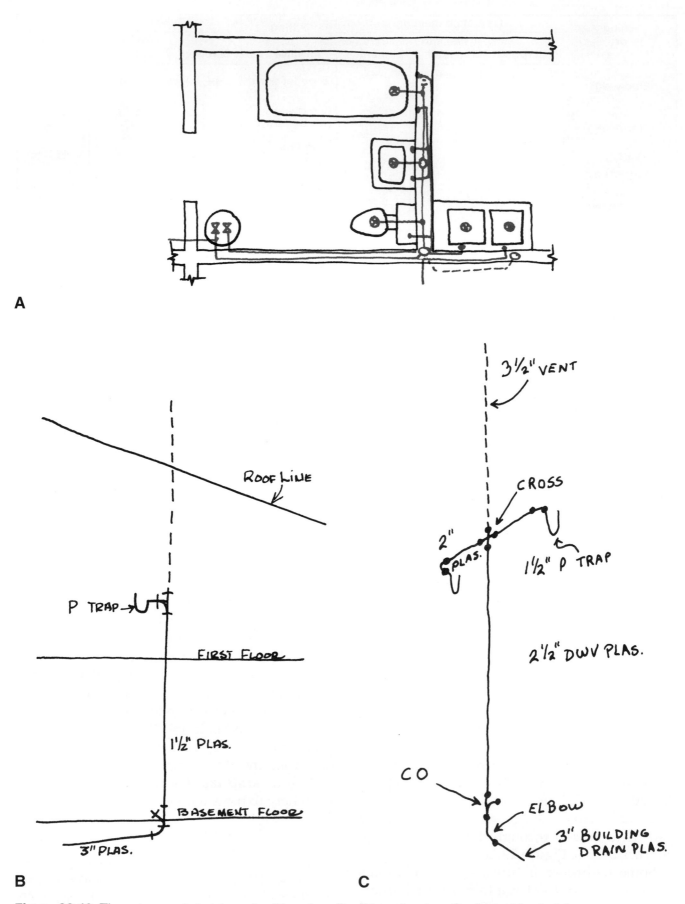

A

B

C

Figure 29-19. Three types of sketches. A—Plan view. B—Riser diagram. C—Isometric sketch.

Figure 29-20. The open spaces in truss joists simplify plumbing runs in a basement or crawl space.

Figure 29-21. This tub/shower unit was installed before the steel studs for the interior partition were erected.

29.8.1 Connecting Copper Pipe

Sweat soldering copper plumbing is relatively easy, but must be carefully done to avoid leaking joints. You will need a gas torch, lead-free solder, paste soldering flux, and a small pad of steel wool or a tube-cleaning brush.

Figure 29-22. Water supply and DWV stub-ins are ready for installation of a bathroom lavatory.

Stub-in: Termination point of plumbing pipes that are to be connected to fixtures.

Vent stack: The part of a DWV system that is open to the atmosphere to allow air into the system, preventing siphoning of traps.

Sweat soldering: Joining process for copper pipe in which the fitting and pipe are heated with a gas flame to draw solder into the joint by capillary action.

PROCEDURE

Soldering copper pipe

1. Ream the cut end of the tubing to remove metal burrs.
2. Clean the end of the copper tubing and the inside of the fitting with steel wool or a cleaning brush. This removes oxidation and reveals bright metal to help achieve a good solder bond.
3. Using a brush, apply paste flux to the cleaned end of the tubing and the inside of the fitting.
4. Slip the fitting over the tubing and heat the joint. It is best to heat the tubing first. Allow the inner cone of the flame to touch the metal. Heat the joint until the flux is smoking.
5. Add solid core solder by touching the end of the solder wire to the joint area at the edge of the fitting, **Figure 29-23.**
6. Feed the solder into the joint as the flame is moved to the center of the fitting. This movement will evenly draw the solder into the fitting, making a leakproof joint. Never try to melt the solder in the flame.
7. Once the solder is drawn into the joint, promptly remove the heat.
8. Wipe excess solder off the joint with a piece of burlap or denim cloth.

Figure 29-23. Sweat soldering copper tubing. The heated tubing, not the flame, must melt the solder. (Jack Klasey)

29.8.2 Bending and Unrolling Flexible Copper Tubing

Flexible copper tubing can be easily bent to change its direction during an installation. To avoid crimping, which may lead to a leak, slide a tubing spring over the section to be bent. Bend the tube and spring just beyond the desired angle, then back to the correct angle. See **Figure 29-24.** Remove the spring by turning the flared end counterclockwise.

Copper tubing comes in coils. Be careful when uncoiling. A procedure that works well is to lightly place a foot on one end while holding the coil upright. Slowly roll the coil away from the secured end. Move the holding foot along until the proper length of tubing has been unrolled.

Compression or *flare fittings* used with copper tubing require no soldering. The fittings are designed to tightly fit against the tubing and mating parts.

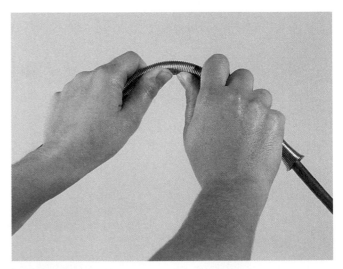

Figure 29-24. Use a tube spring to prevent kinks when bending copper tubing.

Flare fittings: Fittings that require no soldering. They are designed to fit tightly against the tubing and mating parts.

PROCEDURE

Making compression joints

1. Slide the flare nut over the tubing, tapered end facing away from the end of the tubing.
2. Select the correct size hole on the flaring tool and slip the tube into it. The hole should be slightly smaller than the outside diameter of the tubing.
3. With the tubing sticking out 1/32″ to 1/16″ above the bar of the tool, tighten the wing nut so the tubing is tightly held in the bar.
4. Slide the ram over the tube and screw the ram into the end of the tubing. Tighten the ram until a 45° angle flare is made in the tube, **Figure 29-25.**
5. Press the tapered end of the other half of the fitting into the flared end of the tubing.
6. Slide the nut to the tapered fitting and hand-tighten.
7. Using two tubing wrenches or open end wrenches, tighten the assembly until a solid feel is apparent.
8. Test the completed fitting for leaks.

29.8.3 Installing Rigid Plastic Supply Pipe

Rigid plastic pipe is lightweight, low in cost, easy to cut, and easy to fit. Two types are available for indoor supply lines. PVC is suited for cold water only. CPVC can be used for both hot and cold water. Either is sold in 10′ or 20′ length with sizes 1/2″, 3/4″, and 1″ being most common. The fittings slide over the outside of the pipe and are secured with quick-curing solvent cement. Connections for larger-diameter PVC or ABS plastic DWV piping are made in the same way as the connections for plastic water supply piping.

> ### Working Knowledge
>
> To cut plastic pipe with a hacksaw or back saw, use a miter box to secure a square cut. If a pipe cutter is used, install a special cutter wheel designed for plastic. A power "chop saw" with a fine-tooth blade is the fastest and most-accurate cutting method.

PROCEDURE

Installing push-on plastic fittings

1. Remove burrs from the end of the pipe with a knife or sandpaper.
2. Wipe the end with a clean rag to remove any soil. Also clean the inside of the fitting.
3. Remove the glossy finish on the outside of the pipe and the inside of the fitting with sandpaper or a special primer.
4. Slide the fitting onto the pipe until properly seated.

(Continued)

A

B

Figure 29-25. Flaring copper tubing. A—The tubing end should extend 1/32″ to 1/16″ above the surface of the flaring tool. B—The finished flare.

(Continued)

5. Rotate the fitting until it is properly aligned for the run.
6. Draw an alignment mark across the fitting and the pipe.
7. Remove the fitting.
8. Apply a liberal coat of the correct solvent-cement to the outside of the pipe and to the inside of the fitting. Usually, the container has an applicator attached to the cap, **Figure 29-26.** If not, use any soft-bristled brush.
9. Quickly make the connection, giving a quarter turn to evenly spread the cement, **Figure 29-27.** Speed is essential or the connection will be spoiled.
10. Promptly align the marks previously made on the parts. The cement will bond within seconds.

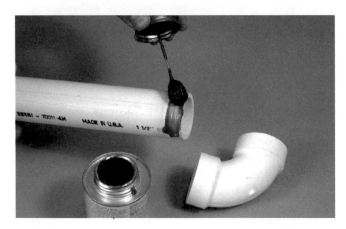

Figure 29-26. Applying solvent to the plastic pipe end.

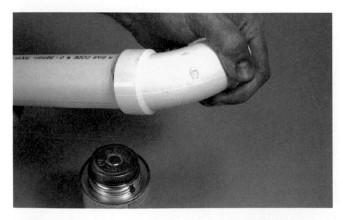

Figure 29-27. Slip the fitting onto the pipe and give it a quarter turn to spread the solvent evenly. Align the fitting with the mark on the pipe.

Safety Note

Solvent-cement is highly flammable. Keep it away from flame. The fumes can also cause lung damage. Provide adequate ventilation and avoid breathing the fumes.

29.8.4 Installing Galvanized Pipe

Galvanized plumbing pipe is still used, though it is not as popular as copper and plastic. It comes in nominal diameters of 1/4″ to 2 1/2″ and in 10′ and 21′ lengths. Both ends are threaded. Galvanized connections are one of the most forgiving if you make a mistake in joining. Threaded joints can be taken apart without destroying the parts.

PROCEDURE

Connecting galvanized pipe

1. Check lengths to ensure accuracy. A "dry fit" assembly can be made to determine this.
2. Make sure threads are clean. Use a wire brush to remove any bits of metal or other debris.
3. Apply pipe compound or Teflon® tape to exterior threads, **Figure 29-28.**
4. Screw on the fittings finger tight.
5. Final tightening should be made with pipe wrenches, **Figure 29-29.** In long runs, tightening of all joints can be made by starting with the last joint in the run. As each joint is tightened, the force is transferred to the next joint until all are tightened.

29.9 Replacing Plumbing Parts

Parts of a plumbing system eventually wear out or leak. This is especially true of faucets and sink drains. Then, repairs or replacement must be made.

Figure 29-28. Wrap Teflon® tape onto the pipe threads in the same direction that will be used to thread on the fitting.

Figure 29-29. Pipe wrenches have teeth to grip the fitting or pipe as it is tightened. For final tightening, a second wrench is often used to grip the pipe and prevent rotation.

29.9.1 Repairing or Replacing Faucets

Many compression faucets still in use today have soft, rubber-like washers that are rotated

---

Washerless: Type of faucet that uses replaceable cartridges, discs, or other methods to control water flow.

against a brass seat to stop water flow. A leaking faucet of this type usually means that the washer, and possibly its seat, has deteriorated and must be replaced. Newer types of faucets are *washerless,* using replaceable cartridges, discs, or other methods to control water flow. Washerless faucet repair is usually fairly simple. Follow the manufacturer's directions.

To repair faucet, first turn off the water supply. Usually there are shutoff valves located on the supply lines under the sink. If not, shut off the main valve for the whole building. If during the repair you find that the valve seat is built into the body of the valve and cannot be removed, it is necessary to use a valve seat dresser.

PROCEDURE

Replacing compression faucet washers and seats

1. With the water supply off, remove the faucet handle, cap, and stem.
2. Inspect the washer at the end of the stem. If it is marred, grooved, or cracked it must be replaced with a new one, **Figure 29-30.**
3. Use a screwdriver to carefully remove the brass screw on the end of the stem.
4. Remove the washer and replace it with a new identical one. If the washer is beveled, make sure to install the beveled side facing outward.
5. Before reinstalling the stem, inspect the seat, **Figure 29-31.** If it is pitted, it needs to be replaced or resurfaced.
6. On most compression faucets, the seat can be removed. Insert a valve seat wrench into the seat and turn it counterclockwise to back it out.
7. Install an identical new seat, but first lubricate its threads with pipe compound.
8. If the seat cannot be removed, insert the valve seat dresser into the valve body until it rests on the seat. Make certain it is upright. Some dressers thread into the valve body for proper positioning. Then, rotate the handle clockwise until the pitting is ground away.
9. Reinstall the stem after lubricating the threads with pipe compound. Replace the cap and test.

Figure 29-30. Washers for compression faucets are made from neoprene or similar compressible material to form a seal against the seat. They can easily be replaced when worn.

Figure 29-31. Badly pitted seats of compression faucets can often be removed and replaced with new ones. If they are not removable, they must be resurfaced with a special tool.

29.9.2 Fixing Sink Drain Problems

A sink drain includes several parts. A seat or flange sits inside the bowl and provides a seal. Its lower end has internal threads for connection to the drain body. A drain body threads into the seat and extends to the tailpiece. A rubber gasket slides onto the tailpiece beneath the bowl to provide a seal. A metal washer fits between the seal and a lock nut. A lock nut secures the tailpiece to the bowl and prevents leakage past the flange and the seal. A pivot rod assembly controls the stopper. Slip couplings fasten the parts of the drain together and make the system leakproof. They are essentially rings with interior threads and are used with washers. A slip coupling and its washer seal the tailpiece to the drain body.

A P-trap connects the tailpiece to the drainpipe and provides a seal to block sewer gases. The trap is designed to remain filled with water so that toxic and disagreeable sewer odors do not enter the house, **Figure 29-33.** As a result of such usage, it is prone to clogging, corrosion, and leaks. To unclog or replace a trap, loosen the couplings that secure it to the tailpiece and drainpipe. Slide the tailpiece down into the P-trap until it clears the sink drain. Rotate the assembly to one side and remove the trap. Clean out the trap or install a new trap. Reverse the procedure to install the trap. Tighten the couplings and test for leaks.

PROCEDURE

Installing a faucet

To install a faucet, follow the instructions provided by the manufacturer, since there are many types on the market. **Figure 29-32** illustrates these typical installation steps:

1. Clean and dry the mounting surface of the sink or lavatory.
2. Slide the deck gasket over the supply piping and mounting studs until it is seated in the underside of the faucet body.
3. Slip the assembly through the openings in the deck.
4. Install the U-shaped clamp bar (channel up) and secure it with the mounting nuts.
5. Carefully position the faucet and gasket on the deck and tighten mounting nuts.
6. Connect flexible supply lines to the hot and cold supply lines.

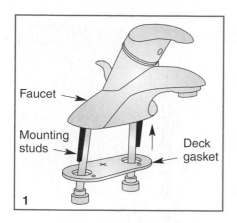

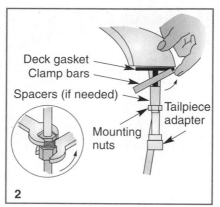

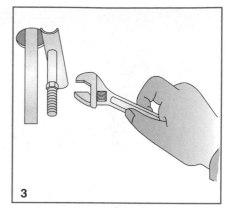

Figure 29-32. Replacing a lavatory faucet. (Moen Incorporated)

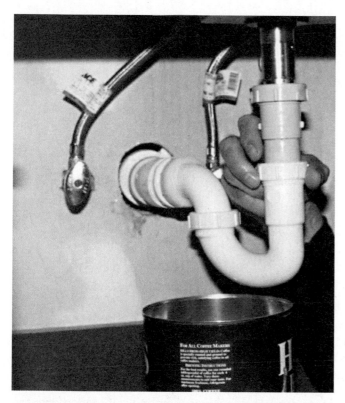

Figure 29-33. P-traps are the most frequent site of clogging and leaking in the DWV system. They are always full of water. Keep a container handy to catch spills when they must be disconnected from the drain.

PROCEDURE

Replacing a drain body

1. Remove the P-trap, as described in the preceding section.

(Continued)

(Continued)

2. Use a spud wrench to loosen the locknut that secures the seat and the drain body to the bowl, **Figure 29-34.**
3. Tap upward on the drain body to break the seal and remove the assembly.
4. Check the manufacturer's literature for drawings that show assembly details. See **Figure 29-35.**
5. Scrape away any debris around the drain hole in the bowl. Carefully clean the area, as shown in **Figure 29-36.**
6. Knead a lump of plumber's putty to soften it. Apply it to the underside of the new seat, **Figure 29-37.**
7. Wrap Teflon® tape onto the top threads of the drain body. Always wrap in the same direction as parts will be threaded on.
8. Thread a mounting nut, along with a washer and a gasket, onto the drain body.
9. From beneath the bowl, slip the drain body through the drain opening. Screw the seat onto the drain body.
10. Align the drain body so the pivot rod opening faces the rear of the bowl.
11. Tighten the mounting nut until snug.
12. Apply Teflon® tape to the threaded end of the tailpiece.
13. Slip the tailpiece into the P-trap and position the assembly under the drain body.
14. Thread the tailpiece into the drain body.
15. Firmly tighten the connection by hand.
16. Tighten all of the mounting nuts and test for leaks.
17. Drop in the drain plug and install the pivot rod according the manufacturer's instructions.

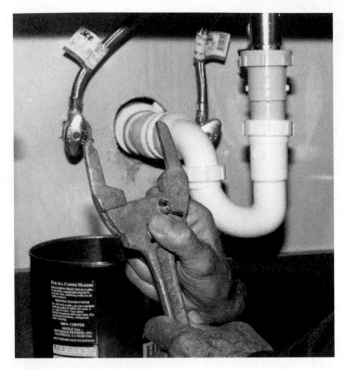

Figure 29-34. A spud wrench adjusts to accommodate many different sizes. They are generally used to loosen or tighten lock nuts on sink drains that are too large for a normal adjustable wrench.

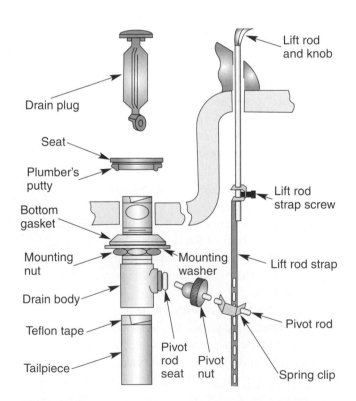

Figure 29-35. Manufacturer's literature should be carefully studied before beginning an installation. The appropriate steps are normally well illustrated. (Moen Incorporated)

Figure 29-36. Carefully clean the drain opening before installing new drain piping.

Figure 29-37. Apply plumber's putty to the underside of the seat. The seat was temporarily threaded onto the drain body to make this step easier. This type of drain body must be installed in the bowl from the underside.

29.10 Other Drain Problems

The homeowner or plumber uses special tools to clear clogged drains. For slow drains, a chemical cleaner may be used. Plungers, snakes, and closet augers are used for more serious clogs. Plungers apply pressure to the clog, while snakes and augers reach and dislodge or retrieve the material causing the obstruction.

Leaking or malfunctioning toilet tanks or stools are common service problems, **Figure 29-38.** Service may involve replacing the ballcock assembly or a flush valve, adjusting the float level, or renewing the linkage between the flush lever and the rod controlling the flush ball. A toilet leaking at its base must be removed to replace the wax ring or rubber seal.

Working Knowledge

When replacing a ballcock assembly, be careful when placing the water refill tubing in the overflow tube. The refill tubing may have to be shortened. If its end is below the water line, water will slowly siphon into the bowl. This could lead to an incorrect diagnosis and unnecessary replacement of the flush valve.

29.11 Wells and Pumps

Many buildings located beyond the lines of municipal water systems have their own wells. Wells require pumps and pressure tanks to maintain a constant water supply.

Drilled wells: Wells used where great depths are needed to reach a suitable water table.
Bored wells: Wells that are made with an earth auger.
Casing: The lining in a well shaft.
Submersible pump: A pump that is installed in a well.

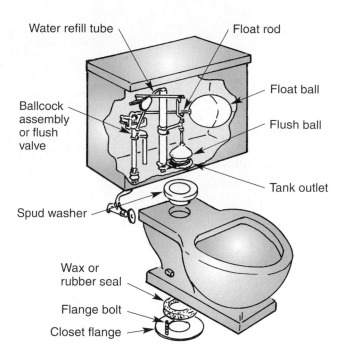

Figure 29-38. Mechanisms in a toilet frequently need adjustment or repairs.

Wells are of different types, **Figure 29-39.** Dug wells were once common. However, most wells today are either drilled or bored.

Drilled wells are used where great depths are needed to reach a suitable water table. There are two drilling methods. The percussion method uses a chisel-shaped bit that is raised and lowered by a cable. This creates a pounding action that breaks up the subsurface material. The rotary method requires a tall mast and more complicated drilling equipment along with a steady flow of water. The slurry created is pumped to the surface and discarded. This method is used when the water table is far below the surface.

Bored wells are made with an earth auger. The auger cuts and lifts subsurface material in much the same way as a twist drill bit cuts and lifts particles.

The diameter of the well must be large enough to accept a casing. The *casing* is a heavy metal tube that keeps contaminants out of the well water. Galvanized or plastic pipe is installed into the well and a pump is connected to this piping. A *submersible pump* may be installed in the well or an above-ground pump may be placed near the well or in the building.

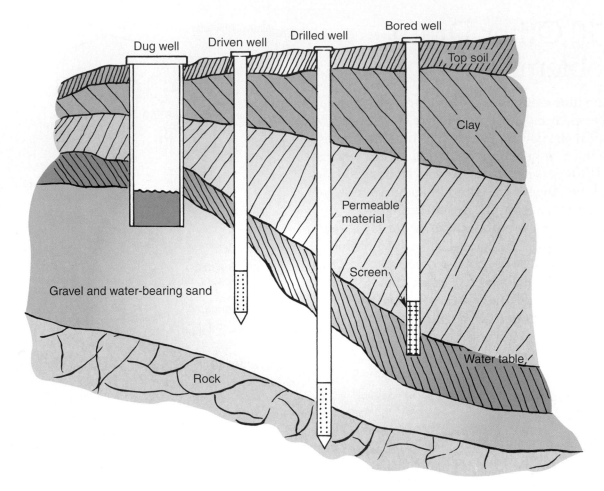

Figure 29-39. Four types of wells. Most wells today are drilled or bored.

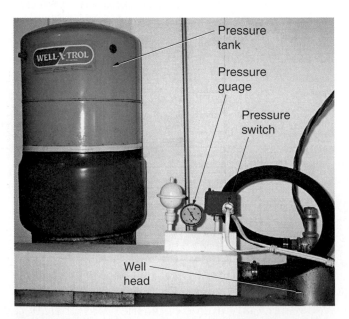

Figure 29-40. A typical setup for a well water system. A submersible pump is located in the well.

Also located near the well or the pump is a pressure tank, **Figure 29-41.** Its function is to store a reserve of water under pressure to meet the continuing needs of the building's occupants. The water is under constant pressure so it can be delivered anywhere in the building at sufficient force to operate fixtures and water-using appliances.

A well requires occasional maintenance and repairs, **Figure 29-42.** Seals harden, crack, and leak; check valves wear out; pumps wear and lose efficiency; pipes rust and develop leaks. Problems within the well, pump, or pressure tank are usually the concern of the well driller. Problems with the supply pipes are corrected by a plumber.

A   B

Figure 29-41. Wells require service from time to time. A—A bad seal is being replaced. B—Lengths of rusted, leaking pipe are being replaced. (Mellin Well Service)

ON THE JOB

Plumber

Plumbers form the largest portion of a broad trade classification dealing with tradespeople who install, maintain, and repair many different types of piping systems. Other trades within that classification include pipelayers, pipefitters, steamfitters, and sprinkler fitters.

The plumber is generally responsible for planning, installing, and testing the water supply and waste disposal (drain/waste/vent) piping and plumbing fixtures within residential, commercial, and other types of structures. Pipelayers excavate for and install the large underground concrete, clay, cast iron, or plastic pipes for water mains, municipal sewer systems, and oil or gas lines. Pipefitters work primarily in industrial and large commercial structures, installing high- and low-pressure piping used for industrial processes or heating and cooling systems. Steamfitters work primarily in industrial and power generation settings installing and maintaining piping used to move high-pressure gases and liquids. Sprinkler fitters specialize in installing and maintaining automatic fire sprinkler systems.

Approximately 3/4 of all plumbers work for contractors who do new construction, repair, or main-tenance work. About 10% are self employed, often operating small businesses devoted to plumbing repairs or remodeling work. The remaining 15% of tradespeople in plumbing occupations, primarily pipefitters and steamfitters, work as maintenance personnel in large industrial operations.

Plumbers must be able to read and follow prints, lay out and install water supply and DWV piping, and be skilled in the techniques for cutting and joining pipes made from various materials. They must have the ability to stand for long periods or work in cramped quarters and have the physical strength to lift and carry piping and associated materials. Although the work is hard and often carried out in uncomfortable conditions, plumbers are among the best-paid members of the building trades.

Like sheet metal workers, plumbers serve a long apprenticeship (4–5 year) to prepare for work in the trade. In addition to on-the-job training, apprentices must take at least 144 hours of classroom instruction each year. Unlike most other building trades, plumbers must be licensed by either state or local authorities. Usually, this involves meeting certain education/experience requirements and passing a written examination. The regulations vary from state to state.

Summary

All of the piping and fixtures that supply water for drinking, cooking, bathing, and laundry and the drainage pipes that dispose of wastewater make up the plumbing system. To install rough plumbing, plumbers must cut holes and notches in the building frame. The plumber must closely work with the carpenter to be sure that the holes will not weaken the joists and studs. Plumbing codes govern the installation of plumbing systems. Plumbing tool kits are generally similar to those used by carpenters and electricians. Plumbing supplies include iron, steel, and plastic pipe; copper tubing; valves; faucets; and fixtures such as lavatories or tubs. Specific procedures are used to install and join each type of pipe and tubing. Plumbing repairs typically involve fixing leaks, replacing faucets and drain components, and clearing clogged drains. Homes located beyond municipal water lines must have a well drilled and a pump installed. Safety in plumbing work requires proper protective gear and careful attention to safe work habits.

Test Your Knowledge

Answer the following questions on a separate piece of paper. Do not write in this book.

1. Why must the cutting and notching of framing members be carefully done?
2. A fire within an electrically powered plumbing tool is known as a Class _____ fire.
3. List three reasons why it is important for plumbers to comply with the existing plumbing codes.
4. The _____ system distributes pressurized water throughout the building.
5. Which plumbing system has venting and what is the purpose of venting?
6. List two types of copper tubing.
7. What does CPVC stand for and what is the maximum temperature at which it can be used?
8. Use a(n) _____ when moving large-diameter pipe.
9. The first phase of installing plumbing in a house is called a(n) _____.
10. Which type of piping may be sweat soldered?
11. How are fittings attached to rigid plastic pipe, such as PVC?
12. Explain the purpose of a P-trap.
13. What materials or tools might be used to unclog a drain with a serious clog?
14. List the two major components of a water system that uses a well.
15. List the four types of wells.

Curricular Connections

Language Arts. Make a short instructional video showing how to sweat solder a joint in copper pipe. If possible, describe what you are doing as you perform each step. To effectively do so, write out the script in large letters on sheets of white posterboard. Have a helper hold these up, just out of camera range, so you can read them as you work. If this method will not work for you, record a separate voice-over narration and combine it with the video. Show the finished product to your class.

Social Studies. Use the library or Internet to find out how the flow of the Chicago River was reversed in 1900. What were the reasons for making the river flow out of Lake Michigan, instead of into the lake? What effect, good or bad, did this have on the city and the people who lived there? Write a report with the results of your research.

Outside Assignments

1. Prepare a piping diagram for a bathroom that has a stool, basin, and shower stall.
2. Using a blueprint supplied by your instructor, install a bathroom DWV in a mockup wall and floor. Use any type of DWV piping and the appropriate fastening system.

Heating, Ventilation, and Air Conditioning

30

Learning Objectives

After studying this chapter, you will be able to:

- Identify ways to conserve energy in housing.
- Define EER ratings and list the appliances to which they are applied.
- List the characteristics of different central air conditioning systems.
- Describe the functions of HVAC system components.
- Explain the design and operation of HVAC systems.
- Describe automatic controls for heating and cooling systems.

Technical Vocabulary

Air exchanger
Boiler
Central heating
Condenser
Conduction
Convection
Cut-in point
Cut-out point
Differential
Direct heating systems
Ducts
Electric radiant heat
Electric resistance
  baseboard heating

Energy efficiency
  rating
Evaporator
Forced-air perimeter
  system
Heat exchanger
Hydronic perimeter
  heating system
Hydronic radiant
  heating system
Manifold
One-pipe system
Plenum
Radiation

Register
Second Law of
  Thermodynamics
Thermostat
Two-pipe system
Zones

Installation of heating, ventilation, and air conditioning (HVAC) is an important building trade. Without it, buildings would not be comfortable places for working or living. They would be either too cold or too hot. Even prehistoric humans with limited technology at hand used open fires to heat their dwellings.

Industry has heavily invested in developing heating and cooling systems. Technology allows for automatic control of systems so that buildings are kept comfortable year-round, regardless of how hot or cold the weather might be.

30.1 Heating and Cooling Principles

To understand how furnaces and air conditioners operate requires an understanding of how heat moves. This brings up what is called the *Second Law of Thermodynamics.* Simply

---

Second Law of Thermodynamics: Heat always moves from hot to cold.

stated, this law dictates that "heat always moves from hot to cold." Consider what happens when you place a metal spoon in a bowl of hot soup. If the spoon is left there for any amount of time, it becomes hotter as the soup becomes colder. Why? Heat in the liquid moved to the cooler spoon.

30.1.1 How Heat Travels

Heating and cooling of buildings depends on the fact that heated matter—whether air, liquid, or solid—passes its heat on to cooler matter. It does so by three methods:

- Radiation.
- Conduction.
- Convection.

Radiation moves heat by way of waves similar to light or radio waves. A source, such as the sun, produces the heat and sends it through the air. The air is not heated, but solid objects that the waves strike absorb the heat.

Conduction is the movement of heat through substances from one end to the other. Heat moving through the metal spoon placed in a bowl of hot soup is a good example of conduction.

Convection is the movement of heat by way of a carrier, such as air or water. This carrier may move naturally or, in the case of furnaces and air conditioners, by the action of fans or blowers.

30.2 Conservation Measures

Fuel costs are constantly rising. Before discussing space conditioning (heating or cooling) mechanisms, it is necessary to consider how to increase their efficiency through conservation measures.

New homes are designed to limit the amount of energy needed to heat and cool them. Savings in heating and cooling come about in several ways. Chapter 15, **Thermal and Sound Insulation,** discussed many of these measures.

- Sealing cracks at building joints with caulking and housewrap cuts down on the loss of conditioned air, **Figure 30-1.** It does so by eliminating infiltration of outside air that is either hotter or cooler than the conditioned air.
- Installing insulated ductwork for forced air heating systems, as shown in **Figure 30-2.**

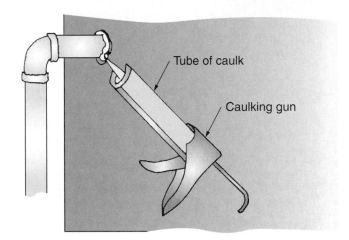

A

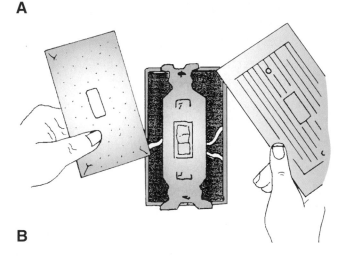

B

Figure 30-1. Cutting down on air infiltration makes a building easier to heat and cool. A—Caulk joints. B—Install seals behind switch and receptacle plates.

Radiation: Transmission of heat by wave motion.

Conduction: The movement of heat through substances from one end to the other.

Convection: Movement of heat through liquids and gases as a result of heated material rising and being replaced by cooler material.

Figure 30-2. Insulating ductwork in an unconditioned space helps conserve energy.

- Increasing the amount of insulation in floors, walls, and ceilings to significantly reduce loss of conditioned air through radiation.

- Using an air exchanger that wrings heat out of the room air before it is exhausted to the outside, thus reducing the heating load on the furnace.

- Installing high efficiency heating and cooling appliances.

The R-rating is an insulating factor. It is not unusual to find new homes with R-ratings double those of twenty years ago. New windows also have much higher R-ratings than earlier designs. Superinsulated, panelized, or modular houses are common in the home market. Panelized systems have wall and roof panels consisting of a 5 1/2″ thick sheet of polystyrene sandwiched between two sheets of plywood or oriented strand board (OSB). Heating and cooling costs have been significantly reduced by the use of such methods.

Furnaces and air conditioners are major users of energy in a home. Assume that the efficiency of a low-efficiency furnace is 80%. For every 100 Btu of fuel energy consumed, the furnace delivers 80 Btu of heat energy. The other 20 Btu of heat energy from the fuel is lost. Much of this lost heat energy escapes up the chimney. Temperatures in a chimney must usually be around 300°F (150°C) for the chimney to draw (pull combustion gases to the atmosphere). High-efficiency furnaces make use of technology to deliver more heat energy from fuel, **Figure 30-3.** Some systems are rated as high as 96% efficient by the U.S. Department of Energy.

A *heat exchanger* is a section of the furnace where heat from combustion gases is transferred by conduction to the circulating air. The heat exchangers in the high-efficiency units are designed to extract more of the heat of combustion. Spent gases leave a high-efficiency furnace at around 100°F (40°C). The furnace does not use a traditional chimney. Instead, the gases are exhausted to the outside through a small plastic pipe. Also, there is no contact between the inside, conditioned air and the outside air used for combustion.

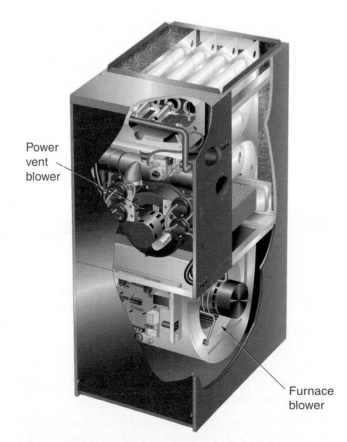

Power vent blower

Furnace blower

Figure 30-3. High-efficiency furnaces are over 90% efficient. (Lennox Industries, Inc.)

Heat exchanger: A series of tubes in a furnace that absorb the heat of combustion and transfer it to the plenum.

Following the energy crisis of 1973, the National Bureau of Standards set up an appliance labeling section. This section sponsors a testing program that assigns efficiency ratings to all appliances. Since 1976, appliance manufacturers have attached *energy efficiency rating* labels, or EER labels, to all of their products. Covered by this rating program are furnaces, central and room air conditioners, water heaters, and other appliances. An EER label contains the following information. See **Figure 30-4.**

- EER of the appliance. The higher the rating, the more efficient the appliance is in its use of electricity, gas, or fuel oil.

- Efficiency range or energy consumption of comparable products on the market.

- Estimate of how much the appliance will cost to operate during the year at various levels of usage.

30.3 Heating Systems

Central heating has all but replaced space heating. Space heating is the use of heating devices, such as a fireplace, in individual rooms. In *central heating,* a single, large heating plant produces the heat. The heat is then circulated throughout the structure by perimeter heating or radiant heating. There are several types of circulating systems used with central heating:

- Forced-air perimeter heating.

- Hydronic perimeter heating.

- Hydronic radiant heating.

30.3.1 Forced-Air Perimeter System

In a *forced-air perimeter system,* air is circulated through large ducts that direct warmed air to every room in the building, **Figure 30-5.** The burner delivers high-temperature combustion gases to the furnace heat exchanger. Heated air from the exchanger collects in a large sheet metal chamber called a *plenum.* A blower forces the heated air in the plenum through *ducts* into the spaces being heated. At the same time, the blower draws cooled air back to the furnace through a series of ducts called the *air return.* An air filter located in the cold air return filters out dust particles picked up from the heated spaces. **Figure 30-3** shows a cutaway view of a high-efficiency gas furnace.

Furnaces may be oil-fired, gas-fired, or electric. Oil-fired units require a storage tank for the fuel. The tank is usually located either underground outside the building or in the basement near the heating unit. For basement tanks, a filler pipe is usually installed that extends through the wall to the outside for easy access during

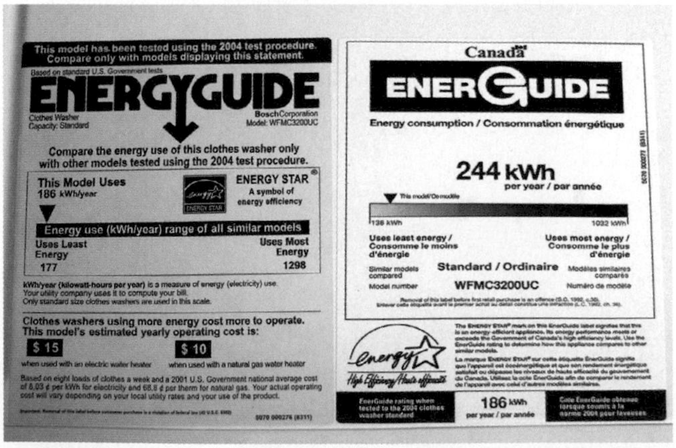

Figure 30-4. A sample of an EER label that all energy-using appliances must display to help customers make purchasing decisions. This appliance has labels for both the United States and Canada.

Energy efficiency rating (EER): A measure of how efficiently an appliance uses energy.

Central heating: A single large heating plant produces the heat, which is then circulated throughout the structure.

Forced-air perimeter system: Air is circulated through large ducts that direct warmed or cooled air to every room in the building.

Plenum: A sheet metal chamber in a furnace that collects heated air in preparation for its transfer to rooms.

Ducts: Tubes or passages used to move heated or cooled air from the furnace or air conditioner to living spaces of the building.

fuel delivery. Gas-fired units may get their fuel supply directly from a gas main supplied by a gas utility or from a storage tank located on the property. Electric furnaces use the resistance of conductors to produce heat, much like an electric stove or oven.

Registers used with forced air furnaces consist of a grille installed over duct openings in the rooms. They usually have shut-offs for controlling airflow. Some have adjustable vanes that control the direction of the airflow as it exits the duct.

Since warm air rises, heating registers are more efficient if located at floor level, **Figure 30-6.** Since cold air sinks, cold air returns should also be located at floor level. Registers should be placed so that air is not discharged directly on room occupants.

Figure 30-6. Warm air registers usually have vanes to direct flow of heated air and a door that can be positioned to reduce or stop airflow.

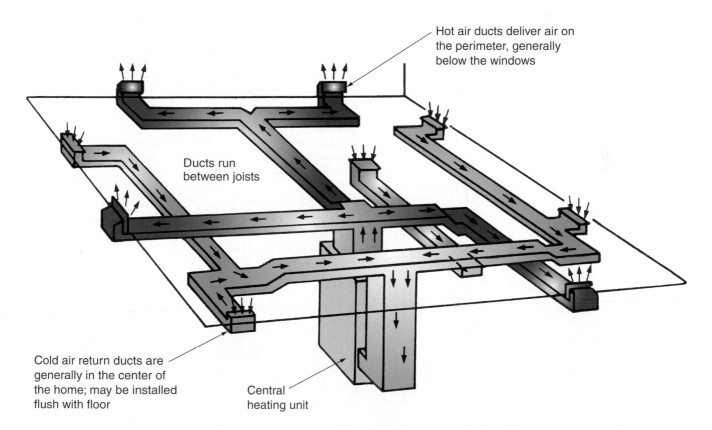

Hot air ducts deliver air on the perimeter, generally below the windows

Ducts run between joists

Cold air return ducts are generally in the center of the home; may be installed flush with floor

Central heating unit

Figure 30-5. A forced-air perimeter heating system. Warm-air ducts deliver heated air to rooms under a slight pressure. The cold air return duct brings air into the furnace for heating each time the circulating blower runs.

Registers: Grilles installed over the ends of heating or air conditioning ducts to direct the heated air.

Warm-air perimeter system

This variation of the forced air system is usually intended for a basement area or for a building with slab-on-grade construction. The ductwork is installed before the slab or the basement floor is poured, **Figure 30-7.** Heated air from the furnace circulates through the duct system embedded in the concrete. This duct system encircles the concrete slab at its outer edges. It is connected to the furnace through a plenum and feeder ducts located beneath the furnace. As in the traditional forced-air system, the warm air is discharged into the rooms

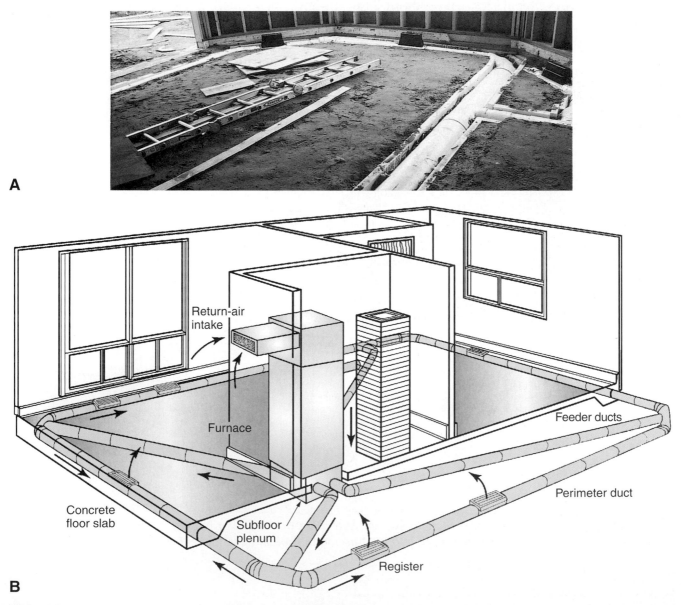

A

B

Figure 30-7. A warm-air perimeter system is being installed to heat a basement. A—Ducts are laid on top of rigid polystyrene insulation. B—A typical installation in slab-on-grade.

through registers. Floor or low sidewall registers are placed along the perimeter of the space. Air returns to the furnace through return-air intakes at high locations on sidewalls or on the sides of the furnace itself.

Ducts may be of sheet metal, vitrified tile, concrete pipe, plastic pipe, or other precast forms. If damp conditions are likely to exist, avoid metal ducts. Metal ducts will corrode and fail over time in damp conditions.

Installing and maintaining forced-air systems

Installation of a forced-air furnace usually follows the installation of the ductwork. The installer is often trained on the job, but probably has some formal instruction in a vocational program. The installer must be able to make mechanical connections and electrical hookups, **Figure 30-8.** The same person may also install the ductwork. This requires a knowledge of certain metal trades as well as a thorough grounding in electrical principles and wiring practice.

Unless the bearings are lubricated for life, motor and blower bearings must be oiled periodically, usually every three months, once the furnace is placed in service. Blower belts should be inspected before the beginning of every heating season. Several times during the heating season, clean or replace filters. Inspect electronic air filters several times during the season and clean as necessary. A trained technician should conduct an annual furnace inspection and servicing.

If there is a humidifier in the system, inspect it at least once a year, usually at the beginning of a heating season. Inspect panels and replace

Figure 30-8. An installer making electrical connections during furnace installation.

them when caked with minerals from the water supply. Check for proper flow of water as well.

30.3.2 Hydronic Perimeter Heating System

A *hydronic perimeter heating system* uses water as a medium to move the heat from the heating unit, **Figure 30-9.** The heating unit is called a *boiler* for the simple reason that it boils water much like a teakettle sitting on a burner. The heated water is piped to radiators that transfer heat to the rooms.

Many hydronic systems include a *manifold*, **Figure 30-9C.** This device balances the delivery of heated water to different *zones* (areas) of the building. The manifold consists of an inlet pipe fitting with several outlets for connecting one pipe with others. Pipes from each outlet are directed to different zones of the building. Each of these pipes has an adjusting valve to regulate

---

Hydronic perimeter heating system: A system that uses water to transfer heat energy from a heat source to radiators located along the outside walls of spaces to be heated.

Boiler: A heating unit for a hydronic heating system.

Manifold: A pipe with several outlets that distributes heated water to different areas.

Zones: Areas of a building that can be separately heated.

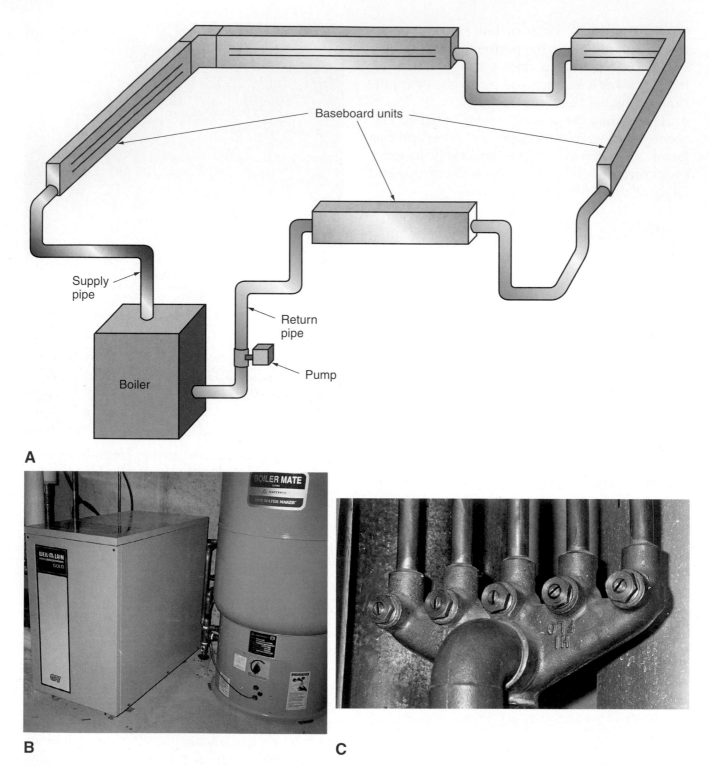

A

B

C

Figure 30-9. A—A basic hydronic system. B—This boiler is teamed with a hot water storage unit. The boiler delivers heated water to the tank until it is used by the hydronic heating system. Domestic hot water is also drawn from the tank. C—This manifold delivers heated water to five different pipes, creating five different heating zones for the house. (United Technologies Carrier) the amount of heated water going through that line.

When the system is completed, the installer places a thermometer in each zone. Then, flow is increased or decreased until the temperature is suitable for the activities in each zone.

Hydronic systems vary somewhat. There are two basic types of piping layout:

- **One-pipe system.** In a *one-pipe system,* a single pipe, called a *main,* supplies the heated water to room heating units. The heated water is pumped through the pipe from radiator to radiator. Finally the water, now cooled, returns to the boiler where it is reheated to repeat the cycle. This type is shown in **Figure 30-9.**

- **Two-pipe system.** In a *two-pipe system,* two mains are used. One main, the *supply main,* simultaneously carries heated water to each room radiator. The other main, known as the *return main,* carries the cooled water from each radiator to the boiler. See **Figure 30-10.**

Hydronic systems have an expansion tank. This acts as a reservoir for the heated water's increase in volume. A pressure-relief valve is also part of the system. It opens to release excess water pressure that might otherwise damage the system. A pressure-reducing valve limits the pressure of the incoming freshwater supply. Air purge valves are used to remove any air that has been introduced into the system. These valves are placed at high points in the piping where air is likely to collect.

Two-pipe systems provide more uniform heating than single-pipe systems. Water enters each radiator at a similar temperature. This even temperature is the result of heated water not having to pass through so many radiators before it returns for reheating.

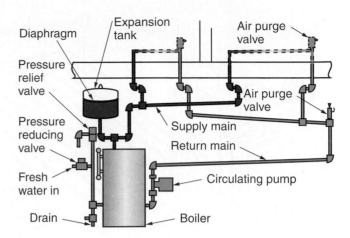

Figure 30-10. A two-pipe hydronic system more efficiently delivers heat than a single-pipe system.

Either system can be designed for greater efficiency by adding zone valves and zone pumps. These allow areas of a building to be heated independent of temperatures in other spaces.

30.3.3 Hydronic Radiant Heating System

A *hydronic radiant heating system* is similar to a hydronic perimeter heating system. However, the piping is installed in a structural part of the building. Hydronic radiant heating systems can be installed several different ways:

- On-grade over a thick underlayment of rigid insulation. Concrete is then poured over the tubing.

- Under subflooring with insulation beneath it. Wood or other types of flooring can be laid over the piping system after provisions are made to prevent damage or crushing of the pipe or tubing. When installing wood or other types of flooring, installers must be careful not to drive fasteners through the heating pipes.

Such systems are usually installed as part of new construction. They also can be part of a remodeling project. Piping is usually copper or flexible plastic laid out in a serpentine pattern that covers the area of installation from side to side, **Figure 30-11.**

One-pipe system: A single pipe supplies heated water to room heating units.

Two-pipe system: The supply main carries heated water to each room unit simultaneously. The return main carries the cooled water to the boiler.

Hydronic radiant heating system: A system that uses water to transfer heat energy from a heat source to tubing located in floors or other structural parts of a building.

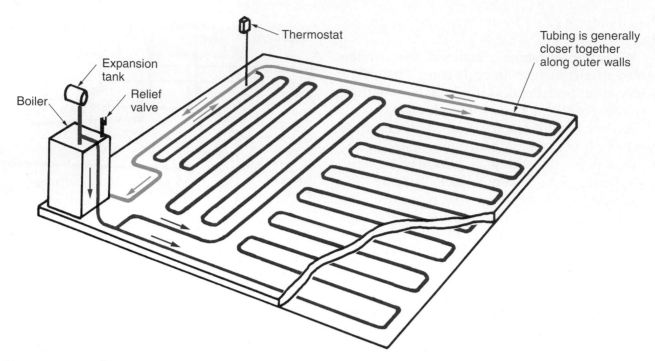

Thermostat

Expansion tank

Boiler

Relief valve

Tubing is generally closer together along outer walls

Figure 30-11. This drawing shows a hydronic radiant heating system as it would be installed in a slab floor. Note the location of the thermostat.

30.3.4 Direct Heating Systems

Direct heating systems are those whose heat source does not use ductwork or a piping system to distribute the heat. This category includes many basic types of heating, such as stoves and fireplaces. It also includes electric resistance baseboard heating and radiant electric heat.

Electric resistance baseboard heating is extremely simple. It consists of a heating element wrapped in a tubular casing. Metal fins spaced along the length of the casing help radiate the heat to the room air. Electric baseboard radiators come in various lengths—from 1' to 12'—and are rated from 100 W/ft to 400 W/ft., **Figure 30-12.** Like the hydronic systems, the radiators are located around the room's outside walls, especially under windows to counteract drafts. This system has some advantages:

- Easy to install.
- Affordable to buy and install.
- Easy to zone heat. Some models have thermostatic controls so that rooms can be individually heated.

Figure 30-12. An electric baseboard unit is sometimes used as auxiliary heat. (Slant/Fin Corp.)

Direct heating system: A heating system that directly gives off heat to objects without benefit of ducts or air system.

Electric resistance baseboard heating: A heating element wrapped in a tubular casing. Metal fins spaced along its length help radiate the heat to the room air.

- Self-contained. Chimneys and piping or storage of fuel are not needed.
- Quiet operating.

The downside is the operating cost. In some areas, electric rates make this type of heating simply too expensive to consider. Consequently, this type of system is usually used only as a backup. In some parts of North America, it is used in combination with wood-burning stoves or gas-fired furnaces.

Another type of electric heating is known as *electric radiant heat.* Such systems are designed to directly heat objects, rather than using radiators and convection currents to heat a room. Heating elements are placed in floors and ceilings in a manner similar to hydronic radiant systems. As the resistance elements heat up, the heat radiates to the concrete or wood floor, wall material, or ceiling material. Ceiling or wall radiant electric heating elements may be encased in plaster. This provides a medium through which the heat can radiate.

Another approach to radiant heating is to install a grid of specially sheathed electrical cable in the floor, **Figure 30-13.** The cables can be directly embedded in concrete or masonry when the floor is poured or laid.

Working Knowledge

To reduce heat loss in radiant installations, be sure to install generous amounts of insulation on the cold side of the heating elements.

30.4 Air Cooling Systems

Cooling systems are basically refrigerators designed to move heated air from a building to the atmosphere. In many cases, the cooling

---

Electric radiant heat: A system that radiates heat by moving it through solid matter, such as a concrete floor.

A

B

Figure 30-13. Electric radiant heat delivers heat directly to solid material such as concrete. A—Heavy insulation is laid on the cold side of heat system. Thick polystyrene sheets are laid over several inches of gravel. A vapor barrier is laid over this. B—A grid of electrical resistance wire is laid on top of the rebar and attached with plastic ties before the concrete is poured. (Kasten-Weiler Construction) system uses the same air distribution system (ductwork) as the heating system, **Figure 30-14.** A system of tubes called an *evaporator* is located in the furnace where return air from the rooms can pass over it.

When the central unit is operating, cooled, liquid refrigerant is pumped through the evaporator. Since it is colder than the room air passing over it, the refrigerant absorbs heat from the air and evaporates. Then, the hot refrigerant gas is drawn back to the outside unit. Here, it passes through another series of coils and fins called a condenser. The *condenser* acts like the radiator of a car and allows the heated refrigerant to pass its heat to the atmosphere. In doing so, the refrigerant returns to a liquid state.

From the condenser, the refrigerant is pumped at a high pressure to the evaporator in the furnace. This completes one cycle of the air conditioner. The cycle is repeated over and over again as the building's air is passed over the cooling coil. **Figure 30-15** shows an air-conditioning unit located outside a home. It contains the compressor, a condenser, and related valves. The compressor is operated by an electric motor.

Figure 30-15. A typical central air-conditioner unit located outside of the house. The condenser totally surrounds the compressor and its related components. This arrangement increases the efficiency of the unit.

30.5 Ducts

Ducts can be constructed of various materials, including metal, rigid plastic, or flexible plastic. Ducts should be as unobtrusive as possible, **Figure 30-16.** Most ducts are prefabricated and then fitted and cut on site. Metal duct sections are joined by the installer on site, **Figure 30-17.**

30.6 Controls

Heating and cooling systems need a device that can sense room temperature. When the temperature falls below or rises above a set point, a signal is sent to the heating or cooling unit. Such a device is called a *thermostat,* **Figure 30-18.**

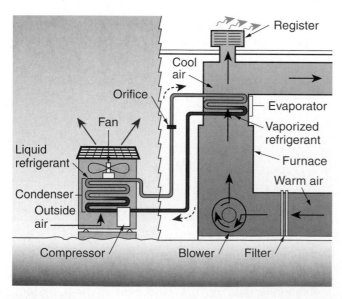

Figure 30-14. A central air-conditioning unit is combined with a forced-air distribution system. A coil located in the furnace receives cold refrigerant. Air forced through the coils is cooled and delivered to rooms by way of the duct system.

Evaporator: The system of coils or tubes in an air conditioning unit that receives the cold refrigerant and allows it to absorb heat and evaporate.

Condenser: The part of an air conditioning unit that receives hot refrigerant and allows it to release heat to the atmosphere and return to a liquid state.

Thermostat: An instrument that automatically controls the operation of heating or cooling devices by responding to changes in temperature.

A   **B**   **C**

Figure 30-16. Ductwork should be concealed wherever possible. A—Round ducts are located between floor joists. B—Ducts located in an attic where they are completely out of sight. C—Ducts running through openings in truss joists.

The temperature setting that causes the thermostat to send a signal is called the *cut-in point.* The temperature setting that causes the thermostat to cease signaling is called the *cut-out point.* The difference between these two points is called the *differential.*

Most thermostats are programmable. They help conserve energy. They can be set to automatically change temperatures at different times of the day. For example, the temperature can be lowered (or raised, if cooling) at night and during daytime hours when occupants are not home. During the hours when people are present, the temperature is increased (or decreased, if cooling). The result can be significant savings in fuel costs.

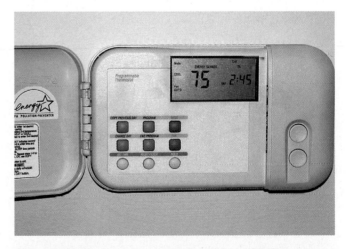

Figure 30-17. Using a metal, strap-type clip called a drive cleat, an installer connects two sections of sheet metal ductwork.

Cut-in point: The temperature setting at which a thermostat signals the furnace or air conditioner to turn on.

Cut-out point: Temperature setting at which a thermostat signals the furnace or air conditioner to turn off.

Differential: The difference between the cut-in point and the cut-out point of a thermostat.

Figure 30-18. A programmable thermostat helps save energy and fuel costs by adjusting heating or cooling levels for different times of the day.

30.7 Air Exchangers

Most newly constructed homes are relatively airtight. They often need a method of replacing indoor air polluted by gases and fumes such as from cooking, cleaning, and other sources. While it would seem that an open window and a fan might be an inexpensive remedy, it is not all that simple. The fresh air might be welcomed, the loss of heat in winter or of cool air in hot weather is not.

An *air exchanger* is designed to draw in fresh outside air, equalize its temperature with that of the inside air, and exhaust the polluted inside air to the atmosphere, **Figure 30-19.** At the same time, the exchanger can filter out most airborne particles, even microscopic particles. Some units also adjust the incoming air for the proper humidity level.

Testing indicates that indoor air should be exchanged with fresh outdoor air every two to three hours. In older, less-tightly-built homes, enough fresh air typically leaks in through cracks around doors, windows, and foundations to meet this requirement.

30.8 Heat Pumps

A heat pump combines heating and cooling in a single unit. The basic system includes the following:

- A pump or compressor for compressing refrigerant.
- An evaporator and condenser.
- A reversing valve to switch the functions of the evaporator and condenser.
- Pipes or tubing connecting the evaporator and condenser to the compressor.

Like an air conditioner, a heat pump uses a refrigerant to collect heat from one place and deliver it to another location. In winter, it collects heat from the outside and delivers it to the inside of a building. In summer, it collects heat from inside of the building and delivers it to the atmosphere.

The secret of heat pump operation is in the reversing valve. By switching the valve, the two

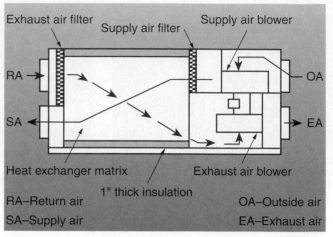

Figure 30-19. An air exchanger is a unit that brings fresh outside air into a building, replacing polluted inside air. A—This drawing shows how an air exchanger works. B—This furnace is set up with an exchanger. (United Technologies Carrier)

chambers are made to switch functions. Thus, in winter, the inside coil becomes a condenser, giving off heat; in the summer, it becomes an evaporator (cooling coil). See **Figure 30-20.**

Air exchanger: A device that exhausts air from a building and draws fresh air in.

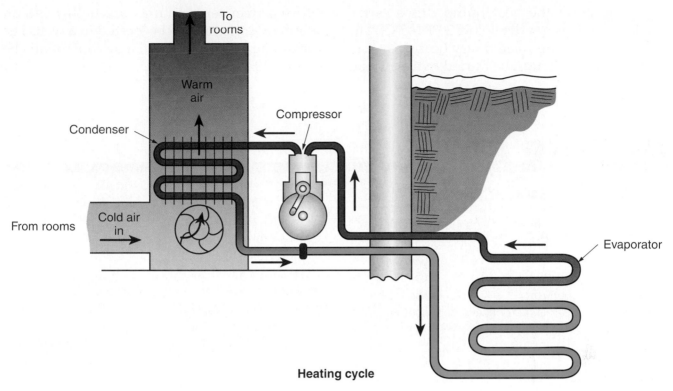

A

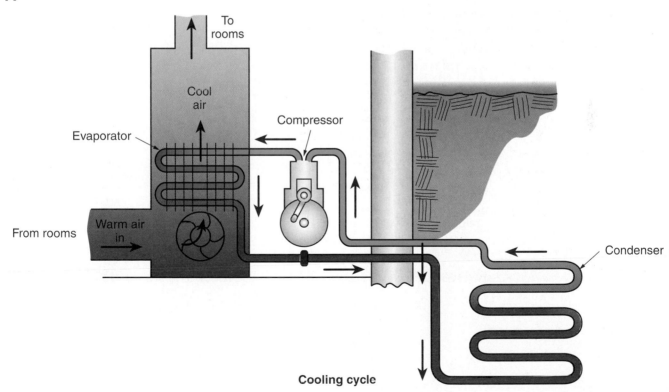

B

Figure 30-20. A—A heat pump in heating mode. B—Through the use of the reversing valve, the heat pump is now in cooling mode.

Efficiency of the heat pump drops as the outside temperatures dip below 20°F (−5°C). The building must have an auxiliary heating system or the outside coil must be buried in the ground where winter temperatures are higher. As an alternative, the coil can be located in a well. The water temperature in a well may consistently be around 40°F (5°C).

ON THE JOB

HVACR Installer

Climate control involves heating, ventilation, and air conditioning of a structure. Without climate control, most homes and commercial buildings would be very uncomfortable or impossible to live and work in. The heating, ventilation, air conditioning, and refrigeration (HVACR) installer is responsible for installing such systems, as well as the refrigeration systems required in many commercial and industrial structures. An installer also often carries out tasks such as diagnosing and repairing system problems or equipment breakdowns.

In larger contracting companies, installers may specialize in one of the subcategories of this trade: heating (furnace) work, air conditioning, or refrigeration systems. Ventilation is common to both heating and air conditioning work. Installers who are self employed or who work for smaller contractors and home improvement stores are most likely to work in any of the areas as required. Typically, an installer physically installs the heating, air conditioning, or refrigeration unit; runs any necessary piping or ductwork; and installs and connects electrical wiring as needed. They then test the operation of the system, making any necessary adjustments. On large scale construction projects where union jurisdictions are often involved, the installer may do only portions of the job. Ductwork, piping, and electrical wiring may have to be installed by members of other trades.

Installers may have to work in confined areas, such as crawl spaces or attics, and may have to deal with uncomfortably high or low temperatures. Like most construction trades, HVACR installers must wear appropriate personal protective equipment and carefully follow safety rules. A particular hazard in air conditioning and refrigeration work is the possibility of suffering frostbite or other damage from mishandled refrigerants.

Since HVACR systems are complex and continue to grow more sophisticated, employers seek installers who have specialized training. High schools, community colleges, and proprietary schools offer training ranging from one semester classes to two year associate degree programs. Formal apprenticeship programs are 3–5 years in length and include both classroom and on-the-job learning. Some installers begin as helpers and acquire their skills over a period of time working with experienced installers and technicians.

Approximately half of the installers and technicians in the HVACR field work for heating and cooling contractors. Most of the rest are employed by industrial plants, institutions, and government agencies. About 15% are self employed.

ON THE JOB

Sheet Metal Worker

The ductwork used to distribute heated and cooled air in most residential and commercial buildings is fabricated and installed by sheet metal workers. Members of the same trade also install metal roofing, fabricate and install metal countertops for residential and commercial kitchens, and sometimes create special ornamental sheet metal installations.

Two out of three sheet metal workers are in construction. The remainder work in industry fabricating products such as cabinets from sheet metal. Construction jobs for sheet metal workers are evenly divided among those with HVAC contractors and those with roofing and sheet metal contractors. Very few sheet metal workers are self employed.

Galvanized steel sheets are the raw material for many sheet metal jobs, especially those in the HVAC area. Copper is used in ornamental work and in roofing and flashing applications. Terne metal (steel coated with lead and tin) is also used in some roofing applications, as is steel that has been factory finished with an abrasion- and weather-resistant coating.

When performing HVAC installations, sheet metal workers often fabricate many of the duct sections in the shop, then move them to the job. Additional work and fabrication is usually needed at the job site. For some projects, virtually all the work is done on the site. Roofers generally do their work entirely on location, cutting and fitting the metal panels as they go.

Careful work habits and good measuring and math skills are needed. Sheet metal workers must plan, lay out, measure, and cut the components they fabricate. Safe working habits are a must, since power tools and sharp metal edges can easily cause injuries. The work is fairly strenuous, since it requires standing for long periods and often involves working overhead. Roofers work outdoors and thus are subject to weather conditions.

While some sheet metal workers learn their skills through informal on-the-job training, most complete an apprenticeship. The sheet metal apprenticeship is one of the most extensive in the building trades, lasting 4–5 years and requiring more than 200 hours of classroom instruction.

Summary

Heating, ventilation, and air conditioning (HVAC) systems keep buildings comfortable for working or living. Heat energy moves from warmer to cooler places by means of radiation, conduction, or convection. Because energy costs continue to increase, homeowners are concerned about keeping energy costs down and take conservation measures. These include sealing air infiltration points, increasing insulation, and installing the most energy-efficient heating and cooling devices. Central heating systems are of three major types: forced air perimeter heating, hydronic perimeter heating, and hydronic radiant heating. Electric resistance heat and electric radiant heating are also used. Cooling systems typically have an outdoor condenser unit to disperse heat and an evaporator to absorb heat from the indoor air. Forced-air heating and cooling systems distribute conditioned air through a network of ducts serving the building. HVAC systems are controlled by thermostats, which turn furnaces and air conditioners on and off at preset points. Programmable thermostats allow different settings for different times of each day. Air exchangers bring in fresh outdoor air and exhaust polluted indoor air to the outside. Heat pumps are single units that handle both heating and cooling tasks. Heating and cooling systems should be checked on a regular basis for efficient and safe operation.

Test Your Knowledge

Answer the following questions on a separate piece of paper. Do not write in this book.

1. Name the three ways that heat travels.
2. List five measures that save energy when heating or cooling a residence.
3. *True or False?* Chimneys are a major source of lost heat energy.
4. What is an EER and what does it cover?
5. List the three types of central heating systems.
6. A(n) _____ heating system is usually installed in a slab-on-grade foundation.
7. What is the main difference between forced-air and hydronic systems?
8. Which heating system has a supply main and a return main?
9. Why is electric resistance baseboard heating not frequently used?
10. *True or False?* In an air-conditioning system, the evaporator is always located where inside air can pass over it.
11. A(n) _____ automatically signals the heating system to deliver more heat.
12. How can a programmable thermostat save energy costs?
13. *True or False?* An air exchanger may also filter dust from incoming air.
14. A(n) _____ is a combination heating and cooling system.
15. What medium is used in a heat pump to collect heat?

Curricular Connections

Science. Demonstrate the heat conduction of different substances. Obtain 1″ × 12″ strips of thin aluminum, copper, steel, wood, and rigid plastic. Grasp one end of a strip between your thumb and forefinger, then hold the other end of the strip against the heat source. An outdoor floodlight lamp can be used as the heat source. Time how long it takes before you can feel heat at the end you are grasping. If no heat is felt after 30 seconds, move on to the next sample. Record the times. Determine which of the samples was the best heat conductor.

Outside Assignments

1. Obtain a set of instructions for installation of a thermostat. Explain to the class how to hook it up to a furnace.
2. Study a book on air-conditioning systems and learn the functions of various components of the system. Present a report to the class.
3. Examine EER rating sheets from two different furnace models. Determine what energy savings are possible by choosing one over the other. Btu ratings should be the same for each furnace in order to have a valid comparison.

Snips and crimping tools are needed for installing ductwork. (Wiss)

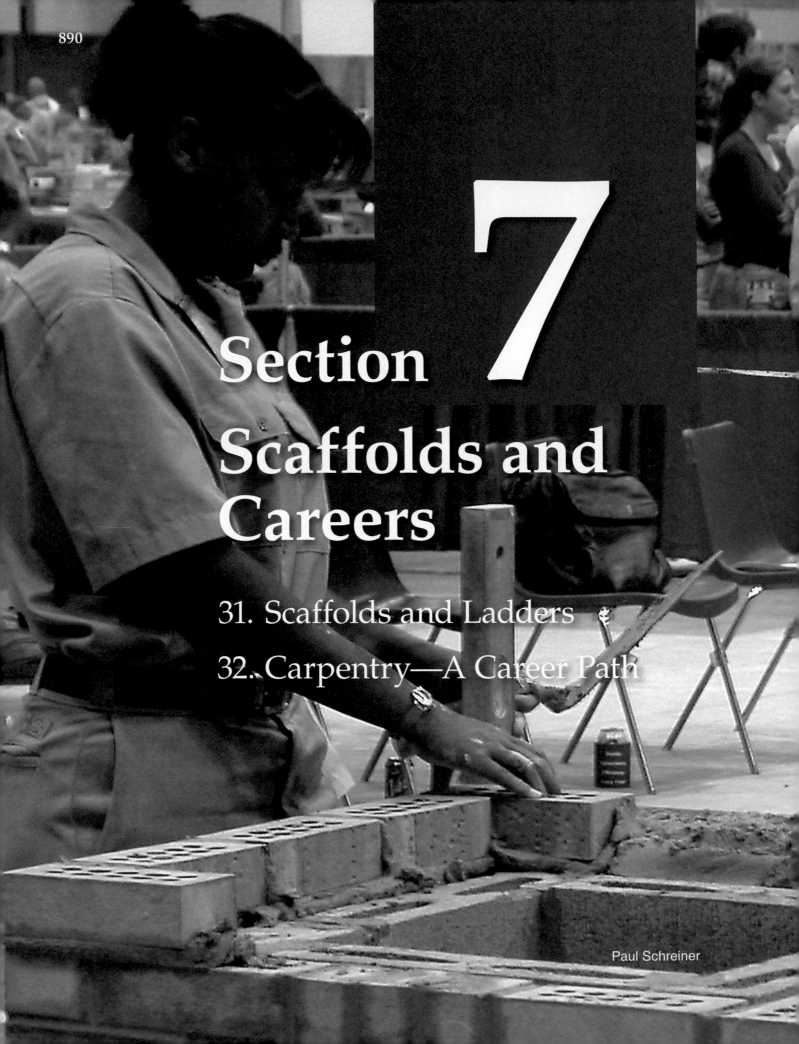

Section 7

Scaffolds and Careers

Paul Schreiner

Scaffolds and Ladders

31

Learning Objectives

After studying this chapter, you will be able to:
- Explain typical designs and construction of manufactured and site-built scaffolding.
- Discuss the types and uses of brackets and jacks.
- List ladder types and maintenance techniques.
- Apply ladder and scaffolding safety rules.

Technical Vocabulary

Bearers
Ladder jacks
Roofing brackets
Scaffolding
Toeboard
Trestle jacks

Carpenters use *scaffolding*, also called *staging*, to work in areas that are out of reach while standing on the ground or floor deck. Scaffolds are temporary raised platforms that support workers, tools, and materials with a high degree of safety, **Figure 31-1.** Their design must help the worker avoid stooping and reaching. The type of scaffold needed for a job depends on how many workers will be on it at one time, the distance it must extend above the ground, and whether it must support building materials as well as people.

31.1 Types of Scaffolding

Scaffolding includes a great variety of designs. In addition to the types manufactured with metal components, there are wood scaffolds that can be built on site by the carpenter crew. Some manufactured types are meant to be assembled and disassembled, so they can be reused at other building sites. One type is completely mobile and can be moved from site to site. Scaffolds of various types are shown throughout this chapter.

31.1.1 Manufactured Scaffolding

Many builders use sectional steel or aluminum scaffolding. It can be quickly and easily assembled from prefabricated frames. The type shown in **Figure 31-2** has light aluminum uprights that are attached to the building with metal braces. The platform is adjusted to various heights by a jack

Scaffolding: A temporary structure or platform used to support workers and materials during building construction. Also called *staging*.

A

B

Figure 31-1. Two types of scaffolding. A—This scaffold system is used for siding work. It has pump-jack units attached to aluminum poles. A foot pedal, as shown, is used to raise the scaffold. B—This system uses metal and wood components. Braces and standards are assembled in many different configurations to meet job requirements. (Alum-A-Pole Corp.)

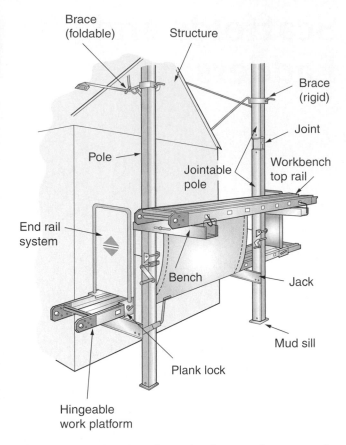

Figure 31-2. Details of an aluminum pole system for scaffolding. (Alum-A-Pole Corp.)

that raises and lowers the brackets supporting the platform. The platform and bench units are also manufactured of aluminum.

Figure 31-3 shows another manufactured type of scaffold. It consists of truss frames and diagonal braces that can be horizontally and vertically assembled to build a scaffold of nearly any safe height or length. **Figure 31-4** illustrates the erection details of this scaffolding system.

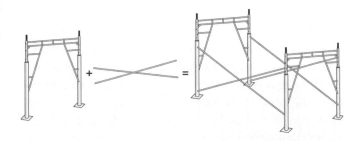

Figure 31-3. Prefabricated truss frames and diagonal braces can be used to rapidly build scaffold sections.

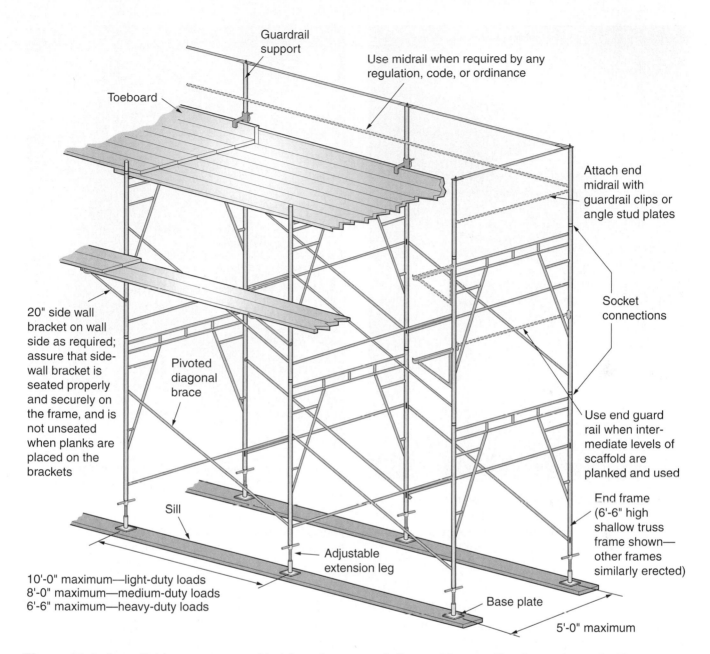

Guardrail support

Use midrail when required by any regulation, code, or ordinance

Toeboard

Attach end midrail with guardrail clips or angle stud plates

20" side wall bracket on wall side as required; assure that side-wall bracket is seated properly and securely on the frame, and is not unseated when planks are placed on the brackets

Pivoted diagonal brace

Socket connections

Use end guard rail when inter-mediate levels of scaffold are planked and used

Sill

End frame (6'-6" high shallow truss frame shown—other frames similarly erected)

Adjustable extension leg

10'-0" maximum—light-duty loads
8'-0" maximum—medium-duty loads
6'-6" maximum—heavy-duty loads

Base plate

5'-0" maximum

Figure 31-4. A scaffold system assembled from frames and diagonal braces like those shown in Figure 31-3. (Patent Scaffolding Co.)

There are many styles of sectional steel scaffolding. Some types have adjustable legs that are attached to the ground-level sections, **Figure 31-5.** Frame sizes range from 2' to 5' wide and from 3' to 10' high. Various lengths of bracing provide frame spacings of 5' to 10'. The basic units are set up and then vertically and horizontally joined. The frames can be equipped with casters when a rolling scaffold is desired.

Another scaffold type is the swinging scaffold. Swinging scaffolds are suspended from the roof or other overhead structures. These are used mainly by painters and should be used only with light equipment and materials.

31.1.2 Mobile Scaffolding

Movable or mobile scaffolding comes assembled and uses either a mechanical or hydraulic system to adjust its height. Although some units are self-propelled, most are mounted on a trailer

Figure 31-5. Adjustable extensions on the legs of scaffolding frames allow leveling for stability and safety. Planks or other support material should be used to prevent legs from sinking into the ground.

Figure 31-6. Mobile scaffolding can be quickly positioned and raised and lowered. This unit is trailer-mounted; others are self propelled.

arrangement that is easily moved from one site to another. This is the major advantage of mobile scaffolding. **Figure 31-6** shows a mobile scaffold raised by a scissors jack that is activated by a hydraulic cylinder. It is set up to be towed behind a truck or other construction vehicle.

31.1.3 Site-Constructed Wood Scaffolding

Where it is not possible or practical to use manufactured scaffolding, carpenters must build scaffolds from available lumber. In constructing wooden scaffolds, the uprights should be made of clear, straight-grained 2 × 4s. The lower ends should be placed on planks to prevent settling into the ground. See **Figure 31-7.**

Bearers, sometimes called *cross ledgers*, consist of 2 × 6s about 4' long. Use at least three 16d nails at each end of the bearers to fasten them to the uprights. For the single-pole scaffold, one end of the bearer should be fastened to a 2 × 6 block securely nailed to the wall. Braces may be made of 1 × 6 lumber fastened to uprights with 10d nails and with 8d nails where they cross. For the platform, 2 × 10 planks without large knots should be used. It is good practice to spike the planks to bearers to prevent slipping. Whether lumber is used for planks or uprights, select lengths with parallel grain, **Figure 31-8.**

Bearers: Horizontal members (usually 2 × 6s) used to connect scaffold uprights. Also called *cross ledgers.*

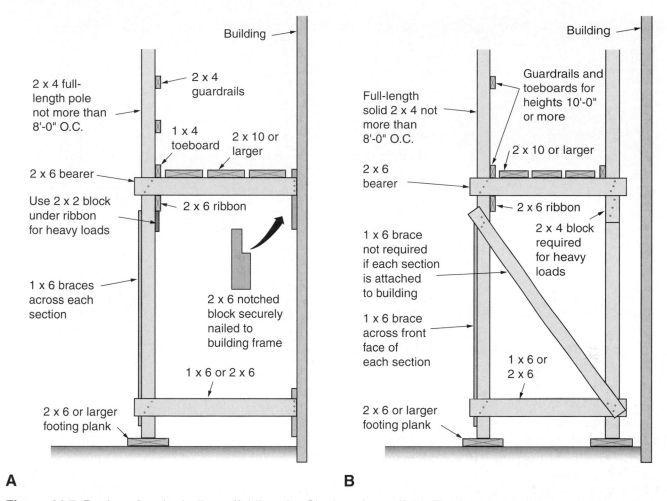

A

B

Figure 31-7. Designs for site-built scaffolding. A—Single pole scaffold. The horizontal distance between bearer sections should never exceed 8′. B—Double-pole scaffold. The structure can be extended upward to form several platforms, but the maximum height should be limited to 18′.

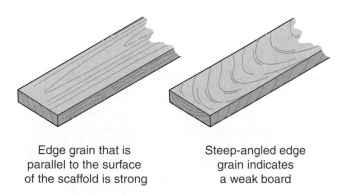

Edge grain that is parallel to the surface of the scaffold is strong

Steep-angled edge grain indicates a weak board

Figure 31-8. For a safe scaffold, use straight-grain lumber for both uprights and platforms.

31.2 Brackets, Jacks, and Trestles

Metal wall brackets are often used for scaffolding in residential construction, **Figure 31-9.** They can be quickly attached to a wall, easily moved from one construction site to another, and installed above overhangs or ledges.

Great care must be used when fastening the brackets to a wall. For light work at low levels, connections with nails may provide enough safety. Use at least four 16d or 20d nails. Be sure the nails penetrate sound framing lumber. Most carpenters prefer the greater safety provided by brackets that hook around wall studs. Holes in the sheathing required to accommodate such

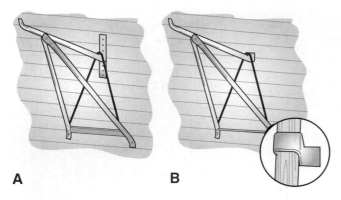

A   **B**

Figure 31-9. Types of metal wall brackets. A—Attached with nails securely set in the building frame. B—Directly hooked to wall studs.

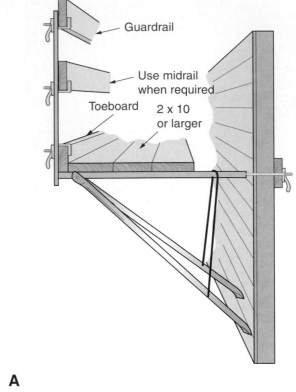

A

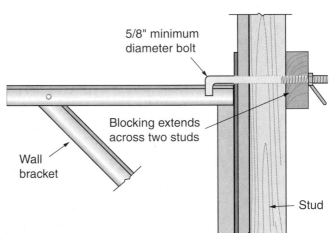

B

Figure 31-10. A—This metal wall bracket is securely bolted through the wall. High scaffolds must be equipped with guardrails and toeboards. B—Detail of the anchoring system. The bolt passes through the wall alongside the stud and is anchored to blocking that must span at least two studs.

brackets must be repaired before the exterior finish is applied.

Some metal wall brackets have posts for holding guardrails and toeboards, **Figure 31-10.** A guardrail protects workers on the scaffold. A *toeboard* helps protect workers beneath from falling tools or other objects. A wire mesh screen is sometimes added on open sides of the scaffold. The scaffold shown in **Figure 31-10** is fastened to the wall with a bolt.

Wall brackets that support a wood platform should not be spaced more than 8′ apart. Metal or reinforced platforms can extend more than 8′ between supports. **Figure 31-11** shows a metal platform supported by metal wall brackets.

Roofing brackets are easy to use and provide safety when working on steep slopes, **Figure 31-12.** One type has an adjustable arm to support the plank. This allows a level platform to be set up on nearly any angle of roof surface. Another type of roofing bracket is fixed at one angle.

Ladder jacks are used to support simple scaffolds for repair jobs. Setup requires two sturdy ladders of the same size and a strong plank. **Figure 31-13** shows a jack that hangs below the

---

Toeboard: A board horizontally fastened to scaffolding slightly above the planking to keep tools and materials from falling on workers below.

Roofing brackets: Provide safety when working on steep slopes.

Ladder jacks: Used to support simple scaffolds for repair jobs; setup requires two sturdy ladders of the same size.

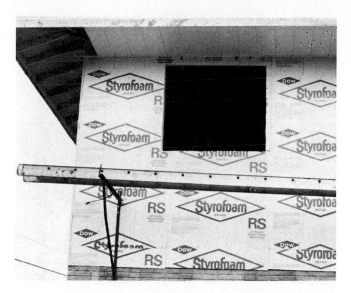

Figure 31-11. Scaffolding formed with metal brackets and a ladder plank. The plank supports a plywood strip and can span nearly twice the maximum distance specified for wood planking. This type of scaffolding should be used at heights of less than 10′. Since it does not include a guardrail, a lifeline should be worn when working on this type of scaffolding.

Figure 31-13. A ladder jack used to provide a low-level scaffold. Two identical ladders must be used. The plank is located inside of the slope of the ladders.

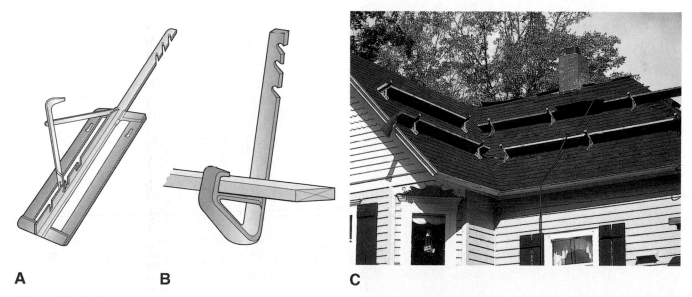

A   B   C

Figure 31-12. Roofing brackets. A—Bracket with slots to slip over nails driven into the roof frame. The adjustable arm is to level planking on roofs of different pitches. B—Bracket with fixed angle for steep roofs. C—Brackets and planks installed for a re-roofing job.

ladder and hooks to the side rails. Another type of jack is shown in **Figure 31-14.**

For interior work, *trestle jacks* are used to support low platforms. They are assembled as

shown in **Figure 31-15.** Follow the manufacturer's guideline for the proper size and weight of jack. Be sure the material used for the *ledger* (horizontal board held by the jacks) is sound and large enough to support the load.

A

B

Figure 31-14. Ladder jacks projecting from the front side of ladders. A—A lifeline should be worn at all times when using a scaffold without a guard rail. B—Ladder jacks used with a metal platform. (Trudeau Construction Co.)

Safety Note

When working on or around scaffolding, observe these rules:

• All scaffolds should be built under the direction of an experienced worker.

• Follow the design specifications listed in local and state codes.

• Inspect scaffolds each day before use.

• Provide adequate pads or sills under scaffold posts.

• Plumb and level scaffold members as each is set.

• Equip all planked areas with proper guardrails, toeboards, and screens when required.

• Power lines near scaffolds are dangerous. Consult the local power company for advice and recommendations.

• Never use ladders or makeshift devices on top of scaffold platforms to increase the height.

• Be certain the planking is heavy enough to carry the load with a safe span length.

• Planking should be lapped at least 12″ and extend 6″ beyond all supports.

• Do not permit planking to extend an unsafe distance beyond supports.

• Before moving a rolling scaffold, remove all materials and equipment from the platform.

• The height of the rolling scaffold platform should not exceed four times the smallest base dimension.

Trestle jacks: Used to support low platforms for interior work.

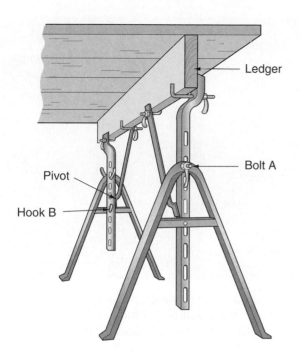

Figure 31-15. Trestle jacks used for low platforms. To raise the jacks, loosen bolt A and move hook B to another slot. Note that the ledger is firmly clamped to the jacks. (Patent Scaffolding Co.)

31.3 Ladders

Types of wood and aluminum ladders commonly used by the carpenter are illustrated in **Figure 31-16.** Stepladders range in size from 4′–20′. A one-piece (single straight) ladder is available in sizes of 8′–26′. Extension ladders provide lengths up to 60′.

Quality wood ladders are made from clear, straight-grained stock. Ladders should be given a clear finish. When reconditioning a wood ladder always apply a clear finish. Never use paint as it may conceal dangerous defects.

Basic care and handling of ladders is illustrated in **Figure 31-17.** Always keep ladders clean. Do not let grease, oil, or paint collect on the rails or rungs. The *rails* are the sides, while the *rungs* are the steps. Keep all fittings tight. On extension ladders, lubricate the locks and pulleys. Replace any frayed or worn rope.

To erect a ladder, place the lower end against a solid base so it cannot slide. Raise the top end. Walk toward the bottom end, grasping and raising the ladder rung by rung as you proceed.

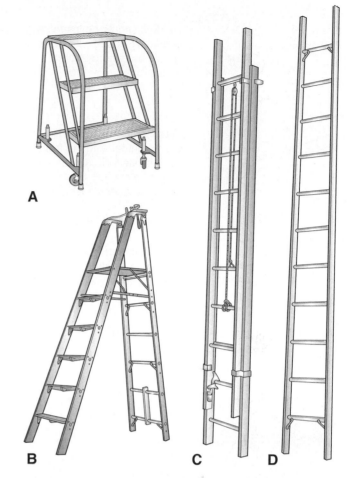

A

B **C** **D**

Figure 31-16. Types of ladders used by carpenters. A—Safety rolling ladder presses against the floor when weight is applied. B—Stepladder with platform and tool holder. C—Extension ladder. D—One-piece (single straight) ladder. (Patent Scaffolding Co.)

When the ladder is vertical, lean it against the structure at the proper angle. Make sure the bottom end of both rails rests on a firm base.

Safety Note

When working on or around ladders, observe these rules:
- Always inspect a ladder before using it.
- Place the ladder so the horizontal distance from its lower end to the vertical wall is 1/4 of the length of the ladder, **Figure 31-18.**
- Before climbing the ladder, be sure both rails rest on solid footing.

(Continued)

Inspection

Ladders should be inspected frequently. Those with defects should be either repaired or destroyed.

Carrying

Always carry a ladder over your shoulder with the front end elevated. Be sure not to drop it or let it fall. Such an impact weakens a ladder.

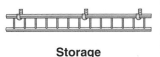

Storage

Store horizontally on supports to prevent sagging. Do not store near heat or out in the weather.

Figure 31-17. Care and handling of ladders.

Figure 31-18. Proper ladder placement. The horizontal distance from the wall to the base of the ladder should be 1/4 of the length of the ladder. (Greco Painting)

(Continued)

- When the ladder is used on surfaces where the bottom might slip, equip the rails with safety shoes, **Figure 31-19.**
- Never place a ladder in front of a doorway if the door can be opened toward the ladder.
- Always face the ladder when climbing up or down.
- Never place a ladder on boxes or any unstable base to get more height.
- Never splice together two short ladders to make a longer ladder.
- Place ladders so work can be done without leaning beyond either side rail.
- Be sure extension ladders have sufficient lap between sections. A 36′ length should lap at least 3′; a 48′, at least 4′.
- When a ladder is used to climb onto a roof, it should extend above the roof at least 3′.
- Keep both hands free to grasp when climbing a ladder. Use a hand line to raise or lower tools and materials.
- Before mounting a stepladder, be sure it is fully open and locked and all four legs are firmly supported.

(Continued)

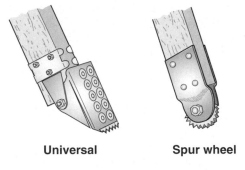

Universal   **Spur wheel**

Figure 31-19. All ladders used on smooth surfaces must be equipped with safety shoes. (Tilley Ladder Co.)

(Continued)

- Do not stand on either of the two top steps of a stepladder.
- Do not leave tools on the top of a stepladder unless the ladder has a special holder.
- Never use metal ladders where contact with electric current is possible.

Summary

Scaffolding is used to work in areas too high to reach from the ground or floor deck. Ladders are used for the same purpose, but are also used to climb to upper portions of the building, such as the roof. Scaffolding may be a manufactured system, with components that vertically and horizontally fit together; mobile units that can be moved from place to place; or site-built structures erected by carpenters. Other types of work platform supports include wall brackets, roof jacks, ladder jacks, and trestle jacks. Since they sometimes involve working at considerable heights above the ground, it is very important to observe all safety precautions while working on scaffolds or ladders.

Test Your Knowledge

Answer the following questions on a separate piece of paper. Do not write in this book.

1. What is the purpose of *scaffolding?*
2. *True or False?* Steel sectional scaffolds should never have casters installed.
3. What is the major advantage of mobile scaffolding?
4. The bearers that support the planking of wood scaffolds should be made with material that has a nominal cross-section of _____.
5. What is a *toeboard?*
6. Metal wall brackets that support a wood platform should not be spaced more than _____ feet apart.
7. The height of the platform on a rolling scaffold unit should not exceed _____ times the smallest dimension of the base.
8. One-piece (single straight) ladders are available in sizes from 8' to _____'.
9. The sides of a ladder are called _____.
10. When a ladder is used to climb onto a roof, it should extend at least _____ feet above the roof edge.

Curricular Connections

Language Arts. The Federal Occupational Safety and Health Act requires the use of fall protection systems on many types of construction projects where people will be working above the ground. Contact the nearest OSHA office or go to the OSHA Web site and find information on the circumstances where fall protection systems must be used, what those systems consist of, and how they work. Present an oral report to the class.

Outside Assignments

1. Obtain descriptive literature from a company that manufactures steel scaffolding. Check with your local building supply dealer or write to the company. From a study of this information, learn about the types and sizes of tubing that are used. Also learn how the various connecting devices operate. Try to find out approximate costs of this type of scaffolding. Prepare carefully organized notes, then write a report.
2. Make a study of the types of ladders used in construction and maintenance work. Deliver an oral report to the class. Include information about the kind and quality of materials used and how the ladders are assembled and finished. Also, include commonly available sizes and list prices. Finally, develop a list of reasons a homeowner may have for buying or renting a ladder. Are these reasons the same as for a builder?

Pump-jack scaffolding being used on condominium construction. This oceanfront building is on a piling foundation that raises it 10′ above the ground.

Carpentry—A Career Path

32

Learning Objectives

After studying this chapter, you will be able to:

- Cite the projected demand for carpenters in coming years.
- List job possibilities for the trained carpenter.
- Describe the sequence of carpentry training and apprenticeship.
- Discuss abilities and characteristics needed by those in the carpentry field.

Technical Vocabulary

Apprenticeship	Operating engineers
Entrepreneurs	Self-employment
Journeyman	SkillsUSA
Manual dexterity	Vocational instructor

Carpentry is a rewarding career. It is ideal for the person who is interested in and has an aptitude for working with tools and materials, **Figure 32-1.** The trade requires the development of manual skills. These skills involve both thinking and doing. Carpentry also requires a thorough knowledge of the materials and methods used in construction work.

Carpenters can expect to work at a variety of construction activities. They may be involved in construction of buildings, highways, bridges, docks, industrial plants, and marine structures. They cut, fit, and assemble wood and other materials. If carpenters work in commercial building construction, they will likely work with metal framing materials. Increasingly, metal framing is found in residential construction.

Carpenters working for general building contractors become involved in many different tasks. These often include laying out building lines, excavating, installing footings and foundations, framing, roofing, building stairs,

Figure 32-1. Potential carpenters enjoy working with tools and on machines. (Montachusett Regional Vo-Tech School)

drywalling, laying flooring, and installing cabinets.

A well-trained carpenter is expected to know the requirements of local building codes. These codes dictate where materials may be used and how a structure is to be built. Failure to follow the codes wastes materials and labor, since inspectors may require work to be redone.

32.1 Economic Outlook for Construction

The population of the United States continues to grow. It surpassed 300 million in the first decade of the 21st century. This continued growth increases the demand for housing and the construction workers who build it. The Bureau of Census projects population growth by regions, with the West and South experiencing the fastest growth. These areas are also the leaders in expected job growth for carpenters. According to state government projections for the 2000–2010 decade, Colorado will experience a 61% increase in carpentry jobs; California will see a 28% increase; North Carolina, 25%; Georgia, 18%, and Louisiana, Mississippi, and Florida will all see 17% growth. Building construction, especially residential structures, continues to increase as the population grows.

Most of this growth will come from the need to improve the nation's infrastructure—roads, bridges, and tunnels. See **Figure 32-2.** Sharing with carpenters in the growth in construction are other building trades workers such as surveyors, electricians, bricklayers, plumbers, cabinetmakers, landscapers, and *operating engineers* (operators of heavy construction equipment). **Figure 32-3** shows a variety of construction-related careers available to anyone who has the interest and aptitude to pursue them.

Operating engineers: A tradesperson who operates heavy equipment, such as cranes, bulldozers, and backhoes.

A

B

Figure 32-2. Infrastructure work, such as road repairs, will be a major source of construction employment in the next 10 years. Next to services, construction is the fastest-growing occupation in the United States.

32.2 Employment Outlook

Carpenters are the largest group of trade workers, with more than 850,000 holding jobs in 2003. Almost one-third of those carpenters (258,000) were working in residential construction; another 140,000 were in nonresidential construction. Almost 250,000 carpenters were listed as working for contractors in the foundation, roofing, drywall, flooring, or related fields. The rest were working for manufacturing firms, government agencies, commercial establishments, or schools.

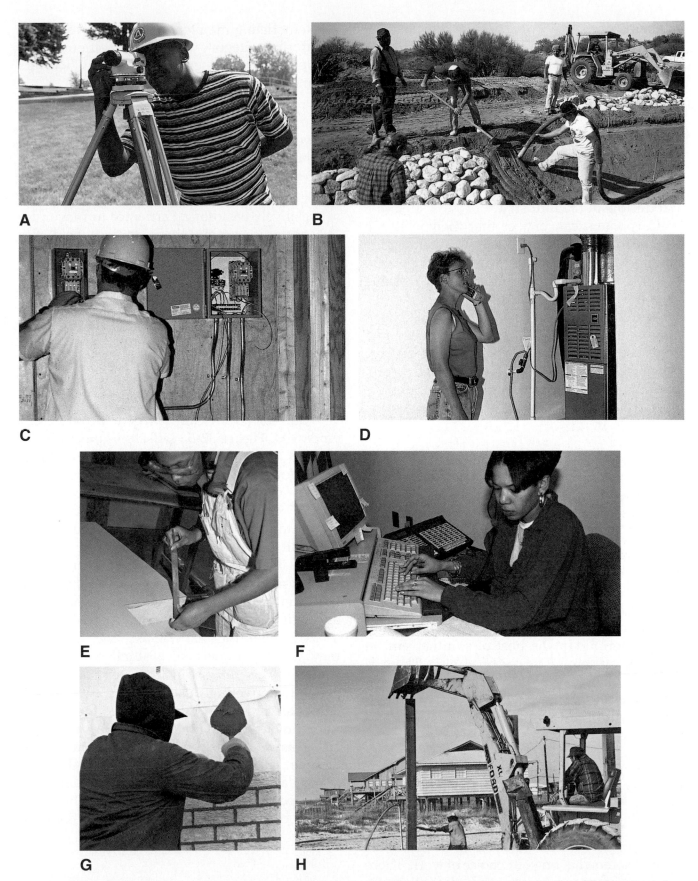

Figure 32-3. Carpentry is related to a wide variety of skilled occupations. A—Surveying. B—Landscaping. C—Electrical trades. D—Construction expediter. E—Cabinetmaking. F—Computer operation (using construction software). G—Brickmason. H—Operating engineer.

Jobs in carpentry are expected to be plentiful in coming years because of increased building demand and the need to replace those who transfer into other jobs or retire.

On the negative side is the prospect of cyclic economic downturns that affect the construction industry. At such times, many people skilled in construction find it hard to get or hold a job. Those who have been in such occupations for a long time have come to expect both growth and decline in the industry.

Carpenters can expect that their jobs will become increasingly technical and automated as time goes on. Thus, apprentices should expect to be learning new tasks throughout their working life.

With experience, new opportunities may come along. In addition to advancement into supervisory positions, carpenters can become architects, educators, engineers, technical representatives for construction materials manufacturers, and union leaders.

32.3 Working Conditions

As with other building trades, carpentry is often hard, physical work. Construction timbers are often heavy and other materials are cumbersome to handle. Climbing, stooping, kneeling, and prolonged standing require stamina. Work has to continue in the heat of summer and in the freezing temperatures of winter. There is also risk of injury from working with sharp tools, power staplers, power nailers, and rough materials. Slips and falls are also a danger and threat.

32.4 Job Opportunities

Job opportunities for you as a carpenter depend somewhat on your choice of work. There are many areas where you can look for a job:
- Building construction—the majority of carpenters work in this area.

- Remodeling, maintenance, and renovations—in this category are many self-employed people, while others work for manufacturers and building management firms.
- Prefabrication of buildings and building components—while this area employs many workers with fewer skills than carpenters, the latter may hold higher-level jobs in this type of fabrication.
- Specialization and related occupations—it is relatively easy for a carpenter to move into specialties within the carpentry field. For example, a general building contractor may subcontract various phases of construction. Carpenters, as a result, may specialize in foundation work, framing, drywalling, insulating, siding, or roofing. **Figure 32-4** shows two of these specialties. A carpenter may also choose to move into related fields such as terrazzo work, heavy equipment operation, bricklaying, or electrical work.

Four of every ten carpenters are self-employed, **Figure 32-5**. *Self-employment* has a number of advantages. Often, profits are higher due to lower overhead. Some carpenters change employers at the conclusion of every construction job. Others alternate between working for a contractor and being the contractor themselves on small jobs.

The income earned by the carpenter is close to that received by other trades. It provides a good living. In addition, the carpenter has pride and satisfaction from doing quality work.

After gaining experience, the carpenter may want to undertake a small construction contract. This may be a first step toward the general contracting business. The experienced carpenter is usually able to handle the work of the estimator. Estimators figure labor and material costs before the contractor bids on a job.

The carpenter who prefers the sales and service aspects of construction work can often find a position with a lumber yard or building supply center. The carpenter can also join the

---

Self-employment: Starting and operating a business of one's own.

A

B

Figure 32-4. Often, carpenters specialize in one particular phase of building. A—This crew specializes in putting in foundations. This involves excavating, building forms, and laying down vapor barrier and reinforcing mat. B—This crew does only framing.

customer service department of a company producing structures in a factory. Customers may need advice on installation or choice of prefabricated units.

32.5 Training

A high school education is highly desirable for a successful career in carpentry or other areas of building construction. This is as true for the aspiring apprentice as for the technician or engineer. Take as many woodworking and building construction courses as possible.

Other courses are also valuable. Drafting is especially useful, as are courses in print reading and career education.

Students interested in the building trades sometimes avoid science and math. This is unfortunate, because these studies are essential if you are to understand the technical aspects of modern methods and materials used in construction work. Include social studies and English because everyone should be prepared to improve our society and be competent in reading, writing, and speaking.

After graduating from high school, you will probably have the skills to directly enter into an apprenticeship training program, **Figure 32-6.** If circumstances permit, you may decide to enroll in a vocational-technical school in your area. There you can take advanced courses in carpentry and related areas. Some high schools offer these vocational courses as a part of the normal programs before graduation. If possible, take classes in concrete work, bricklaying, plumbing, sheet metal, and electric wiring. The carpenter usually closely works with tradespeople in these areas. A basic understanding of their methods and procedures is very valuable. This is especially true if you want to become a supervisor.

Figure 32-5. Self-employed carpenters have lower overhead and no employee problems.

Figure 32-6. An apprenticeship program provides training on the job. Here, a carpenter explains fastening techniques to an apprentice. (United Brotherhood of Carpenters and Joiners of America)

32.5.1 Apprenticeship

Many carpenters learn their trade on the job, taught by their employer or the lead carpenter. However, many learn the trade through apprenticeship or through formal training in vocational training centers or high school programs. *Apprenticeship* is a program that allows a student to both work and attend school to learn a trade—in this instance, carpentry.

Apprenticeship training has its beginning in early times. It started as an arrangement between children and their parents. This system passed the knowledge and skill of a trade on to each succeeding generation. As society and the economic structure became more complex, the trainee often left home and was placed under the guidance and direction of another master of the trade in the community. The student was called an apprentice and learned the trade of this master. During the period of apprenticeship (sometimes as long as seven years) the apprentice often lived in the master's house. There were no wages, but board, room, and clothing were provided.

When the training was complete, the apprentice earned the status of journeyman and could work for wages. The word *journeyman*

arose from the fact that an apprentice who had completed the training period was then free to "journey" to other places in search of employment. The term journeyman still means a tradesperson who is fully qualified to perform all tasks of the trade. To gain more experience, the journeyman worked with other masters as an equal. Then this "seasoned" worker would often set up his or her own business.

This form of apprenticeship rapidly declined during the Industrial Revolution. A new kind of system developed in which the apprentice lived at home and received wages for work done. Under this system, the apprentice was often exploited. Some became only low-paid workers who received little training. Such practices continued in varying degrees until federal legislation was enacted that established standards and specific requirements for apprenticeship training programs. Today, apprenticeship programs are carefully set up and supervised. Local committees consisting of labor and management provide direct control. There is help from schools and state and federal organizations and agencies, **Figure 32-7.**

32.5.2 Apprenticeship Stages

The apprentice works under a signed agreement with an employer. The agreement includes the approval of local and state committees on apprenticeship training.

Applicants must be at least 17 years of age and satisfy the local committee that they have the ability or aptitude to master the trade. Then, they are placed on a waiting list. The waiting period can last from one to five years, depending on local demand.

Apprenticeship: A formal method of learning a certain trade, such as carpentry, that involves instruction as well as working and learning on the job en route to certification as a journeyman.

Journeyman: A tradesperson who has completed an apprenticeship or, through experience, is qualified to work without supervision or further instruction.

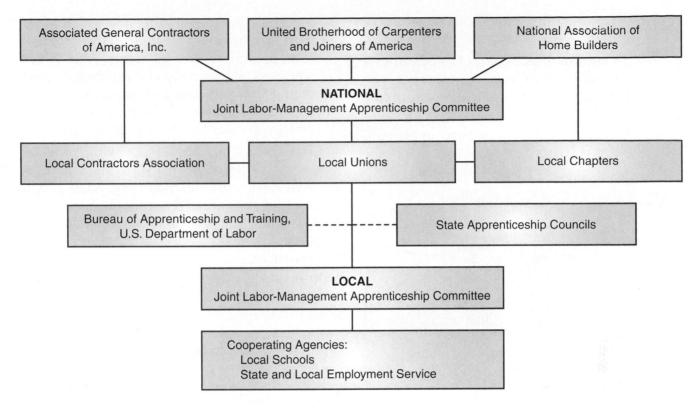

Figure 32-7. The apprenticeship and training system for the carpentry trade is well organized. Additional information may be obtained from the bulletin *National Carpentry Apprenticeship and Training Standards* prepared by the U.S. Department of Labor.

The term of apprenticeship for the field of carpentry is normally four years. This may be adjusted for applicants with significant experience or those who have completed certain advanced courses in vocational-technical schools.

In addition to the instruction and skills learned through regular work on the job, an apprentice attends construction classes in subjects related to the trade. These classes are usually held in the evening and total at least 144 hours per year. They cover technical information about tools, machines, methods, and processes. The classes provide practice in mathematical calculations, print reading, sketching, layout work, estimating, and similar activities. A great deal of study is required to master the technical knowledge needed in carpentry work. See **Figure 32-8.**

The apprentice works and is paid while learning. The wage scale is determined by the local committee. It usually starts at about 50% of the amount received by a journeyman carpenter. This scale is advanced regularly and may approach 90% of a journeyman's pay during the last year.

When the training period is complete and the apprentice has passed a final examination, he or she becomes a journeyman carpenter. A certificate that affirms this status is issued. This document is recognized throughout the country.

32.6 Personal Qualifications

To become a successful carpenter you must:
- Be physically able to do the work.
- Display *manual dexterity.* This means being skilled with your hands and having a talent

---

Manual dexterity: The ability to efficiently accomplish tasks requiring use of the hands.

for working with the tools and materials of the trade.

• Have a sincere interest and enthusiasm that will intensify your efforts as you study and practice the skills and know-how required.

• Possess certain character traits. Honesty in all of your dealings is very important,

B

Figure 32-8. Apprentices begin by working on simpler carpentry tasks. A—This apprentice is nailing wafer-board to a wall frame. B—Before any work is done, the carpenter or another person consults the building plan. Some training in print reading is required for all carpenters. (Henry Brothers Co.; Photography by Larry Morris)

especially in the quality and quantity of work performed. You must show courtesy, respect, and loyalty to those with whom you work. Punctuality and reliability reflect your general attitude and are important not only during your training program, but later when you enter regular employment.

The ability to cooperate with others—students, co-workers, supervisors, and employers—is essential to success. Most people who fail in the carpentry trade do so because of a deficiency in personal characteristics, not because of their skill level.

The hazards associated with carpentry require that you develop a good attitude toward safety. This means that you must be willing to spend time learning the safest way to do your work. You must be willing to follow safety rules and regulations at all times.

Even after you complete your training program, you must continue to perfect your skills and adjust to new methods and techniques, **Figure 32-9.** Each day brings new materials and improved procedures to the construction industry. This presents a special challenge to

Figure 32-9. Changes in materials and techniques require the carpenter to keep up to date. This worker is installing acoustical ceiling tile in a new shopping mall. (USG Acoustical Products Co.)

those in the carpentry trade. You should read and study new books and manufacturer literature, along with trade journals and magazines in the building construction field. Much information can be obtained at association meetings and conventions where new products are exhibited. To be a successful carpenter, you also need to keep informed on code changes, new zoning ordinances, safety regulations, and other aspects of construction work that apply to the local community.

For those who have ability and are willing to work, the field of carpentry is a satisfying, fulfilling, and well-paying occupation. Advancement is limited only by your willingness to try new skills and to seize opportunities as they present themselves.

32.7 Entrepreneurship

Earlier in this chapter, you read that four out of every ten carpenters are self-employed. In other words, they are *entrepreneurs.* This means starting and operating a business of one's own.

There are advantages and disadvantages to operating your own construction company. First, you may find that you gain more satisfaction in certain tasks. As your own boss, you can choose to specialize in these areas and thus offer your services to the community as a subcontractor. You are free to concentrate in the area of your specialty and develop superior skills and competence that allow you to compete with other firms.

Of course, there are disadvantages to owning the company. Failure is possible and you can spend years repaying loans and debts incurred. Your income could be uncertain, depending on how well the business goes or how healthy the economy is. You will have to work longer hours and many of the tasks—paperwork, mainte-

nance, customer relations, setting prices, and organizing your day—are difficult.

An entrepreneur must possess certain characteristics to be successful, **Figure 32-10.** This list summarizes the major characteristics:

* Healthy—long hours and physical labor make heavy demands on the carpenter or construction contractor.

* Knowledgeable—to make a profit, he or she must know all aspects of the trade or trade specialty. In addition, the person must understand the industry and the products being used in the trade.

* Good planner—running a successful business means that nothing is left to chance. She or he must be able to foresee difficulties as well as plan how to take advantage of opportunities.

* Willing to take calculated risks—once a plan has been conceived that takes into account those events likely to occur, the person must have the courage to risk money and future on making the plan work.

* Innovative—successful in finding ways to improve and find ways to produce better work and thus gain the confidence of customers.

* Responsible—willing to accept the consequences of a decision, whether good

Figure 32-10. Entrepreneurs are willing to take risks. This woman left another career to become a general contractor. Here, she inspects a commercial remodeling job with one of her subcontractors and his employee. (RECON—Reconstruction Unlimited, Inc.)

Entrepreneur: A person who starts and operates a new business.

or bad. This includes paying debts, keeping promises, and accepting the responsibility for mistakes of his or her employees.

- Goal-oriented—likes to set goals and works hard to achieve them. In recent years, there has been a large increase in the number of persons over 60 years old. This population group demands many services. Accessible, low-maintenance housing is one of the areas predicted to be in great demand. Thus, a business designed to offer services such as home construction or rehabbing of homes may do well.

32.8 Teaching as a Construction Career

It is not unusual for successful and skilled carpenters to become *vocational instructors.* For some, this is a part-time occupation. Others may leave the construction field to teach their skills to others. This may involve returning to school to take courses leading to accreditation or getting a degree in education. There are also some who, as students, take construction courses to become proficient as preparation for teaching shop courses.

Carpentry instructors not only work in high schools, vocational schools, and universities, but may be employed by tool and equipment manufacturers. In such cases, they may visit educational institutions to instruct other teachers and trainees on how to use the products manufactured by their employers. See **Figure 32-11.**

32.9 Related Occupations

Construction offers a number of other occupations with skills closely allied to carpentry. Opportunities and training are also similar. The next sections discuss several occupations that are related to carpentry.

A

B

Figure 32-11. Teaching as a profession. A—Teaching construction to beginners is a rewarding profession. B—A representative of a tool company conducts training on the proper use of a laser level. (St. Paul Technical College)

32.9.1 Bricklayers and Stonemasons

Bricklayers—or brickmasons, as they are often called—work with manufactured masonry units. They construct walls, floors, partitions, chimneys, and patios. A beginner may work as a hod carrier while learning the trade, **Figure 32-12.** In addition to bringing bricks and other materials to the bricklayer, the hod carrier is expected to mix mortar and set up scaffolding.

Vocational instructor: A teacher specializing in training students for specific occupations, such as carpentry, auto mechanics, or food service.

Figure 32-12. Beginners in the masonry trade work with a journeyman as a helper (called a hod carrier) while learning bricklaying or stonemason skills.

Figure 32-13. Electricians must be highly skilled in electrical theory. They must also keep up-to-date with the National Electrical Code.

Stonemasons construct walls out of stone, set stone for veneer walls on buildings, and lay stone floors. They work with natural stone, like marble, granite, and limestone, as well as with artificial stone made with concrete, marble chips, and other masonry materials. The mason often works with a set of drawings on which each stone is numbered. Helpers may be employed to locate and carry these stones to the mason.

Employment in these occupations is expected to be excellent. Job openings are growing faster than the numbers of persons in training.

32.9.2 Electricians

Electricians normally work from prints as they install, connect, test, and maintain electrical wiring, **Figure 32-13.** Much of their work is in residential construction. However, they also work in factories, office buildings, or exterior wiring installations.

Electric power companies employ some electricians who mostly work as linemen. Their work is always outdoors, dealing with high-voltage electricity. The most common task of a lineman is installing and maintaining electrical service to customers.

Being an electrician is often strenuous, sometimes involving work on scaffolds and ladders. They are also at risk from electrical shock. To avoid injury, they must carefully work, deliberately move about, and follow strict safety procedures.

32.9.3 Plumbers

Plumbers are among the most highly paid construction workers. Employment opportunities should be good for the foreseeable future. The work requires some heavy lifting, much standing, and working in close quarters. Clearing clogged pipes is often disagreeable.

In general, residential plumbers install and repair water, waste disposal, drainage, and gas systems. They also install fixtures—sinks, bathtubs, showers, and toilets—and appliances like dishwashers and water heaters. Some specialize in installing sewers, septic tanks, and drain fields, **Figure 32-14.**

32.9.4 HVACR Technicians

Heating, ventilating, air conditioning, and refrigeration systems involve many mechanical, electrical, and electronic components. Among these are electric motors, pumps, compressors, fans, ductwork, pipes, thermostats, and switches.

Though trained to do both installation and maintenance/repair, HVACR technicians usually specialize in one or the other. Some may narrow their specialization to one type of equipment.

Furnace installers, or heating equipment technicians, install gas, oil, electric, solid-fuel, or multiple-fuel systems. They follow prints or manufacturer's specifications. After placing the units, they install fuel and water supply pipes, air ducts, vents, pumps, and other parts of the system.

Air conditioning technicians, like furnace installers, install and service central air conditioning systems. They must be able to follow manufacturers' instruction manuals, as well as read prints and specifications. After installing all lines and other parts, they charge the system with refrigerant, program the controls, and test for proper operation.

There are many job opportunities in this field. The heating and air conditioning service industry is rapidly expanding. Demand for well-trained technicians should be better than average for all occupations for a number of years. Specialists in service will be in the best position for secure employment.

32.10 Organizations Promoting Construction Training

There are organizations that promote development of excellence in construction. One of them, *SkillsUSA*, holds state and national competitions to encourage students in a variety of occupations, including building trades, **Figure 32-15.**

In the carpentry trade, associations actively promote and support training in construction fields. Two of the most influential are the Associated General Contractors (AGC) and the National Association of Home Builders (NAHB). The major emphasis of the AGC is to represent contractors who do heavy construction work. The NAHB concentrates its efforts on residential and light commercial construction.

The United Brotherhood of Carpenters and Joiners of America is a labor union that looks after the interests of construction workers. While most members are carpenters, a wide spectrum of construction workers also belongs. Membership includes cabinetmakers, pile drivers, millwrights, floor covering workers, and a host of industrial workers employed by factories turning out plywood, lumber, and other construction products. Among their members are thousands of women from many construction occupations.

Information about related occupations is available from several organizations or unions:

- Masonry trade—International Union of Bricklayers and Allied Craftsmen, and the Portland Cement Association.

Figure 32-14. Plumbers must be familiar with and be able to install several different systems and various types of material. This plumber installs septic systems as well as indoor plumbing.

SkillsUSA: An organization that promotes the development of excellence in a variety of occupations, including building trades.

Figure 32-15. Carpentry trainees from all states compete at the SkillsUSA national contest. (SkillsUSA)

- Electrical trade—Independent Electrical Contractors, Inc.; National Electrical Contractors Association (NECA); and the International Brotherhood of Electrical Workers (IBEW).

- Plumbing trade—National Association of Plumbing-Heating-Cooling Contractors, and the Mechanical Contractors Association of America.

- HVACR—The Sheet Metal and Air Conditioning Contractors National Association; Air Conditioning Contractors of America; American Society of Heating, Refrigerating, and Air Conditioning Engineers (ASHRAE); and the National Association of Plumbing-Heating-Cooling Contractors.

ON THE JOB

Construction manager

Sometimes called "constructors," construction managers are in overall charge of planning and coordinating a construction project. The construction manager may work for the general contractor or may be the representative of a separate management firm hired by the developer or owner to oversee the entire project. There are also some self-employed construction managers.

A constructor's duties usually span the entire life of the project, from conceptual development to the building's completion. Working from the project design documents, they oversee all aspects of organization, scheduling and implementation. Very large and complex projects are beyond the managerial abilities of a single person. In such situations, an overall management firm places individual construction managers in charge of specific aspects of the job, such as site preparation, steel erection, concrete work, mechanical systems, plumbing and HVAC, and so on.

Organizational ability, math and computer skills, and a through knowledge of construction processes are vital to success as a construction manager. The manager usually is responsible for creating and adhering to schedules, estimating and projecting costs, and scheduling the work of the various trades-

people. Communication skills are also important. The construction manager must be able to effectively communicate with everyone involved in the project, from the building owner to subcontractors and their workers. They must also communicate with materials suppliers and government representatives, such as building inspectors.

The traditional route to construction management positions led through the building trades, since the work requires solid experience in various aspects of construction. Increasingly, management firms are seeking candidates who combine such experience with a four-year college degree in construction management or civil engineering. Computer experience with software in such areas as CAD, cost estimating, and project management is highly desirable. Certification in this field is becoming more common. Several organizations offer this professional recognition based on education, experience, and a written exam.

Advancement opportunities are best with large management firms, since a manager will be able to assume responsibility for larger and more complex projects with experience. Upper-level managers in such firms may have broad responsibility for a very large project, supervising the work of a number of construction managers.

Summary

There are many employment opportunities for carpenters in residential, commercial, and heavy construction. With continued population growth and demand for new housing, the number of carpentry jobs should continue to grow. Some states expect double-digit growth in the number of construction jobs. However, the construction industry is also subject to seasonal and cyclic downturns. In addition to new construction, carpenters are employed in remodeling and renovating work, manufactured and systems-built home construction in factories, and related construction fields. Four of every ten carpenters are self-employed. Entry into the carpentry field is often through an apprenticeship program, which typically takes four years. When the apprenticeship period is completed, the carpenter has journeyman status. Personal qualifications for carpentry work include manual dexterity, the ability to physically do the work, sound work habits, and good interpersonal skills. Many carpenters are entrepreneurs, operating their own businesses. An additional career path is teaching in a vocational program. Carpenters also shift specialties and move into other building trades areas such as electrical or plumbing work. Organizations such as SkillsUSA, contractors' associations, and labor unions actively promote the carpentry trade.

Test Your Knowledge

Answer the following questions on a separate piece of paper. Do not write in this book.

1. Opportunities for jobs in construction will be greatest in coming years in the _____ and _____ regions of the country.
2. In 2003, more than _____ persons were employed as carpenters in the U. S.
   A. 50,000
   B. 650,000
   C. 850,000
   D. 1,150,000
3. Approximately _____ of all carpenters work in residential construction.
   A. one-fourth
   B. one-third
   C. one-half
   D. three-fourths
4. Carpentry work must continue in the _____ of summer and _____ of winter.
5. Four out of every _____ carpenters work for themselves.
6. What high school courses would be desirable and helpful for persons considering a career in carpentry? List nine.
7. An apprenticeship in carpentry normally covers _____ years.
8. In modern terms, what is a *journeyman*?
9. An *entrepreneur* is someone who _____.
10. What is *SkillsUSA*?

Curricular Connections

Mathematics. Find information on the average hourly pay rate in your area for four construction-related occupations: general laborer, carpenter, brickmason, and plumber. Calculate what a person in each of these occupations would earn (before taxes) in one week, one year, and over an entire working life. Assume that most people retire after a working life of about 45 years. Also assume a rate of inflation of about 3%. Develop a table to show your results. Can you draw any conclusions from the results?

Outside Assignments

1. Find out how to enter a carpentry apprenticeship program in your locality. Contact a carpentry contractor, union local, and state apprenticeship agency. What skills do you have now? What skills must you learn? Report your findings to the class.
2. Interview carpenters to learn their attitudes about what they do for a living. Prepare a list of possible questions to ask.
3. Secure literature from the main associations in construction. Report what they say about opportunities in construction.

Appendix A— Carpentry Math Review

A carpenter or carpenter apprentice must be able to take measurements and make calculations. These skills require an understanding of certain mathematical concepts. This appendix reviews these basic concepts and gives examples of their application to carpentry. Most applications require the ability to make accurate measurements of distance and volume. These basic math skills are needed for two reasons:

- To make the measurements necessary to cut, fit, and locate members in a structure.
- To calculate the quantities of building materials required for a project and the cost.

Using Rules, Tapes, and Squares

One of the first skills a carpenter must master is the accurate reading of rules, tapes, and squares. **Figure A-1** is a portion of a rule showing the

Figure A-1. A carpenter must be able to accurately read a ruler when taking measurements.

various divisions. Note that inch lines go nearly all the way across the rule, while each smaller unit has progressively shorter marks. Each of these marks represents a fraction of an inch. The carpenter must be able to recognize a particular fractional inch by the length of its division mark. In the illustration, the shortest mark represents 1/16″. The next longer marks represent 1/8″ spaces. Still longer marks are used to represent 1/4″ and 1/2″, respectively.

Working with Fractions and Decimals

A *whole* number is an undivided unit—a number without a fractional or decimal portion. It can refer to measurements or quantities of material. Measurements deal with values such as length, area, or volume. There are several units of conventional measure. Inches, feet, and yards are units of length; square inches, square feet, and square yards are units of area; cubic inches, cubic feet, cubic yards, board feet, and gallons are units of volume.

A fraction or a decimal is part of a whole number. For example, 3/4 is three parts of a whole that has been separated into four equal parts; 1/10 or .1 is one part of a whole divided into 10 equal parts.

Fractions

Fractions are always written with one number above the other with a line separating them. The top number of a fraction is called the *numerator;* the bottom number is called the *denominator;* the line separating them is called a *fraction bar.* The denominator tells how many parts the whole is divided into; the numerator tells how many parts are in the fraction concerned. See **Figure A-2.**

An *improper fraction* is a fraction in which the numerator is a larger number than the denominator, such as 11/8. An improper fraction is always equal to a value greater than one. Another way of expressing a value greater than one is as a *mixed number*—a number with a whole number part and a fractional part, such as 1 3/8. To convert a mixed number to an improper fraction:

• Multiply the denominator by the whole number and then add the numerator to find the numerator of the improper fraction.

• The denominator of the improper fraction is the same denominator as the fractional part of the mixed number.

Example:

Convert 2 5/8 to an improper fraction.

Numerator = (8 × 2) + 5 = 16 + 5 = 21
Denominator   = 8
2 5/8   = 21/8

It is easier to visualize mixed numbers than improper fractions. For example, if someone told you that they drove exactly 15 3/16 miles to work, you can easily understand the distance involved. But if the person said they drove 243/16 miles, you would not have an immediate idea of how far they traveled.

However, improper fractions are far easier to work with in calculations. Therefore, when you have a problem and a mixed number is given, convert it to an improper fraction. Then, do the calculation and convert any improper fractions in your final answer to mixed numbers so it is easier to understand.

A whole number can also be expressed as a fraction. The number becomes the numerator and 1 is the denominator. For example, the whole number 7 is the same as the fraction 7/1.

Multiplying Fractions

To multiply fractions, change mixed numbers and whole numbers to improper fractions. Multiply the numerators and then the denominators. The result is called a *product.* Reduce the product to its lowest form.

Example:

Multiply 3/8 × 1/4

Numerator: 3 × 1   = 3
Denominator: 8 × 4 = 32
3/8 × 1/4   = 3/32

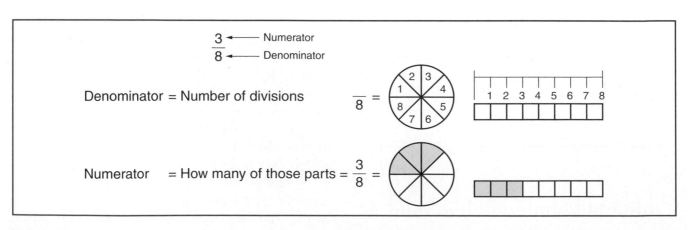

Figure A-2. The denominator tells how many pieces a whole is divided into. The numerator tells how many of those pieces are counted.

Example:

Multiply 3 5/8 × 1/2

Convert the mixed number to an improper fraction:

3 5/8	= 29/8
Numerator: 29 × 1	= 29
Denominator: 8 × 2	= 16
3 5/8 × 1/2	= 29/16
	= 1 13/16

Dividing Fractions

There are three components to a division equation: the *divisor*, the *dividend*, and the *quotient*. These components are shown in **Figure A-3,** along with different ways of writing division equations. As you can see, a fraction is actually a division operation—the numerator is the dividend and the denominator is the divisor.

The process of switching the numerator and denominator of a fraction is called *inversion*. In order to divide fractions, simply invert the divisor (the number you are dividing by) and then multiply the two fractions.

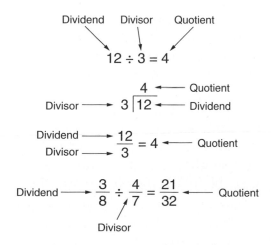

Figure A-3. There are several ways to write a division problem. The dividend divided by the divisor equals the quotient.

Example:

Divide 2/3 by 7/4

Invert the divisor: 7/4 → 4/7
Multiply the dividend by the inverted divisor:

2/3 × 4/7 = 8/21

Changing Denominators

Any fraction can be written in many ways. For example, 1/2 is the same as 3/6 or 4/8 or 6/12. There are situations where it is preferable to change the denominator of the fraction. To convert between equivalent (equal) fractions, simply multiply both the numerator and denominator by the same number.

Example:

Find the equivalent fraction for 1/3 that has 12 as its denominator. Because you multiply 3 by 4 to get 12, multiply both the numerator (1) and denominator (3) by 4 to find the equivalent fraction:

Numerator: 1 × 4	= 4
Denominator: 3 × 4	= 12
1/3	= 4/12

Adding Fractions

Fractions must have identical denominators in order to add them together. They must have a *common denominator*. When fractions have a common denominator, they are added in the following way: the new numerator is obtained by adding the numerators together, the new denominator is the common denominator. When fractions being added have different denominators, one or both of the fractions need to be converted to an equivalent fraction with a common denominator.

Example:

Add 3/8 + 1/8
> Denominators are the same, so add numerators to get new numerator:

> 1+3   = 4

> Keep the same denominator (8)

> 3/8 + 1/8 = 4/8
>   = 1/2

Example:

Add 3/4 + 3/16
> Denominators are different, so one must be changed. Convert 3/4 to an equivalent fraction with 16 as a denominator:

> 3/4 × 4/4   = 12/16
> Add: 12/16 + 3/16 = 15/16

Subtracting Fractions

Subtracting and adding fractions are very similar operations. As with addition, fractions must have common denominators to subtract. The only difference is that you subtract the numerators, rather than adding them.

Example:

Subtract: 3/4 – 1/4

> Denominators are the same, so subtract numerators:

> 3/4 – 1/4 = 2/4
>   = 1/2

Example:

Subtract 4/5 – 3/15
> Multiply 4/5 by 3/3 to obtain a common denominator (15):

> 4/5 × 3/3 = 12/15

> Subtract numerators:

> 12/15 – 3/15 = 9/15
>   = 3/5

Decimals

A *decimal number* has three parts—a whole number part, a decimal point, and a decimal part, **Figure A-4**. The decimal part is like the numerator in a fraction. The denominator in this fraction is 1 followed by the same number of zeros as there are digits after the decimal point. A decimal number is another way of expressing a mixed number.

Example:

Convert 17.31 to a mixed number.

> There are two digits to the right of the decimal point, so there will be two zeros following the 1 in the denominator, with 31 in the numerator.

> 17.31 = 17 31/100

Example:

Convert 2.004 to a mixed number.
> There are three digits to the right of the decimal point, so there will be three zeros following the 1 in the denominator.

> 2.004 = 2 4/1000 = 2 1/250

Figure A-5 is a place value chart. It shows the value of numbers to the left and right of a decimal point. Notice that as numbers are added to the left, their values increase; as numbers are added further to the right, their values decrease.

While zeros have no value of themselves, they are very important to the value of the complete number. The zero is a "place holder." It is used to keep the other numbers in their place. For example, the decimal number .56 is larger than .056, the first being 56/100 and the second

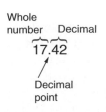

Figure A-4. A non-whole number can be expressed as a decimal.

Ten-thousands place	Thousands place	Hundreds place	Tens place	Ones place	Tenths place	Hundredths place	Thousandths place	Ten-thousandths place
1	2	7	2	6	6	5	7	9
10,000	1,000	100	10	1	1/10	1/100	1/1,000	1/10,000

Figure A-5. Number place chart. Values of each place increases in multiples of ten as you move to the left on the chart.

56/1000—the zero makes the second number smaller than the first number.

Adding and Subtracting Decimal Numbers

To add and subtract decimals, write the numbers in a column with their decimal points aligned. Then, add or subtract as if they were whole numbers. Carry the decimal point down into the answer.

Example:

Add 2.6 + 10.5 + 8.803 + 5 + .0155

Arrange the numbers in a column, aligning decimal points

```
    2.6000
   10.5000
    8.8030
    5.0000
     .0155
   26.9185
```

The number of zeros at the far right end of the decimal portion does not matter, as long as they are at the very end of the number. For example, 45.87 and 45.870000 represent the same number, but 23.002 and 23.0000002 are *different* numbers because there is a non-zero digit to the right of the zeros.

Multiplying Decimal Numbers

Multiplying decimal numbers is a two-step process. First, multiply the numbers, ignoring the decimal points. Then, count the total number of decimal places in the numbers being multiplied and put the same number of decimal places in the product. If there are more decimal places than digits in the product, add zeros to the left to get enough places.

Example:

Multiply 3.511×2.2

Ignoring decimal points, multiply the numbers:

$3511 \times 22 = 77242$

There are four decimal places in the numbers being multiplied, so there must be four decimal places in the product: 7.7242

$3.511 \times 2.2 = 7.7242$

Whenever doing calculations of any kind, be sure to check to see if your answer is reasonable. In the last example, you could round the two factors to the nearest whole numbers (4 and 2) and multiply these together (8) to see if the answer makes sense, which it does.

Dividing Decimal Numbers

Using decimals allows you to divide numbers that will not evenly divide. The division remainder is a fractional part. It can be written as a fraction or the division can be continued for as many decimal places as desired. To divide using decimal places:

- If neither divisor nor dividend contain decimal points and the division does not come out exactly (See 17÷8 in **Figure A-6**):
  A. Place a decimal point to the right of the dividend and after the last number in the answer.

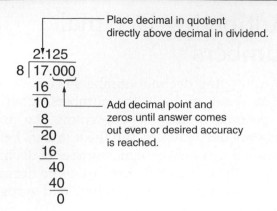

Place decimal in quotient directly above decimal in dividend.

```
    2.125
 8 |17.000
   16
   ___
   10
    8
   ___
   20
   16
   ___
   40
   40
   ___
    0
```

Add decimal point and zeros until answer comes out even or desired accuracy is reached.

Figure A-6. Division problem in which both divisor and dividend are whole numbers.

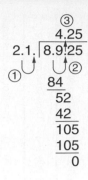

```
        ③
      4.25
 2.1.|8.9.25
  ①    ②
      84
      ___
      52
      42
      ___
      105
      105
      ___
        0
```

① Move decimal point in divisor to far right of number.

② Move decimal point in dividend the same number of places to the right (one, in this case).

③ Decimal point in quotient is located directly above decimal point in dividend.

Figure A-8. Division problem in which the divisor is a decimal.

B. Add zeros to the dividend.

C. Continue dividing to as many decimal places as desired.

- If the dividend contains a decimal point and the divisor does not (See 17.25÷8 in **Figure A-7**):

   A. Divide as usual up to the decimal point.

   B. Place a decimal point to the right of the last number in the answer.

   C. Continue dividing the rest of the dividend. Add zeros to the dividend until the answer is carried out to the number of decimal places desired.

- If the divisor contains a decimal point (See 8.925÷2.1 in **Figure A-8**):

   A. Move the decimal points to the right end of the divisor.

   B. Move the decimal point in the dividend to the right the same number of places.

   C. Divide, being sure to put the decimal point in the quotient directly above the new location of the decimal point in the dividend.

Converting Between Decimals and Fractions

The system for units of measure in the United States uses either fractions or decimals to indicate parts of a whole. While rulers are marked in inches and fractions, micrometers and vernier calipers are marked in decimals. Sometimes it is necessary to convert from one to the other.

Converting Decimals to Fractions

1. Let the number or numbers to the right of the decimal point be the numerator.

2. Count off the places to the right of the decimal (one place = tenths; two places =

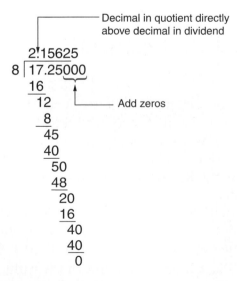

Decimal in quotient directly above decimal in dividend

```
    2.15625
 8 |17.25000
   16
   ___
   12
    8
   ___
   45
   40
   ___
   50
   48
   ___
   20
   16
   ___
   40
   40
   ___
    0
```

Add zeros

Figure A-7. Division problem in which the divisor is a whole number and the dividend is a decimal.

hundredths; three places = thousandths, and so on).

3. Let this number be the denominator.
4. Reduce the resulting fraction to its lowest terms.

Examples:

.5   = 5/10   = 1/2
.50 = 50/100   = 5/10 = 1/2
.175 = 175/1000 = 7/40

Converting Fractions to Decimals

1. Divide the numerator by the denominator.
2. If it does not exactly divide, round off to as many decimal places desired.

See **Figure A-9** for an example.

Knowing and Using Formulas

Often, a carpenter must use formulas. Given a certain problem to solve, the carpenter must know which formula to use and how to work the formula.

Area

The area of square or rectangular shapes can be determined by multiplying the length of two adjoining sides. A carpenter needs to find the area of a rectangle in order to determine the amount of material needed to cover walls, floors, ceilings, and roofs. The product is given in "square" units, such as square feet or square yards.

Often, the two sides of a rectangle are referred to as length and width (L and W, respectively). Then, the formula for area becomes: $L \times W = Area$.

Example:

What is the area of a wall 48′ long and 8′ high?

$48′ \times 8′ = 384$ sq. ft.

When calculating for area, or using any other formula, attention must be paid to the units of measurement involved. When multiplying two measurements with units of feet together, the answer is in square feet. When the measurements are taken in inches, the answer has units of square inches. Measurements with different units cannot be combined—they must be converted to a common unit.

Example:

What is the area of a sidewalk that is 30″ wide and 15′ long?

The measurements are in different units, so they cannot be combined. Change 30″ to 2.5′ and multiply:

$2.5′ \times 15′ = 37.5$ sq. ft.

Example:

How many sheets of 4′ × 8′ plywood are needed to cover a wall 8′ high by 48′ long?

Area of one sheet of plywood: $4′ \times 8′ = 32$ sq. ft.
Area of wall: $8′ \times 48′$ = 384 sq. ft.
Number of sheets: $384 \div 32$ = 12 sheets required

```
                3 ÷ 4

              .75
          4 ⟌ 3.00
              28
              ──
              20
              20
              ──
               0
```

Figure A-9. Converting a fraction (3/4) to a decimal (0.75).

Volume

Volume is a measure of three dimensions: length, width, and height. The product (in dry measure) is given in cubic inches, cubic feet, or cubic yards. To find the volume of a cube, box, or rectangular cylinder, multiply the three dimensions together.

Example:

An excavation is to be dug for the basement of a building. The excavation is to be 78' long by 54' wide by 8' deep. How much dirt will be moved?

Earthwork and concrete quantities are usually measured by cubic yards (1 cubic yard = 27 cubic feet).

Volume = 78' × 54' × 8' = 33,696 cu. ft.

Volume = 33,696 ft. = 1248 cu. yd.
(33,696 divided by 27)

Appendix B— Technical Information

Contents

Standard Abbreviations for Use on Drawings

Above Finished Flooring	AFF	Dormer	DRM	Lattice	LTC
Acoustical Tile	AT or ACT	Double Strength Glass	DSG	Lavatory	LAV
Aggregate	AGGR	Double-Hung Windows	DHW	Length	LGTH
Air Conditioning	AIR COND	Drain	DR	Level	LVL
Air Dried	AD	Drawing	DWG	Light	LT
Alternate	ALT	Dressed and Matched	D&M	Light Switch	LTSW
Alternating Current	AC	Each	EA	Linen Closet	L CL
Aluminum	AL	Edge	EDG	Linoleum	LINO
American Institute of Architects	AIA	Edge Grain	EG	Lintel	LNTL
American Institute of Electrical		Electric Panel	EP	Living Room	LR
Engineers	AIEE	Elevation	EL	Low Voltage	LV
American Society for Testing		Entrance	ENT	Masonry Opening	MO
and Materials	ASTM	Excavate	EXC	Mastic	MSTC
American Standards		Exhaust Vent	EXHV	Material	MATL
Association, Inc.	ASA	Exterior	EXT	Maximum	MAX
Approximate	APPROX	Face Brick	FB	Medicine Cabinet	MC
Architectural	ARCH	Federal Housing Authority	FHA	Minimum	MIN
Asbestos	ASB	Finish	FIN	Miscellaneous	MISC
Asphalt Roof Shingles	ASPHRS	Finished Floor	FNSHFL or FF	Modular	MOD
Basement	BSMT	Fixture	FIX	Molding	MLDG
Batter	BAT	Flashing	FL	Mortar	MOR
Beam	BM	Flat Grain	FG	Nominal	NOM
Better	BTR	Flooring	FLG	Nosing	NOS
Beveled	BEV	Fluorescent	FLUOR	On Center	OC
Blocking Board	BLKG	Flush	FL	Open Web Joint	OWJ
Board	BD	Foot or Feet	FT	Opening	OPNG
Foot	BD FT	Footing	FTG	Paint	PNT
Brick	BRK	Foundation	FDN	Pair	PR
British Thermal Unit	BTU	Furring	FUR	Partition	PTN
Building	BLDG	Fuse	FU	Perpendicular	PERP
Bundle	BDL	Gage	GA	Pilaster	P
Cabinet	CAB	Gallon	GAL	Piping	PP
Carpenter	CPNTR	Galvanize	GALV	Plank	PLK
Casing	CSG	Galvanized Iron	GI	Plaster	PLAS
Cement	CEM	Glass	GL	Plate	PL
Cement Floor	CEM FL	Grade	GR	Plate Glass	PL GL
Cement Mortar	CEM MORT	Gypsum	GYP	Plumbing	PLBG
Center Matched	CM	Hardboard	HBD	Power	PWR
Chimney	CHM	Hardwood	HDWD	Precast	PRCST
Circuit	CKT	Header	HDR	Prefabricated	PREFAB
Closet	CL or CLO	Heat Exchanger	HE	Quart	QT
Column	COL	Herringbone	HGBN	Random	RDM
Common	COM	Horsepower	HP	Receptacle	RCPT
Concrete	CONC	Hose Bibb	HB	Recess	REC
Concrete Block	CONC B	Hot Water	HW	Reference	REF
Conduit	CND	Hundred	C	Refrigerator	REF
Construction	CONST	Insulation	INS	Reinforcing	REINF
Counter	CTR	Interior	INT	Revision	REV
Coupling	CPLG	Iron Pipe	IP	Roll Roofing	RR
Cubic Foot	CU FT	Jamb	JMB	Roof	RF
Cubic Yard	CU YD	Joist	J	Rough	RGH
Cutoff Valve	COV	Keyway	KWY	Rough Opening	RO
Diagram	DIAG	Kiln-dried	KD	Saddle	SDL
Diameter	DIA or DIAM	Kitchen	K	Schedule	SCH
Dimension	DIM	Knee Brace	KB	Screen	SCR
Direct Current	DC	Knife Switch	KNSW	Select	SEL
Ditto	DO	Laminate	LAM	Service	SERV
Door	DR	Lath	LTH	Sewer	SEW

(Continued)

Sheathing	SHTHG	Surfaced One Side and		Valley	VAL
Shelving	SHELV	Two Edges	S1S2E	Vent Stack	VS
Shiplap	S/LAP	Surfaced Two Sides	S2S	Ventilation	VENT
Siding	SDG	Switch	SW or S	Volt	V
Specifications	SPEC	Temperature	TEMP	Voltmeter	VM
Square	SQ	Thermostat	THERMO	Water Closet	WC
Square Feet	SQ FT	Thick	THK	Water Heater	WH
Stairway	STWY	Thousand	M	Weather Stripping	WS
Standard	STD	Timber	TMBR	Weep Hole	WH
Steel	ST or STL	Tongue and Groove	T&G	Weight	WT
Stringer	STGR	Transformer	XFMR	Welded Wire Fabric	WWF
Structural	STR	Truss	TR	Wide Flange	WF or W
Surfaced Four Sides	S4S	Tubing	TBG	With	W/
Surfaced One Side	S1S	Typical	TYP	Without	W/O
		Union	UN	Wood	WD

Slump Testing

While a slump test is normally done in heavy construction work, it is sometimes used in residential construction. A *slump test* measures the consistency, stiffness, and workability of fresh concrete. These characteristics are affected mainly by the amount of water in the mix. Other factors — type of aggregate, air content, admixtures, temperature, and the proportions of all ingredients — also affect slump. Mixing time and standing time also have an effect.

The test is conducted with a slump cone made of metal. It is 12" high with a diameter of 4" at the top and 8" at the bottom. To make the test:

- With the small diameter up, fill the cone in three layers of equal volume, rodding each layer 25 times.

- Strike off the top; then slowly remove the cone with an even motion lasting from 5 to 12 seconds. Do not disturb the mixture or tilt the cone.

- Turn the cone upside-down alongside the mixture and immediately measure the slump with the tamping rod and a rule. The test should take no longer than 1 1/2 minutes.

Slump Range for Different Types of Construction		
Type of construction	**Maximum slump**	**Minimum slump**
Reinforced footings and foundation walls	3"	1"
Plain footings, caissons, and substructure walls	3"	1"
Beams and reinforced walls	4"	1"
Building columns	4"	1"
Pavement and slabs	3"	1"
Mass concrete	2"	1"

Plywood Siding Joints

Vertical wall joints

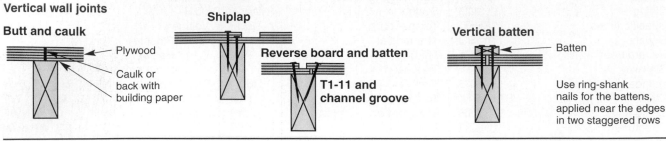

Butt and caulk
- Plywood
- Caulk or back with building paper

Shiplap

Reverse board and batten

T1-11 and channel groove

Vertical batten
- Batten
- Use ring-shank nails for the battens, applied near the edges in two staggered rows

Vertical inside and outside corner joints

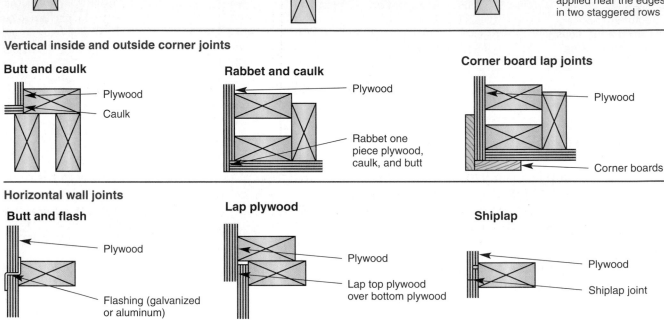

Butt and caulk
- Plywood
- Caulk

Rabbet and caulk
- Plywood
- Rabbet one piece plywood, caulk, and butt

Corner board lap joints
- Plywood
- Corner boards

Horizontal wall joints

Butt and flash
- Plywood
- Flashing (galvanized or aluminum)

Lap plywood
- Plywood
- Lap top plywood over bottom plywood

Shiplap
- Plywood
- Shiplap joint

Horizontal beltline joints

(For multi-story buildings, make provisions at horizontal joints for "settling" shrinkage of framing, especially when applying siding direct to studs.)

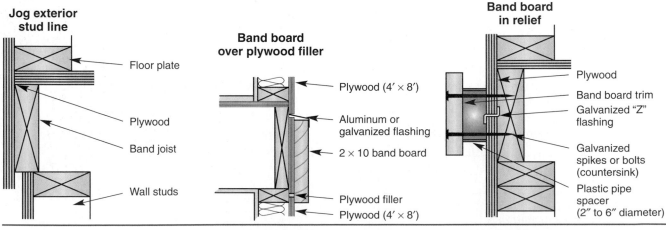

Jog exterior stud line
- Floor plate
- Plywood
- Band joist
- Wall studs

Band board over plywood filler
- Plywood (4′ × 8′)
- Aluminum or galvanized flashing
- 2 × 10 band board
- Plywood filler
- Plywood (4′ × 8′)

Band board in relief
- Plywood
- Band board trim
- Galvanized "Z" flashing
- Galvanized spikes or bolts (countersink)
- Plastic pipe spacer (2″ to 6″ diameter)

Window details

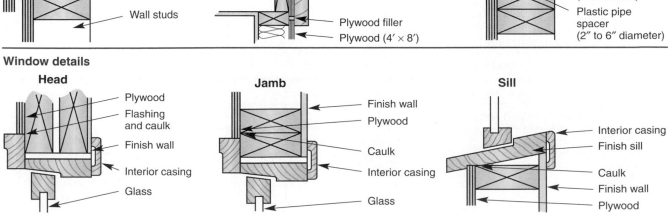

Head
- Plywood
- Flashing and caulk
- Finish wall
- Interior casing
- Glass

Jamb
- Finish wall
- Plywood
- Caulk
- Interior casing
- Glass

Sill
- Interior casing
- Finish sill
- Caulk
- Finish wall
- Plywood

Fire-Resistant Construction

All assemblies shown provide a one-hour fire rating. (APA–The Engineered Wood Association)

One-hour assembly — resilient channel ceiling system

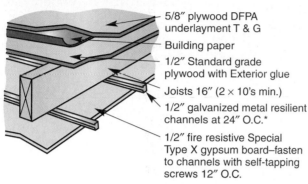

- 5/8" plywood DFPA underlayment T & G
- Building paper
- 1/2" Standard grade plywood with Exterior glue
- Joists 16" (2 × 10's min.)
- 1/2" galvanized metal resilient channels at 24" O.C.*
- 1/2" fire resistive Special Type X gypsum board–fasten to channels with self-tapping screws 12" O.C.

*Channels may be suspended below joists.

One-hour assembly — T-bar grid ceiling system

- 5/8" plywood DFPA underlayment T & G
- Building paper
- 1/2" Standard grade plywood with exterior glue
- Joists 16" O.C. (2 × 10's min.)
- T-bar grid ceiling system
- Main runners 48" O.C.
- Cross-tees 24" O.C.
- 1/2" × 48" × 24" mineral acoustical ceiling panels (install with hold-down clips)

One-hour interior shear wall construction

- 1/2" fire resistive special Type X gypsum board*
- 2 × 4 studs @ 16" O.C.
- 3/8" plywood shear panels

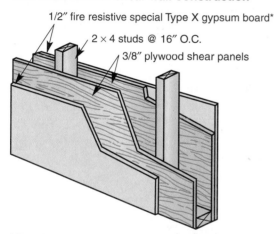

* Regular 1/2" gypsum board may be used when mineral wool or glass fiber batts are used in wall cavity.

Insulation batts in wall cavity also used for sound transmission control.

One-hour exterior wall construction

- 2 × 4 studs @ 16" O.C.
- 3/8" plywood panel* or lap siding
- 1/2" gypsum sheathing
- 5/8" fire resistive Type X gypsum board

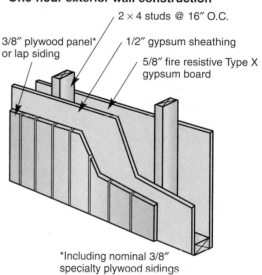

*Including nominal 3/8" specialty plywood sidings

Treated stressed-skin panel construction

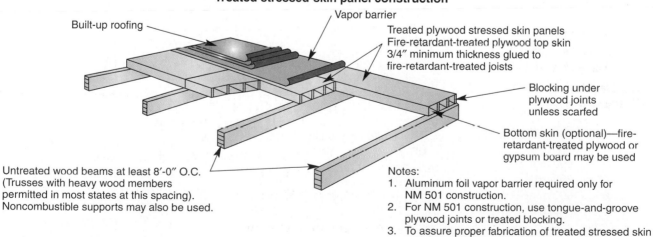

- Built-up roofing
- Vapor barrier
- Treated plywood stressed skin panels Fire-retardant-treated plywood top skin 3/4" minimum thickness glued to fire-retardant-treated joists
- Blocking under plywood joints unless scarfed
- Bottom skin (optional)—fire-retardant-treated plywood or gypsum board may be used

Untreated wood beams at least 8'-0" O.C. (Trusses with heavy wood members permitted in most states at this spacing). Noncombustible supports may also be used.

Notes:
1. Aluminum foil vapor barrier required only for NM 501 construction.
2. For NM 501 construction, use tongue-and-groove plywood joints or treated blocking.
3. To assure proper fabrication of treated stressed skin panels, components bearing the trademark of the Plywood Fabricator Service, Inc. are recommended.

Metrics in Construction

Efficient building construction begins with standard sizes for construction parts. Designs geared for mass production are based on standard modules. Layouts (horizontal and vertical) use one specific size. Whole multiples of the size make up all larger measurements.

The US customary system uses a basic module of 4". The layout grid is further divided into spaces of 16", 24", and 48". Modular design saves material and time.

In countries using the metric system, the 4" module is replaced by a 100 mm module. Smaller sizes (submultiples) include 25 mm, 50 mm, and 75 mm. Large modules measure 400 mm, 600 mm, and 1200 mm, as shown in the drawing below.

The 100 mm module is slightly smaller than the 4" module and the conversion can hardly be detected in small measurements. In a length of 48", however, the difference is about 3/4" and a standard 4' × 8' plywood panel is about 1 1/2" longer than the similar metric size (1200 mm × 2400 mm).

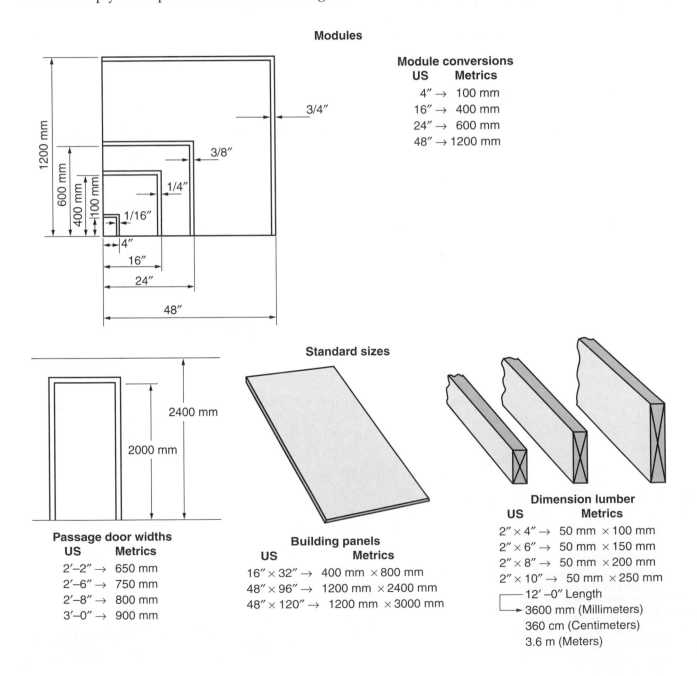

Modules

Module conversions

US		Metrics
4"	→	100 mm
16"	→	400 mm
24"	→	600 mm
48"	→	1200 mm

Standard sizes

Passage door widths

US		Metrics
2'–2"	→	650 mm
2'–6"	→	750 mm
2'–8"	→	800 mm
3'–0"	→	900 mm

Building panels

US		Metrics
16" × 32"	→	400 mm × 800 mm
48" × 96"	→	1200 mm × 2400 mm
48" × 120"	→	1200 mm × 3000 mm

Dimension lumber

US		Metrics
2" × 4"	→	50 mm × 100 mm
2" × 6"	→	50 mm × 150 mm
2" × 8"	→	50 mm × 200 mm
2" × 10"	→	50 mm × 250 mm

12' –0" Length
→ 3600 mm (Millimeters)
360 cm (Centimeters)
3.6 m (Meters)

Millimeter-Inch Equivalents

Inches Fractions	Decimals	Milli-meters	Inches Fractions	Decimals	Milli-meters
	.00394	.1	15/32	.46875	11.9063
	.00787	.2		.47244	12.00
	.01181	.3	31/64	.484375	12.3031
1/64	.015625	.3969	1/2	.5000	12.70
	.01575	.4		.51181	13.00
	.01969	.5	33/64	.515625	13.0969
	.02362	.6	17/32	.53125	13.4938
	.02756	.7	35/64	.546875	13.8907
1/32	.03125	.7938		.55118	14.00
	.0315	.8	9/16	.5625	14.2875
	.03543	.9	37/64	.578125	14.6844
	.03937	1.00		.59055	15.00
3/64	.046875	1.1906	19/32	.59375	15.0813
1/16	.0625	1.5875	39/64	.609375	15.4782
5/64	.078125	1.9844	5/8	.625	15.875
	.07874	2.00		.62992	16.00
3/32	.09375	2.3813	41/64	.640625	16.2719
7/64	.109375	2.7781	21/32	.65625	16.6688
	.11811	3.00		.66929	17.00
1/8	.125	3.175	43/64	.671875	17.0657
9/64	.140625	3.5719	11/16	.6875	17.4625
5/32	.15625	3.9688	45/64	.703125	17.8594
	.15748	4.00		.70866	18.00
11/64	.171875	4.3656	23/32	.71875	18.2563
3/16	.1875	4.7625	47/64	.734375	18.6532
	.19685	5.00		.74803	19.00
13/64	.203125	5.1594	3/4	.7500	19.05
7/32	.21875	5.5563	49/64	.765625	19.4469
15/64	.234375	5.9531	25/32	.78125	19.8438
	.23622	6.00		.7874	20.00
1/4	.2500	6.35	51/64	.796875	20.2407
17/64	.265625	6.7469	13/16	.8125	20.6375
	.27559	7.00		.82677	21.00
9/32	.28125	7.1438	53/64	.828125	21.0344
19/64	.296875	7.5406	27/32	.84375	21.4313
5/16	.3125	7.9375	55/64	.859375	21.8282
	.31496	8.00		.86614	22.00
21/64	.328125	8.3344	7/8	.875	22.225
11/32	.34375	8.7313	57/64	.890625	22.6219
	.35433	9.00		.90551	23.00
23/64	.359375	9.1281	29/32	.90625	23.0188
3/8	.375	9.525	59/64	.921875	23.4157
25/64	.390625	9.9219	15/16	.9375	23.8125
	.3937	10.00		.94488	24.00
13/32	.40625	10.3188	61/64	.953125	24.2094
27/64	.421875	10.7156	31/32	.96875	24.6063
	.43307	11.00		.98425	25.00
7/16	.4375	11.1125	63/64	.984375	25.0032
29/64	.453125	11.5094	1	1.0000	25.4001

The Metric System

Conversion Table: US Customary to SI Metric

When You Know	Multiply By:		To Find
⬇	**Very Accurate**	**Approximate**	**⬇**
Length			
inches	* 25.4		millimeters
inches	* 2.54		centimeters
feet	* 0.3048		meters
feet	* 30.48		centimeters
yards	* 0.9144	0.9	meters
miles	* 1.609344	1.6	kilometers
Weight			
grains	15.43236	15.4	grams
ounces	* 28.349523125	28.0	grams
ounces	* 0.028349523125	.028	kilograms
pounds	* 0.45359237	0.45	kilograms
short tons	* 0.90718474	0.9	tonnes
Volume			
teaspoons		5.0	milliliters
tablespoons		15.0	milliliters
fluid ounces	29.57353	30.0	milliliters
cups		0.24	liters
pints	* 0.473176473	0.47	liters
quarts	* 0.946352946	0.95	liters
gallons	* 3.785411784	3.8	liters
cubic inches	* 0.016387064	0.02	liters
cubic feet	* 0.028316846592	0.03	cubic meters
cubic yards	* 0.764554857984	0.76	cubic meters
Area			
square inches	* 6.4516	6.5	square centimeters
square feet	* 0.09290304	0.09	square meters
square yards	* 0.83612736	0.8	square meters
square miles		2.6	square kilometers
acres	* 0.40468564224	0.4	hectares
Temperature			
Fahrenheit	* 5/9 (after subtracting 32)		Celsius

* = Exact

Conversion Table: SI Metric to US Customary

When You Know ↓	Multiply By:		To Find ↓
	Very Accurate	**Approximate**	
Length			
millimeters	0.0393701	0.04	inches
centimeters	0.3937008	0.4	inches
meters	3.280840	3.3	feet
meters	1.093613	1.1	yards
kilometers	0.621371	0.6	miles
Weight			
grains	0.00228571	0.0023	ounces
grams	0.03527396	0.035	ounces
kilograms	2.204623	2.2	pounds
tonnes	1.1023113	1.1	short tons
Volume			
milliliters		0.2	teaspoons
milliliters	0.06667	0.067	tablespoons
milliliters	0.03381402	0.03	fluid ounces
liters	61.02374	61.024	cubic inches
liters	2.113376	2.1	pints
liters	1.056688	1.06	quarts
liters	0.26417205	0.26	gallons
liters	0.03531467	0.035	cubic feet
cubic meters	61023.74	61023.7	cubic inches
cubic meters	35.31467	35.0	cubic feet
cubic meters	1.3079506	1.3	cubic yards
cubic meters	264.17205	264.0	gallons
Area			
square centimeters	0.1550003	0.16	square inches
square centimeters	0.00107639	0.001	square feet
square meters	10.76391	10.8	square feet
square meters	1.195990	1.2	square yards
square kilometers		0.4	square miles
hectares	2.471054	2.5	acres
Temperature			
Celsius	* 9/5 (then add 32)		Fahrenheit

* = Exact

Standard Sizes, Counts, and Weights

Nominal size is the definition used by the trade, but it is not always the actual size. Sometimes the actual thickness of hardwood flooring is 1/32" less than the so-called nominal size.

Actual size is the mill size for thicknesses and face width. *Counted size* determines the board feet in a shipment. Pieces less than 1" in thickness are considered to be 1".

Nominal	Actual	Counted	Weights (1000 feet)
Tongue and groove—end-matched			
**3/4×3 1/4"	3/4×3 1/4"	1×4"	2210 lbs.
3/4×2 1/4"	3/4×2 1/4"	1×3"	2020 lbs.
3/4×2"	3/4×2"	1×2 3/4"	1920 lbs.
3/4×1 1/2"	3/4×1 1/2"	1×2 1/4"	1820 lbs.
**3/8×2"	11/32×2"	1×2 1/2"	1000 lbs.
**3/8×1 1/2"	11/32×1 1/2"	1×2"	1000 lbs.
**1/2×2"	15/32×2"	1×2 1/2"	1350 lbs.
**1/2×1 1/2"	15/32×1 1/2"	1×2"	1300 lbs.
Square-edge			
**5/16×2"	5/16×2"	face count	1200 lbs.
**5/16×1 1/2"	5/16×1 1/2"	face count	1200 lbs.

Nail Schedule		
Tongue and groove flooring must be blind-nailed		
3/4×1 1/2, 2 1/4 and 3 1/4"	2" machine driven fasteners, 7d or 8d screw or cut nail.	10-12" apart*
3/4×3"** Plank	2" machine driven fasteners, 7d or 8d screw or cut nail.	8" apart into and between joists

*If subfloor is 1/2 inch plywood, fasten into each joist, with additional fastening between.
**Plank flooring over 4" wide must be installed over a subfloor.

Nominal	Actual	Counted	Weights (1000 feet)
Special thickness (T and G—end-matched)			
**33/32×3 1/4"	33/32×3 1/4"	5/4×4"	2400 lbs.
**33/32×2 1/4"	33/32×2 1/4"	5/4×3"	2250 lbs.
**33/32×2"	33/32×2"	5/4×2 3/4"	2250 lbs.
Jointed flooring (i.e., square-edge)			
**3/4×2 1/2"	3/4×2 1/2"	1×3 1/4"	2160 lbs.
**3/4×3 1/4"	3/4×3 1/4"	1×4"	2300 lbs.
**3/4×3 1/2"	3/4×3 1/2"	1×4 1/4"	2400 lbs.
**33/32×2 1/2"	33/32×2 1/2"	5/4×3 1/4"	2500 lbs.
**33/32×3 1/2"	33/32×3 1/2"	5/4×4 1/4"	2600 lbs.

**Special Order Only

Following flooring must be laid on a subfloor		
1/2×1 1/2 & 2"	1 1/2" machine driven fastener, 5d screw, cut steel or wire casing nail.	10" apart
3/8×1 1/2 & 2"	1 1/4" machine driven fastener, or 4d bright wire casing nail.	8" apart
Square-edge flooring as follows, face-nailed through top face		
5/16×1 1/2 & 2"	1 inch 15 gauge fully barbed flooring brad. 2 nails every 7 inches.	
5/16×1 1/3"	1 inch 15 gauge fully barbed flooring brad. 1 nail every 5 inches on alternate sides of strip.	

Tables showing hardwood flooring grades, sizes, counts, and weights make flooring selection easier. Follow the recommended nailing schedule for best results. (National Oak Flooring Manufacturers' Association)

Guide to Hardwood Flooring Grades

Flooring is bundled by averaging the lengths. A bundle may include pieces from 6″ under to 6″ over the nominal length of the bundle. No piece is shorter than 9″. Quantity with length under 4′ held to stated percentage of total footage.

Unfinished oak flooring (Red and white separated)	Beech, birch, and hard maple	Pecan flooring	Prefinished oak flooring (Red and white separated-graded after finishing)
CLEAR (Plain or quarter sawn)** Best appearance. Best grade, most uniform color, limited small character marks. Bundles 1 1/4′ and up. Average length 3 3/4′.	**FIRST GRADE WHITE HARD MAPLE** (Spec. Order) Same as FIRST GRADE except face all bright sapwood. **FIRST GRADE RED BEECH and BIRCH** (Spec. Order) Same as first grade except face all red heartwood. **FIRST GRADE** Best appearance. Natural color variation.	***FIRST GRADE RED** (Spec. Order) Same as FIRST GRADE except face all heartwood. ***FIRST GRADE WHITE** (Spec. Order) Same as FIRST GRADE except face all bright sapwood. **FIRST GRADE** Excellent appearance. Natural color variation, limited character marks, unlimited sap Bdles. 2′. and up. 2 and 3′. bdles. up to 25% footage.	***PRIME** (Special Order Only) Excellent appearance. Natural color variation, limited character marks, unlimited sap. Bundles 1 1/4′ and up. Average length 3 1/2′.
SELECT and BETTER (Special Order) A combination of clear and select grades **SELECT** (Plain or quarter sawn)** Excellent appearance. Limited character marks, unlimited sound sap. Bundles 1 1/4′ and up. Average length 3 1/4′.	**SECOND and BETTER GRADE** Excellent appearance. A combination of FIRST and SECOND GRADES Bdles. 2′. and up. 2 and 3′. Bdles up to 40% footage. (NOTE 5% 1 1/4 bdles. allowed in SECOND and BETTER jointed flg. only. **SECOND GRADE** Variegated appearance. Varying sound wood characteristics of species. Bdles. 2′. and up. 2 and 3′. bdles. up to 45% footage.		**STANDARD and BETTER GRADE** Combination of STANDARD and PRIME. Bundles 1 1/4′ and up. Average length 3 1/2′. **STANDARD GRADE** Variegated appearance. Varying sound wood characteristic of species. A sound floor. Bundles 1 1/4′ and up. Average length 2 3/4′.
NO. 1 COMMON (Red and white may be mixed) Variegated appearance. Light and dark colors; knots, flags, worm holes, and other character marks allowed to provide a variegated appearance, after imperfections are filled and finished. Bundles 1 1/4′ and up. Average length 2 3/4′.		***SECOND GRADE RED** (Special Order Only) Same as SECOND GRADE except face all heartwood. **SECOND GRADE** Variegated appearance. Varying sound wood characteristics of species. Bundles 1 1/4′ and up. 1 1/4′. to 3′. bundles as produced up to 40% footage.	***TAVERN and BETTER GRADE** (Special Order Only) Combination of PRIME, STANDARD, and TAVERN. All wood characteristics of species. Bundles 1 1/4′ and up. Average length 3′. **TAVERN GRADE** Rustic appearance. All wood characteristics of species. Bundles 1 1/4′ and up. Average length 2 1/4′.
NO. 2 COMMON (Red and white may be mixed) Rustic appearance. All wood characteristics of species. A serviceable, economical floor after knot holes, worm holes, checks, and other imperfections are filled and finished. Bundles 1 1/4′ and up. Average length 2 1/4′.	**THIRD and BETTER GRADE** A combination of FIRST, SECOND, and THIRD GRADES. Bundles 1 1/4′ and up. 1 1/4′. and 3′. bundles as produced up to 50% footage. **THIRD GRADE** Rustic appearance All wood characteristics of species. Serviceable, economical floor after filling. Bundles 1 1/4′ and up. 1 1/4′. to 3′. bundles as produced up to 60% footage.	**THIRD GRADE** Rustic appearance. All wood characteristics of species. A serviceable, economical floor after filling. Bundles 1 1/4′ and up. 1 1/4′. to 3′. bundles as produced up to 60% footage.	

* 1 1/4′. SHORTS (Red and white may be mixed) Unique variegated appearance. Lengths 9″–18″. Bundles average nominal 1 1/4′. Production limited.

* NO. 1 COMMON and BETTER SHORTS A combination grade. CLEAR, SELECT, and NO. 1 COMMON, 9″–18″.

* NO. 2 COMMON SHORTS Same as NO. 2 COMMON, except length 9″–18″.

** Quarter sawn—Special order only.
* Check with supplier for grade and species available.
NESTED FLOORING: Random-length tongue and groove, end-matched flooring is bundled end to end continuously to form 8. long (nominal) bundles. Regular grade requirements apply.

NESTED FLOORING: If put up in 8′. nested bundles, 9″–18″ pieces will be admitted in 3/4″ × 2 1/4″ as follows in the species of beech, birch, and hard maple: FIRST GRADE, 4 pcs. per bundle; SECOND GRADE, 8 pcs.; THIRD GRADE, as develops. Average lengths: FIRST GRADE, 42″ SECOND GRADE, 33″; THIRD GRADE, 30″.

PREFINISHED BEECH and PECAN FLOORING
* TAVERN and BETTER GRADE (Special order only) Combination of PRIME, STANDARD, and TAVERN. All wood characteristics of species. Bundles 1 1/4′ and up. Average length 3′.

Common Nail Applications

Joining	Size & type	Placement
Wall framing		
Top plate	8d common 16d common	
Header	8d common 16d common	
Header to joist	16d common	
Studs	8d common 16d common	
Wall sheathing		
Boards	8d common	6" O.C.
Plywood (5/16", 3/8", 1/2")	6d common	6" O.C.
Plywood (5/8", 3/4")	8d common	6" O.C.
Fiberboard	1 3/4" galvanized roofing nail 8d galvanized common nail	6" O.C. 6" O.C.
Foamboard	Cap nail, length sufficient for penetration of 1/2" into framing	12" O.C.
Gypsum	1 3/4" galvanized roofing nail 8d galvanized common nail	6" O.C. 6" O.C.
Subflooring	8d common	10"-12" O.C.
Underlayment	(1 1/4" × 14 ga. annular underlayment nail)	6" O.C. edges 12" O.C. face
Roof framing		
Rafters, beveled or notched	12d common	
Rafter to joist	16d common	
Joist to rafter and stud	10d common	
Ridge beam	8d & 16d common	
Roof sheathing		
Boards	8d common	
Plywood (5/16", 3/8", 1/2")	6d common	12" O.C. and 6" O.C. edges
Plywood (5/8", 3/4")	8d common	12" O.C. and 6" O.C. edges
Roofing, asphalt		
New construction shingles and felt	7/8" through 1 1/2" galvanized roofing	4 per shingle
Re-roofing application shingles and felt	1 3/4" or 2" galvanized roofing	4 per shingle
Roof deck/insulation	Thickness of insulation plus 1" insulation roof deck nail	
Roofing, wood shingles		
New construction	3d –4d galvanized shingle	2-3 per shingle
Re-roofing application	5d –6d galvanized shingle	2-3 per shingle
Soffit	6d –8d galvanized common	12" O.C. max.
Siding*		
Bevel and lap	Aluminum nails are recommended for optimum performance	Consult siding manufacturer's application instructions
Drop and shiplap		
Plywood		
Hardboard	Galvanized hardboard siding nail Galvanized box nail	Consult siding manufacturer's application instructions
Doors, windows, mouldings, furring		
Wood strip to masonry	Nail length is determined by thickness of siding and sheathing. Nails should penetrate at least 1 1/2" into solid wood framing.	
Wood strip to stud or joist		
Paneling		
Wood	4d –8d casing-finishing	24" O.C.
Hardboard	2" × 16 gage annular	8" O.C.
Plywood	3d casing-finishing	8" O.C.
Gypsum	1 1/4" annular drywall	6" O.C.
Lathing	4d common blued	4" O.C.
Exterior projects		
Decks, patios, etc.	8d –16d hot dipped galvanized common	

*Aluminum nails are recommended for maximum protection from staining
NOTE: Usage may vary somewhat due to regional differences and preferences.
(Georgia-Pacific)

Drywall Screws

Description	No.	Length	Applications
Bugle Phillips	1E	6×1	For attaching drywall to metal studs from 25 gage through 20 gage
	2E	6×1 1/8	
	3E	6×1 1/4	
	4E	6×1 5/8	
	5R	6×2	
	6R	6×2 1/4	
	7R	8×2 1/2	
	8R	8×3	
Coarse Thread	1C	6×1	For attaching drywall to 25 gage metal studs, and attaching drywall to wood studs
	2C	6×1 1/8	
	3C	6×1 1/4	
	4C	6×1 5/8	
	5C	6×2	
	6C	6×2 1/4	
Pan Framing	19	6×7/16	For attaching stud to track up to 20 gage
HWH Framing	21	6×7/16	For attaching stud to track up to 20 gage where hex head is desired
	22	8×9/16	
	35	10×3/4	
K-Lath	28	8×9/16	For attaching wire lath, K-lath to 20 gage studs
Laminating	8	10×1 1/2	Type G laminating screw for attaching gypsum to gypsum, a temporary fastener
Trim Head	9	6×1 5/8	Trim head screw for attaching wood trim and base to 25 gage studs
	10	6×2 1/4	

Standard Thicknesses of Insulating Glass

Unit Construction							
Single glass thickness		Overall unit thickness		Air space		Approximate weight	
inches	mm	inches	mm	inches	mm	lb/ft^2	kg/m^2
1/8	3	1/2	12	1/4	6	3.27	16
		3/4	19	1/2	12		
3/16	5	5/8	15	1/4	6	4.90	24
		7/8	22	1/2	12		
		1	25	5/8	15		
1/4	6	3/4	19	1/4	6	6.54	32
		1	25	1/2	12		

(Libby-Owens-Ford Co., Glass Div.)

Welded Wire Fabric Reinforcement

Recommended styles of welded wire fabric reinforcement for concrete		
Type of construction	**Recommended style**	**Remarks**
Barbeque foundation slabs	6×6-W2.0'W2.0 to 4×4-W2.9×W2.9	Use heavier style fabric for heavy, massive fireplaces or barbeque pits.
Basement floors	6×6-W1.4'W1.4, 6×6-W2.0×W2.0, or 6×6-W2.9×W2.9	For small areas (15' maximum side dimension), use 6×6-W1.4×W1.4. As a rule of thumb, the larger the area or the poorer the subsoil, the heavier the gauge.
Driveways	6×6-W2.9'W2.9	Continuous reinforcement between 25' to 30' contraction joints.
Residential foundation slabs	6×6-W1.4×W1.4	Use heavier gauge over poorly drained subsoil or when maximum dimension is greater than 15'.
Garage floors	6×6-W2.9×W2.9	Position at midpoint of 5" or 6" thick slab.
Patios and terraces	6×6-W1.4×W1.4	Use 6×6-W2.0×W2.0 if subsoil is poorly drained.
Porch floors A) 6" thick slab up to 6' span B) 6" thick slab up to 8' span	6×6-W2.9×W2.9 4×4-W4.0×W4.0	Position 1" from bottom form to resist tensile stresses.
Sidewalks	6×6-W1.4×W1.4 or 6×6-W2.0×W2.0	Use heavier gauge over poorly drained subsoil. Construct 25' to 30' slabs as for driveways.
Steps (free span)	6×6-W2.9×W2.9	Use heavier style if more than five risers. Position fabric 1" from bottom form.
Steps (on ground)	6×6-W2.0×W2.0	Use 6×6-W2.9×W2.9 for unstable subsoil.

Gypsum Wallboard Application

Thickness	Approximate Weight lbs square foot	Size	Location	Application Method	Maximum spacing of framing members
1/4"	1.1	4' × 8' to 12'	Over existing walls and ceilings	Horizontal or vertical	
3/8"	1.5	4' × 8' to 14'	Ceilings	Horizontal	16"
3/8"	1.5	4' × 8' to 14'	Sidewalls	Horizontal or vertical	16"
1/2"	2.0	4' × 8' to 14'	Ceilings	Vertical Horizontal	16" 24"
1/2"	2.0	4' × 8' to 14'	Sidewalls	Horizontal or vertical	24"
5/8"	2.5	4' × 8' to 14'	Ceilings	Vertical Horizontal	16" 24"
5/8"	2.5	4' × 8' to 14'	Sidewalls	Horizontal or vertical	24"
1"	4.0	2' × 8' to 12'		For laminated partitions	

Floor Joist Span Data

Species or Group	Grade	2 × 8			2 × 10			2 × 12		
		30 psf live load, 10 psf dead load, deflection <360								
		12" O.C.	16" O.C.	24" O.C.	12" O.C.	16" O.C.	24" O.C.	12" O.C.	16" O.C.	24" O.C.
Douglas Fir and Larch	Sel. Struc.	16-6	15-0	13-1	21-0	19-1	16-8	25-7	23-3	20-3
	No. 1 & Btr.	16-2	14-8	12-10	20-8	18-9	16-1	25-1	22-10	18-8
	No. 1	15-10	14-5	12-4	20-3	18-5	15-0	24-8	21-4	17-5
	No. 2	15-7	14-1	11-6	19-10	17-2	14-1	23-0	19-11	16-3
	No. 3	12-4	10-8	8-8	15-0	13-0	10-7	17-5	15-1	12-4

Species or Group	Grade	2 × 8			2 × 10			2 × 12		
		40 psf live load, 10 psf dead load, deflection <360								
		12" O.C.	16" O.C.	24" O.C.	12" O.C.	16" O.C.	24" O.C.	12" O.C.	16" O.C.	24" O.C.
Douglas Fir and Larch	Sel. Struc.	15-0	13-7	11-11	19-1	17-4	15-2	23-3	21-1	18-5
	No. 1 & Btr.	14-8	13-4	11-8	18-9	17-0	14-5	22-10	20-5	16-8
	No. 1	14-5	13-1	11-0	18-5	16-5	13-5	22-0	19-1	15-7
	No. 2	14-2	12-7	10-3	17-9	15-5	12-7	20-7	17-10	14-7
	No. 3	11-0	9-6	7-9	13-5	11-8	9-6	15-7	13-6	11-0

Species or Group	Grade	2 × 8			2 × 10			2 × 12		
		30 psf live load, 10 psf dead load, deflection <360								
		12" O.C.	16" O.C.	24" O.C.	12" O.C.	16" O.C.	24" O.C.	12" O.C.	16" O.C.	24" O.C.
Southern Pine	Sel. Struc.	16-2	14-8	12-10	20-8	18-9	16-5	25-1	22-10	19-11
	No. 1	15-10	14-5	12-7	20-3	18-5	16-1	24-8	22-5	19-6
	No. 2	15-7	14-2	12-4	19-10	18-0	14-8	24-2	21-1	17-2
	No. 3	13-3	11-6	9-5	15-8	13-7	11-1	18-8	16-2	13-2

Species or Group	Grade	2 × 8			2 × 10			2 × 12		
		40 psf live load, 10 psf dead load, deflection <360								
		12" O.C.	16" O.C.	24" O.C.	12" O.C.	16" O.C.	24" O.C.	12" O.C.	16" O.C.	24" O.C.
Southern Pine	Sel. Struc.	14-8	13-4	11-8	18-9	17-0	14-11	22-10	20-9	18-1
	No. 1	14-5	13-1	11-5	18-5	16-9	14-7	22-5	20-4	17-5
	No. 2	14-2	12-10	11-0	18-0	16-1	13-2	21-9	18-10	15-4
	No. 3	11-11	10-3	8-5	14-0	12-2	9-11	16-8	14-5	11-10

Species or Group	Grade	2 × 8			2 × 10			2 × 12		
		40 psf live load, 10 psf dead load, deflection <360								
		12" O.C.	16" O.C.	24" O.C.	12" O.C.	16" O.C.	24" O.C.	12" O.C.	16" O.C.	24" O.C.
Redwood	Cl. All Heart		7-3	6-0		10-9	8-9		13-6	11-0
	Const. Heart		7-3	6-0		10-9	8-9		13-6	11-0
	Const. Common		7-3	6-0		10-9	8-9		13-6	11-0

Span data is in feet and inches for floor joists of Douglas fir/larch, southern yellow pine, and California redwood. Spans are calculated on the basis of dry sizes with moisture content equal to or less than 19%. Floor joist spans are for a single span.

Ceiling Joist Span Data

20 psf live load, 10 psf dead load, deflection <240													
Drywall ceiling, no future room development, limited attic storage available													
Species or Group	Grade	2 × 4			2 × 6			2 × 8			2 × 10			
		12″ O.C.	16″ O.C.	24″ O.C.	12″ O.C.	16″ O.C.	24″ O.C.	12″ O.C.	16″ O.C.	24″ O.C.	12″ O.C.	16″ O.C.	24″ O.C.
Douglas Fir and Larch	Sel. Struc.	10-5	9-6	8-3	16-4	14-11	13-0	21-7	19-7	17-1	27-6	25-0	20-11
	No. 1 & Btr.	10-3	9-4	8-1	16-1	14-7	12-0	21-2	18-8	15-3	26-4	22-9	18-7
	No. 1	10-0	9-1	7-8	15-9	13-9	11-2	20-1	17-5	14-2	24-6	21-3	17-4
	No. 2	9-10	8-9	7-2	14-10	12-10	10-6	18-9	16-3	13-3	22-11	19-10	16-3
	No. 3	7-8	6-8	5-5	11-2	9-8	7-11	14-2	12-4	10-0	17-4	15-0	12-3

20 psf live load, 10 psf dead load, deflection <240													
Drywall ceiling, no future room development, limited attic storage available													
Species or Group	Grade	2 × 4			2 × 6			2 × 8			2 × 10			
		12″ O.C.	16″ O.C.	24″ O.C.	12″ O.C.	16″ O.C.	24″ O.C.	12″ O.C.	16″ O.C.	24″ O.C.	12″ O.C.	16″ O.C.	24″ O.C.
Southern Pine	Sel. Struc.	10-3	9-1	8-1	16-1	14-7	12-9	21-2	19-3	16-10	26-0	24-7	21-6
	No. 1	10-10	9-4	8-0	15-9	14-4	12-6	20-10	18-11	15-11	26-0	23-2	18-11
	No. 2	9-10	8-11	7-8	15-6	13-6	11-0	20-1	17-5	14-2	24-0	20-9	17-0
	No. 3	8-2	7-1	5-9	12-1	10-5	8-6	15-4	13-3	10-10	18-1	15-8	12-10

Span data are in feet and inches for floor joists of Douglas fir/larch, southern yellow pine, and California redwood. Spans are calculated on the basis of dry sizes with moisture content equal to or less than 19%.

Roof Rafter Span Data

20 psf live load, 10 psf dead load, deflection <240													
Roof slope 3:12 or less, Light roof covering, No ceiling finish													
Species or Group	Grade	2 × 6			2 × 8			2 × 10			2 × 12			
		12″ O.C.	16″ O.C.	24″ O.C.	12″ O.C.	16″ O.C.	24″ O.C.	12″ O.C.	16″ O.C.	24″ O.C.	12″ O.C.	16″ O.C.	24″ O.C.
Douglas Fir and Larch	Sel. Struc.	16-4	14-11	13-0	21-7	19-7	17-2	27-6	25-0	21-10	33-6	30-5	26-1
	No. 1 & Btr.	16-1	14-7	12-9	21-2	19-3	16-10	27-1	24-7	20-9	32-11	29-6	24-1
	No. 1	15-9	14-4	12-6	20-10	18-11	15-10	26-6	23-9	19-5	31-10	27-6	22-6
	No. 2	15-6	14-1	11-9	20-5	18-2	14-10	25-8	22-3	18-2	29-9	25-9	21-0
	No. 3	12-6	10-10	8-10	15-10	13-9	11-3	19-5	16-9	13-8	22-6	19-6	15-11

20 psf live load, 15 psf dead load, deflection <240													
Roof slope 3:12 or less, Light roof covering, No ceiling finish													
Species or Group	Grade	2 × 6			2 × 8			2 × 10			2 × 12			
		12″ O.C.	16″ O.C.	24″ O.C.	12″ O.C.	16″ O.C.	24″ O.C.	12″ O.C.	16″ O.C.	24″ O.C.	12″ O.C.	16″ O.C.	24″ O.C.
Douglas Fir and Larch	Sel. Struc.	16-4	14-11	13-0	21-7	19-7	17-2	27-6	25-0	21-7	33-6	30-5	25-1
	No. 1 & Btr.	16-1	14-7	12-5	21-2	19-3	15-9	27-1	23-7	19-3	31-7	27-4	22-4
	No. 1	15-9	14-3	11-7	20-9	18-0	14-8	25-5	22-0	17-11	29-5	25-6	20-10
	No. 2	15-4	13-3	10-10	19-5	16-10	13-9	23-9	20-7	16-9	27-6	23-10	19-6
	No. 3	11-7	10-1	8-2	14-8	12-9	10-5	17-11	15-7	12-8	20-10	18-0	14-9

Roof Rafter Span Data *(continued)*

		20 psf live load, 10 psf dead load, deflection <240										
		Drywall ceiling, light roofing, snow load										
Species or Group	Grade	2 × 6			2 × 8			2 × 10			2 × 12		
		12″ O.C.	16″ O.C.	24″ O.C.	12″ O.C.	16″ O.C.	24″ O.C.	12″ O.C.	16″ O.C.	24″ O.C.	12″ O.C.	16″ O.C.	24″ O.C.
Southern Pine	Sel. Struc.	16-1	14-7	12-9	21-2	19-3	16-10	26-0	24-7	21-6	26-0	26-0	26-0
	No. 1	15-9	14-4	12-6	20-10	18-11	16-6	26-0	24-1	20-3	26-0	26-0	24-1
	No. 2	15-6	14-1	11-9	20-5	18-6	15-3	25-8	22-3	18-2	26-0	26-0	21-4
	No. 3	12-11	11-2	9-1	16-5	14-3	11-7	19-5	16-10	13-9	23-1	20-0	16-4

		30 psf live load, 15 psf dead load, deflection <240										
		Drywall ceiling, medium roofing, snow load										
Species or Group	Grade	2 × 6			2 × 8			2 × 10			2 × 12		
		12″ O.C.	16″ O.C.	24″ O.C.	12″ O.C.	16″ O.C.	24″ O.C.	12″ O.C.	16″ O.C.	24″ O.C.	12″ O.C.	16″ O.C.	24″ O.C.
Southern Pine	Sel. Struc.	14-1	12-9	11-2	18-6	16-10	14-8	23-8	21-6	18-9	26-0	26-0	22-10
	No. 1	13-9	12-6	10-11	18-2	16-6	13-11	23-2	20-3	16-6	26-0	24-1	19-8
	No. 2	13-6	11-9	9-7	17-7	15-3	12-5	21-0	18-2	14-10	24-7	21-4	17-5
	No. 3	10-6	9-1	7-5	13-5	11-7	9-6	15-10	13-9	11-3	18-10	16-4	13-4

Roof Decking Span Data

With a maximum deflection of 1/240th of the span; live load = 20 psf

Thickness in inches (nominal)	Lumber grade	Simple spans	
		Douglas Fir, Larch, Southern Yellow Pine — Span	Western Red Cedar — Span
2	Construction	9′-5″	8′-1″
2	Standard	9′-5″	6′-9″
3	Select dex.	15′-3″	13′-0″
3	Compl. dex.	15′-3″	13′-0″
4	Select dex.	20′-3″	17′-3″
4	Compl. dex.	20′-3″	17′-3″

Thickness in inches (nominal)	Lumber grade	Random lengths	
		Douglas Fir, Larch, Southern Yellow Pine — Span	Western Red Cedar — Span
2	Construction	10′-3″	8′-10″
2	Standard	10′-3″	6′-9″
3	Select dex.	16′-9″	14′-3″
3	Compl. dex.	16′-9″	13′-6″
4	Select dex.	22′-0″	19′-0″
4	Compl. dex.	22′-0″	18′-0″

Thickness in inches (nominal)	Lumber grade	Comb, simple and two-span continuous	
		Douglas Fir, Larch, Southern Yellow Pine — Span	Western Red Cedar — Span
2	Construction	10′-7″	8′-9″
2	Standard	10′-7″	6′-9″
3	Select dex.	17′-3″	14′-9″
3	Compl. dex.	17′-3″	13′-6″
4	Select dex.	22′-9″	19′-6″
4	Compl. dex.	22′-9″	18′-0″

Manufactured 2″ × 4″ Wood Floor Trusses

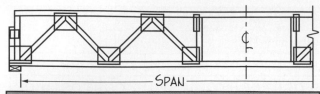 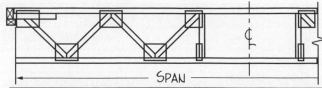

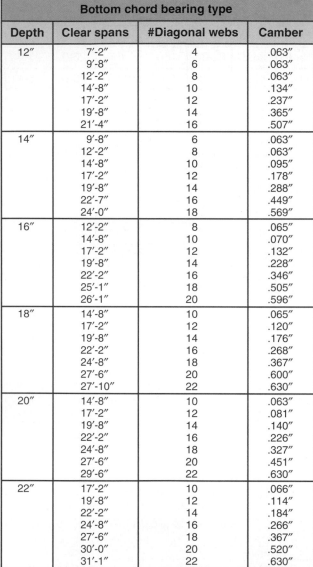

Bottom chord bearing type			
Depth	Clear spans	#Diagonal webs	Camber
12″	7′-2″	4	.063″
	9′-8″	6	.063″
	12′-2″	8	.063″
	14′-8″	10	.134″
	17′-2″	12	.237″
	19′-8″	14	.365″
	21′-4″	16	.507″
14″	9′-8″	6	.063″
	12′-2″	8	.063″
	14′-8″	10	.095″
	17′-2″	12	.178″
	19′-8″	14	.288″
	22′-7″	16	.449″
	24′-0″	18	.569″
16″	12′-2″	8	.065″
	14′-8″	10	.070″
	17′-2″	12	.132″
	19′-8″	14	.228″
	22′-2″	16	.346″
	25′-1″	18	.505″
	26′-1″	20	.596″
18″	14′-8″	10	.065″
	17′-2″	12	.120″
	19′-8″	14	.176″
	22′-2″	16	.268″
	24′-8″	18	.367″
	27′-6″	20	.600″
	27′-10″	22	.630″
20″	14′-8″	10	.063″
	17′-2″	12	.081″
	19′-8″	14	.140″
	22′-2″	16	.226″
	24′-8″	18	.327″
	27′-6″	20	.451″
	29′-6″	22	.630″
22″	17′-2″	10	.066″
	19′-8″	12	.114″
	22′-2″	14	.184″
	24′-8″	16	.266″
	27′-6″	18	.367″
	30′-0″	20	.520″
	31′-1″	22	.630″
24″	17′-2″	12	.063″
	19′-8″	14	.095″
	22′-2″	16	.153″
	24′-8″	18	.235″
	27′-2″	20	.325″
	30′-0″	22	.431″
	32′-6″	24	.630″

Top chord bearing type			
Depth	Clear spans	#Diagonal webs	Camber
12″	6′-10″	4	.063″
	9′-4″	6	.063″
	11′-10″	8	.063″
	14′-4″	10	.122″
	16′-10″	12	.233″
	19′-10″	14	.376″
	21′-4″	16	.507″
14″	9′-5″	6	.063″
	11′-11″	8	.063″
	14′-5″	10	.088″
	16′-11″	12	.167″
	19′-5″	14	.273″
	21′-4″	16	.429″
	24′-0″	18	.569″
16″	12′-0″	8	.063″
	14′-6″	10	.067″
	17′-0″	12	.126″
	19′-6″	14	.219″
	22′-4″	16	.337″
	24′-10″	18	.489″
	26′-1″	20	.596″
18″	14′-6″	10	.063″
	17′-0″	12	.098″
	19′-6″	14	.170″
	22′-0″	16	.260″
	24′-10″	18	.378″
	27′-8″	20	.617″
	27′-10″	22	.630″
20″	14′-6″	10	.063″
	17′-0″	12	.079″
	19′-6″	14	.136″
	22′-0″	16	.221″
	24′-10″	18	.337″
	27′-4″	20	.442″
	29′-6″	22	.630″
22″	17′-1″	12	.065″
	19′-7″	14	.112″
	22′-1″	16	.181″
	24′-10″	18	.275″
	27′-4″	20	.381″
	30′-2″	22	.534″
	31′-1″	24	.630″
24″	17′-1″	12	.063″
	19′-7″	14	.093″
	22′-1″	16	.150″
	24′-7″	18	.231″
	27′-5″	20	.335″
	30′-2″	22	.443″
	32′-6″	24	.630″

Wood floor trusses are typically manufactured from #3 southern yellow pine. Pieces are joined together with 18 and 20 gauge galvanized steel plates applied to both faces of the truss at each joint. Where no sheathing is applied directly to top chords, they should be braced at intervals not to exceed 3′-0″. Where no rigid ceiling is applied directly to bottom chords, they should be braced at intervals not to exceed 10′-0″.

Manufactured wood floor trusses are generally spaced 24″ O.C. and are designed to support various loads. Typical trusses shown here were designed to support 55 psf (live load - 40 psf, dead load - 10 psf, ceiling dead load - 5 psf). A slight bow (camber) is built into each joist so that it will produce a level floor when loaded. Allowable deflection is 1/360 of the span.

Some of the longer trusses require one or more double diagonal webs at both ends. Wood floor trusses are a manufactured product which must be engineered and produced with a high degree of accuracy to attain the desired performance. See your local manufacturer or lumber company for trusses available in your area.

(Courtesy of Trus Joist)

Length of Common Rafters

Feet of run	2-in-12 Inclination (set saw at) 9°28' 12.17″ per foot of run	2 1/2-in-12 Inclination (set saw at) 11°46' 12.26″ per foot of run	3-in-12 Inclination (set saw at) 14°2' 12.37″ per foot of run	3 1/2-in-12 Inclination (set saw at) 16°16' 12.5″ per foot of run	4-in-12 Inclination (set saw at) 18°26' 12.65″ per foot of run	4 1/2-in-12 Inclination (set saw at) 20°33' 12.82″ per foot of run
4	4' 0-11/16″	4' 1-1/32″	4' 1-15/32″	4' 2″	4' 2-19/32″	4' 3-9/32″
5	5' 0-27/32″	5' 1-5/16″	5' 1-27/32″	5' 2-1/2″	5' 3-1/4″	5' 4-3/32″
6	6' 1-1/32″	6' 1-9/16″	6' 2-7/32″	6' 3″	6' 3-29/32″	6' 4-15/16″
7	7' 1-3/16″	7' 1-13/16″	7' 2-19/32″	7' 3-1/2″	7' 4-9/16″	7' 5-3/4″
8	8' 1-3/8″	8' 2-3/32″	8' 2-31/32″	8' 4″	8' 5-3/32″	8' 6-9/16″
9	9' 1-17/32″	9' 2-7/16″	9' 3-11/32″	9' 4-1/2″	9' 5-27/32″	9' 7-3/8″
10	10' 1-23/32″	10' 2-19/32″	10' 3-23/32″	10' 5″	10' 6-1/2″	10' 8-7/32″
11	11' 1-7/8″	11' 2-7/8″	11' 4-1/16″	11' 5-1/2″	11' 7-5/32″	11' 9-1/32″
12	12' 2-1/32″	12' 3-1/8″	12' 4-7/16″	12' 6″	12' 7-13/16″	12' 9-27/32″
13	13' 2-7/32″	13' 3-3/8″	13' 4-13/16″	13' 6-1/2″	13' 8-15/32″	13' 10-21/32″
14	14' 2-3/8″	14' 3-21/32″	14' 5-3/16″	14' 7″	14' 9-3/32″	14' 11-1/2″
15	15' 2-9/16″	15' 3-29/32″	15' 5-9/16″	15' 7-1/2″	15' 9-3/4″	16' 0-5/16″
16	16' 2-23/32″	16' 4-5/32″	16' 5-15/16″	16' 8″	16' 10-13/32″	17' 1-1/8″

Inches of run	These lengths are to be added to those shown above when run involves inches					
1/4	1/4″	1/4″	1/4″	1/4″	1/4″	9/32″
1/2	1/2″	1/2″	1/2″	17/32″	17/32″	17/32″
1	1″	1″	1-1/32″	1-1/32″	1-1/16″	1-1/16″
2	2-1/32″	2-1/32″	2-1/16″	2-3/32″	2-3/32″	2-1/8″
3	3-1/16″	3-1/16″	3-3/32″	3-1/8″	3-5/32″	3-7/32″
4	4-1/16″	4-3/32″	4-1/8″	4-5/32″	4-7/32″	4-9/32″
5	5-1/16″	5-1/8″	5-5/32″	5-7/32″	5-9/32″	5-11/32″
6	6-3/32″	6-1/8″	6-3/16″	6-1/4″	6-5/16″	6-13/32″
7	7-3/32″	7-5/32″	7-7/32″	7-9/32″	7-3/8″	7-15/32″
8	8-1/8″	8-3/16″	8-1/4″	8-11/32″	8-7/16″	8-17/32″
9	9-1/8″	9-3/16″	9-1/4″	9-3/8″	9-15/32″	9-5/8″
10	10-5/32″	10-7/32″	10-5/16″	10-7/16″	10-17/32″	10-11/16″
11	11-5/32″	11-1/4″	11-11/32″	11-15/32″	11-19/32″	11-3/4″

Feet of run	5-in-12 Inclination (set saw at) 22°37' 13.00″ per foot of run	5 1/2-in-12 Inclination (set saw at) 24°37' 13.20″ per foot of run	6-in-12 Inclination (set saw at) 26°34' 13.42″ per foot of run	6 1/2-in-12 Inclination (set saw at) 28°27' 13.65″ per foot of run	7-in-12 Inclination (set saw at) 30°15' 13.89″ per foot of run	7 1/2-in-12 Inclination (set saw at) 32°0' 14.15″ per foot of run	8-in-12 Inclination (set saw at) 33°41' 14.42″ per foot of run
4	4' 4″	4' 4-13/16″	4' 5-11/16″	4' 6-19/32″	4' 7-9/16″	4' 8-19/32″	4' 9-11/16″
5	5' 5″	5'-6″	5' 7-1/8″	5' 8-1/4″	5' 9-15/32″	5' 10-3/4″	6' 0-1/8″
6	6' 6″	6' 7-3/32″	6' 8-1/2″	6' 9-29/32″	6' 11-11/32″	7' 0-29/32″	7' 2-17/32″
7	7' 7″	7' 8-13/32″	7' 9-15/16″	7' 11-9/16″	8' 1-7/32″	8' 3-1/16″	8' 4-15/16″
8	8' 8″	8' 9-19/32″	8' 11-3/8″	9' 1-7/32″	9' 3-1/8″	9' 5-7/32″	9' 7-3/8″
9	9' 9″	9' 10-13/16″	10' 0-25/32″	10' 2-27/32″	10' 5″	10' 7-11/32″	10' 9-25/32″
10	10' 10″	11' 0″	11' 2-7/32″	11' 4-1/2″	11' 6-29/32″	11' 9-1/2″	12' 0-7/32″
11	11' 11″	12' 1-3/32″	12' 3-5/8″	12' 6-5/32″	12' 8-13/16″	12' 11-21/32″	13' 2-5/8″
12	13' 0″	13' 2-13/32″	13' 5-1/32″	13' 7-13/16″	13' 10-11/16″	14' 1-13/16″	14' 5-1/32″
13	14' 1″	14' 3-19/32″	14' 6-15/32″	14' 9-15/32″	15' 0-9/16″	15' 3-31/32″	15' 7-15/32″
14	15' 2″	15' 4-13/16″	15' 7-7/8″	15' 11-1/8″	16' 2-15/32″	16' 6-1/8″	16' 9-7/8″
15	16' 3″	16' 6″	16' 9-5/16″	16' 0-3/4″	17' 4-11/32″	17' 8-1/4″	18' 0-5/16″
16	17' 4″	17' 7-7/32″	17' 10-23/32″	18' 2-13/32″	18' 6-1/4″	18' 10-13/32″	19' 2-23/32″

Inches of run	These lengths are to be added to those shown above when run involves inches						
1/4	9/32″	9/32″	9/32″	9/32″	5/16″	5/16″	
1/2	17/32″	9/16″	9/16″	9/16″	1-9/32″	5/8″	
1	1-3/32″	1-3/32″	1-1/8″	1-1/8″	1-3/16″	1-7/32″	
2	2-5/32″	2-7/32″	2-1/4″	2-9/32″	2-11/32″	2-13/32″	
3	3-1/4″	3-5/16″	3-3/8″	3-13/32″	3-17/32″	3-19/32″	
4	4-11/32″	4-13/32″	4-15/32″	4-9/16″	4-23/32″	4-13/16″	
5	5-13/32″	5-1/2″	5-19/32″	5-11/16″	5-29/32″	6″	
6	6-1/2″	6-19/32″	6-23/32″	6-13/16″	7-1/32″	7-7/32″	
7	7-9/16″	7-23/32″	7-13/16″	7-31/32″	8-1/4″	8-13/32″	
8	8-21/32″	8-13/16″	8-15/16″	9-1/8″	9-7/16″	9-5/8″	
9	9-3/4″	9-29/32″	10-5/32″	10-1/4″	10-19/32″	10-13/16″	
10	10-27/32″	11″	11-3/16″	11-3/8″	11-25/32″	1'0″	
11	11-29/32″	12-3/32″	1'0-5/16″	1'0-1/2″	1'0-31/32″	1' 1-7/32″	

Sizes for Performance-Rated I-Joists

Sizes of performance rated I-joists (APA PRI™)				
Joist series	Letter/number designation	Depth (nominal)	Net depth	Width of flanges
1 × 10	C4 (PRI-15)	10″	9-1/2″	1-1/2″
1 × 10	C6 (PRI-25)	10″	9-1/2″	1-3/4″
1 × 12	C10 (PRI-15)	12″	11-7/8″	1-1/2″
1 × 12	C12 (PRI-25)	12″	11-7/8″	1-3/4″
1 × 14	C14 (PRI-25)	14″	14″	1-3/4″
1 × 14	C16 (PRI-35)	14″	14″	2-5/16″
1 × 16	C18 (PRI-25)	16″	16″	1-3/4″
1 × 16	C20 (PRI-35)	16″	16″	2-5/16″
1 × 10	S2 (PRI-30)	10″	9-1/2″	2-1/2″
1 × 10	S4 (PRI-32)	10″	9-1/2″	2-1/2″
1 × 12	S6 (PRI-30)	12″	11-7/8″	2-1/2″
1 × 12	S8 (PRI-32)	12″	11-7/8″	2-1/2″
1 × 12	S10 (PRI-42)	12″	11-7/8″	3-1/2″
1 × 14	S12 (PRI-32)	14″	14″	2-1/2″
1 × 14	S14 (PRI-42)	14″	14″	3-1/2″
1 × 16	S16 (PRI-32)	16″	16″	2-1/2″
1 × 16	S18 (PRI-42)	16″	16″	3-1/2″

Manufacturing tolerances allowed: flange width, +/-1/32″; depth of joist, +0″, -1/8″

Spans Allowed for Performance-Rated I-Joists

Spans allowed for APA performance-rated I-joists with simple-span O.C. spacing only				
Series and designation	12″ O.C.	16″ O.C.	19.2″ O.C.	24″ O.C.
1 × 10-C4 (PRI-15)*	17′-0″	15′-6″	14′-8″	13′-7″
1 × 10-C6 (PRI-25)	17′-9″	16′-2″	15′-3″	14′-2″
1 × 12-C10 (PRI-15)*	20′-3″	18′-5″	17′-5″	16′-7″
1 × 12-C12 (PRI-25)	21′-1″	19′-3″	18′-2″	17′-3″
1 × 14-C14 (PRI-25)*	24′-0″	21′-10″	21′-0″	19′-8″
1 × 14-C16 (PRI-35)	25′-11″	23′-7″	22′-2″	21′-0″
1 × 16-C18 (PRI-25)	26′-7″	24′-8″	23′-4″	20′-2″
1 × 16-C20 (PRI-35)*	28′-8″	26′-1″	24′-11″	23′-1″
1 × 10-S2 (PRI-30)	18′-0″	16′-3″	14′-10″	13′-3″
1 × 10-S4 (PRI-32)*	19′-0″	17′-4″	16′-4″	15′-4″
1 × 12-S6 (PRI-30)	21′-6″	18′-10″	17′-2″	15′-4″
1 × 12-S8 (PRI-32)*	22′-8″	20′-8″	19′-6″	18′-3″
1 × 12-S10 (PRI-42)	24′-11″	22′-8″	21′-4″	19′-11″
1 × 14-S12 (PRI-32)*	25′-9″	23′-6″	22′-2″	20′-9″
1 × 14-S14 (PRI-42)	28′-3″	25′-9″	24′-3″	22′-8″
1 × 16-S16 (PRI-32)*	28′-7″	26′-1″	24′-7″	23′-0″
1 × 16-S18 (PRI-42)	31′-4″	28′-6″	26′-11″	25′-1″

Notes:
1. Table based on uniform loads (10 psf dead load; 40 psf live load)
2. Clear span allowed here applies only to simple span residential floors.
3. Span limits are for clear distances between support.
4. Minimum end-bearing length shall be 1-3/4″.
5. *These units are the most commonly available.

Clear Spans for Wood I-Beam Rafters

Recommended clear spans for wood I-beam rafters, 1-3/4″ flange (Based on horizontal distances and 16″ O.C. spacing)					
9-1/2″ depth			11-7/8″ depth		
No snow load					
Load (psf) live/dead	Slope less than 4/12	Slope 4/12 to 8/12	Slope more than 8/12	Slope less than 4/12	Slope 4/12 to 8/12	Slope more than 8/12
20/10	22′-6″	20′-6″	17′-11″	26′-11″	24′-4″	21′-4″
20/15	21′-4″	19′-4″	16′-9″	25′-6″	22′-11″	20′-0″
20/20	20′-4″	18′-5″	15′-11″	24′-3″	21′-9″	18′-11″
Snow load					
25/10	21′-4″	19′-4″	17′-4″	25′-7″	23′-1″	20′-8″
25/15	20′-4″	18′-4″	16′-3″	24′-4″	21′-11″	19′-6″
30/10	20′-5″	18′-6″	16′-10″	24′-5″	22′-1″	20′-1″
30/15	19′-6″	17′-8″	15′-11″	23′-4″	21′-1″	19′-0″
40/10	18′-6″	17′-0″	15′-11″	22′-2″	20′-4″	19′-1″
40/15	18′-3″	16′-6″	15′-2″	21′-9″	19′-9″	18′-2″
50/10	17′-1″	15′-8″	13′-3″	20′-5″	18′-9″	18′-2″
50/15	17′-1″	15′-7″	14′-7″	20′-5″	18′-8″	17′-5″

Fink Truss Design

LUMBER—Lumber shall be of a good grade of sufficient quality to permit the following allowable unit stresses:

c = 900psi Compression parallel to grain.
f = 900psi Extreme fiber in bending.
E = 1,600,000psi Modulus of elasticity.

CONNECTORS—Timber connectors shall be 2-1/2" diameter split rings and Trip-L-Grip framing anchors.

BOLTS—Bolts shall be 1/2" diameter machine bolts with 2" × 2" × 1/8" plate washers. 2-1/8" diameter cast or malleable iron washers, or ordinary cut washers.

DIMENSIONS—Dimensions shown will provide approximately 1/2" camber at bottom chord panel points. Utilize full uncut length of bottom chord pieces by increasing the spacing of connectors in the splice.

Span L	Dimensions			Design stresses					
	A	B	C	U_1	U_2	L_1	L_2	V_1	D_1
20'-0"	5'-5"	5'-3 5/8"	2'-7 5/8"	1756	1450	1614	1472	430	430
22'-0"	5'-11 1/2"	5'-10 1/16"	2'-10 13/16"	1932	1595	1775	1619	473	473
24'-0"	6'-6"	6'-4 7/16"	3'-2"	2108	1740	1936	1766	516	516
26'-0"	7'-0 1/2"	6'-10 7/8"	3'-5 1/4"	2282	1885	2097	1913	559	559
28'-0"	7'-7"	7'-5 1/4"	3'-8 7/16"	2459	2030	2258	2060	602	602
30'-0"	8'-1 1/2"	7'-11 11/16"	3'-11 5/8"	2634	2175	2420	2207	645	645
32'-0"	8'-8"	8'-6 1/16"	4'-2 13/16"	2810	2320	2581	2354	688	688

Hardware

No.	Item	Size
11	Split rings	2 1/2" Diam.
2	Trip-L-Grip	Type A
1	Bolt	1/2" × 7 1/2"
2	Bolts	1/2" × 6"
4	Bolts	1/2" × 4"
14	Washers	1/2"

Lumber

Span	2" × 6" No.	2" × 6" Length	2" × 4" No.	2" × 4" Length	Total F.B.M.
20'-0"	2	12'-0"	2	12'-0"	53
			2	10'-0"	
22'-0"	2	14'-0"	2	14'-0"	60
			2	10'-0"	
24'-0"	2	14'-0"	2	14'-0"	63
			2	12'-0"	
26'-0"	2	16'-0"	2	16'-0"	70
			2	12'-0"	
28'-0"	2	16'-0"	2	16'-0"	73
			2	14'-0"	
30'-0"	2	18'-0"	2	18'-0"	79
			2	14'-0"	
32'-0"	2	20'-0"	2	18'-0"	83
			2	14'-0"	

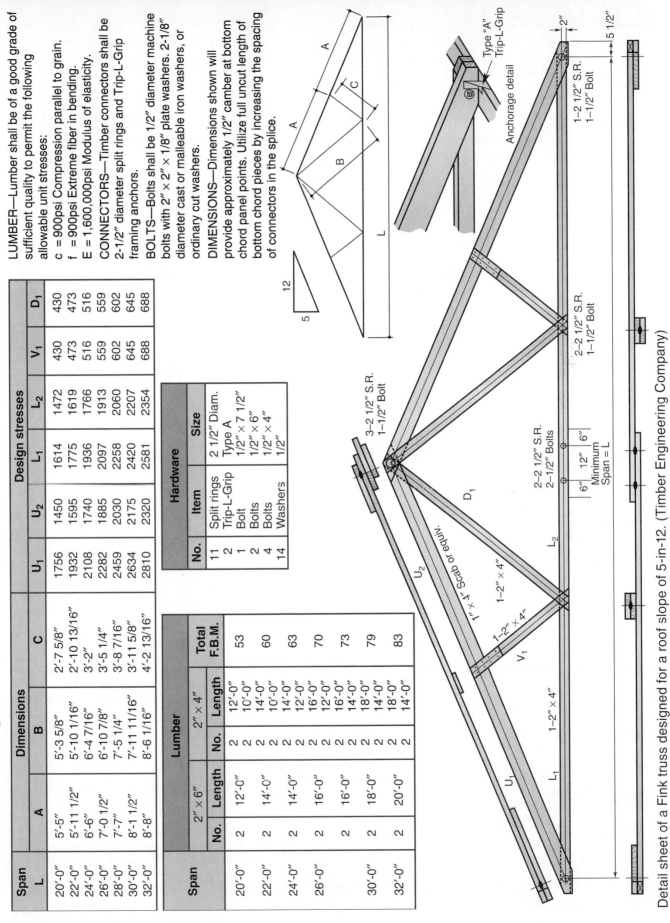

Detail sheet of a Fink truss designed for a roof slope of 5-in-12. (Timber Engineering Company)

Roof Trusses

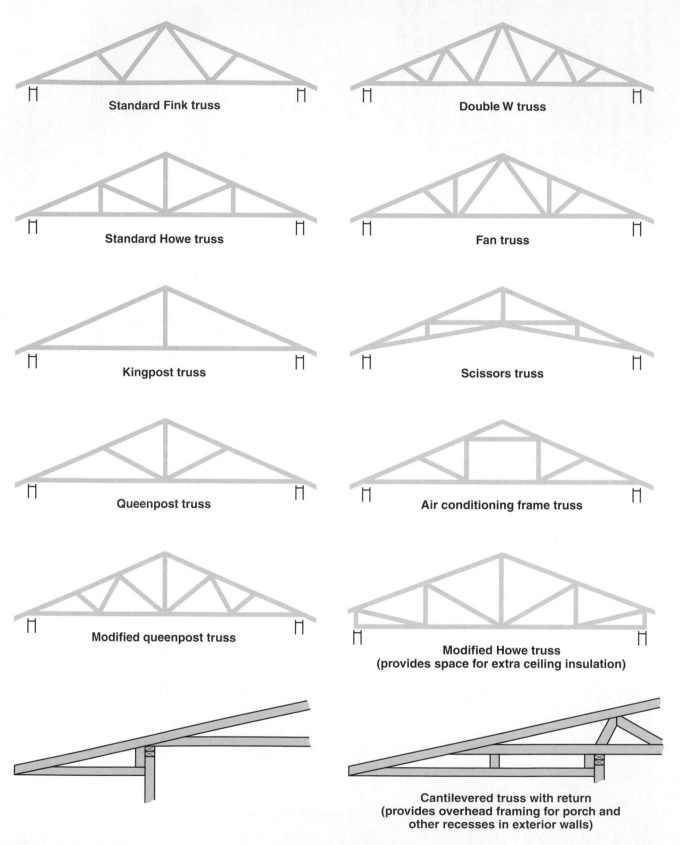

Standard Fink truss

Double W truss

Standard Howe truss

Fan truss

Kingpost truss

Scissors truss

Queenpost truss

Air conditioning frame truss

Modified queenpost truss

Modified Howe truss
(provides space for extra ceiling insulation)

Cantilevered truss with return
(provides overhead framing for porch and
other recesses in exterior walls)

Types of roof trusses and overhangs. The Fink truss is also called a W truss. Any type of truss should be constructed according to designs developed from engineering data.

Ventilation

Better construction methods and materials used by builders today produce houses that are tighter and more draft-free than those built a half century ago. In such snug shelters, moisture created inside the house often saturates insulation and causes paint to peel off exterior walls. The use of a vapor barrier on the warm side of walls and ceilings will minimize this condensation, but adequate ventilation is needed to remove the moisture from the house.

FHA requires that attics have a total net free ventilating area not less that 1/150 of the square foot area, except that a ratio of 1/300 may be provided if:

(a) a vapor barrier having a transmission rate not exceeding one perm is installed on the warm side of the ceiling, or

(b) at least 50% of the required vent area is provided by ventilators located in the upper portion of the space to be ventilated, with the balance of the required ventilation provided by eave or cornice vents.

There are many types of screened ventilators available, including the handy miniature vents illustrated below, which can be used for problem areas or to supplement larger vents.

Free-area ventilation guide Square inches of ventilation required for attic areas												
Width (in feet)	20	22	24	26	28	30	32	34	36	38	40	42
20	192	211	230	250	269	288	307	326	346	365	384	403
22	211	232	253	275	296	317	338	359	380	401	422	444
24	230	253	276	300	323	346	369	392	415	438	461	484
26	250	275	300	324	349	374	399	424	449	474	499	524
28	269	296	323	349	376	403	430	457	484	511	538	564
30	288	317	346	374	403	432	461	490	518	547	576	605
32	307	338	369	399	430	461	492	522	553	584	614	645
34	326	359	392	424	457	490	522	555	588	620	653	685
36	346	380	415	449	484	518	553	588	622	657	691	726
38	365	401	438	474	511	547	584	620	657	693	730	766
40	384	422	461	499	538	576	614	653	691	730	768	806
42	403	444	484	524	564	605	645	685	726	766	806	847
44	422	465	507	549	591	634	676	718	760	803	845	887
46	442	486	530	574	618	662	707	751	795	839	883	927
48	461	507	553	599	645	691	737	783	829	876	922	968
50	480	528	576	624	672	720	768	816	864	912	960	1008

(Length (in feet) is the vertical column.)

Using length and width dimensions of each rectangular or square attic space, find one dimension on vertical column, the other dimension on horizontal column. These wil intersect at the number of square inches of ventilation required to provide 1/300th.

Installing minature vents

1-Inch
Install in paneling used in basement rooms. Can be painted over to match panel finish.

2-Inch
Used to ventilate stud space. For best results, install in top and bottom of each space.

2 1/2-Inch
Made especially to plug the 2 1/2-in. hole cut to blow insulation between studs.

3-Inch
For ventilating rafter space in flat-roof buildings, or other jobs requiring fairly large free area.

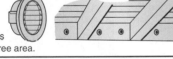

4-Inch
For hard-to-reach spots needing large ventilating area. Large enough for venting soffits.

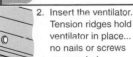

1. Drill or cut a hole the same size as the ventilator.

2. Insert the ventilator. Tension ridges hold ventilator in place... no nails or screws are needed.

3. Tap into place with a hammer, using a wood block to protect the margin. The louvers are recessed so there's no danger of damage during installation.

Ventilators and vent applications.

Ridge vent provides 18 sq. in. of net free area per lineal foot. Installed quickly over a 1 1/2" gap in the sheathing at the ridge.

Roof vents fit over openings cut between rafters to pull hot air out of attic.

Triangle vent fits snugly under the roof gable to provide large vent areas at the highest point of the gable end.

Attic vents are installed in the gable end, usually above the level of the probable level of a future ceiling should the attic be finished later.

Soffit vent replaces a portion of the soffit material to provide continuous ventilation along its entire length.

Brick vents are exactly the size of a brick, can be laid in any brick wall. Screened to meet FHA specs.

Cement block vents are designed to be mortared into the same space as an 8″ × 16″ cement block. At least four should be used to vent crawl space.

Steps for Installing Insulating Glass

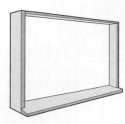

1 – Select a good grade of softwood lumber like ponderosa pine. For a medium-sized frame, use a 1 1/2" thickness. Make the frame slightly larger than the insulating glass. After assembly, apply a coating of water-repellent preservative.

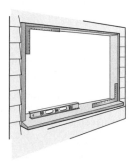

2 – Install frame in opening. Place wedge blocks under sill and at several points around perimeter of frame. When sill is level and frame perfectly square, nail the frame securely to structural members of the building.

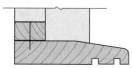

3 – Prepare inside stop members according to recommended sizes. Use miter joints in corners and nail down stops. Also prepare outside stops.

4 – Using recommended grade of glazing sealant, apply thick bed to stops, and top, sides, and bottom of frame. Apply enough material to fill space between frame and glass unit when installation is made.

5 – Install neoprene setting blocks on bottom of frame or edge of glass unit as shown. Blocks (use only two) should be moved in from the corners a distance equal to one-quarter of frame width.

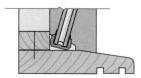

6 – Place bottom edge of insulating glass unit on setting blocks as shown. Then carefully press unit into position against stops until proper thickness of glazing sealant is secured around entire perimeter. Small gauge blocks may be helpful.

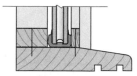

7 – Nail on outside stop. Additional sealant may be required to fill joint. Remove excess sealant from both sides of unit and clean surfaces with an approved solvent.

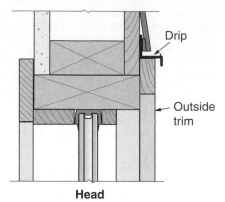

Drip

Outside trim

Head

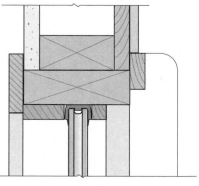

Jamb

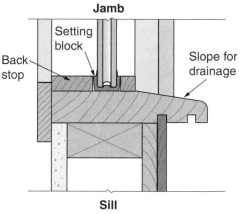

Setting block

Back stop

Slope for drainage

Sill

Details for Constructing a Window Wall

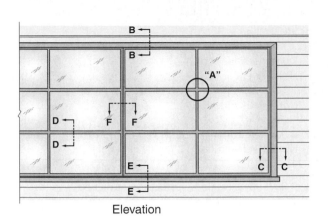

Elevation

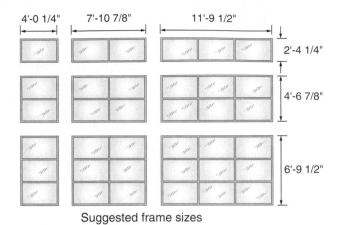

4'-0 1/4" 7'-10 7/8" 11'-9 1/2"

2'-4 1/4"

4'-6 7/8"

6'-9 1/2"

Suggested frame sizes

Glazed metal or wood ventilating units can be installed in any opening as desired without changing the window wall frame. Standard thermopane glass size for window wall construction is 45 1/2" x 25 1/2" for non-ventilating units.

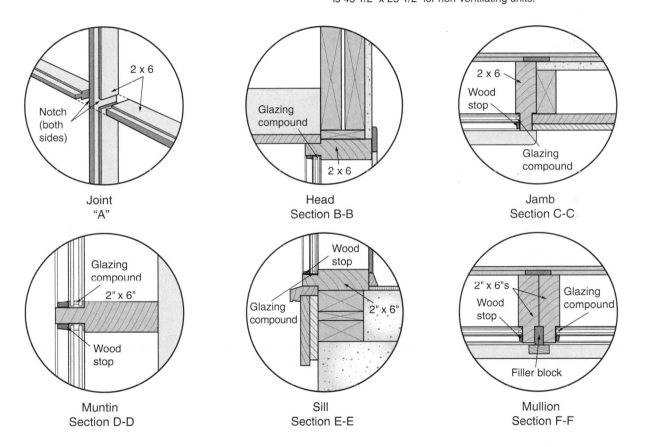

Joint "A"
- 2 x 6
- Notch (both sides)

Head Section B-B
- Glazing compound
- 2 x 6

Jamb Section C-C
- 2 x 6
- Wood stop
- Glazing compound

Muntin Section D-D
- Glazing compound
- 2" x 6"
- Wood stop

Sill Section E-E
- Wood stop
- Glazing compound
- 2" x 6"

Mullion Section F-F
- 2" x 6"s
- Wood stop
- Glazing compound
- Filler block

Insulation Values

Exterior materials

Wood bevel siding, 1/2 × 8, lapped R-0.81
Wood bevel siding, 3/4 × 10, lapped R-1.05
Wood siding shingles, 16″, 7 1/2″ exposure R-0.87
Aluminum or Steel, over sheathing,
hollow-backed R-0.61
Stucco, per inch R-0.20
Building paper R-0.06
1/2″ nail-base insulating board sheathing R-1.14
1/2″ insulating board sheathing, regular density R-1.32
25/32″ insul. board sheathing, regular density R-2.04
Insulating-board backed nominal 3/8″ R-1.82
Insulating-board backed nominal 3/8″ foil backed . . R-2.96
Plywood 1/4″ R-0.31
Plywood 3/8″ R-0.47
Plywood 1/2″ R-0.62
Plywood 5/8″ R-0.78
Hardboard 1/4″ R-0.18
Hardboard, medium density siding 7/16″. R-0.67
Softwood board, fir pine and similar softwoods
   3/4″ R-0.94
   1 1/2″ R-1.89
   2 1/2″ R-3.12
   3 1/2″ R-4.35
Gypsum board 1/2″ R-0.45
Gypsum board 5/8″ R-0.56

Masonry materials

Concrete blocks, three oval cores
   Cinder aggregate, 4″ thick R-1.11
   Cinder aggregate, 12″ thick R-1.89
   Cinder aggregate, 8″ thick R-1.72
   Sand and gravel aggregate, 8″ thick R-1.11
   Sand and gravel aggregate, 12″ thick R-1.28
   Lightweight aggregate (expanded clay,
   shale, slag, pumice, etc.), 8″ thick R-2.00
Concrete blocks, two rectangular cores
   Sand and gravel aggregate, 8″ thick R-1.04
   Lightweight aggregate, 8″ thick R-2.18
   Common brick, per inch R-0.20
   Face brick, per inch R-0.11
   Sand-and-gravel concrete, per inch R-0.08

Insulation

Fiberglass 2″ thick R-7.00
Fiberglass 3 1/2″ thick R-11.00
Fiberglass 6″ thick R-19.00
Fiberglass 12″ thick R-38.00
Foam Board 3/4″ thick R-4.05
Foam Board 1″ tongue and groove R-5.40

Roofing

Asphalt shingles. R-0.44
Wood shingles, plain and plastic film faced. R-0.94

Surface air films

Inside, still air
Heat flow *up* (through horizontal surface)
   Nonreflective. R-0.61
   Reflective R-1.32
Heat flow *down* (through horizontal surface)
   Nonreflective. R-0.92
   Reflective R-4.55
Heat flow *horizontal* (through vertical surface)
   Nonreflective. R-0.68

Outside
Heat flow any direction, surface any position
   15 mph wind (winter). R-0.17
   7.5 (mph wind (summer). R-0.25

Glass

U-Values	Glass Only (Winter)
Single-pane glass	1.16
Double-pane 5/8″ insulating glass (1/4″ air space)	.58
Double-pane × 1″ insulating glass	.55
Double-pane 1″ insulating glass (1/2″ air space)	.49
Double-pane × 1″ insulating glass with combination (2″ air space)	.35

Sample calculation

(to determine the U-value of an exterior wall)

Wall construction	Insulated wall resistance
Outside surface (film), 15 mph wind	0.17
Wood bevel siding, lapped.	0.81
1/2″ ins. bd. sheathing, reg. density	1.32
3 1/2″ air space	—
R-11 insulation	11.00
1/2″ gypsum board.	0.45
Inside surface (film)	0.68
Totals	14.43

For insulated wall, U = 1/R = 1/14.3 = 0.07

Temperature correction factor

Correction factor is an ASHRAE standard to be applied for varying outdoor design temperatures. As follows:

If design temperature is:	−20	−10	0	+10	+20
Then correction factor is:	0.778	0.875	1.0	1.167	1.40

Insulation values for common materials. The method used to calculate U- and R-values for a wall can also be used for ceilings and floors. Temperature correction factors are used by designers of heating and cooling systems. (Andersen Corp.)

Design Temperatures and Degree Days

Design Temperatures and Degree Days (Heating Season)			
State	City	Outside Design Temperature (°F)	Degree Days (°F-Days)
Alabama	Birmingham	19	2,600
Alaska	Anchorage	−25	10,800
Arizona	Phoenix	31	1,800
Arkansas	Little Rock	19	3,200
California	Los Angeles	41	2,000
California	San Francisco	35	3,000
Colorado	Denver	−2	6,200
Connecticut	Hartford	1	6,200
Florida	Tampa	36	600
Georgia	Atlanta	18	3,000
Idaho	Boise	4	5,800
Illinois	Chicago	−3	6,600
Indiana	Indianapolis	0	5,600
Iowa	Des Moines	−7	6,600
Kansas	Wichita	5	4,600
Kentucky	Louisville	8	4,600
Louisiana	New Orleans	32	1,400
Maryland	Baltimore	12	4,600
Massachusetts	Boston	6	5,600
Michigan	Detroit	4	6,200
Minnesota	Minneapolis	−14	8,400
Mississippi	Jackson	21	2,200
Missouri	St. Louis	4	5,000
Montana	Helena	−17	8,200
Nebraska	Lincoln	−4	5,800
Nevada	Reno	2	6,400
New Hampshire	Concord	−11	7,400
New Mexico	Albuquerque	14	4,400
New York	Buffalo	3	7,000
New York	New York City	12	5,000
North Carolina	Raleigh	16	3,400
North Dakota	Bismarck	−24	8,800
Ohio	Columbus	2	5,600
Oklahoma	Tulsa	12	3,800
Oregon	Portland	21	4,600
Pennsylvania	Philadelphia	11	4,400
Pennsylvania	Pittsburgh	5	6,000
Rhode Island	Providence	6	6,000
South Carolina	Charleston	23	2,000
South Dakota	Sioux Falls	−14	7,800
Tennessee	Chattanooga	15	3,200
Texas	Dallas	19	2,400
Texas	San Antonio	25	1,600
Utah	Salt Lake City	5	6,000
Vermont	Burlington	−12	8,200
Virginia	Richmond	14	3,800
Washington	Seattle	28	5,200
West Virginia	Charleston	9	4,400
Wisconsin	Madison	−9	7,800
Wyoming	Cheyenne	−6	7,400

This list of U.S. cities, with their outside design temperatures and degree days, is a useful resource for computer-aided energy analysis. (ASHRAE)

Sound Insulation for Walls and Floors

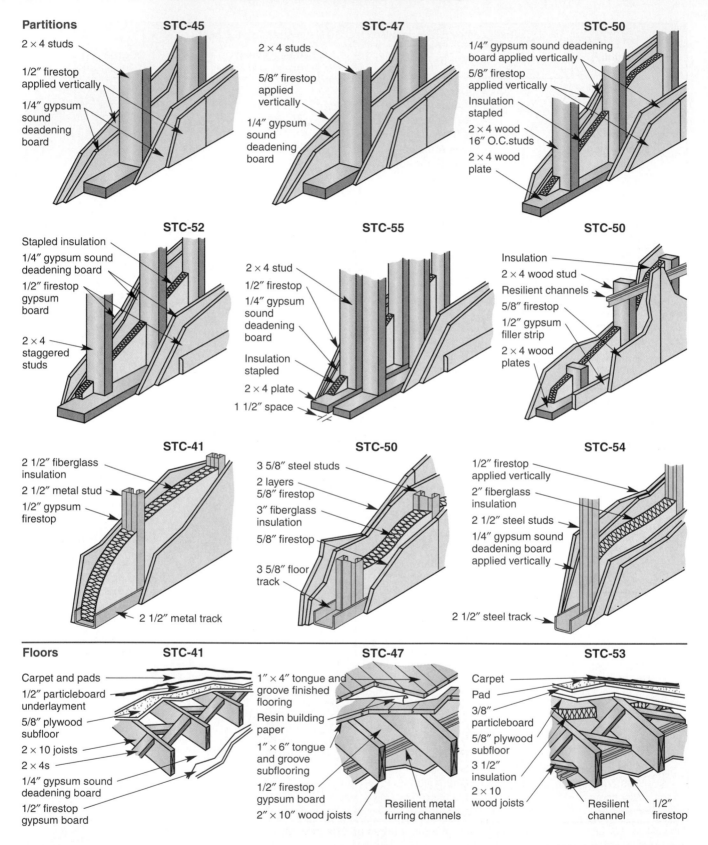

Partitions

STC-45
- 2 × 4 studs
- 1/2″ firestop applied vertically
- 1/4″ gypsum sound deadening board

STC-47
- 2 × 4 studs
- 5/8″ firestop applied vertically
- 1/4″ gypsum sound deadening board

STC-50
- 1/4″ gypsum sound deadening board applied vertically
- 5/8″ firestop applied vertically
- Insulation stapled
- 2 × 4 wood 16″ O.C.studs
- 2 × 4 wood plate

STC-52
- Stapled insulation
- 1/4″ gypsum sound deadening board
- 1/2″ firestop gypsum board
- 2 × 4 staggered studs

STC-55
- 2 × 4 stud
- 1/2″ firestop
- 1/4″ gypsum sound deadening board
- Insulation stapled
- 2 × 4 plate
- 1 1/2″ space

STC-50
- Insulation
- 2 × 4 wood stud
- Resilient channels
- 5/8″ firestop
- 1/2″ gypsum filler strip
- 2 × 4 wood plates

STC-41
- 2 1/2″ fiberglass insulation
- 2 1/2″ metal stud
- 1/2″ gypsum firestop
- 2 1/2″ metal track

STC-50
- 3 5/8″ steel studs
- 2 layers 5/8″ firestop
- 3″ fiberglass insulation
- 5/8″ firestop
- 3 5/8″ floor track

STC-54
- 1/2″ firestop applied vertically
- 2″ fiberglass insulation
- 2 1/2″ steel studs
- 1/4″ gypsum sound deadening board applied vertically
- 2 1/2″ steel track

Floors

STC-41
- Carpet and pads
- 1/2″ particleboard underlayment
- 5/8″ plywood subfloor
- 2 × 10 joists
- 2 × 4s
- 1/4″ gypsum sound deadening board
- 1/2″ firestop gypsum board

STC-47
- 1″ × 4″ tongue and groove finished flooring
- Resin building paper
- 1″ × 6″ tongue and groove subflooring
- 1/2″ firestop gypsum board
- 2″ × 10″ wood joists
- Resilient metal furring channels

STC-53
- Carpet
- Pad
- 3/8″ particleboard
- 5/8″ plywood subfloor
- 3 1/2″ insulation
- 2 × 10 wood joists
- Resilient channel
- 1/2″ firestop

How to build wall partition and floor structures with high STC (sound transmission class) ratings. Ratings are based on sound tests conducted according to ASTM-E90. (Georgia-Pacific)

Branch Pipe Sizes

Fixture	Pipe size	Fixture	Pipe size
Bathtub	1/2″	Shower	1/2″
Dishwasher	1/2″	Urinal – flush valve	3/4″
Drinking fountain	3/8″	– flush tank	1/2″
Hose bibb	1/2″	Washing machine	1/2″
Kitchen sink	1/2″	Water closet – flush valve	1″
Laundry tray	1/2″	– flush tank	3/8″
Lavatory	1/2″	Water heater	1/2″

Plastic Pipe

Type plastic	Uses	Sizes/grades available	Joining methods	Rigid/flexible
ABS	Drain/waste/vent	1 1/4″–6″ Schedule 40 (Sch. 40)	Solvent cement	Rigid
	Sewer lines	3″–8″ Standard dimension ratio (SDR)		
	Water supply	1/2″–2″ SDR		
PB	Hot and cold water distribution	1/8″–1″ CTS	Insert fittings	Flexible
		3/4″–2″ IPS	Compression fittings	
			Instant connect fittings	
PVC	Water distribution	1/4″–12″ Sch. 40 & 80 and SDR	Solvent welding	Rigid
	DWV, sewers, & process piping		Sch. 80: threading & flanges	
CPVC	Hot and cold water distribution	1/2″–3/4″	Solvent welding	Rigid
	Process piping	1/2″–8″ Sch. 40 & 80	Threading & flanges	
PE	Water distribution	1/2″–2″ IPS	Compression & flange fittings	Flexible
	Natural gas distribution Oil field piping	1/2″–6″ IPS		
	Water transmission, sewers	3″–48″ SDR	Fusion welding	
SR	Agricultural field drains Storm drains	3″–8″ SDR	Solvent welding	Rigid

Copper Pipe

Type	Color Code	Application	Straight lengths	Coils (soft temper only)
K	Green	Underground and Interior service	20 ft. in diameters including 8 in. Hard and soft temper	60 ft. and 100 ft. for diameters including 1 in.
L	Blue	Above-ground service	20 ft. in diameters including 10 in. Hard and soft temper	(same as K)
M	Red	Above-ground water supply drainage, waste, and vent	20 ft. in all diameters Hard temper only	Not available
DWV	Yellow	Above-ground drain, waste, and vent piping	20 ft. in all diameters 1 1/4 in. and greater Hard temper only	Not available

Insulated Wire Coverings

Covering types	Letter designation
Rubber	RH, RHH, RHW
Thermoplastic Compound	TW, THW, TBS, THHN

Conductor Applications

Wire type	Apply where	°C	Temperature maximum °F
RH	Dry/damp	75	167
RHH	Dry/damp	90	194
RHW	Dry/wet	75	167
TW	Dry/wet	60	140
THHN	Dry/damp	90	194
THW	Dry/wet	90	194 Under special conditions
THWN	Dry/wet	75	167
XHHW	Dry/damp Wet	90 75	194 167
MI	Dry Wet	90 250	194 Under special conditions 482

Conductor Colors

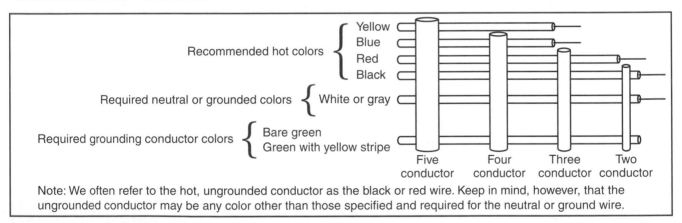

Recommended hot colors { Yellow, Blue, Red, Black

Required neutral or grounded colors { White or gray

Required grounding conductor colors { Bare green, Green with yellow stripe

Five conductor   Four conductor   Three conductor   Two conductor

Note: We often refer to the hot, ungrounded conductor as the black or red wire. Keep in mind, however, that the ungrounded conductor may be any color other than those specified and required for the neutral or ground wire.

Electrical Switches

Type	Purpose
Single-pole, single-throw switch	Controls light or outlet from single location.
Three-way switch	Controls light or outlet from two locations.
Four-way switch	Controls light or outlet from other locations in between pair of three-way switches.
Dimmer switch	Like single-pole, single-throw switch, but also contains rheostat or voltage regulating device which allows all or only portion of electrical energy to outlet or light fixture.
Pilot-lighted switch	Used to control light or outlet which is not in sight. It indicates, by use of pilot light, whether power to device is on or off.
Time-delay switch	Used where delayed shut-off is desirable.
Others: Toggle switch, pushbutton switch, pull chain switch, photoelectric switch, knife switch, tap switch, plate switch, locking switch.	

Glossary of Technical Terms

32 mm construction system: Frameless construction system for cabinetry that has a set of 5 mm holes and spaced 32 mm on center on panels designed to be used on various case types and sizes.

A

Acoustical ceiling tile: Ceiling tile designed to reduce sound by absorbing sound waves.

Acoustical materials: Types of tile, plaster, and other materials that absorb sound waves. Generally applied to interior wall surfaces to reduce reverberation or reflection of sound waves.

Acoustical plaster: Plaster generally used where an unlined or plain surface wall or ceiling is desired, or where curved or intricate planes make the use of sheet materials impractical.

Active solar construction: Construction includes a system that uses mechanical means to transfer a medium heated by solar energy.

Adhesive: A substance capable of holding material together by surface attachment. Also, a general term that includes glue, cement, mastic, and paste.

Admixtures: Chemicals added to concrete to change the characteristics of the mix.

Affective behavior: Attitude and interpersonal relationships.

Aggregate: Materials such as sand, rock, and gravel used to make concrete.

Air conditioning: Control of temperature, humidity, movement, and purity of air in buildings.

Air exchanger: A device that exhausts air from a building and also draws fresh air in. The temperature of the exhausted air is transferred to the incoming fresh air.

Air return: A series of ducts in an HVAC system that return unconditioned air to the air handler.

Air vents: Devices used to purge air that has entered a hydronic heating system.

Air-dried lumber: Lumber seasoned by exposure to the atmosphere without artificial heat. Stacks of lumber may be either covered or uncovered.

Air-entraining agents: Chemical additives that trap tiny air bubbles in concrete, improving its workability and freeze-thaw durability.

Alkyd: A paint in which the vehicle (binder) is an alkyd resin (a type of synthetic resin).

All-hard blades: Heat-treated hacksaw blades that are very brittle and easily broken if misused.

Alligatoring and checking: Paint defects resulting from application of relatively hard coatings over primers that are softer.

Alternating current (ac): An electric current that regularly reverses direction; this type of electric current is used in homes.

Altitude: The distance a rafter extends upward above the wall plate.

Aluminum siding: Metal siding with a factory-finished, baked-on enamel. Can be applied over wood, stucco, concrete block, and other structurally sound surfaces.

Ambient noise: Existing noise in a habitable room that tends to mask the sound entering the room.

Ammoniacal copper quaternary ammonium compound (ACQ): Nontoxic preservative used on lumber. Can cause galvanic corrosion of some types of fasteners, posing possible structural dangers.

Anchor bolts: Bolts embedded in concrete to hold structural members in place.

Anchor straps: Strap fasteners that are embedded in concrete or masonry walls to hold sills in place.

Anchor: A device, usually of metal, that is used to fasten structural building members together.

Angle of repose: The greatest slope at which loose material, such as excavated soil, will stand without sliding.

Annular rings: Rings or layers of wood that represent one growth period of a tree. In cross section, the rings may indicate the age of the tree.

Anodizing: A method of coating metal objects for protection or decoration.

Apprenticeship: A formal method of learning a certain trade, such as carpentry, that involves instruction as well as working and learning on the job. After mastering prescribed tasks over a specified time period, the apprentice is certified as a journeyman and can work without supervision.

Apron: A piece of horizontal trim applied against the wall immediately below the stool of a window.

Architectural drawings: Plans that show the shape, size, and features of a building.

Ash dump: A metal frame with pivoted cover provided in the rear hearth for clearing ashes, if there is space below for an ash pit.

Asphalt: A residue produced from evaporated petroleum used for waterproofing roof coverings, exterior wall coverings, and flooring tile. Insoluble in water, but soluble in gasoline, and melts when heated.

Asphalt shingles: The most common type of roofing material used today. They are manufactured as strip shingles, interlocking shingles, and large individual shingles.

Attached deck: A deck that has a supporting ledger physically attached to the building wall.

Awning window: A window type with a sash that is hinged at the top, allowing the bottom to swing outward.

B

Backfill: The soil replaced around foundations after excavating.

Backfilling: The process of replacing soil that was removed during excavation for the building foundation.

Backing board: In a two-layer drywall system, the base panel of gypsum drywall. It has gray liner paper as facing and is not suitable as a top surface. Also referred to as *backer board*.

Backing sheet: Laminate sheet that is like the regular laminate without the decorative finish and is usually thinner.

Backsaw: Saw with a thin blade reinforced with a heavier steel strip along the back edge. The teeth are small (14–16 points), producing fine cuts. Used mostly for interior finish work.

Backsplash: The vertical surface, usually several inches tall, at the back of a countertop.

Balloon framing: A type of building construction with upright studs that extend from the foundation sill to the rafter plate. A ribbon supports second floor joists. Not commonly used in new construction.

Baluster: Vertical stair member that supports the stair rail.

Balustrade: A railing consisting of a series of balusters resting on a base, usually the treads, that supports a continuous stair or handrail.

Band joist: A joist set perpendicular to the floor joists and to which the ends of the floor joists are attached. Also called *rim joist* or *header*.

Banjo: Taping tool used to apply joint compound and tape at the same time. Has a reservoir to hold the compound and a reel to hold the tape.

Bar or pipe clamps: Clamps useful for holding larger pieces while they are receiving glue or mechanical fasteners.

Baseboard: A finishing board that covers the joint where a wall intersects with the floor.

Baseboard electric resistance heating: An electric heating element wrapped in a tubular casing with metal fins spaced along its length that help radiate the heat to the room air.

Base flashing: The part of chimney flashing that is attached to the roof.

Base level line: A level line that is snapped on the wall from the high point around the wall as far as the cabinets will extend.

Basement: The part of a house that is partly or completely below grade.

Base shoe: Small narrow molding used around the perimeter of a room where the baseboard meets the finish floor.

Batten: A strip of wood placed across a surface to cover joints. In roofing, a horizontal strip placed across the roof in rows to provide a support for clay or cement roofing tile.

Batter: The slope, or inclination from vertical, of a wall or other structure (or a portion of a structure).

Batter boards: A temporary framework of stakes and ledger boards used to locate corners and building lines when laying out a foundation.

Batt insulation: Insulation made of the same fibrous material as blankets, but cut into shorter lengths.

Bay: One of the intervals or spaces into which a building plan is divided by columns, piers, or division walls.

Bay window: A window or group of windows extending beyond the main wall of a building, usually supported on a foundation or by cantilevered joists.

Beam: A horizontal structural member used between posts, columns, or walls to support other beams or joists. May be attached to the building walls or rest on posts or columns. Also called *girders.*

Bearers: Horizontal members (usually 2 × 6s) used to connect scaffold uprights. Also called *cross ledgers.*

Bearing wall: A wall that supports a vertical load in addition to its own weight. Also called a *bearing partition.*

Bedding: A layer of mortar, putty, or other material used to secure a firm bearing.

Bed molding: A molding applied where two surfaces come together at an angle. Commonly used in cornice trim, especially between the soffit and frieze.

Belt sander: Type of portable sander. Its size is determined by the width of the belt.

Benchmark: A mark on a permanent object fixed to the ground from which grade levels and elevations are taken for construction of a building. Sometimes officially established by government survey.

Bevel siding: Used as finish covering on the exterior of a structure. Usually manufactured by diagonally sawing boards to produce two wedge-shaped pieces.

Bevel: To cut to an angle other than a right angle, such as the edge of a board or door.

Bid: An offer to supply, at a specified price, materials, supplies, equipment, an entire structure, or sections of a structure.

Bill of materials: A list of all materials corresponding to a drawing or set of drawings for a project. Also called a *lumber list* or *mill list.*

Binder: Ingredient of paint that remains behind after the oils or resins have dried.

Bird's mouth: A notch cut on the underside of a rafter to fit the top plate.

Blade: The longer, 24″ arm of a framing square. Also called the *body.*

Blemish: Any defect, scar, or mark that tends to detract from the appearance of wood.

Blind nailed: Nailing concealed by installation of another strip of wood; used in tongue-and-groove flooring and in securing door frames to wall framing.

Blind stop: A member applied to the exterior edge of the side and head jamb of a window to serve as a stop for the top sash and to form a rabbet for storm sash, screens, blinds, and shutters.

Blistering and peeling: Paint defect caused by water collecting under the paint film.

Block flooring: Wood flooring cut in square blocks.

Block plane: A small surfacing plane that produces a fine, smooth cut, making it suitable for fitting and trimming work. The blade is mounted at a low angle and the bevel of the cutter is turned up.

Blocking: Solid bridging lumber placed between joists to distribute the floor load. Also, solid pieces fastened between studs in a wall, usually to serve as a base for fastening other materials.

Blue stain: A stain caused by a fungus growth in unseasoned lumber, especially pine. Does not affect the strength of the wood.

Board and batten siding: Siding applications designed around wide, square-edged boards spaced about 1/2" apart.

Board foot: The equivalent of a board 1' square and 1" thick.

Board: Lumber less than 2" (nominal) thick.

Board-on-board siding: A variation of board and batten siding with, in effect, wider battens.

Boiler: A heating unit or heat source for a hydronic heating system.

Booking: A method of pasting and folding wall coverings that allows the paste to soak into the wall covering.

Bored wells: Wells that are made with an earth auger.

Boring: Forming holes in wood that are larger than 1/4" in diameter.

Bow: Warping where the side of the lumber is curved from one end to the other.

Bow window: Similar to a bay window, but may be supported by cables.

Box: Enclosure in an electrical system that houses device and electrical connections.

Box beam: A beam made of one or more plywood webs connected to lumber flanges or built from four dimensional lumber members nailed together to form a box-type header.

Box cornice: A construction that encloses the rafters at the eaves.

Box covers: Covers used to conceal and protect devices, such as switches and receptacles, that are placed in boxes.

Bracket: A projecting support for a shelf or other structure.

Branch circuit: A single circuit, with a fuse or circuit breaker, running from a distribution panel.

Brick construction: A type of construction in which the exterior walls are bearing walls made of brick.

Brick moulding: A moulding for window and exterior door frames. Serves as the boundary moulding for brick or other siding material and forms a rabbet for the screens, storm sash, or combination door.

Brick veneer construction: A type of construction in which a wood-frame structure has an exterior surface of single thickness of brick.

Bridging: Wood or metal pieces fitted in pairs from the bottom of one floor joist to the top of adjacent joist to distribute the floor load.

British thermal unit (Btu): The amount of heat needed to raise the temperature of 1 lb. of water 1°F.

Brown coat: A second layer of plaster that is applied on top of the scratch coat to form a base for the finish coat.

Buck: A temporary wooden form installed within the forms for poured concrete walls to provide an opening or void for later installation of features such as windows or doors.

Builder's level: An optical device used by builders to determine grade levels and angles for laying out buildings on a site. Also called a *dumpy level* or *optical level.*

Building code: A collection of rules and regulations for construction established by organizations and based on experience and experiment.

Building lines: The lines marking where the walls of a structure will be located.

Building paper: Paper and felt combination that prevent infiltration of air when applied to the outside walls of a building, while allowing vapor to pass through.

Building permit: A document issued by building officials, certifying that plans meet the requirements of the local building code. The issuing of a permit allows construction to begin.

Building site: The place where a building is to be constructed.

Build-up strips: Narrow wood pieces that are used to attach the countertop to the base cabinets.

Built in: A piece of furniture (normally cabinets, cupboards, and other large pieces) that is attached to the building frame and, therefore, cannot be readily moved.

Built-up roof: A roof covering composed of several layers of rag felt or jute saturated with coal tar, pitch, or asphalt finished with crushed slag or gravel. Generally used on flat or low-pitched roofs.

Built-up stringer: A stringer to which blocking has been added to form a base for adding treads and risers.

Burnisher: Tool used to form the hook on the edge of a scraper.

Butt: Type of door hinge where one leaf is fitted into a space routed into the door frame jamb and the other into the edge of the door.

Butterfly roof: An inverted gable roof with the low ends joined.

Butt gauge: Layout tool used to lay out the gain (recess) for hinges.

C

Cabinet: Case or box-like assembly consisting of shelves, doors, and drawers. Used primarily for storage.

Cabinet drawer guide: A wood or plastic strip used to guide the drawer as it slides in and out of its opening.

Cabinet drawer kicker: Wood cabinet member placed immediately above and generally at the center of a drawer to prevent the drawer from tilting down when it is pulled out.

Cabinetwork: A term that refers to built-in kitchen and bathroom storage and, in a general way, to such work as closet shelving, wardrobe fittings, desks, bookcases, and dressing tables.

Cable layout: A print to which an architect or electrician has added drawings and symbols to show conductor sizes, number of conductors in a run, and path of conductors as a guide to the electrician or electricians making the installation.

CADD-CAM: The combination of computer-aided design and computer-assisted manufacture.

Camber: A slight arch in a beam or other horizontal member that prevents it from bending into a downward or concave shape due to its weight or load.

Cambium: The layer of wood cells located beneath the bark of a tree.

Cant strip: A strip of wood, triangular in cross section, used under shingles at gable ends or under the edges of roofing on flat decks.

Cantilevered: Horizontally extending beyond a supporting surface.

Carpenter's level: Tool used for checking for level and plumb for members.

Cased opening: An interior opening without a door that is finished with jambs or trim.

Casein glue: An adhesive of casein and hydrated lime that is suitable for gluing oily woods and for laminating wood with a high moisture content.

Casement: A window in which the sash pivots on its vertical edge, so it may be swung in or out.

Casing: The trim around either the outside or inside of a door, window, or other opening. Also, the finished lumber around a post or beam.

Caulk: Flexible material used to seal and waterproof cracks and joints, especially around window and exterior door frames.

C-clamp: Clamping tool that adapts to a wide range of assemblies where parts need to be held together. Available in sizes from 1"–12".

Ceiling frame: The system of support for all components of the ceiling.

Ceiling joists: Main ceiling framing members.

Cement: Binding material that forms concrete when combined with water and aggregate.

Cement board: Fireproof, moisture-resistant, fiber-reinforced cement panels used as a base for finishing materials on walls, floors, and countertops.

Cementitious: Made of cement.

Center guides: A strip of wood or plastic centered between front and rear rails of a cabinet frame to keep a drawer aligned.

Central heating: A single heating plant produces heat that is then circulated throughout the structure by perimeter heating or radiant heating.

Chain saw: Gasoline- or electric-powered saw used to cut heavy timbers, posts, or pilings. Blade lengths are 10"–20".

Chair: Small metal fixture used to hold reinforcing bars away from the ground prior to the pouring of concrete slabs.

Chair rail: Interior moulding applied along the wall of a room to prevent the back of a chair from marring the wall.

Chalk line: Tool used to mark long, straight lines. A thin, strong cord that is covered with powdered chalk is held tight and close to the surface. Then it is snapped, driving the chalk onto the surface and forming a distinct mark.

Chamfer: Corner of a board beveled at a 45° angle. Two boards butt-joined with chamfered edges form a V-joint.

Channel rustic siding: Horizontal wood siding that is usually 3/4″ thick; 6″, 8″, or 10″ wide; and has shiplap-type joints.

Chase: A wood frame jutting from an outside wall that supports a prefabricated chimney. Also, a channel inside of a building for running mechanical lines.

Check: A lengthwise split in a piece of lumber caused by more rapid drying than the rest of the piece.

Checkrails: Meeting rails of a double-hung window that are made thicker to fill the opening between the top and bottom sash. They are usually beveled. Also called a *meeting rail.*

Chevron paneling: Strip paneling diagonally installed in a chevron pattern.

Chimney: A vertical, hollow shaft that exhausts smoke and gases from heating units and incinerators.

Chlorinated polyvinyl chloride (CPVC): A buff-colored thermoplastic used in plumbing.

Chord: The top or bottom member of a floor or roof truss.

Chromated copper arsenate (CCA): Wood preservative treatment once widely used for lumber, but no longer available for residential use due to toxic emissions.

Circuit: A path for electrical power provided by wires (conductors) between a power source and the power-using device (load).

Circuit breaker: Device in a circuit that opens the circuit, preventing the flow of electricity, so that it will not be damaged by an overload or short.

Circulator: A factory-built, metal unit used to increase the heating capacity of a fireplace by circulating cool air from the room around the firebox and exhausting it into the room.

Class A fire: Result from burning wood and debris.

Class B fire: Involve flammable liquids such as gasoline, oil, paints, and oil soaked rags.

Class C fire: Result from faulty electrical wiring and equipment.

Claw hammer: A hammer having a head with one end forked for removing nails.

Cleat: A strip fastened to a wall to support a shelf, fixture, or other objects.

Closed panels: Factory-built housing wall panels that are finished on both sides.

Closed porch: One that is framed with studs and sided like the rest of the house. It is more like an extension of the house than a transitional area.

Closed-cut valley: Method of shingling valleys by carrying each course across the valley onto the adjoining slope. When the adjoining slope is shingled, the overlapping shingles are cut at the valley.

Closet pole: A shaft installed in closets to support hangers.

Closure block: The last concrete block to be placed in any course.

Coatings: All types of finishes, including paints, stains, varnishes, and various synthetic materials.

Coefficient of thermal conductivity (k): The amount of heat, in Btu, transferred in one hour through 1 sq. ft. of a given material that is 1″ thick and has a temperature difference between its surfaces of 1°F.

Cognitive behavior: Knowledge of any sort; in this case, carpentry work.

Coil: An iron core wrapped in conducting wires in which a magnetic field is produced when current is run through the wrapping.

Collar tie: A beam connecting rafters, located considerably above the wall plate. Also called *rafter tie.*

Color wheel: A circular diagram in which related colors are next to each other and complementary colors are opposite. It is useful in picking colors that go well together.

Column: Vertical supporting member in a building.

Combination square: Tool that serves a similar purpose to a try square, and is also used to lay out miter joints. Its adjustable sliding blade allows it to be conveniently used as a gauging tool. It can be used as a marking gauge to make parallel lines and as a square for 90° and 45° angles.

Commercial standard: A voluntary standard that establishes quality, methods of testing, certification, rating, and labeling of manufactured items. It provides a uniform base for fair competition.

Common difference: The difference in length between any two adjacent studs when laying out a gable end frame.

Common rafter: A rafter connected to both the ridge and the wall plate.

Complementary colors: Hues opposite of each other on a color wheel.

Composite board: A board consisting of a core of wood between veneered surfaces.

Composite decking: Material made with a blend of wood fibers and plastic resins. It is worked like wood and creates a low maintenance deck surface.

Composite panels: Panels made up of a core of reconstituted wood with a thin veneer on either side.

Compound miter: An angled cut made across the width and through the thickness of a workpiece.

Compression fittings: Connectors that seal joints by compressing a ring of soft material, such as brass.

Compression valves: Valves that make use of a tapered plug or a washer for full seating to stop water flow or partial seating to control the flow.

Computer-aided drafting and design (CADD): Using computers to create a set of plans and perspective drawings.

Computer-assisted manufacture (CAM): A combination of design and manufacturing activities controlled by a computer.

Concealed header: A header that is butted against the ends of the joists it is supporting.

Concrete: Building material formed by combining cement, water, and aggregate.

Concrete blocks: Blocks made from a mixture of portland cement and aggregates such as sand, fine gravel, or crushed stone. They weigh 40 lb.–50 lb.

Condensation: Water vapor changed to liquid by being cooled; also the collecting of this water in droplets on a surface.

Condenser: The part of an air conditioning unit that receives heated coolant and releases its heat to the atmosphere.

Conductance (C): The amount of heat (Btu) that will flow through the material in one hour per sq. ft. of surface with 1°F of temperature difference.

Conduction: Method of heat transfer in solid material with movement of heat from one end to the other.

Conductor: A material, such as a copper wire, that carries an electric current.

Conduit: A tube of metal or plastic in which wiring is installed.

Conifers: Needle-bearing trees, such as evergreens. Softwoods come from these trees.

Connectors: Fasteners that connect two or more conductors so that current may pass from one to the other.

Contact cement: Neoprene-rubber-based adhesive that instantly bonds on contact of the parts being fastened.

Continuity monitor: A device that checks whether grounding conductors in electrical cords are connected to ground (have continuity).

Control point: A reference point for determining the elevation of footings, floors, and other parts of a building; the highest elevation on the perimeter of an excavation. Also called *high point*.

Convection: Movement of heat through liquids and gases as a result of heated material rising and being replaced by cooler material. Also, the movement of heat by way of a carrier, such as air or water.

Convection current: Airflow produced when heated air becomes lighter and rises.

Convenience outlet: Electrical outlet into which portable equipment such as lamps may be plugged. Also called a *receptacle*.

Convex: An outward-curving shape, like that of a magnifying lens.

Cooling coil: The system of coils or tubes in an air conditioning unit that receives the high-pressure liquid refrigerant and allows it to expand and vaporize, causing the refrigerant to absorb heat.

Cope: To cut or shape the end of a wood moulding member so it will fit the contour of an abutting piece of moulding.

Coping saw: Saw with a thin, flexible blade that is pulled tight by the saw frame. With a blade that has 15 teeth per inch, this saw can make very fine cuts. It is occasionally used for cutting curves.

Copper azole (CA): A copper-based nontoxic wood preservative. Requires the use of stainless steel or heavily galvanized fasteners.

Corbel: To extend outward from the surface of a masonry wall one or more courses to form a supporting ledge.

Corbel block: Concrete block shaped and dimensioned to form a shelf in a wall.

Core: Innermost layer(s) of plywood or the interior area between the face veneers of a flush door.

Corner bead: Moulding used to protect corners. Also, a metal reinforcement placed on corners before plastering.

Corner braces: Diagonal braces let into studs to stiffen and reinforce frame structures.

Corner guide: Device that keeps drawers in place and allow them to slide in and out.

Corner posts: Finish trim pieces used for inside and outside corners in a vinyl siding installation.

Cornice: Exterior trim of a structure at the meeting of the roof and wall; usually consists of panels, boards, and mouldings. Also, an inclusive term for all the parts that enclose or finish off the overhang of a roof at the eaves.

Counterboring: Boring a larger hole partway through stock to recess a bolt head.

Counterflashing: Flashing used on chimneys at the roofline to cover base flashing and prevent leaks. Also called *cap flashing*.

Course: One row of building material such as block, brick, some siding, and roofing.

Course pole: A straight edge, often made from a piece of lumber, marked up to check the height of each course of concrete block.

Cove base: A flexible moulding glued in place to hide the joint between a tile floor and a wall.

Cove moulding: Moulding with a concave profile, used primarily where two members meet at a right angle.

Coverage: The amount of weather protection provided by the overlapping of shingles.

Cracking and scaling: Paint defects resulting from application of paint lacking elasticity.

Crawling: Coating problem in which the finish gathers up into small globules and does not cover the surface.

Crawl space: Unheated space directly above the ground; space between the ground and the floor joists is only a few feet.

Cripple jack: A rafter that intersects neither the wall plate nor the ridge and is terminated at each end by hip and valley rafters. Also called a *cripple rafter*.

Cripple stud: A stud used above or below a wall opening. Extends from the header to the top plate or from sole plate to rough sill. Also called *jack stud*.

Cripple wall: A short wall section between the foundation sill and first floor framing. Also called *underpinning*.

Cross bridging: Set diagonally between the joists to form an X to keep the joists in a vertical position and transfer the load from one joist to the next. Also called *herringbone bridging*.

Cross bands: Inner layers of plywood between the face plies and the core.

Crosscut saw: Saw with pointed teeth designed to cut across the wood grain.

Crown: A slight warping in lumber. Also called a *crook*.

Curing: Preventing loss of moisture in fresh concrete to strengthen it and reduce shrinkage, which also reduces cracking.

Current: Flow of electrons through electrical wiring.

Curtains: Coating problem caused by too-heavy an application.

Curtain wall: A wall, usually nonbearing, between piers or columns.

Cut: The process of removing material to achieve the desired grade.

Cut-in point: The temperature setting at which a thermostat will signal the furnace to turn on.

Cut-out point: Temperature setting at which a thermostat will signal the furnace to turn off.

Cut-out stringer: A stair stringer into which the rise and run are cut.

Cutting in: Painting into the corners formed by ceiling and wall.

D

Dado: A rectangular groove cut in wood across the grain. Also, rectangular horizontal grooves cut in stair stringers to support the treads.

Damper: A venting device in fireplaces used to control combustion, prevent heat loss, and redirect downdrafts.

Darby: A long flat tool used to level the stucco brown coat.

Datum: An established reference point for determining elevations in an area. Also called *benchmark* or simply the "beginning point."

Dead load: The weight of permanent, stationary construction and equipment included in a building.

Deadbolt: Special door security consisting of a hardened steel bolt and a lock. The lock is operated by a key on the outside and by either a key or handle on the inside.

Decay: Disintegration of wood substance due to action of wood-destroying fungi.

Decibel: The unit of measurement used to indicate the loudness or intensity of sound.

Decibels reduction: An expression used to indicate the sound insulating properties of a wall or floor panel.

Deciduous: Broadleaf trees that lose their leaves at the end of the growing season.

Deflection: Bending downward at the center of a joist.

Degree day: Method of measuring the harshness of climate for insulation and heating purposes. One degree day is the product of one day and the number of degrees the mean temperature is below 65°F.

Detail drawing: In working drawings, an enlarged view of a part of the structure that cannot be clearly explained in elevation views or general plan drawings.

Dewpoint: Temperature at which air is sufficiently cooled for water vapor to condense.

Diagonal paneling: Strip paneling installed at a 45° angle.

Differential: The difference between the cut-in and cut-out points of a thermostat.

Dimension lines: Lines on a drawing that show distances and sizes.

Dimension lumber: Lumber 2"–5" thick and up to 12" wide.

Dimensional stability: The ability of a material to resist changes in its dimensions due to temperature, moisture, and physical stress.

Direct current (dc): An electric current flowing in one direction, such as that provided by a battery.

Direct gain system: Passive solar construction in which the sun directly shines into living space to heat it.

Direct heating system: A heating system that directly gives off heat to objects without benefit of ducts or air system.

Disodium octaborate tetrahydrate (DOT): A borax-based wood preservative that is nontoxic and does not cause galvanic corrosion to fasteners.

Door casing: Trim applied to each side of the door frame to cover the space between the jambs and the wall surface.

Door pulls: Hardware that provides a means of grasping and opening a door.

Door frame: An assembly of wood parts that form an enclosure and support for a door.

Door stop: A moulding nailed to the faces of the door frame jambs to prevent the door from swinging through.

Dormer: A framed structure extending from and fastened to a roof slope and having a roof of its own. It provides a wall surface for installing windows.

Double layer: Gypsum board construction used where strength, insulation qualities, and improved fire resistance are important.

Double coursing: A method of wall siding with shingles that involves applying a second layer over each course.

Double-hung window: Window that consists of two sashes that slide up and down past each other in the window frame.

Downdraft: A flow of air down a chimney.

Draft excluder: A threshold or felt weatherstop.

Drainage, waste, and venting (DWV): The nonpressure part of a plumbing system that carries away wastewater and solid waste.

Draw: To pull combustion gases into the atmosphere.

Drawer guides: Wood strips or metal devices that support drawers at the sides or lower corners.

Drawer pulls: Hardware that provides a means of grasping and opening a drawer.

Drawn to scale: The reduced drawing is in exact proportion to the actual size.

Drilled wells: Wells used where great depths are needed to reach a suitable water table.

Drip cap: A moulding that directs water away from a structure to prevent seepage under the exterior facing material. Applied mainly over window and exterior door frames.

Drip edge: A metal edging placed along the rake and eaves before installing shingles.

Drip groove: Semicircular groove on the underside of a drip cap or the lip of a windowsill that prevents water from running back under the member.

Drop siding: Siding usually 3/4″ thick and 6″ wide and machined into various patterns with tongue-and-groove or shiplap joints.

Dry rot: A term loosely applied to many types of decay. When the decay is in an advanced stage, the wood may be easily crushed to a dry powder.

Drywall: Sheet material consisting of a uniform layer of gypsum sandwiched between facings of paper; used as an interior wall covering that can be painted or covered with decorative paper or paneling. Also called *sheetrock* or *gypsum wallboard*.

Drywall screw shooter: Tool used for drywall installation. Drywall screws are inserted one at a time into the nose. Locators control depth below the face of the drywall. Also called *screw gun*.

Drywood termites: Airborne termites that do not need moisture or contact with the ground. They bore into wood structures and furniture.

Ducts: Tubes or passages used to move heated or cooled air from the furnace or air conditioner to living spaces of the building.

Dust mask: A woven covering worn over the mouth and nose to filter out dust and larger particles of airborne debris; not effective in trapping very small particles, such as paint mists.

E

Eased edge: A corner rounded or shaped to a slight radius.

Eaves: The lower part of a roof that projects over an exterior wall. Also called the *overhang*.

Eaves flashing strip: Roof flashing installed at the eaves where outside temperatures drop to 0°F (–17°C) or colder and there is a possibility of ice forming along the eaves.

Eaves trough: A waterway built into the roof surface over the cornice.

Edge-grained lumber: Boards sawed so that the annular rings form an angle with the face of greater than 45°.

Electrical wiring: The wires, boxes, and devices that control the distribution and use of electrical current in a building.

Electric current: Flow of electrons generated when a conductor, such as copper, silver, or aluminum wire, is passed through a magnetic field.

Electric drill: A portable tool powered by electricity and primarily used for drilling holes. May be corded and cordless (battery-powered).

Electric moisture meter: Determines the moisture content of wood based on electrical resistance or capacitance, which varies with change in moisture content.

Electric radiant heat: A system that radiates heat by moving it through solid matter, such as a concrete floor.

Electron: A negatively charged atomic particle.

Elevation: The height of an object above grade level. Also, a type of drawing that shows the front, rear, or side of a building.

Emissivity: The ability of a material to emit heat by radiation.

Energy efficiency rating (EER): A measure of how efficiently an appliance uses energy. The EER must be displayed on a label for all new energy-using appliances.

Engineered lumber products (ELP): Lumber that has been altered by manufacturing processes: wood I-beams, glue laminated beams, laminated strand lumber, laminated veneer lumber, and parallel strand lumber.

Entrance panel: Where service ends inside of the dwelling; houses a main breaker and circuit breakers or fuses for all of the home's circuits.

Entrepreneur: A person who starts and operates a new business.

Equilibrium moisture content (EMC): The moisture content at which wood neither gains nor loses moisture when surrounded by air at a given relative humidity and temperature.

Escutcheon: Protective metal plate surrounding a knob or a keyhole of a door.

Excessive chalking: Loosely-bound pigment on the surface of finished wood caused by ultraviolet rays breaking down oils in the finish coat.

Expanded metal lath: Plaster base made from a copper alloy sheet steel slit that is expanded to form openings for keying the plaster.

Expanded polystyrene foam: Material made from a petroleum by-product and recycled polystyrene; impermeable to moisture. Pentane is the blowing agent used to expand the foam and does not deplete the ozone layer.

Expansion joint: A bituminous fiber strip used to separate blocks or units of concrete to prevent cracking and buckling due to thermal expansion.

Expansion tank: A reservoir in a hydronic heating system that accommodates increased water volume caused by expansion due to heating.

Exposure: The amount of material exposed to the weather in siding or shingles.

Extended rake: A gable overhang.

Exterior finish: All exterior materials of a structure. Generally refers to siding, roofing, cornice, and trim members around openings.

Exterior insulation and finish systems (EIFS): Multilayered exterior wall system consisting of a layer of insulation board attached to the wall sheathing, a water-resistant base coat with a reinforcing mesh, and a finish coat that resembles stucco or stone.

Exterior plywood: Plywood that is bonded with waterproof glues; can be used for siding, concrete forms, and other construction where it will be exposed to the weather or excessive moisture.

Exterior trim: Parts of the cornice and rake structure that are exposed to view.

Extruded polystyrene foam: Material made from a petroleum by-product and recycled polystyrene; impermeable to moisture; has a higher R-value than expanded polystyrene foam. The blowing agent slightly depletes ozone.

F

Facade: The main or front elevation of a building.

Face: The side of the framing square with the manufacturer's name on it. The scale on this side is divided into 8ths and 16ths. Also, the outer plies of plywood.

Face frame construction: A framed cabinet in which the frame is covered by lightweight panels. Also called *frame-with-cover construction*.

Face nailing: A fastening method in which the nail is driven perpendicular to the surface of a piece and through its face.

Face plate: An assembly of solid lumber that is attached to the front edges of cabinets to give a finished appearance and provide support for doors. Also called a *face frame*.

Facing strips: Finish material (as stiles and rails) applied to the front of the cabinet frame.

Factory and shop lumber: A grade of lumber intended to be cut up for use in further manufacture. It is graded on the basis of the percentage of the area that will produce a limited number of cuttings of a specified size and quality.

Fascia: A wood member nailed to the ends of the rafters and lookouts that is used for the outer face of a box cornice.

Fascia backer: A 1″ or 2″ wood nailing strip attached to the rafter ends in a box cornice to support the fascia.

Faucets: Devices that deliver water, usually to a fixture, hose, appliance, or bucket.

Fence: A guide for making rip cuts on a table saw.

Fiberboard: A broad term used to describe sheet material; manufactured from wood, cane, or other vegetable fibers. Fiberboard is available in widely varying densities, depending on the material used for its manufacture.

Fiber-cement siding: Strong, damage-resistant siding material made from cement reinforced with cellulose fiber.

Fiber saturation point: The stage in the drying or wetting of wood at which the cell walls are saturated and the cell cavities are free from water. It is assumed to be 30% moisture content based on oven-dry weight and is the point below which shrinkage occurs.

Field: The middle area of a sheet of wallboard.

Fill: The process of adding material, such as dirt, to achieve the desired grade. Also, the material added.

Fillers: Heavy-bodied liquids used to fill open grain in wood before the application of stains or topcoats.

Finger joints: A joint made in a mill to assemble short lengths of wood into long pieces. The design maximizes the gluing surface of the pieces.

Finish carpentry: A specialty in carpentry that deals with installation of exterior and interior finish.

Finish coat: A third layer of plaster that is applied when the brown coat has set and is somewhat dry.

Finish flooring: The final floor covering.

Finishing sander: Tool used for final sanding where only a small amount of material needs to be removed. It is also used for cutting down and rubbing finishing coats.

Fink truss: Roof truss that can be used for spans up to 50′. Also called *W truss.*

Firestop: A block or stop used in a building wall between studs to prevent the spread of fire and smoke through the air space.

Fireclay: Heat-resistant clay material used for flue linings in masonry chimneys.

Firewall: A wall that subdivides a building to restrict the spread of fire.

Firsts and seconds (FAS): Best grade of hardwood lumber. Pieces are not less than 6″ wide by 8′ long and must yield at least 83 1/3% clear cuttings.

Fish tape: A long, flat, flexible steel tape used to pull electrical wires through conduit.

Fixed anchor: Metal strap connector used to bind intersecting concrete block walls together.

Fixed window: A window that is not designed to open for ventilation.

Fixtures: Devices that receive water from the water supply system and have a means for discharge into the DWV system.

Flare fittings: Fittings that require no soldering; designed to tightly fit against the tubing and mating parts.

Flashing: Sheet metal or other material used in roof and wall construction (especially around chimneys and vents) to prevent rain or other water from entering.

Flat-grained lumber: Softwood lumber cut so the annular rings form an angle of less than 45° with the board's surface.

Flat ICF wall: A solid concrete wall formed by sheets or planks of polystyrene rigid foam filled with concrete.

Flat roof: A roof that is either level or pitched only enough to provide for drainage.

Flexible insulation: Insulation made of loosely felted mats of mineral or vegetable fibers, such as glass fiber, rock wool, slag, wood fiber, and cotton; manufactured in blankets (or quilts) and batts.

Flexible-back blades: Hacksaw blades that are hardened only around the teeth; usually preferred for use on metals.

Flexible measuring tape: Easily carried measuring tape made of fabric or steel.

Flight of stairs: The steps going in the same direction between landings or floors.

Flitch: A steel plate placed between two pieces of 2″ lumber to form a reinforced header.

Float glass process: Glass manufacturing process where molten glass flows onto a flat surface of molten tin.

Floating: Smoothing the surface of wet concrete using a large, flat tool (float).

Floor plans: Line drawings that show the size and outline of the building and its rooms, including dimension lines to show the location and size of inside partitions, doors, windows, and stairs.

Flue: The passage in a chimney through which smoke, gas, and fumes rise.

Flue lining: A noncombustible sleeve, usually made of fireclay, used to provide a smooth surface inside of a masonry chimney.

Flush: Adjoining surfaces that are even, or in the same plane.

Flush door: A door that fits into the opening and does not project outward beyond the frame. In cabinetry, a door with a surface that is perfectly flat.

Flush drawer: A drawer whose front fits flush in its opening.

Foamed-in-place insulation: Insulation materials that can be installed only in open cavities since they expand in volume 100 times beyond the applied thickness. They provide superior insulating quality, seal up gaps and cracks, and are nearly impermeable to moisture.

Folding wood rule: Measuring tool with a standard length of 6'. Some have a ruled metal slide for taking inside measurements.

Footing: An enlargement (usually of poured concrete) at the base of a foundation wall, pier, or column that supports the structure while distributing its weight to the surrounding soil. Also, the entire trench footing up to grade.

Footprint: Size and shape of a building shown on a plot plan.

Forced-air perimeter system: Air is circulated through large ducts around the perimeter of the building that direct warmed air to every room in the building.

Fore and jointer plane: Standard surfacing plane, 18"–24" long.

Form: Wood, metal, or foam structures that support poured concrete until it hardens.

Form ties: Wood or plastic devices that keep concrete wall forms a uniform distance apart.

Foundation: The supporting portion of a structure; located below the structure and supported only by soil or rock.

Foundation plans: Drawings that show the layout of the foundation; often combined with basement plans.

Four-inch method: A shingling pattern for three-tab shingles where the cutouts break joints on thirds. This pattern starts the second course with a strip shortened by 4", the third by 8", and the fourth course starts with a full strip.

Frame and trim saw: A saw that is supported on a pair of overhead shafts or guides. The support rotates left and right a little more than 45° to make crosscuts and miter cuts.

Frame construction: Cabinet construction method in which solid lumber provides a frame that is not covered with panels.

Frameless construction: Cabinet construction method in which heavier panels provide support, rather than the usual framework of narrow pieces of solid wood.

Frame-with-cover construction: A framed cabinet in which the frame is covered by lightweight panels. Also called *face-frame construction*.

Framing plans: Drawings showing the size, number, and location of the structural members of a building's frame.

Framing square: A square designed for carpenters with a short section (called the *tongue*) attached at a right angle to the longer section (called the *blade*). A number of tables are imprinted on its sides.

Framing: The structure of a building that gives it shape and strength. Framing includes interior and exterior walls, floor, roof, and ceilings.

Free-standing deck: An independent structure that is not attached to the building. Also called a *grade-level deck*.

Frequency: Rate at which sound-energized air molecules vibrate; the higher the rate, the more cycles per second (cps).

Frieze: A boxed cornice wood trim member attached to the structure used where the soffit and wall meet.

Frostline: The depth that frost penetrates in any region. Footings must be located below the frostline, unless prescribed methods of insulating the ground are used.

Full-adhesive bonded: Type of sheet vinyl flooring that must be fastened to the substrate with a coat of adhesive covering the entire surface.

Furring: Narrow strips of wood spaced to form a nailing base for another surface; used to level surfaces, form an air space between two surfaces, or give a thicker appearance to the base surface.

Fuses: Protective devices that shut off electrical power when an overload occurs.

G

Gable end: The triangular portion of a wall contained between the slopes of a double-sloped roof or that portion contained between the slope of a single-sloped roof and a line horizontally projected through the lowest elevation of the roof construction.

Gable roof: A roof consisting of a single ridge with slopes in both directions; entirely made of common rafters.

Gain: Recess or mortise cut to receive the end of another structural member, hinge, or other hardware.

Galvanic corrosion: Eating away of metal due to a chemical reaction between dissimilar metals in the presence of moisture; occurs when plain steel or lightly galvanized fasteners are used with lumber that contains a copper-rich preservative.

Gambrel roof: A roof slope formed as if the top of a gable (triangular) roof were cut off and replaced with a cap having a less-steep slope. This cap has a ridge in the center.

Gate valves: Valves that control water with a flat disk that slides up and down perpendicular to the supply pipe. They are never used to restrict flow; they are either fully open or fully closed.

Girder: A principal beam used to support other beams or the ends of joists.

Glass block: Two formed pieces of glass fused together to leave an insulating airspace between.

Glazing: The process of installing glass into sashes and doors. Also refers to glass panes inserted in various types of frames.

Glazing compound: A soft, plastic, putty-like substance used in glazing lights (glass panes) in sashes and doors.

Glue block: A wood block, triangular or rectangular in shape, which is glued into place to reinforce a right angle butt joint. Sometimes

used at the intersection of the tread and riser in a stair.

Glued floor system: The subfloor panels are glued and nailed to the joists.

Glue-laminated lumber: Large beams fabricated by gluing up smaller dimension lumber.

Glulams: Beams made by gluing and then applying heavy pressure to a stack of four or more layers of 1 1/2″ thick stock.

Grade: Quality of lumber. Also, the height or level of a building site.

Grade beam foundation: Similar to a spread foundation footing but differs in that it is reinforced with rebar and rests on pilings.

Grade leveling: Finding the difference in the grade level between several points or transferring the same level from one point to another.

Gravel stop: Fabricated from galvanized sheet metal. Attached to the roof deck to serve as a trim member and to help keep the mineral surface and asphalt in place.

Greenhouse effect: Heating effect of solar energy that can easily pass through glass and creates heat that cannot easily pass back through the glass.

Ground fault circuit interrupter (GFCI): An electrical safety device able to detect a short circuit and automatically shut off power; reacts more rapidly than a fuse or circuit breaker and provides better protection against electrical shock.

Grounding: A system used for electrical safety in which an electrical wire runs from the exposed metal of a power tool to a third prong on the power plug. When used with a grounded receptacle, this wire directs harmful currents away from the operator.

Grounds: Strips of wood installed at the floor line and at openings in a wall as guides to strike off plaster.

Ground-supported slab: Slab that is directly supported on top of the ground. Also called *slab-on-grade.*

Grout: Thin mortar consisting of cement, sand, and water that is used in masonry work to bond units.

Gusset: A panel or bracket of either wood or metal attached to the corners or intersections of a frame to add strength and stiffness.

Gutter: Wood, metal, or plastic trough attached to the edge of a roof to collect and conduct water from rain or melting snow.

Gypboard: Another term for *drywall.*

Gypsum: A powdery, chalk-like, rock material that is the basic ingredient of plaster.

Gypsum lath: A common plaster base consisting of a rigid gypsum filler with a special paper cover. A standard gypsum panel measures 16″ × 48″ and is horizontally applied to the framing members of the structure.

Gypsum wallboard: Wall covering panels consisting of a gypsum core with facing and backing of paper. Also called *drywall* or *sheetrock.*

H

Hacksaw: Saw that can be used to cut nails, bolts, other metal fasteners, and metal trim. Most have an adjustable frame permitting the use of several sizes of blades.

Half round: A moulding whose face forms a half circle. Also, a window whose top forms a half circle.

Half story: That part of a building situated wholly or partially within the roof frame and finished for occupancy.

Hand screws: Wood clamps with broad jaws that distribute the pressure over a wide area. Sizes are designated by the width of the jaw opening and range from 4″–24″. Also called *wood clamps.*

Handrail: A pole installed above and parallel to stair steps to act as a support for persons using the stairs.

Hanger: Connecting hardware used to attach beam ends to other building frame members.

Hardboard: A board material manufactured of wood fiber formed into a panel; has a density of approximately 50–80 lb. per cubic foot.

Hardboard siding: A siding material manufactured by treating ground-up wood with steam and pressure to form it into strips or sheets.

Hard-coat systems: Polymer-modified exterior insulation and finish systems (EIFS).

They are about 1/4″ thick and mechanically attached to the structure.

Hard hat: Protective headgear that is able to resist breaking when struck with an 80 lb. ball dropped from a 5′ height.

Header: Horizontal structural member that supports the load over an opening, such as a window or door. Also called a *lintel.* Also, joists that sit on the sill and to which other floor joists are butted and attached.

Head jamb: Part of an interior door frame that fits between the side jambs, forming the top of the frame.

Head joint: The end of the block that will butt against the previously laid block.

Head lap: The distance in inches from the butt of an overlapping shingle or roll roofing to the top edge of the shingle or roll roofing beneath.

Headroom: The clear space between the floor line and ceiling, as in a stairway.

Hearth: The part of a fireplace that holds the fuel, contains the fire, and projects into the floor area.

Heartwood: The wood extending from the pith or center of the tree to the sapwood. Heartwood cells no longer participate in the life processes of the tree.

Heat exchanger: A series of tubes in a furnace that absorb the heat of combustion and transfer it to the plenum.

Heat pump: An appliance that can provide either heating or cooling.

Heat transmission coefficient: Hourly rate of heat transfer for 1 sq. ft. of surface when there is a temperature difference of 1°F between the air on the two sides of the surface.

Heel: The wide end of a trowel.

Herringbone: Paneling or flooring installed in an alternating diagonal pattern.

High point: The highest elevation on the perimeter of the excavation. Also called the *control point.*

Hip jack rafter: A short rafter connecting to the wall plate and a hip rafter.

Hip rafter: A rafter that runs from the wall plate to the ridge, always at a 45° angle.

Hip roof: A roof that rises from all four sides of a building.

Hip-valley cripple jack: Rafter that intersects neither the plate nor the ridge and is terminated at each end by hip and valley rafters. Also called *cripple jack, valley cripple jack,* or *cripple rafter.*

Holes: Defects in lumber that are openings caused by handling equipment or boring insects and worms.

Hollow back: Removal of a portion of the wood on the unexposed face of a wood member to more properly fit any irregularity in the bearing surface.

Hollow-core door: Flush door with a core assembly of strips or other units that support the outer faces.

Honeycombing: Separation of wood fibers inside the tree, but not always visible in the lumber.

Hopper window: A type of window with a sash hinged at the bottom that swings inward.

Horizontal sliding window: Window with two or more sashes at least one of which horizontally moves within the window frame.

Horn: The extension of a stile, jamb, or sill.

Hose bibb: A water faucet that is threaded so a hose connector can be attached.

Housed stringer: A stair stringer where the edges of the steps are covered with a board.

House wrap: Plastic sheets used to seal exterior walls against air infiltration.

Hub: Bell-shaped end of a cast-iron soil pipe.

Hue: The class of a particular color, such as red or blue.

Hydration: Heat-producing chemical reaction occurring between cement and water to form concrete.

Hydronic perimeter heating system: A system that uses water to transfer heat energy from a heat source to radiators located along the outside walls of spaces to be heated.

Hydronic radiant heating system: A system that uses water to transfer heat energy from a heat source to tubing located in floors or other structural parts of a building.

Hypotenuse: The side of a right triangle opposite of the right angle. In roof framing, the length of the rafter is the hypotenuse.

I

I-beam: A beam made of a single, thin, vertical web connecting two horizontal flanges.

Ice-and-water barrier: A self-sealing, waterproof covering installed at eaves and in valleys before installing roofing.

Impact sounds: Sounds that are carried through a building by the vibrations of the structural materials themselves.

Incinerator: A device that consumes household waste by burning it.

Indirect gain system: Passive solar construction in which solar heat is stored in structures of masonry, water, or another medium and then passed along to the living space by radiation, conduction, or convection.

Induction: A process by which an electric current is produced when a conductor cuts through a magnetic field.

Infiltration: Movement of air into an enclosed space through cracks and other openings.

Insulated concrete form (ICF): A lightweight, rigid, polystyrene form material that becomes a part of the wall providing an insulating factor.

Insulated glass: Multiple glazing with a dead air space or inert gas between the panes.

Insulating fiberboard lath: Plaster base that comes in a 3/8″ and 1/2″ thickness with a shiplap edge; has considerable insulation value and may be used on ceilings or walls adjoining exterior or unheated areas.

Insulation: Any material high in resistance to heat transmission that is placed in structures to reduce the rate of heat flow. Also, the nonconducting covering on electrical conductors.

Insulation board: Rigid insulation attached to the outside surface of the foundation to insulate floors or foundations.

Interior finishing: The installation of cover materials to walls and ceilings.

Interior plywood: Plywood manufactured with either interior or, more typically, exterior glue. A standard panel size is 4′ wide by 8′ long and common thicknesses are 1/4″, 3/8″, 1/2″, and 3/4″.

Interior trim: General term for all of the moulding, casing, baseboard, and other trim items applied by finish carpenters within the building.

Intermediate color: A color that is a combination of a primary color and a secondary color already containing that primary color, such as red-violet, blue-violet, blue-green, yellow-orange, and red-orange.

Internal heat: Heat gain in a home from appliances and occupants.

International Building Code: Building code encompassing previous codes and endorsed by code-writing organizations; adopted by many states.

Isolated gain system: Passive solar construction in which generated heat is stored in a separate sun space and transported to the living space by mechanical means.

Isometric drawing: Drawing in which three surfaces of an object can be viewed. The base of each surface is drawn at a 30° angle from the horizontal.

J

Jack plane: Surfacing plane, 14" long.

Jack posts: Adjustable steel posts used in place of studs for shoring.

Jack rafter: A short rafter framing between the wall plate and a hip rafter or a hip or valley rafter and ridge board.

Jack studs: Studs used above or below a wall opening. They extend from the header to the top plate or from sole plate to rough sill. Also known as *cripple studs.*

Jalousie window: A window with a series of small, horizontal, overlapping glass slats held together by an end metal frame attached to the faces of window frame side jambs or the door stiles and rails. The slats or louvers simultaneously rotate to open.

Jamb: The top and two sides of a door or window frame that contact the door or sash; top jamb and side jambs.

Jamb clips: Steel clips used to hang an interior door, eliminating the need to use wedging.

Jamb extension: Addition of narrow strips of wood to a window jamb to increase its width.

Jig: A device used to position or guide material for accurate cutting or assembly.

Joinery: A term used by woodworkers when referring to the various types of joints used in a structure.

Joint: The point where two members meet and are fastened. Also, the process of straightening the edges of boards by running them over the cutter of a jointer.

Jointer: Tool used to dress the edges and ends of boards, but may also be used for planing a face, or for cutting a rabbet, bevel, chamfer, or taper. The cutter head is cylindrical and usually has three or four knives. As the cylinder rotates at high speed, the knives cut away small chips from the wood.

Jointing: The height of the saw teeth is evenly struck off.

Joist: One of a series of parallel framing members used to support floor and ceiling loads and, in turn, are supported by larger beams, girders, or bearing walls.

Joist hanger: A metal stirrup used to support the end of a joist.

Journeyman: A tradesperson who has completed an apprenticeship or, through experience, is qualified to work without supervision or further instruction.

K

Kerf: The width of cut made by a saw blade.

Kerfing: A longitudinal saw cut or groove made on the unexposed faces of millwork members to relieve stress and prevent warping; members are also kerfed to facilitate bending.

Keyway: A groove formed in the top of a spread footing to tie in the poured-concrete foundation wall.

Kicker: A strip of wood centered between the rails above a drawer to keep it level as it is opened. Also, a board that is added to the rim joist or band to keep a wall from sliding off the platform as it is being raised.

Kiln dried: Wood seasoned in a kiln by means of artificial heat, controlled humidity, and air circulation.

King post truss: Roof truss with top and bottom chords and a vertical center post; used on shorter spans up to 22'.

King stud: A full-length stud nailed alongside the trimmer at each end of the header.

Kneewall: A wall of less than full height.

Knocked down: Refers to structural units requiring assembly after being delivered to the job.

Knot: Lumber defect caused by a branch or limb embedded in the tree and cut through during lumber manufacture.

Kraft paper: A brown building paper that resists puncturing. Kraft paper is used to face some blanket insulation materials.

L

Lacquer, varnish, and sealers: Clear finishes used to protect and enhance the appearance of wood.

Ladder jacks: Used to support simple scaffolds for repair jobs. Setup requires two sturdy ladders of the same size and a strong plank.

Lally column: A cylindrical steel member used to support beams and girders; sometimes filled with concrete.

Laminate: Process of building up and adhering thin layers of wood or other materials. Examples include plastic laminates used as coverings and wood laminates like laminated-strand lumber and laminated-veneer lumber.

Laminated-strand lumber (LSL): Lumber laid up from 1/32" × 1" × 12" strands of wood bonded with polyurethane adhesive. Available in thicknesses of 1 1/4" and 3 1/2". Depths vary up to a maximum 16" and lengths vary up to 35'. Uses include: door and window headers, rim joists in floors, and core stock for flush doors with veneer overlays.

Laminated strip flooring: Flooring manufactured in three layers of hardwood veneer with the face layer prefinished. Edges are tongue-and-groove.

Laminated-veneer lumber (LVL): Lumber produced much like plywood, except all of the veneer panels have their grain running in the same direction.

Landing: A platform between flights of stairs.

Laser level: Leveling instrument that emits a level laser beam over 360°. A receiver attached to a rod and attended by a rod holder makes a sound when the laser beam strikes it.

Laser: Generates a narrow beam of concentrated light and used in construction leveling instruments for layouts.

Lath: A building material of wood, metal, gypsum, or insulating board, fastened to the frame of the building to act as a plaster base.

Lazy Susan: A circular revolving shelf used in corner kitchen cabinet units.

Leader: A vertical pipe that carries rainwater from the gutter to the ground or a drain. Also called a *downspout*.

Ledger board: Horizontal components of batter boards that support the lines set up to locate building lines and corners.

Ledger strip: A 1" or 2" ribbon horizontally nailed along the wall to support the lookouts.

Let in: Refers to any kind of notch in a stud, joist, block, or other structural member that holds another member. The member housed in the notch is said to be "let in."

Level: Device used for laying out vertical and horizontal lines.

Leveling rod: Long rod marked off with numbered graduations that is used along with a builder's level or transit to sight differences in elevation over long distances.

Level-transit: A surveying instrument used to check both level and plumb on a construction site. The telescope tube can swing in a vertical arc to check for plumb or in a horizontal plane for comparing readings at any horizontal position.

Light: A pane of glass or its frame.

Light construction: Construction generally restricted to conventional stud walls, floor and ceiling joists, and rafters; primarily residential in nature, although it does include small commercial buildings.

Light tube: Type of skylight, typically about 12" in diameter, that consists of a clear dome and a tube with a reflective lining that concentrates the light.

Lignin: Substance in wood that binds cell walls together.

Lineal foot: A measurement of length only, pertaining to a distance of 1' long as distinguished from a 1 sq. ft. or 1 cu. ft.; also called *linear foot.*

Line of sight: A straight line that does not dip, sag, or curve. Any point along a level line of sight is at the same height as any other point.

Lintel: A horizontal structural member that supports the load over an opening, such as a door or window. Also called a *header.*

Lipped door: A cabinet door rabbeted (partially recessed) along its edges. The resulting lip conceals the joint.

Lipped drawer: Drawer front that has a lip covering the joint between the drawer and the cabinet's faceplate.

Live load: The total of all moving and variable loads that may be placed on a building.

Lock block: A block of wood that is joined to the inside edge of the stile of a hollow core door and to which the lock is fitted.

Longitudinal beams: Roof beams that run parallel to the supporting side walls and ridge beam to support roof planking or panels.

Lookout: Structural member running between the lower end of a rafter and the outside wall to support the soffit.

Loose-fill insulation: Insulation made from such materials as rock wool, glass, slag wool, wood fibers, shredded redwood bark, shredded paper, granulated cork, ground or macerated (softened by soaking in a liquid) wood pulp products, vermiculite, perlite, powdered gypsum, sawdust, and wood shavings; used in bulk form and supplied in bags or bales.

Lots: Small tracts of land broken up into smaller parcels.

Louver: An opening for ventilation that is protected from the elements by horizontal angled slats that allow the passage of air.

Low-E glass: Window glass with multiglazing that has been coated to screen out UV rays and reflect heat to increase its insulating value.

Lug: Part of a doorjamb extending upward beyond the head jamb.

Lugged tile: Tile with projections at all edges to maintain proper spacing between individual tiles.

Lumber: Logs that have been sawed into usable boards, planks, and timbers. Some planing and matching of ends and edges may be included.

Lumber core: Hardwood plywood core that is easier to cut, the edges are better for shaping and finishing, and they hold nails and screws better than a veneer core.

Lumber list: List that includes all of the materials and assemblies needed to build the structure. Also called a *list of materials, bill of materials,* or *mill list.*

M

Magnetic starter: Safety device on a power tool that automatically turns the switch to the off position in case of a power failure.

Main: In a one-pipe hydronic heating system, the section of the pipe that moves hot water from the boiler to the rooms being heated.

Major module: A unit of measure for modular construction. In the US customary system of units, 48" is the length of a major module. In the metric system, a major module is 1200 mm long.

Manifold: In hydronic heating systems, a pipe with several outlets that distributes heated water to different areas.

Mansard roof: A type of roof in which the pitch of the upper portion of a sloping side is slight and that of the lower portion is steep. The lower portion is usually interrupted by dormer windows.

Manual dexterity: The ability to efficiently accomplish tasks requiring the use of the hands.

Manufactured home: Defined by manufacturers as a trailer longer than 28' and heavier than 4500 pounds; finished and fully equipped at the factory.

Marking gauge: Device used to lay out parallel lines along the edges of material. May also be used to transfer a dimension from one place to another and check sizes of material.

Marquetry: In carpentry, creating patterns in flooring materials by the use of different-colored woods.

Masking sounds: The normal sounds within habitable rooms that tend to hide or mask some of the external sounds entering the room.

Mason's line: Reference line (cord) used to keep masonry units aligned as they are laid up.

Masonry: Stone, brick, hollow tile, concrete block, tile, or other similar materials bonded together with mortar to form a wall, pier, buttress, or other structural element.

Master layout: Full-size drawing of each cabinet member on cardboard or plywood showing all clearances that may be required.

Master stud pattern: Similar to a story pole, but has information marked on it for only a portion of the wall.

Mastic: Thick adhesive used in construction as a bonding agent for tile and paneling.

Matched lumber: Lumber that is edge-dressed and shaped to make a close tongue-and-groove joint at the edges or ends. Also, generally includes lumber with rabbeted edges.

Measuring tape: Tape that comes in lengths of 50′ to 300′ and used for measurements and layouts involving long distances. Also called *steel tape.*

Mechanical cores: Prefabricated building modules that contain one or more of the following utilities: electrical, plumbing, heating, ventilating, and air conditioning. Floor, ceiling, and wall framing are fully formed at the factory and modules are joined at the building site.

Mechanical equipment: In architectural and engineering practice, all equipment included under the general heading of plumbing, heating, air conditioning, gas fitting, and electrical work.

Mechanical systems: Installations in a building including plumbing, electrical wiring, heating, ventilating, and air conditioning.

Medallion: A raised decorative piece, sometimes used on flush doors or on ceilings.

Meeting rail: The bottom rail of the upper sash and the top rail of the lower sash of a double-hung window. Also called a *checkrail.*

Metal sheeting: Roof covering made of sheet metal; often used in areas where heavy snow occurs so that snow tends to slide off before acquiring deep accumulations.

Metal strap bracing: Galvanized steel straps used to help exterior walls resist lateral (sideway) loads; 2″ wide in 18 or 20 gage.

Metal studs: Studs made of steel and used with either metal plates or wood plates.

Mildew: A fungus that attacks organic and nonorganic materials, especially in the presence of moisture.

Mill list: List that includes all of the materials and assemblies needed to build the structure. Also called a *list of materials, bill of materials,* or *lumber list.*

Millwork: Term used to describe products that are primarily manufactured from lumber in a planing mill or woodworking plant including moldings, door frames and entrances, blinds and shutters, sash and window units, doors, stairs, kitchen cabinets, mantels, cabinets, and porch work.

Minor module: A unit of measure for modular construction. In the US customary system of units, 24″ is the length of a minor module. In the metric system, a minor module is 610 mm long.

Miter box: Slots or holders that accurately guide the saw blade when forming miter cuts for various types of joints.

Miter cut: An end cut made at an angle of less than 90°.

Miter gauge: An adjustable guide on a table saw for making square and miter cuts.

Mobile home: Defined by manufacturers as a trailer longer than 28′ and heavier than 4500 pounds; now referred to by industry as *manufactured home.*

Model codes: A single, national family of building codes.

Modular: Term referring to units assembled from parts having standard sizes.

Modular construction: Use of a standard grid for construction based on 4″ squares.

Modular homes: Homes consisting of two or more three-dimensional, factory-built units that are assembled on the building site.

Module (mod): Three-dimensional assembled housing unit built in a factory and transported to a building site to be combined with other modules.

Moisture content: The amount of water contained in wood, expressed as a percentage of the weight of oven-dry wood.

Moisture meter: An electronic device that uses electrical resistance or capacitance to measure moisture content of wood.

Moisture-resistant (MR) wallboard: A type of gypsum wallboard processed to resist the effects of moisture and high humidity; used as a base under ceramic tile and other nonabsorbent finishes in showers and tub alcoves.

Molder: A woodworking machine designed to create mouldings and other wood members with regular or irregular profiles. Also called a *sticker.*

Monolithic: Term used for concrete construction poured and cast in one unit without joints.

Mortar: A combination of cement, sand, lime, and water used to bind masonry blocks.

Mortise: Recessed cavity in a piece of wood, intended to receive hardware or another piece of wood.

Mortise and tenon joint: A type of joint in which a tongue (tenon) on the end of one wood member is inserted into a pocket or slot (mortise) cut into another member. Joints may be glued or fastened with wood pegs or mechanical fasteners.

Mosaic tile: Small pieces of ceramic tile of different colors laid to form a pattern or picture.

Moulding: A relatively narrow strip of wood, typically shaped to a curved profile throughout its length, used to accent and emphasize the ornamentation of a structure and to conceal surface or angle joints.

Movable insulation: Sheet or blanket insulation that can be placed over windows temporarily to insulate against solar energy.

Mudsill: Part of the floor frame that rests on the foundation wall.

Mullion: A slender bar or pier forming a division between units of windows, screens, or similar generally nonstructural frames.

Multi-level deck: Deck structure with connected sections built to different vertical elevations.

Multiple-use window: A single outswinging sash designed so it can be installed in either a horizontal or vertical position.

Muntins: Vertical members between two panels of the same piece of panel work. The vertical and horizontal sashbars separating the different panes of glass in a window.

N

Nail claw: Tool used to pull a nail above the surface of the wood where it can be easily gripped.

Nailer: Tool used for driving nails. Also, lumber, such as 1×6 or 2×4, added as a backing at inside corners.

Nailer boards: Wood strips installed on edge at hips and ridges to support trim tiles.

Nailing strips: Wooden strips cast into a concrete foundation around a window or door opening and used to attach the door or window frame.

Nail puller: A specialty tool with a fixed jaw and a movable jaw and lever. A sliding driver on the end of the handle drives the jaws under the nail head. Leverage tightens the jaw's grip on the nail, raising it above the wood's surface.

Nail set: Tool designed to drive the heads of casing and finishing nails below the surface of the wood. Tips range in diameter from 1/32″–5/32″ in increments of 1/32″.

National Electrical Code (NEC): Collection of rules developed to control and recommend methods of installing electrical systems.

Net floor area: The gross floor area less the area of the partitions, columns, stairs, and other floor openings.

Newel: The main post of the railing system for balconies, stairways, and landings. It also supports the handrails.

No. 1 common: Low grade of lumber which is expected to yield 66 2/3% clear cuttings.

Noise: Unwanted sound.

Noise reduction coefficient (NRC): The sound absorption of acoustical materials is expressed as the average percentage absorption at the four frequencies that are representative of most household noises: 250, 500, 1000, and 2000 hertz.

Nominal size: As applied to timber or lumber, the ordinary commercial size by which lumber is known and sold. The nominal size is normally slightly larger than the actual size.

Nonbearing partition: A partition extending from floor to ceiling that supports no load other than its own weight.

Nosing: The part of a stair tread that projects beyond the riser, or any similar projection. Also, a term applied to the rounded edge of a board.

O

Oakum: Asphalt-saturated, rope-like material used as a packing in cast iron soil pipe joints prior to pouring molten lead.

Ogee: A moulding with a face shaped in an S curve.

Oil-base paints: Coatings with a large percentage of pigment suspended in a vegetable-oil base. Thinning and cleanup must be done with a solvent, such as turpentine or mineral spirits.

Oil stain: Interior stain that is either penetrating or pigmented. Penetrating stains are brushed on and the excess is removed with a dry cloth. Pigmented stains are designed to be applied with either a brush or a cloth pad.

Oilstone: Stone lubricated with oil and used for honing edge tools.

On center (O.C.): A method of indicating the spacing of framing members by stating the measurement from the center of one member to the center of the succeeding one.

One-pipe system: Radiant heating system where a single pipe supplies heated water to room heating units and returns cooled water to the boiler.

Open cornice: A cornice with no enclosing parts.

Open-grain woods: Woods with large pores, such as oak, ash, chestnut, and walnut.

Open porch: An attached structure that has no railing—just a deck, roof, and supporting columns.

Open time: Amount of time between spreading an adhesive and when the parts must be clamped.

Open valley: Method of roofing in which the roof covering is set back from the middle of the flashed valley.

Operating engineers: Those who operate heavy equipment, such as cranes, bulldozers, and backhoes.

Oriel window: A window that projects from the main line of an enclosing wall of a building and is carried on brackets, corbels, or a cantilever.

Orientation: How the building or a wall is located on the site in relation to a direction, such as north or south.

Oriented strand board: A formed panel consisting of layers of compressed, strand-like wood particles arranged at right angles to each other.

O-rings: Rings used on faucets to produce leak-proof seals between the faucet body and internal parts.

Overlay door: A slab door in cabinetry that covers the cabinet frame.

P

Panel door: Style of door that has a frame and separate panels of plywood, hardboard, or steel set into the frame.

Panel saw: Saw used for cutting large sheets of paneling or plywood.

Panelized homes: Homes built in sections in a factory. Components, such as wall sections and roof sections, are then shipped to the building site and erected by carpentry crews.

Parallel lamination: Laminated lumber product in which all of the veneer panels have their grain running in the same direction.

Parallel strand lumber (PSL): Engineered lumber made up of strands of wood up to 8' long that are bonded together using adhesives, heat, and pressure.

Parapet: A low wall or railing along the edge of a roof, balcony, or bridge; the part of a wall that extends above the roofline.

Parquet flooring: Squares of flooring made up of wood pieces, usually forming intricate, decorative patterns.

Particleboard: A formed panel consisting of wood flakes, shavings, slivers, etc., bonded together with a synthetic resin or other added binder.

Parting strip: Thin, wood strip vertically separating the upper and lower sashes of double-hung windows.

Partition: A wall that subdivides space within any story of a building.

Passive solar construction: A building design that uses solar energy as a heat source with no mechanical means to transport the resulting heat.

Paver tile: Concrete-based masonry units used for finish flooring and walkways.

Pencil compass: Tool used to draw arcs and circles or to step off short distances.

Penny (d): Term used to indicate nail size; applies to common, box, casing, and finishing nails.

Pergola: A structure with an open roof, typically of evenly spaced wood members, that provides partial shade from the sun's rays; may be part of a deck or a free-standing structure.

Perimeter bonded: Flooring that is fastened down only around the edges and at seams.

Perm: Unit of measure for water vapor movement through a material.

Permanent wood foundation (PWF): A special building system that uses treated wood in place of masonry for the foundation.

Permeability: Tendency to absorb water.

Permissible exposure limit (PEL): The maximum amount or concentration of a chemical or other material that a worker may be exposed to without causing harm.

Phloem: Outside area of the cambium layer of a tree. It develops cells that form the bark.

Photosynthesis: The chemical process that plants use to store the sun's energy. Trees use sunlight to convert carbon dioxide and water into leaves and wood.

Pictorial sketch: Three-dimensional drawing much like a photo that shows how a project will look when built.

Pier: A column of masonry used to support other structural members. Also, a concrete base used to support a deck structure. Piers should rest on a concrete footing located below the region's frost line

Pigment: Ingredient of paint which provides color and opacity.

Pilaster: A part of a wall that projects not more than one-half of its own width beyond the outside or inside face of a wall to add strength to the wall, but may also be decorative.

Pile: A heavy timber or pillar of metal forced into the earth or pillar of concrete cast in place to form a foundation member.

Piling and girder foundations: Foundation system usually found in warm climates where freezing of supply piping and drains is not a problem.

Pipe compound: Used to seal threaded connections between pipe and other parts.

Pitch: Inclination or slope, as in roofs or stairs. Rise divided by the span yields the pitch.

Pitch pockets: Cavities in lumber that contain or have contained pitch in solid or liquid form.

Plain footings: Footings that carry light loads and usually do not need reinforcing.

Plain sawed: Hardwood lumber that is cut so that the annular rings form an angle of less than 45° with the surface of the board.

Plan: A drawing representing any one of the floors or horizontal cross sections of a building, or the horizontal plane of any other object or area.

Plank: Floor and roof decking material that can be anywhere from 2″ to 4″ thick, depending on the span. Edges may be tongue-and-groove or grooved for a spline joint that can be assembled into a tight, strong surface.

Plank flooring: Usually hardwood material cut into narrow strips, tongued and grooved, and laid in random pattern of end joints; bored and plugged at the factory to simulate the wooden pegs used to fasten the strips in colonial times.

Plaster: Mixture of gypsum and water that can be troweled wet onto interior walls and ceilings.

Plaster grounds: Wood strips as thick as the plaster installed before plastering begins and used as a gauge for thickness.

Plastic laminate: Material commonly used as a surface for counters and tops; offers high resistance to wear and is unharmed by boiling water, alcohol, oil, grease, and ordinary household chemicals.

Plastic wood®: Wood filler consisting of a combination of wood powder and a plastic hardener.

Plat: A map, plan, or chart of a city, town, section, or subdivision, indicating the location and boundaries of individual properties.

Plate: A horizontal structural member placed on a wall or supported on posts, studs, or corbels to carry the trusses of a roof or to directly carry the rafters. Also, a sole or base member of a partition or other frame.

Plate joiner: A portable tool for cutting half-moon slots in the edges of lumber for insertion of wood plates or biscuits. Often referred to as a *biscuit cutter.*

Platform: A horizontal section between two flights of stairs. Also called a *landing.*

Platform framing: A system of framing a building where the floor joists of each story rest on the top plates of the story below (or on the foundation wall for the first story) and the bearing walls and partitions rest on the subfloor of each story.

Plenum: A sheet metal chamber in a furnace that collects heat in preparation for its transfer to rooms.

Plinth: The decorative trim at the bottom end of cased openings; usually wider than the base.

Plot plan: A drawing of the view from above a building site that shows distances from a structure to property lines. Also called a *site plan.*

Plugs: Wooden pieces shaped to fit and forced into a defect.

Plumb: Exactly perpendicular or vertical; at a right angle to the horizon or floor.

Plumb bob: Device that establishes a vertical line when attached to and suspended from a line. Its weight pulls the line in a true vertical position for layout and checking.

Plumber's putty: Putty used on sink rims, strainers, faucets, and toilet bowls to produce a seal.

Plumbing: The work or business of installing pipes, fixtures, and other apparatuses for bringing in the water supply and removing liquid and waterborne wastes. This term is used also to denote the installed fixtures and piping of a building.

Plumbing stack: A general term for the vertical main of a system of DWV piping.

Plunge cutting: An interior cut made with a portable saw without first boring holes.

Pocket door: A sliding door suspended on a track. When opened, it slides into a cavity prepared in the wall. Also called a *recessed door.*

Polyisocyanurate foam: Moisture-resistant, blown-in insulation material. It uses a blowing agent that slightly depletes the ozone. Its R-value is higher than other rigid foam products.

Polystyrene panels: Rigid insulation sheets manufactured from expanded beads of plastic.

Polyurethane adhesive: A single-component, moisture-catalyzed adhesive that bonds with wood of up to 25% moisture content.

Polyvinyl chloride (PVC): Plastic plumbing pipe used for long runs because of its lower thermal expansion.

Polyvinyl resin emulsion adhesive: Wood adhesive, intended for interiors, made from polyvinyl acetates, which are thermoplastic and not suited for temperatures over 165°F (75°C). Also called *white glue.*

Portable tools: Lightweight tools that are intended to be carried around by the carpenter.

Portico: A porch or covered walk consisting of a roof supported by columns; a porch with a continuous row of columns.

Post anchor: Supports the bottom of a post above the floor, protecting the wood from dampness.

Post-and-beam construction: Construction consisting of large framing members—posts, beams, and planks. Because of their great strength, these members may be spaced farther apart than conventional framing members.

Post-and-beam framing: Framing method in which loads are carried by a frame of posts connected with beams, eliminating the need for load-bearing walls.

Post-and-beam ICF wall: Poured-concrete wall contained by insulated forms.

Potable: Fit to drink.

Pour strip: Wood strip attached inside of one of the form walls to indicate the height of the concrete pour.

Powder-actuated tool: A striking tool that uses the force of an exploding cartridge to drive fasteners into concrete and steel.

Power block plane: Small plane that is electrically operated.

Power miter saw: The motor and blade are supported on a pivot; the saw is used to make miter and compound miter cuts.

Power plane: An electrically operated tool that produces finished wood surfaces with speed and accuracy.

Power stapler: Electrically or pneumatically powered tool that drives staples.

Preacher: A small block made from 3/8" or 1/2" hardwood and notched to fit over the siding and used to mark the siding where it is to be cut off. Also called a *siding gauge.*

Precut home: Structure for which lumber is cut, shaped, labeled, and shipped to the building site; reduces labor and saves time on the building site.

Prefabricated construction: Type of construction with a minimum of assembly at the site, usually comprising a series of large units manufactured in a plant.

Prehung door unit: A door frame with the door already installed.

Premium grade: A special grade of hardwood known as *architectural* or *sequence matched.* Usually requires an order to a plywood mill for a series of matched plywood panels.

Preservative: Substance that will prevent the development of wood-destroying fungi, borers of various kinds, and other harmful insects that deteriorate wood.

Pressure-feed spray gun: Spray gun that diverts some air into the paint container. The pressure created forces finishing material into the air stream. This gun is intended for spraying heavy-bodied paints.

Pressure-treated lumber: Wood that has been treated with chemical preservatives to protect it from rot and insects.

Primary colors: Red, blue, and yellow.

Primer: A coating applied to an unpainted surface to provide a suitable base for top coatings.

Prints: Copies of original drawings.

Programmable thermostat: Thermostat that can be set to automatically change temperatures at different times of the day.

Property lines: The boundaries of a site.

Pry bar: Tool designed for removing larger nails or spikes from lumber of larger dimensions.

Psychomotor skills: Eye and hand coordination.

Pull bar: Tool used with a hammer when driving planks together in tight quarters, such as near a wall.

Purlin: Horizontal roof members used to support rafters between the plate and ridge board.

Purlin beams: Longitudinal roof beams.

Push stick: A pole or strip used to push a workpiece when cutting with power saws, jointers, and other power tools. Pushing a board by hand is usually unsafe.

Putty: A plastic, dough-like material used to fill holes and large depressions in wood surfaces. It is also used to fill nail holes before painting.

Q

Quarry tile: Ceramic tile used in construction as finish flooring and countertops.

Quarter round: Moulding with a cross section of one-fourth of a circle.

Quarter sawed: Hardwood cut so the annual rings form an angle of more than 45° with the surface of the board.

Quick square: Triangular measuring tool that can be used to mark any angle for rafter cuts by aligning the desired degree mark on the tool with the edge of the rafter being cut. Also called a *speed square* or *super square.*

R

Rabbet: A rectangular shape consisting of two surfaces cut along the edge or end of a board.

Rabbet plane: Plane designed to form dados, grooves, and rabbets.

Raceway: Pipes or enclosed channels through which conductors are run.

Racking: In flooring, laying out seven or eight rows of strip flooring prior to nailing them in place.

Radial arm saw: A power saw in which the movable motor and blade are carried by an overhead arm and the stock being cut is supported on a stationary table.

Radiant energy: Energy that moves through space in waves. When the waves strike a solid object, the energy is absorbed by the solid matter.

Radiation: Transmission of heat by wave motion, much like light movement.

Rafter: One of a series of structural members of a roof designed to support roof loads. The rafters of a flat roof are sometimes called *roof joists.*

Rafter square: A square specially designed for the carpenter. The blade (or body) is 2″ wide and 24″ long. The tongue is at a right angle to the blade and is 1 1/2″ wide and 16″ long.

Rail: Cross or horizontal members of the framework of a sash, door, blind, or other assembly.

Rake: The trim members that run parallel to the roof slope and form the finish between the roof and wall at a gable end.

Ramp: Inclined plane connecting separate levels.

Rasp: Tool used for trimming, shaping, and smoothing wood.

Rebar: Steel reinforcing rod placed in concrete.

Reciprocating saw: Portable power tool that has linear (back-and-forth) cutting action.

Reflective insulation: A metal foil or foil-surfaced material used to provide thermal insulation. The number of reflecting surfaces, not the thickness of the material, determines its insulating value.

Registers: Grilles installed over the ends of heating or air conditioning ducts to direct the conditioned air.

Reinforced concrete construction: A type of construction in which the principal structural members—floors, columns, and beams—are made of concrete poured around steel bars or steel meshwork in such a manner that the two materials act together to resist force.

Reinforced footings: Footings containing steel rebar for added strength.

Related colors: Hues that are side-by-side on the color wheel.

Relative humidity: Ratio of amount of water vapor in air to the total amount it could hold at the same temperature expressed as a percentage.

Relief valve: Valve that opens an outlet when a set pressure is reached; used to protect pipes and equipment from being damaged by over-pressure.

Resilience: The ability of a material to withstand temporary deformation and assume its original shape when the stresses are removed.

Resistance (R): The reciprocal (opposite) of conductivity or conductance; a measure of insulating value.

Retaining wall: Any wall subjected to lateral pressure other than wind pressures; for example, a wall built to support a bank of earth.

Return main: The part of a double-pipe hydronic heating system that carries cooled water back to the boiler.

Reveal: Amount of setback of a window or door casing from the inside edge of the jamb.

Reverberation sounds: Airborne sounds that continue after the actual source has ceased. They are caused by sound reflecting from floors, walls, and ceilings.

Ribbon: A narrow board let into a stud or other vertical member of a frame, thus adding support to joists or other horizontal members.

Ribbon coursing: Coursing in which both layers are oriented in a straight line. The top course is raised roughly an inch above the under course.

Ridge: The horizontal line at the junction of the top edges of two roof surfaces.

Ridge board: Horizontal part of the roof frame; placed on edge at the ridge and supports the upper ends of the rafters.

Rigid insulation: Insulation manufactured in rigid panels of expanded or extruded polystyrene foam.

Rim joist: Joist that sit on the sill and to which other floor joists are butted and attached. Also called *header* or *band joist.*

Ring-shank nails: Fasteners with annular rings formed around the shaft to provide increased holding power in wood.

Ripping bar: Tool used to strip concrete forms, disassemble scaffolding, or other rough work involving prying, scraping, and nail pulling. Also called *wrecking bar.*

Ripsaw: Saw with chisel-shaped teeth that cut best along the grain.

Rise: In roof framing, the distance a rafter extends upward above the wall plate. In stairs, the vertical distance of the flight.

Riser: The vertical stair member between two consecutive stair treads. In plumbing, a vertical pipe.

Roll roofing: Sheet of asphalt-saturated felt or fiberglass with mineral granules on the top surface. Roll roofing is the uncut form of mineral-surfaced shingle material.

Roof truss: Engineered and prefabricated rafter assemblies. The lower chords also serve as ceiling joists.

Roofing: The materials applied to the structural parts of a roof to make it waterproof.

Roofing brackets: Provide safety when working on steep slopes by allowing planking to be horizontally placed on the roof.

Roofing tile: Tile manufactured from concrete or molded from hard-burned shale or mixtures of shale and clay.

Rosette: Decorative, carved block set at left and right top corners of window and door trim. Use of a rosette eliminates the need to make mitered corners.

Rotary cut veneer: Veneer cut on a lathe that rotates the log against a broad cutting knife. The veneer is cut in a continuous sheet much the same as paper is unwound from a roll.

Rotary hammer drill: Heavy-duty tool for drilling large holes in concrete and other masonry materials.

Rough flooring: The subfloor.

Roughing in: The process of building the structure or framework with all framing members, including wall sheathing, roof sheathing, and subflooring. Also, installing mechanical systems such as plumbing, duct-work, and electrical wiring systems. Always performed before installing drywall.

Rough lumber: Lumber that has been cut to rough size with saws, but that has not been dressed or surfaced.

Rough opening: An opening formed by framing members to receive and support windows or doors.

Router: Tool used to cut irregular shapes and to form various contours on edges.

Router plane: Plane designed to form dados, grooves, and rabbets.

Run: Any horizontal distance. The horizontal distance of rafters from wall plate to ridge. Also, the horizontal distance traveled by a flight of stairs.

Runner guides: U-shaped channels that hold edges of metal soffit material and are attached to the fascia and walls.

Run of stairs: A series of steps that is a continuous section without breaks formed by landings or other constructions.

Runs: Coating problem caused by too-heavy of an application.

R-value: A number related to the efficiency of a thermal insulating material.

S

Saber saw: Portable power saw with a reciprocating cutting action that is useful for a wide range of light work. Also called a *portable jig saw.*

Saddle: A small gable roof placed in back of a chimney on a sloping roof to shed water and debris. Also known as a *cricket.*

Saddle beam: Beam positioned above the joists it supports.

Safety factor: Capable of carrying a load a specified number of times (for example, four) in excess of the expected maximum load.

Safety glasses: Eye protection with a safety lens that must withstand the blow of a 1/8" diameter steel ball dropped from a height of 50".

Sags: Coating problem caused by too-heavy of an application.

Sanded grout: Cement to which sand is added.

Sandwich panels: Panels with cores of such material as rigid polystyrene or paper honeycomb. Skins can be made from a wide variety of sheet materials, including plywood, hardboard, plastic laminates, and aluminum. They are used for curtain walls and various installations where the major support is carried by other components.

Sapwood: The layers of wood next to the bark that are actively involved in the life processes of the tree; usually lighter in color than the heartwood. More susceptible to decay than

heartwood, but sapwood is not necessarily weaker or stronger than heartwood of the same species.

Sash: The framework that holds the glass in a window.

Sash doors: Doors similar to panel doors in appearance and construction but have glass lights in place of one or more wood panels.

Saturated felt: Dry felt soaked with asphalt or coal tar and used under shingles for sheathing paper and as laminations for built-up roofs.

Scaffold: A temporary structure or platform used to support workers and materials during building construction. Also called *staging.*

Scale: A term that specifies the size of a reduced-size drawing. For example, a plan is drawn to 1/4″ scale if every 1/4″ represents 1′ on the real structure.

Scantling: Lumber with a cross section ranging from 2 × 4 to 4 × 4.

Scarfing: Joining the ends of stock together with a sloping lap joint so they appear to be a single piece.

Scarf joint: A mitered-lap joint used where a straight run of baseboard must be joined.

Schedule: A part of the bill of materials that covers windows and doors as to quantity, size of opening, description, and sometimes the name of the manufacturer.

Scissors truss: Truss used for buildings having a sloping ceiling. In general, the slope of the bottom chord is half of the slope of the top chord.

Scotia: A concave molding consisting of an irregular curve. Used under the nosing of stair treads and for cornice trim.

Scraper: Tool with a hooked edge that produces thin, shaving-like cuttings.

Scratch awl: Tool used to scribe lines on the surface of the material. It is also used to mark points and to form starter holes for small screws or nails.

Scratch coat: First layer of plaster that has its surface roughened to provide tooth for succeeding layers.

Screed: A tool used in concrete work to level and smooth a horizontal surface. Also, wood strips attached to or embedded in the concrete surface to serve as a nailing base for the flooring material; also called *sleepers.*

Screeding: The process of leveling off concrete slabs or plastering on interior walls using a screed.

Screened-grid ICF wall: Poured concrete wall using insulated forms; resulting wall forms a grid similar to a window screen.

Scribe: To lay out woodwork to fit flush against an irregular surface.

Scuttle: An opening in a ceiling to provide access to the attic or floor to provide access to the crawl space.

Sealers: Clear coatings applied over wood to protect the wood surface and enhance the beauty of the grain and color.

Seasoning: Removing moisture from green wood to improve its serviceability.

Secondary colors: Colors produced when primary colors are combined, such as violet, green, and orange.

Second growth: Timber that has grown after the removal of a large portion of the previous stand.

Second law of thermodynamics: States that heat always moves from hot to cold.

Section drawing: A type of drawing that shows how a part of a structure would look if cut along a given plane. Also called *section view.*

Selects: The hardwood grade below first and seconds.

Self-adhering tile: Floor tile applied by removing protective paper from the adhesive backing of the tile, positioning the tile on the floor, and pressing it down.

Self employment: Starting and operating a business of one's own.

Self-tapping drywall screws: Screws used to attach wall surface material, such as drywall or paneling, to metal studs.

Selvage: The part of the width of roll roofing that is smooth. For example, a 36″ width has a granular surfaced area 17″ wide and a 19″ wide selvage area.

Semi-enclosed porch: A type of porch with either an ornamental balustrade or a low wall enclosure. Some designs use removable screens and/or storm panels to adapt their use to changing seasons.

Semihoused stringer: A stringer that is partially open.

Semi-transparent stains: Lightly pigmented stains that color the wood, but allow the grain to show through.

Septic tank: A sewage settling tank intended to retain the sludge in immediate contact with the sewage flowing through the tank for a sufficient period to secure satisfactory decomposition of organic sludge solids by bacterial action.

Service: The conductors that bring power from a transformer to a building.

Set of plans: Many sheets of drawings bound together. They include plans, drawings of the mechanical systems, elevation drawings, section drawings, and detail drawings.

Setback: The distance between the building and the property lines.

Setting block: A wood block placed in the glass groove or rabbet of the bottom rail of an insulating glass sash to form a base or bed for the glass.

Setup line: A chalk line the width of two flooring strips (plus 1/2") away from the partition.

Setup time: Time it takes for an adhesive or concrete to stiffen.

Sexual harassment: Any unwelcome sexual advances or requests for sexual favors. It includes any verbal or physical conduct of a sexual nature under specific conditions.

Shakes: Handsplit shingles.

Sheathing: Boards or prefabricated panels that are attached to the exterior studding or rafters of a structure.

Sheathing paper: A building material used in wall, floor, and roof construction to resist the passage of air.

Shed roof: A single-slope roof. Also called a *lean-to roof.*

Sheetrock: A laminated material with a gypsum core and paper covering on either side. Commonly called *drywall* or *gypsum wallboard.*

Sheetrock-drywall saw: Saw with large, specially designed teeth for cutting through paper facings, backings, and the gypsum core of drywall.

Shelves: Fixed or adjustable horizontal dividers in cabinets designed to provide storage.

Shim: A thin strip of wood, sometimes wedge-shaped, for plumbing or leveling wood members. Especially helpful when setting door and window frames.

Shingle butt: Lower, exposed portion of a shingle.

Shoring: Bracing used to provide temporary support.

Short circuit: Accidental grounding of a current-carrying electrical conductor. Also called a *ground fault.*

Shutter: An assembly of stiles and rails to form a frame that encloses panels used in conjunction with door and window frames. Also may consist of vertical boards cleated together.

Side-and-end matched: Strip flooring that is tongue-and-groove both on the sides and ends.

Side guides: Grooves cut in the drawer side before it is assembled and matching strips fastened to the side panels. Used in drawer openings where there is no lower frame.

Siding: The finish covering of the outside wall of a frame building. Many different types are available.

Side jambs: Parts of an interior door frame that are dadoed to receive the head jamb.

Side lap: The overlap distance for side-by-side elements of roofing.

Sill: The lowest member of the frame of a structure, usually horizontal, resting on the foundation and supporting the uprights of the frame. Also, the lowest member of a window or outside door frame.

Sill sealer: Strip of insulation between the sill plate and the foundation.

Single coursing: Applying shingle siding in single layers, as in shingling a roof.

Sistering: Reinforcing rotted studs by nailing another stud alongside the rotted member.

Site preparation: Excavation and other activities needed to ready the ground for foundation work and building activities.

Six-inch method: A shingling pattern for three-tab shingles where the cutouts break joints on every other course. Cutouts are centered over the tab in the course directly below.

SkillsUSA: Organization that promotes development of excellence in construction. It holds state and national competitions to encourage students in a variety of occupations, including building trades.

Skip sheathing: Board sheathing on a roof spaced to allow shingles to dry from the underside.

Skylight: Glazing framed into a roof.

Slab-on-foundation: Slab foundation usually found in commercial construction.

Slab-on-grade: Slab foundation that is directly supported on top of the ground. Also called *ground-supported slab.*

Sleeper: A timber laid on or near the ground or on concrete to support floor joists and other structures above. Also, wood strips laid over or embedded in a concrete floor to which finish flooring is attached; also called *screeds.*

Sliding doors: Cabinet doors often used where the swing of regular doors would be awkward or cause interference.

Slip joint washers: Washers used to seal P-trap connections beneath sinks.

Slip sheet: A sheet of heavy wrapping paper placed over the base surface when adhering laminates.

Slope: Incline of a roof.

Slump test: A procedure to test the moisture content of concrete.

Smoke chamber: Fireplace chamber located between the smoke shelf and the entrance to the flue.

Smoke shelf: The horizontal shelf located adjacent to the damper in a fireplace.

Smooth plane: Surfacing plane, 8"–9" long.

Snub cornice: Cornice with no overhang. Rafters end flush with the sidewalls.

Soffit: The underside of the members of a building, such as staircases, cornices, beams, and arches. Relatively minor in size as compared with ceilings. Also called *drop ceiling* and *furred-down ceiling.*

Soft-coat system: Exterior insulation and finish system (EIFS) that is typically thin (1/8"), flexible, and attached to the structure with adhesives.

Softwoods: Trees that have needle or scalelike leaves and are, for the most part, evergreen (cypress, larch, and tamarack are exceptions). The term has no reference to the actual hardness of the wood. Softwoods

are often referred to as conifers; botanically, they are called gymnosperms.

Soil stack: A general plumbing term for the vertical main vent of a DWV system that extends through the roof.

Solar collectors: Box-like structures with clear glazing to trap solar energy.

Solar furnace: Another name for a thermosiphon.

Solar orientation: Placement of a structure on a building site with respect to exposure to sunlight.

Solar radiation: Radiation from the sun that travels through glass almost as well as it travels through the atmosphere.

Sole plate: The lowest horizontal strip in wall and partition framing. The sole plate is supported by a wood subfloor, concrete slab, or another closed surface.

Solid bridging: Solid pieces of 2" lumber installed between joists. Often installed above a supporting beam to keep the joist vertical, but also adds rigidity to the floor. Also called *blocking.*

Solid stains: Stains that are much like paint, with a heavy pigmentation that provides an opaque color finish.

Sound: A vibration or wave motion that can be heard. It usually reaches the ear through air.

Sound absorption: The capacity of a material or object to reduce sound waves by absorbing them.

Sound transmission class (STC): A single number that represents the minimum sound-deadening performance of a wall or floor at all sound frequencies.

Sound transmission loss (STL): The number of decibels that sound loses when transmitted through a wall or floor.

Span: The distance between structural supports.

Span rating: Maximum distance a structural member can extend without support when the long dimension is at right angles to its supports.

Specific gravity: The ratio of the weight of a body to the weight of an equal volume of water.

Specification: A written document stipulating the type, quality, and sometimes the quantity of materials and work required for a construction job.

Specks in surface: Coating problem resulting from hardened particles in the container of finish, dirty brushes, improperly prepared surfaces, or airborne dust in the room.

Speed square: Triangular measuring tool that can be used to mark any angle for rafter cuts by aligning the desired degree mark on the tool with the edge of the rafter being cut. Also called a *quick square* or *super square*.

Spirit stain: Interior stain with a base material that is either alcohol or acetone. It is generally limited to spray applications.

Splash block: A small masonry block laid with the top near the ground to receive roof drainage and carry it away from the building.

Splay: The rear face of the hearth that slopes toward the front.

Spline: A thin strip of wood that fits into mortises or grooves machined into boards that are to be joined.

Splits and checks: Separations in wood that run along the grain and across the annular growth rings.

Spread foundation: Poured or block foundation resting on a footing twice as wide as the foundation wall.

Spreader: In formwork for poured concrete, a strip of wood used to keep forms apart a certain distance.

Square: A unit of measure—100 square feet— usually applied to roofing material and to some types of siding.

Staggered coursing: Butts of alternating shingles are offset below the line. Offsets are never more than 1" for 16" and 18" shingles or 1 1/2" for 24" shingles.

Stains: Finishing liquids that impart color to wood. Spirit stains have a base either of alcohol or acetone and dry rapidly. Pigmented oil stains often use linseed oil as a base. Penetrating oil stains use turpentine, naphtha, benzol, and other light oils as a base. Water stains are a mixture of powders and water.

Stairwell: The framed opening that receives the stairs.

Stapler: Tool used to insert staples into wood to attach various products, such as insulation, roofing material, underlayment, and ceiling tile. May be powered by hand, electric, or pneumatic means.

Starter strip: Strip of mineral-surfaced material placed beneath the first course of shingles to cover the gaps between shingle tabs.

Starter track: Channels used at the top and bottom of walls and around openings in an EIFS application.

Stationary tools: Heavy equipment mounted on benches or stands. Usually called *machines*.

Station mark: The reference point where a builder's level or transit is set up when laying out building lines or finding grade levels and elevations. It must be a type of marker that will not be disturbed by any activity of construction.

Steel-frame construction: A type of construction in which the structural members are steel or are dependent on a steel frame for support.

Steel framing members: Manufactured in various widths and gages and used as studs, joists, and truss rafters.

Steel posts: Used to support girders and beams; most popular method.

Steel tape: Tape that comes in lengths of 50′ to 300′ and used for measurements and layouts involving long distances. Also called *measuring tape.*

Stepped footing: A footing that changes grade levels at intervals to accommodate a sloping site.

Stickers: Strips of wood used to separate the layers in a pile of lumber so air can circulate.

Stick shellac: Hardened, colored shellac used as filler to repair defects in furniture finishes and surfaces.

Stick wax: A colored wax that closely resembles a crayon. It is rubbed over a wood defect to force the wax into the defect.

Stile: The upright or vertical outside pieces of a sash, door, blind, screen, or face frame.

Stirrups: Hangers made of metal that support joists.

Stock plan: A set of plans that is mass produced to be offered for sale to many clients.

Stool: An interior trim member serving as a sill cap for a sash or window frame. Stools may be beveled or rabbeted to receive the window frame sill.

Stoop: A small porch, veranda, platform, or stairway outside of an entrance to a building.

Storm sash: A second pane of glass in a window used to increase the R-value.

Story: A single floor of a building.

Story pole: A strip of wood used to lay out and transfer measurements for door and window openings, siding and shingle courses, and stairways

Straightedge: A strip of wood or metal with one edge perfectly straight; used to lay out straight lines or check straightness of surfaces.

Straight run: A stairway that does not change direction.

Stressed skin panel: Structural panel with two facings, each glued to the opposite sides of an inner structural framework to form a panel. Facings may be of plywood or other suitable material.

Strike plate: A metal piece mortised into or fastened to the face of a door frame side jamb to receive the latch or bolt when the door is closed.

Stringer: A sloping member that supports the risers and treads of stairs. Also called a *carriage*.

Strip flooring: Usually hardwood material cut into narrow strips, tongued and grooved, and laid in random pattern of end joints.

Strongback: An L-shaped wooden support attached to tops of ceiling joists to strengthen them, maintain spacing, and bring them to the same level.

Structurally supported slab: Slab that can be used with other elements such as walls, piers, and footing.

Structural sheathing: Type 2 plywood or oriented strand board, normally applied to both sides of the wall frame. Used for permanent bracing of load-bearing walls.

Structured wiring: An integrated system of low-voltage wiring and centralized control units used to provide a home with a variety of communication, automation, and security services.

Stub in: Termination point of plumbing pipes that are to be connected to fixtures. Also called *rough in*.

Stucco: An exterior wall finish system consisting of a plaster made from one part portland cement, three parts sand, and hydrated lime. It is applied over metal lath attached to the wall.

Stud: One of a series of vertical structural members in walls and partitions.

Subfloor: Boards or structural panels directly laid on floor joists over which a finished floor is laid.

Submersible pump: A water pump that is installed in a well.

Subterranean termites: The most destructive type of termite. They build earthlike shelter tubes that give them passage over foundation walls, through cracks in walls, on pipes, or other supports leading into buildings.

Suction-feed spray gun: Spray gun designed for applying light-bodied materials such as shellacs, stains, varnishes, lacquers, and synthetic enamels.

Sun space: A separate room with large areas of its south wall glazed to provide solar heat. Sometimes the room has thermal mass built into the inside wall or the floor.

Super square: Triangular measuring tool that can be used to mark any angle for rafter cuts by aligning the desired degree mark on the tool with the edge of the rafter being cut. Also called a *speed square* or *quick square*.

Supply main: The part of a double-pipe hydronic heating system that routes heated water to rooms.

Surfaced lumber: Lumber that is dressed or finished by running it through a planer.

Suspended ceiling: Ceiling finishing system composed of metal framework designed to support tile or panels.

Sweat sheet: A strip of felt or ice-and-water barrier.

Sweat soldering: Joining process for copper pipe in which the fitting and pipe are heated with a gas flame and solder is drawn into the joint by capillary action.

Switches: Devices that control current to lights and appliances.

Symbols: Used in plans to represent materials and other items.

System: Methods and materials of construction. The term is used in connection with floors, ceilings, and roofs as well as walls.

Systems-built housing: Housing built of components designed, fabricated, and assembled in a factory.

T

Table saw: A basic cutting machine used in finish work such as interior trim. Also called a *circular saw.*

Tackers: Hand-powered tool used to insert staples into wood to attach various products, such as insulation, roofing material, underlayment, and ceiling tile.

Tail beam: A relatively short beam or joist supported by a wall on one end and a header on the other.

Tail joist: Short joist.

T-bevel: Tool with an adjustable blade making it possible to transfer an angle from one place to another. It is useful in laying out cuts for hip and valley rafters.

Teflon® tape: Tape used to seal threaded connections between pipe and other parts.

Telegraph: When irregularities in the floor show through on the surface of resilient flooring materials.

Tension bridging: Use of metal connectors to transfer load from one joist to another.

Termination top: The part of the chimney that extends above the roof.

Termite shield: A shield, usually made of sheet metal, placed in or on a foundation wall or around pipes to keep termites out of the structure.

Terne metal roofing: Copper-bearing steel sheets dip-coated with an alloy of lead and tin.

Terrazzo flooring: A floor produced by embedding small chips of marble or colored stone in concrete and then grinding and polishing the surface.

Thermal mass: Often a thick masonry or concrete wall directly located behind large windows or sliding glass doors to absorb solar radiation.

Thermoplastic: Material that softens under heat.

Thermoset adhesives: Adhesives that are more water- and heat-resistant than the polyvinyl adhesives. In this category are polyurethanes, urea formaldehydes, and resorcinal formaldehydes.

Thermosiphon: A solar collector in which heated air (or fluid) is discharged at the top and cool air (or fluid) is drawn in at the bottom.

Thermosiphoning: Movement of heat through a fluid by the action of heated material rising in a narrowed space.

Thermostat: An instrument that automatically controls the operation of heating or cooling devices by responding to changes in temperature.

Three-way switch: A switch designed to operate in conjunction with a similar switch to control one outlet or light from two locations.

Threshold: A member, beveled or tapered on each side, used to close the space between the bottom of a door and the sill or floor underneath. Sometimes called a *saddle.*

Threshold of pain: Painfully loud sounds of over 130 decibels.

Throat: In a fireplace, the narrowed passage above the hearth and below the damper.

Tie beam: A beam so situated that it ties the principal rafters of a roof together and prevents them from thrusting the plate out of line. Also called *collar beam.*

Timbers: Lumber 5″ or larger in least dimension.

Tin snips: Tool used to cut asphalt shingles and light sheet metals, such as flashing.

Toe: The pointed end of a trowel.

Toeboard: A board horizontally fastened to scaffolding, slightly above planking, to keep tools and materials from falling on workers below. The board should be at least 4″ wide.

Toenailing: Driving a nail at a slant with the initial surface so it will penetrate into a second member.

Toe space: A recessed space at the floor line of a base kitchen cabinet or other built-in unit. Permits one to stand close without striking the vertical space with the toes.

Tongue: The shorter arm (16″) of a framing square.

Top plates: Wall-framing members used in conventional platform construction.

Total heat transmission (U): Represented in Btu per sq. ft. per hour with 1°F temperature difference for a structure (wall, ceiling, and floor) that may consist of several materials or spaces.

Total rise: Vertical distance from one floor to another, or total height from the floor or ceiling joist to the peak of the roof.

Total run: In a roof, half of the span (width) of the building. In stairs, the horizontal distance occupied by the stairs measured from the foot of the stairs to a point directly beneath where the stairs rest on a floor or landing above.

Tracheids: Long narrow tubes or cells that make up wood. Also called *fibers*.

Track: Steel framing member, formed into a U-shaped channel, that is used at the top and bottom of a stud wall. The steel studs fit into the track and are secured by screws or welds.

Transformer: A device for changing the voltage characteristics of an electric current.

Transit: Optical leveling instrument commonly used for building layout.

Transom: A small opening above a door separated by a horizontal member (transom bar). Usually contains a sash or a louver panel hinged to the transom bar.

Transverse beams: Roof beams that are similar to exposed rafters on wide spacings.

Trap: A plumbing fitting designed to provide a liquid trap seal to prevent sewer gases from passing through and entering a building.

Tread: Horizontal walking surface on a stair.

Trestle jacks: Used to support low platforms for interior work.

Trim: The finish materials in a building, such as mouldings applied around openings (window trim, door trim) or at the floor and ceiling of rooms (baseboard, cornice, picture moulding).

Trimmer: The beam or floor joist into which a header is framed; adds strength to the side of the opening.

Trimmer stud: A stud that supports the header for a wall opening. The stud extends from the sole plate to the bottom of the header and is parallel to, and in contact with, a full-length stud that extends from sole plate to top plate.

Triple wall: A type of chimney flue made with three metal pipes nested inside of each other. The concentric arrangement provides safety from fire while its light weight makes installation easy.

Trombe wall: Thick wall of masonry placed next to exterior glazing to store solar energy in passive solar construction. Named for French physicist Félix Trombe.

Troweling: Producing a dense, smooth surface on a concrete floor using a trowel.

Truss: A structural unit consisting of such members as beams, bars, and ties arranged to form triangles. Provides rigid support over wide spans with a minimum amount of material. Used to support roofs and joists.

Try square: Square available with blades 6″ to 12″ long. Handles are made of wood or metal. These are used to check the squareness of surfaces and edges and to lay out lines perpendicular to an edge.

Two-pipe system: Hydronic heating system that has a supply main to carry heated water to each room unit simultaneously, while a return main carries the cooled water to the boiler.

U

Undercut: Cutting a wood member so that the back of the member is slightly shorter than the front surface.

Underlayment: A thin sheet material laid down to provide a smooth, even surface for the finish flooring. Also, a thin cover of asphalt-saturated felt or other material that protects sheathing from moisture until shingles are laid, provides weather protection, and prevents direct contact between shingles and resinous areas in the sheathing.

Underpinning: A short wall section between the foundation sill and first floor framing. Also called a *cripple wall.*

Unicom system: A term taken from "uniform manufacture of components." The system uses modules with sizes that are multiples of a standard unit.

Unit rise: The number of inches a common rafter rises for every foot of run. In stairs, the riser height is calculated by dividing the total rise by the total number of risers.

Unit run: In roof framing, one unit of horizontal distance, based on 12".

Unprotected-metal construction: A type of construction in which the structural parts are metal without fireproofing.

Urea-formaldehyde resin glue: Moisture-resistant glue that hardens through chemical action when water is added to the powdered resin.

Utility knives: Tools with very sharp blades useful for trimming wood, cutting veneer, hardboard, particle board, vapor barrier, tar paper, and house wrap.

V

Valley: The internal angle formed by the two slopes of a roof.

Valley cripple jack: A rafter that intersects neither the wall plate nor the ridge and is terminated at each end by hip and valley rafters. Also called a *cripple jack* or *cripple rafter.*

Valley jack: The same as the upper end of a common rafter, but intersects a valley rafter instead of the plate.

Valley rafter: A rafter that forms the intersection of an internal roof angle.

Value: The lightness or darkness of a color.

Valves: Devices that control flow of water in a water supply system.

Vapor barrier: A watertight material used to prevent the passage of moisture or water vapor.

Varnish: Clear penetrating coating applied to wood to enhance appearance and provide protection.

Vehicles: Oils or resins that make the paint a fluid and allow it to be applied to and spread over a surface.

Veneer: A thin sheet or layer of wood or other material.

Veneer core: Hardwood plywood core construction in which the core is made of layers of thin wood; inexpensive, fairly stable, and warp resistant.

Veneer plaster: Interior wall covering consisting of a gypsum lath base and a surface of 1/8" gypsum plaster.

Veneer wall: A masonry facing, such as a single thickness of brick, applied to a frame building wall. The veneer is nonload bearing.

Ventilation: The process of supplying and removing air by natural or mechanical means. Such air may or may not have been conditioned.

Venting: The part of the drainage piping that permits air to circulate in the pipes.

Vent stack: The part of a DWV system that is open to the atmosphere to allow air into the system, preventing siphoning of traps.

Vernier scale: Device on a transit that measures vertical angles.

Vinyl siding: Siding manufactured from a rigid polyvinyl chloride compound that is tough and durable.

Vocational instructor: A teacher specializing in training students for specific occupations, such as carpentry, auto mechanics, or food service.

W

W truss: Roof truss that can be used for spans up to 50'. Also called *Fink truss.*

Waffle-grid ICF wall: Insulated concrete wall system that forms a grid similar to the appearance of a waffle.

Wainscot: A lower interior wall surface (usually 3'–4' above the floor) that contrasts with the wall surface above. May consist of solid wood or plywood.

Waler: A horizontal member used in concrete form construction to stiffen and support the walls of the form; used to keep the form walls from bending outward under the pressure of poured concrete. Also called *wale.*

Wallboard: Wood pulp, gypsum, or other materials made into large rigid sheets that may be fastened to the frame of a building to provide a surface finish.

Wallpapering: Hanging wallcoverings.

Wall pocket: An intentional void left in a concrete wall to support a large timber or I-beam.

Wall rail: In closed stairs, the support rail that is attached to the wall with special metal brackets.

Wall tie: Metal strip or wire used to bind tiers of masonry in cavity wall construction, or to bind brick veneer to a wood frame wall.

Wane: Bark or absence of wood on the edge of lumber.

Warp: Any variation from a true or plane surface. Warp includes any combination of bow, crook, cup, and twist.

Washerless: Type of faucet that uses replaceable cartridges, discs, or other methods to control water flow.

Water-base paints: Paints that can be thinned or cleaned up with water.

Water repellent: A solution, primarily paraffin wax and resin in mineral spirits that penetrates wood, retarding changes in the wood's moisture content.

Water stains: Wood stain that is a mixture of powders and water.

Water supply system: System of pipes that distribute water under pressure to kitchens, bathrooms, and laundry areas.

Water table: A ledge or slight projection at the bottom of a structure that carries the water away from the foundation. Also, the level of water underground.

Water vapor: The water contained within air.

Wave: A variation that transfers energy progressively from point to point in a medium, like the movement of water in a lake or ocean.

Weathering: The mechanical or chemical disintegration and discoloration of the surface of wood or other material. It can be caused by exposure to light, the action of dust and sand carried by winds, and the alternate shrinking and swelling of the surface fibers that comes with the continual variation in moisture content brought by changes in the weather. Weathering does not include decay.

Weatherstrip: Narrow strips of metal, vinyl, plastic, or other material that retard the passage of air, water, moisture, or dust around doors or windows.

Web: Material between the chords in solid-web trusses.

Web-type studs: Studs with a special metal edge with a gap into which nails can be driven.

Weephole: A small hole to drain water to the outside. Commonly used at the lower edges of masonry cavity walls.

Welding bead: The thickened area of metal forming a joint between two pieces that have been melted together by the welding arc.

Wet edge: A paint band narrow enough so that the next band can be completed before the first has dried, eliminating lap marks.

Wet wall: An interior wall finish surface usually consisting of 3/8" gypsum plaster lath and 1/2" gypsum plaster.

Wide flange: Type of steel beam generally used in residential construction. Also called *W-beam*.

Wind: A term used to describe the warp in a board when twisted ("winding"). If laid on a perfectly flat surface, it will rest on two diagonally opposite corners.

Winders: Curved stringers in winding stairs.

Window unit: Unit that consists of a combination of the frame, window, weatherstripping, and sash activation device. May also include screens and/or storm sash. All parts are assembled as a complete operating unit.

Wing: A lateral extension of a building from the main portion; one of two or more coordinate portions of a building that extend from a common junction.

Wing dividers: Tools that are similar to a pencil compass, except that they have no pencil attached to one leg; used to transfer measurements.

Wire glass: Glass with a layer of meshed wire incorporated in the center of the sheet.

Wood bleaching: Applying bleach to remove some of the natural color from the wood.

Wood chisel: Tool used to trim and cut away wood or composition materials to form joints or recesses.

Wood clamps: Wood clamps with broad jaws that distribute the pressure over a wide area. Sizes are designated by the width of the jaw opening and range from 4"–24". Also called *hand screws*.

Wood I-beams: Beams consisting of flanges (also called *chords*) of structural composite lumber, such as laminated veneer lumber, or sawn lumber and a vertical web of plywood or oriented strand board 3/8" or 7/16" thick. Also called *solid-web trusses*.

Wood lath: Plaster base consisting of thin, narrow strips of soft wood nailed to studs.

Wood putty: A mixture of wood and adhesive in powder form.

Wood shakes: Durable, hand-split roofing material.

Wood shingles: Individual wooden pieces with a wedge-shaped cross section, used for both roofing and sidewall applications. Wood shingles are made from cedar or other rot-resistant woods.

Working time: The period during which joint compound or adhesive can be used before it sets.

Woven valley: Running shingles across valleys.

Wrecking bar: Tool used to strip concrete forms, disassemble scaffolding, or other rough work involving prying, scraping, and nail pulling. Also called *ripping bar.*

X

X-bracing: Steel straps in metal framing that diagonally extend from the top to the bottom of the wall. They are attached to the tracks and studs with screws.

Xylem: Inside area of the cambium layer that develops new wood cells.

Z

Zero clearance: A quality of a prefabricated fireplace that allows the unit to be placed adjacent to combustible material such as wood framing.

Zones: Areas of a building that can be separately heated and controlled by individual thermostats.

Index